Lecture Notes on Multidisciplinary Industrial Engineering

Series Editor

J. Paulo Davim, Department of Mechanical Engineering, University of Aveiro, Aveiro, Portugal

"Lecture Notes on Multidisciplinary Industrial Engineering" publishes special volumes of conferences, workshops and symposia in interdisciplinary topics of interest. Disciplines such as materials science, nanosciences, sustainability science, management sciences, computational sciences, mechanical engineering, industrial engineering, manufacturing, mechatronics, electrical engineering, environmental and civil engineering, chemical engineering, systems engineering and biomedical engineering are covered. Selected and peer-reviewed papers from events in these fields can be considered for publication in this series.

More information about this series at http://www.springer.com/series/15734

R. Ganesh Narayanan · Shrikrishna N. Joshi ·
Uday Shanker Dixit
Editors

Advances in Computational Methods in Manufacturing

Select Papers from ICCMM 2019

Springer

Editors
R. Ganesh Narayanan
Department of Mechanical Engineering
Indian Institute of Technology Guwahati
Guwahati, Assam, India

Shrikrishna N. Joshi
Department of Mechanical Engineering
Indian Institute of Technology Guwahati
Guwahati, Assam, India

Uday Shanker Dixit
Department of Mechanical Engineering
Indian Institute of Technology Guwahati
Guwahati, Assam, India

ISSN 2522-5022 ISSN 2522-5030 (electronic)
Lecture Notes on Multidisciplinary Industrial Engineering
ISBN 978-981-32-9071-6 ISBN 978-981-32-9072-3 (eBook)
https://doi.org/10.1007/978-981-32-9072-3

This Springer imprint is published by the registered company Springer Nature Singapore Pte Ltd.
The registered company address is: 152 Beach Road, #21-01/04 Gateway East, Singapore 189721, Singapore

Conference Organizing Committee

Patron

Prof. Gautam Biswas, Director, IIT Guwahati

Chairman

Prof. S. K. Dwivedy, Department of Mechanical Engineering, IIT Guwahati

Executive Chairman

Prof. Uday Shanker Dixit, Department of Mechanical Engineering, IIT Guwahati

Organizing Secretary

Dr. R. Ganesh Narayanan, Department of Mechanical Engineering, IIT Guwahati

Joint Organizing Secretary

Dr. Shrikrishna N. Joshi, Department of Mechanical Engineering, IIT Guwahati

Treasurer

Dr. Swarup Bag, Department of Mechanical Engineering, IIT Guwahati

Members

Dr. P. S. Robi, IIT Guwahati
Dr. Pankaj Biswas, IIT Guwahati
Dr. S. Pal, IIT Guwahati
Dr. Manas Das, IIT Guwahati
Dr. Deepak Sharma, IIT Guwahati
Dr. A. K. Das, IIT Guwahati
Dr. Vinayak Kulkarni, IIT Guwahati
Dr. S. S. Gautam, IIT Guwahati
Dr. Prasenjit Khanikar, IIT Guwahati
Dr. S. Kanagaraj, IIT Guwahati
Dr. Nelson Muthu, IIT Guwahati
Dr. S. Senthilvelan, IIT Guwahati
Dr. Anil Birru, NIT Manipur
Dr. Sajan Kapil, IIT Guwahati

International Advisory Board

Amitabha Ghosh, IIT Kanpur, India
Amitava De, IIT Bombay, India
D. Banabic, Technical University of Cluj-Napoca, Romania
T. A. Dean, University of Birmingham, UK
J. S. Gunasekera, Ohio University, USA
P. Hodgson, Deakin University, Australia
P. K. Jain, IIT Roorkee, India
V. K. Jain, IIT Kanpur, India
S. S. Joshi, IIT Bombay, India
S. V. Kailas, IISc Bangalore, India
Kornel F. Ehmann, Northwestern University, USA
S. M. Mahajan, BHEL, India
S. S. Pande, IIT Bombay, India
J. Paulo Davim, University of Aveiro, Portugal
D. T. Pham, Cardiff University, UK
M. Rahman, NUS, Singapore
N. Ramesh Babu, IIT Madras, India

K. Ramesh Kumar, DRDO, India
A. Roy, Loughborough University, UK
K. P. Rao, City University of Hong Kong, Hong Kong
Santosh Kumar, IIT BHU, India
H. S. Shan, Chandigarh, India
M. S. Shanmugam, IIT Madras, India
Sumitesh Das, Tata Steel, India
Surjya Kanta Pal, IIT Kharagpur, India
S. Venugopal, NIT Nagaland, India
L. Vijayaraghavan, IIT Madras, India

Preface

Manufacturing is the backbone of the economy of a nation. It is one of the oldest activities of the human civilization. In the twenty-first century, the emphasis is on enhancing the efficacy and efficiency of manufacturing processes and systems by using the computational power. Industry 4.0 aims to make use of digital manufacturing, cloud computing and manufacturing, Internet of things, and big data analytics. It is essential to understand the manufacturing processes scientifically, develop models, and optimize them. In December 2011, the 1st International Conference on Computational Methods in Manufacturing (ICCMM 2011) was organized at the Indian Institute of Technology Guwahati. The conference was a huge success with encouraging participation from industry and academia. After a gap of 8 years, the 2nd International Conference on Computational Methods in Manufacturing (ICCMM 2019) was organized on March 8–9, 2019.

The objective of this conference was to promote the use of computational techniques in manufacturing by providing a platform for sharing ideas and expertise. We wanted to bring practicing engineers, experimentalists, and computational experts together. We received contributed papers from the researchers covering all major areas of computational manufacturing and allied fields. A total of 125 full papers were received, which were peer-reviewed. The authors have revised their papers based on the review comments. Finally, 90 papers were selected for inclusion in the book of proceedings.

This book of proceedings covers the following broad topics: material forming, joining and welding, casting, additive manufacturing, material fabrication, subtractive manufacturing, process modeling, simulation, analysis, optimization, and design. Although the focus of the conference is on computational techniques, most of the authors have invariably included experimental results as experiments are the inseparable part of modeling of manufacturing.

We hope that the book of proceedings will be a valuable document for the delegates, scientists, engineers, and research scholars. Springer has taken the responsibility of publishing and distributing it to a wider audience. We wish to thank Ms. Swati Meherishi, Ms. Avni, and other staff members of Springer. We acknowledge the support provided by the reviewers, members of advisory and organizing committees, and colleagues at IIT Guwahati. Any feedback to the editors regarding this collection will be highly appreciated.

Guwahati, India

R. Ganesh Narayanan
Shrikrishna N. Joshi
Uday Shanker Dixit

Contents

About the Editors

R. Ganesh Narayanan is an Associate Professor at the Department of Mechanical Engineering, Indian Institute of Technology (IIT) Guwahati, India. He received his Ph.D. from the IIT Bombay, India. His research areas include metal forming and joining. He has contributed many research articles in reputed journals and international conferences. He has edited few books including 'Sustainable Material Forming and Joining' published by CRC Press USA, 'Strengthening and Joining by Plastic Deformation' published by Springer Singapore, 'Advances in Material Forming and Joining' published by Springer India, and 'Metal Forming Technology and Process Modeling' published by McGraw Hill Education, India. He has also edited special issues of journals including 'Advances in Computational Methods in Manufacturing' in the International Journal of Mechatronics and Manufacturing Systems, and 'Numerical Simulations in Manufacturing' in the Journal of Machining and Forming Technologies. He has organized three international conferences at IIT Guwahati namely International Conference on Computational Methods in Manufacturing (ICCMM) in 2011, the 5th International and 26th All India Manufacturing Technology, Design and Research (AIMTDR) Conference in 2014, and International Conference on Computational Methods in Manufacturing (ICCMM) in 2019. He has also organized a GIAN course on 'Green Material Forming and Joining' at IIT Guwahati in 2016.

Shrikrishna N. Joshi is an Associate Professor in the Department of Mechanical Engineering, Indian Institute of Technology Guwahati, India, where he joined after completing his doctoral studies from IIT Bombay in 2010. His research interests include laser based manufacturing, ultraprecision machining, computer aided design and manufacturing (CAD/CAM) and, manufacturing process modeling & optimization. He has authored/co-authored about 50 research articles in international journals/conferences and six book chapters. He has also co-edited 2 books on "Laser based manufacturing" with Springer.

Uday Shanker Dixit is a Professor in the Department of Mechanical Engineering, Indian Institute of Technology Guwahati. He completed his BE in Mechanical Engineering from the University of Roorkee (now the Indian Institute of Technology Roorkee) in 1987, and his MTech and PhD in Mechanical Engineering from the Indian Institute of Technology Kanpur (IIT Kanpur) in 1993 and 1998, respectively. Professor Dixit has previously worked on designing machines for industry, and as the Officiating Director of the Central Institute of Technology, Kokrajhar. Professor Dixit has been actively engaged in research in various areas of design and manufacturing for the last 25 years. He has authored/co-authored more than 200 research papers, 27 book chapters and 7 books on mechanical engineering. He has also co-edited 7 books on manufacturing. Presently, he is an associate editor of the Journal of the Institution of Engineers (India), Series C and Vice-President of the AIMTDR conference.

Part I
Material Forming

Chapter 1
Numerical Simulation of Micro Forming of Bio-Absorbable AZ80 Magnesium Alloy

D. Rajenthirakumar, N. Srinivasan and R. Sridhar

Abstract The characteristics of grains during the micro forming process are unpredictable. It reflects in the material behavior, interfacial friction and surface finish due to size effects. This paper describes the numerical analysis of size effects and frictional effects during the plastic deformation of magnesium AZ80 alloy in the micro extrusion process. On the basis of true stress–strain curve, the material model is used for numerical investigation. The results reveal that the flow stress increases with decreasing grain size due to increasing interfacial friction. The results are then validated with the micro extrusion experiments conducted. Further, the friction generation and load prediction can be achieved through numerical simulation is the primary criteria.

Keywords Micro forming · Extrusion · Finite element method · Size effect

1.1 Introduction

Miniaturization technology has transformed the very risky assignment into a smart task not only in electronic field but also in medical devices such as pacemakers, implantable defibrillators, catheters, stents, and neuromodulator device. The metals and glasses are replacing the polymers due to its higher strength in minimum grain size. During the transformation from macro scale to micro scale the phenomena called size effect governs the micro plastic deformation of material and its behavior [1]. When scaling down the characteristics of grains are unpredictable pertaining to material behavior. During the micro forming process, surface to volume ratio increases considerably and grain size effect ruins the flow stress of the material [2]. Hence, different methods of experiments and numerical analysis are conducted by

D. Rajenthirakumar (✉) · N. Srinivasan · R. Sridhar
Department of Mechanical Engineering, PSG College of Technology, Coimbatore 641004, India
e-mail: rajentkumar@gmail.com

R. G. Narayanan et al. (eds.), *Advances in Computational Methods in Manufacturing*, Lecture Notes on Multidisciplinary Industrial Engineering,
https://doi.org/10.1007/978-981-32-9072-3_1

researchers worldwide. The primary focus of the research is to control the size effect and minimize the interfacial friction.

Piwnik and Mogielnicki simulated the influence of container roughness on material flow process in micro extrusion. They concluded the significant impact of a wave roughness on material flow in the process [3]. The authors also simulated the interfacial friction towards the improvement of surface roughness [4]. The strategies to estimate grain properties and model grain size effect are developed and the grain size effect on deformation behavior and the scatter of flow stress are modeled. The results are then validated with experiments [5]. So to eliminate the time taken and expenses for experimenting, numerical simulation is employed to predict the behavior in several aspects during the experiments.

In this work, a numerical simulation of AZ80 Mg alloy is performed and the results obtained are then validated with physical experiments. The AZ80 Mg has a huge application in biomedical field [6]. The paper mainly focuses on the numerical analysis of micro extrusion by which the prediction of size effects and friction at interfaces has been achieved. Initially, the material model is obtained by micro compressions test, which is given as input for numerical simulation. The second stage is the FEM simulation. The last stage is the validation of numerical analysis with experiments.

1.2 Micro Compression Test and Size Effects

The material for the micro compression experiment is AZ80 Mg alloy of different grain sizes such as 20, 36, and 84 μm in quasi-static condition. The strain rate is 0.1 mm/s which is kept constant. Figure 1.1 shows the true stress–strain curves for

Fig. 1.1 True stress–strain curve

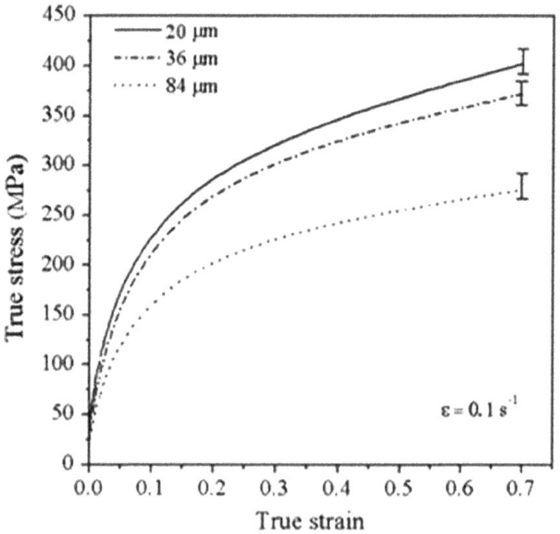

Fig. 1.2 FEM—2D axisymmetric Model

specimens in 1.2 mm diameter and three different grain sizes. The curves illustrate that the flow stress increases with decrease in grain size.

The effects of scaling down with decreasing flow stress are discussed through so-called surface model [7]. The resistance to deformation gets varied from surface layer to inner grains of the material. The difference is due to the hardness variation in surface layer to inner grains of the material. The occurrence of interfacial friction effect can be seen in flow stress curves. The existence of larger grains count possesses maximum resistance to plastic deformation. As a result, the smaller the grains, the larger the friction generated in tool-billet interfaces can be noticeable.

1.3 FEM Model

AFDEX is a metal forming simulator, meets the requirements for bulk metal-forming and simulation. The AFDEX_V17R00 is utilized for predicting the nonlinear behavior of material in 2D and 3D. Based on the previous results in literature [8], 0.3 μm is selected as coefficient of friction as it is a dry condition, as

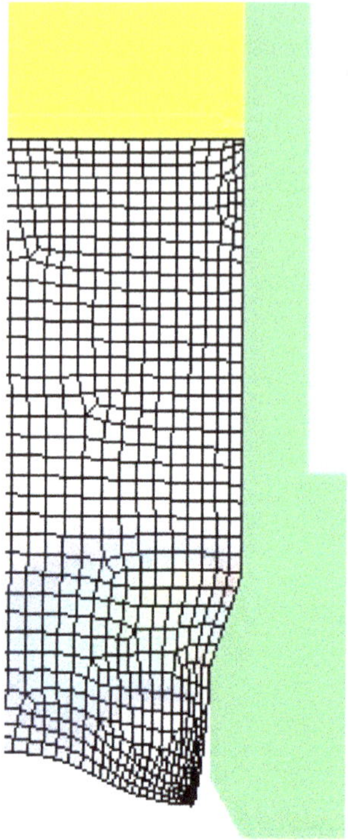

Fig. 1.3 Meshed model

an assumption to compare the frictional effect on forming load. The micro extrusion tooling system is modeled as rigid bodies while the forming material (micro pin) is an elastoplastic body. Simulations are performed with the most widely used von Mises yield criterion as it gives equal weight to all principal stresses. Considering the symmetry in both geometry and loading conditions, a single half experimental setup is modeled (see Fig. 1.2). The meshing in Fig. 1.3 is done with 8200 four nodes 2D axisymmetric quadrilateral element. These elements have maintained the accuracy in stress analysis. The adaptive meshing technique is selected to retain the mesh quality. To obtain a suitable mesh, a local element size has to be applied. This requires some trial and error method.

1.4 Physical Experiment

In order to show the effects of grain size and investigate the interfacial friction, an experimental setup is designed and micro extrusion tests with AZ80 magnesium billets of different grain sizes have been performed. Figure 1.4 shows the exploded view of micro extrusion die assembly and Fig. 1.5 shows the schematic view of micro extrusion die assembly.

Fig. 1.4 Micro extrusion—die assembly

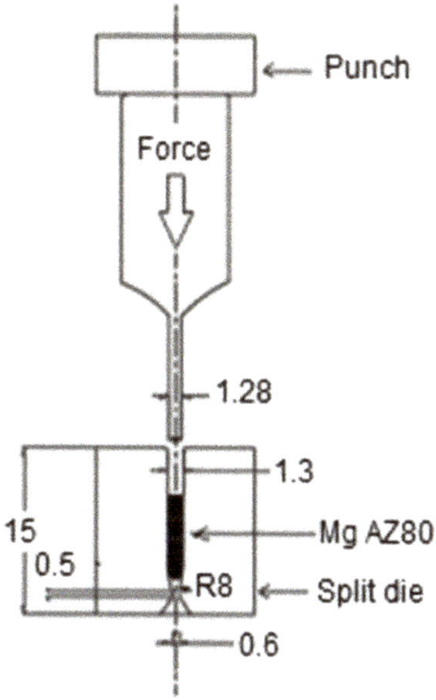

Fig. 1.5 Schematic view of microextrusion assembly

1.5 Results and Discussion

The numerical simulation of micro forward extrusion is performed to evaluate the interfacial friction. The 20 μm grain size is selected for simulations it possesses high grains count per unit volume. The Von-Mises stress distribution and strain distribution of the micro forward extrusion are shown in Figs. 1.6 and 1.7, respectively. The material undergoes the severe shear deformation at the die-billet contact surfaces. The differences in friction coefficient are observed in stress and strain distribution. The interfacial friction observed during the simulation (see Fig. 1.8) is reduced with coefficient of friction.

The experimental results are then compared with numerical simulation for validation purpose. The micro extruded parts are shown in Fig. 1.9. The billet-die interfaces are exposed to friction which increases with the decrease of specimen size. This is due to the increase of the fraction of contact area and so, the peak is observed initially in forming load. It started decreasing due to increase in proper

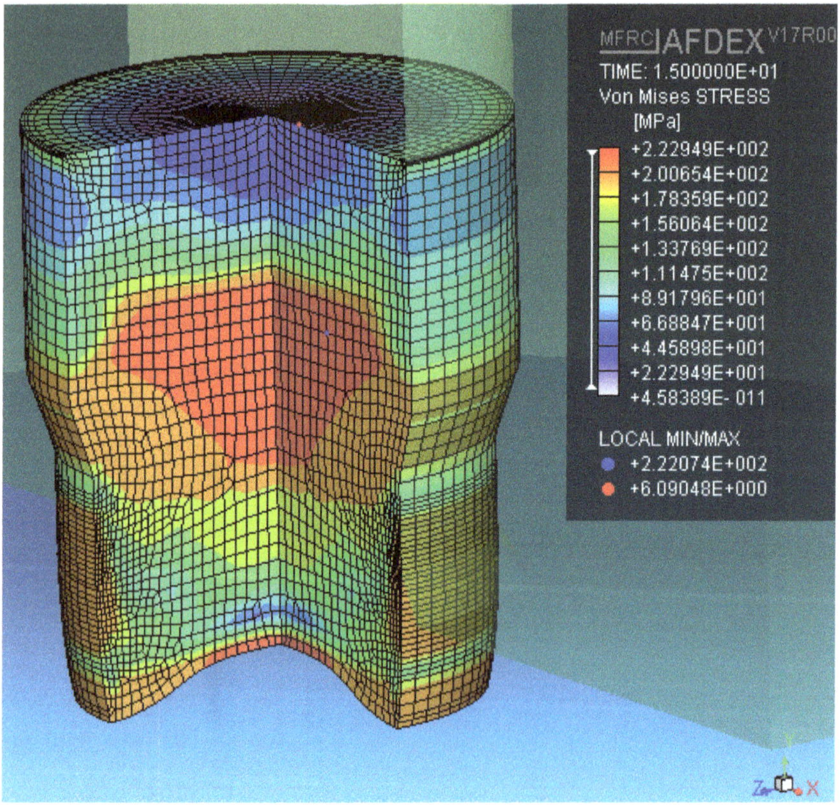

Fig. 1.6 Stress distribution (20 μm)

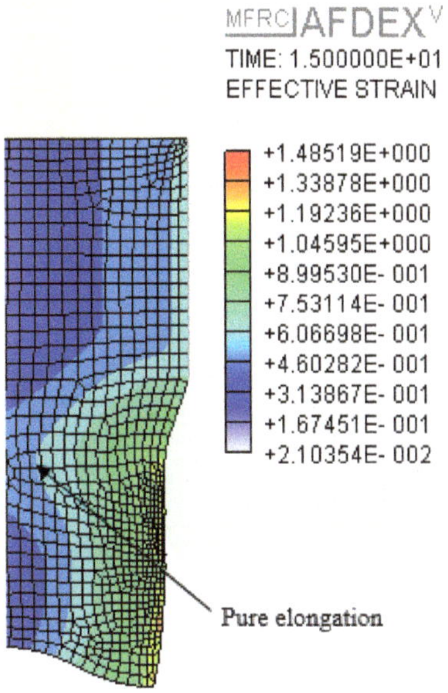

Fig. 1.7 Strain distribution (20 μm)

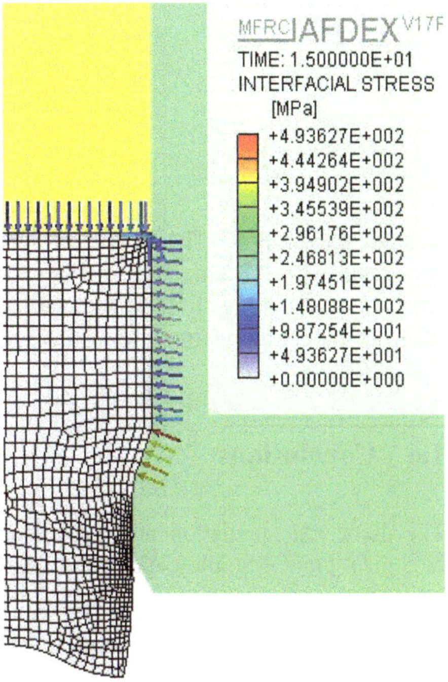

Fig. 1.8 Interfacial stress (20 μm)

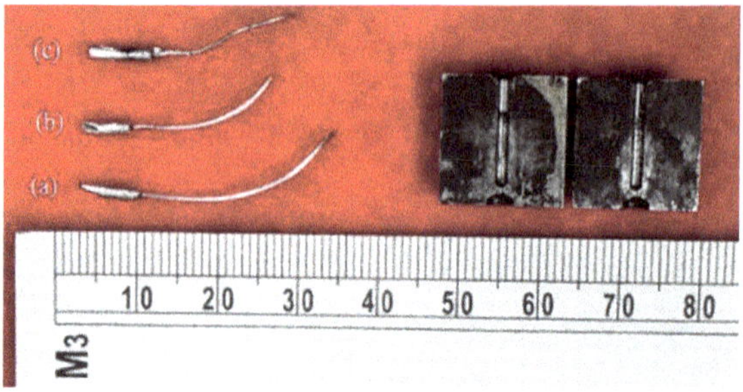

Fig. 1.9 Micro extruded parts, **a** 20 μm, **b** 36 μm and **c** 84 μm

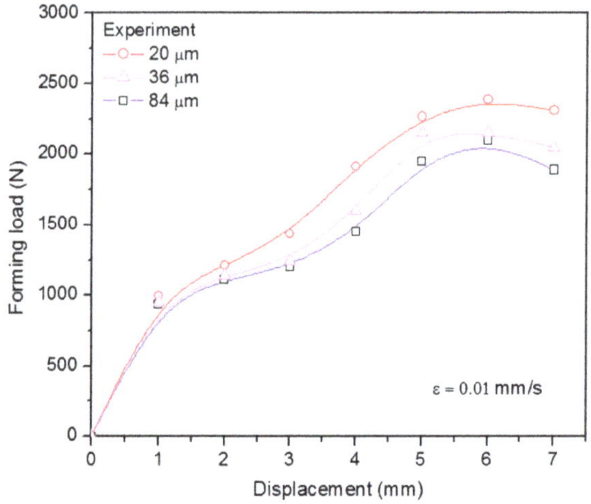

Fig. 1.10 Load versus displacement curve of AZ80 Mg

material plastic flow. The numerical simulation of the micro forward extrusion performed with 0.3 μm. The load–displacement response obtained from experiments is shown in Fig. 1.10. The maximum load consumed by 20 μm is due to high grains count per unit volume which possesses high strain resistance to deformation.

1.6 Conclusions

The numerical simulation and experimental validation has been conducted for biological grade Mg alloy AZ80 and the following conclusions are made:

- The interfacial friction increases with decrease of grain size. It is proved by observing high interfacial stress of 20 μm than the deformation of other grain size Mg AZ80 alloy.
- The flow stress of AZ80 Mg alloy increases with reduction in grain size.
- The deformation load increases with reduction in grain size.

- The numerical simulation results with AFDEX shows the in-depth friction condition effects in the micro forward extrusion process. When the friction coefficient increases the deformation load increases correspondingly.

The numerical simulation has concealed significant benefits such as load prediction for several friction conditions, existence of interfacial friction and minimizing the time and number of experiments. Even though the numerical analysis shows some deviation with experiments, it is likely to be considered for prediction.

Acknowledgements The authors gratefully acknowledge the financial support of Science and Engineering Research Board, Government of India and PSG College of Technology, Coimbatore, Tamil Nadu, India.

References

1. Engel, U., Eckstein, R.: Microforming—from basic research to its realization. J. Mater. Process. Technol. **126**, 35–44 (2002)
2. Qiu, T.T., Hou, Y.K., Cao, H.L.: Review of research progress on size effect in micro-forming. Key Eng. Mater. **717**, 118–121 (2016)
3. Piwnik, J., Mogielnicki, K.: Experimental and FE analysis of aluminium alloy plastic flow in the forward microextrusion processes. Arch. Metall. Mater. **59**, 521–525 (2014)
4. Piwnik, J., Mogielnicki, K.: The friction influence on stress in micro extrusion. Arch. Foundry Eng. **10**, 451–454 (2010)
5. Chan, W.L., Fu, M.W., Lu, J., Liu, J.G.: Modeling of grain size effect on micro deformation behavior in micro-forming of pure copper. Mater. Sci. Eng. A **527**, 6638–6648 (2010)
6. Zheng, Y.: Magnesium alloys as degradable biomaterials. CRC Press, Boca Raton (2015)
7. Chan, W.L., Fu, M.W., Yang, B.: Study of size effect in micro-extrusion process of pure copper. Mater. Des. **32**, 3772–3782 (2011)
8. Krishnan, N., Cao, J., Dohda, K.: Study of the size effects on friction conditions in microextrusion—part I: microextrusion experiments and analysis. J. Manuf. Sci. Eng. **129**, 669 (2007)

Chapter 2
Investigation on Tensile Testing of Tailor Welded Blanks (TWBs)—A Simulation Study

Shalin Marathe and Harit Raval

Abstract The goal of better fuel economy can be fulfilled by the use of technology like TWBs (Tailor Welded Blanks). TWB is a product of joining two or more than two materials before any forming operation. The limitations associated with TWBs are the weld line movement and reduced formability. The formability reduction of TWB is due to the thin/weak material involved in it and therefore the tensile study of the TWBs becomes an important aspect. In the present study, investigation of tensile testing of Similar Material Dissimilar Gauge TWB and Dissimilar Material Similar Gauge TWB has been done numerically. For the simulation purpose, ABAQUS/Explicit has been used. The results indicate that the thin/weak material governs the strength of the TWB because the failure or necking is observed in that particular side only. Also, it has been observed that the failure locations are found to be at the boundary of thin/weak base material and weld zone.

Keywords Tailor welded blanks (TWBs) · Friction stir welding (FSW) · ABAQUS/explicit · Simulation · Tensile testing

2.1 Introduction

Nowadays, automobile, aerospace, transportation, and shipbuilding industries are aiming at weight reduction of the components without compromising the strength and Tailor Welded Blanks (TWBs) have been found to fulfill these requirements. TWB is a product of joining two or more than two material prior to any forming operation. TWB involves joining of two or more than two materials having different properties, thicknesses and/or coatings by any of the conventional joining processes. As it involves joining of materials having different densities the overall weight of the product decreases [1]. Also, TWBs are subjected to forming after the

S. Marathe (✉) · H. Raval
Department of Mechanical Engineering, Sardar Vallabhbhai National Institute of Technology, Surat 395007, India
e-mail: shalin.marathe89@gmail.com

R. G. Narayanan et al. (eds.), *Advances in Computational Methods in Manufacturing*, Lecture Notes on Multidisciplinary Industrial Engineering,
https://doi.org/10.1007/978-981-32-9072-3_2

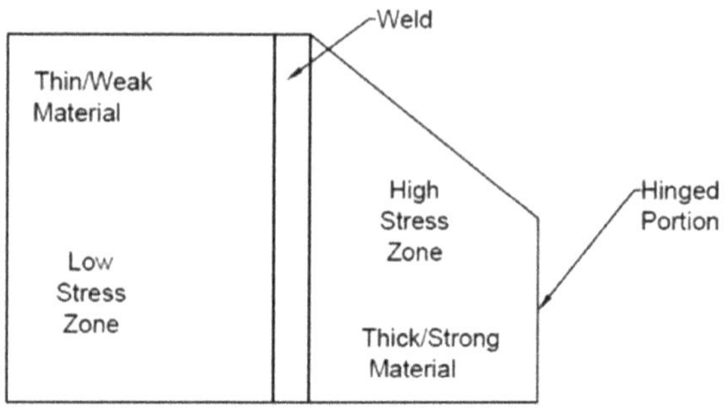

Fig. 2.1 Schematic representation of TWB used at inner door panel of an automobile body

joining has been done and because of that the number of dies required for the forming decreases [1] which also leads to decrease the product lead time and cost as well. Figure 2.1 indicates the schematic of TWB which is used at the inner door panel of an automobile body.

Despite of such tremendous advantages, the application of TWBs has been not up to the mark as limitations like formability reduction [2] and weld line movement [3] are associated with TWBs. In TWBs, as there is a presence of weld between the two parent metals, the whole product becomes inhomogeneous and when TWBs are subjected to forming operations these inhomogeneities result in uneven strain [2] and hardness distribution [2, 4, 5] across the weld. This uneven distribution of hardness and strain leads to failure of the TWBs resulting in the low formability of the component. A lot of research has been done to investigate the formability behavior of TWBs during the conventional forming [2] operations and it has been observed that the formability of TWBs is found to be decreasing in comparison to the parent metal participating in it. Now in order to improve the formability of the TWBs lots of efforts have been done by researchers.

TWBs are generally developed from the materials like steel [6–8], aluminum and its alloys [4, 9] and titanium as more application is found in automobile, shipbuilding, aerospace and transportation industries. The application of steel to develop TWBs has been reduced as the aluminum material gives a higher amount of weight saving [10]. TWBs involving steel as parent material are generally developed using the laser welding [6, 9, 11–13] processes and if aluminum and its alloys are used as parent material than solid-state welding processes like FSW [9, 14–16] is found more suitable. As aluminum has low laser absorption capacity and it is very reflective in nature, it is quite difficult to weld using laser welding and other conventional welding techniques.

In order to decide the quality of the TWBs, there are many testing methods available and tensile testing is one of them. So, in this present work tensile testing of TWBs has been performed using a simulation tool ABAQUS/Explicit. The simulation methods save important resource like time, material and cost. So in present work, testing of TWBs made from similar material having different thickness/gauge and dissimilar materials having similar thickness/gauge are investigated using a simulation tool. In this work, TWB is considered to be made from FSW process which is known as Friction Stir Tailor Welded Blank (FSTWB). Results like plastic strain, Von-misses stresses, Load versus Displacement diagram, Stress–Strain behavior and True strain distribution across the weld are extracted from the simulation and compared with published literatures.

2.2 Numerical Conditions

In this present study, all the simulations are carried out using ABAQUS/Explicit [17]. This study has been divided into two sections. First section contains the simulation study on similar material dissimilar gauge TWBs made from FSW process and second section contains the same investigation for dissimilar material similar gauge TWBs. For all the simulations "Weld Zone" approach is adopted to define the weld properties in the tensile test specimen.

2.2.1 Section-I (Similar Material Dissimilar Gauge TWBs)

In this section, as a Base Material AA 5754 of 1.06 and 1.66 mm is used. The different material properties used during the simulation for the mentioned material is given in Table 2.1.

The material has been modeled as 3 dimensional deformable. The dimensions of the tensile test specimen and the cross-section of the welded joint are shown in Fig. 2.2a, b, respectively. It was assumed that the base metal and weld metal were following the power-law equation.

Figure 2.3a indicates the cross-section (isometric view) and Fig. 2.3b indicates dissimilar gauge tensile test specimen (isometric view) modeled in the ABAQUS environment. The entire simulation was completed in total two steps namely Initial and step-1. In the initial step of the simulation one end of the specimen was encastre

Table 2.1 Properties considered for simulation [18]

Property	Base material	Weld metal
Density (ρ) (kg/m^3)	2600[a]	2600[a]
Young's modulus G (N/m^2)	71×10^9	68×10^9
Yield stress (σ_y) (N/m^2)	132×10^6	146×10^6
Poisson's ratio (μ)	0.33[a]	0.33[a]
Strength coefficient (K) (N/m^2)	551.5×10^6	453.5×10^6
Strain hardening index (n)	0.33	0.23

[a]Assumed

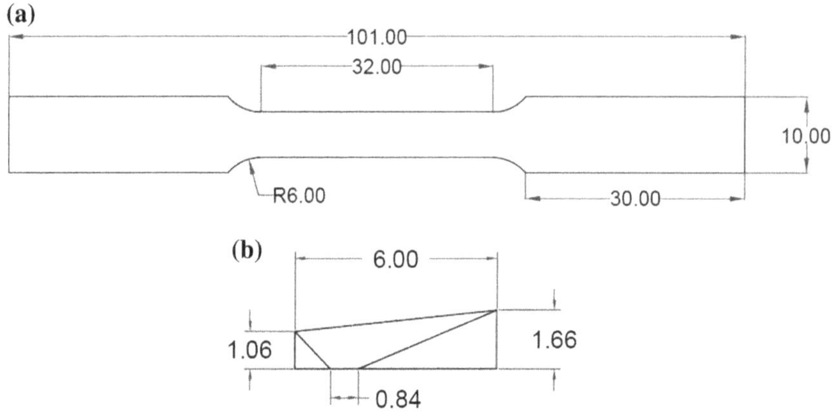

Fig. 2.2 Details of the tensile test specimen, **a** dimensions of the specimen, **b** cross-section of the welded joint (all the dimensions are in mm) [18]

(All DOF (Degree of Freedom) constrained) and in step-1 another end was given the displacement in the X direction only (transverse to the weld).

As shown in Fig. 2.4 thin base metal was given encastre boundary condition and thick base metal was given displacement in the X-direction. Meshing was done using linear tetrahedral elements. Table 2.2 indicates the meshing details used during the simulations.

2.2.2 Section-II (Similar Gauge Dissimilar Material TWBs)

In this section, tensile testing of TWB made from dissimilar material having similar gauge is performed using the ABAQUS/Explicit tool. Figure 2.5 indicates the tensile specimen modeled in the ABAQUS environment. The tensile testing was done as per the ASTM-E8 standards using a sub-sized tensile test specimen. For this section, the material properties were used as per Table 2.3.

In this investigation, one end of specimen was given encastre boundary conditions and another was given the displacement. The meshing details used for the simulation purpose are tabulated in Table 2.4. Figure 2.6 indicates the meshed model of tensile test specimen in the ABAQUS environment. Similar to case-I, this simulation was also completed using total 2 number of steps and materials were assumed to follow the Power law equation.

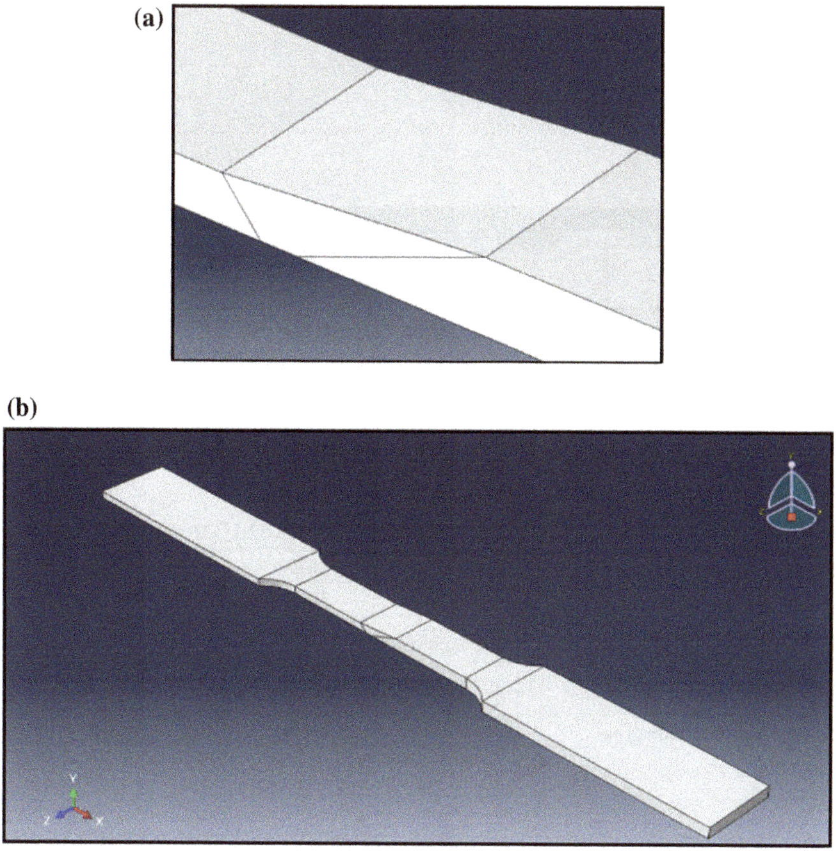

Fig. 2.3 Details of the tensile test specimen, **a** cross-section, **b** overall tensile test specimen

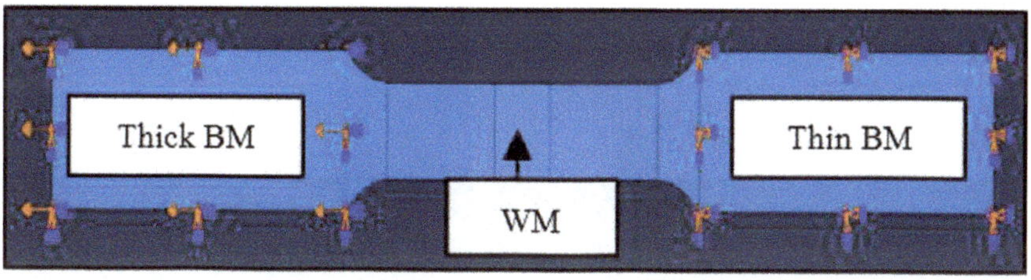

Fig. 2.4 Boundary Conditions applied to the tensile specimen (WM = Weld Metal, BM = Base Metal)

Table 2.2 Mesh statistics

Total number of nodes	3613
Total number of elements	13,355
Linear tetrahedral elements of type	C3D4

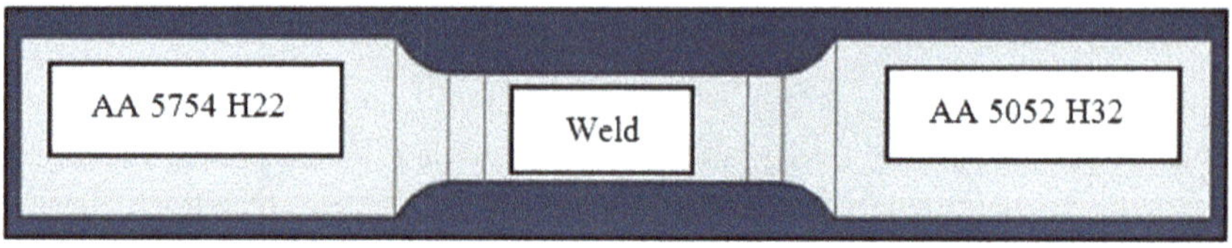

Fig. 2.5 TWB tensile specimen modeled in ABAQUS

Table 2.3 Properties used during simulation [2]

Property	AA 5754 H22	AA 5052 H32	Weld[a]
Density[a] (ρ) (kg/m^3)	2670	2680	2670
Young's modulus[a] G (N/m^2)	70.3×10^9	70.3×10^9	70.3×10^9
Yield stress (σ_y) (N/m^2)	122×10^6	140×10^6	131×10^6
Poisson's ratio[a] (μ)	0.33	0.33	0.33
Strength coefficient (K) (N/m^2)	252.04×10^6	258.7×10^6	255.37×10^6
Strain hardening index (n)	0.112	0.125	0.1185

[a]Assumed

Table 2.4 Mesh statistics

Total number of nodes	3390
Total number of elements	2016
Linear tetrahedral elements of type	C3D8R

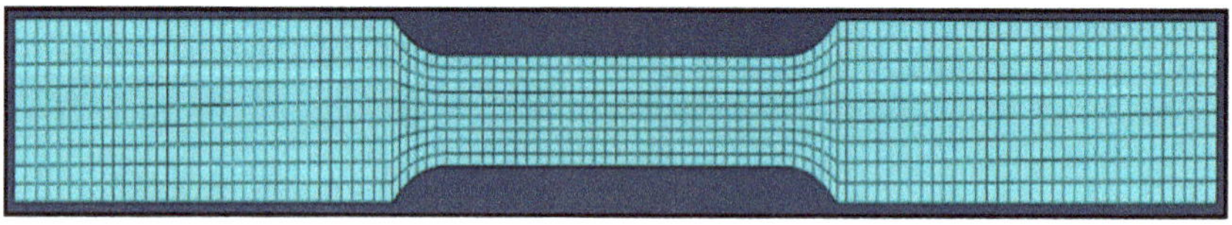

Fig. 2.6 Meshed model of the specimen

2.3 Results and Discussions

In the upcoming sub-sections, the results extracted from the simulation study have been discussed and compared with the same of available literatures. Responses like stress–strain curve, load versus displacement curve, failure location in the specimen and strain developed in the tensile test specimen are compared.

2.3.1 Section-I (Similar Material Dissimilar Gauge TWBs)

For the given boundary conditions, load versus displacement graph was extracted from the simulation and plotted as shown in Fig. 2.7 which also compares it with the published literature.

It can be observed that the natures of both the mentioned results are in good agreement with each other which indirectly authenticates the methodology adopted for the simulation purpose. From the simulations, results of plastic strain were also extracted as indicated in Fig. 2.8.

The maximum plastic strain was observed in the thin material as indicated in Fig. 2.8 and it was found at the boundary region of the thin base material and the weld zone. The value of maximum plastic strain predicted by the present study of simulation was 0.2519 and it was 0.2004 by the published study [18] which seems to be in very good agreement with each other.

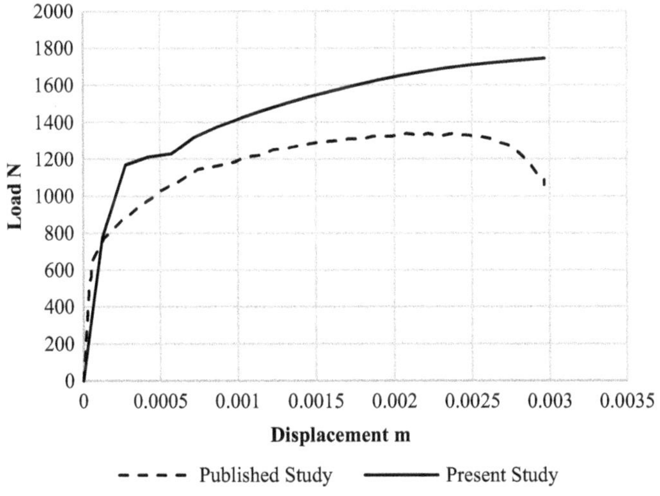

Fig. 2.7 Comparison of load versus displacement graph between published [18] and present study

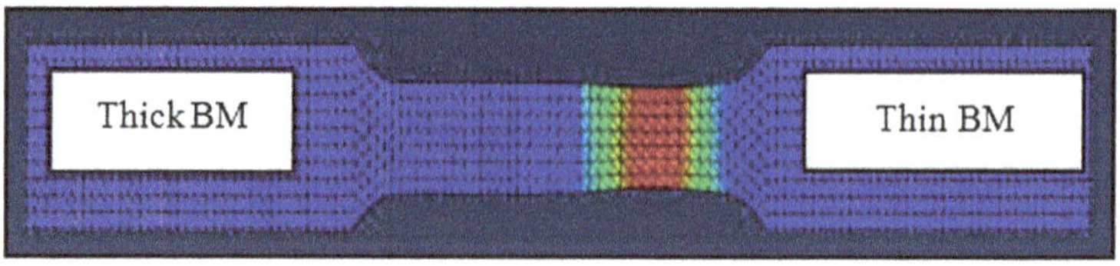

Fig. 2.8 Plastic strain (in X direction only) PE11 for present study

2.3.2 Section-II (Dissimilar Material Similar Gauge TWBs)

From the given boundary conditions in the section-II, the results like failure/necking locations, strain distributions across the weld and stress–strain behavior of TWB were extracted and compared with the published literature. Figure 2.9 indicates the present simulation result for necking location in the tensile test specimen. The failure location from weld center was observed at 11 mm in experimental study of published literature [2] and that was observed at 13 mm for the present simulation study. So, it can be stated that the results were in reasonably good agreement with each other.

The engineering stress–strain behavior of TWB was also plotted and compared with the published results as indicated in Fig. 2.10. From Fig. 2.10, it can be seen that the yield point predicted by present study and published study are in good agreement but for the plastic deformation part of the graph, a slight mismatch is found. This disagreement can be reduced by providing more appropriate boundary conditions and material properties during the simulations.

One more result of strain distribution across the weld was extracted from the simulation and compared with the published results as indicated in Fig. 2.11. The maximum strain value is observed at 13 mm from weld center for the present simulation study and that is the failure/necking point in the tensile test specimen of TWB. There was a disagreement between the present and published study regarding strain distribution which can be resolved using better boundary conditions and material properties.

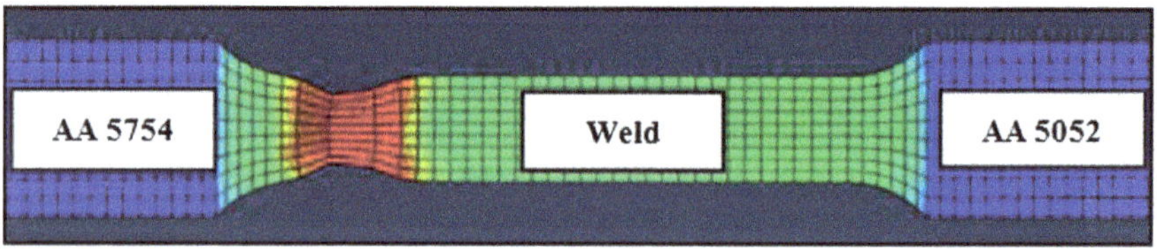

Fig. 2.9 Thinning/necking location for present simulation study (at 13 mm from weld centre)

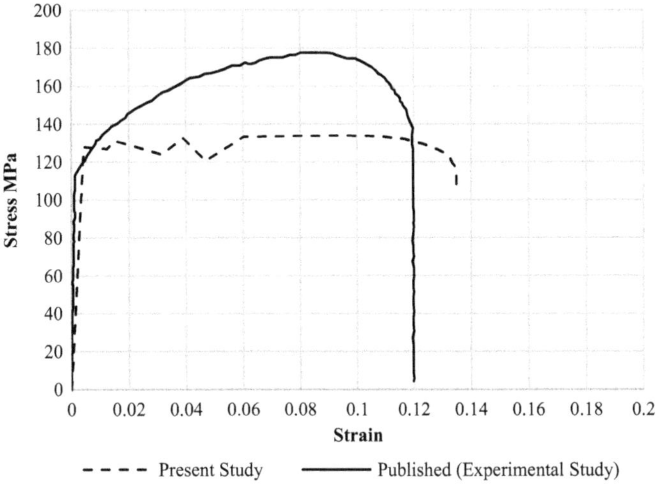

Fig. 2.10 Comparison of engineering stress–strain curve between present (simulation) and published [2] experimental results

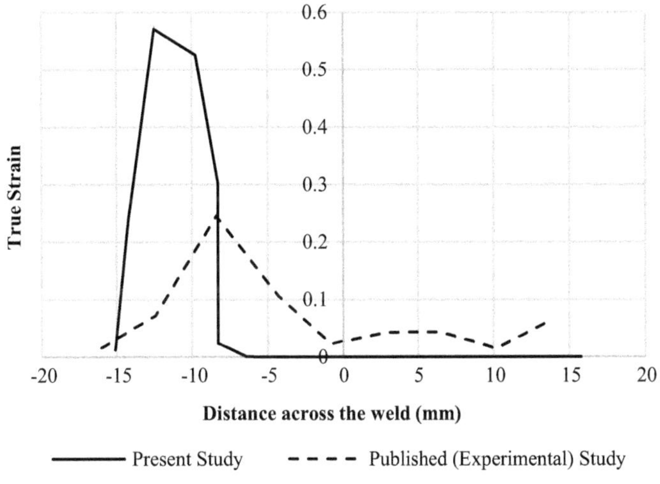

Fig. 2.11 Comparison of strain distribution across the weld for published [2] and present study

2.4 Conclusions

In the present research work, study of tensile behavior of the TWBs through FE analysis has been carried out. Following points can be concluded from the reported study:

- The "Weld Zone" approach used to define the weld properties during the simulation gives reasonably good results.
- The strength of the TWBs depends upon the weak/thin base material participating in it.
- To improve the strength and quality of the TWBs, strain distribution across the weld should be as even as possible.
- The boundary region between the weld zone and the base material is found to be the weakest.
- The simulation methodology not only saves the time, cost and resources but it also leads to reasonably good results for material behavior. The only limitation associated with the simulation methodology is that it demands for the accurate and proper input data in terms of boundary conditions and material properties.
- It is not necessarily required to perform the experimental/physical testing for accurate results. As represented in results of section-II, accurate results can be obtained to predict the material behavior during any deformation process by adopting appropriate approach for the simulation which is "Weld Zone" in this case.

References

1. Parente, M., Safdarian, R., Santos, A., Loureiro, A., Vilaca, P., Jorge, R.: A study on the formability of aluminum tailor welded blanks produced by friction stir welding. Int. J. Adv. Manuf. Technol. **83**, 2129 (2016)
2. Kesharwani, R., Panda, S., Pal, S.: Experimental investigations on formability of aluminum tailor friction stir welded blanks in deep drawing process. J. Mater. Eng. Perform. **24**(2), 1038 (2015)
3. Panda, S.K., Kumar, D.R.: Improvement in formability of tailor welded blanks by application of counter pressure in biaxial stretch forming. J. Mater. Process. Technol. **204**(1–3), 70–79 (2008)
4. Zadpoor, A.A., Sinke, J., Benedictus, R.: The effects of friction stir welding on the mechanical properties and microstructure of 7000 series aluminium tailor-welded blanks. Int. J. Mater. Form. **1**, 1311–1314 (2008)
5. Marathe, S.P., Mistry, H.J., Raval, H.K.: Parametric study of Friction Stir Welding (FSW) of AA 6061 using Taguchi method. In: Proceedings of National Workshop on Industrial Problems on Machines and Mechanical Similar Materials: "Challenges in Manufacturing". NIT, Nagpur (2016)
6. Song, Y., Hua, L., Chu, D., Lan, J.: Characterization of the inhomogeneous constitutive properties of laser welding beams by the micro-Vickers hardness test and the rule of mixture. Mater. Des. **37**, 19–27 (2012)
7. Song, Y.L., Hua, L.: Formability of tailor welded blanks based on the improved rule of mixture considering the welding residual strain and stress. Appl. Mech. Mater. Trans Tech Publ. **152**, 1406–1412 (2012)
8. Jia, Q., Guo, W., Li, W., Zhu, Y., Peng, P., Zou, G.: Microstructure and tensile behavior of fiber laser-welded blanks of DP600 and DP980 steels. J. Mater. Process. Technol. **236**, 73–83 (2016)

9. Ebrahimzadeh, P., Baseri, H., Mirnia, M.J.: Formability of aluminum 5083 friction stir welded blank in two-point incremental forming process. Proc. Inst. Mech. Eng. Part E J. Process Mech. Eng. **232**, 267 (2017)
10. Shibata, K., Iwase, T., Sakamoto, H., Kasukawa, M., Chiba, K., Saeki, H.: Welding of aluminium tailored blanks by Nd: YAG lasers. Weld. Int. **17**(4), 282–286 (2003)
11. Chatterjee, S., Saha, R., Shome, M., Ray, R.K.: Evaluation of formability and mechanical behavior of laser-welded tailored blanks made of interstitial-free and dual-phase steels. Metall. Mater. Trans. A **40**(5), 1142–1152 (2009)
12. Reis, A., Teixeira, P., Duarte, J.F., Santos, A., da Rocha, A.B., Fernandes, A.A.: Tailored welded blanks—an experimental and numerical study in sheet metal forming on the effect of welding. Comput. Struct. **82**(17), 1435–1442 (2004)
13. Rojek, J., Hyrcza-Michalska, M., Bokota, A., Piekarska, W.: Determination of mechanical properties of the weld zone in tailor-welded blanks. Arch. Civil Mech. Eng. **12**(2), 156–162 (2012)
14. Shalin, M., Hiten, M.: Experimental analysis on effect of tool transverse feed, tool rotational speed and tool pin profile type on weld tensile strength of friction stir welded joint of AA 6061. Mater. Today: Proc. **5**(1), 487–493 (2018)
15. Tronci, A., McKenzie, R., Leal, R.M., Rodrigues, D.M.: Microstructural and mechanical characterisation of 5XXX-H111 friction stir welded tailored blanks. Sci. Technol. Weld. Join. **16**(5), 433–439 (2011)
16. Zadpoor, A.A., Sinke, J., Benedictus, R.: Finite element modeling of transition zone in friction stir welded tailor-made blanks. AIP Conf. Proc. AIP **908**(1), 1457–1462 (2007)
17. Cheng, C.H., Chan, L.C., Tang, C.Y., Chow, C.L.: Determination of true stress-strain curve for the weldment of aluminum laser-welded blanks. J. Laser Appl. **17**(3), 159–170 (2005)
18. Garware, M., Kridli, G.T., Mallick, P.K.: Tensile and fatigue behavior of friction-stir welded tailor-welded blank of aluminum alloy 5754. J. Mater. Eng. Perform. **19**(8), 1161–1171 (2010)

Chapter 3
Formability Evaluation of EN-10149-2 (S700mc) Steel Under In-Plane Plane Stretching Condition

Dereje H/Georgis, Perumalla Janaki Ramulu and Habtamu Beri

Abstract The current effort is intended to evaluate the forming behavior of the EN-10149-2 (S700mc) steel of 2.42 mm thickness under in-plane plane-stretching (IPPS) condition at ambient conditions. EN-10149-2 (S700mc) steel has good mechanical properties and lightweight that increase the application in automotive, aerospace, and industrial structural design compared to other metals. For which, the tensile and plain stretching test samples were prepared in three rolling directions 0°, 45°, and 90°. The mechanical properties such as the yield strength, tensile strength, uniform elongation, total elongation, hardening exponent (n-value), material strength coefficient (K) and plastic strain ratio (R-value) of EN-10149-2 (S700mc) steel are evaluated from the tensile testing. The microstructure and hardness of the material were also evaluated. IPPS evaluation test has been carried out by stamping circular grids at the middle of the specimen by which measured the manor and major strain using thickness gradient necking criterion (TGNC). The obtained results from experiments found that the maximum limit strain is appeared for the rolling direction of 45° and followed by 90° and 0°. This indicates that the maximum formability is in 45° rolling direction during the IPPS test.

Keywords Sheet metal forming · Forming limits strain · Mechanical properties · IPPS · TGNC · Circular grids · Major strain · Minor strain

---

D. H/Georgis · H. Beri
Program of Mechanical Design and Manufacturing Engineering, School of Mechanical, Chemical and Materials Engineering, Adama Science and Technology University, Adama, Ethiopia
e-mail: haleluyaa.haile1@gmail.com

H. Beri
e-mail: habtamu_beri@yahoo.com

P. J. Ramulu (✉)
Program of Mechanical Design and Manufacturing Engineering & CoEAME, School of Mechanical, Chemical and Materials Engineering, Adama Science and Technology University, Adama, Ethiopia
e-mail: perumalla.janaki@astu.edu.et

R. G. Narayanan et al. (eds.), *Advances in Computational Methods in Manufacturing*, Lecture Notes on Multidisciplinary Industrial Engineering,
https://doi.org/10.1007/978-981-32-9072-3_3

3.1 Introduction

Basically, metalworking processes are grouped into two classes, namely, bulk-forming and sheet metal forming. The sheet metal blank is plastically deformed into a three-dimensional shape without any critical changes in the thickness of the sheet materials. In other deformation cases, elastic recapturing or springback may be significant because the amounts of plastic and elastic deformations are equivalent. Some of the important sheet metals forming processes are deep drawing, stretching, bending and rubber pad forming [1]. Moreover, numerous car businesses vehicles composed under the thought of lightweight, crash execution, vitality sparing, and ecological perspective are of unique intrigue which is much essential to drive the vehicles' material selection process. Among all the materials existed in the market, steel material has a significant impact on many automotive parts where strength is more important. In this needy, steel industries are developing new grades of advanced high strength steels (AHSS) with high strength to weight ratio as lightweight structures [2]. Likewise, EN-10149-2 (S700mc) steel sheet has many application areas in modern manufacturing industries. It has also good mechanical properties and lightweight that increases the application in automotive, aerospace, and industrial structural design compared to other metals [3, 4]. In-plane tensile tests were conducted to produce strain paths from uniaxial tension to plane strain in mild steel and high strength steel. An in-plane forming limit test that was independent of tooling geometry and friction was created. Results showed that a wide sample with a wide but short gauge section needed to produce forming limit strains close to plane strain. The sample had sufficient area for clamping to avoid fracture outside the plane strain region in the center of the sample was suggested [5]. There were other studies on the in-plane plane-stretching behavior of different materials and their limiting strains were evaluated [6–11].

Although many studies have been investigated the formability of high strength steel low alloy (HSLA) both experimentally and numerically, the formability prediction of HSLA under the in-plane plane stretching condition not been studied much. Researchers published so far described the formability prediction based on the strain path diagrams, yield criteria. The detailed description of the formability of EN-10149-2 (S700mc) steel with 2.42 mm thickness under in-plane plane-stretching (IPPS) condition has not discussed. Based on the applications of the EN-10149-2 (S700mc) steel at different level industries, the present work generates the data for forming characteristics of this particular material which can be used directly without further formability tests.

3.2 Materials and Methodologies

3.2.1 *Materials*

The material selected for this study was high strength low alloy or micro-alloyed steel, EN 10149-2 steel sheet which has 2.42 mm thickness. Chemical composition of material is shown in Table 3.1.

3.2.2 *Methodologies*

Mechanical Properties and Micro-Structure Evaluation of EN-10149-2

The microstructure of EN-10149-2 steel sheet was studied by optical microscope of model Huvitz HR-300 series. Hardness was measured using Vickers's hardness testing machine. The mechanical properties of EN-10149-2 steel sheet were measured using tensile tests as per ASTM- B557 M by cutting sheet in different rolling directions such as 0°, 45°, and 90° and evaluated the mechanical properties and plastic strain ratio (R) as per ASTM- E517 through tensile tests. Tensile tests were performed at a nominal cross-head speed of 1 mm/min, at room temperature. On an ISO 68982-1 universal testing machine and repeated thrice for each set to check the reproducibility. Failure from grip or slippage during testing was not observed. Figure 3.1 shows the base metal tensile specimen. The mechanical properties like yield strength, ultimate tensile strength, uniform elongation, strain hardening coefficient (n) and strength coefficient (K) were evaluated as per the standard procedure after testing the samples till failure. The load-stroke behavior obtained during testing and converted into engineering stress–strain and true stress–strain

Table 3.1 EN-10149-2 (S700mc) steel chemical composition (in wt%)

C	Si	Mn	P	S	Cr	Mo	Ni	Cu	W	Al	Nb	Ti	Fe
0.06	0.04	1.46	0.003	0.002	0.01	0.02	0.03	0.01	0.04	0.05	0.03	0.10	98.15

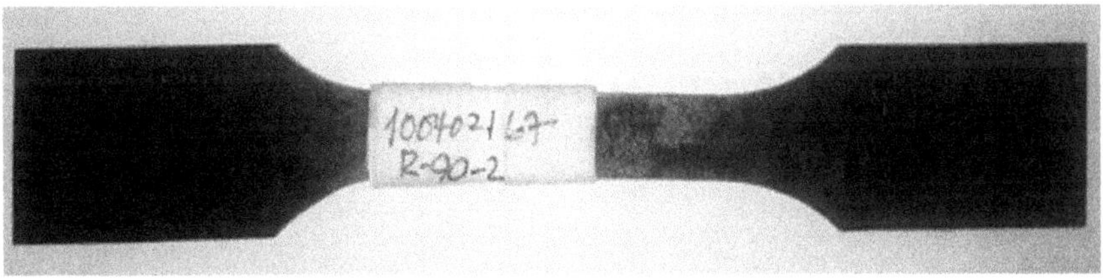

Fig. 3.1 Specimens prepared for tensile tests

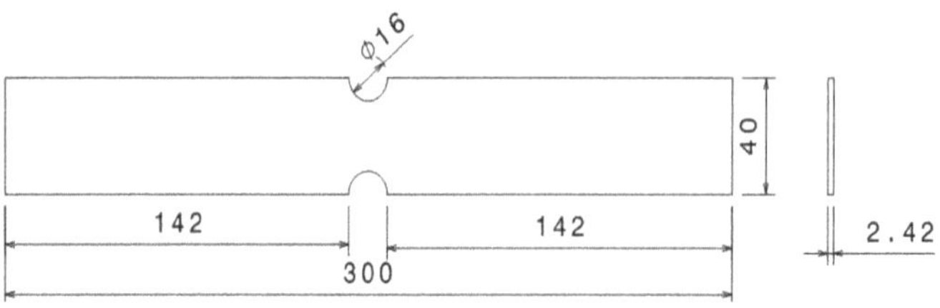

Fig. 3.2 Dimensions for plane–strain tension test

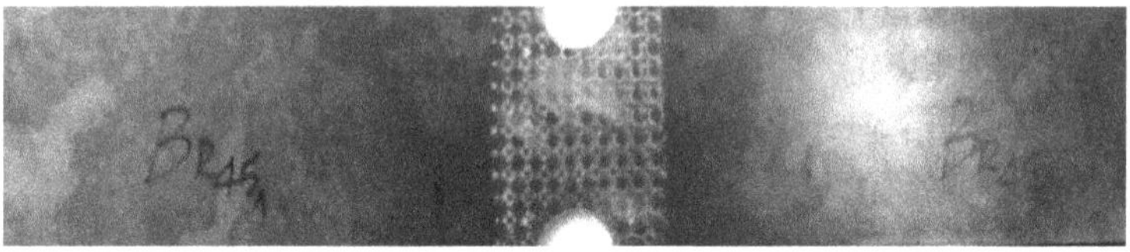

Fig. 3.3 Typical plane–strain tension specimen before test

plot for evaluating the mechanical properties. The plastic strain ratios at different rolling directions were obtained after testing the base material to 10% plastic strain.

Plane–Strain Tension Testing

In this study, plane stretching testing samples were prepared based on the width constraints method. Figures 3.2 and 3.3 show the schematic and original in-plane plane-stretching test specimens with circular grids of 3 mm in diameters on experiment specimen.

3.3 Results and Discussion

3.3.1 Mechanical Properties of Sheet Metal

Figures 3.4 and 3.5 are shown the material EN-10149-2 steel uniaxial tensile test specimen before and after the testing, respectively. Figure 3.5 representing the tested specimen at different rolling direction 0°, 45°, and 90°.

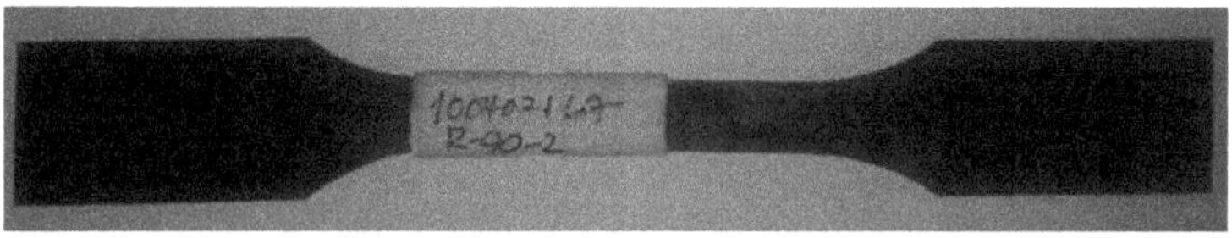

Fig. 3.4 EN-10149-2 specimen before tensile test

Fig. 3.5 EN-10149-2 specimen after tensile test

The mechanical properties of the EN-10149-2 sheet metal are evaluated from the uniaxial tensile tests tabulated in Table 3.2. The evaluated mechanical properties like yield stress, uniform elongation, tensile stress, material strength coefficient (*K*) and strain hardening coefficient (*n*). The base material engineering stress–strain behavior in different rolling directions is shown in Fig. 3.6. Table 3.2 shows the tensile properties of all three reputations and Table 3.3 summarizes the mechanical properties of EN-10149-2 base material. The yield strength and ultimate tensile strength are found to be within 972–1020 MPa and 1018–1054 MPa, respectively. The uniform elongation and strain hardening exponent are about 6.8% and 0.22 respectively.

Table 3.4 indicates summarized mechanical properties of EN-10149-2 (S700mc) sheet steel. The mechanical properties and anisotropy value are shown in Table 3.4.

Table 3.2 Tensile test results of EN-10149-2 sheet metal

Rolling direction	Trial	YS (MPa)	Uniform elongation (%)	Tensile strength (MPa)	Material strength coff. (*K*) (MPa)	Strain hardening exponent (*n*)
0°	1	951.5	5.4	999.3	1512.6	0.142
	2	997.2	7.6	1021.3	1744.5	0.211
	3	983.0	7.6	1036.1	1763.6	0.209
45°	1	927.1	5.2	973.8	1425.7	0.126
	2	968.3	8.2	1023.6	1776.1	0.222
	3	1020.7	6.5	1060.9	1603.1	0.1436
90°	1	1058.8	6.8	1090.5	1628.7	0.1378
	2	1009.2	6.8	1039.62	1796.1	0.215
	3	994.3	6.9	1034.6	1676.1	0.1804

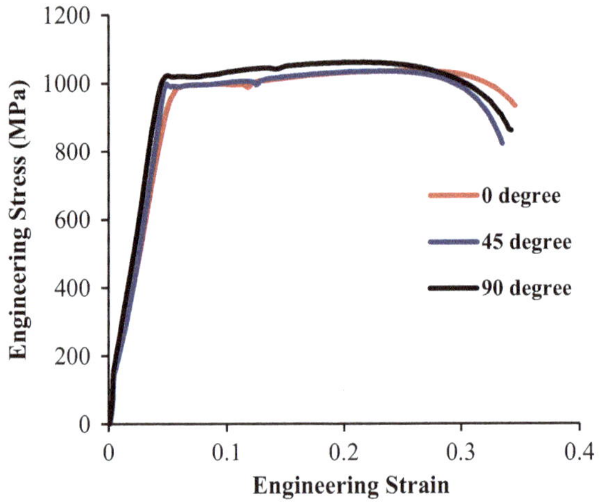

Fig. 3.6 Tensile behavior of base metal in different rolling directions

Table 3.3 Average results of mechanical properties of EN-10149-2 material

Rolling direction	YS (MPa)	Uniform elongation (%)	Tensile strength (MPa)	Material strength coff (*K*) (MPa)	Strain hardening exponent (*n*)
0°	977.3 ± 20	6.87 ± 10	1018.9 ± 20	1673.6 ± 50	0.187 ± 0.014
45°	972.03 ± 30	6.63 ± 3	1018.4 ± 31	1601.6 ± 70	0.164 ± 0.048
90°	1020.7 ± 20	6.8 ± 0.1	1054.9 ± 40	1700.3 ± 70	0.178 ± 0.037

3.3.2 Microstructure of EN-10149-2

Metallographic investigations were carried out on the sample of base metal. Etching with Nital 2% for 10 s; this etchant was used to reveal the banded structure, and the grain boundaries and carbide containing phases. A typical microstructure of EN-10149-2 is shown in Fig. 3.7.

From the microstructure of base metal in Fig. 3.7 indicates the ferrite, pearlite structures along with carbon content. Ferrite and pearlite structure is responsible for the better mechanical properties.

3.3.3 Evaluation of Hardness

Hardness was measured using Vickers hardness (VH) testing machine at five different locations with 10 kg load. The average hardness number of the sample is noted as 124.94 VHN.

Table 3.4 Mechanical property and Plastic Anisotropy (r-value) of EN-10149-2 sheet metal

Rolling direction	YS (MPa)	Uniform elongation (%)	Tensile strength (MPa)	Material strength coff. (*K*) (MPa)	Strain hardening exponent (*n*)	R_0	R_{45}	R_{90}
0°	977.3 ± 20	6.87 ± 10	1018.9 ± 20	1673.6 ± 50	0.187 ± 0.014	0.992	1.287	1.268
45°	972.03 ± 30	6.63 ± 3	1018.4 ± 31	1601.6 ± 70	0.164 ± 0.048			
90°	1020.7 ± 20	6.8 ± 0.1	1054.9 ± 40	1700.3 ± 70	0.178 ± 0.037			

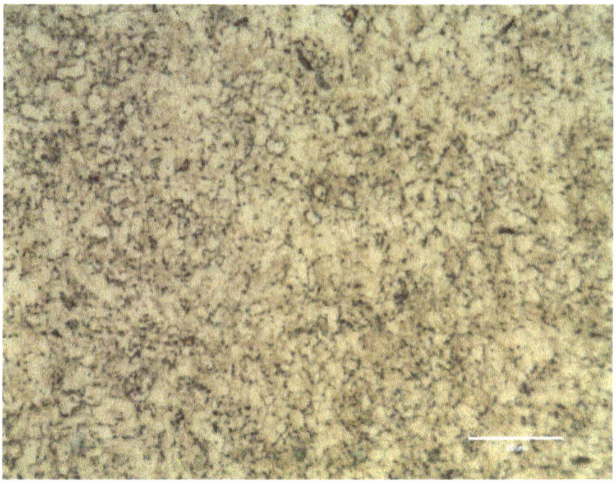

Fig. 3.7 Microstructure of EN-10149-2 at room temperature

3.3.4 In-Plane Plane Stretching (IPPS) Test

The samples were prepared in three rolling directions namely, 0°, 45° and 90° in number two per each rolling direction. The limit strains are evaluated by measuring the length of the major and minor axes of deformed circular grids at the failure region of EN-10149-2 sheet blanks. Figure 3.8 shows circular grids printed on the surface of base materials before deformation and Fig. 3.9 grids elongation after deformation, during IPPS formability tests. The digitized caliper two digits decimal was used to measure the circular grids before and after deformation of base materials for evaluating true major and minor limit strains.

Failure Criteria used for Limit Strains Evaluation

Thickness gradient-based necking criterion (TGNC)

Thickness gradient-based necking criterion is used to obtain forming limit strains of EN-10149-2 sheet blanks [12–15]. Necking can be understood as a thinning phenomenon in any sheet metal during the deformation process. It mostly distinguishes necking as a localized thinning phenomenon, i.e., a localized region within the deforming sheet where a steep gradient in thickness develops. A local critical

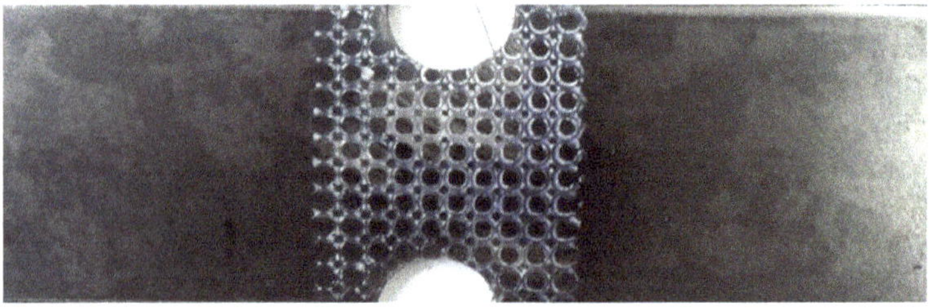

Fig. 3.8 Method of grid measurement before deformation

Fig. 3.9 Method of grid measurement after deformation

thickness gradient (Rc) must exist at the onset of a locally visible neck. The maximum major and minor strain in the thinner element is then reported as the forming limit strain. It is found from the earlier experiments that necking occurs in the sheet metal when the thickness gradient (ratio of the thickness of neighboring elements or circles) falls below 0.92 [12–15].

$$R_{\text{thickness gradient}} = \frac{\text{thickness}_{\text{Necking element}}}{\text{thickness}_{\text{Neighboring element}}} = \frac{t_n}{t_{n-1}} \text{ or } \frac{t_n}{t_{n+1}} \leq R_{\text{Cri}} \Rightarrow \text{material failure or necking}$$

$$R_{\text{Critical}} = 0.92$$

Table 3.5 shows the limit strain results using TGNC from in-plane plane-stretching tested samples in three different rolling directions. Figure 3.10

Table 3.5 Limit strains results using TGNC from in-plane plane-stretching tested

Orientations	Trials	Measured circle diameter (mm)	Measured value- original diameter (mm)	$R_{\text{thickness gradient}}$	Major strain	Minor strain
0°	1	3.124	0.124	0.9193	0.124	0.104
		3.114	0.114			
	2	3.231	0.231	0.9166	0.252	0.114
		3.252	0.252			
45°	1	3.572	0.572	0.91783	0.57	0.541
		3.525	0.525			
	2	3.424	0.424	0.91973	0.464	0.192
		3.461	0.464			
90°	1	3.451	0.451	0.92011	0.451	0.353
		3.416	0.416			
	2	3.294	0.294	0.91588	0.321	0.264
		3.321	0.321			

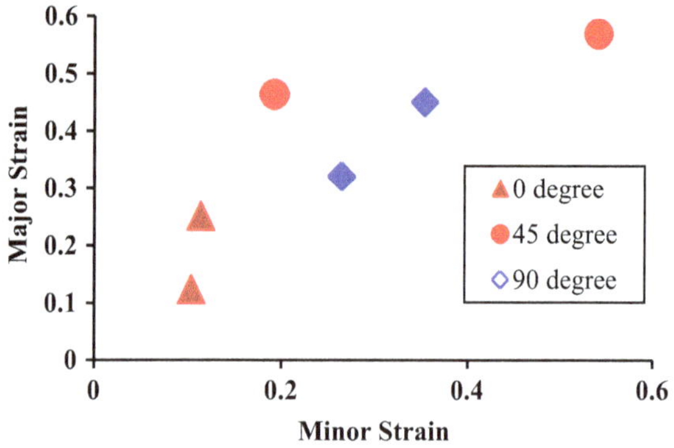

Fig. 3.10 Comparison of limit strain of in-plane plane-stretching test on three rolling directions with two reputations

shows the limit strain comparison of three (0°, 45°, 90°) different rolling direction with two reputations. Limit strain is evaluated using the TGNC after the experiment on three rolling directions. The maximum limit strain is appeared for the rolling direction of 45° and followed by 90° and 0°. This indicates that the maximum formability is in 45° rolling direction during the in-plane plane-stretching test.

3.4 Conclusions

In this work, in-plane plane stretching (IPPS) test is conducted in order to determine limit strains of the EN-10149-02 steel material. From the results, the following conclusions are drawn as follows:

- EN-10149-02 steel material had the highest yield and tensile strength in the transverse direction (90°) and a diagonal direction (45°). The strain handing exponent and total elongation is more at 0° and decreases in the diagonal direction (45°).
- The microstructure result shows that the material constitutes more pearlite matrix grain structure with small carbide boundaries which leads to material more ductile and formable. Vickers's hardness of EN-10149-02 steel material value is 122.24 VHN.
- In-plane plane stretching experimental tests the limit strain evaluated using the TGNC after the experiment on three rolling directions. The maximum limit strain is appeared for the rolling direction of 45° and followed by 90° and 0°. This indicates that the maximum formability is in 45° rolling direction during the in-plane plane-stretching test.

References

1. Handbook, A.M.: Forming and forging. ASM Int. 14 (1993)
2. Panich, S., Barlat, F., Uthaisangsuk, V., Suranuntchai, S., Jirathearanat, S.: Experimental and theoretical formability analysis using strain and stress-based forming limit diagram for advanced high strength steels. Mater. Des. **51**, 756–766 (2013)
3. Garcia, C.I.: High strength low alloyed (HSLA) steels. In: Automotive Steels, pp. 145–167 (2017)
4. Ozturk, F., Lee, D.: Experimental and numerical analysis of out-of-plane formability test. J. Mater. Process. Technol. **170**(1–2), 247–253 (2005)
5. Holmberg, S., Enquist, B., Thilderkvist, P.: Evaluation of sheet metal formability by tensile tests. J. Mater. Process. Technol. **145**(1), 72–83 (2004)
6. Keeler, S., Kimchi, M.: Advanced high-strength steels application guidelines V5. World Auto Steel (2015)
7. Agnew, S.R., Duygulu, Ö.: Plastic anisotropy and the role of non-basal slip in magnesium alloy AZ31B. Int. J. Plast. **21**(6), 1161–1193 (2005)
8. Drewett, L., Bremer, S., Liebeherr, M., De Waele, W., Martin-Meizoso, A., Brozda, J., Gubeljak, N.: HIPERC: a novel, high performance, economic steel concept for line pipe and general structural use. Publications office of the European Union (2010)
9. Janaki Ramulu, P., Satish Babu, A., Ganesh Narayanan, R., Deva Prasad, S., Srinivasa Rao, P.: The behavior of friction stir welded (FSW) sheets of AA6061-T6 during in-plane stretching test. Proced. Eng. **64**, 862–867 (2013)
10. Kumar, M., Kailas, S.V., Ganesh Narayanan, R.: In-plane plane-strain formability investigation of friction stir welded sheets made of dissimilar aluminium alloys. Appl. Mech. Mater. **446–447**, 301–305 (2014)
11. Satheesh Kumar, V., Ganesh Narayanan, R.: In-plane plane strain formability of adhesive-bonded steel sheets: influence of adhesive properties. Int. J. Adv. Manuf. Technol. **76**, 993–1009 (2015)
12. Janaki Ramulu, P., Ganesh Narayanan, R.: Weld zone representation during the formability prediction of friction stir welded blanks with similar thickness sheets. J. Strain Anal. Eng. Des. **46**, 456–477 (2011)
13. Ramulu, P.J., Narayanan, R.G.: Weld zone representation methods during the stretching of friction stir welded blanks with dissimilar sheet thickness: a study using numerical simulations. Materialwiss. Werkstofftech. **43**(3), 241–252 (2012)
14. Ramulu, P.J., Narayanan, R.G.: Experimental evaluation and prediction of forming limit of FSW blanks made of AA 6061 T6 sheets at different weld orientations and weld locations. Materialwiss. Werkstofftech. **44**(6), 527–540 (2013)
15. Ramulu, P.J., Kailas, S.V., Narayanan, R.G.: Formability of friction stir welded sheets made of AA 6061–T6 at different weld orientations and weld locations. Int. J. Mater. Prod. Technol. **50**(2), 147–160 (2015)

Chapter 4
Stretch Formability Prediction of AA6023-T6 Alloy Sheet Under Two Different Heating Conditions

Manaye Mathewos Handiso, Perumalla Janaki Ramulu and G. Somasundaram

Abstract Sheet metal forming is a process widely used in the manufacturing industry. There are several sheet metals forming processes are existing including stretch forming process. In the present work, stretch formability of AA6023-T6 sheet of 2 mm thickness at room temperature and annealed sheet at 400 °C has been performed. For which mechanical properties were evaluated through tensile test for all the conditions. For stretching operations, simulations were performed using limit dome height (LDH) test using PAM STAMP 2G software. During LDH test, the maximum value of LDH for annealed sheets of AA6023-T6 at 400 °C and minimum value for sheets at room temperature is observed. Uniform thickness distribution for annealed blanks at 400 °C is observed compared to room temperature. The room temperature stretch formability of AA6023-T6 alloy sheet has 39.5 maximum dome height and annealed sheet at 400 °C has 43 maximum dome heights. By which, annealed AA6023-T6 sheet has good stretch formability compared to room temperature stretch formability.

Keywords Mechanical properties · Formability · Forming limit diagram · Limit dome height · Thickness distribution

M. M. Handiso · G. Somasundaram
Program of Mechanical Design and Manufacturing Engineering, School of Mechanical, Chemical and Materials Engineering, Adama Science and Technology University, Adama, Ethiopia
e-mail: manayehandiso@gmail.com

G. Somasundaram
e-mail: somasundarram@gmail.com

P. J. Ramulu (✉)
Program of Mechanical Design and Manufacturing Engineering and CoEAME, School of Mechanical, Chemical and Materials Engineering, Adama Science and Technology University, Adama, Ethiopia
e-mail: perumalla.janaki@astu.edu.et

R. G. Narayanan et al. (eds.), *Advances in Computational Methods in Manufacturing*, Lecture Notes on Multidisciplinary Industrial Engineering,
https://doi.org/10.1007/978-981-32-9072-3_4

4.1 Introduction

Sheet metal forming is one of the most used processes in the manufacturing industry. There are many sheet metals forming processes including stretch forming process. In stretch forming process the sheet metal is clamped around its edge and stretched over a die or forming block, which moves upward, downward or sideways, depend up on the machine used. Stretch forming is applicable to produce aircraft-wing skin panel, automotive doors panel and window frames etc. The design of components needs the analysis of instability and determining of possible deformation of sheet metal. The determination of possible forming limits and analysis of instability of sheet metal was predicted from forming limit diagrams (FLDs) [1]. The applied load on sheet metal causes the plastic deformation when it was greater than yield strength of the metal. The stress induced during the process was greater than yield strength, but less than fracture strength of the material [2]. Another important factor which has influence on sheet metal forming is anisotropy. It is the directional variation of mechanical properties. In other words, the reaction of materials to applied stress at different direction has different reaction to the applied stress. The strength and formability of the materials are the main properties which are very important for industrial applications of sheet metals [3]. The forming behavior of a certain sheet metals always depends up on the forming temperature during forming process. For instance, aluminum series of 6xxx alloys had quite less level of formability at room temperature than aluminum series of 5xxx alloys [4, 5]. The amount of strain hardening in the material decided the formability of the materials. The tensile test was one of the mechanical properties' evaluation process in which the strain hardening exponent of the sheet metal evaluated. If the tensile tests of the sheet metal had maximum percentage of elongations, it can be predicted as the sheet metal with a good formability [6]. Forming temperature had significant influence on formability. Many researches were developed on the effect of temperature on the formability of different sheet metals. For example, Nolan et al. [7] studied the formability of a 7000 series alloy under different forming temperature. They performed the tensile tests and deep drawing tests to evaluate the forming behavior. The experiment was taken at both elevated and room temperature. In the drawing process of the metal, the formability of the given sheet metal increased with the increase of forming temperature. This warm forming behavior of the sheet metal indicated that the condition of formability of the alloy improved with that of the forming temperature. For the warm forming condition of 7000 series alloy in general the forming temperature observed in the range of 200–600 °C. The forming condition of stretch flange has a good formability at warm forming condition than cold forming process. Tokita et al. (2017) [8] studied the forming conditions of the stretch formability of two different high strength steels. The forming was preferred on warm forming by spherical stretch forming tests at various forming temperature including room temperature up to 600 °C. Their experiment showed the suitable forming temperature for stretch formability of steels. The observations from the experiments indicated the value of maximum dome height related to the maximum formability indications. There was maximum dome height for test of steel at elevate temperature and minimum dome height at room temperature forming process. High initial forming temperature was one of the affecting factors which causes the loss of quality in the forming process and also the coefficient of friction has a significant influence on the forming behavior of the alloy. Ali and Jumah [2] investigated the forming behavior of annealed aluminum alloy at different forming temperature. The formability of alloy increased with the increasing of the forming temperature. When temperature was at maximum level, the formability was significantly good but the drawback at this forming temperature was the excessive ductility. In this case, the application of alloy limited to its strength. In general, elongation increased, with increase of temperature of AA6061-T6 alloy during tension test. Moreover, at elevated temperature greater or equal to 280 °C, the elongation increases with decreasing strain rate. With spreading of the deformation area, the deformation was more uniform than that at high strain rate; elongation increase gradually [9]. The increase or reduction of strain and plastic anisotropies related to texture intensity and texture distribution should be the main factors for good stretch formability. Refining strain (ductility) of the materials also influence the stretch formability [10, 11]. Forming limit diagram (FLD) was used to predict the materials formability behavior under different forming condition. It played a great role in industries, because it was also the most applicable method to evaluate the formability of sheet metals [12–15]. The value for these parameters calculated with respect to the specimen just before fracture. Forming limit diagrams (FLDs) are a suitable and often used mechanism for classification of the formability and investigation of sheet metal forming process. Forming limit diagram (FLD) also used to indicate the forming limit of a given sheet metals at different state of deformation. For different materials the upper limit curve (ULC) of the forming limit diagram provides breaking, and they calculated with respect to specimens just before breaking. The increase in local necking causes the increase in deformation of a certain specimen at stress concentration area. While the local necking increased further there was no change in stress to the rest of part of the specimen which was not subjected to the necking [16]. The effects of punch and die temperature on the formability of a cup shaped AA3003 aluminum alloy sheet. The formability improves as the temperature difference between the punch and the die increases. Also, the forming speed does not affect localized necking for the given punch velocities in the study. The simulations were captured for both punch force and failure location. The important parameters to control the thickness reduction in the blank were identified as friction and blank holder force. The finite element analysis model was validated with the selected experimental cases. Stretch formability of sheet metals depends on temperature of sheets. Increase in sheet metals temperature increase the stretch formability, but ductility of sheet metals was observed at elevated sheet temperature from the existing literature. To overcome the excessive ductility of sheet metal, annealing in between elevated temperature and room temperature was needed [17]. Based on the above observations, the present study objective is to predict the stretch formability of AA6023-T6 sheet of 2 mm thickness at room temperature and annealed at 400 °C.

4.2 Materials and Methodologies

4.2.1 Base Material

The Material selected for this study was AA6023-T6 alloy sheet which has 2 mm thickness. AA6023-T6 alloy sheet. Table 4.1 shows the nominal chemical composition of this material.

4.2.2 Tensile Test

The mechanical properties of base metal were measured using tensile tests as per ASTM- B557 M. The base metal was cut in different rolling directions like 0°, 45° and 90° and the mechanical properties and plastic strain ratio (R) were evaluated through tensile tests as per ASTM- E517. Tensile tests were performed at a nominal cross-head speed of 1 mm/min, at room temperature and annealed at 400 °C specimens on an ISO 68982-1 universal testing machine tensile testing machine and repeated thrice for each set to check the reproducibility. Failure from grip or slippage during testing was not observed. Figure 4.1 shows base metal tensile specimen. The mechanical properties of base material like yield strength, ultimate tensile strength, total elongation, uniform elongation, strain hardening coefficient (*n*) and strength coefficient (*K*) were evaluated as per the standard procedure after testing the samples till failure. The load-stroke behavior obtained during testing was converted into engineering stress–strain and true stress–strain plot for evaluating the mechanical properties. The plastic strain ratios at different rolling directions were obtained after testing the base material to 10% plastic strain.

Table 4.1 Chemical composition of AA6023-T6 alloy sheet (in wt%)

Chemical composition								
Si	Fe	Cu	Mn	Mg	Bi	Sn	Residuals	Al
0.57	0.5	0.19	0.24	0.41	0.3	1.2	0.59	96

Fig. 4.1 Typical specimen for tensile tests

4.2.3 Simulation Details

Finite element method was used for numerical analysis of sheet metal forming operations. The numerical analysis of stretch forming process was done by PAM STAMP 2G software. The input model consists of die, blank, blank holder, draw and punch were constructed in pre-processor. The punch, die and blank holder had considered as rigid body. The fine meshing was done on the surface of die, blank holder and punch, since, the blank has been considered as a deformable body. In limit dome height (LDH) test, the blank with 2 mm thickness is deformed by a hemispherical punch (diameter: 101.6 mm) inside opening of a die with 105.9 mm diameter. For simulations process, five tests with different size of blanks (100 mm × 200 mm, 125 mm × 200 mm, 150 mm × 200 mm, 175 mm × 200 mm, and 200 mm × 200 mm) were used for each (0°, 45° and 90°) orientation of samples. The tools required for LDH test, punch, die, blank holder, draw bead, and blank generated in solid works, a CAD packages, and meshed using delta mesh facility in PAM STAMP 2G. Both Hollomon's strain hardening law and Hill's 1948 isotropic hardening yield criterion was used as the plasticity model. The base material comprised quadrilateral shell elements of Belytschko–Tsay formulation with five through-thickness integration points. A uniform mesh size of 1 mm was used for all the simulations. The friction coefficient m was assumed to be 0.12 and is kept constant throughout the study. The blank was as the active body and the punch force used was 80 kN to avoid the drawing and tearing near draw beads during forming prediction. For prediction, the forming limit strain thickness gradient necking criterion (TGNC) was used [18–21].

4.3 Results and Discussion

4.3.1 Mechanical Properties of Base Material

The base material true stress–strain behavior at different rolling directions of two different heating conditions are shown in Figs. 4.2 and 4.3. Tables 4.2 and 4.3 summarize the mechanical properties of AA 6061 base material obtained in the present work. The yield strength and ultimate tensile strength are found to be within 180–200 MPa and 193–206 Mpa, respectively.

Heat treatment of the aluminum alloys causes change in mechanical and anisotropy values. Table 4.3 indicates the value of mechanical properties and plastic anisotropy of AA6023-T6 alloy sheet after annealing at 400 °C with one and half hours soaking time.

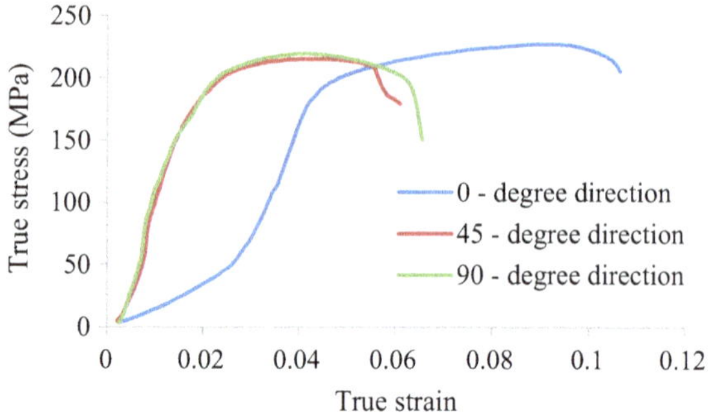

Fig. 4.2 True stress—true strain curve for samples at room temperature

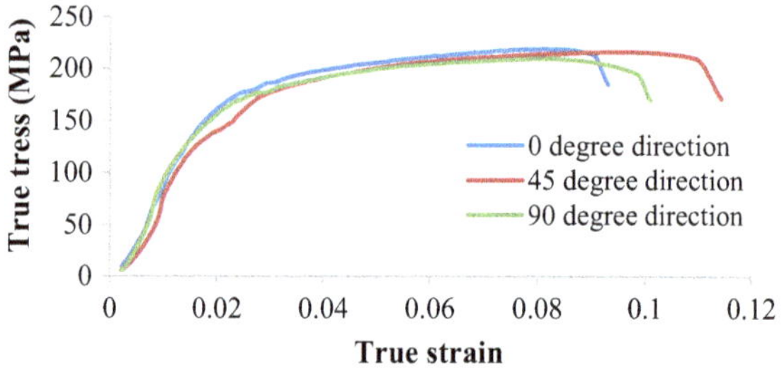

Fig. 4.3 True stress—true strain curve for annealed samples at 400 °C

Table 4.2 Mechanical property of AA6023-T6 alloy sheet at room temperature

Orientation of samples	UE (mm)	UTS (MPa)	σ_{ys} (MPa)	K (MPa)	n	R_0	R_{45}	R_{90}
0°	0.38	200 ± 9	180 ± 3	313 ± 40	0.132	0.49	0.68	0.56
45°	0.54	208 ± 4	190 ± 18	338 ± 20	0.134			
90°	0.52	212 ± 12	200 ± 12	358 ± 20	0.133			

Table 4.3 Mechanical property of AA6023-T6 alloy sheet after heat treatment

Orientation of samples	UE (mm)	UTS (MPa)	σ_{ys} (MPa)	K (MPa)	n	R_0	R_{45}	R_{90}
0°	0.19	206 ± 13	170 ± 17	326 ± 33	0.149	0.43	0.58	0.47
45°	0.21	197 ± 2	168 ± 5	323 ± 16	0.162			
90°	0.25	193 ± 6	166 ± 7	315 ± 28	0.155			

4.3.2 Evaluation of Forming Limit Diagram

The forming limit diagram indicates the stretch formability characteristics of the sheet metal. The forming limit diagram is also the forming limit curve, which is one of necessary value during practical assessment of sheet stretch forming process. Figures 4.4 and 4.5 show the comparison between the form limit curves of AA6023-T6 alloy sheets at room temperature and annealed alloy sheets and Table 4.4 indicate major and minor strain obtained by thickness gradient-based necking criterion at LDH.

Figure 4.4 shows the form limit curves of a sheet that predicts the stretch formability of AA6023-T6 alloy sheet at room temperature. The distribution of the curves of blanks with different dimension to the upper limit curve (ULC) is scattered. There are some blanks lie above the upper limit curve ULC which indicates fracture before the necking. The stretch formability of these blanks which have fracture earlier than the forming limit is not acceptable. The earlier formation of fracture caused due to the cold forming condition. The maximum limit dome height LDH observed in the stretch formability of blanks under room temperature forming

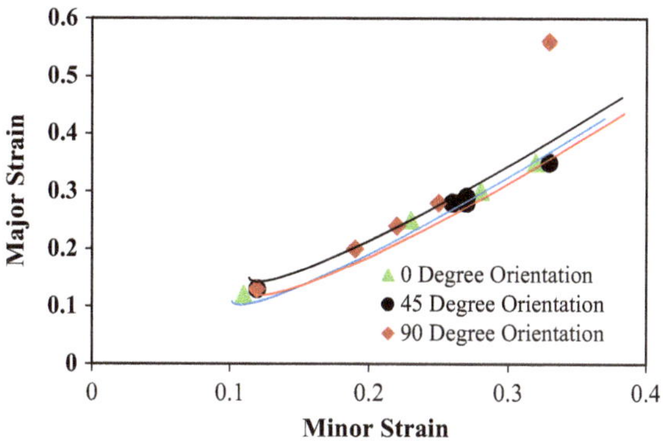

Fig. 4.4 Form limit curve of AA6023-T6 alloy sheet at room temperature forming condition

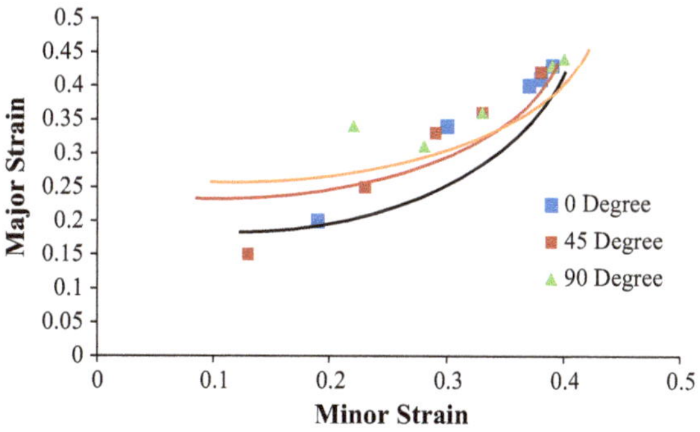

Fig. 4.5 Form limit curve of AA6023-T6 alloy sheet after annealing condition

Table 4.4 Major and minor strain at LDH for room temperature blanks

	0° oriented					45° oriented					90° oriented				
Blanks	1	2	3	4	5	1	2	3	4	5	1	2	3	4	5
Major strain	0.12	0.25	0.30	0.30	0.35	0.13	0.29	0.35	0.28	0.28	0.13	0.24	0.56	0.20	0.28
Minor strain	0.11	0.23	0.28	0.28	0.32	0.12	0.27	0.33	0.27	0.26	0.12	0.22	0.33	0.19	0.25
LDH	26	37.2	39	39	39	29.5	38.5	39	39.5	39.5	32	39	39.5	39.5	39

condition for blanks with dimension of 150 mm × 200 mm, 175 mm × 200 mm and 200 mm × 200 mm is 39.5 mm. From the value of maximum limit dome height LDH, these blanks have good formability compared to the others blanks with minimum value of limit dome height. Table 4.5 indicates major and minor strain obtained by thickness gradient-based necking criterion at LDH for annealed blanks.

Figure 4.5 shows the form limit curves of sheet that predicts the stretch formability of annealed AA6023-T6 alloy sheet. The distribution of the curves of blanks with different dimension to the upper limit curve is scattered. The distribution of the curves of the blanks is under limit of the upper forming limit diagram. This indicates there is no earlier fracture occurred before necking. The stretch formability of these blanks which are scattered blow the upper forming limit curve and near to the upper limit curve is recommended blanks for attaining a good formability. The maximum limit dome height LDH observed in the stretch formability of blank which are formed after annealing condition with dimension of 150 mm × 200 mm, 175 mm × 200 mm and 200 mm × 200 mm is 43 mm. From the value of maximum limit dome height LDH, these blanks have good formability compared to the others blanks with minimum value of limit dome height. The comparison between the maximum limit dome heights of the blanks at room temperature annealed alloy sheet shows the stretch formability of AA6023-T6 alloy sheets. The average maximum limit dome height LDH for room temperature forming condition of the blanks is 39.5 mm and the average maximum limit dome height LDH is 43 mm. Therefore, annealed AA6023-T6 alloy sheet has a good formability at annealing temperature of 400 °C and 90 min soaking time than sheet at room temperature.

Table 4.5 Major and minor strain at LDH for annealed blanks

	0° oriented					45° oriented					90° oriented				
Blanks	1	2	3	4	5	1	2	3	4	5	1	2	3	4	5
Major strain	0.20	0.34	0.40	0.43	0.41	0.15	0.25	0.33	0.36	0.42	0.34	0.31	0.36	0.43	0.44
Minor strain	0.19	0.30	0.37	0.39	0.38	0.13	0.23	0.29	0.33	0.38	0.22	0.28	0.33	0.39	0.40
LDH	31	40.5	41.5	42	41.5	32	42.5	43	43	43	30.5	41.5	43	43	43

4.4 Conclusions

From the present work, the following conclusions are drawn.

- LDH from simulation justified the stretch formability behavior of AA6023-T6 alloy sheet. A blank at room temperature with dimensions of 150 mm × 200 mm, 175 mm × 200 mm and 200 mm × 200 mm have 39.5 mm maximum dome height and blank with dimension of 100 mm × 200 mm and 125 mm 200 mm at room temperature have 26 mm minimum dome height.
- Annealed blanks at 400 °C with 150 mm × 200 mm, 175 mm × 200 mm and 200 × 200 dimensions have maximum dome height compared to 100 mm 200 mm and 125 mm × 200 mm blanks at room temperature. The maximum dome height observed in simulation is 43 mm and minimum 31 mm. By which, the blanks with dimension of 150 mm × 200 mm, 175 mm × 200 mm and 200 mm × 200 mm have good stretch formability compared to blanks with dimension of 100 mm × 200 mm and 125 mm × 200 mm at room temperature.
- The comparison between stretch formability of AA6023-T6 alloy sheet at room temperature and annealed sheet at 400 °C shows, AA6023-T6 alloy sheet at room temperature has 39.5 mm maximum dome height and annealed sheet at 400 °C has 43 mm maximum LDH, therefore, annealed sheet has good stretch formability.

References

1. Stanislav, N.A.S., Emil, E.C.S.: Forming limit diagrams as an important indicator of progressive high-strength steel sheets forming. Transfer inovácií 17–20 (2013)
2. Ali, W.J., Th Jumah, O.: Warm forming of aluminum alloy 2024 at different temperatures. Al-Rafadain Eng. J. **20**(2) (2012)
3. Cardoso, R.P.R., Adetoro, O.B.: A generalisation of the Hill's quadratic yield function for planar plastic anisotropy to consider loading direction. Int. J. Mech. Sci. **128**, 253–268 (2017)
4. Mohamed, M.S., Ismail, A.: Review on sheet metal forming process of aluminium alloys. In: 17th International Conference on Applied Mechanics and Mechanical Engineering (2016)
5. Klos, A., Kahrimanidis, A., Wortberg, D., Merklein, M.: Experimental and numerical studies on the forming behavior of high strain Al–Mg–Si (–Cu) sheet alloys. Proced. Eng. **183**, 95–100 (2017)
6. Abovyan, T., Kridli, G.T., Friedman, P.A., Ayoub, G.: Formability prediction of aluminum sheet alloys under isothermal forming conditions. J. Manuf. Process. **20**, 406–413 (2015)
7. Nolan, R.A., Kulas, M.A., Prangnell, P.B., Quinta da Fonseca, J.: The effect of temperature on the formability of a high strength aluminium automotive alloy. University of Manchester, Manchester (2014)
8. Tokita, Y., Nakagaito, T., Tamai, Y., Urabe, T.: Stretch formability of high strength steel sheets in warm forming. J. Mater. Process. Technol. **246**, 77–84 (2017)
9. Chen, Z., Fang, G., Zhao, J.Q.: Formability evaluation of aluminum alloy 6061-T6 sheet at room and elevated temperatures. J. Mater. Eng. Perform. **26**(9), 4626–4637 (2017)

10. Suzuki, K., Chino, Y., Huang, X., Yuasa, M., Mabuchi, M.: Enhanced room-temperature stretch formability of Mg–0.2 mass% Ce alloy sheets processed by combination of high-temperature pre-annealing and warm rolling. Mater. Trans. **56**(7), 1096–1101 (2015)
11. An, Y.G.: Strain path change effects on stretch formability of 6082 aluminium alloy. Mater. Sci. Technol. **17**(3), 258–263 (2001)
12. Chiba, R., Takeuchi, H., Nakamura, R.: Forming-limit prediction of perforated aluminium sheets with square holes. J. Strain Anal. Eng. Des. **50**(6), 391–404 (2015)
13. Lopez, A.M., van den Boogaard, A.H.: Formability limit curves under stretch-bending. Res. Progr. Mater. Innov. Inst. M2i 1–2 (2011)
14. Shao, Z., Bai, Q., Li, N., Lin, J., Shi, Z., Stanton, M., Dean, T.: Experimental investigation of forming limit curves and deformation features in warm forming of an aluminium alloy. Proc. Inst. Mech. Eng. Part B. J. Eng. Manuf. **232**(3), 465–474 (2018)
15. Butuc, M.C., Teodosiu, C., Barlat, F., Gracio, J.J.: Analysis of sheet metal formability through isotropic and kinematic hardening models. Eur. J. Mech. A/Solids **30**(4), 532–546 (2011)
16. Holmberg, S., Enquist, B., Thilderkvist, P.: Evaluation of sheet metal formability by tensile tests. J. Mater. Process. Technol. **145**(1), 72–83 (2004)
17. Bagheriasl, R.; Formability of aluminum alloy sheet at elevated temperature (2012)
18. Janaki Ramulu, P., Ganesh Narayanan, R.: Weld zone representation during the formability prediction of friction stir welded blanks with similar thickness sheets. J. Strain Anal. Eng. Des. **46**(6), 456–477 (2011)
19. Ramulu, P.J., Narayanan, R.G.: Weld zone representation methods during the stretching of friction stir welded blanks with dissimilar sheet thickness: A study using numerical simulations. Materialwiss. Werkstofftech. **43**(3), 241–252 (2012)
20. Ramulu, P.J., Narayanan, R.G.: Experimental evaluation and prediction of forming limit of FSW blanks made of AA 6061 T6 sheets at different weld orientations and weld locations. Materialwiss. Werkstofftech. **44**(6), 527–540 (2013)
21. Ramulu, P.J., Kailas, S.V., Narayanan, R.G.: Formability of friction stir welded sheets made of AA 6061–T6 at different weld orientations and weld locations. Int. J. Mater. Prod. Technol. **50**(2), 147–160 (2015)

Chapter 5
FE Analysis of Single Point Incremental Forming (SPIF) for Tailor Welded Blanks (TWBs)

Shalin Marathe and Harit Raval

Abstract Fuel economy along with cost reduction is the ultimate target of the automobile industries and in order to achieve that, reduction in weight of the automobile body parts should be done. The homogeneous blanks are replaced by Tailor Welded Blanks (TWBs) while manufacturing the components of an automobile vehicle in order to reduce the weight and cost of the components. TWB is the outcome of joining two or more than two materials having different properties. Despite of having extremely important advantages, technology of TWB is associated with the limitation of weld line shift and formability reduction during the forming process. During the conventional forming process, formability of the TWBs is found to be decreasing and weld line shift is observed to be on higher side. In order to improve the formability of homogeneous blanks, Single point incremental forming process (SPIF) is found to be the excellent solution. However, the combination of SPIF and TWBs technology is not much explored by the researchers. In this present work, simulation of SPIF process for TWBs has been attempted. Effect of different tool initial position on the responses like weld line shift, Plastic Equivalent strain (PEEQ) is investigated. It has been found that the tool dragging effect and deformation of weak material is responsible for the weld line shift during the SPIF process. Particularly for SPIF process, nature and magnitude of weld line shift is affected by the tool initial position.

Keywords Tailor welded blanks (TWBs) · Weld line shift · Formability · Finite element analysis

S. Marathe (✉) · H. Raval
Department of Mechanical Engineering, Sardar Vallabhbhai National Institute of Technology, Surat 395007, India
e-mail: shalin.marathe89@gmail.com

R. G. Narayanan et al. (eds.), *Advances in Computational Methods in Manufacturing*, Lecture Notes on Multidisciplinary Industrial Engineering,
https://doi.org/10.1007/978-981-32-9072-3_5

5.1 Introduction

The automobile industries are aiming at vehicles with more fuel economy along with reduction in the cost of manufacturing the automobile body parts. Reduction in weight without compromising the strength of the component is possible by adopting technology like TWBs. TWB is a product of joining of two or more than two material having different properties which are subjected to forming operations. The joining of material with different densities will lead to the reduction in the overall weight of the component. A schematic representation of TWB used in inner door panel of a car body is shown in Fig. 5.1.

Automobile, aerospace, transportation, and shipbuilding industries are aiming at weight reduction of the components without compromising the strength and TWBs have been found to fulfill these requirements. TWB is a product of joining two or more than two materials prior to any forming operation. TWB involves joining of two or more than two materials having different properties, thicknesses and/or coatings by any of the conventional joining processes. As it involves joining of materials having different densities the overall weight of the product decreases [1]. Also, TWBs are subjected to forming after the joining has been done and because of that the number of dies required for the forming decreases [1] which also leads to decrease the product lead time as well. Figure 5.1 indicates the typical example of TWB.

In conventional technologies, forming of two different materials was done separately and then joining was performed. In the case of TWBs, joining is followed by forming of welded blank using a single die setup and that leads to the manufacturing cost reduction [1, 2]. The joining method adopted for development of TWB depends upon the parent material involved in it. As the application of aluminum instead of steel leads to the 20% reduction of weight [3], use of aluminum is increasing in the automobile applications. For welding of aluminum, friction stir welding process is found to be producing sound and defect-free welds [4].

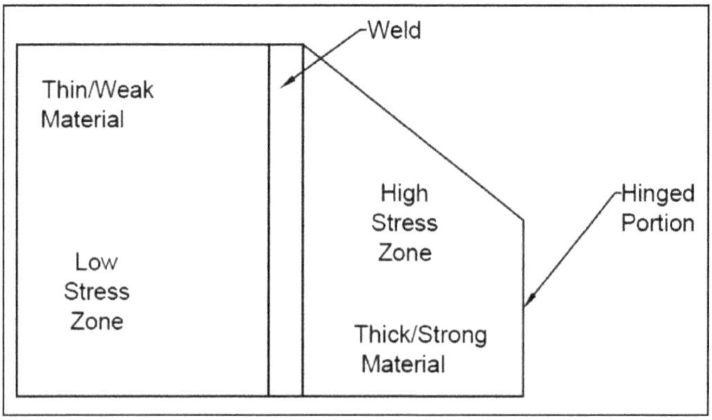

Fig. 5.1 Schematic representation of TWB used in inner door panel of a car body

Despite of having such tremendous advantages, the application of TWBs is limited as they are associated with the limitation of formability reduction [5, 6] and weld line shift [7, 8] during the forming process. In order to improve the forming behavior of TWBs during forming process, many efforts have been made. Such effort was made by Suresh et al. [8] in which the selective heating method was used to heat the strong/thick material to reduce the weld line shift during the stamping process. During the conventional forming process like deep drawing, the weld line shift is observed towards thick/strong material side [5] and formability of TWBs is found to be decreasing [5] than the same of parent material participating in it. However, for homogeneous blank, SPIF results in the better formability compared to the conventional forming processes [9]. So, the combination of SPIF and TWBs technology should be explored which is done by very few researchers.

So, in this present simulation study, FE analysis of SPIF process is carried out for the TWBs developed from the friction stir welding process. The effect of different initial position of tool on the responses like weld line shift, Plastic Equivalent Strain (PEEQ) and formed geometry is investigated.

5.2 Simulation Strategy

5.2.1 Challenges in Simulation of TWBs

For the simulation of TWBs, the major challenge is the consideration of the weld metal properties. During simulation, if the properties of the weld metal are not considered then results of simulation and experimental work are found to be deviating from each other [6, 7, 10]. In order to consider the weld metal properties, testing of welded joint is required to be performed and from that, different properties are extracted and used during the simulation. Rule of Mixture (ROM) can also be adopted to determine the weld metal properties [11]. During simulation of TWBs, if the appropriate properties of the weld metal are considered than actual behavior of the TWBs during that particular process can be predicted [12–14]. In order to consider the weld metal properties during simulation, an approach called "Weld Zone" is used [15]. In this approach, a different zone for weld metal is modeled and separate properties are assigned to it (see Fig. 5.2). In present simulation study, weld zone approach is used during the simulation in order to define the weld zone properties.

FSW results into the different zones in the welded blanks like WN (Weld Nugget), HAZ (Heat Affected Zone), TMAZ (Thermo Mechanically Affected Zone) and Base Metal. In the present simulation WN, HAZ and TMAZ are combined and considered as Weld Metal/Weld Nugget/Weld Zone. The properties of all

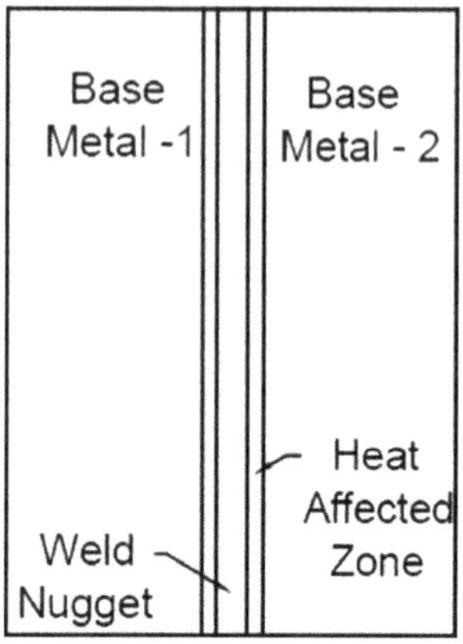

Fig. 5.2 Schematic representation of weld and heat affected zones

the mentioned welding zones can be considered during simulations but for that mechanical testing is required. However, consideration of such properties will lead to more accurate results.

5.2.2 Simulation Methodology

In this work, ABAQUS/Explicit is used for the simulation purpose. In the modeling module, two instances are modeled namely blank and tool. In this present work, it is assumed that two blanks are welded using FSW process in order to develop a TWB. TWB is having dimensions of 70 × 70 mm^2 and 2 mm of thickness. The width of weld zone is assumed to be of 5 mm. Particularly, for FSW, the range of welding zones depends upon the diameter of the FSW tool shoulder. However, this present study is focused on the forming parameters and its effect of the weld line movement. In order to weld the plate of 2 mm thickness, FSW tool having 2 mm pin diameter is required. Considering that, shoulder diameter will be in the range of 4–6 mm [16]. So, the HAZ produced after welding will be in the same range. So, in present simulation study, 5 mm width of weld zone/material is considered.

The tool is considered as analytical rigid body and the blank is considered as shell type deformable body (see Fig. 5.3). The blank material is considered as isotropic in nature for the ease of simulation. However anisotropic properties can be considered and accuracy of results can be further improved.

The material properties used during simulations are represented in Table 5.1. The properties of the weld metal were assumed. The TWB is considered to be fabricated from AA 5754 H22 and AA 5052 H32 using the FSW process. Now onwards AA 5754 H22 and AA 5052 H32 is considered as weak base metal/weak material and strong base metal/strong material respectively.

The material having high value of yield strength is considered as strong material and material having low value of yield strength is considered as weak material.

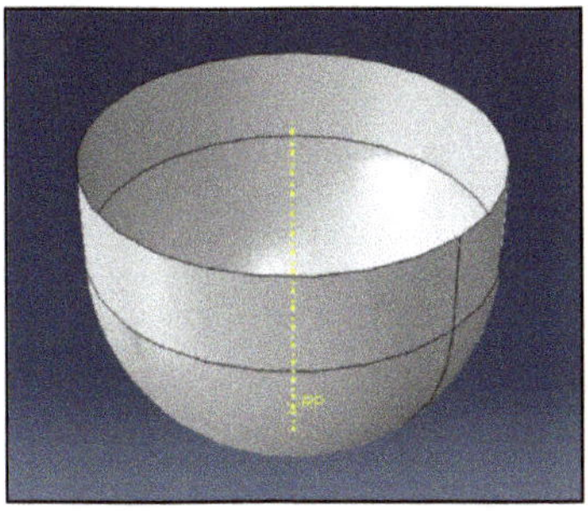

(a) Hemispherical tool (rigid body, 6 mm radius)

(b) TWB – shell deformable body ($70*70\ mm^2$,2 mm thick)

Fig. 5.3 Instances modeled in ABAQUS/explicit

Table 5.1 Properties used during simulation [5]

Property	AA 5754 H22	AA 5052 H32	Weld
Density[a] (ρ) (kg/m^3)	2670	2680	2670
Young's modulus[a] G (N/m^2)	70.3×10^9	70.3×10^9	70.3×10^9
Yield stress (σ_y) (N/m^2)	122×10^6	140×10^6	131×10^6
Poisson's ratio[a] (μ)	0.33	0.33	0.33
Strength coefficient (K) (N/m^2)	252.04×10^6	258.7×10^6	255.37×10^6
Strain Hardening index (n)	0.112	0.125	0.1185

[a]Indicates assumed properties

During simulation, material of TWB is assumed to follow the Power-Law equation as indicated below.

$$\sigma = K\varepsilon^n \tag{1}$$

In Eq. 1, σ and ε indicates true stress and true strain respectively. After the modeling, assembly of the SPIF tool and the blank is done. After assembly, interaction between SPIF tool and blank is provided in terms of friction. A surface to surface contact is defined between tool and blank and friction coefficient of 0.1 [17] is defined in order to consider the effect of friction (see Fig. 5.4).

The flat TWB is deformed using SPIF process and converted on to the target geometry frustum of pyramid. The dimensions of the pyramid are presented in Table 5.2.

During the simulation, the edges of the blank were fixed by defining encastre boundary (see Fig. 5.5) condition. The encastre boundary condition ensures pure stretching of the blank during forming operation and it eliminates the chance of drawing of material.

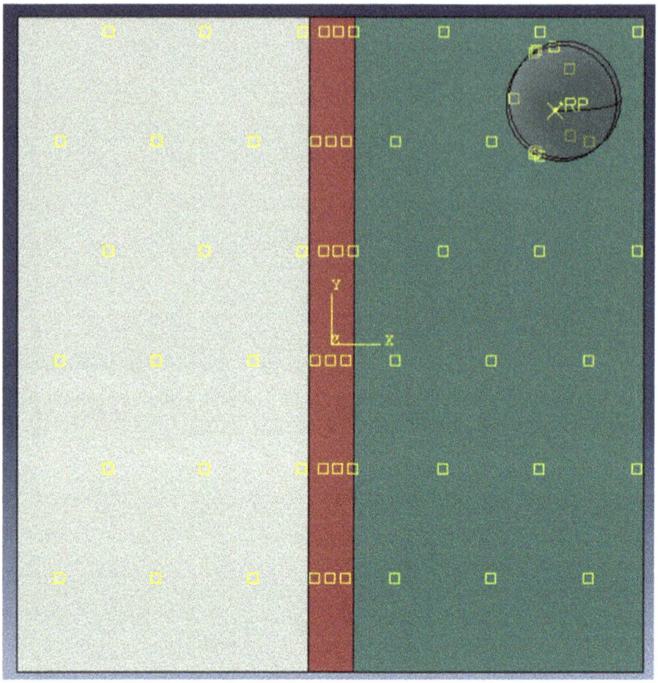

Fig. 5.4 Assembly of tool and blank and interaction between tool and blank

Table 5.2 Dimensions of the target geometry/frustum of pyramid

S. No.	Particular	Unit	Magnitude
1	Wall angle	Degree	45
2	Height of the pyramid	mm	20
3	Top dimensions	mm^2	50 × 50
4	Bottom/base dimensions	mm^2	10 × 10

The tool was given boundary condition in terms of movement in order to produce the target geometry (see Fig. 5.5). Tool starts the movement and goes through the path 1–2–3–4 and then it deforms the blank in the direction of thickness (perpendicular to plane of this paper) by 0.5 mm in step 5. The magnitude of movement of tool in step 5 depends upon the wall angle of target geometry. This cycle is repeated till the desired geometry is produced. Total 205 different boundary conditions were defined along with total 43 different amplitudes for the tool in order to produce the target geometry. The values of SPIF parameters adopted during the simulation are represented in Table 5.3.

After successful implementation of boundary conditions, meshing of blank is done with the help of S4R shell elements by keeping mesh size of 0.5 mm. For the present simulation study, mesh sensitivity analysis was carried out and based on

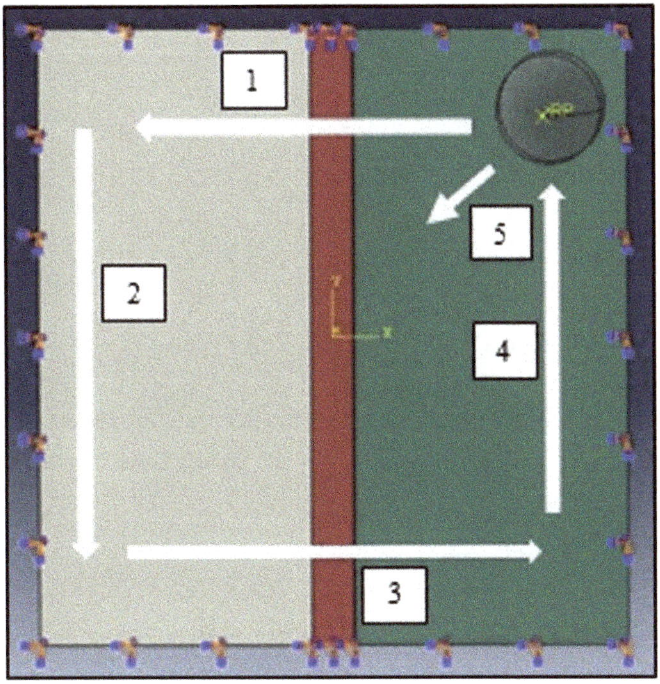

Fig. 5.5 Encastre boundary condition to the edges of the blank and direction of tool movement

Table 5.3 Constant SPIF process parameters for present work

S. No.	Particular	Unit	Magnitude
1	Tool feed	mm/sec	10
2	Tool rotation	RPM	0
3	Tool path	–	Inward tool path
4	Tool radius compensation	mm	Not considered
5	Incremental depth	mm	0.5

computational time for one complete simulation and considering the available computational resources 0.5 of mesh size was selected. The meshing process is followed by the submission of job for the analysis.

The main aim of the present simulation study is to check the effect of different tool starting position on weld line shift during forming. In order to study this, two starting position of tools are considered as indicated in Fig. 5.6.

If the tool is starting its movement from strong material it will be considered as SFS condition (see Fig. 5.6a) and if it is starting from weak material it will be

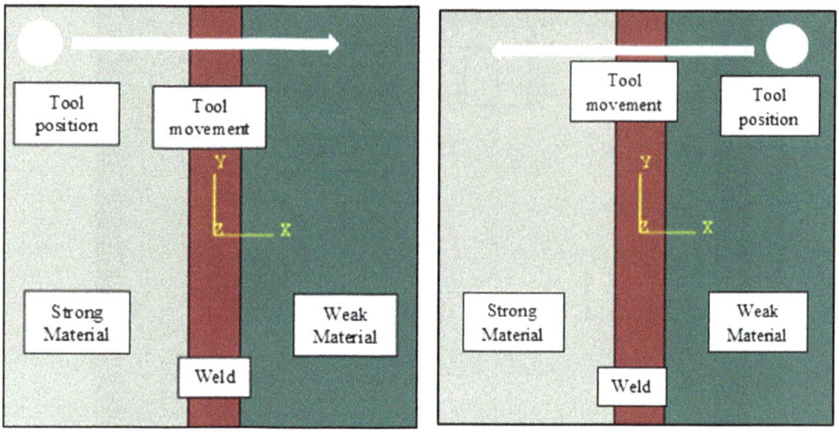

(a) Starting From Strong (SFS) material

(b) Starting From Weak material (SFW)

Fig. 5.6 Two different starting position of SPIF tool

considered as SFW condition (see Fig. 5.6b). In the next sections, the effect of tool starting position on different responses is discussed in detail.

5.3 Results and Discussion

After the successful completion of simulations, results like weld line shift, PEEQ and formed geometry are studied and discussed below.

5.3.1 Formed Component

The formed component is analyzed for the weld line shift. Figure 5.7a, b indicates the top view and side view of the component formed using the SPIF process respectively. It can be observed from Fig. 5.7 that the weld line is shifted towards strong material by marginal amount. In order to investigate the weld line shift in detail, a node path was generated on the weld line and displacement of each node was recorded.

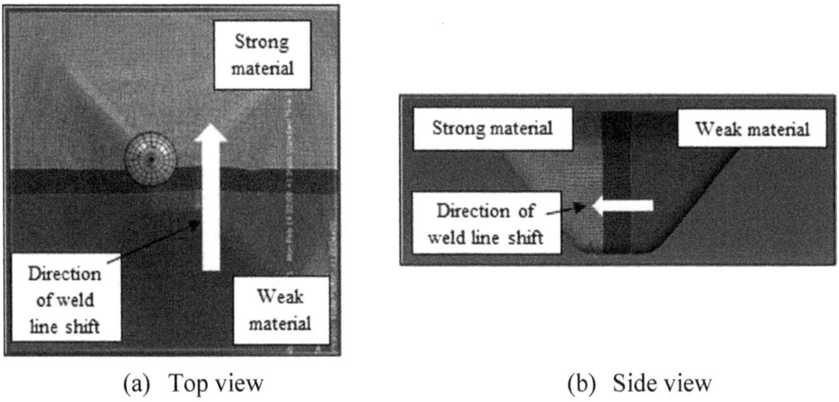

Fig. 5.7 Formed component—frustum of pyramid

5.3.2 Weld Line Shift

During any forming process, same amount of deformation is applied by tool on strong, weak and weld material. As the deformation resistance capacity of the weak material is less compared to strong material, more deformation is experienced by the weak material compared to strong material. Due to this, weld line shifts towards strong material during any forming operation.

As the SPIF process involves local deformation of material due to tool movement, the weld line shift depends upon the initial position of tool. In order to understand the mechanism of weld line shift particularly for SPIF process, consider Fig. 5.8.

Figure 5.8a represents the overall view of the blank and position of weld line shift before and after forming process for SFS case. Figure 5.8b represent the magnified view with different sections. Section B can be called as untouched section, as there is no interaction between tool and blank in it. Section A and C represent the weld line shift in the wall of the pyramid. It should be observed that the weld line shift in section A is less compared to the same in section C. In SFS case, the strong material is above the mean line and weak material below the mean line. When the tool comes from section B to section C, deformation of weak material is supported by the tool dragging effect. Due to which more weld line shift is observed in section C. Now, in section A, deformation of weak material is opposed by the tool dragging and because of that the amount of weld line shift observed in section A is less. The weld line shift observed in section B is only due to section A and C.

Figure 5.9 represents the magnified view of the weld line shift observed for the case SFW. The sections represented in the figure are same as Fig. 5.8. The nature and magnitude of weld line shift of SFW case are different from the same of SFS case. It clearly indicates that the magnitude and nature of weld line shift depends

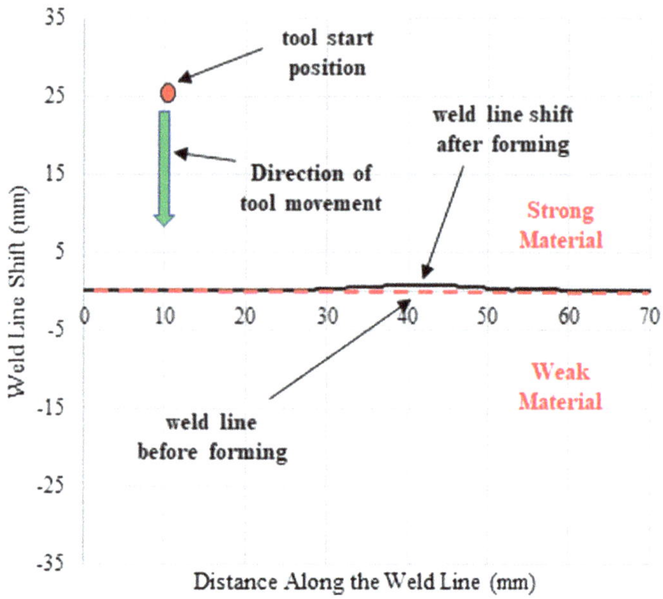

(a) Weld line shift overall view

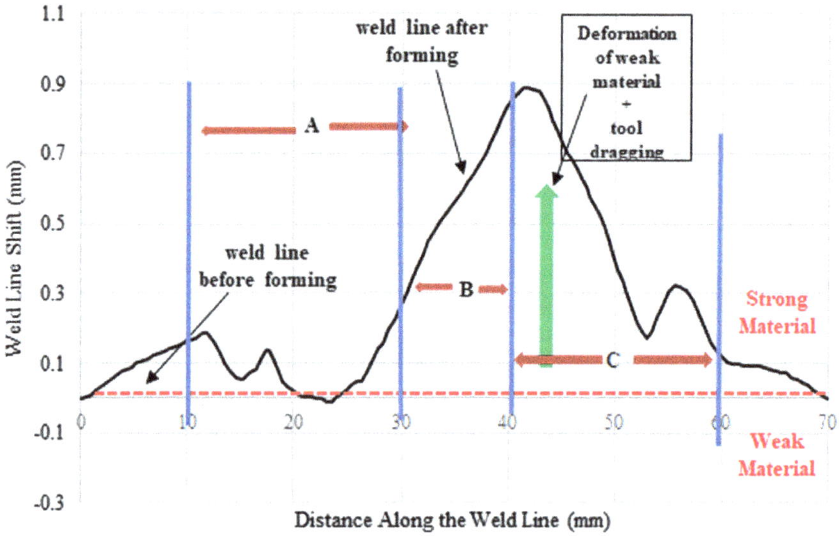

(b) Magnified view of weld line shift

Fig. 5.8 Weld line shift (max. 0.88 towards strong side) for case SFS (Starting From Strong)

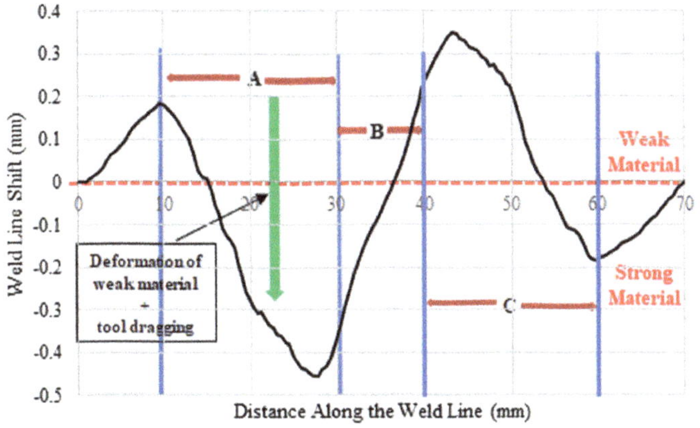

Fig. 5.9 Weld line shift (max. 0.45 towards strong side) for case SFW (starting from weak)—magnified view

upon on the initial position of the tool particularly for the SPIF process. In Fig. 5.9, on comparison of section A and C, it can be observed that the magnitude of weld line shift in section A is higher than the same of section C. For SFW case, the weak material is above the mean line and strong material is below the mean line. In section A, weld line shift is supported by the tool dragging effect and in section C same is opposed by the tool dragging, due to that, section A is experiencing more weld line shift.

In SPIF process, the deformation of the blank is localized unlike other conventional forming processes. Due to this localized deformation mechanism, the weld line shift of TWB during the SPIF process depends upon the initial position of the forming tool.

5.3.3 Plastic Equivalent Strain (PEEQ)

PEEQ result represents the permanent strain present in the body after the forming has been completed. Figure 5.10a, b represent the results of PEEQ for the case SFS and SFW, respectively. These results are based on the von-mises stress criteria.

For the SFS case, more plastic strain is observed in weak material compared to strong material. Due to this, for SFS case, weld line shift is observed only on one side (see Fig. 5.8). More strain leads to excessive thinning and that may lead to the failure of the blank during the forming. Due to in-homogeneity near the weld material, more value of PEEQ is observed there as well.

For the SFW case, value of PEEQ on strong and weak side is observed to have nearly same value. In Fig. 5.10b, two sections are highlighted which are responsible for the weld line shift in section A and C. Due to that, the nature of weld line shift is found to be sinusoidal in nature for SFW case (see Fig. 5.9).

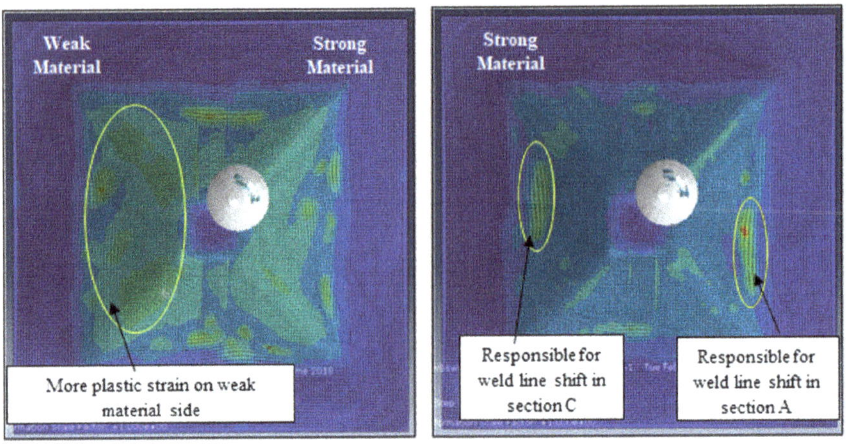

(a) SFS – Max. 4.96 (b) SFW – Max. 6.58

Fig. 5.10 PEEQ in the formed component

5.3.4 Comparison with the Target Geometry

The formed geometry is compared with the target geometry for SFS and SFW case in Fig. 5.11a, b, respectively. It can be observed that the formed geometry is deviating from the target geometry. In order to resolve this effect, proper tool radius compensation should be considered. For both SFS and SFW case, it should be noted that strong material is experiencing more spring back compared to the weak material because strong material is having high yield strength compared to weak material.

5.4 Conclusions

In the present research work, FE analysis of SPIF process is attempted for the TWBs. The effect of two different initial positions of tool on weld line shift, PEEQ and target geometry has been investigated. From the reported simulation study, the following points can be concluded:

- FE analysis can be adopted to study the forming behavior of the TWBs during SPIF process.
- Other than process parameters, weld line shift is affected by the initial position of the tool for the SPIF process. Initial position of tool affects the nature and magnitude of the weld line shift during the SPIF process.
- For the presented simulation conditions SFW results into the less weld line shift compared to the SFS case.

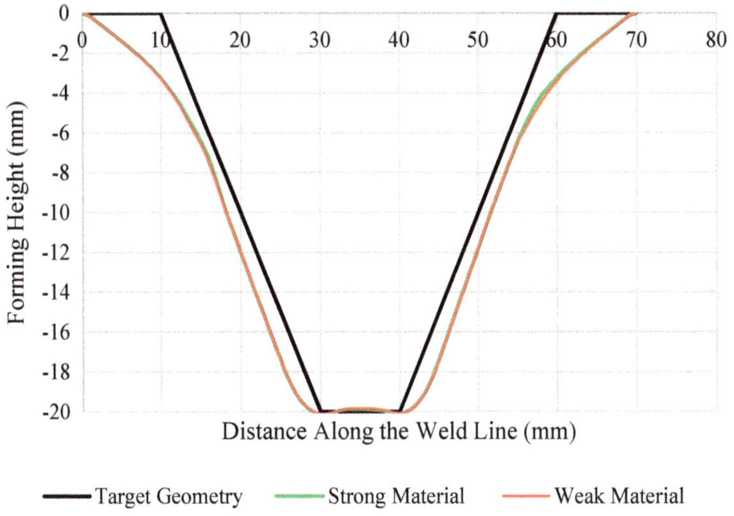

(a) For SFS case

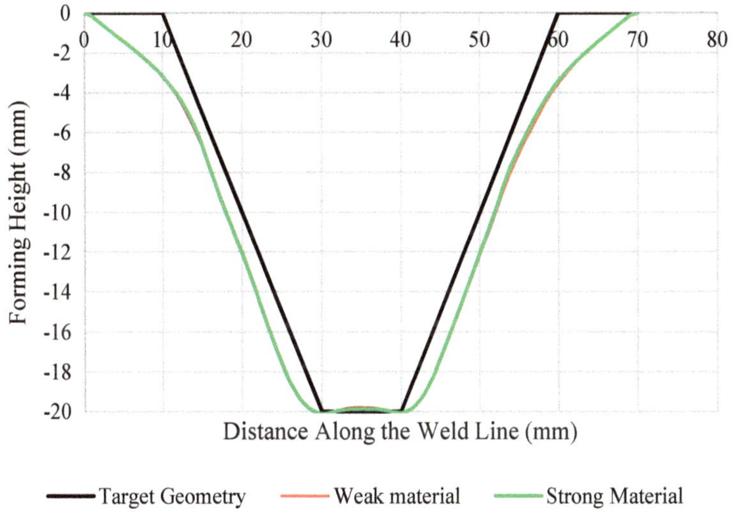

(b) For SFW case

Fig. 5.11 Comparison with target geometry

- More weld line shift is observed in the wall section of the pyramid and that is due to the high amount of strain produced in it.
- If the deformation of weak material is supported by the tool dragging during the forming process, it leads to the maximum weld line shift.

- Formed geometry is nearly matching with the target geometry. However, the gap between them can be reduced by adopting proper tool radius compensation during forming.

References

1. Parente, M., Safdarian, R., Santos, A., Loureiro, A., Vilaca, P., Jorge, R.: A study on the formability of aluminum tailor welded blanks produced by friction stir welding. Int. J. Adv. Manuf. Technol. **83**. https://doi.org/10.1007/s00170-015-7950-0
2. Riahi, M., Amini, A.: Effect of different combinations of tailor-welded blank coupled with change in weld location on mechanical properties by laser welding. Int. J. Adv. Manuf. Technol. **67**, 1–9 (2013). https://doi.org/10.1007/s00170-012-4620-3
3. Shibata, K., Iwase, T., Sakamoto, H., Kasukawa, M., Chiba, K., Saeki, H.: Welding of aluminium tailored blanks by Nd: YAG lasers. Weld. Int. **17**(4), 282–286 (2003)
4. Shalin, M., Hiten, M.: Experimental analysis on effect of tool transverse feed, tool rotational speed and tool pin profile type on weld tensile strength of friction stir welded joint of AA 6061. Mater. Today: Proc. **5**(1), 487–493 (2018)
5. Kesharwani, R., Panda, S., Pal, S.: Experimental investigations on formability of aluminum tailor friction stir welded blanks in deep drawing process. J. Mater. Eng. Perf. **24**(2), 1038 (2015). https://doi.org/10.1007/s11665-014-1361-5
6. Panda, S.K., Kumar, D.R.: Improvement in formability of tailor welded blanks by application of counter pressure in biaxial stretch forming. J. Mater. Process. Technol. **204**(1), 70–79 (2008). https://doi.org/10.1016/j.jmatprotec.2007.10.076
7. Panda, S.K., Baltazar Hernandez, V.H., Kuntz, M.L., Zhou, Y.: Formability analysis of diode-laser-welded tailored blanks of advanced high-strength steel sheets. Metall. Mater. Trans. A **40**(8), 1955–1967 (2009). https://doi.org/10.1007/s11661-009-9875-4
8. Suresh, V.S., Regalla, S.P., Gupta, A.K.: Combined effect of thickness ratio and selective heating on weld line movement in stamped tailor-welded blanks. Mater. Manuf. Process. **32**(12), 1363–1367 (2017). https://doi.org/10.1080/10426914.2016.1257128
9. Micari, F., Ambrogio, G., Filice, L.: Shape and dimensional accuracy in single point incremental forming: state of the art and future trends. J. Mater. Process. Technol. **191**(1–3), 390–395 (2007). https://doi.org/10.1016/j.jmatprotec.2007.03.066
10. Reis, A., Teixeira, P., Duarte, J.F., Santos, A., da Rocha, A.B., Fernandes, A.A.: Tailored welded blanks—an experimental and numerical study in sheet metal forming on the effect of welding. Comput. Struct. **82**(17), 1435–1442 (2004). https://doi.org/10.1016/j.compstruc.2004.03.039
11. Abdullah, K., Wild, P.M., Jeswiet, J.J., Ghasempoor, A.: Tensile testing for weld deformation properties in similar gage tailor welded blanks using the rule of mixtures. J. Mater. Process. Technol. **112**(1), 91–97 (2001). https://doi.org/10.1016/S0924-0136(01)00555-6
12. Song, Y., Hua, L., Chu, D., Lan, J.: Characterization of the inhomogeneous constitutive properties of laser welding beams by the micro-Vickers hardness test and the rule of mixture. Mater. Des. **37**, 19–27 (2012). https://doi.org/10.1016/j.matdes.2011.12.029
13. Jia, Q., Guo, W., Li, W., Zhu, Y., Peng, P., Zou, G.: Microstructure and tensile behavior of fiber laser-welded blanks of DP600 and DP980 steels. J. Mater. Process. Technol. **236**, 73–83 (2016). https://doi.org/10.1016/j.jmatprotec.2016.05.011
14. Chu, G.N., Liu, G., Liu, W.J., Yuan, S.J.: An approach to improve thickness uniformity within tailor-welded tube hydroforming. Int. J. Adv. Manuf. Technol. **60**(9–12), 1247–1253 (2012). https://doi.org/10.1007/s00170-011-3792-6

15. Zadpoor, A.A., Sinke, J., Benedictus, R.: Finite element modeling of transition zone in friction stir welded tailor-made blanks. AIP Conf. Proc. **908**(1), 1457–1462 (2007)
16. Elangovan, K., Balasubramanian, V.: Influences of tool pin profile and tool shoulder diameter on the formation of friction stir processing zone in AA6061 aluminium alloy. Mater. Des. **29** (2), 362–373 (2008). https://doi.org/10.1016/j.matdes.2007.01.030
17. Saidi, B., Boulila, A., Ayadi, M., Nasri, R.: Prediction of the friction coefficient of the incremental sheet forming SPIF. In: The 6th International Congress Design and Modelling of Mechanical Systems CMSM (2015)

Chapter 6
Joining Concentric Tubes by End Forming: A Finite Element Simulation Study

E. Premananda and R. Ganesh Narayanan

Abstract Concentric tubes are used in heat exchangers, chemical and bioreactors, air conditioners, etc. for transmitting fluid from one location to another. Butt joining of tubes is generally performed and this is accomplished by welding and mechanical fastening. The joining of concentric tubes is mostly performed by fusion welding processes. In the present work, an alternative approach to join two concentric tubes using end forming is proposed. A die groove is designed using finite element (FE) analyses to optimize the process parameters for successful joining. FE code ABAQUS/CAE is used for the FE simulations. The tubes are joined by end forming within the die groove. Such FE simulations helped in practical demonstration of the joining process as well.

Keywords Tube joining · End forming · Finite element simulation

6.1 Introduction

Tubes form an integral part of automotive industries, chemical factories, air conditioning system, railway structures, food processing sectors, construction and shipbuilding industries. Concentric tubes are used in heat exchangers, chemical and bioreactors, air conditioners, etc. for transmitting fluid from one location to another. They act as a structural member and hence they are joined and supported suitably to form a mechanical system. Butt joining of tubes is generally performed to take care of length requirements and this is accomplished by welding and mechanical fastening. The joining of concentric tubes is seen in certain applications, in which the ends are joined for sealing and supporting purposes. This is mostly done by

E. Premananda
Department of Production Engineering, VSSUT, Burla, Sambalpur 768018, India

R. Ganesh Narayanan (✉)
Department of Mechanical Engineering, IIT Guwahati, Guwahati 781039, India
e-mail: ganu@iitg.ac.in

R. G. Narayanan et al. (eds.), *Advances in Computational Methods in Manufacturing*, Lecture Notes on Multidisciplinary Industrial Engineering,
https://doi.org/10.1007/978-981-32-9072-3_6

fusion welding processes viz., arc welding and flame welding. Green and sustainable manufacturing and fabrication of parts, as a theme and at a process level, plays a vital role in the development of new materials and processes. One can quote many examples for this. Such a development is meant for cleaner manufacturing and healthy environment. Some examples in which 'welding' is replaced by 'forming or plastic deformation' to accomplish the joining operation are discussed below.

Agrawal and Narayanan [1] proposed a method to join a steel to a thin sheet by end forming of tube. FE simulations were conducted to optimize the parameters. Some criteria were suggested to select the successful joints for fabrication and testing. Alves et al. [2] proposed a method of joining a tube to a sheet using end forming of tube. The forming component involves two localized bends in the tube between which the sheet is placed. FE simulations were conducted to check the possibility of the joining method. A practical example of joining a lever to a fulcrum in an automotive braking system has been demonstrated using the proposed method. A similar method of localized bending of tubes to join two tubes was presented by Alves et al. [3]. In this, joining of carbon steel welded tube (n = 0.06; k = 616.4 MPa) with Aluminium 6060T6 tube was demonstrated. The slenderness ratio and angle of inclination of chamfered edges are optimized to yield a successful product. Silva et al. [4] presented an environmental friendly joining of tubes involving localized bending of tubes. This is similar to tubular lap joints that have a significant overlay with the counterfacing surfaces of the coupling tubes to be joined. The proposal has been demonstrated experimentally and through numerical simulations. The joints performed well in tensile, bending, torsion, and water tightness destructive tests. A new method of fabricating gas and liquid storage reservoirs using nosing of a tubular preform has been demonstrated by Alves et al. [5]. The cases with and without mandrel have been compared. The case in which a mandrel is used for fabrication has been successful. Zhang et al. [6] had used rotary swaging technology for joining two tubes, in which a concave arc region has been formed to take care of the joint efficiency. The ratio of tensile load of the joined tubes to the maximum tensile load of the single tube can reach up to 68% for the inner tube and 47% for the outer tube, depicting the acceptable strength of the joint for practical applications. In separate work, Zhang et al. [7] demonstrated that copper and aluminium tubes can be joined by rotary swaging. The length of the overlapped region plays a vital role in deciding the joint strength. Many such examples can be quoted in the context.

From the discussion, it is imperative that the end forming of tubes is either used to fabricate tubular components or used to join tubes or to join tubes to sheets. By adopting these processes, usage of filler materials, shielding gases, flux in fusion welding processes are ruled out. Thus, joining by end forming can be categorized under green and sustainable joining processes.

In this context, the main objective of the present work is to propose and analyze a simple manufacturing method for joining of concentric tubes having dissimilar properties using end forming. An experimental setup having a die groove has been designed using FE simulations in Abaqus/CAE, FE code. The tubes undergo plastic deformation at their ends in the die groove and get mechanically interlocked to form a joint at the tube end. The influence of three important parameters is studied through FE simulations and suggestions on the sensitivity of those parameters are delivered in the article. Such parameters are optimized within chosen range for practical demonstration.

6.2 Proposed Joining Method

Figure 6.1 shows the proposed method of joining concentric tubes. The tubes of dissimilar qualities are located just above the die groove. The punch is given downward displacement such that the tubes move along the die groove and plastically deform through end curling. During tube end curling, the ends get mechanically interlocked to form a joint at the ends. The joining of concentric tubes by end forming is studied with the inner tube of length 70 mm, internal diameter 37.8 mm, and thickness 1.64 mm, and an outer tube of internal diameter 42.8 mm, and thickness 1.92 mm. The study is carried out to verify process feasibility as a function of gap between the tubes 'd', ratio of length of tubes 'h_1/h_2', and Coulomb's friction coefficient, μ. The input parameters are given in Table 6.1. The range for these parameters was selected after several FE simulation trials.

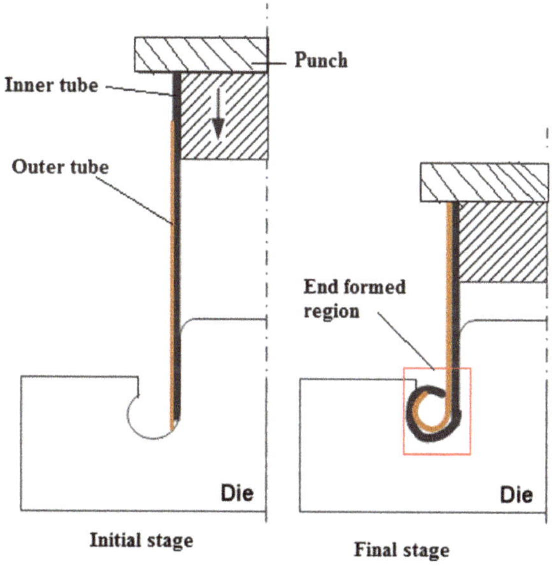

Fig. 6.1 Schematic of joining of concentric tubes by end forming

Table 6.1 List of varying input parameters for concentric tubes joining

Coulomb's friction coefficient (μ)	Ratio of length of tubes (h_1/h_2)	Gap between the tubes (d) in mm
0.05	70/44	0.10
0.08	70/46	0.50
0.10	70/48	0.86
0.15	70/50	1.00
0.20	70/52	1.50

h_1 = length of internal tube referred to as tube1
h_2 = length of external tube referred to as tube2

6.3 FE Simulation and Experiment Trial

The numerical simulation of joining of concentric tubes by end forming is carried out by using the FE code Abaqus/CAE, version 6.7. The selected mesh size for punch and tube is 4 mm and for die is 3.5 mm. The element type of die and punch (rigid bodies) are C3D4 and tube (deformable) is C3D8R. The element size has been chosen after sensitivity analyses and for minimum computational time. The objective of this study is to join the concentric tubes by end forming process. By using the simple die which is used in case of single tube, the locking of tubes did not take place. So a die having groove is designed to fabricate the joint. The entire process can be divided into three stages, i.e., stage 1, stage 2 and stage 3.

Stage 1: As both the tubes are of different length, i.e., internal tube is larger than the external tube, when axial compression takes place the internal tube starts to deform at the die surface. Since there is no contact between the punch and external tube, it rests at the bottom portion without having any deformation which is shown in Fig. 6.2.

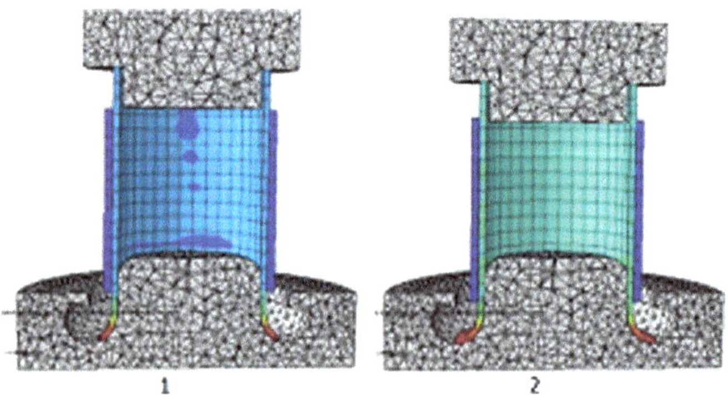

Fig. 6.2 Stage 1 of the proposed joining method

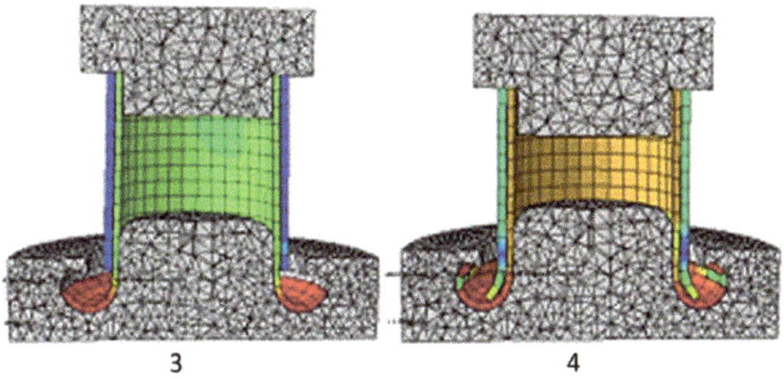

Fig. 6.3 Stage 2 of the proposed joining method

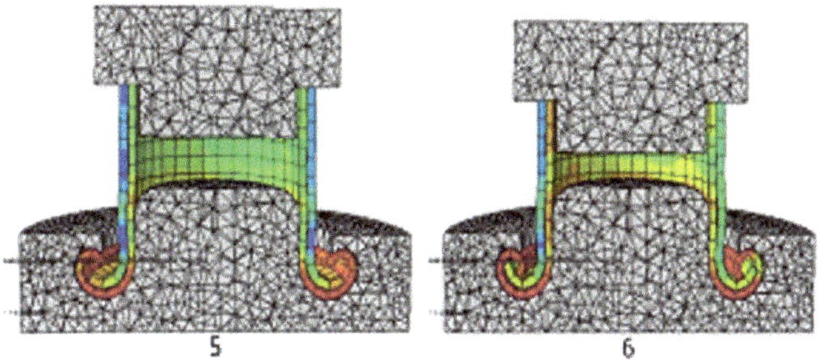

Fig. 6.4 Stage 3 of the proposed joining method

Stage 2: In this stage, the external tube comes in contact with the punch and it also starts deforming above the internal tube. By the time, the internal tube keeps on deforming and acquires the shape of groove of the die. The process of stage 2 is shown in Fig. 6.3.
Stage 3: As the upper portion of the die groove is heading inward, the leading edge of internal tube starts deforming in the inward direction. After a particular stage, the tubes get mechanically locked as shown in Fig. 6.4.

Some experimental trials were conducted to demonstrate the joining method at lab scale. Figure 6.5 shows the experimental setup clamped in a universal testing machine. The tubes were displaced in the downward direction at a uniform cross-head speed of 1 mm/min.

Fig. 6.5 Experimental setup for the proposed joining method

6.4 Material Properties Evaluation

The uniaxial tensile experimental tests were performed in order to study the mechanical properties of the tubes by means of universal testing machine. The specimen used as internal tube is mild steel tube of internal diameter 37.8 mm and thickness of 1.64 mm, the outer tube is of internal diameter 42.8 mm and thickness is 1.92 mm. The mechanical properties obtained for both the tubes from the tensile test are presented in Table 6.2. The internal tube is referred to as tube 1 and the outer tube is referred to as tube 2. The other material properties assumed are density = 7850 kg/m^3, Poisson's ratio = 0.303. These were obtained from tensile tests. The true stress-strain data which is used in the simulation process was fit to the Hollomon's power hardening law defined as, $\sigma = K\varepsilon^n$, where K and n are the strength coefficient and the strain hardening exponent, respectively. For FE simulations, the tubes are considered isotropic, and von-Mises yield function is followed.

Table 6.2 Mechanical properties of the tubes

Tube	Elastic modulus (GPa)	Yield stress (MPa)	Ultimate tensile strength (MPa)	Uniform elongation (%)	n	K (MPa)
Tube 1	84.6	339	406	12.3	0.15	646
Tube 2	106.3	305	342	9.9	0.14	526

6.5 Results and Discussion

Length ratio of both tubes plays a major role in the locking of both the tubes. When the simulation was carried out for tubes having equal length the external tube started to buckle in the mid-region when they were forced to deform along die groove region. Therefore, tubes having different length were introduced in order to get the desired locking of concentric tubes. The length of internal tube is kept fixed and the length of external tube was changed. Figure 6.6 shows the load-displacement curve for different length ratios. The curves have been fit to the data by a polynomial.

It can be seen that for higher and lower length ratios the process is unsuccessful. As the length of external tube increases, the load required for the joint formation also increases. The result of the process for different length ratios, whether it is successful or unsuccessful, is listed in Table 6.3. In case 1, the locking of tubes was not proper. The internal tube strikes the external tube before it moves inside. In cases 2, 3 and 4 the locking was good. In case 5, both the tubes start to buckle before joining.

In forming processes friction plays a key role. The friction between the internal tube and die is considered as frictionless and the friction between both the tubes is

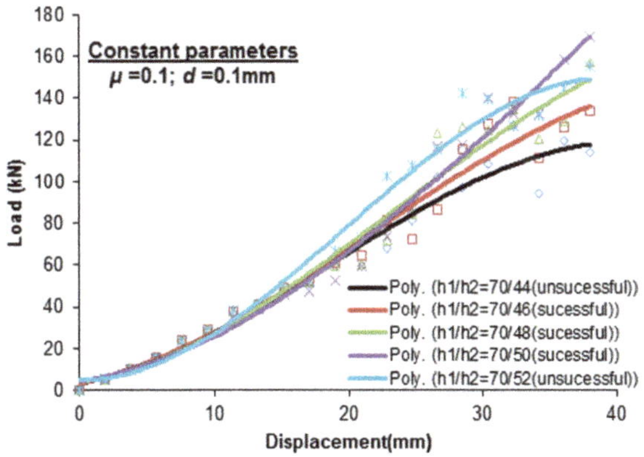

Fig. 6.6 Load evolution for various tubes length ratios

Table 6.3 Effect of length ratio of tubes in concentric tube joining

Cases	h_1/h_2	Results
Case 1	70/44	Unsuccessful
Case 2	70/46	Successful
Case 3	70/48	Successful
Case 4	70/50	Successful
Case 5	70/52	Unsuccessful

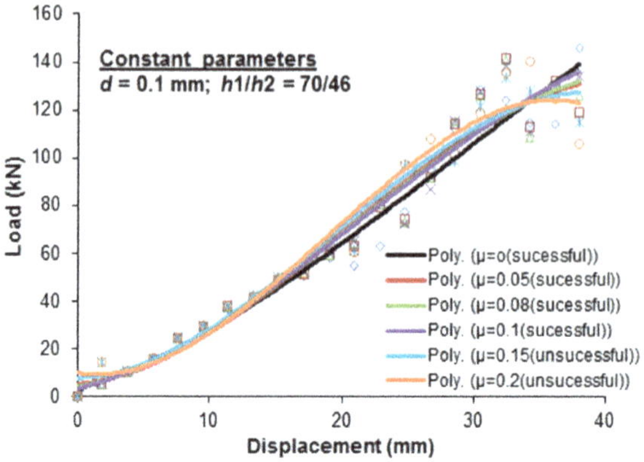

Fig. 6.7 Load evolution for various Coulomb's friction coefficient

Table 6.4 Effect of Coulomb's friction coefficient in concentric tube joining

Cases	μ	Results
Case 1	0	Successful
Case 2	0.05	Successful
Case 3	0.08	Successful
Case 4	0.10	Successful
Case 5	0.15	Unsuccessful
Case 6	0.20	Unsuccessful

varied to observe the effect. As the friction coefficient increases, the load required for the process also increases (Fig. 6.7). For very high value of friction coefficient, the process is unsuccessful. The results for different μ values are listed in Table 6.4. The locking of tubes was acceptable for cases 1–4. In case 5 and 6, tubes started to buckle from the upper portion.

The gap between the tubes was varied in order to evaluate its effect on the forming process. It is observed that for no gap or 0.1 mm gap the process is feasible. For all values other than 0.1 mm, the process is unsuccessful. For all the unsuccessful cases, the load drops when the tubes start buckling (Fig. 6.8). Moreover, in all the unsuccessful cases, the tube buckled when the external tube made contact with the internal tube at die groove region.

Some experimental attempts were made as shown in Fig. 6.9 to join concentric tubes. Though tubes underwent forming in the die groove as observed in simulations, the inner tubes failed before joining is fully accomplished. Such a failure pattern was not observed in FE simulations as instability criterion was not implemented.

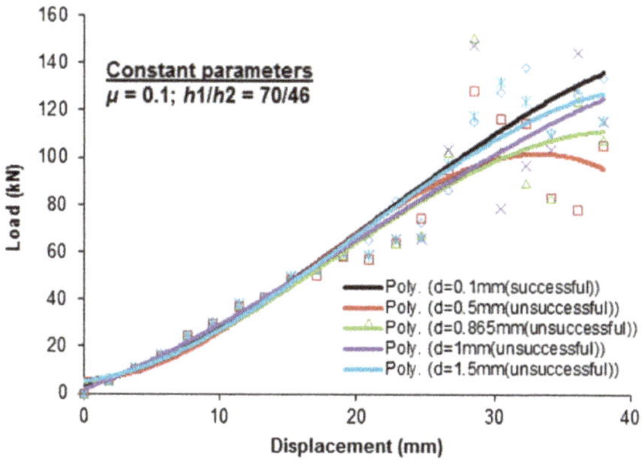

Fig. 6.8 Load evolution for different tube gaps

Fig. 6.9 Joints made by end forming

6.6 Conclusions

An attempt has been made to join concentric tubes by means of end forming operation. The input parameters such as length ratio of tubes, Coulomb's friction coefficient and gap between the tubes have significant effects on the feasibility of the process. The entire analyses have been carried out by FE simulations. The parameters are optimized and experimental demonstration is partially successful. During experimentation, the inner tubes failed due to circumferential stretching before joint formation occurred.

References

1. Agrawal, A.K., Narayanan, R.G.: Joining of a tube to a sheet through end curling. J. Mater. Process. Technol. **246**, 291–304 (2017)
2. Alves, L.M., Dias, E.J., Martins, P.A.F.: Joining sheet panels to thin-walled tubular profiles by tube end forming. J. Clean. Prod. **19**, 712–719 (2011)

3. Alves, L.M., Silva, C.M.A., Martins, P.A.F.: End-to-end joining of tubes by plastic instability. J. Mater. Process. Technol. **214**, 1954–1961 (2014)
4. Silva, C.M.A., Nielsen, C.V., Alves, L.M., Martins, P.A.F.: Environmentally friendly joining of tubes by their ends. J. Clean. Prod. **87**, 777–786 (2015)
5. Alves, L.M., Pardal, T.C.D., Martins, P.A.F.: Nosing thin-walled tubes into axisymmetric seamless reservoirs using recyclable mandrels. J. Clean. Prod. **18**, 1740–1749 (2010)
6. Zhang, Q., Lin, K., Mu, D.: Tube/tube joining technology by using rotary swaging forming method. J. Mater. Process. Technol. **214**, 2085–2094 (2014)
7. Zhang, Q., Zhang, Y., Cao, M., Ben, N., Ma, X., Ma, H.: Joining process for copper and aluminum tubes by rotary swaging method. Int. J. Adv. Manuf. Technol. **89**, 163–173 (2017)

Chapter 7
Optimization of the Forming Parameters in U-Bending for Punch Force and Springback Using Taguchi Method

P. Venkateshwar Reddy, D. Mohana Krishnudu, U. Pranavi and P. Janaki Ramulu

Abstract Springback is a crucial factor that influences the feature of sheet metal in the sheet metal forming (SMF). In SMF operations, springback of the component during unloading mostly determines whether the component confirms to the design dimensions and tolerances. The aim of the current work is to analyze the importance of forming parameters on the responses: punch force and springback in U-Bending of SS 304. Strip length, punch speed and lubricant with three levels each have been considered in the current work as the forming parameters. The effects of different process parameters on U-bending of sheet metal have been investigated by conducting experiments on SS 304. Experiments have been conducted as per Taguchi's L_9 orthogonal array. The optimum conditions have been determined based on their effect on punch force and springback of the sheet metal.

Keywords Springback · U-bending · Optimization · Taguchi · Forming

7.1 Introduction

The sheet metal forming method includes a mixture of elastic–plastic bending and stretch deformation of the sheet. These deformations might prompt a lot of springback of the formed sheet metal part. Springback represents a challenge for

P. Venkateshwar Reddy · D. Mohana Krishnudu
Department of Mechanical Engineering, G Pulla Reddy Engineering College, Kurnool 518007, AP, India

U. Pranavi
Department of Mechanical Engineering, Vardhaman College of Engineering, Hyderabad 501218, T.S, India

P. Janaki Ramulu (✉)
Program of Mechanical Design and Manufacturing Engineering & CoEAME, School of Mechanical, Chemical and Materials Engineering, Adama Science and Technology University, Adama, Ethiopia
e-mail: perumalla.janaki@astu.edu.et

R. G. Narayanan et al. (eds.), *Advances in Computational Methods in Manufacturing*, Lecture Notes on Multidisciplinary Industrial Engineering,
https://doi.org/10.1007/978-981-32-9072-3_7

manufacturers who desires to meet specific dimensions. It is wanted to anticipate and diminish springback with the goal that the last part dimensions can be controlled however much as could reasonably be expected. Significance of the springback calculation is noteworthy from the economical point [1]. Ayres [2] proposed the utilization of various advance procedure to diminish springback in stamping tasks. Liu [3] proposed to shift the holding pressure amid the forming procedure in the way of giving tensile pre-loading to lessen the springback in the formed component. Chou and Hung [4] considered the impact of process factors of the strategies utilized to lessen springback by an optimization procedure for U channel parts. Chan et al. [5] investigated on the springback effect on sheet metal bending by considering various die-punch parameters like punch angle, punch corner radius and die-shoulder radius by performing FEM analysis. Reche et al. [6] studied on air bending tests by considering the effect of bending angle, sheet thickness and loading conditions on spring back. A lot of efforts have been carried out to eliminate the springback, most of them mainly focused on the optimization technique by considering draw-beads, bend curvature radius, punch forces and blank holder force [7, 8]. Maia et al. [9] investigated the U-rail to compare various optimization techniques and prediction tools, with an aim of finding capable strategies for springback compensation. Kakandikar and Nandedkar [10] investigated the process parameters which influences the springback by using optimization technique on automotive component. Vasudevan et al. [11] investigated on the springback of electrogalvanised steel sheet by considering many process parameters and also coating of the materials along with its geometrical parameters and concluded that the punch speed, die opening along with sheet orientation has high influence on the springback. Buang et al. [12] investigated on the stainless steel sheet by considering the effect of the die and punch radii on the springback of V-die bending process and concluded that the springback decreased by decreasing the die and punch radii. Maske and Sawale [13] studied the optimized parameters on the springback effect on aluminum sheet considering the effect of temperature and gauge of the sheet and concluded that the temperature affects the springback phenomenon and their optimized condition is found out. Gautam et al. [14] compared both the numerical and experimental results of springback in 3 ply cladded sheet metal by considering orientation of the sheet metal and concluded that the specimens oriented along the transverse direction have a greater springback when compared to the perpendicular direction.

Only very few of the earlier works have focused on the optimization of forming parameters in sheet metal U-Bending of SS 304. Thus, the aim of the current work is to optimize the forming parameters so as to minimize the punch force and springback using Taguchi Method.

7.2 Materials and Method

7.2.1 Tool and Workpiece Material

Strips made up of SS 304 with thickness of 1 mm were used for the current study. Tables 7.1 and 7.2 shows the chemical composition and mechanical properties of SS 304 respectively. Tools such as punch and die are made of tool steel and their dimensions are shown in Table 7.3. The schematic illustration of tools used in the present study is shown in Fig. 7.1. The fabricated experimental setup is attached to the Universal Testing Machine (UTM) is as shown in Fig. 7.2. Figure 7.3 represents the deformed images of the sheet metal after measuring of springback angle.

7.2.2 Design of Experiments Using Taguchi Method

Taguchi method is used for various problems with multiple parameters due to its practicality and robustness [15, 16]. In Taguchi method, optimization refers to the purpose of best levels of control factors; these best levels are that which maximizes the Signal-to-Noise (S/N) ratios. In this method, the experiments are conducted based on Orthogonal Arrays to conclude the best levels. Strip length, punch speed, and lubricant are the three factors were considered for the experimental outline.

Table 7.1 Chemical composition of SS 304 in wt%

C	Mn	Si	Cr	Ni	P	S	N
0.08	2.0	1.0	18–20	8–10.05	0.045	0.03	0.1

Table 7.2 Mechanical properties of SS304 sheet metal [20]

Property	E (GPa)	ν	σ (MPa)	K (MPa)	n	R_0	R_{45}	R_{90}
Value	210	0.3	420	1069	0.263	1.09	0.95	1.05

Table 7.3 U-bending tool dimensions

Tool	Geometry	Value
Punch	Width	35 mm
	Nose radius	4 mm
	Die width	35 mm
Draw die	Die opening length	37 mm
	Die corner radius	6 mm
Strip	Length	75, 95 and 115
	Width	35 mm

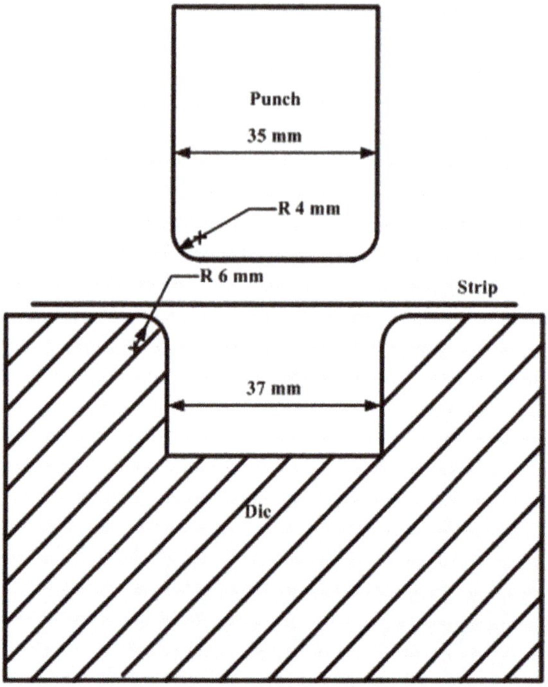

Fig. 7.1 Schematic illustration used in the study for U-bending

Fig. 7.2 Experimental setup for U-bending

Each factor has three levels. So L_9 Orthogonal Array (OA) was shaped in light of the three factors three-level design. The decision of a suitable OA relies upon the measure of control parameters and its levels. To appraise the principle parameter impacts, the design should be Orthogonal and Balanced. Taguchi has arranged some orthogonal arrays that fulfills the over two properties for various number of parameter-level combinations [17]. Inferable from cost and other test confinements, in the present case, a fractional factorial plan with nine runs has been chosen out of

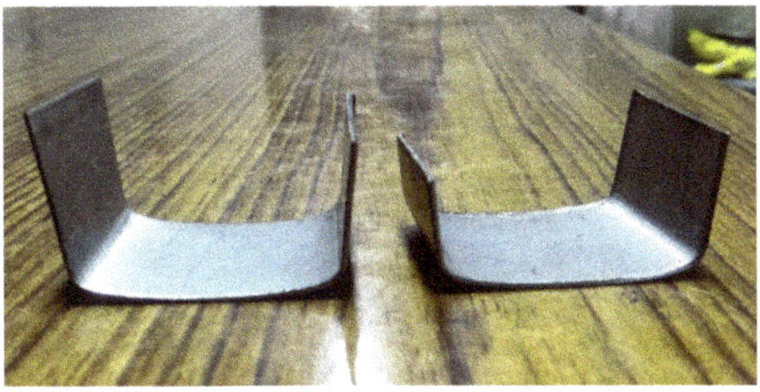

Fig. 7.3 Deformed images after forming the sheet metal

Table 7.4 Considered parameters and their levels

Parameters	Level 1	Level 2	Level 3
Strip length (mm)	75	95	115
Punch speed (mm/min)	5	10	15
Lubricant	Engine oil	Mineral oil	Grease

the full factorial design, $3^3 = 27$ runs. The combination of factors and their levels are recorded in Table 7.4. The recorded methodology of the Taguchi appears in Fig. 7.4. However, this technique helps in deciding the feasible combinations of factors and recognizing the best combination. This, in turn, implies that the resources (materials and time) required for the experiments are also minimum.

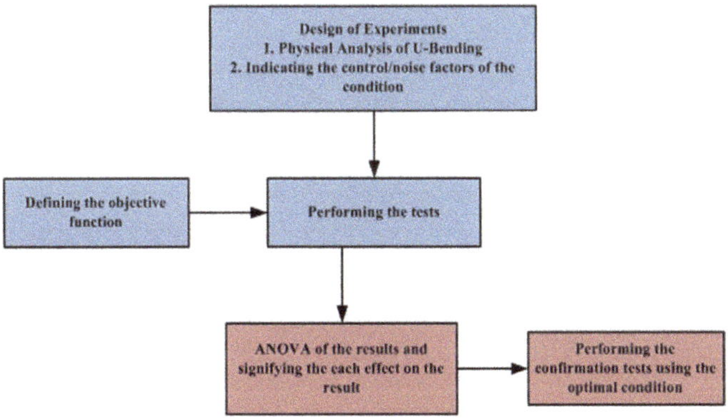

Fig. 7.4 Robust design flowchart for Taguchi method

7.3 Results and Discussion

The main aim of the present study is to optimize the forming parameters like strip length, punch speed and lubricant to achieve low punch force and springback. U-Bending test is analyzed to get a better combination of forming parameters using Taguchi technique for a different set of combinations. Table 7.5 shows the experimental test conditions and its responses.

7.3.1 Effect of Forming Parameters on Punch Force

The influence of forming parameters on Punch force is shown in Fig. 7.5. The punch force increases with increase in strip length since the punch has to travel more distance along with strip into the die. For both punch and die the contact area is increased as the length of the strip increased, thereby the punch force increased as the length of the strip increased. Whereas there is only a minute variation with punch speed and lubrication on punch force. Similar results were observed from the literature work carried out by the authors [18]. In their work lubricant does not affect much on the punch force as the same results were observed in the present study.

Table 7.6 represents the response for mean of means on punch force and it is observed that the most influencing factor on the punch force is the strip length followed by the punch speed and the lubricant. Since the punch travel is more for longer strip length with more contact time between the tools, so the punch force required is more for longer strips. Works carried out by different authors in the literature obtained almost the same results.

Table 7.5 Experimental test conditions and their response values

Test No.	Strip length	Punch speed	Lubricant	Punch force (kN)	Springback angle
1	75	5	Engine oil	10.5	4.6
2	75	10	Mineral oil	11.5	4.5
3	75	15	Grease	10.9	4.9
4	95	5	Mineral oil	17.5	5.1
5	95	10	Grease	16.9	5.5
6	95	15	Engine oil	18.2	5.2
7	115	5	Grease	21.6	5.4
8	115	10	Engine oil	20.6	5.1
9	115	15	Mineral oil	22.1	4.8

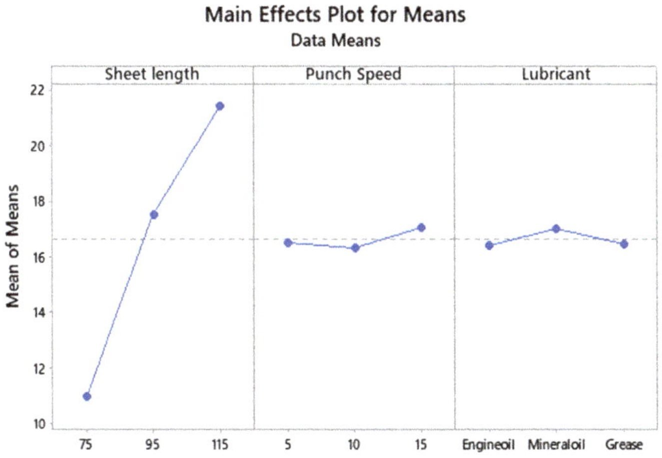

Fig. 7.5 Parameters affect plots on punch force

Table 7.6 Response for mean of means on cutting force

Level	Strip length	Punch speed	Lubricant
1	10.97	16.53	16.43
2	17.53	16.33	17.03
3	21.43	17.07	16.47
Delta	10.47	0.73	0.60
Rank	1	2	3

Table 7.7 represents the ANOVA results for punch force. It is observed that the strip length is the most influencing factor that affects the punch force and the percentage of contribution is about 99%. Meanwhile, the percentage contribution of punch speed and lubricant together is 1%.

Table 7.7 Table representing the ANOVA of punch force

Source	DF	Adj SS	Adj MS	*F*-value	*P*-value
Strip length	2	167.882	83.9411	172.09	0.006
Punch speed	2	0.862	0.4311	0.88	0.531
Lubricant	2	0.682	0.3411	0.70	0.588
Error	2	0.976	0.4878		
Total	8	170.402			
Model summary					
S	*R*-sq	*R*-sq(adj)	*R*-sq(pred)		
0.698411	99.43%	97.71%	88.41%		

7.3.2 *Effect of Forming Parameters on Springback*

The influence of forming parameters on springback is shown in Fig. 7.6. The springback increases with increase in strip length (18.5%) upon a length of 95 mm and then decreases a little (7.2%) for a strip length of 115 mm. Whereas there is only a minute variation with punch speed and also there are some interesting facts observed with the lubricants on springback with variations for different lubricants. Similar results have been observed by the authors in the literature [18]. Lubricants of different viscosity had an impact on the springback in their work similarly same results observed in the present work. The effect of punch speed on springback is not much observed by the literature studies also [19]. Only a minute variation of springback is observed in the present study also similar to the literature study.

Table 7.8 represents the response for mean of means on Springback and it is observed that the most influencing factor on the springback is the strip length and lubricant and only a minute influence by punch speed is observed. Similar results were also observed by the literature studies.

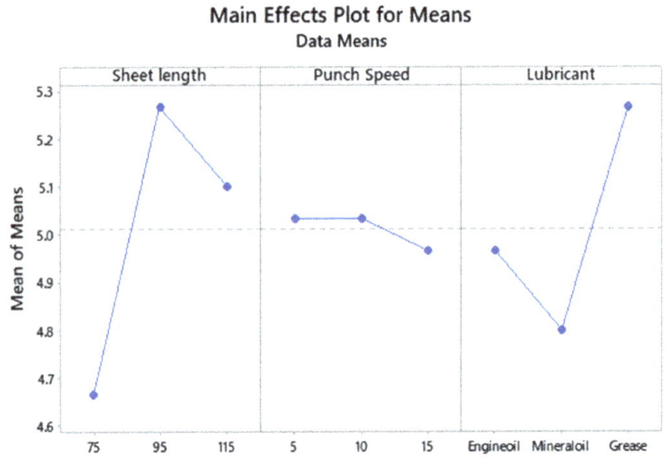

Fig. 7.6 Parameters affect plot on springback

Table 7.8 Response for mean of means on springback

Level	Strip length	Punch speed	Lubricant
1	4.667	5.033	4.967
2	5.267	5.033	4.800
3	5.100	4.967	5.267
Delta	0.600	0.067	0.467
Rank	1	3	2

Table 7.9 ANOVA table of springback angle

Source	DF	Adj SS	Adj MS	*F*-value	*P*-value
Strip length	2	0.575556	0.287778	64.75	0.015
Punch speed	2	0.008889	0.004444	1.00	0.500
Lubricant	2	0.335556	0.167778	37.75	0.026
Error	2	0.008889	0.004444		
Total	8	0.928889			
Model summary					
S	*R*-sq	*R*-sq(adj)	*R*-sq(pred)		
0.0666667	99.04%	96.17%	80.62%		

Table 7.9 represents the ANOVA results for springback. It is observed that the strip length and lubricant are the most influencing factors that affect the springback and the percentage of contributions is about 62.5 and 36.4% respectively. Similar results were observed with the literature studies, all the obtained results are validated with the works carried out by different authors in their works.

7.4 Conclusions

Sheet metal U-Bending was conducted on SS 304 and the forming parameters were analyzed in terms of punch force and springback by varying the strip length, punch speed, and lubricant. Using Taguchi method its optimum machining conditions were determined. Using ANOVA significance parameters were defined. From the present work, the following conclusions were drawn.

- From the statistical analysis, it is observed that strip length is the most influencing parameter that affects both punch force and springback.
- From Taguchi, the optimum forming parameters for punch force were identified as Strip length of 75 mm, Punch speed of 10 mm/min and engine oil as lubricant and are validated by the experiment. The punch force for the optimum conditions is observed as 10.4 kN.
- From Taguchi, the optimum forming parameters for springback were identified as Strip length of 75 mm, Punch speed of 15 mm/min and mineral oil as lubricant and are validated by the experiment. The springback angle for the optimum conditions is observed as 4.5°.
- Obtained results of both punch force and springback are compared to the literature works and a good agreement among the results is observed.

References

1. Gan, W., Wagoner, R.H.: Die design method for sheet springback. Int. J. Mech. Sci. **46**(7), 1097–1113 (2004)
2. Ayres, R.A.: SHAPESET: a process to reduce sidewall curl springback in high-strength steel rails. J. Appl. Metalwork. **3**(2), 127–134 (1984)
3. Liu, Y.C.: The effect of restraining force on shape deviations in flanged channels. J. Eng. Mater. Technol. **110**(4), 389–394 (1988)
4. Chou, I.N., Hung, C.: Finite element analysis and optimization on springback reduction. Int. J. Mach. Tools Manuf. **39**(3), 517–536 (1999)
5. Chan, W.M., Chew, H.I., Lee, H.P., Cheok, B.T.: Finite element analysis of spring-back of V-bending sheet metal forming processes. J. Mater. Process. Technol. **148**(1), 15–24 (2004)
6. Reche, D., Besson, J., Sturel, T., Lemoine, X., Gourgues-Lorenzon, A.F.: Analysis of the air-bending test using finite-element simulation: application to steel sheets. Int. J. Mech. Sci. **57**(1), 43–53 (2012)
7. Yoshihara, S., Manabe, K.I., Nishimura, H.: Effect of blank holder force control in deep-drawing process of magnesium alloy sheet. J. Mater. Process. Technol. **170**(3), 579–585 (2005)
8. Marretta, L., Ingarao, G., Di Lorenzo, R.: Design of sheet stamping operations to control springback and thinning: a multi-objective stochastic optimization approach. Int. J. Mech. Sci. **52**(7), 914–927 (2010)
9. Maia, A., Ferreira, E., Oliveira, M.C., Menezes, L.F., Andrade-Campos, A.: Numerical optimization strategies for springback compensation in sheet metal forming. In: Computational methods and production engineering, pp. 51–82 (2017)
10. Kakandikar, G., Nandedkar, V.: Springback optimization in automotive shock absorber cup with genetic algorithm. Manuf. Review. **5**, 1 (2018)
11. Vasudevan, D., Srinivasan, R., Padmanabhan, P.: Effect of process parameters on springback behaviour during air bending of electrogalvanised steel sheet. J. Zhejiang Univ. Sci. A. **12**(3), 183–189 (2011)
12. Buang, M.S., Abdullah, S.A., Saedon, J.: Effect of die and punch radius on springback of stainless steel sheet metal in the air v-die bending process. J. Mech. Eng. Sci. **8**, 1322–1331 (2015)
13. Maske, N.A., Sawale, J.K.: Taguchi approach for investigation of springback effect in aluminum sheet. Int. J. Mech. Eng. Rob. Res. **2** (2013)
14. Gautam, V., Sharma, P., Kumar, D.R.: Experimental and numerical studies on spring back in U-bending of 3-ply cladded sheet metal. Mater. Today Proc. **5**(2), 4421–4430 (2018)
15. Fei, N.C., Mehat, N.M., Kamaruddin, S.: Practical applications of Taguchi method for optimization of processing parameters for plastic injection moulding: a retrospective review. In: ISRN Industrial Engineering (2013)
16. Reddy, P.V., Reddy, B.V., Rao, P.S.: A numerical study on tube hydroforming process to optimize the process parameters by Taguchi method. Mater. Today Proc **5**(11), 25376–25381 (2018)
17. Logothetis, N.: Managing for total quality: from Deming to Taguchi and SPC. Prentice Hall (1992)
18. Narayanasamy, R., Padmanabhan, P.: Influence of lubrication on springback in air bending process of interstitial free steel sheet. J. Mater. Eng. Perform. **19**(2), 246–251 (2010)
19. Choi, M.K., Huh, H.: Effect of punch speed on the amount of springback in U-bending process of auto-body steel sheets. In: ICTP (2014)
20. Reddy, P.V., Ramulu, P.J., Madhuri, G.S., Govardhan, D., Prasad, P.R.: Design and analysis of deep drawing process on angular deep drawing dies for different anisotropic materials. IOP Conf. Ser. Mater. Sci. Eng. **149**, 012 (2012)

Chapter 8
Experimental Analysis on Springback of JIS 3302 Grade Steel Sheet Material Under Various Heat-Treated Conditions

Kemal Ramato, Perumalla Janaki Ramulu and N. R. R. Anbusagar

Abstract JIS 3302 grade steel sheet material has a wide range of applications in automotive industries. Any kind of sheet metal has to undergo different forming test before a real-time application. There are many forming processes are existed to check the formability of sheet materials. Among all the forming process, bending process is one of the formability tests had a drawback interms of springback. The springback behavior is a measurable quantity for any material after the test. Based on the springback nature of sheet material end product shape finalization will be done. This has a dictum for the present work on springback effect analysis of JIS 3302 steel sheet material is focused. And also, the effect of various heat treatments on springback characteristics of JIS 3302 steel sheet metal studied using experimental and numerical analysis. For the experimental work, sheet was cut with the dimensions of specimen 150 mm × 40 mm × 2 mm in three different rolling directions such as 0°, 45°, and 90° and bending test conducted on V-bending die with 90°. For the tensile testing specimens were cut as per ASTM standard in three rolling directions. The results showed that before heat treatment, the workpiece has taken more load for bending and recorded higher value of springback, whereas, the annealed workpiece has the lowest load consumption to bend and lower springback value.

Keywords Springback · Sheet metal forming · Heat treatment · V-bending · JIS 3302 grade steel · Annealing · Normalizing

K. Ramato · N. R. R. Anbusagar
Program of Mechanical Design and Manufacturing Engineering, School of Mechanical, Chemical and Materials Engineering, Adama Science and Technology University, Adama, Ethiopia
e-mail: kemalramato63@gmail.com

N. R. R. Anbusagar
e-mail: asr2775@gmail.com

P. J. Ramulu (✉)
Program of Mechanical Design and Manufacturing Engineering & CoEAME, School of Mechanical, Chemical and Materials Engineering, Adama Science and Technology University, Adama, Ethiopia
e-mail: perumalla.janaki@astu.edu.et

R. G. Narayanan et al. (eds.), *Advances in Computational Methods in Manufacturing*, Lecture Notes on Multidisciplinary Industrial Engineering,
https://doi.org/10.1007/978-981-32-9072-3_8

8.1 Introduction

Plastic deformation is concerning about metal forming process which has huge applications in various industries. Their main attribute is high efficiency, repeatability, and economy; simultaneously keeping the proper surface condition in mass production. The most frequently used techniques for sheet metal forming are bending, stretching and stamping [1]. Bending is one of the most commonly used methods beyond the sheet metal forming process. The common products obtained with the aid of bending are automobile panels, supermarket shelves, automotive component, aerospace vehicles, pressure vessels, medical equipment, electrical component, drums and housing utensils, etc. [2]. During the forming process force applied when the force is removed, the metal tries to coming back to its original shape and results in a phenomenon named "springback". This is due to the elastic deflection that occurs in the material after the bending process takes place [3]. Ramulu et al. [4] studied springback analysis on AA 6061 Aluminum alloy sheets with seven different rolling directions of sheet with the combination of U and V die shape. In order to make bending at a desired angle the process parameters considered were different load, displacement, and die angle. Finally, concluded that, the effect of rolling direction on springback effects was very sharp and prominent. Main factors that change the springback effect were indicated as sheet thickness, material properties and tool geometry [5–12]. Based on these literatures, the present work aimed to investigate the effect of springback on JIS 3302 steel sheet metal of 2 mm thickness by experimental and numerical work with different heat-treating conditions of the sheet.

8.2 Methodology and Materials

8.2.1 Materials

JIS 3302 grade steel sheet of 2 mm thickness material was used for the present work. This grade material has a wide-ranging of applications and economic considerations in the industries. Table 8.1 shows the chemical composition of JIS 3302 steel material before heat treatment.

Table 8.1 JIS 3302 grade steel chemical composition (in wt%)

C	Si	Mn	P	S	Cr	Mo	Ni	Cu	W	Fe
0.05	0.0065	0.19	0.078	0.035	0.041	0.003	0.014	0.02	0.04	99.5

8.2.2 Experiment: Bending Process

The specimens were prepared in the form strips at different rolling directions (0°, 45°, and 90°) of sheet. This specimen was cut by hydraulic shearing machine from 1000 mm × 2500 mm × 2 mm sheet metal with specimen dimensions of 150 mm × 40 mm × 2 mm as shown in Fig. 8.1. In total, 9 experiments were planned in which 3 tests at ambient condition, 3 tests at full annealing condition and other three tests at normalizing condition in three rolling directions. In a whole, 27 specimens were prepared for three repetitions for each. The full annealing and normalizing heat treatment methods were performed at 900 °C. This 900 °C temperature was considered due to virtually near to recrystallization temperature where phase changes. When phage changes, change in mechanical properties is inevitable.

The V-bending process was used for the bending tests. The experiment set-up was composed of a punch, a die and guide pins which are given in Fig. 8.2. This experiment was performed on hydraulic press brakes machine. The movement of the punch was hydraulically powered and applied enough force on the blank. Experimental setup was not disturbed for all experiments and the tests were conducted under the same operating conditions. The punch stroke was fixed and not moved up to the bottom of the die. A gap of 2 mm, which was equal to sheet thickness was maintained between the final position of punch and the die to avoid crushing of the sheet material at the bending profile. The angle of the die was 90°.

8.2.3 Tensile Test Procedure

The tensile specimens were made from sheet at rolling direction of 0°, 45° and 90° as per ASTM standard E-8 M as shown in Fig. 8.3. Tensile tests were performed at

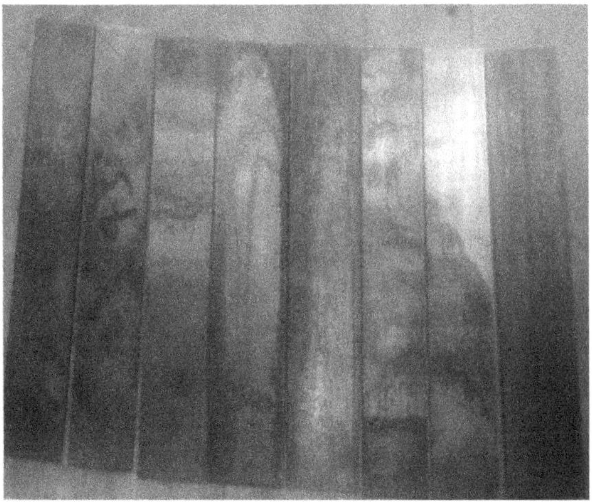

Fig. 8.1 Specimen for bending test with 150 mm × 40 mm × 2 mm dimensions

Fig. 8.2 Experimental setup for V-bending

Fig. 8.3 Specimen prepared for tensile test

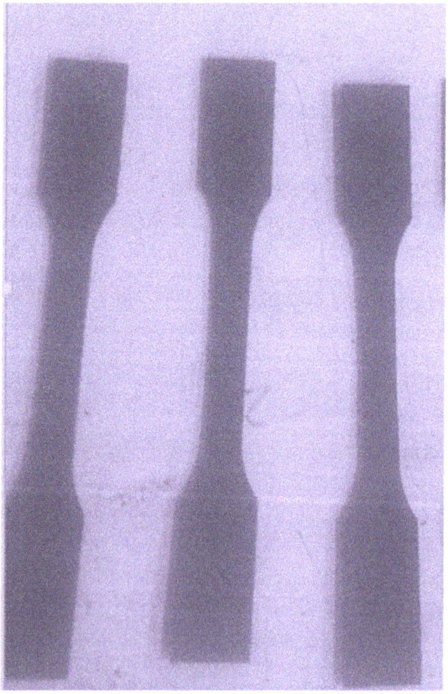

a nominal cross-head speed of 1 mm/min, of samples at room temperature and annealed at 900 °C conditions on an ISO 68982-1 universal testing machine. Tests were repeated three times for each condition. The mechanical properties of JIS 3302 grade steel sheet material like yield strength, ultimate tensile strength, total elongation, uniform elongation, strain hardening coefficient (n) and strength coefficient (K) were evaluated as per the standard procedure after testing the samples till failure. The obtained load-stroke behavior during testing was converted into engineering stress–strain and true stress–strain plot for evaluating the mechanical properties. The plastic strain ratios at different rolling directions were obtained after testing the base material to 10% plastic strain.

8.3 Results and Discussion

8.3.1 Tensile Test Result

The results of tensile tests are shown in Table 8.2 such as ultimate stress (Tensile stress), yield stress, total elongation, strain hardening coefficient (n) and material strength (K). The ultimate stress and yielding stress for annealed specimens are the lowest when it is compared with the results from normalized and before heat treating specimen as presented in Table 8.2.

8.3.2 Springback Results from the Experiments

The experiments are conducted on JIS 3302 grade steel sheet of 2 mm thickness with different rolling directions. For illustration few of the deformed sheets are shown in Fig. 8.4. For each test, three repetitions are followed. In each repetition, springback angles are measured with equal time interval. All the experimental data is tabulated in Table 8.3. Among all the specimens, the major springback has seen without heat-treated specimen as 3.5° in 90° rolling direction and minimum is in annealed sample as 0.5° in 0° rolling direction (Fig. 8.5).

The effect of springback and resulting properties of mechanically formed steel plates have been presented and discussed. It is found that the 2 mm V-bended sheet metal exhibited the largest value of springback 3.060 for experimental. Due to the springback errors, the desired angle of 90° was not achieved. From this of point, it can be concluded that springback causes geometric inaccuracies in sheet metal bending.

Table 8.2 Mechanical properties of JIS 3302 grade steel sheet material

Condition of specimen	Rolling direction	Trials	Mechanical properties of the specimens				
			Tensile stress (MPa)	Yield stress (MPa)	% elongation	n	K (MPa)
Annealing	0°	1	350.3	278.4	52	0.35	723.3
		2	361	214.3	60	0.855	869.3
		3	408.4	330.2	60	0.53	971.0
		Average	373.2	274.3	57.33	0.578	854.5
	45°	1	300	212	58.375	0.32	720.5
		2	299.6	174.1	56	0.4	680.4
		3	320.3	275	54	0.52	856.1
		Average	386.5	220.4	56.125	0.413	763.9
	90°	1	395.5	132.4	55.6	0.49	824.4
		2	397.1	152.4	48.1	0.376	801.7
		3	405.4	289.8	60.7	0.4354	885.3
		Average	399.3	191.5	54.8	0.4338	837.1
Normalizing	0°	1	382.8	376	84.6	0.49	877.4
		2	424.9	352.7	61.5	0.491	977.4
		3	451.9	395.1	61.5	0.3758	883.6
		Average	419.8	308.2	69.2	0.452	912.8
	45°	1	485.4	267.8	73.1	0.5	945.1
		2	402.9	348.1	60	0.51	956.2
		3	444.7	380.2	55.2	0.47	934.6
		Average	431.3	332.0	62.77	0.493	945.9
	90°	1	492.2	421.8	53.8	0.51	1154.5
		2	395.5	334	61.5	0.5325	942.0
		3	440.1	372.3	53.8	0.5412	1051.4
		Average	442.6	176.9	56.37	0.5279	1049.3
Without heat treatment	0°	1	572.5	508.5	44	0.258	1049
		2	562.6	501.8	60	0.0038	363.5
		3	572	529.5	48	0.246	1030.8
		Average	569.0	513.3	50.7	0.169	814.4
	45°	1	474.7	387.3	46.7	0.2864	903.6
		2	534.7	447.8	48.2	0.34	765
		3	584.7	476.65	48	0.53	876
		Average	531.4	437.25	46.3	0.385	848.2
	90°	1	562.8	501.2	46.2	0.153	912.1
		2	549.6	482.3	38.5	0.262	1014.0
		3	521.5	124.3	38.5	0.3	1024.4
		Average	544.6	369.27	41.07	0.238	983.5

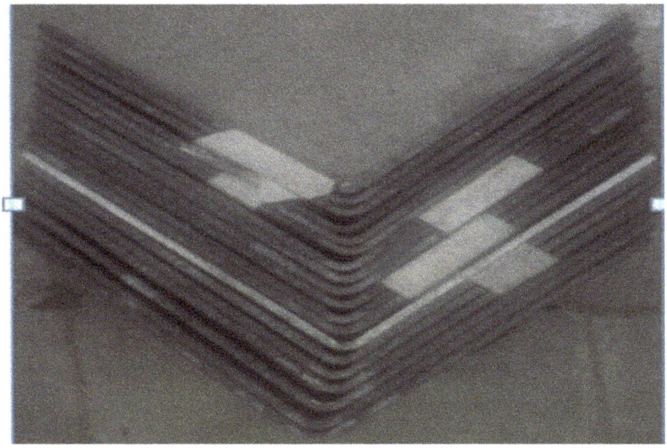

Fig. 8.4 Specimens after V-bending at different heating conditions

Table 8.3 Experimental results of measure springback

Heat treat and with heat treat specimen	Cutting angle of specimen	Experimental result
Annealing	0°	1
	45°	1.43
	90°	1.4
Normalizing	0°	1.5
	45°	1.4
	90°	1.43
Without heat treat	0°	2.6
	45°	3.06
	90°	2.067

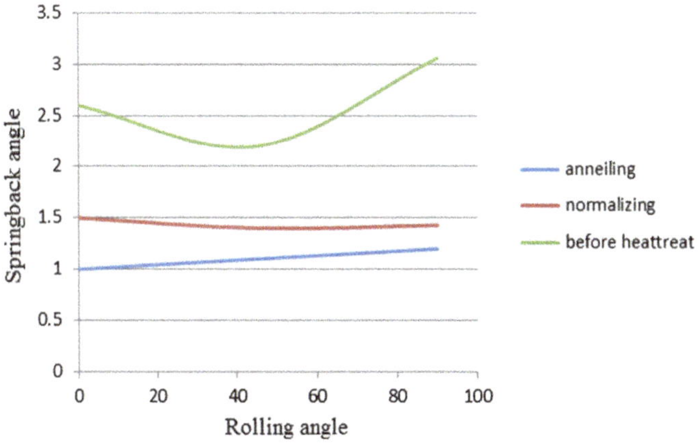

Fig. 8.5 Springback comparison with rolling angle

8.4 Conclusions

In this study, several bending tests are conducted to determine the effect of heat treatment on the springback. For this work, two different types of heat treatment were applied to the JIS 3302 grade steel and the amount of springback is determined under three circumstances. Moreover, the effect of rolling angle to the springback is also studied individually. Hence, the general conclusions attained in this study are as follows:

- The specimen used have the highest yield strength without heat treatment condition compared to annealing and normalizing and it also observed that higher springback obtained in the specimen compared to others.
- It observed that when rolling direction increase the elongation of specimen is decreases.
- In heat treatment condition, required punch force to bend the sheet metal was higher than the material in normalizing condition where the lowest in annealing condition.

References

1. Alghtani, A.H.: Analysis and optimization of springback in sheet metal forming. Ph.D. Diss., University of Leeds (2015)
2. Zulkafli, N.: Experimental and analytical evaluation of bending for aluminum. PhD diss., UMP (2012)
3. Wan-Nawang, W.A., Qin, Y., Liu, X.: An experimental study on the springback in bending of w-shaped micro sheet-metal parts. In MATEC Web of Conferences 2 (2015)
4. Ramulu, P.J., Rao, P.S., Yimer, W.: Springback analysis on AA 6061 aluminum alloy sheets. AIP Conf. Proc. **1769**(1), 200023 (2016)
5. Saikiran Reddy, P., Janaki Ramulu, P., Durga Srinivas, A., Viswa Teja, Y., Pavani, S.: Springback studies on different grades of steel sheets. Appl. Mech. Mater. **813**, 629–633 (2015)
6. Krinninger, M., Opritescu, D., Golle, R., Volk, W.: Experimental investigation of the influence of punch velocity on the springback behavior and the flat length in free bending. Procedia Cirp. **41**, 1066–1071 (2016)
7. Mustafa, O.: Mathematical modeling of the effect of different parameters on spring back in sheet metal formability process. Am. J. Eng. Res. **6**(10), 198–205 (2017)
8. Osman, M.A., Shazly, M., El-Mokaddem, A., Wifi, A.S.: Springback prediction in V-die bending: modelling and experimentation. J. Achie. Mater. Manuf. Eng. **38–2**, 179–186 (2010)
9. Wagoner, R.H., Lim, H., Lee, H.G.: Advanced issues in springback. Int. J. Plast **45**, 3–20 (2013)

10. Toros, S., Alkan, M., Ece, R.E., Ozturk, F.: Effect of prestraining on the springback behavior of the AA5754-O alloy. Mater. Tehnol. **45**(6), 613–618 (2011)
11. Soualem, A.: A detailed experimental study and evaluation of springback under stretch bending process. World Acad. Sci. Eng. Technol. Int. J. Mech. Aerosp. Ind. Mechatron. Manuf. Eng. **8**(6), 1128–1131 (2014)
12. Aso, T., Iizuka, T.: FEM analysis of springback control with lump-punch penetration after V-bending. J. Phys. Conf. Ser. **734**, 032068 (2016)

Chapter 9
Experimental and Numerical Assessment of Residual Stresses in AA6061 After Surface Treatment by Deep Cold Rolling

Rakesh Kumar, Gulshan Kumar, Vijay Gautam and Subhajit Konar

Abstract Deep cold rolling is a mechanical method of surface treatment used to adapt the desired surface properties in machine components for aerospace industries. In this process, localized plastic deformation is rendered on the surface by rolling a form tool wheel on the specimen to condense high dislocation density which imparts high compressive residual stresses resulting in increased tensile and fatigue strength. The main focus of the present study is to characterize the distribution of residual stresses on the surface and subsurface of an 8-mm thick plate made of an aluminium alloy AA6061-O, after deep cold rolling operation to different depths of 0.5, 1, 1.5 and 2.0 mm. The observed residual stresses are correlated with the microstructural features in the deformed depths of the specimens. The residual stress and grain size in the deformed specimens are examined by using grazing incidence X-ray diffraction and electron backscatter diffraction technique, respectively. It is observed that deep cold rolling leads to significant grain refinement, resulting in high values of residual stresses through thickness. The compressive residual stresses are found to be higher on the surface of the specimen and reduce gradually with the penetration depth. The experimental results are compared with the results obtained by FE analysis using ABAQUS. A good agreement is observed between the experimental and the simulated values of residual stresses and the trends of stress distribution pattern.

Keywords AA6061 · Deep cold rolling · Residual stress · Grazing incidence X-ray diffraction · Electron backscatter diffraction · Finite element modelling

R. Kumar (✉) · G. Kumar
Thapar Institute of Engineering and Technology, Patiala, Punjab 147004, India
e-mail: rmnp71@gmail.com

V. Gautam · S. Konar
Delhi Technological University, New Delhi 110042, India

R. G. Narayanan et al. (eds.), *Advances in Computational Methods in Manufacturing*, Lecture Notes on Multidisciplinary Industrial Engineering,
https://doi.org/10.1007/978-981-32-9072-3_9

9.1 Introduction

Residual stresses are the stresses preserved in a material in the absence of any external force. These stresses can originate from phase transformation, thermal origin and defects like vacancies, dislocation, voids, cracks, etc. [1]. Residual stresses are mainly categorized into micro and macro stresses. Micro stresses are developed in grains due to existence of inclusions, dislocations, faults, etc. The macro stress is the average of stresses generated in each grain [2]. The stresses, compressive in nature, are beneficial in fatigue designing of components under dynamic load, whereas the tensile residual stresses are detrimental for the components subjected to fatigue and must be avoided. Undesirable residual stresses can lead to increased chances of crack propagation, distortion and premature failures in the components [3]. There are several methods to impart useful compressive residual stresses like deep cold rolling (DCR), shot peening, burnishing, surface hardening, etc. on the surface of the component to improve its fatigue strength [4]. Although shot peening is generally adopted for most components subjected to fatigue, DCR extends several benefits over the shot peening process. DCR imparts a deeper layer of compressive residual stress with remarkably good surface finish, but this process has a limitation in applying it for different geometrical profiles of the components [5]. This technique is widely used in automotive parts (e.g. crankshafts and shaft fillets), turbine blades, wheel flange, turbine disk, rotors, compressor fan blade, railway axles, etc. [6].

Subsequent paragraphs, however, are indented. DCR improves the fatigue strength, wear resistance and surface finish and decreases crack propagation and stress corrosion cracking by inducing a compressive residual stress on the surface [7]. This process is carried out using a roller shape or a ball-type deforming tool. As the rigid blunt roller or ball tool deforms the material underneath to a predetermined depth, a longitudinal groove is created along the length of the tool path resulting in an elasto-plastic zone. On removal of the deforming forces, the recovery of elastic region develops a zone of large compressive residual stress on the surface [5].

Wong et al. [5] explored the feasibility of DCR on titanium test specimens using three different cold rolling tool designs. The effects of process variables and design variables on yielding depth and residual stresses set-up were investigated. It was concluded that the rolling pressure affects the residual stress profile the most. Bozdana et al. [8] developed a new ultrasonic-assisted DCR, using static and dynamic forces of ultrasonic vibrations and investigated the effects of process parameters on titanium alloy test coupons. The process proved to be advantageous in reduced deforming pressure treating thin components without deteriorating the shape and a magnitude of 900 MPa of residual stress could be set-up at a depth of 0.2 mm. Zhu et al. [9] investigated the fatigue strength of titanium alloy specimens treated and untreated by ultrasonic-assisted DCR. Scanning electron microscopy of the fractured specimens was also done. It was concluded that the new process was able to generate a deep layer of residual stress with high surface finish and resulted in enhanced fatigue strength of the fatigue test specimens. Xie et al. [14] studied the effects of ultrasonic-aided DCR process on microstructure of 30CrMnSiNi2A steel using nano-hardness tester, x-ray diffraction (XRD), scanning electron microscope (SEM), electron backscatter diffraction (EBSD) and transmission electron microscope (TEM). It was observed that severe plastic deformation by ultrasonic-aided DCR produced nano-crystallization, grain refinement, strain-induced martensite transformation, shear bands, intense shear texture and rolling texture in the surface and subsurface of the test specimens. Liu et al. [10] studied the mechanical properties of the surface prepared by ultrasonic surface rolling process, and nano-indentation tests were adopted to evaluate the force and depth of penetration. The magnitude of yield strength and strain hardening exponent were evaluated and the results were compared with the results obtained by FE simulations and found to be in good agreement. Juijerm et al. [11], surface treated the aluminium alloys AA5083 and 6110 using deep rolling at room temperature. It was observed that deep rolling enhanced the fatigue strength of aluminium alloys as compared to the polished condition due to near surface compressive residual stresses as well as work hardening and increased hardness induced by deep rolling.

Aluminium alloys due to its light weight, non-magnetic in nature, resistance to corrosion, high specific strength and better ductility are widely used in various industrial applications (aerospace, automotive, marine, building structure, etc.) [12], and offer lot of potentials to be surface treated by DCR. The development of a surface with high magnitude of compressive residual stress in high strength aluminium alloys depends on many factors like feed rate, roller speed, roller feed, applying force by roller, etc. Feed of roller affects the upper surface of the component, but does not affect the depth of penetration. The optimum combination of rolling force and rolling feed affects the surface hardness the most, whereas, if rolling force is maximum and feed is minimum then maximum surface hardness is achieved. The set-up for the DCR can be easily installed with conventional milling and lathe machines [13].

In the present work, the feasibility study of DCR on aluminium alloy AA6061 is carried out. The main focus of the study is to characterize the distribution of residual stresses on the surface and subsurface of 8-mm-thick plate made of an aluminium alloy AA6061-O after deep cold rolling operation to different depths of 0.5, 1, 1.5 and 2.0 mm. The observed residual stresses are correlated with the microstructural features in the deformed depths of the specimens. The residual stress and grain size in the deformed specimens are examined by using grazing incidence X-ray diffraction and electron backscatter diffraction technique, respectively.

9.2 Methodology

9.2.1 Experimental Work

Material Selection. In the present study, a wrought aluminium alloy AA6061 in annealed state is selected on the basis of its wide applications and high potential to be worked at room temperature by DCR. AA6061 is a precipitation hardening alloy, containing magnesium and silicon as its major alloying elements. The alloying elements enable the material to possess excellent tensile strength coupled with ductility and corrosion resistance and offer high resistance to fracture toughness. Some of the typical examples of applications of this alloy are heavy duty structures, marine structures, rail road, pipe lines, etc. The material is procured in the form of a plate with dimension 300 mm × 100 mm × 8 mm. The chemical composition of the alloy obtained by spark-based spectrometer is given in Table 9.1.

Tensile Properties. The tensile samples are prepared as per the ASTM standard by wire-cut electric discharge machining (WEDM) as shown in Fig. 9.1. A minimum of four samples of AA6061 is tested to obtain the scatter in results. The tensile specimens are tested on a 50 kN UTM, and load versus elongation data are procured using a dedicated Horizon software. The engineering stress–engineering strain curves are plotted to obtain the various tensile properties of the material. The engineering stress–strain plots are converted into true stress–true strain data and ln true stress versus ln true strain data between yield strength and UTS is used to determine the strain hardening exponent and strength coefficients.

Experimental set-up for DCR and sample preparation. Deep rolling set-up is installed on a horizontal milling machine and shown in Fig. 9.2. The milling cutter is replaced with a properly designed roller wheel of 75 mm in diameter, 5 mm thick

Table 9.1 Chemical composition of AA6061 in wt%

AA6061	Mg	Si	Cu	Cr	Fe	Ti	Zn	Co
	0.98	0.61	0.23	0.18	0.21	0.15	0.02	0.15

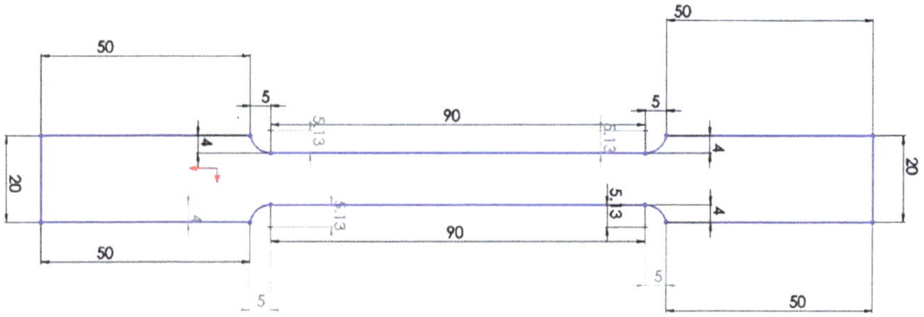

Fig. 9.1 Tensile sample details as per ASTM-E8M (all dimensions are in mm)

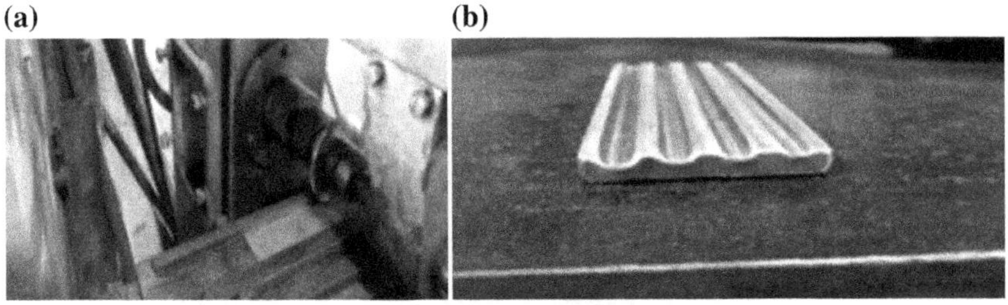

Fig. 9.2 a Deep rolling set-up on a milling machine (*Courtesy* Thapar University, Patiala) and b Deep cold-rolled aluminium specimen

and a nose radius of 2.5 mm made of D2 steel, hardened and tempered to deep roll the aluminium alloy at room temperature. The deep rolling process is carried out on the samples rigidly secured to the table using a fixture and the spindle is rotated at 10RPM to achieve the desired plastic deformation to different depths of 0.5, 1.0, 1.5 and 2 mm.

DCR-processed samples are carefully partitioned longitudinally by a diamond abrasive wheel cutter and polished in different stages from finishing with sub-micron SiC 1200 mesh to electropolishing to perform the microstructure characterization of different depth-rolled samples. The electropolishing is done on prepared samples so as to remove dust and scratches and to get a mirror surface grove of the samples. Electropolishing is done with perchloric acid 20% and methanol 80% at 27 V and 4.5 A current for a soaking time of 10 s.

Electron back scatter diffraction (EBSD), as shown in Fig. 9.3, is performed to measure the grain orientation, texture, grain size and grain boundary in plastically deformed regions of the processed samples to correlate the microstructure with residual stress.

Grazing Incidence X-ray diffraction. The grazing incidence X-ray diffraction (XRD) is a rapid Panalytical make, non-destructive system for analysing a large range of alloys, including metals, polymers and semiconductors as shown in Fig. 9.4. This technique is used to determine the residual stresses on a surface at an omega angle. In this case, the longitudinal stress is measured (σ_{11}) which is residual stress. Residual stress measurements on four samples processed by DCR to different depths of 0.5, 1.0, 1.5 and 2 mm are diagnosed on the surfaces at an omega angle of 0.1°, 0.5° and 0.8°.

9.2.2 *Finite Element Modelling (FEM)*

A three-dimensional, elasto-plastic finite element model is developed for DCR. The FE simulations were conducted using ABAQUS 6.13-1 software. In the simulations, an elastic plastic deformable rectangular aluminium alloy (Al6061) plate

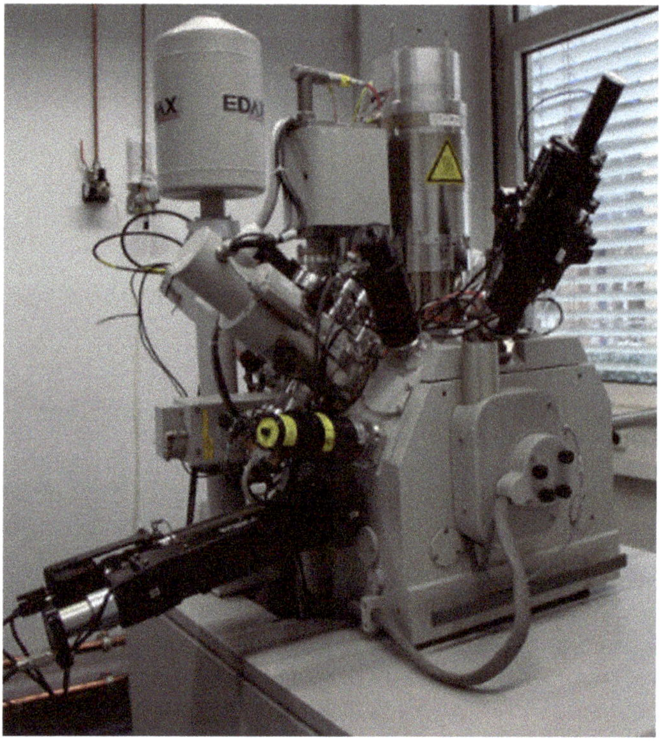

Fig. 9.3 Electron backscatter diffraction instrument to characterize grain structure (*Courtesy* IIT, Bombay)

of dimension 150 mm in length, 100 mm in width and 8 mm thickness is modelled and encastered along the edges to restrict the movement in any direction. The material model is designed to be isotropic in nature with modulus of elasticity (E) and Poisson's ratio (v) as 68.9 GPa and 0.33, respectively. The plastic data between 0.2% offset yield strength and ultimate tensile strength are selected from true stress–true strain plots to simulate the annealed behaviour of the material. The geometric model is using continuum stress/displacement and 8-nodded linear brick C3D8R elements. Element size plays an important role in attaining accurate results. A large number of elements can increase the computational time and a very small number of elements can lead to the inaccurate results. A structured mesh is used in this model. ABAQUS/Explicit has three different options for C3D8R elements, i.e. average strain kinematic, orthogonal and centroid formulations. The average strain kinematic formulation has been used in this study. Though, average strain kinematic formulation required more computational time compared to other two formulation techniques. The Lagrangian approach has been used in the present paper. In this approach, the initial configuration for the time $t = 0$ and the final state for the time $t + \Delta t$ are considered. Dynamic explicit method is used for the simulations.

Fig. 9.4 GIXRD Panalytical™ set-up (*Courtesy* IIT, Bombay)

Mesh density is a significant metric used to control accuracy and a high-density mesh will produce results with high accuracy. However, if a mesh is too dense, it will require a large amount of computer memory and long runtimes, especially for multiple-iteration runs that are typical of nonlinear analyses. The basic fundamental and accurate method for evaluating mesh quality is to refine the mesh until von Mises stress value doesn't change significantly with each refinement.

In this study, 56,250 elements and 58,420 nodes with 15 elements in thickness direction and 250 elements are used in the width direction. To save the time, a biased mesh is generated in the deformation region using double biasing with a bias ratio of 3.

A roller wheel with 75 mm in diameter, 5 mm in thickness, 2.5 mm as nose radius is modelled as discrete rigid. The wheel is provided to have a uniform rotational speed of 10RPM. The simulations were conducted for penetration depths of 0.5, 1.0, 1.5 and 2.0 mm to deform the material longitudinally along the length of the specimen as shown in Fig. 9.5.

Coulomb friction was given to the contact surfaces and it is important to note that adequate friction is required to contact between the sheet and deep roller. Frictional forces at the interface of the roller and the sheet are responsible for pulling the sheet through the roller. A residual stress gradient has been observed at the surface of the deep-rolled sheet.

Boundary conditions are divided into two steps: an initial step and the analysis step. In the initial step, the sheet was given an initial velocity of 10 mm/s in

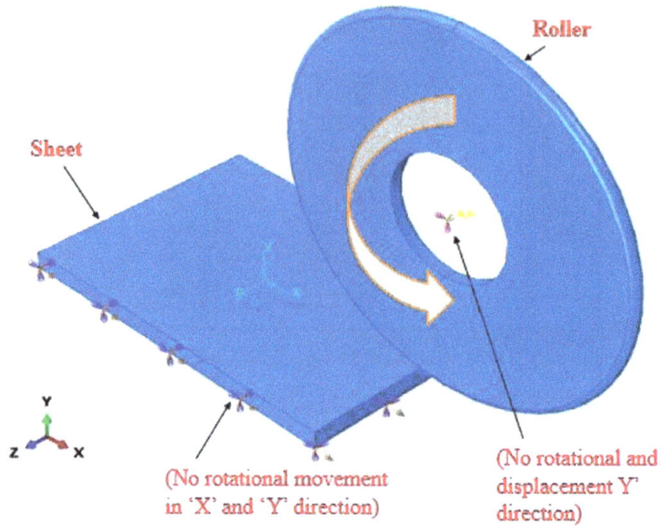

Fig. 9.5 A 3D finite element model DCR

'x'-direction. The interaction between the roller and the upper surface of the sheet is defined in this step. A displacement/rotation boundary condition is defined to constrain the motion of the sheet. The side faces of the sheet are made zero in 'y'- and 'z'-direction. Similarly, the boundary condition of the roller was made such that the roller could rotate only about their respective axis of rotation. All other translational and rotational degrees of freedom were constrained in 'x'- and 'z'-directions. In the analysis step, the rollers were given with an equal and opposite angular velocity. The friction model used in this study, allows the relative motion between the two contact surfaces. On the basis of literature survey on DCR, a frictional coefficient value of 0.2 is used in all the simulations.

9.3 Result and Discussion

9.3.1 Tensile Properties

The results obtained from the uniaxial tension tests conducted at room temperature are plotted and depicted in Fig. 9.6a and ln(true stress) versus ln(true strain) are shown in Fig. 9.6b. It is observed that the yield strength of annealed aluminium alloy AA6061 is 69 MPa and tensile strength is 132 MPa. The ductility of the alloy is excellent and found to be 42%. The strain hardening exponent is 0.3 on an average basis indicating good uniform stretchability.

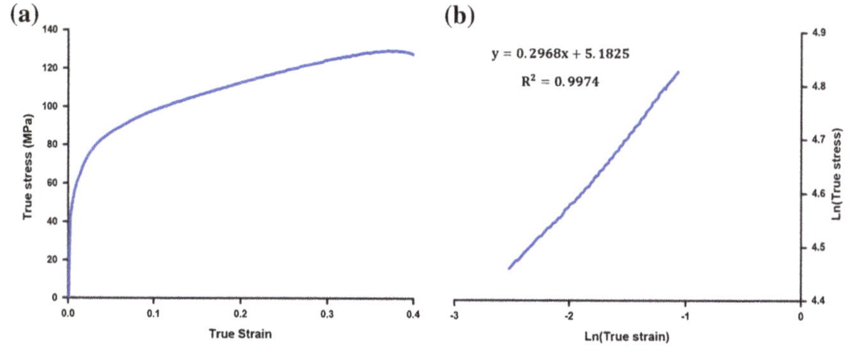

Fig. 9.6 **a** True stress versus true strain plot, **b** ln (true stress) versus ln (true strain) plot for tested tensile specimens of AA6061

9.3.2 Microstructure Characterization

Electron backscatter diffraction (EBSD) is used to characterize the microstructure of the deformed zones of AA6061 specimens. EBSD analysis revealed elongated grains in as-received parent material with a coarse grain size of 135.89 μm as shown in Fig. 9.7a. The various results of EBSD scan obtained at different DCR depths of 0.5, 1.0, 1.5 and 2.0 mm are shown in Fig. 9.7b–e, respectively. It is observed that the grain size with a deformation depth of 0.5 mm is 3.12 μm but the fine grains are segregated in a zigzag manner as shown in Fig. 9.7b. A fairly homogeneous ultrafine-grained microstructure is revealed at a deformation depth of 1.0 mm and the grain size is 1.95 μm. A similar ultrafine homogeneous grain size and distribution is observed with penetration depth of 1.5 and 2.0 mm without any remarkable difference in the average grain size.

9.3.3 Residual Stresses by Experiments and Simulations

The GIXRD method is used to determine the residual stress generated by DCR at the surface and subsurface of the specimens and the results obtained are plotted and shown in Fig. 9.8. It is interesting to find that the specimens of AA6061-O as procured has the range of residual stress ranging from −160 MPa(compressive) to 20 MPa (tensile) at the surface and at a depth of 0.6 mm below surface, respectively. After the first surface treatment by DCR with a tool penetration depth of 0.5 mm, there is a drastic increase in the values of residual stresses ranging from −230 to −160 MPa at surface and 0.6 mm below surface, respectively. All the generated residual stresses are compressive in nature. The similar trend is observed with tool penetration depth of 1, 1.5 and 2.0 mm but the magnitude of residual

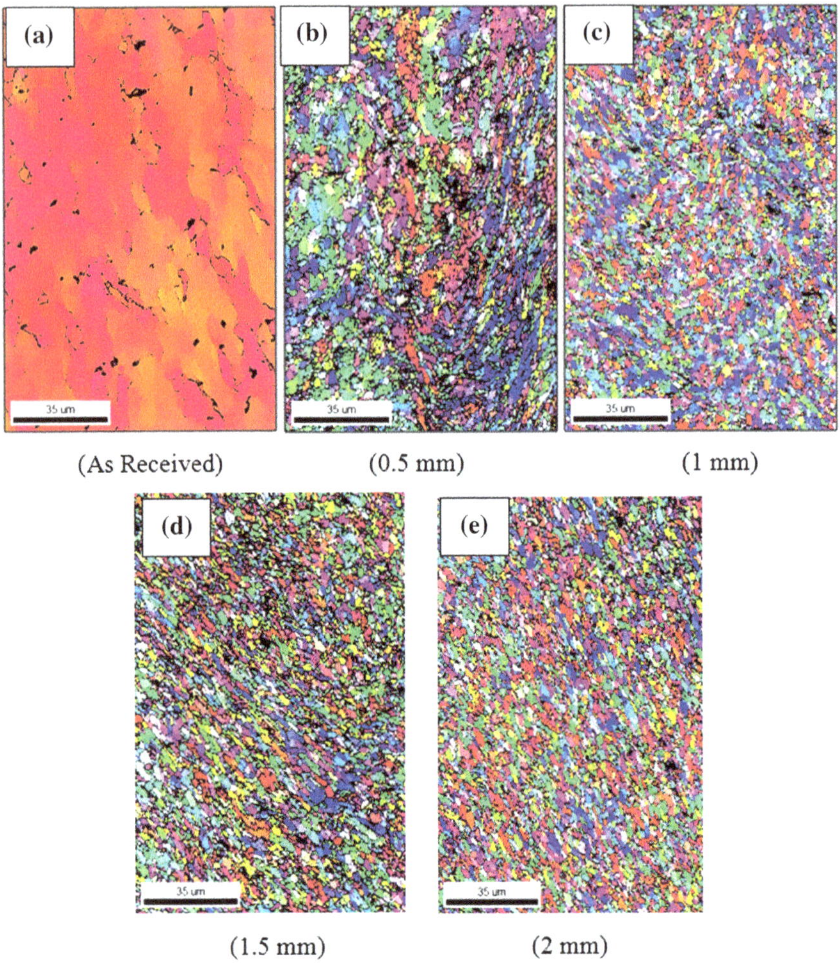

Fig. 9.7 EBSD scans after surface treatment by DCR

stresses is highest with higher depth of penetration by DCR. The values of residual stress on the surface of the material deformed is comparatively higher than the values generated over the subsurface, and the values of compressive residual stress is a function of depth of penetration by DCR, i.e. the stresses are higher on the surface and starts decreasing with increasing distance in depth. This could be attributed to the higher localized deformation at the surface directly under tool and lower concentration of deformation underneath the surface.

The results of residual stresses obtained by FE simulations are captured and plotted between longitudinal stress versus penetration depth for different tool depths of 0.5, 1.0, 1.5 and 2.0 mm and are shown in Fig. 9.9. It is clearly observed that the FE simulation results are in good agreement with the results obtained by

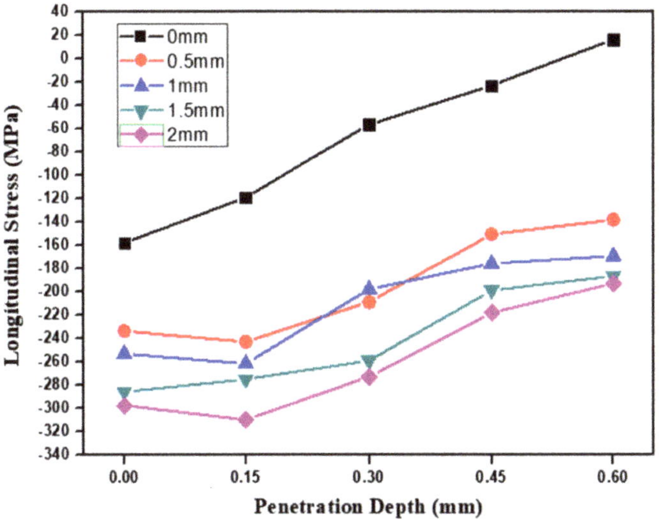

Fig. 9.8 Experimental results of residual stress versus penetration depth

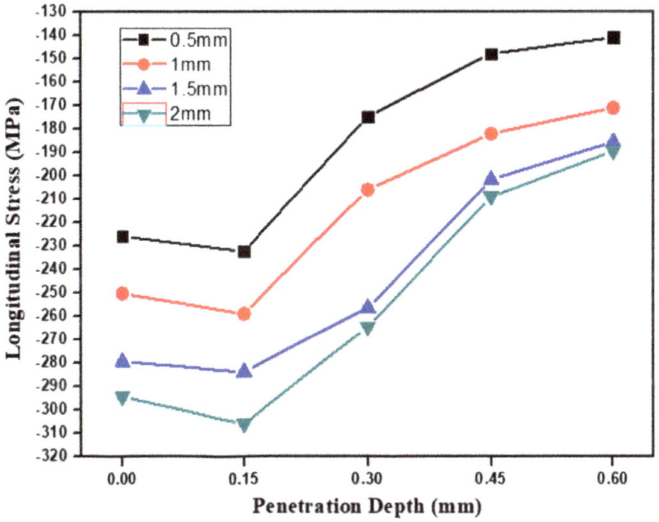

Fig. 9.9 FE simulation results of residual stress gradient in longitudinal direction (LD)

experiments. It is depicted from Table 9.2 that the percentage error in all cases between experimental and simulation for various DCR depth surfaces with their respective penetration depth are below 5%.

It is also observed that up to the depth of 0.15 mm, the values of residual stress increase and then reduce gradually with increasing distance beneath the surface.

Table 9.2 Chemical composition of AA6061 in wt%

DCR depth (mm)	% error				
	Penetration depth (mm)				
	0.00 (%)	0.15 (%)	0.30 (%)	0.45 (%)	0.60 (%)
0.50	3.29	4.36	4.98	1.72	2.10
1.00	1.02	1.03	4.04	3.70	1.20
1.50	2.13	3.20	1.00	1.46	0.48
2.00	1.07	1.29	2.93	3.99	1.76

This could be attributed to the higher plastic deformation resulting in higher dislocation density at a distance of 0.15 mm below the surface, and the material is getting plasticised between the tool and the material itself due to resistance to deformation from the strength offered by the material below the subsurface. A similar trend in the results of residual stresses is observed with different tool penetration depths by DCR of 0.5, 1.0, 1.5 and 2.0 mm. The contour of longitudinal residual stress obtained from simulations is shown in Fig. 9.10. It is found that the stresses on the surface of the specimens in the un-deformed region are tensile in nature whereas the stresses on the deformed region are compressive. This could be attributed to self-equilibrating nature of residual stress, i.e. the net sum of residual stress is zero.

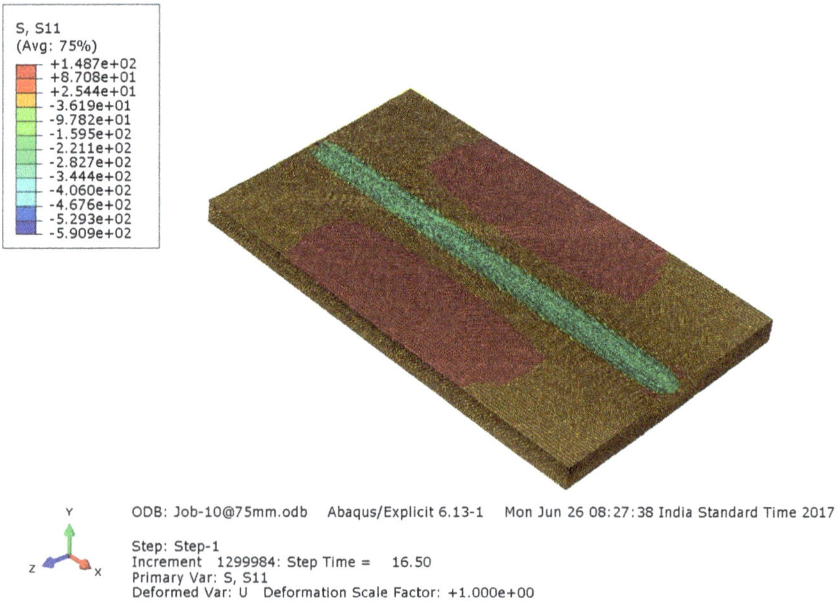

Fig. 9.10 FE simulation results of residual stress by DCR to 0.5 mm tool depth

9.4 Conclusions

Aluminium plate of AA6061 with 8 mm thickness is deep-rolled to different depths of 0.5, 1.0, 1.5 and 2.0 mm. On the basis of experimental work and FE simulations, the following conclusions can be drawn:

- The yield strength of aluminium alloy AA6061 in annealed state is 69 MPa and tensile strength is 132 MPa. The ductility of the alloy is excellent and found to be 42%.
- EBSD analysis revealed elongated grains in as-received parent material with a coarse grain size of 135.89 μm. The grain size with a deformation depth of 0.5 mm is 3.12 μm but the fine grains are segregated in a zigzag manner. A fairly homogeneous ultrafine-grained microstructure is revealed at a deformation depth of 1.0 mm and the grain size is 1.95 μm. A similar ultrafine homogeneous grain size and distribution are observed with penetration depth of 1.5 and 2.0 mm without any remarkable difference in the average grain size.
- The values of residual stress on the surface of the material deformed are comparatively higher than the values generated over the subsurface and is a function of depth of penetration by DCR, i.e. the stresses are higher on the surface and starts decreasing with increasing distance in depth. This could be attributed to the higher localized deformation at the surface directly under tool and lower concentration of deformation underneath the surface.
- The FE simulation results are in good agreement with the results obtained by experiments.

References

1. Understanding stress and strain. Material science and Engineering, http://home.iitk.ac.in/~anandh/E-book.htm. Last accessed 2017/05/23
2. Chatterjee, K., Venkataraman, A., Garbaciak, T., Rotella, J., Sangid, M.D., Beaudoin, A.J., Pilchak, A.L.: Study of grain-level deformation and residual stresses in Ti-7Al under combined bending and tension using high energy diffraction microscopy (HEDM). Int. J. Solids Struct. **94**, 35–49 (2016)
3. Resudual stress info. Proto manufacturing, http://www.protoxrd.com/residual-stress-info.html . Last accessed 2016/11/23
4. Majzoobi, G.H., Azadikhah, K., Nemati, J.: The effects of deep rolling and shot peening on fretting fatigue resistance of Aluminum-7075-T6. Mater. Sci. Eng. A **516**(1), 235–247 (2009)
5. Wong, C.C., Hartawan, A., Teo, W.K.: Deep cold rolling of features on aero-engine components. Procedia CIRP **13**, 350–354 (2014)
6. Liouand, T.I., Wardany, E.I.: Finite element analysis of residual stress in Ti-6Al-4V alloy plate induced by deep rolling process under complex roller path. Int. J. Manuf. Eng. http://dx.doi.org/10.1155/2014/786354
7. Altenberger, I.: deep rolling—the past, the present and the future (2005)

8. Bozdana, A.T., Gindy, N.N., Li, H.: Deep cold rolling with ultrasonic vibrations-a new mechanical surface enhancement technique. Int. J. Mach. Tools Manuf. **45**(6), 713–718 (2005)
9. Zhu, Y.L., et al.: Evaluation of an ultrasound-aided deep rolling process for anti-fatigue applications. J. Mater. Eng. Perform. **18**(8), 1036 (2009)
10. Liu, Y., Zhao, X., Wang, D.: Determination of the plastic properties of materials treated by ultrasonic surface rolling process through instrumented indentation. Mater. Sci. Eng. 21–31 (2014)
11. Juijerm, P., et al.: Fatigue lifetime improvement of aluminium alloys by mechanical surface treatment. In: Proceeding of Thailand Materials Science and Technology Conference
12. http://www.constellium.com/technology-center/aluminium-alloy-properties. Last accessed 2017/07/13
13. Beghini, M., Bertini, L., Monelli, B.D., Santus, C., Bandini, M.: Experimental parameter sensitivity analysis of residual stresses induced by deep rolling on 7075-T6 aluminium alloy. Surf. Coat. Technol. **254**, 175–186 (2014)
14. Xie, J., Zhu, Y., Huang, Y., Bai, C., Ye, X.: Microstructure characteristics of 30CrMnSiNi2A steel after ultrasound-aided deep rolling. J. Mater. Eng. Perform. **22**(6), 1642–1648 (2013)

Chapter 10
Analysis of Hot Workability of Inconel Alloys Using Processing Maps

Gauri Mahalle, Nitin Kotkunde, Amit Kumar Gupta and Swadesh Kumar Singh

Abstract Understanding deformation behavior and workability is key prerequisite to optimize process parameters to improve the material processing conditions and confirming safe performance during forming. The deformation characteristics of Inconel 718 and 625 alloys have been investigated using hot tensile tests at temperature range (room temperature to 973 K) at interval of 100 K and at quasi-static strain rate of 0.0001, 0.001, and 0.01 s^{-1}. The experimental findings confirmed that tensile flow stresses are considerably subtle to strain deformation temperature and strain rate. From tensile flow stress behavior, processing maps have been developed using dynamic material model approach at different value of strains. It was observed that the temperature range ($T < 550$ K) for Inconel 718 and 600–900 K for Inconel 625 for lower strain rate (0.0001 s^{-1}) efficiency were maximum at different considered strain values. The optimum working parameters for both the alloys have been identified.

Keywords Inconel alloys · Deformation behavior · Workability · Hot tensile test · Processing maps

10.1 Introduction

High-strength Inconel alloys are special grade of precipitation-hardenable Ni-Cr-based superalloys with higher concentration (by wt%) of nickel (51–61%), chromium (18–21%), and iron (18.5–4%) in addition with relative higher contents of niobium, molybdenum, titanium, aluminum and carbon [1]. Because of the excellent combination of alloying elements, Inconel alloy possesses superior

G. Mahalle · N. Kotkunde (✉) · A. K. Gupta
Department of Mechanical Engineering, BITS Pilani Hyderabad Campus, Hyderabad 500078, Telangana, India
e-mail: nitink@hyderabad.bits-pilani.ac.in

S. K. Singh
Department of Mechanical Engineering, GRIET, Hyderabad 500072, Telangana, India

R. G. Narayanan et al. (eds.), *Advances in Computational Methods in Manufacturing*, Lecture Notes on Multidisciplinary Industrial Engineering,
https://doi.org/10.1007/978-981-32-9072-3_10

mechanical properties like high yield (700–1250 MPa) and ultimate tensile strengths (950–1650 MPa), good creep and rupture strengths, and high resistance to fatigue and corrosion at elevated temperature. This results in broad applicability in aerospace, nuclear, marine, and chemical industries [2, 3]. Because of constricted forming temperature assortment, more deformation resistance and intricate microstructures, extensive investigation of flow behavior at high temperature of Inconel alloy is very important [4, 5]. Moreover, Inconel alloys are difficult to machine to aimed shape due to extreme toughness, poor thermal conductivity, work hardening characteristics, and presence of hard abrasive carbide at elevated temperature [6].

Understanding deformation behavior is key necessity to optimize process parameters for improving the material processing conditions and confirming safe performance during various manufacturing process. Significant investigations attempted to improve constitutive expression to define complex deformation characteristics by strain, temperature, and strain-rate effects [7, 8]. Tan et al. [9] examined the deformation characteristics for Inconel 718 superalloy tested in hot compression experiments and fitted Arrhenius constitutive equation. Further, optimal processing parameters were determined. Cai et al. [10] stated that strain does not influence the processing map in the test temperature (1050–1180 °C) and developed the deformation equation to describe necessity of peak stress on tested temperature and strain rate. Zhang et al. [11, 12] observed the flow behavior, microstructure evolution, and dislocation behavior with processing maps of the Ni-based superalloy. Peak efficiency in the process of deformation is highly influenced by dislocation density in instability domain with low DRX. Authors investigated effect of the early aging time and microstructure on the processing map of precipitation-hardenable Ni-based superalloy [13, 14]. Though, no substantial effort has been reported in studied literature on the processing plots of Inconel alloys at warm temperatures and quasi-static range strain rates. This paper focuses on the processing map for Inconel alloys at warm temperatures and quasi-static strain rate.

10.2 Materials and Experimental Procedures

In present experimental work, Inconel 718 and 625 alloy sheets, of thickness 1 mm with nominal composition (wt%) as mentioned in Table 10.1, are used. Sub-sized *ASTM E08/E8M-11* standard dimensions are used as flat tensile test specimens, as shown in Fig. 10.1a. Normally, strain-rate scale of the static tension test, either with screw driven or hydraulic machine, is measured in the middle of 10^{-5} and 10^{-1} s^{-1} [15]. It is valuable for studying hot forming performed at lower strain rates as Inconel alloys are strain-rate sensitive [16]. Tensile tests are conducted on a UTM with maximum weight carrying capacity of 50 kN. Two-zone split furnace is having heating capacity up to 1500 K with ±3 K accuracy. Tensile experiments are accomplished from RT to 973 K intervals of 100 K and at 0.01, 0.001, and

Table 10.1 Chemical composition (wt%) of received Inconel 718 and 625 alloy sheet

Element	Ni	Fe	Mo	Cr	Nb	Al	Mn	Ti	Si	Others
For Inconel 718	51.463	20.44	2.871	18.279	5.012	0.561	0.062	1.092	0.051	0.169
For Inconel 625	61.495	3.304	9.479	21.739	3.271	0.067	0.123	0.166	0.101	0.104

0.0001 s^{-1} quasi-static strain rates. First, samples are heated at 20 °C/min to their deformation temperature and then heat preservation time is 10 min to achieve uniform temperature. The Experimental procedure for Quasi-static isothermal hot tensile tests is shown in Fig. 10.1b.

Figure 10.2a–d gives representative true stress–strain curves of alloys measured along rolling direction with different temperatures. Decrease in the flow stress and increased fracture strain show that flow stresses are significantly influenced by test temperature and strain rates. First, flow stress rises linearly to yield stress value for small strain values, due to effect of strain hardening. Noticeable observation from true stress–strain curve at lower strain rate is the serrated yielding phenomenon reported over the temperature range 673–973 K. The serration behavior is described by the Portevin-Le Chatelier effect specially, when homogeneous deformation occurs. The serration observed in Inconel alloys is combination of type A and B, the rise in strain rate and test temperature, according to Rodriguez classification of serrated flow [17]. Similar observations were reported previous studies [18, 19]. Differently, confirming the serrated yielding phenomena is by strain-rate sensitivity (*m*) parameter. The negative m value confirms claim of PLC or DSA effect in alloys.

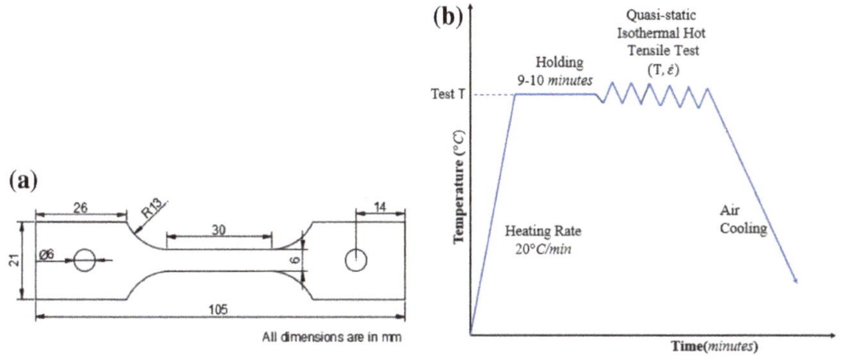

Fig. 10.1 **a** Schematic of used sub-sized ASTM E08/E8M-11 tensile test specimen, **b** Experimental procedure for Quasi-static isothermal hot tensile tests

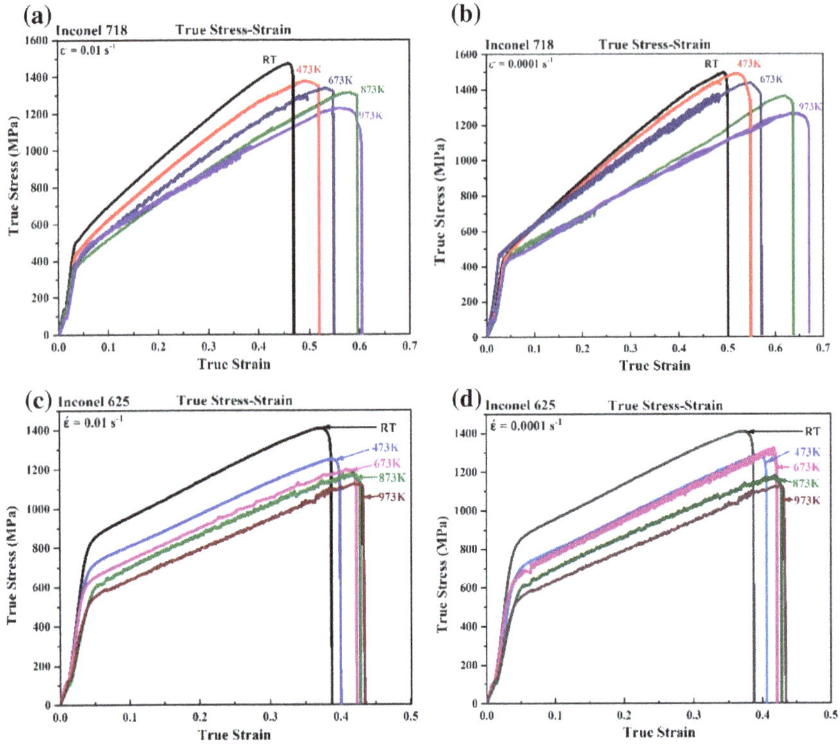

Fig. 10.2 True stress–strain curve of Inconel alloys at various test temperature with strain rate of **a–c** 0.01 s^{-1} and **b–d** 0.0001 s^{-1}

10.3 Processing Map

Processing map is basically combination of the power dissipation efficiency plot and flow instability plot for various temperatures and strain rates. It shows occurrences of damage as a function of processing variable (temperature, strain and strain-rate sensitivity). Processing map is very useful for the selection of deformation processing conditions. Fracture boundaries in processing map are very sensitive to prior thermomechanical history, microstructure, composition of metal, and numerous material parameters (like diffusivity).

At constant temperature and strain, the test metal piece has been considered as energy dissipation system. Considering material as energy dissipation, total input power (*P*) is [20]

$$P = \sigma\dot{\varepsilon} = G + J = \int_0^{\dot{\varepsilon}} \sigma \, \mathrm{d}\dot{\varepsilon} + \int_0^{\sigma} \dot{\varepsilon} \, \mathrm{d}\sigma \tag{10.1}$$

where σ and $\dot{\varepsilon}$ are flow stress and strain rate. G content defines the heat dissipation during plastic deformation and J co-content defines internal energy variation because of the microstructural evolution. While in plastic deformation process, heat dissipated mostly by G content (due to temperature rise) and less amount by J co-content (due to microstructural evolution). Strain-rate sensitivity (m) from flow stress expression is used to determine the power distribution among G content and J co-content. For plastically deforming materials, power law can be represented as

$$\sigma = k(T, \varepsilon, \dot{\varepsilon})\dot{\varepsilon}^{m(T,\varepsilon,\acute{\epsilon})} \tag{10.2}$$

when strain and deformation temperature are considered as constants. The flow stress expression is given as [20, 21]

$$\sigma = k\dot{\varepsilon}^{m} \tag{10.3}$$

where K is strain hardening constant and m is the strain-rate sensitivity, calculated as [20, 21]

$$m = \frac{\partial J}{\partial \sigma} = \frac{\dot{\varepsilon} d\sigma}{\sigma d\dot{\varepsilon}} = \frac{\partial(\ln \sigma)}{\partial(\ln \dot{\varepsilon})} \tag{10.4}$$

Figure 10.3 plotted with ln σ and ln $\dot{\varepsilon}$ values for tested temperatures and fitted with polynomial quadratic fit. m values are determined from slope of these lines. This value varies from 0 to 1 ($m = 0$ represents no power dissipation, $m = 1$ represents ideal linear dissipator, and $m > 0$ represents material with positive strain-rate sensitive).

Diversions of m at 0.15 strain with respect to strain rate and deformation temperature are plotted in Fig. 10.4. From the figures, m value first drops and then rises as deformation temperature increases. It is because of microstructure evolution which needs more heat dissipation for particular test temperature. Even when m is

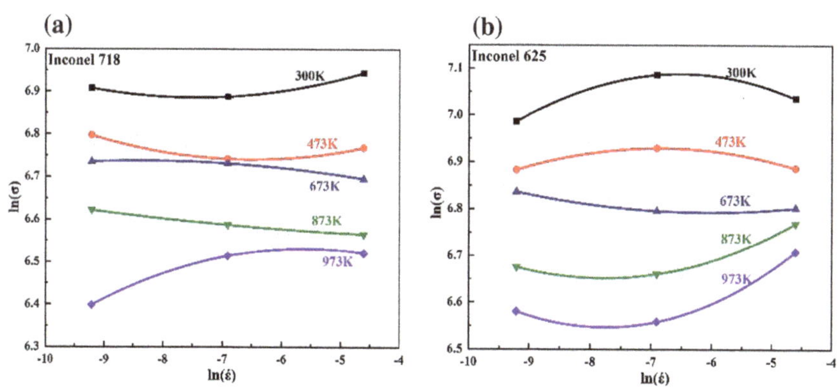

Fig. 10.3 Correlation between the flow stress and strain rate at 0.15 strain of Inconel alloys (**a**, **b**)

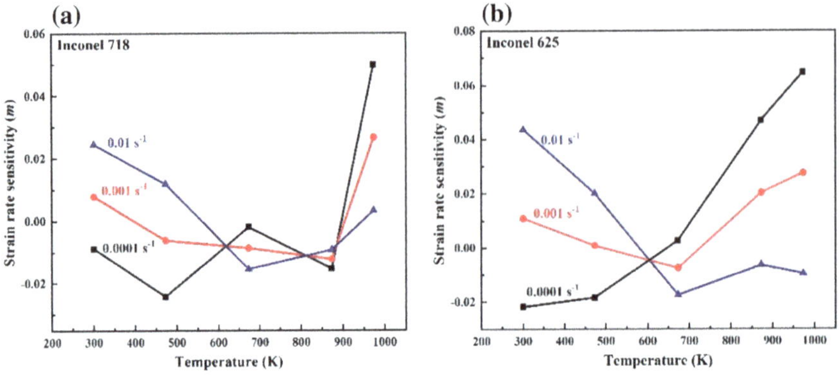

Fig. 10.4 Changes in strain-rate sensitivity (m) exponent at 0.15 strain with temperature for various strain rates of Inconel alloy (**a**, **b**)

negative or zero then also microstructural evolution can occur, this must be evaluated by joined analysis of both efficiency plot and instability plot.

Power dissipation map is one of the representations of the power dissipation efficiency (η) in the microstructural evolution. Dissipation efficiency (η) is expressed as

$$\eta = \frac{J}{J_{\max}} = \frac{2m}{m+1} \tag{10.5}$$

Here,

$$\begin{aligned} J &= \int_0^{\sigma} \dot{\varepsilon} \mathrm{d}\sigma = \frac{m}{m+1} \sigma \dot{\varepsilon} \\ J_{\max} &= \frac{\sigma \dot{\varepsilon}}{2} \end{aligned} \tag{10.6}$$

and $J_{\max}$ represents the ideal linear dissipation (m = 1).

During deformation process, to identify the microstructural instability, number of flow instability criterion has to be reported by researchers [22, 23]. Most commonly used criteria for creation of the processing maps are Prasad instability criterion. Researchers have reported that Prasad instability criterion is more conventional to use than other [24, 25]. Therefore, Conventional Prasad's criterion is centered on maximum rate of entropy creation in the metallurgical system and considered to trace instable regions in processing map. It is represents as

$$\xi = \frac{\partial \ln[m/(m+1)]}{\partial \ln \dot{\varepsilon}} + m \tag{10.7}$$

here, ξ is the dimensionless instability parameter.

Instability plot is divided generally into two regions by instability parameter ξ: region with $\xi < 0$ considered as the instability domain, whereas $\xi > 0$ considered as the stable domain. Contour plot constructed to conclude difference of η at various strain rate and deformation temperature with a fixed strain value. Figure 10.5 signifies contour plot viewing changes of η value considered for Inconel alloys at different strain rate and deformation temperature at strain of 0.15. This plot is called as power dissipation plot. Contour value represents efficiency of the material's power dissipation. From this figure, it is noted that the η value rises with a drop in the strain rate at low test temperature as well as in high test temperature region. Highest for Inconel 718 alloy is 38%, which occurs at a temperature range of 300–480 K and lower stress rate (0.0001 s^{-1}). While the highest η around 35% for Inconel 625 alloy is occurring at 650–700 K temperature range also at lower strain rate of 0.0001 s^{-1}. Generally, except few unpredictable variations (such as unpredictable fracture), high power dissipation efficiency defines better workability of the material [24]. Hence, it is necessary to consider the other aspects (unreliable variations) while deciding the experimental processing conditions of material. The goal of microstructural optimization is not sufficient only on the basis of the highest efficiency [24, 25].

Figure 10.6 represents processing plots by superimposing efficiency plots on instability plots at 0.07, 0.15, and 0.20 strain. The shaded gray region represents the instability ($\xi < 0$). For Inconel 718 alloy, it is perceived that the instability takes place in lower temperature ($T > 600$ K) for strain rates of 0.0001 s^{-1}. This indicates that Inconel 718 alloy possess very poor workability in this range of temperature and strain rates. Whereas for Inconel 625 alloy, instability occurs in high temperature ($T > 700$ K) for all stain rates. The reason for the stability of Inconel 718 alloy is might be due to the formation of precipitate at test temperature which strengthen the alloy. For both the alloys, higher η values are detected in lower strain rates (0.0001 s^{-1}) region. Further, it can be detected that the influence of strain on unstable region is considerable. The total area of the unstable region increases with increase in strain as almost instability is noticed in high strain rate and low-temperature regions. The highest efficiency in the stable region for 0.15 and

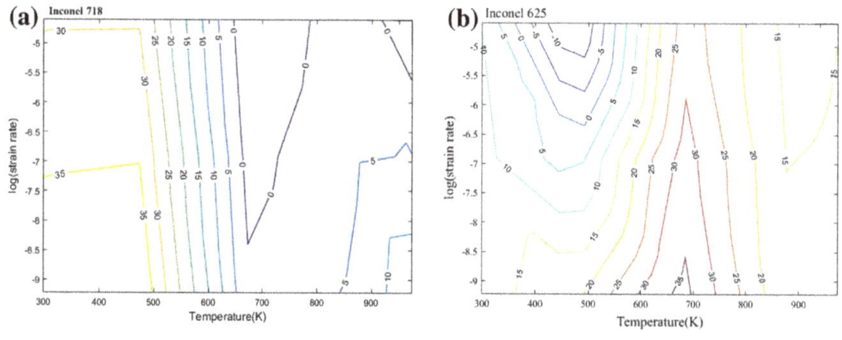

Fig. 10.5 Power dissipation efficiency plot for Inconel alloys (**a**, **b**) at 0.15 strain

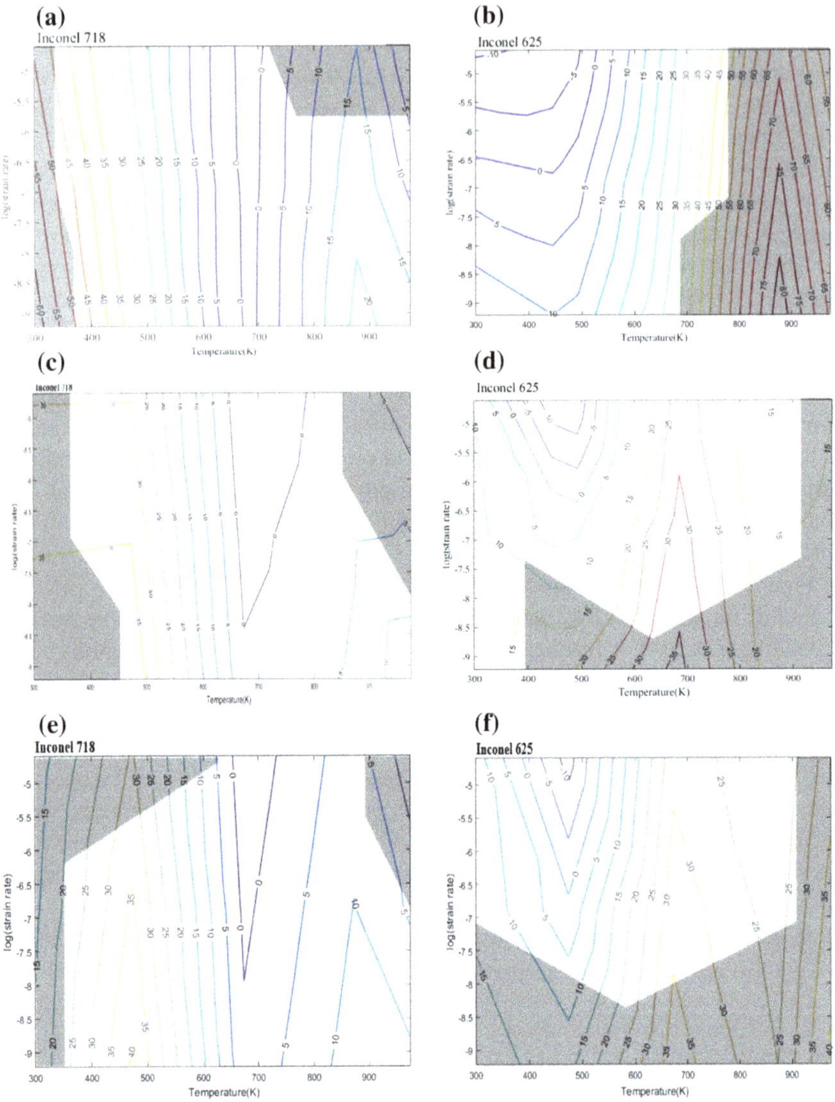

Fig. 10.6 Processing plots for Inconel alloys at **a**, **b** 0.07, **c**, **d** 0.15, and **c** 0.20 strain

0.20 strains is 35 and 38–40%, which is lower than the efficiency at 0.06 strain, i.e. 80 and 60% for Inconel 625 alloy and 718 alloy, respectively. Based on considerations, for the processing of Inconel alloys, 0.06 strain was selected as the optimal parameter condition.

10.4 Conclusions

This study involves workability analysis of Inconel alloy, namely 625 and 718, by processing plots for temperature range of 300–973 K at the strain rates of 0.01, 0.001, and 0.0001 s^{-1}. Following conclusions can be made from the analysis:

- The flow stress and strain hardening behaviors are considerably affected by strain rates and test temperatures in Inconel alloys. The reduction in yield strength observed approx. 31% (From 499.7 to 343.58 MPa) and ultimate strength by approx. 19.8% (from 951.14 to 762.14 MPa), happens with improvement in % elongation from 41.8 to 51.6% from RT to 973 K.
- Processing plots have been established for Inconel alloys at 0.06, 0.15, and 0.20 strains. It is noticed that temperature range ($T < 550$ K) for Inconel 718 and 600–900 K for Inconel 625 for lower strain rate (0.0001 s^{-1}) efficiency were maximum at different considered strain values. Therefore, this condition is considered as optimal parameter processing for the deformation process for Inconel alloys. Further, influence of strain on the instable region is noticed.

Future work includes integration of these established processing plots in finite element analysis of deformation processes of Inconel alloy.

References

1. Special Metals: INCONEL alloy 718. (2007). SMC-066
2. Reed, R.C.: The Superalloys fundamentals and applications. The Superalloys: Fundamentals and Applications. Cambridge University Press (2006). https://doi.org/10.1017/cbo9780511541285
3. Schafrik, R.E., Ward, D.D., Groh, J.R.: Application of alloy 718 in GE aircraft engines: past, present and next five years. In: Superalloys 718, 625, 706 and Various Derivatives (2001). https://doi.org/10.7449/2001/superalloys_2001_1_11
4. Lin, Y.C., Deng, J., Jiang, Y.Q., Wen, D.X., Liu, G.: Effects of initial δ phase on hot tensile deformation behaviors and fracture characteristics of a typical Ni-based superalloy. Mater. Sci. Eng. A **598**, 251–262 (2014). https://doi.org/10.1016/j.msea.2014.01.029
5. Zhang, P., Hu, C., Ding, C.G., Zhu, Q., Qin, H.Y.: Plastic deformation behavior and processing maps of a Ni-based superalloy. Mater. Des. **65**, 575–584 (2015). https://doi.org/10.1016/j.matdes.2014.09.062
6. Rahman, M., Seah, W.K.H., Teo, T.T.: The machinability of Inconel 718. J. Mater. Process. Technol. **63**, 199–204 (1997). https://doi.org/10.1016/S0924-0136(96)02624-6
7. Lin, Y.C., Chen, X.M.: A critical review of experimental results and constitutive descriptions for metals and alloys in hot working. Mater. Des. **32**, 1733–1759 (2011). https://doi.org/10.1016/j.matdes.2010.11.048
8. Liang, R., Khan, A.S.: A critical review of experimental results and constitutive models for BCC and FCC metals over a wide range of strain rates and temperatures. Int. J. Plast. **15**, 963–980 (1999). https://doi.org/10.1016/S0749-6419(99)00021-2
9. Tan, Y.B., Ma, Y.H., Zhao, F.: Hot deformation behavior and constitutive modeling of fine grained Inconel 718 superalloy. J. Alloys Compd. **741**, 85–96 (2018). https://doi.org/10.1016/j.jallcom.2017.12.265

10. Cai, D., Xiong, L., Liu, W., Sun, G., Yao, M.: Development of processing maps for a Ni-based superalloy. Mater. Charact. **58**, 941–946 (2007). https://doi.org/10.1016/j.matchar.2006.09.004
11. Zhang, P., Hu, C., Ding, C., Zhu, Q., Qin, H.: Plastic deformation behavior and processing maps of a Ni-based superalloy. Mater. Des. **65**, 575–584 (2015). https://doi.org/10.1016/j.matdes.2014.09.062
12. Zhang, H., Zhang, K., Lu, Z., Zhao, C., Yang, X.: Hot deformation behavior and processing map of a γ-hardened nickel-based superalloy. Mater. Sci. Eng. A **604**, 1–8 (2014). https://doi.org/10.1016/j.msea.2014.03.015
13. Wen, D.X., et al.: Effects of initial aging time on processing map and microstructures of a nickel-based superalloy. Mater. Sci. Eng. A **620**, 319–332 (2014). https://doi.org/10.1016/j.msea.2014.10.031
14. Sui, F.L., Xu, L.X., Chen, L.Q., Liu, X.H.: Processing map for hot working of Inconel 718 alloy. J. Mater. Process. Technol. **211**, 433–440 (2011). https://doi.org/0.1016/j.jmatprotec.2010.10.015
15. Dieter, G.E., Kuhn, H.A., Semiatin, S.L.: Handbook of Workability and Process Design (2003)
16. Iturbe, A., et al.: Mechanical characterization and modelling of Inconel 718 material behavior for machining process assessment. Mater. Sci. Eng. A **682**, 441–453 (2017). https://doi.org/10.1016/j.msea.2016.11.054
17. Rodriguez, P.: Serrated plastic flow. Bull. Mater. Sci. **6**, 653–663 (1984). https://doi.org/10.1007/BF02743993
18. Caliari, F.R., et al.: Study of the Mechanical Behavior of an Inconel 718 Aged Superalloy Submitted to Hot Tensile Tests. SAE Technical Paper 2011-36-0328, (2011). https://doi.org/10.4271/2011-36-0328
19. Maj, P., et al.: Microstructure and strain-stress analysis of the dynamic strain aging in Inconel 625 at high temperature. Met. Mater. Int. **23**, 54–67 (2017). https://doi.org/10.1007/s12540-017-6264-1
20. Prasad, Y.V.R.K., Rao, K.P., Sasidhara, S.: Hot Working Guide—A Compendium of Processing Maps. ASM International (2015). https://doi.org/10.1016/b978-0-08-033454-7.50019-x
21. Liu, J., Cui, Z., Li, C.: Analysis of metal workability by integration of FEM and 3-D processing maps. J. Mater. Process. Technol. **205**, 497–505 (2008). https://doi.org/10.1016/j.jmatprotec.2007.11.308
22. Prasad, Y.V.R.K.: Processing maps: a status report. J. Mater. Eng. Perform. **12**, 638–645 (2013). https://doi.org/10.1361/105994903322692420
23. Murty, S.V.S.N., Rao, B.N.: On the development of instability criteria during hotworking with reference to IN 718. Mater. Sci. Eng. A **254**, 76–82 (1998). https://doi.org/10.1016/S0921-5093(98)00764-3
24. Sun, Y., Wan, Z., Hu, L., Ren, J.: Characterization of hot processing parameters of powder metallurgy TiAl-based alloy based on the activation energy map and processing map. Mater. Des. **86**, 922–932 (2015). https://doi.org/10.1016/j.matdes.2015.07.140
25. Zhou, G., Ding, H., Cao, F., Zhang, B.: A comparative study of various flow instability criteria in processing map of superalloy GH4742. J. Mater. Sci. Technol. **30**, 217–222 (2014). https://doi.org/10.1016/j.jmst.2013.07.008

Part II
Welding and Joining Technologies

Chapter 11
Investigation on Metallographic Analysis of Electron Beam Ti6Al4V Alloy Welds

Sohini Chowdhury, Yadaiah Nirsanametla and Manapuram Muralidhar

Abstract Electron beam weld quality is evaluated by careful selection of weld parameters and by control of the temporal distribution of electron beam on surface of work piece. Moreover, the integral effect of the electron beam current and weld velocity on weld joint is significant during electron beam welding (EBW) process. In the present work, bead-on-plate EBW of Ti6Al4V alloy plates of 5 mm thickness is carried out to examine the influences of various process parameters with respect to microstructure evolution and weld bead shapes and dimensions. Also, a relationship among input thermal energy, macro- and microstructural characteristics of electron beam Ti6Al4V alloy weldments is established. The experiments were carried out with beam current ranging from 15 to 30 mA and welding velocity from 1200 to 1400 mm min^{-1}. Moreover, a constant focus current and accelerating voltage of 2030 mA and 60 kV are considered. Furthermore, welding modes such as melt-in and keyhole modes are examined with respect to the linear energy and welding process efficiency along with microstructure evolution. In addition, the magnitude of temperature in the heat-affected zone of electron beam welded Ti6Al4V alloy samples is determined using Rosenthal's analytical model at distinct weld regimes. Furthermore, Marangoni and Fourier numbers are employed to investigate weld pool characteristics at distinct weld regimes. Based upon the results, it is observed that electron beam power has a significant influence on weldment profile and penetration level while the welding speed has a noteworthy impact on the solidified structure of fusion zone. Also, three weld regimes were identified under the specified welding conditions. Keyhole mode welding is achieved for linear energy above 77 J mm^{-1}, and melt-in mode welding is attained when linear energy is below 45 J mm^{-1}. Moreover, transition mode welding is identified when the linear energy is in between 56.57 and 75 J mm^{-1}. Furthermore, the efficiency of welding process was assessed by relating electron beam power with fusion zone area. It was recognised that a higher linear energy promoted a higher process efficiency comparative to a lower linear energy. A lower linear

S. Chowdhury · Y. Nirsanametla (✉) · M. Muralidhar
Department of Mechanical Engineering, North Eastern Regional Institute of Science and Technology (NERIST), Nirjuli, Arunachal Pradesh 791109, India
e-mail: ny@nerist.ac.in

R. G. Narayanan et al. (eds.), *Advances in Computational Methods in Manufacturing*, Lecture Notes on Multidisciplinary Industrial Engineering,
https://doi.org/10.1007/978-981-32-9072-3_11

energy promoted a rich domain of acicular α^1 solidified structure in the fusion zone when compared with higher linear energy. Moreover, the integrity of bead-on-plate welds of Ti6Al4V alloy is of sound quality and without any internal weld defects.

Keywords Electron beam welding · Fourier number · Ti6Al4V alloy · Marangoni number · Microstructure evolution · Rosenthal model

11.1 Introduction

The necessity for lightweight materials has substantially escalated in the fabrication industries on account of critical service aspects. Specifically, titanium alloys are currently implemented in all ranges of applications such as automotive, aerospace and nuclear applications. However, the extensive employment of titanium alloys in critical manufacturing domain can be attributed to excellent properties at high temperature, superior fatigue behaviour, corrosion resistance features and high specific strength [1–3]. In addition, the usage of titanium alloys has now been extended to bio-medical devices because of high bio-compatibility property [4]. α-β two-phase Ti6Al4V alloy is the most commonly used titanium-based alloys and is widely implemented in major domain of manufacturing sector. Earlier, several researchers attempted to join titanium assemblies with numerous joining techniques including fusion welding and solid state joining [5–10]. However, an elemental issue of joining titanium and its alloys is its reaction with atmospheric gases at an elevated temperature which results into embrittlement in the welded specimen [11]. Therefore, the EBW process is the most desirable welding technique to join titanium metal assemblies as it provides a controlled atmosphere to operate. Moreover, EBW offers production of wide span of complicated weld geometries along with an enhanced quality level, flexibility and processing capability. Besides, high energy density electron beam produces welds with minimum heat-affected zone (HAZ) and deep penetration which accounts for low level of distortion in weldments. However, the deep penetration mechanism is associated with vapour cavity formation also referred to as keyhole which is enclosed by high-temperature molten weld pool.

Although, the output of welding operation is prominently established upon keyhole characteristics whose integrity is governed by an equilibrium between recoil pressure in the vapour cavity and surface tension of the weld pool. Hence, weld penetration and weld defects are directly related to keyhole stability that implies; if keyhole collapses, a lower penetration is attained. Besides, weld defects in the form of porosities, spiking defects and several others are generated due to keyhole instability and fluctuations. Moreover, several experimental efforts also provided a qualitative understanding of keyhole dynamics as a function of time [12–14]. Also, the keyhole stability is based on optimisation of weld parameters. Akman et al. demonstrated that control of laser output parameters can efficiently control weld penetration and weld bead width in laser welding process

[15]. Squillace et al. [16] reported that the under fill defects in Ti6Al4V alloy laser welds depends on a definite heat input. The authors also determined that the under fill defect is maximum in keyhole mode welds as compared to conduction mode welds. Li et al. studied the influence of electron beam weld parameters on weld profile and weld distortion of titanium-based Ti2AlNb alloy plates [17]. Liu et al. conducted experiments to determine the welding stresses induced after electron beam welding and milling operations from a thick titanium metal plate of 50 mm thickness [18]. Also, Junaid et al. studied the effect of beam parameters on titanium-based Ti5Al2.5Sn alloy sheets with respect to weld bead width and mechanical characteristics. However, from the esteemed reports, it is recognised that EBW of Ti6Al4V alloy is limited. Thus, a comprehensive analysis of EBW of Ti6Al4V alloy can lay a good contribution to the prevailing approaches.

In this work, electron beam welding experiments were performed on 5 mm thick Ti6Al4V alloy plates and weld seams are evaluated in terms of weld geometry and microstructure. The aim of the present work is to establish a relationship between input thermal energy, weld morphology and microstructural characteristics for electron beam welded Ti6Al4V alloy at distinct weld regimes. Moreover, different weld regimes are characterised for Ti6Al4V alloy samples under different welding conditions corresponding to Marangoni number and Fourier numbers. Also, the temperature distribution in HAZ of Ti6Al4V alloy is determined by employing Rosenthal's analytical model.

11.2 Experimental Investigation

In the present work, Ti6Al4V alloy in annealing condition is considered for experimental investigation. The electron beam bead-on-plate welds were conducted on 5 mm thick Ti6Al4V alloy, and the plate dimension (length × width) is 100 mm 50 mm. Prior to welding operations, all the sample surfaces were properly cleaned with acetone and clamped carefully by fixtures to prevent distortion during welding operation. However, to ensure high thermal transfer rate during joining process, the backing plate in fixture is assembled with copper material.

The weld tests were carried out for a suitable combination of weld speed and electron beam power where a stable weld seam is acquired. The different welding conditions that were adapted during EBW of Ti6Al4V alloy are listed in Table 11.1. The experiments were conducted by varying weld speed and electron beam current at fixed accelerating voltage and focusing condition. The accelerating voltage level was maintained constant at 60 kV for all the samples, while the beam current was varied between 15 and 30 mA and weld speed was ranged between 1200 and 1400 mm/min. Furthermore, for evaluation of weld properties, the metallographic examination was conducted. The specimens were sectioned at middle segment across the weld length, grounded and chemically etched with Keller's reagent solution. Thereafter, the weld cross sections and microstructure of the polished weld specimen were acquired with optical microscope.

Table 11.1 Electron beam welding conditions of Ti6Al4V alloy

Weld sample	Accelerating voltage (kV)	Beam current (mA)	Focus current (mA)	Weld speed (mm/min)
1	60	15	2030	1400
2	60	15	2030	1200
3	60	22	2030	1400
4	60	22	2030	1200
5	60	25	2030	1400
6	60	25	2030	1200
7	60	30	2030	1400
8	60	30	2030	1200

To characterise the weld geometry and distinguish weld regimes of Ti6Al4V alloy weldment, linear energy is considered. The linear energy (L_e) which is the ratio of power transferred to the weld speed is represented as

$$L_E = VI/u \tag{11.1}$$

where V, I and u indicate accelerating voltage (kV), electron beam current (mA) and weld speed (mm/s), respectively. Moreover, the process efficiency is quantified by fundamental welding process efficiency (W_E) parameter which correlates the transmitted beam power with fusion zone area. It takes into account the energy required for creating the joining or welding phenomena in the material. Hence, the parameter is represented as

$$W_E = A/L_e \tag{11.2}$$

where A indicates area of the fusion zone (mm^2).

In order to demonstrate thermal and fluid flow characteristics of Ti6Al4V alloy welds, Marangoni number and Fourier numbers are employed. Marangoni number is represented as a ratio of surface tension force to viscous force, and it is mathematically represented as

$$M_a = \frac{d\gamma}{dT} \times \frac{w\Delta T}{\mu\alpha} \tag{11.3}$$

Fourier numbers is defined as a ratio of thermal transmission rate to thermal storage rate and mathematically given as

$$F_o = \frac{\alpha}{ul} \tag{11.4}$$

where $d\gamma/dT$, w, μ, α, ΔT represent surface tension gradient (N m^{-1} K^{-1}), width of molten pool (mm), viscosity (kg m^{-1} s^{-1}), thermal diffusivity (m^2 s^{-1}), temperature gradient (K).

11.3 Results and Discussion

In this section, the macro- and microstructural characteristics of Ti6Al4V alloy weld joint are characterised for a wide range of linear energy varying from 38.57 to 90 J mm^{-1}. The estimation of temperature is also carried out using Rosenthal's analytical model. Moreover, weld pool behaviour is identified by Fourier numbers and Marangoni number at distinct weld regimes.

The effect of process variables such as beam current and welding speed on weld morphology and microstructural characteristics is examined. Figure 11.1 demonstrates the appearance of bead-on-plate Ti6Al4V alloy weld joint corresponding to weld sample #1 at 38.57 J/mm linear energy. It is observed that a smooth, bright, clean weld bead is formed with no oxide formation which indicates the effectiveness of electron beam welding process.

Moreover, the seam morphology of Ti6Al4V alloy weldments corresponding to weld sample #2 and #8 at a constant welding velocity of 1200 mm min^{-1} is presented in Fig. 11.2. Different weld profiles and weld regimes are identified under different linear energy values. The macrograph displays an elliptical profile at linear energy of 45 J/mm (Fig. 11.2a) while a nail head profile (Fig. 11.2b) is attained at linear energy of 90 J/mm. The transition from an elliptical profile to nail head profile is attributed to an increase in linear energy which increases the weld

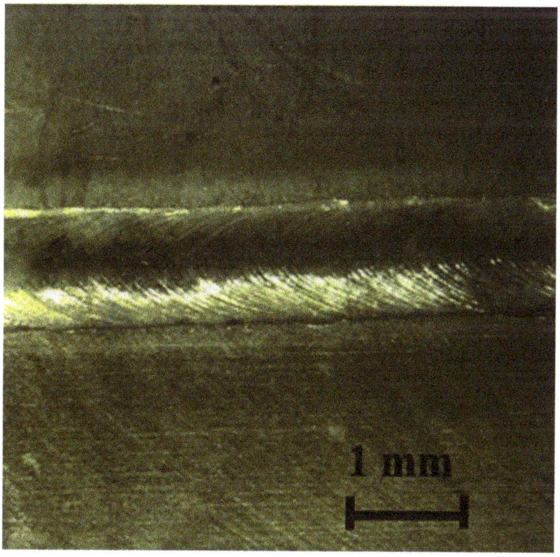

Fig. 11.1 Appearance of the weld surface at linear energy 38.57 J/mm

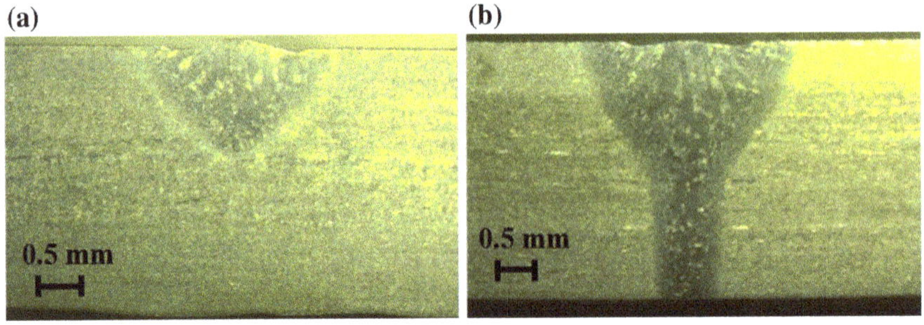

Fig. 11.2 Cross section of weld samples **a** #2 and **b** #8 at constant weld speed of 1200 mm/min

penetration considerably. Also, the cross-sectional macrograph of Ti6Al4V alloy weldments displays a conduction mode welding (Fig. 11.2a) at 45 J/mm and a keyhole mode welding (Fig. 11.2b) at 90 J/mm. This indicates that conduction mode welding is attained for linear energy values at or below 45 J/mm, and keyhole mode welding is attained for linear energy values at or above 77 J/mm. In addition, transition mode welds are also observed between linear energy values of 56.57 and 75 J/mm. Besides, full penetration welds are desirable in most applications due to high joint strength acquired. Moreover, the macrograph reveals no major weld defects such as porosity and spiking defects.

Table 11.2 depicts fusion zone area, linear energy and welding process efficiency values of weld samples. Weld sample 2 reveals a higher welding efficiency than weld sample 1. This is probably due to the fact that the energy spent in the form of heat conduction loss and overheating is higher than energy spent to increase fusion zone area of weld sample 1. Conversely, for weld sample 1, more energy is spent on increase in fusion zone area than thermal losses. Therefore, it can be concluded that lower weld speed or higher linear energy promotes higher process efficiency than higher weld speed or lower linear energy at a constant electron beam power. Figure 11.3 illustrates the cross-sectional macrographs of weld samples under varying linear energy conditions. Figure 11.3a, b corresponds to weld sample

Table 11.2 Efficiency parameters of electron beam welded Ti6Al4V alloy

Weld sample	Weld area (mm^2)	Linear energy ($J\ mm^{-1}$)	Welding process efficiency ($mm^3\ J^{-1}$)
1	3.53	38.57	0.078
2	5.972	45	0.154
3	8.58	56.57	0.135
4	9.43	66	0.166
5	10.422	64.28	0.138
6	11.2	75.75	0.174
7	12.541	77.14	0.139
8	14.25	90	0.184

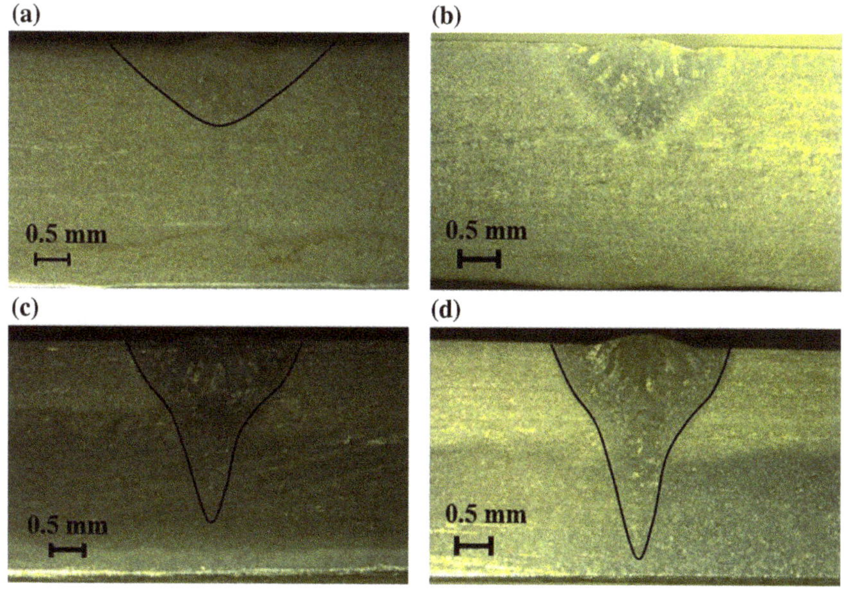

Fig. 11.3 Cross-sectional macrographs of weld samples corresponding to linear energy, **a** 38.57 J mm^{-1}, **b** 45 J mm^{-1}, **c** 75.75 J mm^{-1} and **d** 64.28 J mm^{-1}

1 and 2, respectively, which refers to a decrease in weld speed from 1400 to 1200 mm/min at constant beam current of 15 mA. Similarly, Fig. 11.3c, d corresponds to weld sample 5 and 6, respectively, which refers to a decrease in weld speed from 1400 to 1200 mm/min at a constant beam current of 25 mA. The reduction in weld penetration and an increase in weld width are identified at higher weld speed. Moreover, with an increase in beam current, penetration level in weldment is increased remarkably.

The heat transfer mode and material flow characteristics in the molten metal during welding operation determine the weld joint quality. Moreover, the prediction of temperature using experimental methods is critical during the welding process since molten weld pool width is quite small and there is a possibility of plasma formation. Hence, an analytical model is employed in the present work for estimation of the temperature field in the HAZ analytically. Also, with the purpose of addressing the issue of heat transfer mode in electron beam weldments, Marangoni number [19] and Fourier numbers [20] are employed.

In the present work, Rosenthal's analytical model is employed, since it takes into account the integral effect of processing parameters and thermal properties of Ti6Al4V alloy for prediction of temperature analytically [21, 22]. The positions for estimation of peak temperature in transverse direction of welding line are shown in Fig. 11.4a. The estimated temperature values corresponding to welding conditions are given in Table 11.1 and presented in Fig. 11.4b. It can be identified that the magnitude of temperature in HAZ of weld plates is much lower than liquidus

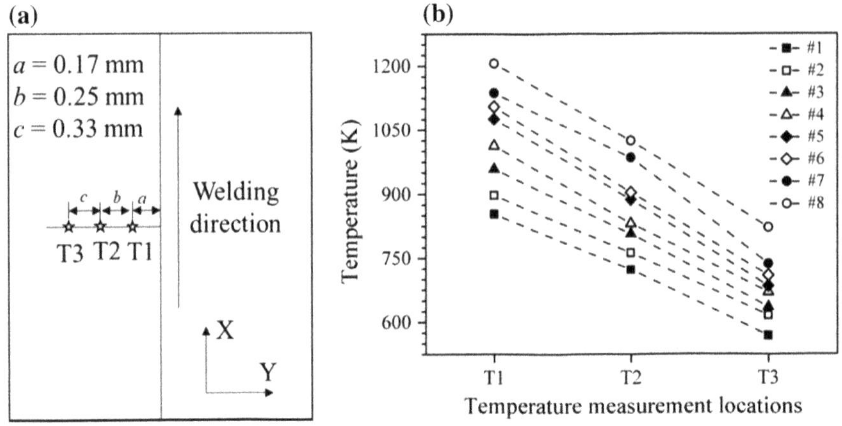

Fig. 11.4 **a** Temperature measurement locations along the welding line and **b** estimated temperature values corresponding to processing conditions of Table 11.2

temperature. Also, temperature at location '*T*1' is determined to be higher than '*T*2' and '*T*3' locations. Since '*T*1' corresponds to a location near FZ while, '*T*2' and '*T*3' correspond to the locations away from the melt zone. Moreover, it is determined that conduction weld (#2) exhibited a lower temperature value when compared with keyhole weld (#7). Furthermore, it is determined that the percentage difference in magnitude of temperature at '*T*1' location for conduction and keyhole mode is 26.0.

Figure 11.5 presents the variation of Marangoni number (Ma) and Fourier numbers as a function of linear energy corresponding to electron beam Ti6Al4V alloy welds. It is observed that *Ma* and *Fo* increases and reduces linearly with increases in magnitude of linear energy, respectively. Moreover, three distinct weld regimes are separated with respect to *Ma* and *Fo* values. Fundamentally, higher *Ma* value indicates higher molten metal velocity at the surface and convective heat

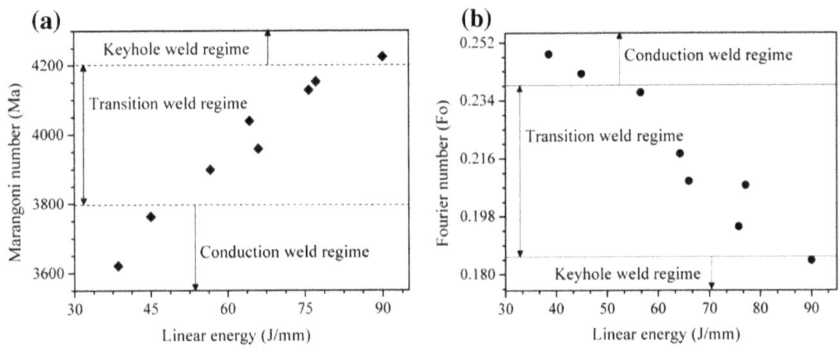

Fig. 11.5 Influence of linear energy on **a** Marangoni number and **b** Fourier number

transmission in the weld pool dictates the process. From Fig. 11.5a, it can be observed that the conduction weld regime exhibits lower Marangoni number when compared with transition and keyhole weld regime. This reflects that conduction welds are subjected to low molten metal velocity at the surface and produces a more stable weld bead with lower aspect ratio. The Fourier numbers is associated with solidification rate of crystals at the FZ which eventually accounts for thermo-mechanical damage in the material. Higher Fourier numbers is always preferable in weldments, since it indicates lower thermal storage rate and lower characteristics length of the molten pool on account of high solidification rate. From Fig. 11.5b, it is observed that conduction weld regime exhibited higher Fourier numbers than transition and keyhole welds. Thus, conduction welds are subjected to high solidification rate and low thermo-mechanical damage relative to transition and keyhole welds.

The microstructure of parent metal of Ti6Al4V alloy primarily consists of body-centred cubic (BCC) β phase distributed in an area of hexagonal close-packed α phases. However, during welding operation, due to rapid heating and cooling process, transformation from α to β phase and decomposition of β to α phase occurs at the fusion zone, respectively, in a short duration of time. This entire heating and cooling operation depends on heating and cooling rate of Ti6Al4V alloy. In case cooling rate is greater than critical cooling rate of Ti6Al4V alloy (around 623 K/s), then microstructure evolution from β to α^1 martensite structure is promoted in the weld zone. Similarly, at lower cooling rate, secondary lamella growth is promoted due to diffusion-controlled nucleation [23]. However, β phases can be traced out in the fusion zone irrespective of cooling rate of Ti6Al4V alloy weldments. The microstructure of fusion zone region of weld samples #1 (Fig. 11.6a) and #8 (Fig. 11.7a) displays similar microstructure at the weld zone comprising of

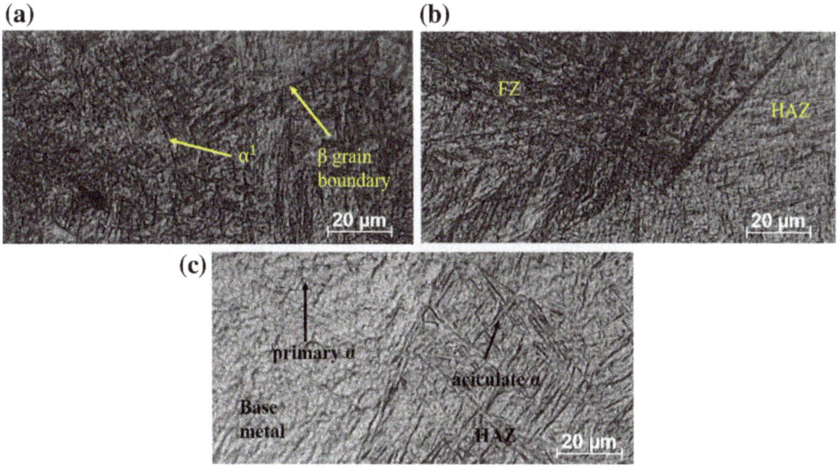

Fig. 11.6 Illustartion of microstructure at different zones for Ti6Al4V alloy weldment at 38.57 J/mm

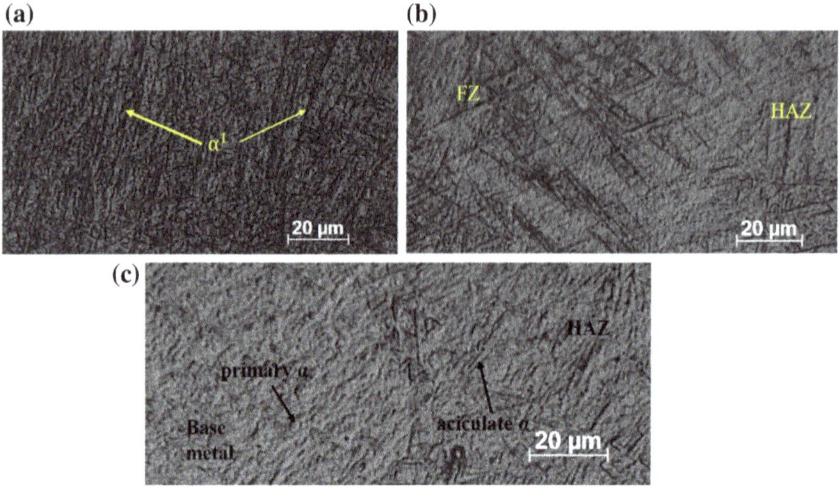

Fig. 11.7 Illustartion of microstructure at different zones for Ti6Al4V alloy weldment at 90 J/mm

martensitic α^1 solidified structure within the β grains. The solidified microstructure α^1 is evolved when β phase is quenched above beta transus temperature at a cooling rate above the critical cooling rate of Ti6Al4V alloy [24]. Moreover, the length and distribution of martensitic structures in the FZ vary in accordance with the solidification rate. It has been determined that the size of α^1 grains is higher in keyhole weld regime relative to the conduction weld regime. However, the percentage content of martensitic structure reduces with increase in linear energy. Therefore, it can be inferred that the fusion zone of weld samples #1 and #8 does not differ much by cooling rate. Although, the amount, distribution and size of acicular α^1 solidified structure at each weld zone differ comparatively corresponding to weld speed. Also, the cooling rate of Ti6Al4V alloy weldment is comparatively enhanced due to implementation of copper as backing plate in fixture [25].

Figures 11.6c and 11.7c display the different weld zones of Ti6Al4V alloy corresponding to weld samples #1 and #8 at 1400 mm/min and 1200 mm/min weld speed, respectively. The heat-affected zone (HAZ) displays a much more complex microstructure as it consists of primary α grains, martensitic α^1 and β grains. Moreover, due to temperature gradient in the Ti6Al4V alloy specimen, the microstructure in the HAZ varied from a rich α^1 (near fusion zone) to a poor α^1 (far from fusion zone) [26]. However, the microstructure in distinct weld zones is directly proportional to the linear energy. A higher linear energy leads to lower cooling rate which induces diffusion-controlled phase transformation. Conversely, a lower linear energy promotes a rich domain of acicular α^1 solidified structure in the fusion zone.

11.4 Conclusion

This paper has presented detailed experimental results of electron beam welding process of 5 mm thick Ti6Al4V alloy specimen. Based on the present investigation, the following conclusions are derived.

- The linear energy which is expressed as power transmitted to weld speed is an influential parameter to determine weld profiles and weld regimes of Ti6Al4V alloy specimen.
- A lower weld speed promotes higher process efficiency than higher weld speed.
- The temperature distribution in HAZ of Ti6Al4V alloy weldments is lower in conduction weld regime than keyhole weld regime.
- The fusion zone of electron beam Ti6Al4V alloy weldments primarily consists of martensitic α^1 structures with small amount of β grains.
- The distribution, amount and size of martensitic α^1 and β grains are influenced by heat input and cooling rate.
- Also, a higher linear energy leads to lower cooling rate which induces diffusion-controlled phase transformation.
- The conduction weld regime exhibited lower martensitic size than keyhole weld regime. However, the density of martensitic structures in the FZ is determined to be more in the conduction weld regime.
- The magnitude of *Ma* number is determined to be lower in conduction weld regime. However, *Fo* number value is relatively higher in conduction weld regime than keyhole weld regime.

Acknowledgements This work was financially supported by second phase of Technical Education Quality Improvement Programme (TEQIP), India and North Eastern Regional Institute of Science and Technology (NERIST). We also thank Hindustan Aeronautics Limited, Koraput, Odisha, India for providing experimental facility.

References

1. Veiga, C., Davim, J.P., Loureiro, A.J.R.: Properties and applications of titanium alloys: a brief review. Adv. Mater. Sci. **32**, 133–148 (2012)
2. Henriques, V.A.R.: Titanium production for aerospace applications. J. Aerosp. Technol. Manag. **1**(1), 7–17 (2009)
3. Peters, M., Kumpfert, J., Ward, C.H., Leyens, C.: Titanium alloys for aerospace applications. Adv. Eng. Mater. **5**(6), 419–427 (2003)
4. Elias, C.N., Lima, J.H.C., Valiev, R., Meyers, M.A.: Biomedical applications of titanium and its alloys. Biol. Mater. Sci. Overv. **60**(3), 46–49 (2008)
5. Saresh, N., Gopalakrishna, P.M., Mathew, J.: Investigation into the effects of electron beam welding on thick Ti-6Al-4V titanium alloy. J. Mater. Process. Technol. **192–193**, 83–88 (2007)
6. Lathabai, S., Jarvis, B.L., Barton, K.J.: Comparison of keyhole and conventional gas tungsten arc welds in commercially pure titanium. Mater. Sci. Eng.: A **299**, 81–93 (2001)

7. Greenwell, T., Ramulu, M., Labossiere, P.: Characterization of tensile behavior in friction stir-welded titanium alloy, Ti-6Al-4V. In: Proceedings of the SEM Annual Conference and Exposition on Experimental and Applied Mechanics, pp. 2659–2668 (2009)
8. Mironov, S., Sato, Y.S., Kokawa, H.: Development of grain structure during friction stir welding of pure titanium. Acta Mater. **57**(45), 19–28 (2009)
9. Balasubramanian, M., Jayabalan, V., Balasubramanian, V.: Effect of pulsed gas tungsten arc welding on corrosion behavior of Ti–6Al–4V titanium alloy. Mater. Des. **29**, 1359–1363 (2008)
10. Li, R., Li, Z., Zhua, Y., Ronga, L.: A comparative study of laser beam welding and laser–MIG hybrid welding of Ti–Al–Zr–Fe titanium alloy. Mater. Sci. Eng.: A **528**, 1138–1142 (2011)
11. Donachie. M.J.: Titanium, A. Technical Guide, 2nd edn. ASM International, Metals Park (2000)
12. Tong, H., Giedt, W.H.: Radiographs of the electron beam welding cavity. Rev. Sci. Instrum. **40**(10), 1283–1285 (1969)
13. Mara, G. L.: A study of the penetration mechanism of electron beam welding and the spiking phenomenon. Ohio State University, Master Thesis (1971)
14. Arata, Y., Matsuda, F., Murakami, T.: Some dynamic aspects of weld molten metal in electron beam welding. Trans. Join. Weld. Res. Inst. 152–161 (1973)
15. Akman, E., Demir, A., Canel, T., Sınmazcelik, T.: Laser welding of Ti6Al4V titanium alloys. J. Mater. Process. Technol. **209**, 3705–3713 (2009)
16. Squillace, A., Prisco, U., Ciliberto, S., Astarita, A.L.: Effect of welding parameters on morphology and mechanical properties of Ti-6Al-4V laser beam welded butt joints. J. Mater. Process. Technol. **212**, 427–436 (2012)
17. Li, Y., Zhao, Y., Li, Q., Wu, A., Zhu, R., Wang, G.: Effects of welding condition on weld shape and distortion in electron beam welded Ti2AlNb alloy joints. Mater. Design **114**, 226–233 (2017)
18. Chuan, L., Jianxun, Z., Bing, W., Shuili, G.: Numerical investigation on the variation of welding stresses after material removal from a thick titanium alloy plate joined by electron beam welding. Mater. Des. **34**, 609–617 (2012)
19. Robert, A., Debroy, T.: Geometry of laser spot welds from dimensionless numbers. Metall. Mater. Trans. B **32**, 941–947 (2001)
20. Mukherjee, T., Manvatkar, V., De, A., DebRoy, T.: Dimensionless numbers in additive manufacturing. J. Appl. Phys. **121**, 064904 (2017)
21. Rosenthal, D.: Mathematical theory of heat distribution during welding and cutting. Weld. J. **20**(5), 220–234 (1941)
22. Rosenthal, D.: The theory of moving sources of heat and its application to metal treatments. Trans. Am. Soc. Mech. Eng. **68**, 849–866 (1946)
23. Mayuri, B., Swarup, B.: Influence of heat Input in microwelding of titanium alloy by microplasma arc welding. J. Mater. Process. Technol. **231**, 100–112 (2016)
24. Fan, Y., Yang, Z., Cheng, P., Egland, K., Yao, Y.L.: Effects of phase transformations on laser forming of Ti-6Al-4V Alloy. J. Appl. Phys. **98**(1), 013518 (2005)
25. Ahmed, T., Rack, H.J.: Phase transformations during cooling in $\alpha + \beta$ titanium alloys. Mater. Sci. Eng.: A **243**, 206–211 (1998)
26. Giuseppe, C., Michelangelo, M., Sabina, L.C.: Ytterbium fiber laser welding of Ti6Al4V alloy. J. Manuf. Process. **20**(1), 250–256 (2015)

Chapter 12
Multi-objective Optimization of FSW Process Variables of Aluminium Matrix Composites Using Taguchi-Based Grey Relational Analysis

Subramanya R. B. Prabhu, Arun Kumar Shettigar, Mervin A. Herbert and Shrikantha S. Rao

Abstract Successful joining of aluminium alloys using friction stir welding (FSW) opens a new window research in extending this technique to join aluminium matrix composites (AMCs). Current research is focused on optimization of process variables for multiple responses simultaneously. Experiments were performed using tool pin profile, rotational speed (RS) and welding speed (WS) as ideal process variables for multi-objective optimization in FSW of AMCs. Tensile strength, macro-hardness and elongation are considered as multi-response behaviour. Grey relational grade for the chosen multiple responses are obtained using grey analysis. Analysis of variance was utilized to understand the influence of process variables on the grey relational grade. Analysis reveals that RS and WS were the most influencing process variables on the output responses. Confirmation experiments were performed at optimized process variables to validate the present study. Predicted values were in good agreement with the experimental results.

Keywords Friction stir welding · Grey relational analysis · Tensile strength · Macro-hardness · Elongation

---

S. R. B. Prabhu (✉) · M. A. Herbert · S. S. Rao
Department of Mechanical Engineering, National Institute of Technology Karnataka, Surathkal, India
e-mail: subramanya.prabhu@gmail.com

S. R. B. Prabhu · A. K. Shettigar
Department of Mechatronics Engineering, Manipal Institute of Technology, Manipal Academy of Higher Education, Manipal, India

R. G. Narayanan et al. (eds.), *Advances in Computational Methods in Manufacturing*, Lecture Notes on Multidisciplinary Industrial Engineering,
https://doi.org/10.1007/978-981-32-9072-3_12

12.1 Introduction

FSW is a novel technique invented by The Welding Institute (TWI, UK) in 1991 to join difficult to weld materials like aluminium and its alloys [1]. FSW is a solid-state welding process where materials are joined by plasticizing and forging action rather than melting and recasting [2]. FSW offers several benefits over conventional fusion welding process. A non-consuming rotating tool alters the material flow and creates a defect-free weld during FSW. The nature of material flow is affected by the tool pin profile as well as process variables [3]. Cavaliere et al. [4] performed FSW of AA6082 alloys and studied the effect of process variables on microstructure and mechanical properties. Rajakumar et al. [5] analysed the influence of process variables such as rotational speed (RS), welding speed (WS), axial load, tool pin profiles and hardness of tool on the mechanical properties of FS-welded AA 7075 alloy. Malopheyev et al. [6] worked on optimization of FSW of 6061-T6 aluminium alloy processing variables and stated that increasing the welding speed hindered precipitation coarsening which improved after ageing. Literature study reveals that most of the study focused on the influence of tool profile and process variables on microstructure and tensile properties. Studies indicate that the quality of the weld mainly depends on various process variables such as shoulder diameter, axial load, RS and WS. To predict the mechanical properties of FS-welded AA 6061, Elangovan et al. [7] created a mathematical model by utilizing process variables like WS, RS, tool geometry and axial load. Murugan and Ashok Kumar [8] also developed similar model to predict tensile strength of friction stir welded stir cast AA6061-T6/AlNp composite incorporating RS, WS, axial load and tool profile as process variables. Artificial neural network (ANN) was used by Ukuyucu et al. [9] to get correlation among mechanical properties of FS-welded aluminium plates and FSW process variables. Subramanya et al. [10] studied multi-response optimization of friction stir welding process variables, using TOPSIS approach to optimize FSW of aluminium matrix composites. Multi-response base on orthogonal array (OA) coupled with grey relational analysis was used by Vijayan et al. [11] to optimize the FSW process variables to join aluminium alloy AA5083 by FSW. Deepandurai and Parameshwaran [12] performed multi-response optimization of FSW variable for AA7075/SiCp using response surface methodology and fuzzy-based grey relational analysis. Comprehensive literature findings indicate that most of the reports allied to the optimization of FSW variables have utilized one response at a time. However, meagre study had been performed on the influence of variables on the multiple responses simultaneously. The present work is focused on optimization of variables while FSW of AMCs for macro-hardness, tensile strength and % elongation using grey relational analysis.

12.2 Experiments

12.2.1 Process Variables Selection

FSW was performed on a conventional vertical milling machine (Make: Bharat Fritz Werner). The material used in this study was AA6061-4.5(wt%) Cu-5(wt%) SiC composite. These composites were prepared by stir casting technique. Composite plates having size of 100 × 50 × 6 mm were prepared from the stir cast composite by machining. FSW tool was prepared using M2 steel with a hardness of 60 HRC. Three types of tool geometries were used in the present study, namely square (SQ), threaded cylindrical (TC) and combined threaded cylindrical and square (CTS). Tool pin geometries are shown in Fig. 12.1.

Set-up used for FSW process is shown in Fig. 12.2. Process was carried out by altering the process variable values. RS, WS and tool pin geometries are chosen as process variables to optimize the FSW of AMCs. Other variables such as tool shoulder diameter of 18 mm and tool tilt angle of 2.50 are maintained constant in order to limit the study. Working range of process variables is fixed by conducting the trial experiments which gives defectless joints. It has been observed that process variables outside the working range yields defects in the joints, such as crack and tunnel hole. The number of experiments for the present study was considered based on Taguchi's design of experiment method. As per Taguchi's method, the degree of

Fig. 12.1 FSW tools with **a** threaded cylindrical (TC), **b** combined threaded cylindrical and square (CTS) and **c** square (SQ) profiled pin

Fig. 12.2 FSW set-up

Table 12.1 FSW process variables with their levels

Factors	Process variables	Level 1	Level 2	Level 3
A	Tool geometry	TC	SQ	CTS
B	Rotational speed (RS)	700	1000	1300
C	Traverse speed (WS)	60	70	80

freedom (DOF) of selected OA must be equal to or higher than the total DOF required for the experiment. In the present work, L9 OA was used to carry out the experiments. Process variables used in the present study and their three levels were shown in Table 12.1.

12.2.2 Evaluation of FSW Performances

Specimens for tensile tests were prepared as per ASTM E8 standard from both welded and the base material. Specimens were cut from the workpiece in a direction normal to the weld path. Three specimens were prepared from each of the welded parts and average of these values was considered as parametric values. Ultimate tensile strength and % elongation values were measured by performing test on universal testing machine.

The hardness test was performed across the weld at the middle region at 3 mm gap on both sides of the weld path. Test was performed on Vickers macro-hardness tester by choosing load of indentation of 5 kg for 15 s. Table 12.2 lists the FSW process variables used in the study and corresponding output responses.

Table 12.2 FSW process variable values and output responses

Expt. No.	Process variables			Output responses		
	Tool profile	RS (RPM)	WS (mm/min)	UTS (MPa)	Vickers hardness (Hv)	% Elongation
1	TC	700	60	118	102	8.6
2	TC	1000	70	158	118	7.4
3	TC	1300	80	135	112	8.1
4	SQ	700	70	144	114	8.8
5	SQ	1000	80	166	123	7.9
6	SQ	1300	60	129	107	9.1
7	CTS	700	80	145	116	8.4
8	CTS	1000	60	150	118	8.2
9	CTS	1300	70	146	126	8.6

12.3 Grey Relational Analysis

The Taguchi method is an important tool for enhancing the productivity and quality during research and development. Taguchi offers meticulous way of design and analysis of experiment to improve the productivity. Most of the Taguchi techniques were on optimization of single responses. However, combining grey relational analysis (GRA) with the Taguchi methods enables the optimization of process variables for two or more than two responses simultaneously [13]. In GRA, grey relational coefficients for various responses are computed and average of these are called grey relational grade (GRG). These GRGs are considered as single-output response for the Taguchi experiment. In the present study, GRA-based Taguchi method is used to optimize the FSW variables for the output responses namely macro-hardness, tensile strength and % elongation.

12.3.1 Response Pre-processing

Since the unit and range of one response differ from the others, pre-processing of responses were necessary in GRA. When the scatter range of response is too large or when target direction of the response is different, then also response pre-processing is to be carried out in GRA. Response pre-processing is a process of converting original responses into comparable values. Responses are to be normalized in the range of zero to one to make them comparable values. In the present study, all the three responses namely tensile strength, macro-hardness and % elongation, 'larger the better' criteria were selected, as quality of the weld was represented by higher values of these responses. The original values of responses were normalized as follows for the 'larger the better' criteria.

$$p_i^*(k) = \frac{p_i(k) - \min p_i(k)}{\max p_i(k) - \min p_i(k)} \tag{12.1}$$

where $p_i(k)$ and $p_i^*(k)$ are the values before and after response pre-processing, where i represents the experiment number and k represents the response. Normalized values after pre-processing were listed in Table 12.3.

Further, $\Delta_{oi}(k)$ represents the deviation of the normalized value $p_i^*(k)$ from the reference value $p_0^*(k)$. Calculation was done for all the experiments and responses and listed in Table 12.4.

$$\Delta_{oi}(k) = \left|p_0^*(k) - p_i^*(k)\right| \tag{12.2}$$

Table 12.3 Normalized value of responses after pre-processing

Expt. No.	UTS (MPa)	Vickers hardness (Hv)	% Elongation
1	0.0000	0.0000	0.7059
2	0.8333	0.6667	0.0000
3	0.3542	0.4167	0.4118
4	0.5417	0.5000	0.8235
5	1.0000	0.8750	0.2941
6	0.2292	0.2083	1.0000
7	0.5625	0.5833	0.5882
8	0.6667	0.6667	0.4706
9	0.5833	1.0000	0.7059

Table 12.4 Deviation sequence of each response

Expt. No.	UTS (MPa)	Vickers hardness (Hv)	% Elongation
1	1.0000	1.0000	0.2941
2	0.1667	0.3333	1.0000
3	0.6458	0.5833	0.5882
4	0.4583	0.5000	0.1765
5	0.0000	0.1250	0.7059
6	0.7708	0.7917	0.0000
7	0.4375	0.4167	0.4118
8	0.3333	0.3333	0.5294
9	0.4167	0.0000	0.2941

12.3.2 Calculating Grey Relational Coefficient (GRC) and Grey Relational Grade (GRG)

A grey relational coefficient (GRC) was calculated using the values obtained after response pre-processing. GRC formulates the relationship among the actual normalized response values with the ideal values. The GRC can be calculated as follows.

$$\delta_i(k) = \frac{\Delta_{\min} + \delta \cdot \Delta_{\max}}{\Delta_{oi}(k) + \delta \cdot \Delta_{\max}} \tag{12.3}$$

where $\Delta_{oi}(k)$ is the deviation of the normalized value $p_i^*(k)$, and from the reference value $p_0^*(k)$, δ is a coefficient of identification. The value of δ is taken as 0.33 by considering the equal preference for all the variables. The GRG for each experiment were calculated and tabulated in Table 12.5.

After calculating GRC, grey relational grade was obtained by taking the average of GRC corresponding to each experiment. The multiple response optimization of the process variables are based on the GRG.

$$\beta_i = \frac{1}{n}\sum_{k=1}^{n} \delta_i(k) \tag{12.4}$$

where $\delta_i(k)$ is the GRG for the *i*th experiment which vary from 1 to 9 and n is the number of responses. Table 12.4 lists the GRG for each experiment. The experiment which is nearer to the ideal normalized value shows higher GRG. From the Table 12.4, Experiment 5 has best multi-response value, since it has the highest GRG. From the present work, it could be inferred that the optimization of multiple responses of FSW of AMCs can be done by converting multiple responses into a grey relational grade.

Table 12.5 Grey relational coefficient (GRC) and grey relational grade (GRG)

Expt. No.	Grey relational coefficient (GRC)			Grey relational grade (GRG)
	UTS	Vickers hardness	Elongation	
1	0.2481	0.2481	0.5287	0.3417
2	0.6644	0.4975	0.2481	0.4700
3	0.3382	0.3613	0.3594	0.3530
4	0.4186	0.3976	0.6516	0.4893
5	1.0000	0.7253	0.3186	0.6813
6	0.2998	0.2942	1.0000	0.5313
7	0.4300	0.4420	0.4449	0.4389
8	0.4975	0.4975	0.3840	0.4597
9	0.4420	1.0000	0.5287	0.6569

Table 12.6 Grey relational grade (GRG) response table

Factors	Process variable	Grey relational grade			Main effect
		Level 1	Level 2	Level 3	
A	Tool profile	0.3882	0.5673[a]	0.5185	0.1791
B	Rotational speed (RS)	0.4233	0.5370[a]	0.5137	0.1137
C	Welding speed (WS)	0.4442	0.5387[a]	0.4911	0.0945

Total mean value of the GRG = 0.4913
[a]Optimum GRG levels

The orthogonal experimental design enables to separate out the influence of each process variables on the GRG at different level. For instance, the hardness at levels 1, 2 and 3 can be computed by averaging the GRG for the Experiments 1–3, 4–6 and 7–9, respectively. The average of GRG for each level of the other responses namely tensile strength and elongation were calculated and listed in Table 12.6. The average of GRG for the nine experiments was computed and shown in Table 12.6. When the GRG is larger, the quality of the product will be nearer to the ideal value. Hence, higher GRG is required for optimum performance.

Based on higher GRG, the optimal process variables for better hardness, tensile strength and % elongation is Level 2 of each process variables as given in Table 12.6.

From the analysis, the optimal process variables are RS of 1000 RPM, WS of 70 mm/min and tool pin profile at Level 2, i.e. square pin profiled tool.

12.3.3 Validation Test

To confirm the enhancement of output responses while FS welding of aluminium composite, validation test was performed. Validation experiments were performed by choosing the process variables at optimum levels. Composites were welded using tool with square profiled pin, choosing RS as 1000RPM and WS as 70 mm/min. Welded samples were tested for tensile strength, macro-hardness and % elongation. Samples show better mechanical properties as given in Table 12.7 in comparison with the sample obtained from 5th experiment which has higher GRG. This implies that Taguchi-based GRA can be successfully employed for optimization of multiple responses simultaneously.

Table 12.7 Comparison of responses for experiment No. 5 and experiment with optimum process variables

Expt. No.	Process variables			Output responses		
	Tool profile	RS (RPM)	WS (mm/min)	UTS (MPa)	Vickers hardness (Hv)	% Elongation
5	SQ	1000	80	166	123	7.9
Optimum variables	SQ	1000	70	170	126	8.4

12.4 Result and Discussion

Amount of heat generated at this region depends on the RS, WS and type of tool, which controls the plasticization and flow of material. Frictional heat generated at the weld region and stirring action of the FSW tool influences the recrystallization of the grain, grain growth and grain orientation. All these factors control the mechanical properties of the joint [14–20]. From the microstructural study, it is found that SiC particles distributed within the matrix dendrites with no definite grain orientations as reported in literature [14]. Figure 12.3 depicts the various zones of weld area. Based on the grain size, the weld area is classified into four zones. (a) Stir zone (SZ) is a dynamically recrystallized zone due to the severe stirring action of the tool. The frictional heat produced at the tool workpiece interface is sufficient to plasticize the material and develops a new set of fine equiaxed grains. (b) Thermo-mechanically affected zone (TMAZ), next to SZ, is not in direct contact with the FSW tool. The heat produced at the SZ and the stirring action of the tool in the SZ alters the size of the grain and its orientation in this region. Material plastically deforms due to the shear stress resulted from the material flow. The grins are elongated and inclined in the direction of shear stress.

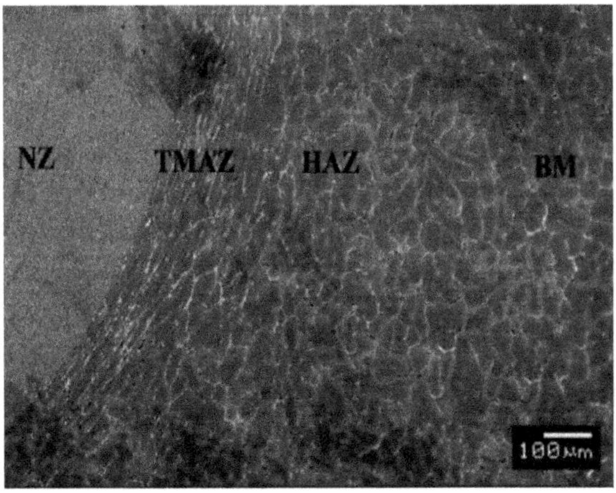

Fig. 12.3 Various zones of friction stir welded joint

(c) Heat affected zone (HAZ), is a zone next to TMAZ, does not undergo plastic deformation but, heavily influenced by the thermal cycle. The heat generated at the weld zone is sufficient to growth of the grains at HAZ. The grain structure at the HAZ is similar to the base material but of bigger size due to grain growth. (d) Base material (BM), region which is neither affected by the thermal cycle nor with the mechanical action of the tool.

Plasticization of material and the material flow in the weld region largely depends on the heat generated at the tool–workpiece interface and the amount of heat supplied to the weld region. Amount of heat generated at the tool–workpiece interface was dependent on the RS and heat supplied to workpiece was on WS [14–16]. As the RS increases, the quantity of heat generates at the weld region increases. So, at low speed, the heat generated at the region is less which affects the plasticization of material which in terms affects the weld quality. RS higher than the optimum speed generates more heat at the weld region, thereby creating turbulent material flow in the region [17]. Optimum heat generation is must for plasticization and proper material flow in the weld region. WS controls the transfer of generated heat at tool–workpiece to the workpiece [18]. Lower WS leads to higher heat input and results in the grain growth and larger heat-affected zone (HAZ). This phenomenon reduces the tensile strength and hardness. Higher WS leads to lower heat input to the weld region, thereby affecting the plasticization of material. Also, higher WS increases the cooling rate which affects the solidification of the plasticized materials in the weld region. Optimum WS is necessary to control the cooling rate which in term controls the grain growth in the weld region [19]. Mechanical properties of the welded samples were dependent on the grain structure and thus optimum RS and WS are necessary to obtain better mechanical properties.

Material flow around the tool pin is affected by the type of pin profile [15, 21]. The factor which varies with different type of tools was the ratio of swept volume to static volume. Conventional unthreaded cylindrical tool has this ratio as one, where both volumes are equal. Presence of thread on the cylindrical pin slightly increases this ratio and it has value of 1.02 in the present study. Polygon-shaped tools have higher swept volume compared to the static volume, results in higher ratio values. Square-shaped pin used in the present study has ratio of 1.57 and combined threaded cylindrical and square pin has ratio of 1.30. As the ratio increases, the pulsating effect of the tool also increases which results in improved material flow around the pin. Proper flow of material results in uniform distribution of particles, finer grain structure, thereby improving the mechanical properties of the welded parts. The present study supports the above phenomenon, indicating FSW with square profiled pin exhibits better mechanical properties.

12.5 Conclusions

The GRA has been used successfully to optimize the FSW process variables in joining AMCs for multiple responses namely macro-hardness, tensile strength and % elongation. Taguchi's L9 orthogonal array has been utilized for the experimental work. The FSW process variables are optimized for multiple responses, namely macro-hardness, tensile strength and % elongation. The present study indicates that complicated multiple response optimizations simultaneously can be significantly simplified by the Taguchi and GRA approach. From the study, it is inferred that to obtain better responses simultaneously, the optimum level of process variables to weld AMCs are square profiled pin with rotational speed of 1000 rpm and WS of 70 mm/min.

References

1. Thomas, W., Nicholas, E., Needham, J., Murch, M., Temple-Smith, P., Dawes, C.: Friction Stir Butt Welding. International Patent No. PCT/GB92/02203, GB Patent No. 9125978.8, 1991, U.S. Patent No. 5,460,317, 1995 (1991)
2. Mishra, R.S., Ma, Z.Y.: Friction stir welding and processing. Mater. Sci. Eng. R **50**, 1–78 (2005)
3. Subramanya, P., Amar, M., Arun, S., Mervin, H., Shrikantha, R.: Friction stir welding of Aluminium matrix composites—a review. In: MATEC Web of Conferences, vol. 144, p. 03002 (2018)
4. Cavaliere, P., De Santis, A., Panella, F., Squillace, A.: Effect of welding parameters on mechanical and microstructural properties of dissimilar AA6082–AA2024 joints produced by friction stir welding. Mater. Des. **30**, 609–616 (2009)
5. Rajakumar, S., Muralidharan, C., Balasubramanian, V.: Influence of friction stir welding process and tool parameters on strength properties of AA7075-T6 aluminium alloy joints. Mater. Des. **32**, 535–549 (2011)
6. Malopheyev, S., Vysotskiy, I., Kulitskiy, V., Mironov, S., Kaibyshev, R.: Optimization of processing-microstructure-properties relationship in friction-stir welded 6061-T6 aluminum alloy. Mater. Sci. Eng. A **662**, 136–143 (2016)
7. Elangovan, K., Balasubramanian, V., Babu, S.: Predicting tensile strength of friction stir welded AA6061 aluminium alloy joints by a mathematical model. Mater. Des. **30**, 188–193 (2009)
8. Murugan, N., Ashok Kumar, B.: Prediction of tensile strength of friction stir welded stir cast AA6061-T6/AlNp composite. Mater. Des. **51**, 998–1007 (2013)
9. Ukuyucu, H., Kurt, A., Areakliuglu, E.: ANN application to FSW of Al plates. Mater. Des. **28**, 78–84 (2007)
10. Prabhu, S.R., Shettigar, A., Herbert, M., Rao, S.: Multi response optimization of friction stir welding process variables using TOPSIS approach. In: IOP Conference Series: Materials Science Engineering, vol. 376, p. 012134 (2018)
11. Vijayan, S., Raju, R., Rao, S.R.K.: Multiobjective optimization of friction stir welding process parameters on aluminum alloy AA 5083 using Taguchi-based grey relation analysis. Mater. Manuf. Process. **25**(11), 1206–1212 (2010)
12. Deepandurai, K., Parameshwaran, R.: Multi response optimization of FSW parameters for cast AA7075/SiCp composite. Mater. Manuf. Process. **31**(10), 1333–1341 (2016)

13. Ghetiya, N.D., Patel, K.M., Kavar, A.J.: Multi-objective optimization of FSW process parameters of aluminium alloy using Taguchi-based grey relational analysis. Trans. Indian Inst. Met. **69**, 917 (2016)
14. Shettigar, A.K., Prabhu, S.R., Malghan, R., Rao, S., Herbert, M.A.: Application of neural network for the prediction of tensile properties of friction stir welded composites. Mater. Sci. Forum **880**, 128–131 (2017)
15. Prabhu, S.R., Shettigar, A.K., Rao, K., Rao, S., Herbert, M.A.: Influence of welding process parameters on microstructure and mechanical properties of friction stir welded aluminium matrix composite. Mater. Sci. Forum **880**, 50–53 (2017)
16. Li, Y.Z., Wang, Q.Z., Xiao, B.L., Ma, Z.Y.: Effect of welding parameters and B_4C contents on the microstructure and mechanical properties of friction stir welded B4C/6061Al joints. J. Mater. Process. Technol. **251**, 305–316 (2018)
17. Vijay, S.J., Murugan, N.: Influence of tool pin profile on the metallurgical and mechanical properties of friction stir welded Al–10 wt% TiB2 metal matrix composite. Mater. Des. **31**, 3585–3589 (2010)
18. Bozkurt, Y., Uzun, H., Salman, S.: Microstructure and mechanical properties of friction stir welded particulate reinforced AA2124/SiC/25p–T4 composite. J. Compos. Mater. **45**, 2237–2245 (2011)
19. Dinaharan, I., Murugan, N.: Effect of friction stir welding on microstructure, mechanical and wear properties of AA6061/ZrB2 in situ cast composites. Mater. Sci. Eng. A **543**, 257–266 (2012)
20. Kalaiselvan, K., Dinaharan, I., Murugan, N.: Characterization of friction stir welded boron carbide particulate reinforced AA6061 aluminum alloy stir cast composite. Mater. Des. **55**, 176–182 (2014)
21. Bist, A., Saini, J.S., Sharma, B.: A review of tool wear prediction during friction stir welding of aluminium matrix composite. Trans. Nonferr. Met. Soc. China **26**(8), 2003–2018 (2016)

Chapter 13
Tool Durability and Weldability in Hybrid Friction Stir Welding of High-Strength Materials

Amit Chaudhary, Chiranthan Ramesh, Viswanath Chinthapenta and Murshid Imam

Abstract Challenges that prevent broad industry adoption for friction stir welding of high-strength materials are (i) premature tool failure, (ii) selection of tool material and design, and (iii) weldability. To address these typical issues, a finite element-based study for laser-assisted friction stir welding process is carried out in ABAQUS, commercially available FE software. Experimentally, it has been shown that for welding of thick plate, laser assistance ahead of the tool is sought to be beneficial. However, the introduction of laser heating adds another complexity in modeling aspects. It is carried out in twofold. Firstly, the laser preheating is modeled to obtain the thermal map on the sample. Next, the regular FSW process is carried out on the already preheated sample. The important findings achieved are improving weld productivity and satisfactory tool life.

Keywords Friction stir welding · High-strength material · Laser heating · Thermal diffusivity · Finite element modeling

13.1 Introduction

Friction stir welding is a solid-state joining process which is developed by W. M. Thomas from TWI, Cambridge, England in 1991 [1]. This process was particularly used for the material which was unweldable by traditional welding technologies. In FSW, a specially designed rotating tool plunged into the abutting edges of the sheet to be welded and then transverse along the joint line. In this process, heat is generated by friction and severe plastic deformation allowing strong metallurgical bonds at the abutting interface. As compared to fusion welding technologies, there is no liquid weld pool formation, and hence excellent metallurgical

A. Chaudhary (✉) · M. Imam
Indian Institute of Technology Patna, Bihta, India
e-mail: ac3aug@gmail.com

C. Ramesh · V. Chinthapenta
Indian Institute of Technology Hyderabad, Sangareddy, India

R. G. Narayanan et al. (eds.), *Advances in Computational Methods in Manufacturing*, Lecture Notes on Multidisciplinary Industrial Engineering,
https://doi.org/10.1007/978-981-32-9072-3_13

properties such as fine microstructures, elimination of cracks in the weld zone, high depth of penetration is , comparatively less energy requirements, no requirement of shielding gas, and absence of consumable electrode [2]. As the capability of the FSW process expands, its demand in joining high-strength material also increases.

During welding of high-strength material, it is observed that the toolpin bend or deform due to high bending and torsional stresses. To overcome this problem, laser preheating of the sheet/plate is carried out ahead of the tool. Laser preheating softens the material so that bending and torsional stresses acting on the tool can be decreased. The present work aims to develop a finite element model for the laser-assisted FSW process. The proposed finite element has the two important industrial applications: (i) it helps to widen the process window by improving the weldability and (ii) improving tool durability by reducing the stresses, especially when FSW of high-strength material with larger sheet thickness needs to be conducted. A passively 3D FE coupled thermo-mechanical model has been developed by considering the thermal and mechanical aspects in ABAQUS/Explicit and to control the heavy distortion, arbitrary Lagrangian-Eulerian (ALE) formulation has been used. For incorporating the moving body heat source, DFLUX user-defined subroutine code was used.

13.2 Finite Element Method

A three-dimensional coupled Eulerian-Lagrangian thermo-mechanical FSW model is developed in ABAQUS/Explicit. The whole simulation is carried out in a dynamic, temp-displacement, explicit step. Stable time increment estimator is used for time marching. In the Lagrangian model, as the material deforms, mesh also deforms with it. In the Eulerian model, the mesh does not change as the material deforms. A large amount of heat is generated due to large plastic deformation and friction. The governing and material constitutive equations are given in Fig. 13.1. The workpiece material used in the present is 2024-T3 aluminum alloy. The dimensions and mesh density of the workpiece are shown in Fig. 13.1b, c. The mechanical and thermal properties are chosen in the simulation is based on the reported work by Mandal et al. [3]. The workpiece is modeled as three-dimensional deformable bodies, reference point and inertia are provided to the tool. Due to large deformation and mesh distortion, arbitrary Lagrangian-Eulerian (ALE) meshing has been considered in the simulation. A C3D8RT element has been chosen, which is a multi-material thermally coupled 8-noded element. Tool–workpiece interaction is defined by using the surface-to-surface contact formulation with tangential force on penalty basis.

Laser beam heating is solved in coupled temperature displacements, a pre-step to plunge. The Gaussian distribution is used to explain the shape of the laser beam heat source. The double-ellipsoidal heat source describes the shape of the body heat source. The laser beam intensity exponentially changes with a penetration depth of

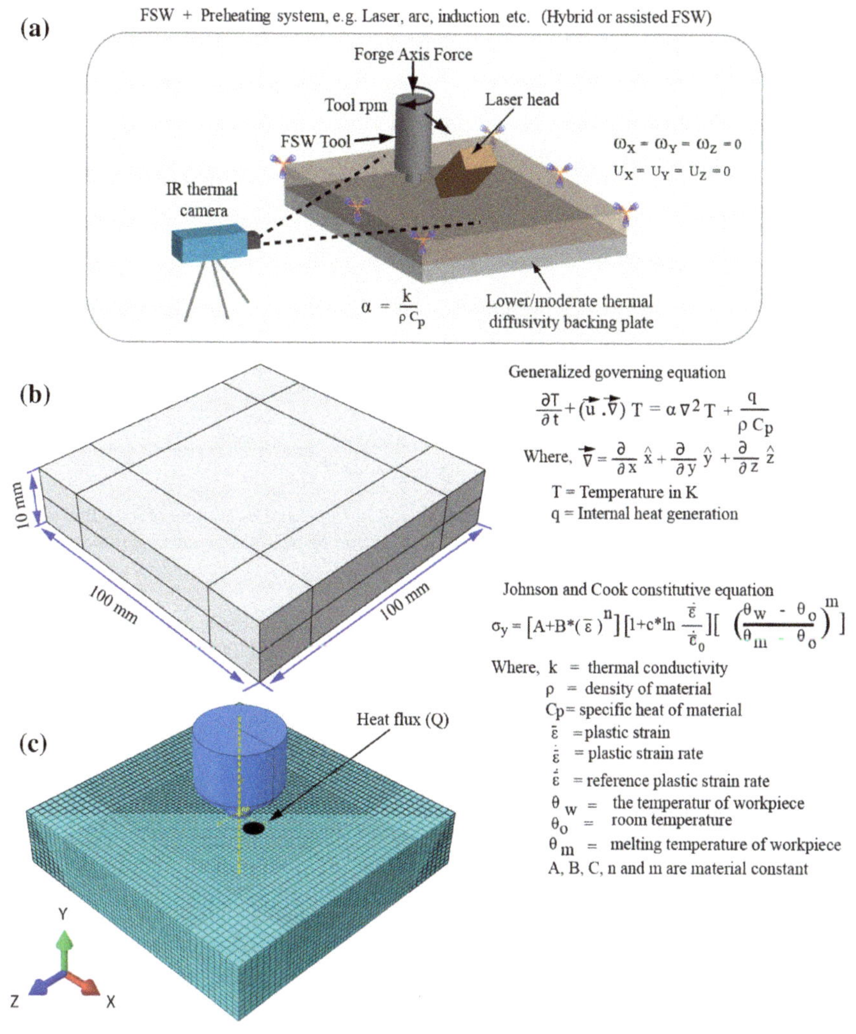

Fig. 13.1 **a** Schematic diagram of FSW with the preheating system, **b** dimensions of the deformable workpiece, and **c** FEA model with laser heat flux and governing equation of material model and heat transfer

welded element. The moving heat source parameters are controlled by changing DFLUX user-subroutine code which is written in FORTRAN programming language. The plunge stage of the tool is modeled in the next step based on the temperature-dependent properties from the laser pre-heating with dynamic, time–temperature explicit analysis. In the present work, FEM formulation is done by using ABAQUS commercial package. Goldak's model is used for simulating the laser heat source. Figure 13.2 shows the schematic of double ellipsoid Goldak heat source model. Double-ellipsoidal heat source Goldak's model was used to preheat

$$q_1(x,y,z) = \frac{6\sqrt{3}f_1Q}{abc_1\pi\sqrt{\pi}}\exp\left(-3\frac{x^2}{a^2} - \frac{y^2}{b^2} - 3\frac{z^2}{C_1^2}\right)$$

$$q_2(x,y,z) = \frac{6\sqrt{3}f_2Q}{abc_1\pi\sqrt{\pi}}\exp\left(-3\frac{x^2}{a^2} - \frac{y^2}{b^2} - 3\frac{z^2}{C_2^2}\right)$$

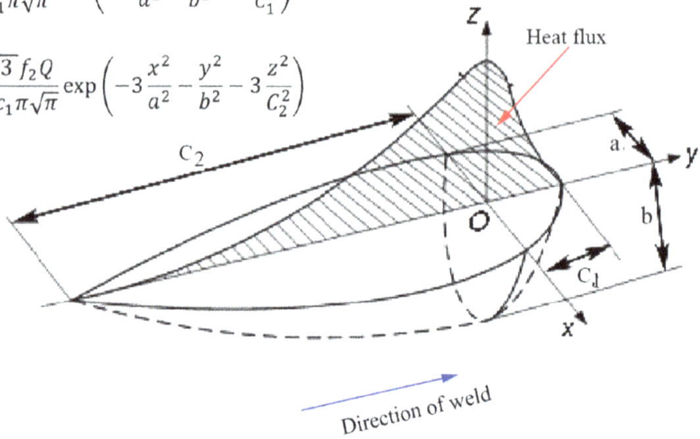

Fig. 13.2 Double ellipsoid Goldak heat source model a, b, c_1, and c_2 are sets of axes that define front and rear ellipsoid, and f_1 and f_2 represent distribution of heat source energy, respectively

the workpiece. Two-half ellipses are connected to single semi-axis. Power distribution of this double-ellipsoidal model heat source is given by the equations as shown in Fig. 13.2. Note that the resultant distribution of heat source is given by $q(x, y, z) = q_1(x, y, z) + q_2(x, y, z)$.

13.3 Results and Discussion

Figure 13.3a shows the relationship between welding parameters and a shift in weld productivity. Note that in this figure, the data for selection of control process parameters for plate thickness >19 mm are compared with the process window of 4-mm-thick sheet [4–8]. It can be cleary seen from this figure that there is a shift in weld productivity of FSW of thicker plate when compared with the smaller sheet thicknesses (<4 mm). The reason for the shift in the weld productivity can be understood as follows: (i) selection of process control parameters requires a more physical/scientific understanding of the process for welding thicker plates. To weld thicker plate, higher heat energy per unit length is needed. The control heat input parameters are tool rotational speed, axial force, and welding speed. In the case of thicker plate, large magnitude of axial force is required to ensure the optimum material mixing along the weld plate thickness. Therefore, the coupling of the tool rotational speed and axial force needs to be adjusted in such a way that it avoids the overheating of the material at the tool shoulder/workpiece interface. Otherwise, the higher degree of thermal softening will make the process unstable and the excess flash formation, (ii) the present industrial practice to avoid such a situation is the weld-flip-weld approach. The first weld pass is made with the shorter tool probe length and then workpiece is flipped for welding from the other side. This approach involves cleaning of the bead obtained after the first weld pass, the additional time for the readjustment of the fixtures and requirement of large clamping forces for the adjustment of the workpiece with the fixtures because of the distortion of the workpiece after the first weld pass. In addition, the tool probe experiences the larger bending and shear stress in the case of thicker plates. The intricate tool probe profile design is detrimental for the tool life. To achieve better welding condition, the tool has been provided with probe surface which acts as a notch and the tool became weaker at that section and may lead to failure. Therefore, the tool material must be strong enough to withstand the large bending and torsion stresses acting on the tool when the welding speed is high. In this context, Laser pre-heating ahead of the tool has been found to be beneficial to solve such problems for welding of thicker plates and for welding of high-strength alloys. From Fig. 13.3b, it can be observed that for the high-strength material (mild- and high-strength steels) the yield strength does

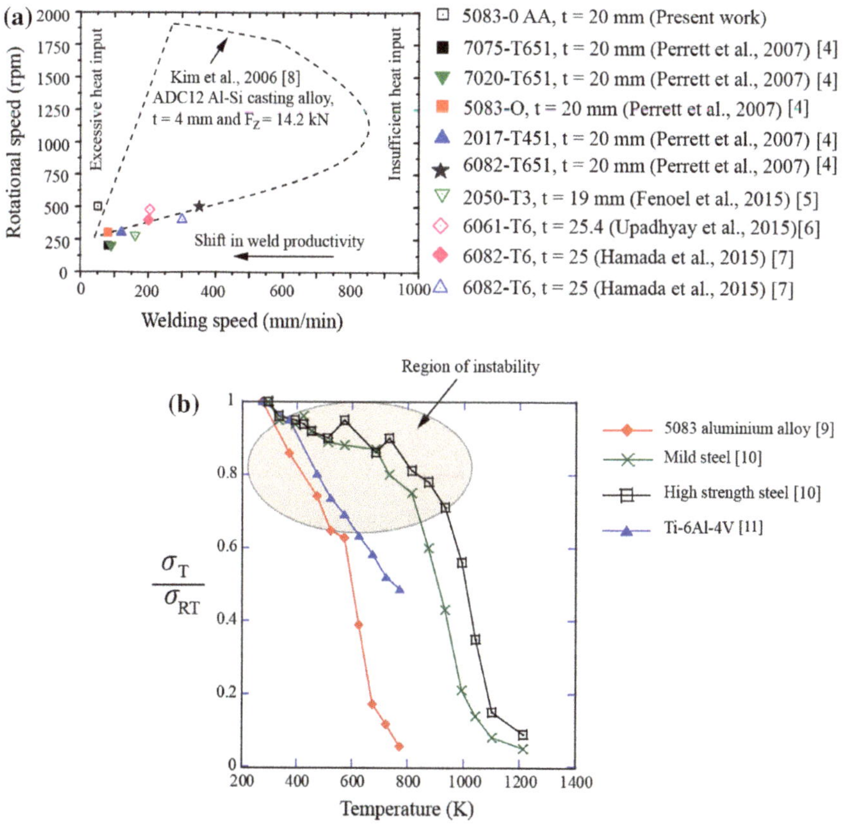

Fig. 13.3 **a** Process map for weld productivity in thick aluminum welds and **b** instability region for carrying out friction stir welding of high-strength materials

not vary significantly for the temperature range from 300 to 800 K (for details see the Fig. 13.3b). Note that σ_T/σ_{RT} is the ratio of the yield strength of the materials at elevated temperatures and room temperature, respectively. At the time of tool plunging, the temperature is not sufficiently high to cause the thermal softening of the work material and thereby leads to the tool failure. Adding the laser heating helps to solve this problem. The finite element simulation with laser pre heating has been carried out to compare the results and benefits of pre-heating over without pre-heating. From the predicted model, it can be observed that stress developed on the tool has been decreased significantly.

Figure 13.4a shows the comparisons of maximum principal stress values at three different locations during the plunging stage. Here, the maximum principal stress values are compared for both the cases (with and without preheating). The three locations (location 1, location 2, and location 3) shown here is for the tool plunging

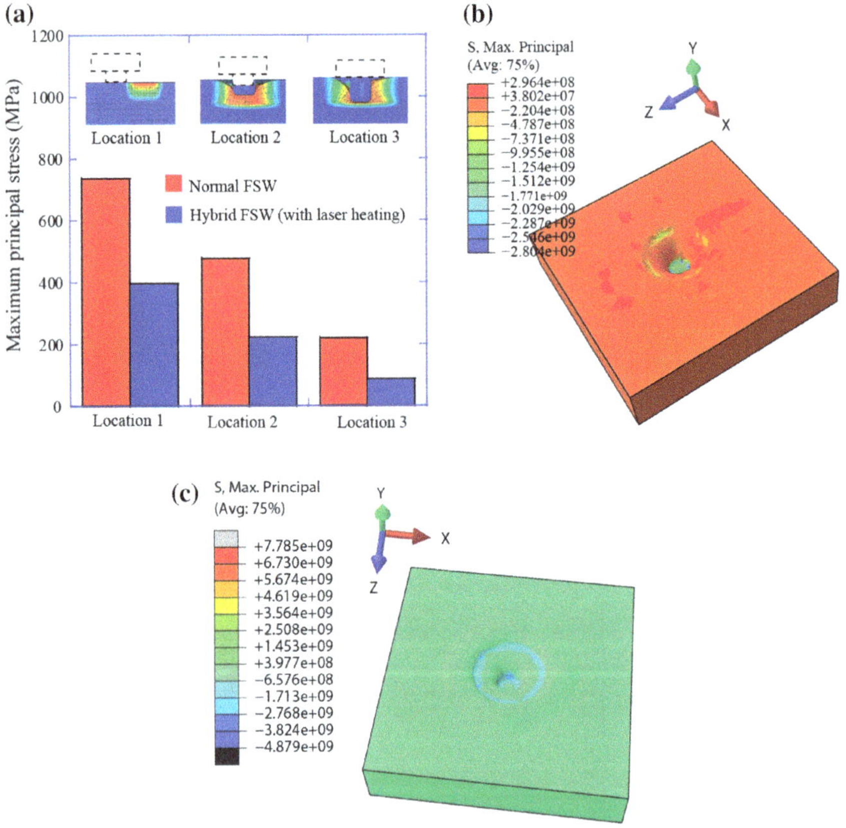

Fig. 13.4 **a** Comparisons of stress values at plunging locations 1, 2, and 3, respectively, **b** finite element simulation result showing the maximum principal stress value for normal FSW process at location 3, and **c** finite element simulation result showing the maximum principle stress value for Hybrid FSW process at location 2

distance of 0.6, 6, and 12 mm, respectively. Note that Fig. 13.4 b, c shows the predicted maximum principal stress on the workpiece at location 3 and location 2 for the case of normal FSW process and laser-assisted FSW. The stresses generated on the tool can be reduced in a very much amount in the case of laser heating as compared to the normal FSW process. Thus, it can be clearly understood that the laser heating is beneficial for causing the thermal softening of the material, particularly during the plunging. On similar lines, Yaduwanshi et al. [9] and Fei et al. [10] pointed out that the laser preheating is beneficial for the welding of dissimilar materials, particularly when aluminum and other high-strength materials (Steel, Titanium, etc.) are involved [11–13].

13.4 Conclusion

Laser-assisted friction stir welding has a potential to lower the stresses acting on the FSW tool by causing the thermal softening of the workpiece material. The finite element model suggests that the tool durability can be greatly improved. The addition of laser heating can also widen the process window, and hence better productivity can be achieved. The development of hybrid FSW system paves the way to establish this inexpensive and simple technique in heavy metal industries where metallurgical problems are still the main challenges to overcome.

References

1. Chena, H.-B., Yan, K., Lin, T., Chen, S.-B., Jiang, C.-Y., Zhao, Y.: The investigation of typical welding defects for 5456 aluminum alloy friction stir welds. Mater. Sci. Eng. A **433**, 64–69 (2006)
2. Imam, M., Sun, Y., Fujii, H., Aoki, Y., Ninshu, M.A., Tsutsumi, S., Murakawa, H.: Friction stir welding of thick aluminium welds—challenges and perspectives. In: Conference Paper, Friction Stir Welding and Processing IX, pp. 119–124. Springer (2017)
3. Mandal, S., Rice, J., Elmustafa, A.A.: Experimental and numerical investigation of the plunge stage in friction stir welding. J. Mater. Process. Technol. **203**(1), 411–419 (2008)
4. Perrett, J.G., Martin, J., Threadgil, P.L., Ahmed, M.M.Z.: Recent developments in friction stir welding of thick section aluminium alloys. In: Proceedings of 6th World Congress, Aluminium 2000, pp. 13–17, Florence, Italy (2007)
5. Avettand-Fenoel, M.-N., Taillard, R.: Heterogeneity of the nugget microstructure in a thick 2050 Al friction stirred weld. Metall. Mater. Trans. A **46**, 300–303 (2015)
6. Upadhyay, P., Reynolds, A.: Effect of backing plate thermal property on friction stir welding of 25 mm-thick AA6061. Metall. Mater. Trans. A **45**, 2091–2100 (2014)
7. Hamada, A.S., Jarvenpaa, A., Ahmed, M.M.Z., Jaskari, M., Wynne, B.P., Porter, D.A., Karjalainen, L.P.: The microstructural evolution of friction stir welded AA6082-T6 aluminium alloy during cyclic deformation. Mater. Sci. Eng. A **642**, 366–376 (2015)
8. Kim, Y.G., Fujii, H., Tsumura, T., Komazaki, T., Nakata, K.: Three defect types in friction stir welding of aluminum die casting alloy. Mater. Sci. Eng. A **415**, 250–254 (2006)

9. Yaduwanshi, D.K., Bag, S., Pal, S.: Numerical modeling and experimental investigation on plasma-assisted hybrid friction stir welding of dissimilar materials. Mater. Des. **92**, 166–183 (2016)
10. Fei, X., Ye, Y., Jin, L., Wang, H., Lv, S.: Special welding parameters study on Cu/Al joint in laser-heated friction stir welding. J. Mater. Process. Technol. **256**, 160–171 (2018)
11. Tello, K.E., Gerlich, A.P., Mendez, P.F.: Constants for hot deformation constitutive models for recent experimental data. Sci. Technol. Weld. Join. **15** (2010)
12. Chen, J., Young, B., Uy, B.: Behaviour of high strength structural steel at elevated temperatures. J. Struct. Eng. **132** (2006)
13. Kim, J., Kim, K.-H., Kwon, D.: Evaluation of high-temperature tensile properties of Ti-6Al-4V using instrumented indentation testing. Met. Mater. Int. **22**, 209–215 (2016)

Chapter 14
Swarm-Intelligence-Based Computation for Parametric Optimization of Electron Beam Fabrication

Sanjib Jaypuria, Amit Kumar Das and Dilip Kumar Pratihar

Abstract In the present paper, two recently developed intelligent swarm optimization algorithms were used for optimization of input variables in the electron beam welding process. The computational intelligence-based optimization algorithms used in this study were bonobo optimizer (BO) and cricket algorithm (CA), which are swarm-intelligence-based algorithms developed by mimicking self-organization, co-evolution, and learning of multiple agents in that particular swarm of the population. Bead-on-plate welding of CuCrZr alloy plate was carried out by electron beam. The welding runs were designed according to central composite design, and regression analysis was conducted to establish input–output dependency. Accelerating voltage, beam current, scanning speed, and focusing distance were considered as input variables of the process, whereas bead geometry parameters and microhardness of weld zone were regarded as the responses of the electron beam welding process. An optimization problem was formulated with an aim to minimize the weldment area without sacrificing the bead penetration and microhardness of the fusion zone. Penalty function approach was also used in this study to optimize the unconstrained problem. Three optimization algorithms were used in this study to solve the optimization constrained problem designed to optimize the input variables of the process yielding proper responses of the weld region. The performances of proposed approaches were evaluated based on percent deviation from that of the experimental result. Bonobo optimizer outperformed cricket algorithm and genetic algorithm for optimizing the process.

Keywords Bonobo optimizer · Cricket algorithm · Electron beam welding · Swarm intelligence · optimization

S. Jaypuria (✉) · A. K. Das · D. K. Pratihar
Soft Computing Lab, Department of Mechanical Engineering, Indian Institute of Technology Kharagpur, Kharagpur 721302, India
e-mail: sanjibjaypuria@iitkgp.ac.in

A. K. Das
e-mail: amit.besus@gmail.com

D. K. Pratihar
e-mail: dkpra@mech.iitkgp.ac.in

R. G. Narayanan et al. (eds.), *Advances in Computational Methods in Manufacturing*, Lecture Notes on Multidisciplinary Industrial Engineering,
https://doi.org/10.1007/978-981-32-9072-3_14

14.1 Introduction

Manufacturing processes are always associated with trial and error method prior to the actual experimentation. It is always not beneficial to conduct these trial experiments due to the wastage of resources and time. Therefore, these days' industries are shifting toward the automation of manufacturing processes, which will reduce the lead time and consumption of resources [1]. In addition to input–output modeling, manufacturing industries also require optimization tools to enhance productivity by ensuring quality-based products. Most of the manufacturing processes deal with multi-input–output systems with complex interdependency of the variables; so, optimization is needed to get a suitable set of input parameters, which ensure a high-quality product. Electron beam welding is such a fusion welding process, where multiple process parameters are involved along with a certain degree of automation of the process. Electron beam welding has gained diverse applications in the fields like nuclear, energy, aerospace, and automotive in the recent era. The wide application of this process is because of low heat input, high aspect ratio, a minimal area of heat-affected zone and capability to weld reactive and reflective metals [2]. Here, electron beam acts as a heat source and fuses the joint by providing intense heat energy, which is being converted from the high-speed electron beam's kinetic energy. The quality of the joint is also decided by the vacuum environment of the welding [3].

Computational intelligence is a part of soft computing tools or metaheuristic algorithms, which are promising tools for modeling, optimization, and prediction of response in material and manufacturing engineering [4]. The most frequently used tools include artificial neural network (ANN), genetic algorithm (GA), fuzzy logic (FL), particle swarm optimization (PSO), and others. A few researchers studied the applicability of metaheuristic algorithms in the different manufacturing process and welding methods.

Benyounis and Olabi [5] reviewed linear regression, response surface method, artificial neural network (ANN), Taguchi method, and design of experiment (DoE) techniques used in welding process for optimization and modeling. It had been concluded from their work that GA and other metaheuristic algorithms are powerful optimization tools, especially in the irregular experimental region. The ability of these evolutionary approaches to operate in a vast set of search points to find optimal point was well appreciated in this work. Sathiya et al. [6] performed a laser welding of AISI 904L and investigated the relationship between laser input parameters and laser-welded joint characteristics. They first established the input–output relationships by artificial neural network and then, the developed model was used for optimization of responses using genetic algorithm. Jha et al. [7] modeled electron beam welding of reactive zircaloy by artificial neural network. The developed neural networks were optimized by GA, PSO, and BP algorithm. These developed bidirectional models had shown good prediction capability of the model. BPNN had proved the better performance as compared to the other two approaches. Park and Rhee [8] performed experiments on the laser welding of automotive grade

AA5182 aluminum alloy and studied the tensile strength of the welded joints. A neural network model tuned by genetic algorithm was proposed to predict the tensile strength using the experimental data. Dey et al. [9] conducted electron beam welding of stainless steel and experiments were designed according to central composite design (CCD). The objective of their study was to minimize the weld area without hampering the maximum bead penetration. Penalty function approach was used to obtain the suitable process parameters of the electron beam welding process. Kanigalpula et al. [10] studied the spiking behavior of partially penetrated ETP copper plates welded using electron beam. The constrained optimization problem for achieving the minimum standard deviation of penetration was solved using genetic algorithm, particle swarm optimization, and desirability function approach. The predicted results with various proposed approaches were verified with that of real experiments and a good agreement between them was achieved. In addition to this, there were a very few available literatures, where swarm-intelligence-based optimization algorithms were implemented on the manufacturing process and system for optimization and modeling.

It is observed from the literature that metaheuristic approaches are mostly used in the optimization of many manufacturing processes in addition to traditional tools because of their flexibility and accuracy. In addition to this, the modeling tool like neural network is also fine-tuned with these evolutionary approaches for the better efficiency of the developed neural network architecture. In most of the available literature, optimization tools like genetic algorithm and particle swarm optimization are used for optimization and modeling of the manufacturing process. It is also found from the research that there are only a few available works of literature, where swarm-intelligence-based approaches were used for optimization of electron beam fabrication.

Therefore, here, two recently developed, swarm-intelligence-based algorithms, e.g., Bonobo optimizer (BO) and cricket algorithm (CA), were used to optimize the process parameters of electron beam welding. The approaches were intended to minimize the weldment area and maximize the bead penetration and weld zone microhardness. The performance comparison and working principle were discussed in this work. In addition to this, genetic algorithm (GA), which is a bio-inspired tool, was also used to optimize the constrained problem. The performances of these developed models were evaluated in terms of absolute percentage deviation and convergence rate.

14.2 Experimentation and Data Collection

Precipitation hardened CuCrZr alloy plates of dimension 100 mm × 65 mm × 10 mm were used for electron beam welding in bead-on-plate configuration. The procured metal had 0.80 wt% Cr, 0.06 wt% Zr, and 0.13% impurities and remaining as copper. The primary input parameters of electron beam welding, namely accelerating voltage (*V*), beam current (*I*), welding speed (*S*), and focusing distance

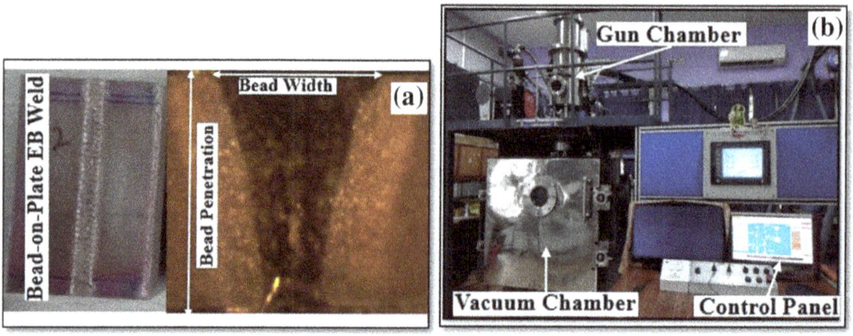

Fig. 14.1 Photograph of **a** weld bead geometry and **b** electron beam welding setup

(B) were taken as process variables, whereas microhardness of fusion zone and bead geometrical parameters, like bead penetration (BP), bead width (BW), microhardness of fusion zone (μ_{FZ}), was considered as the responses to be measured. Figure 14.1a indicates various geometrical features of the weld bead. The photograph of electron beam welding setup along with their principal components is given in Fig. 14.1b. A few trial runs were conducted to select the ranges of input variables. The actual experiments for this study were prepared by considering the method of central composite design (CCD) with three central points. A total of 27 combinations of input parameters were generated, and 81 experiments were conducted by considering three replicates of the original CCD [11]. The actual and test cases experimental data were taken from [11]. A section of the weld (15 $\times$ 10 10 mm) was cut from the obtained weld for measuring the weld responses. The polished and etched samples were examined using optical macroscope to measure bead penetration and width. Microhardness of fusion zone (FZ) was measured for each sample in Vickers microhardness tester at an applied load of 50-gf and dwell time of 15 s.

14.3 Method of Analysis: Proposed Approaches

The mathematical relationships between the inputs and output were established by nonlinear regression before solving the optimization problem. The proposed metaheuristic approaches, which were used as optimization algorithms in the developed problem, were briefly discussed below.

14.3.1 Regression Analysis

A nonlinear statistical regression analysis was conducted using Minitab15 software to establish input–output relationships of the process, as given below in Eq. (14.1).

$$y = \beta_0 + \sum_{i=1}^{k} \beta_i x_i + \sum_{i=1}^{k} \beta_{ii} x_i^2 + \sum_{i=1}^{k} \sum_{i<j}^{k} \beta_{ij} x_i x_j + \varepsilon \quad (14.1)$$

where y is the response parameter; $\beta_0, \beta_i, \beta_{ii}$ are the coefficients to be computed using the least squares method. Here, ε denotes the error in fitting. The nonlinear regression equations for the responses were determined, as shown in Eqs. (14.2)–(14.4).

$$\begin{aligned} \mathrm{BW} = & -172.103 + 2.16103\,V + 0.827001\,I - 0.0522837\,S \\ & + 0.479534\,B + 0.0104345\,V^2 - 0.00000530401\,I^2 \\ & - 0.00000157596\,S^2 + 0.000117136\,B^2 - 0.00057152\,VI \\ & - 0.0000567146\,VS - 0.00867625\,VB - 0.0000572391\,IS \\ & - 0.00193298\,IB + 0.000171340\,SB \end{aligned} \quad (14.2)$$

$$\begin{aligned} \mathrm{BP} = & \ 5857.98 + 8.02798\,V + 0.0920434\,I + 0.0623438\,S \\ & - 33.0775\,B - 0.0313765\,V^2 - 0.00146853\,I^2 \\ & + 0.00000207134\,S^2 + 0.0456195\,B^2 \\ & + 0.0000113958\,VI - 0.0002530645\,VS \\ & - 0.0113766\,VB - 0.000145435\,IS \\ & + 0.00100352\,IB - 0.000114410\,SB \end{aligned} \quad (14.3)$$

$$\begin{aligned} \mu_{\mathrm{FZ}} = & \ 435.876 + 41.1517\,V - 6.48847\,I - 0.289883\,S \\ & - 6.03015\,B - 0.285728\,V^2 + 0.00680864\,I^2 \\ & + 0.000101836\,S^2 + 0.00893827\,B^2 \\ & + 0.00541667\,VI + 0.000987500\,VS - 0.0291667\,VB \\ & + 0.000322917\,IS + 0.013\,IB + 0.000104167\,SB \end{aligned} \quad (14.4)$$

14.3.2 Problem Formulation

A constrained optimization problem was formulated to maximize the weld penetration and microhardness of the weld zone without increasing the weld area.

Minimize

$$A = 0.5 \times (\mathrm{BW} \times \mathrm{BP}) \quad (14.5)$$

subject to

$$\mathrm{BP} \geq 0.7 \times t,$$
$$|\mu_{\mathrm{PM}} - \mu_{\mathrm{FZ}}| \leq 59.0$$

and

$$50 \leq V \leq 60$$
$$70 \leq I \leq 110$$
$$600 \leq S \leq 1000$$
$$365 \leq B \leq 375$$
$$\mu_{\mathrm{PM}} = 153$$
$$t = 10$$

Here, t and μ_{PM} indicate the thickness of the specimen in mm and microhardness of base or parent metal in VHN, respectively. A denotes the weldment area, which had to be minimized for the reduction of bead width and large heat-affected area. Therefore, the constrained optimization problem was designed to minimize weld area without scarifying the bead penetration and microhardness of the fusion zone. The above functional constraint was set by observing the microhardness of FZ for different combinations of input parameters lying within their respective ranges.

14.3.3 Genetic Algorithm (GA)

Genetic algorithm (GA) is a bio-inspired metaheuristic algorithm working based on Darwin's principle of natural selection and evolution. Therefore, concisely, a GA initiates the operation by selecting a random population of solutions and then improves the fitness level through iterative utilization of selection, crossover, and mutation operators. This iterative process stops when the stopping criterion is reached. Interested readers may refer to Pratihar [12] for a detailed discussion of the working principle of GA. The results of GA had been obtained from the literature for comparison purpose [11]. The working flowchart is given in Fig. 14.2a.

14.3.4 Bonobo Optimizer (BO)

Very recently, a new metaheuristic algorithm, named as bonobo optimizer, has been developed [13]. This newly proposed optimization algorithm imitates the social behavior and reproductive strategies of the bonobos. Generally, bonobos have been found to adopt a fusion–fission social practice, and moreover, they have four types of reproductive strategies, like promiscuous, restrictive, consortship, and extra-group mating. Furthermore, in the algorithm, two-phase conditions, namely

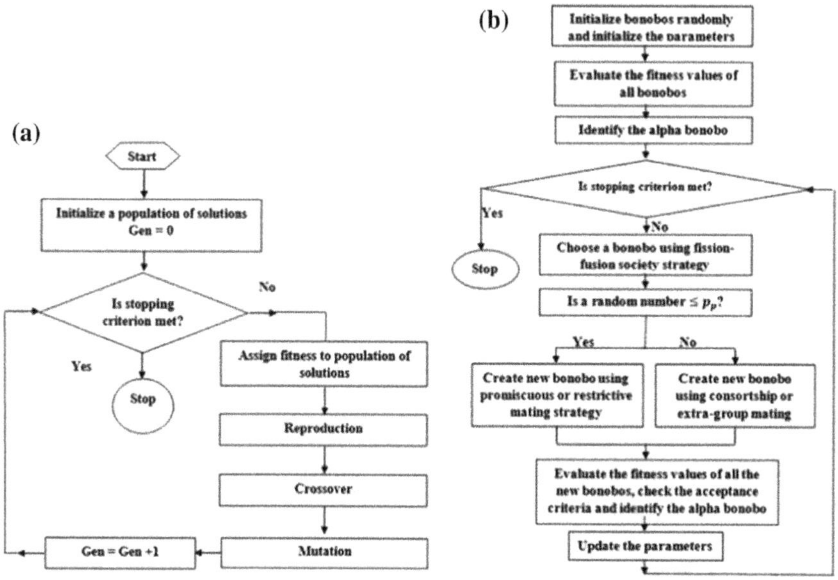

Fig. 14.2 Flowcharts of **a** genetic algorithm and **b** bonobo optimizer

positive phase and negative phase, have been considered and depending upon the phase, its associated phase probability (p_p), either exploitation or exploration phenomenon, is promoted. These strategies have been artificially modeled in the developed BO to make a more efficient optimization technique. Nevertheless, the parameters of the BO have been made adaptive to reach the globally optimum solution with the faster convergence rate. The flowchart of this BO has been shown in Fig. 14.2b.

14.3.5 Cricket Algorithm (CA)

Motivated from the nature and characteristics of an insect, namely cricket, a novel cricket algorithm had been proposed in recent time [14]. This algorithm was developed considering the functional aspects of some swarm-intelligence-based optimization methods, like bat algorithm, firefly algorithm, particle swarm optimization (PSO), etc. Also, a few peculiar characteristics of the crickets, such as forecasting the weather by courtesy of the number of fluttering of crickets and the sound frequency in the temperature, are artificially modeled to solve the optimization problems [15]. It had been applied to solve several standard benchmark functions and various real-world optimization problems. In those experiments, CA had shown its supremacy in performance compared to a few other optimization algorithms.

14.4 Results and Discussion

Sixteen test cases were used to validate this mathematical model. Figure 14.3 shows the absolute percentage deviation in the prediction of responses. The mean absolute percent error (MAPE) in the prediction of BW, BP, and μ_{FZ} were found to be 2.44, 4.84, and 1.14%, respectively. Similarly, the root mean squared error in prediction of BW, BP, and μ_{FZ} were found to be equal to 0.58, 1.03, and 4.47, respectively. In addition to this, *R*-squared value for each response in the regression model was also determined to check the closeness of data points to a fitted regression line. The values were found to be 97.95, 76.15, and 97.77% for BW, BP, and μ_{FZ}, respectively.

A real-coded genetic algorithm was utilized to solve the optimization problem in MATLAB 2015b environment. The minimum value of the fitness function (here the weldment area) was the objective of the study. A penalty function approach was adopted to handle constrained optimization problem, where infeasible solutions were penalized for being removed from the population of solutions. As the performance of the GA is dependent on its parameters, the parametric study was carried out to decide the set of optimal GA parameters by varying one parameter at a time and keeping the others unaltered [12]. The best fitness was obtained for the following GA parameters: $P_c = 0.9$; $P_m = 0.009$; $N = 100$, and $G_{max} = 150$. The GA could yield the optimal solution as follows: $V = 56.08$ kV; $I = 88.6$ mA; $S = 600$ mm/min, and $B = 365$ mm. The optimized value of the area of the weldment, as obtained by the GA, was 12.14 mm^2. A detailed comparison of various algorithms suggested results and is given in Table 14.1.

The common parameters, such as population size ($N = 100$) and maximum number of iterations ($iter_{max} = 150$), are taken the same for each of the optimization

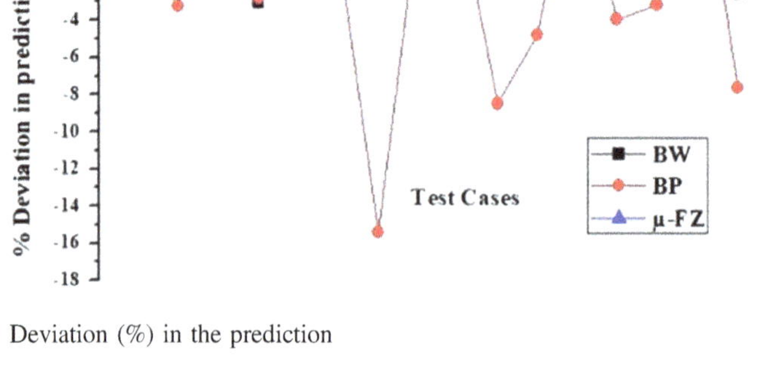

Fig. 14.3 Deviation (%) in the prediction

Table 14.1 Comparison of algorithm suggested results and experimental results

Proposed approaches	V (kW)	I (mA)	S (mm/min)	B (mm)	Area (mm^2)	BW (mm)	BP (mm)	μ_{FZ} (VHN)
CA	57	84	620	375	12.10	3.46	7.00	99.19
EXP	57	84	620	375	11.74	3.40	6.91	100.26
% error					3.093	1.70	1.36	1.066
BO	56	84	607	375	11.99	3.43	7.00	100.76
EXP	56	84	607	375	11.84	3.34	7.09	100.33
% error					1.544	2.54	0.97	0.428
GA [11]	56.08	88.6	600	365	12.14	3.45	7.04	99.66
EXP	56	88	600	365	12.65	3.61	7.01	96.87
% error					4.031	4.43	0.42	2.880

techniques. In addition, the algorithm-specific parameters used for BO are as follows: Initial probability for extra-group mating, $p_{xgm\_initial} = 0.03$; sharing coefficient for alpha bonobo and the selected bonobo are found to be equal to 1.3 and 1.4, respectively; rate of change of phase probability, rcpp = 0.0036; maximum value for the temporary subgroup size factor, $tsgs_{factor\_max} = 0.02$. The BO obtained the optimal solution as follows: $V = 56$ kV; $I = 84$ mA; $S = 607$ mm/min; and $B = 375$ mm. The optimized value of area of weldment as obtained by the BO was 11.99 mm^2. For CA algorithm, the following parameters were taken while it is running: minimum frequency, $Q_{min} = 0$; $\beta_{min} = 0.2$. The CA predicted the optimal solution as $V = 57$ kV; $I = 84$ mA; $S = 620$ mm/min; and $B = 375$ mm. The optimized value of area of weldment as obtained by the CA was 12.10 mm^2.

From Table 14.1, it is observed that BO could yield the best result compared to others for this optimization problem. Therefore, it can be concluded that BO could outperform GA and CA in this study.

The convergence plots of BO and CA algorithms are given in Fig. 14.4. The graph indicates that BO could yield a better solution with faster convergence rate

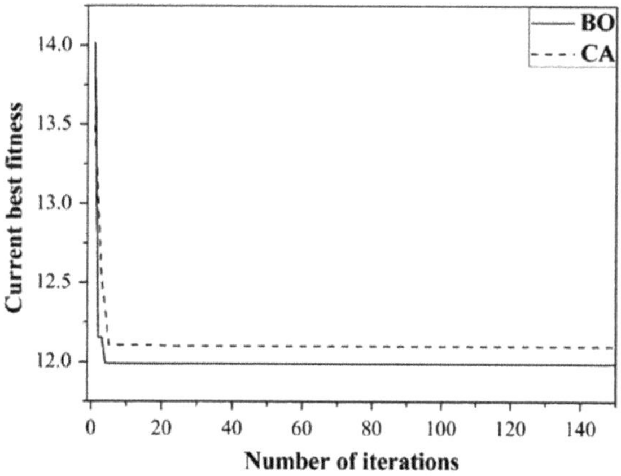

Fig. 14.4 Convergence plot for BO and CA

compared to CA. It can be inferred from the result of absolute percentage deviation and the result of the convergence rate that BO emerges as the winner for the optimization problem attempted in this study. In BO, the parameters of the algorithm are found to be adaptive in nature. According to the need of the situation, either exploitation or exploration phenomenon is promoted, or a proper balance between these two is maintained. In the case of a positive phase, the solutions are updated either using a promiscuous or restrictive type of reproductive strategy, and in this phase, the searching agents are observed to follow the current best solution. On the other hand, a more diversified search is found using either a consortship or extra-group mating strategy during a negative phase. Furthermore, the selection procedure of the mating partner based on the fission–fusion social behavior of the bonobos is also responsible for maintaining a right balance between the selection pressure and population diversity. Due to these reasons, BO is found to have an efficient searching mechanism to reach the globally optimum solution and hence, it has better performance compared to others.

14.5 Conclusion

A constrained optimization problem was formulated to yield minimum weld area without violating the requirement of high penetration and microhardness of the fusion zone. Mathematical relationship among input–output was established using regression, and the obtained equation was validated with experimental results. The following conclusions can be made from the present work.

- The formulated optimization problem was solved using BO and CA, which are swarm intelligence-based algorithm. In addition to this, the solution obtained using the evolutionary genetic algorithm was also taken from available literature for comparison of results among various approaches. The optimal results from different approaches were validated with that of real experiments. The maximum absolute percentage deviation in the prediction of optimal input parameters in all algorithms were found to be less than 4%, which indicates a closer match between algorithms suggested results and experimental results. The absolute percent deviation for all responses was within 2%, and the convergence rate was also faster in BO as compared to that of other approaches. Therefore, it could be concluded that BO performs better than GA and CA based on convergence and accuracy.
- It is advisable to maintain mid-level of voltage and current, a minimum level of welding speed and maximum level of focusing distance to achieve the minimum weld cross section along with a desirable level of penetration and microhardness in weld zone during the electron beam welding of copper alloy.

References

1. Zhong, R.Y., Xu, X., Klotz, E., Newman, S.T.: Intelligent manufacturing in the context of industry 4.0: a review. Engineering **3**(5), 616–630 (2017)
2. Jaypuria, S., Doshi, N., Pratihar, D.K.: Effects of welding parameters on mechanical properties in electron beam welded CuCrZr alloy plates. IOP Conf. Ser.: Mater. Sci. Eng. **338** (1), 012013 (2018)
3. Jaypuria, S., Meher, J., Kanigalpula, P.K.C., Pratihar, D.K.: Electron beam butt welding of Cu-Cr-Zr alloy plates: experimental investigations, studies on metallurgical and mechanical properties. Fusion Eng. Des. **137**, 209–220 (2018)
4. Casalino, G.: Computational intelligence for smart laser materials processing. Opt. Laser Technol. **100**, 165–175 (2018)
5. Benyounis, K.Y., Olabi, A.G.: Optimization of different welding processes using statistical and numerical approaches—a reference guide. Adv. Eng. Softw. **39**(6), 483–496 (2008)
6. Sathiya, P., Panneerselvam, K., Jaleel, M.A.: Optimization of laser welding process parameters for super austenitic stainless steel using artificial neural networks and genetic algorithm. Mater. Des. **1980–2015**(36), 490–498 (2012)
7. Jha, M.N., Pratihar, D.K., Bapat, A.V., Dey, V., Ali, M., Bagchi, A.C.: Knowledge-based systems using neural networks for electron beam welding process of reactive material. J. Intell. Manuf. **25**(6), 1315–1333 (2014)
8. Park, Y.W., Rhee, S.: Process modeling and parameter optimization using neural network and genetic algorithms for aluminum laser welding automation. Int. J. Adv. Manuf. Technol. **37** (9–10), 1014–1021 (2008)
9. Dey, V., Pratihar, D.K., Datta, G.L., Jha, M.N., Saha, T.K., Bapat, A.V.: Optimization of bead geometry in electron beam welding using a genetic algorithm. J. Mater. Process. Technol. **209** (3), 1151–1157 (2009)
10. Kanigalpula, P.K.C., Jaypuria, S., Pratihar, D.K., Jha, M.N.: Experimental investigations, input-output modeling, and optimization of spiking phenomenon in electron beam welding of ETP copper plates. Measurement **129**, 302–318 (2018)
11. Kanigalpula, P.K.C., Pratihar, D.K., Jha, M.N., Derose, J., Bapat, A.V., Pal, A.R.: Experimental investigations, input-output modeling and optimization for electron beam welding of Cu-Cr-Zr alloy plates. Int. J. Adv. Manuf. Technol. **85**(1–4), 711–726 (2016)
12. Pratihar, D.K.: Soft Computing: Fundamentals and Applications. Narosa Publishing House, New Delhi (2013)
13. Das, A.K., Pratihar, D.K.: A new bonobo optimizer (BO) for real-parameter optimization. In: 2019 IEEE Region 10 Symposium. IEEE, Kolkata, India (2019)
14. Canayaz, M., Karci, A.: Cricket behaviour-based evolutionary computation technique in solving engineering optimization problems. Appl. Intell. **44**(2), 362–376 (2016)
15. Canayaz, M., Karcı, A.: Investigation of cricket behaviours as evolutionary computation for system design optimization problems. Measurement **68**, 225–235 (2015)

Chapter 15
Effect of Pulsation on Temperature Distribution of Laser-Welded Dissimilar Joint of Thin Sheet

Bikash Kumar and Swarup Bag

Abstract Dissimilar joints are characterized by microstructural change and compositional gradient which evolve large variation in physical and mechanical properties across the welded joints. The combination of dissimilar material is challenging goal for development of parts with locally optimized parameter especially for Ti-SS joints. Due to high cost (like Ti-alloy), joining of dissimilar material (Ti-SS) helps to reduce the cost incurred by eliminating the otherwise unavoidable usage of costly and rare metal. The present study includes the pulsed laser welding of dissimilar joint of Ti-alloy and AISI 304. EDX analysis is performed on welded specimen for investigation of morphology attained by different zone. It is reported that fusion zone of dissimilar joint contains very fine-grained morphology because grains are not allowed to grow more at high heat input for shorter time span (pulsation). EDX analysis conveys the presence of intermetallic compounds which is majorly composed of Fe, Ti, and Cr. Cooling rate plays important role to determine microstructural variation. But it is difficult to measure it experimentally, so three-dimensional thermal model has been developed and validated with experimental result. Furthermore, average cooling rate is estimated for Ti and SS at different locations across the weld zone. Significant difference in thermo-physical properties like specific heat capacity and conductivity are major parameter due to which large difference has been found in the estimated cooling rate between Ti-alloy and stainless steel alloy at same location at same instant of time.

Keywords Dissimilar joint · Laser welding · EDX analysis · Pulsed thermal modelling · Cooling rate

---

B. Kumar (✉) · S. Bag
Department of Mechanical Engineering, Indian Institute of Technology Guwahati, Guwahati 781039, Assam, India
e-mail: bikash.kumar@iitg.ac.in

R. G. Narayanan et al. (eds.), *Advances in Computational Methods in Manufacturing*, Lecture Notes on Multidisciplinary Industrial Engineering,
https://doi.org/10.1007/978-981-32-9072-3_15

15.1 Introduction

Owning to massive difference in chemical and physical characteristics, welding of dissimilar metal configuration is, however, a challenging task. In order to seize the privilege of certain attributes of each component to enhance the potential of product, dissimilar welding is accomplished. Particularly, dissimilar joints of Ti-alloy and stainless steel are used in petrochemical, chemical, electronics, nuclear and power generation industries [1]. The application of different alloy and metal for components provides greater flexibility to the designer and production engineer and often results in economic and technical advantage over products manufactured from single material. Costly metal (i.e. Ti-alloy) with eminent properties (i.e. low density, high strength, corrosion resistance and lightweight) can be used in critical location with relatively cheaper material (i.e. steel alloy) in order to improve productivity, production flexibility, weld quality and manufacturing opportunity.

In agreement with Fe–Ti diagram, the solubility of iron (Fe) in titanium is very low at room temperature beyond which intermetallic phases (i.e. Fe_2Ti, FeTi) are accounted [2]. These intermetallic morphologies are brittle in nature and leads to spontaneous crack formation during conventional fusion-welded joint. Mismatching of stress generated by thermal cycle (i.e. thermal stress) between parent metal resembles crack formation. To resolve and minimize the undesirable issue arises due to application of conventional process, low energy input process i.e. laser welding process can be efficiently utilized. Rapid heating and cooling cycle, accurate weld bead position, precise and localized beam, process flexibility, low residual stress and distortion generation are its principle features. Therefore, laser beam process is effective tool to control the brittle intermetallic formation at some magnitude.

Unlike continuous, pulse mode welding is characterized by a series of overlapping spot weld. The resulting pulsed weld is one where subtle variations in the pulsing parameters can lead to dramatic differences in quality of weld [3]. However, continuous waved welding process is accomplished of quasi-steady-state temperature distribution during fusion. Modern Nd: YAG laser process has capacity to shape the flux profile for each pulse at wide range of frequency and introduction of several additional parameters like pulse shaping, pulse energy, pulse duration, average power, etc. Owing to its good coupling efficiency, more flexible beam delivery, particularly pulsed mode Nd: YAG laser is a more convenient process for thin sheet welding as compared to continuous mode of CO_2 laser [4]. A precise amount of beam power can be applied for a short period of time causing small heat-affected zone (HAZ) and low distortion is possible. For the same average power, pulsed Nd: YAG laser can provide comparatively much higher penetration than continuous mode.

Matsunawa et al. [5] reported that Ti-alloy is susceptible towards pore formation due to impingement of hydrogen and other gases in melt pool. To overcome this issue, researcher implemented overlapping factor in pulse mode welding process.

Besides, it is shown that the number of porosity can be minimized with increasing overlapping factor, and even defect-free weld joint could be possible for more than 75% overlapping factor [6]. Few literatures have been found which demonstrate and characterize dissimilar joint of Ti-SS alloy. Hiraga et al. [7] investigated lap welding of Ti and AISI 304 of thickness 0.5 mm with a pulsed Nd: YAG laser to produce vacuum container used as thermo bottles. It revealed that when AISI 304 is used as upper sheet, then joint could be produced within a narrow range of welding condition; however, when Ti is used as upper sheet, joint had been found within wide range. Chen et al. [8] examined the dissimilar joint of Ti-alloy and stainless steel (SS304) of thickness 1 mm by using CO_2 laser and characterized the impact of laser beam offset on the microstructural morphology and mechanical properties of joints. Offsetting of laser beam towards steel results more durable joint having intermetallic compounds like FeTi + α-Ti, FeTi + Fe_2Ti + $Ti_5Fe_{17}Cr_5$ but when laser beam was offset towards Ti-alloy, spontaneous fracture happened. It is also reported that tensile strength of joint was higher when laser beam was offset towards SS304. However, characterization of bead geometry found for dissimilar joint based on EDX analysis is still hazy.

A number of FEM-based numerical models have been developed and investigated by different researcher for the simulation of temperature distribution [9, 10]. To determine the temperature distribution numerically by considering temperature-dependent domain properties and, latent heat of fusion and solidification, author has developed quasi-steady-state heat transfer model [11] for GTA welding. Moreover, the influence of welding velocity on double-ellipsoidal volumetric model has been suggested by another researcher [12]. However, effect of pulsation on the temperature distribution of dissimilar joint for thin sheet (thickness: 1 mm) by implementing volumetric heat source model is not focused so far.

The present study focuses on the primary investigation of experimentally observed macrostructural morphology and EDX analysis for dissimilar joint of Ti–6Al–4V alloy and AISI 304 alloy. Weld zone observed from experimental investigation has been compared with FEM-based 3D thermal (numerical) model. The influence of pulse-on temperature distribution during welding has been considered in the present analysis. Cooling rate plays important role for prediction of mechanical properties and evaluation of microstructure for laser welding process. So, based on numerically extracted temperature–time data, average cooling rate for different zones of each component (Ti-alloy and SS304) for post-welded samples have also been estimated. Furthermore, relation is established between microstructure evolved and cooling rate.

15.2 Experimental Investigation

The plate of 304 grade of stainless steel ($60 \times 50 \times 1$ mm^3) and Ti–6Al–4V alloy ($60 \times 50 \times 1$ mm^3) is chosen as specimen for pulsed Nd: YAG laser welding (capacity: 500 W) in butt joint configuration. Figure 15.1a depicts the welding

(a)

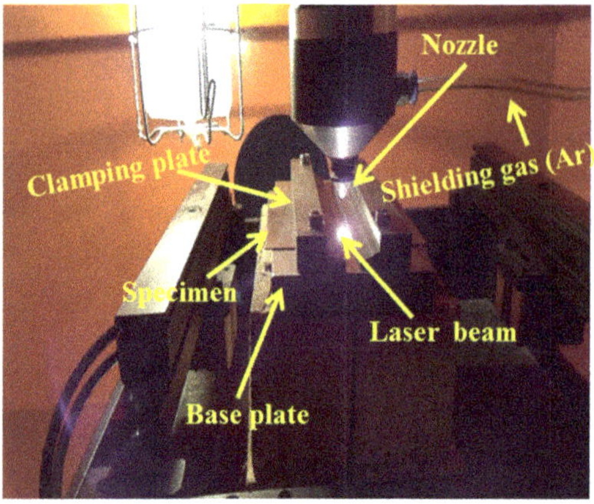

(b)

Process variable (optimized)	value
Peak power (kW)	1.8
Average power (W)	360
Pulse energy (J)	7.25
Heat input (J/mm)	80
Frequency (Hz)	50
Pulse width (ms)	4
Flow rate (L/min)	8
Weld speed (mm/s)	4.5
Offset distance (mm)	5

Fig. 15.1 **a** Schematic diagram of pulsed Nd: YAG laser welding set-up; **b** optimized parameter for dissimilar weld

set-up which is used to perform experiments. It is composed of fixed laser source, while the specimen along with fixture moves at a constant speed with auto-guided work table. The surfaces of specimens (i.e. Ti-alloy, SS304) are carefully pre-processed or polished to remove oxidized or coating layer and then clamped in fixture. As stable welding exists within a narrow range of parameter so heat input is considered as controlling parameter for present analysis to identify optimum process variables. Specimen is welded at an average power of 360 W with weld velocity of 4.5 mm/s (optimized parameter shown in Fig. 15.1b) for a constant pulse repetition rate of 50 Hz.

Welded specimens are cut across the welding line for metallographic analysis by using wire-electrodischarge machine (W-EDM) to avoid alteration of residual stress during cutting. Macro-graph of weld bead geometry is observed under upright optical microscope (*LEICA-S6D*). In order to investigate the morphology of joint and elemental composition, scanning electron microscope (SEM) outfitted with an energy dispersive X-ray spectroscopy (EDX) is used.

15.3 Theoretical Background

15.3.1 Thermal Analysis

A finite element (FE)-based three-dimensional heat transfer model for dissimilar laser welding has been developed by using ABAQUS commercial software. Temperature-dependent thermo-physical properties such as specific heat capacity, conductivity and constant density, latent heat of fusion are considered for the simulation. A full-axisymmetric geometrical domain has been considered in which half geometry is assumed as Ti–6Al–4V alloy, whereas another half geometry is considered as stainless steel (304). Due to coherent and collimate nature of laser beam (spot dia. ~1 mm), the weld line is so narrow that suitably fine mesh is necessary to capture steep temperature. Thermal effects due to solidification of melt pool are simulated by considering the latent heat of fusion for both the material (Ti-alloy and SS). The temporal and spatial diversification of pulsed moving heat source is implemented through DFLUX user subroutine. The Gaussian form of heat flux is assumed over volume in such a way that flux density decreases in thickness (depth) direction and pulse attribute is also included over time domain. Double-ellipsoidal volumetric heat source model is considered for the real representation of laser beam [12]. For the inclusion of pulse characteristics, the temporal variation of flux intensity is assumed as 1 for pulse-on time and considered as 0 for pulse-off time. The average absorptivity of Ti-alloy and SS 304 are taken as 0.34 [13] and 0.17 [14], respectively. The nonlinear heat conduction equation based on energy conservation principle is given by

$$\rho_m C_p \left(\frac{\partial T}{\partial t} - V \frac{\partial T}{\partial x_1} \right) = \frac{\partial}{\partial x_i} \left(k_{ij} \frac{\partial T}{\partial x_j} \right) + h_g \tag{15.1}$$

where x_i for $i = 1, 2, 3$ represents x-, y- and z-axis, ρ_m is the density for Ti-alloy and stainless steel, T is the temperature, C_p is the specific heat, h_g is the internal heat generation or power generation under controlled volume, k_{ij} is the component of thermal conductivity and V is the welding velocity along x-axis. The boundary interaction associated with heat transfer between substrate and surrounding environment during fusion welding are implemented for the solution of conduction equation (shown in Fig. 15.2a).

a. Dirichlet boundary interaction, at time

$$t = 0, T(x, y, z, 0) = T_i \, 300 \text{ K} \tag{15.2}$$

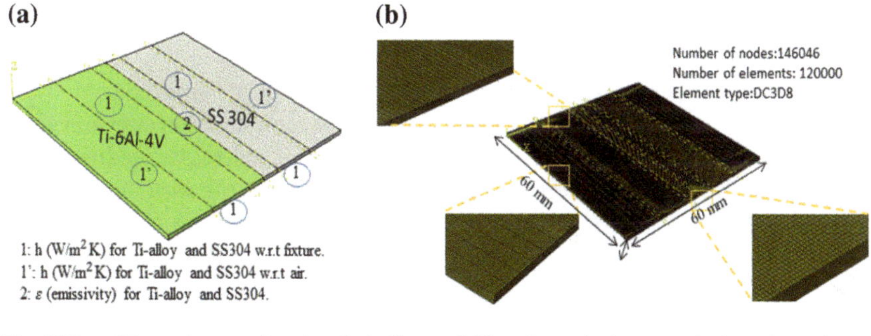

Fig. 15.2 **a** Thermal constraints for dissimilar modelling; **b** meshed geometrical configuration of thermal model

b. Neumann boundary condition; mathematically, heat interactions at all the surfaces can be expressed as

$$k_t \frac{\partial T}{\partial n} - q_f + \varepsilon_r \sigma_r \left(T^4 - T_i^4\right) + h_t (T - T_i) = 0 \tag{15.3}$$

where T_i is ambient temperature, k_t is isotropic thermal conductivity of the material, q_f is Gaussian distributed heat flux impinges on substrate, ε_r is the emissivity, σ_r denoted as Stefan–Boltzman constant (5.67×10^{-8} Jm^{-2} s^{-1} K^{-4}), h_t is the convective heat transfer coefficient between workpiece and ambient surrounding.

The geometrical configuration of dissimilar welded joints with optimum uniform mesh discretization at fusion zone (FZ) and non-uniform meshing at far away distance for each component is shown in Fig. 15.2b.

15.4 Result and Discussions

15.4.1 EDS Analysis

Energy dispersive X-ray (EDX) spectroscopic analysis is performed to determine the elemental distribution and EDX-based scanning electron microscopic (SEM) investigation is carried to understand morphology attained by dissimilar joint of Ti-SS component. Figure 15.3 shows SEM images of different zones of welded specimen. Figure 15.3a depicts the fusion zone (FZ) which is composed of very fine-grained morphology and its grain boundary. Several small pores are also visible on the surface of grain is due to inclusion of hydrogen gas entrapment at high temperature in Ti-alloy. However, basic microstructure of austenitic stainless (304) is consists of equiaxed, austenitic twinned structure as shown in Fig. 15.3b. The austenitic structure completely transformed into delta ferrite upon heating. During welding, the liquid phase is nucleated at triple networks and grain boundary depending upon interface energy at solidus temperature. Upon further heating, the liquid phase is distributed along the grain boundary and generates a continuous junction along the grain edge as clearly visible in Fig. 15.3a.

Due to pulse phenomena, components experience higher temperature for very small interval of time due to which weld pool lasts for short duration at fusion zone. As time spent by the substrate at high temperature domain decreases, grains are not allowed to grow more during cooling which resembles very fine-grained microstructure in FZ. EDS analysis as spectrum 12 (as shown in Fig. 15.4) confirms that fusion zone is majorly composed of elements like Fe (55%), Cr (15%), C (12%), Ti (6%) and Ni (6%). There is possibility that these elements form intermetallic compound like $FeTi + Fe_2Ti + Ti_5Fe_{17}Cr_5 + Ni_3Ti$ and some form of

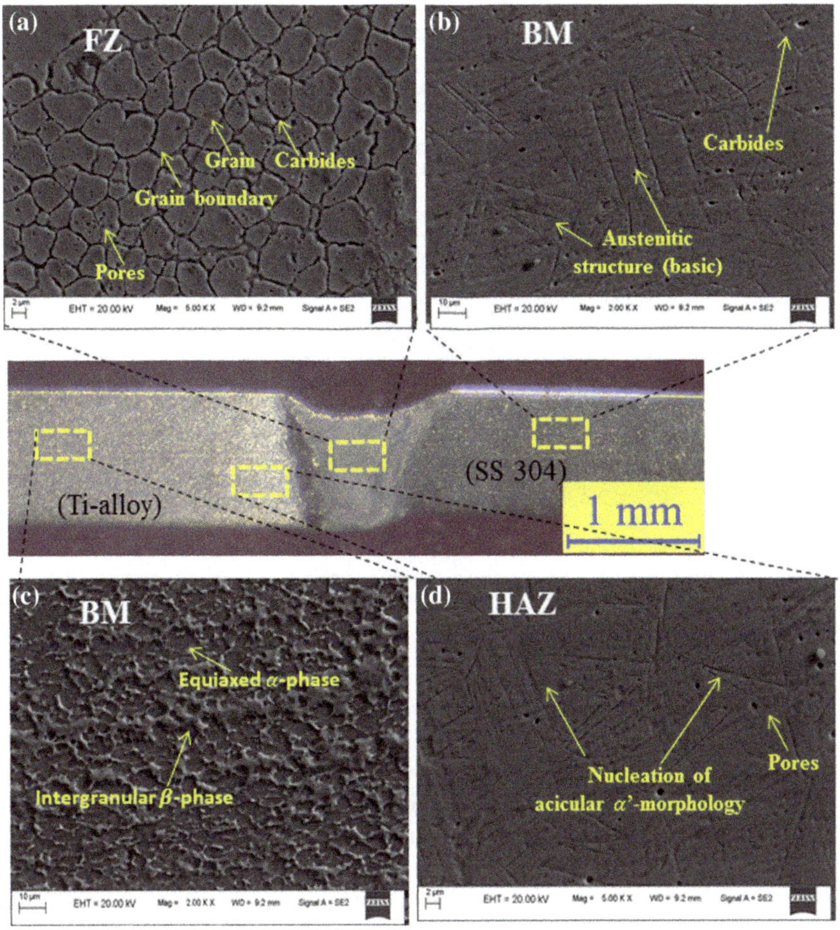

Fig. 15.3 SEM image of **a** FZ; **b** base material (stainless steel); **c** base material (Ti-alloy); **d** HAZ (Ti-alloy)

carbides due to diffusion of Ti-alloy with SS 304 at elevated temperature. Figure 15.3c depicts the basic microstructure of Ti–6Al–4V, consists of equiaxed α-phase (dark region) and line-like white region (light region) as intergranular β-phase. Figure 15.3d signifies the heat-affected zone at the verge of molten zone at Ti-side which clarifies the nucleation of acicular type α' brittle martensitic microstructure.

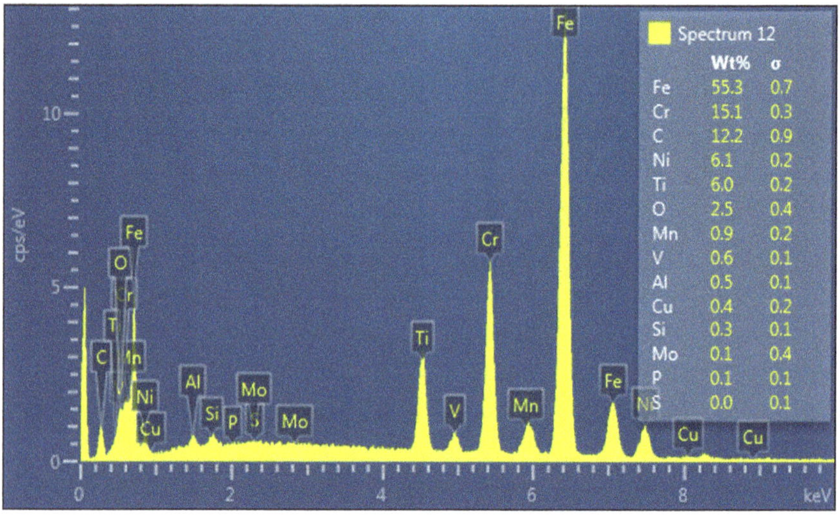

Fig. 15.4 EDS-based spectrum for fusion zone (FZ) of dissimilar weld joint

15.4.2 Numerical Result

FEM-based three-dimensional thermal modelling is undertaken to predict weld bead geometry and to study variation in nodal temperature in both components of dissimilar weld during pulsed laser welding. The numerical simulation of temperature distribution is carried out corresponding to welding speed 4.5 mm/s and average power of 360 W for the heat input of 80 J/mm. Figure 15.5a manifests the comparison between experimental and numerical weld bead profile for dissimilar joint of Ti-alloy and AISI 304. It reveals that macrostructural characteristics (weld bead profile, HAZ) of simulated domain are of similar pattern as anticipated by experimental result. Heat affected a zone which manifests solid-state phase transformation represented by region between blue and yellow colour band, mushy zones are characterized by liquidus (Ti-alloy ~ 1933 K; SS ~ 1723 K) and solidus temperature (Ti-alloy ~ 1877 K; SS ~ 1673 K) is designated as yellow band, whereas FZ is represented by red colour layer for both components. The contour plot of spatial and temporal movement of heat flux in pulsed off and on condition is expressed in Fig. 15.5b. It shows that the red colour band is only visible in pulse-on condition due to attainment of higher temperature for short period which is attributed to entrapment of direct laser beam on the substrate. Experimentally obtained bead profile reveals that the evaporation phenomena also involved for the present optimized process parameter. However, numerical result did not exhibit evaporation occurrence since the maximum temperature attained here is well below of boiling temperature (3500 K). Therefore, it can conclude that conduction mode of welding is prevailing over keyhole model for developed thermal model.

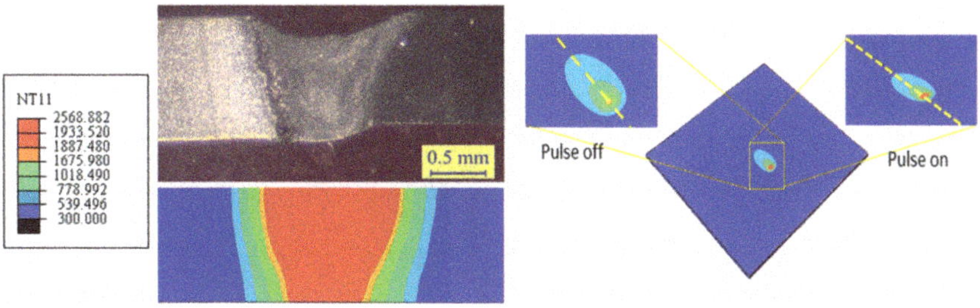

Fig. 15.5 **a** Comparison of experimental and numerical weld bead profile; **b** contour plot of spatial movement of laser bear at pulse-on condition and pulse-off condition

Figure 15.6 depicts the temperature–time variation in Ti-alloy and stainless steel part during pulsed welding. It shows the needle-type structure at peak temperature for both materials. Pulse-on and pulse-off appearances lead to instantaneous temperature gradient at weld line or at closer distance from weld line which is attributed to formation of peak and valley for very short time duration. It is obvious from the figure that peak and valley existence gradually decreases away from the weld line for both the material. Figure 15.6a reveals the attained temperature across the weld line for Ti-alloy side while on stainless steel side is shown in Fig. 15.6b. For both cases, achieved temperature is gradually decreasing as the distance increases from the weld line, and this is due to the fact that at weld line, substrate is in direct contact with laser beam; however, as the distance increases, less amount of heat conduction occurs across the molten zone.

The temperature attained by Ti-alloy is about 2350 K at 0.5 mm distance of weld line, whereas SS 304 achieved 2000 K at same distance across the weld line. The higher diffusive heat transfer coefficient and absorptivity of Ti-alloy as compared with SS304 resemble higher amount of heat diffusion throughout the surface unlike stainless steel. Figure 15.7a conveys that temperature distribution in the

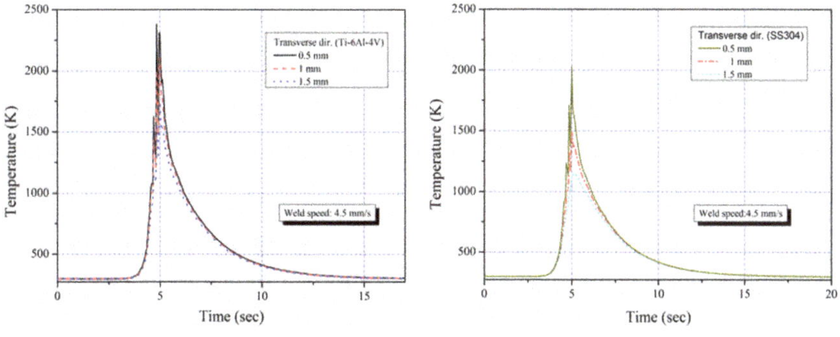

Fig. 15.6 Temperature distribution at different location across the weld zone, **a** Ti–6Al–4V alloy; **b** SS 304

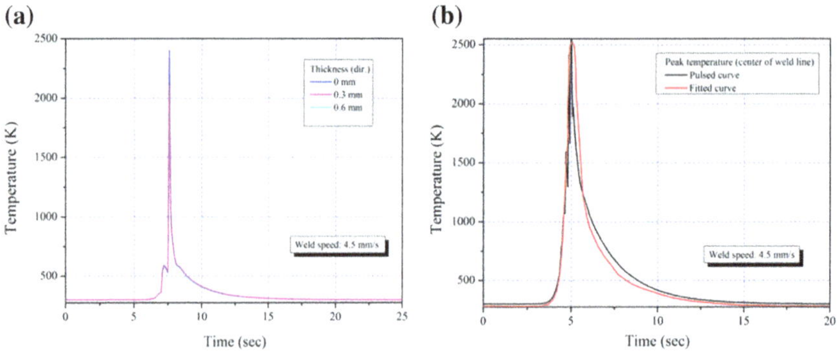

Fig. 15.7 **a** Temperature profile in thickness direction at weld line; **b** temperature profile obtained at mid of weld line in pulsed condition and fitted curve by Gaussian function

component in thickness direction at centre point of weld pool. At the top of the surface, attained temperature is the same as that of peak temperature ($\sim$2500 K). Drop of the temperature throughout the thickness is very low up to 0.5 mm thereafter it increases but not so much because of the thinness of the weldments.

15.4.3 Cooling Rate Estimation

Researchers have already described that cooling rate corresponding to laser welding of thin sheet plays significant role for the prediction of different microstructures attained by weldments and its associated mechanical properties. Senkov et al. [15] investigated the effect of controlled cooling rate on the morphology and mechanical properties of Ti-alloy. The weld thermal cycle mainly consists of continuous heating and followed by cooling it up to ambient environment. It is unsuitable or crucial to determine the attained temperature–time history during welding at fusion zone due to involvement of melting phenomena. Hence, simulation-based time–temperature data are implemented for the estimation of cooling rate at different location of both components of weldments.

In the present work, pulsed mode of welding is considered for the investigation so the temperature profile exhibits instantaneous temperature variation with respect to time at higher temperature. For the sake of simplicity, the achieved temperature profile at weld zone is curve fitted by using Gaussian basic function which provided satisfactory continuous curve. Fitted continuous curve is implemented for the calculation of cooling rate in the present study as depicted in Fig. 15.7b. The temperature at which remarkable phase transformation occurs during cooling cycle is considered for the estimation of cooling rate here. Kumar et al. [16] reported the alpha-dissolution and beta-transus temperature for the dual-phase Ti-alloy as 940 and 1275 K. Mathematical formulation for the estimation of cooling rate is also revealed in the literature. The phase transformation temperature range for stainless steel is assumed as 773 and 1073 K.

Figure 15.8 shows the temperature distribution at different location across the weld line within assumed phase transformation temperature range for Ti–6Al–4V and steel alloy. These curves reveal that temperature profile within this range is approximately linear for both components. Therefore, it is obvious that slope of these curve can predict fairly accurate average cooling rate. Figure 15.8a depicts that the time spent for the attainment of phase transformation temperature at a distance of 0.5 mm (closer to weld line) is comparatively more than (i.e. 1.4 s) that of 1 mm or 1.5 mm across the weld line for steel alloy. Due to direct contact of weld line with heat flux, nearby region of fusion zone experiences high heat input which sustain weld pool for comparatively longer period of time. Therefore, it takes higher solidification time. Hence, average cooling rate is lower (i.e. 220 K/s) at this location, and at 1 and 1.5 mm location, cooling rate are 258 and 277 K/s, respectively.

Figure 15.8b conveys the temperature distribution within alpha-dissolution temperature and beta-transus temperature range for Ti-alloy. It shows that 0.84 s of total time is elapsed for the attainment of phase transformation temperature at the location of 0.5 mm across the weld centre while somewhat less time is taken at the location of 1 and 1.5 mm across the weld line. So, the estimated cooling rate at closer (weld line) distance is comparatively less (i.e. 398 K/s), whereas cooling rate attained at 1 and 1.5 mm distance are 420, 424 K/s, respectively.

It is observed that the estimated cooling rate at same location of two different materials is having remarkable difference, i.e. 220 K/s for stainless steel and 398 K/s for Ti–6Al–4V at 0.5 mm distance from weld centre. Because, cooling rate eminently associated with the thermal behaviour of material. Accomplishment of temperature depends upon the thermo-physical properties like specific heat, diffusion coefficient, thermal conductivity, etc. So, it can deduce that significant difference in these properties of Ti-alloy and SS 304 is responsible for great difference in cooling rate.

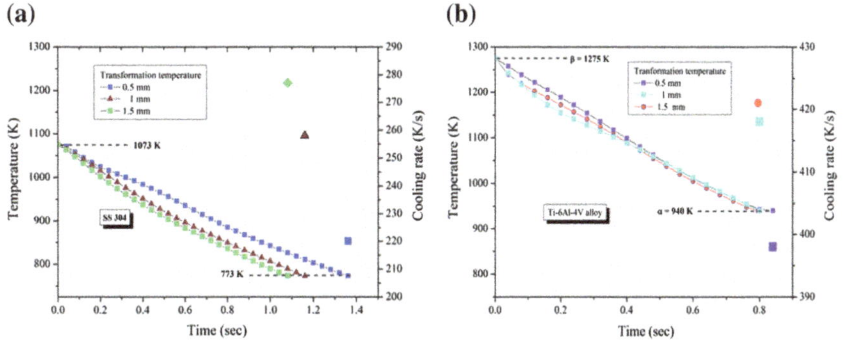

Fig. 15.8 Temperature profile within the range of phase transformation and cooling rate, **a** stainless Steel 304; **b** Ti–6Al–4V alloy

It is already reported in the literature that microstructural evolution can be determined with the help of appropriate cooling rate calculation. It is well established fact that 400 K/s is critical cooling rate for dual-phase Ti-alloy, up to which massive diffusional nucleated secondary lamellar structure can be obtained and beyond which hexagonal acicular α' martensitic structure is possible. Widmanstatten weave-basket-like morphology can be produced at very low cooling rate in dual-phase Ti-alloy. EDS analysis reveals that the nucleation of acicular α' martensitic structure (Fig. 15.3d) has been found at heat-affected zone (nearby region of fusion zone) and it is also observed that the estimated cooling rate at HAZ location is approximately 420 K/s which is well higher that critical cooling rate as reported in the literature. Therefore, it can be concluded that the estimated average cooling rate based on numerical model is fair and accurate and showing good agreement with experimentally observed morphology in the present study.

15.5 Conclusion

Pulsed laser butt welding of titanium alloy to stainless steel has been performed at optimized process variable. EDS-based morphological study and FEM-based temperature distribution have been studied in the present work. Based on this study, conclusions are summarized as following:

- After solidification of weld zone, fine-grained morphology is revealed. Because, due to pulsation effect, high temperature attained at weld zone is only for small period of time that lasts weld pool for shorter period. So, nucleated grain during re-solidification from liquid to solid phase is not allowed to grow more, whereas nucleated acicular α' martensitic structure is observed in HAZ of Ti-alloy side.
- Energy dispersive X-ray (EDX) spectroscopic analysis confirms that FZ majorly contains Fe, Cr, C and Ti elements that resemble formation of intermetallic compound like $FeTi + Fe_2Ti + Ti_5Fe_{17}Cr_5 + Ni_3Ti$ during fusion of both components (Ti-alloy and SS 304) at high temperature.
- Good agreement has been observed between experimentally and numerically welds bead profile. However, evaporation of material at top surface is clearly shown in experimental result but attained maximum temperature of weld zone is well below the boiling temperature so an evaporation phenomenon is neglected in 3D modelling.
- The maximum temperature attained at Ti-side is higher than that of stainless steel side. Significant difference in diffusion coefficient and absorptivity is the possible reason for the same.
- As it is difficult to measure temperature–time data at weld zone experimentally, so cooling rate is estimated at different location (i.e. at a distance of 0.5, 1 and 1.5 mm across the weld line) by using numerical simulated result. Maximum cooling rate has been found at 1.5 mm distance for both of the components of weldments, i.e. 424 K/s for Ti-alloy and 277 K/s for stainless steel.

- It is revealed that remarkable difference in thermo-physical properties such as specific heat capacity, conductivity, diffusivity, etc. are the feasible reason for the large difference in cooling rate of both of the material at the same location.
- Estimated average cooling rate for HAZ of Ti-alloy has predicted similar type of morphology as observed in experimental investigation and also reported by the other researchers.

Acknowledgements The authors gratefully acknowledge Central Instrument Facility (CIF) and Central workshop of IITG for providing experimental facility.

References

1. Sun, Z., Ion, J.C.: Laser welding of dissimilar metal combinations. J. Mater. Sci. **30**(17), 4205–4214 (1995)
2. Mousavi, S.A.A.A., Sartangi, P.F.: Effect of post-weld heat treatment on the interface microstructure of explosively welded titanium–stainless steel composite. Mater. Sci. Eng. A **494**(1-2), 329–336 (2008)
3. Lapsanska, H., Chmelickova, H., Hrabovsky, M.: Effect of beam energy on weld geometric characteristics in Nd: YAG laser overlapping spot welding of thin AISI 304 stainless steel sheets. Metall. Mater. Trans. B **41**(5), 1108–1115 (2010)
4. Tzeng, Y.-F.: Parametric analysis of the pulsed Nd: YAG laser seam-welding process. J. Mater. Process. Technol. **102**(1-3), 40–47 (2000)
5. Matsunawa, A., et al.: Dynamics of keyhole and molten pool in laser welding. J. Laser Appl. **10**(6), 247–254 (1998)
6. Gao, X.-L., et al.: Porosity and microstructure in pulsed Nd: YAG laser welded Ti6Al4V sheet. J. Mater. Process. Technol. **214**(7), 1316–1325 (2014)
7. Hiraga, H., et al.: Nd: YAG laser welding of pure titanium to stainless steel. Weld. Int. **16**(8), 623–631 (2002)
8. Chen, S., et al.: Microstructures and mechanical property of laser butt welding of titanium alloy to stainless steel. Mater. Des. **53**, 504–511 (2014)
9. Bag, S., et al.: Efficient estimation of volumetric heat source in fusion welding process simulation. Weld. World **56**(11-12), 88–97 (2012)
10. Baruah, M., Bag, S.: Influence of heat input in microwelding of titanium alloy by micro plasma arc. J. Mater. Process. Technol. **231**, 100–112 (2016)
11. Bag, S., De, A.: Development of a three-dimensional heat-transfer model for the gas tungsten arc welding process using the finite element method coupled with a genetic algorithm–based identification of uncertain input parameters. Metall. Mater. Trans. A **39**(11), 2698–2710 (2008)
12. Yadaiah, N., Bag, S.: Effect of heat source parameters in thermal and mechanical analysis of linear GTA welding process. ISIJ Int. **52**(11), 2069–2075 (2012)
13. Akbari, M., et al.: Experimental and numerical investigation of temperature distribution and melt pool geometry during pulsed laser welding of Ti6Al4V alloy. Opt. Laser Technol. **59**, 52–59 (2014)
14. Zhu, X.K., Chao, Y.J.: Numerical simulation of transient temperature and residual stresses in friction stir welding of 304L stainless steel. J. Mater. Process. Technol. **146**(2), 263–272 (2004)

15. Senkov, O.N., et al.: Effect of cooling rate on microstructure of Ti-6Al-4V forging. Mater. Sci. Technol. **18**(12), 1471–1478 (2002)
16. Kumar, B., Kebede, D., Bag, S.: Microstructure evolution in thin sheet laser welding of titanium alloy. Int. J. Mechatron. Manuf. Syst. **11**(2-3), 203–229 (2018)

Chapter 16
Activated TIG Welding of AISI 321 Austenitic Stainless Steel for Predicting Parametric Influences on Weld Strength of Tensile Test—Experimental and Finite Element Method Approach

S. Mohan Kumar, N. Siva Shanmugam and K. Sankaranarayanasamy

Abstract Tungsten inert gas (TIG) welding is the most common manufacturing process used to join materials like stainless steel, titanium and aluminium alloys due to their high-quality and inexpensive welds. Furthermore, the difficulty in welding plates having a thickness of 4 mm or more is overcome by employing activated TIG (A-TIG) welding process which uses an activating flux for high depth of penetration in a single pass. Uni-axial tensile test is conducted to assess the tensile strength of the material by experimentation. In this study, a uni-axial tensile test of base metal (BM) and weld metal (WM) samples are simulated using ABAQUS to evaluate the accuracy of finite element (FE) simulation results with the experimental results for predicting the tensile strength. The results show that the stress–strain values predicted by the FE analysis agree with experimental results. Also, the fracture behaviour of experimentation and FE simulation is identical with ductile mode of fracture. The fracture location of the sample in FE analysis is found very similar to experimental fractured samples. The results of ferrite measurement indicate that concentration of delta-ferrite in the WM (5.9 FN) is higher than BM (1.2 FN) content and show better mechanical behaviour in the A-TIG weldments. Also, scanning electron microscope (SEM) shows that the failure of BM and WM resembled to ductile mode-type fracture.

Keywords A-TIG welding · AISI 321 · Tensile test · Finite element analysis · Fractography

---

S. Mohan Kumar · N. Siva Shanmugam (✉) · K. Sankaranarayanasamy
Department of Mechanical Engineering, National Institute of Technology Tiruchirappalli, Tiruchirappalli 620015, Tamil Nadu, India
e-mail: nsiva@nitt.edu

R. G. Narayanan et al. (eds.), *Advances in Computational Methods in Manufacturing*, Lecture Notes on Multidisciplinary Industrial Engineering,
https://doi.org/10.1007/978-981-32-9072-3_16

16.1 Introduction

AISI 321 material has excellent resistance to corrosion [1]. Widely used in automotive industries, jet aircraft components, welded pressure vessels in the defence sector and high-speed impact on firewalls, the encounter of static and dynamics loads on AISI 321 is a common problem. The demand for the controlled weld shape and deeper penetration in TIG welding led to the novel modification of the well-known TIG process, called as A-TIG welding. A-TIG welding was introduced during the 1960s by the E.O. Paton Institute of Electric Welding [2]. The A-TIG joining process uses a thin layer of flux which is spread over the welding region before welding [3]. Tensile testing provides engineers with a wealth of data about the mechanical behaviour of the metals, elastic modulus, tensile strength, yield strength, the percentage of elongation and the stress–strain relationship of the materials. Hence, it is essential to understand the mechanical behaviour and metallurgical aspects that affect the results of the tensile test [4]. Reisgen et al. [5] conducted an FE analysis of uni-axial tensile test to evaluate the tensile strength of laser-welded TRIP700 steel sheet. Garcia-Garino et al. [6] carried out an FE simulation of tensile test for a metal specimen and found good agreement with the analytical and experimental results. Patel et al. [7] studied the effect of activated fluxes like SiO_2 and TiO_2 on the mechanical behaviour of AISI 321 using the TIG welding. However, research works related to FE simulation of tensile test of A-TIG-welded samples are few. Hence, in the present study, tensile test was carried out on base- and A-TIG-welded butt joint of AISI 321 and compared with the experimental and FE analysis results for validation.

16.2 Experimentation Procedure

Bead on plate trials is performed to choose optimum welding parameter. Selection of perfect welding current, arc length and welding speed defines the requirement of the good weld bead. The material selection in the present study is commercially available type AISI 321 austenitic stainless steel. Tables 16.1 and 16.2 show the chemical composition and the physical properties of AISI 321, respectively. Welding equipment model Fronius MagicWave 4000 is used for joining AISI 321 with experimental set-up shown in Fig. 16.1. Particularly during A-TIG joining process, a very thin film of activating flux is painted to the material top surface

Table 16.1 Elemental composition of AISI 321 (wt%)

Ni	Cr	Mn	Si	C	P	S
9.356	17.560	1.007	0.349	0.032	0.031	0.016
Fe	W	Ti	Cu	Mo	Al	Nb
70.775	0.036	0.356	0.356	0.113	0.006	0.003

Table 16.2 Physical properties of AISI 321

Density (kg/m^3)	Young's modulus (GPa)
8090	193

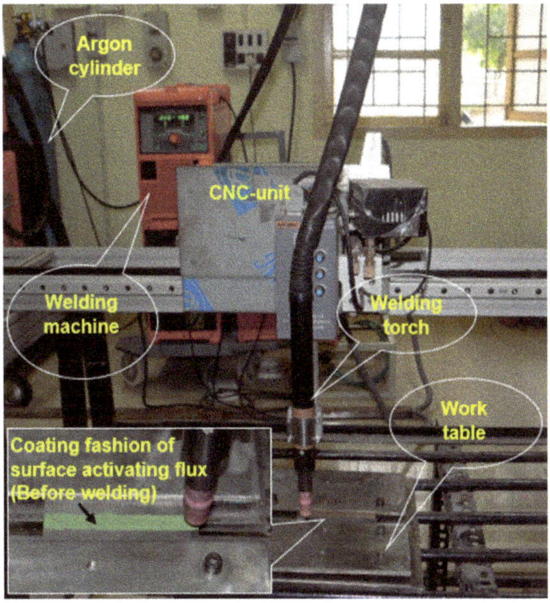

Fig. 16.1 Experimental set-up for A-TIG welding

before joining so as to increase the depth of penetration (DOP). Based on our previous study [8], the commercial Edison Welding Institute (EWI) flux is used in this work, and corresponding ingredients of activated flux are shown in Table 16.3. This increase in DOP is attributed to the mechanisms of arc constriction and Marangoni reversal flow. Before joining the AISI 321 plate of 6 mm thickness was cut into strips of 100 × 100 mm and roughly polished with sheets of silicon carbide to remove the impurities in surface and further scrubbed with acetone. The input weld parameters selected after examinations from the trials of the bead on the plate are shown in Table 16.4. Figure 16.2 represents the welded plate with butt joint configuration. Further, Fig. 16.2b shows the macrostructure of the butt joint with achieved full depth of penetration. The sample for the tensile test is cut from the plate specimen following ASTM E8/E8M-16a standard as shown in Fig. 16.2c.

Table 16.3 Composition of EWI flux

Flux ingredients	Weight (%)
Silicon oxide (SiO_2)	0–20
Chromium III oxide (Cr_2O_3)	0–50
Titanium oxide (TiO_2)	20–60

Table 16.4 A-TIG welding input process parameters

Process parameters	Values	Units
Welding current	220	Amps.
Welding speed	120	mm/min
Arc length	3	mm
Welding voltage	14.9	V
Welding heat input	1.4751	kJ/mm
Electrode diameter	2.4	mm
Shielding gas flow rate (constant—argon purity 99.99%)	20	l/min

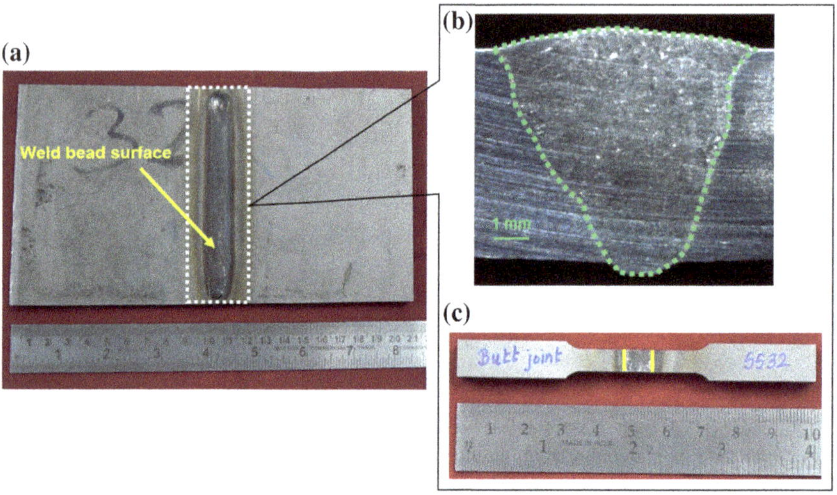

Fig. 16.2 **a** Butt weld geometry made of A-TIG welding, **b** macrostructure of butt joint and **c** sample for the tensile test

16.3 Results and Discussion

16.3.1 Experimental and FE Simulation of the Tensile Test

The AISI 321 tensile specimens of BM and WM were prepared as per standard. The wire-cut EDM process was used to section the specimens. Tensile tests were carried out using Tinius Olsen 50 KN load cell at room temperature with a loading rate of 1 mm/min until failure.

FE analysis of tensile test was performed using ABAQUS—a commercial available FE analysis software. Brief illustration of FEM procedure is shown in Fig. 16.3. The part was modelled in ABAQUS, and physical material properties such as density, Young's modulus and Poisson's ratio assigned are 8090 kg/m^3,

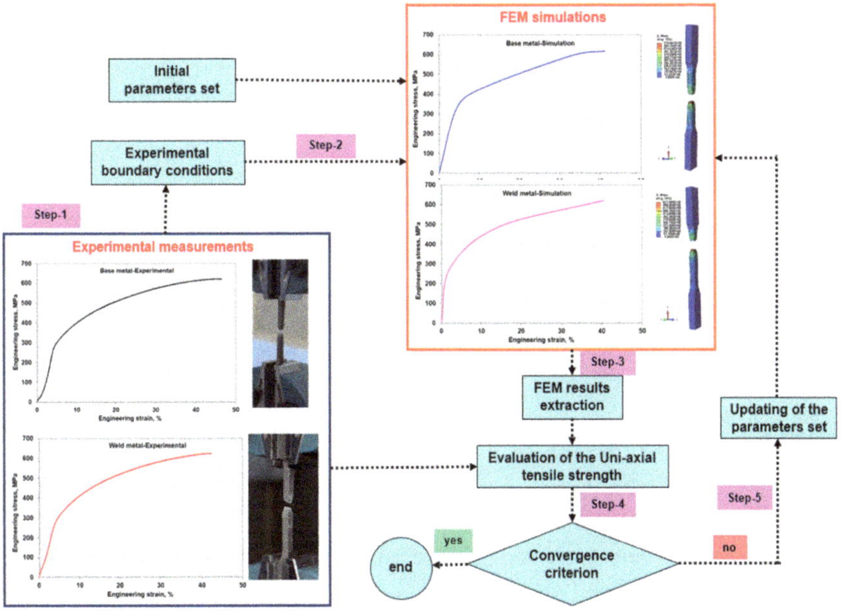

Fig. 16.3 Flow chart of FE method for uni-axial tensile test

Table 16.5 Tensile properties of AISI 321

Part	Yield stress (MPa)	Yield strain (%)	Tensile stress (MPa)	Tensile strain (%)
BM	213	3.536	621	46.56
WM	215	3.088	624	42.4

193 GPa and 0.3, respectively. From uni-axial tensile test, mechanical properties obtained for BM and WM specimens are assigned for plasticity as shown in Table 16.5.

The dynamic explicit analysis was conducted in consideration of nonlinear geometry. Suitable damage material model is necessary to initiate fracture and subsequent failure in the FE model. It is assumed that the material behaviour is isotropic. Hence, the Johnson–Cook model is utilized for analysing the failure. Johnson–Cook (JC) material model is one such model used in the present analysis for material failure, considering the effects of pressure, temperature, equivalent stress and strain rate. JC model's damage parameters assigned for ductile damage in AISI 321 are shown in Table 16.8 [9]. Based on the critical literature review, it is understood that the behaviour of both AISI 316L and AISI 321 austenitic grade stainless steels is almost same. For confirming the similarity in mechanical properties, a uni-axial tensile test has been conducted for both AISI 316L and AISI 321. From the uni-axial tensile test results (Fig. 16.4), the following data are extracted,

Table 16.6 Uni-axial tensile test results of AISI 321 and AISI 316L

Material name	Yield stress (MPa)	Tensile stress (MPa)	Tensile strain (%)
AISI 321	213	621	46.56
AISI 316L	208	594	40.96

Table 16.7 Physical properties

Material name	Density (kg/ m^3)	Young's modulus (GPa)
AISI 321	8090	193
AISI 316L	8000	193

Table 16.8 JC damage parameters assigned for AISI 321

D1	D2	D3	D4	D5
0.05	3.44	2.12	0.002	0.61

and the same have been presented in Table 16.6. The density and Young's modulus are also similar in nature (refer Table 16.7).

Owing to the similarities between the two grades of ASS, the JC damage parameters of AISI 316L material are utilized in the present study for performing the tensile test simulation of AISI 321.

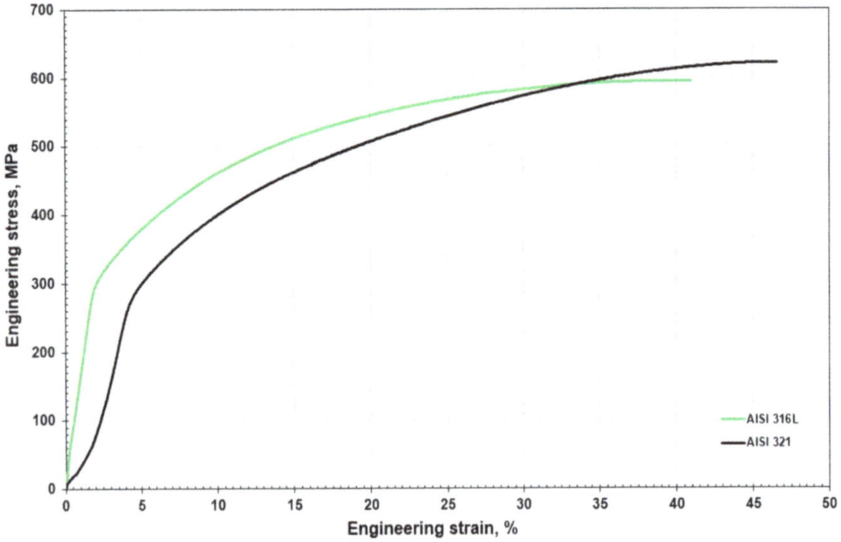

Fig. 16.4 Stress-versus-strain plot for AISI 321 and AISI 316L

The predicted results from the simulation of the uni-axial tensile test also confirm with an error % of 0.51 and 0.48 for BM and WM of AISI 321 steel in comparison with experimental tensile test. JC material model is mathematically expressed by Eqs. 16.1 and 16.2.

The failure is expected when parameter $D = 1$

$$D = \sum \frac{\Delta\varepsilon}{\varepsilon_f} \tag{16.1}$$

$$\bar{\varepsilon}_f = [D_1 + D_2 \exp D_3\sigma^*][1 + D_4 \ln \dot{\varepsilon}^*][1 + D_5 T^*] \tag{16.2}$$

where

$\Delta\varepsilon$ is the increment of effective plastic strain during an increment in loading.
$\bar{\varepsilon}_f$ is the equivalent strain to fracture.
$\dot{\varepsilon}^*$ is the dimensionless strain rate.
T^* is the homologous temperature.
σ^* is the average of the three normal stresses to the von Mises equivalent stress assumed $\sigma^* \leq 1.5$.
D_1, D_2, D_3, D_4 and D_5 are material-dependent constants.

The experimental setting provides boundary conditions for the tensile specimen. One end of the specimen is ENCASTERED (U1 = U2 = U3 = UR1 = UR2 = UR3 = 0), where movement is restricted in all degrees of freedom, while another end is allowed to move in the *X*-direction for BM and WM (refer Fig. 16.5a, b).

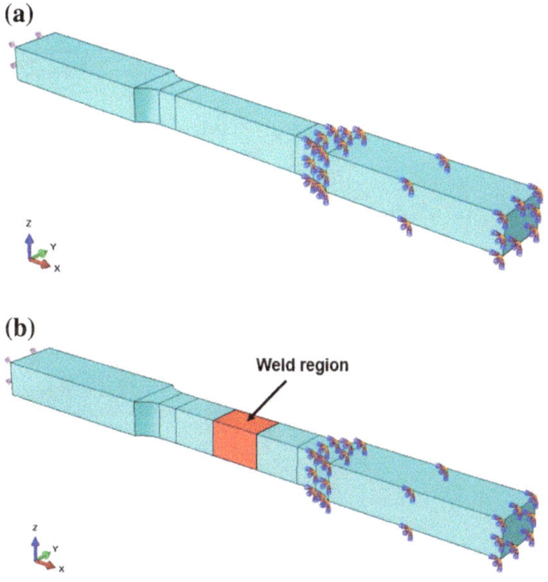

Fig. 16.5 Boundary conditions for the modelled tensile test specimens, **a** BM and **b** WM

Table 16.9 Meshing parameters used for uni-axial tensile test simulation

Part	Shape of element	Meshing technique	Elements generated	Element type
BM and WM	Hexagonal	Structured	6672	C3D8R (an eight-node linear brick, reduced integration with hourglass control)

The meshing of the model is conducted by employing fine mesh at the gauge length section and coarse mesh at the gripping section (to reduce the computational time) as results are expected only at gauge length section. The parameters used for meshing in FE analysis are shown in Table 16.9. Further, the meshed tensile test model of BM and WM is described in Fig. 16.6a, b separately.

Figure 16.7a shows the experimental tensile test depicting the fracture behaviour of the BM. Similarly, a tensile test conducted using the FE model is shown in Fig. 16.7b with a subsequent fracture. A good match can be observed in fracture at the centre of the specimen in both the analyses. Also, the fracture behaviour of the BM and WM samples was observed with the ductile mode of fracture and found similar with FE analysis as shown in Figs. 16.8 and 16.10. Likewise, for WM, the fracture region obtained after performing experimental and FE analysis is shown in Fig. 16.9a, b, respectively. From our earlier investigations [8], the mechanical properties of HAZ were better than the BM using A-TIG process. From the results

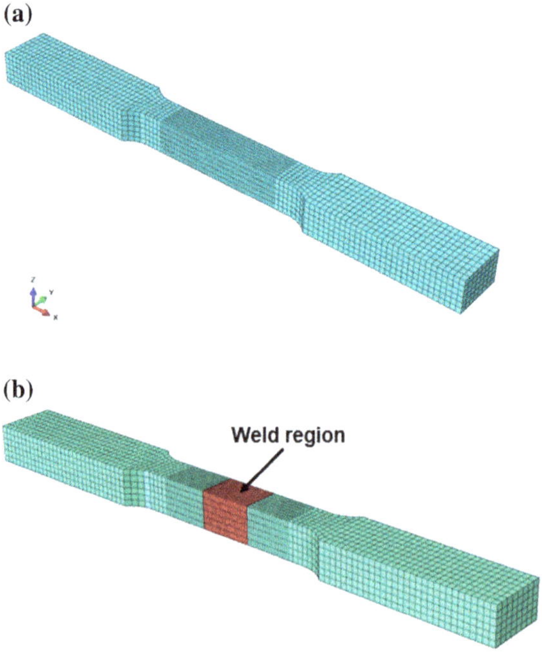

Fig. 16.6 Modelled tensile test specimens with mesh, **a** BM and **b** WM

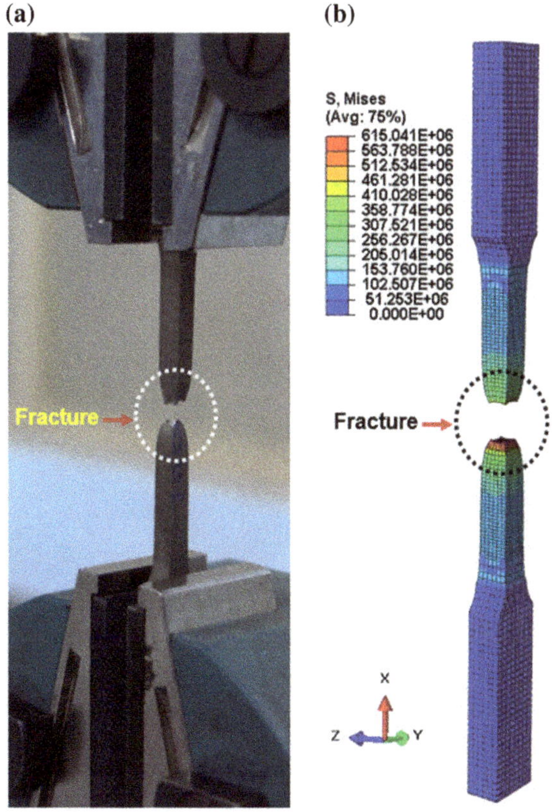

Fig. 16.7 Comparison of the BM tensile test between experimental (**a**) and FE analysis (**b**)

of uni-axial tensile test on both BM and WM, it is evident that the fracture occurred in the BM regions. Hence, the effect of heat-affected zone in the FE simulation is not considered. Excellent agreement is revealed through visual comparison as fracture occurred always in the BM, i.e. ductile mode of failure as shown in Fig. 16.10 between experimental and FE analysis.

The stress–strain curves determined from the tensile test are shown in Fig. 16.11. The tensile strength for the BM and WM tested experimentally at 621 and 624 MPa, respectively, while the FE simulation tensile strength for the BM and WM is 617.8 and 621 MPa, respectively. Table 16.10 presents the tensile strength and the crack position for BM and WM specimens. It was observed that welded specimen has slightly higher tensile strength compared to base, and percentage of elongation of weld specimen decreased considerably when compared to the BM specimen in experimental and FE analysis. From Table 16.10, it is evident that excellent result is obtained in terms of tensile strength for base and weld metal with the percentage of error 0.51 and 0.48%, respectively. Results of ferrite measurement (Fischer FERITSCOPE FMP30) in BM and WM zones of AISI 321 are 1.2 and 5.9 FN, respectively. It is confirmed that the increase in tensile strength and decrease in

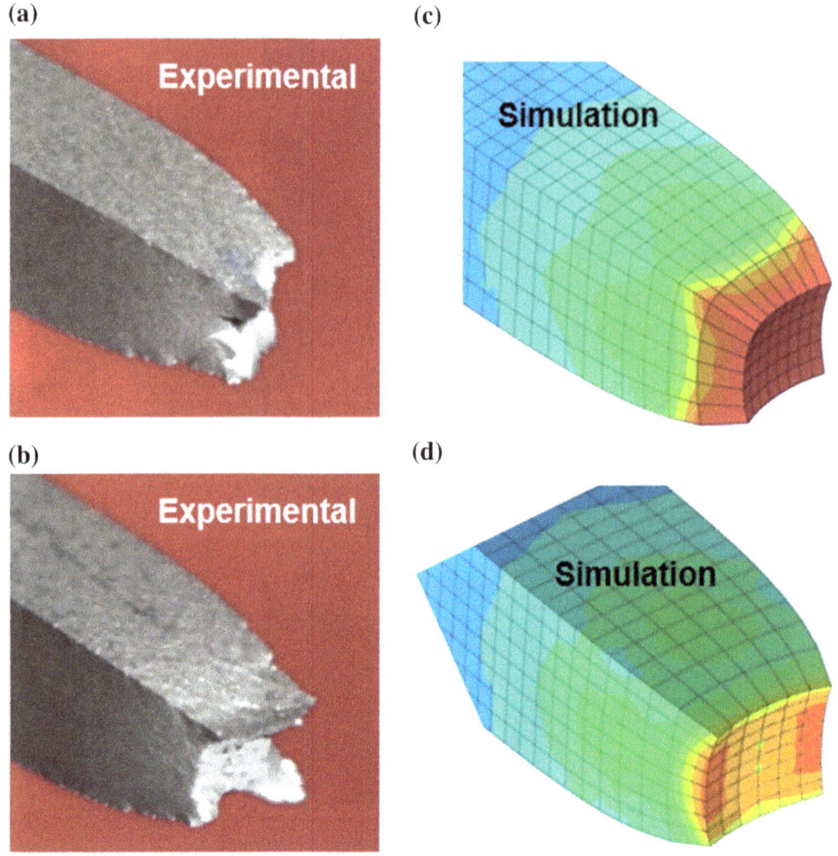

Fig. 16.8 Comparison of BM tensile test fracture surface mode between experimental (**a** and **b**) and FE analysis (**c** and **d**)

ductility are attributed to the hike in the delta-ferrite content. The tensile test results can corroborate that the A-TIG WM tensile strength is better than BM [10]. The experimental and FE simulation joint efficiency of the WM is 101.1 and 100.51%, respectively. The fractured specimens also showed a ductile fracture.

16.3.2 Fractography Examination

The fractured tensile specimen is cleaned ultrasonically and examined using SEM [11]. The tensile strength of the WM is higher than that of the BM only due to the influence of the welding heat input leading to rearrangement of the molecular orientation. Further, as is clear from Figs. 16.12 and 16.13, it has characteristics

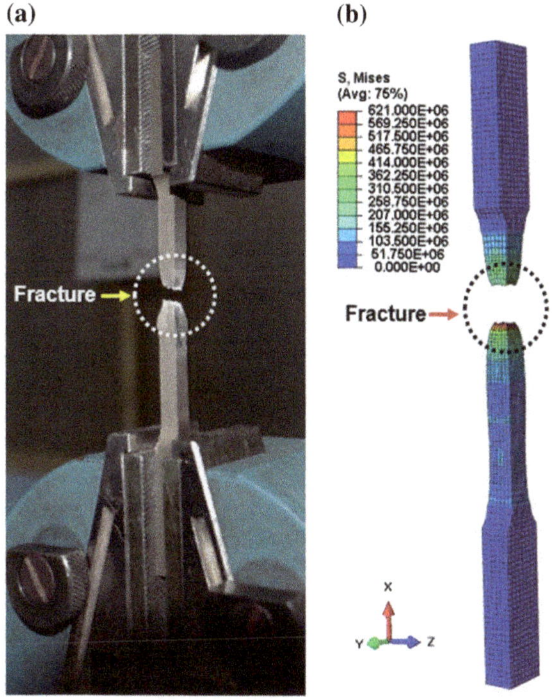

Fig. 16.9 Comparison of the WM tensile test between experimental (**a**) and FE analysis (**b**)

similar to those of the BM and the WM. It is understood from the fracture surface of the WM sample that there are microvoids and fine dimples as compared with the BM and such fine dimples contribute to the improvement of the load-carrying capacity of the WM. Also, on the fractured surface of the WM and the BM, it has similar characteristics, that is, the evolution of a honeycomb structure with voids and dimples confirming the ductile failure.

16.4 Conclusions

The following most important inferences are summarized from the current study.

- A-TIG welding was performed on AISI 321 with optimal input parameters achieving good depth of penetration and bead width for butt joint configuration.
- Experimental tests performed on BM and WM were evaluated for mechanical strength. The uni-axial tensile test showed higher strength for WM—624 MPa compared to BM—621 MPa. The increase in weld strength and reduction in ductility are due to the increase in the delta-ferrite content.
- The prediction by FE analysis was observed with a tensile strength of 621 and 617.8 MPa for WM and BM, respectively.

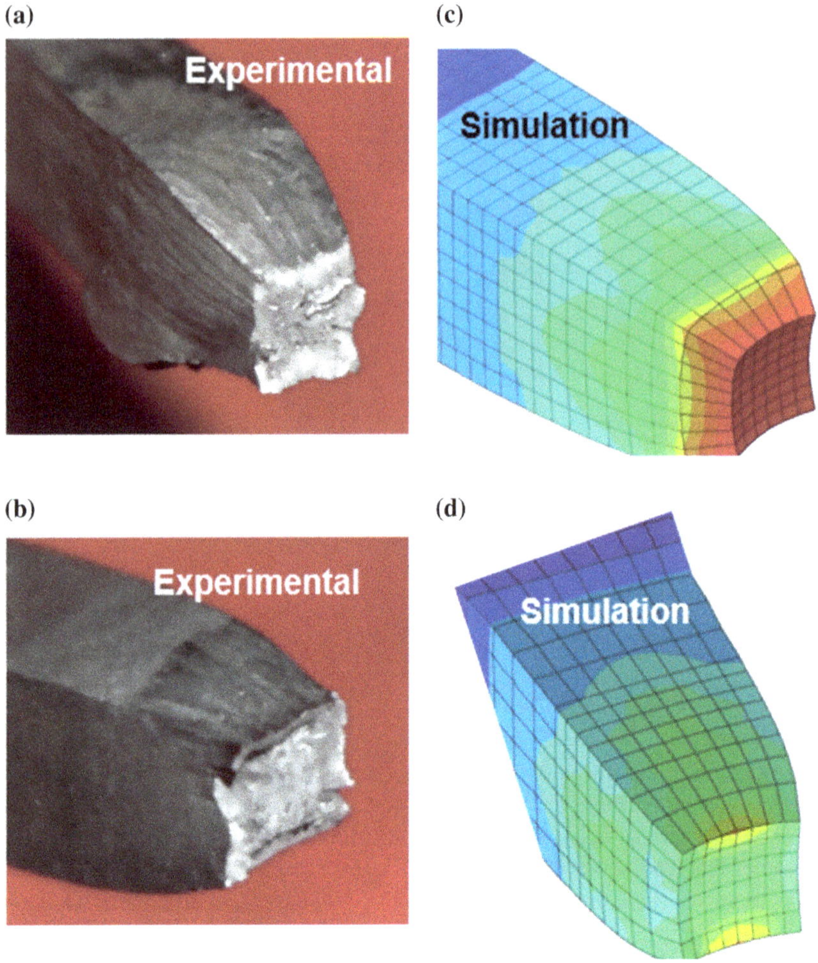

Fig. 16.10 Comparison of WM tensile test fracture surface mode between experimental (**a** and **b**) and FE analysis (**c** and **d**)

- The fracture position of the FE analysis tensile test is found very similar to fractured samples.
- The proper validation by calculating the percentage of error for the tensile test with an error is less than 1%.
- The observation of fractography reveals that voids and dimples are found in the fractured area of the tensile test sample and confirms ductile mode of failure.

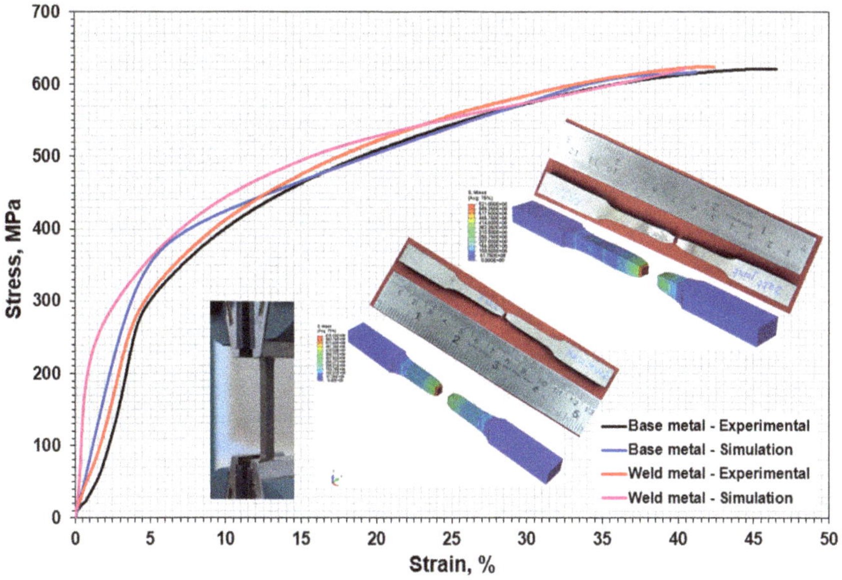

Fig. 16.11 Stress-versus-strain plot for BM and WM

Table 16.10 Uni-axial tensile test results

Sample type	Experimental tensile strength (MPa)	Simulation tensile strength (MPa)	% of error	Position of fracture
BM	621	617.8	0.51	Centre
WM	624	621.0	0.48	BM

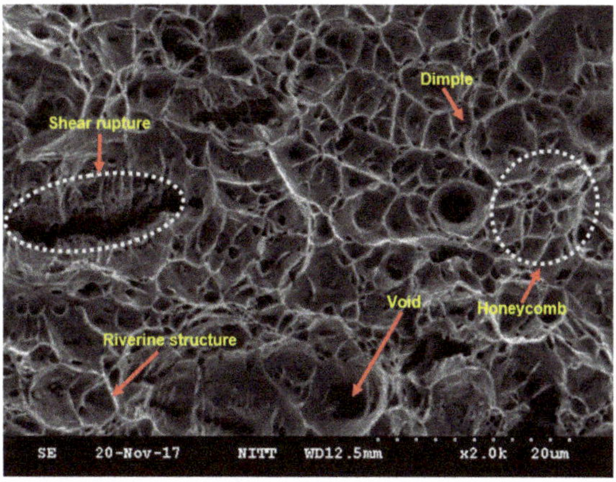

Fig. 16.12 SEM photomicrograph of fractured zone in BM tensile test sample

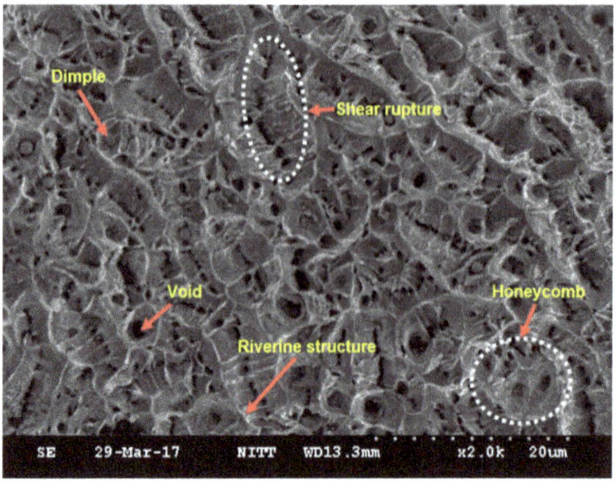

Fig. 16.13 SEM photomicrograph of fractured zone in WM tensile test sample

References

1. Nakhodchi, S., Shokuhfar, A., Iraj, S.A., Thomas, B.G.: Evolution of temperature distribution and microstructure in multipass welded AISI 321 stainless steel plates with different thicknesses. J. Press. Vessel Technol. **137**(6), 061405-1-15 (2015)
2. Fujii, H., Sato, T., Lu, S., Nogi, K.: Development of an advanced A-TIG (AA-TIG) welding method by control of Marangoni convection. Mater. Sci. Eng. A **495**, 296–303 (2008)
3. Vidyarthy, R.S., Dwivedi, D.K.: Activating flux tungsten inert gas welding for enhanced weld penetration: J. Manuf. Process. **22**, 211–228 (2016)
4. Joun, M., Choi, I., Eom, J., Lee, M.: Finite element analysis of tensile testing with emphasis on necking: Comput. Mater. Sci. **41**, 63–69 (2017)
5. Reisgen, U., Schleser, M., Mokrov, O., Ahmed, E.: Numerical and experimental investigation of tensile behavior of laser beam welded TRIP700 steel. ISIJ Int. **51**(3), 429–434 (2011)
6. Garcia-Garino, C., Gabaldon, F., Goicoleab, J.M.: Finite element simulation of the simple tension test in metals. Finite Elem. Anal. Des. **42**, 1187–1197 (2006)
7. Patel, A.B., Patel, S.P.: The effect of activating fluxes in TIG welding by using Anova for SS 321. Int. J. Eng. Res. Appl. **4**, 41–48 (2014)
8. Mohan Kumar, S., Siva Shanmugam, N.: Studies on the weldability, mechanical properties and microstructural characterization of activated flux TIG welding of AISI 321 austenitic stainless steel. Mater. Res. Express **5**, 106524, 1–48 (2018)
9. Sawakar, N., Boob, G.: Finite element based simulation of orthogonal cutting process to determine residual stress induced. Int. J. Comput. Appl. 0975–8887 (2014)
10. Liu, G.-H., Liu, M.-H., Yi, Y.-Y.: Activated flux tungsten inert gas welding of 8 mm-thick AISI 304 austenitic stainless steel. J. Cent. South Univ. **22**, 800–805 (2015)
11. Pouranvari, M., Asgari, H.R., Mosavizadch, S.M., Mariachi, P.H., Goodarzi, M.: Effect of weld nugget size on overload failure mode of resistance spot welds. Sci. Technol. Weld. Join. **2**, 217–225 (2007)

Chapter 17
Design Optimization of Welding Parameters for Multi-response Optimization in TIG Welding Using RSM-Based Grey Relational Analysis

S. C. Moi, R. Rudrapati, A. Bandyopadhyay and P. K. Pal

Abstract This paper presents a research work on optimization of input process parameters using response surface methodology (RSM) along with grey relational analysis to enhance the weldment quality responses in tungsten inert gas welding of 316L austenitic stainless steels. The process control parameters selected for the investigation are welding current, welding speed, and gas flow rate. The experiments have been performed using Box–Behnken design of RSM with three factors, three levels each. Experiments have been performed and output responses: ultimate tensile strength (UTS) and weld width (WW) have also been measured. Grey relational analysis technique has been applied on experimental data to convert multi-output problem into single-response problem. Mathematical model has been developed correlating the process parameters with grey relational grade. The effects of welding input parameters on multi-performance characteristics have been studied graphical contour plots which made from the developed mathematical model. Optimal process parametric has been obtained at higher grey relational grade value. The results of ANOVA and contour plots indicating that output responses, UTS and WW, are significantly influenced by welding input parameters. The concluding remarks have been drawn from the study.

Keywords TIG welding · 316L austenitic stainless steel · Response surface methodology · Grey relational analysis · Welding performance

---

S. C. Moi · A. Bandyopadhyay · P. K. Pal
Mechanical Engineering Department, Jadavpur University, Kolkata, India
e-mail: sc_moi1@rediffmail.com

A. Bandyopadhyay
e-mail: asishbanerjee@yahoo.com

P. K. Pal
e-mail: pradippal54@yahoo.com

R. Rudrapati (✉)
Mechanical Engineering Department, Hawassa University, Awasa, Ethiopia
e-mail: rameshrudrapati@gmail.com

R. G. Narayanan et al. (eds.), *Advances in Computational Methods in Manufacturing*, Lecture Notes on Multidisciplinary Industrial Engineering,
https://doi.org/10.1007/978-981-32-9072-3_17

17.1 Introduction

The selection of the material for any industrial application is based on its mechanical properties like strength, ductility resistance to oxidation and/corrosion, sensitization, etc. Variety of austenitic stainless steel materials which are used for many industrial purposes, type 316L is usually considered over other austenitic steel materials as a structural material, because, of its higher corrosion resistance and excellent mechanical properties both at low and high temperature [1]. Austenitic stainless steel (type 316L) is one of the most consumable materials in orthopedic implants [2], and it is considered as a major structural part for sodium cooled fast reactor applications [3], etc. TIG welding is very useful for similar or dissimilar welding of metals compared to laser welding, because, of high welding quality, high stability in welding of variety of industrial applications. Tungsten inert gas (TIG) welding with and without use of filler material is generally applied welding methods for joining of stainless steels for the industrial applications like pipe, automotive exhaust gas system, chemical industry-based equipment, power plants [3], etc. Welding of 316L austenitic stainless steel (ASS) is important area of research [4] due to corrosion and crack formation resistance [5].

In TIG welding, correct selection of welding process parametric setting is very important to decide the quality of the weld specimen. The quality levels of weldments is highly depending on many interactive input variables such as, the gap maintained between the joined parts, evaporation of parts, torch angle, quantity of the shielding gases present, pulse frequencies and the electrode tip angle selected, etc. when conduction of welding operation. It is very important to select the optimal welding process parametric combination to obtain required weld performances [6]. The systematic analysis and optimization approaches are required to produce desired quality welding and to enhance the quality responses of joining parts [7] but it required to study and understand the welding process extensively because many factors control the welding process [8].

The experimental analysis and optimization of any welding process are difficult job due to many interlinked parameters involved while solving it. The experimental analysis of welding is often a very costly as well as time-consuming process [9]. With a view to achieving the desired quality characteristics in TIG welding of 316L austenitic stainless steels, statistically designed experiments based on the Box–Behnken design of RSM have been used to conduct experimental study. The Box–Behnken experimental design-based RSM is very useful to study system or process effectively to obtain the required information about the direct and interaction effects of input process variables on performance measures with only fifteen number of experimental runs. The input parameters considered in the present work are current, speed, and gas flow rate. The welding performance characteristics measured after the experimentation are ultimate tensile strength (UTS) and weld width. Solving the multi-performance characteristics is not an easy task. Ghosh et al. [10], Wahule and Wasankar [11], and Bahar et al. [12] stated that grey relational analysis (GRA) is efficient for solving multi-objective problems. In the present research investigation,

GRA is utilized to solve the multi-responses by converting multi-output parameters into single response. RSM has been used to make mathematical equation between the input factors and grey relational grade. Two-dimensional contour plots have been made RSM to analyze the factor effects on output response(s).

17.1.1 Literature Survey

Recently, some papers reported in literature to establish a new experimental optimization approach using Taguchi method to analyze TIG welding with minimum number experimental runs.

Durgutlu [13] had investigated the microstructure, penetration, and mechanical properties of 316L ASS in TIG welding operation. They tested the effects of shielding gas (i.e., hydrogen in argon) while welding of 316L ASS. They found highest tensile strength of sample at 1.5% H_2-Ar from their study. Soltani and Tayebi [5] had conducted a comparative analysis of welding of 304L and 316L stainless steels using laser and TIG welding operations. They considered yield strength, elongation, hardness, and weld width as output responses. Authors stated from their study that TIG welding has produced better weld results compared to laser welding. Thomas et al. [4] had been investigated the fatigue crack growth rate in stainless steel (type 316LN) sheets joined by using traditional multi-pass TIG and activated TIG welding processes. Researchers had been found that better results while welding of multi-pass welding of 316LN stainless steel compared to activated TIG welding. Cheng et al. [14] studied the significances of TIG flux on the quality responses of dissimilar welds between mild steel G3131 and austenitic stainless steel 316L. The experimental runs were performed to investigate the effects of CaO, Fe_2O_3, Cr_2O_3, and SiO_2. They found residual slag on TIG weld surface due to oxide flux. Improvements in weld joint penetration and increased weld depth to width ratio were obtained by using SiO_2 powder. Reddy [15] conducted experiments to weld AISI 4140 and AISI 316 with and without using filler metal in TIG welding. The quality responses were evaluated from the study.

Ghosh et al. [10] made an investigation to visualize the welding factor effects on multi-performance characteristics of weld joint in metal inert gas (MIG) welding of ferritic stainless steel to austenitic stainless steel with the use of grey relational analysis and Taguchi method. Authors mentioned that welding responses were highly depend on the proper selection input parametric combination. Researchers had also been stated that grey relational analysis is efficient to deal with multi-responses problems. Wahule and Wasankar [11] had been carried out experimental research work to optimize the multi-outputs such as tensile strength (TS) and % elongation (PE) in TIG welding of SS304 and Fe-410 materials. They considered grey relational analysis technique to solve multi-objective problem (TS and PE) into single objective problem. Researchers used analysis of variance tool to study the significances of input parameters on both the output responses. They stated from their work that input welding parameters are significantly affecting the mechanical properties of welded joint. They optimized multi-objectives: TS and PE simultaneously by using grey relational analysis. Bahar et al. [12] had conducted analysis on TIG welding of SS316 and MS1020 materials to enhance the welding performance. Authors selected current, welding speed, and gas flow rate as input parameters and hardness, percentage elongation, and UTS as output responses. GRA had been used to optimize the multi-output variables. Factor effects on all the responses were analyzed by analysis of variance and graphical main effect plots. Optimum input combination had been obtained by grey based Taguchi method. Authors had stated that GRA was very useful to optimize multi-objectives simultaneously.

From the literature review, one can identify that various aspects of TIG welding had been investigated by the researchers. These aspects include parametric studies related to enhance the welded joint quality characteristics as well. However, research is still being continued. This indicates the need for further research. The effects of weld process parameters on weld joint quality characteristics like UTS and weld width though studied by some investigators, conclusive relationship between process parameters, and quality characteristics—explaining the effects will only emerge through more extensive research.

17.2 Response Surface Methodology (RSM) and Grey Relational Analysis (GRA)

RSM is used to determine the relationships between output response(s) and a set of quantitative input parameters. RSM is very useful to predict the improved product quality over the quality achieved. In the present work, RSM's Box–Behnken experimental table with three factors, three levels each, and fifteen runs (Table 17.1) are chosen. The mathematical equation is then postulated that illustrates the relationship between the input process variables and response characteristic. The quantitative form of a mathematical model to postulate the relationships between input and output variables is given as follows:

$$Y = f(\text{current, speed, gas flow rate}) \tag{17.1}$$

where Y is the output response. Here, it is assumed that the independent control variables are continuous and controllable by experiments with negligible errors. The complete explanation of the process behavior can be obtained through second-order polynomial equation is given below.

$$\begin{aligned} Y = {} & \beta_0 + \beta_1(\text{current}) + \beta_2(\text{speed}) + \beta_3(\text{gas flow rate}) + \beta_{11}(\text{current} * \text{current}) \\ & + \beta_{22}(\text{speed} * \text{speed}) + \beta_{33}(\text{gas flow rate} * \text{gas flow rate}) + \beta_{12}(\text{current} * \text{speed}) \\ & + \beta_{13}(\text{current} * \text{gas flow rate}) + \beta_{23}(\text{speed} * \text{gas flow rate}) \end{aligned} \tag{17.2}$$

Table 17.1 Box–Behnken design, output responses, and grey relational grade and ordering

Run	Input parameters			Output responses		Grey relational analysis	
	Current (A)	Speed (cm/min)	GFR (l/min)	UTS (MPa)	Weld width (mm)	Grey relational grade	Grey relational order
1	100	12	9	612.13	7.61	0.418719	13
2	150	12	9	616.42	6.78	0.542807	11
3	100	18	9	608.48	6.22	0.711446	7
4	150	18	9	630.37	6.35	0.76222	5
5	100	15	6	621.26	6.81	0.561772	9
6	150	15	6	612.91	6.64	0.559544	10
7	100	15	12	580.61	7.37	0.354331	15
8	150	15	12	607.69	6.82	0.498924	12
9	125	12	6	624.15	6.83	0.575156	8
10	125	18	6	616.24	6.18	0.770095	4
11	125	12	12	592.42	7.44	0.371909	14
12	125	18	12	615.88	6.24	0.729702	6
13	125	15	9	639.42	6.55	0.782465	3
14	125	15	9	642.64	6.54	0.832558	1
15	125	15	9	639.67	6.56	0.782788	2

where Y = output response and the betas are coefficients of linear, quadratic, and interaction of welding process parameters. The term β_0 is the intercept term; β_1, β_2, and β_3 are the linear terms; β_{11}, β_{22}, and β_{33} are the squared terms; and β_{12}, β_{13}, and β_{23} are the interactions between the input factors. This empirical mathematical model is helpful to identify the optimum welding parametric setting to obtain desired weld performance.

GRA is mainly utilized to solve the multi-attribute decision-making problems. It considers multi-objectives simultaneously to identify the best alternative. Data preprocessing steps of GRA can be expressed as:

Step 1: In the GRA, first step is to preprocess the experimental data because the factor goals and directions may be different in each other. The preprocessing of the data is called grey relational generation. The equations for calculating higher-the-better and lower-the-better are given in Eqs. 17.3 and 17.4, respectively. Higher-the-better:

$$X_i(k) = \frac{Y_i(k) - \min Y_i(k)}{\max Y_i(k) - \min Y_i(k)} \tag{17.3}$$

where $X_i(k)$ is the value after grey relational generation while min $Y_i(k)$ and max $Y_i(k)$ are, respectively, the smallest and largest values of $Y_i(k)$ for the kth response. Lower-the-better:

$$x_i(k) = \frac{\max y_i(k) - y_i(k)}{\max y_i(k) - \min y_i(k)} \quad (17.4)$$

where $x_i(k)$ is grey relational generation; min $y_i(k)$ and max $y_i(k)$ are the smallest and largest values of $y_i(k)$ for the *k*th response.

Step 2: Grey relation coefficient

By normalizing, grey relational coefficient (GRC) is calculated by using Eq. 17.5 which is given below

$$\xi_i(k) = \frac{\Delta \min + r\Delta \max}{\Delta o_i k + r\Delta \max} \quad (17.5)$$

where $\Delta o_i = \| x_0(k) \_ x_i(k) \|$ is variance of the absolute value between $x_0(k)$ and $x_i(k)$ and here "*r*" is distinguishing coefficient used to adjust the difference of the relational coefficients; Δ min and Δ max are minimum and maximum values of Δo_i for the *k*th response; "*r*" is within the set {0, 1}. $\xi_i(k)$ is grey relation coefficient.

Step 3: Determination of the grey relational grade (GRG) for each experiment combination

$$\alpha_i = \sum_{k=1}^{n} \xi_i(k)/n \quad (17.6)$$

where "*n*" is the number of performance characteristics; $\xi_i(k)$ = grey relation coefficient of each response; α_i is grey relational grade.

Step 4: Grey relational ordering: In GRA, final step is to provide the grades of GRG an ascending order to all factor combinations. Optimal parametric setting can be obtained at which parametric combination possess highest grade value.

17.3 Experimental Procedure

As already mentioned, that present work is planned to obtain the better weld qualities on 316L austenitic stainless steel material in TIG welding operation by using RSM and grey relational analysis. The selected dimensions of base material, i.e., 316L austenitic stainless steel are: 50 mm × 80 mm × 3 mm, and 1.6 mm diameter of ER316L austenitic stainless steel filler rod has been selected for the work. Juang and Tarng [16] and Ugur et al. [17] had mentioned that correct selection of process parameters had highest influence on the weld pool geometry as well as mechanical properties of welded joint in TIG welding operation. The important process parameters: welding current, gas flow rate, and welding speed have been selected for the present experimentation investigation. The selected input parametric settings and their levels are current = 100, 125, and 150 A; speed = 12, 15, and 18 cm/min; and gas flow rate = 6, 9, and 12 l/min. The experiments have been performed as per

Fig. 17.1 Experimental setup

Box–Behnken design of response surface methodology. Figure 17.1 shows the photographic view of a semi-automatic TIG welding machine (Model-SUPERGEN320, M/C SR. No-12111097670 and Make-Ador Welding Ltd.). Box–Behnken design matrix has been shown in Table 17.1. After TIG welding operation, visual and X-ray radiographic tests have been conducted on welded specimen to check the defects in it. However, visual and X-ray radiographic test results have not been discussed in the present paper. Output responses, ultimate tensile strength (UTS) and weld width (WW), have been measured and given in Table 17.1.

17.4 Converting Multi-response into Single-Response Problem

Grey relational analysis has been conducted on experimental data to translate multi-responses into single objective problem by following the steps mentioned above. Present work, UTS and WW, must be maximized to enhance the mechanical properties and WW needs to be minimized to reduce the material requirement. As mentioned above, in GRA, first step is to preprocess the data. Here, Eqs. 17.3 and 17.4 have been used to preprocess the data of UTS and WW, respectively. Equation 17.5 has been used to calculate the grey relational coefficient. Finally, Eq. 17.6 has been used to determine the grey relational grade (GRG) by averaging the grey relational coefficients (GRC) of both the responses, UTS and WW. Grey relational ordering has been made to grey relational grade values. The GRG values and ordering are given in Table 17.1.

17.5 Results and Analysis

The experiments have been made and output responses are measured and shown in Table 17.1. GRA is used to translate multi-objectives, i.e., UTS and WW into single response, grey relational grade (Table 17.1). RSM is applied on grey relational grade to postulate the relationships between welding factors: current, speed and gas flow rate, and grey relational grade. Contour plots have been made from developed mathematical equation by using RSM application. Optimum welding condition obtained highest grey relational grade.

17.5.1 Mathematical Modeling and Contour Plots

Developing mathematical model from the experimental data can create a more scope to study and analyze the experimental results. In the present investigation, RSM has been utilized to develop the mathematical relationships between the welding process parameters and grey relational grade (GRG) as given in Table 17.1. Here, GRG is developed from the output responses: UTS and PE as already mentioned above. RSM is one of the empirical mathematical modeling tools used to carry out to develop mathematical relationships between input and output parameters. General polynomial quadratic model of is already given in Eq. 17.2. The constant values of β_0, β_1, β_2, β_3, β_{11}, β_{22}, β_{33}, β_{12}, $\beta_{13,}$ and β_{23} as given in Eq. 17.3 are calculated from experimental data (i.e., GRG in the present case) as shown in Table 17.1 using least square. The developed models are shown in Eq. 17.7.

$$\begin{aligned} Y = & -6.33747 + 0.0486566 * \text{current} + 0.404600 * \text{speed} \\ & + 0.212709 * \text{gas flow rate} - 1.98029\text{E} - 04 * \text{current} * \text{current} \\ & - 0.0127831 * \text{speed} * \text{speed} - 0.0103304 * \text{gas flow rate} * \text{gas flow rate} \\ & + 2.32751\text{E} - 05 * \text{current} * \text{speed} + 5.16623\text{E} - 05 * \text{current} * \text{gas flow rate} \\ & - 0.00207718 * \text{speed} * \text{gas flow rate} \end{aligned} \tag{17.7}$$

where Y is output response.

The contour plots are two-dimensional plots which are very useful to study the main and interaction effects of two factors on outcome in a single plot [18]. In the present work, contour plots have been drawn from the developed mathematical model (Eq. 17.7) of grey relational grade (GRG) by RSM application from the MINITAB 16.1 software. The relationship between GRG with the welding input parameters has been demonstrated graphically by these plots. Each line in the contour plots is a constant—response line. The contour lines in a plot, comparatively straighter lines, indicate lesser or no interaction effect on the output variable.

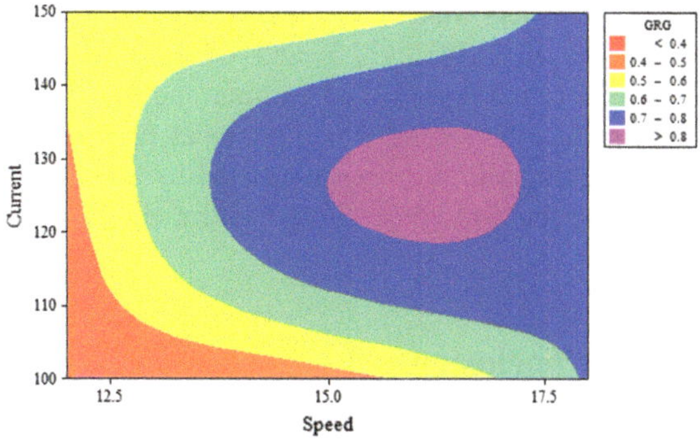

Fig. 17.2 Contour plot showing combined effects of current and speed on GRG

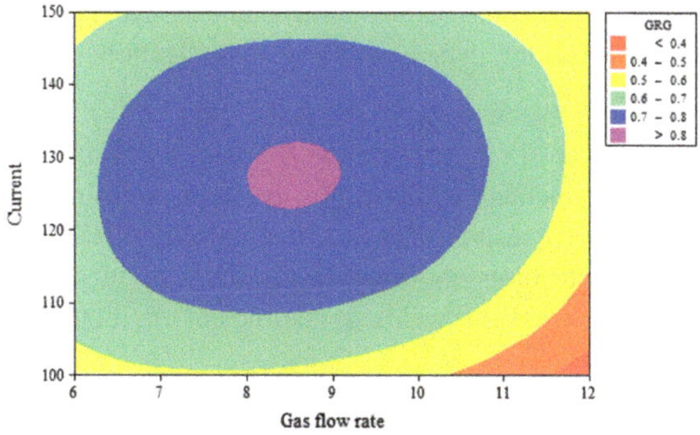

Fig. 17.3 Contour plot showing combined effects of current and gas flow rate GRG

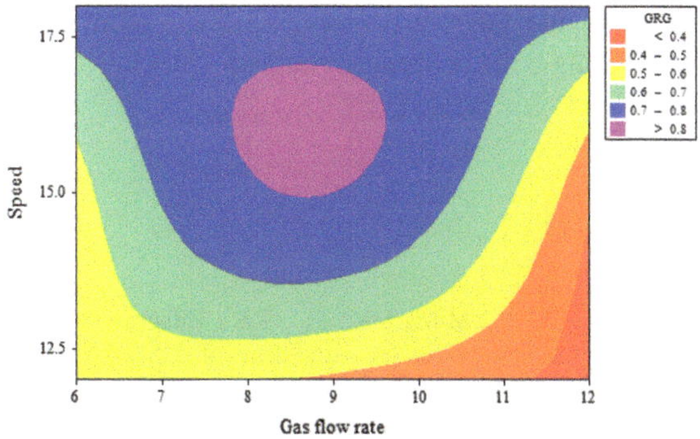

Fig. 17.4 Contour plot showing combined effects of speed and gas flow rate on GRG

The elliptical nature or bent/circular contours in a plot shows that input parameters are significant on the response. The contour plots for GRG is shown in Figs. 17.2, 17.3, and 17.4.

From contour plots of grey relational grade (GRG) as given in Figs. 17.2, 17.3, and 17.4, it is noted that all the welding input variables are most significant for UTS and weld width. From the contour plots, one can identify the optimum parametric setting where better outcome obtained. In a contour plot, each shade has set of two input parameters which projecting certain output values. From the all contour plots shown in Figs. 17.2, 17.3 and 17.4, it is observed that input parameters current = 125 A, speed = 15 cm/min and gas flow rate = 9 l/min, is one of optimum parametric setting which produces higher grey relational grade. Same parametric setting has also been suggested by grey relational analysis (Table 17.1) for predicting multi-response characteristics combinedly.

17.6 Conclusions

The experiments have been conducted to study, model, and optimize the weld responses: ultimate tensile strength (UTS) and weld width (WW) combinedly in TIG welding on 316L austenitic stainless steel plates. Following are the conclusions drawn from the present research investigation:

1. Multi-objective problem (UTS and WW) is converted into single-response problem (i.e., grey relational grade) by GRA approach
2. Mathematical relationships between welding parameters and grey relational grade by response surface methodology (RSM)
3. Contour plots are drawn using RSM application
4. The effects of welding parameters on both the responses are studied from contour plots
5. Input welding parameters are most significant for both the responses
6. Optimum welding condition: current = 125 A, speed = 15 cm/min, and gas flow rate = 9 l/min is obtained by GRA
7. RSM and GRA are very useful to optimize the welding conditions of TIG welding 316L austenitic stainless steel.

References

1. Samanta, S.K., Mitra, S.K., Pal, T.K.: Effect of rare earth elements on microstructure and oxidation behavior in TIG weldments of AISI 316L stainless steel. Mater. Sci. Eng., A **430**, 242–247 (2006)
2. Dadfar, M., Fathi, M.H., Karimzadeh, F., Dadfar, M.R., Saatchi, A.: Effect of TIG welding on corrosion behavior of 316L stainless steel. Mater. Lett. **61**, 2343–2346 (2007)
3. Sakthivel, T., Vasudevan, M., Laha, K., Parameswaran, P., Chandravathi, K.S., Mathew, M. D., Bhaduri, A.K.: Comparison of creep rupture behaviour of type 316L(N) austenitic stainless-steel joints welded by TIG and activated TIG welding processes. Mater. Sci. Eng., A **528**, 6971–6980 (2011)
4. Thomas, M., Raghu, V., Prakash, S., Ganesh, S.R., Vasudevan, M.: High temperature fatigue crack growth rate studies in stainless steel 316L(N) welds processed by A-TIG and MP-TIG welding. In: 12th International Fatigue Congress (2018)

5. Soltani, H.M., Tayebi, M.: Comparative study of AISI 304L to AISI 316L stainless steels joints by TIG and Nd:YAG laser welding. J. Alloys Compd. (2018). https://doi.org/10.1016/j.jallcom.2018.06.302
6. Konkol, P.J., Koons, G.F.: Optimization of parameters for two wire AC - ACSAW. Weld. J. **27**(12), 367s–374s (1978)
7. Tarng, Y.S., Tsai, H.L., Yeh, S.S.: Modeling, optimization and classification of weld quality in tungsten inert gas welding. Int. J. Mach. Tools Manuf. **39**, 1427–1438 (1999)
8. Sudhakaran, R., VelMurugan, V., Sivasakthivel, P.S.: Effect of process parameters on depth of penetration in gas tungsten arc welded (GTAW) 202 grade stainless steel plates using response surface methodology. TJER **9**(1), 64–79 (2012)
9. Correia, D.S., Gonçalves, C.V., da Cunha Jr., S.S., Ferraresi, V.A.: Comparison between genetic algorithms and response surface methodology in GMAW welding optimization. J. Mater. Process. Technol. **160**, 70–76 (2005)
10. Ghosh, N., Pal, P.K., Goutam, N.: Investigation on dissimilar welding of AISI 409 ferritic stainless steel to AISI 316L austenitic stainless steel by using grey based Taguchi method. Adv. Mater. Process. Technol. (2018). 2374068X.2018.1451182
11. Wahule, A., Wasankar, K.: Multi-response optimization of process parameters of TIG welding for dissimilar metals (SS-304 and Fe-410) using grey relational analysis. Int. Res. J. Eng. Technol. **5**(6), 986–993 (2018)
12. Bahar, D., Venkatesh, A., Vinay, B.K., Siva, T.R.K.P.: Multi objective optimization of TIG welding parameters for joining dissimilar materials. Int. J. Eng. Sci. Res. **8**, 26–33 (2018)
13. Durgutlu, A.: Experimental investigation of the effect of hydrogen in argon as a shielding gas on TIG welding of austenitic stainless steel. Mater. Des. **25**, 19–23 (2004)
14. Cheng, H.K., Kuang, H.T., Chang, P.C.: Effect of activated TIG flux on performance of dissimilar welds between mild steel and stainless steel. Key Eng. Mater. **479**, 74–80 (2011)
15. Reddy, M.P.: Assessment of mechanical properties of AISI 4140 and AISI 316 dissimilar weldments. In: 7th International Conference on Materials for Advance Technology. Elsevier, Amsterdam (2014)
16. Juang, S.C., Tarng, Y.S.: Process parameter selection for optimizing the weld pool geometry in the tungsten inert gas welding of stainless steel. J. Mater. Process. Technol. **122**, 33–37 (2002)
17. Ugur, E., Melih, B, Yugut, K., Sueda, O.: Optimization of weld bead geometry in TIG welding process using grey relation analysis and Taguchi method. MTAEC9 **43**(3), 143–149 (2009)
18. Moi, S.C., Rudrapati, R., Bandyopadhyay, A., Pal. P.K.: Determination of tungsten inert gas welding input parameters to attain maximum tensile strength of 316 l austenitic stainless steel. J. Mech. Eng. Strojnicky Casopis **68**, 231–248 (2018)

Chapter 18
Subjective Factors Consideration in the Selection of Welding Technique for Welded Tube Manufacturing

Ravindra S. Saluja and Varinder Singh

Abstract In choosing the right welding process for an application like welded tube manufacturing, the manufacturing engineers have to take into account a large number of selection criteria. Choosing the right welding process for any application is difficult not only due to multiple factors affecting the decision but also due to a number of emerging new welding techniques becoming available as alternatives. Some of the alternative welding processes for manufacturing tubes have been Fretz-Moon process (FM), low-frequency (LF) as well as high-frequency (HF) alternating current resistance welding, direct current (DC)-based resistance welding in addition to gas metal arc welding (GMAW), and submerged arc welding (SAW) in some specific applications. A systematic examination of the decision situation may lead us to an appropriate choice rather than basing the decision on one factor or the other in stand-alone manner. This paper presents an analytic hierarchy process (AHP)- based decision analysis for the selection of welding process in a case of manufacturing small diameter steel tubes.

Keywords Steel tubes · Welding process · Decision analysis · AHP

18.1 Introduction

Tube manufacturing is one of the major industries supporting almost all other sectors dealing with engineering products. Most of the tubes are manufactured by a technique involving roll bending a sheet or strip and welding the ends in a continuous operation to give it a tube shape. This type of manufacturing technique is preferred one in comparison to seamless tube manufacturing as it offers significant

R. S. Saluja (✉) · V. Singh
Department of Mechanical Engineering, K. K. Birla Goa Campus, BITS Pilani, Sancoale 403726, Goa, India
e-mail: rsaluja@goa.bits-pilani.ac.in

V. Singh
e-mail: vsingh@goa.bits-pilani.ac.in

R. G. Narayanan et al. (eds.), *Advances in Computational Methods in Manufacturing*, Lecture Notes on Multidisciplinary Industrial Engineering,
https://doi.org/10.1007/978-981-32-9072-3_18

advantages on economics and production front. Tubes manufactured by this technique are commonly called welded tubes and are able to satisfy general quality requirements. By subjecting such products to post-processing operations such as cold drawing and some other operations, even stringent quality requirements are met easily. The choice of welding technique plays a major role in defining the cost and quality of such tubes. In general, the techniques of welding deployed for manufacturing different tubes are decided based on stand-alone technical considerations. In the past, many researchers [1–4] have pointed out the inadequacies of such approach while deciding welding techniques for different applications. The application of multiple criteria decision tools for systematic consideration of varied factors has been a topic of active research in the recent past wherein such investigations have resulted in well thought out selection processes for applications as trivial as butt joints for aluminum parts to hard facing of carbon steels by welding harder material.

For commercially important tube manufacturing industry, no systematic treatment is yet reported in literature for consideration of multiple alternatives available and the multiple attributes affecting the decision. An appropriate choice of welding for a mass scale industry of tube manufacturing may make a significant difference in the competitiveness of the industry in terms of cost and quality in addition to the some benefits even for general society in terms of resource sustainability. In the modern times, a vast amount of technological research is resulting in the advent of newer techniques in welding technology. The present research may also pave way for consideration of newer alternatives which may result in fruitful and commercially exploitable innovations. Thus, a systematic decision-making process is very much needed that can take the multiple attributes as well as multiple alternatives into consideration for selection of the most appropriate welding technique for welded tube manufacturing. Different multi-attribute decision-making (MADM) tools may be used for analyzing this decision. Analytic hierarchy process (AHP) is one of the popular ones used by a number of researchers for decision making in different applications [5]. Technique for Order Preference by Similarity to Ideal Solution (TOPSIS) is another multi-attribute method that has been used to deal with a number of multi-attribute decision problems [1, 6–8]. A hybrid AHP-TOPSIS approach has also been in use to provide decision support in many applications. Similarly, there are many more MADM tools, all with the aim of resolving the multi-attribute decision situation in a systematic manner while following different algorithms. On the other hand, the fuzzy scales have also been adopted by a number of researchers within AHP, TOPSIS, and different other MADM methodologies [9–11]. Researchers [11] have also discussed that the use of different MADM tools do not create significant difference in results and the choice of the tool may be based on minimizing the computational load. AHP in its basic form also holds significant promise and is widely used to address decision situations involving subjective judgments in pairwise comparison of attributes as well as alternatives in the domain of the chosen attributes of decision making.

The present paper investigates the choice of welding technique for small diameter tube manufacturing considering multiple subjective attributes. AHP method is employed as it offers a simple way to systematically arrive at the user preference to meet the goal of selection process.

18.2 Methodology and Implementation

The overall methodology used in this paper for investigating decision making in welding process selection for small diameter tube manufacturing while considering subjective factor is graphically presented in Fig. 18.1.

18.2.1 AHP Model

For calculating the weights and user preference scores based on the subjective factors, AHP method is used. The general steps in AHP are described below which are based on established literature sources [3–5].

(a) Index the welding processes that engineers can take into consideration for potential pick.
(b) Find the attributes with measurable values that could be used in decision making.

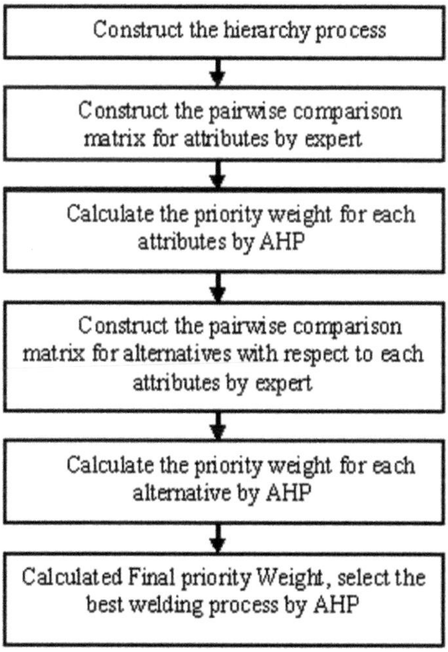

Fig. 18.1 Welding process selection based on the subjective factor by using an AHP methodology for tube manufacturing

(c) Next, the pairwise comparison matrix of set of attributes is completed by capturing relative importance as per expert views.
(d) Once the pairwise comparison has been captured in a matrix for attributes, the normalized values are estimated. The values of each column are normalized by dividing each value in column by the sum of the values in the same column. Then, the average of each row is calculated to provide priorities weights (PW) of the criteria.
(e) Similar to steps (c and d), the pairwise comparisons of all the alternatives is used to obtain priority weights (PW) of each alternative with respect to different attributes.
(f) Also, before weights are calculated in steps (c, d, and e), it is necessary to check the pairwise comparisons are consistent. For that purpose, a measure known as consistency ratio (CR) is obtained as below.

$$\mathrm{CR} = \mathrm{CI}/\mathrm{RI}, \tag{18.1}$$

where consistency index,

$$(\mathrm{CI}) = (\lambda_{\max} - n)/(n - 1) \tag{18.2}$$

$\lambda_{\max}$ = Maximum eigen value
n = Dimension of matrix.
And RI is a random index based on a table in reference [5].
So as long as CR < 0.1, the evaluations arrived at are considered consistent.
(g) For AHP-based selection, the weighted normalized decision matrix is calculated by summations of the products of attribute weights and alternative weights to obtain the final rankings.
(h) Compute the weighted priority scores for each alternative.
(i) In the end, the ranks are assigned to alternatives having the highest weighted priority score.

18.2.2 Implementation of Decision-Making Model

The case of manufacturing welded tube with the following design specifications has been considered for this study.

Material: Low carbon steel.
Diameter: 15 mm
Thickness: 2 mm

Based on the design specifications and the guidelines of the Society of Manufacturing Engineering [12, 13], four welding processes have been shortlisted for this case for further study namely Fretz-Moon method (FM), direct current electric resistance welding (DC), low-frequency electric resistance welding (LF), and high-frequency electric resistance welding (HF). Many research papers [1–4] dealing with selection of welding processes for different applications use many different attributes.

The hierarchical structure of the decision problem is presented in Fig. 18.2 where the top level indicates the overall goal, which is to bring the best welding process available for tube manufacturing on mass scale. As the attributes important for the current decision making, namely setup time requirement (SR), worker skill requirement (WSR), occupational factor (OF), ease of automation (EA), and quality (QY), are explained in Table 18.1, at the second level, such attributes and their relative importance toward achievement of the goal of selecting the best alternative are analyzed through pairwise comparison. At the next level, the four potential alternatives, namely Fretz-Moon process (FM), direct current electric resistance welding (DC), low-frequency electric resistance welding (LF), and high-frequency electric resistance welding (HF), are considered for pairwise comparison in the direct of each attribute.

For pairwise comparison of different attributes as well as alternatives, the scale of comparison is presented in Table 18.2. It is used to fill the pairwise comparison elements for each attribute in Table 18.3. It is also used to fill pairwise comparison for each alternative with respect to all considered attribute in Table 18.4.

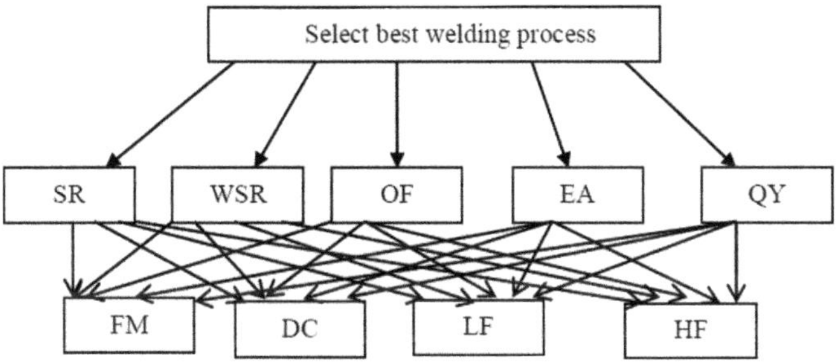

Fig. 18.2 Schematic diagram of analytic hierarchy process

Table 18.1 Description of attributes considered [3, 4]

S. No.	Subjective attributes	Description
1	Setup time requirement (SR)	Welding parameters setting, cleaning the base metal, and clamping joint in fixtures
2	Worker skill requirement (WSR)	Refer to the skill level of worker required
3	Occupational factor (OF)	Arc glare, smoke, and fumes
4	Ease of automation (EA)	Manual, semi-automatic, and fully automatic
5	Quality (QY)	Weld bead appearance and surface finish

Table 18.2 Strength of preference for pairwise comparison [3–5]

Strength of importance	Verbal judgement
1	Equally preferable
3	Moderately preferable
5	Strongly preferable
7	Very strongly preferable
9	Extremely strongly preferable
2, 4, 6, 8	Intermediate values
1/2, 1/3, 1/4 etc.	Reciprocal of above values

18.3 Results and Discussion

The pairwise comparison of attributes for this case of tube manufacturing is presented in Table 18.3. As evident from the priority weight scores in Table 18.3, the quality of the welding process is considered as the most important attributes followed by occupational factor and so on.

Similar way, the pairwise comparison of welding processes with regard to each of the considered attributes and values of priority weights (PW) is presented in Table 18.4. It may be noted that the calculated consistency ratio by using Eq. (18.1) in Tables 18.3 and 18.4 is found to be less than 0.1 as per the requirement of AHP procedures.

The results of Table 18.4 are summarized in Table 18.5 and overall composite priority weights of alternatives are calculated according to step (f) and step (g) as mentioned in Sect. 18.2.1 and are reported in Table 18.6. From the overall priority weights for different alternatives as available in Table 18.6, the highest value for direct current (DC) resistance welding is observed.

18.3.1 Discussion

It may be observed that DC resistance welding is standing high in the order of preference. The reason for this outcome is that the quality of weld in terms of clean

Table 18.3 Comparison of subjective attributes

Attributes	SR	WSR	OF	EA	QY	Priority weight
SR	1.000	0.111	0.143	0.200	0.111	0.030
WSR	9.000	1.000	0.333	0.500	0.333	0.138
OF	7.000	3.000	1.000	3.000	0.333	0.254
EA	5.000	2.000	0.333	1.000	0.250	0.138
QY	9.000	3.000	3.000	4.000	1.000	0.439
CR = 0.094						

Table 18.4 Pairwise comparison matrix of processes with respect to the attributes

Attributes	Process	FM	DC	LF	HF	Priority weights	Consistency ratio
Setup time requirement (SR)	FM	1.00	0.20	0.20	0.14	0.05	0.02
	DC	5.00	1.00	1.00	0.33	0.21	
	LF	5.00	1.00	1.00	0.33	0.21	
	HF	7.00	3.00	3.00	1.00	0.52	
Worker skill requirement (WSR)	FM	1.00	3.00	3.00	3.00	0.50	0
	DC	0.33	1.00	1.00	1.00	0.17	
	LF	0.33	1.00	1.00	1.00	0.17	
	HF	0.33	1.00	1.00	1.00	0.17	
Occupational factor (OF)	FM	1.00	0.20	0.20	0.14	0.05	0.024
	DC	5.00	1.00	1.00	0.33	0.21	
	LF	5.00	1.00	1.00	0.33	0.21	
	HF	7.00	3.00	3.00	1.00	0.52	
Ease of automation (EA)	FM	1.00	0.20	0.20	0.20	0.06	0
	DC	5.00	1.00	1.00	1.00	0.31	
	LF	5.00	1.00	1.00	1.00	0.31	
	HF	5.00	1.00	1.00	1.00	0.31	
Quality (QY)	FM	1.00	0.14	0.50	0.25	0.06	0.08
	DC	7.00	1.00	5.00	3.00	0.59	
	LF	2.00	0.20	1.00	0.33	0.11	
	HF	4.00	0.33	3.00	1.00	0.23	

Table 18.5 Summary of priority weights of alternative welding processes

Process	SR	WSR	OF	EA	QY
FM	0.05	0.50	0.05	0.06	0.06
DC	0.21	0.17	0.21	0.31	0.59
LF	0.21	0.17	0.21	0.31	0.11
HF	0.52	0.17	0.52	0.31	0.23

Table 18.6 Ranking of welding processes by AHP model

Process	SR	WSR	OF	EA	QY	Composite priority weight	Ranking
FM	0.002	0.069	0.013	0.008	0.028	0.12	4
DC	0.006	0.023	0.054	0.043	0.260	0.39	1
LF	0.006	0.023	0.054	0.043	0.048	0.17	3
HF	0.016	0.023	0.133	0.043	0.103	0.32	2

weld is of highest importance in the present case of small diameter tube manufacturing on mass scale, as the post weld flash removal is extremely difficult. The DC resistance welding is otherwise a poorer choice in comparison to a more popular alternative of high-frequency alternating current resistance welding as is evident from Table 18.4. For medium diameter tubes, the cleaner quality weld may acquire lesser importance and the choice may tilt toward high-frequency alternating current resistance welding as flash removal is easier owing to larger diameter. The utility of such decision analysis is that it considers different practical aspects and uses the provided information systematically, which help the design engineer and manufacturing engineer to choose the most appropriate process considering several attributes simultaneously.

18.4 Conclusions and Future Scope

In this work, the AHP model has been used for investigating the issue of selecting appropriate welding process for a case of small diameter tube manufacturing on mass scale while considering multiple attributes, namely setup time requirement, worker skill requirement, occupational factor, ease of automation, and quality of weld. The conclusions drawn from the study and the future scope are presented below:

i. For the case of manufacturing a welded tube with the following design specifications: material—carbon steel, diameter—15 mm and thickness—2 mm, the direct current (DC)-based resistance welding is found to be the most appropriate process while considering multiple attributes.
ii. AHP-based analysis is found to be simple for adaptation by experts while systematically collecting expert opinion on different factors in the otherwise complex problem of selection involving multiple attribute and choices as well.
iii. In future, there is scope to consider emerging welding processes for multiple attribute-based analysis before actual commercial deployment. Also, there is scope to consider multiple MADM methods as well as fuzzy approach due to subjective judgments involved.

References

1. Mostafa, J., Vahdat, S.E.: A fuzzy multi-attribute approach to select the welding process at high pressure vessel manufacturing. J. Manuf. Process. **14**, 250–256 (2012)
2. Darwish, S.M., Tamimi, A.A., Habdan, S.A.: A knowledge base for metal welding process selection. Int. J. Mach. Tools Manuf. **3**, 1007–1023 (1997)
3. Ravisankar, V., Balasubramanian, V., Muralidharan, C.: Selection of welding process to fabricate butt joints of high strength aluminum alloys using analytic hierarchic process. Mater. Des. **27**, 373–380 (2006)

4. Balasubramanian, V., Varahamoorthy, R., Ramachandran, C.S., Muralidharan, C.: Selection of welding process for hard facing on carbon steels based on quantitative and qualitative factors. Int. J. Adv. Manuf. Technol. **40**, 887–897 (2009)
5. Saaty, T.L.: Analytic Hierarchy Process. McGraw Hill Publications, New York (1980)
6. Singh, V., Agrawal, V.P., Deb, P.: A decision making method for selection of finish process. In: Proceedings of IEEE International Conference on Industrial Engineering and Engineering Management, pp. 38–42, Macao, China, 7th–10th December (IEEM-2010). https://doi.org/10.1109/ieem.2010.5674417
7. Rao, R.V.: Machinability evaluation of work materials using a combined multiple attributes decision-making method. Int. J. Adv. Manuf. Technol. **28**, 221–227 (2006)
8. Rao, R.V., Davim, J.P.: A decision-making framework model for material selection using a combined multiple attributes decision-making method. Int. J. Adv. Manuf. Technol. **35**, 751–760 (2008)
9. Chang, D.Y.: Applications of the extent analysis method on fuzzy AHP. Eur. J. Oper. Res. **95** (3), 649–655 (1996)
10. Deng, H.: Multi-criteria analysis with fuzzy pair-wise comparison. Int. J. Approximate Reasoning **21**(3), 215–231 (1999)
11. Sabzi, H.Z., King, J.P., Gard, C.C., Abudu, S.: Statistical and analytical comparison of multi-criteria decision-making techniques under fuzzy environment. Oper. Res. Perspect. **3**, 92–117 (2016)
12. Baheti, M.: Everything About Welded Tube Technology. DeeTee Industries Ltd., Indore, India (2016)
13. Veilleux, R.F., Petro L.W.: Tool and manufacturing engineer's handbook. In: Manufacturing Management, vol. 5, 4th edn. Society of Manufacturing Engineers (1998). ISBN 978-0-872633-06-3

Chapter 19
GMAW Investigation of AISI 201 Stainless Steel and Industry Need Optimization Using Genetic Algorithm

Krity Lata, M. Chandrasekaran, S. K. Tamang, R. Ramesh and N. K. Rana

Abstract In this work, the welding investigation of AISI 201-grade stainless steel (120 mm × 60 mm with 4 mm thickness) using gas metal arc welding (GMAW) has been carried out. Four weld parameters, viz. wire feed rate, welding voltage, nozzle-to-plate distance and welding speed, are used to investigate different weld bead characteristics [i.e. penetration (*P*), bead width (*W*) and bead height (*H*)]. The welding experiments are performed using Taguchi L9 experimental design, and each run is completed in a single pass. The predictive models are developed to predict weld bead geometry, and the performance of the model is validated. The increase in wire feed rate and voltage increases penetration and bead width, while an increase in nozzle-to-plate distance decreases the value of penetration and bead width. The welding parameters are optimized using genetic algorithm to determine the optimal combinations of welding parameters for better quality of component having maximum *P* with minimum *W* and *H* at which the maximum weld quality be achieved.

Keywords AISI 201-grade stainless steel · GMAW · Weld bead · Genetic algorithm

19.1 Introduction

Nitrogen-strengthened austenitic stainless steel (200 series) has found more applications in modern industries compared to conventional AISI 300 series of stainless steel components [1]. The price of nickel is relatively higher compared to other austenite formation elements such as manganese, and changing so, it becomes

K. Lata · M. Chandrasekaran (✉) · S. K. Tamang · N. K. Rana
Department of Mechanical Engineering, NERIST (DU), Nirjuli 791109, Arunachal Pradesh, India
e-mail: mchse1@yahoo.com

R. Ramesh
PSG College of Technology, Coimbatore, Tamil Nadu, India

R. G. Narayanan et al. (eds.), *Advances in Computational Methods in Manufacturing*, Lecture Notes on Multidisciplinary Industrial Engineering,
https://doi.org/10.1007/978-981-32-9072-3_19

necessary to increase such a material like AISI 201, as an option, with the same weight of chromium 18%, and instead of nickel, 7 wt% of manganese is added to stabilize the austenitic phase, and weight of nickel is decreased from 8 to 3 wt% in stainless steel of AISI 201 grade [2]. Hence, AISI 201 SS is a high Cr–Mn–N stainless steel which is free of nickel. This material is useful in manufacturing of battle tanks in defence applications possessing low cost, better corrosion resistance and excellent mechanical properties. This material can be used for conventional construction applications such as nuclear reactors, thermal power plants, civil construction, vessels and heat exchangers and for multiple industrial applications such as food and dairy processing equipment and kitchen equipment for consumer use.

Gas metal arc welding (GMAW) is a renowned soldering process used for various ferrous and non-ferrous materials. This is a cheaper and cost-effective process, without flaws, without faults, with proper appearance and with easy mechanizing [3]. In this process of welding, an arc is struck between a continuous consumable wire electrode and the workpiece metal. To provide a good-quality weld having economic production of desired weld bead geometry, optimum selection of welding parameters is important. A number of researches were conducted on welding investigation and parametric optimization in different welding processes on joining various materials. Choudhury and Chandrasekaran [4] reviewed previous research works published in the areas of welding investigation and optimization on aerospace materials. In gas tungsten arc welding (GTAW), Juang and Tarng [5] studied the selection of optimal process parameters to obtain optimum welding bead geometry. The effects of hardness and penetration depth were investigated by Ibrahim et al. [6] on various MIG welding process parameters. The optimum parameter combination was assessed by Datta et al. [7], using the Taguchi approach and the Grey relational analysis, in order to reach an optimum penetration depth and minimum bead height, bead height and depth of the welds. They concluded that Taguchi–Grey method adopted achieves an improved weld quality. The effect of weld speed on microstructure, mechanical characteristics and corrosion studies of AISI 201 GTA welded stainless steel sheets was studied by Chuaiphan and Srijaroenpramong [2]. Kim et al. [8] studied gas metal arc welding process parameters modelling and optimizing with GA and RSM. Sathiya et al. [9] have examined laser welding processes of the AISI 904L super austenitic stainless steel materials' assemblies and process parameters for optimization with the help of soft computing methods such as ANN and GA. The author has attempted to investigate the effect of various weld parameters on weld bead characteristics during the AISI 201 SS GMAW process on beads on plates in the present investigation. The experimental results are analysed and weld parameters are optimized using GA to get better penetration with minimum bead width and bead height. The result provides important guidelines for the selection of weld parameters for particular application in fabrication industries.

19.2 GMAW Experimentation Procedure

The experiments designed were based on Taguchi L9 OA having 4 factors and 3 levels. The welding parameters were chosen as wire feed rate (*W*), welding voltage (*V*), nozzle-to-plate distance (*N*) and welding speed (WS), and response parameters were chosen as penetration (*P*), bead width (*W*) and bead height (*H*). Experiments were done on bead on plate (AISI 201 stainless steel of size 120 mm × 60 mm, 4 mm thick) as per design matrix using TransPuls Synergic 3200 MIG welding machine. The experimental set-up was prepared as shown in Fig. 19.1. GMAW machine was connected to shielding gas (98% Ar + 2% O_2) cylinder, and gas flow rate was controlled electronically using this machine. AISI 316L wire rod having diameter 1.2 mm was selected as welding electrode and was attached to the welding torch which was further connected to the set-up table and fixed at a torch angle of 90°. DCEP electrical polarity was used as it concentrated more heat on melting the filler material in the welding area. The variations in welding voltage automatically changed the current in this experimental set-up. Before welding, the workpiece was cleaned, degreased and clamped in the experimental set-up. After the complete configuration, the tests were performed randomly using the design matrix so that any systemic error was avoided. By conducting test runs, the working range of all the selected factors was determined by changing one factor while the remaining factor was kept as constant values. Thus, when inspecting the weld bead for a smooth appearance, no visible defects such as undercut and surface porosity were obtained in various test experiments and given in Table 19.1, and the range of parameters of process variables was defined.

Figure 19.2 shows welded specimens. After welding, the specimen was cut and grinded in a grinder disc and polished using nylon polishing cloth with levigated polishing alumina solution. The specimen was etched with etchant solution (15% HCL + 10% HNO_3 + 2–3 drop glycerine) and then washed with alcohol solution.

Fig. 19.1 GMAW experimental set-up

Table 19.1 GMAW parameters and their levels

Parameter	Unit	Symbol	Level 1	Level 2	Level 3
Wire feed rate	m/min	WFR	6	7	8
Welding voltage	V	*V*	20	22	24
Nozzle-to-plate distance	mm	*N*	16	20	24
Welding speed	mm/min	WS	220	260	300

Fig. 19.2 Weld runs

The bead profile parameters such as penetration, bead width and bead height were measured using 'stereo microscope' to get bead profile of experiments. The measured weld profiles are shown in Fig. 19.3.

19.2.1 Effect of Welding Parameters On the Performance Measures

The experimental results obtained from GMAW experiments were analysed through graphical figure. It shows and studies the effects of each parameter on responses. The effect of wire feed rate (WFR) on different geometries of weld beads, such as penetration, bead width and bead height, is shown in Fig. 19.4. The increase in wire feed rate, penetration and bead width is observed. For all wire feed values, bead height is constant.

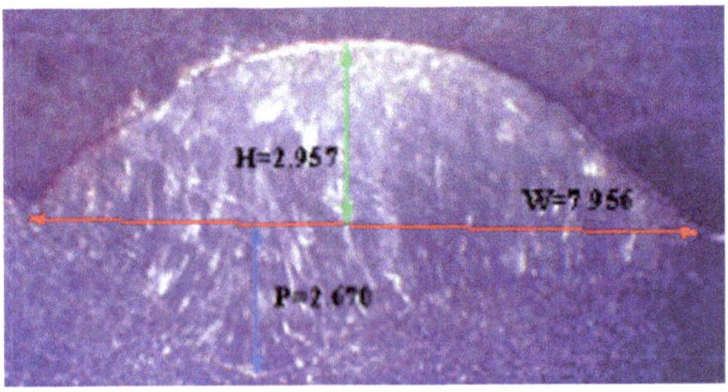

Fig. 19.3 Microstructure measurement of a weld bead profile

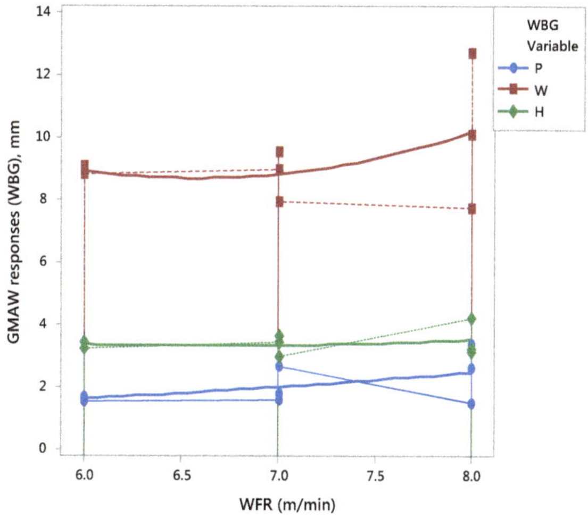

Fig. 19.4 Effect of wire feed rate on *P*, *W* and *H*

Similarly, the effect of other weld parameters is studied. It is observed that increase in welding voltage increases penetration and bead width, whereas bead height decreases very slightly. The increase in the value of the nozzle-to-plate distance lowers the penetration value and the width of the bead.

19.3 Development of RSM Modelling

The experimental results are used to develop predictive modelling of different weld characteristics using response surface methodology (RSM). RSM is statistical and empirical method being widely used for the development of predictive models. Three different models are developed for predicting *P*, *W* and *H* characteristics of weld bead geometry.

In this work, functional correlation between different weld bead characteristics and process parameters is found using regression analysis. The linear models with interaction terms show better correlation. In general, linear model with interaction terms is given in Eq. 19.1.

$$y = \beta_0 + \sum_{i=1}^{k} \beta_i x_i + \sum_{j=i+1}^{k} \sum_{i=0}^{k} \beta_{ij} x_i x_j + \varepsilon, \tag{19.1}$$

where y is the response and x_i (1, 2, …, k) is the level of k variable. β_0 is constant, β's are model coefficients, and ε is modelling error, respectively. The regression model is developed using statistical software Minitab 17®. The predictive model equations obtained for *P*, *W* and *H* are given in Eqs. 19.2, 19.3 and 19.4, respectively. The correlation coefficient (R^2) values of each model equation are above 95%.

$$P = 34.11 - 5.093\,\text{WFR} - 1.358\,V - 0.427\,N + 0.00254\,\text{WS} + 0.2176\,\text{WFR} \times V + 0.0379\,\text{WFR} \times N + 0.0057\,V \times N, \quad (19.2)$$

(R^2 = 99.96%, R^2 = 99.67%)

$$W = 187.6 - 23.6\,\text{WFR} - 9.24\,V - 0.58\,N + 0.0124\,\text{WS} + 1.180\,\text{WFR} \times V - 0.074\,\text{WFR} \times N + 0.062\,V \times N, \quad (19.3)$$

(R^2 = 95.03%, R^2 = 60.23%)

$$H = -14.6 + 3.46\,\text{WFR} + 0.896\,V - 0.185\,N - 0.00489\,\text{WS} - 0.163\,\text{WFR} \times V + 0.112\,\text{WFR} \times N + 0.0057 \times N, \quad (19.4)$$

(R^2 = 97.15%, R = 77.23)

Table 19.2 presents predictive performance of the model comparing with the experimental results, and it shows an average error of 0.621, 4.780 and 1.89% for *P*, *W* and *H*, respectively. Also, the maximum error percentage is well within 5.5%, except one data set for *W*. The established models can be used with considerable accuracy to determine penetration, bead width and bead height.

Table 19.2 Model predictions

S. No.	Experimental results			Model prediction			% Error		
	P	*W*	*H*	*P*	*W*	*H*	*P*	*W*	*H*
1	1.686	8.863	3.427	1.693	8.984	3.383	0.42	1.360	1.280
2	1.600	9.105	3.415	1.575	8.504	3.236	1.58	1.600	5.240
3	1.542	8.851	3.224	1.549	9.016	3.181	0.46	1.860	1.330
4	1.554	8.976	3.403	1.571	9.360	3.401	1.09	4.270	0.060
5	1.764	9.546	3.607	1.781	9.952	3.600	0.96	4.250	0.200
6	2.670	7.956	2.957	2.685	8.344	2.950	0.56	4.870	0.230
7	1.452	7.751	4.190	1.447	7.656	4.095	0.35	1.220	2.260
8	2.581	10.103	3.216	2.575	9.992	3.117	0.24	1.090	3.070
9	3.378	12.736	3.139	3.372	14.968	3.04	0.18	17.520	3.150
Average percentage error (APE)							0.621	4.780	1.890

19.4 Parametric Analysis

The effect of weld parameters on various weld measures is analysed through contour figures. Three weld bead characteristics, viz. penetration, bead width and bead height, are studied over the range of parameters under consideration. The effect of responses is investigated by two different parameters, while other parameters are kept constant at middle level. As the wire feed rate and voltage increase, the penetration and bead width increase, while the value of penetration and bead width decreases with increase in the value of nozzle-to-plate distance. In this contour figures, you can better understand how weld parameters interact with various weld bead responses.

19.4.1 Contour Figures for Penetration (P)

From the results, it is clear that the welding voltage (*V*), interaction terms WFR × *V* and wire feed rate (WFR) are the most significant parameters that affect the penetration. Four figures, viz. *P* versus WFR and *V*, WS and *V*, WFR and *N*, and *V* and *N*, are figured, and it is shown in Fig. 19.5. The dark green colour indicates maximum penetration, and dark blue colour indicates minimum penetration.

The variation in *V* and WFR on *P* is shown in Fig. 19.5a. It shows that *V* and WFR bears a direct relationship with penetration. Increased value of *V* and WFR increases *P* in a range of 1.5–3 mm. The Fig. 19.5b also shows that *V* and *N* direct variation in *P*, and Fig. 19.5c shows that as the value of WFR increases, the value of *P* increases. A minimal effect of *N* is observed in case of penetration. Figure 19.5d shows the variation of *P* with *N* and *V* that shows direct relationship. The deeper penetration needs high values of input factors.

19.4.2 Contour Figures for Bead Width (W)

Figure 19.6 shows the contour figures for bead width (*W*) versus WFR and *V*, WFR and *N*, *V* and WS, *and V* and *N*. From the Fig. 19.6a, it is observed that WFR and *N* holds a direct relationship with bead width, and as WFR and *N* increases, the value *W* also increases. Figure 19.6b shows that with an increase in *V*, the value of *W* decreases. From the Fig. 19.6c, *W* increases with increase in the value of *N* and WS, and in Fig. 19.6d, low value of *N* shows minimum *W*, and W slightly increases with increase in *N*. *W* is almost unaffected with an increase in *V*.

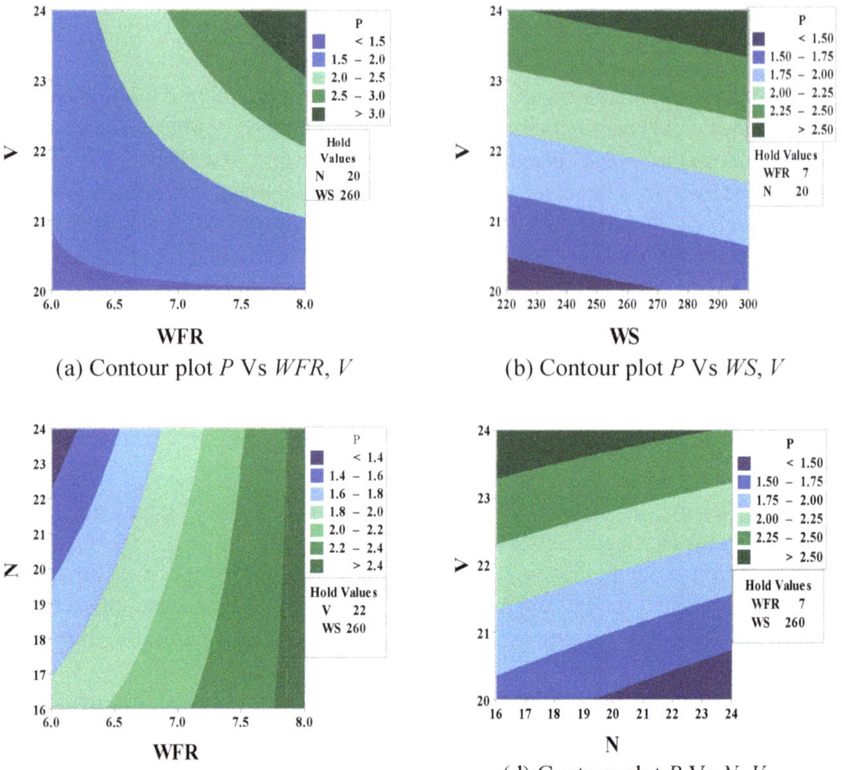

(a) Contour plot *P* Vs *WFR*, *V*

(b) Contour plot *P* Vs *WS*, *V*

(c) Contour plot *P* Vs *WFR*, *N*

(d) Contour plot *P* Vs *N*, *V*

Fig. 19.5 Contour figures for penetration

*19.4.3 Contour Figures for Bead Height (*H*)*

Contour figures of *H* versus WFR and *V*, WFR and WS, *and V* and *N* shown in Fig. 19.7 provide better understanding of weld parameter on *H*. From the Fig. 19.7a, it is observed that the value of height (*H*) increases with increase in wire feed rate (*W*) and remains unaffected by *V* value. From Fig. 19.7b, it is also observed that with an increase in WFR, bead height increases and *H* decreases with an increase in welding speed (WS). From the Fig. 19.7c, *H* is maximum at low value of *V* and minimum when the welding voltage (*V*) increases and *H* increases with increase in nozzle-to-plate distance (*N*).

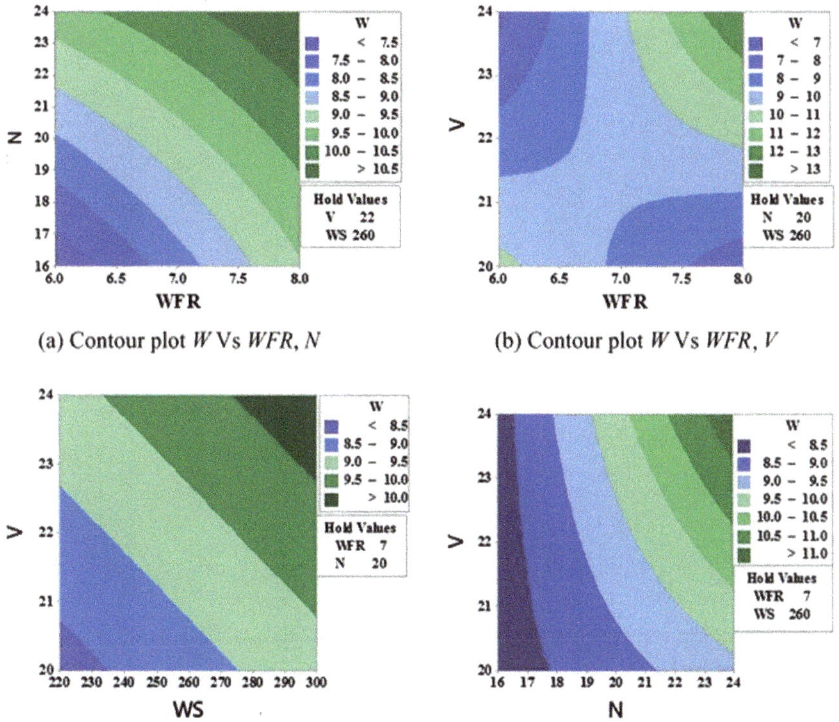

(a) Contour plot *W* Vs *WFR, N*

(b) Contour plot *W* Vs *WFR, V*

(c) Contour plot *W* Vs *WS, V*

(d) Contour plot *W* Vs *N,V*

Fig. 19.6 Contour figures for bead width

19.5 Optimization of GMAW Welding

19.5.1 Single-Objective Optimization Using GA

In single-objective optimization, optimum process parameters are obtained to maximize penetration and minimize bead width and bead height individually. The optimization problem is formulated with fitness function derived from regression model equations and is optimized using GA toolbox available in MATLAB 14. The function *P* (penetration) is maximized and function *W* (bead width) and *H* (bead height) are minimized subjected to variable bounds of the factors. The range of the level of each factor decides the upper and lower bounds of the variables. The GA optimization is carried out for both single and multiple objectives. The problem is formulated for single-objective optimization as follows:

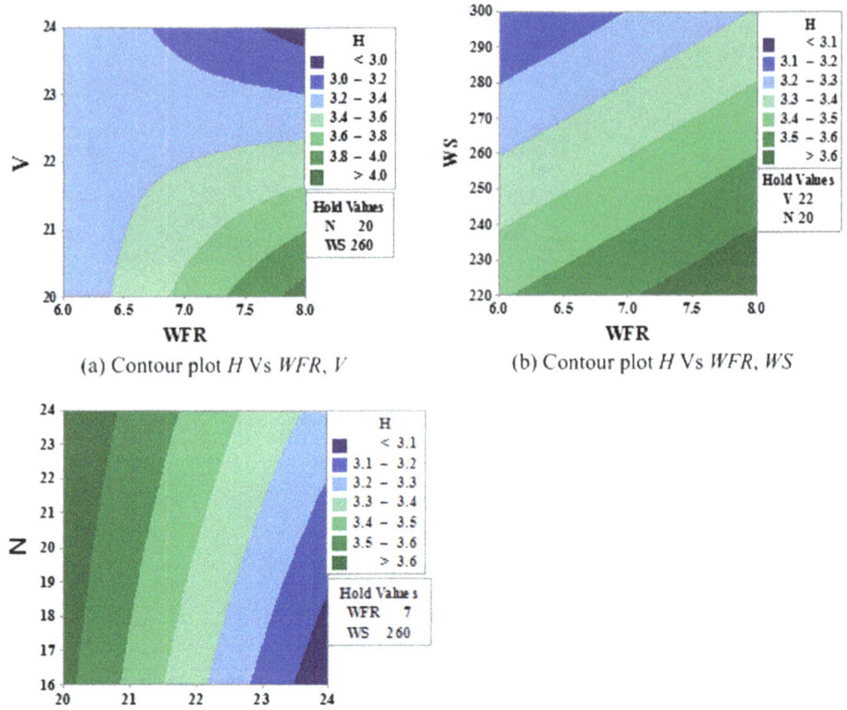

(a) Contour plot H Vs WFR, V

(b) Contour plot H Vs WFR, WS

(c) Contour plot H Vs V, N

Fig. 19.7 Contour figures for bead height

Objective 1

$$\textbf{Maximize}\, \boldsymbol{P}\,(\boldsymbol{WFR}, \boldsymbol{V}, \boldsymbol{N}, \boldsymbol{WS}) = -(34.11 - 5.093\,\mathrm{WFR} - 1.358\,V - 0.427\,N + 0.00254\,\mathrm{WS} + 0.2176\,\mathrm{WFR} \times V + 0.0379\,\mathrm{WFR} \times N + 0.00570\,V \times N) \tag{19.5}$$

Objective 2

$$\textbf{Minimize}\, \boldsymbol{W}\,(\boldsymbol{WFR}, \boldsymbol{V}, \boldsymbol{N}, \boldsymbol{WS}) = 187.6 - 23.6\,\mathrm{WFR} - 9.24V - 0.58N + 0.0124\,\mathrm{WS} + 1.180\,\mathrm{WFR} \times V - 0.074\,\mathrm{WFR} \times N + 0.062\,V \times N \tag{19.6}$$

Objective 3

$$\textbf{Minimize}\, \boldsymbol{H}\,(\boldsymbol{WFR}, \boldsymbol{V}, \boldsymbol{N}, \boldsymbol{WS}) = -14.6 + 3.46\,\mathrm{WFR} + 0.896\,V - 0.185\,N - 0.00489\,\mathrm{WS} - 0.163\,\mathrm{WFR} \times V + 0.0112\,\mathrm{WFR} \times N + 0.0057\,V \times N \tag{19.7}$$

Table 19.3 GA results for single-objective optimization

S. No.	Optimization objective	No. of iterations	Optimum process parameters				Results obtained (mm)
			WFR (m/min)	*V* (V)	*N* (mm)	WS (mm/min)	
1	Maximize *P*	121	8	24	20	258	3.468
2	Minimize *W*	134	6	24	16	220	4.312
3	Minimize *H*	109	8	24	16	300	2.483

The variable process is bounded by its lower bound (LB) and upper bound (UB). The LB and UB are given as $6 \leq \text{WFR} \leq 8$; $20 \leq V \leq 24$; $16 \leq N \leq 24$; $220 \leq \text{WS} \leq 300$. The problem can be optimized with the MATLAB 14 GA toolbox. Table 19.3 shows GA results for single optimization.

19.5.2 Simultaneous Optimization

In this work, multi-optimization is performed for two different optimization problems. Each optimization problem provides 18 Pareto-optimal fronts that provide different values of optimum parameters. Each combination of optimum parameter provides the corresponding value of maximum penetration (*P*), minimum bead width (*W*) and minimum bead height (*H*) obtained in the weld bead geometry. The problem is formulated as follows:

Multi-optimization objective 1

$$\textbf{Maximize}\,\boldsymbol{P}\,\textbf{and minimize}\,\boldsymbol{W}(\boldsymbol{WFR}, \boldsymbol{V}, \boldsymbol{N}, \boldsymbol{WS}) \tag{19.8}$$

Multi-optimization objective 2

$$\textbf{Maximize}\,\boldsymbol{P}\,\textbf{and minimize}\,\boldsymbol{H}(\boldsymbol{WFR}, \boldsymbol{V}, \boldsymbol{N}, \boldsymbol{WS}) \tag{19.9}$$

subjected to $6 \leq \text{WFR} \leq 8; 20 \leq V \leq 24; 16 \leq N \leq 24; 220 \leq \text{WS} \leq 300$.

The above multi-objective optimization problem of GMAW of AISI 201 SS is carried out using GA toolbox in MATLAB 14. Figure 19.8 shows the non-dominated Pareto front figure being a number of optimal solutions for objectives 1 and 2, respectively.

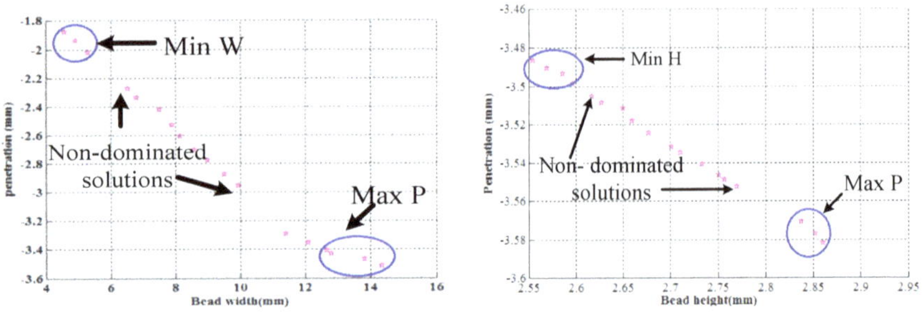

(a) For *P* and *W* (b) For *P* and *H*

Fig. 19.8 Pareto-optimal solutions

19.6 Industrial Application Optimization

Manufacturing industries require production of better quality of components having maximum penetration (*P*) with minimum bead width (*W*) and bead height (*H*) for economic production of quality components. In reference to optimization problem 1, GA solutions are given in Table 19.4. For the industrial use of this optimization

Table 19.4 Technology table (*P* and *W*)

Multi-optimization (problem 1)						
S. No.	Optimum parameters				Optimum responses	
	WFR (m/min)	*V* (V)	*N* (mm)	WS (mm/min)	*P* (mm)	*W* (mm)
1	6.001	23.998	16.310	228.799	1.852	4.571
2	6.001	23.998	16.310	228.799	1.852	4.571
3	6.058	23.961	16.530	232.176	1.888	4.954
4	6.192	23.996	16.448	230.687	1.992	5.328
5	6.265	23.991	16.796	229.464	2.025	5.727
6	6.364	23.985	16.351	240.986	2.151	6.022
7	6.576	23.990	16.359	228.908	2.278	6.612
8	6.602	23.977	16.577	242.833	2.323	6.975
9	7.051	23.986	16.627	237.872	2.649	8.490
10	7.405	23.994	17.120	246.260	2.936	10.004
11	7.557	23.991	16.632	242.902	3.047	10.315
12	7.679	23.992	16.823	240.726	3.134	10.778
13	7.735	23.989	17.895	248.331	3.198	11.420
14	7.853	23.977	18.105	251.684	3.297	11.916
15	7.928	23.928	17.484	270.552	3.377	12.144
16	7.976	23.988	18.390	264.639	3.439	12.595
17	7.977	23.885	20.216	283.906	3.460	13.269
18	7.999	23.982	20.669	286.882	3.541	13.661

results, the value of penetration *P* is arranged in ascending order. Considering optimum parameters to be obtained for a weld design having $P \geq 2$, the GA provides 14 possible optimal solutions (i.e. S. No. 5–18). For minimum bead width, S. No. 5 is found to be optimum providing minimum bead width *W* satisfying the penetration (*P*) in the desired level. If $P \geq 3$, GA provides 8 possible optimal solutions (i.e. S. No. 11–18) and S. No. 11 is the optimal solution providing minimum bead width (*W*) satisfying penetration (*P*).

Similarly, in reference to optimization problem 2, GA solutions are given in Table 19.5. Considering optimum parameters to be obtained for a weld design having $P \leq 3.5$, the GA provides 5 possible optimal solutions (i.e. S. No. 1–5). For minimum bead height, S. No. 1 is found to be optimum providing minimum *H* satisfying the *P* in the desired level. If $P \geq 3.5$, GA provides 13 possible optimal solutions (i.e. S. No. 6–18) and S. No. 6 is the optimal solution providing minimum *H* satisfying *P*.

Table 19.5 Technology table (*P* and *H*)

Multi-optimization (problem 2)						
S. No.	Optimum parameters				Optimum responses	
	WFR (m/min)	*V* (V)	*N* (mm)	WS (mm/min)	*P* (mm)	*W* (mm)
1	7.99	23.995	17.332	281.302	3.493	2.631
2	7.99	23.996	17.385	281.316	3.494	2.633
3	7.99	23.999	17.483	281.315	3.495	2.637
4	7.99	23.998	17.584	281.323	3.497	2.641
5	7.99	23.989	17.757	281.333	3.499	2.648
6	7.99	23.991	17.983	281.312	3.502	2.658
7	7.99	23.999	18.141	281.311	3.504	2.664
8	7.99	23.993	18.222	281.317	3.505	2.667
9	7.99	23.991	18.321	281.324	3.506	2.672
10	7.99	23.992	18.391	281.313	3.507	2.675
11	7.99	23.994	18.474	281.319	3.508	2.678
12	7.99	23.997	18.831	281.320	3.513	2.693
13	7.99	23.991	18.937	281.322	3.514	2.697
14	7.99	23.995	19.030	281.317	3.515	2.701
15	7.99	23.996	19.121	281.325	3.517	2.704
16	7.99	23.994	19.317	281.331	3.519	2.713
17	7.99	23.998	19.429	281.333	3.520	2.716
18	7.99	23.999	19.430	281.332	3.521	2.717

19.7 Conclusion

This study investigates the influence of process parameters of gas metal arc welding (GMAW) like wire feed rate, welding voltage, nozzle-to-plate distance and welding speed on weld bead geometry such as penetration, bead width and bead height during the welding process of stainless steel of AISI 201 grade. Parametric analysis and development of predictive model are carried out. The most important parameters for the geometry of weld beads are welding voltage and wire feed rates. The increase in wire feed rates and voltage increases penetration and bead width, while increasing the distance between nozzles decreases the penetration value and bead width. The deeper penetration needs high values of input factors. The developed model having R^2 value greater than 95% and predictive performance with an average error of less than 5% confirms the effectiveness of the model. The single optimization using GA provides maximum P of 3.468 mm, minimum W of 4.312 mm and minimum H of 2.483 mm corresponding to different combinations of weld parameters. The simultaneous optimization of both (i) penetration (P) and bead width (W); (ii) penetration (P) and bead height (H) using GA gives 18 combinations of Pareto-optimal solutions. Optimized results can be used as guidelines for manufacturing industry to select optimum welding parameters to produce desired penetration value (P) with either minimum width or height of weld bead. The approach is found to be an effective and have minimum effort.

Acknowledgements The authors acknowledge the financial support received from NERIST, Arunachal Pradesh, under TEQIP-II scheme and Advance Welding Training and Research Centre in Ajay Kumar Garg Engineering College, Ghaziabad, Uttar Pradesh, in carrying out the experimentation work.

References

1. David, S.B., Cheryl, A.B.: AL 201HPTM (UNS S20100) alloy A high performance, lower nickel alternative to 300 series alloy. Reprinted with the permission of KCI Publishing, Netherlands (3) 17–20 (2005)
2. Chuaiphan, W., Srijaroenpramong, L.: Effect of welding speed on microstructures, mechanical properties and corrosion behaviour of GTA-welded AISI 201 stainless steel sheets. J. Mater. Process. Technol. **214**, 402–408 (2014)
3. Houldcroft, P.T.: Welding Process Technology, 1st edn. Cambridge University Press, Cambridge (1977)
4. Choudhury, B., Chandrasekaran, M.: Investigation on welding characteristics on aerospace materials—a review. Mater. Today Proc. **4**, 7519–7526 (2017)
5. Juang, S.C., Tarng, Y.S.: Process parameters selection for optimizing the weld pool geometry in the tungsten inert gas welding of stainless steel. J. Mater. Process. Technol. **122** 33–37 (2002)
6. Ibrahim, I.A., Mohamat, S.A., Amir, A., Ghalib, A.: The effect of gas metal arc welding (GMAW) processes on different welding parameters. Procedia Eng. **41**, 1502–1506 (2012)

7. Datta, S.V., Bandyopadhyay, A.B., Kumar, P.: Grey-based Taguchi method for optimization of bead geometry in submerged arc bead-on-plate welding. Int. J. Adv. Manuf. Technol. **39**, 1136–1143 (2008)
8. Kim, D., Rhee, S., Park, H.: Modelling and optimization GMA welding by genetic algorithms and response surface methodology. Int. J. Prod. Res. **40**, 1699–1711 (2002)
9. Sathiya, P., Paneerselvam, K., Abdul Jaleel, M.Y.: Optimization of laser welding process parameters for super austenitic stainless steel using artificial neural networks and genetic algorithm. Mater. Des. **36**, 490–498 (2012)

Chapter 20
Finite Element Modelling and Experimental Verification of Dissimilar Joining Between Inconel 718 and SS 316L by Micro-plasma Arc Welding

Ajit Kumar Sahu and Swarup Bag

Abstract In the present research work, an attempt is made to successfully weld Inconel 718, a nickel-based superalloy with SS 316L, an austenitic stainless steel in autogenous mode by using constant current micro-plasma welding process. A finite element method (FEM)-based, three-dimensional (3D) thermal model is developed for butt welding configuration between the selected bimetallic combinations by using a double-ellipsoidal volumetric heat source model. A good consistency is found between the numerically obtained and experimentally obtained weld bead measurements. The numerical model is further used to extract peak temperatures, time–temperature profiles along with average cooling rates of the welding processes. Obtained cooling rates are correlated with the weld micro-structures and mechanical properties. Micro-structural study by SEM analysis has shown coarse columnar dendritic structure in the weld interior for the higher heat input welding case, whereas lower heat input during welding leads to fine weld micro-structure due to high cooling rate. An improvement in the tensile strength and hardness value is observed with increase in weld cooling rate and low heat input during the joining process.

Keywords Inconel 718 · Finite element modelling · Dissimilar welding · Cooling rate · Micro-structure

A. K. Sahu (✉) · S. Bag
Department of Mechanical Engineering, Indian Institute of Technology Guwahati, Guwahati 781039, Assam, India
e-mail: ajit.sahu@iitg.ac.in

S. Bag
e-mail: swarupbag@iitg.ac.in

R. G. Narayanan et al. (eds.), *Advances in Computational Methods in Manufacturing*, Lecture Notes on Multidisciplinary Industrial Engineering,
https://doi.org/10.1007/978-981-32-9072-3_20

20.1 Introduction

Welding of dissimilar materials is always a challenging task due to the differences in material composition and physical properties which need several trials to optimize successful welding parameters. Dissimilar material welding has been in demand for various industries to reduce material cost and to improve flexibility in design. It not only satisfies the service condition but also reduces component cost by reducing the amount of high-cost materials in a component [1]. Inconel 718 is a nickel–chromium-based precipitation hardened superalloy which is strengthened mostly by γ'' (Ni_3Nb) precipitates [2]. Excellent corrosion and oxidation resistance along with extensive mechanical properties at elevated temperature makes Inconel 718 suitable for various critical condition applications. Higher resistance to strain age cracking during the welding makes it one of the most successful high temperatures nickel–chromium alloy ever designed. It is widely used to make hot section components of aero engine, liquid rocket cryogenic engines, and gas turbines blades. It is also widely used in chemical industries, process industries, and nuclear reactors in various severe working conditions [3]. Similarly, austenitic steels are widely used in petrochemical, aerospace, and power generating sector in the form of plates, coils, sheets, bars, and tubing. AISI 316L contains lower amount of carbon which makes this austenitic steel immune to the sensitization at elevated temperatures [4]. AISI 316L exhibits higher creep and tensile strength as compared to other nickel–chromium austenitic steels over a wide temperature range. It also possesses excellent toughness even at the cryogenic temperature which makes it distinct from other chromium–nickel austenitic stainless steels.

The aforesaid features of Inconel 718 and AISI 316L make them suitable to use in the form of bimetallic joint for various engineering applications to make various components in aerospace engine and gas turbine engine. This bimetallic combination is used by ALSTOM Power Sweden for making compressor rotor where high pressure compressor stages are made by Inconel 718 and low pressure stages are made of austenitic based stainless steel [5]. Fusion welding is one of the most preferred joining techniques for this bimetallic combination to establish a good metallurgical bond between the mating surfaces. Although fusion welding of Inconel 718 is not utterly free from welding defects, it can be minimized by taking required precautions during welding process. Defects like solidification cracking and micro-fissuring in the fusion zone (FZ) and heat affected zone (HAZ) are major problem associated with the Inconel 718 and AISI 316L [6]. The generation of intermetallic Laves phase by segregation of niobium (Nb) in the interdendritic region during weld metal solidification is the main cause of above welding defects in Inconel 718 [7], whereas formation of 3–9% of δ ferrite reduces tendency of hot cracking, but higher percentage of ferrite content reduces creep life along with corrosion resistance of the AISI 316L joint [8]. A higher weld cooling rate has resulted in minimization of these defects which can be achieved by using a heat source of higher power density [9]. Thavamani et al. [10] have used gas tungsten arc welding (GTAW) process for welding Inconel 718 at three different heat input of 530, 670 and 800 J/mm and reported the size of dendrites and crack sensitivity of the weld increasing with increase in heat input. Lower value of heat input improves the weld cooling rate, whereas higher value slows down the weld cooling rate which leads to segregation of various useful alloying element into the interdendritic regions, leading to deterioration of mechanical properties of the welded component. Bansal et al. [11] have used microwave welding to join Inconel 718 and AISI316L with Inconel 718 powder as interfacing layer between the two surfaces and reported good joint strength for the bimetallic welded joint. Ramkumar et al. [12] have used multi-pass GTAW process to weld this bimetallic combination of material and reported superior mechanical properties for the joint as compared to the AISI 316L base material. Ramkumar et al. [13] inspected the dissimilar welding of Inconel 718 and SS 316L with the help of constant and pulsed current mode GTAW process. They found weld prepared by pulsed current mode has superior tensile, hardness and toughness properties as compared to constant current mode weld which is mainly due to the control of Laves phase formation due to higher cooling rate, obtained during pulsed current mode of welding. Kumar et al. [4] also employed CCGTAW and PCGTAW to weld Inconel 625 and AISI 316L and stated presence of various secondary phases in the HAZ are minimal for PCGTAW condition. From the above literature review, it can be concluded that the scope and application of joining this bimetallic combination are wider and required through investigation in terms of joint performance at ambient temperature. The selection of proper welding techniques and process condition is an important aspect of welding dissimilar materials. By using advanced welding techniques, full penetration can be achieved comparatively at a lower heat input; as a result, the weld cooing rate can be enhanced and formation of various deleterious secondary phases can be constricted [14]. From the open literature, it was evident most of the reported work were limited to GTAW process for joining this bimetallic combination. Hence, in the present study, micro-plasma arc welding technique which has higher power intensity among all the arc welding processes is used to join this bimetallic combination [15]. μ-PAW also possesses good arc stability even at very low range of current as compared to GTAW which makes this process more suitable for joining thin sheets. The present study is mainly focused on weldability, mechanical and metallurgical analysis of these bimetallic joints obtained at different condition of continuous current μ-PAW process. Another objective of the present study is to correlate the effect of heat input/cooling rate on fusion zone micro-structure and Laves phase formation by different welding condition.

Direct experimental investigations during welding are extremely expensive and often impossible because of the complexity of the process and presence of high temperature arc. Hence, a three-dimensional finite element (FE)-based numerical model was developed using ABAQUS to obtain the temperature isotherm in the fusion zone along with time–temperature history of μ-PAW welding process. The conductive heat transfer-based numerical modelling is done by using a moving double-ellipsoidal heat source that is accountable for complex physical phenomena like conduction, convection along with radiation heat transfer from the joining surfaces to the ambient. The double-ellipsoidal heat source is used in the current study due to its unique capability to incorporate steep gradient of temperature in the front of heat source and a gentle gradient in the trailing edge of weld pool [16]. Temperature-based physical properties of the materials were considered in the model. Thereafter, the time–temperature profile of the fusion zone from the numerical model is accounted for estimation of the cooling rate. The influence of weld cooling rate/solidification rate on weld micro-structure and corresponding outcome on mechanical properties are reported in the present study.

20.2 Experimental Procedure

As received Inconel 718 and AISI 316L stainless steel sheets of thickness 0.7 mm were sheared into 100 × 60 mm coupons with welding direction normal to the rolling direction, the chemical composition of the base materials were analysed by energy dispersive X-ray (EDX) technique and results are shown in Fig. 20.1. Based on EDX analysis, the resulted composition (wt%) of Inconel 718 and AISI 316L are given in Table 20.1.

The welding surfaces were cleaned with acetone prior to the welding process for removing oxides and other foreign impurities. Sheets were clamped in a restrain fixture made up of copper to achieve high cooling rate. The welding torch of micro-plasma set-up is kept fixed and the workpiece is allowed to move at a constant velocity. Commercially available argon gas (99% pure) is used as plasma gas as well as used as a shielding gas for the protection of weld pool. Butt configuration weld joints were prepared by constant current (CC) mode using micro-plasma arc welding set-up with a DC current range of 0.05–50 A. The plasma gas and shielding gas flow rate are maintained at 0.5 and 5 L/min, respectively, with a nozzle standoff distance of 2 mm. The other welding parameters employed for making defect free welds are depicted in Table 20.2 which is obtained by a wide range of welding trials.

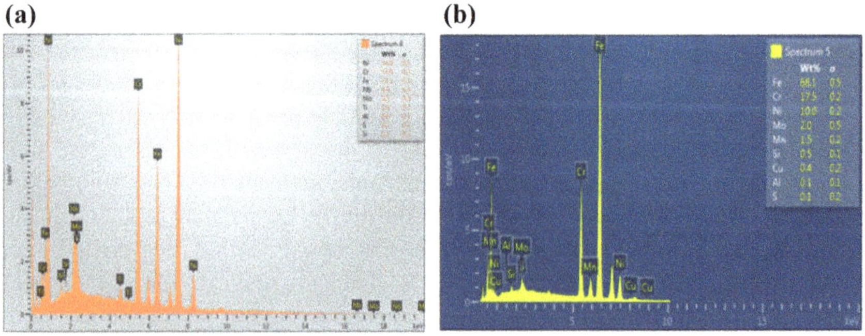

Fig. 20.1 EDX elemental analysis of **a** Inconel 718 and **b** AISI 316L base material

Table 20.1 Chemical composition (wt%) of Inconel 718 and AISI 316L

Elements	Ni	Cr	Nb	Mo	Ti	Al	Si	S	Fe
Inconel 718	54.0	18.6	4.8	2.5	1.0	0.6	0.2	0.2	Bal.
Elements	Ni	Cr	Mo	Mn	Si	Cu	Al	S	Fe
AISI 316L	10.0	17.5	2.0	1.5	0.5	0.4	0.1	0.1	Bal.

Table 20.2 Process parameters for μ-PAW of Inconel 718 and SS 316L weldments

Sample	Current (A)	Voltage (V)	Welding speed (mm/s)	Heat input (J/mm)
S1	14	24	3.4	99
S2	15	24	3.2	112

The welded samples were sheared normal to the welding direction and polished for macro-structural analysis. Then, the mirror polished samples were etched in a mixture of HCl, HF and CH_3COOH in the ratio of 1.5:1:1, respectively, for the micro-structural analysis. The micro-structural analyses were done by the help of optical microscope and scanning electron microscope. Vickers's micro-hardness measurements were done across the weld bead for each specimen to analyse the hardness behaviour. Tensile test of the welded specimens and base materials are conducted at room temperature to analyse the yield strength, ultimate tensile strength and percentage elongation. For each condition, two sub-size transverse specimens were prepared according to ASTM E8 M standard with a gauge length of 25 mm. The tests were conducted by a computer controlled servo-hydraulic universal tensile testing machine with a cross-head speed of 1 mm/min.

20.3 Numerical Model

Experimental investigations have certain limitations due the involvement of high temperature heat source during the welding process. Hence, a finite element-based conduction heat transfer model is developed for the micro-plasma arc welding process to obtain temperature distribution of the entire weldment. The moving heat source is incorporated in the model by using a DFLUX subroutine with the numerical model. To approximate the influence of convective transport of heat in weld pool, a Gaussian distributed double-ellipsoidal volumetric heat source is used to distribute the heat flux into the joining surfaces. Temperature-based thermo-physical properties for SS 316L and Inconel 718 were considered in the present analysis [17, 18]. For simplification, following assumptions were considered in the model: Top surface of the fusion zone is assumed as flat and neglecting the effect of plasma arc and shielding gas. The initial temperature of workpiece was taken as uniform ambient temperature (T_0) of 300 K. The latent heat of melting and cooling is incorporated, whereas material evaporation from the fusion zone is neglected. The surface of the base plates which is exposed to heat flux and other surfaces is subjected to heat losses due to convection and radiation. The surfaces which are directly exposed to air were assigned a convective heat transfer coefficient (h) of 35 W/m^2 K [16]. However, surfaces which are in direct contact with the copper fixture and backing plate, which have high thermal conductivity, are assigned with a high heat transfer coefficient of 1000 W/m^2 K [19]. The fundamental governing equation of heat transfer or conservation of energy in 3D Cartesian coordinate is expressed as

$$\rho C\left(\frac{\partial T}{\partial t}-v\frac{\partial T}{\partial x}\right)=\frac{\partial}{\partial x}\left(k\frac{\partial T}{\partial x}\right)+\frac{\partial}{\partial y}\left(k\frac{\partial T}{\partial y}\right)+\frac{\partial}{\partial z}\left(k\frac{\partial T}{\partial z}\right)+\dot{q} \tag{20.1}$$

where (x, y, z) is the coordinate system associated with the moving heat source. ρ, k and C refer to density, thermal conductivity and specific heat of the base material, respectively. T and t represents temperature and time variable, respectively. $\dot{q}$ is the internal heat generation per unit time and per unit volume and 'v' is considered as the velocity of the plasma arc in welding direction (x-axis) or moving coordinate axis. The meshing arrangement of base plate is shown in Fig. 20.2. Very fine meshing is provided in the fusion zone area to capture thermal history precisely as it is exposed to a very concentric heat source where the temperature gradient becomes steep. While coarser meshing was generated in the area far from the heat flux to reduce computational cost, Continuum solid eight nodded brick elements (DC3D8 type) were selected in the thermal analysis for diffusive heat transfer.

The double-ellipsoidal parameters like width (b) and depth (c) of the double-ellipsoidal heat source were directly taken from the experimental results. The length of front (c_f) and rear (c_r) quadrant were estimated as a function of velocity [20]. In dissimilar welding, the fusion zone becomes a mixture of both base materials which will produce various intermetallic compounds during the

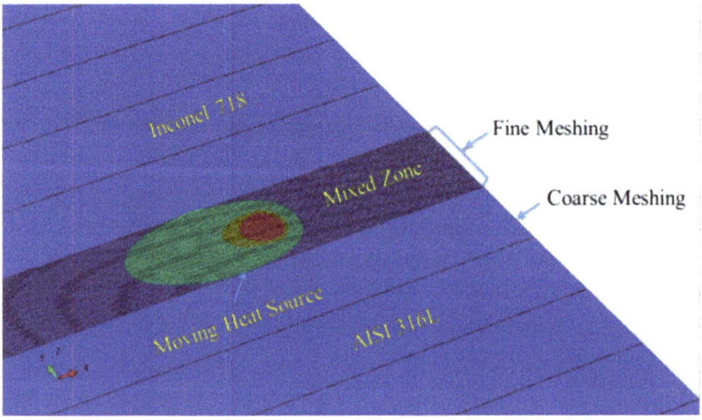

Fig. 20.2 Finite element meshing of solution domain

solidification and will have an impact the joint efficiency. Therefore, to correlate the weld cooling rate with intermetallic phase formation precisely, the temperature dependent thermo-physical properties of fusion zone were considered as the average of both base materials which is named as mixed zone as shown in Fig. 20.2.

20.4 Results and Discussion

20.4.1 Temperature Field Characteristics

Temperature distribution corresponding to S1 welding condition along with the moving heat source is shown in Fig. 20.3. A peak temperature of 2169 K is achieved at the weld centre line. The fusion zone (FZ) is defined by 1730 K which is the liquidus temperature of AISI 316L, i.e. maximum between the base materials [1]. And the heat affected zone (HAZ) was defined below 1547 K which is the solidus temperature of Inconel 718, i.e. minimum between the base materials [21]. The region between these two isotherm contours is considered as mushy zone where liquid and solid phase coexists together.

The comparison between simulated and experimentally estimated weld macrographs in transverse cross section for both welding conditions is shown in Fig. 20.4, where FZ, mushy zone, and HAZ are well defined. The shape and size of the numerically simulated outcomes were found to be in good agreement with the experimentally obtained weld macrographs. With increase in welding heat input per unit length, the size of weld bead increases and the fusion zone becomes wider. The weld cross section of S2 welding condition is little bit of straight type as compared

Fig. 20.3 Temperature distribution for S1 welding condition

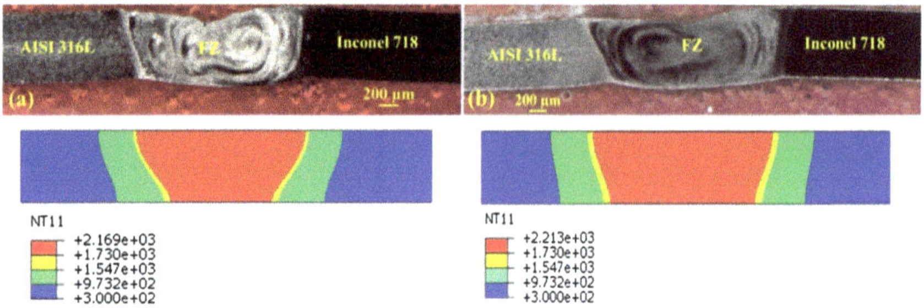

Fig. 20.4 Comparison of weld macrograph for **a** S1 and **b** S2 welding condition

to S1 welding condition which is due to higher heat input during welding process. The maximum error found to be 6% in the weld bead width comparison at the cap and root between experimentally and numerically measured value (Table 20.3).

Figure 20.5 shows the time–temperature profiles of S1 and S2 welding condition at weld centre along with the average weld cooling rate. The temperature rises rapidly as the heat source approaches and drops steadily to ambient temperature when the heat source passes away from the node. The quality of weld is greatly dependent on heat input per unit length during the welding process, peak

Table 20.3 Weld bead dimensions at different location

Location	S1		S2	
	Experimental	Numerical	Experimental	Numerical
Weld cap (mm)	2.46	2.38	2.6	2.44
Weld root (mm)	1.91	1.8	1.96	1.94

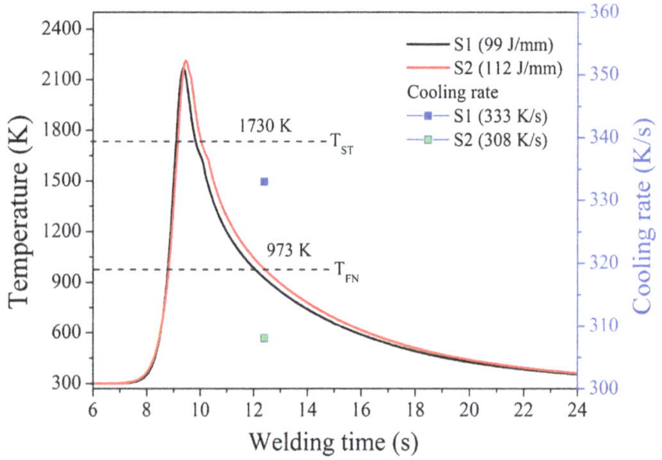

Fig. 20.5 Time–temperature profile and average cooling rate of S1 and S2 welding conditions

temperature, and cooling rate. The peak temperature of S1 and S2 welding conditions were found to be 2169 and 2213 K, respectively. Higher peak temperature for S2 welding condition is attributed to the higher heat input during welding process. The average weld cooling rate was calculated by taking the slope of time–temperature curve between transformation start (T_{ST}) and finish temperature (T_{FN}). Higher liquidus temperature among Inconel 718 and SS 316L was considered as T_{ST} as below this temperature phase transformation initiated. Minimum value of solid phase transformation end temperature among the bimetallic is taken as T_{FN} as below this temperature phase transformation does not occurs. The calculated average cooling rate for S1 and S2 welding condition were 333 and 308 K/s. When the weld peak temperature increases, weld pool lasts for longer period of time resulted in lower average cooling rate.

20.4.2 *Effect of Cooling Rate on Weld Micro-structure and Mechanical Properties*

Formation of various intermetallic phases during weld cooling period is the major challenge of fusion welding process. Segregation of Mo and Nb in the interdendritic region leads to formation of deleterious Laves phases in Inconel 718 which promotes liquation cracks and micro-fissuring in the FZ and HAZ [14]. In AISI 316L, formation of higher percentage of δ ferrite in interdendritic region has deleterious effect on mechanical properties and corrosion resistance of the joint. The scale of solidification structure is dominated by the cooling rate; higher cooling rate will produce finer dendrites, whereas lower value will produce coarse dendrites. In Fig. 20.6, higher magnification SEM micro-structure of weld interior for both the welding conditions is shown. Higher cooling rate of S1 weld condition (333 K/s) leads to fine micro-structure as compared to S2 weld condition (308 K/s). Coarse columnar dendrites were observed in the weld interior for S2 welding condition, whereas fine interconnected equiaxed type of micro-structure was observed in S1 welding condition.

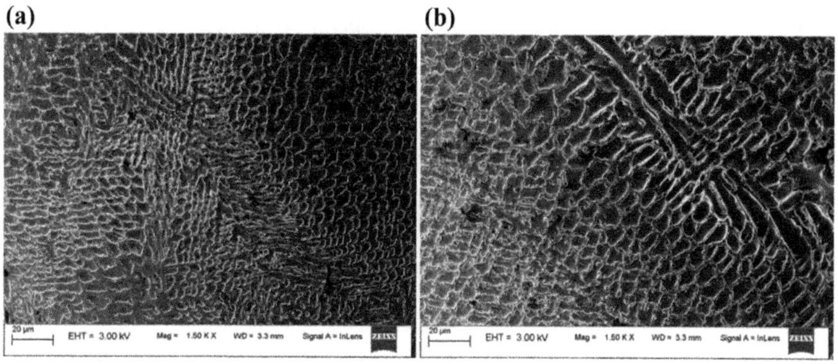

Fig. 20.6 Fusion zone micro-structure of **a** S1 and **b** S2 welding condition

Table 20.4 Results of tensile test

	Inconel 718 (base material)	AISI 316L (base material)	S1 (weld joint)	S2 (weld joint)
YS (MPa)	594	408	389	337
UTS (MPa)	878	679	653	596
Elongation (%)	23	39	26	21

The weld cooling rate plays a crucial role in controlling interdendritic segregation and formation of various intermetallic phases in the fusion zone. Lower heat input assures a higher cooling rate during the solidification that hinders the segregational tendency of solutes in the interdendritic regions, resulting in a fined micro-structure in the fusion zone. Formation of various intermetallic phases consumes a notable amount of beneficial alloying elements from the base material which results deterioration of weld quality. Hence, control of intermetallic Laves and δ-ferrite phases in the solidified zones improves mechanical properties of the joint. The tensile properties of both base and welded materials are given in Table 20.4. S1 welded samples were failed in the AISI 316L base material sides, whereas S2 welded samples were failed in the fusion zone as shown in Fig. 20.7. The tensile strength of S1 weld metal were found superior to the S2 weld material which is mainly due to the fined micro-structure of the fusion zone that reduces segregation and formation of intermetallic in the grain boundary. S1 welded sample has a tensile strength of 96% to that of the AISI 316L base material and a tensile strength of 74% to that of Inconel 718. The improvement in tensile strength of S1 welded condition is mainly attributed to the fine weld micro-structure which is mainly due to the higher cooling rate achieved during the welding process.

Cross-weld micro-hardness profile of the S1 and S2 weldments are shown in Fig. 20.8. The average micro-hardness values of base material are 219 HV and 134 HV for Inconel 718 and AISI 316L, respectively. Fusion zone of S1 welding condition has an average micro-hardness value of 154 HV, whereas for S2 welding condition, it has a lower value of 148 HV. Higher hardness of S1 fusion zone is mainly due to the fine micro-structure because of higher cooling rate. The hardness is found to be less in the HAZ region as compared to the FZ and BM, which is mainly attributed to the grain coarsening from FZ towards BM.

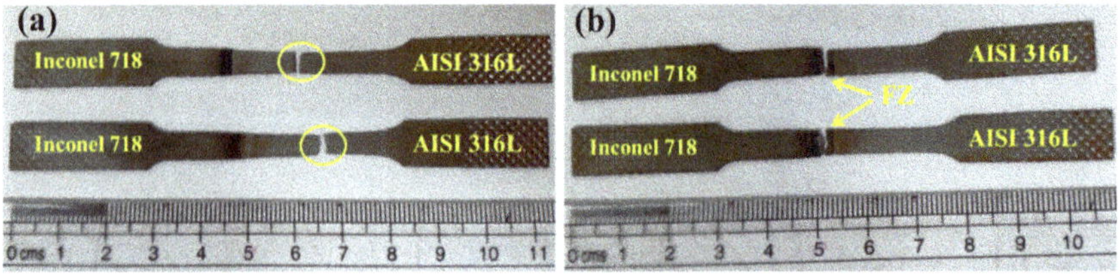

Fig. 20.7 Location of tensile fracture for **a** S1 and **b** S2 welding conditions

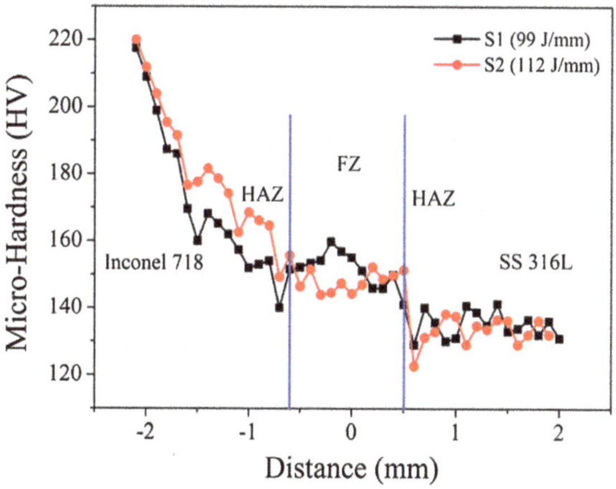

Fig. 20.8 Cross-weld micro-hardness distribution

20.5 Conclusion

In the present study micro-plasma arc welding between Inconel 718 with AISI 316L sheet materials with butt joint configuration has been undertaken. A three-dimensional FE-based thermal model is developed to investigate the thermal cycle during the welding process. The following conclusions were drawn from the current study

- A good agreement was obtained between the shape and size of computed and experimentally obtained weld bead.
- Full weld penetration for the bimetallic combination was achieved at a heat input of 99 J/mm.
- A maximum cooling rate of 333 K/s was calculated from the numerical model for the low heat input of 99 J/mm.
- Refinement in weld micro-structure is observed at 333 K/s cooling rate as compared to 308 K/s which results in improved joint efficiency (96.17% of AISI 316L base material) and ductility (26%) of the welded joint.
- A maximum micro-hardness difference of 6 HV is reported between the FZ of 99 and 112 J/mm welding condition.

Acknowledgements The authors would like to acknowledge the Department of Mechanical Engineering, Central Instrument Facility (CIF), and Central Workshop of IIT Guwahati, India for providing all facilities to carry out the experimental and numerical work.

References

1. Lee, H.T., Jeng, S.L., Yen, C.H., Kuo, T.Y.: Dissimilar welding of nickel-based Alloy 690 to SUS 304L with Ti addition. J. Nucl. Mater. **335**, 59–69 (2004)
2. Ye, X., Hua, X., Wang, M., Lou, S.: Controlling hot cracking in Ni-based Inconel-718 superalloy cast sheets during tungsten inert gas welding. J. Mater. Process. Technol. **222**, 381–390 (2015)
3. Huang, C.A., Wang, T.H., Lee, C.H., Han, W.C.: A study of the heat-affected zone (HAZ) of an Inconel 718 sheet welded with electron-beam welding (EBW). Mater. Sci. Eng., A **398**, 275–281 (2005)
4. Kumar, K.G., Ramkumar, K.D., Arivazhagan, N.: Characterization of metallurgical and mechanical properties on the multi-pass welding of Inconel 625 and AISI 316L. J. Mech. Sci. Technol. **29**, 1039–1047 (2015)
5. Henderson, M.B., Arrell, D., Larsson, R., Heobel, M., Marchant, G.: Nickel based superalloy welding practices for industrial gas turbine applications. Sci. Technol. Weld. Joining **9**, 13–21 (2004)
6. Radhakrishnan, B., Thompson, R.G.: A model for the formation and solidification of grain boundary liquid in the heat-affected zone (HAZ) of welds. Metall. Mater. Trans. A **23**, 1783–1799 (1992)
7. Janaki Ram, G.D., Venugopal Reddy, A., Prasad Rao, K., Reddy, G.M., Sarin Sundar, J.K.: Microstructure and tensile properties of Inconel 718 pulsed Nd-YAG laser welds. J. Mater. Process. Technol. **167**, 73–82 (2005)
8. Kar, J., Roy, S.K., Roy, G.G.: Effect of beam oscillation on microstructure and mechanical properties of AISI 316L electron beam welds. Metall. Mater. Trans. A **48**, 1759–1770 (2017)
9. Radhakrishna, C.H., Rao, K.P.: The formation and control of Laves phase in superalloy 718 welds. J. Mater. Sci. **32**, 1977–1984 (1997)
10. Thavamani, R., Balusamy, V., Nampoothiri, J., Subramanian, R., Ravi, K.R.: Mitigation of hot cracking in Inconel 718 superalloy by ultrasonic vibration during gas tungsten arc welding. J. Alloy. Compd. **740**, 870–878 (2018)
11. Bansal, A., Sharma, A.K., Das, S., Kumar, P.: On microstructure and strength properties of microwave welded Inconel 718/stainless steel (SS-316L). Proc. Inst. Mech. Eng., Part L: J. Mater. Des. Appl. **230**, 939–948 (2016)
12. Ramkumar, T., Selvakumar, M., Narayanasamy, P., Begam, A.A., Mathavan, P., Raj, A.A.: Studies on the structural property, mechanical relationships and corrosion behaviour of Inconel 718 and SS 316L dissimilar joints by TIG welding without using activated flux. J. Manuf. Process. **30**, 290–298 (2017)
13. Ramkumar, K.D., Patel, S.D., Praveen, S.S., Choudhury, D.J., Prabaharan, P., Arivazhagan, N., Xavior, M.A.: Influence of filler metals and welding techniques on the structure–property relationships of Inconel 718 and AISI 316L dissimilar weldments. Mater. Des. (1980–2015) **62**, 175–188 (2014)
14. Manikandan, S.G.K., Sivakumar, D., Rao, K.P., Kamaraj, M.: Effect of weld cooling rate on Laves phase formation in Inconel 718 fusion zone. J. Mater. Process. Technol. **214**, 358–364 (2014)
15. Kou, S.: Welding Metallurgy. Wiley, New Jersey, USA (2003)
16. Goldak, J., Chakravarti, A., Bibby, M.: A new finite element model for welding heat sources. MTB **15**, 299–305 (1984)
17. Rahman Chukkan, J., Vasudevan, M., Muthukumaran, S., Ravi Kumar, R., Chandrasekhar, N.: Simulation of laser butt welding of AISI 316L stainless steel sheet using various heat sources and experimental validation. J. Mater. Process. Technol. **219**, 48–59 (2015)
18. Dye, D., Hunziker, O., Roberts, S.M., Reed, R.C.: Modeling of the mechanical effects induced by the tungsten inert-gas welding of the IN718 superalloy. Metall. Mater. Trans. A **32**, 1713–1725 (2001)

19. Baruah, M., Bag, S.: Influence of heat input in microwelding of titanium alloy by micro plasma arc. J. Mater. Process. Technol. **231**, 100–112 (2016)
20. Yadaiah, N., Bag, S.: Effect of heat source parameters in thermal and mechanical analysis of linear GTA welding process. ISIJ Int. **52**, 2069–2075 (2012)
21. Aktaş Çelik, G., Polat, Ş., Atapek, Ş.H., Haidemenopoulos, G.N.: Microstructural and mechanical characterization of solidified austenitic stainless steels. Arch. Foundry Eng. **17**, 163–167 (2017)

Chapter 21
Numerical Simulation and Experimental Validation of A-TIG Welding of 2.25Cr–1Mo Steel

A. R. Pavan, B. Arivazhagan, S. Arun Kumar, M. Vasudevan and S. Mahadevan

Abstract Localized heating and cooling during welding introduces residual stress in the weld joints due to the thermal gradients. The microstructure formation in 2.25Cr–1Mo steel is very much sensitive to the rate of cooling and eventually, the austenite may transform to pearlite/bainite/martensite. A variant of Tungsten Inert Gas (TIG) welding called Activated-TIG (A-TIG) welding was made use of in this study. Here, we have tried to develop a numerical model based on the finite element method to predict the thermo-mechanical behavior of 6 mm thick 2.25Cr–1Mo steel weld joints, with accounting solid-state phase transformation. The heat energy supplied to the workpiece is determined and calibrated in the heat source model. Simulated thermal cycle showed a peak temperature of 1377 °C at 8 mm distance away from the weld centerline, whereas, the experimental result showed 1371 °C. A *k*-type thermocouple was used to measure the temperature distribution during welding. The simulated thermal distributions were sequentially coupled to mechanical analysis. The evolution of stress and ultimately the locked-in residual stresses were determined. X-ray diffraction studies showed the peak residual stress near heat affected zone of 592 MPa and the vertical height gauge was used to measure the distortion before and after welding. The predicted residual stresses and distortion showed good agreement with the experimental measurements.

Keywords 2.25Cr–1Mo steel · A-TIG welding · FEM · Solid-state phase transformation · Welding residual stress and distortion

A. R. Pavan · B. Arivazhagan · S. Arun Kumar · M. Vasudevan (✉) · S. Mahadevan
Metallurgy and Materials Group, Indira Gandhi Centre for Atomic Research, Kalpakkam, Tamil Nadu, India
e-mail: dev@igcar.gov.in

A. R. Pavan · B. Arivazhagan · M. Vasudevan · S. Mahadevan
Homi Bhabha National Institute, Indira Gandhi Centre for Atomic Research, Kalpakkam, Tamil Nadu, India

R. G. Narayanan et al. (eds.), *Advances in Computational Methods in Manufacturing*, Lecture Notes on Multidisciplinary Industrial Engineering,
https://doi.org/10.1007/978-981-32-9072-3_21

21.1 Introduction

A higher level of welding residual stress or locked-in stress in the weld joint is a life-limiting factor under service conditions. Steep thermal gradient and shrinkage of sizeable molten metal volume generate tensile stress near the weld joint and is inevitable. This, in turn, leads to stress corrosion cracking, causes brittle fracture and premature failure of engineering components [1]. But, compressive residual stresses are beneficial in certain circumstances [2]. Therefore, by knowing the magnitude of residual stress, one can estimate the remaining life of weld joints with safety measures. However, the locked-in stresses can be controlled and reduced by proper choice of the welding process and use of restraints. Thus, the safety of the structure and design can be increased.

The 2.25Cr–1Mo steel has been most widely used in petrochemical, modern power generation industry in the form of boilers, heaters, heat exchangers, reactors, and hydrocrackers usually built as heavy wall pressure vessels in very high temperature range of 350–550 °C because of its excellent strength at elevated temperatures [3, 4]. The excellent resistance to creep and corrosion, high toughness and good weldability makes it as a candidate material in service conditions. Alongside, these materials have good physical properties such as a low thermal expansion coefficient and a high thermal conductivity. Hence, Cr–Mo steels are promising to operate at high temperatures and under low stresses. Moreover, the integrity of the weld joint decides the performance of any components. Further, the mechanical properties of 2.25Cr–1Mo steel depend on the microstructure, which is sensitive to the cooling rate which in turn depends on heat input during welding [5]. Normally air-cooled steel forms bainitic microstructure at room temperature. This type of steel is usually air cooled after austenitization for service in power plant applications [6]. For 2.25Cr–1Mo joint fabrication, Shielded Metal Arc Welding (SMAW) and TIG welding have been widely adopted [7]. Moreover, TIG has lower productivity while joining the thick (>3 mm) components due to a large number of weld passes utilized to create a joint which eventually introduces high tensile residual stress near the joint with very high distortion. Recently, the variant of TIG welding known as A-TIG welding is widely applied due to its higher productivity [7], low cost, lower micro inclusion, low residual stress and distortion and also because of autogenous welding. An activating flux containing multi-component oxides, applied on the joint alters the surface tension of the weld pool and controls the arc constriction during TIG welding and eventually enhances the depth of penetration up to 300% compared to conventional TIG welding process [8].

However, experimentally assessing the evolution of welding residual stress and optimizing the welding parameters to achieve better weld geometry is time-consuming and costly. Thus, a computational approach is utilized to envisage the thermo-mechanical behavior of the weld joint. The welding thermal cycle, residual stress and distortion could be determined by simulation and appropriate measures or changes in the welding process and conditions could be adopted during the fabrication to reduce the level of residual stresses and distortion in the actual component.

The computational simulation tool has been used from the past few decades to understand the evolution of welding residual stress and distortion in the welded structures. The commonly used approach is one way coupled approach, viz. sequential coupling of thermal history to mechanical analysis. Various studies [9–13] are available on the prediction of welding residual stress and distortion. Further, the complexity in the numerical model is, to consider the solid-state phase transformations during welding. Many studies have successfully incorporated phase transformation in the numerical model [6, 14, 15]. In addition, very few literatures predicted the welding residual stress and distortion in 2.25Cr–1Mo steel [16–18]. However, there is a limited number of literatures on understanding the evolution of welding residual stress and distortion in A-TIG welded 2.25Cr–1Mo steel joints. Therefore, the motivation of this research work is to predict welding residual stress and distortion by considering solid-state phase transformation in 2.25Cr–1Mo steel weld joint fabricated by autogenous A-TIG welding process by numerical simulation.

21.2 Finite Element Modeling

The dimension of 2.25Cr–1Mo steel plate is 300 × 120 × 6 mm^3. Straight-sided square butt joint configuration was used for A-TIG welding process. Numerical simulations were performed using commercially available SYSWELD software. The optimized meshing is necessary to resemble the exact experimental result in the numerical model. Hence, optimization of mesh size, with the right choice of mesh dimensions is required for improving the accuracy of the simulation. In this work, around 30 mm near to the weld line, fine mesh to a size of 1 mm × 0.8 mm (quadrilateral) was chosen, since the effect of the temperature gradient is more near the weld region and heat affected zone as shown in Fig. 21.1.

Goldak double ellipsoidal heat source model was used for defining a heat source [19]. The thermomechanical analysis is done by sequential coupling of thermal and mechanical analysis. Here the thermal and phase transformation strains are coupled

Fig. 21.1 Weld pad model

with non-linear mechanical analysis to understand the evolution of stress and distortion in the weld joint. It was assumed that there is no influence of any structural changes on the thermal analysis part.

21.2.1 Material Properties

The residual stresses and the dimensional change in the weld joint originate owing to the changes in thermo-physical, mechanical properties and microstructural characteristics of materials. Therefore, to get good accuracy in the model, it is necessary to choose appropriate material properties. For 2.25Cr–1Mo steel, the material database was taken from the literature and SYSWELD database [16] as shown in Fig. 21.2. Heat transfer due to convective flow in molten metal is considered in the model by increasing the thermal conductivity value by 2–3 times above the melting temperature.

The heat source fitting tool is used to calibrate the double ellipsoidal heat source and compared with the experimental weld bead macrograph. In addition, the experimental and predicted thermal profiles were compared to know significant heat supply into the material. Further, the obtained thermal stains are sequentially coupled to the mechanical analysis.

21.2.2 Governing Equations

Welding introduces significant heat energy in a localized region. Thus, the heat transfer during welding is primarily due to conduction mode. Heat conduction can be described as shown in the following Eq. (21.1).

$$\nabla(k\nabla T) + Q = \rho C(\partial T/\partial t) \tag{21.1}$$

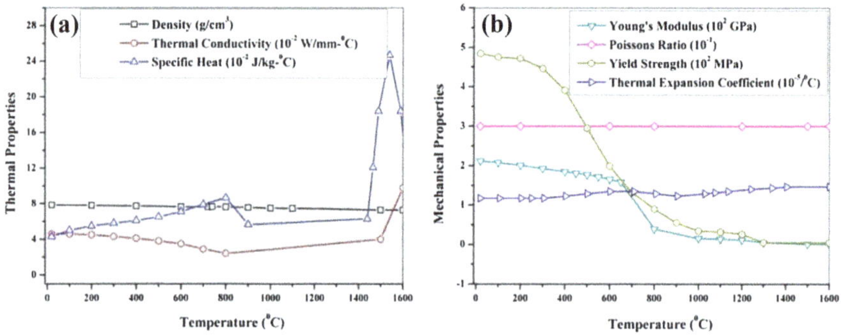

Fig. 21.2 2.25Cr–1Mo steel material properties **a** thermal physical and **b** mechanical properties

where k is the thermal conductivity (W/m °C), Q is the volumetric heat flux (W/m^3), ρ is the density (kg/m^3), and C is the specific heat (J/kg °C).

After welding, the heated material starts to cool to ambient temperature due to convection and radiation. The rate of convection heat transfer is governed by Newton's law of cooling as Eq. (21.2).

$$Q_c - h_c(T_S - T_0) = 0 \tag{21.2}$$

From, Stefan–Boltzmann law the radiation heat transfer is expressed as Eq. (21.3),

$$Q_r - \varepsilon\sigma(T_S^4 - T_0^4) = 0 \tag{21.3}$$

where h_c is the film coefficient for convection [~25 W/(m^2 K)], T_S is the surface temperature (K), T_0 is the ambient temperature (K), ε is the emissivity (0.8), and σ is the Stefan–Boltzmann constant (5.67×10^{-8} W/m^2 K^4). The total heat transfer can be written as,

$$Q_t = Q_c + Q_r \tag{21.4}$$

$$Q_t = h_c(T_S - T_0) + \varepsilon\sigma\,(T_S^4 - T_0^4) \tag{21.5}$$

By solving Eq. (21.5)

$$\left.\begin{aligned} Q_t &= h_c(T_S - T_0) + \varepsilon\sigma(T_S - T_0)(T_S + T_0)(T_S^2 + T_0^2) \\ Q_t &= [h_c + \varepsilon\sigma(T_S + T_0)(T_S^2 + T_0^2)](T_S - T_0) \end{aligned}\right\}$$

$$h_{eq} = h_c + \varepsilon\sigma(T_S + T_0)\,(T_S^2 + T_0^2) \tag{21.6}$$

$$Q_t = h_{eq}(T_S - T_0) \tag{21.7}$$

where Q_t is the total surface heat transfer and h_{eq} is the overall heat transfer coefficient. To account the phase transformation from parent phase to austenite and austenite to bainite phase transformation in the analysis, Leblond's equation and Johnson–Mehl–Avrami (JMA) were used (Eqs. 21.8 and 21.9).

$$\mathrm{d}z/\mathrm{d}t = \left(z_{eq}(T) - z/\tau(T)\right)f(\dot{T}) \tag{21.8}$$

$$p = 1 - \exp(-kt^n) \tag{21.9}$$

where z_{eq} refers to austenite proportion at phase equilibrium, z is austenite proportion, p is the fraction of bainite, k and n are kinetic parameters and t is time. The parameters of the JMA equation should be obtained from the Continuous Cooling Transformation (CCT) diagram. Here, we have assumed in the model that there is no formation of martensite upon cooling in the weld metal.

In addition, after the completion of the thermal analysis, by solving Eqs. (21.1) and (21.7) the obtained temperature distributions are coupled to a mechanical analysis by using Eqs. (21.10) and (21.11). In simple terms, the thermal analysis is sequentially coupled to mechanical analysis. Hence, the possible total strain induced on the material during welding is by elastic, thermal, plastic and due to phase transformation as shown in Eq. (21.10).

$$\Delta\varepsilon^{\text{total}} = \Delta\varepsilon^{\text{elastic}} + \Delta\varepsilon^{\text{thermal}} + \Delta\varepsilon^{\text{plastic}} + \Delta\varepsilon^{\text{phase}} \tag{21.10}$$

For, mechanical analysis, the force equilibrium is given as Eq. (21.11)

$$\sigma_{ij} + p_j = 0 \tag{21.11}$$

where σ_{ij} is stress tensor and p_j is body force at any point within the boundary. For the finite element formulation, the basic equilibrium equation can be used to achieve the virtual induced strain and displacement.

21.2.3 Boundary Condition

During the experiment, the weld joint was clamped to avoid vertical movement. Therefore, to match with experimental conditions, the apt external constraints were provided in the model. The rigid clamps were made use of to avoid rotational motion and translational motion of the weld joint. The elastic clamp with a finite stiffness of 10,000 N/mm is used to achieve the proper deformation. Isotropic hardening model was used to consider elastic-plastic material behavior.

21.3 Experimental

As received 2.25Cr–1Mo plates with normalized at 950 °C for 20 min and tempered at 730 °C for 60 min was used with dimension 300 mm × 120 mm × 6 mm having a straight square but joint configuration. Weld plate is fabricated by using A-TIG welding with heat input 2.38 kJ/mm. The transient temperature during the welding process was captured by Chromel-Alumel thermocouples placed 8 mm away from the center of the weld as shown in Fig. 21.3. Vertical height gauge was used to obtain the difference in height of the plate before and after welding (least count 0.006 mm) to estimate the level of distortion.

X-ray diffraction was utilized to estimate the magnitude of welding residual stress with Cr–Kα radiation. The $\sin^2\psi$ method was performed to evaluate the magnitude of residual strain and later strain is converted into stress. Initially, the

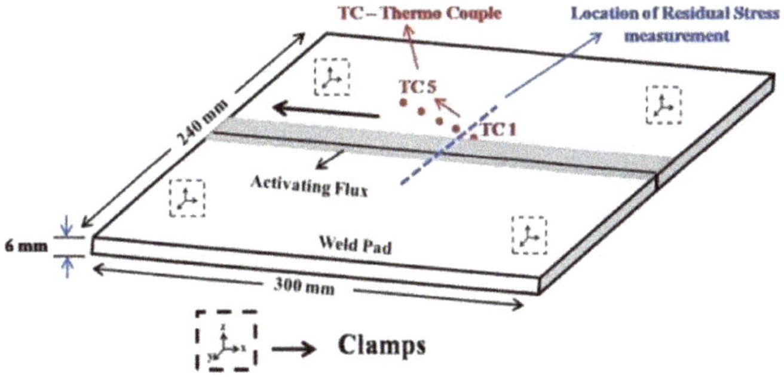

Fig. 21.3 Weld pad configuration

change in *d*-spacing is characterized at different ψ angles. The slope of a linear plot of *d*-spacing versus $\sin^2\psi$ gives the residual strain in the material, where, *d* represents the lattice spacing and ψ is the angle between the surface normal and the bisector of the incident and diffracted beam which is normally varied in the angular range of 0° to 45°.

21.4 Results and Discussion

21.4.1 Thermal Analysis

The thermo-mechanical simulation was carried out by utilizing FEM based SYSWELD software. The heat source fitting tool is employed to calibrate the heat source by comparing with the experimental weld macrograph to ensure the significant amount of heat energy supplied into the material (Fig. 21.4). Initially, heat energy supplied into the weld joint was calibrated using double ellipsoidal heat

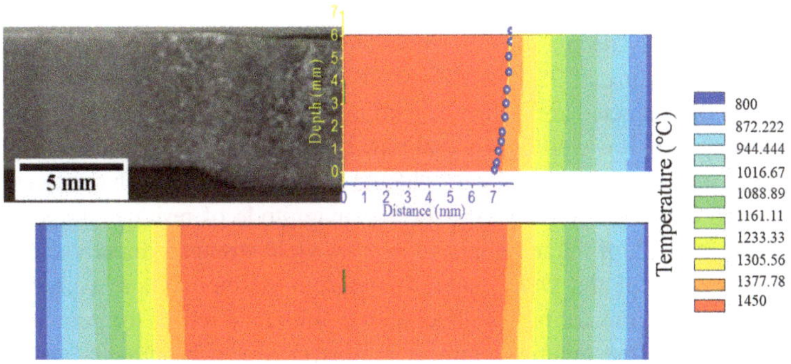

Fig. 21.4 Heat source fitting

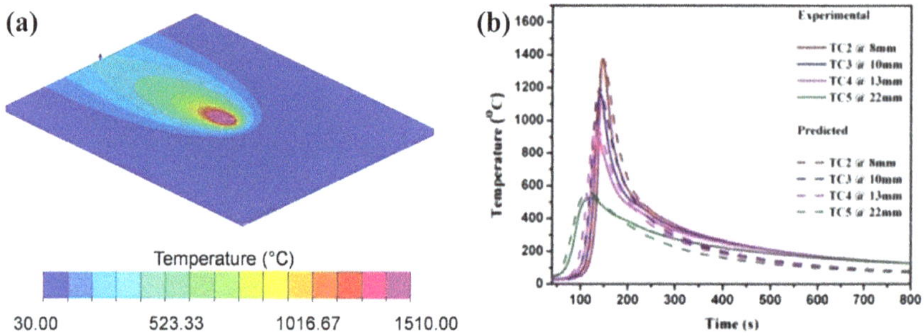

Fig. 21.5 **a** Thermal analysis of weld plate **b** thermal cycles near the weld center

source function parameters as shown in Eqs. (21.12) and (21.13) and fitted with the experimental result as shown in Fig. 21.5b.

Heat input, $Q = \left(\frac{I \times V \times \eta}{v}\right)$

$$q_{\mathrm{f}}(x, y, z, t) = \frac{6\sqrt{3}f_{\mathrm{r}}Q}{a_{\mathrm{f}}bc\pi\sqrt{\pi}} \mathrm{e}^{-3x^2/a_{\mathrm{f}}^2} \mathrm{e}^{-3y^2/b^2} \mathrm{e}^{-3[z+v(\tau-t)]^2/c^2} \tag{21.12}$$

$$q_{\mathrm{r}}(x, y, z, t) = \frac{6\sqrt{3}f_{\mathrm{r}}Q}{a_{\mathrm{r}}bc\pi\sqrt{\pi}} \mathrm{e}^{-3x^2/a_{\mathrm{r}}^2} \mathrm{e}^{-3y^2/b^2} \mathrm{e}^{-3[z+v(\tau-t)]^2/c^2} \tag{21.13}$$

I is the current, V is voltage, η is the efficiency of the welding process and v is the torch velocity in m/s. Efficiency is assumed to be 80% for A-TIG welding. Where a_f and a_r is the front and rear length of the weld pool, respectively, c is the weld penetration depth and b is half width of the weld bead. f_f and f_r ($f_r + f_f = 2$) is the front and rear fractions of the double ellipsoidal heat source, which is considered as 0.4 and 1.6 respectively. q_r and q_f are power densities in W/m^3. The first thermocouple fixed near (8 mm) to the weld bead shows highest peak temperature of 1371 °C and the fifth thermocouple is fixed away (22 mm) from the center of the weld showed lowest peak temperature of 520 °C. This change in peak temperature shows the Gaussian heat distribution, where, the heat flux density is high at the center and gradually decreases away from the center.

During the welding of 2.25Cr–1Mo steel, the microstructures in the weld zone and near the heat affected zone alters according to the applied heat input and mainly due to cooling rate. It is well known that phase transformation affects the magnitude of residual stress due to volume change (body-centered cubic structure to face-centered cubic structure). However, the magnitude of residual stress depends on the yield strength of the material and in turn, yield strength, directly depends on the microstructure of the weld material. Hence it is important to consider yield strength and phase transformation effect on the numerical model for good accuracy. Figure 21.6 shows the granular bainitic microstructure of base metal and bainitic lath structure of as-welded sample; this displayed the typical characteristics of the

(a) (b)

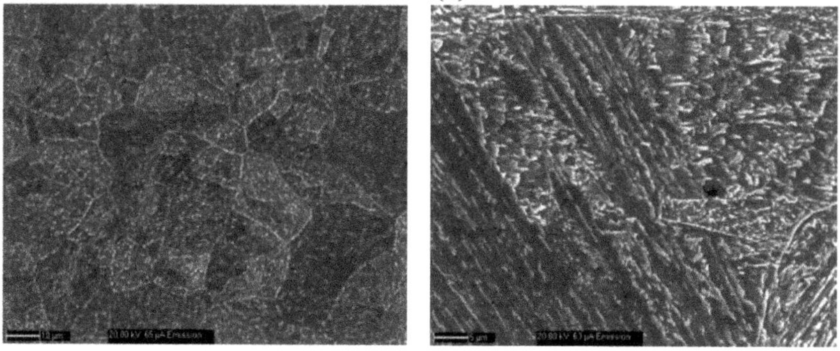

Fig. 21.6 Microstructure a base metal b weld metal

bainitic microstructure. To consider ferrite/bainite to austenite and austenite to bainite phase transformation, Eqs. (21.4) and (21.5) were used in the model respectively.

21.4.2 Mechanical Analysis

The temperature histories were then sequentially coupled for the mechanical analysis to determine the level of welding residual stress, shrinkage effect and distortion. The strains produced by the thermal energy are coupled to mechanical analysis. The thermal strains, plastic and elastic strains along with volume strains produced due to solid-state phase transformation effect, induce stress in the material. To consider all these factors in the numerical model, temperature dependent material properties must be incorporated, that is, Young's modulus and Poisson's ratio was taken into account for elastic strains. The coefficient of thermal expansion was used for achieving thermal strains. For plastic and phase transformation induced strains, temperature dependent stain hardening and change in volume viz., phase transformation properties were considered in the material database, respectively. A dip in the residual stress distribution across the fusion zone indicates volume relaxation of the weld metal due to phase transformation effect as shown in the contours (Fig. 21.7a, b) of longitudinal residual stress in 2.25Cr–1Mo weld joints.

Figure 21.7 shows the comparison of measured and predicted longitudinal welding residual stress profiles at the mid-cross-section of the weld joint. Measured peak tensile residual stress is 592 MPa while the predicted value was 568 MPa. The weld-induced residual stress has high magnitude at heat affected zone (HAZ) and at the weld center; there was a relaxation in the stress value. This is mainly due to localized heating and cooling, the only a small area near the joint is affected by the

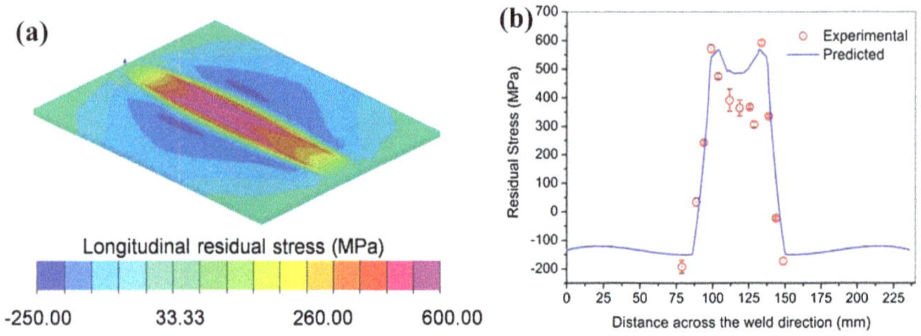

Fig. 21.7 Longitudinal residual stress **a** contour image **b** validation

intense heat source which leads to the steep thermal gradient. Hence, the contraction of the molten metal pool is hindered by the cooler region. Therefore, usually in a weld metal peak tensile stress near the interface (heat affected zone) can be found (Fig. 21.7b). Further, volume change during phase transformation relieves the built-up stress and eventually leads to the stress relaxation at the weld center. The predicted and measured tensile residual stresses near the vicinity of the joint are of the order of yield strength of the material. Thus, the excess generation of stress will drive the deformation in the structure, called distortion (Fig. 21.8a, b).

The formation of stress and eventually the residual stress during heating and cooling cycles at a spot can be understood from Fig. 21.9a, b. As the heating starts, compressive stress increases to a peak value and the stress is limited to materials yield strength. Further, compressive stress starts to decrease as the metal starts to soften. Eventually drops down to zero at the melting point of the material. During cooling, tensile stress starts to build slowly due to thermal strains and at the bainitic start temperature (Bs (T)) there was small stress relieving effect due to the phase transformation from austenite to bainite and also due to change in yield strength. For 2.25Cr–1Mo steel, the Bs (T) varies from 490 to 520 °C. Also, it can be predicted by using the below equation [7].

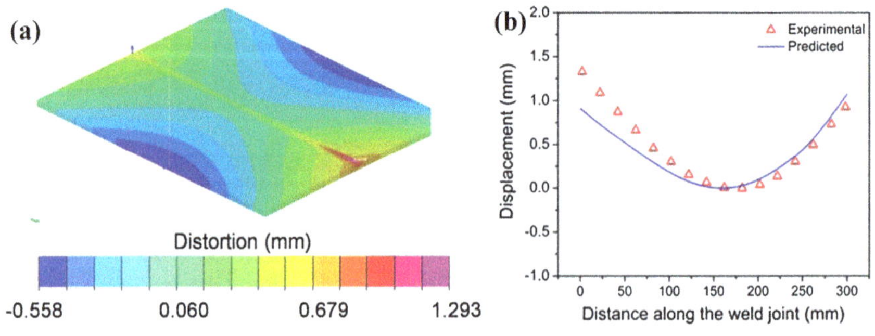

Fig. 21.8 Distortion **a** contour image **b** validation

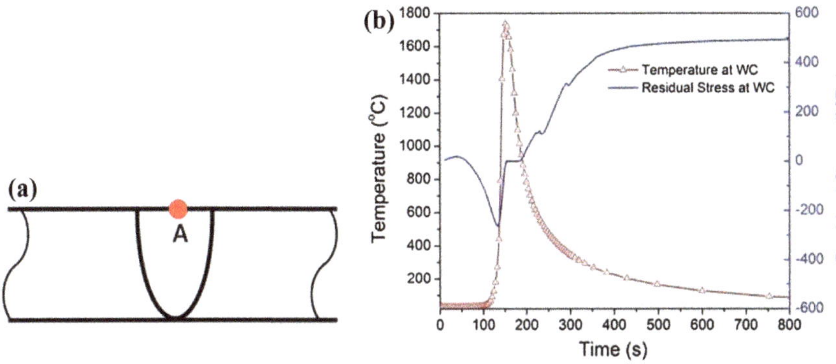

Fig. 21.9 **a** Schematic sketch showing the location of a spot for predicting stress evolution. **b** Evolution of stress during welding thermal cycle at a spot

$$Bs(^\circ C) = 630 - 45\,Mn - 40\,V - 35\,Si - 30\,Cr - 25\,Mo - 20\,Ni - 15\,W. \tag{21.14}$$

Meanwhile, cooling beyond bainitic start temperature causes the generation of very high tensile stress because of the dominating effect of thermal strains over volume strain. Eventually, the tensile stress reaches the maximum value of the order of yield strength of base/weld metal.

21.5 Conclusions

Thermo-Mechanical analysis of A-TIG welding of 2.25Cr–1Mo steel has been successfully carried out by the numerical model. The predictions of the numerical model are validated by experimental tools. The following conclusions are drawn from this study:

1. The peak temperature at 8 mm distance was measured as 1371 °C using thermocouples and the thermal cycle exhibited a sharp temperature gradient. Numerical prediction showed 1377 °C at the same spot and therefore, there was good agreement between the predicted and measured temperatures.
2. The predicted and the measured peak tensile residual stress were found to be 592 and 568 MPa respectively near the heat affected region. The welding residual stress profile across the weld joint displayed a typical "M" shape curve which implied that austenite to bainitic phase transformation had an influence on the final residual stress.

3. The observed microstructure at room temperature showed typical bainitic microstructure. Also, the model successfully predicted the distortion with good accuracy. The predicted thermal cycle, residual stress and distortion showed good agreement with the experimentally measured values.

References

1. Mochizuki, M.: Control of welding residual stress for ensuring integrity against fatigue and stress–corrosion cracking. Nucl. Eng. Des. **237**(2), 107–123 (2007)
2. Vasantharaja, P., Maduarimuthu, V., Vasudevan, M., Palanichamy, P.: Assessment of residual stresses and distortion in stainless steel weld joints. Mater. Manuf. Processes **27**(12), 1376–1381 (2012)
3. Na, H., Lee, S., Kang, C.: Effect of micro-segregation on impact toughness of 2.25 Cr-1Mo steel after post weld heat treatment. Metals **8**(6), 373 (2018)
4. Trautwein, A., Mayer, H., Gysel, W., Walser, B.: Structure and mechanical properties of 2¼Cr-1Mo cast steel for pressure components with wall thicknesses up to 500 mm. In: Application of 2¼Cr-1 Mo Steel for Thick-Wall Pressure Vessels, ASTM International (1982)
5. Gojic, M., Kosec, L., Matkovic, P.: The effect of tempering temperature on mechanical properties and microstructure of low alloy Cr and CrMo steel. J. Mater. Sci. **33**(2), 395–403 (1998)
6. Wada, T., Eldis, G.: Transformation characteristics of 2¼Cr-1Mo steel. In: Application of 2¼Cr-1 Mo Steel for Thick-Wall Pressure Vessels, ASTM International (1982)
7. Arivazhagan, B., Vasudevan, M.: Studies on A-TIG welding of 2.25 Cr-1Mo (P22) steel. J. Manuf. Processes **18**, 55–59 (2015)
8. Mills, K., Keene, B., Brooks, R., Shirali, A.: Marangoni effects in welding. Philos. Trans. R. Soc. Lond., Ser. A Math. Phys. Eng. Sci. 911–926 (1998)
9. Friedman, E.: Thermomechanical analysis of the welding process using the finite element method. J. Pressure Vessel Technol. **97**(3), 206–213 (1975)
10. Lindgren, L.-E.: Numerical modelling of welding. Comput. Methods Appl. Mech. Eng. **195** (48–49), 6710–6736 (2006)
11. Hyde, T., Yaghi, A., Tanner, D., Bennett, C., Becker, A., Williams, E., Sun, W.: Current capabilities of the thermo-mechanical modelling of welding processes. J. Multiscale Model. **1**, 451–478 (2009)
12. Deshpande, A., Xu, L., Sun, W., McCartney, D., Hyde, T.: Finite-element-based parametric study on welding-induced distortion of TIG-welded stainless steel 304 sheets. J. Strain Anal. Eng. Des. **46**(4), 267–279 (2011)
13. Lindgren, L.-E.: Computational Welding Mechanics. Elsevier (2014)
14. Leblond, J., Devaux, J.: A new kinetic model for anisothermal metallurgical transformations in steels including effect of austenite grain size. Acta Metall. **32**(1), 137–146 (1984)
15. Heinze, C., Pittner, A., Rethmeier, M., Babu, S.: Dependency of martensite start temperature on prior austenite grain size and its influence on welding-induced residual stresses. Comput. Mater. Sci. **69**, 251–260 (2013)
16. Deng, D., Murakawa, H.: Finite element analysis of temperature field, microstructure and residual stress in multi-pass butt-welded 2.25 Cr–1Mo steel pipes. Comput. Mater. Sci. **43**(4), 681–695 (2008)

17. Vakili-Tahami, F., Daei-Sorkhabi, A., Saeimi-S, M., Homayounfar, A.: 3D finite element analysis of the residual stresses in butt-welded plates with modeling of the electrode-movement. J. Zhejiang Univ. Sci. A **10**(1), 37–43 (2009)
18. Deng, D., Tong, Y., Ma, N., Murakawa, H.: Prediction of the residual welding stress in 2.25 Cr-1Mo steel by taking into account the effect of the solid-state phase transformations. Acta Metall. Sin. (English Letters) **26**(3), 333–339 (2013)
19. Goldak, J., Chakravarti, A., Bibby, M.: A new finite element model for welding heat sources. Metall. Trans. B **15**(2), 299–305 (1984)

Chapter 22
Finite Element Modeling of Hybrid Laser-TIG Welding of Type 316L(N) Stainless Steel

M. Ragavendran, M. Vasudevan and M. Menaka

Abstract In the present study, thermomechanical analysis of autogenous hybrid laser-TIG welding of type 316L(N) austenitic stainless steel has been carried out. It has been realized that the residual stress due to welding plays a vital role in the mechanical properties of the joint and its performance during service. Therefore, it is important to study the residual stress of type 316L(N) stainless steel weld joint produced by hybrid welding process. First, the thermal analysis was carried using a hybrid heat source consisting of conical and double ellipsoidal models. The heat source was calibrated by comparing the simulated weld bead profile with that of the experimentally obtained weld bead profile. Then, the obtained temperature distribution was sequentially coupled to the mechanical analysis. The simulated thermal cycle was validated by temperature measurements using noncontact real-time IR thermography. The simulated residual profile was validated by an ultrasonic technique employing critically refracted longitudinal waves. The ferrite content of the weld metal was measured using ferrite scope. Distortion in the weld joint is measured using digital vertical height gauge. Optical microscopy is employed for microstructural characterization of the weld joint. The weld metal exhibited a peak tensile residual stress value of 320 MPa. The predicted distortion values were very low. There was a good agreement found between the predicted and experimentally measured thermal cycles, residual stresses, and distortion.

Keywords Hybrid laser-TIG welding · 316(L)N stainless steels · Thermomechanical analysis

M. Ragavendran (✉)
Homi Bhabha National Institute, Indira Gandhi Centre for Atomic Research, Kalpakkam, Tamil Nadu, India
e-mail: mragavendran@igcar.gov.in

M. Vasudevan
Metallurgy and Materials Group, Indira Gandhi Centre for Atomic Research, Kalpakkam, Tamil Nadu, India

M. Menaka
Health Safety and Environmental Group, Indira Gandhi Centre for Atomic Research, Kalpakkam, Tamil Nadu, India

R. G. Narayanan et al. (eds.), *Advances in Computational Methods in Manufacturing*, Lecture Notes on Multidisciplinary Industrial Engineering,
https://doi.org/10.1007/978-981-32-9072-3_22

22.1 Introduction

Type 316L(N) austenitic stainless steel is the main structural material used in the construction of nuclear reactors. It is the preferred nuclear grade structural material for reactor due to its good mechanical properties, corrosion, and irradiation resistance [1]. Welding is the predominant fabrication method employed during their construction. Therefore, the utilization of an advanced welding process for welding of type 316L(N) stainless steel is of high importance. Laser welding has high-energy density and the advantages are lower residual stress and distortion that are due to low heat input, high welding speeds, and ease of automation. But for welding high-thickness plates using a laser, keyhole instability occurs which produces a lot of porosity in the welds. Arc welding is most widely used, but it has a lot of limitations namely high residual stress and distortion due to high heat input. Slow welding speed and need of groove preparation limit the productivity.

For welding, heavier components, an advanced welding process such as hybrid laser-TIG welding process could be used. In our earlier study, hybrid laser-TIG welding process parameters were optimized for welding of type 316L(N) austenitic stainless steel using RSM and GA [2, 3]. The advantages of hybrid welding are good gap bridging ability and enhanced productivity; high penetration capability with faster welding speeds due to more energy delivered into the material [4]. The process improves the weld quality with reduction of porosity and cracks due to slow cooling rate and enhances ductility [5]. The accurate measurement of temperature during hybrid welding is difficult using conventional thermocouples. It cannot measure the weld pool temperature; it will measure only the nearby base metal temperature. Therefore, in this study, these limitations will be overcome by using noncontact real-time IR thermography for precise measurement of temperature. It is also employed for authenticating, fine-tuning, and validating the FEM model.

The novelty in this investigation was the use of IR thermography to measure the temperature distribution in the weld pool and its surroundings to validate the FEM predictions. Further, the thermal model was sequentially coupled to the mechanical analysis for determination of residual stress. The predictions of the numerical analysis presented in this investigation on the temperature distribution, residual stress, and distortion were confirmed by measurements by the use of an experimental tool.

22.2 Experimental

The 316L(N) plates of size 300 mm × 125 mm × 5.6 mm were used in straight square butt joint configuration. The setup is given in Fig. 22.1. The laser-TIG hybrid welding was carried out with the determined optimal parameters with a heat input of 0.466 kJ/mm. Weld joint was made autogenously. After the weld joint was fabricated, the residual stress studies were made by utilizing an ultrasonic technique

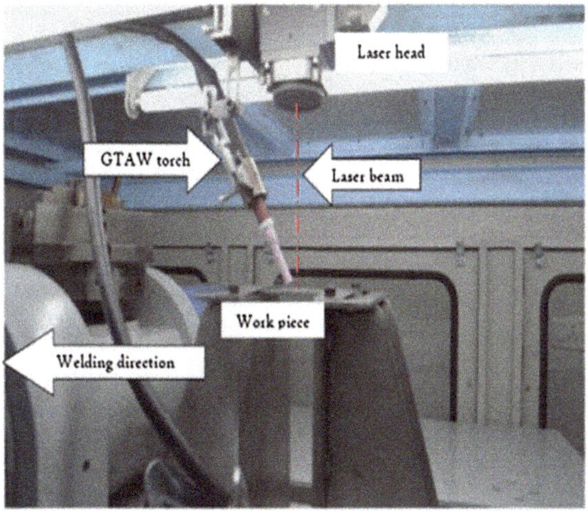

Fig. 22.1 Experimental setup

Table 22.1 Laser welding heat source specification

TRUMPF laser welding specification	
Max average power	530 W
Pulse power	0.5–10 kW
Pulse duration	0.3–50 ms
Max pulse energy	60–100 J
Max rep rate	0.1–833 Hz
λ of laser light	1064 nm

using L_{CR} waves. The weld joint was cross-sectioned, polished, and etched electrolytically using equal amounts of concentrated nitric acid and distilled water. The temperature was measured in real time using IR thermography. Distortion was measured using digital height gauge. Tables 22.1 and 22.2 show the heat source

Table 22.2 Welding parameters

Welding parameters	
Leading heat source	TIG
Welding speed	20 cm/min
Defocusing	−2 mm
Gap between heat sources	3 mm
Torch angle	45°
Arc gap	1 mm
Shielding gas/flow rate	Argon, 10 L/min
Power	3750 W
Pulse frequency	12 Hz
Pulse duration	10 ms
TIG current	120 A

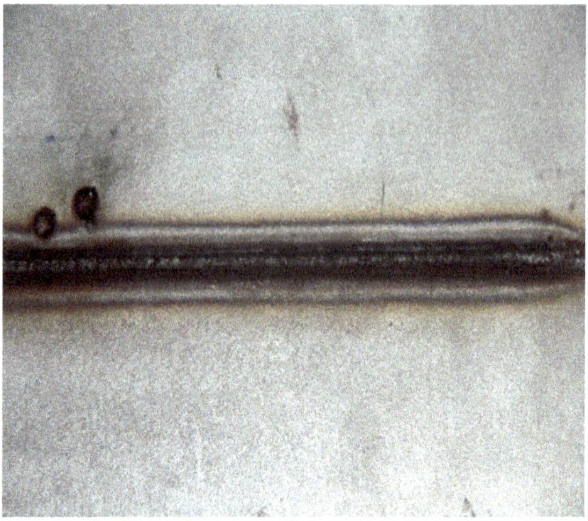

Fig. 22.2 Weld joint fabricated by hybrid laser-TIG welding process

details and welding parameters used in hybrid laser welding. Figure 22.2 shows the joint fabricated by the hybrid welding process.

22.3 Numerical Modeling

The thermal analysis of hybrid laser-TIG welding simulation was executed using commercially available SYSWELD software which is exclusively used for the modeling and simulation of fusion welding processes. The hybrid heat source model consisting of double ellipsoid for the arc welding and three-dimensional conical for the laser welding is used in the analysis. The geometrical parameters of the heat source models are taken from the experimental weld bead. Figure 22.3 schematically represents the combined heat source model employed in this study.

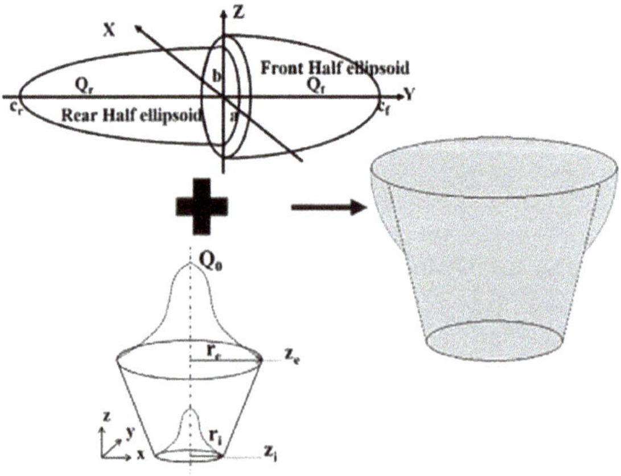

Fig. 22.3 Coupled double ellipsoidal with three-dimensional conical heat source model

22.3.1 Governing Equations

The heat transfer due to confined heating while welding is mainly due to conduction mode. The amount of heat conduction relies on the materials thermophysical properties, the amount of weld metal volume produced, and weld morphology. Heat conduction defined by finding the solution for the following equation.

$$\nabla(k\nabla T) + Q(x, y, z, t) = \rho \mathrm{Cp}(\partial T/\partial t) \tag{22.1}$$

where Q = Heat flux volume (W m^{-3}), k = Thermal conductivity (W m^{-1} °C^{-1}), ρ = Density (kg m^{-3}), Cp = Specific heat capacity (J kg^{-1} °C^{-1}), and T = Transient temperature (°C), t = time (s). The heat loss due to radiation and conduction throughout the welding is given by Stefan–Boltzmann relation and Newton's law of cooling.

22.3.2 Boundary Condition

The hybrid laser-TIG hybrid was conducted using various welding parameters. The initial temperature was 30 °C. All the heat losses namely radiation heat loss which is predominant in the weld pool, convective and conductive heat losses that are prevalent away from the weld pool are taken into account. In the model, convective and radiative losses were considered on all its surfaces except the symmetry plane. Similar to the experimental conditions, the plates to be welded are clamped appropriately to avoid the translational and rotational movements.

22.3.3 Finite Element Modeling

The dimension of the weld plate used in finite element modeling was 300 × 125 × 5.6 mm. The butt joint configuration with a square edge preparation, without a groove, was used for welding. To improve the precision and effectiveness of the simulation, optimum mesh size should be used. The effect of heat input due to welding will be more in the weld center and heat-affected zone (HAZ). So to get the accurate weld bead morphology, finer mesh size was used in these zones and the other regions are given little coarser mesh to reduce the computation time. The heat source moves in *Y*-axis. The model was formulated such as the welding direction was along the *Y*-axis and the model was about the *XY* plane. The thermophysical properties of the type 316L(N) austenitic stainless steel were considered from room temperature to melting point are taken from IGCAR databook [6].

22.4 Results and Discussion

22.4.1 *Temperature Distribution and Weld Bead Comparison*

Temperature distribution in the transverse direction is plotted from the weld centerline. The induced peak temperature is almost the same near the regions next to the weldment. The distance far away from the weldment is not affected by two heat sources. The peak temperature changes if the heat source traverses in its path throughout the welding. All the nodes are heated till its peak temperature because the heat source strikes a corresponding node and it begins to cool when it travels further. The profiles are obtained in the quasi-state condition. The weld bead produced by the hybrid welding process is perfectly matching with the predicted finite element model using SYSWELD. If the predicted weld bead geometry is matching well with experiments, then the results given by the predicted model are reliable and almost accurate. Figure 22.4 shows the weld macrograph which is compared with the FEM model.

The thermal heat source profiles of IR thermography are compared with FEM model. Temperature profile matches well in the quasi-state condition. There was a good agreement found within the measurements and the predicted temperature using the model. The peak temperature measured also almost found equal to that of the predicted by the model. The variations seen may be due to assumptions during the material modeling beyond the melting point of the material and experimental constraints. These deviations are reduced by improving the numerical modeling analysis. Figure 22.5 shows the IR temperature distribution profile which is compared with that of the FEM model. Figure 22.6a presents the numerical modeling results, in which the induced peak temperature due to welding which keeps on

Fig. 22.4 Weld macrograph compared with FEM model

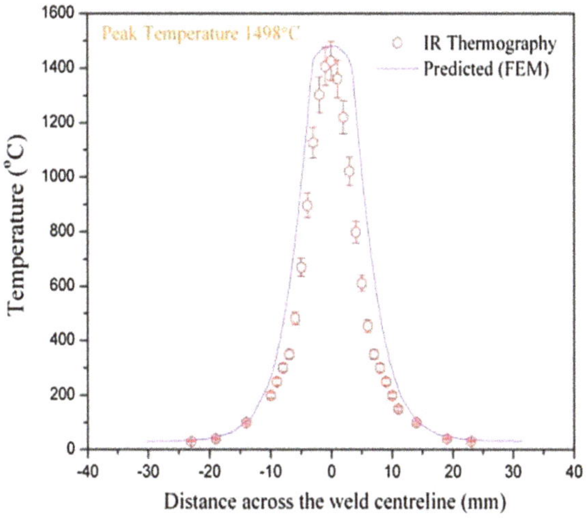

Fig. 22.5 Temperature measured by IR thermography compared with FEM at 50 s

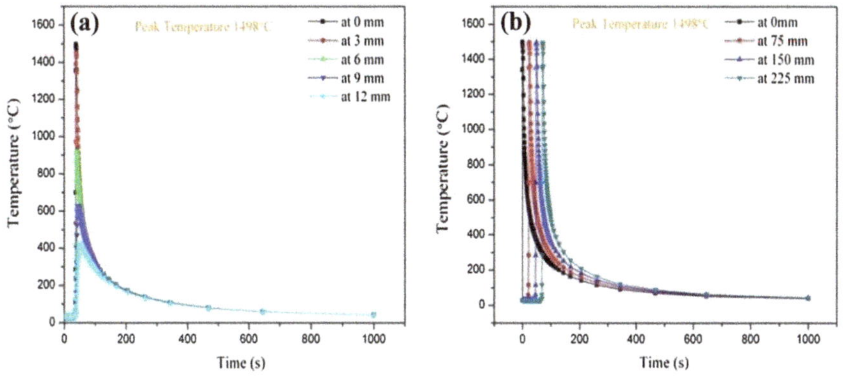

Fig. 22.6 Thermal cycle from welding centerline: **a** transverse direction **b** longitudinal direction

decreasing in the transverse direction. Thermal cycle along the weld center line for various locations is shown in Fig. 22.6b.

22.4.2 Residual Stress Analysis

The residual stress measurement was conducted by employing ultrasonic testing. The predicted residual stress profiles were confirmed by longitudinal residual stress measurements taken at the center of the weld joints. The residual stress values evaluated are compared with the model and it is presented in Fig. 22.7a.

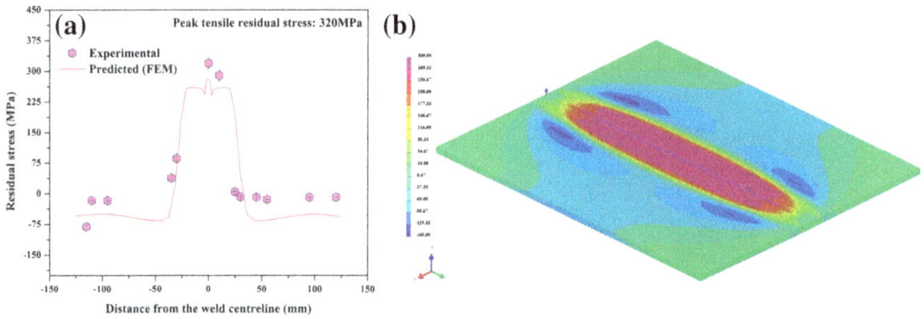

Fig. 22.7 Residual stress **a** comparison of FEM versus ultrasonic testing **b** distribution in welded plate

Figure 22.7b shows the residual stress distribution in the welded plate. The peak tensile stress is at the weld center and the value is around 280 MPa as predicted by the model. The maximum tensile stress measured by the ultrasonic technique is 320 MPa. The flat region observed next to the peak tensile residual stress is due to the autogenous hybrid heat sources. That is, if the filler wire is used, heat input will be utilized for melting the filler wire. But in the case of autogenous welding, the heat input is spread over a more area next to weld metal. Moreover, the type 316L (N) austenitic stainless steel has low thermal conductivity, so it retains high temperature over a more extended period due to which the residual stress profile is flat near to the weldment. There was a good agreement found within the predicted model and the experimentally calculated residual stress.

The evolution of stress and finally the residual stress during heating and cooling cycles due to welding can be explained by spot analysis given in Fig. 22.8a, b as a function of time. When welding starts, compressive stress rises to a peak value and the stress is restricted to materials yield strength. Then, the compressive stress begins to decline as the metal commences softening. Finally, stress falls to zero at the melting point of the base metal. During cooling, the tensile stress rises to

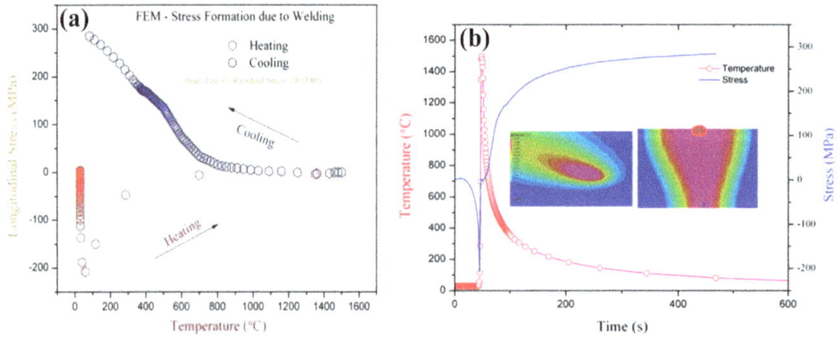

Fig. 22.8 **a** Stress evolution during welding—spot analysis **b** stress evolution as a function of time

3. Chandrasekhar, N., Ragavendran, M., Ravikumar, R., Vasudevan, M., Murugan, S.: Optimization of hybrid laser–TIG welding of 316LN stainless steel using genetic algorithm. Mater. Manuf. Processes **32**(10), 1094–1100 (2017)
4. Bagger, C., Olsen, F.O.: Review of laser hybrid welding. J. Laser Appl. **17**(2), 2–14 (2005)
5. Ribic, B., Palmer, T.A., DebRoy, T.: Problems and issues in laser-arc hybrid welding. Int. Mater. Rev. **54**(4), 223–244 (2009)
6. Data Book (1999) Material Properties for Design. IGCAR
7. Suutala, N.: Effect of solidification conditions on the solidification mode in austenitic stainless steels. Metall. Trans. A **14**(1), 191–197 (1983)
8. Gao, M., Zeng, X.Y., Hu, Q.W.: Effects of welding parameters on melting energy of CO_2 laser–GMA hybrid welding. Sci. Technol. Weld. Joining **11**(5), 517–522 (2006)

Chapter 23
Numerical Study on Electromagnetic Crimping of Aluminum Tube and Steel Profiled Rod

Getu Tilahun Areda and Sachin Dnyandeo Kore

Abstract Electromagnetic crimping (EMC) is one of the advanced manufacturing technologies used to join lightweight material including metals with metallurgical and thermal property differences. This paper investigates EMC process numerically to deform aluminum tube (Al6061-T6) to the steel profiled rod using the electromagnetic (EM) module of LS-DYNA^{TM}. Electromagnetic crimping methods can join multiple material systems without melting. A longitudinal groove was designed on steel rod for mechanical interlocking to resist the torsional load. Groove geometrical parameters like groove length, depth, edge radius, and width were kept constant. Effects of voltage on an effective plastic strain, resultant impact velocity, resultant displacement, and temperature were studied in detail. Moreover, the simulation was carried out also to analyze the effect of capacitance on process parameters, and results are discussed in detail. The developed model is validated and found in a good agreement with experiment. Based on the results found, it is possible to predict essential process parameters for enhanced joint strength.

Keywords Electromagnetic crimping (EMC) · Numerical modeling · Groove geometrical parameters · Maxwell's equation · LS-DYNA^{TM} · Johnson-Cook

23.1 Introduction

A new era of a manufacturing process is looking for developing advanced techniques for joining lightweight materials that can overcome problems associated with the conventional approach. Joining methods like fusion welding has several

G. T. Areda (✉)
Mechanical Engineering Department, Indian Institute of Technology Guwahati, Guwahati 781039, Assam, India
e-mail: a.getu@iitg.ac.in

S. D. Kore
Mechanical Engineering Department, Indian Institute of Technology Goa, Farmagudi 403401, Goa, India

R. G. Narayanan et al. (eds.), *Advances in Computational Methods in Manufacturing*, Lecture Notes on Multidisciplinary Industrial Engineering,
https://doi.org/10.1007/978-981-32-9072-3_23

limitations like the creation of intermetallic compound, crack, voids, etc. that needs to be addressed. An advanced technological solution like solid-state joining becomes best options to overcome the existed limitations. Joining metals which have high aspect ratio for lightweight applications requires a method which can provide enhanced joint strength without being affected by thermal property differences between the mating parts. EMC process is a punchless, an environmentally friendly, green, and able to complete the work within a microsecond. This method becomes best suited to resolve defects associated with a thermal-based joining technique like arc welding. Automotive and space frame industries are exceedingly demanding the advantage of this process.

Thermal, mechanical, and electromagnetic are phenomenon's which needed to be addressed using numerical analysis in case of an electromagnetic forming (EMF) process. LSTC^{TM} has introduced an electromagnetic module to simulate the EMF process [1]. In the electromagnetic module of LS-DYNA^{TM}, magnetic field, Lorentz force, current in the conductor, etc. are calculated. The mechanical solver deals with the structural analysis of the deformation. The electromagnetic module solves Maxwell's equations. The thermal solver calculates internal energy, generation of heat and its flow. As the process involves high strain rates, it has all advantages of high velocity forming like improved FLD, minimized spring back, and less wrinkling [2]. Metal forming is one of the LS-DYNA^{TM} main applications with capabilities that allow one to simulate conventional and high-speed forming process [3].

In this chapter, the electromagnetic crimping of an aluminum tube on profiled steel rod was modeled and simulated with 3D finite element method. Simulations are carried out using LS-DYNA^{TM}. Effects of voltage and capacitance on process parameters of electromagnetic crimping were studied. Resultant velocity, resultant displacement, effective plastic strain, and temperature are predicted, and the effect of high voltage on deformation was analyzed using the EM module of LS-DYNA^{TM} software. The compression of the aluminum tube on profiled carbon steel rod has been predicted with 3D finite element simulation at various voltage and capacitance levels.

Capacitor bank in an electromagnetic forming machine is used to store the required amount of energy that will be discharged into the coil, while the high current switch is closed. Induced current starts to flow on the surface of the tube which is caused by a current in the coil. The magnetic field generated due to primary and induced current results radial force known as the Lorentz force used to launch the workpiece away from the coil. The impact that produced due to high speed will make the workpieces to deform beyond its elastic limit for different EMF applications like joining, cutting, welding, and shearing. The total duration of the process takes no longer than 100 microseconds. Electromagnetic crimping process has several applications such that medical (pharmaceutical glass bottles), automotive, space frame structures, wire crimping, double seaming, hemming, and clinching [4, 5]. Capacitors, resistors, and inductors are the main components of the EMF system as shown in Fig. 23.1. The electrical properties of the considered workpieces are exceedingly significant in increasing or decreasing deformation

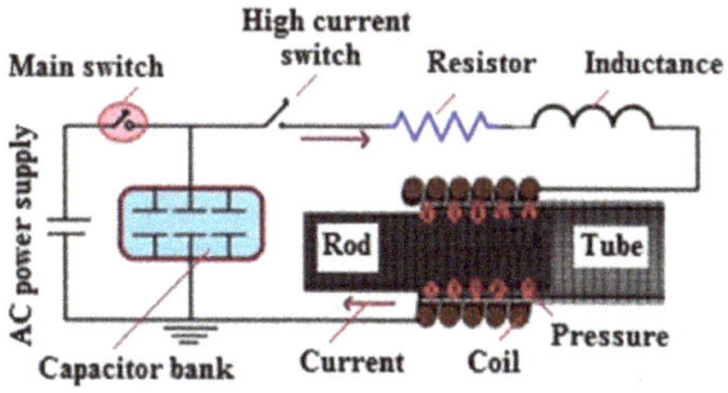

Fig. 23.1 Circuit diagram of EMC process

levels. Direct application of electromagnetic energy can deform metals like magnesium and aluminum which has low formability in the conventional forming process. Applying magnetic energy directly to deform lower electrically conductive metals like stainless steel is too difficult.

23.2 Governing Equations of EMF Process

The required electrical energy which is stored in the bank of the capacitor is given by Eq. 23.1, and the resulting current *I*(*t*) by Eq. 23.2.

$$E_c(t) = \frac{1}{2}CV^2(t) \tag{23.1}$$

$$I(t) = V_0 \sin(wt)\mathrm{e}^{-\beta t} \tag{23.2}$$

where *V*(*t*) voltage across the capacitor, *C* capacitance, ω frequency, β damping coefficient, and *t* time. Set of Maxwell's equations form the basis of numerical computation in any FEM software. The generation of magnetic pressure due to the interaction of two opposite magnetic field is computed using Maxwell's equations. The resultant Lorentz force which is a function of current density and magnetic field density also calculated in the same approach. Differential forms of Maxwell's equations (Eqs. 23.3–23.6) are shown below.

$$\nabla \times B = \mu_0(J + \varepsilon_0 \partial E/\partial t) \tag{23.3}$$

$$\nabla \times E = -\partial B/\partial t \tag{23.4}$$

$$\nabla \cdot E = \rho/\varepsilon_0 \tag{23.5}$$

$$\nabla \cdot B = 0 \tag{23.6}$$

where *E* electric flux intensity, *B* magnetic field density, μ_0 magnetic permeability, *J* current density, ρ net charge density, ε_0 permittivity of vacuum, and *t* is for time.

23.3 Geometrical Dimensions of Workpiece and Material Properties

Geometrical dimensions tube, copper coil, and steel profiled rod are shown in Fig. 23.2a–c. Total of four longitudinal grooves was designed, and its geometrical parameters kept constant for all simulation. Axisymmetric coil with dimension and shape as shown in Fig. 23.2b was used.

Table 23.1 shows the properties of the aluminum tube, copper oil, and steel rod used in this study.

23.4 3D Finite Element Modeling

Simulations were carried out for different voltages levels using identical coil geometry which is modeled as a rigid body to make it simple for calculation in the boundary element method. Hence, deformation was analyzed only on the tube at the joining zone. The ranges of voltage were varied from 1.5 to 4 kV with an increasing rate of 0.5 kV. The arrangement of workpieces and tool coil for simulation is shown in Fig. 23.3. The model parts have meshed with 3290, 6580, and 2394 number of elements for the coil, rod, and tube, respectively.

Electromagnetic crimping is a high-speed joining process using a magnetic pulse, and the material model which take care of strain rate hardening and softening due to temperature rise is Johnson cook (JC). In order to evaluate the constitutive response of the tube, the flow stress which is determined as a function of the plastic

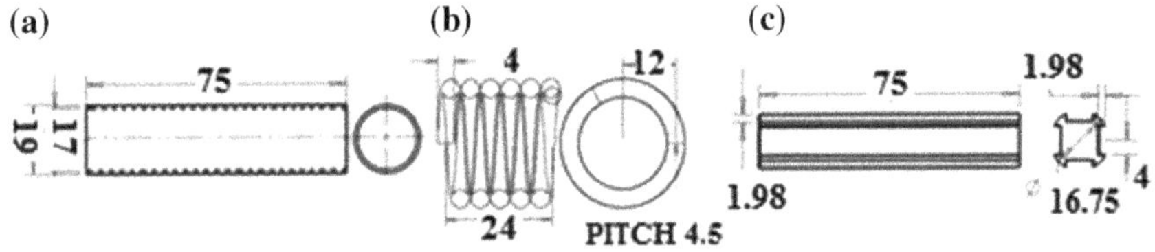

Fig. 23.2 Geometrical dimension of **a** tube **b** coil, and **c** rod (dimensions are all in mm)

Table 23.1 Material properties of tube, rod, and coil used in the simulation [6, 7]

Material	ρ (kg/m^3)	G (GPa)	E (GPa)	γ	C (J/kg K)	K (W/m k)	σ (MS/m)	T_m (K)
Al6061-T6	2700	26	68.9	0.31	896	167	25	925
Steel	7850	–	207	0.29	–	–	–	–
Cu1001	8960	–	124	0.27	–	–	59.6	–

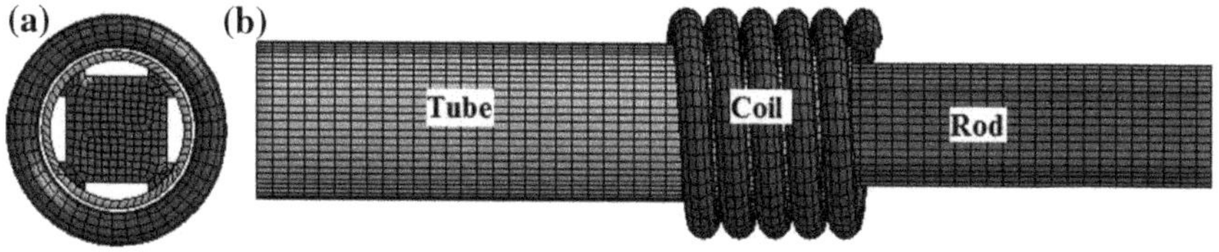

Fig. 23.3 Arrangement of rod, coil, and tube **a** side and **b** front view

Table 23.2 Parameters of JC strength model for Al6061-T6 and steel [9, 10]

Materials	A (MPa)	B (MPa)	C	n	m	T_m (K)	$\dot{\varepsilon}^*$ (s^{-1})
Al6061-T6	324	114	0.002	0.42	1.34	925	1
4340 Steel	729.2	5.095	0.014	0.26	–	1793	–

Table 23.3 Parameters for linear polynomial EOS for tube [10]

C_0	C_1	C_2	C_3	C_4	C_5	C_6	E_0	V_0
0	74.2 GPa	60.5 GPa	36.5 GPa	1.96	0	0	0	1

strain, strain rate, and temperature is related in JC constitutive equation as discussed by Schwer [8].

$$\sigma_y = \left(A + B\,\bar{\varepsilon}^{p^n}\right)\left(1 + c\ln\dot{\varepsilon}^*\right)\left(1 - \left(\frac{T - T_R}{T_m - T_R}\right)^m\right) \tag{23.7}$$

where A is an initial yield strength, B and C are material constant, $\bar{\varepsilon}^{p^n}$ reference plastic strain, $\dot{\varepsilon}^*$ equivalent plastic strain rate of the workpiece, n strain hardening index, m thermal softening index, T_R room temperature, and T_m melting temperature of the workpiece, T operating temperature.

Parameters of JC strength model for Al6061-T6 and steel and parameters for linear polynomial EOS for Al6061-T6 are shown in Tables 23.2 and 23.3.

23.5 Results and Discussion

Numerical results are discussed and analyzed in detail as follows. Effects of voltage and capacitance on other process parameters were explained based on the selected element in the joining zone. Moreover, the transient magnetic field is determined for different values of capacitance.

23.5.1 Effect of Voltage on Effective Plastic Strain

Effective plastic strain result at different voltage levels is shown in Fig. 23.4. Maximum of 1.038 and minimum of 0.0127 effective plastic strain were obtained using 4.0 and 1.5 kV for an element (H-24494) which is found in the deformation zone. These results were obtained from the element found at maximum deformation which is parallel to the groove surface. There were rebounding effects when we considered voltage more than 4 kV that leads to having minimum radial displacement and effective plastic strain.

23.5.2 Effect of Voltage on Resultant Displacement and Velocity

Figure 23.5a shows the resultant displacement of an element (H-24494) at different voltage levels. Keep increasing voltage levels causes the resultant displacement to reach its maximum due to increasing discharge energies. Increasing voltage to 4.0 kV leads an element to displace by 2.29 and 0.085 mm was found while using 1.5 kV voltage. Resultant displacement reaches a peak at a different time for each voltage level but almost equal to for more than 2.5 kV with small variations. If we go beyond 4.0 kV, radial displacements will significantly decrease due to rebounding effect. Measuring parameters like a radial or resultant displacement during an experiment is a challenging task due to the involving speed, but it is possible to predict using 3D simulation as shown in Fig. 23.5a. Besides that, resultant velocity is one of the critical process parameters needs to be determined by the electromagnetic crimping process. Increasing voltage from 1.5 to 4.0 kV causes the velocity to rise from 15.075 to 438.44 m/s. The percentage of increment of

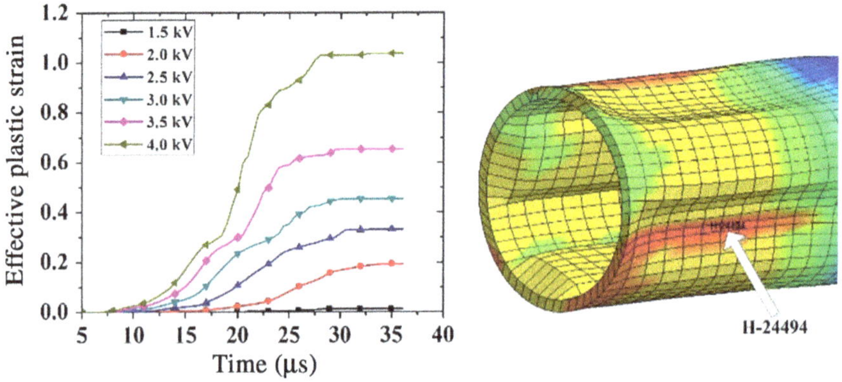

Fig. 23.4 Effect of voltage on effective plastic strain

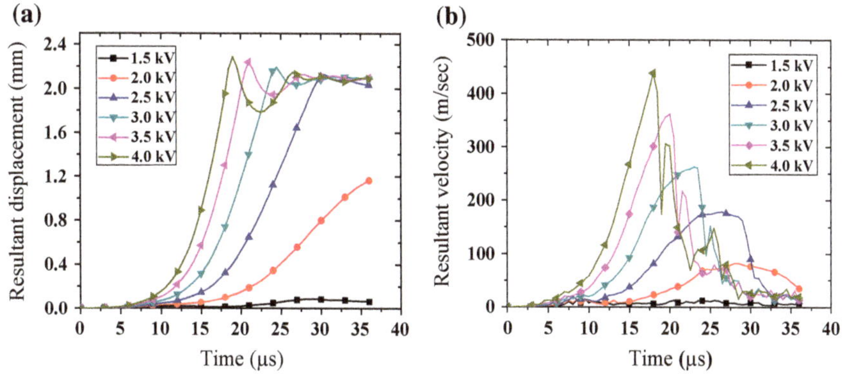

Fig. 23.5 Effect of voltage on **a** resultant displacement and **b** resultant velocity

velocity compared to variation in voltage shows some difference, but it has direct influence as shown in Fig. 23.5b. Minimum velocity of 15.075 m/s was found while applying a voltage of 1.5 kV.

23.5.3 Effect of Voltage on Temperature and Current

As a solid-state process, increments of temperature up to a melting point of any of the metal under consideration were not expected. The rising of temperature in electromagnetic crimping process is not sufficient to cause phase change during deformation, but it may soften the joining zone. Maximum of temperature 568.19 K was obtained using 4.0 kV as shown in Fig. 23.6a. The result shows an apparent variation of temperature caused by different voltage level. Minimum increment of 27.15 K from room temperature was found while using 1.5 kV compared to

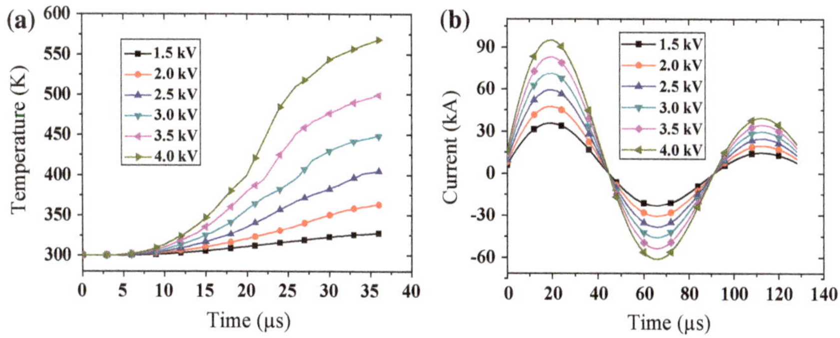

Fig. 23.6 Effect of voltage on **a** temperature and **b** current

different voltage levels. Variations of current at different voltage levels for identical RLC circuit are shown in Fig. 23.6b. The amplitude of the current curve for all voltage levels reaches a maximum at 20 microseconds. Maximum current was achieved while using 4.0 kV that is 95 kA, and a minimum of 35.6 kA was found while using 1.5 kV. The direct proportionality between voltage and current obtains the pattern as shown in Fig. 23.6b for different voltage levels. Current tends to increase as voltage increases, and variation of 0.5 kV has a significant effect on increasing or decreasing its amplitude.

23.5.4 Effect of Capacitance on Process Parameters (V and ε)

Simulation result obtained using 3.0 kV was referred to analyze the effect of capacitance on resultant velocity and effective plastic strain. Increasing 22.68% of capacitance from 426 to 551 μF leads to an increment of a 10.66% resultant velocity. This increment is depicted graphically in Fig. 23.7a. It is required to have a significant difference in capacitance to obtain great variation in resultant velocity. Stored energy will be high if the capacitance is increased and that leads to having an increased effective plastic strain as shown in Fig. 23.7b.

An element (H-24494) on the outside surface of the tube and at the middle of the coil was selected to measure the strain. The effective plastic strain was found uniformly increased as capacitance varies from 426 to 551 μF. Maximum and minimum effective plastic strain of 0.530 and 0.454 were obtained while rising values of capacitance from 426 to 551 μF. It is essential to increase capacitance significantly to attain a meaningful variation in resultant velocity and plastic strain.

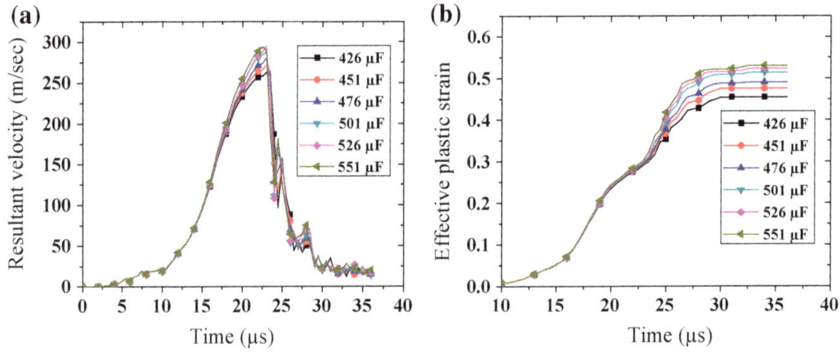

Fig. 23.7 Effect of capacitance on **a** resultant velocity and **b** effective plastic strain

23.5.5 *Effect of Capacitance on Resultant Displacement and Temperature*

Increasing capacitance by 22.6% leads the resultant displacement to increase only by 3.22% as demonstrated in Fig. 23.8a. The proportional variation between capacitance and resultant displacement shows great differences. Beside that Fig. 23.8b shows how temperature varies due to different levels of capacitance where inductance, resistance, and voltage kept constant. A temperature increment of 4.9% was obtained by comparing the maximum and the minimum result found for increasing capacitance by 22.6%. Room temperature was considered as simulation condition which is 300 K.

23.5.6 *Effect of Capacitance on Magnetic Field*

Magnetic field increases as capacitance increases by keeping other circuit parameters constant. The minimum magnetic field of 10.7 T was found using 426 μF, and a maximum of 14.62 T obtained using 551 μF capacitance. Lorentz force which is a function of the magnetic field also increases accordingly. The result revealed that increasing the capacitance by 29.34% leads to increasing the magnetic field by 36.6%. It is vital to add capacitance of the EMF (electromagnetic forming) system based on the required level of deformation, which gives a significant difference in the resulted magnetic field.

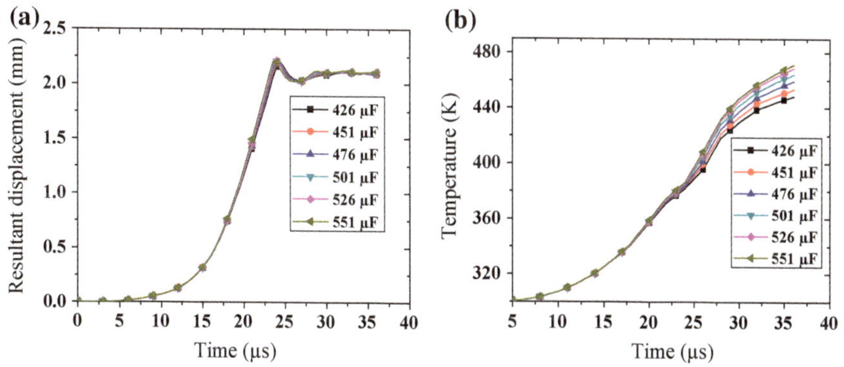

Fig. 23.8 Effect of capacitance on **a** resultant displacement and **b** temperature

23.5.7 Failure Prediction

Figure 23.9 shows von Mises stress distribution on the joining zone at a different time value using 4 kV. The magnitude of stress is found maximum at groove edge that possibly leads to generate necking and thinning which cause joint failure. One of the critical geometrical parameters which can directly affect the joint strength by acting as a cutting edge is the groove edge radius which is 0.5 mm in this study. Moreover, Fig. 23.10 confirms the failure region by showing where shear strain, von Mises stress, and shear stress become maximum.

Results are taken at the end of 36 microseconds for different energy levels. A minimum and maximum von Mises stress of 341.6 and 382.6 MPa were found at the various voltage levels. A better groove filling is observed while using 4 kV that leads to resist higher torque load. Therefore, 4 kV is found optimum voltage for the given material system and geometrical parameters.

23.5.8 Experimental Work

The experiment is carried out using a setup shown in Fig. 23.11 which has a similar geometrical parameter to the FE model. The same geometrical and process parameters are used to conduct the experiment. Discharge energy generated using

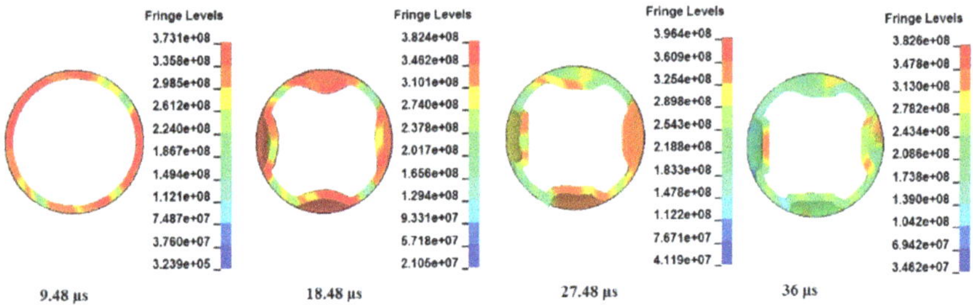

Fig. 23.9 Distribution of von Mises stress on joining zone for 4 kV at the different time value

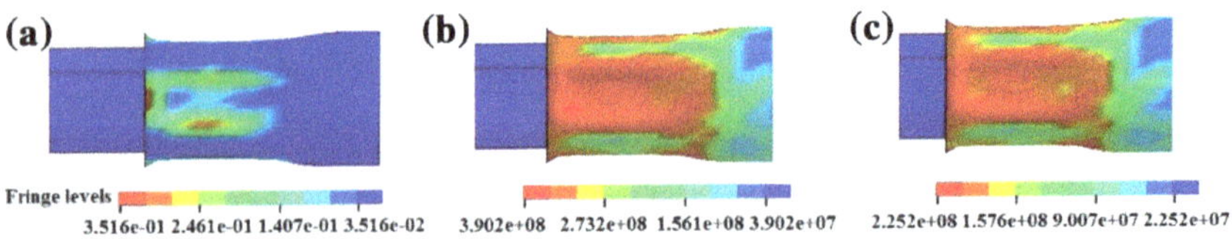

Fig. 23.10 Contours of **a** maximum shear strain **b** von Mises stress, and **c** Tresca (maximum shear stress)

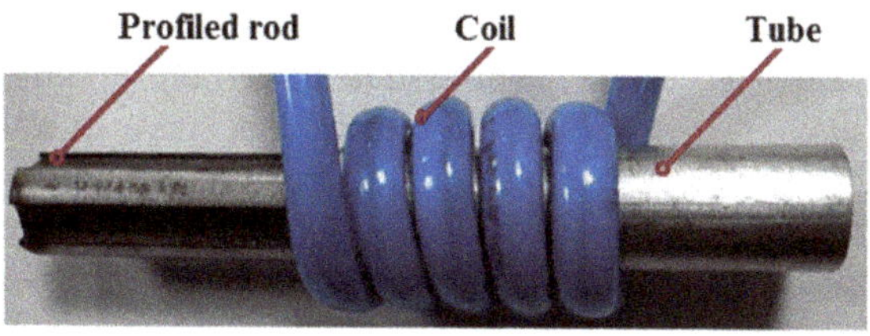

Fig. 23.11 Experimental setup

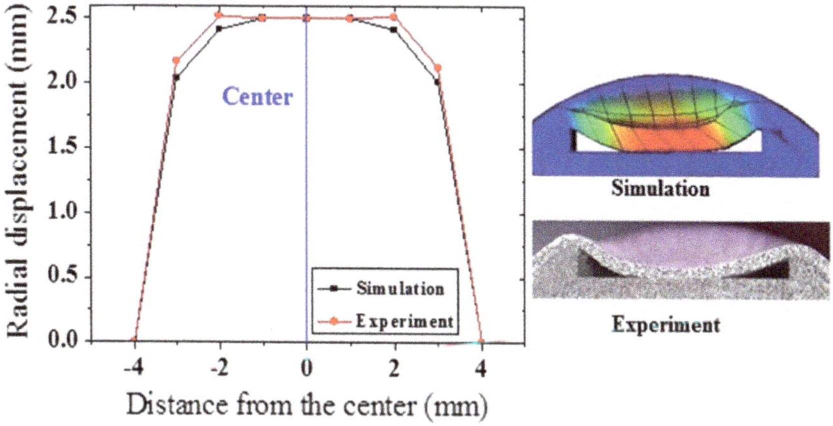

Fig. 23.12 Radial displacement in simulation and experiment

4 kV is applied, and the radial displacement is measured to compare with the simulation.

The radial displacement was measured using cross section of crimped sample and compared with simulation as depicted in Fig. 23.12. Measured gap between the tube and groove surface at the deformation zone was found in good agreement with the simulation. The deformation at the center of the coil is higher that leads to having the maximum radial displacement. The radial displacement was measured with 1 mm gap starting from the center to ends of the groove.

23.6 Conclusions

In this study, numerical modeling and simulations of electromagnetic crimping for transferring torsional load were demonstrated. The aluminum tube gets crimped on the groove of the steel rod successfully, while electromagnetic energy is applied. Process parameters which are difficult to measure and the possible failure region during the experiment were predicted successfully using the EM module of

LS-DYNA^{TM}. The increasing voltage causes the effective plastic strain, radial displacement, resultant velocity, temperature, and output current to increase accordingly. Moreover, the effects of increasing capacitance on resultant velocity, effective plastic strain, radial displacement, temperature, and magnetic field were found less significant. The result revealed that increasing the capacitance by 29.34% leads to increasing the magnetic field by 36.6%. Further, the resultant displacement is getting decreased as voltage increases beyond 4 kV due to re-bouncing effect. Therefore, 4 kV is found optimum voltage for the given material system and geometrical parameters. A mean percentage error which is less than 7% was found while comparing radial displacement obtained from simulation and experiment. The developed model was validated and shows a good agreement with experiment.

References

1. L'Eplattenier, P., Çaldichoury, I.: Update on the electromagnetism module in LS-DYNA. In: 12th LS-DYNA Users Conference, Detroit (2012)
2. LS-DYNA Theory Manual, LSTC
3. L'Eplattenier, P., Cook, G., Ashcraft, C., Burger, M., Shapiro, A., Daehn, G., Seth, M.: Introduction of an electromagnetism module in LS-DYNA for coupled mechanical-thermal-electromagnetic simulations. Steel Res. Int. **80**(5), 38–56 (2009)
4. Psyk, V., Risch, D., Kinsey, B.L., Tekkaya, A.E., Kleiner, M.: Electromagnetic forming—a review. J. Mater. Process. Technol. **211**(5), 787–829 (2011)
5. Zittel, G.: A historical review of high speed metal forming. In: Proceedings of the 4th International Conference on High Speed Forming—ICHSF, pp. 2–15 (2010)
6. Patel, C., Kore, S.D.: Effect of frequency on electromagnetic expansion of thin tubes. J. Mach. Form. Technol. **7**(1/2), 91(2015)
7. Rajak, A., Kore, S.: Electromagnetic hemming of aluminum sheets using FEM. In: ICMMD-2016 Advances in Intelligent Systems Research, pp. 77–82. Atlantis Press (2017)
8. Schwer, L.: Optional strain-rate forms for the Johnson-Cook constitutive model and the role of the parameter. Epsilon_0, LS-DYNA Anwenderforum, LS-DYNA Frankenthal, pp. 1–14 (2007)
9. Johnson, G.R., Cook, W.H.: A constitutive model and data for metals subjected to large strains, high strain rates and high temperatures. In: Proceedings of the 7th International Symposium on Ballistics, vol. 2, pp. 541–547 (1983)
10. Schwer, L.E., Windsor, C.A.: Aluminum plate perforation: a comparative case study using Lagrange with erosion, multi-material ALE, and smooth particle hydrodynamics. In: 7th European LS-DYNA Conference, May (2009)

Chapter 24
Temperature Prediction During Self-pierce Riveting of Sheets by FEA-ANN Hybrid Model

Deepak Mylavarapu, R. Ganesh Narayanan and Manas Das

Abstract Nowadays, the automobile industries are more concerned about reducing automobile weight for improving fuel efficiency and reduced vehicle emission. Hence, there is a need for relatively lighter weight materials like aluminium alloy replacing steel parts for automobiles. However, joining of these lightweight materials is very difficult using conventional spot welding technique, which is a slow process also. In the present study, an alternative high-speed advanced mechanical fastening technique, namely self-pierce riveting is used for joining of these lightweight sheets of different materials. In this process, a semi-tubular rivet is pressed by a punch into two or more sheets which are supported on a die. Due to the special design of die shape, the rivet flares inside bottom sheet to form a mechanical interlock. Experimental measurement of temperature inside the sheets during the process is a difficult task, as the process takes place for a fraction of a second within a span of 10 mm. In the present study, the temperature rise during SPR process is predicted by finite element simulation using Abaqus® for any combination of sheet materials, process conditions and tool dimensions to understand the process. In addition, an ANN model is developed to predict the temperature rise during joint formation in sheets for any combination of process parameters. The ANN model accurately predicted the temperature evolution during the process. Hence, with the hybrid ANN-FEM strategy one can design and optimize the process, tool and material conditions, efficiently, which otherwise will be time and resource-intensive.

Keywords Self-pierce rivet · Temperature · Neural network · Finite element simulation

D. Mylavarapu · R. Ganesh Narayanan (✉) · M. Das
Department of Mechanical Engineering, IIT Guwahati, Guwahati 781039, India
e-mail: ganu@iitg.ac.in

M. Das
e-mail: manasdas@iitg.ac.in

R. G. Narayanan et al. (eds.), *Advances in Computational Methods in Manufacturing*, Lecture Notes on Multidisciplinary Industrial Engineering,
https://doi.org/10.1007/978-981-32-9072-3_24

24.1 Introduction

The overall weight of an automobile governs the fuel efficiency and vehicle emissions. Polmear [1] pointed out that the fuel consumption can be reduced by 5.5% for each 10% reduction in weight. This is possible when lightweight structures and lightweight materials are used in automotive sectors. Use of aluminium alloys in place of steel is one of the possible solutions. Hirsch [2] highlighted that, as part of Super Light Car project, about 53% of the parts contribution is from Aluminium alloy in the form of sheet panels, extrusion in front rails, bumper, crash parts, in the rear underbody rail, and in wheelhouse structure as high-pressure die-cast. The formability of Aluminium sheets is also acceptable and can be improved by materials processing. Joining of lightweight sheets including Aluminium alloys is sometimes difficult with traditional welding processes. In this context, friction stir welding, adhesive bonding, mechanical clinching, laser welding and some hybrid joining methods are used successfully with support from numerical simulations and parametric optimization [3].

Self-pierce riveting (SPR) is a high-speed cold joining method used for joining similar and dissimilar materials including non-metals and metals. In SPR, a rivet is made to deform the sheets arranged in lap configuration above a die with groove. A punch is used to displace the rivet downwards. Depending on the die design, the sheets are mechanically locked. In the process of joining, the upper sheet is pierced and the rivet is flared, while the lower sheet is just deformed.

The stages in SPR along with the schematic of load evolution are shown in Fig. 24.1. During clamping, the rivet is pushed by a punch perpendicular to the sheet surface. In this stage, the rivet clamps the sheets. In piercing stage, the rivet causes severe plastic deformation of the upper sheet resulting in sheet piercing. The lower end of the rivet flares in the third stage and follows the contours of the die. Sheet overlap and interlock distance that are responsible for the required mechanical resistance are achieved during the stage. In the compression stage, the punch continues to push the rivet inside the sheets to be joined. The punch movement stops

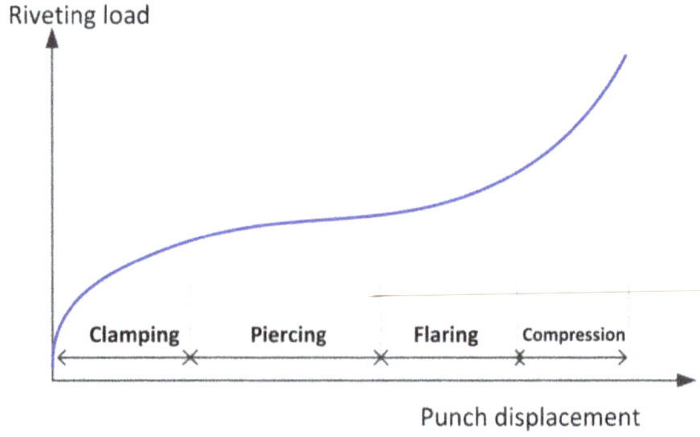

Fig. 24.1 Load evolution in SPR

when it reaches a predetermined load or displacement. The rivet head surface is levelled with the sheet surface at this stage.

Some important contributions in SPR of variety of materials including those involving modelling is presented here. Xu [4] studied the influence of three important factors, namely die type (height of projected region is varied), rivet length and sheet combination on the SPR quality indices like bottom sheet thickness, undercut and rivet spread. The relation between SPR variables and outputs is not unique. The die requirements change with the outputs monitored. The rivet length has negative effect on bottom thickness, but positive effect on undercut and spread. Some of these parameters have interaction effect as well. The experimental analyses by Sun et al. [5] showed superiority of SPR joints (made of aluminium and steel grade sheets) as compared to resistance spot welded joints. The adhesive bonding also improved the static and fatigue behaviour, specifically in lap shear mode. In a similar attempt by Mori et al., it has been observed that the static and fatigue strengths of SPR joints are better than that of clinched joints and resistance spot welded joints. There exists a slight slip at the joint interface in the case of SPR and clinched joints that have improved the fatigue behaviour as compared to resistance spot welding [6]. Johnson et al. [7] attempted to evaluate the joint quality by computer image processing method. With this system, one can identify the number of rivets present in the machine jaw and the material thickness after joining. A comparative analysis of SPR joints made by steel rivet and aluminium rivet by mechanical testing revealed that the rivet material has no influence on the initial stiffness of the mechanical response [8]. But the maximum force and the displacement at maximum force get affected significantly. It is also observed that the SPR joint made by steel rivet can tolerate larger normalized maximum load as compared to the one made by aluminium rivet at varying loading angle.

Several authors have attempted to study the SPR process through numerical simulations. Mori et al. [9] through finite element analyses (FEA) optimized the die shape to produce a successful SPR joint made of multi-layered steel-Al sheets. Porcaro et al. [10] performed numerical simulations using a 2D axisymmetric model in LS-DYNA for joining two sheets using rivet and tools. The effect of mesh size, adaptive time interval, friction and material failure on the joint prediction is studied. Bouchard et al. [11] investigated the SPR process using Forge 2005, an FE code. A Lemaitre-coupled damage model has been applied to predict failure during the process. In the same manner, Casalino et al. [12] implemented fracture conditions through plastic strain threshold during SPR numerical simulations to understand the piercing of rivet leg into the sheets. He et al. [13] through numerical simulations and experiments showed that the SPR joints and hybrid SPR joints (SPR + adhesive bonded) performed better than adhesive bonded joints. The article on recent developments in FE simulations of SPR process presented by He et al. [14] is noteworthy. Mylavarapu et al. [15] predicted critical thinning during SPR through FE simulations. They proposed a strain mapping method to evaluate the thinning of sheets. Later, they used a trained artificial neural network (ANN) model for which FE simulations were conducted to predict the temperature evolution during SPR. The friction coefficient, upper sheet materials properties and lower sheet material properties like strain hardening exponent and strength coefficient are the input parameters for temperature prediction [16]. The present work is an extension of the previous attempt made by the authors.

The main aim of the present work is to predict the temperature evolution during SPR by FEA-ANN hybrid model. For this case, ten different SPR parameters are considered for temperature prediction. SPR process is modelled and simulated using a FE code ABAQUS® using explicit-temperature-displacement module. Temperature prediction during SPR is not attempted until now. Such analyses will help in selecting the materials and process conditions appropriately during SPR.

24.2 FE Simulation of SPR

FE simulation of SPR is performed using a 2D-axisymmetric model that is generated in ABAQUS® using explicit-temperature-displacement module. Four-node 2D elements with four Gauss points and hourglass control are selected for modelling. An average element size of 0.5 mm × 0.5 mm is used for meshing. The element type of CPS4RT, which is a four-node plane-strain quadrilateral element with reduced integration along with temperature DOF, is used. Reduced integration corresponds to lowering of the order of integration as compared to that of full integration. The meshing in the sheets and the blank holder is done in structured form, but the meshing of the rivet, die and the punch are done in free meshing form. This causes the rivet to have varying element dimensions. The FE model contains a rivet, bottom plate, top plate, punch, blank holder and die. Among this, the punch, blank holder and die are assumed as rigid parts. The rivet and sheets are considered plastically deformable during riveting process and are modelled as elastic–plastic materials. The rivet geometry is assumed to have a nominal diameter of 5 mm and is made of high-strength steel. The punch is of 5 mm diameter with 2 mm as its corner radius. The blank holder has inner radius of 10 mm and outer radius of 20 mm and is modelled as an annular ring. The die geometry has a projection at the centre that controls the rivet flaring. The blank holder and punch geometries are modelled such that sufficient clearance is provided between the surfaces. The dimensions are based on standards available practically. Figure 24.2 shows the FE model for SPR analyses and one of the simulated cases.

The sheet material properties represent a 5xxx Al alloy, and the stress–strain behaviour of the material is shown in Fig. 24.3. The stress–strain curve was obtained from uniaxial tensile tests following standard procedure. The rivet material was considered much stronger than the sheet. Since its material properties were not evaluated, a multiplying factor was considered with respect to the sheet to obtain its stress–strain behaviour. However, both are modelled as elasto–plastic materials. The other tools (die, punch and the blank holder) were modelled as rigid by considering properties of tool steel which has very high elastic modulus. The other material properties required to simulate the FE explicit-temperature-displacement model are given in Table 24.1.

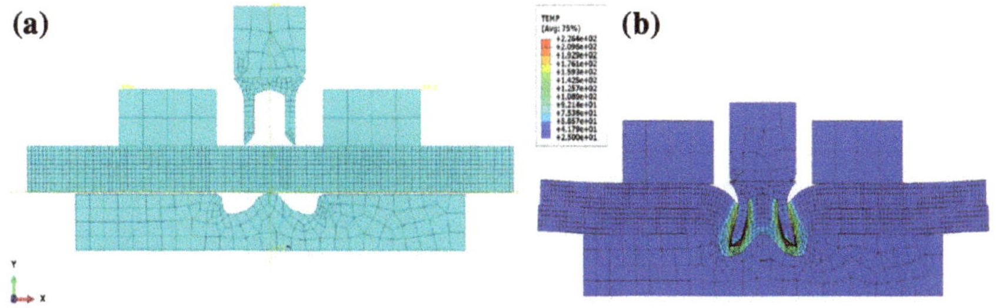

Fig. 24.2 FE model of SPR **a** initial stage for 0.21 tip height, **b** FE simulation of an SPR case

Fig. 24.3 True stress–strain behaviour of 5xxx Al sheet

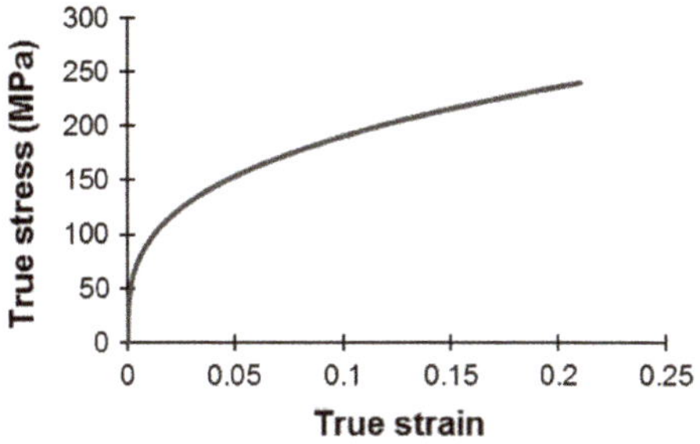

Table 24.1 Material properties used for FE simulations

Material	Properties					
	Density (kg/m^3)	Specific heat (J/kg K)	Thermal conductivity (W/m K)	Inelastic heat factor	Elasticity (GPa)	Poisson's ratio
5xxx sheet	2700	900	205	0.9	68	0.33
Rivet	7830	490	50.2	0.9	180	0.3
Tool	7830	490	50.2	0.9	210	0.3

In the FE model, the interactions are created between the rivet and the sheets, the upper and the lower sheets, the blank holder and the upper sheet, the rivet and the punch, and the die and the lower sheet. The contact was modelled using surface-to-surface penalty formulation. Constant Coulomb's friction coefficient was set as per the levels decided (given later). The sheets are clamped between the die and the blank holder with a holding force of about 5 kN. The die and the blank holder are fixed during SPR simulations. The punch is provided with a prescribed displacement of 6.5 mm to push the rivet through the sheets for joint formation. The rate of punch movement is fixed such that the total displacement occurs in 1 s in real practice. After the completion of each simulation, temperature at an element that is in contact with the rivet tip was evaluated with respect to punch displacement.

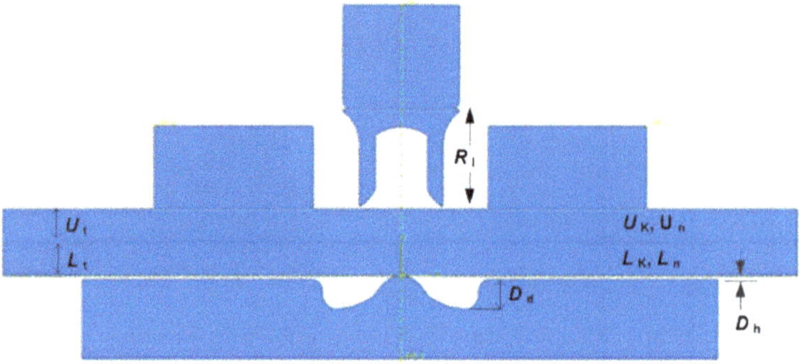

Fig. 24.4 Definition of SPR parameters

24.3 SPR Parameters

The SPR parameters considered for the ANN modelling are Coulomb's coefficience of friction (μ), die tip height (D_h), die depth (D_d), rivet length (R_l), upper sheet thickness (U_t), lower sheet thickness (L_t), upper sheet K (U_K), upper sheet n (U_n), lower sheet K (L_K), lower sheet n (L_n). Here, K and n are the strength coefficient and strain hardening exponent of the sheet materials. The parameters are defined in Fig. 24.4. The levels for these parameters (shown in Table 24.2) were selected based on available SPR data and material properties. The range was chosen to satisfy the practical range available for these parameters.

24.4 ANN Model

In the present work, ANN model was developed to predict the temperature evolution during SPR. A total of 440 FE simulations were carried out for ten parameters (Table 24.2) with several levels for each parameter. To predict the temperature evolution during SPR, the temperature-displacement curves from FE simulations were divided into segments at equal intervals. For example, in Fig. 24.5, the plot for four combinations out of 440 combinations is shown with the segments made at different displacements. Here, for 6-mm-punch displacement, 10 segments are considered. Hence, for every 0.6 mm a partition line is drawn. For one partition line, the curves intersect at four different points. These four temperatures are considered as output for a particular displacement at four different SPR conditions. Similar exercise has been followed for all the segments resulting in large data set for temperature evolution. A sample table for four combinations at one partition line is given in Table 24.3. The eleven different temperature values for a condition will be trained to generate an ANN model.

Table 24.2 Levels of SPR variables used for FE simulations and ANN model

P	L																
	1	2	3	4	5	6	7	8	9	10	11	12	13	14	15	16	17
μ	0.1	0.12	0.14	0.16	0.18	0.2	0.22	0.24	0.26	0.28	0.30	0.32	0.38	0.4	0.43	–	–
U_K	100	125	150	175	200	225	250	275	300	310	330	350	370	390	410	430	460
U_n	0.1	0.14	0.16	0.17	0.21	0.23	0.25	0.27	0.29	0.31	0.33	0.34	0.37	0.39	0.4	0.43	0.47
L_K	100	125	150	175	200	225	250	275	300	310	330	350	370	390	410	430	460
L_n	0.1	0.14	0.16	0.17	0.21	0.23	0.25	0.27	0.29	0.31	0.33	0.34	0.37	0.39	0.4	0.43	0.47
D_h	0.21	0.1	0	−0.1	−0.2	−0.29	−0.4	−0.5	−0.6	−0.79	−1	−1.5	−2	–	–	–	–
D_d	1.8	1.85	1.9	1.95	2	2.05	2.15	2.2	–	–	–	–	–	–	–	–	–
U_t	1	1.25	1.5	1.75	2	2.25	2.5	2.75	3	–	–	–	–	–	–	–	–
L_t	1	1.25	1.5	1.75	2	2.25	2.5	2.75	3	–	–	–	–	–	–	–	–
R_l	5.5	5.8	6	6.25	6.5	6.75	7	–	–	–	–	–	–	–	–	–	–

L: Levels; P: Parameters; U_K, L_K in MPa; D_h, D_d, U_t, L_t and R_l in mm

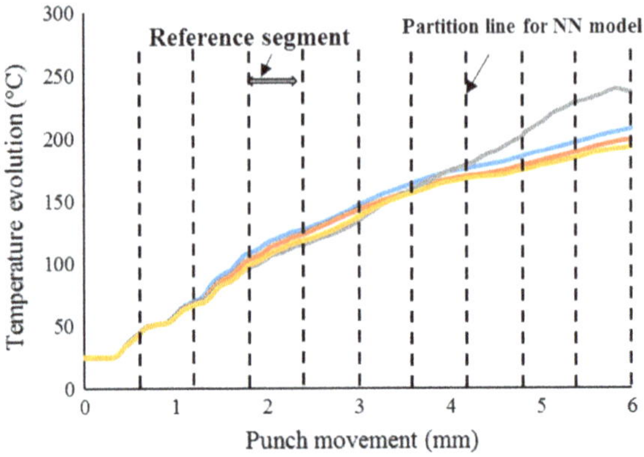

Fig. 24.5 Temperature evolution for four combinations of parameters

Table 24.3 Sample input table for ANN model

SPR combinations	SPR parameters										Temperature (°C)
	U_K	U_n	L_K	L_n	μ	D_h	D_d	U_t	L_t	R_l	
Combination 1	390	0.31	390	0.31	0.47	0.21	2	2	2	6	227
Combination 2	390	0.22	390	0.31	0.10	0.21	2	2	2	6	196
Combination 3	390	0.26	390	0.31	0.10	0.21	2	2	2	6	188
Combination 4	390	0.31	390	0.31	0.17	0.21	2	2	2	6	183

The ANN model was trained by using back-propagation algorithm using an in-house C++ code. The algorithm is based on the error-correction learning rule. In the current application, the objective is to use the network to learn mapping between input and output patterns. The components of the input pattern consist of the control variables of the SPR operation (Table 24.2), whereas the output pattern components represent the temperature rise during the process. The nodes in the hidden layer are necessary to implement the nonlinear mapping between the input and output patterns. Each neuron is connected to all the other neurons in the adjacent layer through the weighted connections. In the forward pass, the synaptic weights remain unaltered throughout the network and the function signals of the network are computed on a neuron-by-neuron basis.

Training and prediction are performed separately for each output temperature, i.e., for each partition line. Thus, the eleven outputs are trained separately, generating eleven ANN architectures. The final ANN architecture contains five input neurons, one hidden layer with ten neurons, and one output neuron corresponding to temperature at each displacement. The final architecture has been decided based on numerous trial exercises. For the first eight displacement levels, tan-sigmoid transfer function has been used between input and hidden layer, while log-sigmoid between hidden layer and output layer. However, for the last three displacement levels, tan-sigmoid transfer function has been used between input and hidden layer and hidden layer and output layer. The coefficient of transfer function between the input and hidden neurons and the coefficient of transfer function between the hidden neurons and output, learning constant, momentum constant are decided based on trial and error and accuracy of the network. The trained ANN has been validated with the FE simulation results at 44 intermediate levels (Table 24.4) for checking the adequacy of the ANN model.

Table 24.4 Validation cases for checking the adequacy of ANN model

Validation cases	U_K	U_n	L_K	L_n	μ	D_h	D_d	U_t	L_t	R_l
Case 1	390	0.31	390	0.31	0.17	0.21	2	2	2	6
Case 2	390	0.31	390	0.31	0.25	0.21	2	2	2	6
Case 3	390	0.31	390	0.31	0.39	0.21	2	2	2	6
Case 4	**390**	**0.31**	**390**	**0.31**	**0.47**	**0.21**	**2**	**2**	**2**	**6**
Case 5	112	0.31	390	0.31	0.1	0.21	2	2	2	6
Case 6	167	0.31	390	0.31	0.1	0.21	2	2	2	6
Case 7	**237**	**0.31**	**390**	**0.31**	**0.1**	**0.21**	**2**	**2**	**2**	**6**
Case 8	350	0.31	390	0.31	0.1	0.21	2	2	2	6
Case 9	390	0.12	390	0.31	0.1	0.21	2	2	2	6
Case 10	390	0.26	390	0.31	0.1	0.21	2	2	2	6
Case 11	390	0.35	390	0.31	0.1	0.21	2	2	2	6
Case 12	**390**	**0.22**	**390**	**0.31**	**0.1**	**0.21**	**2**	**2**	**2**	**6**
Case 13	**390**	**0.31**	**112**	**0.31**	**0.1**	**0.21**	**2**	**2**	**2**	**6**
Case 14	390	0.31	167	0.31	0.1	0.21	2	2	2	6
Case 15	390	0.31	237	0.31	0.1	0.21	2	2	2	6
Case 16	390	0.31	350	0.12	0.1	0.21	2	2	2	6
Case 17	390	0.31	390	0.29	0.1	0.21	2	2	2	6
Case 18	390	0.31	390	0.35	0.1	0.21	2	2	2	6
Case 19	**390**	**0.31**	**390**	**0.22**	**0.1**	**0.21**	**2**	**2**	**2**	**6**
Case 20	112	0.31	390	0.31	0.15	0.21	2	2	2	6
Case 21	167	0.31	390	0.31	0.25	0.21	2	2	2	6
Case 22	390	0.12	390	0.31	0.15	0.21	2	2	2	6
Case 23	390	0.26	390	0.31	0.25	0.21	2	2	2	6
Case 24	390	0.31	112	0.31	0.15	0.21	2	2	2	6
Case 25	390	0.31	167	0.31	0.25	0.21	2	2	2	6
Case 26	390	0.31	390	0.29	0.15	0.21	2	2	2	6
Case 27	390	0.31	390	0.35	0.25	0.21	2	2	2	6
Case 28	**390**	**0.31**	**390**	**0.31**	**0.1**	**−0.3**	**2**	**2**	**2**	**6**
Case 29	390	0.31	390	0.31	0.1	−0.7	2	2	2	6
Case 30	390	0.31	390	0.31	0.1	0.05	2	2	2	6
Case 31	390	0.31	390	0.31	0.1	−0.05	2	2	2	6
Case 32	390	0.31	390	0.31	0.1	−0.15	2	2	2	6
Case 33	**390**	**0.31**	**390**	**0.31**	**0.1**	**0.21**	**1.97**	**2**	**2**	**6**
Case 34	390	0.31	390	0.31	0.1	0.21	2.07	2	2	6

(continued)

Table 24.4 (continued)

Validation cases	U_K	U_n	L_K	L_n	μ	D_h	D_d	U_t	L_t	R_l
Case 35	390	0.31	390	0.31	0.1	0.21	2.18	2	2	6
Case 36	390	0.31	390	0.31	0.1	0.21	2.02	2	2	6
Case 37	390	0.31	390	0.31	0.1	0.21	2	1.12	2	6
Case 38	**390**	**0.31**	**390**	**0.31**	**0.1**	**0.21**	**2**	**2.12**	**2**	**6**
Case 39	390	0.31	390	0.31	0.1	0.21	2	2.07	2	6
Case 40	390	0.31	390	0.31	0.1	0.21	2	2	1.12	6
Case 41	**390**	**0.31**	**390**	**0.31**	**0.1**	**0.21**	**2**	**2**	**2.12**	**6**
Case 42	390	0.31	390	0.31	0.1	0.21	2	2	2.07	6
Case 43	**390**	**0.31**	**390**	**0.31**	**0.1**	**0.21**	**2**	**2**	**2**	**6.12**
Case 44	390	0.31	390	0.31	0.1	0.21	2	2	2	5.92

24.5 ANN Prediction

The trained ANN model has been validated for 44 intermediate cases (given in Table 24.4) that were not part of the training set. Out of these, validation for ten cases, namely case 4, case 7, case 12, case 13, case 19, case 28, case 33, case 38, case 41, case 43 (indicated in Table 24.4), is shown in Fig. 24.6. It is observed that the temperature evolution predictions from ANN agree well with that of from FE simulations. There is about 5% deviation. Similar accuracy levels are observed in all other cases as well. Figure 24.6k shows the accurate ANN predictions for all 44 SPR cases. With the FEA-ANN hybrid model, one can design and optimize the SPR process, tool and material conditions, efficiently, which otherwise will be time and resource intensive.

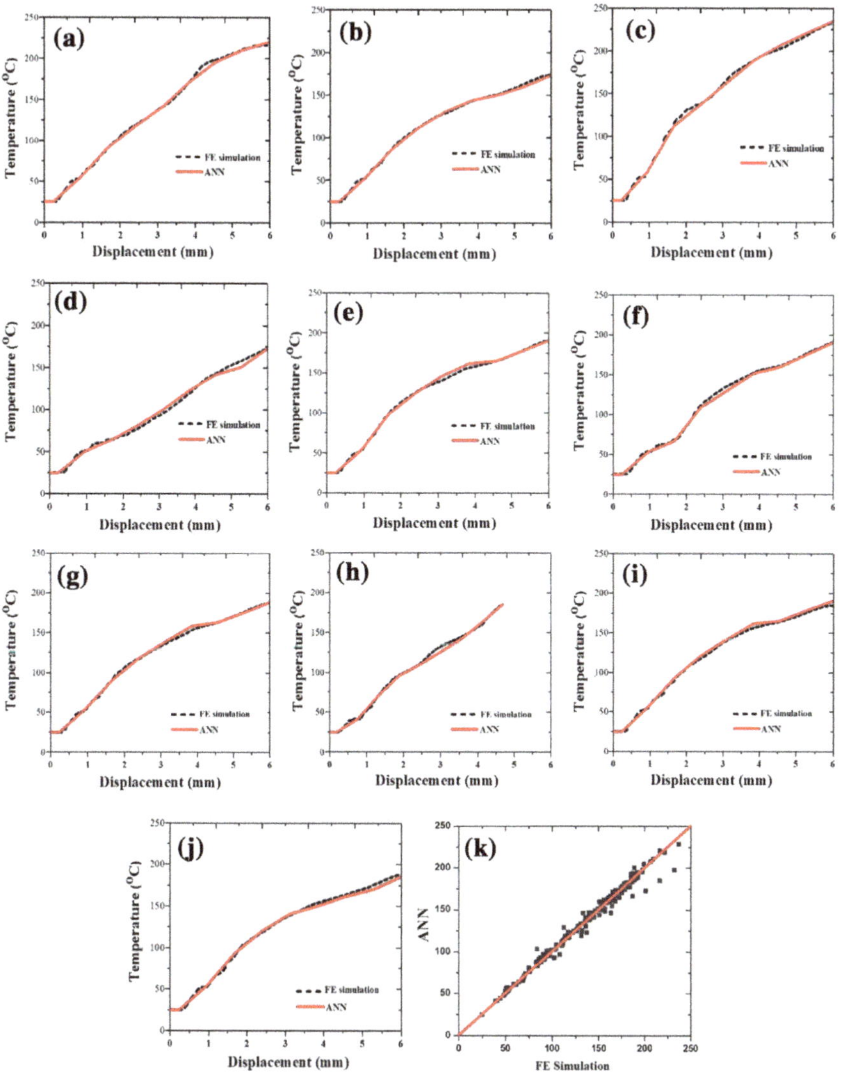

Fig. 24.6 ANN model validation with FE simulation results **a** case 4, **b** case 7, **c** case 12, **d** case 13, **e** case 19, **f** case 28, **g** case 33, **h** case 38, **i** case 41, **j** case 43 (indicated in Table 24.4), **k** ANN versus FE simulation results for all 44 cases

24.6 Conclusions

The temperature measurement during SPR is challenging and difficult task to perform in shop floor. The temperature has to be monitored in the deforming region of the sheet which is inside the die with a diameter of about 10 mm and the process occurs for a small duration. The present work demonstrates that the temperature rise during SPR process can be predicted using a robust FEA-ANN strategy for any combination of sheet materials, process conditions and tool dimensions. The ANN model is sufficiently accurate to predict the temperature evolution during SPR and hence can be used as a 'virtual machine' for analyses.

References

1. Polmear, J.: Light Metals: Metallurgy of the Light Metals, vol. 3, 3rd edn. Wiley, USA (1995)
2. Hirsch, J.: Recent development in aluminium for automotive applications. Trans. Nonferrous Met. Soc. China **24**, 1995–2002 (2014)
3. Mori, K., Bay, N., Fratini, L., Micari, F., Tekkaya, A.E.: Joining by plastic deformation. CIRP Ann. **62**(2), 673–694 (2013)
4. Xu, Y.: Effects of factors on physical attributes of self-piercing riveted joints. Sci. Technol. Weld. Joining **11**(6), 666–671 (2006)
5. Sun, X., Stephens, E.V., Khaleel, M.A.: Fatigue behaviors of self-piercing rivets joining similar and dissimilar sheet metals. Int. J. Fatigue **29**, 370–386 (2007)
6. Mori, K., Abe, Y., Kato, T.: Mechanism of superiority of fatigue strength for aluminium alloy sheets joined by mechanical clinching and self-pierce riveting. J. Mater. Process. Technol. **212**, 1900–1905 (2012)
7. Johnson, P., Cullen, J.D., Sharples, L., Shaw, A., Al-Shamma'a, A.I.: Online visual measurement of self-pierce riveting systems to help determine the quality of the mechanical interlock. Measurement **42**, 661–667 (2009)
8. Hoang, N.H., Porcaro, R., Langseth, M., Hanssen, A.G.: Self-piercing riveting connections using aluminium rivets. Int. J. Solids Struct. **47**, 427–439 (2010)
9. Mori, K., Abe, Y., Kato, T.: Self-pierce riveting of multiple steel and aluminium alloy sheets. J. Mater. Process. Technol. **214**, 2002–2008 (2014)
10. Porcaro, R., Hanssen, A.G., Langseth, M., Aalberg, A.: Self-piercing riveting process: an experimental and numerical investigation. J. Mater. Process. Technol. **171**, 10–20 (2006)
11. Bouchard, P.O., Laurenta, T., Tollier, L.: Numerical modeling of self-pierce riveting-From riveting process modeling down to structural analysis. J. Mater. Process. Technol. **202**, 290–300 (2008)
12. Casalino, G., Rotondo, A., Ludovico, A.: On the numerical modelling of the multiphysics self-piercing riveting process based on the finite element technique. Adv. Eng. Softw. **39**, 787–795 (2008)
13. He, X., Xing, B., Zeng, K., Gu, F., Ball, A.: Numerical and experimental investigations of self-piercing riveting. Int. J. Adv. Manuf. Technol. **69**, 715–721 (2013)
14. He, X., Gu, F., Ball, A.: Recent development in finite element analysis of self-piercing riveted joints. Int. J. Adv. Manuf. Technol. **58**, 643–649 (2012)
15. Mylavarapu, D., Narayanan, R.G., Das, M.: Prediction of critical thinning during self-pierced riveting of sheets. In: Proceedings of 6th International & 27th All India Manufacturing Technology, Design and Research Conference (AIMTDR-2016), pp. 1017–1020, College of Engineering, Pune, India (2016)
16. Mylavarapu, D., Das, M., Narayanan, R.G.: Prediction of temperature evolution during self-pierced riveting of sheets. In: Handbook of Research on Manufacturing Process Modeling and Optimization Strategies (Chapter 18), pp. 381–398, IGI Global, USA (2017)

Part III
Subtractive Manufacturing

Chapter 25
Determination of Optimal Cutting and Tool Geometry Parameters for Better Surface Integrity of Hard Turned AISI 52100 Steel-Hybrid GRA-PCA

P. Umamaheswarrao, D. Ranga Raju, K. N. S. Suman and B. Ravi Sankar

Abstract In the present work, AISI 52100 steel hard turning has been performed using PCBN tools. Cutting speed, feed, depth of cut, nose radius and negative rake angle are the input parameters, and the measured responses are surface roughness and workpiece surface temperature. Experiments are planned as per central composite rotatable design (CCD) of response surface methodology (RSM). The effect of input parameters and their interactions are discussed with main effects plot and response surface plots. Further, the multi-objective optimization scheme is proposed by adopting grey relational analysis (GRA) coupled with the principal component analysis (PCA). Results demonstrated that responses considerably affected by speed followed by nose radius, feed, depth of cut and negative rake angle. Cutting speed 1000 rpm, feed 0.04 mm/rev, depth of cut 0.4 mm, nose radius 1 mm and negative rake angle 15° are the obtained optimum cutting parameters.

Keywords Hard turning · AISI 52100 steel · Surface roughness · Workpiece surface temperature

---

P. Umamaheswarrao (✉) · B. Ravi Sankar
Department of Mechanical Engineering, Bapatla Engineering College,
Bapatla 522102, AP, India
e-mail: maheshponugoti@gmail.com

D. Ranga Raju
Department of Mechanical Engineering, Srinivasa Institute of Engineering
and Technology, Amalapuram 533216, AP, India

K. N. S. Suman
Department of Mechanical Engineering, College of Engineering,
Andhra University, Visakhapatnam 530003, AP, India

R. G. Narayanan et al. (eds.), *Advances in Computational Methods in Manufacturing*, Lecture Notes on Multidisciplinary Industrial Engineering,
https://doi.org/10.1007/978-981-32-9072-3_25

25.1 Introduction

Optimum machining parameter's determination in manufacturing industry leads to the reduction in production cost and time, increases the production rate, improves the product quality and reduces environment impact which in turn leads to productivity enhancement [1, 2]. The comparison of conventional machining versus hard turning is shown in Fig. 25.1. [3]. Higher material removal rate, less machining time, less set-up time and greater flexibility are the merits in hard machining in contrast to grinding. Complex part fabrication using hard turning saves 30% manufacturing costs [4]. During AISI 52100 steel hard turning with PCBN tool, feed rate has a significant influence on the surface finish, while cutting speed and depth of cut had marginal effect [5, 6]. Ravi Sankar et al. [7] revealed surface roughness greatly influenced by the speed and nose radius in AISI 52100 hard turning with PCBN tools.

Suhail Adeel et al. [8, 9] obtained better surface roughness at higher the workpiece surface temperature for AISI 1020 steel. The cutting performance could be effectively sensed and controlled by the workpiece surface temperature, while AISI D2 steel hard turning cutting speed, depth of cut followed by feed were the major contributors to workpiece surface temperature development [10]. During turning of EN-9 steel, the workpiece surface temperature was considerably affected by feed subsequently cutting speed and depth of cut [11]. Multi-objective

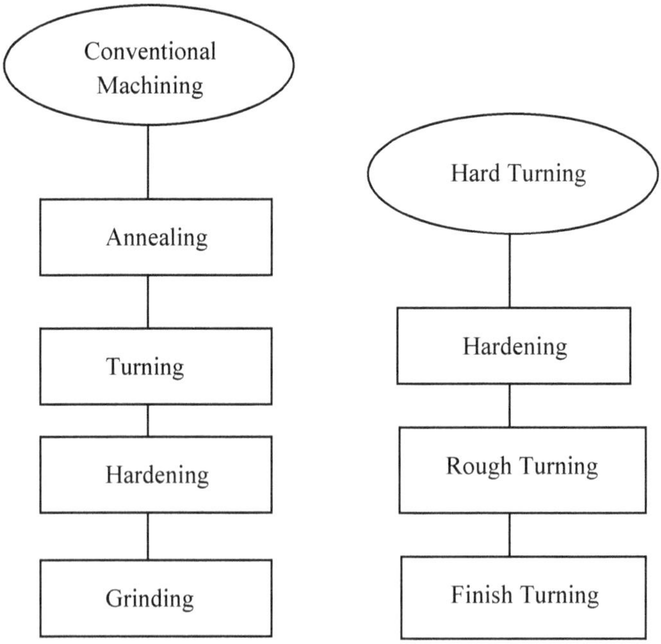

Fig. 25.1 Conventional machining and hard turning comparison [4]

optimization of parameters was performed by deploying GRA coupled with PCA and gained wider accomplishment [12–14].

From the literature, it was elucidated that hard turning was the best alternative to grinding owing to its merits. Much emphasis was made on AISI 52100 steel hard tuning by several researchers due to its applications in various parts of the industry. The past studies made a large amount of interest to investigate the effect of cutting parameters on the responses. Scanty literature was available on the effect of tool parameters on the output responses, and the much interest was focused on nose radius. Most of the studies focused on the machining forces developed and surface roughness during hard turning.

However, the workpiece surface temperature reported to be the sensible and controlling factor for turning performance. Hence, the present study was aimed to conduct hard turning of AISI 52100 steel using PCBN tools with cutting speed, feed rate and depth of cut as cutting conditions whereas nose radius and negative rake angle as tool geometry parameters. Surface roughness and workpiece surface temperature (also termed as surface temperature) were considered as responses. Further, multi-objective optimization was performed deploying GRA coupled with PCA for optimum cutting conditions.

25.2 Experimental Work

Experiments are designed using central composite rotatable design (CCD) of second order which is found to be the most efficient tool in response surface method (RSM). Five input parameters were varied at five levels throughout the experiment, and their influences on responses like surface roughness and workpiece surface temperature were examined. Levels and their factors are depicted in Table 25.1.

Experiments were conducted using variable speed and feed drive lathe in dry condition. Experimental set-up is depicted in Fig. 25.2. Round bars of 48 mm diameter and 500 mm length made of AISI 52100 steel were used for the experiment with a hardness of 57 HRC. Machining length and diameter of workpiece

Table 25.1 Factors and levels

S. No.	Factors	Units	Notation	Levels				
				−2	−1	0	1	2
1	Speed	rpm	v	200	400	600	800	1000
2	Feed	mm/rev	f	0.02	0.04	0.06	0.08	0.1
3	Depth of cut	mm	d	0.4	0.5	0.6	0.7	0.8
4	Nose radius	mm	r	0.4	0.6	0.8	1	1.2
5	Negative rake angle		α	−5	−15	−25	−35	−45

Fig. 25.2 Experimental set-up

were fixed at 30 and 48 mm respectively for every experimental run. In the present work, PCBN inserts were used with different nose radii such as *r* = 0.4, 0.6, 0.8, 1 and 1.2 mm with different rake angles (−5, −15, −25, −35 and −45). Inserts are mounted on a tool holder of ISO Designation PSBNR2525 M12.

Mitutoyo make surface roughness tester (SJ-210) was used to measure the surface roughness of the turned samples. Workpiece surface temperature was measured using an infrared thermometer manufactured by AMPROBE (range: −50 to 1550 °C, Model: IR 750). Experimental matrix with results is depicted in Table 25.2.

Table 25.2 Experimental results

Expt. No.	Speed (rpm)	Feed (mm/rev)	Depth of cut (mm)	Nose radius (mm)	Negative rake angle (°)	Surface roughness (μm)	Workpiece surface temperature (°C)
1	400	0.04	0.5	0.6	35	0.525	57.43
2	800	0.04	0.5	0.6	15	0.465	74.4
3	400	0.08	0.5	0.6	15	0.453	65.19
4	800	0.08	0.5	0.6	35	0.545	77.68
5	400	0.04	0.7	0.6	15	0.552	71.96
6	800	0.04	0.7	0.6	35	0.507	82.88
7	400	0.08	0.7	0.6	35	0.539	70.27
8	800	0.08	0.7	0.6	15	0.471	65.48
9	400	0.04	0.5	1	15	0.485	66.3
10	800	0.04	0.5	1	35	0.401	66.3
11	400	0.08	0.5	1	35	0.507	84.38

(continued)

Table 25.2 (continued)

Expt. No.	Speed (rpm)	Feed (mm/rev)	Depth of cut (mm)	Nose radius (mm)	Negative rake angle (°)	Surface roughness (μm)	Workpiece surface temperature (°C)
12	800	0.08	0.5	1	15	0.502	80.11
13	400	0.04	0.7	1	35	0.508	67.07
14	800	0.04	0.7	1	15	0.408	80.3
15	400	0.08	0.7	1	15	0.604	68.76
16	800	0.08	0.7	1	35	0.498	76.5
17	200	0.06	0.6	0.8	25	0.559	66.61
18	1000	0.06	0.6	0.8	25	0.456	82.73
19	600	0.02	0.6	0.8	25	0.468	70.85
20	600	0.1	0.6	0.8	25	0.53	74.88
21	600	0.06	0.4	0.8	25	0.45	71.39
22	600	0.06	0.8	0.8	25	0.48	74.2
23	600	0.06	0.6	0.4	25	0.514	67.18
24	600	0.06	0.6	1.2	25	0.485	76.05
25	600	0.06	0.6	0.8	5	0.484	70.32
26	600	0.06	0.6	0.8	45	0.509	73.68
27	600	0.06	0.6	0.8	25	0.507	68.6
28	600	0.06	0.6	0.8	25	0.518	74.94
29	600	0.06	0.6	0.8	25	0.52	71.41
30	600	0.06	0.6	0.8	25	0.512	66.36
31	600	0.06	0.6	0.8	25	0.488	76
32	600	0.06	0.6	0.8	25	0.522	69

25.3 Methodology Adopted

Hybrid GRA-PCA

Hybrid GRA-PCA is adopted to obtain optimum parametric combinations, and the methodology is shown in Fig. 25.3.

25.4 Results and Discussion

The experiments were conducted as per the CCD plan of response surface method. The experimental runs with rank, GRC and GRG are depicted in Table 25.3.

Better multiple performance characteristics are obtained at larger GRG. A larger delta value indicates the higher significance of the parameter in controlling the response.

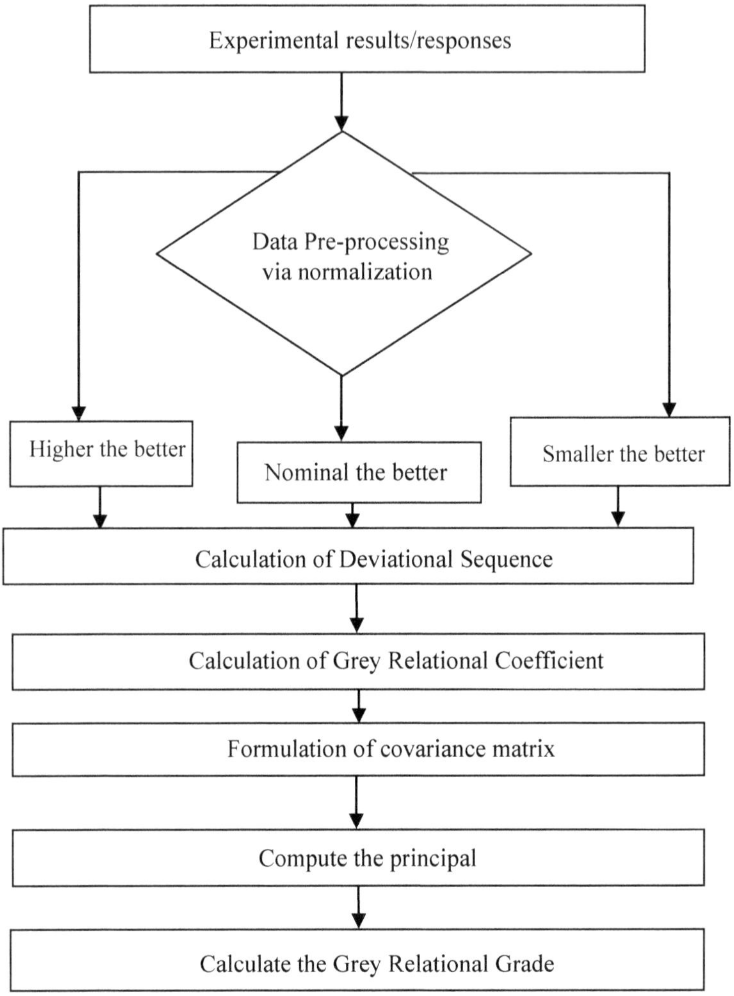

Fig. 25.3 Methodology for hybrid GRA-PCA

ANOVA is used to estimate each machining parameters percentage of contribution on multi-objective optimization. From the ANOVA analysis, it is clear that speed contribution is highest followed by nose radius, feed, depth of cut and negative rake angle as depicted in Table 25.4.

Quadratic equation for GRG after eliminating insignificant terms for 95% confidence level is given below

Table 25.3 Grey relational coefficient, grey relational grade and rank of the surface roughness and workpiece surface temperature

Expt. No.	Grey relational coefficient		Grey relational grade	Rank
	Surface roughness	Workpiece surface temperature		
1	0.45011	0.333333	0.391714	31
2	0.61329	0.498178	0.555725	7
3	0.66123	0.39276	0.52699	12
4	0.41344	0.550829	0.482126	19
5	0.40198	0.465106	0.433535	29
6	0.48915	0.661699	0.575416	5
7	0.42379	0.444661	0.434222	28
8	0.59183	0.395397	0.493607	16
9	0.54716	0.403041	0.475096	20
10	1	0.403041	0.701507	2
11	0.48915	0.702486	0.595809	4
12	0.50123	0.597622	0.549418	8
13	0.48681	0.410493	0.448643	24
14	0.93548	0.601618	0.768536	1
15	0.33333	0.427856	0.380587	32
16	0.51133	0.530653	0.520984	14
17	0.39113	0.406008	0.398565	30
18	0.64856	0.657879	0.653208	3
19	0.60237	0.451472	0.526913	13
20	0.44034	0.505246	0.472787	21
21	0.67441	0.458004	0.5662	6
22	0.56232	0.495291	0.528799	11
23	0.47319	0.411580	0.442378	26
24	0.54716	0.523343	0.535246	9
25	0.55013	0.445240	0.497678	15
26	0.48448	0.487940	0.486204	17
27	0.48915	0.426149	0.457644	23
28	0.46453	0.506143	0.485328	18
29	0.46031	0.458249	0.459274	22
30	0.47764	0.403612	0.440621	27
31	0.53846	0.522543	0.530492	10
32	0.45617	0.430441	0.443302	25

Table 25.4 ANOVA table for GRG

Source	SS	DF	MS	*F*	*P*	% C
Speed	0.09	1	0.09	140.82	<0.0001	39.63012
Feed	0.0094	1	0.0094	14.68	0.0028	4.139146
Depth of cut	0.0037	1	0.0037	5.77	0.0351	1.629238
Nose radius	0.0224	1	0.0224	35.01	0.0001	9.863496
Negative rake angle	0.0001	1	0.0001	0.2045	0.6599	0.044033
Speed * Feed	0.0346	1	0.0346	54.06	<0.0001	15.23558
Speed * Depth of cut	0.0082	1	0.0082	12.84	0.0043	3.610744
Speed * Nose radius	0.0064	1	0.0064	10	0.009	2.818142
Speed * Negative rake angle	0.0013	1	0.0013	1.96	0.1896	0.572435
Feed * Depth of cut	0.0114	1	0.0114	17.83	0.0014	5.019815
Feed * Nose radius	0.0067	1	0.0067	10.49	0.0079	2.950242
Feed * Negative rake angle	0.0025	1	0.0025	3.84	0.0759	1.100837
Depth of cut * Nose radius	0.0021	1	0.0021	3.28	0.0973	0.924703
Depth of cut * Negative rake angle	0.0016	1	0.0016	2.53	0.1399	0.704535
Nose radius * Negative rake angle	0.003	1	0.003	4.72	0.0526	1.321004
Speed * Speed	0.0058	1	0.0058	9.07	0.0118	2.553941
Feed * Feed	0.0017	1	0.0017	2.62	0.134	0.748569
Depth of cut * Depth of cut	0.0111	1	0.0111	17.38	0.0016	4.887715
Nose radius * Nose radius	0.0007	1	0.0007	1.05	0.3266	0.308234
Negative rake angle * Negative rake angle	0.0009	1	0.0009	1.43	0.2575	0.396301
Residual	0.007	11	0.0006			3.082343
Lack of fit	0.0013	6	0.0002	0.1885	0.967	0.572435
Pure error	0.0057	5	0.0011			2.509908
Total	0.2271	31				

$$
\begin{aligned}
\mathrm{GRG} = {} & 0.122501Xv - 0.039557Xf - 0.024805Xd + 0.061082Xr - 0.004668X\alpha \\
& + 0.056249XvXv + 0.077862XdXd - 0.185917XvXf + 0.090597XvXd \\
& + 0.079974XvXr - 0.106758XfXd - 0.081885XfXr + 0.469492
\end{aligned} \tag{25.1}
$$

$$S = 0.02529\ \mathrm{R{-}Sq} = 96.9\%\ \mathrm{R{-}Sq(adj)} = 91.3\%$$

From the main effects plot (Fig. 25.4) shows that grey relational grade increases significantly with an increase in speed which indexes the occurrence of optimum surface roughness and workpiece surface temperature. This is due to more thermal softening of material with an increase in speed [15, 16]. As feed increases, grey relational grade reduces due to the more feed marks which lead to higher surface

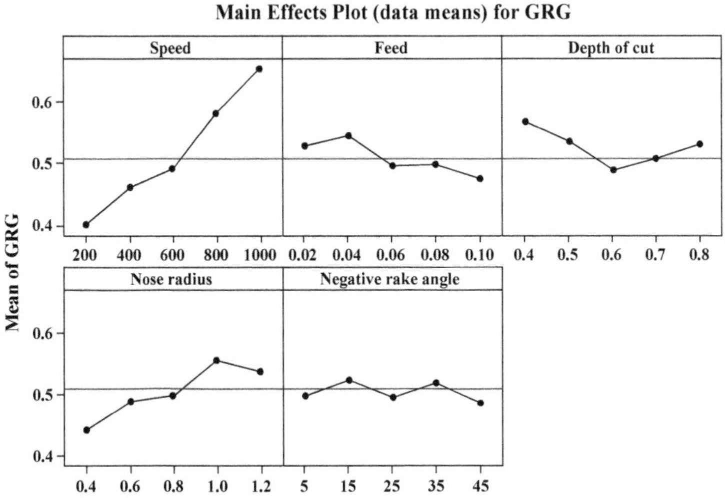

Fig. 25.4 Main effects plot for GRG

roughness and ploughing effect [17]. As the depth of cut increases, grey relational grade decreases and then increases. This is attributed to the increased cutting resistance and cutting force since chipping does not take place at nose radius which has a notable influence on the surface roughness as well as workpiece surface temperature [18]. Grey relational grade increases with an increase in nose radius. This is attributed to the increased nose radius leads to a reduction in surface roughness from general principles of metal cutting. Grey relational grade varies marginally with an increase in negative rake angle. This is due to the increased edge strength of the tool with an increase in negative rake angle causing ease in machining [18, 19]

From the main effect plot (Fig. 25.4), it is observed that the optimistic overall grey relational grade can be achieved with speed = 1000 rpm, feed = 0.04 mm/rev, depth of cut = 0.4 mm, nose radius = 1 mm and negative rake angle = 15°, respectively. GRG for the obtained optimum combination of parameters was 0.80887 estimated from Eq. 25.2 and was 5.24% more than the highest GRG in Table 25.3. Hence, the obtained values were optimum.

$$\gamma = \gamma_m + \sum_{i=1}^{q} (\overline{\gamma_j} - \gamma_m) \tag{25.2}$$

The interaction among the parameters is discussed with the help of response surface plots shown in Fig. 25.5. The optimum grey relational grade is noticed at the combinations of upper limits of speed due to thermal softening [15, 16], lower limits of feed due to less feed marks and upper limits of nose radius due to more

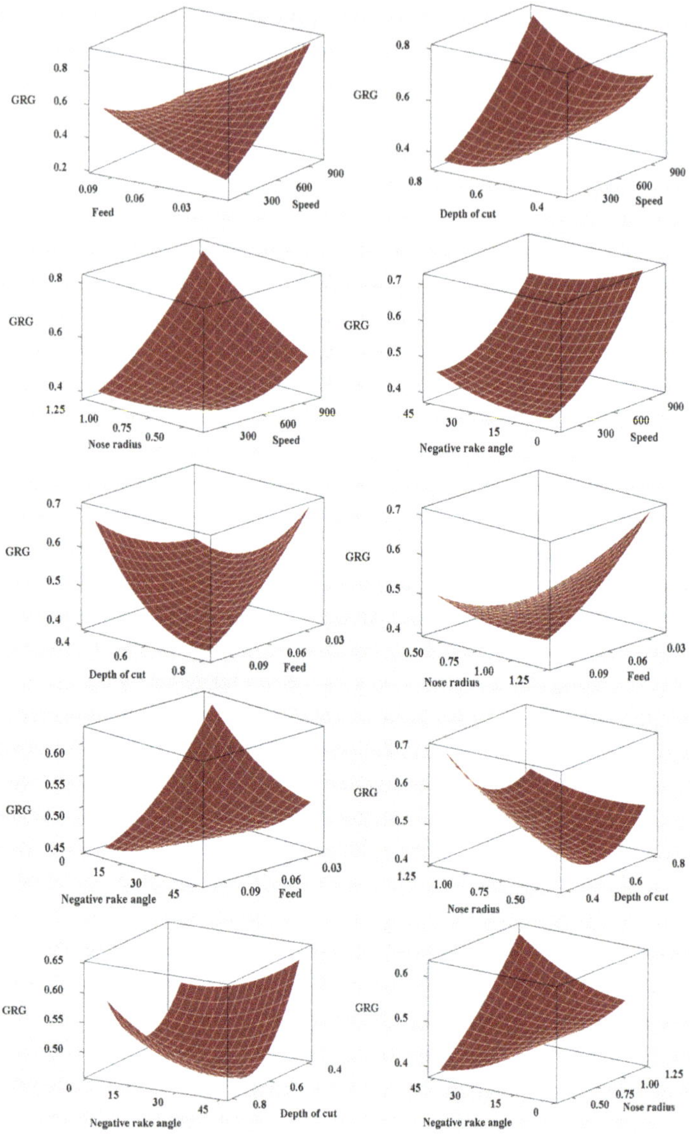

Fig. 25.5 Response surface plots [hold values were speed 600 rpm, feed 0.06, depth of cut 0.6, nose radius 0.8 and negative rake angle 25]

ploughing energy [20]. However, the interaction of depth of cut and negative rake angle with other parameters is observed to be significant, due to the variation of the developed cutting resistance and edge strength.

25.5 Conclusions

Multi-objective optimization for AISI 52100 hardened steel was performed using hybrid GRA-PCA

- The obtained optimum combination of parameters and their levels were A5B2C1D4E2, i.e. (speed = 1000 rpm, feed = 0.04 mm/rev, depth of cut = 0.4 mm, nose radius = 1 mm and negative rake angle = 15°).
- Responses were notably affected by speed followed by nose radius, feed, depth of cut and negative rake angle.
- Higher GRG is noticed during the interaction at upper limits of speed, lower limits of feed and higher limits of nose radius. However, the depth of cut and negative rake angle exhibit considerable interaction with other parameters.
- An increase in the value of predicted weighted GRG from 0.768536 to 0.80887 confirms the improvement in the performance of hard turning.

References

1. König, W., Hochschule, T., Komanduri, R., Schenectady, D., Tönshoff, H.K.: Machining of hard materials. Ann. CIRP **33**(2), 417–427 (1984)
2. Tönshoff, H.K., Arendt, C., Amor, R.B.: Cutting of hardened steel. Ann. CIRP **49**(2), 547–566 (2000)
3. Dogra, M., Sharma, V.S., Sachdeva, A., Suri, N.M., Dureja, J.: Tool wear, chip formation and work piece surface issues in CBN hard turning: a review. Int. J. Precis. Eng. Manuf. **11**(2), 341–358 (2010)
4. Huang, O.Y., Chou, Y.K., Liang, Y.S.: CBN tool wear in hard turning: a survey on research progresses. Int. J. Adv. Manuf. Technol. **35**, 443–453 (2007)
5. Bagawade, A.D., Ramdasi, P.G., Pawade, R.S., Bramhankar, P.K.: The effects of cutting conditions on chip area ratio and surface roughness in hard turning of AISI52100 steel. Int. J. Eng. Res. Technol. **1**(10) (2012)
6. Benga, G., Savu, D., Olei, A.: Influence of the cutting parameters on the surface roughness when machining hardened steel with ceramic and PCBN cutting tools. Adv. Eng. Forum **13**, 19–22 (2015)
7. Ravi Sankar, B., Umamaheswarrao, P., Nawaz Sharief, S.K., Suresh, T., Raju, R.: Parametric investigations on surface roughness of hard turned AISI 52100 steel. Appl. Mech. Mater. **867**, 171–176 (2017)
8. Suhail Adeel, H., Ismail, N., Wong, S.V., Abdul Jalil, N.A.: Optimization of cutting parameters based on surface roughness and assistance of work piece surface temperature in turning process. Am. J. Eng. Appl. Sci. **3**(1), 102–108 (2010)
9. Suhail Adeel, H., Ismail, N., Wong, S.V., Abdul Jalil, N.A.: Surface roughness identification using the grey relational analysis with multiple performance characteristics in turning operations. Arab. J. Sci. Eng. **37**(4), 1111–1117 (2012)
10. Das, S.R., Nayak, R.P., Dhupal, D.: Optimization of cutting parameters on tool wear and work piece surface temperature in turning of AISI D2 steel. Int. J. Lean Thinking **3**(2), 140–156 (2012)
11. Singh, H., et al.: Study of cutting parameters on turning using EN9. Int. J. Adv. Ind. Eng. **1**(2), 40–42 (2013)

12. Ravi Sankar, B., Umamaheswarrao, P.: Optimisation of hardness and tensile strength of friction stir welded AA6061 alloy using response surface methodology coupled with grey relational analysis and principle component analysis. Int. J. Eng. Sci. Technol. **7**(4), 21–29 (2015)
13. Ravi Sankar, B., Umamaheswarrao, P., Srinivasulu, V., Kishore Chowdari, G.: Optimization of milling process on jute polyester composite using Taguchi based grey relational analysis coupled with principle component analysis. Mater. Today Proc. **2**, 2522–2531 (2015)
14. Pradhan, M.K.: Estimating the effect of process parameters on MRR, TWR and radial overcut of EDMed AISI D2 tool steel by RSM and GRA coupled with PCA. Int. J. Adv. Manuf. Technol. **68**(1–4), 591–605 (2013)
15. Ebrahimi, A., Moshksar, M.M.: Evaluation of machinability in turning of microalloyed and quenched tempered steels: tool wear, statistical analysis, chip morphology. J. Mater. Process. Technol. **209**, 910–921 (2009)
16. Lin, H.M., Liao, Y.S., Wei, C.C.: Wear behavior in turning high hardness alloy steel by CBN tool. Wear **264**, 679–684 (2008)
17. Shihab, S.K., Khan, Z.A., Mohammad, A.A.S., Siddiquee, A.N.: Optimization of surface integrity in dry hard turning using RSM. Sadhana **39**(5), 1035–1053 (2014)
18. Suresh, R., Basavarajappa, S., Gaitonde, V.N., Samuel, G.L., Paulo Davim, J.: State-of-the-art research in machinability of hardened steels. Proc. Inst. Mech. Eng. Part B J. Eng. Manuf. **227** (2), 191–209 (2013)
19. Qian, L., Hossan, M.R.: Effect on cutting force in turning hardened tool steels with cubic boron nitride inserts. J. Mater. Process. Tech. **191**(1–3), 274–278 (2007)
20. Chou, Y.K., Song, H.: Tool nose radius effects on finish hard turning. J. Mater. Process. Technol. **148**, 259–268 (2004)

Chapter 26
Study of Cutting Edge Radius Effect on the Cutting Forces and Temperature During Machining of Ti6Al4V

Siddharam Mane, Shyamprasad Karagadde and Suhas S. Joshi

Abstract The cutting-edge radius of the tool has significant effects on the machining process, as it influences the cutting forces, stresses, and temperature at the tool–chip interface. These parameters ultimately affect the tool life and surface integrity of the finished workpiece. The presence of cutting-edge radius in the tools protects them from easily chipping off during the cutting process. A finite element based ABAQUS™ model is used to evaluate the effect of cutting-edge radius for 20, 40, and 60 μm on the cutting forces for orthogonal cutting of Ti6Al4V alloy at different cutting parameters. It was observed that the cutting-edge radius influences both the cutting and thrust forces. An increase of 4–8% and 12–14% in the cutting force and thrust force was observed when the cutting-edge radius changes from 20 to 60 μm. The temperature in the tool was increased with increasing cutting-edge radius.

Keywords Machining · Ti6Al4V · Cutting-edge radius · Finite element modeling

26.1 Introduction

In any machining process, a cutting tool with rounded cutting edges provides an enhanced tool life. The optimum cutting-edge radius for a machining process depends on the work material, tool material, and machining process parameters. Tool-edge geometry has significant effects on the cutting process, as it affects cutting forces, stresses, temperatures, the coefficient of friction, and tool–chip contact length. Nasr et al. [1] showed increasing cutting-edge radius increases both the residual stresses and temperature in the workpiece for AISI 316L steels. Liu and Melkote [2] have studied the tool-edge radius effect on the microstructure of the hard turned of AISI 52100 steel. They observed that large edge honed tools produce substantial subsurface

S. Mane · S. Karagadde · S. S. Joshi (✉)
Department of Mechanical Engineering, Indian Institute of Technology Bombay, Mumbai 400076, India
e-mail: ssjoshi@iitb.ac.in

R. G. Narayanan et al. (eds.), *Advances in Computational Methods in Manufacturing*, Lecture Notes on Multidisciplinary Industrial Engineering,
https://doi.org/10.1007/978-981-32-9072-3_26

plastic flow. Wyen and Wegener [3] conducted the experimental study for orthogonal turning on Ti6Al4V with different cutting-edge radii. They found that the coefficient of friction is influenced by the cutting-edge radius of the tool.

Kim et al. [4] carried out numerical and experimental analysis considering the effect of the tool edge on the cutting process. They confirmed that a major cause for the size effect is the tool-edge radius and increase in the tool-edge radius that causes a change in the temperature distribution in the tool and changes the location of maximum temperature. Endres and Kountanya [5] showed that increasing the cutting-edge radius reduces the flank wear of the tool. Increasing cutting-edge radius increases mechanical stability, but the increasing contact length on the flank face due to the increase in cutting edge results in different thermo-mechanical loads on the cutting edge [6]. Zhao et al. [7] experimentally carried out hard turning of AISI 52100 steel under the cutting-edge radius of 20, 30, and 40. They found that the flank wear decreases with increasing radius.

Use of the cutting tool with some radius protects the edge from chipping off during machining, hence increases tool life. Considering the effect of cutting edge on the cutting forces, temperature profile, the coefficient of friction and contact length, an attempt is made to understand the effect of cutting-edge radius on the cutting forces and temperature in the machining of difficult to machine alloy like Ti6Al4V. This is done by developing a numerical model for two cutting-edge radii of 20 and 40, and 60 μm for different cutting speeds.

26.2 Numerical Modeling

A finite element model using ABAQUS™ is used for modeling and simulation of the tool and the workpiece. A 2D dynamic coupled temperature displacement model is used for simulating orthogonal cutting as shown in Fig. 26.1.

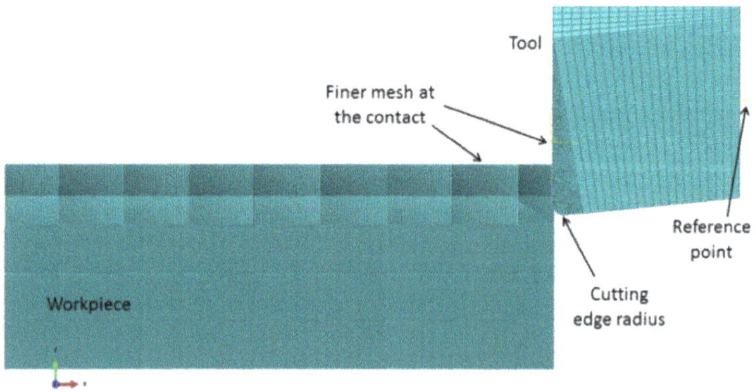

Fig. 26.1 Numerical modeling of orthogonal cutting

Table 26.1 Numerical model cutting parameters

Parameters	Value
Workpiece	Ti6Al4V
Tool	Tungsten carbide
Feed (mm/rev)	0.1
Cutting speed (m/min)	30, 50, 70, 90
Depth of cut (mm)	0.1
Rake angle	0
Cutting-edge radius (μm)	20, 40, 60

Table 26.2 Thermophysical properties of the workpiece and tool [8, 9]

Parameter	Workpiece	Tool
Specific heat (J/kg k)	670	178
Thermal conductivity (W/m k)	6.6	24
Density (kg/m^3)	4430	15700
Young's modulus (GPa)	110	705
Poisson's ratio	0.33	0.23

Table 26.3 Johnson–Cook parameter for Ti6Al4V alloy [10]

A (MPa)	B (MPa)	n	C	m	T_m	T_0
1098	1092	0.93	0.014	1.1	1630	20

The element type used for both the tool and workpiece refers to CPE4RT elements. The tool is assumed to be a rigid body since there will not be any deformation in the tool. A reference point is located at the tip of the tool for providing velocity to the tool. 20, 40, and 60 μm are the three different cutting-edge radii of the tool used in the simulation. The workpiece used is a rectangular block of 3 × 5 × 15 mm. Finer mesh size is used near the tool–workpiece contact region, as shown in Fig. 26.1. A total number of elements in the tool and the workpiece are 2392 and 39140, respectively. The cutting parameters for the modeling are listed in Table 26.1. The thermophysical properties of the tool and the workpiece are listed in Table 26.2.

Since machining is a dynamic process, there will be high strains, strain rate, and temperature involved. To accommodate this, a Johnson–Cook flow stress model was used for Ti alloy shown in Table 26.3. The Johnson–Cook fracture model is given in Table 26.4. The thermoplastic behavior of titanium alloy by Johnson–Cook model is given by,

Table 26.4 Johnson–Cook fracture model parameters for Ti6Al4V alloy [10]

D_1	D_2	D_3	D_4	D_5	ε_0
−0.09	0.25	−0.5	0.014	3.87	1

$$\sigma = [A + B\varepsilon^n]\left[1 + C\ln\frac{\dot{\varepsilon}}{\varepsilon_0}\right]\left[1 - \left(\frac{T - T_0}{T - T_\mathrm{m}}\right)^m\right] \tag{26.1}$$

where σ is the equivalent stress, ε is the equivalent plastic strain, $\dot{\varepsilon}$ is the equivalent strain rate, and ε_0 is the reference strain rate. T, T_0, and T_m are the current temperature, room temperature, and melting temperature of the workpiece, respectively. The parameters A, B, n, C, and m are material constants, and they refer to initial yield stress, hardening modulus, strain hardening coefficient, strain rate dependency coefficient, and thermal softening coefficient, respectively.

26.3 Results and Discussion

Figure 26.2 shows an example of simulation at a cutting speed of 90 m/min. The stresses observed in the model are in the range of 1.7–1.9 GPa, which is generally observed in machining conditions.

26.3.1 *Forces Evaluation*

Forces generated in machining plays an important role in the heat flow and effectively the temperature distribution along the rake face of the tool. Hence, the estimation of cutting forces is important in order to understand the heat flow and temperature.

Fig. 26.2 Stresses (Pa) obtained in the finite element model

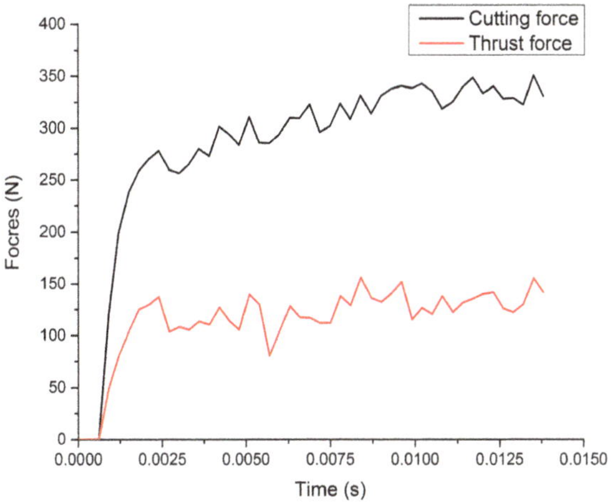

Fig. 26.3 Variation in the cutting and thrust forces with time

Figure 26.3 shows the variation in the cutting and thrust force with time. The cutting forces are large as compared to the thrust forces. These forces are evaluated for all four cutting speeds and three cutting-edge radii, as shown in Table 26.1.

The average cutting forces for cutting conditions are plotted in Fig. 26.4. As can be seen, with the change in cutting speed, change in the cutting force is around 4–5% only, whereas changing cutting edge from 20 to 60 μm changes cutting force and feed force on an average by 4–8 and 12–14%, respectively. For a radius of 20 μm, the cutting forces and thrust force are in the range of 250 and 80 N, respectively. However, for a larger radius of 40 μm, the cutting force shows some variation with cutting speeds, but the thrust force remains more ever in the range of 70–80 N. Overall, changing cutting edge influences the forces more as compared to the cutting force. The contact area at the tip of the tool is more in case of 60 μm as compared to the 20 μm. This increase in contact area increases the resistance the tool tip, and hence, the cutting forces observed in the case of 60 μm are higher.

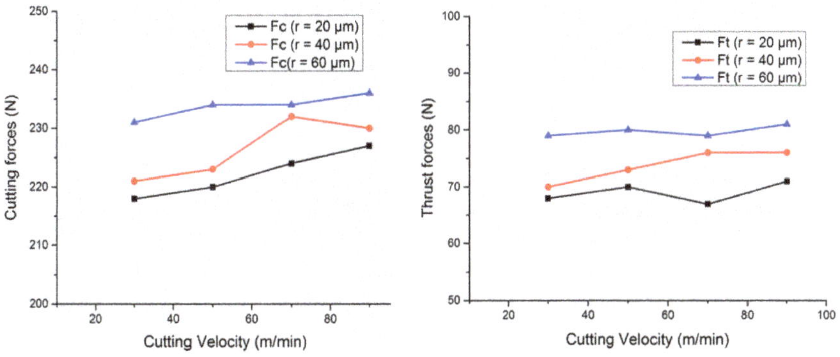

Fig. 26.4 Cutting and thrust forces for $f = 0.1$ mm/rev

26.3.2 Maximum Temperature in the Tool

Figure 26.5 shows the location of the maximum temperature developed around the tool tip. For a smaller cutting-edge radius of 20 μm, the maximum temperature location was away from the tool tip at all cutting speeds except 90 m/min where the maximum temperature was observed at the tool tip. In the case of 40 μm radius, two peaks of temperature are observed at all cutting speeds. These peaks are at the tool tip and around 1.2–1.8 mm from the tip of the tool, see Fig. 26.5b. The reason for observing the second peak on the rake face is due to the friction at the tool–chip interface. It is observed that for a larger radius of 60 μm, the maximum temperature location shifts away from the tool tip which is more toward the flank side of the tool, such a trend is observed because, in the case of larger cutting edge, the contact between the flank face of the tool and workpiece is more, can be seen in Fig. 26.5c. This causes more friction between the tool and workpiece away from the tool causing shifting of the maximum temperature location away from the tool tip. Whereas, in the case of 20 μm cutting-edge radius, the contact area was less, hence, less temperature around tool tip (Fig. 26.5a).

The temperature developed for the cutting process in the case of all the three cutting-edge radii is plotted in Fig. 26.6. The maximum temperature is observed away from the tool tip in the case of low edge radius (Fig. 26.6a), whereas, for an intermediate radius of the tool, i.e., 40 μm, the temperature obtained at the tool tip are higher as compared to 20 μm cutting radius. The maximum temperature obtained in the simulation is around 450 and 500°C for 20 and 40 μm at a cutting speed of 90 m/min, respectively. Also, the spread of the temperature is observed to be more in case of 40 and 60 μm as compared to the 20 μm cutting edge because of the large contact area available between the tool and workpiece for 40 and 60 μm edges. The maximum temperature of 550 °C was observed in the case of 60 μm

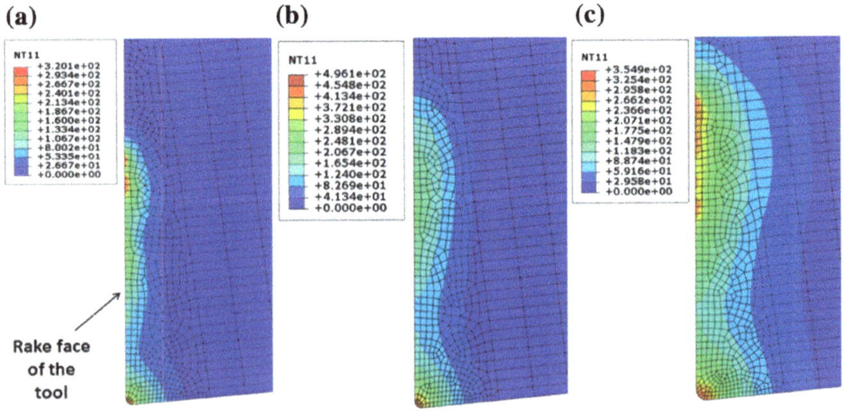

Fig. 26.5 Location of the maximum temperature generated along the rake face of the tool, **a** r = 20 μm, **b** r = 40 μm, **c** r = 60 μm

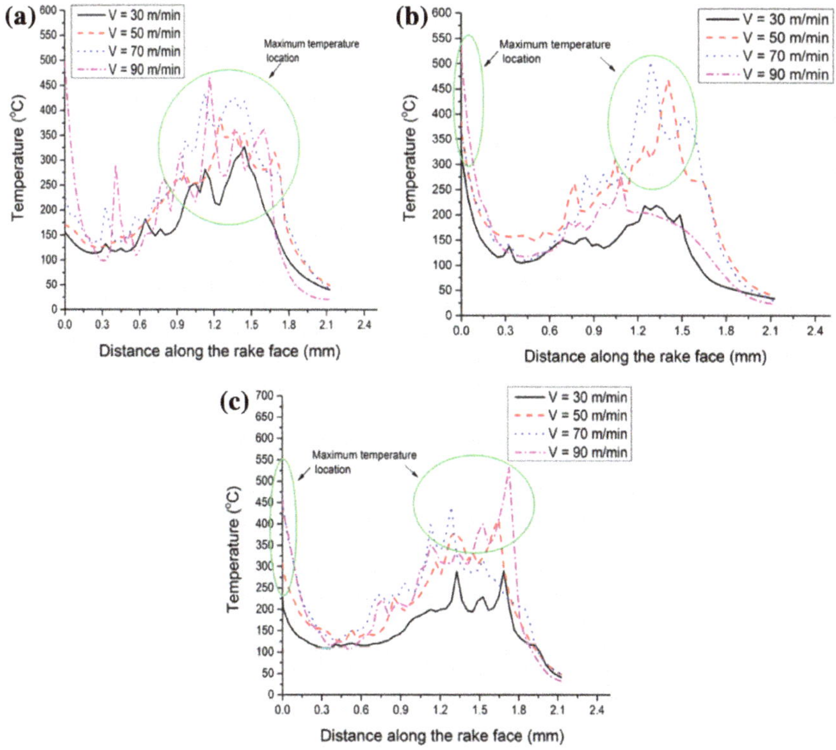

Fig. 26.6 Temperature along the rake face of the tool in the case of **a** r = 20 μm, **b** r = 40 μm, **c** r = 60 μm

cutting-edge radius at a speed of 90 m/min. Hence, the cutting-edge radius has an influence on not only the maximum induced temperature but also over the zone of temperature.

26.4 Conclusions

A FEM-based model is developed for predicting the cutting forces and temperature in the machining of Ti6Al4V alloy. It is seen that the cutting edge of the tool plays a major role in the induced temperature of machining and cutting forces. The cutting force and thrust force are seen to be increased by 4–8 and 12–14%, for a cutting-edge radius of 20 and 60 μm, respectively. The obtained temperature contours show that the maximum temperature location changes along the rake face of the tool as the cutting-edge radius changes. Although the effect of changing tool-edge radius on the forces is small, its effect on the tool temperature influences significantly. Selecting a suitable cutting edge according to cutting conditions and material properties should be selected to improve the surface integrity and enhancing the cutting tool life.

References

1. Nasr, M., Ng, E., Elbestawi, M.: Modelling the effects of tool-edge radius on residual stresses when orthogonal cutting AISI 316L. Int. J. Mach. Tools. Manuf. **47**(2), 401–411 (2007)
2. Liu, K., Melkote, S.: Finite element analysis of the influence of tool edge radius on size effect in orthogonal micro-cutting process. Int. J. Mech. Sci. **49**(5), 650–660 (2007)
3. Wyen, C., Wegener, K.: Manufacturing technology influence of cutting edge radius on cutting forces in machining titanium. CIRP Ann. Manuf. Technol. **59**(1), 93–96 (2010)
4. Kim, K., Lee, W., Sin, H.: Finite-element analysis of machining with the tool edge considered. J. Mater. Process. Technol. **86**(1–3), 45–55 (1999)
5. Endres, W., Kountanya, R.: The effects of corner radius and edge radius on tool flank wear. J. Manuf. Process. **4**, 89–96 (2002)
6. Denkena, B., Biermann, D.: Manufacturing technology cutting edge geometries. CIRP Ann. Manuf. Technol. **63**(2), 631–653 (2014)
7. Zhao, T., Zhou, J., Bushlya, V., Stahl, J.: Effect of cutting edge radius on surface roughness and tool wear in hard turning of AISI 52100 steel. Int. J. Adv. Manuf. Technol. **91**(9–12), 3611–3618 (2017)
8. Welsch, G., Boyer, R., Collings, E.: Materials Properties Handbook: Titanium Alloys. ASM International, Ohio (1993)
9. Pierson, H.: Handbook of Refractory Carbides and Nitrides, pp. 100–117. William Andrew Publishing, Westwood, NJ (1996)
10. Leseur, D.: Experimental investigations of material models for Ti-6A1-4V and 2024-T3. Technical report (1999)

Chapter 27
Comparative Study on Effect of Inclusions and Machinability Aspects of Alloy Steels

V. Krishnaraj, K. Thillairajan and R. Rajeshshyam

Abstract In recent decades, the noticeable amount of research has been carried out in the role of non-metallic inclusions and their relationship to the machinability of various steels. This paper compares the performances of EN alloy steels for varied machinability aspects throughout the machining under dry cutting condition. Cutting speed, feed, and depth of cut were the major ruling parameters affecting the machinability of materials. The present works review the machining of different alloy steels and the role of non-metallic inclusions in them. More precisely, the effects of composition and morphology of inclusions on machinability factors such as cutting force and chip size and shape are discussed and summarized. EN 1A Pb&Te alloy steel provides lower cutting force, and good curl or breakdown chips thus provides the better machinability.

Keywords Alloy steel · Non-metallic inclusions · Machinability · Chip morphology

27.1 Introduction

Several alloy steels, which can be machined at higher cutting speeds and show longer tool life by adding additives, have been developed and used for making the screws, fasteners, and many automobile components. The performance of steels is becoming increasingly stringent. Consumer's expectation is high on the quality and performance of alloy steels. Microstructure, composition, segregation, and cleanliness are the dominant criteria in determining the properties of steel. At present, most of the steel manufacturers are unable to attain completely pure steel without any non-metallic inclusions. These are present in steel greater or lesser according to their mixture and conditions of production. Usually, these non-metallic inclusions

V. Krishnaraj (✉) · K. Thillairajan · R. Rajeshshyam
PSG College of Technology, Coimbatore 641004, Tamilnadu, India
e-mail: vkr.prod@psgtech.ac.in

R. G. Narayanan et al. (eds.), *Advances in Computational Methods in Manufacturing*, Lecture Notes on Multidisciplinary Industrial Engineering,
https://doi.org/10.1007/978-981-32-9072-3_27

are the foreign substances; they affect the homogeneity of the structure and influence the machinability. Inclusions are compound materials, embedded inside the steel at the time of the manufacturing process, and it has a distinctive chemical composition and properties. Non-metallic inclusions influence the properties of steels like machinability, formability, and toughness and corrosion resistance.

In general, machinability can be represented as an optimal combination of the following factors that are low cutting forces, high metal removal rates, low tool wear rates, excellent surface finish, and good curl or breakdown of the chip. Tanaka et al. [1] investigated the machinability study of BN free-machined steel. They identified that BN free-machining steel exhibits slightly lower cutting temperature and smaller cutting forces. Jiang et al. [2] studied titanium diboride particles on machinability and machining criteria optimization during machining of TiB_2/Al MMCs. They revealed that TiB_2 particles influence on machining force, residual stress, and surface roughness. Lalbondre et al. [3] experimented the machinability of AISI 51100 and AISI 52100. They revealed that machinability of AISI 51100 is superior to AISI 52100. Anmark et al. [4] proposed that machinability of clean steels is significantly improved by resulfurization. The better machinability of the standard steel grade R is partly linked to its high content of sulphur (0.028 wt% S) which results in higher amount of MnS inclusion in the steel matrix. This MnS inclusion acts as stress risers in the primary shear region during machining. They concluded that MnS particles improved the chip formation process and prolong the cutting life. Alizadeh [5] examined the influence on the machinability of sintered steels and reported that addition of 0.3 MnS resulted in higher machinability. Joseph [6] revealed the effects of inclusions in carbon and alloy steels in their mechanical properties and cause of failure. Addition to that, inclusions, such as oxides and manganese sulphides, affect the anisotropy of these materials with respect to rolling condition. MnS particles particularly resist the anisotropy because of its elongated morphology after hot rolling.

Gupta et al. [7] quantified the micro-scale effects of inclusions on the steel sheets. They identified that the size, properties (hard and soft), and shape of the inclusions have a substantial effect on stress concentration and peak stress in a steel sheet in which inclusions are ingrained. Das et al. [8] described the comparative study on machinability of alloy steel using uncoated carbide and coated cermet inserts during hard turning. They experienced coated cermets attained a lower cutting force, minor flank wear, and reduced workpiece surface temperature. The revealed the depth of cut was the influential parameter for feed force and radial force.

The objective of the present work is on the comparative analysis of three different alloy steels. The machining operation was conducted in a dry environment based on Box–Behnken design using three factors and three levels. The machinability of the workpiece is investigated based on the output reacts like machining force and chip morphology.

27.2 Experimental Details

A complete testing procedure was carried out to examine the effect of inclusions on machinability of alloy steels when turning at several cutting conditions without coolant supply. The work materials were EN 8M free-machining steel, EN 1A alloy, and EN 1A Pb&Te alloy steel. Figure 27.1 represents the cylindrical bars of 28 mm diameter and 440 mm length made of different alloying elements which are used for the experiments.

Three parameters such as cutting speed, feed, and depth of cut are discrete at three levels during the experiments and their influences on output parameters such as cutting force and chip morphology are examined. The experimental runs are designed using Box–Behnken design without compromising the accuracy of results. Before getting into the actual machining, the rough layers were removed from the specimens.

27.2.1 Chemical Composition

A portion of the workpiece was cut using the parting tool and was polished using different grades of abrasive sheets and was subjected to the spectroscopy tests. The chemical composition obtained and major constitutions are summarized in Table 27.1.

27.2.2 Hardness and Microstructure

Hardness was determined by tests conducted using Vickers hardness testing machine wherein a square-based pyramid-shaped diamond indenter was used at 300 g load for 10 s. The diagonal of the indentation was measured and Vickers

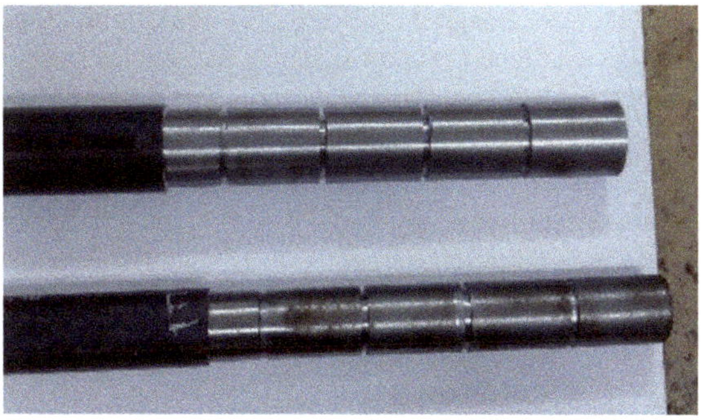

Fig. 27.1 Workpiece material

Table 27.1 Chemical composition of alloy steels (in weight %)

Element (in mass%)	EN 8M steel	EN 1A steel	EN 1A Pb&Te
C	0.37	0.07	0.077
Si	0.21	0.057	0.057
Mn	1.27	1.28	1.25
S	0.14	0.23	0.24
Cr	0.06	0.038	0.04
Pb	–	–	>0.38
Te	–	–	0.001

Table 27.2 Vickers hardness of the alloy steels

Specimens	EN 8M steel	EN 1A steel	EN 1A Pb&Te
Vickers hardness	193	168	161

hardness was determined. The hardness values of the three steels are listed in Table 27.2.

The piece was then observed under an optical microscope after application of natal etchant. EN 8M has medium carbon steel 40% ferrite and 60% pearlite formation, EN 1A has low carbon steel with 70–75% ferrite and 25–30% pearlite, and EN 1A Pb&Te has almost ferrite structure with pearlite of 5–10%.

27.2.3 Inclusions

Inclusions are non-metallic compounds embedded in steel during the manufacturing process. It has a different chemical origin and gives different mechanical properties to steel. Inclusions have domination on several properties of steel, such as formability, toughness, and machinability and corrosion resistance. The workpieces were then observed under an optical microscope to study the distribution of inclusions on the surface as shown in Figs. 27.2, 27.3 and 27.4.

The above image indicates the Type B alumina inclusions. Literature has shown that the existence of non-metallic inclusions [particularly the hard ones such as alumina (Al_2O_3)] and complex oxides and voids has adverse effects on mechanical properties of steels. The inclusions are distributed in a particular area, and it looks like a clustered formation as shown in Fig. 27.2a. Clusters of inclusions are also unfavourable since they may result in a local drop of mechanical properties such as toughness and fatigue strength. These inclusions tend to increase the hardness of the workpiece while simultaneously increasing strength to a particular limit. This increase in hardness makes it difficult to machine.

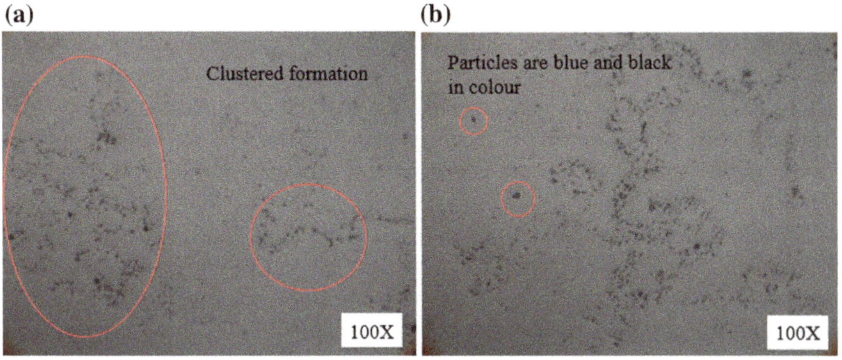

Fig. 27.2 **a** Type B–alumina type thin series 2.5 (fine size 2–9 μm). **b** Type B alumina thin series 1.5 inclusions in EN 8M free-machining steel

Fig. 27.3 Type D globular oxide thin series 3 (size 3–8 μm) inclusions in EN 1A steel

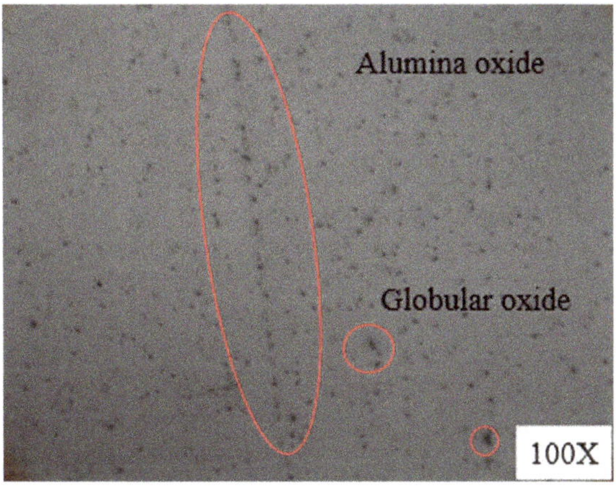

Fig. 27.4 Type D globular oxide thin series 3 inclusions in EN 1A Pb&Te steel

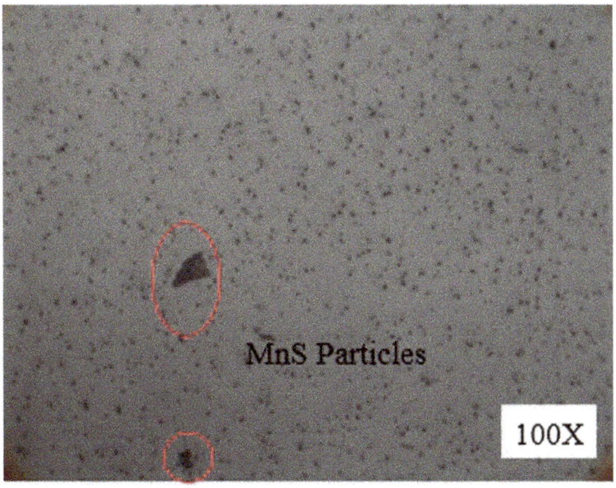

Figure 27.3 shows the type of inclusions formed in EN 1A Alloy steel. Globular shape of inclusions is desirable since their effect on the mechanical properties of steel is moderate. The spherical shape of globular inclusions has shown their formation in a liquid state at low content of aluminium, as indicated in Fig. 27.3. Examples of globular inclusions are manganese sulphide and oxysulphide formed during solidification in the spaces between the dendrite arms, iron aluminates, and silicates.

The MnS inclusions marked are shown in Fig. 27.4. MnS inclusions have a positive influence on the machinability, as they are soft in comparison with steel matrix and act as voids. As a result, the chips are smaller in length which leads to easier removal and declines the total power consumption during the machining.

27.2.4 Experimentation

In the present work, the multilayer CVD-coated inserts are used. The insert grade of CNMG 120408 MT TT5100 was procured from Taegu Tec. The ISO Range is (P20-P35) and shape of the insert is square (90° point angle) without any specific type of chip breaker geometry. This insert is mounted on a tool holder of ISO designation ECLNR-2020K12 as shown in Fig. 27.5. The inserts utilized in these experiments contain multilayer coatings (TiCN–Al_2O_3–TiN). Titanium nitride coatings were given at outer layer, alumina oxide, and titanium carbon nitride are as an intermediate and bottom layer of the insert.

The various control parameters adopted during the experiment are shown in Table 27.3.

Figure 27.6a, b shows the experimental setup. The turning operation was executed on a lathe of Pinacho SC 200x750 with spindle power of 4 kW and speed range of 40–2800 rpm. For measuring the cutting force, three-dimensional Kistler tool dynamometer was used.

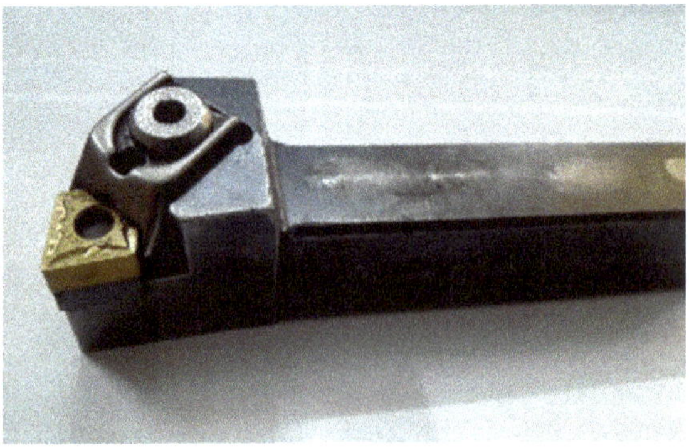

Fig. 27.5 Cutting inserts and tool holder

Table 27.3 Control parameters and their levels for Box–Behnken design

S. No	Parameter	Unit	Symbol	Low level	Medium level	High Level
1	Cutting speed	m/min	*v*	125	162.5	200
2	Feed	mm/rev	*f*	0.1	0.15	0.2
3	Depth of cut	mm	*d*	0.5	0.75	1

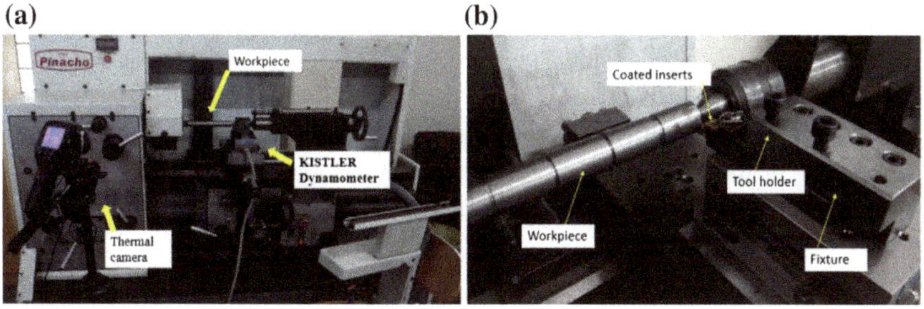

Fig. 27.6 Experimental setup

27.3 Result and Discussion

From Fig. 27.7, it was observed that the resultant cutting force of EN 8M steel was higher when compared to other two materials. When comparing to these three steels, EN 1A Pb&Te steel, accomplished the lower cutting forces at all cutting conditions owing to their low hardness and effects of lead and tellurium content in the steel.

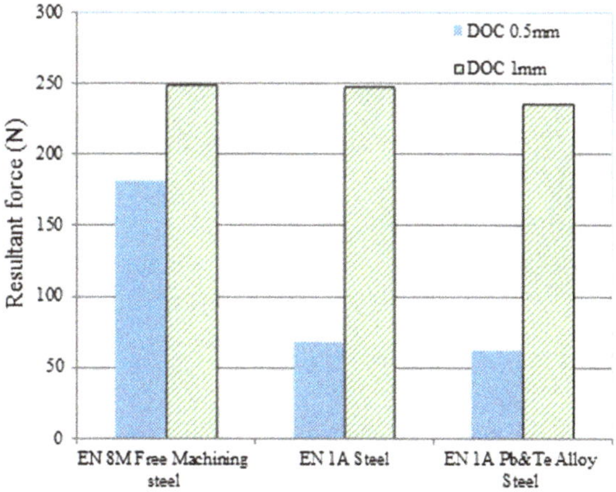

Fig. 27.7 Comparison of resultant forces for three different steels (when cutting speed 162.5 m/min, feed–0.1 mm/rev)

27.3.1 Chip Morphology

All cases which occur in a material during machining are depicted in the chip character, for example, in shape and density of chip morphology. Machining of EN 8M free machining steel obtained helical and continuous chips at higher cutting speed. On the other hand, chips were obtained blue in colour because of carbon content, which is 0.37%. During machining, an adequate amount of heat is expelled from the workpiece. Figure 27.8 shows the chips generated during machining of EN 8M steel. EN 1A alloy steel has produced the chips with light golden in colour and helical continuous one at higher cutting speed as shown in Fig. 27.9. The Lead and Tellurium alloy steel obtained small and discontinuous chips in all cutting conditions. The high content of sulphur (0.24%) results in an increased amount of MnS inclusions in the steel matrix. It is due to the effect of MnS inclusions and leads which predominantly improves the machinability. MnS inclusions act as a stress raiser in the primary shear region during the machining. As a result, they improve the chip formation process as shown in Fig. 27.10.

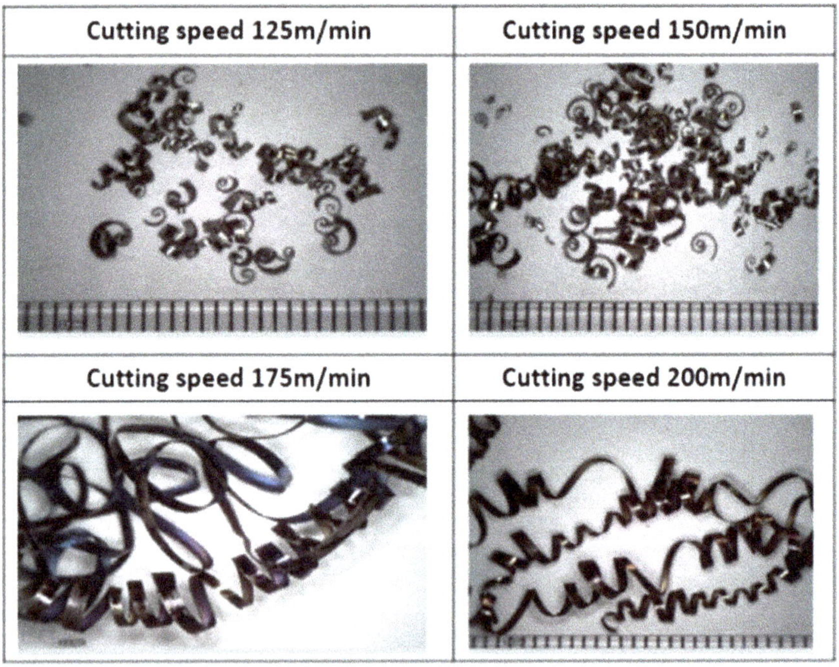

Fig. 27.8 Chips formation of EN 8M free-machining steel

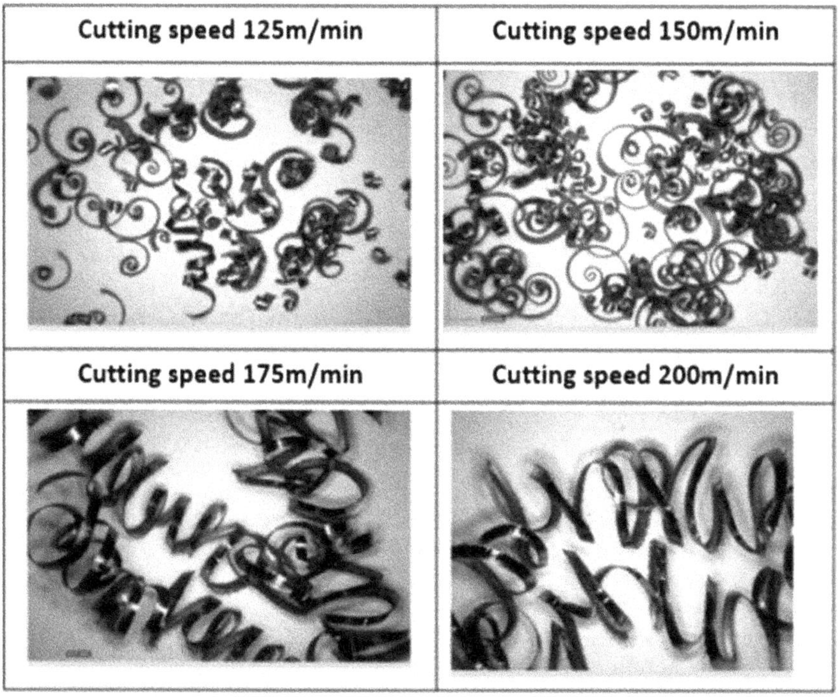

Fig. 27.9 Chip formation of EN 1A alloy steel

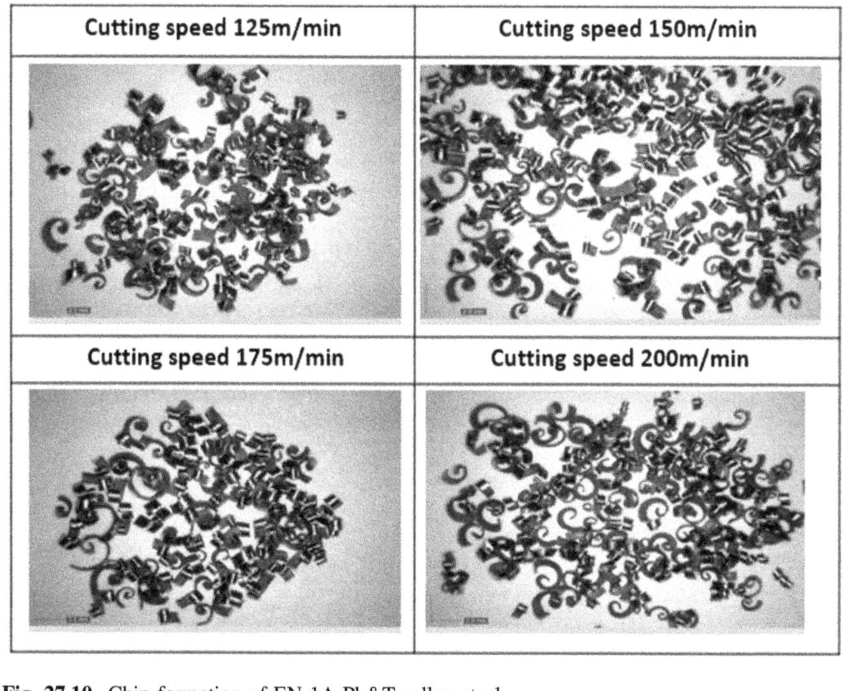

Fig. 27.10 Chip formation of EN 1A Pb&Te alloy steel

27.3.2 *Discussion*

In this overall experimentation, dynamic cutting forces under different cutting speeds and chip morphology were measured during dry machining of different alloy steels. Due to the chemical composition and effect of inclusions, the results were significantly varied. EN 8M free-machining steel is the hardest one because it has 40% ferrite and 60% pearlite, so it generated the highest cutting forces. The chips were segmented associated with low cutting speed, and it is lengthy and continuous at higher cutting speed.

EN 1A has a significant amount of manganese and sulphur, due to this effect, it has a lower level of cutting forces were observed. Chips were discontinuous at lower cutting speed, and it was longer and continuous at higher cutting speed.

EN 1A Pb&Te alloy has an almost ferrite structure and only 5–10% pearlite structure with large amount of inclusions (MnS Particles). Hence, it has generated the least amount of cutting forces; moreover, it has Pb content of >0.38% and Mn 1.25 with S 0.24. These three elements improve machinability. It has produced discontinuous chips in all cutting conditions when compared to other alloy steels.

27.4 Conclusion

The following outcomes are drawn out from the results gained from the current work.

- An influence of carbon content in EN 8M free-machining steel has experienced higher cutting force and obtained long and continues chips at higher cutting speeds.
- Manganese combined with sulphur (MnS) provides better machinability. The tendency of MnS inclusions in EN 1A alloy steel has generated cutting forces moderately and chips were seen curled and continues at higher cutting speed.
- As manganese, sulphur, and lead contents are increased, cutting forces were significantly decreased. Particularly in EN 1A Pb&Te alloy, lower cutting force was seen and discontinuous chips were found. This provides better machinability.

Acknowledgements We would like to thank JSW Steels Pvt. ltd. Salem, for funding this project. We also thank the faculty members of Metallurgical department and Centre for advanced tooling and precision dies, PSG Institute of Advanced Studies, Coimbatore.

References

1. Tanaka, R., Yamane, Y., Sekiya, K., Narutaki, N., Shiraga, T.: Machinability of BN free machining steel in turning. Int. J. Mach. Tools Manuf. **47**(12–13), 1971–1977 (2007)
2. Jiang, R., Chen, X., Ge, R., Wang, W.: Influence on TiB_2 particles on machinability parameter optimization of TiB_2/Al MMCs. Chin. J. Aeronaut. **31**(1), 187–196 (2017)
3. Lalbondre, R., Krishna, P., Mohankumar, G.C.: An experimental investigation on machinability studies of steels by face turning. Procedia Mater. Sci. **6**, 1386–1395 (2014)
4. Anmark, N., Karasev, A., Jönsson, P.G.: The Influence of Microstructure and Non-Metallic Inclusions on the Machinability of Clean Steels. KTH Royal Institute of Technology, WILEY-VCH, Weinheim (2015)
5. Alizadeh, E.: Factors influencing the machinability of sintered steels. Powder Metall. Met. Ceram. **47**(5–6), 304–315 (2008)
6. Joseph, M.: The effeects of sulfide inclusions on mechanical properties and failures of steel components. J. Fail. Anal. Prev. **15**(2), 169–178 (2015)
7. Gupta, A., Goyal, S., Padmanabhan, K.A., Singh, A.K.: Inclusions in steel: micro-macro modelling approach to analyse the effects of inclusions on the properties of steel. Int. J. Adv. Manuf. Technol. **77**(1–4), 565–572 (2014)
8. Das, A., Mukhopadhyay, A., Patel, S.K., Biswal, B.B.: Comparative assessment on machinability aspects of AISI 4340 alloy steel using uncoated carbide and coated cermets insets during hard turning. Arab. J. Sci. Eng. **41**, 4531–4552 (2016)

Chapter 28
Process Parameters Optimization of Electrical Discharge Machining of Al7075/SiC/WS2 by Using MCDM

Rakesh Kumar Patel and M. K. Pradhan

Abstract This paper determines the optimum process parameters of the non-conventional electrical discharge machining. The performance of the EDM machine depends upon the process parameter used. In this analysis, weight percentage, pulse current (I_p), discharge voltage (V), and pulse duration (T_{on}) are used as process parameters. The optimized output parameters are MRR, TWR, surface roughness, and radial overcut. By using face-centered composite design, nine trials were conducted on the workpiece which is made up of Al7075/SiC/WS2 hybrid composite. The trial results obtained were used in decision-making method correlation coefficient and standard deviation (CCSD) integrated approach. These results give useful information on how to control the machining parameters and accuracy of the components produced from EDM. Decision-making method used is simple, and results obtained are confirmed by conducting confirmation experiments.

Keywords Electrical discharge machining (EDM) · Material removal rate (MRR) · Tool wear rate (TWR) · Correlation coefficient and standard deviation (CCSD)

28.1 Introduction

In this article, investigation has been done on the effect of EDM parameters on machining characteristics of EDM machine for Al7075/SiC/WS_2 hybrid composite and the Multi-criteria decision-making (MCDM) method, Correlation coefficient and standard deviation (CCSD) integrated approach used for determining the weights of attributes. CCSD can be used in a multi-objective problem in various fields. EDM is a non-conventional manufacturing process, which is widely used to

R. K. Patel (✉) · M. K. Pradhan
Maulana Azad National Institute of Technology, Bhopal 462003, India
e-mail: rakesh1993patel@gmail.com

M. K. Pradhan
e-mail: mohanrkl@gmail.com

R. G. Narayanan et al. (eds.), *Advances in Computational Methods in Manufacturing*, Lecture Notes on Multidisciplinary Industrial Engineering,
https://doi.org/10.1007/978-981-32-9072-3_28

make more precise, accurate, and attractive components in modern manufacturing industries. In this study, the main focus was to increase material removal rate and minimize surface roughness, tool wear rate, and radial overcut for composite. The Al7075/SiC/WS_2 hybrid composite has been selected to analyze this optimization method. The various EDM input parameters such as discharge voltage (V), discharge current (I_p), pulse-on time (T_{on}), and wt% of reinforcement materials have applied to machining of a composite. The machining of particulate metal matrix composites presents a significant challenge since a number of reinforcement materials are significantly harder than the commonly used high-speed steel (HSS) and carbide tools. The reinforcement phase causes rapid abrasive tool wear, and therefore, the widespread usage of particulate metal matrix composites is significantly impeded by their poor machinability and high machining costs. Although often categorized as difficult to machine, MMCs are actually readily machinable. They form short cutting chips and require moderate cutting forces, and the range of machining parameters at which they can be machined is quite wide. However, MMCs are highly abrasive, and tools can wear rapidly.

To avoid these problems, EDM was used due to its non-contact machining characteristic. More optimum machining parameters in EDM machine are optimized to get a desirable magnitude of EDM responses on particulate metal matrix composites.

Even many attempts have been made by the experimenters for the enhancement of the MRR and improvement in accuracy, but association of CCSD method for obtaining optimal setting on EDM of Al7075/SiC/WS_2 hybrid composite has never been attempted; thus, in this research, the aforesaid unique combination has been attempted.

28.2 Literature Review

A significant number of research papers are reviewed to know the approach of yielding optimal EDM performance measures of high MRR, low TWR, and acceptable ROC. This section provides a study of each of the performance measures and the scheme for their enhancement. Roy et al. [1] used dielectric fluid mixed with Al powder and conducted experiment in EDM machine, and to analyze the experimental results, response surface methodology (RSM) is used. The input process parameters used are pulse current, pulse-on time, and concentration of Al powder on kerosene dielectric and found that due to increase in Al powder on dielectric fluid reduced the MRR and increases surface finish. Pradhan and Biswas [2] have implemented successfully by employing RSM and investigated the individual effect of the input parameters on the responses like MRR and surface roughness (R_a). Further, Pradhan and Biswas [3] have explored the effects of I_p, T_{on}, duty cycle (T_{au}), and V on various response parameters using two neuro-fuzzy and one neural network model. Mandal et al. [4] used an artificial neural network model and multi-objective optimization approach for obtaining the best process parametric combinations for a better productivity and fine surface finish. Dvivedi et al. [5] used Al6063 SiCp metal matrix composite to check machineability in EDM. They investigated the optimum value of MRR and TWR by considering the input variables T_{on}, T_{off}, and I_p. Kanagarajan et al. [6] had investigated the EDM performance of tungsten carbide/cobalt cemented carbide by considering the input parameters such as electrode rotation, I_p, T_{on}, and flushing pressure. Jaharah et al. [7] found that on machining of tool steel on EDM, MRR increases and TWR decreases with increase in the I_p and T_{on}. Dhar et al. [8, 9] used Al–4Cu–6Si alloy—10 wt% SiCp composites on EDM machine—to find the effect of I_p, T_{on}, and V on MRR, TWR, and gap on EDM. To develop the relationship between the machining parameters used a second-order nonlinear mathematical model and found that MRR, TWR, and gap on EDM increase with increase in I_p and T_{on}.

In past, substantial efforts are carried out to increase output parameters such as MRR, TWR, and accuracy of EDM process. The important issue is to pick the process parameters such as I_p, T_{au}, T_{on}, V, flushing pressure, dielectric fluid, and polarity in such a way that increases MRR and accuracy, and simultaneously ROC, TWR, and surface roughness should diminish.

28.3 Description of the Experiments

The experiments were conducted under various parameter settings of discharge current, pulse-on time, and discharge voltage. The Minitab 18 software was used for Taguchi's method to modeling alternatives and attributes of EDM parameters. Finally, the soft computing techniques were employed for modeling of MRR, TWR, radial overcut, and R_a.

Calculate the responses of EDM machine

$$\text{MRR}\left(\frac{\text{mm}^3}{\text{min}}\right) = \frac{\text{Weight Loss (gm)} \times 60}{\text{Density of Sample}\left(\frac{\text{gm}}{\text{mm}^3}\right) \times \text{Machining Time (s)}}$$

$$\text{TWR}\left(\text{mm}^3/\text{min}\right) = \frac{\text{Weight Loss(gm)} \times 60}{\text{Density of Tool}(\text{gm}/\text{mm}^3) \times \text{Machining Time (s)}}$$

$$\text{Radial Over Cut (ROC)} = \frac{D_1 - D_2}{2}$$

where D_1 = right side reading on digital micrometer and D = left side reading on digital micrometer. The Toolmaker microscope was used to calculate ROC. The digital surface roughness meter is used to calculate surface roughness of surface of cavity on composite surface generated by EDM machine (Table 28.1), where 0.75, 1, and 1.5 represent the wt% of WS_2 in composite with constant 10 wt% of SiC (Fig. 28.1).

Table 28.1 Experimental results

Run	Wt. (%)	I_p (A)	T_{on} (μs)	V (V)	MRR (mm^3/min)	TWR (mm^3/min)	ROC (μm)	R_a (μm)
1	0.75	8	75	50	38.144	0.146	5.646	7.100
2	0.75	10	100	60	17.764	0.091	5.706	11.328
3	0.75	12	150	70	56.070	0.119	5.678	12.510
4	1	8	100	70	18.332	0.292	5.693	8.041
5	1	10	150	50	19.132	0.378	5.626	10.381
6	1	12	75	60	39.220	0.363	5.637	8.844
7	1.5	8	150	60	18.033	0.231	5.625	10.323
8	1.5	10	75	70	18.287	0.047	5.678	7.120
9	1.5	12	100	50	39.716	0.102	5.581	9.723

Fig. 28.1 Samples machined on EDM machine

28.4 Analysis Method

Wang et al. [10] used CCSD multi-criteria decision making method. Suppose there are n decision alternatives A_1, …, A_n to be evaluated in terms of m attributes $O_1, O_2, \ldots, O_m$, it forms a decision matrix which can be represented as $X = (x_{ij})$ $n \times m$ where x_{ij} is the performance value of A_i with respect to O_j. The m attributes can be categorized into two groups: beneficial and non-beneficial. The beneficial attributes are those whose values are always larger the better. The non-beneficial attributes are those whose values are smaller the better. Due to the attributes in different units, the decision matrix $X = (x_{ij})n \times m$ needs to be normalized to eliminate its dimensional units. The commonly used normalization method is:

for beneficial attributes

$$z_{ij} = \frac{x_{ij} - x_j^{\min}}{x_j^{\max} - x_j^{\min}} \tag{28.1}$$

for non-beneficial attributes

$$z_{ij} = \frac{x_j^{\max} - x_{ij}}{x_j^{\max} - x_j^{\min}} \tag{28.2}$$

z_{ij} Normalized value of output parameter.

Now remove attribute O_j one by one from the set of attributes one at a time, and see its impact on decision making. When O_j is removed, the overall assessment value of each alternative can be calculated as

$$d_{ij} = \sum_{k=1,k\neq j}^{m} z_{ik} w_k \quad i = 1, 2, \ldots, n$$

The correlation coefficient between the attributes and the score values can be expressed as

$$R_j = \frac{\sum_{i=1}^{n} \left(z_{ij} - p_j\right)\left(d_{ij} - q_j\right)}{\sqrt{\sum_{i=1}^{n} \left(z_{ij} - p_j\right)^2 \left(d_{ij} - q_j\right)^2}}$$

where

$$p_j = \frac{1}{n}\sum_{i=1}^{n} z_{ij}, \quad q_j = \frac{1}{n}\sum_{i=1}^{n} d_{ij}$$

The attribute given more weightage which have bigger standard deviation value and smaller weightage which have lower standard deviation value in comparison.

From the above calculations, the weights of attributes can be determined as

$$w_j = \frac{\sigma_j \sqrt{1 - R_j}}{\sum_{j=1}^{m} \sigma_j \sqrt{1 - R_j}} \quad j = 1, 2, \ldots, n \tag{28.3}$$

R_j Correlation coefficient between the values of O_j.
σ_j Standard deviation of the values of O_j.

The score of each decision alternative can be determined by

$$d_{ij} = \sum_{j=1}^{m} z_{ij} w_j \quad i = 1, 2, \ldots, n \tag{28.4}$$

28.5 Results and Discussion

To get the optimum result of the machining parameters and to reduce the manufacturing cost of composite product, the CCSD method was used to optimization of EDM parameters for machining of hybrid composite (Al7075 + SiC + WS2) and find the rank of alternatives. According to the CCSD integrated approach, the weight of the attributes is 0.2528, 0.2518, 0.2118, and 0.2737 for the attribute MRR, TWR, ROC, and R_a, respectively, which was calculated by using Eq. (28.3).

The overall assessment value of each decision alternative can be computed by

$$d_i = \sum_{j=1}^{m} z_{ij} w_j, \quad i = 1, 2, \ldots, n$$

From Table 28.2, values shown are normalized value which was calculated by using Eq. (28.1) for beneficial and Eq. (28.2) for non-beneficial attribute, and score was calculated by using above Eq. (28.4) (Fig. 28.2).

From the above graph, alternative 9 has highest score, so it is best suitable alternative according to MCDM. EDM of hybrid composite Al7075/SiC/WS2 with input value of alternative 9 gives optimum value of output parameters.

Table 28.2 Overall score assessment and ranking

Alternatives	MRR (mm^3/min)	TWR (mm^3/min)	ROC (mm)	R_a (μm)	Score	Rank
1	0.5320	0.7013	0.4784	1	0.6913	2
2	0	0.8680	0	0.2184	0.287	8
3	1	0.7819	0.2192	0	0.5009	4
4	0.0148	0.2593	0.1	0.8260	0.3171	7
5	0.0357	0	0.6384	0.3935	0.2553	9
6	0.5601	0.0469	0.5528	0.6776	0.4555	5
7	0.0070	0.4454	0.6464	0.4042	0.3698	6
8	0.0136	1	0.2192	0.9963	0.5835	3
9	0.5730	0.8346	1	0.515	0.7197	1

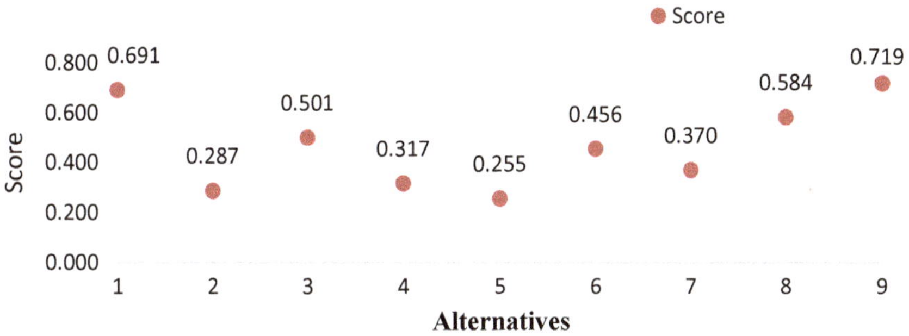

Fig. 28.2 Score of the alternatives

28.6 Conclusion

This study proposes optimization of the EDM input process parameters. CCSD method shows that the alternative 9 is the best alternative. The optimal condition out of the given alternatives determined by the CCSD methods is as follows: pulse current 12 A, pulse-on time 100 μs, weight percentage 1.5, and discharge voltage 100 V. These results will help in reducing the machining cost, error, and time consumption in the machining process and also enhance the surface quality and efficiency. This study gives the optimum input process parameter of EDM to achieve desired MRR, TWR, surface roughness, and radial overcut, in the die sinking of Al7075/SiC/WS2 hybrid composite.

References

1. Roy, C., Syed, K.H., Kuppan, P.: Machinability of Al/10% SiC/2.5% TiB2 metal matrix composite with powder-mixed electrical discharge machning. Procedia Technol. **25**, 1056–1063 (2016)
2. Pradhan, M.K., Biswas, C.K.: Neuro-fuzzy and neural network-based prediction of various responses in electrical discharge machining of AISI D2 steel. Int. J. Adv. Manuf. Technol. **50** (5–8), 591–610 (2010)
3. Pradhan, M.K., Biswas, C.K.: Investigating the effect of machining parameters on EDMed components a RSM approach. J. Mech. Eng. **7**(1), 47–64 (2010)
4. Mandal, D., Pal, S.K., Saha, P.: Modeling of electrical discharge machining process using back propagation neural network and multi-objective optimization using non-dominating sorting genetic algorithm-II. J. Mater. Process. Technol. **186**(1–3), 154–162 (2007)
5. Dvivedi, A., Kumar, P., Singh, I.: Experimental investigation and optimization in EDM of al 6063 SiCp metal matrix composite. Int. J. Mach. Mach. Mater. **3**(3–4), 293–308 (2008)
6. Kanagarajan, D., Karthikeyan, R., Palani kumar, K., Sivaraj, P.: Influence of process parameters on electric-discharge machining of WC/30%Co composites. Proc. Inst. Mech. Eng. Part B J. Eng. Manuf. **222**(7), 807–815 (2008)

7. Jaharah, A.G., Liang, C.G., Wahid, S.Z., Ab, R.M., Che, H.C.: Performance of copper electrode in electrical discharge machining (EDM) of AISI H13 harden steel. Int. J. Mech. Mater. Eng. **3**(1), 25–29 (2008)
8. Dhar, S., Purohit, R., Saini, N., Sharma, A., Kumar, G.H.: Mathematical modeling of electric discharge machining of cast Al–4Cu–6Si alloy–10 wt.% SiCP composites. J. Mater. Process. Technol. **194**(1–3), 24–29 (2007)
9. Purohit, R., Krishna, C.M., Rana, R.S., Kumar, V., Patel, K.K.: Optimization of electric discharge machining parameters for Al–3 weight% nano SiCp composites using copper electrode. Trends Ind. Mech. Eng. **105** (2016)
10. Wang, Y.M., Luo, Y.: Integration of correlations with standard deviations for determining attribute weights in multiple attribute decision making. Math. Comput. Model. **51**(1–2), 1–2 (2010)

Chapter 29
CNC Turning and Simulation of Residual Stress Measurement on H13 Tool Steel

Ravi Butola, Qasim Murtaza and Ranganath M. Singari

Abstract This paper deals with the measurement of residual stress produced during turning operation. CNC turning is a machining process to remove unwanted material from the workpiece to produce desired rotational parts. Turning is a widely used cutting process in the manufacturing industry, and thus, continuous research is going on to optimize it because of cutting high-performance dynamics, workpiece materials, and cutting tools. Turning operation was performed on H13 tool steel workpiece dimensions 110 × Ø22 mm on CNC Lathe machine tools. H13 tool steel is used for manufacturing of friction stir welding and processing tool. The main aim of the current study is to determine the residual stress using PulsetecμX-360n portable stress analyzer setup, and Abaqus 6.14 software was used to simulate the turning process. Experimental tests were performed on H13 tool steel materials which are in form of FSW/FSP tool, which consists of different pin profiles. The experimental outcomes are then compared with the outcomes obtained from the simulation. Experimental results validate the simulation results for both the tool pin profiles which are lying within the acceptable range.

Keywords Turning · H13 tool steel · Simulation

R. Butola · Q. Murtaza (✉) · R. M. Singari
Department of Mechanical, Production and Automobile Engineering, Delhi Technological University, Delhi 110042, India
e-mail: qasimmurtaza@dce.ac.in

R. Butola
e-mail: ravibutola33855@gmail.com

R. M. Singari
e-mail: ranganath@dce.ac.in

R. G. Narayanan et al. (eds.), *Advances in Computational Methods in Manufacturing*, Lecture Notes on Multidisciplinary Industrial Engineering, https://doi.org/10.1007/978-981-32-9072-3_29

29.1 Introduction

Turning operation was performed on H13 tool steel workpiece on CNC Turning Center Lathe. The dimension of workpiece was 110-mm-long cylinder with 22 mm diameter. The cutting tool insert used was uncoated tungsten carbide CNMG 120404–THM for turning operation, and it was a dry turning since no cutting fluid was used. In turning processes, large plastic deformation takes place; the prediction and control of the material fracture is a critical issue. In order to surface finish and integrity of the new component, the damage and fracture of the materials should be predicted. H13 tool steel is used for manufacturing of friction stir welding and processing tool which are used for making welding joints and surface composite. Friction stir processing, a development based on friction stir welding (FSW), is a solid-state processing technique for microstructure modification [1]. FSW/FSP tool consists of shoulder and different pin profiles, i.e., cylindrical and taper cylindrical pin profiles have been used to process the aluminum-based surface composite at different tool rotation and tool traveling speeds. Cylinder pin profile is simplest in construction considering the machining operations performed in making the cylindrical pin profile. The parameters considered for the tool design are shoulder diameter, pin diameter, and length of the pin. Huang et al. [2] studied the impact of residual stress at the initial stage and turning-induced stresses after deformation of aluminum and found that turning-induced residual stresses are the primary cause of distortion. Plate deformation is increased by coupling of compressive residual stress and machining-induced residual stress while deformation is decreased by the coupling action of tensile initial residual stress and machining-induced residual stress. Wang et al. [3] carried out a study on residual stress in machined surface of monocrystalline silicon based on molecular dynamics simulation. They found that on the increasing depth of cut, the maximum tensile residual stress decreases while compressive residual stress increases underneath the surface and same happened when the tool rake angle changed from positive to negative. Fergani et al. [4] predicted the residual stress in multi-pass milling machining using thermomechanical stress model using Neumann–Duhamel law. A regeneration algorithm was proposed that considers the combination of initial stresses and material cyclic hardening in residual stress calculations. Sharma and Butola [5] reviewed Taguchi's technique, response surface methodology, and gray fuzzy algorithm for optimal values of machining parameters in the CNC process of turning. They found that material removal rate increased with increase in feed rate and depth of cut, while the increase in cutting speed decreased it. Huang et al. [6] investigated the influence of residual stress on the monolithic component deformation using FEM as well as experimental work and found that initial residual stress in the blank was the main factor of deformation for three-frame monolithic beam and deformation is increased by coupling of initial and turning residual stresses. Johnson and Cook [7] have been developed that the flow stress is defined as a function of strain, strain rate, and temperature. Pres et al. [8] have been reported a measurement, simulation, and the comparison between the cutting force and geometrical feature of the chip. Wang et al. [9] have been presented the frequency and cutting force for Ti6Al4V after that they compare their simulation and experimental result, and they found good agreement between them. Umbrello et al. [10] develop a hardness flow stress simulation model of turning processes based on Johnson–Cook equation. Chaudhary et al. [11] have reviewed that different FSP tool pin is used at different parameter. FSP can be widely used in metals such as Al alloy that is lightweight and can be used extensively in various applications. The significance of this research will have a major and robust impact on the manufacturing industry. The prediction of residual stresses by using Abaqus software could become an opportunity to undertake by the tool manufacturers by changing or improving the cutting tool design prior to manufacturing and same supported by experimental data using PulsetecμX-360n portable stress analyzer setup.

The objective of this research is to the measurement of residual stress using PulsetecμX-360n portable stress analyzer setup. After CNC turning, the process on H13 tool steel materials which are in form of FSW/FSP tool has manufactured two different tool profiles, i.e., cylindrical and taper cylindrical pin and shoulder. To simulate the turning of H13 tool steel workpiece and determine stresses, Abaqus 6.14 software was used.

29.2 Experimental Method

29.2.1 Selection of Tool Materials

Each material is characterized by a unique set of physical, mechanical, and chemical properties which can be treated as attributes of a specific material. The selection of material is primarily dictated by the specific attributes that are required for manufacturing friction stir processing tool. H13 tool steel is used for manufacturing of friction stir welding and processing tool which are used for making welding joints and surface composite. The tool requisite for friction stir processing must resist thermal fatigue cracking under hot working operations. H13 tool steel is versatile chromium–molybdenum hot work steel that is widely used in manufacturing of friction stir welding and processing tool. The workpiece in form H13 tool steel rod of diameter 22 mm and length 110 mm, processed by Mahindra and Mahindra Steel, was used in this investigation. The chemical composition of these materials in weight percentage as supplied by the producer is given in Table 29.1.

Table 29.1 Chemical composition of tool material-H13 tool steel

C	Mn	Si	Cr	Mo	V
0.40	0.40	1.00	5.25	1.35	1.00

29.2.2 Machining Methodology

Machining methodology consists of machining techniques and equipment used for machining or shaping the tools. In this research, two workpieces were made with different pin profiles like cylindrical and taper cylindrical. For machining the tool material into the desired tool pin profile, CNC Lathe machine tool was used. Primarily, turning was used for making the shoulder of the tool. For a cylindrical pin, the further turning operation was performed. In the first step, turning reduced the diameter from 22 to 19.95 mm for the entire length of the tool because 19.95 mm diameter is the maximum diameter that can accommodate the FSW machine in our laboratory. In the second step, turning reduced the diameter from 19.65 to 7.22 mm for pin length of 3.50 mm. The taper cylindrical pin has generated a taper of length 3.50 mm with diameter linearly varying from 7.22 to 3.50 mm, and tool pin length depends upon the thickness of the workpiece to be processed (Fig. 29.1; Table 29.2).

29.2.3 Measurement Condition of H13 Tool Steel

Measurement of residual stress in cylindrical pin and taper cylindrical pin was performed on the following conditions: pitch 50 μm, X-ray irradiation time (setup) 30 s, X-ray irradiation time (Meas.) 30 s, X-ray irradiation time (Max) 101 s, X-ray

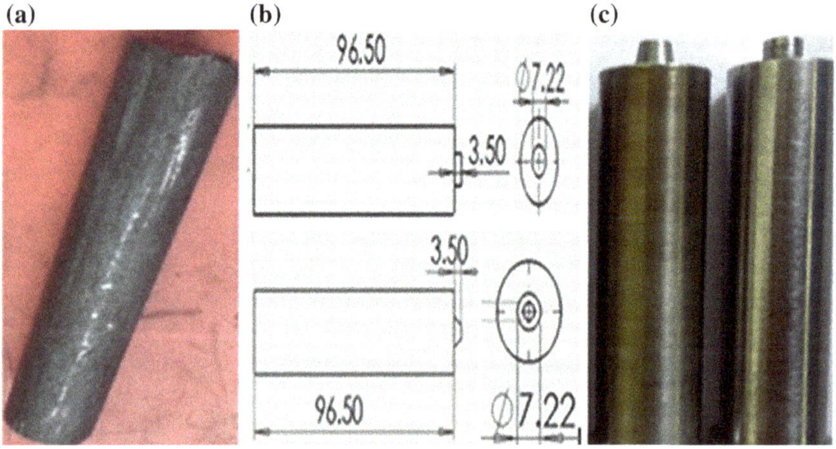

Fig. 29.1 **a** H13 tool steel rod, **b** dimension of tool, and **c** after turning

Table 29.2 Input machining parameter

Cutting speed (S)	Feed rate (F)	Depth of cut
1500 (Rpm)	0.10 (mm/rev)	1 (mm)

tube current 1.00 mA, X-ray tube voltage 30.00 kV, X-ray incidence angle 35.0°, offset of alpha angle 0°, X-ray wavelength (K-Alpha) 2.29093 A (Cr), X-ray wavelength (K-Beta) 2.08480 A (Cr), total measurement count 6631, oscillation count 7, detection sensitivity 26.1% (224,879), peak strength (Ave) 209 k, and temperature 31.56 °C.

29.2.3.1 Sample Information

The basic idea was to measure residual stresses on the machined workpieces of H13 tool steel as follows: wavelength K-Alpha, diffraction angle $(2\theta) = 156.396°$, diffraction lattice angle $(2\eta) = 23.604°$, interplanar spacing $(d) = 1.170$, diffraction plane $(h, k, l) = 2, 1, 1$, crystal structure BCC, Young's modulus $(E) = 224.000$ GPa, Poisson's ratio $(v) = 0.280$, sigma (x) stress constant $(K) = -465.097$ GPa, tau (xy) stress constant (K) 380.985 GPa, and sigma (y) stress constant $(K) = -2091.661$ GPa.

29.2.3.2 Measurement of Residual Stresses

Residual stress measurement was performed in Precision Manufacturing Lab at Delhi Technological University. After turning operation of H13 tool steel, the next step is to the measurement of residual stress has been done on the specimen using PulsetecμX-360n portable stress analyzer setup. Two specimens were performed for measurement of residual stress, i.e., cylindrical and taper cylindrical profiles. Mechanical residual stresses are generally produced because of the plastic deformation in the non-uniform manner caused during manufacturing operations. Residual stresses can be made naturally during the operation or may be produced in order to generate the desired stress profile in the component [12] (Fig. 29.2).

29.3 Modeling of Cylindrical and Taper Cylindrical Tool

The following steps will illustrate in different stages involved in the FEA of the turning process to create a cylindrical and taper cylindrical tool profile discussed below

- Standard and explicit model was selected upon opening the Abaqus 6.14 package. This allows for explicit integration scheme to solve highly nonlinear systems with many complex contacts under transient loads.
- In the Part module, a 3-D deformable part with solid extrusion feature is chosen. This part is the H13 tool steel workpiece.
- Figure 29.3a shows a circular sketch of 22 mm radius which is drawn. This was the raw dimension of the workpiece. Figure 29.3b shows the part is extruded to

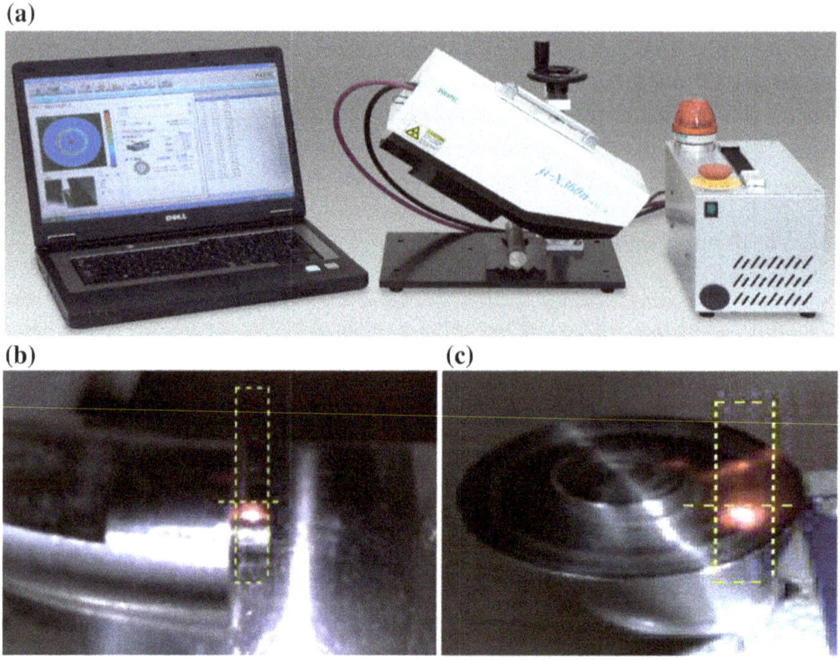

Fig. 29.2 **a** PulsetecμX-360n portable stress analyzer set up, **b**, **c** image captured by stress analyzer

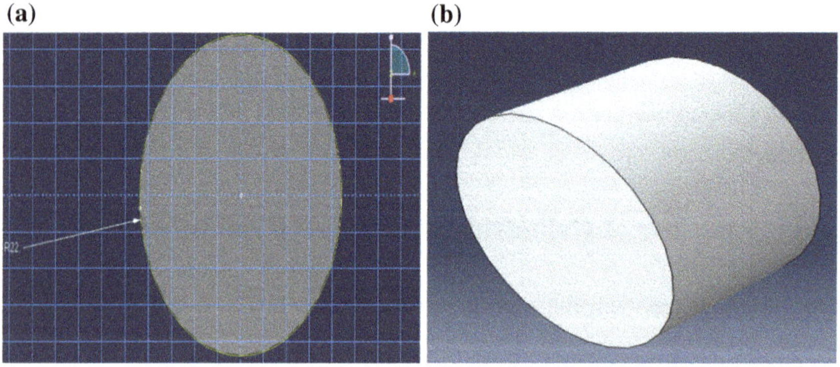

Fig. 29.3 **a** Geometry for workpiece with R 22 mm and **b** shown 30-mm-extruded cylindrical workpiece

30 mm depth. Note that it is shorter than the actual length of the workpiece. This is because we are only interested in the region most affected by the turning process.

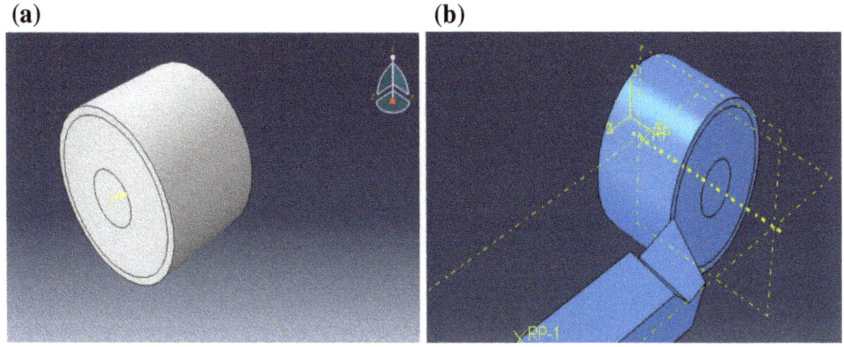

Fig. 29.4 **a** Partitioned regions in the workpiece and **b** assembly

- Figure 29.4a shows the part is partitioned into three different regions. It will help in improving the mesh generation. Since there is step cutting involved, different regions can be meshed to be of finer/coarser quality.
- In the Materials module, the material is defined by entering the properties of H13 tool steel under the *Material Behaviors* section. Properties like conductivity, ductile damage, shear damage with damage evolution, density, elastic, expansion, and plastic behavior are defined. Johnson-Cook damage criterion was followed.
- Figure 29.5a shows the Mesh module, 3D stress family of explicit, and linear geometric order, and Hex element type is assigned to the workpiece. Element deletion is chosen since we are not interested in chip flow and only the stresses on the workpiece surface. In the interaction module, friction contact with the coefficient of 0.4 is used. In the load module, boundary conditions are added. Here, the angular velocity to the workpiece, the depth of cut, and feed rate are given to the tool.

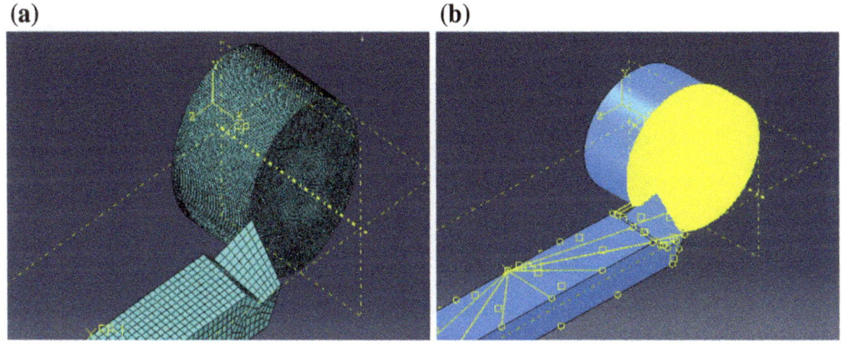

Fig. 29.5 **a** Meshing and **b** constraints added to the workpiece and tool

The procedure followed for the stress analysis of the cylindrical tool pin was similar to the procedure described in the previous section. The only exception lies in the load module, where the boundary conditions are modified so that the tool travels in a straight path parallel to the workpiece axis.

29.4 Results and Discussion

As these stresses play a very important role in the working performance of the component, it becomes quite necessary to examine the magnitude and nature of the residual stresses contained in the component. Stresses present can be tensile or compressive in nature but generally tensile nature of residual stresses is undesired as it can because quench cracking, stress-induced corrosion cracking, and fatigue failure. Compressive nature is desired one as it gives opposite effect of the tensile stresses on the component.

The developed FE analysis model was validated by comparing the simulation of the turning process with the experimental data. Figures 29.5 and 29.6 show the simulation of turning process by using Abaqus 6.14 software. Due to advancement in technology and new methodology, now residual stresses due to turning operation can be visualized using computer simulation. Experiments have done by using PulsetecμX-360n portable stress analyzer machine. Residual stress name is given to the stress which is left in the component after removal of the load. It remains almost in every rigid structure generated due to every operation performed on it such as metallurgical, thermal, and mechanical during manufacturing. Simulation of residual stress is an important process when H13 tool steel is used for CNC turning operation. Measurement of residual stresses due to CNC turning can now be visualized via computer simulations due to advancements in technology.

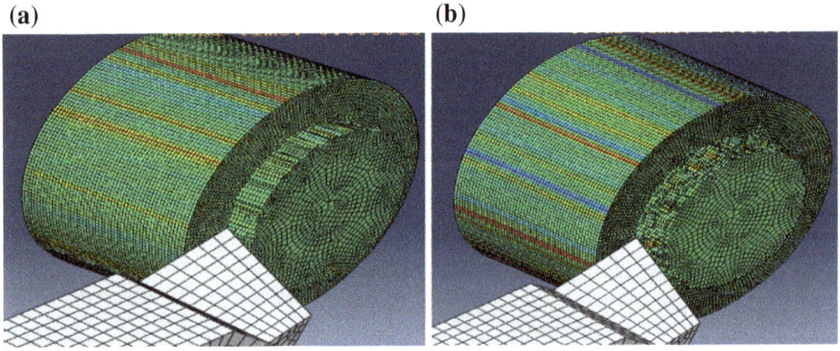

Fig. 29.6 **a** Stress contours of cylindrical tool pin and **b** stress contours of taper cylindrical tool pin

Unfortunately, turning processes have not yet been simulated to perfection in software due to a large number of deformations and variables involved. To overcome this problem, FE analysis is carried out in Abaqus/CAE 6.14. Experimental data is obtained from by using a PulsetecμX-360n portable stress analyzer setup. Experimental and simulation results show the percentage error lying within the acceptable range for both the cylindrical and taper cylindrical tool of H13 tool steel (Figs. 29.7, 29.8, 29.9 and 29.10; Tables 29.3 and 29.4).

The outcome of the residual stress on the periphery of a cylindrical pin and shoulder was observed to be compressive 166 and 280 MPa, respectively. The standard deviation was 16, 19 MPa, respectively. The outcome of the residual stress on the periphery of the taper cylindrical pin and shoulder was observed to be compressive 301 and 444 MPa, respectively. The standard deviation was 17 and 18 MPa, respectively.

After the measurement of residual stress, an experimental data and simulation data for cylindrical tool pin were found to be compressive 223 MPa and compressive 240 MPa, respectively, so the percentage error was observed that 7.62%. For taper cylindrical tool pin, an experimental data and simulation data were found to be compressive 372 MPa and compressive 398 MPa, respectively, so the percentage error observed was 6.98%.

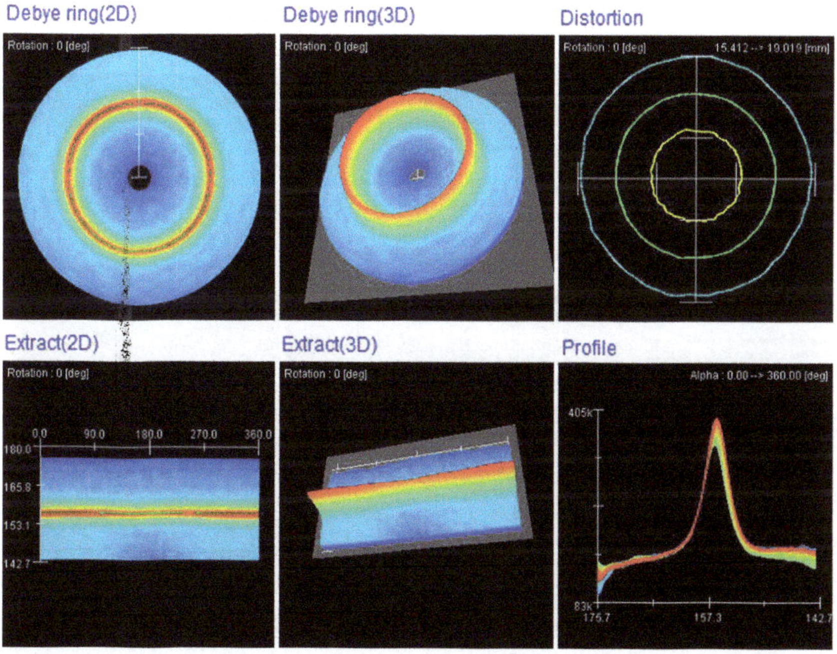

Fig. 29.7 Cylindrical tool pin. **a** Debye ring 2D and 3D, **b** distortion, **c** extract 2D and 3D, and **d** profile

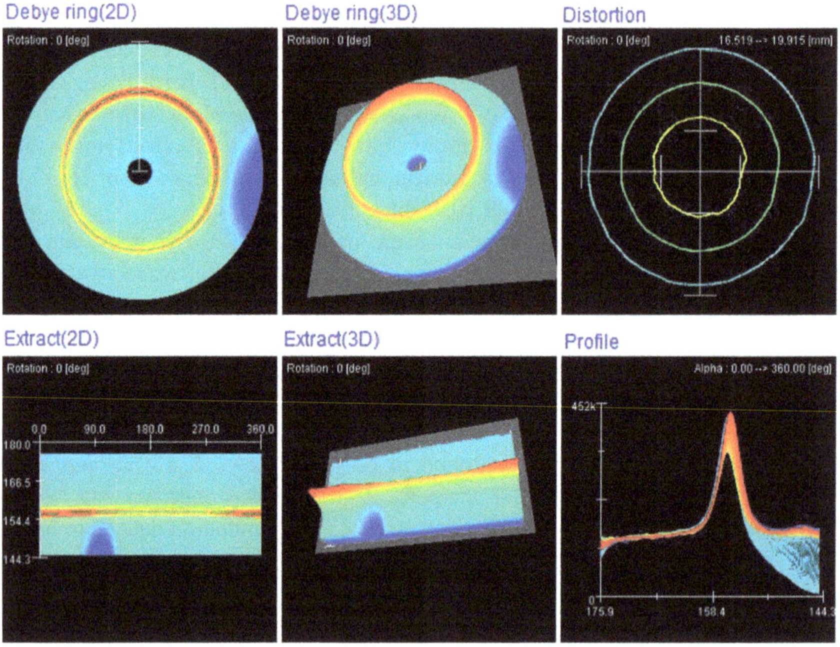

Fig. 29.8 Cylindrical tool shoulder. **a** Debye ring 2D and 3D, **b** distortion, **c** extract 2D and 3D, and **d** profile

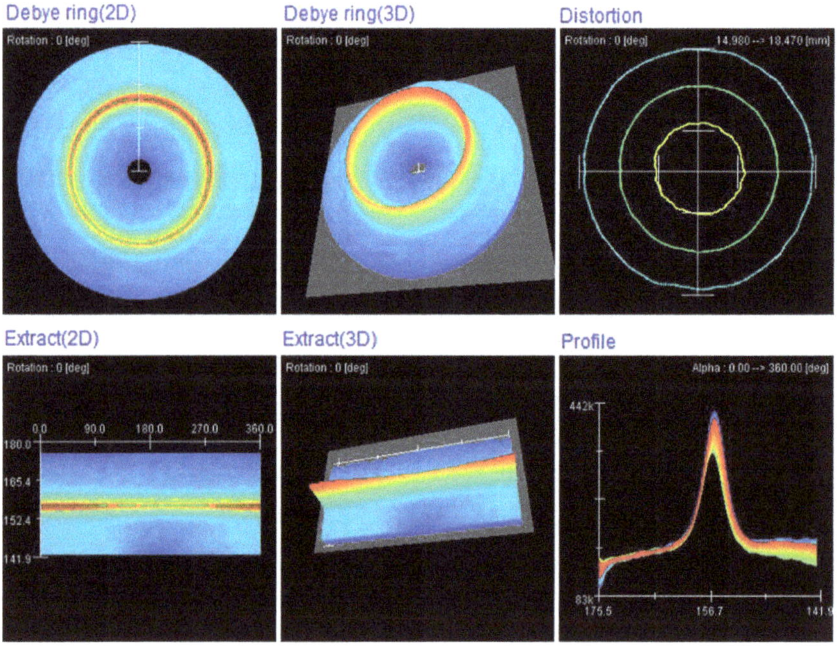

Fig. 29.9 Taper cylindrical tool pin. **a** Debye ring 2D and 3D, **b** distortion, **c** extract 2D and 3D, and **d** profile

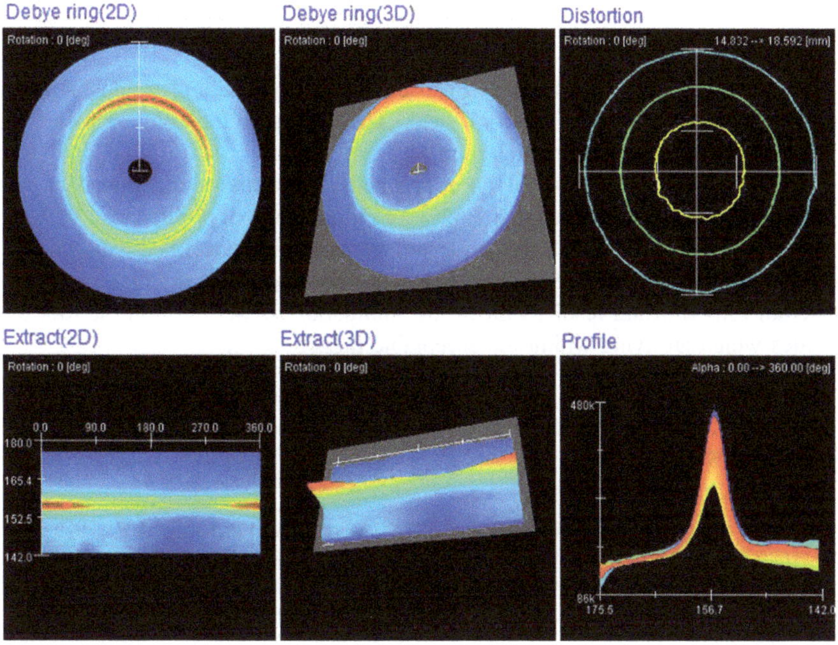

Fig. 29.10 Taper cylindrical tool pin. **a** Debye ring 2D and 3D, **b** distortion, **c** extract 2D and 3D, and **d** profile

Table 29.3 Measurement of residual stress in different tool pin and shoulder

S. No.	Pin profile	Residual stress [MPa] (Absolute)	Average stress [MPa] (Absolute)
1	Cylindrical pin	166	223
2	Cylindrical shoulder	280	
3	Tapered cylindrical pin	301	372
4	Tapered cylindrical shoulder	444	

Table 29.4 Summary of results and error %

S. No.	Tool pin profile	Residual stress [MPa] (Absolute) from simulation	Average stress [MPa] (Absolute) from experiment	Error %
1	Cylindrical pin	240	223	7.62
2	Tapered cylindrical pin	398	372	6.98

29.5 Conclusions

In the current research, the study was carried out on simulation of residual stresses stimulated by turning of H13 tool steel. Two different tool pin profiles, i.e., cylindrical and taper cylindrical is used.

After analyzing the result of an experimental data and simulation data, the percentage error was observed that 7.62% for cylindrical tool profile and 6.98% for taper cylindrical tool profile and the percentage error can be rectified with the help by improving the mesh size quality that could be helpful for production of new material. Experimental results validate the simulation results for both the tool pin profiles which are lying within the acceptable range.

References

1. Mishra, R.S., Mahoney, M.W., McFadden, S.X., Mara, N.A., Mukherjee, A.K.: High strain rate super plasticity in a friction stir processed 7075 Al alloy. Scr. Mater. **42**, 163–168 (2000)
2. Huang, X., Sun, J., Li, J.: Effect of Initial residual stress and machining-induced residual stress on the deformation of aluminium alloy plate. J. Mech. Eng. **61**(2), 131–139 (2015)
3. Wang, Y., Shi, J., Ji, C.: A numerical study of residual stress induced in machined silicon surfaces by molecular dynamics simulation. Appl. Phys. A **115**, 1263–1279 (2014)
4. Fergani, O., Jiang, X., Shao, Y., Welo, T., Yang, J., Liang, S.: Prediction of residual stress regeneration in multi-pass milling. Int. J. Adv. Manuf. Technol. **83**, 1153–1160 (2016)
5. Sharma, V., Butola, R.: Optimization of machining parameters in CNC turning of hybrid metal matrix composites using different techniques: a review. IJARI **5**, 78–82 (2017)
6. Huang, X., Sun, J., Li, J.: Finite element simulation and experimental investigation on the residual stress-related monolithic component deformation. Int. J. Adv. Manuf. Technol. **77**, 1035 (2015)
7. Johnson, G.R., Cook, W.H.: A constitutive model and data for metals subjected to large strains, high strain-rate, and temperatures. In: Proceedings of the 7th International Symposium on Ballistics, The Hague, The Netherlands, pp. 1–7 (1983)
8. Preś, P., Skoczyński, W., Jaśkiewicz, K.: Research and modeling workpiece edge formation process during orthogonal cutting. Arch. Civ. Mech. Eng. **14**(5), 622–635 (2014)
9. Wang, B., Liu, Z.: Investigations on the chip formation mechanism and shear localization sensitivity of high-speed machining Ti6Al4V. Int. J. Adv. Manuf. Technol. **75**(5), 1065–1076 (2014)
10. Umbrello, D., Rizzuti, S., Outeiro, J., Shivpuri, R., M'Saoubi, R.: Hardness-based flow stress for numerical simulation of hard machining AISI H13 tool steel. J. Mater. Process. Technol. **199**, 64–73 (2008)
11. Chaudhary, A., Kumar, D.A., Goel, A., Butola, R., Ranganath, M.S.: The mechanical properties of different alloys in friction stir processing: a review. Mater. Today Proc. **5**, 5553–5562 (2018)
12. Brien, E.O.: Residual Stress—Friend or Foe A Civil Aircraft Perspective, Vol. 1, pp. 13−21. ICRS-6, Oxford, UK, 10–12 July (2000)

Chapter 30
Small-Scale Machining Simulations

Anuj Bisht, Anish Roy, Uday S. Dixit, Satyam Suwas and Vadim V. Silberschmidt

Abstract Molecular dynamics (MD) and single-crystal plasticity finite-element method (CP FEM) are approaches used to simulate the micro-machining process. At such small-length scales, anisotropic behaviour of material becomes important; the two methods can essentially capture it. Therefore, it is important to understand the fundamental principle behind these methods as well as their capabilities and limitations in order to select the scheme to simulate the micro-machining process. In this paper, the fundamentals of MD and CP FEM are introduced in brief. The applicability of the respective method is further illustrated with the help of examples from the literature. Thereafter, the two methods are compared and discussed in terms of their various aspects and capabilities. This discussion should enrich the reader and assist their choice of an appropriate simulation method.

Keywords Micromachining · CP FEM simulation · Molecular dynamics

30.1 Introduction

There is a push for miniaturization of devices with improved performance in electronic and sensor industry. Precision micromachining in such components is unavoidable often to facilitate assembly of parts. The behaviour of such devices under the process of micro-machining is not completely understood. This is primarily due to the change in behaviour of material at lower length scales, which in turn depends on intrinsic and extrinsic material parameters. Firstly, grain size is the most important intrinsic material parameter, which influences the materials

A. Bisht · A. Roy (✉) · V. V. Silberschmidt
Loughborough University, Loughborough LE11 3TU, UK
e-mail: a.roy3@lboro.ac.uk

U. S. Dixit
Indian Institute of Technology Guwahati, Guwahati 781038, India

S. Suwas
Indian Institute of Science, Bangalore 560012, India

R. G. Narayanan et al. (eds.), *Advances in Computational Methods in Manufacturing*, Lecture Notes on Multidisciplinary Industrial Engineering,
https://doi.org/10.1007/978-981-32-9072-3_30

behaviour. The strength of the material increases with the decrease in grain size from micron to submicron and nanometre regime, known as "Hall–Petch (HP) effect" [1, 2]. The increase in strength upon grain size reduction is mainly attributed to the increase in grain boundary area, which acts as barrier for dislocation motion [3]. Secondly, the size of the sample is an extrinsic material factor, and the strength of the material increases in decreasing the sample dimension to micron and submicron scale [4–6], an effect similar to the HP effect. This phenomenon is mainly attributed to the starvation of dislocation (plasticity carriers) on reducing the sample dimension. Apart from the mentioned phenomena, material is observed to drift from the conventional behaviour when it is machined at micron and nanoscale. Similar behaviour is also observed under indentation at various length scales, known as "indentation size effect (ISP)" [7–9]. Due to the unconventional behaviour of material showing scale effects, there is a need to study the process to better understand the phenomenon occurring during micro-machining process.

Experimental studies are instrumental in reflecting the actual events. However, due to the limitations of method involved in experiments, the phenomena can be studied only after its completion and the evolution of defects in the material is speculated based on the findings. In this regard, simulation methods are quite instrumental to study the evolution of material. Finite-element methods are used conventionally to simulate various material processes [10–12], which includes deformation behaviour, using bulk phenomenological-based models. They are good in simulating phenomenon where the scales are larger compared to the inherent dislocation slip phenomena. Johnson–Cook model is a popular model used and is for isotropic material. However, materials are crystalline and, hence, inherently anisotropic in nature [13]. The anisotropic nature becomes prominent under the event of limited small-scale deformation process, where the deformation length scale becomes comparable to deformation features like slip bands due to dislocation slip activity. To account for the crystal anisotropy, there are two simulation methods adopted for simulating the micro-machining processes, namely single-crystal plasticity (CP) finite-element method (FEM) [14–16] and molecular dynamics (MD) [17–19]. The two mentioned simulation methods are based on different principles.

The CP FEM modelling takes into account the anisotropy in material in a FEM framework. On the other hand, MD is an atomic-level simulation and is primarily based on Newton's second law of motion to describe the trajectory of atoms with time. Atomic potential function is used to describe the interatomic interaction between atoms. This essentially means the scale of the two simulations is very different.

The aim of the present paper is to introduce to the reader the two methods used for simulation of micro-machining process. They are described using two examples from the literature. The principle behind the two methods is introduced in brief in their respective section. The example also focuses on the information, which can be extracted from the respective simulation.

30.2 Simulation Techniques

Here, we discuss the basics of the two simulation methods, namely molecular dynamics (MD) simulation and finite-element (FE)-based crystal plasticity (CP) numerical models. In this section, we briefly discuss the two techniques, which are commonly used to predict machining outcomes in the small-length scales.

30.2.1 Molecular Dynamics Simulation

Molecular dynamics (MD) is a versatile simulation technique, which finds applications in the simulation of material behaviour at very small-length scales (scale of molecules). In MD, the behaviour of an ensemble of molecules of the material is studied when subjected to an externally imposed thermo-mechanical load. It proceeds by solving the many body problem involving atoms at each time step. The principle relation in MD is based on Newton's second law of motion. Newton's equation of motion for an atom i in the system is given by

$$f_i = m_i a_i, \tag{30.1}$$

where f_i is the net force exerted on the atom i, m_i is its mass and a_i is the acceleration of atom i. The net force f_i on atom i is the summation of individual forces (f_{ij}) exerted by atoms j on atom i.

The net force f_i is given by the gradient of the potential energy (V):

$$f_i = -\nabla_i V. \tag{30.2}$$

Potential functions are used to describe the interacting forces (f_{ij}) among the atoms i and j. Lennard-Jones (LJ) potential [20, 21], analytical bond-order potential (ABOP) [22], embedded atom method (EAM) [23] and Tersoff potential [24] are important potential functions to name a few. Specific potential functions were developed for specific bond types including ionic bond, metallic bond, covalent bond, valence bond and others. The choice of an appropriate potential function depends on the material and problem at hand [25]. For example, the EAM potential is useful for describing the interaction among interacting metallic atoms, while ABOP is trustable for describing covalent materials.

Once the force is obtained at the current time step t, the position (r_i) and velocity (v_i) of atom i can be obtained for the next time step $t + \delta t$ by time integration step. Various time integration schemes are available to march ahead in time. Velocity Verlet integrator [26] and rRESPA multi-timescale integrator [25] are popular time integration schemes. The position (r_i) and velocity (v_i) of atom i at time step $t + \delta t$ as given be Velocity Verlet algorithm are

$$x_i(t+\delta t) = x_i(t) + v_i(t)\Delta t + \frac{1}{2}a_i(t)\delta t^2, \tag{30.3}$$

$$v_i(t+\delta t) = v_i(t) + \frac{a_i(t) + a_i(t+\delta t)}{2}\delta t. \tag{30.4}$$

30.2.2 Crystal Plasticity Finite-Element Simulation

The classical crystal plasticity theory is discussed in the present section. When a material is deformed plastically, it experiences both elastic and plastic strain. A deformation gradient ($\boldsymbol{F}$), in a continuum sense, can be decomposed into an elastic part ($\boldsymbol{F}_e$) and plastic part ($\boldsymbol{F}_p$):

$$\boldsymbol{F} = \boldsymbol{F}_e\boldsymbol{F}_p. \tag{30.5}$$

On differentiating the deformation gradient ($\boldsymbol{F}$), we get:

$$\dot{\boldsymbol{F}} = \dot{\boldsymbol{F}}_e\boldsymbol{F}_p + \boldsymbol{F}_e\dot{\boldsymbol{F}}_p. \tag{30.6}$$

Now, the velocity gradient ($\boldsymbol{L}$) is defined as

$$\boldsymbol{L} = \dot{\boldsymbol{F}}\boldsymbol{F}^{-1} \tag{30.7}$$

and, thus, $\boldsymbol{L}$ can be decomposed into

$$\boldsymbol{L} = \dot{\boldsymbol{F}}_e\boldsymbol{F}_e^{-1} + \boldsymbol{F}_e\left(\dot{\boldsymbol{F}}_p\boldsymbol{F}_p^{-1}\right)\boldsymbol{F}_e^{-1} = \boldsymbol{L}_e + \boldsymbol{L}_p. \tag{30.8}$$

Plastic deformation in metals is attributed to the cumulative motion of individual dislocations on their respective slip planes, which leads to shear deformation along the slip planes in a crystal. Following this, the plastic velocity gradient ($\boldsymbol{L}_p$) is defined as the sum of shear rates due on all slip systems:

$$\boldsymbol{L}_p = \sum_{i=1}^{N} \dot{\gamma}^\alpha \boldsymbol{s}^\alpha \otimes \boldsymbol{m}^\alpha, \tag{30.9}$$

where $\dot{\gamma}^\alpha$ is the shear slip rate on slip system α; unit vectors $\boldsymbol{s}^\alpha$ and $\boldsymbol{m}^\alpha$ define the slip direction and slip plane; and N is the total number of slip systems. The velocity gradient can further be decomposed in terms of rate of stretching $\boldsymbol{D}$ and rate of spin $\boldsymbol{W}$, for both elastic and plastic component. From Eqs. (30.8) and (30.9), the elastic and plastic part of velocity gradient $\boldsymbol{L}$ is expressed as

$$L_e = D_e + W_e = \dot{F}_e F_e^{-1}, \quad L_p = D_p + W_p = \sum_{i=1}^{N} \dot{\gamma}^{\alpha} s^{\alpha} \otimes m^{\alpha}. \tag{30.10}$$

The relationship between the elastic part of the rate of stretching ($\boldsymbol{D}_e$) and the Jaumann rate of Cauchy stress ($\boldsymbol{\Sigma}$) as a constitutive law is given by Huang [27] as:

$$\overset{\nabla}{\boldsymbol{\Sigma}} + \boldsymbol{\Sigma}(\boldsymbol{I} : \boldsymbol{D}_e) = \boldsymbol{C} : (\boldsymbol{D} - \boldsymbol{D}_p) \tag{30.11}$$

where $\boldsymbol{C}$ is the fourth-order elastic stiffness tensor and $\boldsymbol{I}$ is the second-order unit tensor. The Jaumann rate of stress evolution is

$$\overset{\nabla}{\boldsymbol{\Sigma}} = \dot{\boldsymbol{\Sigma}} - \boldsymbol{W}_e \boldsymbol{\Sigma} + \boldsymbol{\Sigma} \boldsymbol{W}_e. \tag{30.12}$$

The calculation of plastic velocity gradient $\boldsymbol{L}_p$ is based on the shear strain rate $\dot{\gamma}^{\alpha}$ experienced by individual slip system α and is typically expressed as a power law [28] in terms of resolved shear stress τ^{α} as

$$\dot{\gamma}^{\alpha} = \dot{\gamma}_0 \left| \frac{\tau^{\alpha}}{g^{\alpha}} \right|^n \operatorname{sgn}(\tau^{\alpha}), \tag{30.13}$$

where $\dot{\gamma}_0$ is the reference shear strain rate, τ^{α} is the strain rate sensitivity and g^{α} is the slip resistance of the slip system α. The resolved shear stress τ^{α} on a slip system α is obtained using Schmid's law

$$\tau^{\alpha} = \operatorname{sym}(\boldsymbol{s}^{\alpha} \otimes \boldsymbol{m}^{\alpha}) : \boldsymbol{\Sigma}. \tag{30.14}$$

and the slip resistance $\dot{g}^{\alpha}$ is given by

$$\dot{g}^{\alpha} = \sum_{i=1}^{N} h_{\alpha\beta} |\dot{\gamma}^{\beta}|, \tag{30.15}$$

where $h_{\alpha\beta}$ is the hardening modulus. The hardening in a slip system α can be due to dislocation present on the same slip system α (given by $h_{\alpha\alpha}$) or due to the presence of dislocation on other slip system β (given by $h_{\alpha\beta}$). Modified version of the hardening relation proposed by Asaro [29] is used in the case studied [14] and is given by,

$$h_{\alpha\alpha} = (h_0 - h_s) \operatorname{sech}^2 \left(\frac{(h_0 - h_s)\gamma}{\tau_0 - \tau_s} \right) + h_s,$$

$$h_{\alpha\beta} = q h_{\alpha\alpha}(\alpha\beta), \qquad \gamma = \sum_{\alpha} \int_0^t |\dot{\gamma}^{\alpha}| dt, \tag{30.16}$$

where τ is the shear stress, h is the hardening modulus and q is the latent hardening ratio. The subscripts "0" and "s" represent the initial and saturated condition. γ is the accumulative shear strain over all the slip systems.

The CP formulation is implemented in a general purpose finite-element software ABAQUS/Explicit via a material subroutine (VUMAT). Details of the implementation are available elsewhere [14]. Material removal in a single crystal is non-trivial. To this end, a new approach was proposed, which has capability in accounting for shear deformation in individual slip systems. This is given by [14].

$$\max\left(\gamma - \gamma_{cr}, \gamma_{sl,min} - \gamma_{sl,cr}\right) \geq 0, \tag{30.17}$$

$$\gamma = \min(\gamma^{\alpha}), \ \alpha = 1, 2, \ldots N, \tag{30.18}$$

where subscripts "sl, cr" and "cr" correspond to a critical value for a single slip system and for all slip systems, respectively. The criterion accounts for the total slip (from all slip systems) at a spatial location as well as the induced slip from individual slip systems. Once the mentioned criteria were satisfied, the element was either deleted or converted into smoothed-particle hydrodynamics (SPH) particles.

30.3 Case Studies

In this section, we illustrate two case studies for simulation of machining processes at lower length scales, each for MD and CP FEM. For MD, we refer to the recent work of Goel et al. [19] on the cutting behaviour of silicon (Si). For CP FEM, the work of Liu et al. [14] on the micromachining of single-crystal metal is referred.

30.3.1 MD: Influence of Microstructure on Cutting Behaviour of Silicon [19]

Goel et al. [19] carried out the MD simulation of nanometric cutting of silicon in order to study the difference in the cutting behaviour of single crystalline (SC) and polycrystalline (PC) Si. The MD simulation was performed using a large-scale atomic/molecular massively parallel simulator (LAMMPS) [30]. The simulation set-up at the start of cutting simulation after equilibration is shown in Fig. 30.1. The atoms in the workpiece (Si) and diamond cutting tools (*C*) were categorized into three zones according to their functions. The Newton atoms are atoms in the region, which were directly involved in the tool–workpiece interaction. The thermostatic atoms are a thin boundary layer of atoms, which were subjected to a thermostat in order to mimic the heat dissipation in the cutting region due to coolant during the machining operation. The boundary atoms were subjected to rigid boundary condition. A sharp cutting edge was used to ensure the brittle-regime machining in

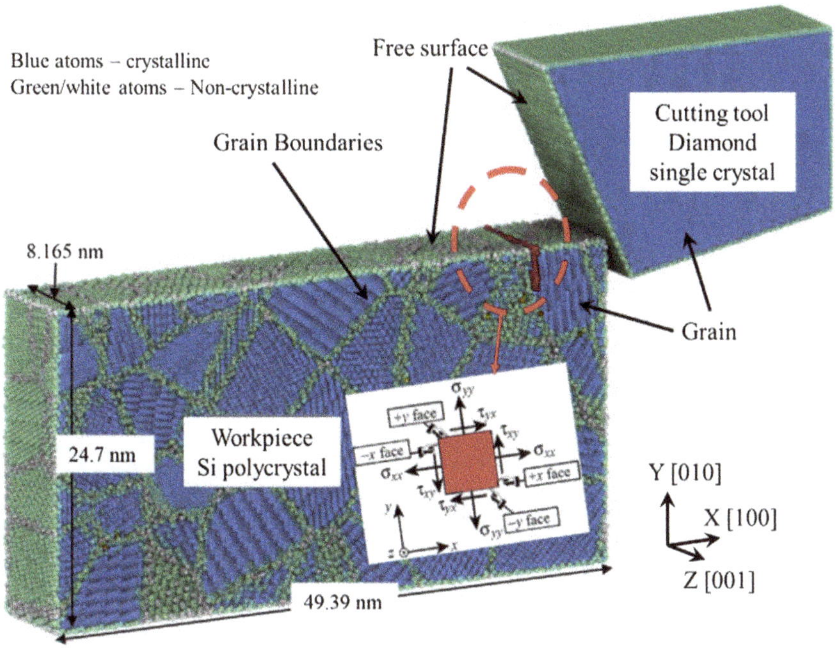

Fig. 30.1 A schematic diagram of the MD simulation model of nanometric cutting of polycrystalline silicon. The blue atoms are diamond atoms while white and green atoms are atoms at grain boundaries and free surface [19]

order to study such a process. To simulate the atomic interactions, ABOP potential was found to be suitable and advantageous as it describes the interaction between carbon and silicon.

The SC silicon required a larger cutting force during machining when compared to PC silicon. The chip for machining SC silicon was more curled (Fig. 30.2a) than the chip obtained from machining PC silicon (Fig. 30.2b). The chip formation in PC silicon takes place with plastic deformation concentrated along the grain boundaries via amorphization of the grain boundaries (Fig. 30.2d). Chunks of grains are pulled off as chips with no defect activity (Fig. 30.2f), indicating minimal plastic activity within the grains. The grain boundaries had large residual stresses (Fig. 30.2b) which help to reduce the critical stress to activate plastic deformation mediated along grain boundary, and, thus, resulting in lower cutting force for PC silicon.

An interesting observation in machining of SC silicon is the periodic occurrence of nano-grooves on the machined surface (Fig. 30.2c). The nano-grooves are regions of amorphous silicon indicating brittle cracking. Such nano-grooves were also observed in experimental studies of nanomachined silicon [25] (inset in Fig. 30.2c). These grooves form along (010) orientation and were inclined 45°–55° to the cutting direction.

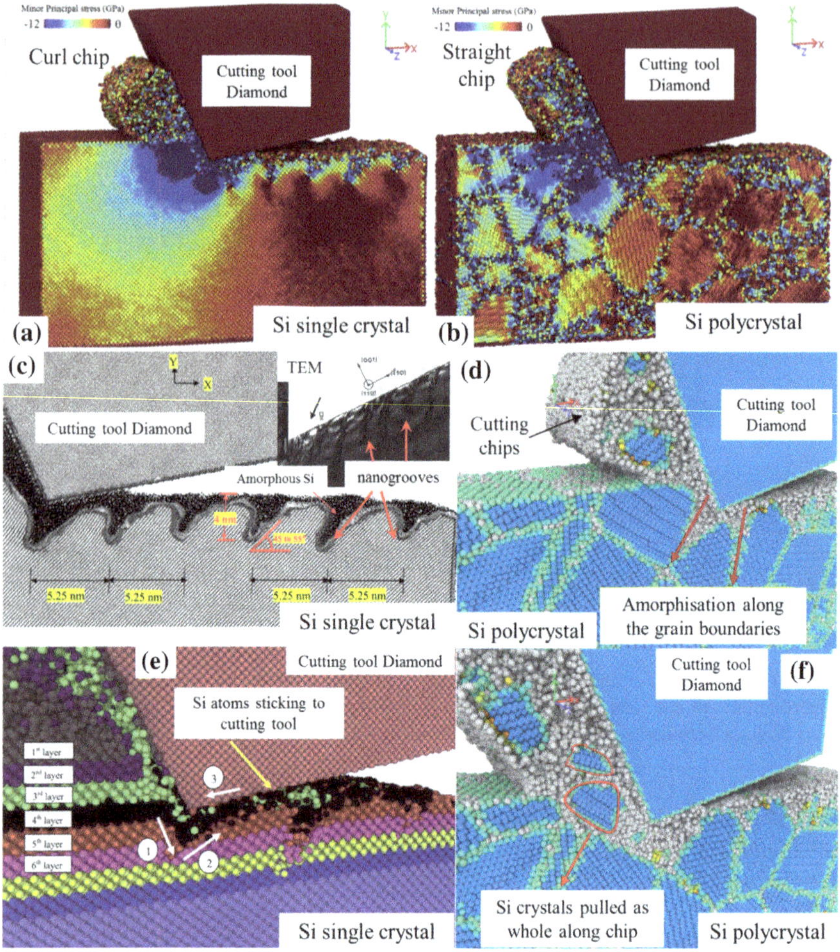

Fig. 30.2 Variation in the minor principal stress (compressive) during cutting of single-crystal (**a**) and polycrystalline (**b**) silicon [19]. (**c**) Formation of nano-grooves in the cutting zone. The inset in (**c**) is TEM image of diamond-turned specimen along (010) orientation [25]. (**e**) The path of vorticity as observed from displacement of atoms from different coloured layers [19]. Images (**d**) and (**f**) are the snapshot at various stages of cutting showing amorphization and grain pulled-outs along the chip [19]

The authors further investigated the grooving mechanism by visualization. For this, the layers of atoms of 0.54 nm thickness before machining are coloured differently. A snapshot of the machined SC silicon shows the grooving process (Fig. 30.2e). On machining, the material layer above the point of stagnation (layer 4 black Fig. 30.2e) was sheared and eventually separated from the workpiece as chips, while the layer beneath was under pure compression and slipped under the bottom the tool face. This process of flow of silicon under the tool creates a vortex.

This vorticity phenomenon leads to the flow of material (black layer) below the cutting tool and downwards in the wake of the cutting edge [19]. Sites formed by such process have abundant amorphous silicon.

One surprising finding in this study as reported by the authors [19] was the more or less absence of dislocation formation upon cutting. A burst of dislocations are normally observed under nanoindentation simulations [31]. Only a few ¼ <111> partial dislocations were observed in a selective case in the sub-surface after machining. Plasticity was mainly via amorphization and flow of silicon and was responsible for the ductile behaviour [19].

30.4 Crystal Plasticity Simulation of Micromachining of Single-Crystal Metal: Methodology and Analysis [14]

A finite-element model for single-crystal copper (FCC structure) was used to study micro-scratching (Fig. 30.3a) [14]. The micro-scratching was performed in two stages using a rigid frictionless cutting tool with 60° wedge angle and 6.25° clearance angle. The two-stage micro-scratching consisted of a linearly increasing cutting depth up to 18 μm (engagement stage) followed by a constant cutting depth of 18 μm (full-engagement stage), as shown in Fig. 30.3b. A cutting velocity of 10 mm/min was used. The model was meshed using eight-node brick elements with reduced integration (C3D8R), and a finer mesh was employed near the cutting zone for better accuracy. The copper single-crystal model is oriented such that (110) plane is parallel to the cutting plane. In the study, the cut was performed along [1–10] direction (0°) (symmetric) and at 45° to [1–10] direction (asymmetric).

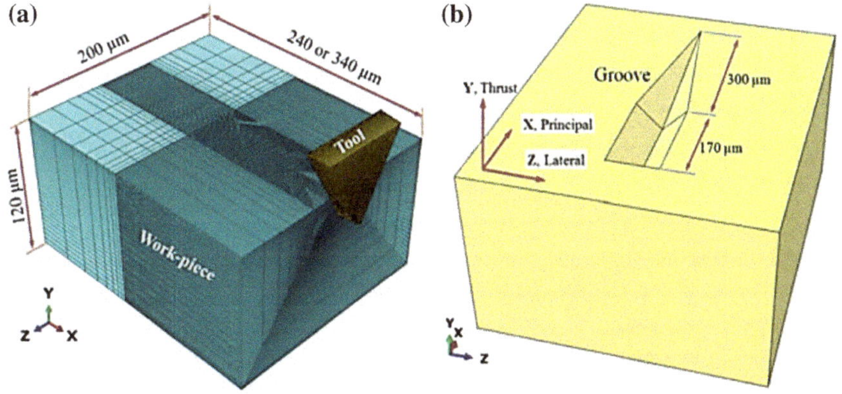

Fig. 30.3 **a** Finite-element model for micro-scratching simulation. **b** Schematic of the two-step micro-scratching experiment [14]

Different material removal models (element deletion, ALE and SPH) adopted resulted in similar cutting and reaction force in the full-engagement stage. However, the time for simulation was significantly different for the three methods. The element deletion was the most computationally efficient. ALE and SPH had higher computational costs, around 7 and 10 times that of element deletion, respectively. Thus, in the study, all further simulations were performed using element deletion.

The cutting forces during the two-step micro-scratching process for 0° and 45° direction are shown in Fig. 30.4a and b, respectively. The principal force matches well with the experimental results. The principal force for the 45° direction was 40% higher than that for 0° direction and shows the anisotropy in the material. In comparison, the predictions for thrust forces are relatively poor. It was observed to be not significantly dependent on the cutting direction.

The chip morphology for 0° direction and 45° direction from simulations (Fig. 30.5a and c) and experiments (Fig. 30.5b and d) is shown in Fig. 30.5. The simulations are capable of capturing the chip morphology and compare well with experiments. For 0° direction, the chip is symmetric, while for 45° direction, it is asymmetric. This is due to the crystal symmetry/asymmetry along the cutting direction. The crystal possesses a twofold symmetry along [1–10] direction (0° direction); hence, activated slip systems were symmetrical with respect to the cutting direction 0°. This results in a symmetrical chip formation (Fig. 30.5a) and stress field. However, no crystal symmetry exists along the cutting 45° direction, which explains the asymmetrical chip (Fig. 30.5c).

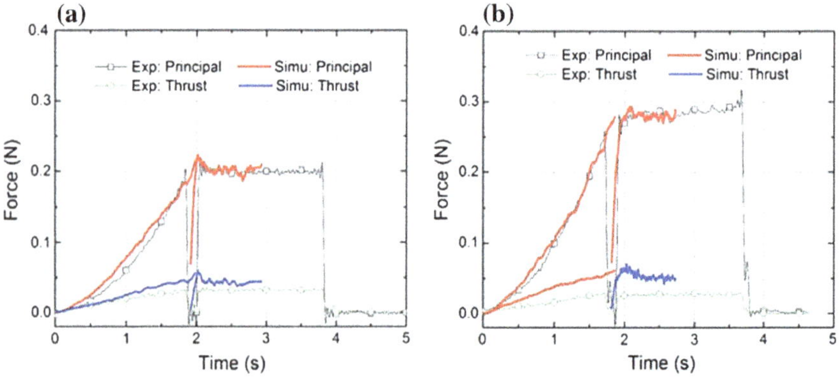

Fig. 30.4 Comparison of cutting force for FE simulations and experimental data in micro-scratching of single-crystal copper along 0° (**a**) and 45° (**b**) direction [14]

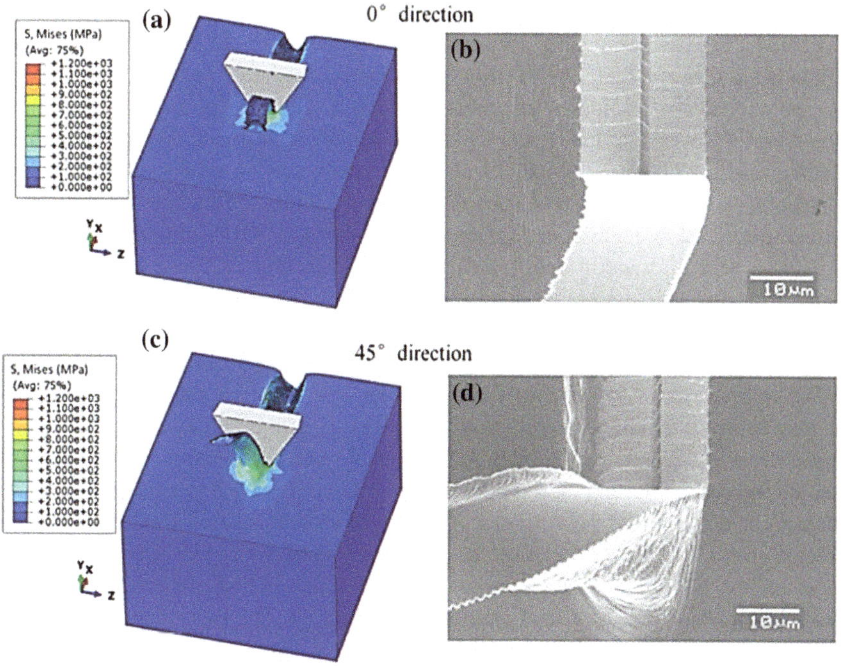

Fig. 30.5 Comparison of chip morphology on micro-scratching of single-crystal copper simulation (**a**) and experiment along 0° (**b**) direction; (**c**) simulation along 45° direction; (**d**) experiment along 45° direction [14]

30.5 Comparison of the Two Methods

The two methods for simulating micromachining processes, namely MD and CP FEM, are versatile techniques with several advantages and disadvantages. A right choice of simulation technique depends on several factors including the workpiece material, processing condition, size of machining cut to name a few.

Molecular dynamics (MD) is very helpful in situations when the underlying deformation mechanism is not well understood. Since potential function is the only necessary information needed to run a simulation. The results obtained provide fundamental insight into the nuances of the deformation processes. Depending on the boundary conditions and interaction among atoms, the system evolves in time. No prior knowledge of how the system will evolve is required. However, we note that the results obtained are dependent on an accurate potential function being defined for the material of interest. This is because a wide variety of interactions, such as ionic, metallic and covalent, are possible among atoms. This particularly becomes important as the number of interacting atom types increases in the system, which leads to an increase in the complexity of the potential and simulation computational cost. Nevertheless, a great depth of information ranging from vacancy cluster, dislocation type, surface defects to phase transformation can be extracted from the simulation. Timescale and domain size are other factors, which often limit the applicability of MD simulations. With an increase in size, the computational cost increases exponentially. Even with the advances in computational facility available today, the simulation length scale for an MD simulation is limited to a few hundreds of nanometre. MD usually needs a time step of a few fem to seconds for numerical convergence. This again limits the time available for the simulation to a few nanoseconds. Due to this limitation, MD simulations generally involve a very high strain rate, typically 10^8 s^{-1} and above. These strain rates are unrealistic for real-world applications and merely provide an insight into the nuances of deformation processes.

Single-crystal plasticity-based FEM (CP FEM) uses phenomenological constitutive material laws to account for the anisotropic material deformation mechanisms. From a practical standpoint, CP FEM is computationally efficient and possesses the capability of modelling reasonable spatial domains. Various deformation modes including slip systems [14, 32] and twin systems [33] can be incorporated in the CP FEM model. The choice of deformation modes to be considered for CP FEM simulation is principally based on the active dominant mechanisms. Slip-based gradient plasticity is dominant in FCC materials, especially those having high stacking fault energy like Al and Ni. The twin-induced plasticity is dominant primarily in FCC materials with low stacking fault energy (brass and silver) and in HCP materials (Mg, Ti, Zr). By incorporating multiple modes in simulation, the activity of each system during the deformation can be studied via CP FEM simulations. Moreover, the deformation modes changes with sample size and temperature. Such dependency of deformation modes on size [32] and temperature [33] can be incorporated with the help of experimental data, which widens the applicability of CP FEM modelling to a large-scale scenario. Furthermore, the effect of individual deformation mode can be investigated via CP FEM by only considering a particular mode during the simulation, which helps in better understanding the role of individual deformation system. Nevertheless, it should be noted that CP FEM simulation is based on phenomenological constitutive material laws and, thus, for new materials initial experimental investigation for identification of deformation modes and calibration of material model is required.

30.6 Conclusion

In this paper, two methods, namely molecular dynamics (MD) and single-crystal plasticity (CP) FEM, used for simulation of micro-machining process are described in brief with the help of two examples from the literature. Both methods capture the anisotropic behaviour of the material. The basics of the two methods along with their capability are illustrated. Various aspects of the two methods are compared to give the reader a better understanding of respective advantages and deficiencies. MD is more suitable when the underlying deformation mechanisms are not well understood. On the other hand, CP FEM is apt for realistic simulations and can be executed for a wide range of loading conditions. It is clear that even the type of data obtained from the two simulation methods is very different. Thus, the appropriate choice of simulation method depends primarily on the investigation at hand.

References

1. Chokshi, A.H., Rosen, A., Karch, J., Gleiter, H.: On the validity of the hall-petch relationship in nanocrystalline materials. Scr. Metall. **23**, 1679–1683 (1989). https://doi.org/10.1016/0036-9748(89)90342-6
2. Petch, J.N.: The cleavage strength of polycrystals. J. Iron Steel Inst. **174**, 25–28 (1953)
3. Siegel, R.W., Fougere, G.E.: Mechanical properties of nanophase metals. Nanostruct. Mater. **6**, 205–216 (1995). https://doi.org/10.1016/0965-9773(95)00044-5
4. Suzuki, H., Ikeda, S., Takeuchi, S.: Deformation of thin copper crystals. J. Phys. Soc. Japan. **11**, 382–393 (1956). https://doi.org/10.1143/JPSJ.11.382
5. Bei, H., Shim, S., George, E.P., Miller, M.K., Herbert, E.G., Pharr, G.M.: Compressive strengths of molybdenum alloy micro-pillars prepared using a new technique. Scr. Mater. **57**, 397–400 (2007). https://doi.org/10.1016/J.SCRIPTAMAT.2007.05.010
6. Uchic, M.D., Dimiduk, D.M., Florando, J.N., Nix, W.D.: Sample dimensions influence strength and crystal plasticity. Science **305**, 986–989 (2004). https://doi.org/10.1126/science.1098993
7. Nix, W.D., Gao, H.: Indentation size effects in crystalline materials: a law for strain gradient plasticity. J. Mech. Phys. Solids **46**, 411–425 (1998). https://doi.org/10.1016/S0022-5096(97)00086-0
8. Ma, Q., Clarke, D.R.: Size dependent hardness of silver single crystals. J. Mater. Res. **10**, 853–863 (1995). https://doi.org/10.1557/JMR.1995.0853
9. Fleck, N.A., Muller, G.M., Ashby, M.F., Hutchinson, J.W.: Strain gradient plasticity: theory and experiment. Acta Metall. **42**, 475–487 (1994). https://doi.org/10.1016/0956-7151(94)90502-9
10. Misra, A., M. Pandey, P., Dixit, U.S.: Modeling of material removal in ultrasonic assisted magnetic abrasive finishing process. Int. J. Mech. Sci. 131–132, 853–867 (2017). https://doi.org/10.1016/j.ijmecsci.2017.07.023
11. Dutta, P.P., Kalita, K., Dixit, U.S.: Electromagnetic-force-assisted bending and straightening of AH36 steel strip by laser irradiation. Lasers Manuf. Mater. Process. **5**, 201–221 (2018). https://doi.org/10.1007/s40516-018-0062-6
12. Misra, A., Pandey, P.M., Dixit, U.S.: Modeling and simulation of surface roughness in ultrasonic assisted magnetic abrasive finishing process. Int. J. Mech. Sci. **133**, 344–356 (2017). https://doi.org/10.1016/j.ijmecsci.2017.08.056
13. Hall, R.C.: Single crystal anisotropy and magnetostriction constants of several ferromagnetic materials including alloys of NiFe, SiFe, AlFe, CoNi, and CoFe. J. Appl. Phys. **30**, 816–819 (1959). https://doi.org/10.1063/1.1735247
14. Liu, Q., Roy, A., Tamura, S., Matsumura, T., Silberschmidt, V.V.: Micro-cutting of single-crystal metal: Finite-element analysis of deformation and material removal. Int. J. Mech. Sci. **118**, 135–143 (2016). https://doi.org/10.1016/j.ijmecsci.2016.09.021
15. Demiral, M., Roy, A., Sayed, T. El, Silberschmidt, V.V.: Numerical modelling of micro-machining of f.c.c. single crystal: Influence of strain gradients. Comput. Mater. Sci. **94**, 273–278 (2014). https://doi.org/10.1016/j.commatsci.2014.05.058
16. Liu, Q., Dodla, S., Roy, A., Silberschmidt, V.V.: Crystal-plasticity simulation of micromachining of single-crystal metal: methodology and analysis, 1st edn. Springer, Singapore (2016)

17. Goel, S., Luo, X., Reuben, R.L., Rashid, W.: Bin: atomistic aspects of ductile responses of cubic silicon carbide during nanometric cutting. Nanoscale Res. Lett. **6**, 1–9 (2011). https://doi.org/10.1186/1556-276X-6-589
18. Goel, S., Luo, X.C., Reuben, R.L., Bin Rashid, W., Sun, J.N.: Single point diamond turning of single crystal silicon carbide: molecular dynamic simulation study. Key Eng. Mater. **496**, 150–155 (2011). https://doi.org/10.4028/www.scientific.net/KEM.496.150
19. Goel, S., Kovalchenko, A., Stukowski, A., Cross, G.: Influence of microstructure on the cutting behaviour of silicon. Acta Mater. **105**, 464–478 (2016). https://doi.org/10.1016/j.actamat.2015.11.046
20. Jones, J.E., Ingham, A.E.: On the calculation of certain crystal potential constants, and on the cubic crystal of least potential energy. Proc. R. Soc. A Math. Phys. Eng. Sci. **107**, 636–653 (1925). https://doi.org/10.1098/rspa.1925.0047
21. Jones, J.E.: On the determination of molecular fields. II. From the equation of state of a gas. Proc. R. Soc. A Math. Phys. Eng. Sci. **106**, 463–477 (1924). https://doi.org/10.1098/rspa.1924.0082
22. Erhart, P., Albe, K.: Analytical potential for atomistic simulations of silicon, carbon, and silicon carbide. Phys. Rev. B. **71**, 035211 (2005). https://doi.org/10.1103/PhysRevB.71.035211
23. Daw, M.S., Foiles, S.M., Baskes, M.I.: The embedded-atom method: a review of theory and applications. Mater. Sci. Reports. **9**, 251–310 (1993). https://doi.org/10.1016/0920-2307(93)90001-U
24. Tersoff, J.: Modeling solid-state chemistry: interatomic potentials for multicomponent systems. Phys. Rev. B. **39**, 5566–5568 (1989). https://doi.org/10.1103/PhysRevB.39.5566
25. Becker, C.A., Tavazza, F., Trautt, Z.T., Buarque de Macedo, R.A.: Considerations for choosing and using force fields and interatomic potentials in materials science and engineering. Curr. Opin. Solid State Mater. Sci. **17**, 277–283 (2013). https://doi.org/10.1016/j.cossms.2013.10.001
26. Verlet, L.: Computer "experiments" on classical fluids. I. Thermodynamical properties of Lennard-Jones molecules. Phys. Rev. **159**, 98–103 (1967). https://doi.org/10.1103/physrev.159.98
27. Huang, Y.: A user-material subroutine incorporating single crystal plasticity in the ABAQUS finite element program, Harvard (1991)
28. Hutchinson, J.W.: Bounds and self-consistent estimates for creep of polycrystalline materials. Proc. R. Soc. A Math. Phys. Eng. Sci. **348**, 101–127 (1976). https://doi.org/10.1098/rspa.1976.0027
29. Asaro, R.J.: Crystal Plasticity. J. Appl. Mech. **50**, 921 (1983). https://doi.org/10.1115/1.3167205
30. Plimpton, S.: Fast parallel algorithms for short-range molecular dynamics. J. Comput. Phys. **117**, 1–19 (1995). https://doi.org/10.1006/JCPH.1995.1039
31. Goel, S., Cross, G., Stukowski, A., Gamsjäger, E., Beake, B., Agrawal, A.: Designing nanoindentation simulation studies by appropriate indenter choices: case study on single crystal tungsten. Comput. Mater. Sci. **152**, 196–210 (2018). https://doi.org/10.1016/j.commatsci.2018.04.044
32. Liu, Q., Roy, A., Silberschmidt, V.V.: Size-dependent crystal plasticity: from micro-pillar compression to bending. Mech. Mater. **100**, 31–40 (2016). https://doi.org/10.1016/j.mechmat.2016.06.002
33. Liu, Q., Roy, A., Silberschmidt, V.V.: Temperature-dependent crystal-plasticity model for magnesium: a bottom-up approach. Mech. Mater. **113**, 44–56 (2017). https://doi.org/10.1016/j.mechmat.2017.07.008

Chapter 31
Process Parameters Optimization of Cylindrical Grinding Machining Using Taguchi Method

T. Nancharaiah and M. Subramanyam

Abstract The main objective of this work is to study the effect of material hardness, workpiece speed and depth of cut on surface finish and material removal rate in cylindrical grinding process. For three process parameters at three different levels, L9 orthogonal array (OA) is selected. Experiments were conducted, and roughness values and material removal rates are calculated. S/N ratio is performed to get the optimal process parameters for surface roughness and material removal rate. ANOVA is carried out, and the percentage of contribution of each selected parameter was found.

Keywords Cylindrical grinding · Surface roughness · Material removal rate

31.1 Introduction

Machining involves removing materials using cutting tools for getting rid of unwanted materials from workpiece and converting into the desired shape. There are four movements in cylindrical grinding. They are the work must revolve, the wheel must revolve, the work must pass the wheel and wheel must pass work. Surface finish is the important output parameter in cylindrical grinding. Many researchers worked in this area, a few are cited here for references.

Athreya and Venkatesh [1] illustrated the parameter design of the Taguchi method and optimizing the process parameters. Rupesh [2] applied Taguchi method to decide the optimal process parameter for surface roughness. Prasad [3] measured the vibration and temperature rise. Habrat [4] revealed that grinding with the use of resin bond grinding wheel provides significantly lower grinding force components

T. Nancharaiah (✉)
Department of Mechanical Engineering, Bapatla Engineering College, 522102 Bapatla, India
e-mail: tata_nr@rediffmail.com

M. Subramanyam
Department of Mechanical Engineering, Bapatla Engineering College, Bapatla 522102 AP, India

R. G. Narayanan et al. (eds.), *Advances in Computational Methods in Manufacturing*, Lecture Notes on Multidisciplinary Industrial Engineering,
https://doi.org/10.1007/978-981-32-9072-3_31

during the process. Mohite et al. [5] focused on the ability of a grinding wheel measured in terms of surface of workpiece during the subsequent grinding operation.

31.2 Experimentation

31.2.1 Grinding Machine Specifications

See (Table 31.1).

31.2.2 En Alloy Steels

Mild steel of different compositions is considered for analysis. EN 19, EN 24 and EN 31 of different grades of the mild steels are turned on a lathe and then cylindrical grinding operation is performed as per L9 orthogonal array. The chemical composition of EN 19, EN 24 and EN3 1 is as shown in Tables 31.2, 31.3 and 31.4, respectively.

Table 31.1 Specifications of grinding machine

Model	DEVCO UC-150
Center distance	160 mm
Center height	102 mm
Angle of swiveling	±9 °C
Speed of transverse	140-260-370-700 mm/min
In feed of hand wheel	0.01 Division
Grinding wheel size	250*25*76.2 mm
Grinding wheel speed	2300 rpm
Work head single spindle speed	90-210-360 rpm
Spindle in taper	MT-3
Total power required	3.5HP (MAX)

Table 31.2 Chemical composition of EN 19 steel

Carbon	0.36–0.44%
Silicon	0.10–0.35%
Manganese	0.45–0.70%
Sulfur	0.040 Max
Phosphorus	0.035 Max
Chromium	1.00–1.40%
Molybdenum	0.20–0.35%

Table 31.3 Chemical composition of EN 24 steel

Carbon	0.36–0.44%
Silicon	0.10–0.40%
Manganese	0.70–1.00%
Sulfur	0.040 Max
Phosphorus	0.035 Max
Chromium	0.90–1.20%
Molybdenum	0.25–0.35%

Table 31.4 Chemical composition of EN 31 steel

Carbon	0.90–1.20%
Silicon	0.10–0.35%
Manganese	0.30–0.75%
Sulfur	0.040%
Phosphorus	0.040%
Chromium	1.00–1.60%

31.2.3 Input Process Parameters and Output Variables

The aim of grinding process is to get a better surface finish and higher rate of production. Currently there is a need for precise and accurate components in every field, so grinding operation is used to achieve this goal.

31.2.3.1 Process Parameters

In this study, the process parameters at their levels are given in Table 31.5.

31.2.3.2 Output Parameters

The material removal rate (MRR) and surface roughness of the workpiece have been measured from machined surfaces. To have minimum MRR and minimum roughness value, the optimal process parameters should be chosen.

Table 31.5 Selection of different levels of input process parameters

Parameter	Level 1	Level 2	Level 3
Hardness	40	47	55
Speed (rpm)	100	214	340
Depth of cut (microns)	10	20	30

31.2.4 Experimental Setup

The experiments were performed on DEVCO UC-150 cylindrical grinding machine. Depth of cut (*D*), speed (*N*) and hardness (*H*) is set at different levels, and the effects of these parameters on material removal rate and surface roughness are studied using Taguchi method. Surface roughness values are measured using SURFTEST SJ301.

For better surface roughness, smaller is better S/N ratio was used in this study and is calculated using the formula given below.

$$\mathbf{S/N\ ratio = -10log\ (Ra^2)}$$

In the Taguchi method of the smaller the better method, the smallest S/N value which is calculated is the best optimized running in the system. Smallest S/N ratio has highest surface finishing (Tables 31.6, 31.7, 31.8, 31.9, 31.10 and 31.11).

From the above Ra values, the signal to noise ratio is calculated using the below formula

Table 31.6 Surface roughness values as per L9 orthogonal array

S. No. (microns) (Ra)	Hardness roughness	Speed (rpm)	Depth of cut	Roughness (Ra)
1	40	100	10	0.81
2	40	214	20	0.78
3	40	340	30	1.25
4	47	100	20	1.06
5	47	214	30	1.08
6	47	340	10	1.20
7	55	100	30	1.60
8	55	214	10	1.04
9	55	340	20	1.54

Table 31.7 S/N values for machining the EN 19, 24, 31 workpieces

S. No.	Hardness	Speed (rpm)	Depth of cut (microns)	Ra	S/N ratio
1	40	100	10	0.81	−18.17
2	40	214	20	0.78	−14.83
3	40	340	30	1.25	−17.20
4	47	100	20	1.06	−17.50
5	47	214	30	1.08	−15.98
6	47	340	10	1.20	−21.65
7	55	100	30	1.60	−19.32
8	55	214	10	1.04	−20.34
9	55	340	20	1.54	−20.75

Table 31.8 Experimental data for MRR

S. No.	Depth (microns)	Speed (rpm)	Hardness	Metal removal rate (MRR) (mm^3/min)
1	10	100	40	152
2	10	214	47	163
3	10	340	55	158
4	20	214	55	169
5	20	340	40	176
6	20	100	47	171
7	30	340	47	246
8	30	100	55	224
9	30	214	40	282

Table 31.9 S/N values for machining the EN 19, 24, 31 workpieces

S. No.	Depth (microns)	Speed (rpm)	Hardness	Metal removal rate (MRR) (mm^3/min)	S/N ratio
1	10	100	40	152	−20.81
2	10	214	47	163	−19.89
3	10	340	55	158	−20.52
4	20	214	55	169	−19.42
5	20	340	40	176	−18.53
6	20	100	47	171	−18.86
7	30	340	47	246	−16.18
8	30	100	55	224	−17.27
9	30	214	40	282	−15.36

Table 31.10 ANOVA table for surface roughness

Parameter	Dof	SS	SS_m	% of contribution	F-test
Hardness	2	0.3066	0.1533	47.52	37.85
Speed	2	0.1983	0.0945	30.74	22.59
Depth of cut	2	0.1322	0.0661	20.49	16.32
Error	2	0.0081	0.00405	1.25	–
Total	8	0.6452	–	–	–

$$\mathbf{S/N\ ratio} = -\mathbf{10log}\ (\mathbf{Ra}^2)$$

Lower S/N ratio indicates better surface finish.

From the above MRR values, the signal to noise ratio is calculated using the below formula

Table 31.11 ANOVA table for MRR

Parameter	Dof	SS	SS_m	% of contribution	F-test
Depth of cut	2	15042.8	7521.43	89.30	31.82
Speed	2	748.223	391.11	4.44	1.65
Hardness	2	580.223	280.11	3.44	1.22
Error	2	472.69	236.349	2.806	–
Total	8	16844	–	–	–

$$SN = -10x\log\left[\frac{1}{n}\sum_{i}^{n} y_i^2\right]$$

Higher S/N ratio indicates that maximum MRR value.

31.3 Results and Discussions

31.3.1 Discussions

From Fig. 31.1, it was observed that surface roughness value increases if the depth of cut and speed increases and decreases if hardness increases.

It was observed from Fig. 31.2 that material removal rate increases if depth of cut increases and decreases if hardness and speed increases.

From the ANOVA, it was found that hardness effects on surface roughness are 47.52%, speed are 30.74% and depth of cut are 20.49%, and hardness effects on MRR are 3.44%, speed are 4.44% and depth of cut are 89.30%.

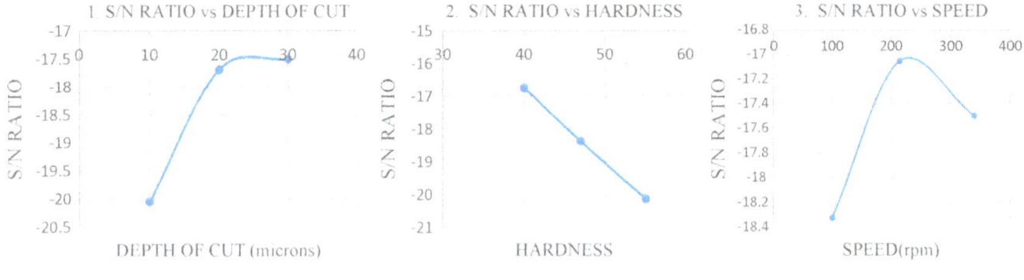

Fig. 31.1 Mean effective plots for surface roughness

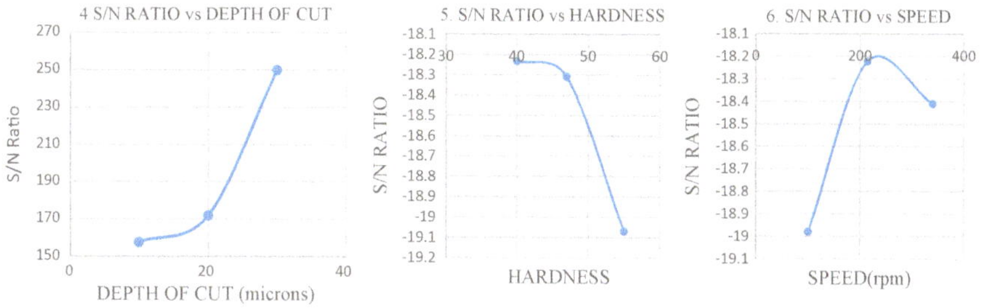

Fig. 31.2 Mean effective plots for material removal rate

31.4 Conclusions

Based on Taguchi method, experiments were conducted using EN alloy steels by varying depth of cut, hardness and speed parameters on cylindrical grinding machine, and the following conclusions were drawn.

(a) Higher depth of cut results MRR increases and surface finish decreases
(b) Increase in hardness, MRR decreases and surface finish increases
(c) Increase in speed, MRR increases and then decreases, surface finish decreases and then increases

From ANOVA, it concludes that hardness is more effective on surface roughness, and depth of cut is more effective on MRR.

References

1. Athreya, S., Venkatesh, Y.D.: Application of Taguchi method for optimization of process parameters in improving the surface roughness of lathe facing operation. Int. Refereed J. Eng. Sci. pp. 13–19 (2012)
2. Karande, R.J.: Optimization of cylindrical grinding machine parameters for minimum surface roughness and maximum MRR. GRD J. Global Res. Develop. J. Eng. **2**(5) (2017, April). ISSN: 2455–5703
3. Prasad, P.P: Design and analysis of grinding wheel of fly ash grinding machine. Int. J. Sci. Eng. Technol. Res. (IJSETR) **5**(8) (2016, August)
4. Habrat, W.F.: Effect of bond type and process parameters on grinding force components in grinding of cemented carbide. In: International Conference on Manufacturing Engineering and Materials, ICMEM 2016, 6–10 June 2016, Nový Smokovec, Slovakia (2016)
5. Mohite, D.D., Patil, K.R., Karande, R.J., Jadhav, V.S. (2017) Modeling and optimization of cylindrical grinding parameters for MRR and surface roughness. Int. J. Eng. Sci. Res. Technol. **6**(4) (2017, April)

Chapter 32
Effect of Sequential Drilling Process on In-situ Bone Temperature

Varatharajan Prasannavenkadesan and Ponnusamy Pandithevan

Abstract In orthopedic surgery, sequential drilling is the process of making multiple holes to facilitate the implant fixation. During the bone drilling process, the generated heat will cause thermal damage that affects the implant fixation strength because of "osteonecrosis," a permanent death of the tissues around the drilling site. In this work, an attempt was made to identify the effects of sequential drilling of cadaveric human femur bone on the heat accumulation. To perform the sequential drilling process, the rotational speed of 1500 rpm and 80 mm/min feed rate was considered with the 4 mm deep holes. Thermal images were captured using Sonel® infrared thermography during the drilling process. It was identified that the accumulated heat increased as the number of sequential holes increased. It was also observed that the distance between the drilled holes influenced the amount of temperature rise. The number of sequential holes drilled and the distance between the holes greatly influenced the temperature rise. This study showed the accumulation of heat during sequential drilling and its consequences in the temperature rise. It is recommended to the surgeons to increase the field of a drilling site in such a way to increase the distance between the subsequent holes to avoid thermal damage to the bone to prevent from further complications.

Keywords Orthopedic surgery · Sequential drilling · Femur bone · In-situ temperature · Osteonecrosis

32.1 Introduction

Bone drilling procedure is one of the most commonly employed surgical procedures in orthopedics, during joint replacement and implant fixations. During the drilling of bone, the interaction between tool and bone; and the contact between the chip

V. Prasannavenkadesan · P. Pandithevan (✉)
Department of Mechanical Engineering, Indian Institute of Information Technology Design and Manufacturing Kancheepuram, Chennai 600127, India
e-mail: ppthevan@iiitdm.ac.in

R. G. Narayanan et al. (eds.), *Advances in Computational Methods in Manufacturing*, Lecture Notes on Multidisciplinary Industrial Engineering,
https://doi.org/10.1007/978-981-32-9072-3_32

and the drill bore, creates friction which in turn induces heat [1]. When the generated temperature exceeded the threshold will lead to a condition called "thermal necrosis," a permanent death of the bone cells and because of that, the strength of the bone will reduce [2]. The phenomenon of necrosis will lead the bone to loss its healing capability and this will become more common in case of drilling, especially in multi-pass or sequential drilling [3]. This kind of thermal damage increases with increase in temperature and also with respect to the exposure time [4]. Apart from the threshold value that causes the thermal necrosis, the time being exposed also decides the severity of the damage. It has been reported in the literature that the necrosis condition will occur if the temperature is at 47 °C for 1 min and if the temperature increased to 53 °C, then the time to cause the failure will be less than a second [5, 6]. Based on the literature, it was identified that the temperature which is greater than or at 70 °C will result in a sudden state of thermal injury, and also even a lesser temperature with prolonged exposure time will provide the same adverse effects. Other than time, the significant parameters that increase the temperature rise include the thickness of the bone considered and also depth of the holes drilled [7]. Even though the temperature generated during drilling can be reduced with the help of irrigation, it is not recommended in some actual clinical cases [4]. Because of the low thermal conductivity of bone as reported in the literature, the dissipation of heat will be difficult and the build-up heat due to the multi-pass drilling will be more [8]. The main objective of this work was to identify the effect of heat accumulation in the sequential drilling with respect to the distance between the drilling sites, as the severity due to the temperature is large when compared with the single-hole drilling of femoral cortical bone. Based on this work, necessary recommendations can be provided to the surgeons to aid the drilling process in clinical trials.

32.2 Materials and Methods

Sequential drilling was performed in a vertical CNC milling machine (MTAB Maxmill®). Sonel® KT-160A infrared thermal imaging camera was used to capture the temperature data and processed with the Sonel ThermoAnalyze® software. Figure 32.1 shows the experimental setup considered. The surgical drill bit of 3.2 mm diameter was used in this study. Sequential drilling was performed at a rotational speed of 1500 rpm and feed rate of 80 mm/min with 4 mm deep holes to study the accumulation of heat.

The most commonly used parameter range for orthopedic surgery was identified through the literature and from surgeon guidance. The combination of rotational speed and feed rate that provides the higher temperature rise was identified through trial experiments. Based on that, the parameters were selected for this work. To facilitate the process of drilling, the femur bone was cut into pieces and mount into polymethylmethacrylate (PMMA). In order to replicate the clinical case, the dwell time of 1 s was provided between each pass. The accumulation of heat during the sequential drilling is due to the adjacent holes. In general, the thermocouple is

Fig. 32.1 Experimental setup used to conduct the sequential drilling in bone

employed to capture the data during single-hole drilling. But, in the case of sequential hole drilling, thc multiple placements of thermocouple require several holes to house the thermocouples, which affects the integrity of the bone [9]. Because of this, the thermocouple is not employed in this study. A 2 × 3-hole array was considered and its schematic representation is shown in Fig. 32.2. The scheme considered for the sequential drilling is based on the distance from margin to margin (4 mm) of the drilled hole. Two points A and B were selected to study the accumulation of heat with respect to the drilling (Fig. 32.2). Six holes were drilled and two reference points “A” and “B” were considered as an analyzing region to identify the heat accumulation phenomenon.

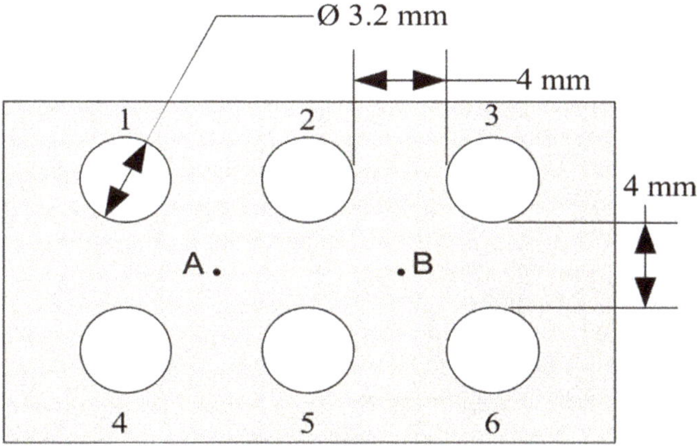

Fig. 32.2 Drilling sequence considered in this work with the reference points A and B

32.3 Results and Discussion

The variation in temperature (T) with respect to the time (t) for each drilling action is shown in Fig. 32.3. A representative thermal image and the corresponding thermal profile (Temperature, T vs. Distance, d) captured are shown in Fig. 32.4 for illustrative purpose. Based on the considered feed rate (80 mm/min), to achieve the full depth (4 mm), 3 s was taken. The dwell period provided was 1 s and thus a total time period of 4 s was taken to complete a single hole. The upright line, shown in Fig. 32.3 represents the hole number and the corresponding abscissa provides the time (t).

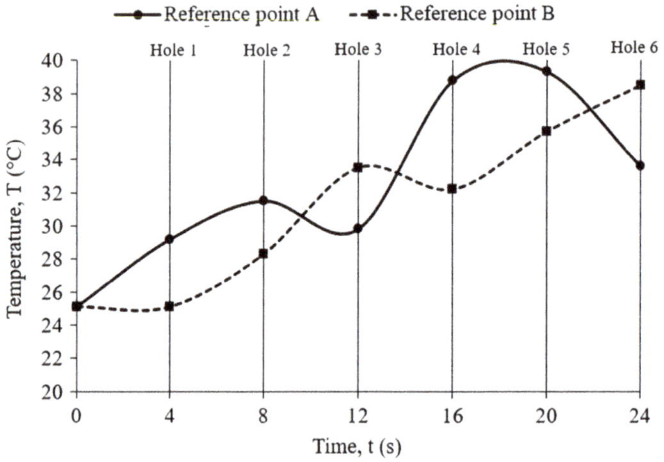

Fig. 32.3 Captured temperature data with respect to the time during sequential drilling

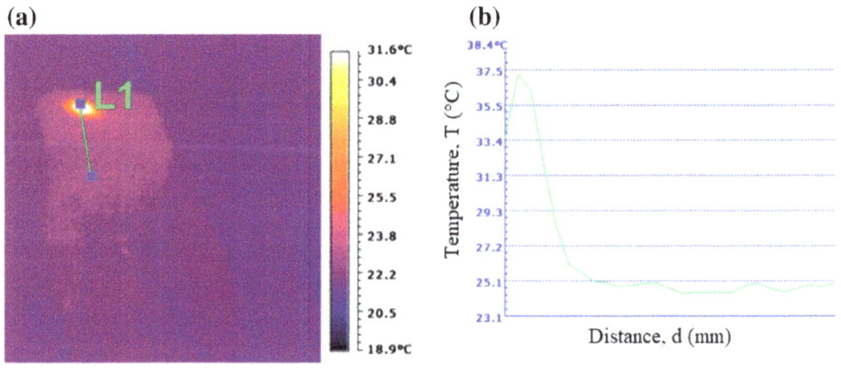

Fig. 32.4 Representative temperature data, **a** thermal image captured during the drilling with its temperature contour and **b** temperature profile of the processed thermal image

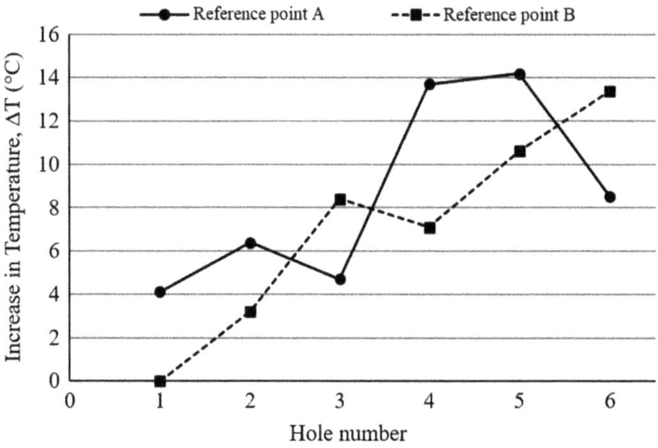

Fig. 32.5 Temperature rise during drilling with respect to the hole number

As the sequential drilling procedure starts, the reference point A (RP_A) responds first and then the reference point B (RP_B). This can be identified through Fig. 32.5, which shows the relation between the increase in temperature rise (ΔT) during the drilling and the hole number.

When drilling was made at the position of the hole number 1, the temperature information was obtained through RP_A and at this particular instance, the absence of heat was observed at RP_B. As the sequential drilling process proceeded through the subsequent passes, the accumulation of heat was clearly visible. The temperature rise was decreased substantially as the position of RP_A was far from the hole number 3. When the drilling site is in proximal (hole number 4) to the RP_A, again the temperature was increased drastically, which may be due to the accumulation of heat. Hole number 6 also showed the same trend as that of hole number 3. In the case of RP_A, it may be reported as the heat accumulation was increased for the drilling sites which are proximal to RP and gets decreased when the drilling site was farther. In the case of RP_B, it is noticed from the obtained data (Fig. 32.5) that there was not an initial rise in temperature when drilling was performed at hole number 1. A similar trend was obtained in the case of RP_B, in such a way that the heat accumulation was greater when the RP_B is at nearer to the drill site. Also, it can be noticed from Fig. 32.5 that the temperature rise at RP_A and RP_B follows the opposite trend. It may be due to the reason that the considered reference points A and B are at opposite to each other. Due to this, at a particular instance, for a considered drilled hole higher temperature rise or heat accumulation was observed at the reference point which is nearer to the drilled hole. Thus, the RP_A responds in a greater way to the sites 1, 2, 4, and 5; the RP_B provides the temperature information for the sites 2, 3, 5, and 6. The farther distance might be the reason for the RP_B for not providing any information about hole number 1 at the initial time. As the sequential drilling is involved in the drilling of more than a hole, the heat gets accumulated and gets transferred to the next drilling site. Due to this, more chance for the osteonecrosis condition when the temperature reaches 47 °C. So to avoid that in sequential drilling, the surgeons are advised to select the drilling site farther from the previously drilled hole. In this present work, even though, the sequence of the holes to be drilled is changed since the distance between the drilling sites were maintained at the same distance, the temperature profile will not vary. As a preliminary work, those effects were not included in this study. Based on the conducted experiments, it may be concluded that the temperature rise during multi-pass drilling can be reasonably reduced by increasing the dwell time, and this discussion is in good agreement with the numerical study reported earlier [10]. By increasing the dwell period or the time interval between the passes, more time to disintegrate the heat generated can be obtained so that the accumulation can be reduced.

32.4 Conclusion

In this work, an attempt was made to investigate the accumulation of temperature in the cadaveric human femur bone during sequential drilling with a 3.2 mm surgical drill bit for orthopedic applications. Temperature is higher at the margin of the drilled holes, but the accumulation of heat was analyzed at the considered reference points A and B. Accumulation of heat was found to be increased during the subsequent passes of the drilling. From the investigation made, it may be concluded that the distance which is considered as a datum for comparison has a significant impact over the captured temperature information. Because of this kind of heat accumulation due to the sequential drilling, the thermal damage occurred during the orthopedic surgery will be more. By this preliminary study, it was identified that by increasing the margin to margin distance between the drilled holes, the thermal damage can be decreased through the decrease in the heat accumulation or temperature rise. The main limitation of this work is that the observations provided were based on the very limited experimental study. So more follow-up experiments are need to be conducted to have better repeatability and understanding about the effects of sequential drilling. The postoperative success of the drilling process in orthopedic surgery not only depends on the reduction of temperature rise through the identification of optimal distance or optimal parameters but also the geometry of drill bit design, which reduces the efforts in the process of surgery. The authors tend to travel in further directions of this work in such a way to explore the effects of exposure time and the distance between the holes to the thermal damage caused, to implement this phenomenon in the virtual orthopedic surgery protocol. As the temperature generation is more at the proximity of the drill site, and it is also necessary to optimize the way in which the sequential drills can be made to avoid heat accumulation. So the association between the personnel from engineering and physicians is required to develop the surgical protocol for the automated surgeries.

Acknowledgements This research work was funded by Science and Engineering Research Board (File No. YSS/2015/001053), Department of Science and Technology, New Delhi, Government of India.

Ethical Approval This study was approved by the local ethics committee of the Institute and the participants' next of kin provided informed consent before commencing the present study.

References

1. Hillery, M.T., Shuaib, I.: Temperature effects in the drilling of human and bovine bone. J. Mater. Process. Technol. **92**, 302–308 (1999)
2. Bachus, K.N., Rondina, M.T., Hutchinson, D.T.: The effects of drilling force on cortical temperatures and their duration: an in vitro study. Med. Eng. Phys. **22**(10), 685–691 (2000)
3. Baron, R., Horne, W.C.: Bone resorption: regulation of osteoclast activity. In: Bronner, F., Farach-Carson, M.C. (eds.) Topics in Bone Biology. Springer, London (2005)
4. Eriksson, A.R., Albrektsson, T.: Temperature threshold levels for heat-induced bone tissue injury: a vital-microscopic study in the rabbit. J. Prosthet. Dent. **50**(1), 101–107 (1983)
5. Pandey, R.K., Panda, S.S.: Drilling of bone: a comprehensive review. J. Clin. Orthop. Trauma **4**(1), 15–30 (2013)
6. Heydari, H., Cheraghi Kazerooni, N., Zolfaghari, M., Ghoreishi, M., Tahmasbi, V.: Analytical and experimental study of effective parameters on process temperature during cortical bone drilling. Proc. IMechE Part H J. Eng. Med. **232**(9), 871–883 (2018)
7. Wiggins, K.L., Malkin, S.: Drilling of bone. J. Biomech. **9**(9), 555–559 (1976)
8. Lee, J., Chavez, C.L., Park, J.: Parameters affecting mechanical and thermal responses in bone drilling: a review. J. Biomech. **71**, 4–21 (2018)
9. Palmisano, A.C., Tai, B.L., Belmont, B., Irwin, T.A., Shih, A., Holmes, J.R.: Heat accumulation during sequential cortical bone drilling. J. Orthop. Res. **34**(3), 463–470 (2016)
10. Tai, B.L., Palmisano, A.C., Belmont, B., Irwin, T.A., Holmes, J., Shih, A.J.: Numerical evaluation of sequential bone drilling strategies based on thermal damage. Med. Eng. Phys. **37** (9), 855–861 (2015)

Part IV
Additive Manufacturing and Casting

Chapter 33
FE-Based Heat Transfer Analysis of Laser Additive Manufacturing on Ti–6Al–4V Alloy

Sapam Ningthemba Singh, Sohini Chowdhury, Md. S. Mujaheed Khan, Manapuram Muralidhar and Yadaiah Nirsanametla

Abstract A thorough understanding of laser-based additive manufacturing process and effect of various process variables such as scanning velocity and laser beam power on melt-pool dimensions and temperature variation is a promising task in design and manufacture of an able product. The present work is focused on comprehending the thermal and melt-pool behavior of a high layer thickness five-layer laser additive manufacturing of Ti–6Al–4V alloy quantitatively. A three-dimensional (3D) nonlinear transient thermal model is developed based on a finite element procedure to simulate single- and multi-layer of Ti–6Al–4V alloy and to estimate melt-pool dimensions and thermal cycles. In this work, temperature-dependent material properties and Gaussian distributed 'disk' heat source model are implemented along with actual process boundary and initial conditions. Also, the influence of laser beam power and laser scanning velocity was examined with respect to melt-pool characteristics and thermal cycles. The laser scanning velocity ranges from 200 to 500 mm s^{-1} and laser beam power from 100 to 400 W are examined. It is observed that the temperature rises for successive layers as the laser power supply continues on consecutive layers. Also, it is obvious that with the rise in temperature, melt-pool dimensions also increase. Furthermore, the melt-pool dimensions increase as the number of deposited layers increases. Time–temperature history and melt-pool evolution in different layers with respect to laser beam power and laser scanning velocity are presented. To verify the effectiveness of the developed model, simulated results are compared with experimentally measured melt-pool profiles and dimensions. A fair agreement between experimental results and computed values is achieved.

Keywords Laser additive manufacturing · FEM · Melt-pool dimensions · Thermal cycles and Ti–6Al–4V alloy

---

S. N. Singh · S. Chowdhury · Md.S. M. Khan · M. Muralidhar · Y. Nirsanametla (✉)
Department of Mechanical Engineering, North Eastern Regional Institute of Science and Technology (NERIST), Nirjuli, Arunachal Pradesh 791109, India
e-mail: ny@nerist.ac.in

R. G. Narayanan et al. (eds.), *Advances in Computational Methods in Manufacturing*, Lecture Notes on Multidisciplinary Industrial Engineering,
https://doi.org/10.1007/978-981-32-9072-3_33

33.1 Introduction

Laser-based additive manufacturing (AM) is an advanced manufacturing technique in which an object is fabricated in a layer-wise manner directly from a CAD file. Moreover, it is quite different from conventional machining processes where parent material undergoes subtractive operations to acquire the desired final product [1]. Hence, it gains some significant advantages over conventional machining processes in terms of design freedom, manufacturing flexibility, the ability to manufacture complex and customized shapes. Also, a variety of materials ranging from polymers, metals, alloys to ceramics can be utilized to manufacture desired products either in powder form or the form of the wire material. The main applications of laser-based additive manufacturing processes are in the field of aerospace, automotive, prototyping and healthcare—dental and medical. Ti–6Al–4V alloy is mainly used in the aerospace, automotive and healthcare industries where high strength, high corrosion resistance and light materials are of main necessity [2–4]. Hence, it is of prime interest to investigate and to have good knowledge of laser additive manufacturing of Ti–6Al–4V which will help to predict the nature of the interaction of the laser with the substrate, melt-pool formation and temperature variations during and after manufacturing operation. Data from such analysis can be used to further study mechanical and microstructural properties. This, in turn, will help in designing an efficient process, optimized machines for specific jobs.

Owing to the importance of additive manufacturing of Ti–6Al–4V alloy in different industrial domains, several researchers have carried out experimental investigations to understand the thermo-physics and its effect on final microstructural and mechanical characteristics of the built product. Yadroitsev et al. [5] reported selective laser melting (SLM) of Ti–6Al–4V alloy for single-layer and single-track application to monitor temperature and microstructural changes during the process. The authors implemented a CCD camera to determine temperature distribution and melt-pool dimensions. It was found that the depth and width of melt pool increase with a continuous supply of laser power. Presence of foreign particles or the addition of extra elements in material results in a drastic change in surface tension and shape of the deposited layer during the LAM process. Traces of rare earth elements such as lanthanum hexaboride and boron in the Ti–6Al–4V and slow cooling rate result in very large Ti particles and improved strength. However, the final structure of the material is more hardened due to the presence of these elements [6]. It is possible to maintain constant energy density by adjusting laser intensity, scanning speed, layer thickness, cooling rate, etc. However, changing the process parameters will result in varying shapes and mechanical properties for constant energy density. Bartolomeu et al. [7] examined the influence of heat source (laser power) and scanning velocity on hardness, density and shear strength of Ti–6Al–4V material during a SLM process. Further studies were carried out on the influence of process variables on microstructure. However, they ignored heat transfer and interaction during the whole process. In experimental research on single-track SLM of Ti–6Al–4V material, it is asserted that the process variables can be varied to control melt-pool dimensions. Controlling melt-pool dimensions is necessary to adjust the final microstructure of the object. However, single-track experiments cannot provide complete information on a multi-layer built object. Contrary to the theoretical concept, when layer thickness varies from 20 to 60 μm, it is found that there is a negligible effect on the melt-pool depth and width when the scanning speed and laser power are kept constant [8].

Chandrakanth et al. [9] compared the effects on the melt-pool characteristics of Ti–6Al–4V with respect to laser power and scanning velocity during the SLM process considering when the material is in powder state and no powder state [9].

Conducting experiments corresponding to every AM criterions is not desirable in terms of resources and expenditure. Hence, it is preferable to simulate AM process using available tools and methods to examine the effect of input factors on the final built object. Numerical simulations can be used to compute melt-pool dimensions and time–temperature distribution in the laser additive manufacturing process. Qingcheng et al. [10] developed a finite element model to determine the thermo-mechanical behavior of Ti–6Al–4V alloy for laser engineered net shaping (LENS) process using ellipsoidal heat source model. The computed results were determined to be in good agreement with the experimental data [10]. Over the past few years, several thermal models have been developed to simulate AM processes to examine the thermal and mechanical effects with respect to process variables. A computational model of 40 layers of the Ti–6Al–4V alloy was studied by Xufei et al. [11] during the solid forming process using generic heat source model. However, performing finite element analysis for 40 layers will increase computational cost. They determined that the initial temperature of the substrate and the cooling rate after the fabrication process are significant factors. The SLM model developed by Ali et al. [12] considered absorptivity, laser spot area and laser power for effective evaluation of optical penetration of laser beam. They also found that there is no or little noticeable increase in the maximum temperature after the fourth layer. Hence, taking more number of the layer to study the thermal analysis and melt-pool formation is not required. Kurian et al. [13] carried out a detailed comparison of FEM result with experimental data during the SLM process for single-track and single-layer stainless steel 316L material [13]. Moreover, the computational model yielded good results when compared with experimental data, and it is within an acceptable range. Similarly, Jun et al. [14] developed a FEM model to analyze the distortion and residual stresses in electron beam-based additive manufacturing of Ti–6Al–4V alloy. In their work, the authors have implemented a uniform distribution heat source model to simulate the AM process. It was found that the distortion level increases initially for 2–3 layers and then decreases as the layers accumulate [14]. However, the heat source in the real world is not uniform. Thermal distribution along the layer can be applied for determining minimum scanning velocity. A minimum energy density should be maintained to avoid incomplete melting and to prevent balling effect [15]. Jihong et al. [16] analyzed that the heat-affected zone (HAZ) is a significant output parameter in the SLM process [16]. However, several investigators neglected HAZ in their analysis [17–22].

The aim of the present work is to develop a FE model for a five-layer thin-walled Ti–6Al–4V alloy during the laser additive manufacturing process to study the effect of laser power and scan velocity. Using the developed model, five different layers are examined with respect to time–temperature history and melt-pool formation using moving Gaussian distributed disk heat source model. To verify the correctness of the developed model, predicted results are compared with experimental values and found in good agreement.

33.2 Theoretical Background

A transient finite element-based model is developed for five-layer SLM process of Ti–6Al–4V to determine the nonlinear, time–temperature history and melt-pool dimensions at each point of the five layers. SLM setup consists of a laser source, a powder feeding mechanism, a base or a building platform, roller scrapper and other auxiliary parts [16]. The transient heat conduction equation for a three-dimensional heat flux can be represented as

$$\boldsymbol{k}\left(\frac{\partial^2 \boldsymbol{T}}{\partial \boldsymbol{x}^2}+\frac{\partial^2 \boldsymbol{T}}{\partial \boldsymbol{y}^2}+\frac{\partial^2 \boldsymbol{T}}{\partial \boldsymbol{z}^2}\right)+\dot{\boldsymbol{Q}}=\boldsymbol{\rho c_p}\left(\frac{\partial \boldsymbol{T}}{\partial \boldsymbol{t}}+\boldsymbol{v}\frac{\partial \boldsymbol{T}}{\partial \boldsymbol{y}}\right), \quad (33.1)$$

where ρ, c_p, T, v, k and $\dot{Q}$ are the density of the material, specific heat, velocity vector, temperature variable, thermal conductivity and heat generation rate (internal) per unit volume. The first term represents the heat conduction process in Cartesian coordinates system along with internal heat generation rate in the substrate. The initial temperature distribution at time $t = 0$ is given by

$$T(x, y, z)|_{t=0} = T_0, \quad (33.2)$$

where T_0 is the ambient temperature and is assumed to be 298 K.

The thermal transfer within the melt pool is largely due to the latent heat of conduction. Heat is dissipated from the layers as well from the substrate to the atmosphere by convection and radiation modes. The necessary boundary conditions are expressed by Newtonian convection (Newton's Law of Cooling) and gray-surface body behavior [23–29], and mathematically represented as

$$k\frac{\partial T}{\partial n} - q + q_{\mathrm{c}} + q_{\mathrm{r}} = 0, \quad (33.3)$$

where q_{c} and q_{r} are heat loss due to convection and radiation mode, and q is the rate of heat influx. The basic equation of convection heat loss and radiation heat loss is

$$q_c = h(T - T_0), \quad (33.4)$$

$$q_r = \sigma\epsilon(T^4 - T_0^4), \quad (33.5)$$

where $h\sigma$ is the coefficient of convective heat transfer, is the Stephan–Boltzmann constant (5.67 J/m^2 s K^4) and ϵ is melt-pool emissivity, T_0 is initial temperature and T is the instantaneous temperature.

The Gaussian distributed disk heat source model that is utilized in the present analysis is given by [30].

$$q = \frac{3\eta P}{\pi r^2}\exp\left(\frac{-3(x^2+y^2)}{r^2}\right), \quad (33.6)$$

where q is the heat flux, η is the laser absorption coefficient, P is the laser power, r is the effective laser beam radius, and x and y are the respective position of the laser beam at a given time.

A computational model was developed using FE-based ANSYS 14.5 software. The dimensions of the substrate are 10 mm × 31 mm × 4 mm and five layers with layer thickness 1 mm as shown in Fig. 33.1. Also, temperature-dependent material properties such as thermal conductivity, density and specific heat are used [31]. The scanning pattern of the laser on successive layers is shown in Fig. 33.2. The process variables used in the present work are given in Table 33.1. Six different data sets are utilized to simulate laser additive manufacturing process and to analyze melt-pool dimensions and temperature distribution.

The deposition of layers is simulated using the element birth and death technique in ANSYS APDL. When the first layer is exposed to the laser, the elements in the other four layers are classified as death elements. After the first layer, the second layer is activated using element birth feature while keeping the last three layers deactivated, and this process is repeated till the fifth layer.

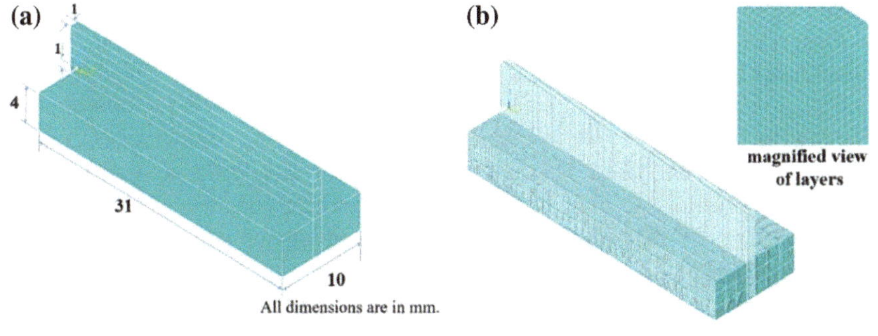

Fig. 33.1 **a** Geometry of the present work, **b** geometry of the present work after meshing

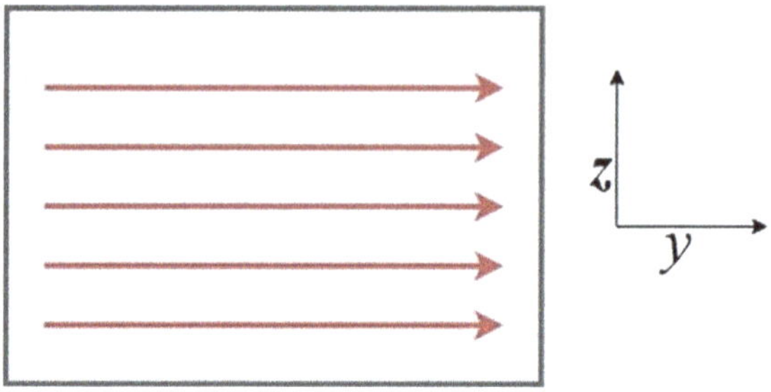

Fig. 33.2 Scanning pattern of the laser on successive layers

Table 33.1 Process parameters used in the present work

Data set no.	1	2	3	4	5	6
Power (W)	100	300	200	200	300	400
Speed (mm s^{-1})	200	200	200	300	400	500

33.3 Results and Discussions

In this section, the influence of laser beam power and scanning velocity on the melt-pool profile and time–temperature distribution at selected points on each layer is presented along with the validation of model predictions with experimental results. Variation of melt-pool dimensions as well as the effect of laser power variation on peak temperature magnitude and overall temperature values is presented. Also, the effect of scanning velocity at constant laser power is also analyzed. To verify the effectiveness of the model, it is calibrated and compared with the experimental data.

For validation of the numerical model results, experimental work is considered from the literature [8]. An SLM experiment was conducted by Yang et al. (2016) which consists of IPG YLR-500 fiber laser, a powder delivery system, a support and building platform along with a process control computer system using Ti–6Al–4V alloy powder with a range of 30–50 μm in an inert environment of Argon. Table 2 depicts the process parameters used in the experimental investigation. From the experimental investigation, it is observed that the melt-pool width (~70 μm to ~160 μm) and depth (~10 μm to ~200 μm) increase as the power is increased from 100 to 500 W keeping the other parameters unchanged. But the melt pool decreases when the scanning speed increases from 100 to 1000 mm s^{-1} keeping the laser power and layer thickness constant.

A finite element-based heat transfer model is developed to examine the melt-pool formation and thermal cycles. To confirm the viability of the developed model, the single-layer FEM model is developed initially of 1 mm layer thickness. The single-layer model is compared and validated with the experimental data from literature [8]. The melt-pool depth corresponding to the computed value and the experimental result of data set #3 of Table 33.1 are determined to 0.12 and

0.13 mm. Moreover, a percentage difference in melt-pool depth is estimated to be 5.38% for all processing conditions of Table 33.1 and which is under acceptable range. Comparison of computed (left) and experimental (right) melt-pool shape corresponding to data set #3 of Table 33.1 is presented in Fig. 33.3, and the red color zone represents the melt zone.

With the application of laser power, a sudden rise in temperature is observed with a high-temperature gradient. Figure 33.4 shows the 3D transient temperature distribution at different positions and at different layers corresponding to process variables given in Table 33.1. Figure 33.4a shows the location of the laser source

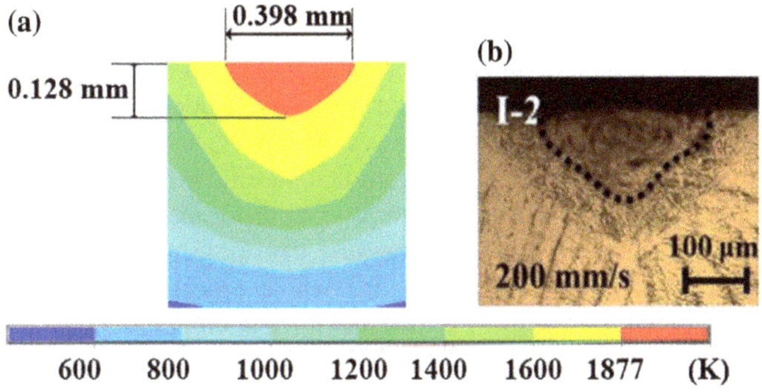

Fig. 33.3 Comparison of computed (left) and experimental (right) melt-pool shape and size corresponding to data set #3 of Table 33.1 of for a single-layer laser additive manufacturing

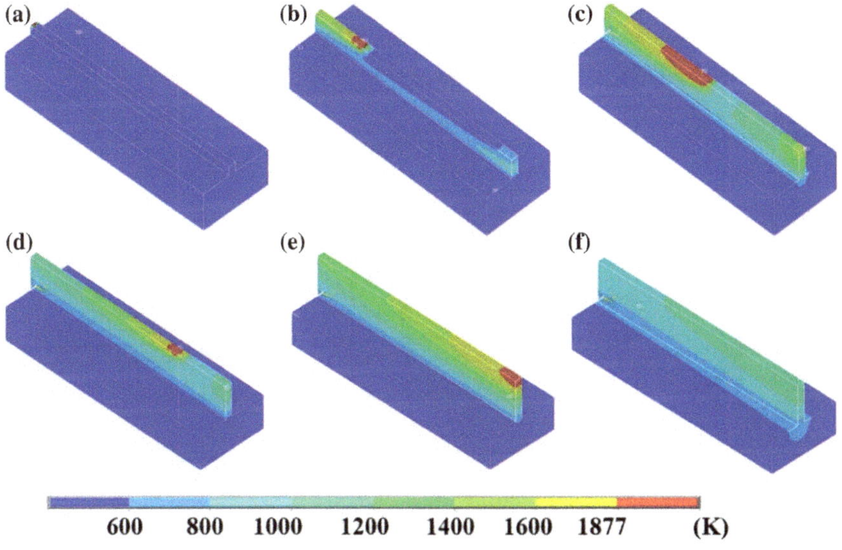

Fig. 33.4 3D transient temperature distribution at different positions and different layers during five-layer laser additive manufacturing of Ti–6Al–4V alloy

which is at the beginning of the first layer corresponding to data set #6. Figure 33.4 (b) represents the location of the laser source which is at a distance of 7.75 mm along the scanning direction of the second layer corresponding to data set #5. Similarly, Fig. 33.4c demonstrates the location of the laser source which is at the center of the third layer corresponding to the data set #4. Figure 33.4d represents the location of the laser source which is at a distance of 23.25 mm along the scanning direction of the fourth layer corresponding to data set #3. Figure 33.4e shows the location of the laser source at the end of the fifth layer corresponding to data set #2. Figure 33.4f represents the cooling nature of laser additive manufacturing of Ti–6Al–4V alloy. It is observed from Fig. 33.4c, d that the melt zone and heat-affected zone are wider for high laser beam power and low scanning velocity.

Figure 33.5 depicts the time–temperature history of each layer at a distance of 7.75 mm along the scanning direction. This figure shows a rapid rise in temperature followed by a gradual decrease in temperature. The sudden rise in temperature is due to sudden impingement of laser power at a point, and a gradual decrease is because of laser source moving away from that specific point. Theoretically, there should be five different peaks for the first layer, four for the second layer and three for the third layer and likewise. This is because laser power passes above a specific point for five, four, three times during the whole process. However, the graph reveals only three visible peaks. It shows that for a point below three layers where the laser is striking, there is little effect of laser power on that point for specific data set. However, other data set represents five peaks during the whole process. This reflects that high laser power affects the temperature at a point, even though layers are away from it. However, the overall temperature of the point still increases as the number of layers increases. After the manufacturing process, there is a rapid decrease in the temperature due to cooling up to 2.5 s, after that the cooling takes place slowly. This is due to the temperature difference between the ambient and builds parts decreases, and heat transfer due to convection decreases sharply.

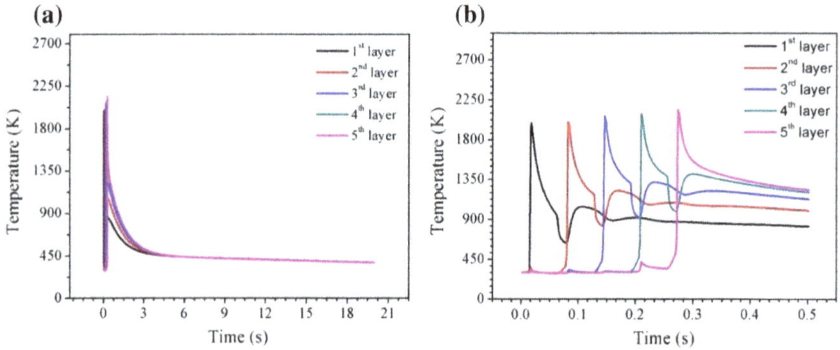

Fig. 33.5 Time–temperature history of each layer during a five-layer laser additive manufacturing of Ti–6Al–4V alloy for data set # 6 of Table 33.1

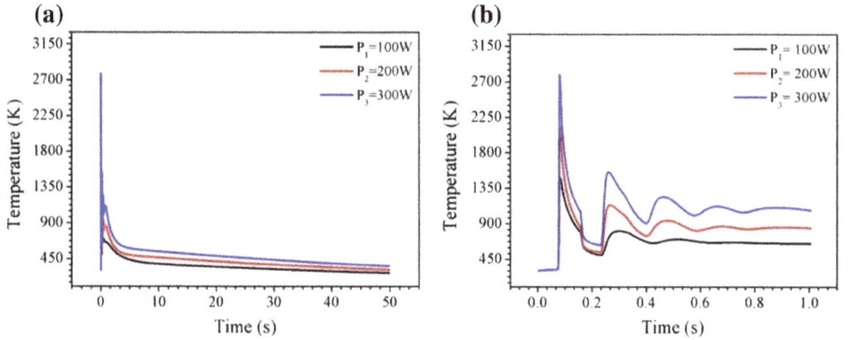

Fig. 33.6 Thermal cycles at constant scanning velocity of 200 mm s^{-1} and different laser beam power: **a** full view and **b** magnified view

The effect of laser power on the time–temperature history of Ti-alloy material is graphically represented in Fig. 33.6. For constant laser scanning velocity of 200 mm s^{-1}, the laser power is varied from 100 to 300 W. Correspondingly, the peak temperatures are 1484, 2113 and 2788 K at the top of the first layer and a distance of 23.25 mm along the scanning direction corresponding to the laser power 100, 200 and 300 W, respectively. There are five peaks for 300 and 200 W laser power; however, for 100 W there are only three peaks. Figure 33.7 shows the increased melt-pool dimensions when laser power varies from 100 to 200 W at constant scanning velocity. It can be identified that there is an incomplete melting and sintered which is represented by the yellow and light green region. This results in a change of microstructure and mechanical properties in the final product.

Figure 33.8 depicts the temperature variation at constant laser power and increasing laser scanning velocity. At constant laser power of 200 W (Fig. 33.8a) and an increasing scanning velocity from 200 to 300 mm s^{-1}, the temperature reduces from 2113 to 1738 K at the top of the first layer and a distance of 23.25 mm along the scanning path. A similar trend is observed when power is

Fig. 33.7 Effect of laser power on the melt-pool formation at a constant scanning velocity of 200 mm s^{-1} in Ti–6Al–4V alloy during laser additive manufacturing

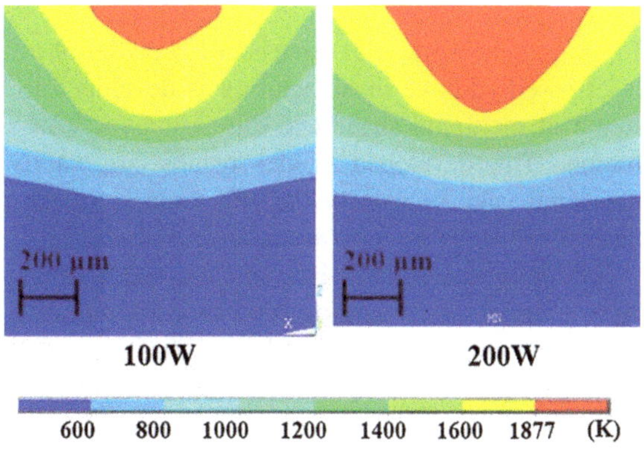

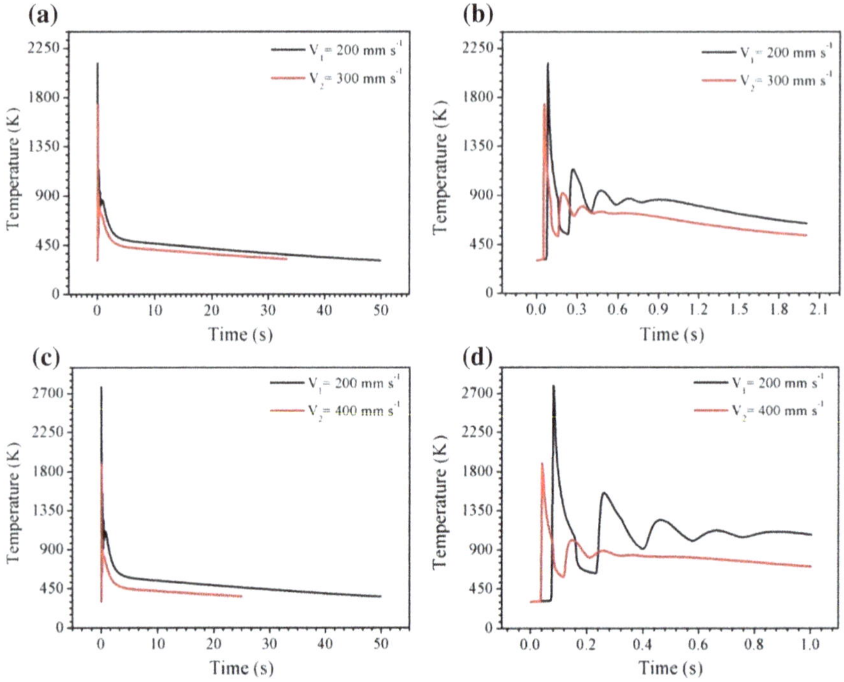

Fig. 33.8 Effect of scanning speed at constant laser beam power **a**, **b** 200 W and **c**, **d** 300 W

300 W (Fig. 33.8c) and scanning velocity is increased from 200 to 400 mm s^{-1}, and temperature decreases from 2788 to 1895 K. The temperature distribution and peak temperatures were estimated using developed model along with melt-pool formation.

33.4 Conclusion

In the present work, a 3D transient FE model of a large layer thickness five-layer laser-based additive manufacturing is developed to understand the complex evolution of melt-pool shape and dimensions as well as transient temperature distribution. The developed model results are validated with the experimentally measured values to verify the effectiveness of the process model. The model is determined to be in good accordance with the experimental data.

The influence of laser beam power and scanning velocity examined five-layer laser additive manufacturing process using a developed model. It is observed that the temperature of subsequent layers rises with an increase in layer number. In the first layer, the substrate played an important part in conducting the heat away from the layers. The results also showed that for low power or high speed, there is the negligible effect of temperature on the first layer when the laser source passes through the fourth layer. However, if the laser power is high and the scanning speed is low, it is still a significant influencing factor for the first layer even when the laser source passes on the fifth layer. It is also found that the temperature increases up to 2788 from 1484 K when laser beam power increases from 100 to 300 W at the top of the first layer for the same scanning velocity, 200 mm s^{-1}. Moreover, the temperature decreases from 2788 to 1895 K when the laser power is 300 W, and scanning velocity is changed from 200 to 400 mm s^{-1}. The future scope of the present work is the development of a mechanical model to estimate the residual stresses and mechanical properties using the present thermal model. Also, this model will provide a base to develop a process model for a multi-layer, multi-tracks during additive manufacturing.

Acknowledgements The authors thank the TEQIP II grant for financially supporting this work at North Eastern Regional Institute of Science & Technology, Itanagar, Arunachal Pradesh, India.

References

1. International Organization for Standardization: Standard terminology for additive manufacturing—general principles—part 1: terminology, 2015, (52900:2015 (E))
2. Bandyopadhyay, A., Espana, F., Balla, V.K., Bose, S., Ohgami, Y., Davies, N.M.: Influence of porosity on mechanical properties and in vivo response of Ti–6Al–4V implants. Acta Biomater. **6**, 1640–1648 (2010)
3. Vrancken, B., Thijs, L., Kruth, J.P., Humbeeck, J.V.: Heat treatment of Ti–6Al–4V produced by selective laser melting: Microstructure and mechanical properties. J. Alloy. Compd. **541**, 177–185 (2012)
4. Sun, J., Yang, Y., Wang, D.: Parametric optimization of selective laser melting for forming Ti–6Al–4V samples by Taguchi method. Opt. Laser Technol. **49**, 118–124 (2013)
5. Yadroitsev, I., Krakhmalev, P., Yadroitsava, I.: Selective laser melting of Ti–6Al–4V alloy for biomedical applications: temperature monitoring and microstructural evolution. J. Alloy. Compd. **583**, 404–409 (2014)
6. Bermingham, M.J., McDonald, S.D., Dargusch, M.S.: Effect of trace lanthanum hexaboride and boron additions on microstructure, tensile properties and anisotropy of Ti–6Al–4V produced by additive manufacturing. Mater. Sci. Eng., A **719**, 1–11 (2018)
7. Bartolomeu, F., Faria, S., Carvalho, O., Pinto, E., Alves, N., Silva, F.S., Miranda, G.: Predictive models for physical and mechanical properties of Ti6Al4 V produced by selective laser melting. Mater. Sci. Eng., A **663**, 181–192 (2016)
8. Jingjing, Y., Jie, H., Hanchen, Y., Jie, Y., Ming, G., Zemin, W., Xiaoyan, Z.: Role of molten pool mode on formability, microstructure and mechanical properties of selective laser melted Ti–6Al–4V alloy. Mater. Des. **110**, 558–570 (2016)
9. Chandrakanth K.: The effect of laser power and scan speed on melt pool characteristics of pure titanium and Ti–6Al–4V alloy for selective laser melting. Wright State University, MSME Thesis, 1510 (2016)
10. Qingcheng, Y., Zhang, P., Cheng, L., Zheng, M., Chyu, M., Albert, C.T.: Finite element modelling and validation of thermo-mechanical behaviour of Ti–6Al–4V in directed energy deposition additive manufacturing. Addit. Manuf. **12**, 169–177 (2016)

11. Xufei, L., Xin, L., Chiumenti, M., Miguel, C., Li, J., Liang, M., Wei, L., Yunlong, H., Weidong, H.: Finite element analysis and experimental validation of the thermomechanical behaviour in laser solid forming of Ti–6Al–4V. Addit. Manuf. **21**, 30–40 (2018)
12. Ali, F., Mohsen, B., Ehsan, F., Golabi, S.: Finite element simulation of selective laser melting process considering optical penetration depth of laser in powder bed. Mater. Des. **89**, 255–263 (2016)
13. Kurian, A., Arivazhagan, N., Senthilkumaran, K.: Numerical and experimental investigations on laser melting of stainless steel 316L metal powders. J. Manuf. Process. **16**(3), 345–355 (2014)
14. Jun, C., Michael, A.G., Philip, N.: Finite-element analysis and experimental validation of thermal residual stress and distortion in electron beam additive manufactured Ti–6Al–4V build plates. J. Mater. Process. Technol. **237**, 409–419 (2016)
15. Patcharapit, P., Recep, O., Shi-Chune, Y.: Numerical and experimental investigations of micro and macro characteristics of direct metal laser sintered Ti–6Al–4V products. J. Mater. Process. Technol. **240**, 262–273 (2017)
16. Jihong, Y., Shoujin, S., Milan, B., Wenyi, Y.: Experimental investigation and 3D finite element prediction of the heat affected zone during laser assisted machining of Ti–6Al–4V alloy. J. Mater. Process. Technol. **210**(15), 2215–2222 (2010)
17. Peyre, P., Aubry, P., Fabbro, R., Neveu, R., Longuet, A.: Analytical and numerical modelling of the direct metal deposition laser process. J. Phys. D Appl. Phys. **41**(2), 025403 (2008)
18. Gockel, J., Beuth, J., Taminger, K.: Integrated control of solidification microstructure and melt pool dimensions in electron beam wire feed additive manufacturing of Ti–6Al–4V. Addit. Manuf. **1–4**, 119–126 (2014)
19. Fachinotti, V.D., Cardona, A., Baufeld, B., Van der Biest, O.: Finite-element modelling of heat transfer in shaped metal deposition and experimental validation. Acta Mater. **60**(19), 6621–6630 (2012)
20. Romano, J., Ladani, L., Razmi, J., Sadowski, M.: Temperature distribution and melt geometry in laser and electron-beam melting processes: A comparison among common materials. Addit. Manuf. **8**, 1–11 (2015)
21. Labudovic, M., Hu, D., Kovacevic, R.: A three dimensional model for direct laser metal powder deposition and rapid prototyping. J. Mater. Sci. **38**(1), 35–49 (2003)
22. Wang, L., Felicelli, S., Gooroochurn, Y., Wang, P.T., Horstemeyer, M.F.: Optimization of the LENS® process for steady molten pool size. Mater. Sci. Eng., A **474**(1), 148–156 (2008)
23. Beauchamp, M.J., Nordin, G.P., Woolley, A.T.: Moving from mill fluidic to truly microfluidic sub-100-μm cross-section 3D printed devices. Anal. Bioanal. Chem. **409**, 4311–4319 (2017)
24. Monaghan, T., Harding, M.J., Harris, R.A., Friel, R.J., Christie, S.D.: Customisable 3D printed microfluidics for integrated analysis and optimisation. Lab Chip **16**, 3362–3373 (2016)
25. Sweet, E.C., Rudra, R., Mehta, R.L., Lin, L.: Finger-powered 3D printed microfluidic pumps. In: 19th International Solid-State Sensors, Actuators and Microsystems Conference Transducers, pp. 1766–1769, IEEE, Taiwan (2017)
26. Bose, S., Robertson, S.F., Bandyopadhyay, A.: Surface modification of biomaterials and biomedical devices using additive manufacturing. Acta Biomater. **66**, 6–22 (2018)
27. Bandyopadhyay, A., Bose, S.: Additive Manufacturing. CRC Press, US (2015)
28. Bose, S., Ke, D., Sahasrabudhe, H., Bandyopadhyay, A.: Additive manufacturing of biomaterials. Prog. Mater Sci. **93**, 45–111 (2018)
29. Puleo, D.A., Nanci, A.: Understanding and controlling the bone–implant interface. Biomaterials **20**(23), 2311–2321 (1999)
30. Yadaiah, N., Bag, S.: Development of egg-configuration heat source model in numerical simulation of autogenous fusion welding process. Int. J. Therm. Sci. **86**, 125–138 (2014)
31. Boyer, R., Collings, E.W., Welsch, G.: Materials Properties Handbook: Titanium Alloys. ASM International, US (1994)

Chapter 34
Material Perspective and Deformation Pattern of Micro-Sized Metallic Particle Using Cold Gas Dynamic Spray

Abdul Faheem, Faisal Hasan and Qasim Murtaza

Abstract The cold gas dynamic spray is a solid powder deposition phenomenon of coating and repairing with a broad materials range. This deposition technology gained attention not only in institutions but also in industrial aspects. This coating technology is a surface engineering technology which is presently replacing electron, laser beam type of technologies for additive manufacturing. This study provides the pattern of deformation of both particle and substrate under the impact of high particle velocity. An ABAQUS/Explicit, a finite element methodology was used to forecast the coating behaviour of different engineering materials. A FEM approach was used for the simulation of the metallic impact of substrate and particle. A wide range of impact velocities was used to finding the threshold or critical velocity for the proper deposition, phenomenon of rebounding and also to divulge the mechanism of governance, i.e. adiabatic shear instability (ASI). This study focused and revealed how engineering materials that are commonly used in aerospace engineering behaved under different impact conditions using kinetic spray coating technology.

Keywords Cold gas dynamic spray · Numerical simulation · Critical velocity · Finite element methodology

34.1 Introduction

Cold gas dynamic spray or cold spray (CS) or kinetic spray is a new material deposition technology which is procuring attention of surface engineering. This is a newly gaining noticed coating process used for a wide range of materials, i.e. ductile, brittle, ceramics and composite materials, etc., [1] without noteworthy heating of micro- or

A. Faheem · F. Hasan
Department of Mechanical Engineering, AMU, Aligarh 202002, India

Q. Murtaza (✉)
Department of Mechanical Engineering, DTU, New Delhi 110042, India
e-mail: qasimmurtuza@dce.ac.in

R. G. Narayanan et al. (eds.), *Advances in Computational Methods in Manufacturing*, Lecture Notes on Multidisciplinary Industrial Engineering,
https://doi.org/10.1007/978-981-32-9072-3_34

nano-solid particles. A kinetic spray, firstly, at the Russian's Institute of Applied and Theoretical Mechanics of Science was discovered more than two decades ago and the first patent on it done in 1994 [2]. In CS, adhesion between the particle and substrate occurs due to the impact of the high velocity of micro- or nano-size of particles [3]. This high velocity not only effects the deformation of both of it but also equated a large amount of plastic deformation [4]. The acceleration of particles is obtained by the convergent–divergent nozzle under high pressure with a Mach number in the range of 3 for the impact phenomenon [5]. Apart from heating, the dimension instability and surface protection also be done by alternate technology, i.e. kinetic spray [6]. This metal joining process is similar to cladding in which the fusion of parts occurs to fasten the dissimilar metals [7, 8] (Fig. 34.1).

The major disadvantage of high-pressure cold spray system is the erosion at the throat of the nozzle due to erosion particle which not only affects the working of the nozzle but also the deposition quality [9]. (2) Low-pressure cold spray is a system in which powder is fed at that point in the nozzle where the gas expanded the least pressure. This system is commonly used for metal alloys and ceramics. LPCS systems are generally portable and smaller and also workable on a limited range of velocity (300–600 m/s). Due to the portability of LPCS, it does not need pressurizer feeder (Fig. 34.2).

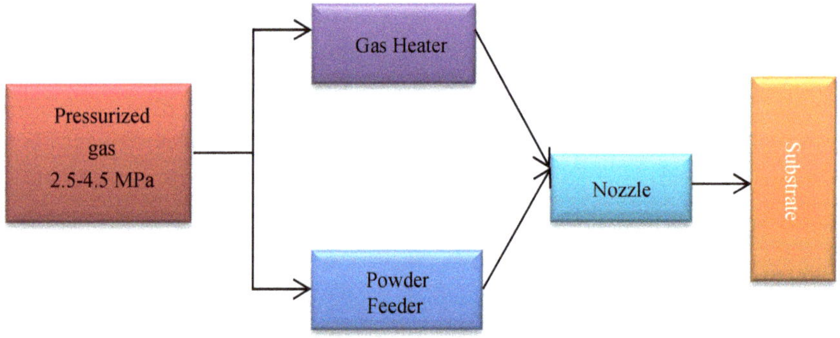

Fig. 34.1 High-pressure cold gas dynamic spray system

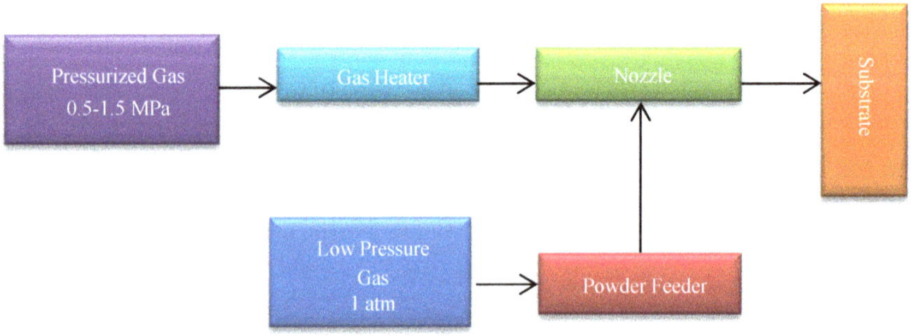

Fig. 34.2 Low-pressure cold gas dynamic spray system

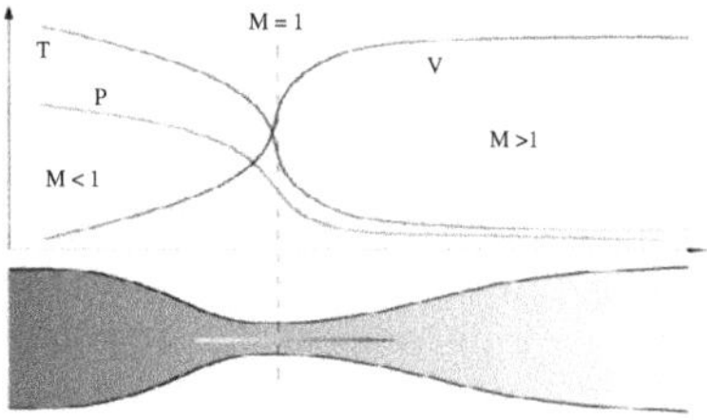

Fig. 34.3 Pressure, velocity and temperature variation in convergent–divergent nozzle

Generally, the pressure of working gas is between 0.5 and 1 MPa, and the flow rate is about 0.5–2 m^3/min. In LPCS, the design of the nozzle at exit is limited to Mach number less than 3, and inlet pressure is also less 1 MPa. If it exceeds, then atmospheric pressure is not able to supply powder particles into the nozzle. Due to this restriction, only low velocities particle reached into the nozzle. There are different variants of the standard apparatus of cold spray like kinetic metallization, pulsed gas dynamic spraying and vacuum cold spray.

KM is a variant of cold gas dynamic spraying that uses a convergent nozzle to achieve Mach number 1 at the exit of the gas velocity and small divergence for the compensation of friction impact. The deposition efficiency of micro- or nano-sized particles either preheated or at normal temperature solely depends on the mechanical and thermal properties of the engineering materials [10] (Fig. 34.3).

34.2 Numerical Methodology

Modelling of high velocity has been done by ABAQUS/Explicit using the code of finite element analysis (FEM). FEM has been commonly adopted for impact analysis in the kinetic spray. Adiabatic condition has been assumed for all kind of impact phenomenon. Particle is in range of 10–20 μm with the axis-symmetric condition for the impacting single particle on the substrate. Aluminium, copper and titanium have been used for the high-velocity impact simulation. Both the particle and substrate have not been preheated, and the impact problem was assumed to be adiabatic. Elements for the geometries are four-node bilinear quad elements and also with reduced integration and hourglass control (CAX4R) [11]. Reference strain rate for all the materials is taken as 1. Johnson–Cook (J-K) plasticity approach has been used which deals with thermal softening, strain and strain rate hardening. The elasticity response assumed to be linear for both the materials. Considerations of gravity, friction and rebounding force have been neglected in this study. The damage parameters are not considered in this study. For the J-K approach, the values of equivalent plastic strain (PEEQ) solely depend at the integration points. An isotropic condition has been assumed for both thermal and mechanical properties of the material. The boundary conditions for constraining the substrate and particle motion have been shown in Fig. 34.4. For both the particle and substrate, axis-symmetric boundary condition has been used. Fixed boundary condition is used for the right side of the substrate's surface. J-K plasticity approach has been used for the plastic deformation of the particle as well as the substrate. This model gives the rate dependency of material during plastic deformation. Thermal softening, strain hardening and strain rate sensitivity are also defined for the model. The equivalent flow stress is given by as

$$\sigma_e = \left(A + B\epsilon_p^n\right)(1 + C\ln\epsilon^*)(1 - (T^*)^m) \quad (34.1)$$

where A and B are yield stress and work hardening parameter of the material. T^* is the homologous temperature. Different boundary conditions were used from ABAQUS/Explicit, i.e. symmetry about a plane where X is constant (XSYMM) and ENCASTRE where $U1 = U2 = U3 = 0$ (Table 34.1).

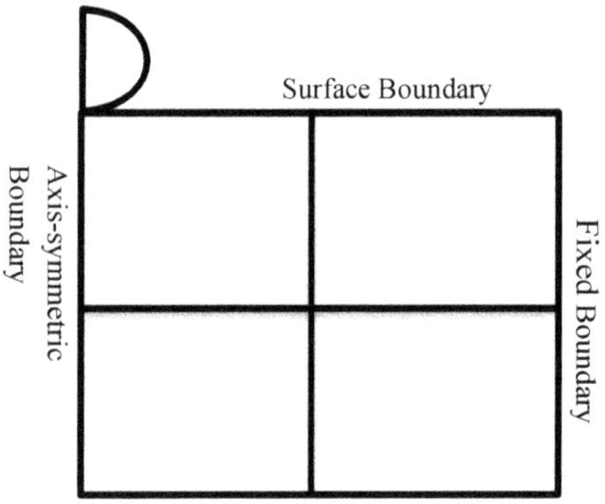

Fig. 34.4 Boundary conditions for the axis-symmetric single particle impact

Table 34.1 Properties of different materials

Material	Cu	Al	Ti
Density (kg/m^3)	8960	2700	7870
Heat capacity (J/kgK)	383	920	528
Young's modulus (GPa)	124	65.762	116
Poisson's ratio	0.34	0.3	0.34
A (MPa)	90	148.361	806.57
B (MPa)	292	345.513	481.61
N	0.31	0.183	0.319
C	0.025	0.001	0.0194
M	1.09	0.859	0.655
Melting point (K)	1356	930	1923
Reference temperature (K)	298	298	298

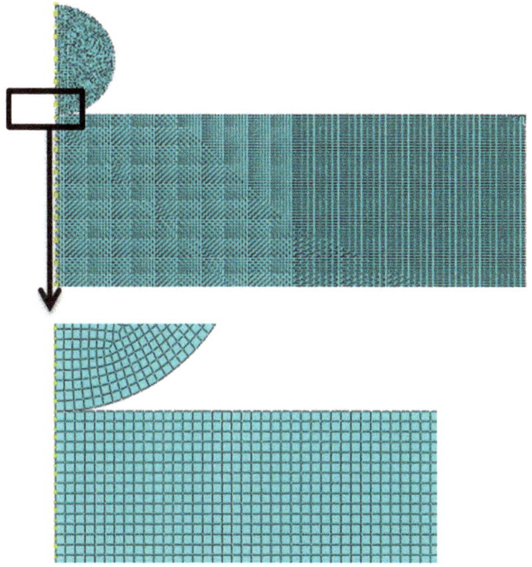

(a) Mesh Distribution in axis-symmetry Single particle impact. Number of elements and nodes are 12840 and 13075 respectively for 20 micrometer copper particle.

(b) Magnified view of Mesh at the contact zone of particle and substrate.

Fig. 34.5 Meshing of axis-symmetric impact behaviour of cold spray

Meshing plays a very crucial role in every computational problem. It also plays a vital role in the estimation of computational cost. Different meshing has been used like the diameter of particle (d_p)/32, d_p/35 and d_p/40. It was found that after d_p/40, there were no improvements in the results occurred. So, d_p/40 meshing size as shown in Fig. 34.5 has been used for all the computational impact problem in cold gas dynamic spray.

34.3 Results and Discussion

Copper particle having a diameter of 10 μm at a diameter of particle (d_p)/40 meshing at different velocities and also different values of equivalent plastic strain (PEEQ), compression ratio and flattening ratio, etc., were found for checking the adequate bonding between them. Figures 34.6 and 34.7 show that the flattening ratio and compression ratio are totally depended on the velocities and mechanical properties. Figure 34.8 also shows that values of PEEQ solely depend on the velocity of particle and substrate materials. Thermal properties also play a very important role for the deformation of the particle, i.e. if the specific heat of the material is low, then the particle will plastically deform at low velocity.

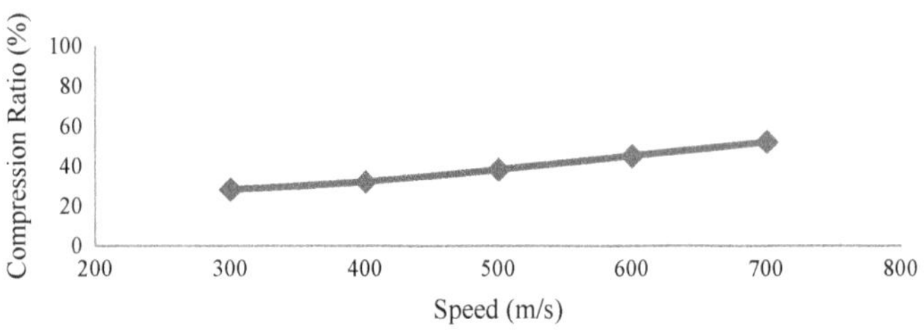

Fig. 34.6 Variation of compression ratio vs increasing the speed of 10 μm copper particle with meshing size $d_p/40$ using the lagrangian approach

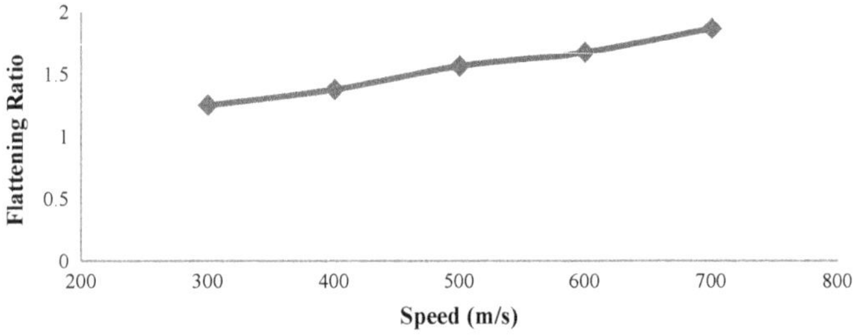

Fig. 34.7 Variation of flattening ratio by increasing the speed of 10 μm copper particle with meshing size $d_p/40$ using the lagrangian approach

34.3.1 Deformation Behaviour of Soft Particle on Hard Substrate

It has been seen from Fig. 34.9 that the deformation behaviour of a soft particle like aluminium when impacted on the hard substrate than splattering phenomenon occured. When micro-sized particle impacts at a certain angle like 30°, 45° and 60° especially when the particle made of soft material, splattering of the particle occurs on the surface of substrate with minute crater.

Formation of the jet by the deformation of the particle also been seen that removes the oxide layer from the surface. As shown in Fig. 34.9, when 20 μm aluminium particle is impacted at a different angle, then deformation behaviour is totally different from the perpendicular impact.

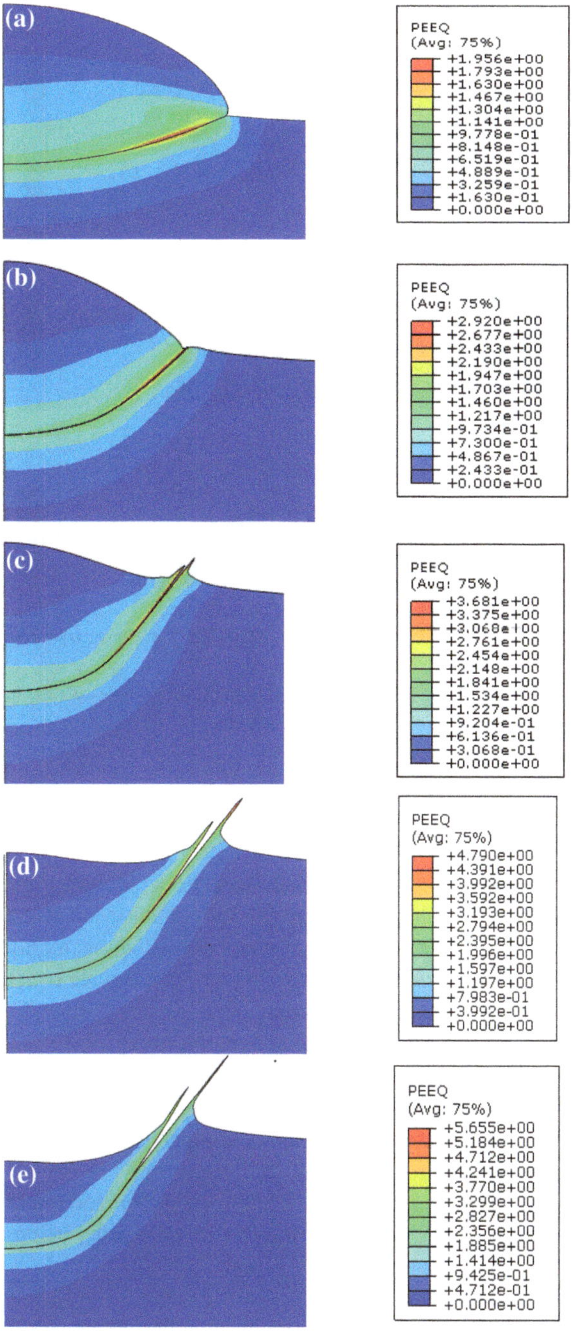

Fig. 34.8 Contour of 10 μm diameter of the copper particle with copper substrate and PEEQ at different velocities **a** 300, **b** 400, **c** 500, **d** 600 and **e** 700 m/s respectively

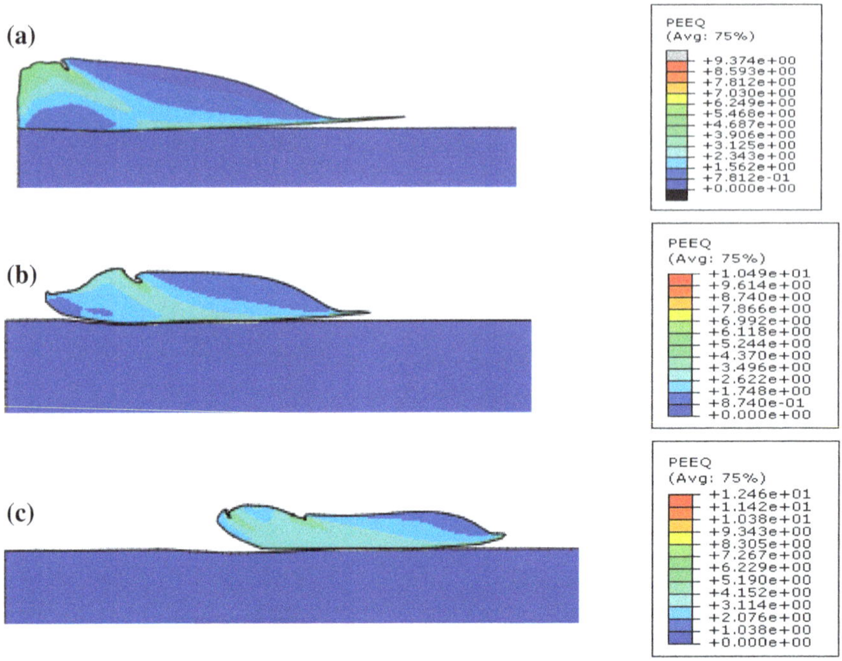

Fig. 34.9 Contour and values of PEEQ for 20 μm aluminium particle on titanium substrate at an angle of **a** 30°, **b** 45° and **c** 60° at 750 m/s, respectively

34.4 Conclusion

Different ranges of micro-sized particles have been probed during the plastic deformation process using kinetic spray technology. It has been investigated that for same particle and substrate, the critical velocity which is required for proper bonding, contour and crater formation and also for the adequate penetration is in the range of 500–560 m/s. Critical velocity has been very important for the adequate metal coating on the substrate by cold gas dynamic spray.

On the contrary, using soft particle and impacting on the hard substrate, it has been probed that huge amount of increment in the critical velocity or adequate penetration requires more particle velocity as compared to same particle/substrate. In this impact condition, homogeneity and uniformity of metal coating have occurred which indicates better coating. Lastly, the inclination angle also played a crucial role in the splattering and uniformity in metal coating using cold spray.

References

1. King, P.C., Bae, G., Zahiri, S.H., Jahedi, M., Lee, C.: An experimental and finite element study of cold spray copper impact onto two aluminum substrates. J. Therm. Spray Technol. **19** (3), 620–634 (2010)
2. Alkhimov, A.P., Kosarev, V.F., Papyrin, A.N.: A method of cold gas-dynamic deposition. Dokl. Akad. Nauk SSSR **35**(12), 1062–1065 (1990)
3. Yin, S., Suo, X., Guo, Z., Liao, H., Wang, X.: Deposition features of cold sprayed copper particles on preheated substrate. Surf. Coat. Technol. **268**, 252–256 (2015)
4. Li, W., Liao, H., Li, C., Li, G., Coddet, C., Wang, X.: On high velocity impact of micro-sized metallic particles in cold spraying. App. Surf. Sci. **253**, 2852–2862 (2006)
5. Yin, S., Wang, W., Li, X., Li, Y.: Numerical study on the effect of substrate size on the supersonic jet flow and temperature distribution within the substrate in cold spraying. J. Therm. Spray Technol. **21**(3–4), 628–635 (2012)
6. Grujicic, M.: Adiabatic shear instability based mechanism for particles/substrate bonding in the cold-gas dynamic-spray process. Mater. Design. **25**, 681–688 (2004)
7. Klassen, T., Kreye, H.: On parameter selection in cold spraying. J. Therm. Spray Technol. **20** (6), 1161–1176 (2011)
8. Yin, S., Liu, Q., Liao, H., Wang, X.: Effect of injection pressure on particle acceleration, dispersion and deposition in cold spray. Comput. Mater. Sci. **90**, 7–15 (2014)
9. Assadi, H., Ga, F., Stoltenhoff, T., Kreye, H.: Bonding mechanism in cold gas spraying. Acta Mater. **51**, 4379–4394 (2003)
10. Klinkov, S.V., Kosarev, V.F.: Measurements of cold spray deposition efficiency. J. Therm. Spray Technol. **15**(9), 364–371 (2006)
11. ABAQUS/CAE User Manual 6.14. Hibbitt, Karlsson and Soerensen, Pawtucket

Chapter 35
Towards an Improved Understanding of Stereolithography Process—A Computational Study

Usharani Rath and Pulak M. Pandey

Abstract Amongst various futuristic manufacturing technologies, additive manufacturing is fast emerging from the lap of laboratory-scale testing to full-scale commercial applications. Stereolithography of polymeric materials has the potential to offer cost-effective and accurate solutions for many industrial sectors such as automotive, aerospace and electronics. However, the process of understanding and comprehending the underlying physics of the liquid additive manufacturing process is still in its infantile stage. In the present work, an attempt was made to computationally model the laser beam and the photosensitive resin interaction in the stereolithography process. The laser source was considered as a Gaussian heat source, and the liquid vat was considered as a homogenous isotropic fluid domain. Finite element modelling approach was utilized to evaluate the thermal changes occurring during the photo-initiated curing process. The resultant process quality indicators such as transient temperature distribution across the liquid vat, conductive and radiative heat flux, laser intensity on the liquid layer and its diminishing impact across the layer thickness were computed and compared. Maximum local temperature as high as 390 K was found on the irradiated location which cools down rapidly due to heat conduction and diffusion. However, the temperature reduces exponentially with the height of resin vat, and it takes around 4 s for the irradiated point to cool down to the ambient temperature and gets cured. Approximately 1.1×10^6 W/m^2 of laser intensity is produced on the bottom surface of laser. Some selected results were validated with contemporary literature and found to be satisfactorily consistent.

Keywords Additive manufacturing · Stereolithography · Finite element modelling

---

U. Rath · P. M. Pandey (✉)
Rapid Prototyping Laboratory, Department of Mechanical Engineering,
Indian Institute of Technology Delhi, New Delhi 110016, India
e-mail: pmpandey@mech.iitd.ac.in

U. Rath
e-mail: usha.rrath@gmail.com

R. G. Narayanan et al. (eds.), *Advances in Computational Methods in Manufacturing*, Lecture Notes on Multidisciplinary Industrial Engineering,
https://doi.org/10.1007/978-981-32-9072-3_35

Nomenclature

ρ	Density of PMMA resin
C_p	Specific heat capacity of PMMA resin
K	Thermal conductivity of PMA resin
Q	Heat generated during photo-curing reaction
I_s	Laser intensity at the bottom surface of resin vat
I_0	Original laser intensity
s	Coordinate of irradiated point
w	Laser spot width
I_Z	Laser intensity at a height z from the bottom surface of resin vat
D_P	Penetration depth
n	Normal direction to boundary 1–5
q	Heat flux due to radiation
ϵ	Emissivity
σ	Stefan Boltzmann's constant
T_{amb}	Ambient temperature
T	Transient temperature

35.1 Introduction

3D printing technology has been a watershed across the manufacturing industry in the last two decades. Some distinct features, such as ability to produce any complex three-dimensional geometries from a CAD file employing a layer-by-layer approach, zero material wastage, nearly net shape production of parts and enhanced automation leading to improved process capability, make these set of additive technologies futuristic and transformational. This solid free-form fabrication technique uses layer-by-layer approach to obtain physical objects from a CAD model by generating material deposition path or laser scanning path.

Stereolithography is one of the pioneers and widely used 3D printing technique utilizing photosensitive liquid resin as the feedstock. The pre-programmed laser cures the liquid resin vat in a layer-by-layer approach resulting from the chemical crosslinking of the monomer resin to produce solid 3D polymeric components as designed by the input CAD model. It produces physical 3D objects with highest precision and accuracy as compared to other commercially available 3D printers such as selective laser sintering or fused deposition modelling.

A good number of research expeditions are reported on stereolithography process and its capabilities. However, the process modelling and analysis of underlying physics of stereolithography process are still in its early stages, and very few research contributions are reported on the same. Huang et al. [1] generated dynamic finite element code for evaluating the shrinkage effects in constrained-surface stereolithography process. They suggested that the surface tension of liquid resin and shape of physical model may give rise to the inaccuracies of 3D-printed parts. Lee et al. [2] utilized artificial neural network to build up a model that can predict the effects of process parameters on the build accuracy of stereolithography process. They also mentioned that understanding the process physics can improve the process performance of the photo-curing process. They concluded that reduced layer thickness and increased hatch spacing improved the part performance. Bartolo [3] compared the phenomenological approach with the mechanistic approach for the stereolithography process. They suggested that the photo-curing process is an exothermic reaction increasing the local temperature of the irradiated region which gradually diminishes through conduction, convection and diffusion. Furthermore, they also implied that the fraction variation contour has similar resemblance as that of laser intensity which is conical. It was also explained that the local temperature and light intensity can lead to increased curing reaction and reduced printing time. Jiang et al. [4] utilized a dynamic finite element-based method to anatomize the shrinkage effects occurred during conventional stereolithography and mask-based stereolithography processes. They could also investigate the presence of residual stresses during the stereolithography process. Emami et al. [5] developed an analytical model for scanning-projection stereolithography process. They evaluated and compared the total laser energy intensity received by projection stereolithography and scanning-projection stereolithography processes during curing of each layer. They also indicated that SPSL process produces better accuracy than the PSL process. Liravi et al. [6] adopted cohesive zone model and conducted finite element simulation to predict the separation force existing between the cured part and the resin tray. Bae et al. [7] proposed a segregation model to determine optimum process parameters for direct ceramic stereolithography. The model was developed from the degree of segregation and the time constants such as writing time and settling time. Bikas et al. [8] recommended that the major cause of poor quality and inaccurate parts in additive manufacturing lies in the limited available modelling and optimization approaches.

In the light of the literature survey, it was realized that the quantum of research on the modelling and simulation of stereolithography process is scanty. However, from the research contributions by Bartolo [3] and Jiang et al. [4], it has been established that modelling of the stereolithography process can lead to better understanding of the underlying physics and improve the process performance. However, none of the research papers reported the transient temperature distribution along the resin surface and along the depth. Furthermore, findings regarding the laser intensity variation along the depth of resin are also very few.

Thus, in the wake of the aforementioned literature explorations, the objective of the present research investigation was formulated to vindicate some of the underlying physics of the stereolithography process. In the present study, the finite element analysis of the bottom-up stereolithography process was carried out for a single-layer curing of resin vat. As an outcome, the transient temperature distribution, laser intensity variation, conductive and convective flux contours, transient residual stress distribution and variation on the resin surface were evaluated and analysed for improved part quality and better process performance.

35.2 Process Modelling

A finite element-based commercial multi-physics software package, COMSOL, which operated under Windows 64-bit operating system, was used for the simulation study. The physical model of the present study primarily consists of a methyl methacrylate resin vat of 10 cm length, 5 cm width and 3 mm height. The fluid properties of resin and specifications of laser used in bottom-up stereolithography set-up are presented in Tables 35.1 and 35.2. The UV laser was considered as a moving heat source which travels across the bottom surface of resin vat. The X-coordinate of laser beam follows a ramp function with a slope of 1/50, so that the laser moves 10 cm distance along X-direction in 5 s.

35.2.1 Assumptions

There are broadly five assumptions considered in the present modelling expedition in line with the spirit of some of the reviewed literature [3, 4] and the contemporary understanding of stereolithography process.

1. The model was considered as a homogenous and isotropic fluid domain.
2. The effect of optical scattering was not considered.
3. Convection effects were taken as negligible as air is quiescent inside the stereolithography chamber.
4. The heat generation was considered to be only due to heat of polymerization as the stereolithography process is governed by photo-curing reaction.
5. The absorption of UV laser intensity was considered to be governed by Beer–Lambert's law.

Table 35.1 Resin fluid properties

Serial number	Properties	Value
01	Density (g/cm^3)	1.12
02	Dynamic viscosity (Pa s)	890
03	Thermal conductivity (W/mK)	0.25
04	Specific heat (J/kgK)	1500
05	Ambient temperature (K)	298.15
06	Emissivity	0.9

Table 35.2 Laser specifications

Serial number	Specifications	Value
01	Power (mW)	120
02	Wavelength (nm)	405
03	Spot size (mm)	0.155
04	Radiation	Ultraviolet

35.2.2 Governing Equations

A set of governing equations were used in the current study to model the physics of heat transfer due to laser radiation. There are two equations governing the three-dimensional transient energy equation and absorption of UV radiation (Beer–Lambert's law) by the liquid resin vat, given by Eqs. 35.1, 35.2 and 35.3, respectively. The governing Equation 35.1 is solved considering the heat source 'Q' as a Gaussian heat source.

$$\rho C_{\mathrm{p}} \frac{\partial T}{\partial t} + \rho C_{\mathrm{p}} u \nabla T + \nabla \cdot (-K \nabla T) = Q \tag{35.1}$$

$$I_{\mathrm{s}} = I_0 \exp[-2 \times (s/w)^2] \tag{35.2}$$

$$I_Z = I_0 \exp\left(\frac{-z}{D_{\mathrm{P}}}\right) \tag{35.3}$$

where ρ, C_{p} and K are the density, specific heat capacity and thermal conductivity of liquid resin used, respectively. Q is the heat generated from the photo-curing exothermic reaction. I_{s} and I_0 are the laser intensity at the resin bottom surface and the original laser intensity, respectively. s and w are the coordinate of irradiated point and laser spot width, respectively. I_Z is the laser intensity at a height z from the bottom surface of resin vat. D_{P} is a material property of the photo-polymeric resin, known as penetration depth.

35.2.3 Beer–Lambert's Law

Beer–Lambert's law is utilized to establish a relation between the depth of penetration of a light source in a material and the attenuation coefficient of the material. The attenuation coefficient of the material is inversely proportional to the material property (D_{P})of the medium through which the material is travelling. A material or medium with higher attenuation coefficient is supposed to have reduced depth of penetration of light source travelling through it. Equations 35.2 and 35.3 govern the Beer–Lambert' law at the resin vat bottom surface and at a specific height from the bottom surface.

35.2.4 Domain and Boundary Conditions

Various boundary and initial conditions were applied in the present model for the UV radiation and subsequent phenomena such as transient heat conduction,

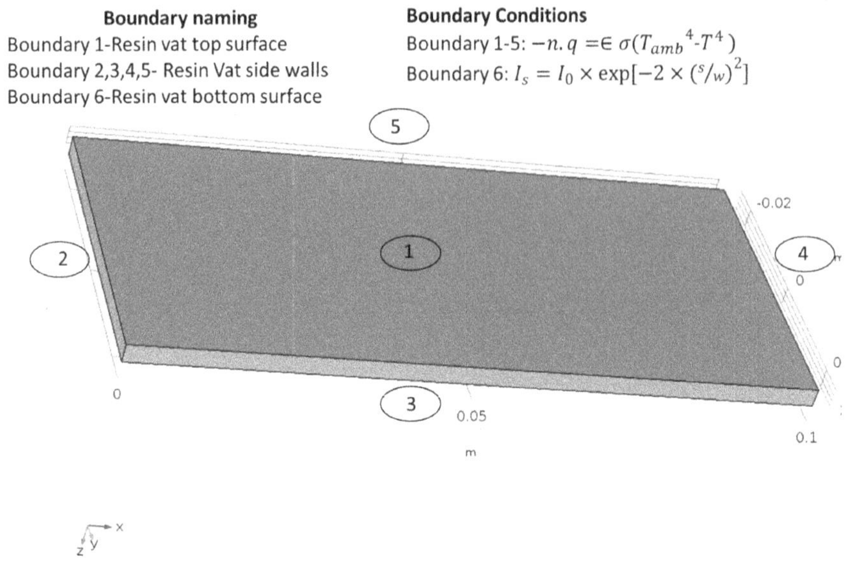

Fig. 35.1 Geometric model with boundary and initial conditions

convection and diffusion to happen. The boundary and initial conditions are clearly depicted from Fig. 35.1. A single rectangular fluid domain was considered for the simplicity of the model with six boundaries. Boundary 1 was considered as the top surface of the resin vat, and boundaries 2, 3, 4 and 5 were modelled as the side surfaces of the resin vat. Boundary 6 was considered as the bottom surface of the resin vat, on which the laser light was irradiated.

35.2.5 Meshing

A customized meshing chronology was used to discretize the three-dimensional geometric model of resin vat. Tetrahedral meshing was done with a maximum element size as 0.001 mm for the entire domain for the convergence of the solution. A total of 536,841 numbers of elements were generated. The model generation and succeeding analysis were done using COMSOL commercial software package. The simulation was performed for an irradiation time of 5 s with 0.1 as the time step and 0.001 as the relative tolerance of the simulation study.

35.3 Results and Discussion

The dynamic temperature variation on the bottom surface of resin vat is represented in Fig. 35.2. The maximum local temperature obtained during the curing process at the point of irradiation was predicted to be 390 K, which diminishes to ambient temperature by heat conduction and diffusion in a very short span of time. Furthermore, it can also be clearly visualized that the temperature rise due to irradiation at each irradiated point is maximum and reduces radially as per the Gaussian distribution. The variation of temperature with the height from the bottom surface of resin vat at a particular irradiated point is represented in Fig. 35.3. The decrease in temperature with height signifies that the absorption of UV radiation on resin vat is according to the Beer–Lambert's principle. Nonetheless, further slight increase in temperature might be caused due to the diminished conduction and diffusion as compared to the bottom surface. The variation of temperature with time at a particular irradiated point (boundary 2) on the resin bottom surface is also shown in Fig. 35.4. The curing profile of the photo-polymeric resin is in accordance with the star-weave build structure, which resembles a conical shape [9]. The laser intensity distribution follows a conical pattern as presented in Fig. 35.5, validating the work done by Bartolo [3]. Moreover, the laser intensity reduces from the bottom to top surface of resin vat in accordance with the Beer–Lambert's principle as represented in Eq. 35.3 and cures the resin irradiated beyond the critical laser intensity. The value of critical laser intensity varies depending on the resin and

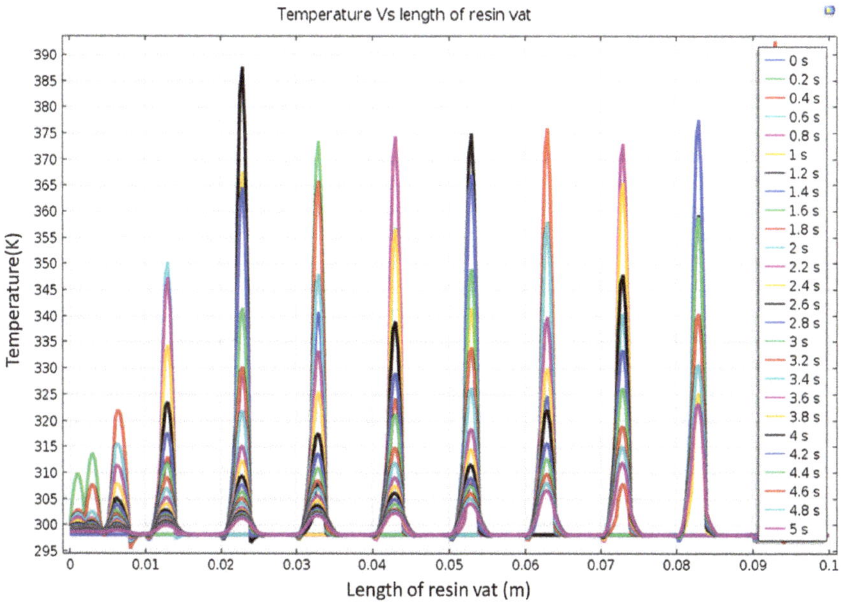

Fig. 35.2 Temperature variation on the bottom surface of resin vat

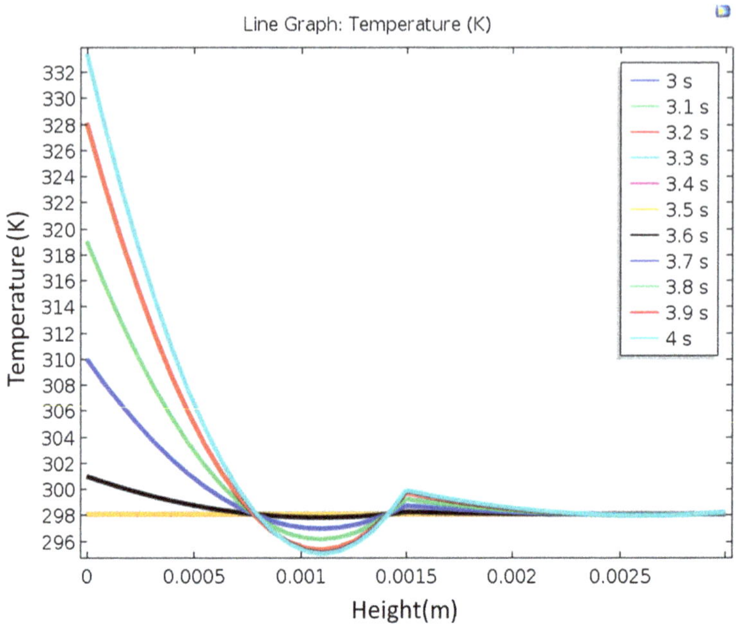

Fig. 35.3 Temperature variation with height of resin vat

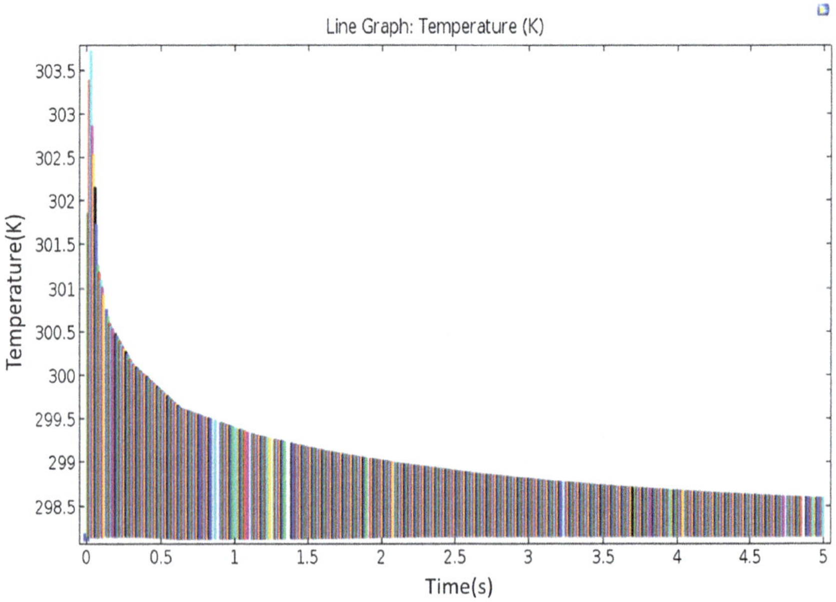

Fig. 35.4 Temperature variation with time

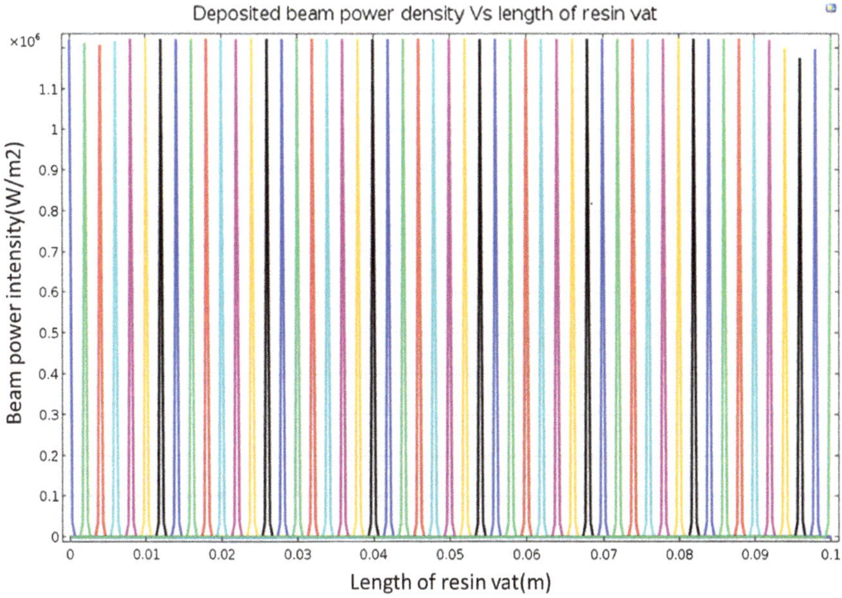

Fig. 35.5 Laser intensity variation on the bottom surface of resin vat

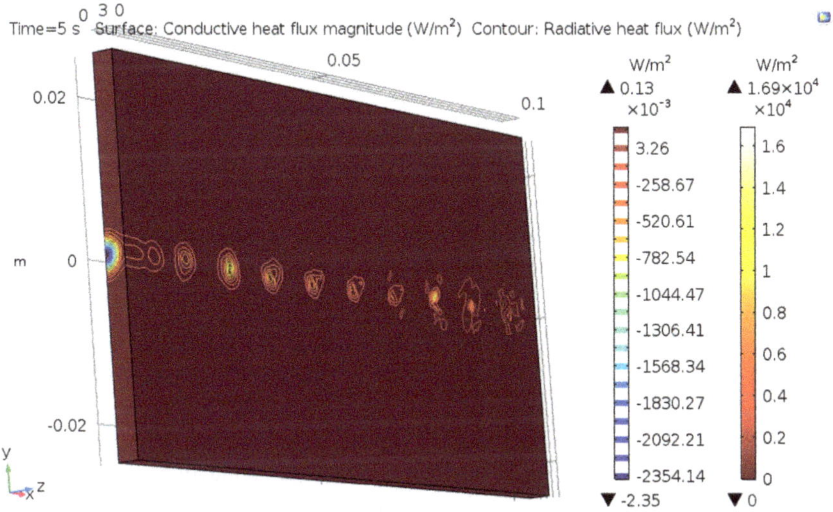

Fig. 35.6 Conductive heat flux contour

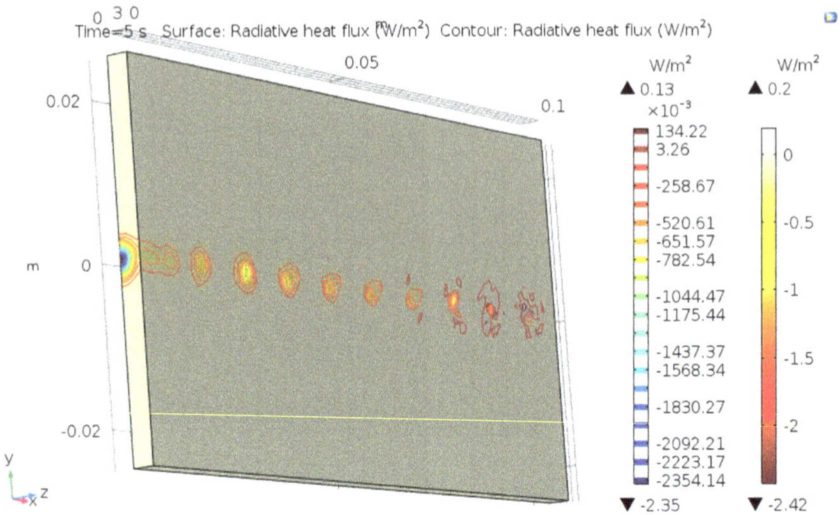

Fig. 35.7 Radiative heat flux contour

machine specifications. It can be evaluated by conducting a good number of experiments and plotting the semi-logarithmic graph between cure depth and logarithm of maximum exposed laser energy. The local heat generation and subsequent temperature rise due to exothermic photo-curing reaction decline expeditiously by heat conduction and diffusion. The conductive heat flux and radiative heat flux contours are depicted from Figs. 35.6 and 35.7, respectively.

35.4 Conclusions

A macroscopic model established by finite element method using the governing equations for the physical changes occurring during the stereolithography process was proposed. The following inferences can be deduced from the research work.

1. The transient temperature history on the surface and along the depth of the liquid resin irradiated by ultraviolet laser light with 405 nm wavelength was analysed.
2. Irradiated zone of resin undergoes a local increase in temperature as high as 390 K, which reduces quickly due to the advent of conduction and diffusion.
3. Laser intensity cures the laser across the depth in a conical manner, resulting in the prediction of fractional conversion contour of liquid resin to the green-state/semi-solid resin as conical [3]. The bottom surface of resin vat was irradiated with a laser intensity of 1.1×10^6 W/m^2.

The simulation approach can be improved further by considering the chemical kinetics of the photo-curing reaction, which involves the monomer and photo-initiator concentration in the liquid resin.

References

1. Huang, Y., Jeng, J., Jiang, C.: Increased accuracy by using dynamic finite element method in the constrain-surface stereolithography system. J. Mater. Process. Tech. **140**, 191–196 (2003)
2. Lee, S.H., Park, W.S., Cho, H.S., Zhang, W.L.M.C.: A neural network approach to the modelling and analysis of stereolithography processes. Proc. Inst. Mech. Eng. Part B J. Eng. Manuf. **215**, 1719–1734 (2001)
3. Bartolo, P.J.: Mechanistic approaches. Ann. CIRP **55**(1) (2006)
4. Jiang, C., Huang, Y.-M., Liu, C.-H.: Dynamic finite element analysis of photopolymerization in stereolithography. Rapid Prototyp. J. **12**(3) (2012)
5. Emami, M., Barazandeh, F., Yaghmaie, F.: An analytical model for scanning-projection based stereolithography. J. Mater. Process. Tech. **219**, 17–27 (2015)
6. Liravi, F., Das, S., Zhou, C.: Computer-aided design separation force analysis and prediction based on cohesive element model for constrained-surface stereolithography processes. Comput. Des. **69**, 134–142 (2015)
7. Bae, C., Halloran, J.W.: A segregation model study of suspension-based additive manufacturing. J. Eur. Ceram. Soc. **38**(15), 5160–5166 (2018)
8. Bikas, H., Stavropoulos, P., Chryssolouris, G.: Additive manufacturing methods and modelling approaches: a critical review. Int. J. Adv. Manuf. Technol. **83**, 389–405 (2016)
9. Rahmati, S., Ghadami, F.: Process parameters optimization to improve dimensional accuracy of stereolithography parts. Int. J. Adv. Des. Manuf. Technol. **7** (2014)

Chapter 36
Experimental Investigations for Effects of Raster Orientation and Infill Design on Mechanical Properties in Additive Manufacturing by Fused Deposition Modelling

Pushpendra Yadav, Ankit Sahai and Rahul Swarup Sharma

Abstract Manufacturing and designing of various functional components in the fields of engineering, automobile and aerospace are being done by additive manufacturing (AM) technology. These activities must be supported by the knowledge of process parameters that may influence mechanical properties of industrial products. Fused deposition modelling (FDM) is one of the most popular AM technologies in which the quality of part depends on selection of process variables. The present work focuses on two input process parameters, i.e. raster orientation and infill density and its effect on the mechanical properties of 3D-printed samples. The PLA samples were prepared in *XY* and *XZ* orientation with infill density of 20, 40, 60, 80 and 100%. The tensile testing is conducted to evaluate the effects of these process parameters on tensile strength, ultimate strength, etc. Further, fractography was performed and it is concluded that dimple size and distribution on fractured surface were affected by the infill density. Printing time was also checked and compared for specimens printed for different parameters, which shows printing time is least for *XY* orientation and maximum for *XZ* orientation. Printing time increases as the infill density increases from 20 to 100%.

Keywords 3D printing · FDM · Tensile testing · Infill density · Fractography

---

P. Yadav · A. Sahai (✉) · R. S. Sharma
Department of Mechanical Engineering, Faculty of Engineering,
Dayalbagh Educational Institute Dayalbagh, Agra 282005, India
e-mail: sahaiankit@dei.ac.in

P. Yadav
e-mail: er.pushpendrayadav@gmail.com

R. G. Narayanan et al. (eds.), *Advances in Computational Methods in Manufacturing*, Lecture Notes on Multidisciplinary Industrial Engineering,
https://doi.org/10.1007/978-981-32-9072-3_36

36.1 Introduction

3D printing is considered as the future of the technology in the world and is a manufacturing phenomenon that will revolutionize industries, factories and mass production in the coming years. With the advent of Industry 4.0, 3D printing has become a buzzword in almost all industrial sectors. Although 3D printing technologies took birth in the 1980s, mainly by the name of rapid prototyping, they got a fresh impetus in this century and have been growing at an exponential rate in last few years [1]. This technology allows designers to quickly and inexpensively produce prototypes without tooling or machining which act as catalysts to students, researchers, scientists and innovators to rapidly produce prototypes based on their imagination and thus 3D printing is helping users across all age groups to enhance product manufacturing by simplification in working. Some economists have considered this a third industrial revolution following mechanization and assembly line production [2]. This industry has grown quickly in recent years to an annual value of over $5 billion. Sales of 3D desktop printers were 160,000 in 2014 and grew to nearly 300,000 in 2015 [3]. Some predictions have shown that the 3D printing total market revenue will be between $7 and $23 billion per year by 2020. Because of this massive expansion and growing significance, FDM 3D printing is being studied in the areas of medical, aerospace, building equipment, materials, preparation techniques, etc. and is still the subject of research [4–8]. 3D printing is far different from traditional manufacturing processes and saves lots of build time, production waste and can be easily worked by ordinary workers as compared to skilled machining, forging, casting, etc. where skilled workers are required to complete the job. But with any manufacturing process, it is important to focus on resource saving and environmental issues and with the new technologies coming up in 3D printing, it is necessary to pay more attention to its sustainability in context to the present scenario before we integrate 3D printing with industrial set-up. The sustainability studies will also help scientists, researchers and other decision makers to understand the 3D printing process and its implications in different applications or products [7]. Therefore, it is extremely important to understand the influence of process variables or parameters on product development. These variables can be classified in: material properties, build specifications, part positioning and orientation and environment [8]. A similar study was developed by Ziemian et al. [9] in which the group evaluated the effect of raster orientation in the direction of the strain and developed the optimal levels for improving the strength of the specimen. The most studied parameters, in a static situation, have been build orientation, raster angle, raster width, raster-to-raster air gap and nozzle diameter or slice height [10, 11]. Some researchers have also analysed the effect of 3D printing in Engineering education which is suitable for classrooms and public spaces and encourages action-based learning, exploration and innovation, to accelerate the acceptance and involvement of 3D printing technologies in the manufacturing sector [12, 13].

Not many people have focused the research on the effect of infill density on the mechanical properties of PLA materials for different raster orientations. Therefore, the present work focuses on the effect of infill density on mechanical properties in *XY* and *XZ* orientation and its fractography analysis using scanning electron microscopy (SEM).

36.2 Fused Deposition Modelling Process

One of the most exciting and widely used additive manufacturing technologies is fused deposition modelling (Fig. 36.1) which is based on the material extrusion process. In FDM process, the polymer filament is fed to the liquefier chamber where it gets heated to a particular temperature depending upon the melting point of the material used for printing. Once the material is heated, it gets softens and melts and after that molten material is pushed through the nozzle from the liquefier with some suitable velocity. The extruded material in molten form is deposited on the platform layer by layer which slowly gets cooled at room temperature. The layer thickness and other parameters depend upon the G-code generated by the slicing software.

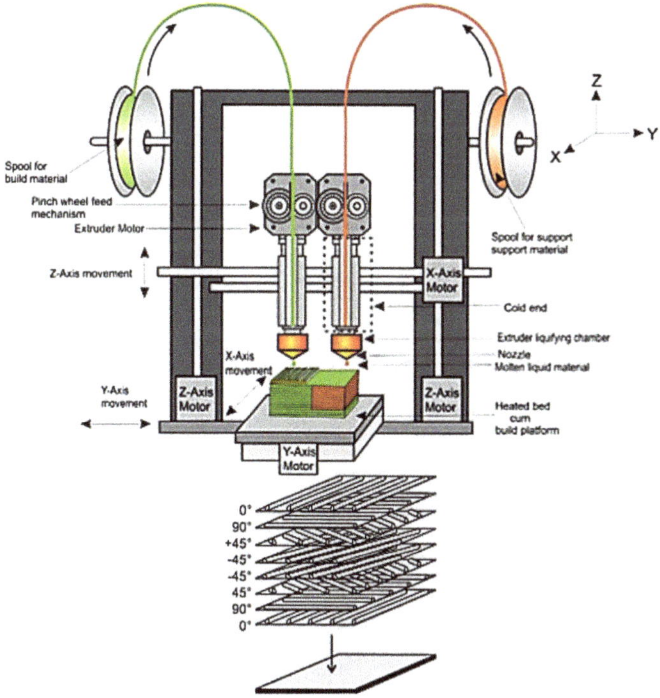

Fig. 36.1 Complete process of FDM process machine set-up

These 3D printers working in FDM technology are easy to design, build and quick to work upon. The students, researchers and even normal household users can easily install and work to produce the desired products using the different parameter combinations and of different shapes and sizes. These printers can be a great learning experience as the users can visualize, model and print in quick succession to get the rapid prototype of the specimen or product.

36.3 Experimental Details

In the present work, raster orientation and infill density are input parameters and based on their different combinations, mechanical properties like tensile strength and yield strength are calculated. The tensile test specimens were prepared based on ASTM D638, standard dimensions are used corresponding to Type-1 specimen in Fig. 36.2, and each sample set consisted of five specimens, Fig. 36.3. The geometry of the 3D-printed specimens was modelled using SolidWorks software exported as an STL file and imported to the 3D printing software. A 50 KN universal testing machine was used for tensile testing with speed of 5 mm/min, and fractography was performed using scanning electron microscope (SEM). The material used for current experiment is PLA.

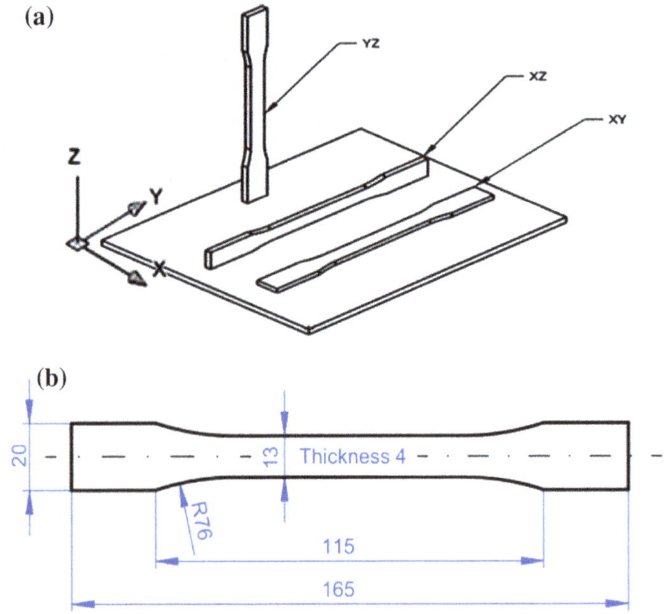

Fig. 36.2 Different raster orientations *XY*, *XZ*, *YZ* and ASTM D638 Type-1 tensile test specimen

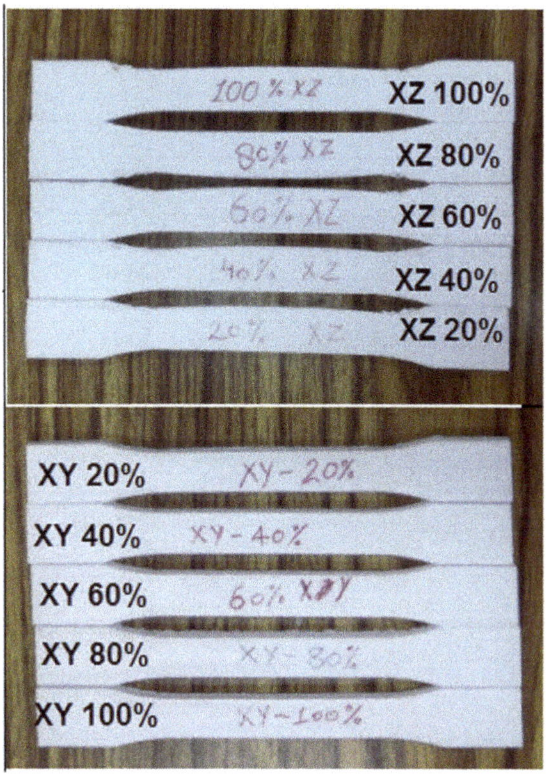

Fig. 36.3 Printed specimens in each orientation of *XY*-and *XZ*-directions

Before performing fractography, tensile tested samples were cut and gold (75%) and palladium (25%) coating was done for better image analysis. During experimentation, mechanical properties were evaluated for raster orientation *XY* and *XZ* as mentioned in Tables 36.1, 36.2 based on certain fixed parameters such as infill pattern, layer thickness, temperature and air gap while infill density and raster orientation were changed, Table 36.3.

Table 36.1 Mechanical properties of 3D-printed PLA in *XY*-raster orientation and different infill densities

Raster orientation—*XY*					
Infill density (%)	20	40	60	80	100
Area(mm^2)	52	52	58.08	52.8	52
Gauge length (mm)	50	50.071	50.062	50.091	50
Peak load (kN)	1.183	1.401	1.424	1.859	2.345
Peak stress (MPa)	22.746	26.946	24.518	35.203	45.089
Yield stress (MPa)	17.583	18.81	19.754	28.476	34.428
Yield load (kN)	0.914	0.978	1.147	1.504	1.79
Elongation at break (%)	7.435	6.922	4.4	3.1	14.678

Table 36.2 Mechanical properties of 3D-printed PLA in *XZ*-raster orientation and different infill densities

Raster orientation—*XZ*					
Infill density (%)	20	40	60	80	100
Area (mm^2)	50.92	54.6	54.8	49.4	52
Gauge length (mm)	50	50	50	50	50
Peak load (kN)	1.5	1.652	1.836	1.939	1.684
Peak stress (MPa)	29.46	30.254	33.496	39.256	32.382
Yield stress (MPa)	22.212	23.492	24.896	31.33	28.028
Yield load (kN)	1.131	1.283	1.364	1.548	1.457
Elongation at break (%)	5.127	6.422	5.586	4.818	3.25

Table 36.3 Printing parameters of PLA 3D printing

Fixed variables		Control factors	
Speed	60 mm/s	Infill density	20%
Layer thickness	0.2 mm		40%
Number of shells	3		60%
Air gap	0		80%
Infill pattern	Rectilinear		100%
Nozzle temperature	210 °C	Raster orientation	*XZ*
Bed temperature	60 °C		*XY*

36.4 Results and Discussion

The experimentation was mainly focussed on the evaluation of mechanical properties with the variation in raster orientation and infill density. The stress–strain curve in Fig. 36.4 indicates that PLA-printed specimen in *XY* plane exhibits the maximum tensile stress 45.09 MPa and elongation as 14.68%. In *XZ* orientation,

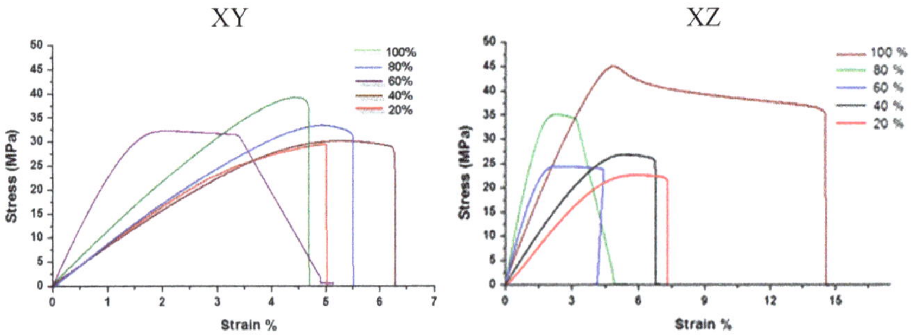

Fig. 36.4 Stress–strain curve for different infill densities in *XY* and *XZ*-raster orientations

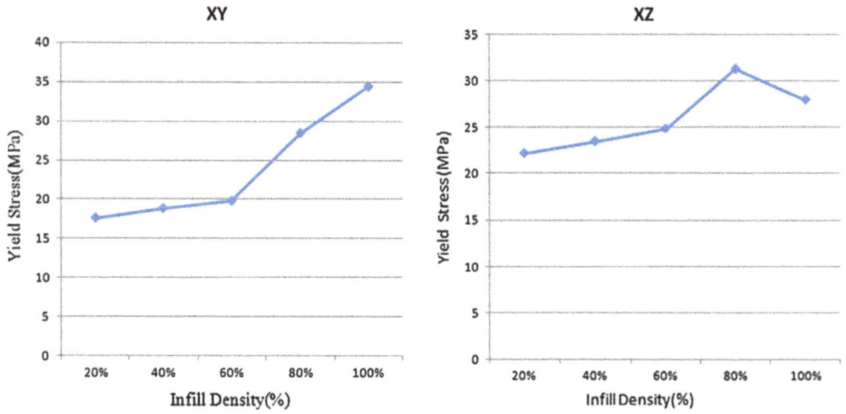

Fig. 36.5 Variation of yield stress for different infill densities and raster orientations

maximum tensile strength is 39.25 MPa for 100% infill density while maximum elongation is 6.3% for 40% infill density. The graph clearly indicates that the value of tensile strength in *XZ* for all infill densities are close to each other while in *XY* orientation the trend is completely different and the values range from 21 to 45 MPa. Figure 36.5 displays the variation of yield strength for *XY* and *XZ* orientation with different infill densities.

36.5 Fractography

In product design and manufacturing, failure analysis is an important part and, therefore, in this work the fractured surface of PLA specimens is also analysed using SEM for different raster orientations and infill densities, Fig. 36.6. It is depicted that ductile fracture is observed in *XZ* orientation as fibre is deformed in different planes leading to trans-layer tensile failure. The fracture gets transferred from one layer to another in *XZ*-direction. Specimens with *XY* orientation displayed intermediate brittle-ductile fracture behaviour. Failure is generally caused by raster pulling and rupturing as clearly depicted in Fig. 36.6. The fractured surface exhibits different fracture modes indicating hairline and major cracks marked with arrows at different location of fractured surface. In *XY* orientation, yield stress and elongation at break are more as compared to *XZ*, thereby exhibiting prolonged necking while applying tensile load as depicted in Tables 36.1 and 36.2. The strain (%) is more in *XY* orientation compared with *XZ* orientation. *XY* orientation shows brittle and ductile fracture which can be attributed to cooling rate behaviour during the solidification of material. The *XZ* orientation displays ductile fracture with dimples visible at different locations of fractured surface.

Figure 36.7 displays the printing time for different infill densities and raster orientations for PLA material. The printing time and number of layers for the tensile

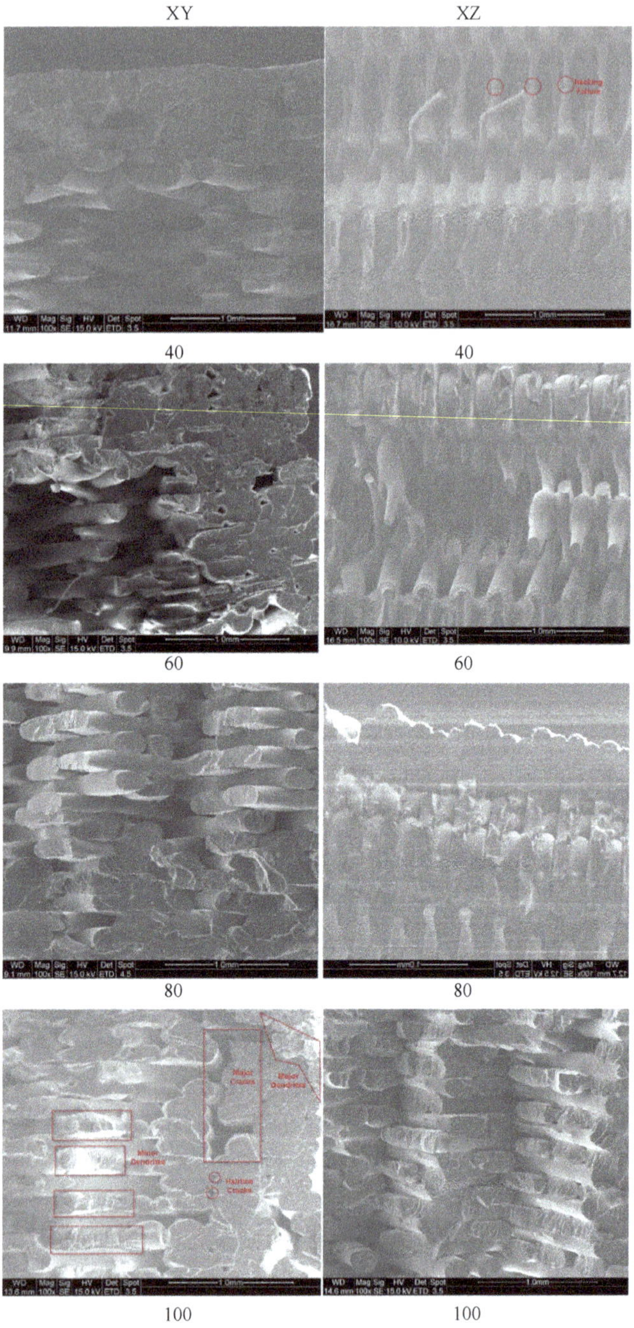

Fig. 36.6 SEM image fractured surface of *XY* and *XZ*-raster orientation and different infill density

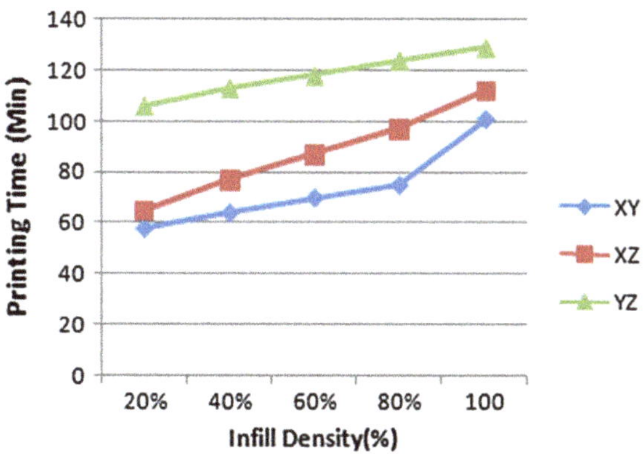

Fig. 36.7 Variation of printing time for various infill densities and raster orientations

samples depend upon the various process parameters such as raster orientation, layer thickness and feed rate. Also, the printing time is directly related to manufacturing cost [9]. Raster orientation *YZ* exhibited the longest printing time as compared to *XY* samples which have the shortest time. Printing time also decreases as feed rate is increased for *XY* orientation. In case of the *YZ* orientation, printing time remained almost constant for high feed rate values.

36.6 Conclusion

The research outlines the following conclusions:

- The maximum tensile stress in *XY* orientation is 45.09 MPa for 100% infill density with elongation as 14.68%, and similarly for the *XZ* orientation maximum tensile stress 39.256 MPa for 80% and elongation as 6.422% for 40% infill density.
- In *XY*-raster orientation, yield stress goes on increasing at different infill densities from 20 to 100% in step of 20. The maximum rate of change of yield stress is **0.958** with reference to 20% infill density. Similarly, in *XZ*-raster orientation, yield stress is increasing but at a smaller rate of change is **0.2618** with reference to 20% infill density. This clearly indicates that from yield stress point of view *XY* orientation is stronger than *XZ* orientation. This is also supported from Tables 36.1 and 36.2 that yield load is more at *XY* orientation as compared to *XZ* orientation.
- *XY* orientation has least printing time as compared to *XZ* and *YZ* orientation, with *YZ* taking maximum printing time.
- Fracture morphologies are affected by the raster orientation direction and infill density with *XY* orientation displayed intermediate brittle-ductile fracture behaviour while *XZ* displayed ductile fracture.
- Failure after tensile loading exhibits hairline and major cracks in *XY* orientation.

Acknowledgements The authors acknowledge the help from ACMS, IIT Kanpur for allowing to perform tensile testing and fractography. Support from TEQIP-III, Faculty of Engineering, DEI and MHRD NMEICT is highly appreciated.

References

1. Anderson, I.: Mechanical properties of specimens 3D printed with virgin and recycled polylactic acid. 3D Print. Addit. Manuf. **4**(2), 110–115 (2017)
2. Huang, S., Liu, P., Mokasdar, A.: Additive manufacturing and its societal impact: a literature review. Int. J. Adv. Manuf. Technol. 67:1191–1203 (2013)
3. Dawoud, M., Taha, I., Ebeid, S.: Mechanical behavior of ABS an experimental study using FDM and injection molding techniques. J. Manuf. Process. **21**, 39–45 (2016)
4. Joshi, S.C., Sheikh, A.A.: 3D printing in aerospace and its long-term sustainability. Virtual Phys. Prototyp. **10**, 175–185 (2015)
5. Lee Ventola, C.: Medical applications for 3D printing: current and projected uses. Pharm. Ther. **39**(10), 704–711 (2014)
6. Ford, S., Despeisse, M.: Additive manufacturing and sustainability: an exploratory study of the advantages and challenges. J. Clean. Prod. **137**, 1573–1587 (2016)
7. Kim, K., Lee, D.S., Kim, Y.C.: Sustainable design process based on 3D printing technology. Indian J. Sci. Technol. **9**(46), (2016)
8. Montero, M., Roundy, S., Odell, D., Ahn, S.H., Wright, P.K.: Material characterization of fused deposition modeling (FDM) ABS by designed experiments. In: Proceedings of Rapid Prototyping and Manufacturing Conference, pp. 1–21. SME, Cincinnati, Ohio, (2001)
9. Ziemian, C.W., Sharma, M.M., Ziemian, S.N.: Anisotropic mechanical properties of ABS parts fabricated by fused deposition modelling. In: Gokcek, M. (ed.) Mechanical Engineering, pp. 159–180. In Tech, New York, NY, USA (2012)
10. Bellini, A., Güçeri, S.: Mechanical characterization of parts fabricated using fused deposition modeling. J. Rapid Prototyp. **9**, 252–264 (2003)
11. Yadav, P., Kapoor S., Sahai A., Sharma, R.S.: Effect of build orientation and infill density on mechanical properties of 3D printed parts by Fused Deposition Modeling. In: 34th National Convention of Mechanical Engineers & National Seminar on. Mechanical Engineering and Emerging Challenges for Manufacturing, pp. 118–124, The Institution of Engineers India, New Delhi (2018). ISBN: 978-81-936149-8-3
12. Sharma R.S., Singhal I., Gupta S.: Innovative training framework for additive manufacturing ecosystem to accelerate adoption of three-dimensional printing technologies. 3D Print. Addit. Manuf. **5**(2), (2018)
13. Yadav, P., Singhal, I., Sahai, A., Sharma, R.S.: Intensifying hands-on learning and experimentation of fused deposition modelling three dimensional printers. In: 7th International and 28th AIMTDR Conference, Anna University, Chennai, India (2018)

Chapter 37
Development of Methods for Machine Intelligence Quotient Measurement for Piston Casting by Die Casting Machines: A Study

Rupshree Ozah and Muralidhar Manapuram

Abstract Pistons are the most important parts of an automobile engine. Pistons are manufactured by various die casting processes like sand casting, gravity die casting, pressure die casting and automatic die casting. Manufacturing assets and processes have different intelligent attributes based on embedded intelligent systems. Not much of work seems to be available for measurement and quantification of machine intelligent quotient (MIQ) for metal casting manufacturing machines. To assess the intelligence of die casting machines, three methods are developed. High MIQ indicates the lower interaction between human and die casting machine, i.e., the machine can take comparatively higher percentage of decisions in machine operations. In the present work, an attempt has been made to develop a stepwise procedure for MIQ measurement by three methods, i.e., human–machine cooperative system, geometric polytope method and multiple perspective analysis methods. MIQ for three die casting machines is computed based on human–machine cooperative method, and it has been found that automatic die casting machine has the highest MIQ (109.31) followed by pressure die casting machine (59.69) and gravity die casting machine (26.44) the lowest.

Keywords MIQ · Die casting machine · Piston casting · Human–machine cooperative system · Geometric polytope method · Multiple perspective analysis method

---

R. Ozah (✉) · M. Manapuram
Department of Mechanical Engineering, North Eastern Regional Institute of Science and Technology, Itanagar, Arunachal Pradesh 791109, India
e-mail: rupshreeozah91@gmail.com

M. Manapuram
e-mail: mm@nerist.ac.in

R. G. Narayanan et al. (eds.), *Advances in Computational Methods in Manufacturing*, Lecture Notes on Multidisciplinary Industrial Engineering,
https://doi.org/10.1007/978-981-32-9072-3_37

37.1 Introduction

Automobile industry is the fast growing and technology-intensive industry. India is the second-largest casting producing country in the world. Globalization has led to increase the pressure on engineers and technologists to increase the efficiency of casting manufacturing processes and assets to edge-over other manufacturing processes and cost effectiveness. With the advent of computers, Internet, intelligent techniques, Industry 4.0 and cyber physical system's decision-making attributes of machines have been increased significantly. But there are no systematic methods to measure MIQ of machines. Moreover, efforts of researchers to embed human intelligent attributes by various software programs and hardware devices enable to enhance MIQ. Various tools [1] used to communicate or to transfer intelligence to the machine are fuzzy logic, genetic algorithm, vision system, charge couple device (CCD) cameras, sensors and actuators.

Intelligence is always studied in humans but has been observed in both non-human animals and plants. Intelligence in the machine is also called as artificial intelligence. Engineers and scientists are trying to impart human intelligence attributes to machines to improve machine intelligence [2, 3]. Not of much work seems to be available for measurement of MIQ. Hence in this paper, a stepwise procedure is developed to measure the machine intelligence quotient. MIQ for three types of die casting machines (gravity die casting machine, pressure die casting machine and automatic die casting machine) for the manufacture of piston casting is also considered by human–machine cooperative system. In addition, two more methods for measuring MIQ have been considered and discussed in brief.

37.2 Machine Intelligent Quotient and Its Sub-systems

Intelligence is always a hard topic to express and there is no clear definition for it. It may be defined as the ability to respond rapidly and successfully to a new situation. Intelligence requires the ability to sense the environment, make decisions and control action [4]. Various attributes of intelligence are adaptability, self-maintenance, communication, autonomy, learning, self-improvement, anticipation, goal seeking, inference and creativity. Intelligent machine is a machine that learns and can solve problems by reasoning and inference. Intelligent controller is any control system that can deliver a degree of human-like response to stimuli. Nowadays, intelligent factories have been evolving with intelligent machine systems along with intelligent human–machine interfaces.

Intelligent systems can be broadly classified into four types viz. biological intelligence, artificial intelligence, machine intelligence and computational intelligence. Intelligent quotient is developed in human beings in the form of intelligence quotient (IQ) which can be defined as the ratio of mental age and chronological age multiplied by hundred.

Machine intelligence is another term used to measure the intelligence of the machine, i.e., theory and development of the computer systems facilitating to perform the tasks normally which requires human intelligence.

Machine intelligence is the process of analyzing, organizing and converting data into knowledge, where (machine) knowledge is defined as the structured information acquired and applied to remove ignorance and uncertainty about a specific task pertaining to the intelligent machine [5]. Machine intelligence consists of two components, i.e., control intelligence and interface intelligence. Interface intelligence indicates the degree of intelligence of human–computer interaction that may be human to machine interaction and vice versa. Machine intelligence quotient can be considered as a union of machine control intelligence and machine interface intelligence [6]. High machine intelligent quotient indicates the number of decisions taken by machines is more than human, i.e., machine to human interaction is more than human to machine interaction. Nowadays, machines are taking complex decisions without any interaction of human in fully automatic process. The taxonomy of intelligence and its meaning and the paradigms are presented in Table 37.1.

Intelligence attributes consist of activity attributes (cognition, memory, logical thinking, creative thinking and evaluation), material attributes (images, symbols, non-verbal perceptions), product attributes (classes, relations, systems, unitary items, transformations and implications) and their combinations. There is a continuous effort by the engineers and the technologies to impart human intelligent attributes to machines and industrial shop-floor systems [7, 8].

The human intelligence has been gradually increasing linearly with time, but the machine intelligence has not been significant during 1950–2000 [9]. In Fig. 37.1, human intelligence and machine intelligence with time are shown. With the advent of computer, artificial intelligence, sensors, actuators, neural networks, the intellectual power of machines has been steeply increased. The point where the machine

Table 37.1 Taxonomy of intelligence and its paradigms

Types of intelligence	Literal meaning	Paradigms of intelligence
Biological intelligence	Naturally grown organism, human knowledge plus sensory input	Human and other developed species brain
Artificial intelligence	Cognitive inspired artificial models, knowledge tidbits and sensor data	Intelligent systems, knowledge systems, decision-making systems
Machine intelligence	Complex machines, PLC controlled machines	Robot, neural network
Computational intelligence	Computational methodologies and software systems, mathematical computation plus sensor data	Expert system, Fuzzy system, robotic nose, robotic arm, intelligent agent-based systems, Internet of things, cloud manufacturing, genetic algorithm

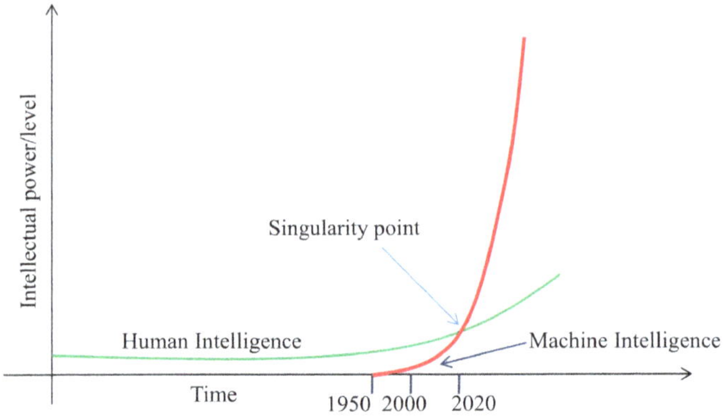

Fig. 37.1 Evolution of human and machine intelligence

intelligence outperforms the human intelligence is termed as singularity. In this century, machine intelligence is likely to supersede human intelligence.

37.3 Die Casting Machines for Piston Manufacturing

Pistons are the short piece of metal that moves up and down under pressure inside a cylinder of an engine and are made gas tight by piston rings. Pistons are usually made of Al–Si alloy and manufactured by different types of die casting machines viz. gravity, pressure and automatic based on the quantity, quality and lead times. The number of automobiles manufactured [10, 11] and the number of pistons required are quantified during 2008–2016 and presented in Table 37.2. The numbers of automobiles and pistons produced during a decade period are 157.1 million and 238.2 million, respectively. The annual requirement of pistons for automobile industry is around 60.12 million.

Al–Si pistons are generally manufactured by die casting processes, i.e., gravity die casting, pressure die casting and automatic die casting process. In die casting machine, molten metal enters the mold cavity under gravity [12]. Pressure die casting is characterized by forcing molten metal under high pressure into a mold cavity. Gravity die casting (GDC) is preferred in medium to large volumes and of a heavier section thickness than high pressure die casting, but thinner sections than sand casting. It has very good dimensional accuracy and also good cast surface finish. Pressure die casting cycle time is less, reliable and cost-effective manufacturing process for the production of high volumes and cast components that are near net-shaped having tight tolerances. In automatic die casting, more operations castings are performed by machines like automatic die coating, automatic pouring and ejection without any human interaction. The tasks of die casting machine,

Table 37.2 No. of automobiles and pistons produced in India during 2008–17

Type of automobile	Existing automobiles plying (2008–16) (Million) (1)	No. of pistons/automobile (2)	Total no. of pistons for existing automobiles (3 = 1 * 2)	10% annual spare parts for existing automobiles (Million) (4 = 3 * 10%)	2017 annual automobile production (Million) (5)	No. of piston for 2017 (Million) (6)	Total no. of pistons required annually (Million) (7 = 4 + 6)
Two wheelers	125.6	1	125.6	12.56	19.9	19.9	32.46
Three wheelers	6.7	2	13.4	1.34	0.8	1.6	2.94
Four wheelers	24.8	4	99.2	9.92	3.7	14.8	24.72
Total	157.1		238.2	23.82	24.4	36.3	60.12

Table 37.3 Die casting machine/human task chart

Task performed by			
Die casting machine	Gravity die casting machine	Pressure die casting machine	Automatic die casting machine
Machine on/off	Human	Human	Human
Die lubrication	Human	Human	Machine
Core insertion	Human	Machine	Machine
Closing mold	Machine	Machine	Machine
Pouring the metal	Human	Human	Machine
Opening the mold	Machine	Machine	Machine
Ejection of the casting	Human	Human	Machine
Extraction of the casting	Human	Human	Machine

sequence of operations, task performed by human or machine for gravity, pressure die and automatic die casting machines are presented in Table 37.3.

37.4 Methods for MIQ Measurement for Die Casting Machines

In the present research, three methods for computation of MIQ of die casting machines have been considered. These include machine–human cooperative system, geometric polytope method and multiple perspective analysis method.

37.4.1 Method for MIQ Measurement for Die Casting Machines by Human–Machine Cooperative System

This system basically estimates the mental workload and data quantity and their interface complexity. Stepwise procedure to estimate MIQ is as follows:

Step 1: To find out the input and output variables of the target machine, i.e., the die casting machines and decide the task performed by either machine or human.
Step 2: To define the die casting intelligent task set,

$$T = \{T_1, T_2, T_3, T_4, T_5, T_6, T_7, T_8\} \tag{37.1}$$

where $T_1, T_2, T_3, T_4, T_5, T_6, T_7, T_8$ are the intelligent tasks of die casting machine performed by machine and human. These are the set of tasks required to control and operate the event.

Step 3: To determine the intelligent task cost,

$$t = \{t_1, t_2, t_3, t_4, t_5, t_6, t_7, t_8\} \tag{37.2}$$

where $t_1, t_2, t_3, t_4, t_5, t_6, t_7, t_8$ are the task intelligent costs. Task intelligent costs t_i are intelligence required to execute the task T_j which is varying from 6 to 22 for the tasks 1–8.

Step 4: To prepare data transfer matrix $D = d_{ij}$, for transferring data from task T_i to T_j.

$$D = d_{ij} = \begin{bmatrix} d11 & d12 & d13 & d14 & d15 & d16 & d17 & d18 \\ d21 & d22 & d23 & d24 & d25 & d26 & d27 & d28 \\ d31 & d32 & d33 & d34 & d35 & d36 & d37 & d38 \\ d41 & d42 & d43 & d44 & d45 & d46 & d47 & d48 \\ d51 & d52 & d53 & d54 & d55 & d56 & d57 & d58 \\ d61 & d62 & d63 & d64 & d65 & d66 & d67 & d68 \\ d71 & d72 & d73 & d74 & d75 & d76 & d77 & d78 \\ d81 & d82 & d83 & d84 & d85 & d86 & d87 & d88 \end{bmatrix} \tag{37.3}$$

Step 5: To prepare the task allocation matrix for gravity die casting machine, pressure die casting and automatic die casting machines. The element of this type of matrix can only have binary numbers, i.e., 0 or 1. If the machine performs a task T_i then a_{i1} is 1, and if human performs the task a_{i2} will be 1. If the task T_i is neither assigned by the machine nor human, then it will be 0. Hence, $a_{i1} + a_{i2} + a_{i3} = 1$ for all $1 \leq i \leq n$ task allocation matrix will be $3 \times n$.

$$A = \begin{pmatrix} a_{i1} & a_{j1} & 0 \\ \vdots & \ddots & \vdots \\ a_{i8} & a_{j8} & 0 \end{pmatrix} \tag{37.4}$$

Step 6: To prepare the task allocation matrix considering the task performed by machine or human. To prepare task allocation table considering the task performed by either human or machine, let us consider the machine–human interface (C_{mh}) and human–machine interface complexity (C_{hm}). The values can be from 0.01 to 0.09 depending on the human–machine interaction. Higher values are assigned where more human interaction is required and vice versa. The interface complexities are to be considered for three die casting machines.

Step 7: To prepare the task allocation graph by considering the set of tasks and their interactions like machine to machine, human to machine and machine to human.

Step 8: To compute the machine control intelligent quotient C, i.e., the sum of all intelligent task costs [10].

$$C = \sum_{i=1}^{n} a_{i1} t_i + \sum_{i=1}^{n} a_{i2} t_i \tag{37.5}$$

where C is the control intelligence quotient (CIQ) of the overall control system containing the human and machine. C is the sum of all task intelligence costs (both machine and human).

$$C = t_1 + t_2 + t_3 + t_4 + t_5 + t_6 + t_7 + t_8 \tag{37.6}$$

Step 9: To compute the human intelligence quotient (H). It is the sum of task intelligence costs allocated to the human and interface intelligence costs. We can obtain MIQ by subtracting H from C.

$$H = \sum_{i=1}^{n} a_{i2} t_i + C_{\mathrm{mh}} \sum_{i=1}^{n} \sum_{j=1}^{n} a_{i1} a_{j2} d_{ij} + C_{\mathrm{hm}} \sum_{i=1}^{n} \sum_{j=1}^{n} a_{i2} a_{j1} d_{ij} \tag{37.7}$$

Step 10: MIQ is divided into two components, i.e., machine control intelligence and interface intelligence.

$$M = M_{\mathrm{i}} + M_{\mathrm{c}} \tag{37.8}$$

$$M_{\mathrm{i}} = -\left\{ C_{\mathrm{mh}} \sum_{i=1}^{n} \sum_{j=1}^{n} a_{i1} a_{j2} d_{ij} + C_{\mathrm{hm}} \sum_{i=1}^{n} \sum_{j=1}^{n} a_{i2} a_{j1} d_{ij} \right\} \tag{37.9}$$

$$M_{\mathrm{c}} = \sum_{i=1}^{n} a_{i1} t_i \tag{37.10}$$

M_{i} values are either negative or zero. It implies that if M_{i} is zero, then it will not have any effect on MIQ. If M_{i} is negative, it means the interface overhead deteriorates the machine intelligence quotient.
Step 11: To compute MIQ, subtract H from C;

$$M = C - H \tag{37.11}$$

37.4.2 Estimation of MIQ by Geometric Polytope Method

This method is used to estimate a relative means of determining how to contrast across different machines for comparative intelligence or autonomy. A stepwise procedure for MIQ measurement or estimation is as follows:

Step 1: To define the axes of the geometric polytope to be relevant to that performance of the attribute (goal) under consideration.
Step 2: To define the scales of each axis of the geometric polytope to be relevant to the attribute or goal of interest.
Step 3: To plot the attribute values (productivity, cost, cycle time, etc.) for different machines (die casting) on the same axis.
Step 4: To consider polytope area resulting will provide a relative comparison of the efficiency of a particular machine to perform certain attributes.
Step 5: To use the relative measures of machine intelligence, the effectiveness to perform specific attributes or goal under various conditions can be determined. It is difficult to get an absolute value of MIQ but the comparison is made based on the area of polytope. Higher the area of polytope, then higher is the MIQ of machine.

37.4.3 Estimation of MIQ by Multiple Perspective Analysis

In this method, three filters are used, i.e., technical filter (T), organizational filter (O), personal filter (P). Technical filter includes

Step 1: To define linguistic variables to take values on the interval between zero and one for measuring machine intelligence.
Step 2: To consider three filters, i.e., technical, organizational and personal and their attributes.
Step 3: To normalize the micro and macroeconomic measures (attributes) like productivity, cost and cycle time, etc., for three die casting machines.
Step 4: To correlate the fuzzified values {T-O-P} to MIQ.

For example, three die machines have {0.4, 0.3, 0.2} for gravity die casting, {0.5, 0.4, 0.3} for pressure die casting and {0.7, 0.6, 0.5} for automatic die casting. Then, automatic die casting machine has got the higher MIQ of {0.7, 0.6, 0.5} on linguistics scale which reflects 0.7 for technical, 0.6 for organizational and 0.5 for personal filter.

37.5 Computation of MIQ by Machine–Human Cooperative System for Die Casting Machines

Computation of MIQ for three die casting machines is given below considering intelligent tasks and costs.

Step 1: To find out the input–output variables of the die casting machines and decide the performance performed by either machine or human.
Step 2: To define die casting machine intelligent task set.

$$T = \{T_1, T_2, T_3, T_4, T_5, T_6, T_7, T_8\}$$

Step 3: To determine intelligent task cost,

$$t_i = \{6, 10, 15, 19, 22, 18, 16, 14\}$$

Step 4: To prepare data transfer matrix,

$$\begin{aligned}
D &= d_{ij} = d_{8\times 8} \\
D_1 &= \{d_{14}, d_{16}, d_{24}, d_{26}, d_{34}, d_{36}, d_{45}, d_{46}, d_{47}, d_{56}, d_{67}, d_{68}\} \\
&= \{25, 25, 15, 17, 19, 15, 30, 12, 20, 20, 20, 22\} \\
D_2 &= \{d_{13}, d_{14}, d_{16}, d_{17}, d_{23}, d_{24}, d_{26}, d_{27}, d_{34}, d_{35}, d_{38}, d_{45}, d_{48}, d_{56}, d_{57}, d_{67}, d_{68}, d_{78}\} \\
&= \{20, 22, 12, 14, 15, 24, 17, 15, 12, 30, 12, 32, 14, 28, 24, 15, 19, 22\} \\
D_3 &= \{d_{12}, d_{13}, d_{14}, d_{15}, d_{17}, d_{23}, d_{28}, d_{34}, d_{38}, d_{45}, d_{48}, d_{56}, d_{58}, d_{67}, d_{68}, d_{78}\} \\
&= \{14, 18, 28, 30, 26, 19, 12, 20, 16, 12, 24, 22, 22, 30, 20, 28, 26\}
\end{aligned}$$

Step 5: Task allocation matrix

$$\begin{bmatrix} 0 & 1 & 0 \\ 0 & 1 & 0 \\ 0 & 1 & 0 \\ 1 & 0 & 0 \\ 0 & 1 & 0 \\ 1 & 0 & 0 \\ 0 & 1 & 0 \\ 0 & 1 & 0 \end{bmatrix} \quad \begin{bmatrix} 0 & 1 & 0 \\ 0 & 1 & 0 \\ 1 & 0 & 0 \\ 1 & 0 & 0 \\ 0 & 1 & 0 \\ 1 & 0 & 0 \\ 1 & 0 & 0 \\ 0 & 1 & 0 \end{bmatrix} \quad \begin{bmatrix} 0 & 1 & 0 \\ 1 & 0 & 0 \\ 1 & 0 & 0 \\ 1 & 0 & 0 \\ 1 & 0 & 0 \\ 1 & 0 & 0 \\ 1 & 0 & 0 \\ 1 & 0 & 0 \end{bmatrix}$$

Gravity die casting machine   Pressure die casting machine   Automatic die casting machine

Step 6: To prepare task allocation table considering the task performed by either human or machine, let us consider the machine–human interface (C_{mh}) and human–machine interface complexity (C_{hm}). The values considered are from 0.02 to 0.05.
Step 7: Task allocation shows the task performed by human, machine and events of three die casting machines. Tasks allocations are shown in task allocation matrices (step 5).
Step 8: To determine the control intelligence quotient for three die casting machines,

$$C_1 = 120,\ C_2 = 120, C_3 = 120$$

(Eq. 37.6)
Step 9: To compute the human intelligent quotient (Eq. 37.7) for three die casting machines,

$$H_1 = 83 + (0.04 \times 94) + (0.05 \times 136) = 93.56;$$
$$H_2 = 52 + (0.03 \times 129) + (0.04 \times 111) = 60.31;$$
$$H_3 = 6 + (0.025 \times 138) + (0.02 \times 62) = 10.69$$

Step 10: To compute die casting machine control intelligence (Eq. 37.10) and interface intelligence (Eq. 37.9),

$$M_{c1} = 37,\ M_{c2} = 68,\ M_{c3} = 114;$$
$$M_{i1} = -10.56,\ M_{i2} = -8.31,\ M_{i3} = -4.69$$

Step 11: To compute MIQ (using Eq. 37.11),

$$M_1 = C_1 - H_1 = 120 - 93.56 = 26.44;$$
$$M_2 = C_2 - H_2 = 120 - 60.31 = 59.69;$$
$$M_3 = C_3 - H_3 = 120 - 10.69 = 109.31$$

The total intelligent costs have been considered 120 for all eight tasks for die casting machines and considered the task costs from 6 to 22 according to the task complexity, and it is shown in Table 37.4. The interface intelligence for all of the casting processes is negative due to the fact that the interface overhead deteriorates the MIQ. The control intelligence stayed top for the automatic die casting machine as this machine has controlled the more intelligence task than other two methods. The computed MIQ values are presented in Table 37.5.

Table 37.4 Task description, tasks intelligent costs

Task element	Task description	Intelligent task description	Task intelligent cost	Task intelligent cost
T_1	Machine on/off	Detect	t_1	6
T_2	Die lubrication	Identify	t_2	10
T_3	Core insertion	Observe	t_3	15
T_4	Closing mold	Interpret	t_4	19
T_5	Pouring the metal	Define tasks	t_5	22
T_6	Opening the mold	Select actions	t_6	18
T_7	Ejection of the casting	Execute	t_7	16
T_8	Extraction of the casting	Evaluate	t_8	14
			Total	120

Table 37.5 Estimation of die casting machines intelligence quotient

Description	Gravity die casting machine	Pressure die casting machine	Automatic die casting machine
Total intelligent cost ($C = H + M$)	120	120	120
Human intelligence quotient (H)	93.56	60.31	10.69
Control intelligence (M_c)	37	68	114
Interface intelligence (M_i)	−10.56	−8.31	−4.69
Machine intelligence quotient (M)	26.44	59.69	109.31

37.6 Conclusions

In the present paper, MIQ and its sub-systems are defined and formulated for computation of machine intelligence. Three methods for measuring MIQ were formulated, and MIQ was calculated by human–machine cooperative process for gravity die casting, pressure die casting and automatic die casting machines. Human–machine cooperative system is more suitable for estimation of die casting machines for the manufacture of pistons. MIQ of automatic die casting machine has the highest and gravity die casting machine lowest. It is also useful for comparison of different machines with different intelligent attributes manufactured by original equipment manufacturers. From the MIQ computation, it was found that automatic die casting machine has the highest MIQ of 109.31 followed by pressure die casting machine (56.69) and gravity die casting machine (26.44).

Acknowledgements The authors gratefully acknowledge the support extended by North Eastern Regional Institute of Science and Technology (NERIST), Itanagar, Arunachal Pradesh-791109, India in the present work.

References

1. Bien, Z., Bang, W.C., Kim, D.Y., Han, J.S.: Machine intelligence quotient: its measurements and applications. Fuzzy Sets Syst. **127**, 3–16 (2002)
2. Muralidhar, M.: Introduction to Methods for Calculation of Machine Intelligent Quotient, Short Term Training Program on Soft Computing in Engineering Application, pp. 87–103, NERIST, Arunachal Pradesh, India (2010)
3. Muralidhar, M.: Development of intelligent induction furnace and measurement of furnace intelligence quotient: a study. In: 69th World Foundry Congress, pp. 0754–0761, Hangzou, PR China (2010)

4. Anthony, A., Jannett, T.C.: Measuring machine intelligence of an agent-based distributed sensor network system. In: Advances and Innovations in Systems, Computing Sciences and Software Engineering, pp 531–535 (2007)
5. Wang, Y.: On Abstract Intelligence: Toward a Unifying Theory of Natural, Artificial, Machinable, and Computational Intelligence, Int. J. Softw. Sci. Comput. Intell. 1(1), 1–17 (2009)
6. Muralidhar, M.: Mathematical computational methods for determination of MIQ for machines. In: International Conference on Mathematics and Applied Mechanics, Bangalore (2007)
7. Mourtzis, D., Milas, N., Vlachou, A.: An internet of things-based monitoring system for shop-floor control. J. Comput. Inf. Sci. Eng. **18**, 0210051-8, (2018)
8. Park, H.J., Kim, B.K., Lim, K.Y.: Measuring the machine intelligence quotient (MIQ) of Human–machine cooperative systems. In: IEEE Trans Syst, Man, and Cybern—Part A: Syst Hum **31**(2), 89–96 (2001)
9. The Machine Intelligence Continuum: https://www.topbots.com/topbots-ai-machine-intelligence-continuum. Last accessed on 2018/11/02
10. Kumar, S., Muralidhar, M.: CTS Estimation in Shock Absorber Assembly Workforce in Indian Automobile Industry: A Study. HWWE, AMU, Aligarh, India (2017)
11. Reddy, A.C., Rajanna, Ch.: Design of gravity die casting process parameters of Al-Si-Mg alloys. J. Mach. Form. Technol. 1–24 (2009)
12. Ashwini, A., Avinash, K.S.: Rejection analysis in Piston manufacturing unit. Int. J. Innov. Res. Sci., Eng. Technol. **4**(3), 1157–1163 (2015)

Chapter 38
Optimization of Porosity in Cast Aluminium Foams

R. Kumar, G. Avinash and Neha Gupta

Abstract Closed-cell ultra-light aluminium foams are often used as structural materials in aerospace, railways and automobile sectors as they provide good strength to weight ratio with high impact strength and corrosion resistance. Melt - based metal foams can be casted using indirect foaming techniques, wherein gas bubbles are created inside the melt which has been pre-treated in a suitable way. The properties of these metal foams are directly affected by nature of porosity, which is difficult to measure and control experimentally. Therefore, theoretical solutions have to be developed to control this porous structure. In the present work, a aluminium foam, prepared using foaming method using blowing agents with porosity $\sim 86\%$, is used, where the effect of melt viscosity and solidification time on gas bubble size and its rising velocity inside melt have been theoretically studied. The calculations showed that rising velocity of hydrogen gas bubbles, aluminium melt viscosity and solidification time are responsible for motion of bubbles in the aluminium melt. Control of this movement of bubbles can result in optimum porosity in the solidified foam. A relationship has been established for viscosity enhancement due to the addition of additives like Ca, Al_2O_3, SiC, TiB_2, SiO_2, BN and their influence on bubble size and its rising velocity during solidification of this foam. Overall, the study suggested amount of additives that are required to be added in a given volume of aluminium melt, to increase melt viscosity and to maximize bubble entrapment. The results also showed that a cooling time of ~ 8 s leads to efficient bubble entrapment with uniform pore structure and hence properties of aluminium foam.

Keywords Aluminium foam · Porosity control · Blowing agents · Solidification

R. Kumar · G. Avinash · N. Gupta (✉)
Department of Metallurgical and Materials Engineering,
NIT Raipur, Raipur (C.G.) 492010, India
e-mail: ngupta.met@nitrr.ac.in

R. G. Narayanan et al. (eds.), *Advances in Computational Methods in Manufacturing*, Lecture Notes on Multidisciplinary Industrial Engineering,
https://doi.org/10.1007/978-981-32-9072-3_38

38.1 Introduction

In recent years, light-weight metal foams have gained a lot of attraction both for the scientific research and industrial applications. Amongst metal foams, aluminium alloys are commercially the most exploited ones due to their high strength to density ratio, good corrosion resistance and competitive cost [1]. The application potential of these materials has already being explored for different fields, namely construction, energy absorption, thermal or sound insulation, etc. where improvement in safety is needed. Several processing methods can be used to produce these structures as casting, powder metallurgy, metal deposition, additive manufacturing and many more [2]. Closed-cell metallic foams can be economically obtained by solidifying melt containing well-dispersed gas bubbles, also known as indirect gas foaming technique. Such foams are commonly used for structural applications as they exhibit properties as enhanced energy absorption with high stiffness to weight ratio [3], thereby offering significant performance gains.

Most of the properties, such as strength, sound insulation, impact energy absorption, etc. of these foams are extremely sensitive and are directly affected by its final pore size and its distribution [1–3]. Hence, a well-controlled closed-cell pore structure is often desired in order to obtain high-quality metal foams, so as to meet the requirements of specific applications. Therefore, it is very much required to have good pore distribution and its control during processing. However, it is a tedious task to experimentally control porosity using foaming techniques due to the complex nature of foaming process which often leads to non-homogeneous distribution of gas bubbles. Pore size, shape and its distribution depend on melt viscosity, bubble size and rising velocity, solidification time, surface tension, fluid buoyancy and drag velocity. Therefore, it is extremely important to control these parameters during foaming process. Only few studies have been reported on theoretical optimization studies related to pores formation in metal foams [4–8]. It is known that low melt viscosity and high bubble rise velocity tends to drain the melt faster leading to rapid escape of gas bubbles before foaming [5]. This leads to fabrication of imperfect foam having lower porosity. Therefore, melt with sufficient viscosity and optimal bubble rise velocity has to be prepared to efficiently stabilize the liquid foam and fabricate structurally sound aluminium foams [6, 7]. Talaria discussed and predicted terminal bubble velocity in a given liquid [5]. His viscosity expression can be easily used to relate with a given bubble and it's rising velocity. Studies by Banhart showed the importance of studies, to understand stabilization mechanisms for metal foams [6]. Akiyama et al. have prepared closed-cell aluminium foams through indirect gas foaming technique which used titanium hydride (TiH_2) as foaming agent and calcium as additive to enhance the viscosity of the melt so as to adjust its foaming properties [7].

In this work, a closed-cell aluminium foam was prepared using similar technique as by Akiyama et al. by adding TiH_2 as foaming agent to create hydrogen bubbles [7]. These bubbles resulted in inhomogeneous pore development in aluminium foam upon solidification. Here, the effect of melt viscosity and cooling conditions on pore formation in aluminium foam has been investigated theoretically. Further, the effect of different additives (Ca, Al_2O_3, TiB_2, SiO_2, SiC and BN) on melt viscosity and its influence on bubble diameter and hence its rising velocity during solidification of this foam was analysed. A relationship has been developed for melt viscosity, solidification time, bubble size and bubble rising velocity for efficient pore entrapment and thereby uniform distribution in aluminium foam.

38.2 Materials and Methods

Aluminium alloy was heated at 720 °C in a Heat Treat Electric furnace for 1.5 h, to which 1.8 wt% additive (calcium) was added as thickening agent under continuous stirring to adjust the melt viscosity. Thereafter, 1.5 wt% TiH_2 was added as blowing agent to this melt to form hydrogen gas bubbles. Continuous stirring (total time = 150 s, speed = 1000 rpm) of the melt resulted in a homogeneous distribution of additive, and blowing agent inside melt till cooling helped to obtain a foam with porosity of $\sim$86%.

Under these well-controlled experimental conditions, a porous structure with uniform pore size and spherical shape was expected. However, a non-uniform porous structure was observed as illustrated in Fig. 38.1, despite taking precautions in the process, which resulted in decrement of foam properties (strength, stiffness, etc.) when compared to theoretical values, as was also observed by Suresh et al. [9]. Efficient control of porosity in the foam is a solution to this. Also, it is known that bubble rising velocity and melt viscosity in the aluminium melt affects the porosity in foam [1–3]. For study, a mould with dimensions $15 \times 15 \times 15$ cm^3 with 20 cm sprue length and 2 cm sprue diameter was used. Here, an effort has been done to solve and study for the in situ pore size and its distribution with respect to melt viscosity during solidification using theoretical calculations.

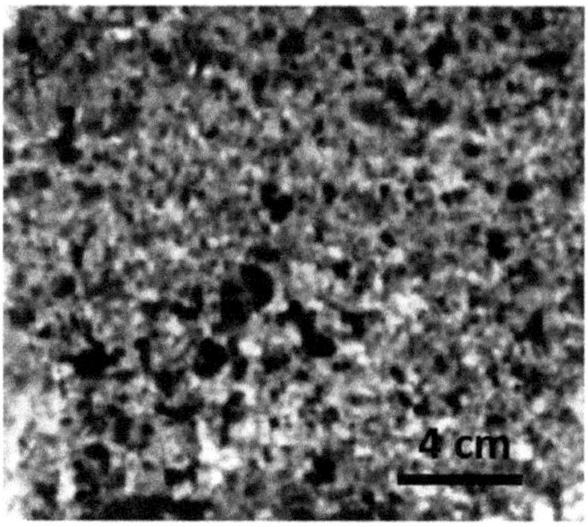

Fig. 38.1 Microstructure of foam showing non-uniform porosity

38.3 Results and Discussion

In continuation of our earlier works [4], we know that pore size can be controlled by controlling the diameter of hydrogen bubble generated inside the aluminium melt. Also, pore size distribution of foam can be controlled by controlling the rising velocity of these hydrogen bubbles. Here, a relationship between bubble size with pore size and bubble rising velocity with pore distribution inside the aluminium foam during solidification have been developed. Using Stokes equation [4, 10], hydrogen gas bubble rising velocity can be calculated, which indicates the movement of bubbles inside aluminium melt leading to pore distribution inside foam upon solidification. It was calculated that the small isolated gas bubbles rose in melt at 720 °C with rising velocity of 0.038 m/s. For this rising velocity, the distance covered by bubbles at solidification start (~615 °C, 30 s) [11] is 1.1241 m. This bubble displacement was too high compared to the height of the mould used; therefore, the gas bubbles get easily escaped resulting in undesired low porosity foam structure. The bubble rising velocity has to be therefore decreased by increasing the melt viscosity so that gas bubbles stay in the melt for longer time, thus resulting in a structurally sound porous structure. The same was confirmed by Akiyama et al., who used calcium as thickening agent to increase the melt viscosity [7].

This velocity is strongly dependant on the aluminium melt viscosity. The relative viscosity (μ_{rel}) due to the addition of additives is [4, 11]:

$$\mu_{\text{rel}} = \frac{\mu_{\text{sus}}}{\mu_{\text{liq}}} \tag{38.1}$$

μ_{sus} Dynamic viscosity of the suspension
μ_{liq} Dynamic viscosity of the aluminium melt (1.38 MPa) [12].

Using Eq. (38.1), it was calculated that after the addition of 1.8 wt% additive, the viscosity of aluminium melt increased from 1.38 to 7.4 MPa and was high enough to hold gas bubbles inside the melt for longer times. This increase in viscosity decreased the bubble rise velocity from 0.038 to 0.007 m/s, which helped in efficient entrapment of gas bubbles in the melt to achieve desired pore structure.

Table 38.1 Volume percentage of different additives that can be used

Additives	Volume %
Ca	2.79
Al_2O_3	7.11
TiB_2	8.14
SiO_2	4.77
SiC	5.78
BN	3.16

Different additives can be used to obtain such increment in melt viscosity. Table 38.1 reports the calculated volume percentages of some of these additives.

However, when TiH_2 was added to the melt, it decomposed and resulted in formation of gas bubbles. These bubbles further expanded as they rose in the melt. The increase in bubble diameter w.r.t. solidification time can be given as [13]:

$$d(t) = 2\left[\frac{K}{\sqrt{D} \pm \sqrt{\frac{K^2}{D} + 2(0.9253K)}}\right] t^{\frac{1}{2}} \quad (38.2)$$

where,

$d(t)$	Bubble diameter (mm)
$K = \sqrt{\frac{3}{\pi}} \times \frac{D(C - C_w)}{\rho_g}$	1.22×10^{-6} m^2/s
C	Gas concentration in the liquid (0.14×10^{-3} kg/m^3)
C_w	Gas concentration within the pore (0.055×10^{-3} kg/m^3)
ρ_g	Bubble density (0.17×10^{-3} kg/m^3)
D	Diffusion coefficient of hydrogen in the Al-melt (2.5×10^{-6} m^2/s)
t	Solidification time (s).

Equation (38.2) showed that gas expansion leads to bubble growth, which further influences its rising velocity. These changes were calculated using Stokes equation and Eq. (38.2) and are reported in Table 38.2. Table 38.2 shows that with increase in solidification time, bubble growth occurs, which also enhances their rising velocities due to expansion. This increase in bubble growth during solidification can lead to increased pore size.

Overall, the study concludes that for a foam size of 15 cm × 15 cm × 15 cm, to have hydrogen gas bubble being effectively entrapped inside aluminium melt, a reduction in solidification time has to be done. As per calculations, the solidification time should be less than 8 s so as to get uniform porosity in foam. Therefore, for

Table 38.2 Influence of solidification time and bubble size on bubble rising velocity

t (sec)	d (mm)	U_∞ (mm/s)
1	0.74	2.33
2	1.05	4.71
3	1.29	7.05
4	1.49	9.40
5	1.67	11.72
6	1.83	14.11
7	1.97	16.46
8	2.11	18.8
9	2.24	21.15
10	2.36	23.52

parameters chosen, along with improvement in melt viscosity by adding 1.8 wt% additives, fast cooling of the melt are required so that an optimal uniform pore size and distribution can be achieved.

38.4 Conclusion

This work was performed to have a better understanding and control over pore size and its distribution in aluminium foam to have better property control. It was observed that the melt viscosity increased on the addition of 1.8 wt% additives like Ca, Al_2O_3, TiB_2, SiO_2, SiC and BN, which helped in holding bubbles in the melt. The viscosity enhanced from 1.38 to 7.4 MPa. This increased melt viscosity helped in entrapping the bubbles in the melt by decreasing bubble rise velocity from 0.038 to 0.007 m/s.

It was also observed that with increment in solidification time, bubble growth occurs due to gas expansion, which affects their rising velocity. This leads to increase in pore size, which is sometimes undesirable. Calculations suggested that reducing solidification time from 30 s to less than 8 s, improved the entrapment of gas bubbles inside the aluminium melt, thereby creating effective porosity in the foam.

Overall, using calculations discussed here, it is expected that a researcher can optimize his/her parameters used to fabricate foams as mould size, melt used, melt viscosity, amount of additives and solidification time to obtain uniform porosity in the metal foam and hence it's properties. This study can help in better designing of the experiments used for development of metal foams by indirect foaming method.

References

1. Banhart, J.: Europhys News (30), 17–20 (1999)
2. Banhart, J.: Manufacture, characterization and application of cellular metals and metal foams Prog. Mater. Sci. **46**, 559–632 (2001)
3. Gibson, L.J., Ashby, M.F.: Cellular Solids: Structure and Properties, Cambridge. Cambridge University Press, UK (1997)
4. Avinash, G., Harika, V., Sandeepika, Ch., Gupta, N.: Pore size control in aluminium foam by standardizing bubble rise velocity and melt viscosity. IOP Conf. Ser.: Mater. Sci. Eng. **338**, 012010 (2018)
5. Talaia, M.A.R.: Terminal velocity of a bubble rise in a liquid column. World Acad. Sci. Eng. Technol. (28), 264–68 (2007)
6. Banhart, J.: Metal foam: production and stability. Adv. Eng. Mater. **8**(9), 781–794 (2006)
7. Akiyama, S., Imagawa, K., Kitahara, A., Nagata, S., Morimoto, K., Nishikawa, T., Itoh, M.: Foamed metal and method of producing same. US Patent 4713277 A (1987)
8. Suresh, C., Vidyashankar, S.: Investigation of the effect of pore size on mechanical properties of open cell aluminium foam structure. International J. Adv. Eng. Manag. Sci. **2**(9), 1641–1644 (2016)

9. Stokes, G.G.: Mathematical and Physical Papers, Sommerfeld. Cambridge University Press, London (1880)
10. Bala, K.C., Khan, R.H.: Rate of solidification of aluminium casting in varying wall thickness of cylindrical metallic moulds. Leonardo J. Sci. (25), 19–30 (2014)
11. Konijn, B.J., Sanderink, O.B.J., Kruyt, N.P.: Experimental study of the viscosity of suspensions: effect of solid fraction, particle size and suspending liquid. Powder Technol. (266), 61–69 (2014)
12. Assael, M.J., Kakosimos, K., Banish, R.M., Brillo, J., Egry, I., Brooks, R., Quested, P.N., Mills, K.C., Nagashima, A., Sato, Y., Wakeham, W.A.: Reference data for the density and viscosity of liquid aluminum and liquid iron. J. Phys. Chem. Ref. Data **1**(35), 285–300 (2006)
13. Yang, C.C., Nakae, H.: The effects of viscosity and cooling conditions on the formability of aluminum alloy. J. Mater. Process. Technol. (141), 202–206 (2003)

Chapter 39
To Predict Surface Roughness and Linear Shrinkage of Die Casting Process by Using of Fuzzy Algorithm Model

Narendra Krishania and Anil Kumar Birru

Abstract This research paper narrates a manually constructed Mamdani based on fuzzy algorithm model for envisaging surface roughness and linear shrinkage of die casting, so that defects can be refrained from the casting in terms of surface finish and dimensions. A set of rules established by the help of mathematical model have been used to derive two fuzzy controllers which are being used in this process. With the help of this fuzzy algorithm, high production rate and high quality of products can be obtained by controlling process parameters. Confirmation experiments reveal that these fuzzy logics are able to attain optimum grouping of the process parameters. Hence, the quality of casted products in die casting process can be improved to a greater extent by this approach. The predicted surface roughness and linear shrinkage by this model had an error of only 3.55 and 6.02%, respectively, which was confirmed by checking the validity of the model developed by performing confirmation experiments. This proposed model can be reasonably utilized by the industries involved in die casting around the world to increase the overall effectiveness of the process and product.

Keywords Fuzzy logic · Orthogonal array · Surface roughness · Linear shrinkage

39.1 Introduction

Metal casting is a vital part of industrial production. It is generally used to develop near net shape component. Nowadays, the theory of direct/net shape manufacturing is attaining great importance in terms of lead time and cost reduction [1].

The die casting is an economical and efficient method offering a wide range of shapes than any other process. During manufacturing a metal part, the shapes of the

N. Krishania (✉) · A. K. Birru
National Institute of Technology Manipur, Imphal 795004, India
e-mail: anilbirru@nitmanipur.ac.in

R. G. Narayanan et al. (eds.), *Advances in Computational Methods in Manufacturing*, Lecture Notes on Multidisciplinary Industrial Engineering,
https://doi.org/10.1007/978-981-32-9072-3_39

metals depend upon the injection temperature and liquid level. Die casting parameters like injection temperature, moisture level, injection pressure, and liquid level by using embedded controller. In the conventional system, the die casting parameters are measured by manual. The process of die casting involves five main stages which are clamping, injection, cooling, ejection, trimming. Basically, non-ferrous alloys are die casted like alloys of aluminium, alloys of copper, alloys of magnesium, and alloys of zinc. Die casting machines used are of two types—one is hot chamber machine (alloys with low melting temperature are die casted like zinc) and cold chamber machine (alloys with high melting temperature are die casted, like aluminium). Several engine components and pump components are manufactured using die casting.

Accuracy of die casting is affected by many factors like injection pressure too high, injection temperature too high, cooling time too short, low pouring temperature, clamp force too low, ejection force too high, non-uniform cooling rate.

To recognize the significant controllable process parameters which influence the quality of the casted products, it is essential to understand and find out the relationship among these controllable process parameters and quality of the casted products. The selection of the suitable process parameters for making qualitative casted products is one of the primary challenges in casting industry. It is very difficult to set appropriate process parameters with consideration of multiple performance characteristics and to depend on large amount of experimental operations. Therefore, it is important to effectively obtain the optimal process parameters to reduce trial and error time and consuming cost in the casting process [2].

Fuzzy reasoning is one of the prevailing tools which is used for modelling and governing undefined multi-criterion conclusion-making problems. It is grounded on the notion to obtain a set of association between inputs and outputs describing a procedure. So, the technique of fuzzy modelling conveys a nonlinear course better than several existing process [3]. Fuzzy logic techniques have been implemented by a number of researchers in their conclusion-making glitches. A fuzzy logic controller for forecasting the level regulator of molten steel in strip casting methods was created by Park and Cho [4]. With the aid of orthogonal array fuzzy linguistic approach, a general optimization system was implemented in the CNC turning operation for surface by Lan [5]. With the help of fuzzy-based Taguchi technique, Hsiang et al. [6] discovered the optimal process parameters that were maximizing the multiple performance characteristic index (MPCI) for hot extrusion of AZ31 and AZ61 magnesium alloy bicycle carriers [7, 8]. So, in the present study an effort has been made to create fuzzy logic controllers for estimation of quality of casting products with some provided set of process parameters like injection temperature, injection pressure, and time. It was originated that the predicted and measured values are in close proximity [9–12]. The closeness of results implies that the fuzzy modelling technique is efficient to estimate the quality of the patterns used for die casting.

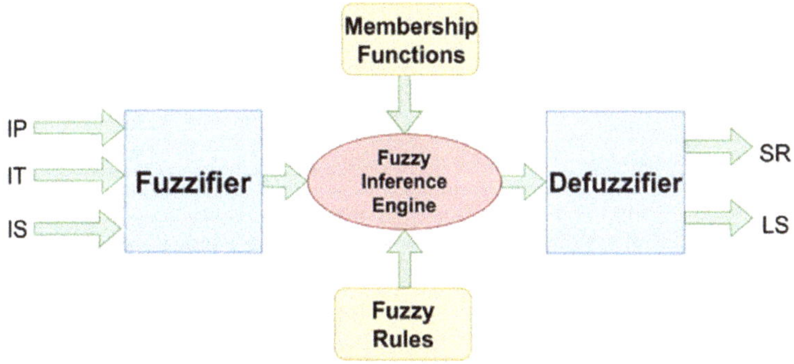

Fig. 39.1 Structure of three inputs two outputs' fuzzy logic

39.1.1 Fuzzy Logic Overview

Mamdani method is the most used fuzzy inference. Initially, one of the first fuzzy processes to regulate a steam engine and boiler was developed by Mamdani and Assilia in [13–15]. A fuzzy logic arrangement consisting of three inputs and two outputs is developed for this study which is shown in Fig. 39.1 [16]. Following steps are followed to execute Mamdani fuzzy inference:

(a) fuzzification of the input variables,
(b) rule output are evaluated and aggregated, and
(c) defuzzification.

Mamdani fuzzy rule is described as:

$$\begin{aligned}&\text{Rule 1: If } y_1 \text{ is } M_1,\ y_2 \text{ is } N_1 \text{ and } y_3 \text{ is } O_1 \text{ then } z_1 \text{ is } P_1 \text{ and } z_2 \text{ is } Q_1 \text{ else}\\&\text{Rule 2: If } y_1 \text{ is } M_2,\ y_2 \text{ is } N_2 \text{ and } y_3 \text{ is } O_2 \text{ then } z_1 \text{ is } P_2 \text{ and } z_2 \text{ is } Q_2 \text{ else}\\&\text{Rule } n\text{: If } y_1 \text{ is } M_n,\ y_2 \text{ is } N_n \text{ and } y_3 \text{ is } O_n \text{ then } z_1 \text{ is } P_n \text{ and } z_2 \text{ is } Q_n \text{ else}\end{aligned}$$

where M_i, N_i, O_i, P_i, and Q_i are termed as fuzzy subsets which are described by the analogous membership function [17–20].

39.2 Experimental Work

By performing the experiment with single variable at a time approach, the span of the particular process parameters is determined. The designed symbols and ranges of the process parameters are provided in Table 39.1. In the present study, the response parameters considered are surface roughness (SR) in μm and linear shrinkage (LS) in percentage.

Table 39.1 Factors and levels used in the experiments

Factor	Description	Unit	Range	Level 1	Level 2	Level 3
A	Pouring temperature	°C	650–750	650	700	750
B	Injector pressure	Bar	120–240	120	180	240
C	Plunger velocity	m/s	1.2–3.8	1.2	2.5	3.8

Table 39.2 L_9 orthogonal array used in the primary experiments

Experiment no.	A	B	C	SR	LS
1	650	120	1.2	48.5541	1.412
2	650	180	2.5	55.8354	1.384
3	650	240	3.8	63.0352	1.204
4	700	120	1.2	58.2841	2.974
5	700	180	2.5	65.3251	2.575
6	700	240	3.8	73.7327	2.045
7	750	120	1.2	61.0512	3.715
8	750	180	2.5	66.0263	3.315
9	750	240	3.8	75.1121	3.012

The primary experiments were conducted using nine test trials; these nine test trials were designed on the basis of L_9 orthogonal array. The reason of choosing L_9 orthogonal array is that shorter arrays are not capable of producing ample data to appropriately analyse the method since they are too simple and longer arrays would create complexity for the trial method. The response of assigned L_9 orthogonal array is shown in Table 39.2. In order to consider the three unlike performance characteristics in Taguchi method, corresponding values of the surface roughness (SR) and linear shrinkage (LS) are processed by the fuzzy logic unit.

39.2.1 Fuzzy Logic Modelling

In this work, three inputs and two outputs are used to evolve a fuzzy controller. Pouring temperature (PT), injector pressure (IP), and plunger velocity (PV) are three input variables, and surface roughness (SR) and linear shrinkage (LS) are two output variables. To make the fuzzy rule base, Mamdani fuzzy "If-Then" declarations for 27 rules has been considered, and to execute inference of max–min operation, Mamdani implication method has been adopted. To change fuzzy data to crisp response values, the centre of gravity method has been used by the defuzzifier. By acquiring triangular membership function, fuzzy division of input and output variables is carried out. Block diagram of fuzzy controller is shown in Fig. 39.2. This has been developed by MATLAB.

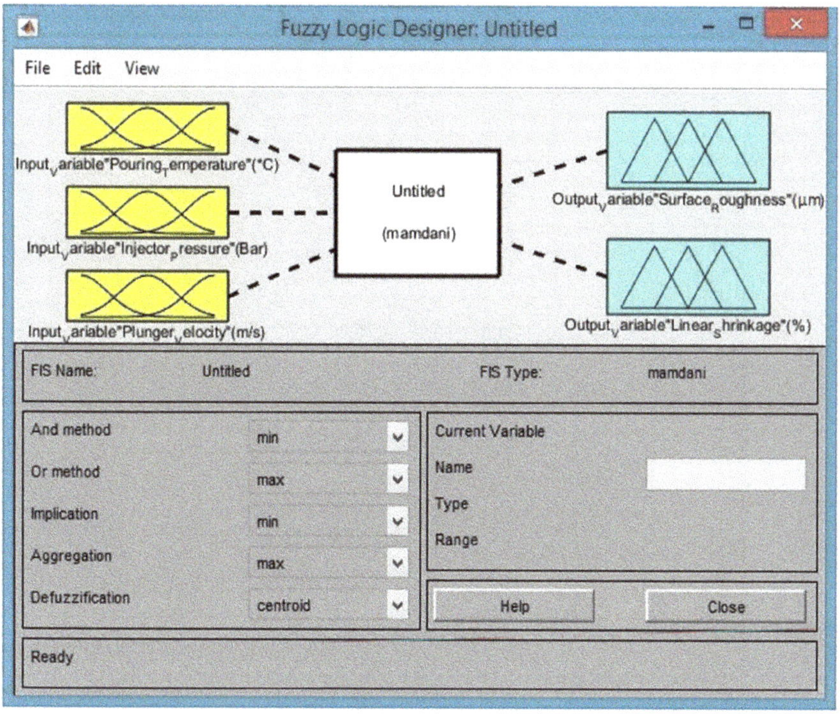

Fig. 39.2 Block diagram of fuzzy controller

Membership functions of the three inputs are small (S), medium (M), and large (L), and membership functions of the two outputs are very small (VS), small (S), medium (M), large (L), and very large (VL). Input and output functions are shown in Figs. 39.3 and 39.4. Table 39.3 shows the fuzzy intervals for input and output for forecasting the response, namely surface roughness (SR) and linear shrinkage (LS) of the casted products in die casting process. By taking into account the max–min combinational operation and with provided set of inputs, experimental results are originated on which these fuzzy rules are established. A non-fuzzy output is provided by the fuzzy reasoning of these rules as shown in Fig. 39.5.

Validation of Fuzzy Model Prediction

Three experiments were performed with different combination of the input process parameters which were not incorporated in the training set (randomly chosen). In Table 39.4, the corresponding experimental responses are measured and listed. From the fuzzy controller, the response values are again acquired for the same set of input process parameters. Then, to find out the extent by which the fuzzy controller is working, the percentage of error was calculated. The experimental and the predicted result obtained from the Mamdani fuzzy model has been plotted on graph as shown in Fig. 39.6. Also, bar graph has been drawn between experimental and fuzzy predicted data as shown in Fig. 39.7.

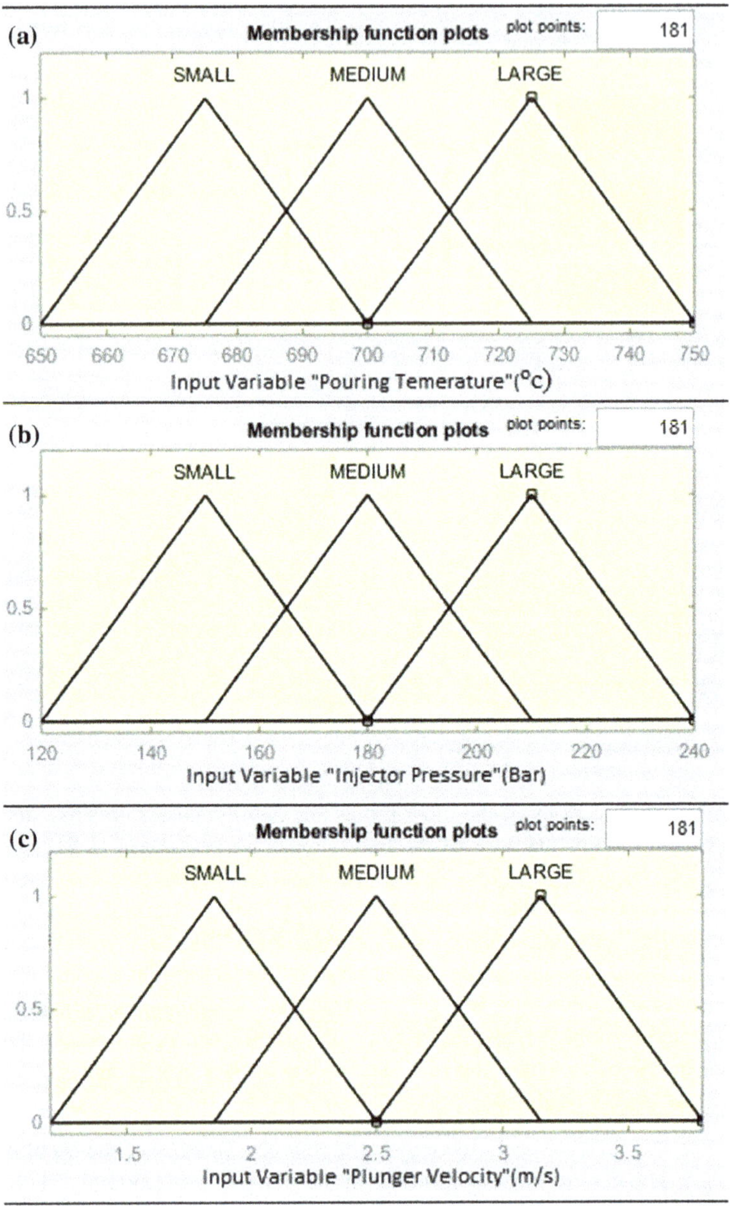

Fig. 39.3 Membership function of input variables: **a** pouring temperature, **b** injector pressure, and **c** plunger velocity

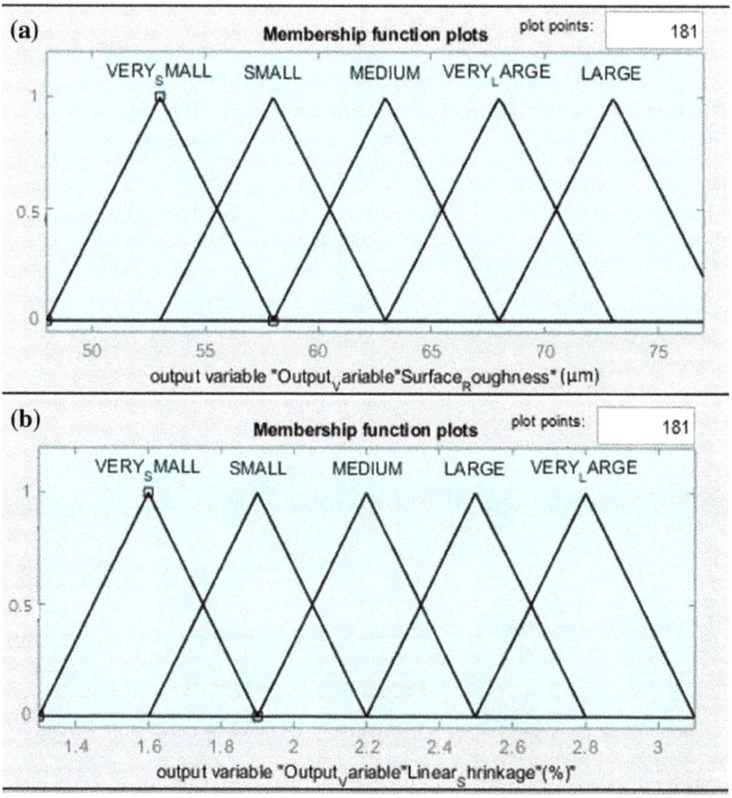

Fig. 39.4 Membership function of output variables: **a** surface roughness and **b** linear shrinkage

Table 39.3 Input output and fuzzy intervals

S. no.	System's linguistic variable		Linguistic values	Fuzzy interval (min–average–max)
1	Inputs	Injection speed	Small	650–675–700
			Medium	675–700–725
			Large	700–725–750
		Injection pressure	Small	120–150–180
			Medium	150–180–210
			Large	180–210–240
		Injection time	Small	1.2–1.85–2.5
			Medium	1.85–2.5–3.15
			Large	2.5–3.15–3.8

(continued)

Table 39.3 (continued)

S. no.	System's linguistic variable		Linguistic values	Fuzzy interval (min–average–max)
2	Outputs	surface roughness (μm)	Very small	48–53–58
			Small	53–58–63
			Medium	58–63–68
			Large	63–68–73
			Very large	68–73–78
		linear shrinkage (%)	Very small	1.3-1.6–1.9
			Small	1.6–1.9–2.2
			Medium	1.9–2.2–2.5
			Large	2.2–2.5–2.8
			Very large	2.5–2.8–3.1

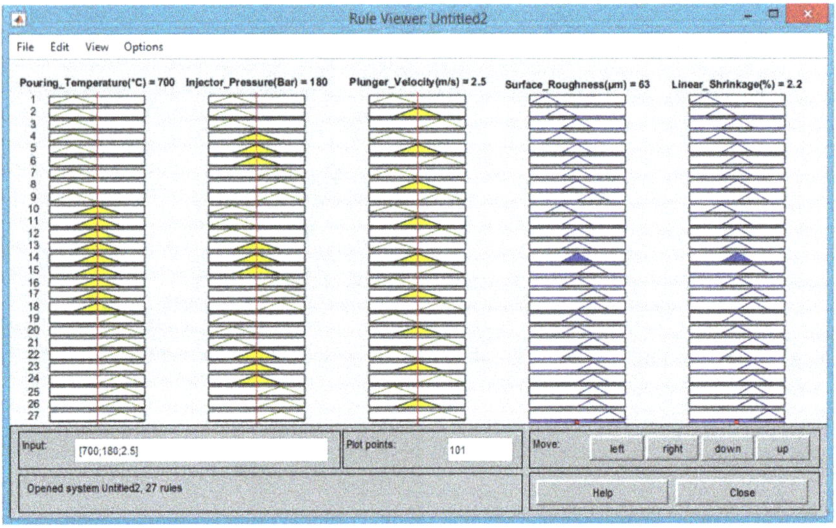

Fig. 39.5 Fuzzy inputs and prediction of outputs

Table 39.4 Comparison of results between the actual and the predicted values

S. no.	Injection process parameter			Experimental values		Fuzzy predicted values		% error	
	PT	IP	PV	SR	LS	SR	LS	SR	LS
1	650	120	1.2	48.554	1.412	52	1.5	7.09	6.23
2	700	180	2.5	65.325	2.075	63	2.2	3.55	6.02
3	750	240	3.8	75.112	3.012	79	3.3	5.17	9.56

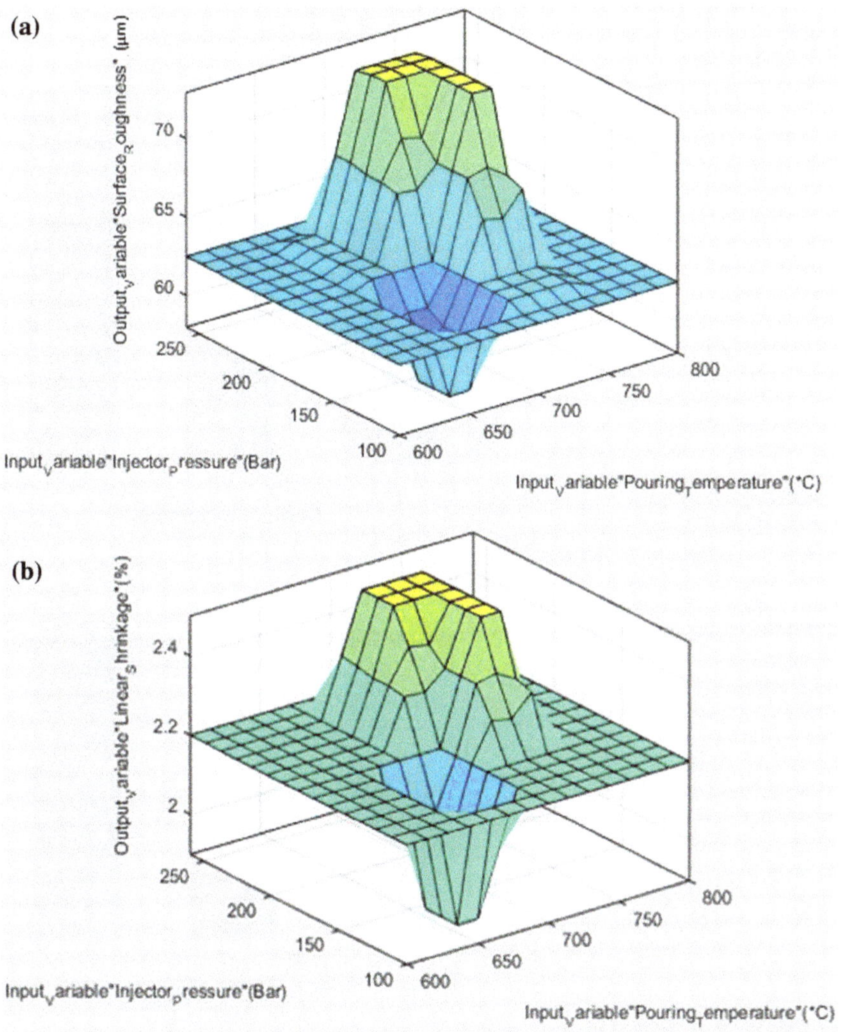

Fig. 39.6 Predicted surface roughness and linear shrinkage by fuzzy logic in relation to parameters change: **a** pouring temperature and injector pressure and **b** pouring temperature and injector pressure

39.3 Discussion

In this work, to predict the quality characteristics, namely surface roughness and linear shrinkage of casted product in die casting process, a fuzzy logic controller using Taguchi orthogonal array design has been developed. Prediction results show that Mamdani inference was on the basis of maximum error calculated. So, for

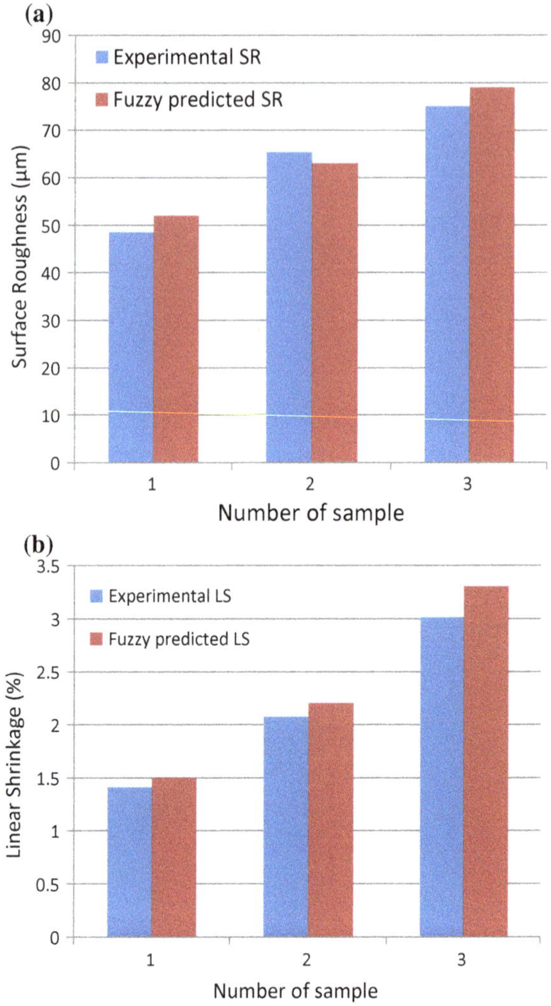

Fig. 39.7 Prediction of **a** surface roughness and **b** linear shrinkage

improving the quality and reliability of the entire process, it has proved to be an effective and powerful tool. By enhancing the number of experiments, the anticipated values of fuzzy output can be further precisely predicted. It is also very useful and can also be used for the optimization of multiple performance quality process parameters and casted product material, to test the ability of the expert systems in prediction of the outputs.

39.4 Conclusion

From the present study, resulting conclusions are summarized as follows:

- The capability of generalization and prediction of pattern characteristics within the range of experimental data such as surface roughness and linear shrinkage of casted product in die casting process have been shown with the help of Mamdani fuzzy inference.
- Industries which are using the experience of skilled workers can use a set of fuzzy rules which further result in improvement in forecasting ability of the process.
- For automation of the process, this methodology seems to be beneficial.

References

1. Campbell, J.: The concept of net shape for casting. Mater. Des. **21**, 373–380 (2000)
2. Mahapatra, S.S., Nanda, S.K., Panigrahy, B.K.: A Cascaded fuzzy inference system for Indian river water quality prediction. Adv. Eng. Softw. **42**, 787–796 (2011)
3. Khrais, S., Al-Hawari, T., Al-Araidah, O.: A Fuzzy logic application for selecting layered manufacturing techniques. Expert Syst. Appl. **38**, 10286–10291 (2011)
4. Park, Y., Cho, H.: A fuzzy logic controller for the molten steel level control of strip casting processes. Control Eng. Pract. **13**, 821–834 (2005)
5. Lan T.S.: Fuzzy linguistic optimization on surface roughness for CNC turning. Math Prob Eng, pp. 1–10, Hindawi Publishing Corporation (2010)
6. Hsiang, S.H., Lin, Y.W., Lai, J.W.: Application of fuzzy-based Taguchi method to the optimization of extrusion of magnesium alloy bicycle carriers. J. Intell. Manuf. (2010)
7. Zimmermann, H.: Fuzzy Set Theory and Its Applications Second. Revised Edition, Springer (1992)
8. Xu, D.: Applications of Fuzzy Logic in Bioinformatics. Imperial College Press, London, UK (2008)
9. Ramesh, S., Karunamoorthy, L., Palanikumar, K.: Fuzzy modeling and analysis of machining parameters in machining titanium alloy. Mater. Manuf. Processes **23**, 439–447 (2008)
10. Lo, S.-P.: An adaptive-network based fuzzy inference system for prediction of workpiece surface roughness in endmilling. J. Mater. Process. Technol. **142**, 665–675 (2003)
11. Jiao, Y., Lei, S., Pei, Z.J., Lee, E.S.: Fuzzy adaptive networks in machining process modeling: surface roughness prediction for turning operations. Int. J. Mach. Tools Manuf. **44**, 1643–1651 (2004)
12. Gupta, A., Singh, H., Aggarwal, A.: Taguchi-fuzzy multi output optimization (MOO) in high speed CNC turning ofAISI P-20 tool steel. Expert Syst. Appl. **38**, 6822–6828 (2011)
13. Mamdani, E.H., Assilia, S.: An experiment in linguistic synthesis with a fuzzy logic controller. Int. J. Man Mach. Stud. **7**, 1–13 (1975)
14. Dussud, M., Galichet, S., Foulloy, L., Simonnin, P.: Fuzzy logic control of continuous casting installations. In: Proceedings of the13th IFAC World Congr., San Fransisco, CA, vol. M, pp. 469–474 (1996)
15. Foulloy, L., Galichet, S.: Typology of fuzzy controllers. In: Nguyen, H.T., Sugeno, M., Tong, R., Yager, R. (eds.) Theoretical Aspects of Fuzzy Control, pp. 65–90. Wiley, New York (1995)
16. Zadeh, L.A.: Fuzzy sets. Inf. Control **8**, 338–353 (1965)

17. Saarlainen, E.J., Lautala, P., Inkinnen, M., Johansson, J.: Steel caster mold level control using fuzzy logic. In: 13th PTD Conference Proceedings 1995, pp. 397–402
18. Klir, G.J., Yuan, B.: Fuzzy Sets and Fuzzy Logic: Theory and Applications. Prentice-Hall, New Delhi (1997)
19. Bouhouch, S., Lahreche, M., Moussaoui, M., Bast, J.: Quality monitoring using principal component analysis and fuzzy logic. Appl. Continuous Casting Process, Am. J Appl. Sci. **4**(9), 637–644 (2007)
20. Bartolini, G., et al.: Development of performance adaptive fuzzy controllers with application to continuous casting plants. In: Sugeno, M. (ed.) In Industrial Applications of Fuzzy Control. Amsterdam, Elsevier, The Netherlands (1985)

Chapter 40
Prediction of Hot Spot and Hot Tear of the Al–Cu Cast Alloy by Casting Simulation Software

Kumara Swamy Pulisheru and Anil Kumar Birru

Abstract The process of casting in foundry is complex and involves large process to get the required part, and such process needs simulation before manufacturing. Casting simulation may save material and gives optimal process of casting in foundry. The defect minimization or prevention is challenging and plays significantly in enhancing the productivity of the foundry. The aim of the present study is to compute hot spots and hot tearing in Al–Cu cast alloy. The originating points of the feed paths represent the hot spots that lead to a shrinkage defect. Some studies have been carried out on hot tear. It was observed that the location of hot tear is at inter-junction of mold cavity to the gate. The developed model is used as bench mark to predict the location of hot spot and hot tear of Al–Cu cast alloy by green sand casting method with casting simulation software.

Keywords Hot spot · Hot tear · Sand casting · Al–Cu cast alloy · Simulation

40.1 Introduction

Poor quality and low yield are major issues for foundries which will increase the rejection rate. Additional melting of molten metal, which directly reflects the quality of the casting and casting yield, highly depend upon solidification of molten metal in the cavity. The direction of molten metal referred to as a feed path which is automatically generated using the vector element method (VEM). The volumetric contraction during solidification compensated from adjacent hotter regions of molten metal. Solidification of casting progressively takes place from the mold boundary which isolates last solidifying points in the mold cavity said to be a 'hot spot' which are the most probable locations for shrinkage related defects. Computer-aided design and simulation have been playing vital role from the last two decades. Casting simulation is to improve the quality and casting yield in

K. S. Pulisheru · A. K. Birru (✉)
National Institute of Technology Manipur, Imphal 795004, India
e-mail: anilbirru@gmail.com

R. G. Narayanan et al. (eds.), *Advances in Computational Methods in Manufacturing*, Lecture Notes on Multidisciplinary Industrial Engineering,
https://doi.org/10.1007/978-981-32-9072-3_40

shorter time in shop floor trial-and-error approach. The effect of feeder location, shape, size, and feed aids might give tangible information by casting simulation software. The minimization of defects may be performed by an appropriate feeding system to ensure directional solidification from thin to thick sections of the casted produced. Sutaria et al. [1] observed that computed feed path and hot spot by combining level set method were the last solidified region in casting by vector notation method. Computed feed paths during solidification represent the mass-less particles along the liquid–solid interface. Sutaria et al. [2] investigated with the application of gradient vector method (GVM) and found that directional solidification and hot spot in the casting observed by pouring three layouts of a multi-junction casting. Sutaria et al. [3] investigated the optimization of casting feeders by vector element method (VEM) using feed paths.VEM resulted in terms of hot spot locations correlated with FEM.

Choudhari et al. [4] it was observed that the optimizing feeder dimensions and increasing feeding efficiency of casting by minimizing of the casting defect. Casting simulation enables visualization of progressive freezing regions inside the casting to locate the hot spot. Choudhari et al. [5] studied that the defect minimization in casting to improve the casting yield was compared with its experimental validation. Similarly, applied simulation-based gradient vector method is used to identify the location of shrinkage porosity.

Jain et al. [6] comparative study was made on sand casting and investment casting to minimize the tooling cost for reduction of production cost. It was found that the tooling cost was more in case of sand casting due to additional allowances to get defect-free casting which gives more errors in dimensional allowances, whereas in case of investment casting get more surface finishing. Hodbe et al. [7] studied the manual design, fabrication of tools, and trial-and-error in process to minimize the defects, said process was repeated to get defect-free casting in foundry industries. The defect-free casting increases the casting yield. Simulation-based trials save the time, loss of material, and other industrial expenditure. Rabindra et al. [8] investigated the solidification behavior and detection of hotspots in aluminum alloy castings: Computer-aided analysis by AutoCAST and compared with experimental validation. Birru and Benny Karunakar [9] investigated minimization of hot tear by adding the grain refiners with different levels of composition Al–2.5, Ti–0.5C and Al–3.5, Ti–1.5C with the variation of pouring temperatures as 700, 750, and 780 °C. Gajbhiye et al. [10] investigated the reducing shrinkage porosity in a sand casting process by simulation in AutoCAST-X software with experimental validation by destructive testing. Hence, much work is not emphasized to compute hot spots and hot tearing in Al–Cu cast alloy, and especially in sand casting, authors made an attempt to predict the hot tearing and hot spots.

40.2 Methodology

40.2.1 Problem Definition

The objective of the study is to identify the defects in Al–Cu cast alloy casting such as hot spot and hot tear by AutoCAST simulation software. By optimizing the defects in casting enables to improve the casting yield and quality of the casting part.

40.2.2 Process Design

The entire study has been divided as section, viz. design of model, numerical simulation using AutoCAST software, and analyzing the results. The entire process of simulation is as shown in Fig. 40.1. Casting material is used as Al–Cu alloy and mold is taken as green sand as per aimed work for the sand casting foundry requirements. The CAD model is developed in Creo software. Design parameters are taken from the design calculation as depicted in Table 40.1. The material properties are mentioned in Table 40.2, restricted to simulation software which is considered as boundary conditions to casting simulation.

Fig. 40.1 Flow chart for the casting simulation in foundry

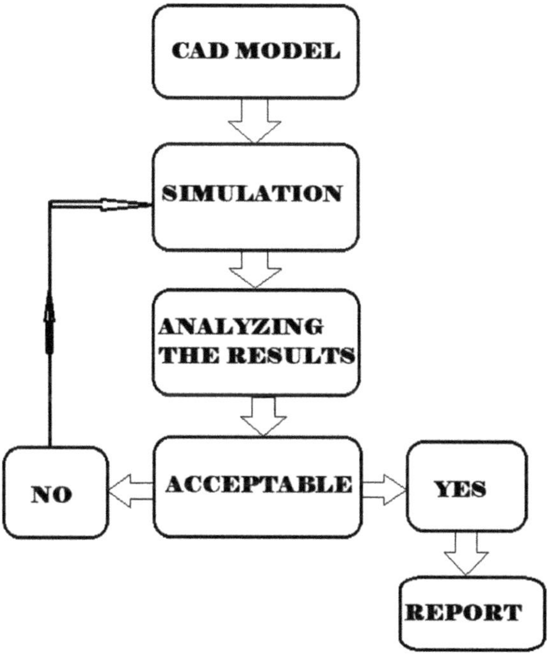

Table 40.1 Design parameters

S. N.	Design parameters	Value (mm)	Cross section
1	Pouring basin, height, length, and width	8, 20, 15	Rectangular
2	Diameter of sprue at top, bottom, and sprue length	10, 8, 23.5	Cylindrical
3	Sprue well diameter and height	10, 5	Cylindrical
4	Ingate dimensions, width, height, and length	8, 4, 52	Rectangular

Table 40.2 Material properties

S. N.	Description	Designated value	Units
1	Density	2800	Kg/m^3
2	Thermal conductivity	121	W/m-K
3	Specific heat	920	J/Kg-K
4	Solidus temperature	570	°C
5	Liquidus temperature	650	°C

The design of pattern helps to obtain the acceptable design mold which leads to get good quality casting in foundry industries. The design of part consists of various elements such as pouring basin, sprue, sprue well, and ingates, and they act as a channel to molten metal to flow into the mold cavity to get the desired shape of the mold cavity.

40.2.3 Numerical Simulation

Numerical simulation describes the applications of AutoCAST software for method design and optimization of casting process to enable to get defect-free casting. In the present investigation, simulation process of the selected CAD part was modeled in (.stl file), imported into AutoCAST for simulation. The mold box dimensions have been considered as 200 × 100 × 80 mm as shown in Fig. 40.2. The simulation was carried out by choosing the green sand casting process with aluminum-based alloy. The mold is subdivided into cubic elements with the element size of 1 mm for internal computation for uniform solidification and mold filling. The pouring temperature is considered 700 °C in casting simulation. It was observed from Fig. 40.3 that the Al–Cu alloys were much prone at TC-1 region and TC-2 region.

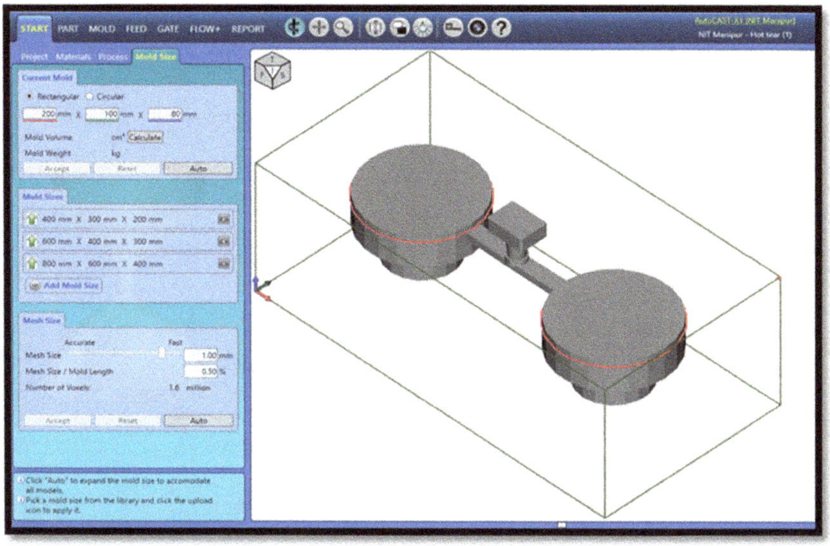

Fig. 40.2 Mold box dimensions for the present investigations

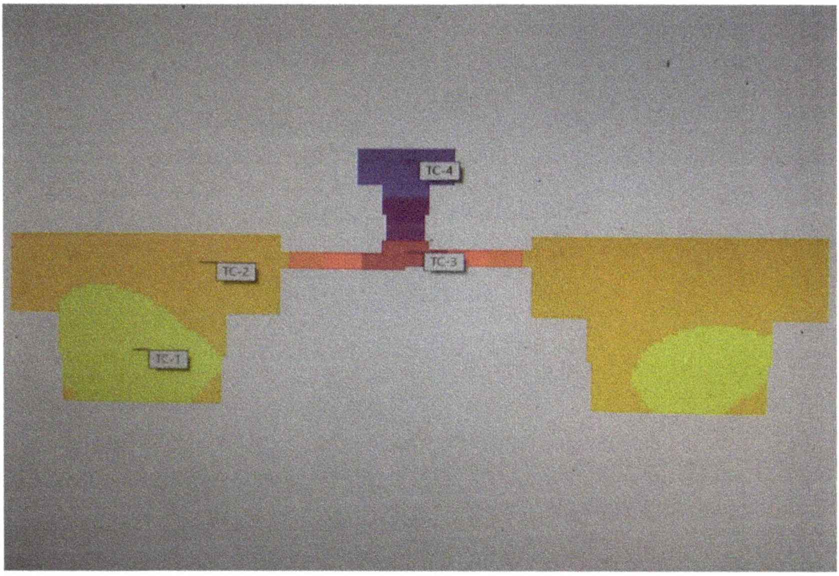

Fig. 40.3 Temperature at TC-1 to TC-4 regions in the casted part

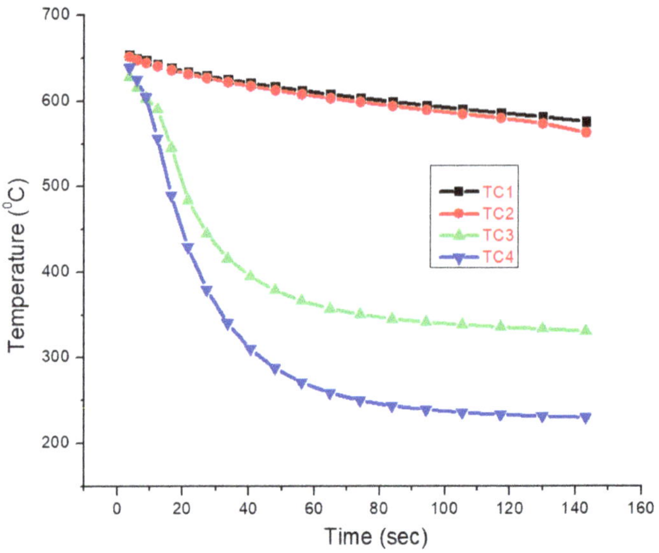

Fig. 40.4 Effect of pouring temperature on different zones in Al–Cu cast alloy

Hot tear was shown at the left end of the casted product, similarity index at the right side of the casted product was observed, TC-3 region the molten metal was distributed at both the sides uniformly, and TC-4 was observed as hard zone of the Al–Cu alloy casted product.

The solidification of molten metal has been compared with the time Vs temperature as shown in Fig. 40.4. An average of four K type thermo-couples at different regions in casting solidification process was observed. TC-1, TC-2 was hot sport and hot tear zones, respectively TC-3 is junction zone were the molten Al–Cu alloy flows, and TC-4 is the hard zone.

40.3 Results and Discussion

In Fig. 40.5, it was observed that the feed metal paths show the solidification direction of Al–Cu alloy casting in 2D and 3D sectioned casting with feed path vector notation as mentioned below

- Progressive Solidification-Cooling
- Directional Solidification-Feed metal paths.

The progressive solidification is represented by isothermal maps (at equal temperature). The directional solidification is represented by feed paths (with temperature gradients).

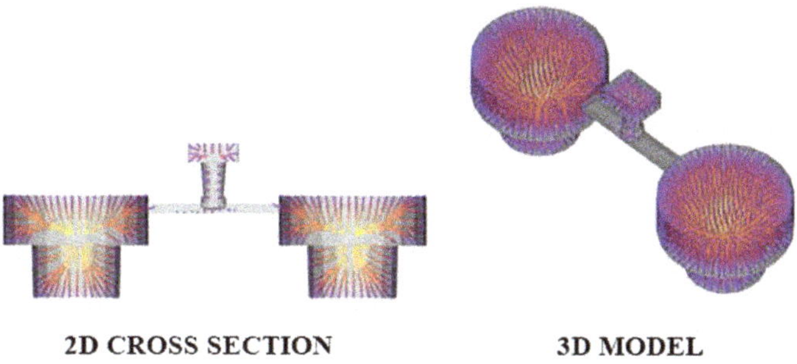

Fig. 40.5 Feed paths by vector notation method of 2D cross section and 3D model

The last solidified region in casting called the hot spot which leads to volumetric shrinkage of the molten metal in the process of solidification. In casting process to compensate the volumetric shrinkage of the molten metal in solidification process the feeder has facilitate to overcome the shrinkage porosity in the casting part. The casting simulation enables to identify the hot spot region to facilitating the feeder for getting the quality casting in foundry.

Figures 40.6 and 40.7 it shows the hot tear occurs where the weakest zone in cavity while in solidification process due to sudden cooling of the molten metal from high temperature to low temperature. Casting simulation enables to identify the weaker zone in mold cavity for the redesign of mold cavity to overcome the hot tears in casting hence, to get the good quality casting part in foundry.

The hot spot and hot tear zones were correlated with Figs. 40.6 and 40.7 for TC-1 to TC-4, the significance of pouring temperature with respect to the solidification time has effect on the Al–Cu cast alloy. In the process of solidification noticed that the TC-3 conditions as the temperature decrease with respect to the time, the flow of material decreased as the time increased and it was also noted that drastically drop down in the temperature was observed at 20–40 s. Similarly, hard zone at TC-4 condition, the temperature drastically drops down at 15–35 s. Hence, it leads to the hot tearing defect in casting solidification. Similarly, for TC1-TC2, the hot spot was observed at the verge of solidification, i.e., TC-1, respectively,

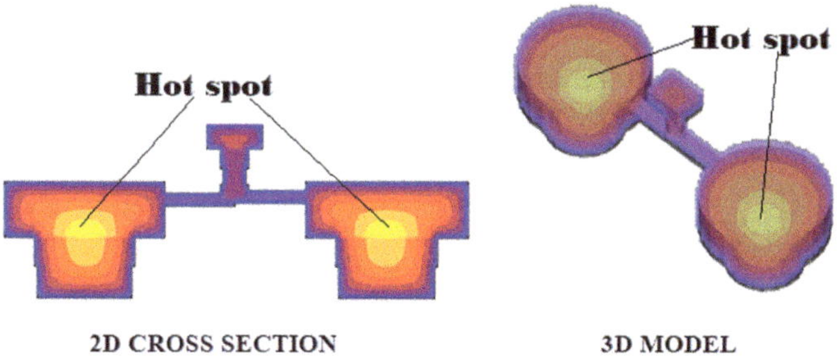

Fig. 40.6 Hot spot region of Al–Cu cast alloy: 2D-cross-sectional view and 3D-isometric view

Fig. 40.7 Hot tear region of Al–Cu cast alloy: 2D-top view and 3D-model

before the formation of the hot spot, the ending of the hot tear starts in Al–Cu cast alloy, and it was further scope to the authors that the grain refinement may also consider for aforesaid alloy.

40.4 Conclusion

Casting simulation was implemented for Al–Cu alloy with green sand mold which enables to predict the problems in casting process in foundry industries. With a motive to minimize the defects in casting especially for green sand molds to acquire the quality of the castings

1. The present study enabled to identify the hot spot and hot tears in casting part by casting simulation.
2. It facilitated optimization of design and casting yield for the preceding process to overcome the hot spot and hot tears by providing feed aids and redesign of the mold cavity.
3. It was concluded that casting simulation enables for trial-and-error to get the accuracy in casting by optimizing the defects in casting, which saves the time and loss of material in foundry which ultimately impacts to enhance the productivity of the industry.
4. It was observed that the Al–Cu alloy in the green sand mold system, the hot tearing and hot spot occurred in the range of 600–700 °C as the solidification time increases at 120–140 s with a minor variation in the hot tearing and hot spot hence, it was concluded that the hot spot increases, respectively, as hot tearing increases vice versa.

Acknowledgements Authors are much thankful for the TEQUIP-III, NPIU-New Delhi, for financial support for purchase of AutoCAST software to carry out the simulation of designed casting.

References

1. Sutaria, M., Gada, Vinesh H., Sharma, A., Ravi, B.: Computation of feed-paths for casting solidification using level-set-method. J. Mater. Process. Technol. **212**, 1236–1249 (2012)
2. Sutaria, M., Ravi, B.: Computation of casting solidification feed-paths using gradient vector method with various boundary conditions. Int. J. Adv. Manuf. Technol. **75**, 209–223 (2014)
3. Sutaria, M., Joshi, D., Jagdishwar, M., Ravi, B.: Automatic optimization of casting feeders using feed-paths generated by VEM. In: Proceedings of the ASME 2011 International Mechanical Engineering Congress & Exposition IMECE-2011, Nov 11–17, 2011, Denver, Colorado, USA
4. Choudhari, C.M., Narkhede, B.E., Mahajan, S.K.: Casting design and simulation of cover plate using AutoCAST software for defect minimization with experimental validation. In: 3rd International Conference on Materials Processing and Characterisation (ICMPC 2014), Procedia Materials Science 6, pp. 786–797 (2014)
5. Choudhari, C.M., Narkhede, B.E., Mahajan, S.K.: Methoding and simulation of LM 6 sand casting for defect minimization with its experimental validation. In: 12th Global Congress on Manufacturing and Management, GCMM 2014, Procedia Engineering 97, pp. 1145–1154 (2014)
6. Jain, P., Kuthe, A.M.: Feasibility study of manufacturing using rapid prototyping: FDM approach. The Manufacturing Engineering Society International Conference, MESIC 2013, Procedia Engineering 63, pp. 4–11 (2013)
7. Hodbe, G.A., Shinde, B.R.: Design and Simulation of Lm 25 sand casting for defect minimization. In: ICMPC 2017, Materials Today: Proceedings 5, pp. 4489–4497 (2018)
8. Rabindra, B., Kayal, S., Sutradhar, G.: Solidification behavior and detection of hot spots in aluminium alloy castings: computer aided analysis and experimental validation. Int. J. Appl. Eng. Res., Dindigul, 1(4), (2011)
9. Birru, A.K., Benny Karunakar, D.: Effects of grain refinement and residual elements on hot tearing of A713 aluminium cast alloy. Trans. Non ferrous Met. Soc. China **26**, 1783–1790 (2016)
10. Gajbhiye, A.K., Choudhari, C.M., Raut, D.N., Narkhede, B.E., Bhandarkar, B.M.: Minimization of shrinkage porosity in A sand casting process by simulation in AUTOCAST software with experimental validation by destructive testing. Int. J. Modern Eng. Res. (IJMER), **4**(5), 18 (2014)

Chapter 41
Influence of Pouring Temperature on Fluidity of Al–Cu Cast Alloy: Analyses Using Casting Simulation Software

Kumara Swamy Pulisheru and Anil Kumar Birru

Abstract The fluidity of molten metal plays a vital role in the casting process when we are confined in manufacturing thinner sections. The aim of the present study is to determine the effect of pouring temperature on the fluidity of Al–Cu cast alloy. The simulation was performed by AutoCAST-X software with altered pouring temperatures of 700, 750, and 800 °C. It was observed that the fluidity was enhanced when the pouring temperature at 800 °C for the aforesaid cast alloy. Similarly, it was found that the volume of fill increased with the increasing pouring temperature and also strong agreement was observed with the solidification time and fill time.

Keywords Sand casting · Fluidity · Al–Cu cast alloy · Pouring temperature · Simulation

41.1 Introduction

In the manufacturing process, casting is one of the most economical processes among all the available process even in the twenty-first century with respect to the volume of the production. Aluminum alloys are one of the most widely used materials from the past few decades and versatile raw material found in the common foundry metal. The ratio of the cast to wrought aluminum alloy products is increasing in a number of castings being used in automotive, aerospace industries, and domestic goods. The advantage of aluminum alloys mainly from their low density, moderate castability, and excellent mechanical properties.

K. S. Pulisheru · A. K. Birru (✉)
National Institute of Technology Manipur, Imphal 795004, India
e-mail: anilbirru@gmail.com

R. G. Narayanan et al. (eds.), *Advances in Computational Methods in Manufacturing*, Lecture Notes on Multidisciplinary Industrial Engineering,
https://doi.org/10.1007/978-981-32-9072-3_41

In manufacturing industry, computer-aided design and simulation play a vital role in forecasting the problems in the manufacturing process for improving of production and quality of the product. The application of casting simulation is increasing in the foundry as it essentially replaces or minimizes the shop-floor trials to achieve the desired internal quality at the highest possible time. Casting solidification simulation is now widely used to visualize the effect of feeder location, shape, size, and feedaids (such as feeder sleeves and mold chills) on the solidification sequence, temperatures, and cooling rates. Based on the earlier researchers, the feeding system design is iteratively modified and simulated until an optimal design the smallest number and volume of feeders to prevent casting shrinkage defects is achieved. In a casting process, the term "fluidity" indicates the distance of the molten metal can flow in a mold at a constant cross-sectional area before it solidifies [1]. The fluidity in casting enhances the casting yield. Over the decades, a considerable amount of experimental and numerical work has been carried out to understand casting solidification and feeding mechanism. Mohan Krishna et al. [2] investigated understanding of the fluidity behavior of narrow and wide freezing range aluminum alloys, respectively (47,100, and 51,300), then extended to cast superalloy CMSX4 and ductile cast iron (GJS-400-18) product. It was found that spiral and complex shaped sand and mullite molds with molten alloys having varied superheats of 25, 50, and 100 °C over the liquid temperatures were simulated using finite element method-based software. Birru et al. [1] investigated the fluidity of Al–Zn alloys, such as the standard A713 alloy with and without scrap addition. Fluidity measurements were performed with a double spiral fluidity test consisting of gravity casting of double spirals in green sand molds with good reproducibility. The influence of recycled alloy on fluidity has been compared with that of the virgin alloy, and the results showed that the fluidity decreased with the increase in recycled alloy at minimum pouring temperatures. Joshi et al. [3] investigated the analytical hierarchy process to the determination of relative weights of various criteria for castability evaluation. The criteria are categorized in a three-level hierarchy, with the middle level comprising mold, feeding, and gating groups. The methodology is illustrated by modifying the product and tooling design of a gray iron bracket casting and estimating the improvement in manufacturability. Manoj et al. [4] studied on casting simulation of a virtual process which ensures casting without defects. It involves computer-aided modeling, pattern design, simulation, and optimization. Fluidity of molten metal plays an important role in producing sound complex castings. The objective of the analysis study is to identify the factors affecting casting fluidity by changing sprue height, pouring temperature, and flow rate on the fluidity of aluminum alloy (AlSi-13).

Ravi et al. [5] studied the desired quality at the least cost without shop-floor trials, with design, process simulation, and quality evaluation and cost estimation carried out. It was found that simulation results are evaluated using three quality indices, moldability, fillability, and feedability. Sutaria et al. [6] investigated the volumetric contraction accompanying solidification of molten metal manifests in defects like shrinkage cavity, porosity, centerline shrinkage, corner shrinkage, and sink. These defects may be minimized by designing an appropriate feeding system to ensure directional solidification from thin to thick sections in the casting, leading to feeders. Ravi et al. [7] identified the location of a feeder, calculating to minimum size creating its solid model, attaching the feeder model to the casting, carrying out solidification simulation, predicting quality and estimating the yield. Naveen et al. [8] studied the castings that serve these specific applications have to achieve the quality requirements of superior mechanical properties and zero porosity. To achieve these objectives within a limited time frame in a product development process, CAD technologies combined with process simulation tools are increased to optimize the form filling and solidification of the cast parts. Hence, the newly developed simulation of flywheel component that was prototyped via sand casting route. Results of casting trials showed a high level of confidence in the simulation tools.

41.2 Methodology

41.2.1 Problem Definition

The objective of the present study on the fluidity of Al–Cu cast alloy with altered pouring temperature by choosing the method of green sand mould process. In which effect of pouring temperature on the casted part is being analyzed to optimize the defects in the casting process by choosing an appropriate method design. The optimization of method design is carried out by casting simulation.

41.2.2 Process Design

The entire study has been divided into sections, viz. design of the model, numerical simulation using AutoCAST software, and lastly analyzing the results. The entire process is as shown in flowchart as in Fig. 41.1. The entire process shows the method design in the foundry.

To study the fluidity of Al–Cu cast alloy with altered pouring temperatures such as 700, 750, and 800 °C considered the model as a rectangular plate with the size of 220 × 140 × 5 mm by Cre-O software as shown in Fig. 41.2 and converted it into **.stl** file to import it into an AutoCAST software for simulation.

Casting material is used as A206 [9], the chemical composition is as shown in Table 41.1, and mold material is taken as green sand with the dimensions of 300 × 100 × 160 as shown in Fig. 41.3, as per the foundry requirement. The design of the gating system consists of various elements like pouring basin, sprue,

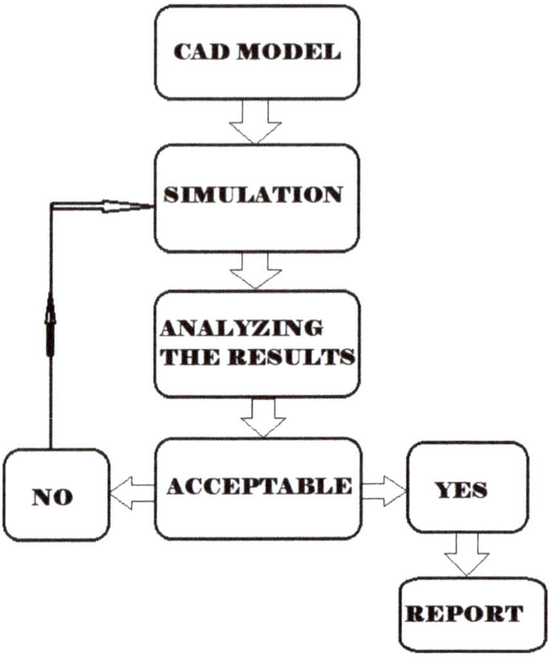

Fig. 41.1 Flowchart for the casting simulation

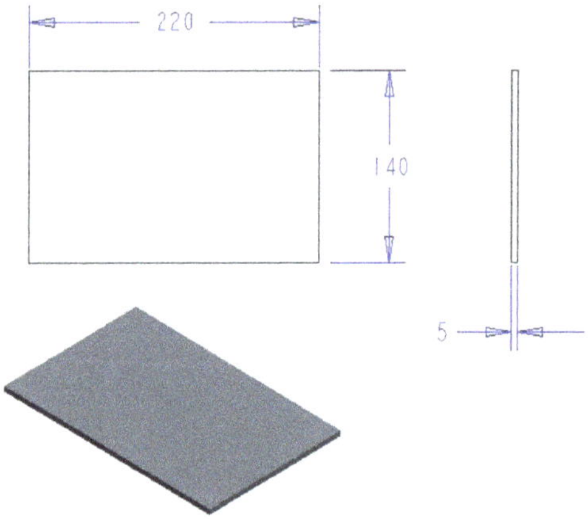

Fig. 41.2 Part model by Cre-O for simulation

Table 41.1 Chemical composition of A206 alloy [9]

Element	Cu	Mn	Mg	Si	Ti	Sr	Al
% of composition	5.0	0.5	0.35	0.05	0.25	0.20	Balance

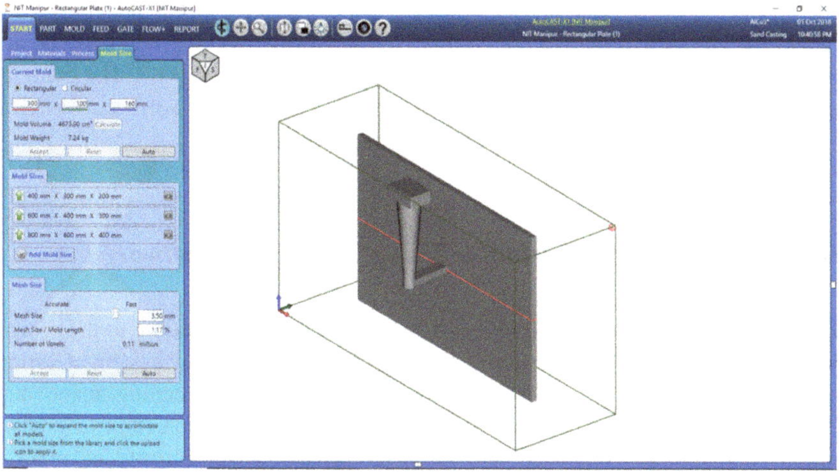

Fig. 41.3 Casting part with mold box

Table 41.2 Design parameters

Sr. no.	Design parameters	Designed value (mm)	Section shape
1	Pouring basin height, length, width	10, 30, 20	Rectangle
2	Diameter of sprue at top, bottom, and sprue length	20, 10	Cylindrical
3	Sprue well: diameter and height	10, 10	Cylindrical
4	Ingrate dimensions: width, height, and length	10, 5, 35	Rectangle

Table 41.3 Material properties

S. no.	Description	Designated value	Units
1	Density	2800	Kg/m^3
2	Thermal conductivity	121	W/m-K
3	Specific heat	920	J/Kg-K
4	Solidus temperature	570	°C
5	Liquidus temperature	650	°C

and sprue well, runner, and ingates. They act as passageways for the flow of molten metal from the ladle to various portions of the mold cavity. Design dimensions are obtained for gating system, and the feeder is mentioned in Table 41.2. The material properties, as mentioned in Table 41.3, restricted to simulation software which is considered as boundary conditions to casting simulation.

41.2.3 Numerical Simulation Using AutoCAST Software

Simulation has been performed in AutoCAST environment according to the design dimensions obtained for a pattern with allowances and gating system. Casting simulation enables to check the part-process compatibility, parting plane and moldability index, and feed metal paths–directional solidification (thin to thicker regions) and to predict the location of shrinkage defects such as porosity and cracks.

Simulation-based trials do not involve wastage of material, energy, and labor and do not hold up regular production. Computer simulation provides a clear understanding of the casting phenomena to identify the location and extent of internal defects, ensuring defect-free castings. Thus, numerical simulation of casting can be considered as an important method to make casting technique change from experience test to science guidance. It requires a standard **.stl** file of the methods layout. However, the application of proper boundary conditions leads to accurate results of the simulation.

Part and Mold box

The prerequisite of this software is to create the part model in CAD software and save it as a standard **.stl** file for importing in AutoCAST. The mold box dimensions have been taken as 300 × 100 × 160 mm as shown in Fig. 41.3. The entire mold containing the casting is automatically subdivided into cubic elements for internal computations such as thickness, solidification, and mold filling. The element size mesh was considered as 3.5 mm.

Simulation with different pouring temperatures

The simulation at different pouring temperatures of 700, 750, and 800 °C was performed. It was observed that the fill time and solidification time increased with respect to the increasing temperature, and the volume of fill is high at higher pouring temperature. Casting simulation between Figs. 41.4 and 41.12 indicates the fill time, solidification time, and percentage of the volume of fill with respect to the altered pouring temperatures, i.e., 700, 750, and 800 °C. In this process, molten metal flow may be visualized at different cross sections of the mold cavity. It enables to visualize the flow of molten metal regions and solidification regions in the casting process (Figs. 41.4, 41.5, 41.6, 41.7, 41.8, 41.9, 41.10, 41.11, and 41.12).

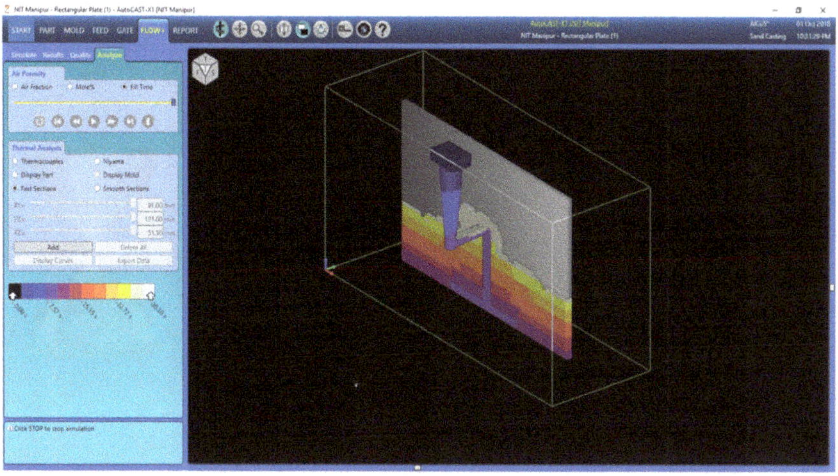

Fig. 41.4 Fill time with the pouring temperature of 700 °C

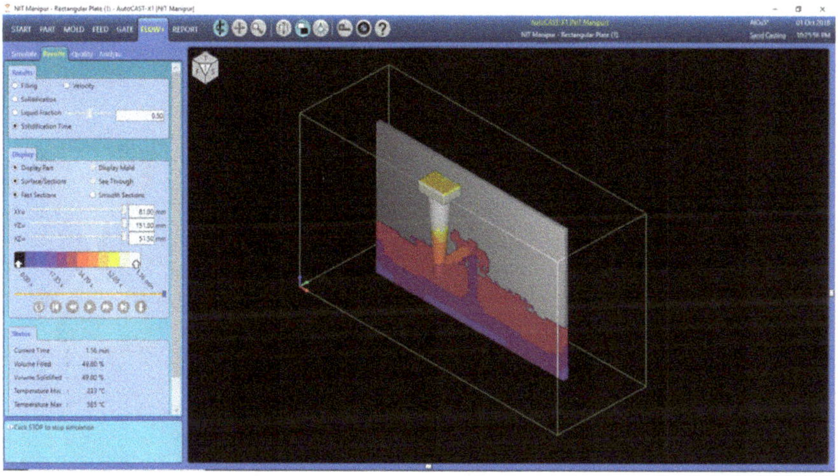

Fig. 41.5 Solidification time with the pouring temperature of 700 °C

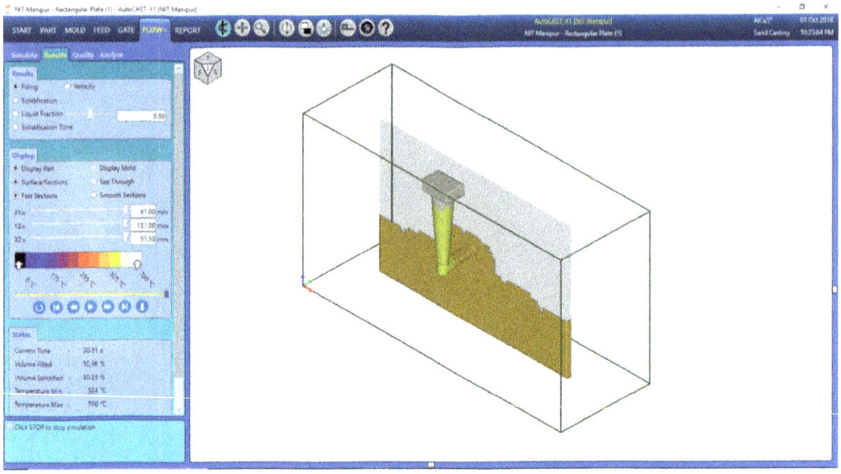

Fig. 41.6 Volume of fill with the pouring temperature of 700 °C

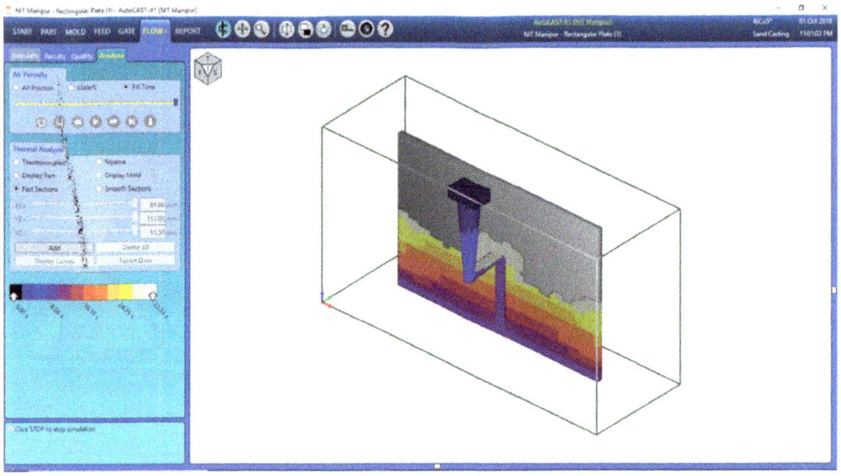

Fig. 41.7 Fill time with the pouring temperature of 750 °C

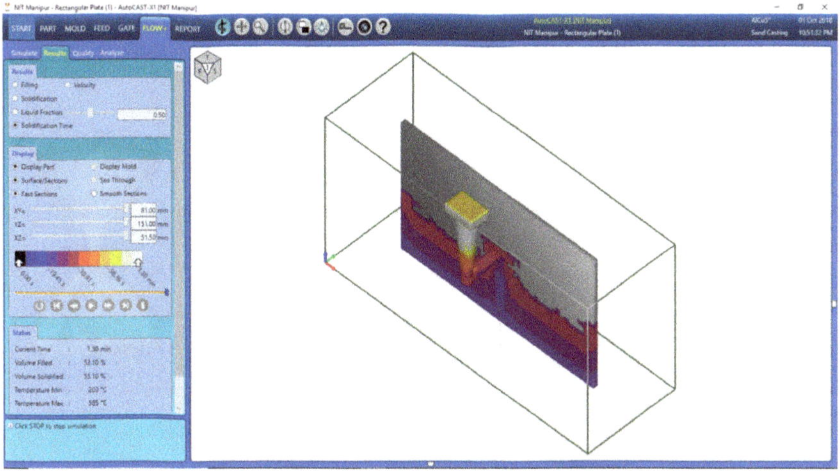

Fig. 41.8 Solidification time with the pouring temperature of 750 °C

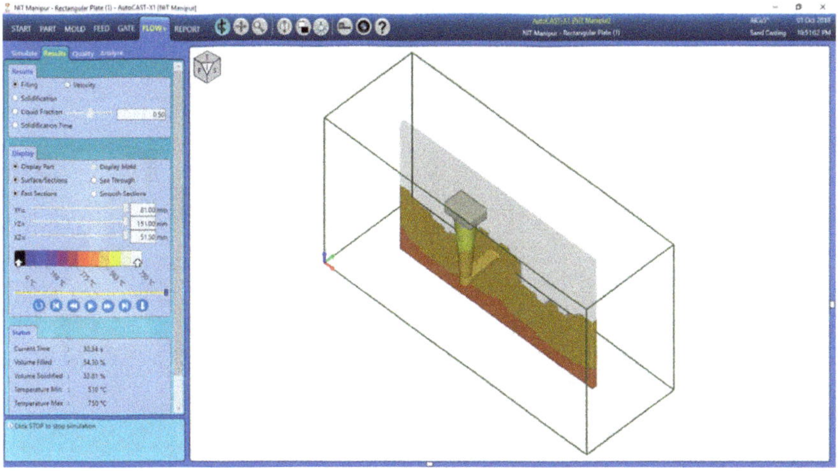

Fig. 41.9 Volume of fill with the pouring temperature of 750 °C

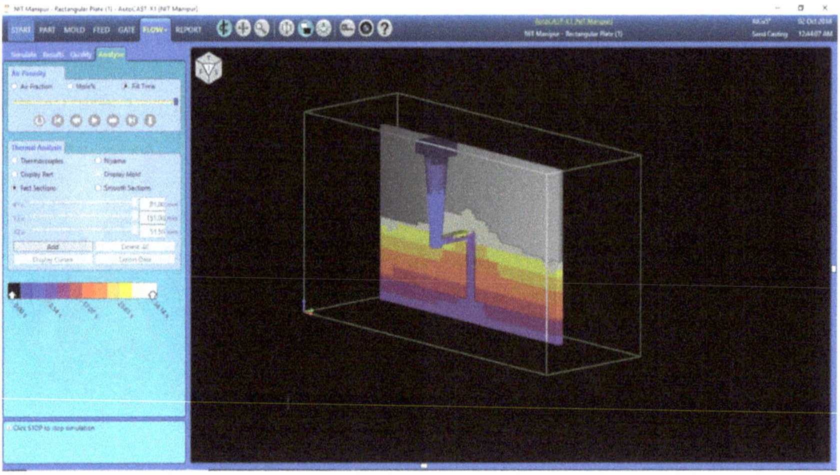

Fig. 41.10 Fill time with the pouring temperature of 800 °C

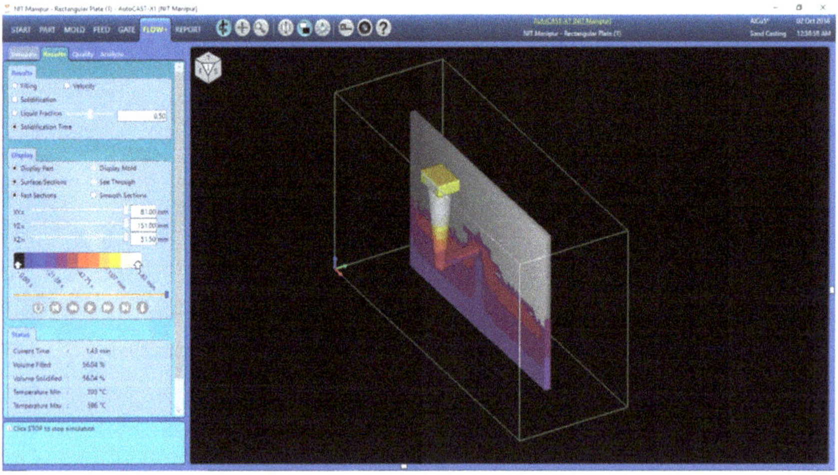

Fig. 41.11 Solidification time with the pouring temperature of 800 °C

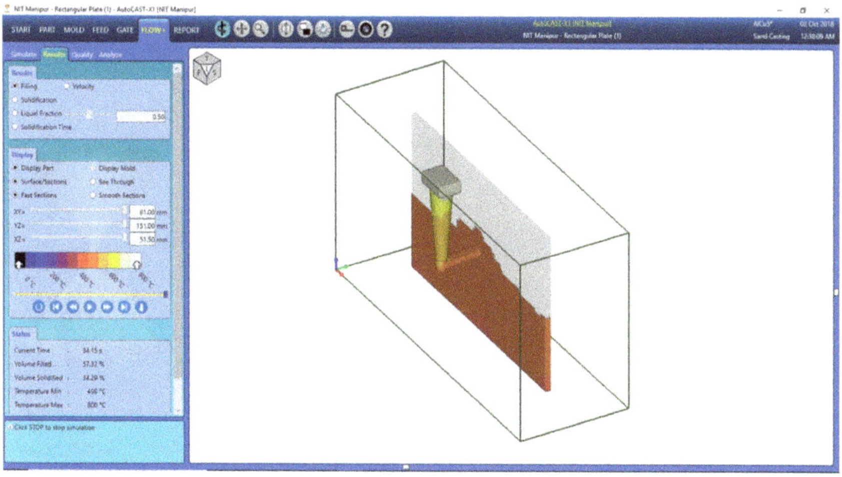

Fig. 41.12 Volume of fill with the pouring temperature of 800 °C

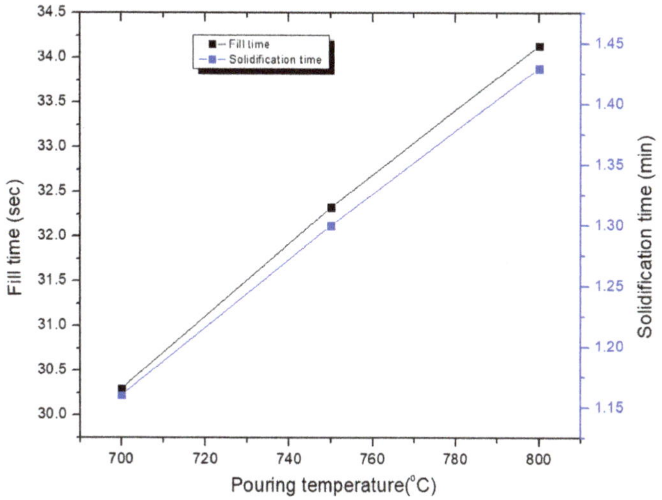

Fig. 41.13 Fill time and solidification time with altered pouring temperatures

41.3 Results and Discussion

Casting simulation enhanced the fluidity of the molten metal with altered pouring temperatures. In Fig. 41.13, it was observed that with the increasing pouring temperature, the fill time and solidification time were increased. In Fig. 41.14, it was observed that when the fill time increased, then the percentage of the volume of fill in the cavity increased. In Fig. 41.15, it was observed that the percentage of the

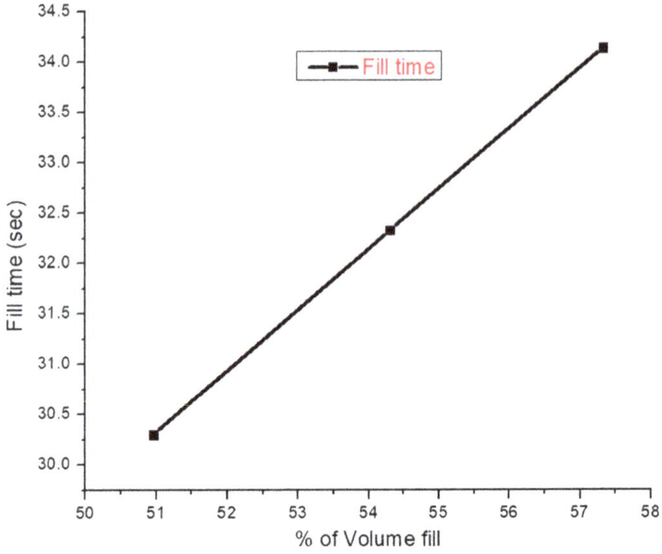

Fig. 41.14 Percentage of volume of fill with respect to fill time

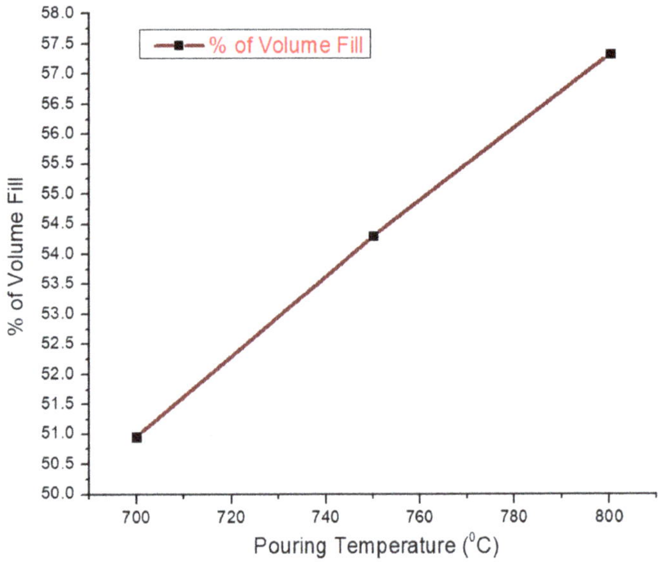

Fig. 41.15 Percentage of volume fill with the altered pouring temperatures

volume of fill increased with the increased pouring temperature. Hence, it was observed that the fluidity increased with the increasing temperature and fill time according as shown in Figs. 41.13, 41.14, and 41.15. Hence, it was observed that fluidity increased with increased pouring temperature.

41.4 Conclusion

In this present investigation, it was observed that solidification simulation enables visualization of freezing inside a casting and identification of the last freezing regions.

1. The effect of pouring temperature enabled the filling time and solidification time; it was observed at with the increasing pouring temperatures, i.e., 700 to 800 °C, the fill time and solidification time were increased.
2. It was observed that the volume of fill increased with increasing temperature and with the increased pouring temperature, the fill time and percentage of the volume of fill increased.
3. In casting simulation, it was observed that the effect of pouring temperature on the fluidity of the Al–Cu cast alloy was increased with the increasing temperature. It was found that the volume of fill increased with the increasing pouring temperature and also strong agreement was observed with the solidification time and fill time.
4. It was concluded that fluidity of Al–Cu cast alloy might be increased with increasing temperature. Hence, it was recommended that the industries have might implement the simulation before manufacturing to economize the product to reach all the levels of the customers.

Acknowledgements Authors are much thankful for the TEQIP-III, NPIU-New Delhi, for financial support for the purchase of AutoCAST-X software to carry out the simulation of designed casting.

References

1. Birru, A.K., Karunakar, D.B.: Fluidity of A713 Cast Alloy with and without scrap addition using double spiral fluidity test: a comparison. Int. J. Mech. Aerosp. Ind. Mechatronic Manufacturing Eng. **6**(1) (2012)
2. Mohan Krishna, D., Reddy, G.S.: Simulation of fluidity in aluminum alloys, super alloy cmsx4, and ductile cast iron. In: 5th International & 26th All India Manufacturing Technology, Design and Research Conference (AIMTDR 2014) December 12th–14th, IIT Guwahati, Assam, India (2014)
3. Joshi, D., Ravi, B.: Early castability evaluation using analytical hierarchy process Int. J. Adv. Manuf. Technol. **50**, 21–36 (2010)
4. Manoj, K., Vaibhav, J., Jitendra, P., Qureshi, F.: Filling analysis of spiral fluidity standard using ProCast simulation. Int. J. Eng. Trends Technol. (IJETT) **46**(8) (2017)
5. Ravi, B., Joshi, D., Kamalesh Singh, D.: Part, tooling and method optimisation driven by castability analysis and cost model. In: 68th WFC—World Foundry Congress 7th–10th February, pp. 261–266 (2008)
6. Sutaria, M., Joshi, D., Ravi, B.: Feeding system design and evaluation using temperature gradient (Feed Path) Maps. American Foundry Society (2014)

7. Ravi, B., Joshi, D.: Feed ability analysis and optimisation driven by casting simulation. Technical Paper submitted to the Indian Foundry Journal (2017)
8. Naveen, H., Sunil, M.: Casting simulation for sand casting of flywheel. IOSR J. Mech. Civil Eng. (IOSR-JMCE), **11**, 37–41 (2014)
9. Birru, A.K., Karunakar, D.B.: A study on hot tearing susceptibility of Al–Cu, Al–Mg, and Al–Zn alloys. Trans. Indian Inst. Met. **65**(1), 97–105 (2012, February)

Part V
Modelling and Analysis

Chapter 42
Analysis and Optimization of Tool Wear Rate in Magnetic Field-Assisted Powder-Mixed Electrical Discharge Machining of Al6061 Alloy Using TLBO

Arun Kumar Rouniyar and Pragya Shandilya

Abstract This paper presents the optimization and analysis of machining parameters on machining of Al6061 alloy using magnetic field-assisted powder-mixed electrical discharge machining (MFAPM-EDM) process. For performance analysis, peak current (I_P), spark on time (S_{ON}), spark off time (S_{OFF}), powder concentration (PC), and magnetic field (MF) are considered as machining parameters and tool wear rate (TWR) as performance measures. The experimental design based on Box Behnken Design is used for conducting the experiments and single objective optimization is performed using teaching–learning-based optimization (TLBO) algorithm. Peak current is observed as the most significant parameters followed by spark on time, magnetic field, and powder concentration. Model to predict TWR is developed in terms of machining parameters. The optimum set of machining parameters for minimum TWR using TLBO algorithm is obtained at I_P-1 A, S_{ON}-30 μs, S_{OFF}-51 μs, PC-10 g/l, and MF-0.45 T.

Keywords MFAPM-EDM · Al6061 alloy · Tool wear rate · TLBO

42.1 Introduction

Aluminum 6061 alloy is light and medium strength alloy with magnesium and silicon as principal element other than aluminum. Al6061 alloy is used in the field of aerospace, automobile, marine industries, and architectural and structural applications. This is mainly due to specific properties such as lightweight, hard, strong and toughness, high corrosion resistance, better machinability and

A. K. Rouniyar (✉) · P. Shandilya
Department of Mechanical Engineering, Motilal Nehru National Institute of Technology, Allahabad, Uttar Pradesh 211004, India
e-mail: rme1602@mnnit.ac.in

P. Shandilya
e-mail: pragya20@mnnit.ac.in

R. G. Narayanan et al. (eds.), *Advances in Computational Methods in Manufacturing*, Lecture Notes on Multidisciplinary Industrial Engineering,
https://doi.org/10.1007/978-981-32-9072-3_42

formability [1, 2]. Conventional machining of Al6061 possesses severe difficulties such as built of edge formation, adhesion between tool and workpiece, poor surface quality, dimensional inaccuracy and burr formation [3, 4]. Thus, to mitigate the above difficulties, the researchers have opted for the non-conventional machining process.

Among non-convectional machining process, electrical discharge machining (EDM) was found suitable for machining intricate profiles on difficult to machine materials irrespective of material hardness with high degree of automation [2, 5]. Material removal mechanism in EDM mainly occurs by melting and evaporation. EDM process was mainly used for tool and die making but at the current scenario, applications are extended to aerospace, military, automobile, and surgical industries. However, the machining with EDM at large scale has certain limitations such as low machining efficiency, poor surface finish, prone to heat and residual stress generated on workpiece material during machining [6–8]. Thus to overcome such problem, hybrid machining process known as magnetic field-assisted powder-mixed EDM (MFAPM-EDM) was developed which can machine complex profiles of conductive materials with higher efficiency and better surface quality. In this process, removal of material from the surface of the workpiece is assisted by powder particle mixed in a dielectric fluid in the presence of magnetic field. The material removal occurs by multiple discharges due to bridging of powder particle under the sparking area instead of single spark combined with Lorentz force developed due to the magnetic field-coupled electric field which assists in removing material. The Lorentz force changes the path of electrolytic ions by reducing the mean free path shifting of debris from the workpiece and tool in the discharge zone as well as accounts for increasing the electron density. Thus, this Lorentz force has a significant impact on the plasma channel which effectively expelling the debris particles leading to increase erosion volume and maintaining uniform discharge waves by gap causing improvement in the MRR, TWR, and better surface integrity [9, 10]. For example, Bhatt et al. [9] machined AISI D3, H13, and AISI D3 steel with tungsten and titanium powder mixed in EDM oil under the influence of magnetic field to examine the effect of process parameters on MRR, overcut, surface roughness, TWR, and microhardness. Krishna Kant [11] used *U*-shaped copper tool to machine EN-19 steel with graphite powder in the dielectric fluid in assisted with a magnetic field to study the effect of process parameters on overcut, MRR and surface roughness. Bains et al. [10] carried out machining of Al–Si MMCs with silicon carbide powder-mixed dielectric fluid under the influence of fixed magnetic field in EDM. Under the influence of the magnetic field at high spark energy, increase in material erosion rate and decrease in recast layer thickness, surface roughness, and microhardness were observed. Hienz et al. [12] performed machining on Ti-6Al-4V alloy under the influence of magnetic field to study the influence on erosion efficiency, plasma temperature, and debris field characterization.

After epitomized scrutiny of literature review, most of the literatures are available in magnetic assisted EDM or powder-mixed EDM process. Few literatures available on machining the difficult to cut conductive materials using

MFAPM-EDM process were improvement in MRR, TWR, and surface integrity was observed. In this present experimental work, MFAPM-EDM process was used to machine Al6061 alloy with aluminum powder mixed in EDM oil under the influence of magnetic field. The effect of machining parameters, namely peak current (I_P), spark on time (S_{ON}), spark off time (S_{OFF}), powder concentration (PC), and magnetic field (MF) on tool wear rate (TWR) were studied. Response surface model to predict TWR was developed, and single objective optimization was performed using teaching–learning-based optimization (TLBO) to determine the optimum machining parameters.

42.2 Experimental Setup and Methods

The experiments were performed on modified conventional die-sinking machine (EMS 5030, Elektra) with fabricated attachment equipped to it as shown in Fig. 42.1. The machine is supplied with ELECTRONIA PRS-20 which can operate peak current up to 35 A. Proper stirring system is designed to solve the problem of powder settling, and a magnetic system is attached for the quick removal of debris for the machining area. Al6061 alloy of dimension 40 mm × 40 mm × 10 mm was machined using 12-mm-diameter electrolytic copper as a tool, aluminum powder (mesh size-200), grade 5 EDM oil as dielectric fluid, and neodymium ring magnet. The machining parameters considered for the experimental work were peak current, spark on time, spark off time, powder concentration, and magnetic field. Box Behnken design (BBD)-based response surface methodology (RSM) was used for designing the experiments [13]. The level and range of machining parameters were decided based on literature review, one parameter at a time and specifications of machine setup. Table 42.1 shows the range and level of experimental machining conditions. For TWR, measurement of the weight of tool before and after

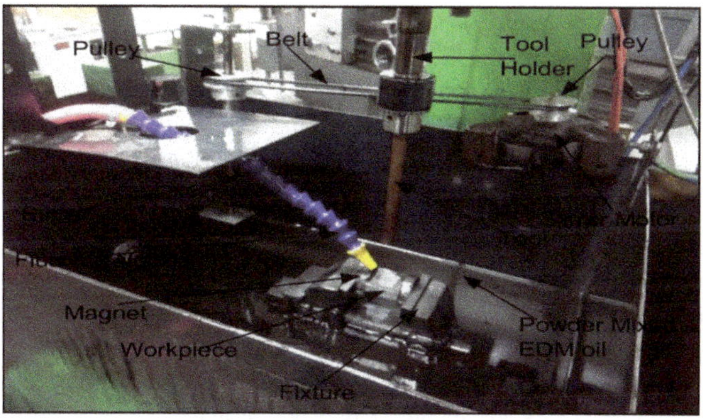

Fig. 42.1 Pictorial view of experimental setup for MFAPM-EDM

Table 42.1 Range and levels of experimental machining conditions

Sl. no.	Machining Parameters	Symbol	Levels		
			1	2	3
1	Peak current (A)	I_P	1	4	7
2	Spark on time (μs)	S_{ON}	30	50	70
3	Spark off time (μs)	S_{OFF}	35	43	51
4	Powder concentration (g/l)	PC	6	8	10
5	Magnetic field (T)	MF	0.15	0.30	0.45

machining was measured using digital precision pan balance (Model- SI-215D, Denver Instruments) with a maximum capacity of 210 grams. TWR was calculated using Eq. (42.1) [14]. The magnetic field was measured using digital gauss meter.

$$\text{TWR}\,(\text{g/min}) = \frac{T_{\text{bm}} - T_{\text{am}}}{t} \tag{42.1}$$

where

T_{bm} Tool weight before machining (g)
T_{am} Tool weight after machining (g)
t Machining time = 10 min.

42.3 Optimization Using TLBO

For this experimental work, teaching–learning-based optimization (TLBO) was employed. TLBO is a single as well as multi-objective optimizing algorithm to obtain solutions for discrete and continuous data problem. TLBO was coined by R. V. Rao for higher consistency which defines the teaching–learning capability of teachers and learners in the given classroom. Two basic modes, i.e., teacher phase and learner phase, are present in this TLBO algorithm. During teacher phase, the teacher tries to give knowledge to the learners and tries to improve the mean results of the classroom. While during learner phase, they try to interact among themselves to increase the knowledge [15]. The following steps are involved to find the optimum solution using TLBO algorithm.

(a) Design variables and its range, objective function, population size, and the number of iterations are defined.
(b) A table containing design variables value, objective function with population size are prepared and mean of all design variables are calculated.
(c) Based on objective function value, the teacher values are calculated.
(d) Other variable values are updated based on the teacher values.

(e) Comparison between new values and corresponding old values of an objective function are performed and the table is updated.
(f) Random values are selected and compared among the learners to improve the values.
(g) Comparison between new values and previous old values of the objective function are performed and the table is updated.
(h) Iterations are performed up to termination criteria.

42.4 Results and Discussion

The BBD design along with experimental results of TWR is shown in Table 42.2. Forty six different experimental combinations run are performed on machining of an Al6061 alloy by the MFAPM-EDM process.

Table 42.2 BBD design with experimental results

STD	I_P	S_{ON}	S_{OFF}	PC	MF	TWR (g/min)
1	1	30	43	8	0.3	0.0013897
2	7	30	43	8	0.3	0.0019722
3	1	70	43	8	0.3	0.0017885
4	7	70	43	8	0.3	0.0023923
5	4	50	35	6	0.3	0.0023388
6	4	50	51	6	0.3	0.0022273
7	4	50	35	10	0.3	0.0021488
8	4	50	51	10	0.3	0.0019249
9	4	30	43	8	0.15	0.0020373
10	4	70	43	8	0.15	0.0022568
11	4	30	43	8	0.45	0.0014504
12	4	70	43	8	0.45	0.0021381
13	1	50	35	8	0.3	0.0017269
14	7	50	35	8	0.3	0.0022719
15	1	50	51	8	0.3	0.0015414
16	7	50	51	8	0.3	0.0021310
17	4	50	43	6	0.15	0.0024057
18	4	50	43	10	0.15	0.0022345
19	4	50	43	6	0.45	0.0022559
20	4	50	43	10	0.45	0.0017867
21	4	30	35	8	0.3	0.0017430
22	4	70	35	8	0.3	0.0022585

(continued)

Table 42.2 (continued)

STD	I_P	S_{ON}	S_{OFF}	PC	MF	TWR (g/min)
23	4	30	51	8	0.3	0.0016038
24	4	70	51	8	0.3	0.0020801
25	1	50	43	6	0.3	0.0018884
26	7	50	43	6	0.3	0.0025449
27	1	50	43	10	0.3	0.0015708
28	7	50	43	10	0.3	0.0021372
29	4	50	35	8	0.15	0.0023692
30	4	50	51	8	0.15	0.0021542
31	4	50	35	8	0.45	0.0019972
32	4	50	51	8	0.45	0.0018848
33	1	50	43	8	0.15	0.0017885
34	7	50	43	8	0.15	0.0023174
35	1	50	43	8	0.45	0.0016011
36	7	50	43	8	0.45	0.0020561
37	4	30	43	6	0.3	0.0018072
38	4	70	43	6	0.3	0.0022951
39	4	30	43	10	0.3	0.0015976
40	4	70	43	10	0.3	0.0020364
41	4	50	43	8	0.3	0.0020480
42	4	50	43	8	0.3	0.0019392
43	4	50	43	8	0.3	0.0019365
44	4	50	43	8	0.3	0.0020703
45	4	50	43	8	0.3	0.0022817
46	4	50	43	8	0.3	0.0020436

42.4.1 Analysis of Tool Wear Rate

The calculated TWR for different experimental conditions was considered for the analysis. Table 42.3 describes the analysis of variance (ANOVA) for TWR carried out using design expert *V* 9.0.5 at 95% confidence level to determine the significant parameters. From Table 42.3, the *F*-value of a model was observed to be 64.36 implying the significance of the model. Likewise, I_P, S_{ON}, S_{OFF}, PC, and MF passed the significant test as $p < 0.05$ was observed. I_P was most influential machining parameters followed by S_{ON}, PC, and MF, while the effect of S_{OFF} was least. However, S_{ON}-PC and PC-MF among interactions and for quadratic terms I_P, S_{ON}, and PC was observed to be significant.

The effect of machining parameters depicted in main effect plot is shown in Fig. 42.2a–e. With increase in peak current from 1 to 7 A, increase in TWR was observed from Fig. 42.2a. This is mainly because of the increment in electrical power which melts the material from tool electrode due to high spark energy which is proportional to peak current. However, increase in TWR was observed with

Table 42.3 ANOVA for tool wear rate (TWR)

Source	SS	DOF	MS	F-value	p-value Prob > F
Model	3.348×10^{-6}	10	3.348×10^{-7}	64.36	<0.0001
A-I_P	1.281×10^{-6}	1	1.281×10^{-6}	246.3	<0.0001
B-S_{ON}	8.302×10^{-7}	1	8.302×10^{-7}	159.6	<0.0001
C-S_{OFF}	1.067×10^{-7}	1	1.067×10^{-7}	20.52	<0.0001
D-PC	3.383×10^{-7}	1	3.383×10^{-7}	65.03	<0.0001
E-MF	3.580×10^{-7}	1	3.580×10^{-7}	68.82	<0.0001
BE	5.480×10^{-8}	1	5.480×10^{-8}	10.54	0.0026
DE	2.220×10^{-8}	1	2.220×10^{-8}	4.27	0.0463
A^2	1.212×10^{-7}	1	1.212×10^{-7}	23.31	<0.0001
B^2	1.758×10^{-7}	1	1.758×10^{-7}	33.79	<0.0001
D^2	3.890×10^{-8}	1	3.890×10^{-8}	7.48	0.0097
Residual	1.820×10^{-7}	35	5.201×10^{-9}		
Lack of fit	1.028×10^{-7}	30	3.427×10^{-9}	0.22	0.9970
Pure error	7.924×10^{-8}	5	1.585×10^{-8}	R-Sq. = 0.9484, Adj. R-Sq. = 0.9337, Pred. R-Sq. = 0.9199	
Cor total	3.530×10^{-6}	45			

SS Sum of Square, *MS* Mean Square, *DOF* Degree of Freedom

increase in S_{ON} from 30 to 70 μs as shown in Fig. 42.2b; this is due to an increase in pulse duration for machining that produces high spark energy which causes melting and evaporation of material from tool surface. Also, decrease in TWR was observed from Fig. 42.2c with increase in spark off time from 35 to 51 μs. It is because of less time the spark gets to be in contact with the workpiece material resulting in the decrease in TWR. From Fig. 42.2d, the decrease in TWR was observed with an increase in powder concentration from 6 to 10 g/l. This was a result of the formation of a narrow channel which transfers the heat to the workpiece materials rather than tool after the addition of powder in EDM oil causing low TWR. From Fig. 42.2e, the decrease in TWR, with an increase in MF from 0.15 to 0.45 T, was attributed to the diamagnetic behavior of electrolytic copper which opposed the magnetic field when increased and the debris particles were repelled away causing low TWR.

42.4.2 *Response Surface Modeling*

A second-order model was developed in terms of significant machining parameters to predict the TWR. The regression coefficient of the model was obtained using experimental data at 95% confidence level. The second-order model for TWR in coded factors is stated in Eq. (42.2).

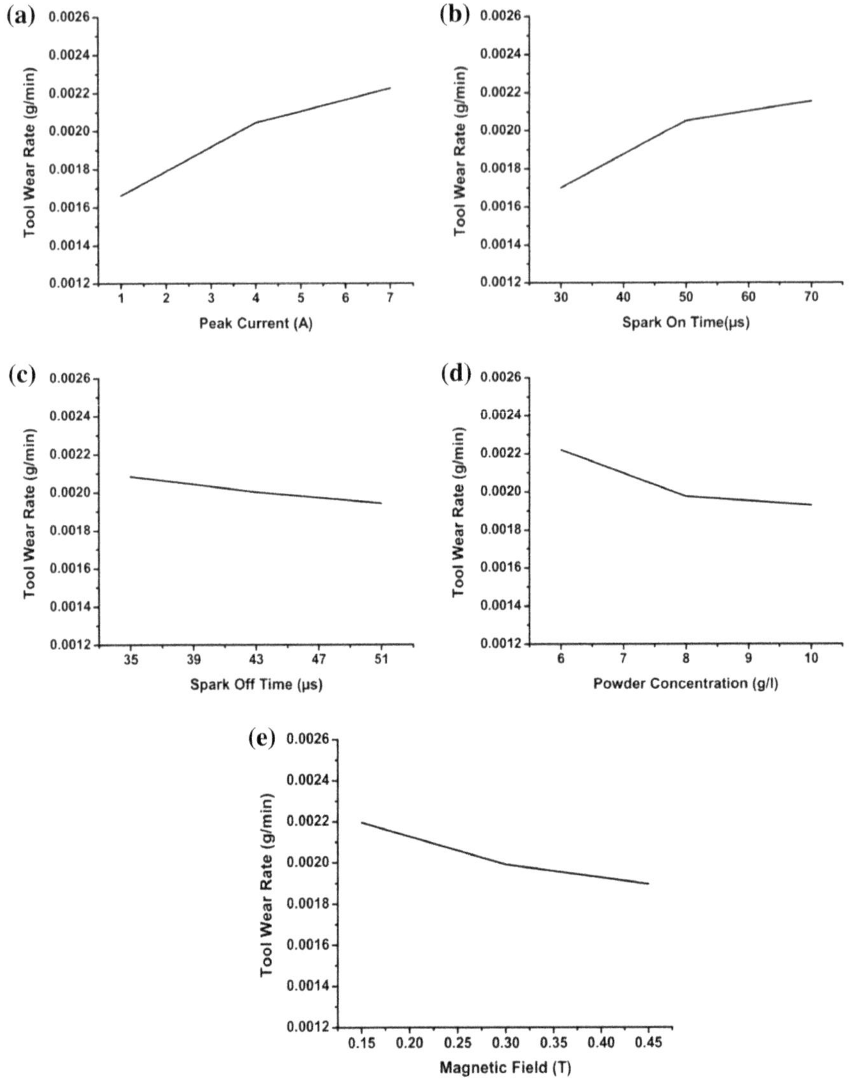

Fig. 42.2 Main effect plot of machining parameters **a** peak current **b** spark on time **c** spark off time **d** powder concentration **e** magnetic field on TWR

$$\begin{aligned} \text{TWR} = &+2.073 \times 10^{-3} + 2.830 \times 10^{-4} \times A + 2.278 \times 10^{-4} \times B - 8.167 \times 10^{-5} \times C \\ &- 1.454 \times 10^{-4} \times D - 1.496 \times 10^{-4} \times E + 1.171 \times 10^{-4} \times B \times E - 7.450 \times 10^{-5} \times D \times E \\ &- 1.108 \times 10^{-4} \times A^2 - 1.334 \times 10^{-4} \times B^2 + 6.274 \times 10^{-5} \times D^2 \end{aligned} \tag{42.2}$$

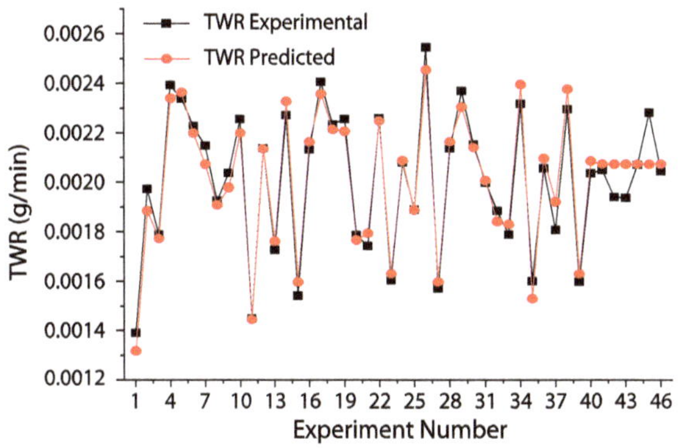

Fig. 42.3 Experimental and predicted TWR against experimental runs

Predicted R-Sq. value of 0.9199 for TWR was in reasonable agreement with the adjusted R-Sq. value of 0.9337. The second-order model developed was found to be congruent with the experimental data as shown in Fig. 42.3.

42.4.3 Optimization Using TLBO

Response surface model developed and experimental data were used in teacher–learner phase for optimizing the machining parameters. Single objective optimization was carried out to determine the optimum set of machining parameters for TWR. For optimization, the computer code was developed with population size = 46 and number of iterations = 20 using MATLAB R2016a. The optimal solution of machining parameters for TWR with the number of iterations is shown in Fig. 42.4, and convergence graph shows the decreasing curve for TWR with increasing the successive iterations. The optimized value of TWR was obtained at eighth iterations, and the optimum set of machining parameters for minimum TWR (0.000813 gm/min) was obtained at I_P-1 A, S_{ON}-30 μs, S_{OFF}-51 μs, PC-10 g/l, and MF-0.45 T, respectively.

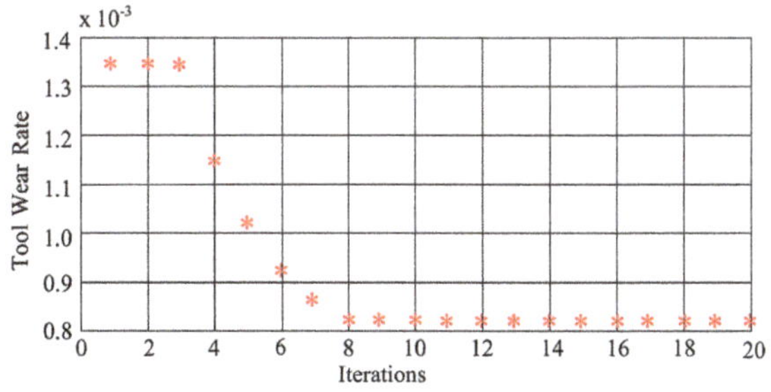

Fig. 42.4 TWR with number of iterations

Table 42.4 Confirmation results for TWR

Input machining parameters					TWR (g/min)	
I_P (A)	S_{ON} (μs)	S_{OFF} (μs)	PC (g/l)	MF (T)	Optimized	Experimental
1	30	51	10	0.45	0.000813	0.000894

42.4.4 *Confirmation Test*

Once the optimum machining parameters were identified using TLBO algorithm for minimum TWR, the same combination of machining parameters was used for the validation by performing the experiments in MFAPM-EDM process. The results obtained with TLBO and experiments for optimum machining parameters are shown in Table 42.4.

42.5 Conclusions

In this paper, magnetic field-assisted powder-mixed EDM was successfully used for machining Al6061 alloy with aluminum powder mixed in EDM oil. The experimental design was based on BBD, and ANOVA was carried out at 95% confidence level to determine the influential machining parameters for TWR. Modeling in terms of coded machining parameters was carried out using response surface modeling, and the optimized set of machining parameters was obtained using TLBO algorithm. The following conclusions were obtained from this work.

1. Peak current was most influential machining parameters followed by spark on time, magnetic field, and powder concentration while spark off time has least effect.
2. Second-order model to predict the TWR was developed using response surface and was found to be congruent with the experimental results.
3. Using TLBO, optimized set of machining parameters was obtained at eighth iteration for I_P-1 A, S_{ON}-30 μs, S_{OFF}-51 μs, PC-10 g/l, and MF-0.45 *T* for minimum TWR.

References

1. Kumar, R., Pradhan, M.K., Kumar, R.: Modeling and optimization of milling parameters on Al-60161 alloy using multi-objective genetic algorithm. In: 5th International and 26th All India Manufacturing Technology, Design and Research Conference, vol. 89, pp. 1–7 (2014)
2. Arooj, S., Shah, M., Sadiq, S., Jaffery, S.H.I., Khushnood, S.: Effect of current in the EDM machining of aluminum 6061 T6 and its effect on the surface morphology. Arabian J. Sci. Eng. **39**(5), 4187–4199 (2014)

3. Sanchez, J.M., Rubio, E., Alvarez, M., Sebastian, M.A., Marcos, M.: Microstructural characterisation of material adhered over cutting tool in the dry machining of aerospace aluminium alloys. J. Mater. Process. Technol. **164–165**, 911–918 (2005)
4. Roy, P., Sarangi, S.K., Ghosh, A., Chattopadhyay, A.K.: Machinability study of pure aluminium and Al–12% Si alloys against uncoated and coated carbide inserts. Int. J. Refract Metal Hard Mater. **27**, 535–544 (2009)
5. Bisaria, H., Shandilya, P.: The machining characteristics and surface integrity of Ni-rich NiTi shape memory alloy using wire electric discharge machining. In: Proceedings of the Institution of Mechanical Engineers, Part C: Journal of Mechanical Engineering Science (2018). https://doi.org/10.1177/0954406218763477
6. Pramanik, A., Basak, A.K., Islam, M.N., Littlefair, G.: Electrical discharge machining of 6061 aluminium alloy. Trans. Nonferrous Metals Soc. China **25**(9), 2866–2874 (2015)
7. Bisaria, H., Shandilya, P.: Experimental studies on electrical discharge wire cutting performance of Ni-rich NiTi shape memory alloy. Mater. Manuf. Process. **33**(9), 977–985 (2017)
8. Bisaria, H., Shandilya, P.: Machining of metal matrix composites by EDM and its variants: a Review. Chapter 23 in DAAAM International Scientific Book, pp. 267–282, ISSN 1726-9687. Vienna, Austria (2015)
9. Bhatt, G., Batish, A., Bhattacharya, A.: Experimental investigation of magnetic field assisted powder mixed electric discharge machining. Part. Sci. Technol. **33**, 1–28 (2015)
10. Bains, P.S., Sidhu, S.S., Payal, H.S., Kaur, S.: Magnetic Field Influence on Surface Modifications in Powder Mixed EDM. Silicon (2018). https://doi.org/10.1007/s12633-018-9907-z
11. Kant, K.: Experimental investigation of magnetic field on EDM process by using Taguchi method on EN-19 tool steel. Global J. Res. Eng. Mech. Mech. **15**(3), 27–35 (2015)
12. Heinz, K., Kapoor, S.G., Devor, R.E., Surla, V.: An investigation of magnetic-field-assisted material removal in micro-EDM for nonmagnetic materials. J. Manuf. Sci. Eng. **133**, 021002-1-9 (2011)
13. Montgomery, D.C.: Design and analysis of experiments, 4th edn. Wiley, New York (1997)
14. Prasanna, J., Rajamanickam, S., Kumar, O.A., Raj, G.K., Narayanan, P.V.V.S.: MRR and TWR evaluation on electrical discharge machining of Ti-6Al-4V using tungsten: copper composite electrode. In: IOP Conference Series: Materials Science and Engineering, vol. 197, pp. 012087-1-7 (2017)
15. Rao, R.V.: Teaching learning based optimization algorithm and its engineering applications. Springer (2016). https://doi.org/10.1007/978-3-319-22732-0

Chapter 43
ANFIS-Based Subtractive Clustering Algorithm for Prediction of Response Parameters in WEDM of EN-31

Kunal Chopra, Aishwarya Payla, Guneet Kaur and Eswara Krishna Mussada

Abstract Wire electrode discharge machining (WEDM) is an accurate but an expensive and time-consuming process. In order to establish a stable connection between input and output variables, implementation of soft computing techniques can be useful. Therefore, the current study focuses on comparing adaptive neuro-fuzzy inference system (ANFIS)-based subtractive clustering algorithm with numerous input combinations as well as multivariate regression models in order to simulate and map the output variables with the process parameters used during experimentations, namely pulse-on time (T_{on}), servo voltage (S_v), wire feed (W_f), and wire tension (W_t). Results show that ANFIS models have the ability to estimate the edge roughness (E_r) and kerf width (K_w) more accurately with 96.2 and 97.3% accuracy. ANFIS model is more reliable, accurate, and productive as it uses the learning of neural networks to predict. Also, the developed model has been used to study and explain the effect of various input variables upon the quality of machining. High pulse-on time directly decreases the quality increasing the edge roughness and kerf width which are both undesirable. Low wire feed has shown to decrease both the response parameters regardless of other input parameters. Wire tension has shown much less significant effect as compared to the other three variables.

Keywords WEDM · ANFIS · Machining · EN-31 · Artificial Intelligence

---

K. Chopra · A. Payla · G. Kaur
SOET, BML Munjal University, Gurgaon 122413, India
e-mail: kunalchopra954@gmail.com

A. Payla
e-mail: aishwaryapayla21@gmail.com

G. Kaur
e-mail: guneet.kaur31297@gmail.com

E. K. Mussada (✉)
Amity Institute of Technology, Amity University, Noida 201313, Uttar Pradesh, India
e-mail: eswar_mussada@yahoo.com; ekmussada@amity.edu

R. G. Narayanan et al. (eds.), *Advances in Computational Methods in Manufacturing*, Lecture Notes on Multidisciplinary Industrial Engineering,
https://doi.org/10.1007/978-981-32-9072-3_43

43.1 Introduction

Manufacturing industry all around the world is focusing on improving quality even for intricate geometries but not at the expense of high cost. The issue of high forces generated in traditional machining processes has been overcome by the advent of non-conventional machining processes which removed the contact between the tool and workpiece and hence reducing contact and frictional stresses. Wire electrode discharge machining (WEDM) is one of the advanced machining process which uses electro-thermal process for machining conductive materials. A thin metal wire as the electrode removes the material by melting and vaporization of workpiece material [1]. Continuously flowing dielectric fluid ensures the flushing of debris thus formed from the workpiece surface. The spark erosion process used for material removal in WEDM makes it possible to machine even the hard materials. Since there is no contact between tool and workpiece surfaces, friction does not come into play and hence residual stresses generated at the cutting zone are low.

Mahapatra et al. [2] through the study of different input variables including pulse-on time (T_{on}), pulse frequency, discharge current, wire speed, wire tension (W_t) and dielectric flow on the surface finish, and kerf width (K_w) found discharge current, pulse duration, and dielectric flow rate to be the most influential parameters for minimization of surface roughness and kerf width in rough cutting operations. Kumar et al. [3] found I_P (peak current), T_{on}, T_{off}, and W_f to have effect on the surface finish in that order while WEDM of high-speed steel (HSS) through zinc-coated wire. Sharma et al. [4] did the experimental study followed by parameter optimization on WEDM and found out that T_{on} majorly affects the response parameters such as surface roughness (S_r) and C_r. Experimental investigation done by Garg et al. [5] focused on machining Al/10% $ZrO_{2(P)}$ metal matric composite by WEDM. They studied the effect of the input parameters on surface roughness (S_r) and cutting velocity (*CV*). Machining parameters such as pulse width, short time pulse and time between pulses were found to affect both the response parameters most. Servo control mean reference voltage was observed to have similar significance on affecting the response parameters. Fard et al. [6] machined AlSiC metal matrix composite to study the effect of T_{on}, T_{off}, W_f, W_t, S_v, and discharge current on S_r and cutting speed by applying ANOVA to the experimental results. In his study, T_{on} and discharge current were found to be the most significant factors influencing the studied targets while W_t had the least impact. Parmanik et al. [7] processed duplex stainless steel in WEDM and observed that increasing T_{on} did not have a linear effect on K_w and craters along with resolidified molten were noticed on the machined surface. Tosun et al. [8] analyzed the crater sizes formed due to 0.25 mm diameter brass wire under different cutting parameters. It was concluded that pulse duration and open-circuit voltage have adverse effect on crater [8]. Chopra et al. [9] found low T_{on} and S_v to be the important parameters to minimize edge roughness (E_r) while machining EN-31 through WEDM.

Regression is a statistical approach to establish a relationship between two sets of variables, dependent variables and the independent variables. Regression can further be multivariate regression or multiple regression. Multivariate regression refers to estimating more than one output variable using single regression model and when multivariate regression consists of more than one predictor variable it refers to multivariate multiple regression. Majumder et al. [10] studied the impact of pulse-on time, pulse-off time and wire feed on kerf width (K_w), $S_{r,}$ and material removal rate (MRR). They developed general regression neural network (GRNN) and multiple regression models for prediction and comparison of studied key machinability aspects. Maher et al. [11] performed experiments to study the impact on CNC WEDM response parameters like cutting speed, surface roughness, and white layer thickness while machining titanium alloy grade 5(Ti6Al4V). Their study involves development of a regression model to study the effect of performance index on the machinability aspects studied. In order to increase the productivity and accuracy of the machining of super alloy, Udimet-L605 done by WEDM, Nain et al. [12] developed linear regression model to understand the relation between input and output parameters followed by application of particle swarm optimization (PSO), concluding T_{on} to be the most impacting factor on C_r.

Artificial neural networks were introduced to simulate the neural structure similar to brain by establishing a relationship between I/O data during training phase which could be recalled during the verification or optimization process [13]. In ANN algorithms, each processing element or neurons can reproduce the biological NN having numerous inputs and a single output considering certain assumptions and constraints. ANN is an effective technique which can approximate with high precision even nonlinear functions [14] but the efficiency is constrained to problem under consideration [15]. Ramakrishnan et al. [16] designed an ANN model using back-propagation algorithm to predict the characteristic parameters, namely MRR and S_r, of CNC WEDM while machining Inconel 718. Optimization was done using multi-response signal-to-noise ratio (MRSN) and analysis of variance (ANOVA) was applied to determine the level of impact of the response parameters.

Adaptive neuro-fuzzy inference system (ANFIS) combines ANN with fuzzy modeling thus capturing benefits of both techniques in one framework only. It provides a means to map the input–output data relations. Yilmaz et al. [17] developed a fuzzy model that allowed precise selection of EDM parameters using if-then. Fuzzification was done by triangular membership function while defuzzification process was then carried out by centroid area method in order to obtain minimum electrode wear and better surface quality. Maher et al. [18] applied ANFIS to predict the value of white layer thickness (WLT) in WEDM and obtained an accuracy of 97.39%. Maher et al. [19] successfully designed ANFIS model for the selection of optimum machining parameters with the objective of achieving higher productivity at the highest possible surface quality and sustainable development at minimum cost.

Present study is conducted with the objective of investigating the effect of four input parameters, pulse-on time, servo voltage, wire tension, and wire feed on the machining performance of die steel EN-31 and to develop both multivariate regression and ANFIS model to predict and improve the output parameters, viz. edge roughness (E_r) and kerf width (K_w). EN-31 alloy steel has a number of applications in manufacturing industry due to its outstanding wear resistance and high load-carrying capacity. There is limited literature available on the prediction and optimization of K_w; however, no study has been carried out for prediction of E_r while machining through WEDM.

43.2 Experimental Work

The experiments were conducted on "Electronica Wire-cut Electric Discharge Machine" which can cut with speeds upto 230 mm^2/min and workpiece of heights up to 250 mm can be held in its clamp. Out of many wires available for cutting operation, brass wire having a diameter of 250 μm was used which provides better integrity of surface while conducting current. Deionized water is a proven dielectric to provide a good conducting sphere of influence while machining and hence was used in our experimentations.

Four machining parameters, namely pulse-on time (T_{on}), servo voltage (S_v), wire tension (W_t), and wire feed (W_f), were selected in this study for investigating the effect on edge roughness and kerf width. These parameters were identified on the basis of literature survey and pilot experimentation to maintain the appropriate spark gap. T_{on} and S_v were scrutinized at three levels while W_f and W_t were scrutinized at two levels on the basis of economic impact of changing the parameters. Duty factor for the machine was kept at 0.7. A general full-factorial design for experimentation was used for studying the effect of each parameter on performance. Table 43.1 shows the control parameters and their levels along with the other parameters that were kept constant throughout the machining operation.

The crests and troughs formed along the edge were the result of craters and irregularities formed due to the rapid melting and re-solidification of material. The accuracy of machined component is a direct measure of kerf width and edge roughness which were measured at several locations along the machined edge by "Olympus optical microscope" as shown in Fig. 43.1 and averaged over all the readings for each experiment. Each machined specimen was cleaned with acetone before observing under the optical microscope.

Table 43.1 Control and constant parameters

Parameters	Level-1 (low)	Level-2 (medium)	Level-3 (high)
T_{on} (μs)	100	110	120
S_v (V)	20	40	60
W_f (m/min)	5	8	–
W_t (g)	660 g	1140 g	–
Wire diameter	250 μm	–	–
Constant machining parameters			
Wire material		Standard brass	
Peak current		12 (fine pulse and 2 power pulses ON)	
Peak voltage		11	
Cutting speed override		25%	
Duty cycle		0.70	
Servo feed		10 mm/min	
Dielectric fluid		De-ionized water	

43.3 Modeling Approach

Regression analysis has been used to map the input and output variables in order to obtain a function which could predict the response measures on the basis of given input values. Since the ranges of our parameters are not consistent and hence weights were required, instead, techniques like feature scaling, normalization, and polynomial feature scaling were applied as a part of data preprocessing step. It is usually done using Min-Max scaling method, which is the simplest method of all and consists of rescaling the range of features in range of [−1, 1]. Equation (43.1) shows the formula used. Platform used for this project is Jupyter Notebook and programming language was Python. Scikit-learn is a popular open-source library used to implement machine learning and perform statistical analysis in Python. It features various classification, regression, and clustering algorithms and is designed to interoperate with the Python numerical and scientific libraries NumPy, SciPy, and Pandas.

$$x' = \frac{x - \min(x)}{\max(x) - \min(x)} \tag{43.1}$$

After that, polynomial feature transformation was applied for both degree 2 polynomial and degree 3 polynomial. It helped find relationship between different dependent variables and their impact on independent variables, E_r and K_w. Cubic polynomial regression helped improve the R score from 67 to 94% for both of our response variables which is more accurate than linear regression. The function obtained for K_w and E_r are depicted in Eqs. (43.2) and (43.3).

$$k_w = (-2.50e-10) + (-192.45)x1 + (-2165.67)x2 + (554.52)x3 + (-7.49)x4 + (-48.36)x1^2 + (6677.44)x1x2 + (-1715.03)x1x3 + (79.42)x1x4 + (-754.35)x2^2 + (-725.72)x2x3 + (-233.69)x2x4 + (554.52)x3^2 + (-159.07)x3x4 + (-7.49)x4^2 + (94.92)x1^3 + (-3708.53)x1^2x2 + (2081.60)x1^2x3 + (-222.51)x1^2x4 + (-125.49)x1x2^2 + (575.63)x1x2x3 + (-854.05)x1x2x4 + (-1950.18)x1x3^2 + (895.14)x1x3x4 + (115.820)x1x4^2 + (439.84)x2^3 + (118.36)x2^2x3 + (-11.80)x2^2x4 + (-35.93)x2x3^2 + (213.71)x2x3x4 + (605.77)x2x4^2 + (554.52)x3^3 + (-210.62)x3^2x4 + (-355.05)x3x4^2 + (-7.49)x4^3 \quad (43.2)$$

$$E_r = (6.68409772e-13) + (44.15)x1 + (95.005)x2 + (47.19)x3 + (-10.25)x4 + (-2.44)x1^2 + (-310.30)x1x2 + (-165.98)x1x3 + (62.73)x1x4 + (36.996)x2^2 + (59.24)x2x3 + (8.78)x2x4 + (47.19)x3^2 + (-33.66)x3x4 + (-10.25)x4^2 + (-48.77)x1^3 + (236.87)x1^2x2 + (207.42)x1^2x3 + (-156.19)x1^2x4 + (-34.74)x1x2^2 + (-127.937)x1x2x3 + (34.075)x1x2x4 + (-198.43)x1x3^2 + (216.10)x1x3x4 + (65.22)x1x4^2 + (-12.088)x2^3 + (8.95)x2^2x3 + (13.79)x2^2x4 + (42.10)x2x3^2 + (-21.76)x2x3x4 + (-28.67)x2x4^2 + (47.19)x3^3 + (-41.62)x3^2x4 + (-55.87)x3x4^2 + (-10.25)x4^3 \quad (43.3)$$

where $x1$, $x2$, $x3$, and $x4$ are the independent variables (T_{on}, S_v, W_f, and W_t, respectively) along with the regression coefficients.

ANFIS model exploits the capability of learning by neural network in order to establish a relationship between input–output parameters which uses fuzzy rules. Due to high variability in the input variables, subtractive clustering algorithm was used for fuzzy rule determination process which checks each and every data point for potential cluster center and the point with highest density in its vicinity is considered as the cluster center with readings outside the range of influence of that center being excluded and are again checked for potential of being a cluster center. It is an iterative process that runs until all points lie within the range of influence of a cluster center. Subtractive clustering approximates the function as a generalized bell curve and determine the number of rules and membership functions. Since the number of centers to be created for the distribution of data was not clear, automated algorithm of subtractive clustering and fuzzy rules were used. Similar to feature scaling, the data were first normalized as ranges for our variables are not same. In normalization, mean is involved when standardizing the data as shown in the equation below:

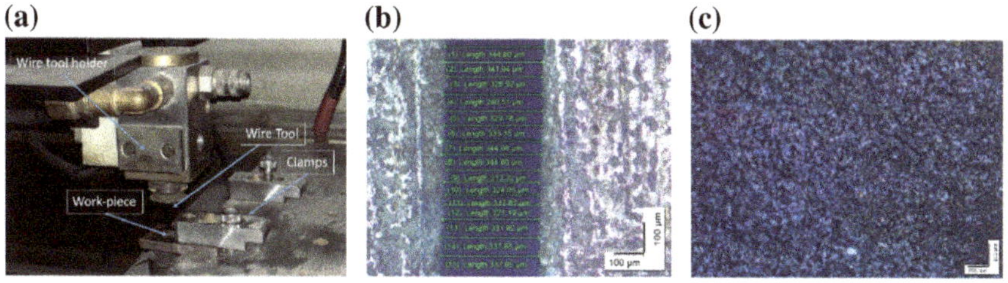

Fig. 43.1 **a** Experimental setup, **b** kerf width measurement, and **c** microscopic image of machined surface at T_{on} = 110 µs, S_v = 60 V

Table 43.2 Information of the trained ANFIS model

Number of nodes	307
Number of linear parameters	150
Number of nonlinear parameters	240
Number of training data pairs (membership functions)	30
Number of fuzzy rules	30

$$x' = \frac{x - \text{average}(x)}{\text{stdev}(x)}. \tag{43.4}$$

The data were then trained using the MATLAB's ANFIS toolkit using subtractive clustering as an FIS technique having 0.5 s the range of influence using the hybrid training algorithm. The structure with four input and single output parameters formed a structure containing 30 input membership functions and fuzzy rules as shown in Fig. 43.1. The membership functions and parameters obtained are shown in Table 43.2. Training error for E_r and K_w came out to be 2.16e–7 and 1.51e–7, respectively. The trained model was used to test the data and predict for randomly selected readings as mentioned in Sect. 43.4.

43.4 Results and Discussion

Effect of machining parameters on E_r and K_w is shown in Figs. 43.2 and 43.3. With the rise in S_v and T_{on} values roughness is found to be increasing. While wire tension at higher S_v values and wire feed has negligible effect on the roughness. Higher values of wire tension reduce the vibration in the wire thus maintaining the spark gap required for the machining process. The reduced vibrations give more interaction time to electrode with the workpiece thus utilizing the energy available for material removal and forming larger number of craters on the surface. At higher gap, voltage energy density is more resulting in more rapid vaporization and hence localized material removal takes place at the cutting edge which leads to higher

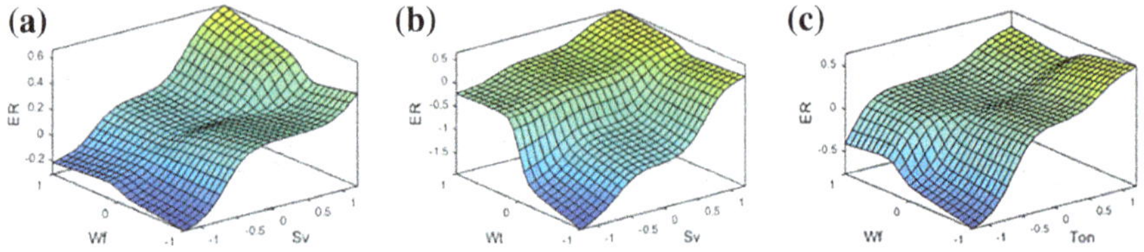

Fig. 43.2 Modeled edge roughness by ANFIS w.r.t **a** wire feed and servo voltage, **b** wire tension and servo voltage, and **c** wire feed and pulse-on time

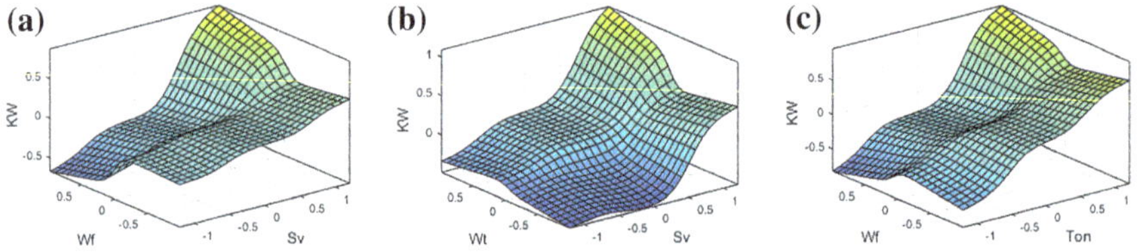

Fig. 43.3 Modeled kerf width by ANFIS in w.r.t. **a** wire feed and servo voltage, **b** wire tension and servo voltage, and **c** wire feed and pulse-on time

roughness. As the pulse duration increases, discharge energy around the wire intensifies. Due to more energy available at the spark zone, deep craters are formed which contribute to the roughness of the cutting edge.

Kerf width depends on the radius of effective spark gap. With the rise in S_v spark energy increases, this increases accounts for a wider spark gap and hence higher kerf width values. Kerf width was found to be increase with T_{on}. As at higher T_{on} values, more electrons from the wire comes in contact with the neutral particles of dielectric fluid. This interaction magnifies the ionization effect which causes more material to vaporize. While wire tension is not affecting the kerf much, wire feed at lower S_v and T_{on} values decreases the kerf. This can be accounted to the inability of wire to maintain least required spark gap for conduction when the vibrations in the wire are more at higher feed. When more discharge energy is available, increase in vibrations results in widening the gap further.

43.5 Model Verification

Six random readings were selected to test the models generated by both the methods. First, they were put into the relation obtained from multivariate regression and each output value was compared with experimental value and a prediction error was calculated for both E_r and K_w as shown in Tables 43.3 and 43.4. Average error, E_{avg} was measured by the following formula:

Table 43.3 Comparison of experimental and predicted responses of E_r for test data

Machining parameters				Response parameters			Predictive error	
T_{on}	S_v	W_f	W_t	Experimental	ANFIS	Regression	ANFIS	Regression
100	40	8	1140	8.09	8.44	6.91	0.043	0.147
100	60	8	660	11.61	10.97	15.09	0.054	0.301
110	40	5	660	10.49	10.79	11.22	0.028	0.069
110	60	5	1140	10.81	10.63	11.43	0.017	0.057
120	20	8	1140	10.12	9.82	10.98	0.030	0.084
120	20	5	1140	9.38	9.90	7.04	0.055	0.249
						E_{avg}	0.038	0.151

Table 43.4 Comparison of experimental and predicted responses of K_w for test data

Machining parameters				Response parameters			Error	
T_{on}	S_v	W_f	W_t	Experimental	ANFIS	Regression	ANFIS	Regression
100	40	8	1140	302.689	299.811	302.689	0.009	0.055
100	60	8	660	292.041	297.377	292.041	0.018	0.002
110	40	5	660	271.998	288.615	271.998	0.057	0.129
110	60	5	1140	325.289	316.444	325.289	0.027	0.033
120	20	8	1140	331.386	325.799	331.386	0.016	0.003
120	20	5	1140	311.654	322.612	311.654	0.035	0.095
						E_{avg}	0.027	0.053

$$E_{avg} = \frac{1}{m}\sum_{i=1}^{m} Ei. \quad (43.5)$$

where E_i = error of sample number i an m is the total number of samples.

$$E_i = \left|\frac{\text{Experimental} - \text{Predicted}}{\text{Experimental}}\right| \quad (43.6)$$

The errors in multivariate regression model came out to be 15.1 and 5.3% for E_r, and $K_{w,}$ respectively. The same six readings were then predicted with the ANFIS model and again a prediction error was measured. ANFIS model predicted the output with an error of 3.8 and 2.7% for E_r and $K_{w,}$ respectively. Figure 43.4 shows the comparison of measured and predicted values of test data in ANFIS model. These low values of average error show that the ANFIS model developed is adequate to predict the studied response parameters, E_r and K_w. However, the average prediction error obtained through multivariate regression model for E_r and K_w are quite high when compared with ANFIS model and is thus, not capable of predicting the machinability aspects accurately.

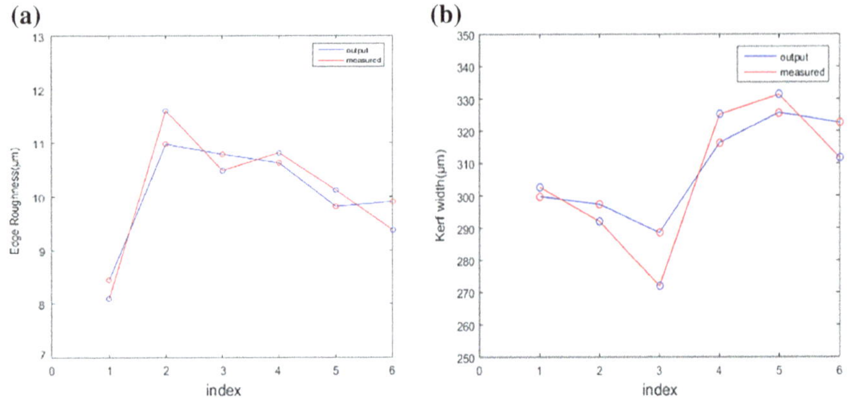

Fig. 43.4 Measured vs. predicted values for **a** edge roughness and **b** kerf width

43.5.1 Analysis of Machining Variables

Analysis of variance (ANOVA) was performed to determine significance of each machining parameters on both the responses. E_r is most influenced by servo voltage followed by W_t, W_f, and T_{on} as can be seen from Table 43.5 below, whereas Table 43.6 shows that T_{on} and W_f are the most and least important parameters, respectively, for affecting the kerf.

Main effects plots for the mean of E_r and K_w were also plotted using MiniTab software. Figure 43.5a shows the relation of E_r with the various machining parameters. Roughness is increasing with the increase in W_t, S_v, and T_{on} but is inversely dependent on the W_f, while K_w is seen to be directly dependent on all the input parameters.

Table 43.5 ANOVA table for mean w.r.t. E_r

Source	DF	Adj SS	Adj MS	*F*-value	*P*-value
T_{on}	2	2.565	1.282	0.3	0.745
S_v	2	56.17	28.087	10.39	0
W_f	1	14.83	14.827	3.86	0.058
W_t	1	1.489	1.489	0.35	0.557

Table 43.6 ANOVA table for mean w.r.t. K_w

Source	DF	Adj SS	Adj MS	*F*-value	*P*-value
T_{on}	2	7344	3671.9	16.83	0
S_v	2	2023	1011.6	2.67	0.084
W_f	1	267.3	267.3	0.64	0.43
W_t	1	729.7	729.7	1.8	0.189

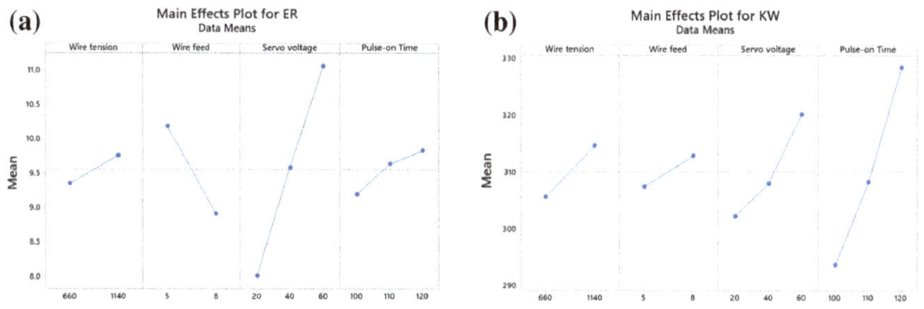

Fig. 43.5 Main effects plot for **a** edge roughness and **b** kerf width

43.6 Conclusion

Manufacturing Industry has a socio-economic responsibility in providing quality products keeping the costs incurred as low as possible. This has driven many researchers to look for best set of machining conditions. A full-factorial design has been used for experimentations upon which predictive models were built through multivariate regression and ANFIS. ANFIS model was able to predict with an accuracy of 96.2 and 97.3% for E_r and $K_{w,}$ respectively, which is higher than that obtained by regression model. Surface plots from ANFIS show that high T_{on} values increase the E_r significantly. High W_t also increased the roughness on edges. Similarly, increasing T_{on} and S_v increased the K_w. Lower wire feed value was able to decrease the K_w even at high S_v and T_{on}. These inputs cannot be kept too low as productivity is again a matter to be considered while production and hence a trade-off is necessary while deciding the input variables for any machining process. Further work can include finding the most optimum input variables which could give highly efficient process.

References

1. Mussada, E.K., Hua, C.C., Rao, A.K.P.: Surface hardenability studies of the die steel machined by WEDM. Mater. Manuf. Process. 1–6 (2018). https://doi.org/10.1080/10426914.2018.1476695
2. Mahapatra, S.S., Patnaik, A.: Optimization of wire electrical discharge machining (WEDM) process parameters using Taguchi method. Int. J. Adv. Manuf. Technol. **34**, 911–925 (2006). https://doi.org/10.1007/s00170-006-0672-6
3. Kumar, K., Agarwal, S.: Multi-objective parametric optimization on machining with wire electric discharge machining. Int. J. Adv. Manuf. Technol. **62**(5–8), 617–633 (2011)
4. Sharma, N., Raj, T., Jangra, KK.: Parameter optimization and experimental study on wire electrical discharge machining of porous Ni40Ti60 alloy. In: Proceedings of the Institution of Mechanical Engineers, Part B: Journal of Engineering Manufacture, 1–15(2015). https://doi.org/10.1177/0954405415577710

5. Garg, S.K., Manna, A., Jai, A.: An investigation on machinability of Al/10% ZrO_2 (P)-metal matrix composite by WEDM and parametric optimization using desirability function approach. Arabian J. Sci. Eng. **39**(4), 3251–3270 (2014). https://doi.org/10.1007/s13369-013-0941-2
6. Babu, K.A., Venkataramaiah, P.: Multi-response optimization in wire electrical discharge machining (WEDM) of Al6061/SiCp composite using hybrid approach. J. Manuf. Sci. Prod. **15**(4), 327–338 (2015). https://doi.org/10.1515/jmsp-2015-0010
7. Pramanik, A., Basak, A.K., Dixit, A.R.: Processing of duplex stainless steel by WEDM. Mater. Manuf. Process. (2018). https://doi.org/10.1080/10426914.2018.1453165
8. Tosun, N., Pihtili, H.: The effect of cutting parameters on wire crater sizes in wire EDM. Int. J. Adv. Manuf. Technol. **21**(10–11), 857–865 (2003). https://doi.org/10.1007/S00170-002-1404-1
9. Chopra, K., Payla, A., Mussada, E.K.: Detailed experimental investigations on machinability of EN31 steel by WEDM. Trans. Indian Inst. Metals, 1–9(2019). https://doi.org/10.1007/s12666-018-1552-0
10. Majumder, H., Maity, K.P.: Predictive analysis on responses in WEDM of titanium grade 6 using general regression neural network (GRNN) and multiple regression analysis (MRA). Silicon **10**(4), 1763–1776 (2018). https://doi.org/10.1007/s12633-017-9667-1
11. Maher, I., Sarhan, A.A.D.: Proposing a new performance index to identify the effect of spark energy and pulse frequency simultaneously to achieve high machining performance in WEDM. Int. J. Adv. Manuf. Technol. **91**(1–4), 433–443 (2017). https://doi.org/10.1007/s00170-016-9680-3
12. Nain, S.S., Gard, D., Kumar, S.: Investigation for obtaining the optimal solution for improving the performance of WEDM of super alloy Udimet-L605 using particle swarm optimization. Eng. Sci. Technol. Int. J. **21**(2), 261–273 (2018). https://doi.org/10.1016/j.jestch.2018.03.005
13. Chojaczyk, AA.: Review and application of artificial neural networks models in reliability analysis of steel structures. Struct Saf, (2014). http://dx.doi.org/10.1016/j.strusafe.2014.09.002
14. Cardoso, J., de Almeida, J.R., Dias, J., Coelho, P.: Structural reliability analysis using Monte Carlo simulation and neural networks. Adv. Eng. Softw. **39**(6), 505–513 (2008)
15. Bucher, C.: Most T. A comparison of approximate response functions in structural reliability analysis. Probab. Eng. Mech. **23**(2–3), 154–63 (2008)
16. Ramakrishnan, R., Karunamoorthy, L.: Modeling and multi-response optimization of Inconel 718 on machining of CNC WEDM process. J. Mater. Process. Technol. **207**, 343–349 (2008). https://doi.org/10.1016/j.jmatprotec.2008.06.040
17. Yilmaz, O., Eyercioglu, O., Gindy, N.N.Z.: A user-friendly fuzzy-based system for the selection of electro discharge machining process parameters. J. Mater. Process. Technol. **172**, 363–371 (2006)
18. Maher, I., Sarhan, A.A.D., Marashi, H., Barzani, M.M., Hamdi, M.: White layer thickness prediction in WEDM-ANFIS modeling. Malaysian Int. Tribology Conf. 16–17, 240–241 (2015). Penang, Malaysia
19. Maher, I., Sarhan, A.A.D., Marashi, H., Barzani, M.M., Hamdi, M.: Increasing the productivity of the wire-cut electrical discharge machine associated with sustainable production. J. Cleaner Prod. (2015). https://doi.org/10.1016/j.jclepro.2015.06.047

Chapter 44
On Estimation of Scallop Height from CNC Part Programs

A. Kukreja and S. S. Pande

Abstract Today, complex freeform surfaces are widely used on parts in automobile, aerospace, and die– mold industries. Multi-axis computer numerical control (CNC) machines are used to manufacture such parts as they need tighter tolerances and high surface finish to meet functional and aesthetic requirements. Scallop height is primarily used to measure and assess the surface quality of such products. The inspection methods used in practice are expensive and need experienced personnel. No simulation techniques exist, as of now, to estimate the part quality apriori. This paper reports the development of an algorithm to estimate the scallop heights for freeform surface machining from the CNC part programs. The developed system takes STL file of the simulated machined component from the VERICUT and slices it along the specified planes. Points so generated are analyzed to estimate scallop heights at various regions on the surface using numerical curve-fitting techniques. The developed system has been validated with the case studies. The system provides a robust and accurate estimation of the part quality.

Keywords Scallop height estimation · Surface topography · CNC machining

44.1 Introduction

Today freeform surfaces are widely used on parts in automobile, aerospace, and die–mold industries. To suit various functional and aesthetic requirements, these parts need to be manufactured with tighter tolerances and good surface quality.

A. Kukreja
CAM Lab, Department of Mechanical Engineering, Indian Institute of Technology Bombay, Powai, Mumbai 400076, Maharashtra, India
e-mail: aman.kukreja1@gmail.com

S. S. Pande (✉)
CAM Lab, Department of Mechanical Engineering, Indian Institute of Technology Bombay, Powai, Mumbai 400076, Maharashtra, India
e-mail: s.s.pande@iitb.ac.in

R. G. Narayanan et al. (eds.), *Advances in Computational Methods in Manufacturing*, Lecture Notes on Multidisciplinary Industrial Engineering,
https://doi.org/10.1007/978-981-32-9072-3_44

Multi-axes (3–5 axis) CNC machines are mainly used to manufacture such complex freeform surface parts [1]. They need a robust, error-free part program to drive the tool for machining.

The literature documents a focused research effort to develop various toolpath planning strategies (iso-planar, iso-scallop, etc.) [2–4] during the past two decades. In the commercial domain, several CAM systems exist which provide facilities for part modeling, toolpath planning, and CNC code simulation to generate and validate CNC part programs. CNC simulators provide the graphical simulation of toolpath and provide an estimation of machining time and code length. However, there is no facility in any CAM software/simulator to estimate the part scallop height which forms a primary measure of the part quality. VERICUT is one such software that provides the simulation of machining and gives the user a tool to measure scallop height at particular points [5]. The "manual" measurement system allows the measurement of scallops one point at a time which is cumbersome, error-prone, and could even leave some of the areas unobserved.

Literature reports very scant work focused on the prediction of surface roughness during milling. Ralph and Loftus [6] proposed a mathematical model for the average height, area, and volume of the scallops in freeform surface machining. However, it agreed poorly with machining trials. Bedi et al. [7] made a comparison of toroidal, ball end, and flat end mill in terms of surface roughness quantified as scallop heights and feed marks. A comprehensive review of the inspection techniques for machined freeform surfaces was presented by Li and Gu [8]. They pointed out that the inspection techniques are costly, complex, and time-consuming. The CNC part programs are thus required to be generated in a sequential design–plan–machine–inspect route which increases the cycle time. No system has been developed that can predict scallop height variation directly from part programs in the simulation. A need, thus, exists to develop an accurate procedure to estimate a priori part quality (scallop height) from the CNC part program during simulation. This paper is an attempt in this direction.

Rest of the paper is organized as follows. Section 44.2 explains the modular diagram and data flow of the system developed. The case studies are discussed in Sect. 44.3, and the conclusion is presented in Sect. 44.4.

44.2 Modular Diagram of the System

Figure 44.1 shows the modular diagram of the system. The developed system takes NC codes as input. It is simulated in VERICUT 8.0 to generate a CAD model of the machined surface in the STL format. The STL model file is exported, and is, in effect, processed by our system. The CAD model (STL file) is sliced by the inspection planes perpendicular to the lay direction of the machined surface. The inter-planar distance is definable by the user to decide a number of slicing planes. The intersection of the inspection (slicing) plane and STL machined model generates the surface profile in the form of points. A curve is fitted to these intersection

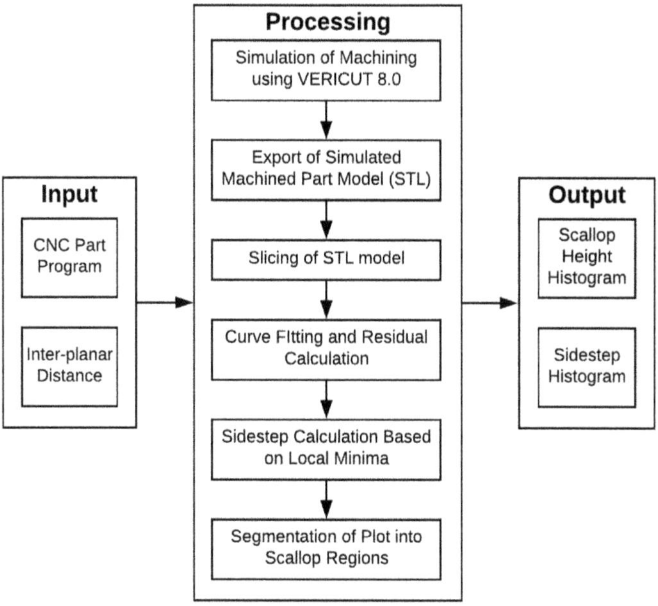

Fig. 44.1 Modular design of the system

points to generate a surface profile in the evaluation plane. Scallop height is computed from the curve by the residual calculation and segmentation at several points throughout the surface. Scallop height and step size estimated are plotted as histograms. In what follows, various system modules are explained.

44.2.1 Slicing of CAD Model

The machined part model (STL) model generated in VERICUT is sliced using parallel planes that are perpendicular to the lay direction (Fig. 44.2a). The intersection of the slicing planes and STL model would give points that form the surface topology pattern with superimposed waviness and roughness profiles (Fig. 44.2b).

Due to the direct slicing of STL facets, the scallop curve therefrom is well defined with less noise. A MATLAB function file “stl_slice_and_plot” was used [9]. The function first finds the triangles that intersect with the slicing plane. The lines created on the triangle face by the intersection plane are identified and processed further. The distance between the slicing planes is user definable, and hence, the user can control the areas to be evaluated and generate data on the desired surface topography.

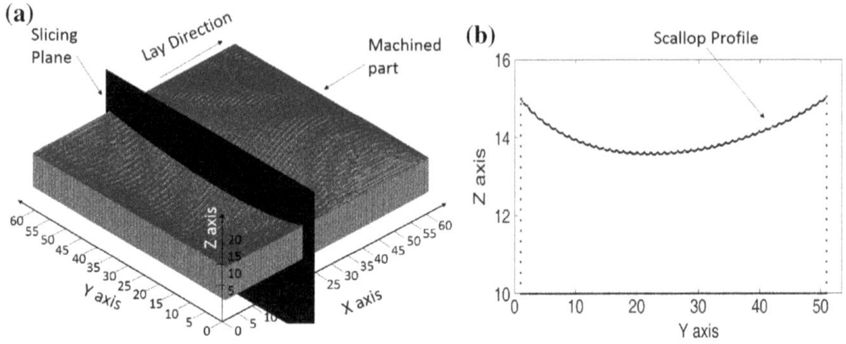

Fig. 44.2 **a** Slicing of the STL model and **b** points generated after slicing

44.2.2 Curve Fitting and Residual Calculation

After getting the point data, the next task was to remove waviness from the profile and calculate scallop heights based on the surface roughness profile. The waviness is removed by fitting a curve and then calculating the residuals of all data points. Residual gives the measure of skewness of the scallop data. Several curve-fitting methods were tried out [10]. For a large dataset, using "Gaussian fitting" and "third-degree polynomial fitting" resulted in poor fitting as the residual values were coming out to be large and were skewed on either side of the curve. After several trials, the ninth-degree polynomial fitting was found to be giving best approximation with residual variation within the desired level. Hence, in the present work, ninth-degree polynomial curve fitting approach has been used. The main advantages of polynomial fits include reasonable flexibility for data that is linear and not too complicated. This makes the fitting process fast, accurate, and computationally simple [11]. Equation (44.1) shows the polynomial model for curve fitting used in this work.

$$z = \sum_{k=0}^{n} p_i y^{n+1-i} \tag{44.1}$$

where $n + 1$ is the order and n is the degree of the polynomial ($1 \leq n \leq 9$). The order gives the number of coefficients to be fitted, and the degree gives the highest power of the predictor variable. The minimum distance (residual) of all the scallop points from the curve points are calculated based on Eq. (44.2).

$$Ɽ(i) = \min(\sqrt{(z_{1i} - z_{2j})^2 + (y_{1i} - y_{2j})^2} \ \forall j\epsilon[1,m]) \ \forall i\epsilon[1,k] \quad (44.2)$$

where Ɽ = minimum distance of scallop point "*i*" from the fitted curve; m is the total number of points taken on the fitted curve; y_1, z_1 signifies the scallop points; *k* is the total number of scallop points; y_2, z_2 signifies the sample points taken on the fitted curve.

44.2.3 Segmentation and Scallop Height Calculation

Scallop pattern can be clearly seen on the profile after the residual calculation (Fig. 44.3b). The generated residual curve is then segmented based on the sidestep while generating the CNC toolpath. The sidesteps are identified by searching for consecutive local minima or local maxima points. The local minima are found by comparing the *z*-values of the preceding and succeeding points with the current point.

The distance of two consecutive local minima (the difference between *y*-values) gives the sidestep variation in the slicing plane, that, in turn, facilitate calculation of scallops. The plot shown in Fig. 44.4 is then segmented as per calculated sidesteps.

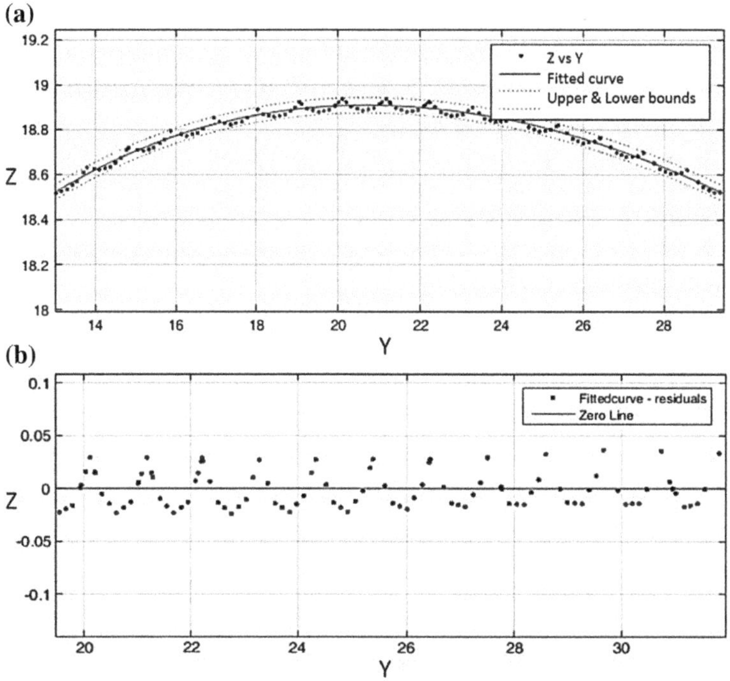

Fig. 44.3 **a** Curve fitting and **b** residual plot

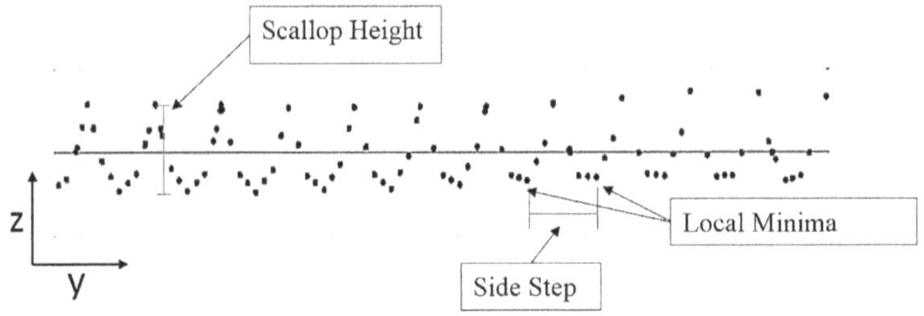

Fig. 44.4 Segmentation and scallop height calculation

The highest and lowest points (z) in the segments give scallop height for each sidestep (Eq. 44.3).

$$\check{S} = \max(z_i) - \min(z_i) \tag{44.3}$$

where $\check{S}$ is scallop height and z_i is the z-coordinate of scallop points. The case studies have been presented in the next section.

44.3 Results and Discussions

The algorithm for scallop estimation was implemented using MATLAB R2015b on Windows 10 platform, Intel Core i5, and 4 Gb RAM. The simulation software VERICUT 8.0 was used to generate a CAD model of the part (in the STL format) produced by the input CNC part program. The validation of the developed system was done using Mastercam X5-based iso-planar toolpath planning strategy. This was followed by extensive testing of the system on different kinds of surfaces produced with CNC programs with various toolpath strategies, like iso-parametric, iso-scallop, and adaptive planar [12].

The following case studies show the results and validation of the approach. The case study 1 refers to a relatively simple geometry with known values of scallops while case study 2 explores a complex bicubic Bezier surface with adaptive sidesteps generated using algorithm proposed by Dhanda and Pande [12] for three-axis CNC machining.

44.3.1 Case Study 1

In this case, the CNC part program was generated using Mastercam X5. The model has concave and convex portions (Fig. 44.5a). The scallop height was set to be 0.05 mm on a flat surface with sidestep 1.091 mm. It can go up to 0.1 mm at 45° to the lay direction.

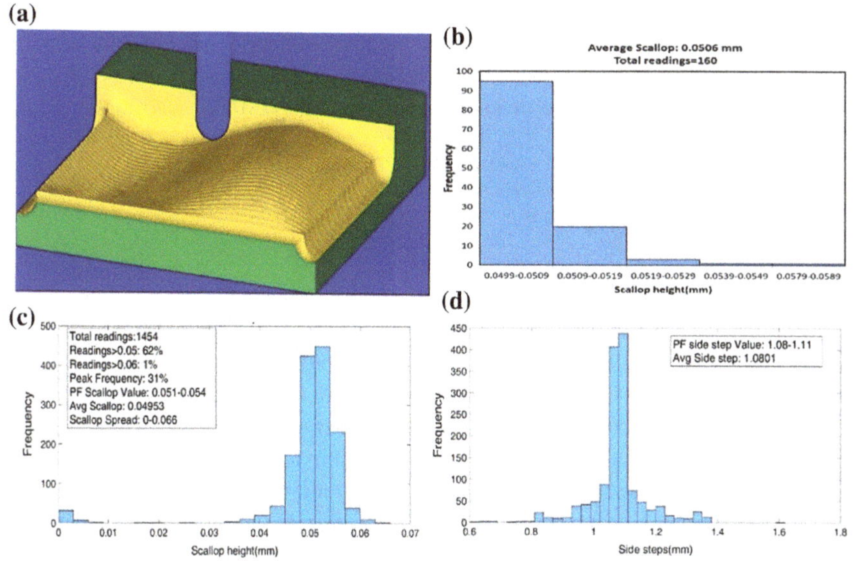

Fig. 44.5 Case 1; **a** Mastercam generated tool paths; **b** scallop estimation using VERICUT; **c** scallop estimation using developed system; and **d** sidestep variation

The scallop option in the VERICUT X-Caliper was used to measure the scallop height formed by two parallel intersecting cylinders chosen manually. It then identifies the cylinders being measured by displaying intersecting circles on the machined part model [5]. Figure 44.5b shows the generated histogram of the values estimated by the VERICUT (Range: 0.0499–0.058 mm). The scallop heights were estimated on 160 points (15 planes) throughout the surface using the X-Caliper tool of VERICUT. The average scallop was found to be 0.0506 mm. It is a known fact that the concave region has larger scallop heights and convex region has smaller scallop heights than the flatter region for the same sidesteps. The values estimated using VERICUT does not show this trend. This is because it does not consider the curvature of the surface while estimating scallops. The measurement/estimation is very approximate with limited points and sometimes unable to cover the whole machined surface.

Figure 44.5c shows the scallop height values estimated using the developed system (Range: 0.04–0.06 mm). The values below 0.05 mm indicate the convex regions and the values above 0.05 mm indicates concave regions. The scallop heights are calculated based on the sidesteps that are found using the method explained in Sect. 44.2.3. A similar trend is seen for the sidestep (Fig. 44.5d). In both the Fig. 44.5c and d, there is a spread but it is quite small. It is seen that the scallop frequency histogram from our system (Fig. 44.5c) has a well-defined symmetric peak around the expected scallop value (0.05 mm).

The spread in the sidestep values is due to the approximation in generating STL model of the machined stock, and the slicing of the STL model which may induce errors (in calculating local minima) as the plane (facet) and curve intersection was solved using numerical methods in the developed system. This error in sidestep values directly affects the scallop height estimation and therefore some values of scallop height (Fig. 44.5c) are outside the tolerance band (0.04–0.06 mm).

44.3.2 Case Study 2

In this case, a complex bicubic Bezier surface was designed (Fig. 44.6a) and a CNC part program was generated using an adaptive planar strategy [12]. The maximum permissible sidesteps in various regions were automatically calculated by the algorithm reported in [12] based on surface curvature while maintaining the surface finish (i.e., tolerance <0.05 mm). The sidesteps were found to be in the range of 0.75–1.5 mm to maintain the required surface quality [12].

Figure 44.6(b) shows the scallop heights frequency histogram estimated using VERICUT (Range: 0.0235–0.0485 mm). The scallop height was estimated at 120 points (9 planes). The average scallop was found to be 0.0372 mm. Similar to case study 1, it is seen that the VERICUT values are not considering the effect of surface curvature. The scallop values are under tolerance but its range is much wider than desired. The developed system shows that most of the values of scallop height are in the range of 0.04–0.05 mm (Fig. 44.6c). About 83% of values were found to be

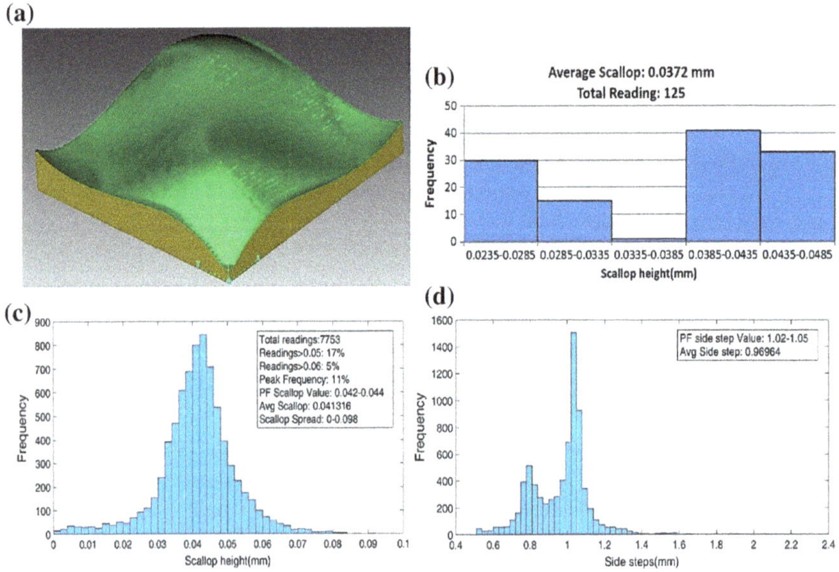

Fig. 44.6 Case 2; **a** Mastercam generated tool paths; **b** scallop estimation using VERICUT; **c** scallop estimation using developed system; **d** sidestep variation

below 0.05 mm with the average scallop of 0.041 mm. Hence, it agrees well with the desired values. According to the adaptive strategy reported in [12], the sidesteps for the concave regions were below 1.091 mm and for convex regions above 1.091 mm, a similar trend can be seen in Fig. 44.6d. The scallop heights were calculated based on these sidesteps values. The out-of-bound values seen in Fig. 44.6d may be again due to inherent error in the approximation done during STL generation and slicing algorithm.

As seen in case studies, the system is robust enough to estimate scallop for any kind of machined surface. The developed system could also be applied to five-axis machined parts. VERICUT cannot analyze these, and it fits the circle on the machined surface to find scallop heights but the surface generated in five-axis machines using flat end mill forms elliptical scallops. In essence, the developed system gives an accurate, fast, and robust estimation of scallop values from the simulation of a CNC part program without performing the actual cutting. It will thus help in significantly reducing product development lead time.

44.4 Conclusions

Despite a lot of research on CNC toolpath planning and simulation, no system exists that can estimate part quality (scallop height) directly from CNC part programs for the entire surface. Simulation software like VERICUT provides a very approximate tool. The system developed in this paper precisely caters to these requirements. The work proposes an algorithm for scallop height estimation from the simulation of machining using CNC part programs. In essence, it analyzes the simulated 3D machined surface, develops a mathematical model for the surface topography, and estimates the scallop and wavelength parameters. The system has been extensively tested for a variety of CNC part programs and surface topographies. It provides an accurate, fast, and robust estimate of part quality from the CNC part programs apriori.

References

1. Choi, B.K., Jerard, R.B.: Sculptured surface machining: theory and applications. Kluwer Academic Publishers, the University of Michigan, USA, Dordrecht (1998)
2. Elbert, G., Cohen, E.: Tool path generation for freeform surface models. Comput. Aided Des. **26**(6), 490–496 (1994)
3. Feng, H.Y., Huiwen, L.: Constant scallop-height tool path generation for three-axis sculptured surface machining. Comput. Aided Des. **34**(9), 647–654 (2002)
4. Park, S.C., Choi, B.K.: Tool-path planning for direction-parallel area milling. CAD Comput. Aided Design **32**(1), 17–25 (2000)
5. Getting Started with VERICUT Introduction. Introduction to VERICUT, https://community.ptc.com/sejnu66972/attachments/sejnu66972/installation_licensing/12339/1/vericut_help.pdf. Last accessed 09 Apr 2018

6. Ralph, W.L., Loftus, M.: Cusp geometry analysis in free-form surface machining. Int. J. Prod. Res. **30**(11), 2697–2711 (1992)
7. Bedi, S., Ismail, F., Mahjoob, M. J., C. Y.: Toroidal versus ball nose and flat bottom end mills. Int. J. Adv. Manf. Tech. 326–332 (1997)
8. Li, Y., Gu, P.: Free-form surface inspection techniques state of the art review. Comput. Aided Des. **36**(13), 1395–1417 (2004)
9. STL slice and create path, https://www.mathworks.com/matlabcentral/fileexchange/62113-slice_stl_create_path-triangles-slice_height. Last accessed 09 Apr 2018
10. MathWorks.: Polynomial curve fitting (polyfit), 1–19 (2017)
11. Mastercam.: Getting started series—basic 3D machining, http://eng.auburn.edu/~griffgj/training/Basic_3D_Machining.pdf. Last accessed 09 Apr 2018
12. Dhanda, M., Pande, S.S.: Adaptive tool path planning strategy for freeform surface machining using point cloud. Camput. Aided Design Appl. **16**(2), 289–307 (2019)

Chapter 45
Influence of Guides on Critical Speeds of Circular Saws

S. Singhania, P. Kumar, S. K. Gupta and M. Law

Abstract This paper investigates the influence of guides on the critical speeds of circular saws. Guides constrain the out-of-plane lateral motion of these rotating saws and have a stabilizing effect by increasing the critical speeds, below which the rotating saw is stable. We present expanded analytical formulations in which guides are modelled as multiple discrete spring-damper elements approximating the distributed nature of the guide pad and saw interactions. We observe that for a given guide pad area, convergence analysis is necessary to understand how many discrete spring-damper elements are actually necessary to approximate the distributed nature of the guide pad and saw interactions. Curiously, we observe that damping in the guides has no significant influence on the critical speed and that it only changes the nature of frequency-speed characteristics of the rotating saw. We also find that critical speeds are sensitive to guides modelled with a distribution of discrete spring elements along the radial and/or circumferential directions. These observations suggest that more generalized formulations that model the guide pads as distributed spring-damper systems, rather than multiple discrete spring-damper elements, are necessary. We expect our findings to instruct and advise the placement of guides on rotating circular saws, such that a preferential increase in critical speeds can be obtained to make possible high speed and productivity circular sawing operations in the wood-working and metal-cutting industries concerned with circular sawing processes.

Keywords Circular saw · Guides · Critical speeds instability

S. Singhania · P. Kumar · S. K. Gupta · M. Law (✉)
Machine Tool Dynamics Laboratory, Department of Mechanical Engineering,
Indian Institute of Technology Kanpur, Kanpur, UP 208016, India
e-mail: mlaw@iitk.ac.in

R. G. Narayanan et al. (eds.), *Advances in Computational Methods in Manufacturing*, Lecture Notes on Multidisciplinary Industrial Engineering,
https://doi.org/10.1007/978-981-32-9072-3_45

45.1 Introduction

Circular sawing is a common cutting operation across the wood-working and metal-cutting industries. It involves feeding a rotating circular saw into the work-piece to cut logs, bars, and tubes to lengths of desired sections. To meet the demands on high productivity in these industries, circular saws are desired to be operated at high speeds allowing more material to be removed per unit time. However, when the saw is rotating, forward- and backward-travelling waves are generated. The natural frequencies of the backward-travelling waves decrease with an increase in saw rotation speed, and the speed at which a natural frequency corresponding to a backward wave becomes zero is known as the critical speed [1]. When rotation speeds correspond to the critical speed, large amplitude vibrations appear, which can destroy the saw, the work material being cut, and the machine tool system. It is hence necessary to understand the factors governing critical speed instability, which is the focus of this paper.

Significant research attention across industries and academia has been paid on characterizing, predicting, and avoiding critical speeds. In some early seminal work, Mote and Nieh [2], and Szymani and Mote [3] studied the vibrations and critical speeds of rotating circular saws. Contemporarily, as a means of increasing the critical speeds of saws, Iwan and Moeller [4] presented formulations for saws constrained by guides. Guides constrain the out-of-plane lateral motion of rotating flexible saws and increase their critical speeds. Since these guides offer resistance to motion and act as a transverse load, Schajer [1], Hutton et al. [5], and Lehmann and Hutton [6], all investigated the influence of guided saws, with guides modelled as lumped spring elements. Hutton [5] observed that multiple guides were necessary to increase the critical speed. Though their [1, 5, 6] formulations allowed for the possibility of incorporating multiple guides, they ignored the influence of damping, if any, offered by guides. Damping in guides was investigated by Mote [7], and he observed that, if not favourably positioned, guides promote instability. However, his analysis was for the case of a stationary saw subjected to a moving load, and not the case we are interested in—a rotating, and constrained saw.

Since guides are actually pads distributed over a sector area constraining the lateral out-of-plane motion of the rotating saw, they need to be modelled as distributed systems. However, all the earlier reported work [1, 5–7] modelled the guide as a spring lumped at the geometric centre of the guide pad. Mohammadpanah and Hutton [8] were the first to investigate the distributed pad–saw interaction by modelling the guide as 16 linear springs distributed equally over the area of the guide pad. However, the effect of distributed damping in the guide was ignored. They also did not report on how many springs are sufficient to describe the distributed guide area, i.e. no convergence analysis of critical speeds to an increase in number of springs was presented.

This paper addresses the above two issues, i.e. understanding through systematic convergence studies, how many spring-damper systems are actually necessary to approximate the distributed pad–saw interaction, and understanding the role of damping in guides on the critical speeds of circular saws. To do so, we systematically modify the formulations for guided rotating circular saws, to include the possibility of incorporating multiple spring-damper systems. These modified formulations are presented in Sect. 45.2. Models are first numerically verified in Sect. 45.3 with results from a commercially available circular sawing stability code, CSAW® [9]. Verification is only provided for the undamped guide modelled as a single spring. Numerically verified models are then subjected to convergence analysis in Sect. 45.4, which includes discussions on the influence of damping on critical speeds, as well as discussion about the influence of modelling discrete spring-damper elements distributed radially, and circumferentially. The paper is concluded in Sect. 45.5.

45.2 Mathematical Model for the Critical Speed of Circular Saws

A circular saw, laterally constrained by a guide pad, is modelled as a flexible disc with a clamped inner radius, r_i, outer radius, r_0, rotating with an angular velocity of Ω in the clockwise direction as shown in Fig. 45.1. Ignoring any external load, the governing equation of motion of the circular saw in the stationary frame of reference with a guide pad modelled as multiple spring-damper systems, distributed equally over the area of guide pad, is modified from [5] to become:

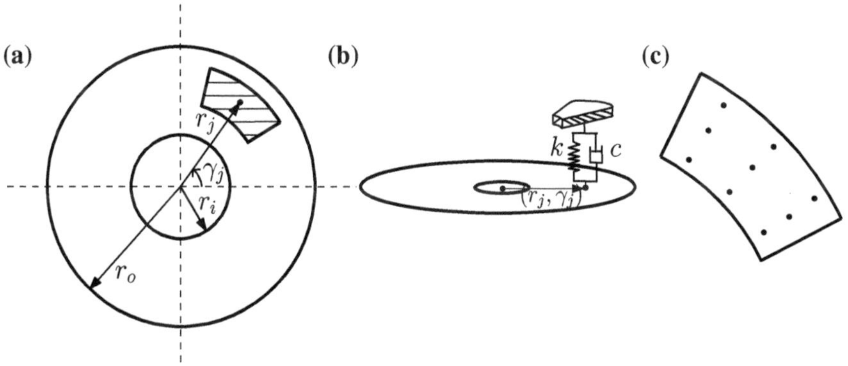

Fig. 45.1 Schematic of **a** circular saw and guide pad **b** the interaction between the guide pad and circular saw and **c** distribution of discrete spring-damper elements over the guide pad

$$
\begin{aligned}
&D\nabla^4 u + \rho h\left(\frac{\partial^2 u}{\partial t^2} + 2\Omega\frac{\partial^2 u}{\partial t\partial\gamma} + \Omega^2\frac{\partial^2 u}{\partial\gamma^2}\right) \\
&\quad - h\left[\left(\frac{1}{r}\right)\left(\frac{\partial}{\partial r}\right)\left(r\sigma_r\frac{\partial u}{\partial r}\right) + \left(\frac{1}{r^2}\right)\left(\frac{\partial}{\partial\gamma}\right)\left(\sigma_\gamma\frac{\partial u}{\partial\gamma}\right)\right] \\
&\quad + \sum_{j=1}^{J} k_j u\left(\frac{1}{r}\right)\delta(r-r_j)\delta(\gamma-\gamma_j) + \sum_{j=1}^{J} c_j\frac{\partial u}{\partial t}\left(\frac{1}{r}\right)\delta(r-r_j)\delta(\gamma-\gamma_j) = 0
\end{aligned}
\tag{45.1}
$$

wherein the first term represents bending stiffness, the second term represents inertial stress, the third term represents in-plane rotational stress, and the fourth and fifth terms represent stiffness and damping due to the interaction between the guide and circular saw, respectively, at the location of r_j, γ_j.

In Eq. (45.1), D is the flexural rigidity $(= Eh^3/12(1-v^2))$, E is the Young's modulus, h is circular saw thickness, v is the Poisson ratio, ρ is the density of circular saw; $\nabla^4\left(=\left[\frac{\partial^2}{\partial r^2} + \frac{1}{r}\frac{\partial}{\partial r} + \left(\frac{1}{r^2}\right)\left(\frac{\partial^2}{\partial\gamma^2}\right)\right]^2\right)$ is a bi-harmonic operator; $\delta(\cdot)$ is the Dirac delta function, k and c represent the stiffness and damping of the guide pad; J is the number of spring-damper elements; $u(r,\gamma,t)$ is the transverse displacement of the circular saw, and r, γ represent the radial and angular coordinates with respect to a fixed frame of reference.

Equation (45.1) is a fourth-order partial differential equation, and since it is difficult to solve it analytically, we use the Galerkin projection method to solve it [5]. To represent the deflection of circular saw due to bending, the transverse displacement $u(r,\gamma,t)$ can be written by a modal expansion. Therefore, the solution of Eq. (45.1) is assumed in the following form [5, 10]:

$$
u(r,\gamma,t) = \sum_{m=0}^{m}\sum_{n=0}^{n}[C_{mn}(t)\cos(n\gamma) + S_{mn}(t)\sin(n\gamma)]R_{mn}(r) \tag{45.2}
$$

wherein $C_{mn}(t)$ and $S_{mn}(t)$ are the unknown functions (modal coordinates), m and n represent the number of nodal circles and nodal diameters, respectively, and $R_{mn}(r)$ are radial functions chosen to satisfy the inner and outer boundary conditions of the circular saw. These $R_{mn}(r)$ can be approximated by polynomial functions as [5]:

$$
R_{mn}(r) = \sum_{i=0}^{4} E_{mni}r^{m+i}, \tag{45.3}
$$

wherein E_{mni} are unknown coefficients determined from the out-of-plane boundary conditions given by Eq. (45.4) and additional normalizing conditions, $R_{mn}(r_o) = 1$. Since the circular saw is clamped at its inner radius and is free at its outer radius, displacement and the slope at the inner radius will be zero, whereas bending moment and shear force at the outer radius will be zero. Therefore, the boundary condition at the inner and outer radius in terms of transverse displacement can be written as:

$$\left.\begin{array}{r} u(r_i,\gamma,t)=0, \frac{\partial}{\partial r}u(r_i,\gamma,t)=0, \\ M_{rr}(r_o,\gamma,t)=0, Q_r(r_o,\gamma,t)-\frac{1}{r_o}\left(\frac{\partial}{\partial\gamma}\right)M_{r\gamma}(r_o,\gamma,t)=0, \end{array}\right\} \tag{45.4}$$

wherein M_{rr} and $M_{r\gamma}$ are bending moments, and Q_r is the shear force [5].

Equation (45.1) can be rewritten with brevity, in terms of a differential operator (L) as:

$$L(u)=0, \tag{45.5}$$

wherein

$$L=D\nabla^4+\rho h\left(\frac{\partial^2}{\partial t^2}+2\Omega\frac{\partial^2}{\partial t\partial\gamma}+\Omega^2\frac{\partial^2}{\partial\gamma^2}\right)-h\left[\left(\frac{1}{r}\right)\left(\frac{\partial}{\partial r}\right)\left(r\sigma_r\frac{\partial}{\partial r}\right)+\left(\frac{1}{r^2}\right)\left(\frac{\partial}{\partial\gamma}\right)\left(\sigma_\gamma\frac{\partial}{\partial\gamma}\right)\right]$$
$$+\sum_{j=1}^{J}k_j\left(\frac{1}{r}\right)\delta(r-r_j)\delta(\gamma-\gamma_j)+\sum_{j=1}^{J}c_j\frac{\partial}{\partial t}\left(\frac{1}{r}\right)\delta(r-r_j)\delta(\gamma-\gamma_j).$$

Since the assumed form of the solution does not satisfy the equation of motion exactly, the residue can be minimized using the Galerkin projection approach, resulting in:

$$\int_0^{2\pi}\int_{r_i}^{r_o}L(u)(R_{ql}\cos(l\gamma))rdrd\gamma=0, \tag{45.6}$$

$$\int_0^{2\pi}\int_{r_i}^{r_o}L(u)(R_{ql}\sin(l\gamma))rdrd\gamma=0, \tag{45.7}$$

wherein $q=0,1,\ldots,m,\ l=0,1,\ldots n$. Solving and simplifying Eqs. (45.6) and (45.7), we get:

$$\sum_{m=0}^{m}\alpha_1\left[\delta^{(1)}_{qml}\frac{d^2C_{ml}}{dt^2}+\delta^{(2)}_{qml}\frac{dS_{ml}}{dt}+\delta^{(3)}_{qml}C_{ml}\right]+\sum_{m=0}^{m}\left[N^{(1)}_{qml}C_{ml}+N^{(2)}_{qml}S_{ml}\right]=0, \tag{45.8}$$

$$\sum_{m=0}^{m}\beta_1\left[\delta^{(1)}_{qml}\frac{d^2S_{ml}}{dt^2}-\delta^{(2)}_{qml}\frac{dC_{ml}}{dt}+\delta^{(3)}_{qml}S_{ml}\right]+\sum_{m=0}^{m}\left[N^{(3)}_{qml}C_{ml}+N^{(4)}_{qml}S_{ml}\right]=0, \tag{45.9}$$

$$\alpha_1=\left\{\begin{array}{ll}\pi, & l\neq 0\\ 2\pi, & l=0\end{array}\right\}, \beta_1=\left\{\begin{array}{ll}\pi, & l\neq 0\\ 2\pi, & l=0\end{array}\right\}.$$

For the sake of brevity, $\delta_{qml}^{(1)}, N_{qml}^{(1)}, N_{qml}^{(2)}, N_{qml}^{(3)}, N_{qml}^{(4)} \delta_{qml}^{(2)}, \delta_{qml}^{(3)}$ are not defined here, and the reader is directed to [5] for details. The above Eqs. (45.8) and (45.9) can be rewritten in a compact form as:

$$[A]\{\ddot{x}\} + [B]\{\dot{x}\} + [C]\{x\} = 0 \tag{45.10}$$

where the vector $\{x\}$ is an array of $\{C_{0,0}, C_{0,1},\ldots, C_{1,0}, C_{1,1},\ldots, C_{m,n}, S_{0,1}, S_{0,2},\ldots, S_{1,1}, S_{1,2}, \ldots, S_{m,,n}\}$. In Eq. (45.10), $[A]$ is the mass matrix, $[B]$ is the gyroscopic matrix, and $[C]$ is the stiffness matrix. Since $[B]$ and $[C]$ depend on the rotational speed of circular saw, dynamics of circular saw depends on rotational speed. As Eq. (45.10) is a second-order linear differential equation, the solution of the equation can be assumed in the form of $x(t) = \{X\}e^{\lambda t}$. On substituting the assumed form of the solution, Eq. (45.10) forms an eigenvalue problem which can be solved for every combination of rotational speed and other parameters. The imaginary part of the eigenvalue, λ of this eigenvalue problem, corresponds to a natural frequency of the circular saw. To reduce the effective number of parameters, we non-dimensionalize Eq. (45.10) as:

$$\Omega_0 = \left[\frac{\rho h r_o^4}{D}\right]^{\frac{1}{2}} \Omega, \zeta = k\left(\frac{r_o^2}{D}\right), \phi = \left[\frac{\rho h r_o^4}{D}\right]^{\frac{1}{2}} \omega, c_1 = \frac{c r_o^2}{\sqrt{\rho h r_o^4 D}}$$

wherein c_1 is the non-dimensional damping, ζ is the non-dimensional stiffness, ω is the natural frequency, ϕ is the non-dimensional natural frequency, and Ω_0 is the non-dimensional rotational speed.

We solve the above set of equations and, at first, verify the formulations against results from CSAW®, as discussed next, in Sect. 45.3, followed by convergence testing and investigations with damping in Sect. 45.4.

45.3 Numerical Verification of Critical Speed Model

As a preliminary check of our proposed formulations, we first verify our results with those obtained from a commercially available circular sawing stability code, CSAW®. Comparisons are made with frequency-speed diagrams, known as the Campbell diagram. We limit our numerical checks for the case of an unguided saw, and for the case of a guided saw, with the guide modelled as an undamped point spring—since CSAW® does not have the facility of including damping. We limit our comparisons for the number of nodal circles, m as 1, and the number of nodal diameters, n as 4. The saw, assumed to be made of steel, is clamped at its inner radius, $r_i = 42.5$ mm and has a free outer radius, $r_o = 142.5$ mm.

Campbell diagrams for the unguided case are compared in Fig. 45.2. As is evident, the critical speed obtained from the proposed analytical model (Fig. 45.2a) occurs at 8497 rpm, whereas the critical speed obtained from CSAW® (Fig. 45.2b)

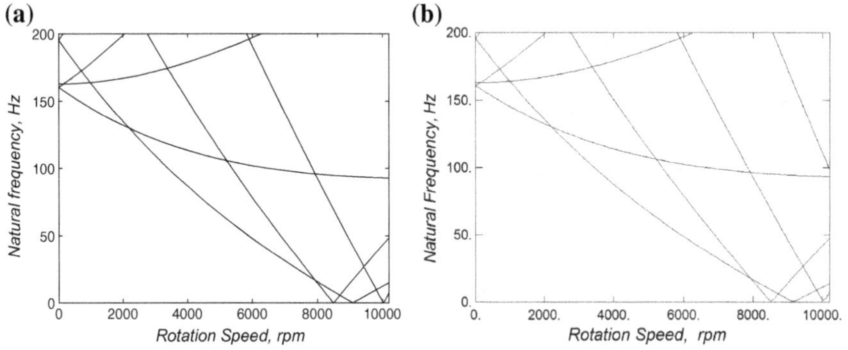

Fig. 45.2 Campbell diagram for unguided circular saw obtained **a** using analytical model and **b** using CSAW®

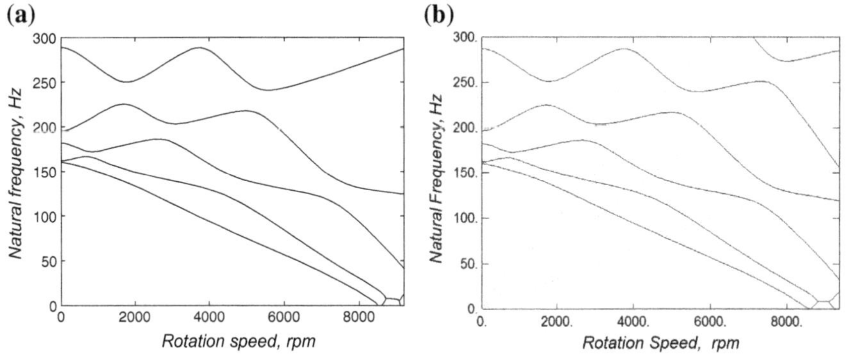

Fig. 45.3 Campbell diagram for guided circular saw obtained **a** using analytical model and **b** using CSAW®

occurs at 8519 rpm, which is approximately same as our analytical model. In the Campbell diagram obtained from CSAW®, there appear four straight lines from the origin. These represent the first four engine order (EO) lines and correspond to imperfections or unbalance in the rotating parts of the system. These faults generate harmonic and sub-harmonic levels of vibrations [8] as shown in Fig. 45.2b.

Campbell diagrams for the guided case are compared in Fig. 45.3. The guide pad has been modelled as a sector of a circular disc with orientation of inner and outer angle as 5° and 36°, respectively, whereas, inner and outer radii of the guide pad are 57 and 142.5 mm, respectively. The spring element, located at the geometric centre of the pad, has a stiffness 7.5×10^8 N/m. We observe that the analytical model predicted response (Fig. 45.3a) is consistent with the results from CSAW® (Fig. 45.3b), and that there is no change in the critical speed with a single guide. This observation is consistent with those reported earlier in [5].

Having numerically verified the analytical model, we use the analytical model to understand how many spring-damper systems are actually necessary to approximate distributed guide and saw interactions, and to understand the role of damping in guides on the critical speeds of circular saws, as is discussed next, in Sect. 45.4.

45.4 Results and Discussion

In this section, we first present convergence analysis for understanding how many spring-damper systems are actually necessary to approximate distributed guide and saw interactions, for a given guide area, followed by analysis for damped guides, followed by analysis on the influence of modelling multiple spring-damper systems radially and circumferentially. Guide pad area for all results presented herein is 4614 mm^2. Guide location is the same as described in Sect. 45.3. All combinations of the spring-damper elements approximating the guide pad are distributed uniformly over the guide pad area. All results presented herein are for the mode corresponding to the first critical speed. All results are presented in the non-dimensional form of the Campbell diagrams.

45.4.1 Convergence Analysis of Discrete Multiple Spring Elements to Model the Guide Pad

For a given guide area, and location, it is, in general, difficult to a prior establish the required number of discrete spring elements, necessary to capture the saw and guide interactions. We neglect the damping in the guide presently, influence of which is discussed separately in Sect. 4.2. We analyse sensitivity of the first critical speed with five combinations of spring elements approximating the same guide area, i.e. with J as 1, 4, 9, 16, and 25. Overall effective stiffness k_j in each case was taken to be as 7.5×10^8 N/m. Results for all cases are shown in Fig. 45.4.

As is evident from Fig. 45.4, an increase in number of springs increases the critical speed. However, beyond 16 springs, the critical speed does not appear to change, suggesting that 16 springs are sufficient to describe the size and location of the guide pad under consideration. Interestingly, we observe that with four discrete springs, the critical speed increases, which is not consistent with the findings reported in [5], in which they found that to increase the critical speed, at least eight discrete uniformly distributed guides modelled as a discrete point spring elements are necessary. This suggests that it is indeed necessary to model the pad–saw interaction as multiple discrete spring elements.

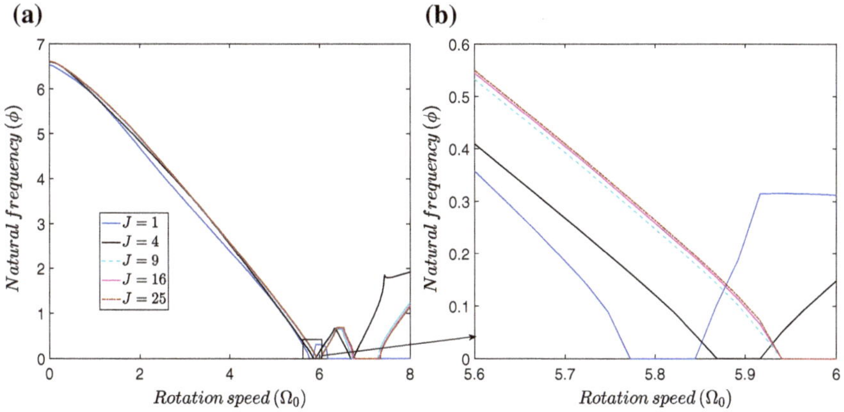

Fig. 45.4 Campbell diagram for multiple discrete spring elements distributed equally over the guide pad corresponding to the lowest mode

45.4.2 Influence of Damping on the Critical Speed

Since guides offer not just stiffness, but also damping, influence of damping on the critical speed is reported herein. To delineate the influence of damping, i.e. without stiffness, we present results for a guide with only damping in Fig. 45.5a and for a guide with stiffness and damping in Fig. 45.5b. For the case of a guide with damping only, we model the guide to have an overall effective damping of $c_j = 17.2\,\mathrm{N/m/s}$, for equivalent 16 damping elements. For the case of a guide modelled with discrete spring-damper elements, we take the number of elements to be 16, with an equivalent overall stiffness and damping of $7.5 \times 10^8\,\mathrm{N/m}$ and $17.2\,\mathrm{N/m/s}$, respectively.

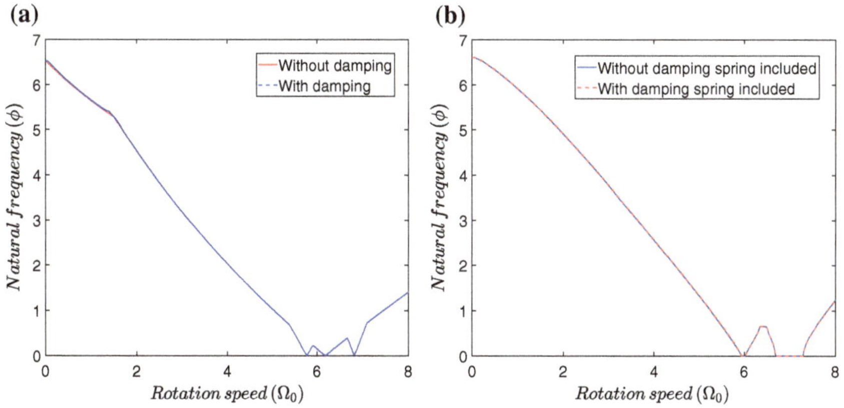

Fig. 45.5 Campbell diagram for circular saw with pad–saw interaction modelled as **a** discrete damper elements **b** discrete spring-damper elements corresponds to the lowest mode

As is evident from Fig. 45.5a, saws with guides with damping only, i.e. without springs, do not influence the critical speed and behave like unguided saws, which are curious. We also observe that damping only affects the frequency-speed characteristics. Furthermore, even for the case of guides modelled with multiple discrete spring-damper systems, the critical speed does not change when compared to the critical speed for the case of guides modelled only as springs, as is evident from Fig. 45.5b. These observations suggest that damping has no influence on the critical speeds of saws for the given guide pad configuration (mentioned in Sect. 45.3), which is unusual and unexpected and needs further investigations.

45.4.3 Influence of Circumferential and Radial Distribution of Spring Elements on the Critical Speed

In this section, we present the influence of orientation of the distribution of discrete spring elements that are to approximate a given guide pad area, on the critical speed. We present results for the guide pad approximated by different number of discrete spring elements, distributed uniformly in the circumferential and radial directions for a given area. Results for a uniform distribution of discrete spring elements along the circumferential direction (with fixed radial location at 99.75 mm) are shown in Fig. 45.6, and results for uniform distribution of discrete spring elements along the radial direction (with fixed circumferential location at 20.5°) are shown in Fig. 45.7. Since damping was shown to not influence the critical speed, we neglect it herein. As before, the overall effective stiffness in each case is taken to be 7.5×10^8 N/m. As is evident from Fig. 45.6, as the number of discrete spring elements increases along the circumferential direction, the critical speed increases and becomes constant after $J = 8$. However, as is evident from Fig. 45.7, for the case of different discrete spring elements along the radial direction

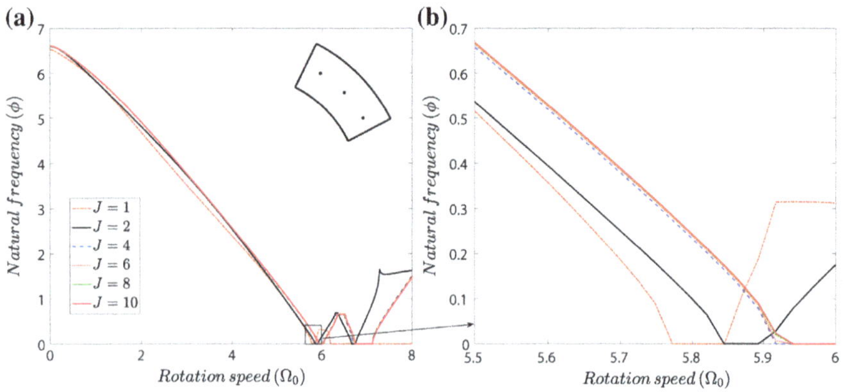

Fig. 45.6 Influence of multiple spring element over the guide pad in circumferential direction

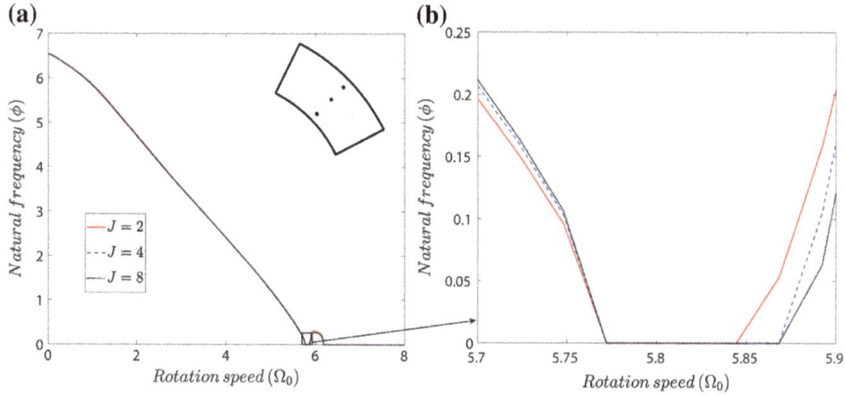

Fig. 45.7 Influence of multiple spring element over the guide pad in radial direction

at a fixed orientation, there is no observable change (increase) in critical speed with an increase in the discrete spring elements. We can hence conclude that for a given guide pad area modelled as discrete springs, an increase in discrete spring elements distributed along the circumferential direction leads to an increase in critical speed, whereas an increase in discrete spring elements distributed along the radial direction does not have any influence on critical speed.

These observations, which are different than those observed in the convergence analysis in Sect. 44.4.1, where we observed the critical speed to not change beyond 16 discrete spring elements approximating the same guide area, suggests that the approach to distributing discrete springs to approximate the guide area has a large bearing on the critical speed. Furthermore, we suspect that these results are not generalizable but need to be investigated separately for each guide pad area, and location, pointing to the need for a more generalized formulation that models the guide pads as distributed spring-damper systems, rather than multiple discrete spring-damper elements. Further investigations are also necessary to understand how these findings might change for the case of a saw with multiple guides.

45.5 Conclusion

This paper reports on the influence of guides on the critical speed of rotating saws. Since guides are known to improve critical speeds, understanding how they are modelled is essential to advise guide placement. Using numerically verified analytical models that include guides modelled as multiple discrete spring-damper elements, we observe that damping has no influence on critical speed, which is confounding, and needs further investigations. We also observe that for guides approximated with multiple discrete spring elements, the number, orientation, and location of the springs govern the increase in the critical speed, if any. We observe that for a given guide area, it is necessary to carry out convergence analysis to understand how many discrete spring elements are actually necessary to approximate distributed guide and saw interactions. We also observe that critical speeds are very sensitive to guides modelled with a distribution of discrete spring elements along the radial direction and/or circumferential direction. These observations suggest that a more generalized formulation that models the guide pads as distributed spring-damper systems, rather than multiple discrete spring-damper elements are necessary. Experimental validation is also necessary and is planned.

Acknowledgements This research was supported by the Government of India's Science and Engineering Research Board's Early Career Research Award—SERB/ECR/2016/000619.

References

1. Schajer, G.S.: Guided circular saw critical speed theory." Proceedings SawTech 1989, Oakland, CA, USA
2. Mote, C.D., Nieh, L.T.: On the foundation of circular-saw stability theory. Wood Fiber Sci. **5**(2), 160–169 (1973)
3. Szymani, R., Mote, C.D.: Principal developments in thin circular saw vibration and control research. Holz als Roh-und Werkst. **35**, 219–225 (1977)
4. Iwan, W.D., Moeller, T.L.: The stability of a spinning elastic disk with a transverse load system. J. Appl. Mech. **43**(3), 485–490 (1976)
5. Hutton, S.G., Chonan, S., Lehmann, B.F.: Dynamic response of a guided circular saw. J. Sound Vib. **112**(3), 527–539 (1987)
6. Lehmann, B.F., Hutton, S.G.: Self-excitation in guided circular saws. J. Vib. Acoust. Stress. Reliab. Des. **110**(3), 338–344 (1988)
7. Mote Jr., C.D.: Moving-load stability of a circular plate on a floating central collar. J. Acoust. Soc. Am. **61**(2), 439–447 (1977)
8. A. Mohammadpanah and S. G. Hutton, "Flutter instability speeds of guided splined disks: an experimental and analytical investigation," Shock Vib. **2015** (2015)
9. Schajer, G.S.: CSAW 4.0 Computer software for optimizing circular saw design. Wood Mach, Institute, Berkeley, CA USA (2007)
10. Hagedorn, P., DasGupta, A.: Vibrations and waves in continuous mechanical systems. Wiley (2007)

Chapter 46
Investigations on the Influence of Radial Run-Out on Cutting Forces for Serrated Cutters

P. Bari, M. Law and P. Wahi

Abstract Serrated cutters with their complex local tool geometries result in continuously changing chip thickness and engagement conditions due to a change in local radius along the serration height. This change in geometry results in a reduction of cutting forces, and hence these cutters are favoured in the rough cutting of difficult-to-cut materials. The change in local radius of the cutter is also influenced by the radial run-out on the tool. Run-out, though undesirable, is inevitable. How this run-out influences the local radius, chip thickness, and forces is not entirely understood for serrated cutters. This paper fills that gap, by presenting a model that factors in the influence of radial run-out on cutting forces for serrated cutters. Investigations with different sets of cutting parameters show that for cutting with a low feed, and high axial and radial depths of cut, the influence of run-out on the cutting forces is negligible. For the case of cutting with high feeds however, we find that run-out adversely impacts cutting forces for any combination of axial and radial depths of cut. We also observe that with certain combinations of cutting parameters and run-outs, serrated cutters lose their advantage over regular end mills with similar levels of run-outs. Our findings suggest that serrated cutters may be used even with high levels of run-outs when the feed is low, and that their use should be avoided for any level of run-out when cutting with high feed rates.

Keywords Serrated cutters · Run-out · Chip thickness · Cutting forces

46.1 Introduction

Serrated cutters with their complex local tool geometries result in continuously changing chip thickness and engagement conditions due to a change in local radius along the serration height. This change in geometry results in a preferential

P. Bari · M. Law (✉) · P. Wahi
Department of Mechanical Engineering, Indian Institute
of Technology Kanpur, Kanpur, Uttar Pradesh 208016, India
e-mail: mlaw@iitk.ac.in

R. G. Narayanan et al. (eds.), *Advances in Computational Methods in Manufacturing*, Lecture Notes on Multidisciplinary Industrial Engineering,
https://doi.org/10.1007/978-981-32-9072-3_46

reduction of cutting forces, and hence these cutters are favoured in the rough cutting of difficult-to-cut materials. The change in local radius of the cutter is also influenced by the radial run-out on the tool. Run-out is characterized as the difference in the geometrical axis of the tool from the spindle's rotational axis. Run-out, though undesirable, is inevitable due to how the tool is clamped in the tool holder and also due to manufacturing tolerances and errors. How this run-out influences the local radius, chip thickness, and forces is not entirely understood for serrated cutters. The main focus of this paper is hence to incorporate run-out in predictive force models for serrated cutters and present systematic investigations on how different levels of run-out influence cutting forces for serrated cutters for different combinations of cutting parameters.

Since run-out changes the nature of cutting forces, contributes to surface roughness errors on machined workpieces, and also contributes to premature tool wear and potential breakage of the tool, a lot of research attention has been paid on modelling and understanding the influence of run-out for milling cutters. Early seminal work by Martellotti [1, 2] reported on the influence of run-out in milling processes. Later, Kline and DeVor [3] presented mathematical models incorporating the influence of run-out on force prediction for regular end mills. Zheng et al. [4] reported on face milling force profile changes with run-outs. Wan et al. [5] proposed new techniques to calibrate cutting force coefficients and run-out parameters in peripheral milling. The effect of run-out on cutting force variation and surface roughness generation was also studied by Krüger and Denkena [6], in which they identified run-out parameters from measured process forces using model-based run-out identification methods. Schmitz et al. [7] explored the effect of milling cutter run-out on surface topography, surface location error, and stability in end milling operations. More recently, Li et al. [8] addressed the issue of run-out in five-axis milling processes. Most literature addressing the influence of run-out in milling has focused only on regular end mills, and even though run-out is an important factor for end mills with serrations, the reported research on serrated end mills has focused more on modelling the serration geometry and its influences than on run-out.

Some classical early work on straight fluted serrated end mills was presented by Tlusty et al. [9], in which it was shown that due to the reduced contact between the tool and the workpiece on account of serrations, cutting forces reduce. Merdol and Altintas [10, 11], and Dombovari et al. [12] reported modelling of cylindrical and tapered serrated end mills with a generalized representation of serration profiles using cubic splines without considering the effect of run-out. Later, Campomanes [13] presented a detailed mechanics-based model that included the influence of helix with sinusoidal serrations, and used an approximate chip thickness model. He also made passing reference to how run-out does not have a significant effect on the cutting forces for serrated end mills at lower feed rates. His work, however, or the work of others before, and/or after him, does not support the claim based on model-based systematic investigations. Since serrated cutters preferentially reduce cutting forces, the influence of run-out on cutting forces, if any, and if run-out negates the positive influence of serrations in reducing forces, needs to be understood comprehensively based on systematic model-based investigations and is the main motivation of the present work.

We present an expanded geometric model for serrated cutters that includes the influence of radial run-out, in Sect. 46.2. We limit our discussions in this paper to serrations only of the sinusoidal kind. We ignore axial run-out presently. We also assume that run-out is measurable on serrated tools. Measuring run-out on regular end mills is possible and has been reported on elsewhere; see, for example, [14–16]. However, measuring run-out on serrated cutters due to constantly changing radius along the serration wavelength is non-trivial and needs to be addressed separately. Assuming run-out to be measurable, we systematically investigate its influence on force profiles for different combinations of cutting parameters in Sect. 46.3. We systematically discuss the influence of run-out for different levels of feed, and axial and radial depths of cut. This is followed by the main conclusions in Sect. 46.4.

46.2 Geometric Model of Serrated Cutter with Run-Out

This section describes the geometry of serrated cutters considering the influence of radial run-out. Models described in this section build on the classical work done by Merdol and Altintas [11], and by Dombovari et al. [12], and also on our own earlier reported work [17], by including the influences of run-out.

46.2.1 Modelling of Serration Profile

A schematic and a cross-sectional view of a sinusoidal serrated cutter with run-out is shown in Fig. 46.1. Due to the run-out, cutter geometrical centre, denoted by O_g, deviates from spindle rotation centre O_r, by a constant radial deviation, ρ.. The run-out angle between direction of radial offset (deviation) and the nearest tooth at the bottom of the tool is denoted by δ. We assume that there is no axial run-out or any cutter geometric axis tilt; i.e. ρ is constant along the cutter height. Due to the run-out, since the whole system rotates about the spindle axis through O_r, it is convenient to model all things with respect to O_r, and hence the xyz coordinate frame is attached to O_r as shown.

The cutter can have N number of flutes (teeth), but as an example, only three (ith, $(i+1)$th, and $(i+l)$th) flutes are shown in Fig. 46.1. As there is a wavy surface along the flute of the serrated cutter, the local geometrical radius changes along the flute and the height. The local geometrical radius which is measured with respect to geometrical axis (dotted line through O_g shown in Fig. 46.1a) which is parallel to z-axis for the ith flute at the height z is defined as:

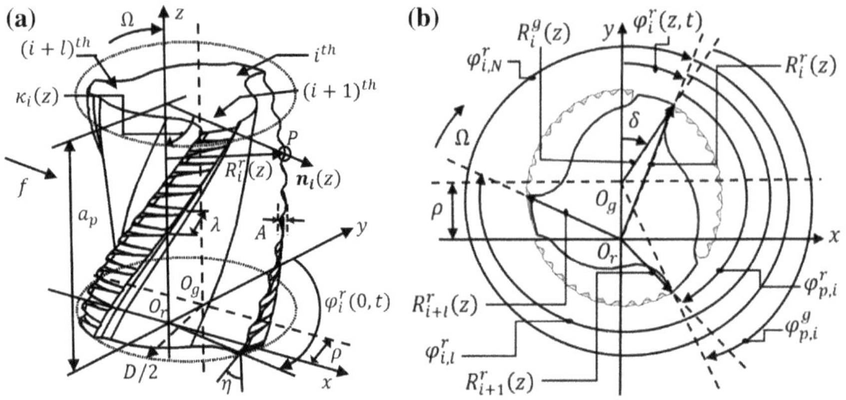

Fig. 46.1 **a** Geometry of the serrated cutter with run-out. Figure is adapted and modified from [17]. **b** Cross-sectional view at height z with run-out

$$R_i^g(z) = \frac{D}{2} - \Delta R_i^g(z) \tag{46.1}$$

where D is shank diameter (nominal diameter) of the cutter, and $\Delta R_i^g(z)$ is the variation in local geometrical radius for the sinusoidal serration profile:

$$\Delta R_i^g(z) = \frac{A}{2} - \frac{A}{2}\sin\left(\frac{2\pi z}{\lambda\cos\eta} - \psi_i + \frac{\pi}{2}\right) \tag{46.2}$$

where A is the serration amplitude, which is half of the peak-to-peak serration height; λ is the wavelength; η is the helix angle; and ψ_i is the phase shift due of the serrations on different flutes, explained in [17]. All geometries are measured with respect to spindle rotational axis which do not align with cutter geometrical axis due to eccentricity between them caused by the run-out. Due to the run-out, we define another local radius called the rotational radius with respect to the z-axis through O_r as follows:

$$R_i^r(z) = \left[\left(\rho + R_i^g(z)\cos\left(\delta - \sum_{k=1}^{i-1}\varphi_{p,k}^g - \frac{2z\tan\eta}{D}\right)\right)^2 + \left(R_i^g(z)\sin\left(\delta - \sum_{k=1}^{i-1}\varphi_{p,k}^g - \frac{2z\tan\eta}{D}\right)\right)^2\right]^{0.5} \tag{46.3}$$

The local geometrical radius $R_i^g(z)$ for a representative four-fluted sinusoidal serrated cutter without and with run-out is shown in Fig. 46.2a and b, respectively. Cutter parameters are as follows: shank diameter D = 16 mm, serration amplitude

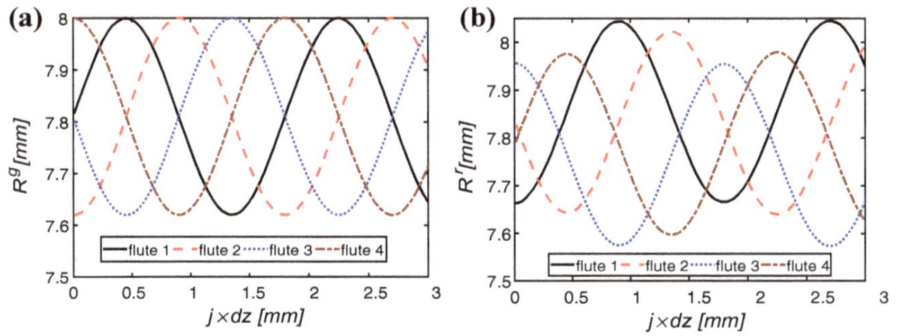

Fig. 46.2 **a** Variation of local geometrical radius, R^g, along height without run-out. **b** Variation of local rotational radius, R^r, along height with run-out: $\rho = 50\,\mu m, \delta = 30°$

and wavelength $A = 0.38$ mm and $\lambda = 1.8$ mm; and, helix angle $\eta = 20°$. Run-out is assumed as $\rho = 50\,\mu$m and $\delta = 30$. Comparing Fig. 46.2a and b, we see that run-out makes the radius irregular.

In addition to causing a change in the local radius, run-out also changes the instantaneous radial immersion angle. The angular position for the ith flute at height z, measured from the y-axis in a clockwise direction considering the run-out, called the instantaneous radial immersion angle, is calculated as follows

$$\varphi_i^r(z,t) = \Omega t + \sum_{k=1}^{i-1} \varphi_{p,k}^r - \frac{2z\tan\eta}{D} \tag{46.4}$$

where Ω is the clockwise spindle speed (rad/sec) and $\varphi_{p,k}^r$ is new pitch angle with respect to O_r due to the run-out. The relationship between the new pitch angle $\varphi_{p,k}^r$ with respect to O_r and the old geometrical pitch angle $\varphi_{p,k}^g$ with respect to O_g is shown in Fig. 46.3a. From Fig. 46.3a, using the cosine triangle formula, $\varphi_{p,k}^r$ is calculated as follows:

$$\varphi_{p,k}^r = \cos^{-1}\frac{\left(R_i^r(z)\right)^2 + \left(R_{i+1}^r(z)\right)^2 - \left(L_i^r(z)\right)^2}{2R_i^r(z)R_{i+1}^r(z)} \tag{46.5}$$

where

$$L_i^r(z) = \sqrt{\left(R_i^g(z)\right)^2 + \left(R_{i+1}^g(z)\right)^2 - 2R_i^g(z)R_{i+1}^g(z)\cos\varphi_{p,k}^g}$$

In addition to a change in the instantaneous radial immersion angle and pitch angles, run-out also changes the axial immersion angle, and that is calculated as detailed in [17]. Furthermore, the changed pitch angle, due to run-out, may result in a delay between formation of the current surface being generated and the previous

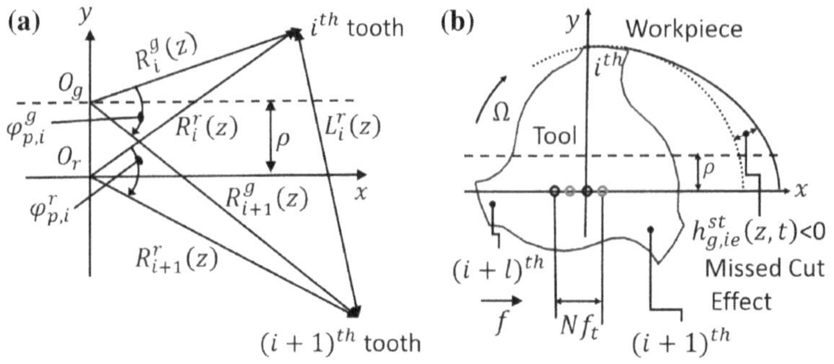

Fig. 46.3 **a** Calculation of the new rotational pitch angle, $\varphi_{p,k}^r$, due to the run-out on serrated cutters. **b** Missed cut effect with run-out for serrated cutter

surface generated, due to which redistribution of multiple delays may occur, and this can be factored in as discussed in [17] and as discussed next in Sect. 2.2.

46.2.2 Local Chip Thickness

Elemental physical static chip thickness is defined as the local distance between previous and current cut surfaces in the direction of normal vector $\boldsymbol{n}_i(z)$ of the flute considering run-out with circular tool path approximation as:

$$h_i^{st}(z,t) = g_i(z,t) \min_{l=1}^{N} \left[\left(R_i^r(z) - R_{i+l}^r(z) \right) + f_{i,l}(z,t) \sin\varphi_i^r(z,t) \right] \sin\kappa_i(z) \quad (46.6)$$

where $f_{i,l}(z,t)$ is the corresponding feed motion during delay time $\tau_{i,l}$, $\kappa_i(z)$ is the axial immersion angle, $g_i(z,t)$ is the screening function due to radial immersion and missed cut (Ref. Fig. 46.3b) effect, explained in [17], and $R_i^r(z)$ is calculated from Eq. (46.3). The missed cut effect considering the run-out creates a non-uniform chip thickness profile as shown schematically in Fig. 46.3(b).

Variation in the chip thickness profile for a four-fluted sinusoidal serrated cutter considering zero run-out is shown in Fig. 46.4a, and considering run-out ($\rho = 50\,\mu\text{m}, \delta = 30°$) is shown in Fig. 46.4b. The cutting conditions chosen are: depth of cut, $a_p = 3$ mm; feed, $f_t = 0.25$ mm/tooth; and speed, 5000 rpm, for slotting, i.e. 100% immersion. Comparing Fig. 46.4a and b, we see that due to the run-out, the flute chip thickness for the first and second flutes increases by 150%, and for the other two flutes, it decreases by 5%. Since chip thickness is different with run-outs, cutting forces which depend on chip thickness are also expected to be different.

The force model presented in [17] is used incorporating Eq. (46.6) to investigate the influence of run-out on force profiles, as discussed next, in Sect. 46.3. Furthermore, since the chip thickness also depends on the instantaneous entry and

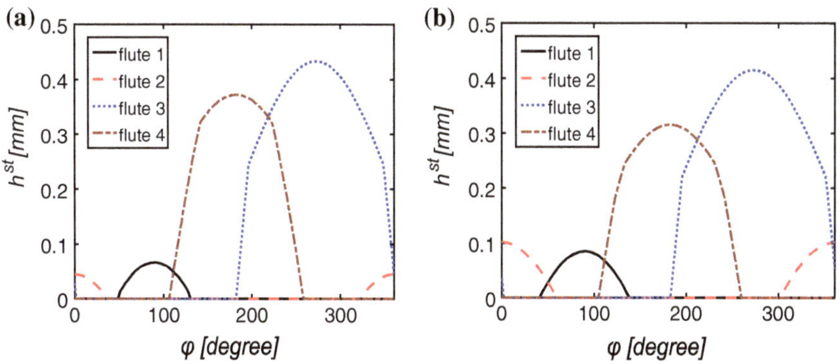

Fig. 46.4 Chip thickness (hst) distribution at tool tip ($z = 0\,\text{mm}$) **a** without run-out and **b** with run-out:$\rho = 50\,\mu\text{m}$, $\delta = 30°$

exit angles, these too will change with run-out, and they will also result in the force model being updated. However, for brevity, those formulations are not reported herein, since we observed that even without the updated entry and exit angle formulations, the force profiles do not change much.

46.3 Effect of Run-Out for Serrated Cutter at Different Operating Conditions

In this section, we discuss the influence of run-out on cutting forces for different combinations of cutting parameters. Our serrated cutting force model without run-out was numerically verified against other reported results and was also experimentally validated in our earlier reported work in [17]. We have also numerically verified our run-out model, albeit for the regular end mill case (i.e. by putting $A = 0$ in force model), using a commercially available advanced machining process simulation software, CutPRO [18], but have not reported those results here for brevity. Herein, we only present numerical investigations for our serrated model with run-out.

Although serrated cutters are more effective at reducing forces with low feeds than at high feeds, the effect of run-out is investigated for all possible operating conditions. Two types of feeds, one low (0.02 mm/tooth) and the other high (0.25 mm/tooth), are chosen. We present results with a low feed in Sect. 3.1 and present a subset of results with a high feed in Sect. 3.2. At each feed, two levels of engagement conditions, one low (10% up milling) and the other high immersion (100% slotting), are taken. Again, for each engagement condition, two levels of depths of cut, one low ($a_p = 3\,\text{mm}$) and the other high ($a_p = 8$ mm), are chosen. Other cutting conditions for all cases are as follows: run-out angle $\delta = 30°$, 5000 rpm, helix $\eta = 20°$. For each case, cutting force is compared for zero run-out, small run-out (5 μm), and large run-out (50 μm), as discussed in subsequent subsections. To also understand, how/if run-out on serrated cutters can potentially negate the positive influence of serrations, which is to otherwise reduce force, we also discuss in Sect. 46.3.3 results for regular end mills with run-outs.

46.3.1 Influence of Run-Out at Low Feed (0.02 mm/tooth) Rates

Effect of run-out on serrated cutting forces (F_x) at low feed (0.02 mm/tooth) rates is shown in Fig. 46.5. From Fig. 46.5, we observe that the ratio of maximum to average force increases with an increase in run-out. Maximum difference occurs between zero run-out and large run-out force profiles. For 10% up milling, at a low depth of cut ($a_p = 3$ mm) (Fig. 46.5a), the increase of the peak amplitude of up to 25% from the case of zero run-out is observed, and at a high depth of cut ($a_p = 8$ mm) (Fig. 46.5b), the increase of the peak amplitude of up to 12% is observed.

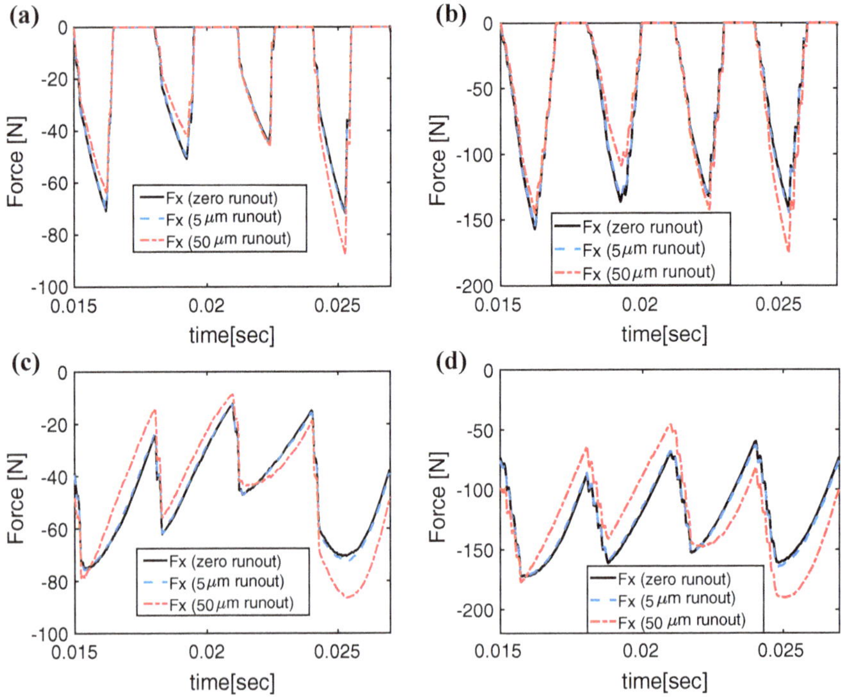

Fig. 46.5 Influence of run-out on cutting force (F_x) for a sinusoidal serrated end mill for low feed rates: **a** $a_p = 3$ mm, at 10% up milling; **b** $a_p = 8$ mm, at 10% up milling; **c** $a_p = 3$ mm, at 100% milling (slotting); and **d** $a_p = 8$ mm, at 100% milling (slotting)

For 100% milling (slotting) at a low depth of cut ($a_p = 3\,\text{mm}$) (Fig. 46.5c), the increase of the peak amplitude of up to 13% is observed, and at a high depth of cut ($a_p = 8\,\text{mm}$) (Fig. 46.5d), the increase of the peak amplitude of up to 10% is observed.

From these investigations, we can conclude that at low feeds, there is no substantial increase in the maximum force, or a significant change in the force profile, even with large values of run-out. This is interesting and is consistent with the insightful observations made in passing by Campomanes [13]. We also see that for the cases investigated, there is no significant change in the mean value of the forces, even with large run-outs on serrated cutters. These observations suggest that serrated end mills may be used even with large run-outs on them at low feeds, and potentially any combination of axial and radial depths of cuts, without resulting in any potentially significant damage to the tool, or the surface profile.

46.3.2 Influence of Run-Out at High Feed (0.25 mm/tooth) Rates

Effect of run-out on serrated cutting forces (F_x) at high feed (0.25 mm/tooth) rates is shown in Fig. 46.6. Here, we report on only a subset of the cutting combinations reported on in Sect. 46.3.1. As before, in this case too, we observe that the ratio of maximum to average force increases with an increase in run-out. For 10% up milling at low depth of cut ($a_p = 3\,\text{mm}$) (Fig. 46.6a), the increase of the peak amplitude of up to 35% is observed. For the 100% milling (slotting) at high depth of cut ($a_p = 8\,\text{mm}$) (Fig. 46.6b), the increase of the peak amplitude of up to 25% is observed.

Contrasting the results for high immersion with those of the low immersion case, we see that with high feed rates and for low immersion cutting, run-out appears to

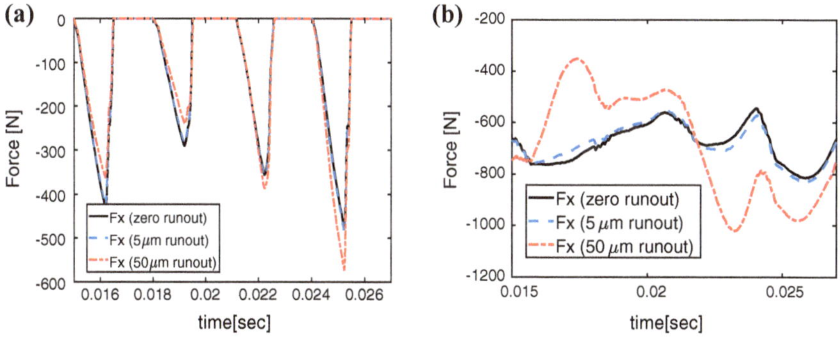

Fig. 46.6 Influence of run-out on cutting force (F_x) for sinusoidal serrated end mill for high feed rates: **a** $a_p = 3\,\text{mm}$, at 10% up milling, and **b** $a_p = 8\,\text{mm}$, at 100% milling (slotting)

have a more severe effect than at the high feed and high immersion case. Moreover, for the high feed and high axial and radial depths of cut, there is also a significant change in the force profile for high levels of run-out as compared to the case of zero run-out. These observations suggest that if serrated cutters have high levels of run-out, they best be avoided to be used with high feed rates. A sudden increase in forces, may cause impact loading on the cutter, which may cause it to prematurely break.

Seeing that serrated cutters with high run-outs are not advisable to use with high feed rates, to understand if/how run-out on serrated cutters used at high feed rates can potentially negate the positive influence of serrations, which is to otherwise reduce forces, we discuss results for regular end mills with run-outs next in Sect. 3.3 and contrast those findings with those observed here.

46.3.3 *Influence of Run-Out at High Feed (0.25 mm/tooth) Rates for Regular End Mills*

Effect of run-out on regular end mill cutting forces (F_x) at high feed (0.25 mm/tooth) rates is shown in Fig. 46.7. Here too, we report on only a subset of the cutting combinations reported on in Sect. 3.1. Results herein are obtained by setting $A = 0$ in the force model, i.e. essentially making it a regular end mill. All other geometries are the same as for the serrated end mill. As observed for serrated end mills, we observe that the ratio of maximum to average force increases with an increase in run-out. As evident in Fig. 46.7, for 10% up milling at a low depth of cut $(a_p = 3\,\text{mm})$ (Fig. 46.7a), the increase of peak amplitude of up to 37% is observed, whereas for 100% milling (slotting) at a high depth of cut $(a_p = 8\,\text{mm})$ (Fig. 46.7b), difference of the peak amplitude up to 46.5% is observed. These observations, as before, suggest that if regular end mills have high levels of run-out, they best be avoided to be used with high feed rates.

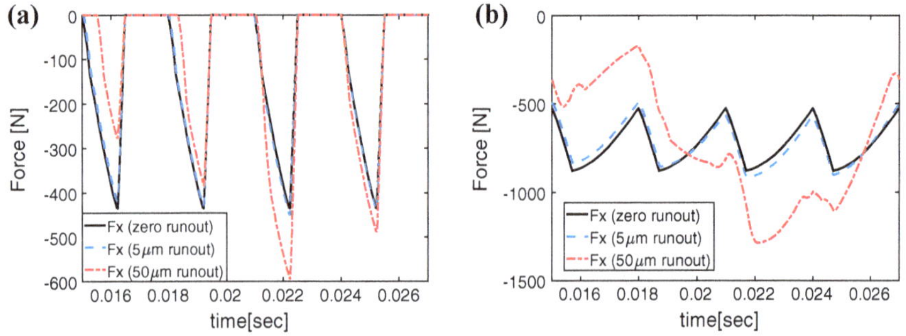

Fig. 46.7 Influence of run-out on cutting forces (F_x) for regular end mill at high feed rates: **a** $a_p = 3\,\text{mm}$, at 10% up milling, and **b** $a_p = 8\,\text{mm}$, at 100% milling (slotting)

Comparing the results for serrated cutters with run-out and with cutting at high feed and low immersion in Fig. 46.6a with those in Fig. 46.7a for regular end mills with run-out and high feed, with also low immersion, it is clear that even in the case of no run-out, the maximum force for the serrated end mill considered is 11% higher than regular end mills, which suggests that serrated cutters are ineffective at high feed rates, since there is no preferential reduction in cutting forces. The trend, however, reverses for the case with large run-out, which is strange, and needs to be investigated further. For the case of serrated end mill cutting with high feed and high immersion (axial, and radial), the peak force from Fig. 46.6b is lower than the case of regular end mills from Fig. 46.7b, without and with large run-outs. These observations are interesting and suggest that at certain combination of cutting and run-out parameters, serrated end mills retain their advantage over regular end mills, while at certain other combinations of cutting and run-out parameters, serrated cutters lose their advantage over regular end mills.

46.4 Conclusions

This paper discussed the influence of run-out on serrated cutters. We expanded the geometric force model for sinusoidal serrated cutters by including the effect of run-out. We systematically investigated how the run-out influences local chip thickness distribution, and how that in turn can potentially change the force profile for serrated cutters with run-out. Comprehensive investigations on the influence of run-out on forces were presented for a combination of feed rates, and axial and radial depths of cut. We observed that high levels of run-out increase the ratio of maximum to average forces and also change the nature of the cutting force profile. Interestingly, we observed that at low feeds, and large axial and radial depths of cut, run-out had no significant effect on the serrated force profiles, suggesting that even for large levels of run-outs, serrated cutters may still be used with low feed and high axial and radial depths of cuts, without worrying too much about if run-out may damage the workpiece quality or cause tool breakage. For cutting with high feed rates however, we observed that run-out adversely impacts cutting forces for any combination of axial and radial depths of cut. We also observed that with certain combinations of cutting parameters and run-outs, serrated cutters lose their advantage over regular end mills with similar levels of run-outs, and hence their use in such cases should be avoided.

Our findings were only for the case of sinusoidal serrations and having neglecting axial run-out. How run-out potentially influences force profiles for other serration types needs to be also investigated. Furthermore, numerical investigations presented herein need experimental validation, which will also include actually measuring run-out on serrated cutters, and these form part of our planned future investigations.

References

1. Martellotti, M.E.: An analysis of the milling process. Trans. Am. Soc. Mech. Eng. **63**, 677 (1941)
2. Martellotti, M.E.: An analysis of the milling process: Part II down milling. Trans. Am. Soc. Mech. Eng. **74**, 233 (1945)
3. Kline, W.A., DeVor, R.E.: The effect of runout on cutting geometry and forces in end milling. Int. J. Mach. Tool Design Res. **23**, 123–140 (1983)
4. Zheng, H.Q., et al.: Theoretical modeling and simulation of cutting forces in face milling with cutter runout. Int. J. Mach. Tools Manuf. **39**, 2003–2018 (1999)
5. Wan, M., et al.: New procedures for calibration of instantaneous cutting force coefficients and cutter runout parameters in peripheral milling. Int. J. Mach. Tools Manuf. **49**, 1144–1151 (2009)
6. Krüger, M., Denkena, B.: Model-based identification of tool run-out in end milling and estimation of surface roughness from measured cutting forces. Int. J. Adv. Manuf. Technol. **65**(5–8) (2013)
7. Schmitz, T.L., et al.: Runout effects in milling: Surface finish, surface location error, and stability. Int. J. Mach. Tools Manuf. **47**, 841–851 (2007)
8. Li, Z., et al.: Mechanistic modeling of five-axis machining with a general end mill considering cutter runout. Int. J. Mach. Tools Manuf. **96**, 67–79 (2015)
9. Tlusty, j., et al.: Use of Special Milling Cutters against Chatter. Proceedings of NAMRC 11, SME (1983)
10. Altintas, Y., Ber, A.: Manufacturing Automation: Metal Cutting Mechanics, Machine Tool Vibrations, and CNC Design (2001)
11. Merdol, S.D., Altintas, Y.: Mechanics and dynamics of serrated cylindrical and tapered end mills. J. Manuf. Sci. Eng (2004)
12. Dombovari, Z., et al.: The effect of serration on mechanics and stability of milling cutters. Int. J. Mach. Tools Manuf (2010)
13. Campomanes, M.L.: Kinematics and dynamics of milling with roughing tools. Metal Cutt. High Speed Mach (2002)
14. Wang, J.-J.J., Zheng, C.M.: Identification of cutter offset in end milling without a prior knowledge of cutting coefficients. Int. J. Mach. Tools Manuf. **43**, 687–697 (2003)
15. Hekman, K.A., Liang, S.Y.: In-process monitoring of end milling cutter run-out. Mechatronics **7**(1), 1–10 (1997)
16. Lee, K.Y., et al.: A run-out measuring method using modeling and simulation in four-fluted end milling. J. Mater. Process. Technol. **187–188**, 207–211 (2007)
17. Bari, P., Wahi, P., Law, M.: Investigations on the influence of serration parameters on cutting forces. Proceedings of the AIMTDR, 2018, India (to appear)
18. CUTPRO © Advanced Machining Process Measurement and Simulation Software (2000)

Chapter 47
Modeling and Optimization of WEDM Parameters Using Design of Experiments

P. Bharathi, G. Srinivasarao and P. G. Krishnaiah

Abstract In the current work, modeling and optimization of input parameters in wirecut electric discharge machine (WCEDM) of titanium [Ti-6A-4V] (Grade 5) were attempted. WCEDM working variables such as pulse-on-time (T_{ON}), pulse-off-time (T_{OFF}), servo-voltage (SV), peak-current (IP), and wire feed were considered to study the responses of cutting speed or speed of cut (CS) and wire wear ratio (WWR). Each input parameter was set at three levels. Experiments were conducted as per central composite face (CCF)-centered design. Based upon the runs' data, model equations for CS and WWR were developed using multiple linear regression. Optimization of practice parameters had been performed with the help of level mean analysis, response graphs, and model equations. From the analysis, it was observed that pulse-on-time was the substantial considerable input variable followed by servo-voltage, peak-current, and pulse-off-time.

Keywords WCEDM · Modeling · Optimization · Cutting speed · Wire wear ratio

47.1 Introduction

The WCEDM is a non-classical thermoelectric machining practice used for cutting newer, harder materials like titanium, composites, and ceramics. Titanium and its alloy are winning materials owing to their unique amalgamation of properties like high hardness and chemical wear resistance to substantial environments and tendency to form a stable, tightly adherent, protective surface film [1]. Due to the excellent properties of titanium alloys, the usage is being increased in applications

P. Bharathi (✉)
NRI Institute Technology Agiripalli, Krishna, AP, India
e-mail: nitbharathi307@gmail.com

G. Srinivasarao
R.V.R & J.C. College of Engineering Chowdavaram, Guntur, AP, India

P. G. Krishnaiah
PVPS Institute of Technology, Kanuru, Krishna, AP, India

R. G. Narayanan et al. (eds.), *Advances in Computational Methods in Manufacturing*, Lecture Notes on Multidisciplinary Industrial Engineering,
https://doi.org/10.1007/978-981-32-9072-3_47

like aerospace, chemical processing, surgical implants, and missile component [2]. Titanium and its alloys are the popular choice for fixation devices and are used to make artificial hips and knee replacement surgeries and other biological implants [3]. [Ti-6A-4V] is an alpha–beta alloy group, in which α-stabilizer possesses excellent creep resistance and β-stabilizers exhibit high hardenability and a higher density [4]. It is a substantial suitable process for machining very complex profiles of conductive hard materials where the high degree of precision and surface finish is the crucial factor. In WCEDM, material removal mechanism is very complex with a series of electric sparks generating the tremendous amount of heat between workpiece material and conductive tool which melts and evaporates minute amount of the workpiece and electrode material. The dielectric medium apart from providing the necessary dielectric strength continuously flushes away the solidified debris and also dissipates the heat from the machining zone. The tool wire (usually less than 0.3 mm diameter) is continuously fed from a supply spool, and the tool electrode is controlled and monitored by the CNC system. Researchers have focused on the input variables, pulse-on-time, pulse-off-time, servo-voltage, and peak-current, to avoid severe problems like the short-circuit of wire, wire fracture, and work exterior surface spoil leading to decrease in the overall efficiency of the process. Ghodsiyeh et al. [5] studied the effect of cutting parameters counting pulse-on-time, pulse-off-time, and peak-current on exterior surface roughness, sparking gap, and material removal rate of titanium (Ti-6Al-4V) alloy. Statistically the optimization model is the best manufacturing tool. Kaojy et al. [6] conducted the input variables' optimization of electric liberation cutting process of Ti-6Al-4V alloy considering multiple recital attribute wire wear ratio, material removal rate [MRR], and exterior surface coarseness using the Taguchi technique and gray relational coefficient grades. The process parameters chosen in this revision are discharge current, open voltage, pulse duration, and duty factor. The optimized process parameters prove an improved electrode wear ratio of 0.15, MRR of 0.12, and surface roughness of 0.19. Krishna et al. [7] built a hybrid replica by artificial neural networks [ANN] and applied genetic algorithm technique to optimize the exterior surface coarseness in electric discharge cutting of [Ti-6Al-4V] alloy by allowing for the concurrent effect of peak-current and voltage.

Xess et al. [8] performed concurrent optimization of the practice accomplishment material removal rate, surface roughness with aid of gray Taguchi's technique for cutting of titanium [Ti-6A-4V] alloy in EDM. Jabbaripour et al. [9] investigated the response distinctiveness of material removal rate [MRR], ratio of tool rapture [TWR], and diverse aspects of integrity of surface for Ti-6Al-4V on electrical discharge machining (EDM) specimens like topography of cutting surface, crack initiation, white layer (recast layer) height, and microhardness. Here, three process machining variables including pulse-current, pulse-on-time, and open-circuit voltage were altered through EDM runs. The effects of MRR and TWR versus process cutting variables were carried out based on foremost and combine outcome plots and too confirmed by analysis of variance results. The practical values exposed that broad aspects of exterior surface reliability for machined specimens are substantially exaggerated by pulse-current and pulse-on-time. Spedding and Wang et al.

[10] developed models through RSM and ANN to characterize the responses in WCEDM ultimate performance obtained by two techniques. Susanta et al. [11] applied weighted principal component [WPC] procedure to optimize the numerous outputs of the WCEDM process. The outcome offered good improvement in overall quality level by the WPC rather than other approaches. Rupesh et al. [12] had conducted studies on pure titanium with input variables such as pulse-on-time, pulse-off-time, peak-current, wire feed, wire tension, and servo-voltage. Dual-response optimization was performed on performance characteristics like cutting speed and surface coarseness. Anish-Kumar et al. [13] investigated on WCEDM using RSM with pure titanium as source material and copper alloy brass electrode. The responses of material removal rate, surface coarseness, dimensional deviation, and wire wear ratio were optimized by using multi-output optimization through desirability approach. Nayak et al. [14] calculated the optimal combination of variables for maximum MRR, minimum SR, and kerf width. Amitesh et al. [15] carried out studies on surface integrity, material removal rate, and wire wear ratio in WCEDM of Nimonic 80 alloys. Pulse-on-time and pulse-off-time were identified as considerable parameters. Further, single- and multi-response optimization were compared and validated. Advanced microscope, i.e., SEM analysis, concluded that the effect of the higher pulse-on-time causes height of recast layer. Sanjeev et al. [16] used the desirability function approach to optimize surface coarseness and velocity of cut while cutting alloys of composite (MMC). It was identified that the pulse width (PW), the time among pulses, small pulse time, and servo manage to mean reference voltage have the individual and interactive influence on the response variables. Srinivasarao et al. [17] conducted experiments on AISI 52100 steel using a CCF design and studied the effect of the input variables like pulse-on-time, pulse-off-time, de-ionized water pressure, and wire feed on surface roughness. Srinivasarao et al. [18] run the titanium Grade 5 material with CCF design with input variables like pulse-on-time, pulse-off-time, peak-current (IP), servo-voltage (SV), and wire feed (WF). Mathematical models were formulated for material subtraction rate and surface coarseness, and multi-objective optimization has been performed using desirability approach. Bharathi et al. [19] focused on some cutting limits T_{ON}, T_{OFF}, WF, and voltage while machining SS304 material for high MRR, lower SR, and kerf width using multi-objective optimization. Nihat et al. [20] investigated on wire wear ratio in WCEDM with the practice parameters like pulse duration, open-circuit voltage, wire speed, dielectric flushing pressure on 'AISI-4140' steel (work material), and brass wire (tool). From this study, it was observed that WWR increased by amplifying pulse duration and open-circuit voltage, while it is decreased by wire speed and non-conductive flushing pressure.

47.1.1 Materials and Processes

The work material used for experimentation is Ti-6A-4V. The annealed brass wire was used as a cutting tool with a diameter 0.25 mm and 1000 MPa tensile strength

Table 47.1 Element composition of the specimen

Element	N	C	Fe	O	Al	V	Ti
Composition	0.1	0.1	0.3	0.2	6.8	4.5	88.08

to manufacture very excellent, exact, and spotless cuts. The chemical composition of the workpiece is shown in Table 47.1 [21]. Workpieces in the form of the rectangle plate are considered for conducting the runs.

The speed of the cutting determined by dividing the cutting length by time is taken for each experiment. Wire wear ratio was determined as follows:

$$\text{wire wear ratio} = \left[\frac{\text{IWW} - \text{FWW}}{\text{IWW}}\right]$$

IWW = wire weight before machining
FWW = wire weight after machining

The tool electrode was weighted before and after the machining, and the weight reduction was determined to get the output response, i.e., WWR. An electronic weighing scale with 0.001 g accuracy was taken for weight dimension. The mean value was used to minimize the error in weighing.

47.1.2 Design of Experiments

Five input variables, specifically pulse-on-time, pulse-off-time, servo-voltage, peak-current, and wire feed, were accounted to revise the responses cutting speed (CS) and wire wear ratio [WWR]. Each input variables was located as three levels which are shown in Table 47.2. Experiments run using CCF design for five factors, and the CCF layout includes 28 experiments in which there are 16 fractional points, 10 axial points, and 2 center-points. All the experiments are carried out in random order in order to avoid bias. The runs along with responses, i.e., cutting speed and wire wear ratio, are shown in Table 47.3.

Table 47.2 Factors and corresponding levels

Factor	Factors	Level 1 (−1)	Level 2 (0)	Level 3 (+1)
T_{ON} (μs)	Pulse-on-time	107	113	119
T_{OFF} (μs)	Pulse-off-time	51	57	63
IP (amp)	Peak-current	10	11	12
SV (V)	Servo-voltage	30	40	50
WF (meter/minute)	Wire feed	3	6	9

Table 47.3 Experimental design with response

S. No	T_{ON}	T_{OFF}	IP	SV	WF	CS (mm/minute)	WWR
1	−1	−1	−1	−1	1	1.2245	0.011
2	−1	−1	−1	1	−1	0.59	0.008
3	−1	−1	1	−1	−1	1.943	0.003
4	−1	−1	1	1	1	0.9891	0.006
5	−1	1	−1	−1	−1	0.6645	0.006
6	−1	1	−1	1	1	0.05	0.004
7	−1	1	1	−1	1	0.805	0.006
8	−1	1	1	1	−1	0.2	0.001
9	1	−1	−1	−1	−1	1.653	0.009
10	1	−1	−1	1	1	1.152	0.003
11	1	−1	1	−1	1	3.22	0.009
12	1	−1	1	1	−1	2.1	0.022
13	1	1	−1	−1	1	1.059	0.077
14	1	1	−1	1	−1	0.741	0.013
15	1	1	1	−1	−1	2.5	0.009
16	1	1	1	1	1	1.179	0.009
17	−1	0	0	0	0	0.531	0.001
18	1	0	0	0	0	1.5	0.024
19	0	−1	0	0	0	2.865	0.011
20	0	1	0	0	0	1.151	0.008
21	0	0	−1	0	0	0.784	0.008
22	0	0	1	0	0	1.228	0.011
23	0	0	0	−1	0	1.9	0.016
24	0	0	0	1	0	1.152	0.006
25	0	0	0	0	−1	1.861	0.009
26	0	0	0	0	1	1.737	0.011
27	0	0	0	0	0	1.1	0.008
28	0	0	0	0	0	1.11	0.004

47.2 Examination of Data and Analysis of the Outcome

The foremost objective of this work is to examine the influence of variables on the response by means of response tables, response plots, and mathematical modeling. The main effects are studied from the factorial segment of the CCF layout which consists of the experiments from 1 to 16. One-factor-at-a-time analysis has been made using the experiments from 17 to 28. The performance of each control variable on the output assumed, i.e., cutting speed has been executed with level mean analysis. A level mean of a factor is the average of the cutting speed value of experiments in which the factor is at a particular level. For the factorial portion of CCF design, the mean value of the cutting speed for the T_{ON} at level '−1' and '+1' can be determined by a mean response for runs 1–8 and 9–16, respectively. The average of the response for every level of the cutting variables can be computed in a similar manner. The control variable with the strongest influence is calculated by the difference between the average values of the factor at large and small levels. For one-factor-at-a-time analysis, the level means' values of T_{ON} are obtained by considering the experimental values for experiments 17, an average of 27 and 28, and 18 (i.e., 0.531, the average value of 1.1 and 1.11, 1.50). For T_{OFF}, the level means' values are obtained by considering the experimental values for experiments 19, an average of 27 and 28, and 20 (i.e., 2.865, the average value of 1.1 and 1.11, 1.151). From one-factor-at-a-time analysis, the change in each parameter is observed when the factor is changing from level '−1' to level '0' and to level '+1'. The level averages obtained from the fractional factorial part of CCF layout and one-factor-at-a-time analysis for cutting speed and wire wear ratio are represented in Tables 47.4, 47.5, 47.6, and 47.7 respectively.

Table 47.4 Level means: The factorial segment of [CCF] layout

Level	T_{ON}	T_{OFF}	IP	SV	WF
−1	0.8082	1.608	0.891	1.633	1.298
1	1.7	0.899	1.617	0.875	1.209
Difference	0.8918	0.709	0.716	0.758	0.089
Rank	1	4	3	2	5

Table 47.5 Level means: individual factor at a time scrutiny

	T_{ON}	T_{OFF}	IP	SV	WF
Level [−1]	0.531	2.865	0.784	1.9	1.861
Level [0]	1.105	1.105	1.105	1.105	1.105
Level [+1]	1.5	1.151	1.228	1.152	1.737
Difference	0.969	1.714	0.444	0.748	0.124
Rank	2	1	4	3	5

Table 47.6 Level means: the factorial part of CCF layout for WWR

Level	T_{ON}	T_{OFF}	IP	SV	WF
−1	0.0056	0.0089	0.0164	0.0163	0.0089
1	0.0189	0.0156	0.0081	0.0083	0.0156
Difference	0.0133	0.0067	0.0083	0.008	0.0067
Rank	1	4	2	3	4

Table 47.7 Level means: single-factor simultaneous analysis for WWR

	T_{ON}	T_{OFF}	IP	SV	WF
[Level. −1]	0.001	0.011	0.008	0.016	0.009
[Level. 0]	0.006	0.006	0.006	0.006	0.006
[Level +1]	0.024	0.008	0.011	0.006	0.011
Difference	0.023	0.003	0.003	0.01	0.002
Rank	1	3	3	2	4

47.2.1 Cutting Speed (CS)

The level means from Table 47.4 show that the T_{ON} is the substantial considerable parameters followed by servo-voltage, peak-current, and pulse-off-time. The wire feed rate is observed as not considerable. The same behavior was observed in the case of single-factor simultaneous analysis, but the pulse-off-time is identified as substantial considerable instead of the pulse-on-time. This might be due to the existence of combined effects (Fig. 47.1).

47.2.2 Wire Wear Ratio (WWR)

The level means from Table 47.6 shows that the T_{ON} is the considerable parameters followed by peak-current, servo-voltage, pulse-off-time, and wire feed rate. From Table 47.7, it can be observed from the single-factor simultaneous analysis that the pulse-on-time is identified as substantial considerable and remaining all factors are having more or less the similar effect. The reverse trend in case of pulse-off-time and peak-current has been identified, i.e., in the factorial portion if the pulse-off-time is escalating from level '−1' to level '+1', the wire wear ratio is increasing, whereas in one-factor-at-a-time analysis, if the pulse-off-time is escalating from level '−1' to level '+1' the wire wear ratio is decreasing. The similar trend can also be observed in the case of peak-current. From factorial portion when the peak-current is growing from starting level '−1' to end level '+1', the wire wear ratio is decreasing, whereas in one-factor-at-a-time analysis, if the peak-current is increasing from level '−1' to level '+1', the wire wear ratio is increasing. This is due to the existence of combined effects. The combined effects are studied by means of model equations and surface graphs.

47.2.3 Development of Mathematical Models

Based upon the experimental data, second-order models are developed for cutting speed and wire wear ratio. The data is given in the coded appearance of factors,

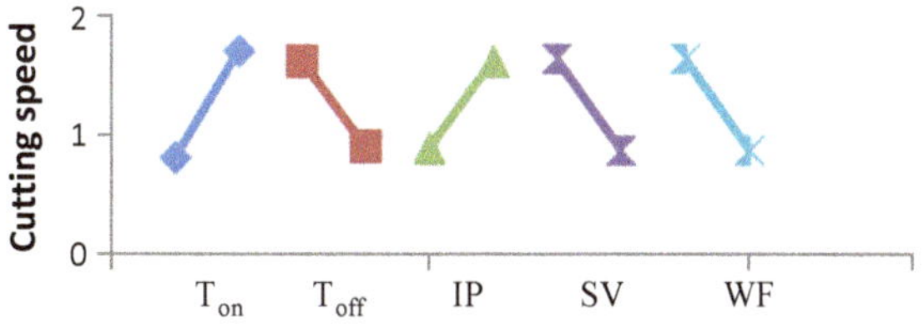

Fig. 47.1 Response plot for fractional factorial of CCF design for CS

i.e., minus one to plus one to SPSS software to estimate the model equation. The model equation for cutting speed is shown below:

$$\begin{aligned} \mathrm{CS} = {} & 1.425 + 0.45T_{\mathrm{ON}} - 0.41T_{\mathrm{OFF}} + 0.347\mathrm{IP} - 0.3795\mathrm{SV} \\ & - 0.482(T_{\mathrm{ON}})^2 + 0.511(T_{\mathrm{OFF}})^2 - 0.491(\mathrm{IP})^2 \\ & + 0.302(\mathrm{WF})^2 + 0.187T_{\mathrm{ON}}\mathrm{IP} - 0.121\mathrm{IPSV} \\ R^2 = {} & 0.928 \end{aligned}$$

The R^2, the value of 0.928, exhibits that variability in cutting speed is 92.8%, described by the model.

Based on the model, the response of the CS decreases as an increase in the T_{OFF} and SV and increases with an increase in T_{ON} and IP. The combination of IP with T_{ON} and SV is also identified as considerable. Further, the presence of square terms of the pulse-on-time, pulse-off-time, peak-current, and wire feed indicates the quadratic nature in output.

Combined effects mean the dual effect of two input variables on the response. The combined effect of pulse-on-time and peak-current on speed of cut gives that CS goes to a maximum value 1.436 mm/minute at larger value of pulse-on-time as well as peak-current at the same time as it reaches minimum value 0.158 mm/minute, where pulse-on-time is smaller level and peak-current is also smaller level as shown in Fig. 47.2. The combined effect of peak-current with servo-voltage on cutting speed gives that cutting speed of maximum value 1.781 mm/minute at the larger level of peak-current and smaller level of servo-voltage, while it reaches the least amount value of 0.845 mm/minute at the smaller level of peak-current and smaller level of servo-voltage. Decreasing SV decreases the gap between the electrode and work material that results in high ionization of dielectric medium and hence generates high discharge energy.

47.2.3.1 Optimum Values of Speed of Cut

The optimum values corresponding to the utmost cutting speed of 3.3719 mm/minute are the pulse-on-time 119 ms μs, pulse-off-time 51 ms μs, peak-current 12 A, servo-voltage 30 V, and wire feed 3 m/minute (Figs. 47.3 and 47.4).

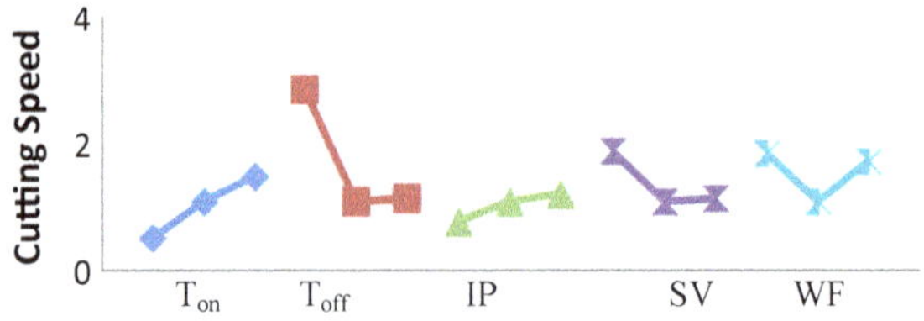

Fig. 47.2 Response plot for one-factor-at-a-time analysis for CS

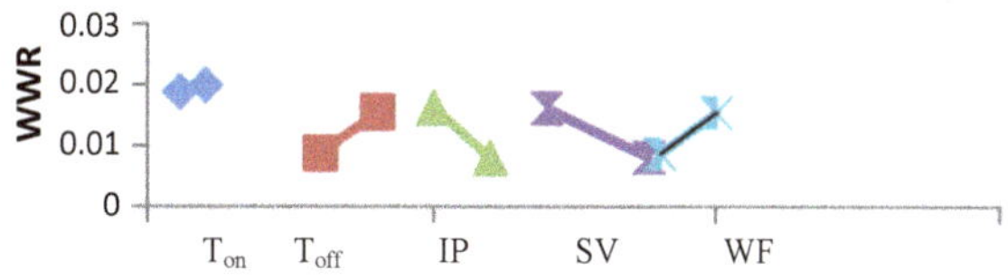

Fig. 47.3 Response plot for factorial of CCF design (WWR)

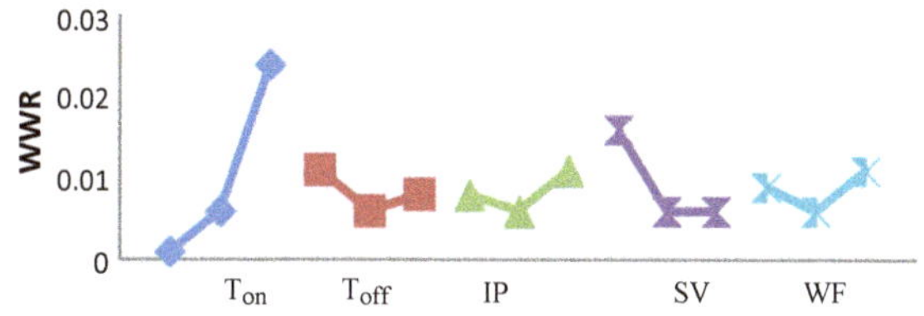

Fig. 47.4 Response plot for one-factor-at-a-time analysis (WWR)

The model equation for wire wear ratio is shown below:

$$\begin{aligned}\text{WWR} = {} & (0.01) + 0.004\,T_{\text{ON}} + 0.003\,T_{\text{OFF}} - 0.003\,\text{IP} - 0.004\,\text{SV} \\ & + 0.003\,\text{WF} + 0.003(T_{\text{ON}})^2 + 0.005\,T_{\text{ON}}\,T_{\text{OFF}} - 0.0\,T_{\text{ON}}\text{IP} - 0.003T_{\text{ON}}\text{SV} + 0.002T_{\text{ON}}\text{WF} \\ & - 0.005T_{\text{OFF}}\text{IP} - 0.005\,T_{\text{OFF}}\text{SV} + 0.005\,T_{\text{OFF}}\text{WF} + 0.005\,\text{IP SV} - 0.004\,\text{IP WF} \\ & - 0.006\,\text{SV WF} - (R^2 - 0.963)\end{aligned}$$

The R^2 value 0.963 implies that 96.3% of the variability in wire wear ratio is described by the model as given in the above equation (Figs. 47.5 and 47.6).

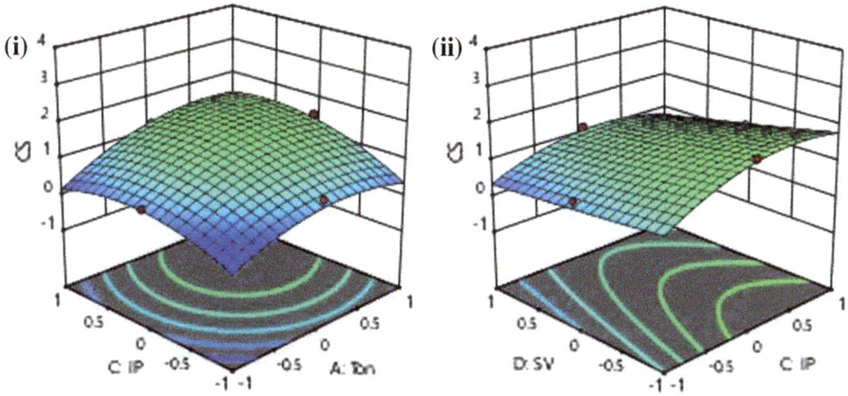

Fig. 47.5 (i) Effect of IP and T_{ON} on cutting speed (T_{OFF} = 0, SV = 0, WF = 0) and (ii) effect of IP and SV on cutting speed (T_{ON} = 0, T_{OFF} = 0, WF = 0)

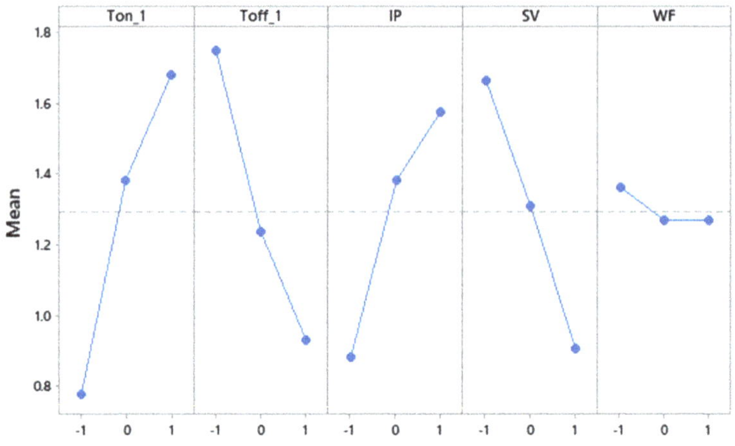

Fig. 47.6 Main effect plots for cutting speed

The combined effect of the pulse-on-time and pulse-off-time on WWR during their interaction gives that WWR goes to a maximum value of 0.028 at the higher value of pulse-on-time same as pulse-off-time, while it reaches the minimum value of 0.008, where the pulse-on-time and pulse-off-time were at lower level. The combined effect of pulse-off-time with peak-current on WWR shows that WWR's maximum value is 0.012 at the higher level of pulse-off-time and lower level of peak-current while it reaches minimum value 0.005 at the smaller level of pulse-off-time as well as lower level of peak-current. The optimum input variables equivalent to the minimum WWR 0.0099 were the pulse-on-time = 107 microseconds; pulse-off-time = 63 microseconds, peak-current = 12 A, servo-voltage = 30 V, and wire feed rate = 3 m/minute (Figs. 47.7 and 47.8).

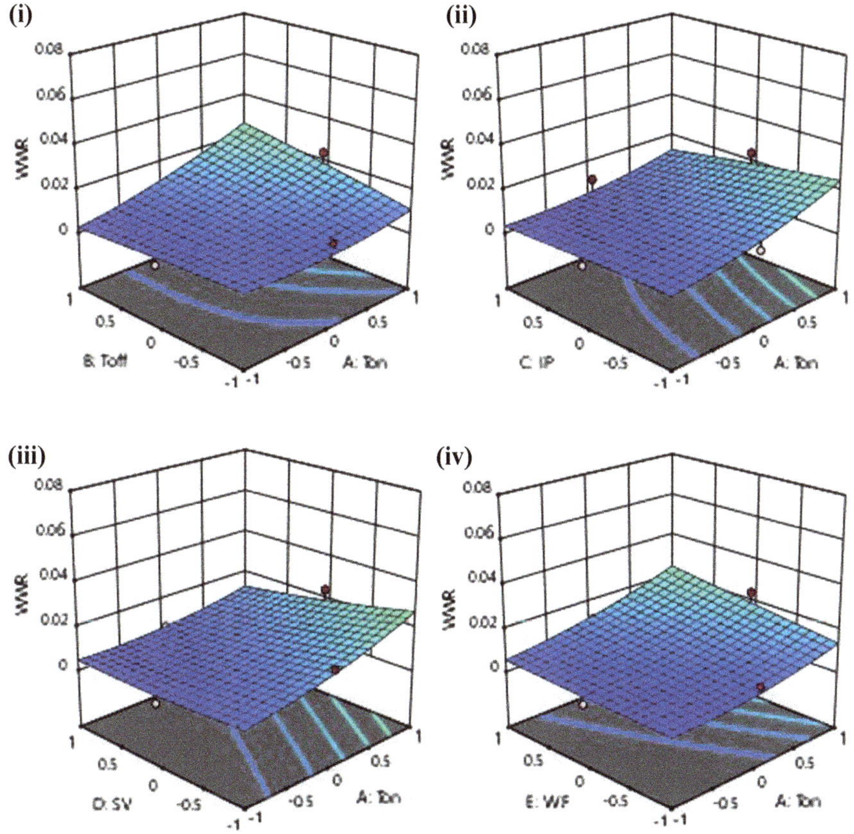

Fig. 47.7 (i) Effect of T_{OFF} and T_{ON} on WWR, (ii) effect of IP and T_{ON} on WWR, (iii) effect of SV and T_{ON} on WWR, and (iv) effect of WF and T_{ON} on WWR

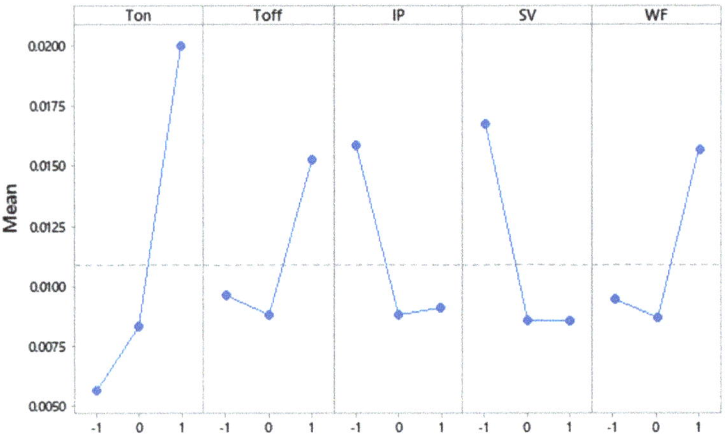

Fig. 47.8 Major outcome plots for wire wear ratio

47.3 Conclusions

Modeling and optimization of input variables have been performed in WCEDM of titanium [Ti-6A-4V] material with annealed brass wire as the electrode. The following observations are made:

1. All input variables having a considerable effect on cutting speed and wire wear ratio.
2. The factors pulse-on-time and peak-current have a positive effect, whereas pulse-off-time and servo-voltage have a negative effect on cutting speed.
3. The factors pulse-on-time, pulse-off-time, and wire feed have a positive outcome, whereas peak-current and servo-voltage have a negative effect on WWR.
4. The importance of interaction and second-order terms which shows the quadratic nature of the response is also observed on both responses.
5. The optimum values corresponding to the high cutting speed of 3.3719 mm/minute are the pulse-on-time 119 μs, pulse-off-time 51 μs, peak-current 12A, servo-voltage 30 V, and wire feed 3 m/minute.
6. The optimum values related to the minimum WWR 0.0099 were pulse-on-time = 107 μs; pulse-off-time = 63 μs, peak-current = 12 A, servo-voltage = 30 V, and wire feed rate = 3 m/minute

References

1. Yang, X., Richard Liu, C.: Machining titanium and its alloys. Mach. Sci. Technol. **3**(1), 107–139 (1999)
2. Myers, J.R., Bomberger, H.B., Froes F.H.: Corrosion behavior and use of titanium and its alloys. J. Metals (1984)
3. Sharma, A., Sharma, M. D., Sehgal, R.: Experimental study of machining characteristics of titanium alloy (Ti–6Al–4V). Arab. J. Sci. Eng. **38**(11), 3201–3209 (2013)
4. Catia S., Lopes, D., Mariana, T., Donato, Ramgi. P.: Comparative corrosion behavior of titanium alloys (Ti15MO and [Ti-6A-4V] for dental implants applications: a Review. Cor-Ros. Prot. Mater. **35**, 5–14 (2016)
5. Ghodsiyeh, D., Lahiji, M. A., Ghanbari, M., Shirdar, M. R., Golshan, A.: Optimizing Material Removal Rate (MRR) in WEDMing Titanium alloy (Ti6Al4V) using the Taguchi method. Res. J. Appl. Sci. Eng. Technol. **4**(17), 3154–3161 (2012)
6. Kao, J.Y., Tsao, C.C., Wang, S.S., Hsu, C.Y.: Optimization of the EDM variables on machining Ti–6Al–4V with multiple quality characteristics. Int. J. Adv. Manuf. Technol. **47**, 395–402 (2010)
7. Rao, K.M., Rangajanardhaa, G., Rao, H.D., Rao, S.M.: Development of hybrid model and optimization of surface roughness in electric discharge machining using artificial neural networks and genetic algorithm. J. Mater. Process. Technol. **209**(3), 1512–1520 (2009)
8. Xess, P.A., Sandhyarani, B., Manoj, M.: Optimization of the EDM variables on machining [Ti-6A-4V] alloy with multiple quality characteristics. Appl. Mech. Mater. **619**, 89–93 (2014)
9. Jabbaripour, B., Sadeghi, M.H., Faridv, S., Shabgard, M.R.: Investigating the effects of EDM variables on surface integrity, MRR and TWR in machining of Ti–6AL–4V. Mach. Sci. Technol. **16**, 419–444 (2012)
10. Spedding, T.A., Wang, Z.Q.: Study on modeling of wire EDM process. J. Mater. Process. Technol. **69**, 18–28 (1997)
11. Susanta, K.G., Shankar, C.: Optimization of multiple response for WCEDM processes using weighted principal components. Int. J. Adv. Manuf. Technol. **40,** 1102–1110 (2009)
12. Chalisgaonkar, R., Kumar, J.: Optimization of WCEDM process of pure titanium with multiple performance characteristics using Taguchi's DOE approach and utility concept. Front. Mech. Eng. **8**(2), 201–214 (2013)
13. Kumar, A., Kumar, V., Kumar, J.: Multi-response optimization of process parameters based on response surface methodology for pure titanium using WEDM process. Int. J. Adv. Manuf. Technol. **68**(9–12), 2645–2668 (2013)
14. Nayak, B.B., Mahapatra, S.S.: Multi-response optimization of WCEDM process variables us ing the AHP an TOPSIS method. **2**, 2319–3182 (2013)
15. Goswami, A., Kumar, J.: Investigation of surface integrity, material removal rate and wire wear ratio for WEDM of Nimonic 80A alloy using GRA and Taguchi method. Eng. Sci. Technol. Int. J. **17**(4), 173–184 (2014)
16. Garg, S. K., Manna, A., Jain, A.: An investigation on machinability of Al/10% ZrO_2 (P)-metal matrix composite by WEDM and parametric optimization using desirability function approach. Arab. J. Sci. Eng. **39**(4), 3251–3270 (2014)
17. Srinivasarao, G., Santhipriya, P., Suneel, D.: Modelling of surface roughness for AISI 52100 steel in WCEDM by Design of Experiments. Int. J. Eng. Res. Technol. **5** (2016)
18. Srinivasarao, G., Suneel, D.: Parametric optimization of WCEDM on α-β titanium alloy using desirability approach. Mater. Today Proc. **5**, 7937–7946 (2018)
19. Bharathi, P., Priyanaka, T.G.L., Srinivasa Rao, G., Nageswara Rao, B.: Optimum WCEDM Process variables of SS304 using Taguchi Method. Int. J. Ind. Manuf. Syst. Eng. **1**,69–72 (2016)
20. Tosun, N., Cogun, C.: An investigation on wire wear in WEDM. J. Mater. Process. Technol. **134**(3), 273–278 (2003)
21. ATI, Technical Data Sheet.: ATI [Ti-6A-4V], Grade 5, version 1, Pittsburg, USA, pp. 1–4 (2012)

Chapter 48
Analysis on Thermal Characteristics of Micro-Drilled Glass Using Microwave Energy at 2.45 GHz

Gaurav Kumar and Apurbba Kumar Sharma

Abstract Use of microwave energy at 2.45 GHz for micromachining offers promising nontraditional machining technique for brittle materials like glass. However, process is yet to be analyzed for better understanding of the process mechanism to minimize the defects associated with the thermal damages in the machined materials. In the present paper, the process was studied using the finite element method, and a 3D model of the microwave drilling setup was developed using COMSOL Multiphysics v5.2 software. Simulation was carried out to study the thermal characteristics of the materials with time while irradiated during the process. The main affecting parameters were analyzed to find out the possible reasons for the defect associated with this process. Simulation of the microwave drilling process was carried out for a combination of graphite concentrator (diameter = 500 μm) and borosilicate glass (thickness = 1.2 mm) at 700 W. Simulation results revealed that the electric field intensity near the concentrator tip was higher (approximately 8.99×10^6 V/m) enough to ionize the air dielectric media which causes the generation of plasma around the tip of concentrator. The maximum temperature in the machining zone was observed approximately 1100 °C in 6 s. The rapid rise in temperature in the specimen induces a very high thermal stress (approximately 85 MPa) which reaches beyond the fracture strength of the borosilicate glass. The experimental investigation of the same concentrator-specimen combination demonstrated the similar pattern of fracture while drilling in the air using the same processing parameters.

Keywords Microwave drilling · Glass · Stress

G. Kumar
NIT Uttarakhand, Srinagar (Garhwal) 246174, Uttarakhand, India

G. Kumar · A. K. Sharma (✉)
Microwave Material Processing Laboratory, IIT Roorkee, Roorkee 247667, Uttarakhand, India
e-mail: akshafme@gmail.com; akshafme@iitr.ac.in

R. G. Narayanan et al. (eds.), *Advances in Computational Methods in Manufacturing*, Lecture Notes on Multidisciplinary Industrial Engineering,
https://doi.org/10.1007/978-981-32-9072-3_48

48.1 Introduction

Microwave drilling is a thermal energy-based advanced machining method in which electromagnetic energy concentrated at the tip of the concentrator is used to ablate/melt the material from the workpiece just beneath the concentrator. Microwave drilling was reported successfully for the first time by Jerby et al. in the year 2002. An extendable monopole antenna was used to concentrate the microwave into a small hot spot with the help of a coaxial waveguide. Due to the creation of hot spot, material became soft and antenna was made to penetrate inside the softened material to create a hole [1–3]. Later on, they reported that microwave drilling takes place due to thermal run-away effect at the tip of metallic drill. Theoretical analysis of thermal run-away effect was done by coupling the Maxwell's and heat equations to understand the temperature variation of hot spot with respect to time. It was reported that temperature of hot spot rises very rapidly [4]. In the year 2015, Lautre et al. performed drilling on various materials using microwave energy in a domestic microwave applicator using gravity feed. Drilling on various materials like polymer, glass, etc. was reported [5, 6]. But, a significant amount of thermal damage was present around the hole. To understand the mechanism, the process was simulated using COMSOL Multiphysics. It was reported that the jamming of the concentrator was taking place due to solidification of glass melt as the upper plasma sphere was relatively cooler. The high temperature of the plasma sphere around the concentrator tip caused it to melt and get blunt. So, as the cylindricity or conicity of the concentrator tip is lost, plasma formation at concentrator tip stops [7]. Later on, precursors were used to reduce thermal shock on the glass specimen but problems of concentrator jamming and blunting were still present [8]. In the year 2018, Kumar and Sharma used dielectrics to reduce thermal damage and avoid problem of concentrator melting and jamming while drilling glass [9].

Finite element-based modeling has been used for predicting the liquid and solid-state material transformation, the transient temperature distribution, and residual stresses induced in the workpiece during thermal energy-based machining processes [10–12]. Dhanik and Joshi studied the effect of inter-electrode gap on plasma temperature. It was found that temperature lies within 8100 ± 1750 K with a change in the inter-electrode gap [11]. Jiao and Wang simulated the laser cutting of glass plate to study the temperature and thermal stress distribution using ANSYS. It was reported that a dual-laser beam method where a laser beam was focused to machine the undesired material from the glass and an off-focused CO_2-laser beam was used to preheat the glass sample helped in reducing the thermal gradients. Due to reduced thermal gradient, the thermal stress became less than the critical fracture strength of the glass [12]. Krötz et al. simulated the single discharge of electrochemical arc machining for analyzing the heat-affected zone. It was found that duration of discharge and heat transfer from discharge to workpiece governs the diameter of heat-affected zone rather than the arc spot diameter [13].

Therefore, the objective of the present study is to simulate the microwave drilling process to understand the mechanism of the process and to find out the reason behind thermal damage induced in the workpiece during machining.

48.2 Modeling and Simulation

Figure 48.1 shows the experimental setup of microwave drilling. The pin vice is made up of structural steel and is used to hold concentrator. The material of the concentrator is also of structural steel. Microwave drilling is achieved by concentrating electric field at the tip of a metallic concentrator which rests just above the top surface of the workpiece (borosilicate glass) with the help of pin vice. As the electric field strength around concentrator tip reaches above the ionization potential of the medium around it, plasma is formed at the concentrator tip as shown in Fig. 48.2. The plasma formed ablates/melts the material from the workpiece just beneath it and the concentrator penetrates inside the workpiece to create a hole. But, it is very difficult to have a clear understanding of the process mechanism of microwave drilling due to the limitation in experimental assessment of distribution of electric field inside the cavity, temperature, and thermal stress along the surface of the workpiece, etc. So, the software tool like COMSOL Multiphysics can help in having a better understanding of the process mechanism.

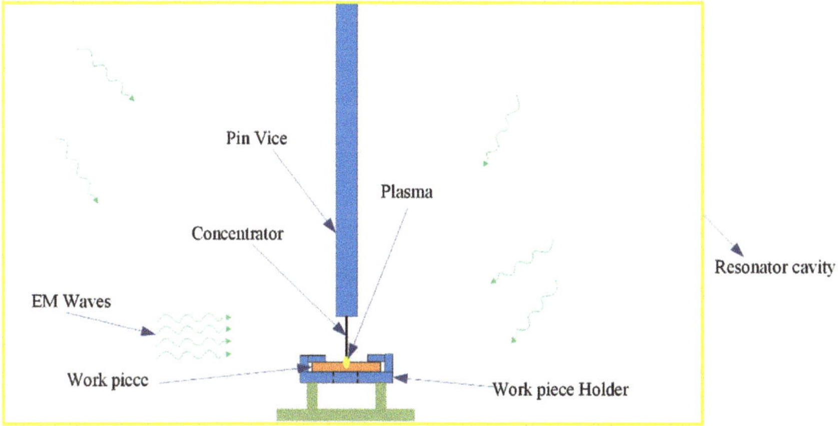

Fig. 48.1 Schematic diagram of the microwave drilling experimental setup

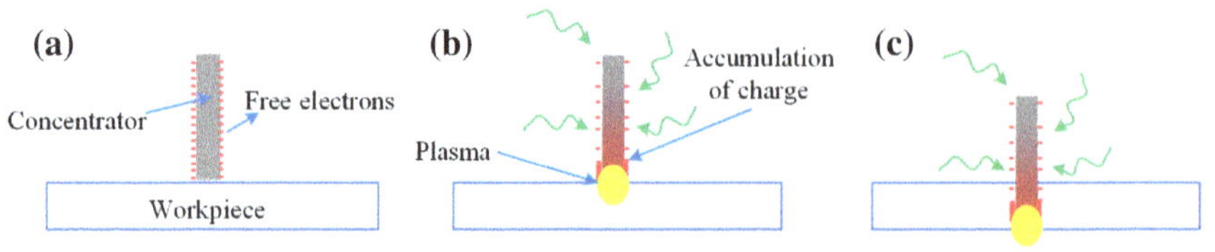

Fig. 48.2 Schematic of: **a** concentrator resting just above workpiece, **b** plasma formation after the microwave oven is switched on, and **c** concentrator penetrating inside the workpiece

48.2.1 Model Development

A 3D model of the experimental setup of microwave drilling as shown in Fig. 48.3 was developed using COMSOL Multiphysics 5.2 software to analyze the process and the behavior of workpiece while machining. The dimensions of the above model are shown in Table 48.1.

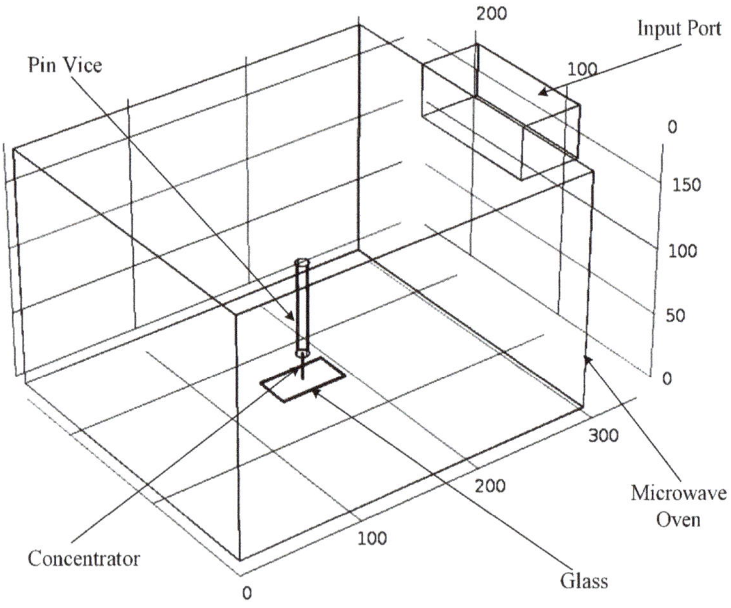

Fig. 48.3 Schematic of the 3D model of the microwave drilling

Table 48.1 Dimensions of the component used to generate the 3D model

Component	Oven	Waveguide	Pin vice	Concentrator	Workpiece
Dimensions (mm)	292 × 256 × 180	110 × 40 × 50	70 × 4.2(ϕ)	20 * 0.5(ϕ)	53 × 28 × 1.2

48.2.2 Meshing

Physics-controlled meshing is used to mesh the 3D model of the microwave drilling. Figure 48.4 shows the meshed 3D model. Fine meshing has been used to mesh the cavity (Fig. 48.4a), and extremely fine meshing has been used to mesh the concentrator and workpiece (Fig. 48.4b). The number of vertex elements, edge elements, triangular elements, and tetrahedral elements is 40, 557, 4504, and 45,401, respectively. The degrees of freedom solved for are 292,505. Figure 48.4c shows the plot of element quality of meshed geometry. The domain element statistics are as follows: average element quality = 0.653; element volume ratio = 1.924×10^{-7}; mesh volume = 1.368×107; average growth rate: 2.077.

48.2.3 Assumptions

The phenomenon of microwave drilling is governed by the formation of plasma at the tip of the concentrator and its subsequent interaction with the workpiece material. The formation of plasma depends upon the coupling of electromagnetic waves with the material of workpiece and concentrator. Besides it, shape of concentrator and its absorption behavior also influence the intensity of plasma formed. Therefore, following assumption has been made in order to simplify the model: (a) The workpiece, concentrator, and concentrator holder material used in the study are homogeneous and isotropic. (b) The atmosphere temperature inside the system is considered to be 27 °C, initially. (c) The air domain and jigs used are not

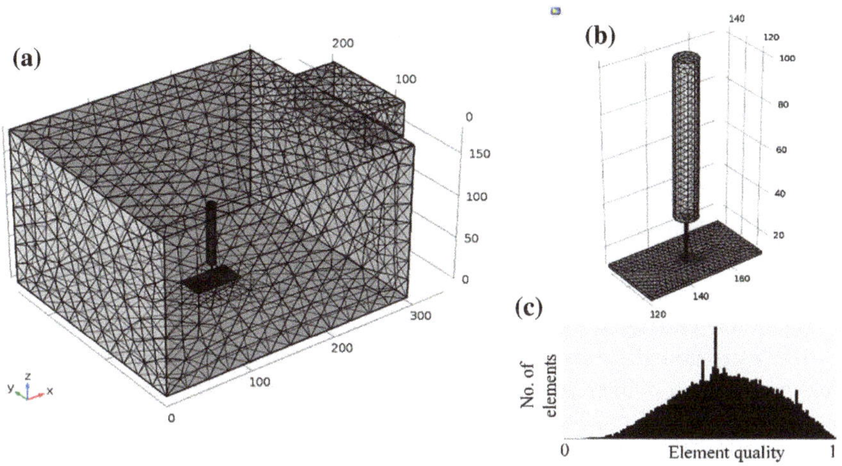

Fig. 48.4 Image showing: **a** the fine-meshed resonating cavity, **b** extremely fine-meshed workpiece and concentrator, and **c** plot of element quality of meshed geometry

Table 48.2 Property of the material of different component of experimental setup

Property	Materials		
	glass	Steel	Copper
Density (kg/m^3)	2210	7850	8960
Thermal Conductivity (W/(m × K))	1.4	44.5	400
Heat Capacity at constant pressure (J/(kg × K))	730	475	385
Relative permeability	1	1	1
Relative permittivity	4.2	1	1
Electrical conductivity (S/m)	1×10^{-14}	4.032×10^6	5.998×10^7
Young's modulus (N/m^2)	70×10^9	200×10^9	110×10^9
Poisson's ratio	0.22	0.30	0.35
Coefficient of thermal expansion(1/K)	3.3	12.3×10^{-6}	17×10^{-6}

considered for thermal analysis as they have zero dielectric losses. (d) The walls of the cavity and waveguide are considered to be made up of copper. (e) The port is excited by a transverse electromagnetic standing wave fields in the Z direction. The property of the material of the different component of the experimental setup is shown in Table 48.2 which has been taken from the material library of the software.

48.2.4 Governing Equations

COMSOL Multiphysics (version 5.2), a finite element-based software has been used to solve the governing equations. The phenomenon of interaction of microwave with the concentrator and glass can be explained with the help of Maxwell's equations. The following equation based on standard Maxwell's EM equations with the stipulated excitation at the rectangular port is solved for electric field (***E***) inside the waveguide and oven:

$$\nabla \times \left(\mu_r^{-1} \nabla \times \boldsymbol{E}\right) - k_0^2 \left(\varepsilon_r - \frac{j\sigma}{\omega \varepsilon_0}\right) \boldsymbol{E} = \boldsymbol{0} \tag{48.1}$$

where μ_r = relative permeability, k_0 = wave number in the free space, ε_r = relative permittivity, j = imaginary unit, ω = angular frequency, σ = conductivity, and ε_0 = permittivity of free space.

Due to the externally applied electromagnetic field, a very intense electric field ($\sim 10^6$ V/m) around the tip of the concentrator gets developed. This high electric field propagates through dielectric materials by ionizing it and creating a hot spot just beneath the tool tip. The heat generation in the concentrator, at the interface of concentrator and workpiece and in the dielectric media, can be expressed by the heat transfer equation as:

$$\rho C_p \frac{\partial T}{\partial t} = \nabla.(k\nabla T) + P(r,T) - P_{\mathrm{Loss}}(T_S) \tag{48.2}$$

where t = time (s), ρ = material density (kg m^{-3}), C_p = specific heat capacity (J kg^{-1} K^{-1}), and k = thermal conductivity (Wm^{-1} K^{-1}), $P(r,T)$ = power absorbed by the dielectric, $P_{\mathrm{Loss}}(T_S)$ = power lost to the surrounding due to convection and radiation respectively.

Due to the creation of hot spot on the surface of the tool tip, a thermal gradient is developed on the surface of workpiece. Due to this thermal gradient, a thermal stress is induced inside the surface of workpiece which is expressed using the following equation:

$$\sigma_{\mathrm{th}} = \frac{E\alpha}{1-\vartheta}(\Delta T) \tag{48.3}$$

where σ_{th} = residual thermal stress (MPa), E = Young's modulus, ϑ = Poisson's ratio and α = coefficient of temperature expansion, ΔT = temperature differential on the surface of the workpiece.

48.2.5 Validation

The result obtained after simulating the above experimental condition was validated experimentally. Figure 48.5a shows the optical image of a hole drilled in air. It is evident from the figure that circular and radial cracks are present around the hole. These cracks originate from the critically stressed zone and propagate through the materials around it. It is evident from the figure that an area of ~4.09 mm^2 is critically stressed. Figure 48.5b shows the simulated result of the distribution of around the hole created. It is evident that an area of 3.878 mm^2 is having Von

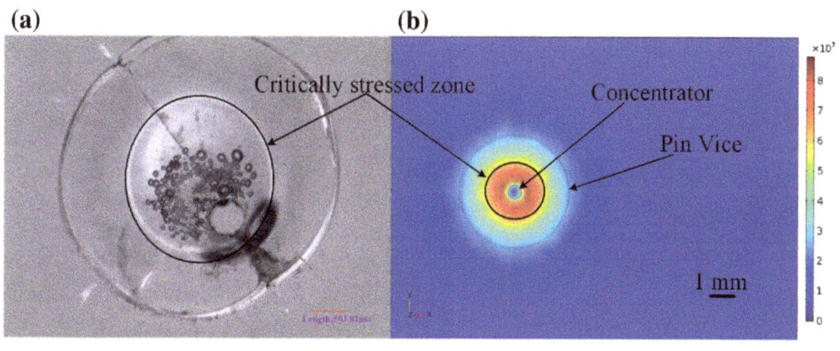

Fig. 48.5 **a** Optical image of the workpiece drilled in air and **b** Top view of the simulated stressed zone around the hole drilled on the borosilicate glass

Mises stress value greater than 60 MPa (critical stress limit for fracture of glass) which is called critically stressed zone. Both the experimental and simulated results are almost close to each other. Thus, the above-proposed model can be used for further study.

48.3 Results and Discussion

Figure 48.6 shows the distribution of electric field around concentrator tip. It can be observed from the zoomed view of electric field distribution around corners of concentrator tip that electric field strength is more than 10^6 V/m. Due to the high strength of electric field around concentrator tip, medium around the concentrator tip is ionized and plasma formation takes place. Consequently, maximum temperature in the surface of the workpiece just beneath the concentrator tip and concentrator rises rapidly up to 1100 °C and 950 °C in 6 s, respectively as shown in Fig. 48.7. It is also evident that maximum temperature in workpiece is higher as compared to the concentrator. This shows that core of the plasma formed, i.e., the zone of maximum temperature is around the surface of the workpiece. This is also evident from Fig. 48.8. It takes around 2 s to reach the softening temperature of borosilicate glass (~820 °C). As the glass melts, the concentrator starts to penetrate inside the workpiece. But, after few seconds, when the temperature is around 1100 °C, phenomena of ablation also takes place. Figure 48.8 shows the distribution of temperature along the different direction of the surface. It can be observed from Fig. 48.8 that temperature drops rapidly along the surface away from concentrator. The temperature just beneath the concentrator is around 950 °C and increases to around 1100 °C up to a distance of 0.4 mm. But after 0.4 mm, temperature decreases rapidly, and after 0.8 mm, the temperature comes below 800 °C. Due to this, the material removal takes place only up to a radial distance of 0.4 mm.

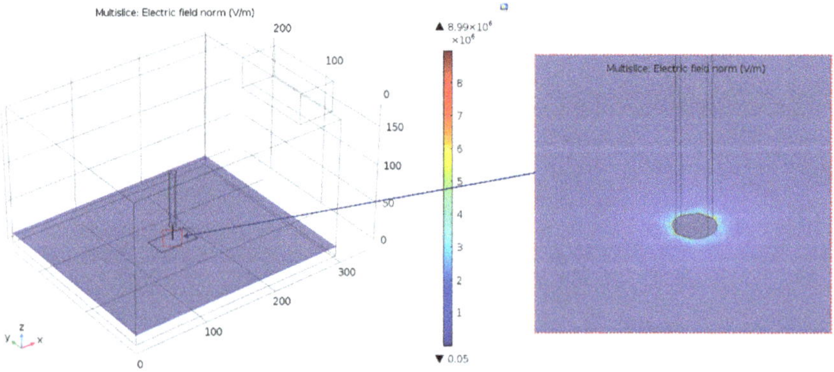

Fig. 48.6 Distribution of electric field around concentrator (inset: zoomed view of electric field distribution around concentrator tip)

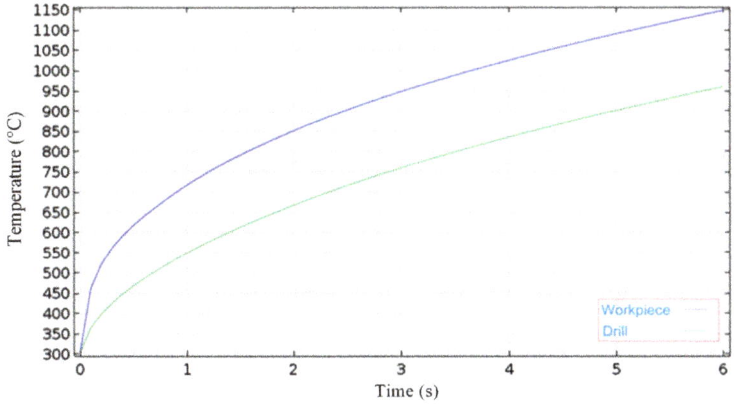

Fig. 48.7 Distribution of maximum temperature in concentrator and workpiece with respect to time

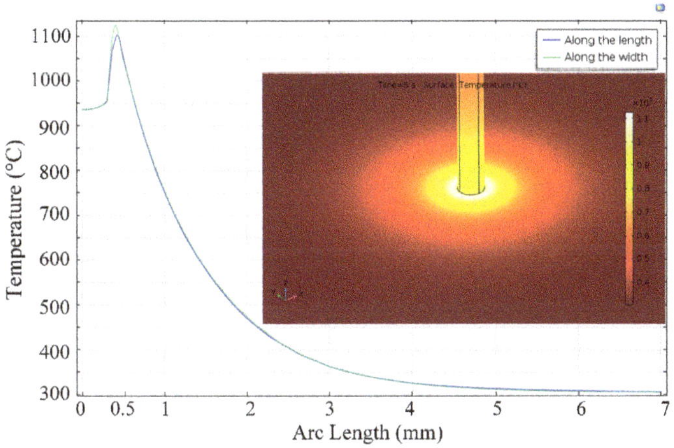

Fig. 48.8 Distribution of temperature along the surface with respect to arc length in different direction

The area after radial distance of 0.4 mm is heat-affected zone and highly stressed. Due to the high thermal gradient after 0.4 mm, the value of Von Mises stress rises above the critical stress limit for fracture of glass (60 MPa) as shown in Fig. 48.9. The region up to a radial distance of ~ 1.3 mm is highly stressed zone. Crack initiates in this zone and propagates radially up to certain distance on the surface of the workpiece and after that circumferentially around the hole as shown in Fig. 48.5a. Due to the high thermal stress, glass often cracks while drilling in air. The hole drilled in glass in air at a controlled feed rate of 1.2 mm/s is shown in

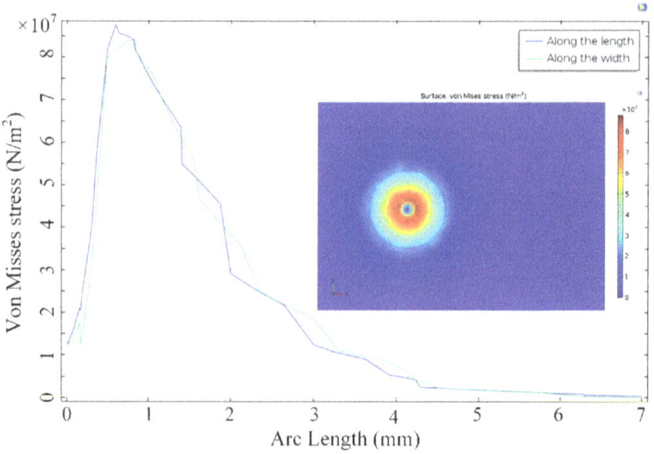

Fig. 48.9 Stress distribution along the surface just below the tip of the concentrator

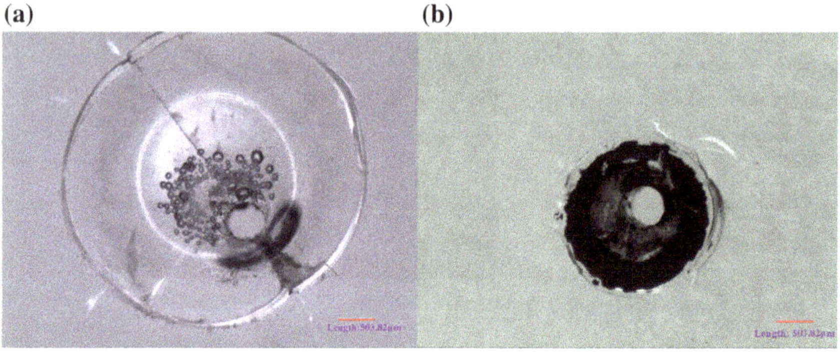

Fig. 48.10 Optical image of the hole drilled in (**a**) air (**b**) dielectric at a feed rate of 1.2 mm/s

Fig. 48.10a. It can be observed that significant amount of cracks is present around hole drilled. But as the medium around the concentrator tip is changed by a dielectric like transformer oil, the cracks and heat-affected zone are reduced significantly as shown in Fig. 48.10b. Dielectrics help in reducing the thermal gradient due to which cracks around the hole are reduced. Dielectric also helps in confining the shape of the plasma which ultimately reduces the thermal damage area as shown in Fig. 48.10b. But, a significant amount of carbon deposit is seen around the hole.

48.4 Conclusion

COMSOL Multiphysics 5.2 was used to simulate the microwave drilling process. Followings are the conclusions based on the simulation study:

- High electric field strength around the concentrator tip ($\sim 8.99 \times 10^6$ V/m) which is more than dielectric breakdown strength of air ionizes the medium around it to produce plasma.
- The maximum temperature on the surface of workpiece just beneath the concentrator tip is ~ 1100 °C which is higher than the ablation temperature of the borosilicate glass.
- The mechanism of material removal from the target area consists of combined action of ablation and melting.
- A very high thermal stress (~ 85 MPa) around the concentrator tip is induced in the surface of the workpiece which is the prime reason behind the crack generation while drilling.
- Dielectric helps in reducing the thermal gradient which ultimately lessens the crack and thermal damage around the hole.

References

1. Jerby, E., Dikhtyar, V., Aktushev, O., Grosglick, U.: The microwave drill. Science **298**, 587–589 (2002)
2. Jerby, E., Dikhtyar V.: Drilling into hard non-conductive materials by localized microwave radiation. In Advances in Microwave and Radio Frequency Processing, pp. 687–694. Springer, Berlin, Heidelberg (2006)
3. Jerby, E., Thompson, A.M.: Microwave drilling of ceramic thermal-barrier coatings. J. Am. Ceram. Soc. **87**(2), 308–310 (2004)
4. Jerby, E., Aktushev, O., Dikhtyar, V.: Theoretical analysis of the microwave-drill near-field localized heating effect. J. Appl. Phys. **97**(3), 034909 (2005)
5. Lautre, N.K., Sharma, A.K., Kumar, P., Das, S.: A photoelasticity approach for characterization of defects in microwave drilling of soda lime glass. J. Mater. Process. Technol. **225**, 151–161 (2015)
6. Lautre, N.K., Sharma, A.K., Kumar, P.: Distortions in hole and tool during microwave drilling of perspex in a customized applicator. In Microwave Symposium (IMS), 2014 IEEE MTT-S International, pp. 1–3 (2014)
7. Lautre, N.K., Sharma, A.K., Pradeep, K., Das, S.: A simulation approach to material removal in microwave drilling of soda lime glass at 2.45 GHz. Appl. Phys. A **120**(4), 1261–1274 (2015)
8. Lautre, N.K., Sharma, A.K., Das, S., Kumar, P.: On crack control strategy in near-field microwave drilling of soda lime glass using precursors. J. Thermal Sci. Eng. Appl. **7**(4), 041001 (2015)
9. Kumar, G., Sharma, A.K.: Role of dielectric fluid and concentrator material in microwave drilling of borosilicate glass. J. Manuf. Process. **33**, 184–193 (2018)

10. Das, S., Klotz, M., Klocke, F.: EDM simulation: finite element-based calculation of deformation, microstructure and residual stresses. J. Mater. Process. Technol. **142**(2), 434–451 (2003)
11. Dhanik, S., Joshi, S.S.: Modeling of a single resistance capacitance pulse discharge in micro-electro discharge machining. J. Manuf. Sci. Eng. **127**(4), 759–767 (2005)
12. Jiao, J., Wang, X.: A numerical simulation of machining glass by dual CO_2-laser beams. Opt. Laser Technol. **40**(2), 297–301 (2008)
13. Krötz, H., Roth, R., Wegener, K.: Experimental investigation and simulation of heat flux into metallic surfaces due to single discharges in micro-electrochemical arc machining (micro-ECAM). Int. J. Adv. Manuf. Technol. **68**(5–8), 1267–1275 (2013)

Chapter 49
A Study on the Effect of Polarity Change on Various Parameters on Ti6Al4V in Powder-Mixed Micro-EDM Using Multi-objective Grey Fuzzy Optimization

Deepak Sharma, Anusha Roohi Siddique, Vishnu Kumar, Shalini Mohanty and Alok Kumar Das

Abstract The study emphasizes on the effect of polarity change while conducting experiments using tungsten disulphide in powder-suspended micro-EDM process on Ti6Al4V alloy workpiece with brass tool electrode. For design of experiment, Taguchi method with L9 orthogonal array has been adopted. For each set of experiments, three levels of factors such as voltage, duty factor and machining time are considered. The responses such as MRR (higher the better) and SR (lower the better) are calculated, and a grey fuzzy logic model is built for optimization of the parameters. ANOVA table shows that the voltage plays the most significant role when machining in straight polarity followed by machining time and duty factor, whereas for reverse polarity machining time is the most significant one followed by voltage and duty factor. In case of straight polarity, confirmatory tests were performed at V = 25, DF = 55 and T = 8 and in reverse polarity at V = 25, DF = 45 and T = 8 and GFRG in case of straight polarity was found to increase by 0.216 and increased by 0.182 for the reverse polarity.

Keywords Micro-EDM · Surface roughness · Grey fuzzy · ANOVA · MRR

D. Sharma · A. R. Siddique · V. Kumar · S. Mohanty (✉) · A. K. Das
Department of Mechanical Engineering, Indian Institute of Technology
(Indian School of Mines), Dhanbad 826004, India
e-mail: shalinimohanty5@yahoo.com

D. Sharma
e-mail: mr.deepaksharma7s7@gmail.com

A. R. Siddique
e-mail: anusha.roohi@gmail.com

V. Kumar
e-mail: vishnu.bit15@gmail.com

A. K. Das
e-mail: eralok@yahoo.co.in

R. G. Narayanan et al. (eds.), *Advances in Computational Methods in Manufacturing*, Lecture Notes on Multidisciplinary Industrial Engineering,
https://doi.org/10.1007/978-981-32-9072-3_49

49.1 Introduction

With the everyday advancement in the field of engineering and technology, one gets to witness novel methods of machining at every step. Now is the time when the researchers and industries are moving towards the non-traditional machining and hybrid machining at a very fast rate. With growing requirements of precision, tolerances and other high aspect ratio micro-features, micro-EDM has served as a very important technique to achieve the former. Micro-EDM is nothing but just an evolution of the process of EDM where the size of the tool and movement of the axis of resolution are small to the level of microns and the discharge energy which is released is very less as compared to the conventional EDM [1].

Various parameters are considered for carrying out experiments in micro-EDM, but polarity change has been a challenging parameter by researchers. The difference in polarities of the workpiece and electrode and comparing them help in comprehending the results better. Generally, the aim of a machining process is to provide better surface finish with optimized parameters. Polarity of the process holds great significance in such a case. To this aspect, Schluze [2] studied that EDM with reverse polarities (tool positive) proves to be important when small material is required to be removed during finishing operations. It can also be used to remove oxide layers on the surface when used in hybrid processes with ECM. Parshuramalu et al. [3] have provided with a methodology for testing and finding out how the different tools and workpiece can be used with alteration of polarity. It is clear that if the polarity is not chosen properly, serious effects on wear and stability can occur. The experiments were carried out on a steel workpiece in a sinking EDM with copper electrode. From the results, it was found that the removal of steel was greatly affected by the polarity and the current settings. Researchers have been trying to improve the EDM process for a long time now. From changing the electrode to improving the dielectric, everything is tried to increase the efficiency of this process. Kumar et al. [4] mixed nanopowder of Al_2O_3 in dielectric. In this process, reverse polarity was used and it was found out that there was a reduction in micro-cracks with enhanced surface finish. Tyagi et al. [5] produced green sintered electrodes along with copper powder in die-sinking EDM. They performed EDC in straight polarity. As a consequence, coating with enhanced thickness and better surface properties were formed. It is evident that research progress with reference to the polarity change has not been much prevalent and hence, in the present work, an effort is done to study the effect of different parameters of the machining process on surface roughness of Ti6Al4V workpiece when machined with mixing powder in dielectric in micro-EDM.

Whenever machining is done, it is very necessary that the parameters at which it is done are optimum. It is an arduous work to ensure that the process parameters are best. Hence, certain optimization methods like principal component analysis, Taguchi methodology and GRA are used and have proved to be of great importance for obtaining desirable results. Various researchers have used GRA for optimization of the performance parameters while experimenting. Gopal and Chakradhar [6] optimized the parametric settings of feed rate, voltage applied and the concentration of the electrolyte for an ECM process. Durairaj et al. [7] made use of the GRA method in WEDM for finding out the vital process parameters which affected the optimal values of kerf width, and SR. Kumar and Kumar [8] determined optimal values of different parameters like pulse on duration (Ton), discharge current and gap voltage for an EDM process using GRA method considering MRR, electrode wear, SR as response values. Natarajan and Arunachalam [9] used Taguchi system and GRA to improve the EDM machining inputs by many regular attributes. Gopalakannan et al. [10] conducted experiments by selecting process parameters like gap voltage, pulse current, pulse on and off duration and used Taguchi-based GRA for EDM process with several outputs like EWR, MRR and SR. They decided that current and pulse on duration were the parameters of highest significance. Dewangan et al. [11] conducted experiments by means of RSM on composites with wire EDM and applied grey fuzzy method. Optimum values of process parameters were acquired, and consequent values of the MRR and SR were established. Ramanan and Shanavas [12] applied Taguchi and grey fuzzy logic to improve parameters of performance in turning of the AISI D2, while Soepangkat et al. [13] used the same for optimization in submerged arc welding.

Though a lot of researchers have used the above-mentioned methodology for optimizing difficult parameters of various non-traditional machining processes and electrical discharge machining in general [14], a optimization of parameters specifically for a micro-EDM has not been done. The purpose of the present study is to highlight the multi-objective optimization of powder-mixed micro-EDM under different polarities using grey fuzzy technique.

49.2 Methodology

49.2.1 Grey Relational Generation

In grey method, some interrelated information is well known and other is unknown. These multiple interrelated performance characteristics can be solved by grey relational analysis. In this analysis by using certain formulas, experimental data (MRR and SR) was normalized in the range of 0 and 1 and these values were used in generating grey relational coefficient which will direct the connection among the desired and actual experimental numbers. According to the type of normalized data required, GRA is of three types, namely lower the better, higher the better and nominal the better. To get normalized higher-the-better MRR values, the following equation was used:

$$X_a(b) = \frac{y_a(b) - \min y_a(b)}{\max y_a(b) - \min y_a(b)} \tag{49.1}$$

Similarly, normalized smaller-the-better SR values can be obtained by:

$$Y_a(b) = \frac{\max y_a(b) - y_a(b)}{\max y_a(b) - \min y_a(b)} \quad (49.2)$$

where a = 1, 2, 3, … k and b = 1, 2, 3, … q; k is the number of experimental data; q is the number of factors; $y_a(b)$ is the actual sequence; $y_a(b)$ is the value post-grey relational generation; min $y_a(b)$ and max $y_a(b)$ are the smallest and largest values of $y_a(b)$, respectively.

Grey relational coefficient (GRC)

$$\epsilon_a(b) = \frac{\Delta\min + \omega\Delta\max}{\Delta oi(k) + \omega\Delta\max} \quad (49.3)$$

where $\epsilon_a(b)$ is the grey relational coefficient; Δ_{oa} is difference among $y_o(b)$ and $y_a(b)$; $y_o(b)$ is the best sequence; $\Delta\max$ is uppermost value of $\Delta_{oa}(b)$; $\Delta\min$ is smallest value of $\Delta_{oa}(b)$; ω is the distinctive coefficient which is in the range of $0 \le \omega \le 1$. To get the results closer to optimal normalized value, highest value of grey relational coefficient has been used.

Grey Relation Grade (GRG)

The average of GRC value corresponding to each criterion is the GRG value.

$$\Gamma_a = \frac{1}{n}\sum_{b=1}^{n} \epsilon_a(b) \quad (49.4)$$

where n is no. of characteristics; greater the value of GRC, more will be its priority. The greater value of GRC suggests that corresponding process parameter is close to the optimal parameter.

49.2.2 Fuzzy Logic

Fuzzy logic is a scheme where the perceptive analysis of certain data by a human brain is represented. Fuzzy logic helps in making certain deductions which are similar to the results from a certain scheme of proposals which show considerable similarity with the initial data. A fuzzy logic division comprises of membership functions, fuzzifier, a fuzzy rule base, the inference engine (Mamdani) and a defuzzifier. The fuzzy interface engine (Mamdani) is shown in Fig. 49.1. Membership functions are fed to the fuzzifier which generates the grey relational coefficients. The membership function of MRR and SR is shown in Fig. 49.2, and membership function of GFRG is shown in Fig. 49.3. The fuzzifier interface engine converts these values to fuzzy values, and at end, defuzzifier alters fuzzy values in the form of grey fuzzy reasoning grade.

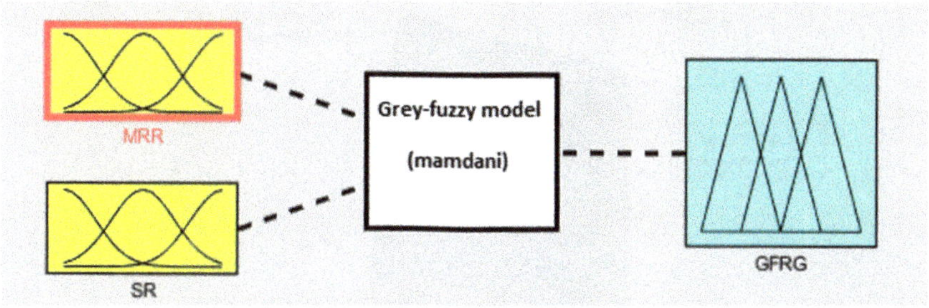

Fig. 49.1 Fuzzy logic model for process responses

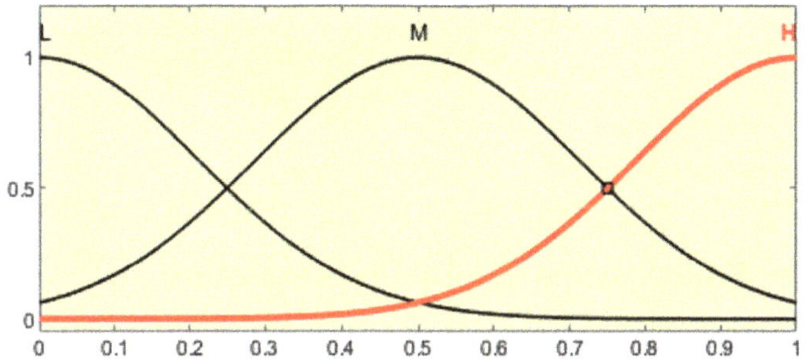

Fig. 49.2 Membership function for MRR and SR

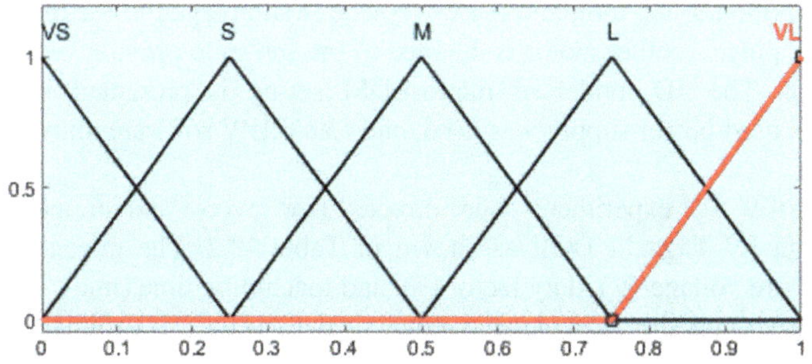

Fig. 49.3 Membership function for grey fuzzy reasoning grade

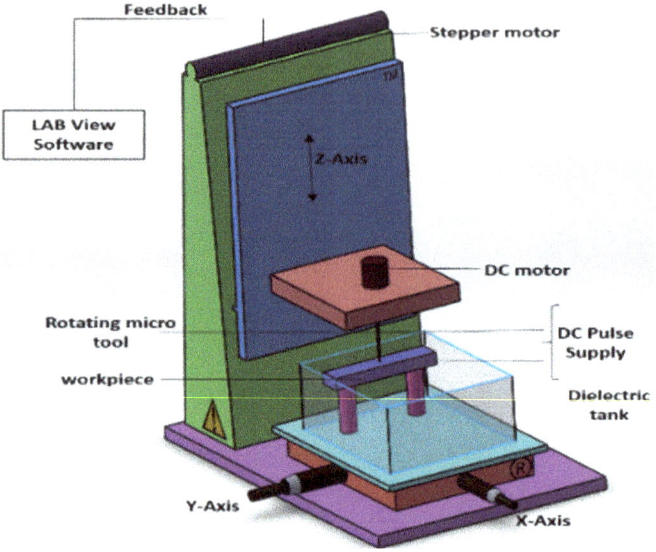

Fig. 49.4 Three-dimensional figure of micro-EDM system

49.3 The Experimental Set-Up and Procedure

Experiment was conducted on Ti6Al4V workpiece of dimensions 10 × 10 × 1.5-mm thickness with brass tool of 1-mm diameter and 30-mm length. De-ionized water was used as dielectric medium and tungsten disulphide as suspended powder particles. To conduct the experiments, a three-axis system fitted out with the micro-EDM set-up is used as shown in Fig. 49.5. XY-stage is used to provide longitudinal and transverse motion to the workpiece. A DC motor is used to impart rotational motion to the tool, and the workpiece is submerged in the dielectric with proper clamping. Another motor is devoted to the set-up to provide vertical motion to the tool. The 3D model of micro-EDM set-up is presented in Fig. 49.4. Transistor-based power supply was used, and LabVIEW software allows the z-axis motion.

A total of 9 × 2 experiments were directed (for reverse and straight polarities) by choosing L9 Taguchi DOE as shown in Table 49.2. The uncontrolled input parameters are voltage (V), duty factor (%) and machining time (min). Tool is fixed to be positive and the workpiece is connected to negative terminal for reverse polarity, and tool is negative and workpiece is positive for straight polarity. Trial experiments were conducted to decide different levels of parameter values, and it was observed that as the difference between the values of parameters for consecutive levels was small it was difficult to differentiate between the performance parameter values so middle-range parameters with sufficient difference were nominated for the experiments as shown in Table 49.1. The performance of

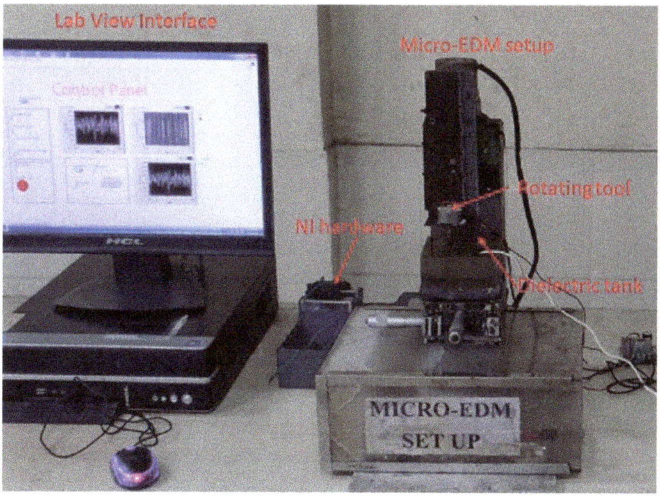

Fig. 49.5 Micro-EDM set-up

Table 49.1 Parameters of the process and their levels

Factors	Parameters	Level 1	Level 2	Level 3
A	Voltage, *V*	25	40	55
B	Duty factor, DF	35	45	55
C	Machining time, *T* (min)	6	8	10

experiments was measured by MRR and SR. Each of these values is considered at once to get GFRG which yield the optimum MRR and SR. Design of experiment is generated by L9 Taguchi array system.

The flow chart shown in Fig. 49.6 gives a concise idea about how the grey fuzzy logic works. It starts with the construction of design of experiment with the help of Taguchi L9 design. Experimental data was then converted to normalized data with the help of formulas. This normalized data was used to calculate grey relational coefficients for both MRR and SR, and the GRCs for both straight and reverse polarities are then fed to the fuzzifier as input membership functions. With the help of certain "if and then" rules, fuzzy interface engine (Mamdani) converts these values to fuzzy values and then the defuzzifier converts them into grey fuzzy reasoning grades. GFRG values decide the optimal solution, and then confirmatory tests were done to verify the optimal solution.

Table 49.2 Design of experiment with responses

Exp. no.	V	DF	T	MRR_{st} (g/s)	MRR_{rev} (g/s)	SR_{st} (μm)	SR_{rev} (μm)
1	25	35	6	0.00101	0.00023	0.998	0.266
2	25	45	8	0.00172	0.00015	1.556	0.177
3	25	55	10	0.00347	0.00032	1.856	0.379
4	40	35	8	0.00077	0.00031	2.085	0.481
5	40	45	10	0.00523	0.00038	2.83	0.75
6	40	55	6	0.00054	0.00006	1.85	0.489
7	55	35	10	0.00282	0.00020	3.248	0.588
8	55	45	6	0.00395	0.00023	2.041	0.575
9	55	55	8	0.00887	0.00038	2.755	0.696

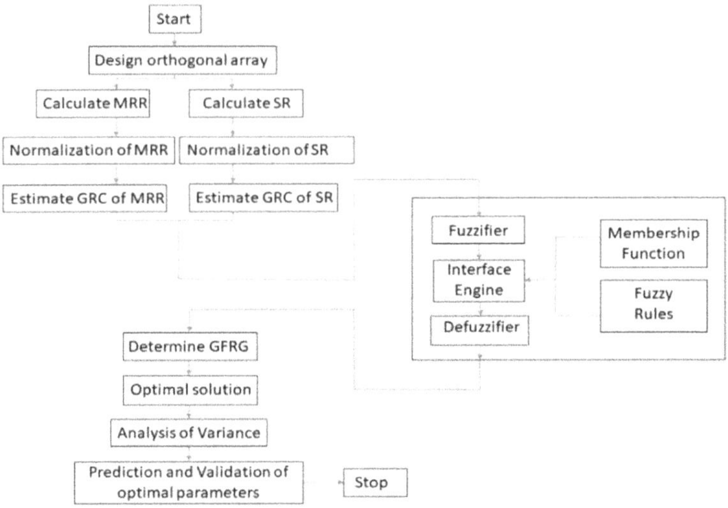

Fig. 49.6 Flow chart of grey fuzzy-based algorithm

49.4 Results and Discussion

49.4.1 The Grey Reasoning Coefficient

For each performance parameter, normalized values were calculated. Table 49.3 displays the calculated normalized value of each quality characteristic, grey relational coefficients and grey relational grade.

Table 49.3 Normalized and GRC table

Exp no.	Normalized value of		Normalized value of		Grey relational coefficient of		Grey relational coefficient of		GRG_{st}	GRG_{rev}
	MRR_{st}	MRR_{rev}	SR_{st}	SR_{rev}	MRR_{st}	MRR_{rev}	SR_{st}	SR_{rev}		
1	0.05664	0.51586	1	0.84467	0.34641	0.50805	1	0.76298	0.673	0.636
2	0.14220	0.25953	0.752	1	0.36824	0.40307	0.66844	1	0.518	0.702
3	0.35161	0.80422	0.61866	0.64746	0.43539	0.71862	0.56732	0.58648	0.501	0.653
4	0.02760	0.77218	0.51688	0.46945	0.33958	0.68699	0.50859	0.48518	0.424	0.586
5	0.56342	0.99647	0.18577	0	0.53385	0.993	0.38045	0.33333	0.457	0.663
6	0	0	0.62133	0.45549	0.33333	0.33333	0.56904	0.47869	0.451	0.406
7	0.27457	0.43255	0	0.28272	0.40802	0.46840	0.33333	0.41075	0.371	0.440
8	0.40921	0.52162	0.53644	0.30541	0.45838	0.51105	0.51891	0.41855	0.489	0.465
9	1	1	0.21911	0.09424	1	1	0.39035	0.35568	0.695	0.678

49.4.2 The Grey Fuzzy Reasoning Analysis

MRR and surface roughness are two inputs, and GFRG is one output of fuzzy logic method. The GRC for MRR and surface roughness are considered as the inputs to the fuzzy logic method. To establish the fuzzy if and then rule as shown in Table 49.4, the language-based membership operators L, M and H were used to indicate the grey relational coefficient of the entered variables. Similarly, the output grey relational grade has been signified through the member functions VS, S, M, L, VL on triangular shape membership function.

The common methods for defuzzification are centre of maximum (COM) and centre of area (COA). The relationship between two inputs and output is developed by certain "if-then" rules that are:

Rule 1: If $m1$ is $V1$ and $n2$ is $W1$ and $p3$ is $X1$ and $q4$ is $Y1$, then output (G) is $Z1$.
Rule 2: If $m1$ is $V2$ and $n2$ is $W2$ and $p3$ is $X2$ and $q4$ is $Y2$, then output (G) is $Z2$.
Rule n: If $m1$ is Vn and $n2$ is Wn and $p3$ is Xn and $q4$ is Yn, then output (G) is Zn.

where Vi, Wi, Xi and Yi are the fuzzy subsections distinct by the equivalent member functions, i.e. μVi, μWi, μXi and μYi, respectively. The inference engine (Mamdani) does fuzzy reasoning on rules defined by fuzzy system while considering max and min inference for generating a specific fuzzy value, $\mu c_o(G)$.

$$\begin{aligned}\mu c_o(G) = {} & (\mu V_1(m_1) \wedge \mu W_1(n_2) \wedge \mu X_1(p_3) \wedge \mu Y_1(q_4) \wedge \mu Z_1(G)) \\ & \vee (\mu V_1(m_1) \wedge \mu W_2(n_1) \wedge \mu X_2(p_2) \wedge \mu Y_2(q_3) \wedge \mu Z_2(G)) \\ & \vee (\mu V_n(m_1) \wedge \mu W_n(n_2) \wedge \mu X_n(p_3) \wedge \mu Y_n(q_3) \wedge \mu Z_n(G)).\end{aligned}$$

where $\vee$ is the maximum operator and $\wedge$ is the minimum operator. The process of defuzzification has been used to convert the fuzzy interface output into exact value of GFRG(G_o).

Table 49.4 Fuzzy if and then rule

I/C criteria	Input (if)		MPCI (then)
	Individual desirability values of MRR	Individual desirability values of SR	
1	L	L	VS
2	L	M	S
3	L	H	M
4	M	L	S
5	M	M	M
6	M	H	L
7	H	L	M
8	H	M	L
9	H	H	VL

$$G_o = \frac{\sum G\mu c_o(G)}{\sum \mu c_o(G)} \tag{49.5}$$

Graphical illustration of fuzzy logic reasoning is shown in Fig. 49.7 (straight polarity) and Fig. 49.8 (reverse polarity). The nine rows represent rules, and three columns represent MRR, SR and GFRG, respectively. Both the graphs are taken at optimal parameters, and in Fig. 49.7 the input values of grey relational coefficient of SR and MRR are 0.39 and 1, respectively, and defuzzified output is 0.666 which shows the combined effect of both the input values. Similarly, in Fig. 49.8 the input values are 0.403 and 1 and corresponding output value is 0.671. The GFRG values with their respective ranks are shown in Table 49.5.

The response analysis in Tables 49.6 and 49.7 determined that for straight polarity the parameter combination $V = 25$, DF = 55 and $T = 8$ gives the optimal value and for reverse polarity the parameter combination $V = 25$, DF = 45 and $T = 8$ gives the optimal value of MRR and SR. Figures 49.9 and 49.10 show the mean of mean, response graph of GFRG for straight and reverse polarities, respectively.

49.4.3 Analysis of Variance

ANOVA is formulated for both straight and reverse polarities using GFRG values to calculate the percentage by which each parameter is affecting the process. The results of ANOVA for straight polarity are given in Table 49.8 and for reverse polarity in Table 49.9. From ANOVA tables, it is justified that in case of straight polarity voltage (27.19%) influences more on alloying process followed by

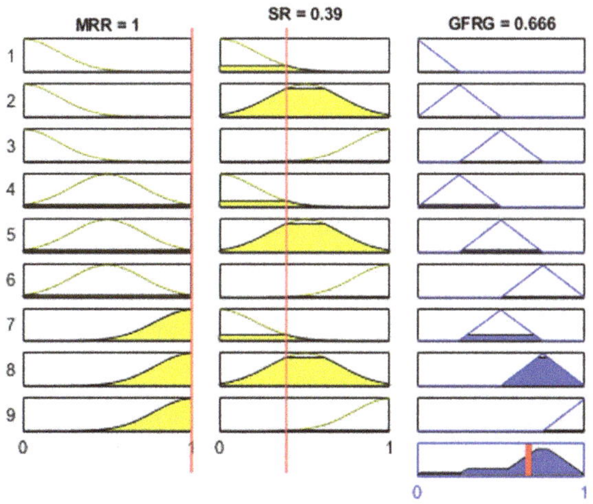

Fig. 49.7 Fuzzy logic rules for straight polarity

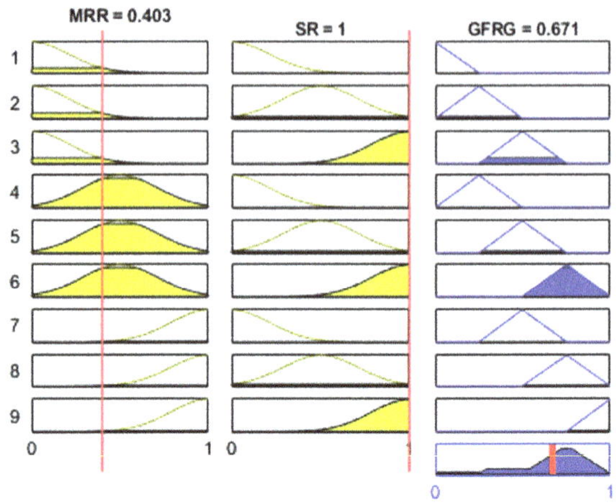

Fig. 49.8 Fuzzy logic rules for reverse polarity

Table 49.5 GFRG with the respective rank

Exp. no.	$GFRG_{st}$	$Rank_{st}$	$GFRG_{rev}$	$Rank_{rev}$
1	0.649	2	0.611	4
2	0.516	3	0.671	1
3	0.501	4	0.604	5
4	0.445	8	0.566	6
5	0.47	6	0.641	3
6	0.46	7	0.435	9
7	0.422	9	0.466	8
8	0.493	5	0.478	7
9	0.666	1	0.653	2

Table 49.6 Response values for the GFRG straight polarity

Factors	Level 1	Level 2	Level 3
V	0.5553	0.4583	0.5270
DF	0.5053	0.4930	0.5423
T	0.5340	0.5423	0.46432

Table 49.7 Response values for the GFRG reverse polarity

Factors	Level 1	Level 2	Level 3
V	0.6287	0.5473	0.5323
DF	0.5477	0.5967	0.5640
T	0.5080	0.6300	0.5703

machining time (20.04%) and duty factor (7.201). Similarly, from table in case of reverse polarity machining time (33.63%) affects more followed by voltage (24.27%) and duty factor (5.62%).

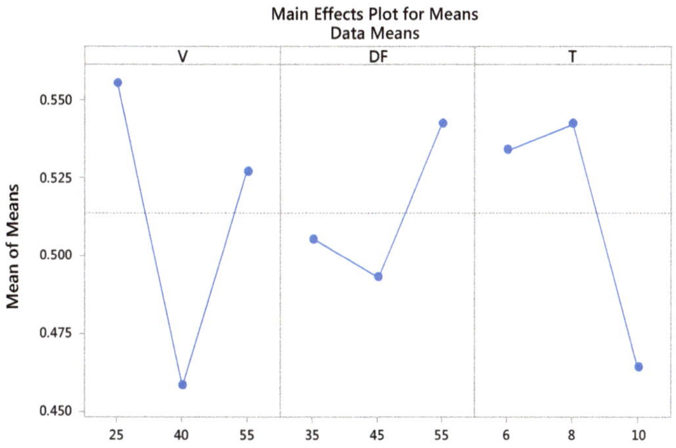

Fig. 49.9 Response graph for straight polarity

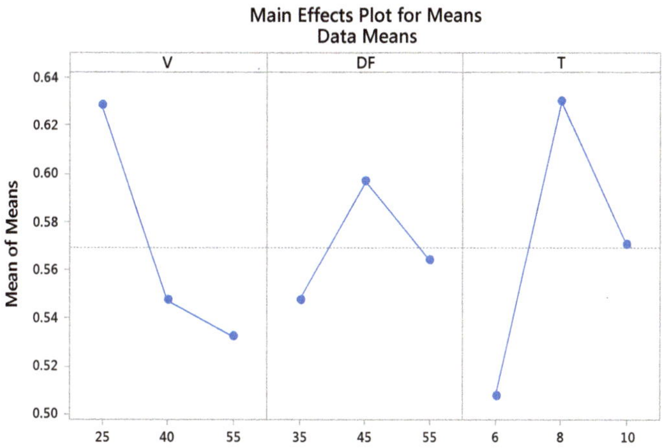

Fig. 49.10 Response graph for reverse polarity

Table 49.8 Analysis of variance of grey fuzzy relation grade for straight polarity

Factor	DOF	Sum of squares	Mean squares	F value	P value	% contribution
V	2	0.01493	0.007463	0.99	0.424	27.19
DF	2	0.003955	0.001977	0.21	0.815	7.201
T	2	0.01101	0.005503	0.67	0.545	20.046
Error	2	0.025027	0.025028	–	–	45.519
Total	8	0.054922	–	–	–	100

Table 49.9 Analysis of variance of grey fuzzy relation grade for reverse polarity

Factor	DOF	Sum of squares	Mean squares	*F* value	*P* value	% contribution
V	2	0.016120	0.008060	1.04	0.409	24.2778
DF	2	0.003735	0.001867	0.19	0.831	5.62516
T	2	0.0223	0.011165	1.67	0.265	33.6305
Error	2	0.024213	0.024213	–	–	36.4664
Total	8	0.066398	–	–	–	100

49.4.4 Theoretical Prediction and Confirmation of Experiments

After obtaining optimal level of design parameters in case of straight and reverse polarities, the last step is to forecast and settle the upgradation of performance features. Table 49.10 shows the contrast of original process parameters and optimal process parameters in case of straight polarity, and as noted the MRR improved from 0.00347 to 0.009225 g/min, surface roughness (SR) decreased from 1.856 to 1.83 μm, and GFRG is improved by 0.216. Similarly, Table 49.11 shows the comparison for reverse polarity, and as noted MRR increased from the value of 0.00034 to 0.00035 g/min and GFRG is improved by 0.182. It is clearly revealed that multiple performance parameters are improved using grey fuzzy optimization technique. MRR is calculated by the formula as given in Eq. 49.6.

$$\mathrm{MRR} = \frac{W_\mathrm{i} - W_\mathrm{f}}{t} \tag{49.6}$$

where W_i and W_f are the weights before and after the experiments and t is the time of machining in seconds.

FESEM images of both straight and reverse polarities at 40 V, 35% duty factor and 8 min machining time are shown in Figs. 49.11 and 49.12, respectively. It is found that the recast layer is more uneven in case of straight polarity whereas a

Table 49.10 Results of process performance by means of the original and optimal factors for straight polarity

	The initial process parameters	The optimal process parameters	
		Prediction	Experiment
Level	A1B3C3	A1B3C2	A1B3C2
MRR	0.00347	0.009005	0.009225
Surface roughness	1.856	1.85	1.83
GFRG	0.501	0.704	0.717
Improvement of GFRG		0.203	0.216

Table 49.11 Results of process performance by means of the original and optimal factors for reverse polarity

	The initial process parameters	The optimal process parameters	
		Prediction	Experiment
Level	A1B3C3	A1B2C2	A1B2C2
MRR	0.00032	0.00034	0.00035
Surface roughness	0.379	0.15	0.177
GFRG	0.604	0.767	0.786
Improvement of GFRG		0.163	0.182

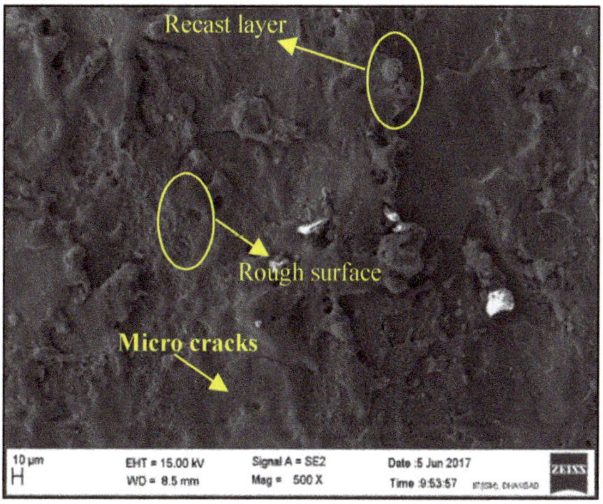

Fig. 49.11 FESEM for straight polarity

comparatively lower surface roughness is seen in reverse polarity. This may be due to less discharge energy available at workpiece in case of reverse polarity as compared to straight polarity. In addition to that, WS_2 is a solid lubricating powder; thus, it plays an important role to improve surface properties of workpiece.

49.5 Conclusions

The combination of optimization method and grey fuzzy logic is used in current research work for concurrently optimizing the numerous performance characteristics like MRR and SR both in straight and reverse polarities. After summarizing the key attributes of this experimental investigation, the following conclusions were drawn:

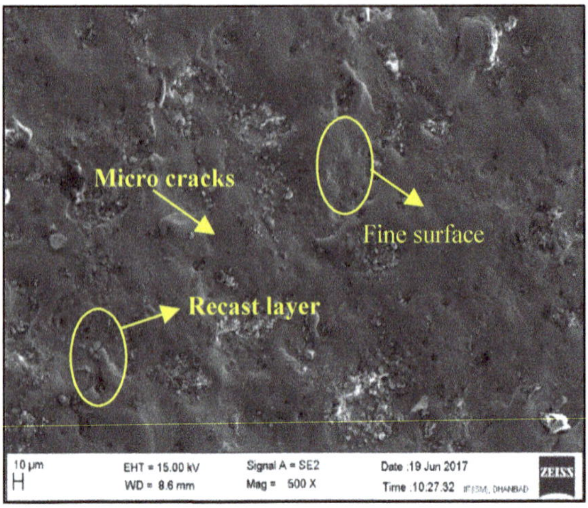

Fig. 49.12 FESEM for reverse polarity

1. The optimal parametric combination in straight polarity is $V = 25$, DF = 55 and $T = 8$ and in reverse polarity is $V = 25$, DF = 45 and $T = 8$ determined by grey fuzzy logic.
2. The ANOVA results determine that in case of straight polarity voltage (27.19%) influences more and in case of reverse polarity machining time (33.63%) affects more.
3. It has been concluded from the results that the lower surface roughness was obtained in case of reverse polarity as compared to what obtained in straight polarity as the craters formed in case of straight polarity were of comparatively greater size due to the direction of current which in turn resulted in high energy flow to the workpiece.
4. It is indeed complicated to convert multiple parametric optimizations into a single GFRG. So, an amalgamated technique like grey fuzzy logic method along with orthogonal array system provides an efficient way for optimizing the process parameters in micro-EDM.

References

1. Masuzawa, T.: State of the art of micromachining. CIRP Ann. Manuf. Technol. **49**(2), 473–488 (2000)
2. Schulze, H.P.: Importance of polarity change in the electrical discharge machining. AIP Conf. Proc. **1896**(1), 050001. AIP Publishing (2017)
3. Parshuramulu, A., Buschaiah, K., Laxminarayana, P.: A study on influence of polarity on the machining characteristics of sinker EDM. Int. J. Adv. Res. Eng. Technol. (IJARET) **4**(3), 158–162 (2013)

4. Kumar, A., Mandal, A., Dixit, A.R., Das, A.K.: Performance evaluation of Al2O3 nano powder mixed dielectric for electric discharge machining of Inconel 825. Mater. Manuf. Processes **33**(9), 986–995 (2018)
5. Tyagi, R., Das, A.K., Mandal, A.: Electrical discharge coating using WS2 and Cu powder mixture for solid lubrication and enhanced tribological performance. Tribol. Int. **120**, 80–92 (2018)
6. Gopal, A.V., Chakradhar, D.: Parametric optimization in electrochemical machining of EN-31 steel based on grey relation approach. Appl. Mech. Mater. **110**. Trans Tech Publications (2012)
7. Durairaj, M., Sudharsun, D., Swamynathan, N.: Analysis of process parameters in wire EDM with stainless steel using single objective Taguchi method and multi objective grey relational grade. Procedia Eng. **64**, 868–887 (2013)
8. Kumar, V.S., Kumar, P.M.: Optimization of cryogenic cooled EDM process parameters using grey relational analysis. J. Mech. Sci. Technol. **28**, 3777–3784 (2014)
9. Natarajan, N., Arunachalam, R.M.: Optimization of micro-EDM with multiple performance characteristics using Taguchi method and Grey relational analysis (2011)
10. Gopalakannan, S., Senthilvelan, T., Ranganathan, S.: Statistical optimization of EDM parameters on machining of aluminium hybrid metal matrix composite by applying Taguchi based grey analysis (2013)
11. Dewangan, S., Gangopadhyay, S., Biswas, C.K.: Multi-response optimization of surface integrity characteristics of EDM process using grey-fuzzy logic-based hybrid approach. Eng. Sci. Technol., Int. J. **18**(3), 361–368 (2015)
12. Ramanan, G., Shanavas, S.: Multi objective optimization of machining parameters for AA7075 metal matrix composite using grey-fuzzy technique. Int. J. Appl. Eng. Res. **12**(8), 1729–1735 (2017)
13. Soepangkat, B.O.P., Soesanti, A., Pramujati, B.: The use of Taguchi-grey-fuzzy to optimize performance characteristics in turning of AISI D2. Appl. Mech. Mater. **315**. Trans Tech Publications (2013)
14. Winarni, S., Indratno, S.W.: Application of multi response optimization with grey relational analysis and fuzzy logic method. J. Phys.: Conf. Ser. **948**(1). IOP Publishing (2018)

Chapter 50
Multi-output Response Optimization for Overall Enhancement of Mechanical Characteristics Using Utility Approach for AISI 316L Austenitic Stainless Steel Using Microplasma Arc Welding

Srikant Prasad, Dibakor Boruah and Pintu Thakur

Abstract This paper presents multi-output advancement for microplasma arc welding (MPAW) process using utility methodology. It is an advanced unconventional welding process in which very thin (usually less than 1 mm) metal surfaces are joined using plasma arc. A successful trial has been made to weld 0.5-mm-thick 316L stainless steel sheet. An attempt has been made to optimize the MPAW parameters for a desired equal-weighted multi-output such as ultimate tensile strength, yield strength, percentage elongation, Young's modulus, and weld hardness at fusion zone and heat-affected zone. The design of experiments L_{27} orthogonal matrix was conducted using levels of several visually successful trials. An analytical tool analysis of variance and interaction has been clubbed with utility approach to get desired aim of experimentation. As per different work application, the different preferable weight can be incorporated into the output. The confirmation test correlated with the best-desired aim of the output parameter. It was observed that overall utility for all mechanical properties and corresponding S/N ratio is enhanced with 0.28 and 0.42 values, respectively. It was found that welding speed with 49.01% and pulse time with 13.21% are the most significant parameters.

S. Prasad (✉)
Mechanical Engineering, Indian Institute of Technology Guwahati,
Guwahati 781039, Assam, India
e-mail: shrikant@iitg.ac.in

D. Boruah
Faculty of Engineering, Environment and Computing, Coventry University,
Coventry CV1 5FB, UK
e-mail: boruahd@uni.coventry.ac.uk

P. Thakur
Girijananda Chowdhury Institute of Management and Technology,
Tezpur 784501, Assam, India
e-mail: pintuthakur07@gmail.com

R. G. Narayanan et al. (eds.), *Advances in Computational Methods in Manufacturing*, Lecture Notes on Multidisciplinary Industrial Engineering,
https://doi.org/10.1007/978-981-32-9072-3_50

The enhancement in the responses confirmed the approach of the feasibility of utility analysis which can help to enhance the product reliability for an industrial application.

Keywords MPAW · Optimization · Overall utility · ANOVA

50.1 Introduction

Microplasma arc welding (MPAW) was invented first by Robert Gage in 1957, which was introduced in 1964 to the welding industry for better control of the arc-welding process in lower current ranges. The plasma gas (the fourth state of matter) is used to generate arc with the help of argon gas. An ionization potential of plasma gas is 15.6 eV [1]. It takes very low currents to generate plasma arc. The argon gas is preferred as plasma gas because its low ionization potential ensures reliable starting and dependable pilot arc [2]. This arc is utilized to weld very thin plates even less than 0.5 mm. Low-alloy stainless steel sheet 316L is the useful engineering materials for industrial applications since they enriched with an excellent combination of properties like high strength-to-weight ratio, corrosion resistance, good toughness, fatigue life, which make it attractive for different applications through advanced joining process like MPAW. During welding, several input parameters which influence the welding properties need to be optimized for the better performance in the field application. The welding input parameters influence weld joint quality during the welding process. Therefore, it can be considered as a multi-input and multi-output process. Recently, a few researchers have reported optimization of 304L and Inconel considering effect of grain size, bead geometry, hardness, and ultimate tensile strength (UTS) using pulse microplasma [3]. The pulsed current microplasma arc-welding parameters were optimized to maximize UTS for AISI 304L sheets material using response surface method [4] and Hooke and Jeeves Algorithm by Shiv Prasad et al. [5]. The latest research work on the optimization of hydroxyapatite synthesis and microplasma spraying of porous coatings onto titanium implants has been reported by Alontsevadalontseva et al. [6]. According to the authors best of knowledge, no literature reported optimization of AISI 316L austenitic stainless steel sheet of using pulse MPAW. The various tests have been conducted to evaluate mechanical properties like tensile test, hardness, and weld bead geometry. As a multi-output response ultimate tensile strength (UTS), yield strength (YS), percentage elongation (%E), young modulus (YM), hardness of fusion zone (H_{FZ}) and hardness of heat-affected zone (H_{HAZ}) were used to study the optimization process using utility approach which helped to quantify the integrated performance of the MPAW. The confirmation experiment was conducted to validate the optimized parameters of the process. The predicted parameters have been correlated with the experimental result of the optimized parameters. The optimal welding conditions help to increase the productivity, minimize the total operating cost and strengthening of weld joint.

50.2 Experimental Work

The welding experiments were conducted using Aircraft plasma MP-50. The stainless steel 316L sheet of size (100 mm × 150 mm), thickness 0.5 mm was welded without filler material. The experiment had been designed as per design of experiment (DOE), using orthogonal array L27 Taguchi fractional factorial design method. The Argon gas was used as an orifice gas and shielding gas. The shielding gas prevents the weld from oxidation during welding and welds remain shiny. The coupling of orifice and shielding gas made with help of "T" joint attached at the end of cylinder exit pipe. The transferred arc mode in which the torch is brought in close to the workpiece, welding current is initiated, and the arc is transferred from the electrode to the workpiece which produces excess temperature up to 1700 °C to meet the welding requirement, then it cooled down to permissible temperature by means of circulating water through the nozzle. The complete experimental setup and welding fixture have been developed at IIT Guwahati central workshop. The setup has been shown in Fig. 50.1.

50.2.1 Weld Direction Determination for Rolled Sheet

To enhance percentage elongation, the weld direction has been determined with special light scattering technique which always shines in the direction perpendicular to the rolling direction. Due to rolling, grains are elongated in the direction of rolling resulting in less percentage elongation and hardness in the rolling direction. Therefore, care has been taken during the experiment and welding direction has been proposed normal to the light scattering direction [2]. Figure 50.2a illustrates the weld direction and rolling direction with the help of light scattering direction.

Fig. 50.1 Microplasma arc-welding (MPAW) setup

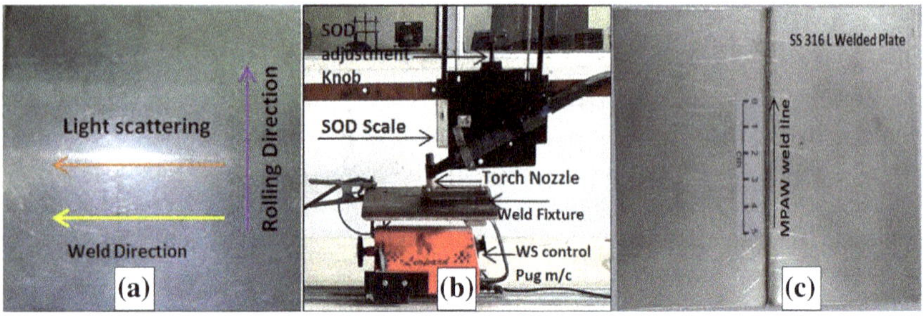

Fig. 50.2 **a** Weld direction and rolling direction, **b** torch and fixture equipped with SOD adjustment, **c** welded plate

Table 50.1 Stainless steel 316L chemical composition

Chemical composition (%) of SS 316L	C	Mn	Si	P	S	Cr	Mo	Ni	N
	0.03	2.0	0.75	0.045	0.03	18.0	3.00–2	10–14	1.0

The yellow arrow shows the rolling direction, and the blue arrow shows the welding direction.

The welding setup, fixture, and standoff distance adjustment features for MPAW setup were fabricated at IIT Guwahati central workshop for accuracy of weld quality, which has been shown in Fig. 50.2b, a welded sample was shown in Fig. 50.2c. The weld fixture was also designed as per dimension of weld sheet and less arc deflection. The interface gap between two weld plates, which is not permissible was ensured properly before welding. The chemical composition of the weld material used in these experiments has shown in Table 50.1. The significant six parameters in three levels which have an impact on weld quality, taken into consideration and shown in Table 50.2.

During post-weld process, tensile samples were cut through CO_2 laser-cutting machine, make: model no: LVD Orion3015, power rated 2.5 kW. The hardness samples (5 mm × 5 mm) and tensile samples as per the dimension of standard

Table 50.2 Details regarding the welding parameters and their levels

Sl. no.	Parameters (unit)	Abbreviations	Level 1	Level 2	Level 3
1	Current (A)	IP	11.9	13.8	15.3
2	Welding speed (mm/s)	WS	2.7	3.7	4.7
3	Plasma gas flow rate (L/s)	PGFR	0.6	0.7	0.8
4	Pulse time (ms)	PT	30	40	45
5	Pulse frequency (Hz)	PF	35	45	50
6	Standoff distance (mm)	SOD	1	1.5	2

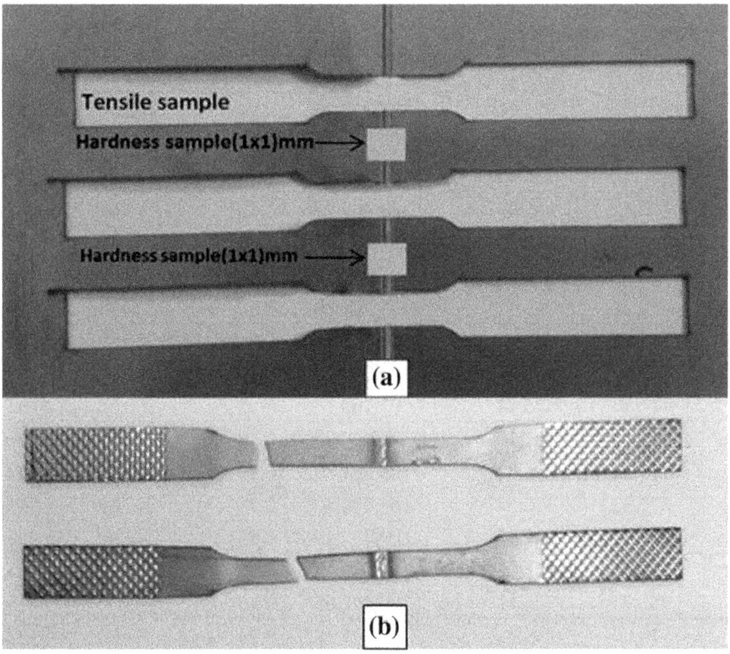

Fig. 50.3 **a** Extraction of tensile, hardness, and microstructure sample from a welded plate, **b** tested tensile samples

ASTM–B557M-10 were extracted. Figure 50.3a shows the extraction of tensile, hardness, and microstructure sample, and Fig. 50.3b shows the tested tensile sample.

The material properties considered in the present investigation are: UTS, YS, YM, %E, $H_{FZ,}$ and H_{HAZ}. The bead geometry is the mirror of the mechanical properties of a weld which depends upon reinforcement height and bead width [5]. The proper reinforcement height is required to get good strength of the welding [6]. All the mechanical output properties have been shown in Table 50.3.

Table 50.3 Experimental data of response characteristics

Exp. no.	UTS (MPa)	Avg. YS (MPa)	YM (GPa)	%E	H_{FZ} (HV)	H_{HAZ} (HV)
1	794.72	385.55	97,061.72	0.62	195.00	207.00
2	768.46	395.00	90,436.35	0.61	212.00	206.00
3	769.07	419.39	10,2576.10	0.68	188.00	208.00
4	743.43	377.71	106,978.70	0.59	240.00	199.00
5	792.52	392.19	100,899.00	0.61	211.00	209.00
6	803.36	375.11	98,579.45	0.59	199.00	212.00

(continued)

Table 50.3 (continued)

Exp. no.	UTS (MPa)	Avg. YS (MPa)	YM (GPa)	%E	H_{FZ} (HV)	H_{HAZ} (HV)
7	577.31	353.26	97,777.82	0.54	190.00	189.00
8	592.03	323.37	95,458.42	0.58	199.00	196.00
9	668.22	136.43	106,095.70	0.35	187.00	199.00
10	722.15	365.03	98,463.28	0.60	194.00	193.00
11	710.71	343.73	88,487.05	0.59	218.00	217.00
12	791.57	365.46	99,683.59	0.59	209.00	196.00
13	756.22	360.67	108,467.80	0.58	222.00	210.00
14	752.82	349.91	100,781.40	0.57	227.00	202.00
15	758.56	355.83	100,782.40	0.69	183.00	198.00
16	714.19	323.44	97,160.93	0.55	200.00	195.00
17	745.25	121.60	97,984.79	0.49	206.00	200.00
18	775.23	336.01	105,061.80	0.52	201.00	203.00
19	677.32	326.22	106,704.80	0.55	222.00	245.00
20	681.98	339.42	109,134.40	0.55	228.00	221.00
21	773.71	362.02	105,543.20	0.55	206.00	201.00
22	751.79	356.41	89,591.68	0.55	235.00	235.00
23	776.90	358.99	97,847.78	0.59	193.00	192.00
24	764.09	353.50	103,777.10	0.55	209.00	199.00
25	485.53	322.49	100,041.60	0.54	240.00	199.00
26	501.92	358.29	101,397.30	0.56	214.00	212.00
27	505.21	328.08	99,956.11	0.54	195.00	198.00
AVG	709.417	340.1891	100,249.3	0.568874	208.2593	205.2222

50.3 Result and Discussion

50.3.1 Optimization Using Utility Approach

Multi-response optimization is the most prominent area in the research for the welding industries to meet satisfaction, reliability, and cost reduction. Nowadays, unconventional welding process like MPAW is a very successful welding technology to weld materials with a very low thickness. Very precise control on parameters is required for its utility. Various process parameters need to be optimized to meet several demandable responses from the customer to manufacturer. The optimization of weld parameters is necessary to get optimal response for all quality characteristics. The multi-output characteristic is needed to scale for a single-output characteristic on a single composite scale index to get a single optimal setting of MPAW parameters [7–9]. The multi-response optimization is a need of today for industrial application where more than one response is coupled simultaneously using utility analysis. Several analytical techniques were explored to handle the need for multi-optimization problem. Previously, some work has been reported for optimization is gray relation analysis, response surface methodology (RSM), and Taguchi methods. A research paper has been published on gray-based Taguchi method for parameters optimization of submerged arc bead on plate-welding process. Till now, no publication has been reported on multi-optimization using utility tool for MPAW process. For single output response Taguchi and RSM methodology are reliable optimization, but for multi-output welding process Taguchi analysis is not feasible [10–13]. Very few works are presented in literature for multi-response optimization problems [14, 15]. In comparison with utility approach, the desirability approach is not efficient up to a level. In the desirability approach, individual desirability is scaled into single composite index, for overall multi-response. Due to the complexity of computational aspects, several desirability values appear with zero value during the calculation which is difficult to analyze with ANOVA. Few publications on multi-output optimization have been reported on bead on plate welding. However, during the experiment, it was observed that the responses of real-welding process are different than the bead on plate welding [16]. Datta et al., have reported on submerged arc bead on plate welding using gray-based Taguchi technique. This research paper has been focused on overall enhancement of weighted mechanical properties for MPAW [17]. The mechanical property of a MPAW has been shown in Table 50.3.

50.3.2 Utility Concept Approach

The utility alludes to the fulfillment of customer; it contrasts from individual to individual, time to time and place, as defined by Prof. Hobson, "Utility is the capacity of a decent to fulfill a need". The business analyst estimated the utility in cardinal (numerical terms), yet there was no standard unit for estimating utility. In this way, the financial analysts utilized a fanciful measure, known as ('Util'). The costumers assess the item based on handiness. Taguchi configuration cannot deal with multi-output responses since every execution trademark have distinctive unit [18]. The general utility is the composite record of all quality as indicated by their value, which is appointed through the weight of the particular responses. A few research papers have been reported on Taguchi-constructed utility approach in light of various welding processes like Laser, EBM, and SAW process. However, no work has been accounted for on 0.5-mm-thin plate Austenitic steel 316L sheet welding through MPAW process. Subsequently, an effort has been made to represent the enhancement using utility-based Taguchi approach for multi-objective MPAW process for a thin austenitic steel sheet of 316L. This method was utilized to decide the ideal and best mix of process parameters for this process specifically welding pulse current, welding speed, pulse width, pulse frequency, plasma gas flow rate and stand of distance to maximize UTS, YS, %E, and hardness of the weld with the weight priority index to meet customer requirement.

50.3.3 Determination of Utility Value

All the output data has been converted into overall utility index to determine the utility for quality characteristics. In this approach, a preference scale has to construct using weight and composite numbers which is called overall utility. There are several preference scales, namely (i) exponential, (ii) logarithmic, and (iii) linear, but in the present research, investigation logarithmic approach has been considered as a preference scale index.

It is the evaluation of diverse quality characteristics in a rational choice by combining to make a composite index which represents the utility of a product. The overall utility index of a weld output is a sum of utilities of individual's quality characteristics.

$$U(X_1, X_2, \ldots\ldots\ldots\ldots, X_n) = \sum_{i=1}^{n} U_n(X_i) \tag{50.1}$$

$$P_i = A \times \log\left(\frac{X_{i,}}{X_{i,}}\right) \tag{50.2}$$

$$A = \frac{9}{\log\left(\frac{X_i^*}{X_{i,}}\right)} \tag{50.3}$$

50.3.3.1 Steps in Calculation of Utility

Step 1: In the utility analysis, the minimum acceptable quality is set numeric index 0 and the best acceptable quality level is set 9. Among the several scales mentioned above, the log scale has been chosen to find the preference number (P_i). The following equation is followed to find the constant value (A).

Step 2: For finding the value (A) using (P_i) value 9 in Eq. (50.3), we have to use (X_i*) as optimal pridicted output, and ($X_{i,}$), as minimum characteristic output among all.

Step 3: We have to find preference number (P_i) using Eq. (50.4). At this stage, we have to incorporate index (A) value and (X_i) individual present output carectersitic value, and (X_i), as minimum charectersitic output among all. The chart repersented below has been prepared using above Eq. 50.4 for all mechanical responcees like UTS-P, YS-P, YM-P, %E-P, $H_{FZ\text{-}P}$, and $H_{HAZ\text{-}P}$, respectively. The equal weightage has been considered for all outputs for predicted mechanical properties which are shown in Table 50.4. Depending on the customer requirements, the attributes to preference weightage may be given priority among all property. The property can be adjusted by incorporating weight (W) into individual utility index. Sum of all weights should not more than 1.

Table 50.4 Overall utility of data processing for predicted properties after assigning weights

Exp. no.	UTS-P W	YS-P W	YM-P W	%E-P W	H_{FZ-P} W	H_{HAZ-P} W
1	0.69	1.98	0.71	1.23	0.1	0.2
2	0.65	2.02	0.17	1.21	0.24	0.19
3	0.65	2.13	1.13	1.44	0.04	0.21
4	0.6	1.95	1.46	1.11	0.43	0.11
5	0.69	2.01	1.01	1.19	0.23	0.22
6	0.71	1.93	0.83	1.12	0.13	0.25
7	0.24	1.83	0.77	0.94	0.06	0
8	0.28	1.68	0.58	1.1	0.13	0.08
9	0.45	0.2	1.39	0	0.03	0.11
10	0.56	1.89	0.82	1.15	0.09	0.05
11	0.54	1.78	0	1.13	0.28	0.3
12	0.69	1.89	0.91	1.14	0.21	0.08
13	0.62	1.87	1.56	1.08	0.31	0.23
14	0.62	1.82	1	1.03	0.34	0.14
15	0.63	1.84	1	1.46	0	0.1
16	0.54	1.68	0.72	0.96	0.14	0.07
17	0.6	0	0.78	0.72	0.19	0.12
18	0.66	1.75	1.32	0.86	0.15	0.15
19	0.47	1.69	1.44	0.98	0.31	0.56
20	0.48	1.76	1.61	0.98	0.35	0.34
21	0.66	1.87	1.35	0.97	0.19	0.13
22	0.62	1.85	0.1	0.97	0.4	0.47
23	0.66	1.86	0.77	1.13	0.09	0.03
24	0.64	1.83	1.22	0.98	0.21	0.11
25	0	1.68	0.94	0.91	0.43	0.11
26	0.05	1.86	1.04	1.02	0.25	0.25
27	0.06	1.7	0.93	0.94	0.1	0.1

$$U(X_1, X_2, \ldots\ldots\ldots, X_n) = \sum_{i=1}^{n} W_i U_n(X_i) \tag{50.4}$$

A methodology of combining Taguchi and utility concept was introduced for predicting the optimal parameter settings of the combined responses. The six quality characteristics, namely UTS, YS, YM, %E, H_{FZ}, H_{HAZ}, during the experimental observation ratio of top width to top reinforcement were considered in this study for best output. The response parameters were individually optimized using Taguchi's technique to obtain the optimal input settings for response characteristics of MPAW, and using corresponding value, the mean for responses was predicted. These results are tabulated in Table 50.5.

Table 50.5 Optimal parameter setting for individual response characteristics

Response characteristics	Optimal level of process parameters	Predicted optimal value quality characteristics
UTS	$IP_2 + WS_2 + PGFR_1 + PT_2 + PF_3 + SOD_1$	919.99
YS	$IP_1 + WS_1 + PGFR_2 + PT_1 + PF_1 + SOD_1$	450.65
YM	$IP_3 + WS_2 + PGFR_3 + PT_2 + PF_3 + SOD_1$	111,895.1
%E	$IP_1 + WS_1 + PGFR_4 + PT_1 + PF_2 + SOD_1$	0.66
H_{FZ}	$IP_3 + WS_2 + PGFR_5 + PT_2 + PF_1 + SOD_3$	242.48
H_{HAZ}	$IP_3 + WS_1 + PGFR_6 + PT_2 + PF_1 + SOD_3$	232.88

Analysis of utility data for optimal parameter settings: the response graphs of overall utility and corresponding S/N ratio graphs plotted are shown below in Fig. 50.4.

The iteration between parameters (WS × SOD) and (PT × SOD) has been considered in the present study. The result reveals that third level of peak current (I_{P3}), first level of welding speed (WS_1), plasma gas flow rate (PGF_1), pulse time (PT_2), and third level of pulse frequency (PF_3) are expected to produce maximum values of utility and S/N ratio within the boundaries experimental zone. The ANOVA of utility data and S/N ratio are given in following Tables 50.6.

Analysis of Variance for mean of overall utility.

Where, SSQ = Sum of squares, MSS = Mean sum of square, DOF = Degrees of freedom, F = Fisher's ratio (the test statistic).

The utility approach reveals that welding speed (49.01%), pulse time (13.21%), and interaction of WS × SOD (12.11%), PT × SOD (11%) and standoff distance (9.57%). Interaction effect of welding speed and stand of distance are high

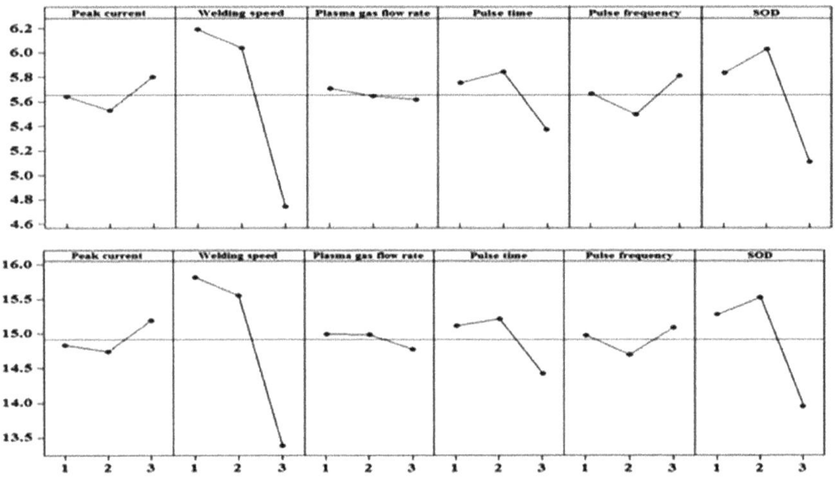

Fig. 50.4 Response graph of overall utility and S/N ratio

Table 50.6 Analysis of variance for mean of overall utility

Source (SS)	DOF	SSQ	MSS	Fisher's ratio	Percent contribution (%)
IP	2	0.23113	0.115565	0.575351	1
WS	2	9.73059	4.865295	24.22234	49
PGF	2	0.15301	0.076505	0.380888	1
PT	2	2.62276	1.31138	6.528831	13
PF	2	0.54734	0.27367	1.362492	3
SOD	2	1.89957	0.949785	4.728596	10
(WS × SOD)	4	2.134469	0.533617	2.656665	11
(PT × SOD)	4	1.326292	0.331573	1.650768	7
Residual error	6	1.205159	0.20086		6
SST	26	19.85032			100

compared with that of other input parameters even though other parameters such as peak current, plasma gas flow rate, and pulse frequency also contribute to the overall enhancement of the response characteristics. The percentage contribution of the input variables on the response characteristics is depicted by means of the pie chart Fig. 50.5.

The variation of overall utility with experiments is given in Table 50.7. It can be observed from the below table that experiment 9 and 17 have the lowest utility value, and experiment 13 and 24 have the highest utility value. The utility values are nearly uniform for experiments 1, 3, 5. From this variation, it can be inferred that the welding process is highly unpredictable. The predicted mean for overall utility has been tabulated based on Eq. (50.5) for the optimal parameter setting, where T_u is the total average of overall utility and μ_m is the predicted mean value.

$$\mu_m = IP_3 + WS_1 + PGFR_2 + PT_2 + FP_3 + SOD_2 - 5T_u \quad (50.5)$$

The predicted value of mean for overall utility obtained from the above formula is 5.6682 and corresponding S/N ratio is 15.0689.

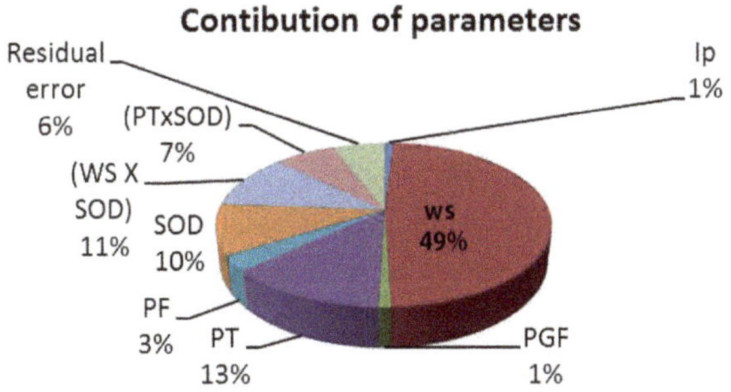

Fig. 50.5 Pie chart showing the relative contribution of process parameters

Table 50.7 Overall utility and their corresponding S/N ratio

Exp. no.	Overall utility	S/N	Exp. no.	Overall utility	S/N
1	4.914012	13.82872	15	5.029823	14.03105
2	4.469039	13.00428	16	4.108907	12.27453
3	5.597911	14.96052	17	2.411887	7.64714
4	5.656226	15.05054	18	4.886665	13.78025
5	5.342404	14.55474	19	5.445744	14.72115
6	4.976322	13.93817	20	5.516175	14.83276
7	3.843378	11.69426	21	5.174645	14.27761
8	3.854209	11.7187	22	4.401555	12.87212
9	2.184868	6.788505	23	4.537717	13.13675
10	4.557094	13.17376	24	5.002038	13.98294
11	4.024767	12.09482	25	4.069355	12.19051
12	4.920941	13.84096	26	4.468833	13.00388
13	5.668243	15.06897	27	3.832711	11.67012
14	4.951167	13.89415			

Maximum Overall utility and S/N ratio are 5.6682 and 15.06897

The obtained values of overall utility and S/N ratios for L_{27} are given in Table 50.7.

50.3.4 *Confirmation Experiment*

The confirmation experiment was performed using predicted value which was calculated from optimization using overall utility approach analysis. The best experiment was considered on the basis of maximum overall utility, not on the basis of particular single properties. The stress–strain curve obtained by this confirmation test which is target aim considering optimization using utility approach, as shown in Fig. 50.6.

The observed values of quality characteristics were correlated with experimental confirmation test results in the given Table 50.8.

The results ensure the existence and reliability of optimization coupling with utility approach for MPAW. It ensures that it is a cost-effective way to minimize the production cost and quality welds. It was observed from the analysis that enhancement in the overall utility is 0.28 and S/N ratio is 0.42, compared with the initial best setting. For the general requirements, UTS is considered dominant properties among all, so UTS properties have been highlighted in Table 50.9.

Hence, the model developed based on overall utility approach using Taguchi is well applicable and reliable to get desired weighted output for all properties of MPAW process.

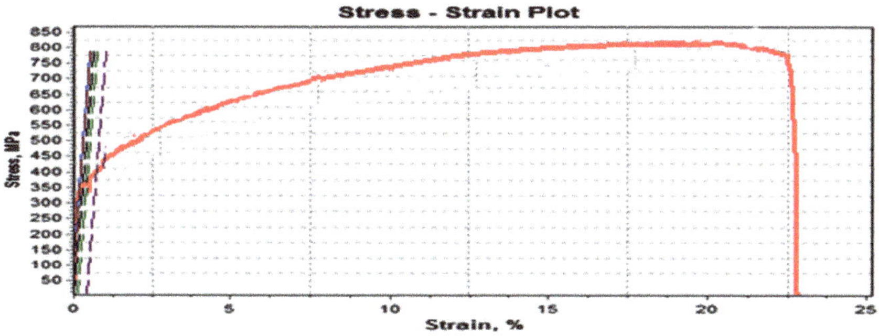

Fig. 50.6 Stress–strain curve of confirmation test

Table 50.8 Observed values of response characteristics for confirmation experiment

Response characteristics	Best factor setting for overall utility (Exp. 13)	Optimal process condition	
	IP_2, WS_2, $PG_{f/r3}$, PT_1, PF_1, SOD_2	Prediction with optimum setting considering interactions (WS × SOD), (PT × SOD)	Confirmation experiment with optimal condition
UTS (MPa)	756.22	IP_3 WS_2 $PGFR_2$ PT_2 PF_1 SOD_1 (Based on mean responce graph)	823.00
YS (MPa)	360.67		552.93
YM (GPa)	108467.8		98956
%E	0.57		0.67
H_{FZ}	222		199
H_{HAZ}	210		212
Initial S/N ratio	15.06897 (Max)	15.9438	15.48
Overall utility for best setting	5.668243 (Max)	5.91000	5.94
Improvement in overall utility = (5.94 − 5.66) = 0.28 Improvement in S/N ratio = 15.48 − 15.06 = 0.42			

Table 50.9 UTS of confirmation test and some highlighted experiments

Sample name	UTS (MPa)
(a) Exp. 25	395.75 MPa (lowest value)
(b) Exp. 6	803.54 MPa (highest value for UTS)
(c) Exp. 13	756.21 MPa (UTS is moderate but overall utility as multi-output has highest significant value)
(c) Base metal	725 MPa (reference value)
(d) The enhanced value	823 MPa (enhanced UTS using utility approach)

50.4 Conclusion

1. The recognition of rolling direction of sheet to choose welding direction has a vital importance in welding, and it improves weld quality such as % elongation.
2. It was observed that enhancement in the overall utility is 0.28 and S/N ratio is 0.42 compared with the initial best setting. Hence, the model developed based on overall utility optimization approach is well applicable to get desired weighted output for the UTS, YS, %E, YM, $H_{FZ,}$ and H_{HAZ} properties of MPAW process.
3. ANOVA coupled with utility approach was performed to determine the significant parameters in this study, and it was found that welding speed with 49.01% and pulse time with 13.21% were most influential factors affecting the weld quality. Even though the contribution of other parameters was minimal but cannot be ignored in the study.
4. The confirmation test projects an enhancement in UTS which is 823 MPa. It is showing improvement in UTS from Exp. 6 and Exp. 13.
5. The optimization work using utility approach helps to enhance the productivity, reliability, quality, and cost reduction for an industrial application of the MPAW process.

References

1. Sanders, F.W., Hurst, G.S., Bortner, T.E.A.: Study of Alpha Particle in Argon Mixtures. Oak Ridge National Laboratory (1959)
2. Foest, R., Schmidt, M., Becker, K.: Microplasma an emerging field of low-temperature plasma science and technology. Int. J. Mass Spectrom. **248**, 87–102 (2006)
3. Shiv Prasad, K., Srinivasa, R.C., Nageswara, R.D.: Optimization of fusion zone grain size, hardness, and ultimate tensile strength of pulsed current microplasma arc welded AISI 304L sheets using genetic algorithm. Int. J. Manuf. Eng. 1–8 (2014)
4. Shiv Prasad, K., Srinivasa, R.C., Nageswara, R.D.: Optimizing pulsed current microplasma arc welding parameters to maximize ultimate tensile strength of AISI 304L sheets using response surface method. Int. J. Eng. Sci. Technol. 226–236 (2011)
5. Shiv Prasad, K., Srinivasa, R.C., Nageswara, R.D.: Optimizing pulsed current microplasma arc welding parameters to maximize ultimate tensile strength of SS304L sheets using Hooke and Jeeves Algorithm. J. Manuf. Prod. 39–48 (2011)
6. Alontsevadalontseva, D.L., Abilev, M.B., Zhilkashinova, A.M., Voinarovych, S.G., Kyslytsia, O.N., Ghassemieh, E., Russakova, A., Łatka, L.: Optimization of hydroxyapatite synthesis and microplasma spraying of porous coatings onto titanium implants, 79–95 (2018)
7. Yang, W.H., Tarng, Y.S.: Design optimization of cutting parameters for turning operations based on the Taguchi method. J. Mater. Process. Technol. **84**(1), 122–129 (1998)
8. Yang, J.L., Chen, J.C.: A systematic approach for identifying optimum surface roughness performance in end milling operations. J. Ind. Technol. **17**(2), 1–8 (2001)
9. Deng, C.S., Chin, J.H.: Hole roundness in deep hole drilling as analysed by Taguchi methods. Int. J. Adv. Manuf. Technol. **25**(5–16), 420–426 (2005)
10. Kumar, P., Creese, R.: Effect of Zircon Sand on Surface Finish of Al-7% Si Alloy V-Process Castings, pp. 95–98. AFS Transactions, USA (1999)

11. Creese, R., Kumar, P.: Effect of Fly Ash (As an Additive to Greensand) on the Surface Roughness of Al-7%Si Alloy Castings Using Taguchi Technique. AFS Transactions, USA (2001)
12. Lee, H.T., Yur, J.P.: Characteristic analysis of EDM surfaces using Taguchi approach. J. Mater. Process. Manuf. **15**(6), 781–806 (2000)
13. Kumar, P., Barua, P.B., Gaindher, J.L.: Quality of V-process method through the Taguchi technique. J. Qual. Reliab. Eng. **12**, 421–427 (1996)
14. Tong, L.I., Su, C.T.: Optimizing multi-response problems in the Taguchi method by fuzzy multiple attribute decision making. J. Reliab. Qual. Eng. Int. **13**, 25–34 (1997)
15. Antony, J.: Multi-response optimization in industrial experiments using Taguchi quality loss function and principle component analysis. J. Reliab. Qual. Eng. Int. **16**, 3–8 (2000)
16. Datta, S., Bandyopadhyay, A., Pal, Kumar P.: Grey-based Taguchi method for optimization of bead geometry in submerged arc bead-on-plate welding. Int. J. Adv. Manuf. Technol. **39**(11–12), 1136–1143 (2008)
17. Kumar, P., Barua, P.B., Gaindhar, J.L.: Quality optimization (multi characteristic) through Taguchi's technique and utility concept. Qual. Reliab. Eng. Int. **16**, 475–485 (2000)
18. Goyal, T., Walia, R.S., Sidhu, T.S.: Multi- response optimization of low-pressure cold-spread coatings through Taguchi method and utility concept. Int. J. Adv. Manuf. Technol. **64**, 903–914 (2013)

Chapter 51
Characterization of Geometrical Features of Ultra-Short Pulse Laser-drilled Microholes Using Computed Tomography

K. Kiran Kumar, G. L. Samuel and M. S. Shunmugam

Abstract High-aspect-ratio high-quality microholes are required in turbine blades to improve cooling performance. These cooling holes are drilled by pulsed laser and hence dimensional as well as geometrical tolerances like circularity and cylindricity are important. The measurement of geometrical features of the microholes is a very challenging task without destroying the components. In the present work, the microholes are produced on Ti6Al4V alloy by ultra-short pulse laser. The geometrical features of microholes are then captured using a non-destructive technique, namely computed tomography. CT-scanned 3D data is directly used for geometrical analysis using open-source software, GOM Inspect. Since algorithms used in the GOM Inspect are proprietary in nature, the extracted coordinate data are also analyzed using the computational methods developed by the authors based on least squares technique. The dimension, circularity, and cylindricity of microholes are compared with the results obtained from GOM Inspect software and a close match is found.

Keywords Microholes · Ultra-short pulse laser · Computed tomography · Geometrical analysis · Least squares technique

51.1 Introduction

The high-aspect-ratio microholes with tight tolerances are in high demand in automotive, aerospace, electronic, and biomedical industries. Specific applications include fuel injection nozzle, turbine blade and printed circuit board (PCB), inkjet printer nozzle and microneedles. The microholes for film cooling are drilled on turbine blade having complex geometry by ultra-short pulse laser. Unlike tool-based machining in which the geometry and axis of the hole are controlled by

---

K. Kiran Kumar · G. L. Samuel · M. S. Shunmugam (✉)
Indian Institute of Technology, Madras, Chennai 600036, India
e-mail: shun@iitm.ac.in

R. G. Narayanan et al. (eds.), *Advances in Computational Methods in Manufacturing*, Lecture Notes on Multidisciplinary Industrial Engineering,
https://doi.org/10.1007/978-981-32-9072-3_51

the tool and its movement, the geometry of hole and its axis are determined by the ablation process in the laser drilling. Therefore, selection of parameters for proper control of laser ablation process is important to produce high-quality holes. It is imperative that the dimensional and form errors of hole produced by laser drilling are measured and characterized. Measurement of the dimensional and geometrical errors of microholes is a difficult task. The geometrical deviations of microhole are characterized as 2D and 3D errors, namely circularity and cylindricity. Different approaches have been reported in the literature to measure the dimensional and geometrical accuracy of microholes. The conventional methods reported in the literature mostly involve sectioning and polishing. These techniques require accurate axis alignment and they are also time-consuming.

It is important to characterize the microholes without destroying the machined holes. Kao and Shih [1] have investigated the form measurements of microholes using coordinate measuring machine (CMM) with a combination of optical and contact sensors. Non-destructive testing method based on optical system has been used for the inspection of cooling holes in turbine blade in aero-engine application [2]. Computed tomography (CT) is the only technology capable of measuring inner and outer geometry of components without destroying the machined surface. CT is able to detect pores or small delaminations within composites and measure small and hidden coolant bores of turbine blade [3]. CT technique has been used for industrial applications in the field of dimensional metrology [4]. Zhang et al., have reported measurement of features of the laser-drilled microholes using micro-CT [5]. 3D geometrical measurements of ablated crater such as volume, depth, and width and cone angle have been reported by Galmed et al. [6]. The characterization of geometrical features of microholes without destroying the holes still remains as a very big challenge.

In order to avoid destruction of machined component, non-destructive technique based on computed tomography is employed in the present work. CT is used to scan an array of 3 × 3 microholes drilled by ultra-short pulse laser micromachining. Further, the CT-scanned volume data is analyzed by using GOM Inspect available as open-source software. Dimensional and geometrical errors of laser-drilled microholes are also evaluated using computation technique developed by the authors based on least squares technique. The results are compared with those obtained from GOM Inspect software and discussed.

51.2 Microhole Drilling

The microhole drilling experiments are conducted by using ultra-short pulse laser. The Yb-doped fiber laser (make: Amplitude Systems; model: Satsuma HP2) operates at a central wavelength of 1030 nm with pulse duration in the range of 350 fs to 10 ps. The maximum average power is 20 W and pulse energy of 40 μJ is delivered at a repetition rate of 200 kHz. The laser system is integrated with the ALPhNOV machine and laser beam is guided with trepanning head of GFH GmbH. Trepanning drilling technique is used to machine an array of 3 × 3 microholes in 1.5-mm-thick Ti6Al4V alloy plate.

51.3 Characterization of Microholes

The characterization of microholes is very challenging task. The conventional methods using sectioning and polishing are time-consuming and inaccurate due to axis misalignment. In order to overcome these problems, non-destructive technique like computed tomography (CT) is used. CT technology is capable of measuring the internal and external geometric features of complex-shaped components.

51.3.1 Computed Tomography

X-ray-based computed tomography technique is used for dimensional metrology of industrial components. The radiation emitted by a point-source X-ray passes through the measured object to a flat detector (sensor) and several 2D images are taken at various angular positions with the help of a rotating stage. 3D data reconstruction generates a 3D voxel model. Post-processing of voxel data using image processing technique detects the edges of the component for further analysis. The schematic diagram of computed tomography technique is shown in Fig. 51.1.

For precise measurement using CT, a high-quality tomogram is extracted from the detected surface. The CT scan image quality is mainly dependent on voxel size and focal spot size. The sharper images can be obtained with smaller X-ray spot size of high-power nanofocus tube. In the present work, nanotomphoenix|x-ray (make: GE, Germany) available at Centre for Non-destructive Evaluation (CNDE), Mechanical Engineering Department, IIT Madras, shown in Fig. 51.2 is used. The nanotom CT works at a tube voltage of 110 kV and current of 80 μA, with a minimum voxel size of <500 nm. The array of 3 × 3 microholes is scanned, and the scanning time is approximately 45 min.

From the generated volume data file, volume graphics (VG viewer) open-source software creates 3D and 2D images for visualization [7]. The 2D and 3D views of the component are shown in Fig. 51.3. In the present work, GOM Inspect open-source software [8] is used in the next step to generate the geometric model as shown in Fig. 51.4.

Computed tomography volume data (.vgl format) file is directly imported in GOM Inspect software, without converting into STL format. This is an easy

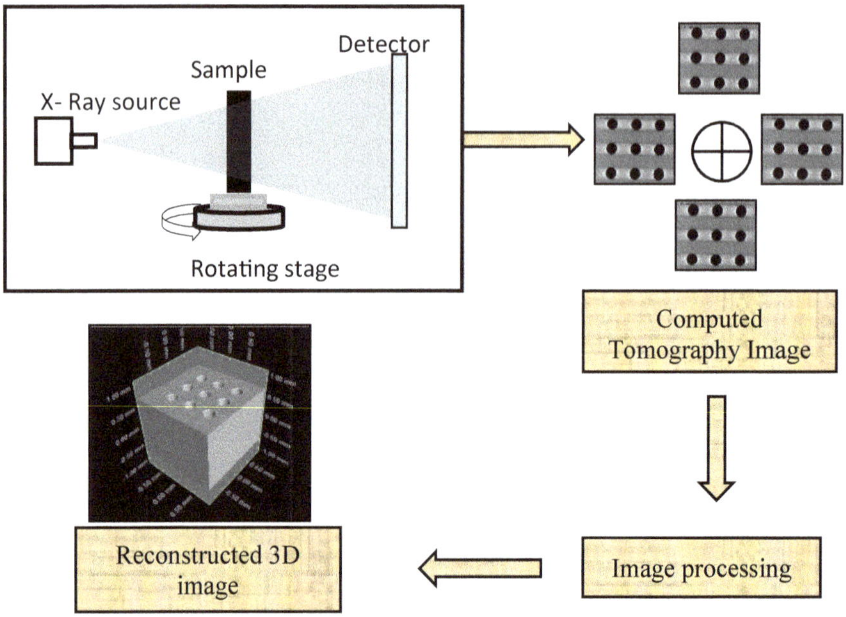

Fig. 51.1 Schematic diagram of computed tomography technique

Fig. 51.2 Computed tomography (CT) system at CNDE, IIT Madras

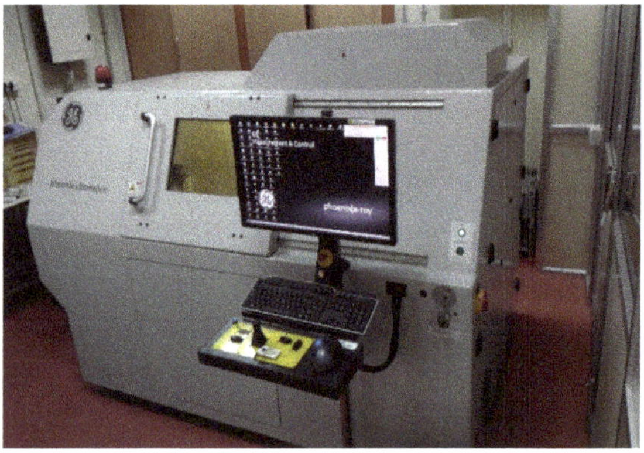

approach for the dimension analysis of CT-scanned complex shape and size. After importing CT volume data in GOM inspect software, it converts into mesh elements. Cylinder surfaces are constructed by selecting the holes. The coordinate transformation is done by using axis alignment with respect to z-axis in vertical direction. From the geometric model, dimensional and form errors of microholes are assessed as outlined in Sect. 51.4.

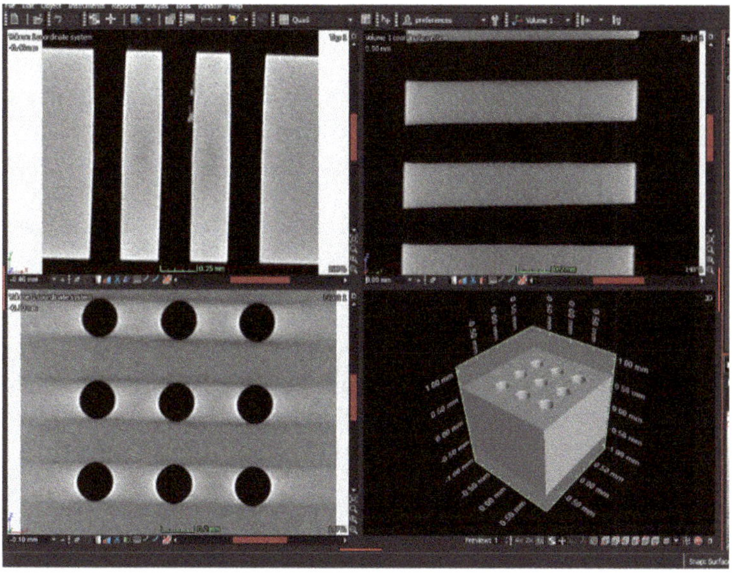

Fig. 51.3 Screenshot of reconstruction of CT-scanned images using volume graphics module

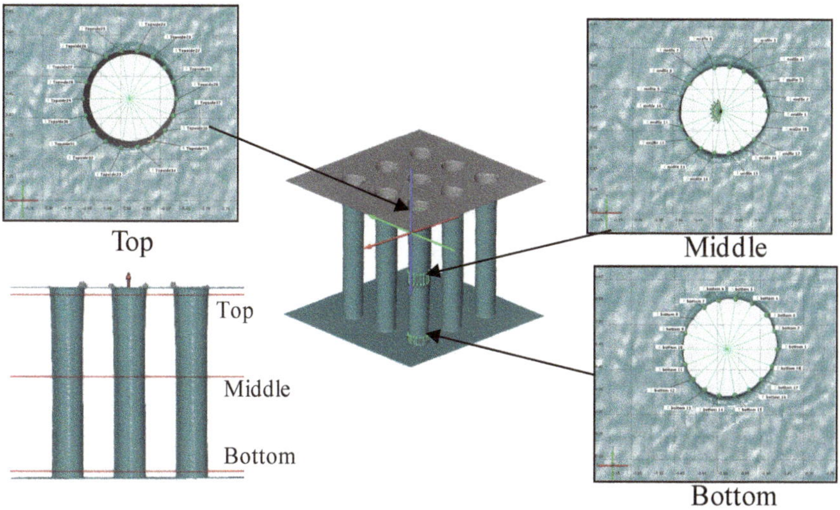

Fig. 51.4 Extracting the xyz coordinates from CAD geometry at three sections (top, middle, and bottom)

51.4 Computational Methods

To compare with the results obtained from GOM Inspect software, least squares technique developed by the authors is used for evaluation of form errors. The coordinates of microholes are extracted using GOM Inspect at selected sections as shown in Fig. 51.4. The *xyz* coordinates are extracted at top, middle, and bottom sections, respectively. The *xyz* coordinates at top surface of one microhole are given in Table 51.1.

The extracted coordinate data of microholes contains both size and form information, and a suitable method has to be implemented to separate the form data and size information. Samuel and Shunmugam [9] have proposed a transformation technique for evaluation of form error from the coordinate data. Accordingly, a reference circle is chosen passing through three selected points and the deviations are computed using Eq. (51.1).

$$d_i = \left[(x_i - x_R)^2 + (y_i - y_R)^2\right]^{1/2} - r_R \tag{51.1}$$

Table 51.1 *xyz* coordinate data (top, middle, and bottom sections)

Sl. no.	z' = 0.056 mm		z' = 0.681 mm		z'= 1.306 mm	
	x mm	*y* mm	*x* mm	*y* mm	*x* mm	*y* mm
1	−0.643	0.446	−0.637	0.446	−0.640	0.446
2	−0.633	0.487	−0.630	0.485	−0.633	0.487
3	−0.612	0.522	−0.609	0.519	−0.613	0.523
4	−0.581	0.548	−0.578	0.543	−0.581	0.548
5	−0.543	0.562	−0.541	0.554	−0.542	0.558
6	−0.502	0.562	−0.504	0.552	−0.503	0.555
7	−0.463	0.548	−0.470	0.538	−0.469	0.539
8	−0.432	0.522	−0.442	0.514	−0.442	0.514
9	−0.412	0.486	−0.423	0.482	−0.425	0.482
10	−0.403	0.446	−0.417	0.446	−0.417	0.446
11	−0.410	0.405	−0.423	0.410	−0.425	0.411
12	−0.430	0.369	−0.442	0.379	−0.443	0.380
13	−0.462	0.342	−0.471	0.357	−0.470	0.356
14	−0.501	0.328	−0.504	0.345	−0.504	0.343
15	−0.543	0.329	−0.540	0.344	−0.540	0.344
16	−0.582	0.343	−0.575	0.356	−0.574	0.356
17	−0.613	0.370	−0.604	0.378	−0.603	0.379
18	−0.634	0.406	−0.627	0.408	−0.627	0.408

Points for establishing reference circle (1, 4, 7) at all three sections where; (x_i, y_i) are coordinates of extracted data. (x_R, y_R) is center of reference circle and r_R is its radius. The form data in the form of (r_i, θ_i) is computed as

$$r_i = d_i - d_{\min} + d_o \tag{51.2}$$

$$\theta_i = \tan^{-1}\left[\frac{(y_i - y_R)}{(x_i - x_R)}\right] \tag{51.3}$$

where; $d_{\min}$ is minimum value of d_i and d_o is taken as 0.001 mm.

In the present work, transformation approach is implemented for evaluating circularity and cylindricity of microholes and the results are discussed in Sect. 51.5. 2D and 3D form errors (circularity, cylindricity) of the microholes are evaluated by using least squares technique [10]. The deviations from the limacon and limacon cylinder are expressed in the linear form as follows.

$$e_i = r_i - (R_o + x_o \cos\theta_i + y_o \sin\theta_i) \tag{51.4}$$

$$e_i = r_i - [R_o + (x_o + l_o z_i)\cos\theta_i + (y_o + m_o z_i)\sin\theta_i] \tag{51.5}$$

where; e_i is linear deviation at ith point, R_o is estimated radius for limacon/limacon cylinder, (r_i, θ_i) are polar coordinates at ith point, (x_o, y_o) are estimated center coordinates and (l_o, m_o) are estimated slope values for limacon cylinder.

The least squares parameter for the limacon and limacon cylinder are estimated from the following equations.

$$\begin{bmatrix} \sum\cos^2\theta_i & \sum\sin\theta_i\cos\theta_i & \sum\cos\theta_i \\ \sum\sin\theta_i\cos\theta & \sum\sin^2\theta_i & \sum\sin\theta_i \\ \sum\cos\theta_i & \sum\sin\theta_i & \sum 1 \end{bmatrix} \begin{bmatrix} x_o \\ y_o \\ R_o \end{bmatrix} = \begin{bmatrix} \sum r_i\cos\theta_i \\ \sum r_i\sin\theta_i \\ \sum r_i \end{bmatrix} \tag{51.6}$$

$$\begin{bmatrix} \sum\cos^2\theta_i & \sum\sin\theta_i\cos\theta_i & \sum z_i\cos^2\theta_i & \sum z_i\sin\theta_i\cos\theta_i & \sum\cos\theta_i \\ \sum\sin\theta_i\cos\theta & \sum\sin^2\theta_i & \sum z_i\sin\theta_i\cos\theta & \sum z_i\sin^2\theta_i & \sum\sin\theta_i \\ \sum z_i\cos^2\theta_i & \sum z_i\sin\theta_i\cos\theta_i & \sum z_i^2\cos^2\theta_i & \sum z_i^2\sin\theta_i\cos\theta_i & \sum z_i\sin\theta_i \\ \sum z_i\sin\theta_i\cos\theta_i & \sum z_i\sin^2\theta_i & \sum z_i^2\sin\theta_i\cos\theta_i & \sum z_i^2\sin\theta_i & \sum z_i\sin\theta_i \\ \sum\cos\theta_i & \sum\sin\theta_i & \sum z_i\cos\theta_i & \sum z_i\sin\theta_i & \sum 1 \end{bmatrix} \begin{bmatrix} x_o \\ y_o \\ l_o \\ m_o \\ R_o \end{bmatrix} = \begin{bmatrix} \sum r_i\cos\theta_i \\ \sum r_i\sin\theta_i \\ \sum r_i z_i\cos\theta_i \\ \sum r_i z_i\sin\theta_i \\ \sum r_i \end{bmatrix} \tag{51.7}$$

51.5 Results and Discussion

The microhole's dimensions are measured at top, middle, and bottom sections by fitting three-point circle at the respective sections in the GOM software. The diameters of microhole are obtained as 240, 220, and 220 μm at top, middle, and bottom sections, respectively, with height of 0.056, 0.681, and 1.306 mm, respectively, as shown in Fig. 51.5. If the practice of quantifying taper based on entry (top) and exit (bottom) diameters is followed, there is a general taper in the microhole with the size narrowing with the depth.

Similarly, the cylindricity error of microhole is measured as 11.2 μm by using Gaussian fit method, which is available in GOM Inspect software.

51.5.1 Comparison of Results with Least Squares Limacon Technique

The microhole geometrical xyz coordinate data given Table 51.1 is transformed to polar coordinates with respect to a reference circle. The reference circle is established by using a circle fitted through three selected points from given data. For example, center coordinates and radius of reference of circle passing through points 1, 4, and 7 are obtained as $x_R = -0.522$ mm; $y_R = 0.443$ mm and radius $r_R = 0.120$ mm, respectively. The deviations of each point with respect to reference circle are computed using Eq. (51.1). From the form data (r_i, θ_i), the form errors are evaluated with reference to a limacon. This is because, the center (x_R, y_R) chosen for transformation is different from the actual center of the circular feature. It has been shown in the literature that, it has been shown that the offset introduces certain distortion in the transformed profile and hence a limacon must be used as a reference for evaluating the circularity error form from data [11].

The transformed data (r_i, θ_i) is not listed in this paper due to space limitation. The data exhibits nonuniform angular spacing, and hence the general solution given by Eq. (51.6) is used to arrive at the limacon parameters. The results are plotted for top, middle, and bottom sections in polar chart as shown in Fig. 51.6. The parameters are given in Table 51.2, along with circularity errors and diameters. The circularity error is obtained as 3.182, 5.670, and 7.699 μm at top, middle, and bottom sections, respectively. From the above results, it observed that the quality of laser-drilled microhole changes with respect to depth. There is a good match between the diameter values obtained.

The cylindricity of microhole is evaluated by considering the height (z-coordinates). The xyz coordinates are extracted from geometric model for top, middle, and bottom sections as shown in Fig. 51.5. Table 51.3 gives value parameters obtained in the cylindricity evaluation. The cylindricity error of microhole is evaluated as

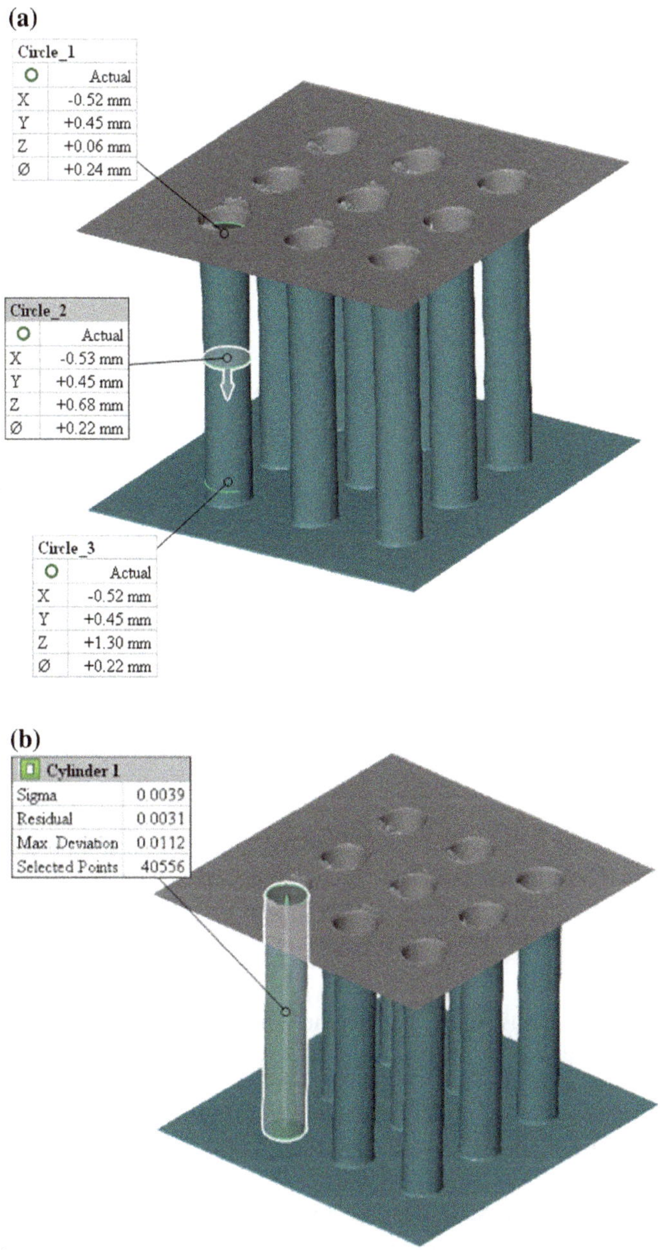

Fig. 51.5 Screen shot of microhole geometry measurement **a** diameter, **b** cylindricity

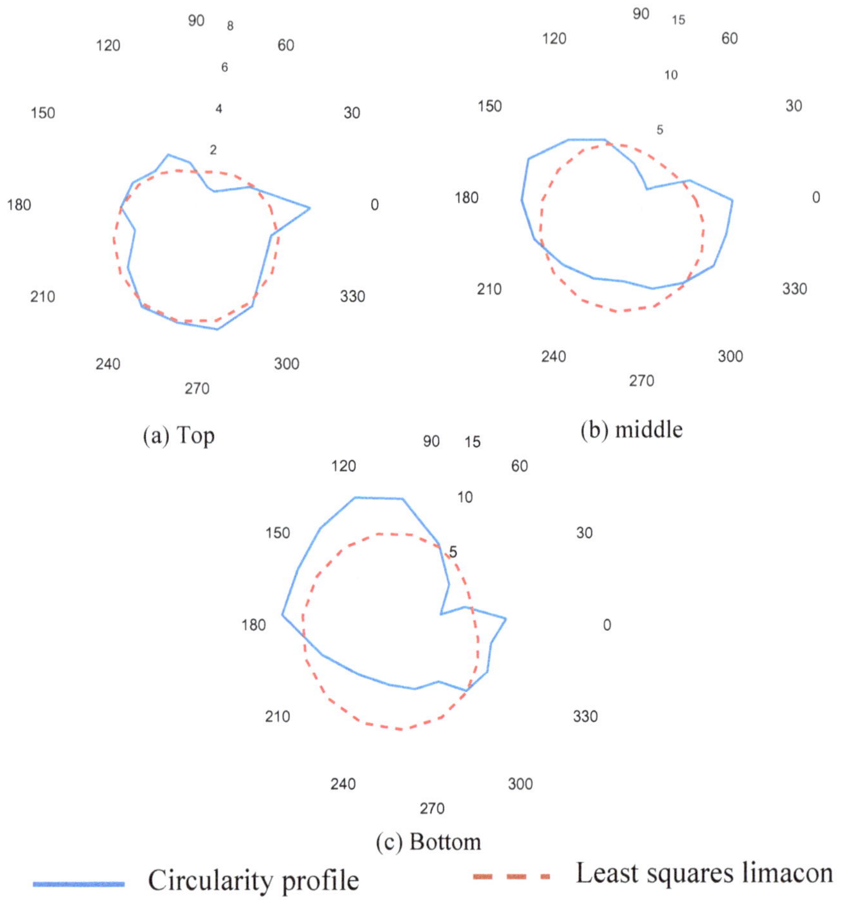

Fig. 51.6 Circularity profile along with least squares limacon at different sections

16.317 μm by using least squares technique. The cylindricity value of 11.2 μm is obtained from Gaussian fit method in GOM Inspect software. The value is lower than that computed from least squares limacon technique. Without the details of Gaussian fit method used in GOM Inspect, it is difficult to give reasons for the difference in the cylindricity errors. Interestingly, the diameter values agree with each other well.

In the above table, subscripts R and o refer to reference cylinder and limacon cylinder, respectively, while x, y represent the intersection of axis with top plane; l, m are the slope of axis; r is the radius of the cylinder.

Table 51.2 Circularity evaluation

	Top	Middle	Bottom
Initial values of reference circle			
x_R (mm)	−0.522	−0.528	−0.532
y_R (mm)	0.443	0.446	0.451
r_R (mm)	0.120	0.108	0.107
After fitting limacon			
x_o (μm)	−0.582	−2.096	−4.042
y_o (μm)	−2.045	−2.466	0.972
r_o (μm)	3.620	6.749	7.751
Circularity error (μm)	3.182	5.670	7.699
Diameter (mm)[a]	0.245	0.227	0.227
GOM inspect			
Circularity error (μm)	Not available in free version		
Diameter (mm)	0.24	0.22	0.22

[a]Diameter = $2\{r_R + [(r_O/1000) - 0.001]\}$

Table 51.3 Cylindricity evaluation (top section taken as $z = 0$ mm)

Initial values of reference cylinder	
x_R (mm)	−0.522
y_R (mm)	0.443
l_R (μm/mm)	−8.0
m_R (μm/mm)	6.4
r_R (mm)	0.107
Values of limacon cylinder	
x_o (μm)	0.321
y_o (μm)	−2.576
l_o (μm/mm)	−3.005
m_o (μm/mm)	2.399
r_o (μm)	10.721
Cylindricity error (μm)	16.317
Diameter (mm)[a]	0.233
GOM inspect	
Cylindricity error (μm)	11.2
Diameter (mm)	0.22

[a]Diameter = $2\{r_R + [(r_O/1000) - 0.001]\}$

51.6 Conclusion

New approach is established in the present work to characterize the geometrical features of the ultra-short pulse laser-drilled microholes. Using computed tomography, it is feasible to characterize the laser-drilled microholes without destroying the sample. The dimensional accuracy of computed tomography volume data is directly analyzed in GOM Inspect software. The diameters of microhole are 240,

220, and 220 μm at top, middle, and bottom sections, respectively, showing a taper in the microhole.

The form errors of microholes are established by least squares limacon technique. The circularity errors are 3.188, 5.670, and 7.699 μm at top, middle, and bottom sections, respectively. Cylindricity of microhole is evaluated as 16.317 μm using least squares limacon technique, which is higher than the result obtained from Gaussian fit in GOM Inspect software. Future work will be extended to evaluate the conicity error of microholes using computational technique.

Acknowledgements Authors would like to thank M/s Amplitude Systems, France, for providing the ultra-short pulse laser-drilled test sample. Authors also thank Aeronautical Research and Development Board (ARDB), Government of India (Project number: ARDB/01/2031768/M/I dated 10 Aug 2015) and CNDE, Mechanical Engineering Department, Indian Institute of Technology, Madras.

References

1. Kao, C.C., Shih, A.J.: Form measurements of micro-holes. Meas. Sci. Technol. **18**(11), 3603–3611 (2007)
2. Shetty, D., Eppes, T., Campana, C., Filburn, T., Nazaryan, N.: New approach to the inspection of cooling holes in aero-engines. Opt. Lasers Eng. **47**, 686–694 (2009). https://doi.org/10.1016/j.optlaseng.2008.12.001
3. Hanke, R., Fuchs, T., Uhlmann, N.: X-ray based methods for non-destructive testing and material characterization. Nucl. Instrum. Methods Phys. Res., Sect. A Accel. Spectrom. Detect. Assoc. Equip. **591**, 14–18 (2008). https://doi.org/10.1016/j.nima.2008.03.016
4. Kruth, J.P., Bartscher, M., Carmignato, S., Schmitt, R., De Chiffre, L., Weckenmann, A.: Computed tomography for dimensional metrology. CIRP Ann. - Manuf. Technol. **60**, 821–842 (2011). https://doi.org/10.1016/j.cirp.2011.05.006
5. Zhang, Y., Wang, Y., Zhang, J., Liu, Y., Yang, X., Zhang, Q.: Micromachining features of TiC ceramic by femtosecond pulsed laser. Ceram. Int. **41**, 6525–6533 (2015). https://doi.org/10.1016/j.ceramint.2015.01.095
6. Galmed, A.H., du Plessis, A., le Roux, S.G., Hartnick, E., Von Bergmann, H., Maaza, M.: Three dimensional characterization of laser ablation craters using high resolution X-ray computed tomography. Spectrochim. Acta - Part B At. Spectrosc. **139**, 75–82 (2018). https://doi.org/10.1016/j.sab.2017.11.011
7. Volume Graphics, https://www.volumegraphics.com/
8. GOM mbH, GOM Inspect Software. http://www.gom.com (2018)
9. Samuel, G.L., Shunmugam, M.S.: Evaluation of circularity and sphericity from coordinate measurement data. J. Mater. Process. Technol. **139**, 90–95 (2003). https://doi.org/10.1016/S0924-0136(03)00187-0
10. Shunmugam, M.S.: On assessment of geometric errors. Int. J. Prod. Res. **24**, 413–425 (1986)
11. Samuel, G.L., Shunmugam, M.S.: Evaluation of circularity from coordinate and form data using computational geometric techniques. Precis. Eng. **24**(3), 251–263 (2000)

Chapter 52
Investigation on Rotary Electrical Discharge Machining (EDM) of EN-25 Material Using Copper Electrode for Improving Geometrical Errors

L. Selvarajan, R. Rajavel, F. Leo Princely, R. Aravind and T. P. Habeeb Masood

Abstract In this analysis, an investigation and testing work were carried out on the spark EDM of EN-25 material using copper electrodes. Considering the machining parameters such as pulse on time, pulse off time, current, dielectric pressure for machining the effect of these following input parameters on output characteristics like metal removal rate (MRR), tool wear rate (EWR), wear ratio (WR), cylindricity (CYL), circularity (CIR), and perpendicularity (PER). The investigation was carried on with L_9 orthogonal array. The effect of each machining parameters on response characteristics was studied independently using trend analysis. From results, circularity has been minimized to decrease of current and increase of dielectric pressure. Cylindricity has been minimized to decrease of current and increase of dielectric pressure. Perpendicularity has been minimized to increase of current and decrease of dielectric pressure.

Keywords EDM · Geometrical tolerances · Circularity · Cylindricity · Perpendicularity

---

L. Selvarajan (✉) · R. Rajavel · T. P. Habeeb Masood
Department of Mechanical Engineering, Mahendra Institute of Engineering and Technology, Salem, India
e-mail: selvalakshmanan86@gmail.com

R. Rajavel
e-mail: catchrajavel@gmail.com

T. P. Habeeb Masood
e-mail: habi1@yahoo.com

F. Leo Princely
Department of Production Engineering, National Institute of Technology, Tiruchirappalli, India
e-mail: leoprincely@gmail.com

R. Aravind
Alstom Transport India Limited, Bengaluru, India
e-mail: aravindrajagopalan96@gmail.com

R. G. Narayanan et al. (eds.), *Advances in Computational Methods in Manufacturing*, Lecture Notes on Multidisciplinary Industrial Engineering,
https://doi.org/10.1007/978-981-32-9072-3_52

52.1 Introduction

As the world of manufacturing is growing at an exponential rate, the focus on surface finish and accuracy of parts produced is becoming highly important. This importance to finish and accuracy is evolving due to modern unconventional machining processes such as electric discharge machining. The shift to nontraditional machining is due to the exogenous nature of machining in case of conventional machining process. Owing to these current industry trends, an experimental study on the EDM of EN-25 is mandatory rather than that of a choice. It is used in automobile components and aerospace components. It is used for Adapters, Axles, Die Holders, Connecting Rods, Ejector Rods and Drill Shanks. Thus, in this investigation, EN-25 steel was used as the workpiece. The experiments were performed with EDM oil as dielectric fluid and 2-mm-copper tool. The input factors considered are current (A), pulse off time (μs) and pulse on time (μs) and dielectric fluid (kg/cm^2). This experimental investigation will increase the depth of knowledge of EDM and hence increase the propensity to success in terms of accuracy and surface finish.

52.2 Literature Review

Selvarajan et al. [1] in this work deal with the review of various papers related to optimization of EDM process. The electrode is chosen from regularly used Cu, CuW, and AgW as copper, which is based on the study by Muttamara et al. [2]. Uno et al. [3] spark EDM has emerged as an outstanding approach for cutting conductive metals that are otherwise highly difficult to machine. Lin et al. [4] in research introduce the multiresponse performance index is used to solve the spark EDM process with multiple responses or output characteristics. The machining parameters (the workpiece polarity, duty factor, pulse on time, and current, dielectric fluid and discharge voltage) are optimized with examinations of the multiple performance characteristics (electrode wear ratio and metal removal rate). Luis et al. [5] in this work, a comparative study of the spark EDM of various conductive ceramics had been carried out. Based on the experience acquired in the field of EDM for other materials, as, when the pulse time is reduced, which is equivalent to improving duty cycle, the MRR value tends to gradually increase. Arivazhaghan and Rajaduri [6] in this paper, illustration of the influence of input machining parameters on the metal removal rate is determined along with performance measurement analysis. The objective of this paper is to study the influence of operating technique parameters of copper electrode on the metal removal rate of EN-19 material followed by optimization. Soni and Chakraverti [7] in this paper describe optimum parameters for Titanium Grade (Ti-6Al-4V) using rotary spark machining process. The response of volume metal removal rate (MRR) and surface roughness (Ra) is considered for enhancing or increasing the machining efficiency.

Selvarajan et al. [8] EN-19 is tensile strength steel so it is widely used in different machinery parts such as shafts, racks, studs, bolts, nuts, rollers, etc. Wire EDM seems a good choice for machining the irregular shapes on medium steel Material. In the research, optimization of EDM parameters to improve geometrical errors by Selvarajan et al. [9] and the various methods to improve tolerance in $MoSi_2$–SiC was done. In another study by Selvarajan et al. [10], the optimization of machining parameters for improving circularity of Si_3N_4–TiN composite was done. Optimization of spark EDM Parameters in machining of $MoSi_2$–SiC composites for enhancing geometrical error was done by Selvarajan et al. [11]. It is evident that in all the above research studies conducted on either different materials or with different parameters. There is a noticeable research gap in terms of research on analyzing the machining of input parameters on geometrical output characteristics such as circularity, cylindricity, and perpendicularity. Thus, the research objective was chosen to be analyzing the individual impact of input parameters on each of these geometrical characteristics. The output parameters such as perpendicularity, cylindricity, and circularity are measured by using CMM, they are found applications in high accuracy.

52.3 Research Plan

Figure 52.1 shows that the research plan at EN-25 material with copper electrode.

52.4 Experimental Methodology

The experimental setup of EDM drilling machine is shown in Fig. 52.2. The workpiece material is shown in Fig. 52.3. The chemical composition of EN-25 steel was shown in Table 52.1

The properties of EN-25 steel are furnished in Tables 52.2 and 52.3 determine the working condition of the EDM drilling used for machining EN-25 steel.

The number of experiments to be conducted is nine and thus the L_9 orthogonal array was found to be suitable as instead of conducting 81 experiments based on factorial setup; this orthogonal array helps to finish the study within nine experiments. The values for each factor at various levels are shown below (Table 52.4).

This is the Taguchi's L_9 orthogonal array table where the experiment is done according to this procedure. The outputs are measured using a CMM whose image is provided in Fig. 52.4.

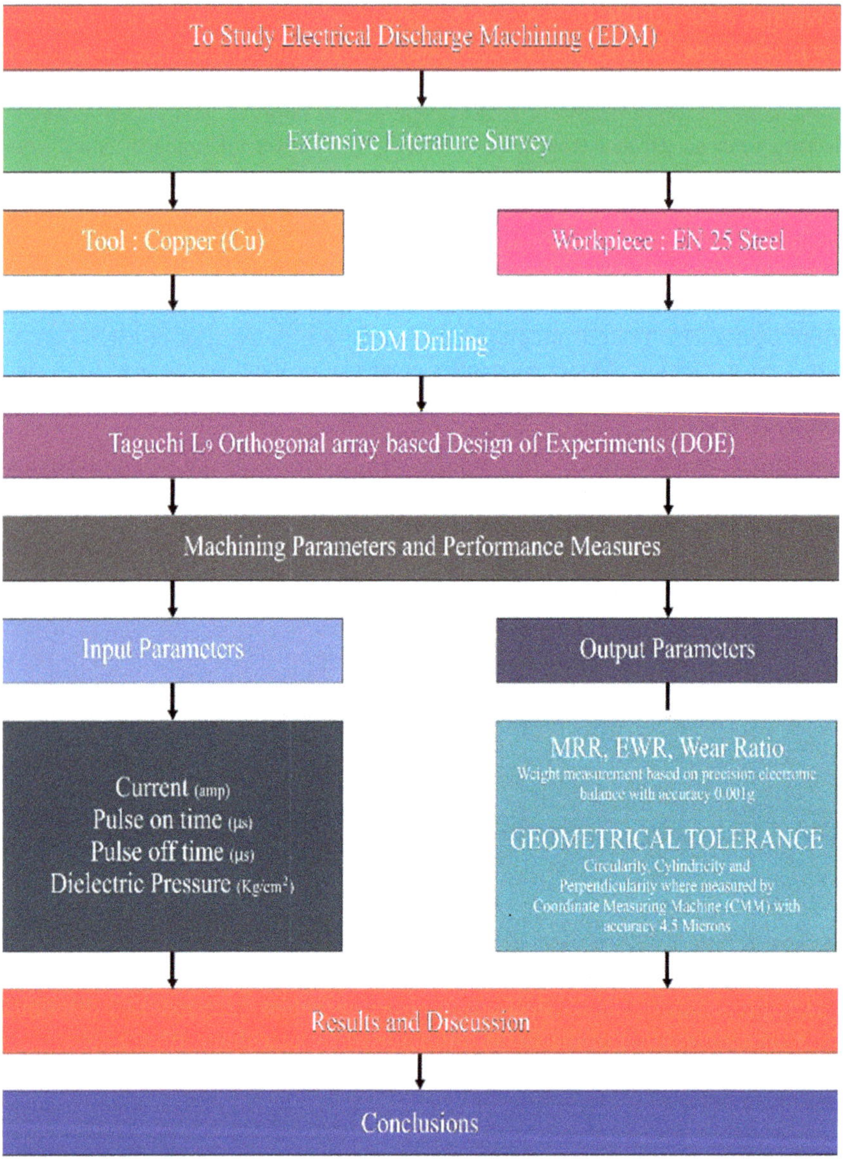

Fig. 52.1 Research plan

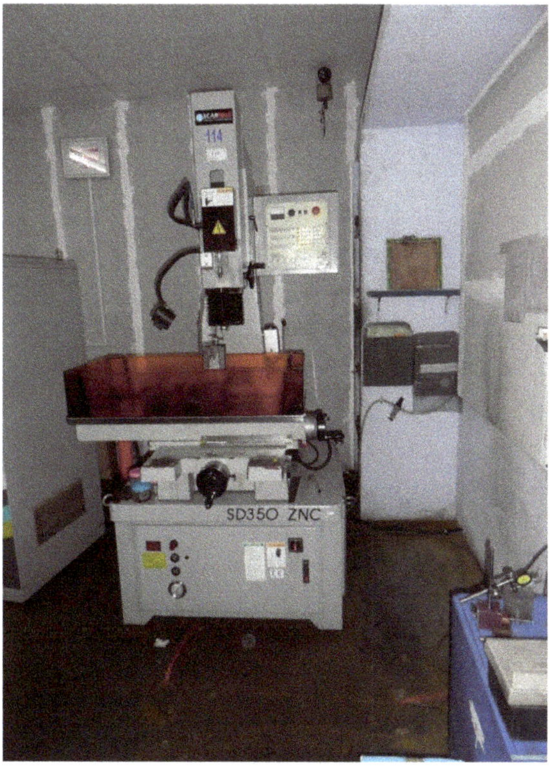

Fig. 52.2 Experimental setup of EDM drilling machine

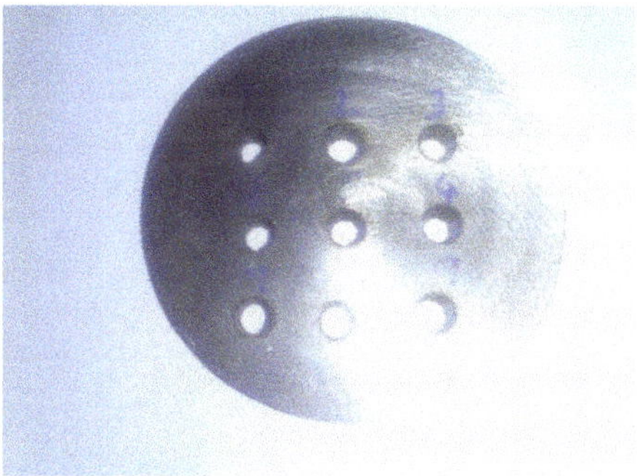

Fig. 52.3 EN-25 workpiece material

Table 52.1 Chemical composition of EN-25

C	Si	Mn	Ni	Cr	Mo	S	P
0.35	0.20	0.62	2.60	0.67	0.58	0.03	0.03

Table 52.2 Properties of EN-25 steel

Properties	EN-25 Steel
Density	7.59 (kg/m^3)
Thermal expansion	12.5 (10^{-6} k)
Melting point	1400 (°C)
Thermal conductivity	48.6 (W/m k)
Specific heat	1500 (J/Kg k)
Electrical resistivity	1200 (10^{-9} W m)
Tensile strength	980 (MPa)
Brinell hardness	627 (kg)

Table 52.3 EDM operating conditions for EN-25

Working conditions	Description
Electrode material	Copper
Diameter of drilling hole	3 mm
Depth of drilling hole	2.5 mm
Electrode polarity	Positive
Workpiece polarity	Negative
Specimen material	EN-25 steel
Type of current	DC power supply
Discharge current (I, A)	4–8
Pulse on time (t_{on}, μs)	20–30
Pulse off time (t_{off}, μs)	4–6
Dielectric fluid	EDM oil
Fluid pressure (kg/cm^2)	24–28

Table 52.4 Machining parameters and levels of EDM drilling

Factor	Parameter	Units	Level 1	Level 2	Level 3
A	Current	A	4	6	8
B	Pulse on time	μs	20	25	30
C	Pulse off time	μs	4	5	6
D	Dielectric pressure	kg/cm^2	24	26	28

52.4.1 MRR and TWR Formula

Measurement of metal rmoval rate:

- MRR = Machined workpiece weight loss (g)/machining time (min)

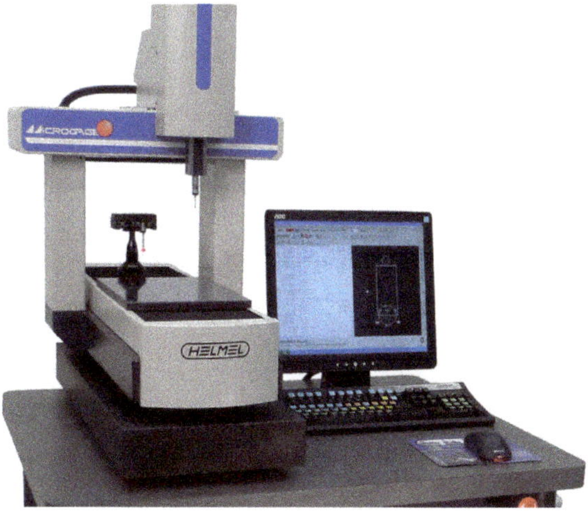

Fig. 52.4 CMM testing machine

Measurement of tool wear rate:

- TWR = Electrode weight loss (g)/machining time (min)

Measurement of percentage wear rate:

- Wear ratio is defined as the ratio of the tool wear rate to the material removal rate.
  %WR = EWR/MRR

52.5 Results and Discussion

Table 52.5 depicts the input parameters, namely current, pulse on time, pulse off time, and dielectric pressure. The weight of the electrode and workpiece before and after machining is taken to calculate MRR and EWR as per formula in Sect. 52.4.1

The machining time, material removal rate, electrode wear rate, wear ratio, circularity, perpendicularity, and cylindricity are shown in Table 52.6.

Now, the relationship between output parameters and each input parameter can be deeply studied. Firstly relationship with respect to current is studied in Fig. 52.5a–c.

The material and electrode wear rate are increased with increasing current. The geometrical tolerances like circularity, perpendicularity, and cylindricity increase with respect to current at a very gradual manner. We can also observe that the wear ratio is decreasing with current, and thus by decreasing current we can decrease wear ratio. Maximum metal removal rate (0.743 g/min) is obtained on the condition of current 7 A, pulse on time 30 μs, pulse off time 5 μs, and dielectric pressure 24 kg/cm^2.

Table 52.5 Input processes parameters at EN-25 material

Exp. no.	Input parameters				Workpiece weight		Electrode weight		Machining time
	A	*B*	*C*	*D*	Before machining	After machining	Before machining	After machining	
	A	μs	μs	kg/cm^2	g	g	g	g	min
1	4	20	4	24	25.168	24.992	19.795	19.758	1′ 12″
2	4	25	5	26	24.992	24.815	19.758	19.721	1′ 01″
3	4	30	6	28	24.815	24.631	19.721	19.687	48″
4	6	20	5	28	24.631	24.455	19.687	19.621	36″
5	6	25	6	24	24.455	24.273	19.621	19.573	42″
6	6	30	4	26	24.273	24.105	19.573	19.522	30″
7	7	20	6	26	24.105	23.947	19.522	19.417	35″
8	7	25	4	28	23.947	23.747	19.417	19.367	24″
9	7	30	5	24	23.747	23.576	19.367	19.314	14″

Table 52.6 Response parameters of EN-25 material use copper electrode

Output parameters						
S. no.	MRR g/min	EWR g/min	WR %	CIR mm	PER mm	CYL mm
1	0.146	0.030	20.55	0.0653	0.0292	0.1134
2	0.422	0.036	8.53	0.0361	0.0384	0.1201
3	0.230	0.042	18.26	0.0365	0.0155	0.0907
4	0.293	0.110	37.54	0.0595	0.0149	0.1274
5	0.260	0.068	26.15	0.0512	0.0359	0.1548
6	0.336	0.102	30.36	0.0497	0.0390	0.1490
7	0.272	0.181	66.54	0.0546	0.0452	0.1505
8	0.500	0.125	25.00	0.0429	0.0642	0.1246
9	0.743	0.230	30.96	0.0603	0.0140	0.1011

The relationship of performance characteristics with respect to pulse on time is shown in Fig. 52.6a–c.

The material and electrode wear rate are high when pulse on time is high and they are at maximum pulse on time of 30 μs. Cylindricity is decreasing with pulse on time, but there is only negligible change in perpendicularity and circularity with pulse on time. The wear ratio of the whole shows an increasing trend with respect to pulse on time. Minimum EWR (0.030 g/min) is acquired at the condition of current 4 A, pulse on time 20 μs, pulse off time 4 μs, and dielectric pressure 24 kg/cm^2.

The relationship between output characteristics and pulse off time is shown in Fig. 52.7a–c.

The material removal rate is showing a decreasing trend with pulse off time, and electrode wear rate has a negligible variation with respect to pulse off time. All geometrical characteristics, namely perpendicularity, cylindricity, and circularity

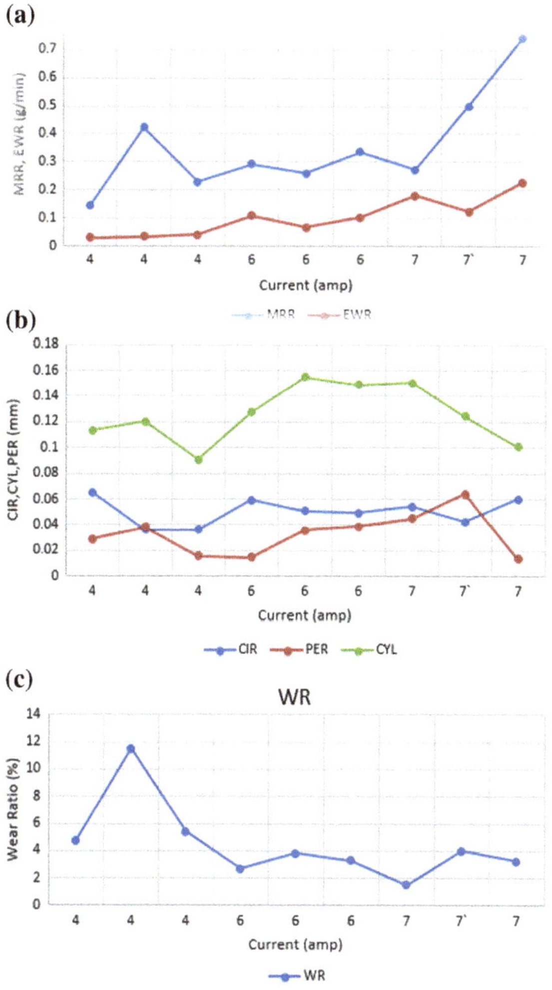

Fig. 52.5 **a** Relationship between current and MRR and EWR. **b** Relationship between current and CIR, PER, and CYL. **c** Relationship between current and wear ratio

show only a negligible change with respect to pulse off time. Whereas a gradual increasing trend is observed with wear ratio and pulse off time. Minimum circularity 0.0361 mm is acquired at the condition of current 4 A, pulse on time 25 μs, pulse off time 5 μs, and dielectric pressure 26 kg/cm^2. Minimum perpendicularity 0.0140 mm is acquired at the condition of current 7 A, pulse on time 30 μs, pulse off time 5 μs, and dielectric pressure 24 kg/cm^2. Minimum cylindricity 0.0907 mm is acquired at the condition of current 4 A, pulse on time 30 μs, pulse off time 6 μs, and dielectric pressure 28 kg/cm^2.

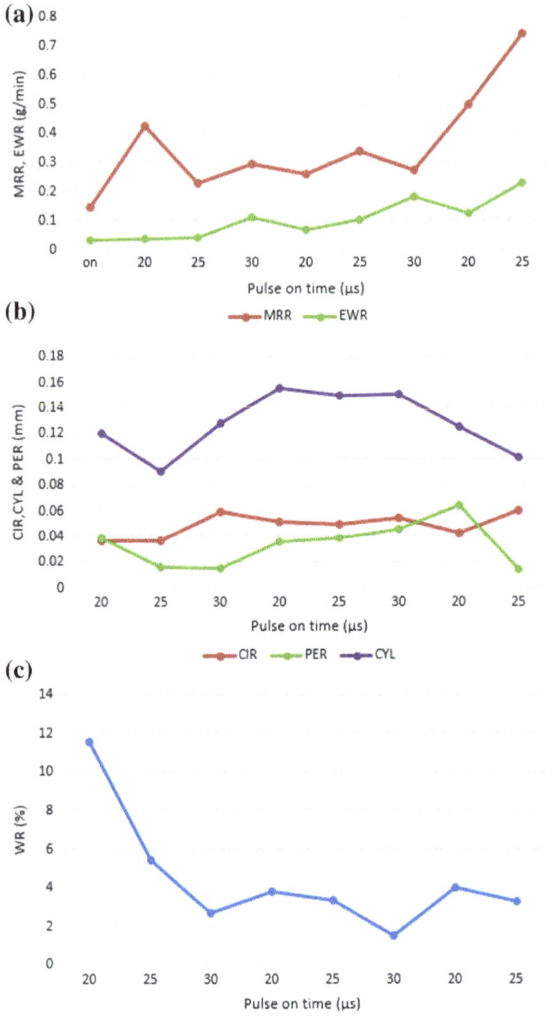

Fig. 52.6 **a** Relationship between pulse on time and MRR and EWR. **b** Relationship between pulse on time and CIR, PER, and CYL. **c** Relationship between pulse on time and wear ratio

The relationship of output characteristics with dielectric pressure is expressed in Fig. 52.8a–c.

There is only negligible change in both MRR and EWR, but at minimum dielectric pressure of (24 kg/cm^2), both are at their maximum. Cylindricity shows the gradual decrease with respect to dielectric pressure, circularity, and perpendicularity have not witnessed only negligible change with respect to dielectric pressure. There is a gradual decrease in wear ratio with respect to dielectric pressure. Minimum wear ratio 8.53% is obtained on the effect of current 4 A, pulse on time 25 µs, pulse off time 5 µs, and dielectric pressure 26 kg/cm^2.

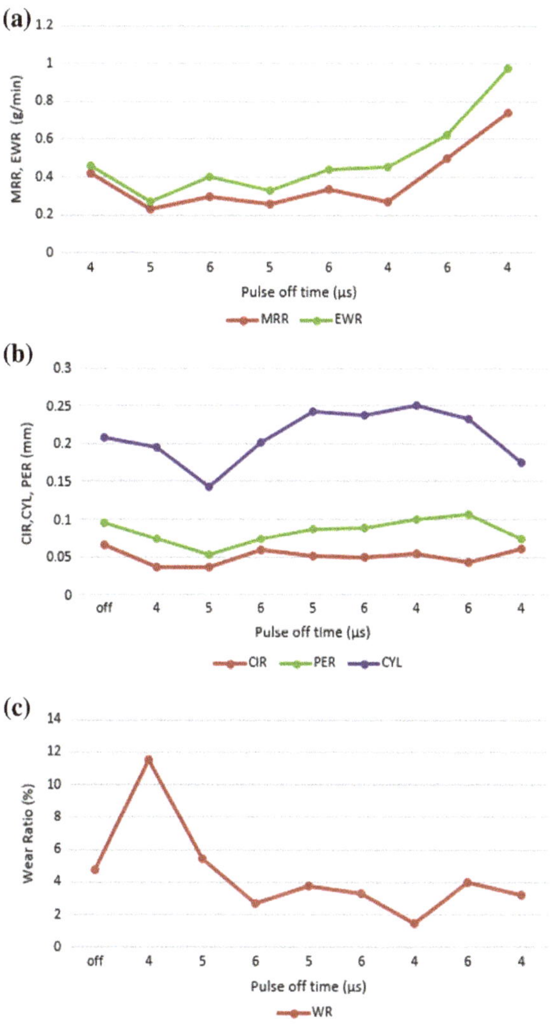

Fig. 52.7 **a** Relationship between pulse off time to MRR and EWR. **b** Relationship between pulse off time to CIR, PER, and CYL. **c** Relationship between pulse off time and wear ratio

52.6 Conclusions

- The three significant factors influencing the value of the MRR are the current (A), pulse on time (μs), and the pulse off time (μs). From results, MRR has been increased with increase current (A), pulse on time (μs), and decrease of pulse off time (μs). Metal removal enhancing increases with increase of pulse off time up to 6 μs then it's tended to decreases.
- The three main outstanding factors affecting the value of the EWR are the current (A), pulse on time, and the pulse off time. From results, EWR has been decreasing with decrease current (A), dielectric pressure (kg/cm^2). The value of

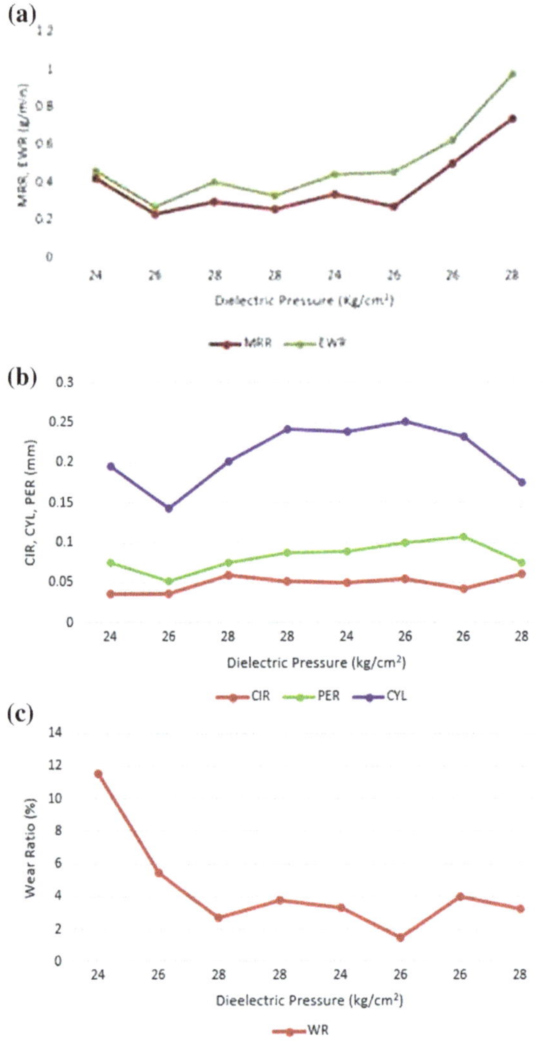

Fig. 52.8 **a** Relationship between dielectric pressure to MRR and EWR. **b** Relationship between dielectric pressure to CIR, PER, and CYL. **c** Relationship between dielectric pressure to wear ratio

EWR decreases with the decrease of pulse on time. However, EWR decreases with increase the pulse off time up to 6 μs.

- The two main notable factors affecting the geometric tolerance of circularity are the current (A) and dielectric pressure (kg/cm^2). From results, circularity has been minimized with decrease of current (A) and increase of dielectric pressure (kg/cm^2).
- The two main important factors affecting the geometric tolerance of perpendicularity are the current (A) and dielectric pressure (kg/cm^2). From results, perpendicularity has been minimized with increase of current (A) and decrease of dielectric pressure (kg/cm^2).

- The two main notable factors affecting the geometric tolerance of cylindricity are the current (A) and dielectric pressure (kg/cm^2). From results, cylindricity has been minimized with decrease of current (A) and increase of dielectric pressure (kg/cm^2).
- From the results of response table and response graph, the optimal level setting of four machining parameters is *A*, *B*, *C*, and *D* for minimizing the electrode wear rate and maximizing material removal rate among the nine experiments by using the Taguchi L_9 orthogonal array.
- The most important factors affecting the EWR and MRR of the spark EDM process have been identified as factor *A* (current), factor *B* (spark on time), factor *C* (spark off time), and factor *D* (dielectric pressure).

References

1. Selvarajan, L., Sathiya Narayanan, C., Jeyapaul, R.: Optimization of machining characteristics in EDM of Si_3N_4-TiN composites by Taguchi grey relational analysis. Appl. Mech. Mater. **592**, 600–604 (2014)
2. Muttamara, A., Fukuzawa, Y., Mohri, N., Tani, T.: Effect of electrode material on electrical discharge machining of alumina. J. Mater. Process. Technol. **209**(5), 2545–2552 (2009)
3. Uno, Y., Okado, A., Itoh, M., Yamaguchi, T.: Electrical discharge machining of a groove with a rotating disk electrode. Int. J. Electr. Mach. **1**, 13–20 (1996)
4. Lin, C.L., Lin, J.L., Ko, T.C.: Optimisation of the EDM process based on the orthogonal array with fuzzy logic and grey relational analysis method. Int. J. Adv. Manuf. Technol. **19**(4), 271–277 (2002)
5. Luis, C.J., Puertas, I., Villa, G.: Material removal rate and electrode wear study on the EDM of silicon carbide. J. Mater. Process. Technol. **164**, 889–896 (2005)
6. Arivazhaghan, P., Rajaduri, A.: Electronic discharge machining of Al/SiC carbide MMCs. In: Proceeding of the International Conference on advanced Mechanics and Industrial Engineering, University of Roorkee, p. 709 (1997)
7. Soni, J.S., Chakraverti, G.: Machining characteristics of titanium with rotary electro-discharge machining. Wear **171**, 51–58 (1994)
8. Selvarajan, L., Katherasan, D., Srivijai, B., Rajavel, R., Ramamoorthi, M.: Experimental analysis of EN 19 alloy material on EDM for improving geometrical errors using copper pentagon shaped electrode. In: International Conference of Materials Processing and Characterization, Materials Today Proceedings, vol. 5, pp. 4508–4514 (2018)
9. Selvarajan, L., Sathiya Narayanan, C., Jeyapaul, R.: Optimization of process parameters to improve form and orientation tolerances in EDM of MoSi2-SiC composites. Mater. Manuf. Processes **30**(8), 954–960 (2015)
10. Selvarajan, L., Sathiya Narayanan, C.S., Jeyapaul, R.: Optimization of EDM parameters on machining Si3N4–TiN composite for improving circularity, cylindricity, and perpendicularity. Mater. Manuf. Processes **31**(4), 405–412 (2016)
11. Selvarajan, L., Sathiya Narayanan, C., Jeyapaul, R.: Optimization of EDM hole drilling parameters in machining of MoSi2-SiC intermetallic/composites for improving geometrical tolerances. J. Adv. Manuf. Syst. **14**(04), 259–272 (2015)

Chapter 53
Computation of End-Cutting-Edge Wear of Single-Point Cutting Tool Using Image Processing

Vishal K. Singh and Shrikrishna N. Joshi

Abstract This paper presents a simple and efficient method for computation of cutting-edge wear of a single-point cutting tool in a plain turning operation. It comprises development of an efficient and accurate image processing-based computer algorithm which compares two images of the cutting tool, i.e., before and after its use in the machining operation. The first image has been used as the reference image. Location of the tool in the image was found out using the developed image processing algorithm. Vertical and horizontal magnifications of the tool image were adjusted such that both the images of the tool have same tool width in terms of pixels. The portion of the images containing tools was extracted and used for the wear computation. The performance of the algorithm was tested by conducting systematic experiments and it was found to be efficient and stable. Only 28 s was noted for the computation of tool wear based on the images captured by a simple digital camera. The algorithm efficiently removes the noise present in the images of worn-out cutting tool used in real-life conditions.

Keywords Plain turning · Tool wear · Cutting-edge wear · Computer vision · Image processing

53.1 Introduction

In the rapidly changing global marketplace, cost-cutting and boosting up productivity are the main goals of the manufacturing industry. With the introduction of manufacturing concepts like JIT, high-speed machining and precision manufacturing have gained more importance. Predicting exact tool life has been a challenge for manufacturing engineers for a long time. Thus, the tool rooms prefer preventive maintenance of the machine tools and cutting tools. However, often the capacity of

V. K. Singh · S. N. Joshi (✉)
Department of Mechanical Engineering, Indian Institute of Technology Guwahati, Guwahati 781039, Assam, India
e-mail: snj@iitg.ac.in

R. G. Narayanan et al. (eds.), *Advances in Computational Methods in Manufacturing*, Lecture Notes on Multidisciplinary Industrial Engineering,
https://doi.org/10.1007/978-981-32-9072-3_53

the tool remains unused and thus increases the cost of production. Therefore, predicting tool life with a fair degree of accuracy is essential to carry out an efficient metal cutting process. Further, the realization of unattended machining processes is a prerequisite for ensuring successful operation of intelligent manufacturing systems (IMSs), flexible manufacturing systems (FMSs), and flexible manufacturing cells (FMCs) [1].

To analyze the performance of cutting tool life, dimensional tool life can be used as a parameter. Astakhov [2] defined the term 'dimensional tool life' as the time period within which the cutting tool assures the required dimensional tolerance and required surface integrity of the workpiece. Authors derived the mathematical equations based on the tribology of metal cutting. Literature reports eminent attempts on development of multisensor models to predict the tool failure [3–7]. Most of them are based on the indirect approach of measurement/monitoring and predict the time of catastrophic failures only [4]. Use of multisensor models is expensive and costly for being employed in real-world daily machining work. On the importance scale, computation of gradual wear ranks higher than catastrophic failure, as it controls the dimensional tolerance of the tool. Direct methods of measurement, particularly computer vision are by nature more suitable for monitoring gradual wear of the tool; but tedious [8]. Therefore, researchers worldwide focused their attention on measuring tool wear between the machining cycles and *ex situ* tool wear measurement.

Use of image processing in the field of tool wear monitoring dates back to the early 90s. Kurada and Bradley in their work proposed a tool wear monitoring system to measure the flank wear. This system was dependent on a He–Ne laser source for illumination [9]. Fadare and Oni proposed a system, which uses two light sources inclined at 45° to horizontal. In addition, two perpendicular views of the tool are captured [10]. Shahabi and Ratnam used workpiece roughness profile, obtained using computer vision for assessment of flank wear and nose radius wear [11].

Recent work in the field of tool wear computation and prediction involves computer vision; use of high-quality charge-coupled device (CCD) sensors and artificial neural network (ANN). ANN accounts for the unknown variables and parameters involved. Most of these research works have been focused on measurement of flank wear and crater wear. Flank wear is the major criterion for measurement of tool life but capturing images during machining is quite a challenging task; a lot of noise in form of chips, dirt, and cutting fluid is present in the background [8]. Tan et al. in their work used sensor placement and optimum fusion of sensors for real-time machine condition monitoring [12].

In the present work, a simple and efficient method for computation of cutting-edge wear of the single-point cutting tool in a plain turning operation is developed. It comprises of development of an efficient and accurate image processing-based computer algorithm which compares two images of the cutting tool, i.e., before and after its use in the machining operation. In what follows, the details of development of algorithm and its experimental validation are presented in the following sections.

53.2 Methodology

A true-color digital image is a 3-dimensional matrix arrangement of primary color pixels with different intensities, whereas a grayscale image is a 2-dimensional arrangement of pixel intensities. Since every digital image or video (series of images varying with time) is a matrix; therefore, two images can be differentiated using the contents of their matrix using mathematical operations. This is the basic idea behind using image processing to determine the tool wear. Use of computer algorithms to detect, improve, or compute features of the image, or the objects in the image is known as image processing. Feature extraction from image requires various steps to be followed which are presented in Fig. 53.1.

53.2.1 Image Acquisition

It is the first step of image processing. It comprises of camera calibration and image capturing.

Camera Calibration. In this step, the parameters of camera's lens and image sensor are estimated. These parameters help to correct the lens distortion, to measure the size of an object in world units, and to determine the location of the camera in the scene. Camera parameters include intrinsic, extrinsic, and distortion coefficients. To estimate the camera parameters, 3-D world points and their corresponding 2-D image points are required. Multiple images of a calibration pattern such as a checkerboard are used. Using the correspondences, camera parameters are estimated. After calibration of the camera, the accuracy of the estimated parameters could be evaluated using reprojection errors and parameter estimation errors. Figure 53.2 shows the developed image acquisition setup for capturing the images of worn-out cutting tool.

Image Capturing Setup. Image capturing step comprises of a click-type camera and LED light source (a set of three numbers of 10 W LED bulbs) placed at the bottom of a transparent and hollow box-shaped tool holder. A white paper sheet placed

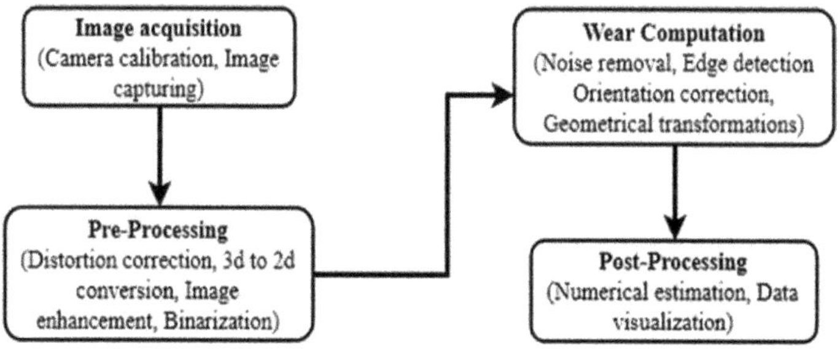

Fig. 53.1 Image processing flow diagram

Fig. 53.2 Image acquisition setup

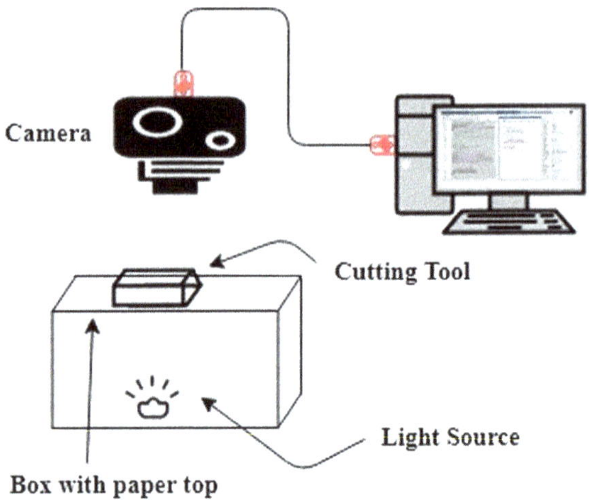

between the tool holder and tool produces a diffused light effect. The diffused light effect helps to minimize the light blobs appearing in the captured images due to the unbalanced distribution of the light from the illumination source. The cutting tool is placed on the marked space on tool holder. An image is captured on the camera which is placed directly above the tool holder. Two images of the tool are captured—one before the start of a machining run carried out for a chosen set of process conditions, and the second image captured after completion of the machining run. The Sony DSC-W570 camera captures a true-color image of size 4608 × 3456 and then, the images are imported into MATLAB™-based image processing utility.

53.2.2 Preprocessing

Distortion Correction. The environmental factors, namely temperature and humidity, affect the quality of images. This causes faulty representation of real-world objects in the image world, which reduces the precision in the measurement of parameters of an image. All the regions in an image cannot be assigned a proportionate number of pixels per world unit.

This problem was solved by using MATLAB™ undistorting function. It was used during camera calibration. Figure 53.3 shows the possible radial camera distortions.

3D to 2D Conversion. In this work, the true-color image [stacking of red (R), green (G), and blue (B) intensity distribution] is converted to a grayscale image. Equation (53.1) is used for conversion of true-color image to grayscale image, I.

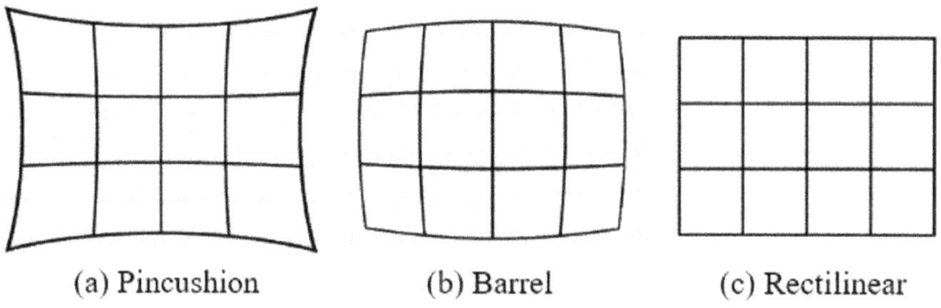

Fig. 53.3 Types of camera distortion

$$I = 0.2989\ \times R + 0.5870 \times G + 0.1140 \times B \tag{53.1}$$

Image filtering. Some of the captured images have inherited noise, namely Gaussian noise and salt and pepper noise. These noises are removed by Gaussian filter and median filter, respectively. Gaussian filtering is achieved by convolving the 2-D Gaussian distribution function with the image matrix.

$$G(x, y) = \frac{1}{\sqrt{2\pi\sigma^2}} \exp\left(-\frac{x^2 + y^2}{2\sigma^2}\right) \tag{53.2}$$

where G is the output of Gaussian operation on the image; x and y are the directions in which the function is operating. The standard deviation σ is the most important of all; it controls the amount of blurring, and its higher value causes more blurring and might lead to the loss or distortion of image features. A median filter is more effective than convolution as it reduces the noise and preserves the edges. Use of smoothening filters causes blurred edges. For proper detection of features and edges, unsharp masking (subtraction of a blurred version of the image from itself) sharpens the images. Image taken has poor contrast between the region of interest (ROI) and background; to tackle this problem 'imadjust,' a MATLAB™ function is employed. It uses the histogram stretching/shrinking to increase or decrease the contrast.

Binarization. The operations performed on the image are combinations of addition, subtraction, multiplication, and matrix inversion. It is convenient to perform these operations on matrix ones (1s) and zeros (0s). Binarization is the conversion of pixel intensities to ones and zeros. Pixels above the threshold are assigned ones and those below the threshold are assigned zeros. Threshold calculated using Otsu's method is more suitable for separation of classes in gray level [11]. In the current work, the effectiveness of the threshold varied from 0.9348 to one during the experiments. If the binarization assigns ones to the background, then inversion (reversing assigned values to pixels) of the image is required to increase the efficiency of computation.

53.2.3 Wear Computation

Noise Removal. Noise occurring in the images is of two types, namely disconnected noise and connected noise. Figure 53.4 shows the various types of noise that can be associated with a camera image. Disconnected noise is present in the image but stays at a distance apart from ROI. Connected noise is attached to ROI or in the vicinity of ROI. In addition, due to their random nature, connected noise cannot be separated using predefined steps. Connected noise occurs due to the shade of the tool as well. The width of such noise happens to be constant if the image acquisition setup does not change, and thus, making it possible to clear them. When the image is binarized, all the disconnected noises are present; there is no difference in the noise blocks and tool in terms of pixels. In the present algorithm, the binarized image was labeled, i.e., provided pixel numbers to each object in the image. Figure 53.5 shows typical binary image matrices. The camera was placed in such a manner that tool gets the maximum area in captured images. This helps in detecting the objects related to cut-chips, as the size of disconnected noise can be in the form of chips, which will not be larger than the tool itself. Then, the identity of the largest object is detected, and the rest of the objects in the image are discarded (Fig. 53.5).

Edge Detection. Orientation correction of the image is resource-consuming operation; it works faster with the edge image of the tool. The Canny operator is more suitable to find the edges in the image by using the local maxima of the image gradient. This method uses two thresholds to detect strong edges (high magnitude

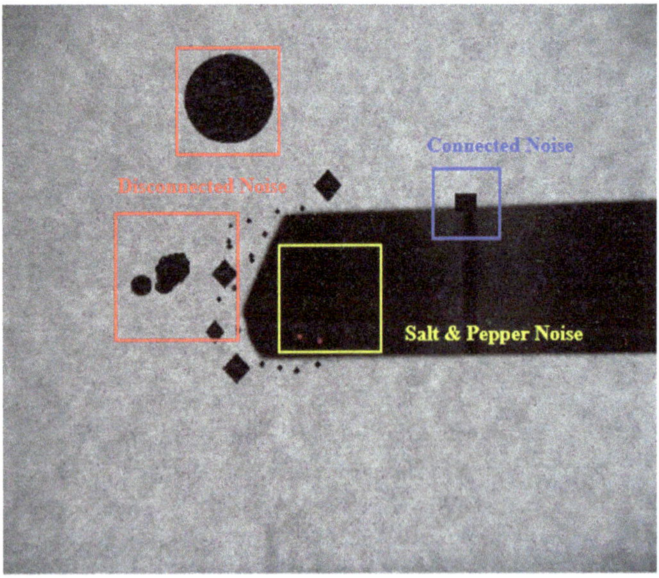

Fig. 53.4 Types of noise, disconnected noise, connected noise, and salt and pepper noise

of gradient) and weak edges (low magnitude of gradient) while preserving the weak edges connected to strong edges. However, the use of two thresholds makes the Canny method less susceptible to noise and helps in faster detection of true weak edges.

Use of automatic anisotropic Gaussian filters has shown improved results for edge detection process which removes the inherent deficiency of missing edges due to nonmaxima suppression associated with the Canny's algorithm [13].

Orientation Correction. The tool images were captured after removing the cutting tool from the lathe machine; and therefore, they are vulnerable to a slight deviation in tool orientation in the horizontal plane. In addition, the amount of deviation is different for any two images. This leads to false area detection. The orientation of images is corrected using Radon transform and is given by

$$\acute{f} = Rf = \int_L f(x, y)\mathrm{d}s \tag{53.3}$$

$$x = p\cos\phi - s\sin\phi \tag{53.4}$$

$$y = p\sin\phi + s\cos\phi \tag{53.5}$$

where ds is the increment in length along L (Eq. 53.3), With reference to the image, the Radon transform is the projection of the image intensity along a radial line oriented at a specific angle, ϕ; and ϕ varies from $\pi/2$ to $\pi/2$. If ϕ is a scalar, $\acute{f}$ is a column vector containing the Radon transform for theta degrees. The angle ϕ corresponding to the maximum value of Radon transform is the orientation angle β for the object in the image. In addition, the accuracy of orientation angle can be tuned to suit the need.

Geometrical Transformations. The cutting tool shown in Fig. 53.4 can have four possible orientations, horizontal with tool base on left or right, vertical with base on top or bottom. By using the geometrical properties, the tip, and the base of the tool are identified. After obtaining the orientation angle of the tool in an image, the image is rotated in a manner (i.e., using rotation angle as $\pm\beta$ or $90^\circ \pm \beta$ or $180^\circ \pm \beta$) that every tool image has the same orientation. Rotation of the image matrix in MATLAB^{TM} increases its size. It is due to image padding (addition of ones and zeros based on user's choice) action of rotation function. In addition, an increase in the size of an image is different for different angles of rotation. To subtract one image from another, the sizes of two images must be same. Therefore, the image cropping accompanied translation is necessary. Figure 53.6 shows the captured images of a typical cutting tool and their corresponding binary images. The centroid of binary image of the worn tool is aligned to the centroid of healthy tool. Then, the subtraction operation is performed to obtain the worn-out area, as shown in Fig. 53.7.

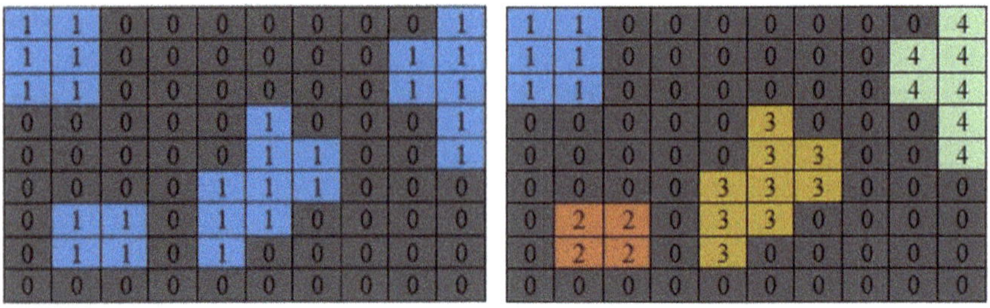

(a) Unlabelled matrix **(b)** Labelled matrix

Fig. 53.5 Binary image matrices

Numerical Estimation. Tool wear data is computed in terms of pixels, and size of pixel is a function of camera's intrinsic properties, standoff distance between camera and objects. Therefore, this data is converted to world units, millimeters. It may happen that worn tool image and healthy tool image are assigned different number of pixels for same actual width. This discrepancy is taken care by resizing both the images such that 700 pixels are assigned to the width of the tool in each image; the number of pixels assigned to tool width for the experimental setup varied around 700.

53.3 Experimental Setup

To validate the performance of the developed algorithm, in-house experiments on plain turning operation were carried out on cylindrical workpiece made up of H13 steel, measuring 25 mm in diameter and 110 mm length. The experiments were performed on 11 kW, high-speed precision lathe, HMT NH26. At start, an image of the cutting tool was captured just after it was ground and the second image was captured after its failure. Machining parameters used for this experiment are listed in Table 53.1. The image for cutting tool was taken in the lab in presence of proper illumination source and away from machining noises like chips, dirt, cutting fluid,

Table 53.1 Machining parameters

Sr. no.	Parameter	Value
1	Tool material	High-speed steel (HSS) (*Miranda Tools make*)
2	Tool dimension	1/4″ × 4″
3	Work material	H13 steel (cylindrical)
4	Spindle speed	598 rpm
5	Feed	0.058 mm/rev
6	Depth of cut	0.5 mm
7	Machining time	2.1 min (stopped due to tool failure)

etc. The details regarding the image acquisition setup are already presented in Sect. 53.2.1. A schematic representation of image acquisition setup is shown in Fig. 53.2.

53.4 Results and Discussion

53.4.1 Image processing

Figure 53.6a, c shows the input true-color images and Fig. 53.6b, d shows the noise removed in corresponding binary images with scale factor one. Tables 53.2 and 53.3 show the properties of tools in corresponding binary images. Table 53.4 shows the numerical value of wear area.

The number of pixels assigned for tools in the image is 687 for the healthy tool and 731 for the worn tool image. Both the images were resized to bring the tool width to 700 pixels. Table 53.2 shows the properties of both the images.

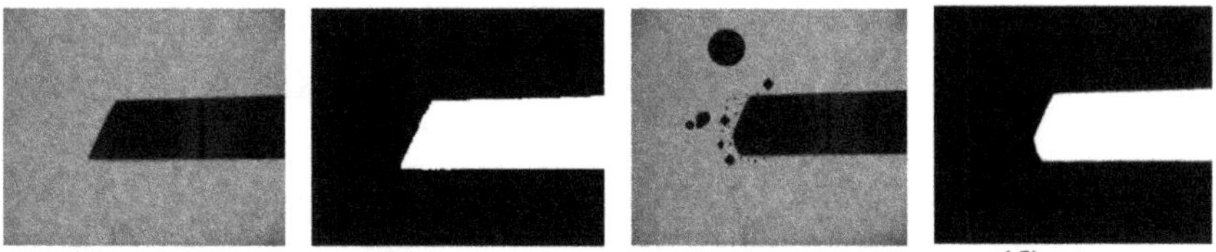

(a) Input 1 **(b)** Binary 1 **(c)** Input 2 **(d)** Binary 2

Fig. 53.6 Captured images of cutting tool and their corresponding binary images

Table 53.2 Properties of binary images of tools

Properties	Values before machining	Values after machining
Area	2,286,601	2,271,953
Perimeter	72,483.3	70,788.2
Equivalent diameter	1706.3	1700.8
Centroid (x, y)	2851.1, 1609.6	2869.8, 1551.5
Euler number	1	1
Initial orientation	−0.5°	−0.3°

Table 53.3 Properties of the binary image of worn-out area in pixels

Sr. no.	Properties	Values
1	Area	24,739
2	Perimeter	815.57
3	Equivalent diameter	177.48
4	Centroid (x, y)	1464.5, 1963.8

Table 53.4 Results: worn-out area

Properties	Values in pixels	Values in world units
Area	24,739	2.03 mm^2
Perimeter	815.57	7.39 mm
Equivalent diameter	177.48	1.41 mm

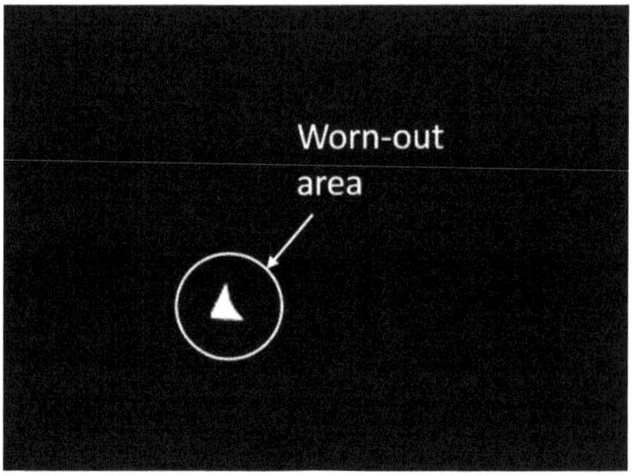

Fig. 53.7 Worn-out area image

Area mentioned in Table 53.2 is in terms of number of pixels. It can clearly be noted that there is reduction in the area of captured image after the machining run. Equivalent diameters were computed by treating the area to be equal to the area of a circle. A difference in centroid coordinates of both images occurs because of two reasons; first, actual position of tool while image capturing and second, due to wear of a portion of the tool.

Figure 53.7 shows the worn-out area image. The original worn-out area image contains some disconnected noise. This disconnected noise occurs, as the alignment of the image tools is not exactly 100%, but very close to it. The size of noise is very small and can be removed by using disconnected noise removal method as mentioned in Sect. 53.2.3.

The width of the tool is 6.35 mm and 700 in terms of pixels. Based on this, the conversion scale was computed as 0.0091 mm/pixel. Since the pixels are square of the data obtained using the current experimental setup, it was assumed that area conversion scale is square of conversion scale. It is equal to 8.23×10^{-5} mm^2/ area in terms of pixels. Table 53.4 shows the worn-out area in terms of pixels and in world units.

In this work, image processing toolbox available in MATLAB™ is used. It has a plethora of predefined functions, some of which can be customized to suit the need of the work. In addition, MATLAB™ provides a seamless integration of its various predeveloped modules which found to be very helpful during the development of algorithm of the present work.

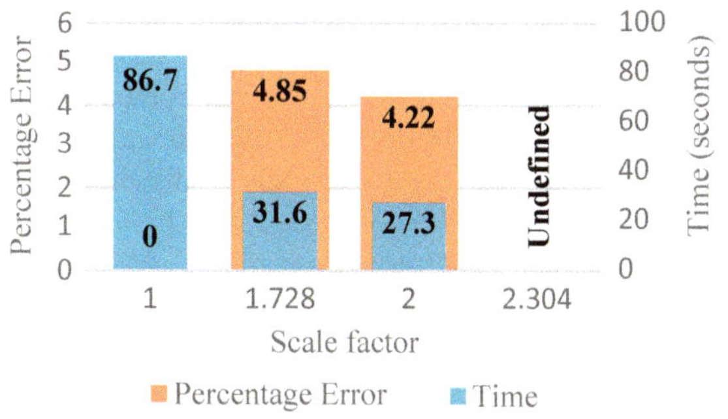

Fig. 53.8 Variation of time and percentage error with different scale factors

53.4.2 Speed Test Results

During the algorithm development, it was noticed that a trial on a large size image, i.e., 4608 × 3456 (original image) consumed a considerable amount of time, 87 s; therefore, a speed test was conducted to test the response of the developed algorithm. In this test, the dimensions of the original image were scaled and the effect of reduction in size of the image was studied. The reduction in size of the image was performed manually using Windows Paint application. The time for single run of the developed algorithm was collected from MATLAB™ profiler. First, the scale factor (by what factor the image size has been reduced) was varied from 1 to 2.304. Then, the errors in wear areas were determined. It was noted that by using the scaled images, the speed and accuracy of the computation were improved. Figure 53.8 shows that a scale factor (reduction) of two performed well and can be used for the further computations.

Overall, this work demonstrates a simple and easy method to compute the tool wear, in which only a simple digital camera is sufficient for computation of tool wear. Camera can be mounted in CNC machines, which captures the images in intervals when tool is not interacting with workpiece. By computing the end-cutting-edge wear, dimensional accuracy of workpiece could easily be improved by applying in-process tool compensation. The present work may also help in monitoring the in-process tool wear.

53.5 Conclusion

In the present work, a simple and efficient image processing algorithm has been developed for computation of wear in cutting edges of single-point cutting tool used in plain turning operation. An in-house image acquisition setup has been developed to capture the images of cutting tools (before and after machining). The captured images were preprocessed using various advanced image processing functions such as distortion correction, conversion from 3D to 2D, filtering using Gaussian and median filters, and binarization. Canny operator was used to properly orient the captured images. The orientation of the images was corrected using Radon transform. Finally, the worn-out area was computed in terms of pixels by using the geometric transformations of captured images. The size of pixel is a function of camera's intrinsic properties, standoff distance between camera and objects. The performance of the developed algorithm has been validated by conducting in-house experiments on plain turning operation, and it was found satisfactory. During the speed test, it was noticed that processing of the original image (4608 × 3456) consumed a considerable amount of time, 87 s. However, the scaled image with reduction factor of two improved the speed of computation as well as accuracy.

The work presented hereby is useful for tool monitoring process and for tool change decision-making process. The presented algorithm has potential and flexibility of being used with fully automatic tool changing process. The whole process involves use of a simple digital camera, which reduces the cost of the operation. This work would be useful for all tool rooms, machine tool manufacturers, and small-scale industry. However, reduction in computation time would further enhance the potential of this work. Integration of artificial neural networks (ANN) would reduce the frequency of image acquisition required for the process.

References

1. Cho, D.W., Jung, M.Y.: On-line monitoring of tool breakage in face milling using a self-organized neural network. J. Manuf. Syst. **14**(2), 80–90 (1995)
2. Astakhov, V.P.: Tribology of Metal Cutting. Elsevier Ltd., Amsterdam (2006)
3. Bahr, B., Motavalli, S., Arfi, T.: Sensor fusion for monitoring machine tool condition. Int. J. Comput. Integr. Manuf. **10**, 314–323 (1997)
4. Balasmo, V., Caggiano, A., Jemielniak, K.: Multi sensor signal processing for catastrophic tool failure detection in turning. Procedia CIRP **41**, 939–944 (2016)
5. Brooks, R.R., Iyengar, S.S.: Multi Sensor Fusion: Fundamentals and Applications with Software. Prentice Hall PTR (1998)
6. Chirssolouris, G., Domroese, M., Beaulieu, P.: Sensor synthesis for control of manufacturing processes. J. Eng. Ind. Trans. ASME **114**, 158–174 (1992)
7. Dimla, D.E., Lister, P.M., Leighton, N.J.: A multi sensor integration method of signals in a metal cutting operation via application of multi-layer perceptron neural networks. In: 5th International Conference on Artificial Neural Networks, pp. 306–311. Cambridge (1997)
8. Dutta, S., Pal, S.K., Mukhopadhyay, S., Sen, R.: Application of digital image processing in tool condition monitoring: a review. CIRP J. Manufact. Sci. Technol. **6**(3), 212–232 (2013)
9. Kurada, S., Bradley, C.: A machine vision system for tool wear assessment. Tribol. Int. **30**(4), 295–304 (1997)
10. Fadare, D.A., Oni, A.O.: Development and application of a machine vision system for measurement of tool wear. ARPN J. Eng. Appl. Sci. **4**(4) (2009)
11. Shahabi, H.H., Ratnam, M.M.: Assessment of flank wear and nose radius wear from workpiece roughness profile in turning operation using machine vision. Int. J. Adv. Manuf. Technol. **43**, 11–21 (2009)

12. Tan, K.K., Er, P.V., Yang, R., Teo, C.S.: Selective precision motion control using weighted sensor fusion approach. In: IEEE International Conference on Mechatronics and Automation, IEEE ICMA 2013, pp. 179–184 (2013)
13. Zhang, W., Zhao, Y., Breckon, T.P., Chen, L.: Noise robust image edge detection based upon the automatic anisotropic Gaussian kernels. Pattern Recogn. **63**, 193–205 (2017)

Chapter 54
Modelling and Analysis of Multi-agent Approach for an IoT-Enabled Autonomous Manufacturing System

K. B. Badri Narayanan and M. Sreekumar

Abstract Industrial Internet of Things (IIoT) is a digital platform to perform machine-to-machine communication, acquire sensor data, analyse, and deliver comprehensive information and also facilitate the manufacturing system to implement smart decisions faster. This paper presents modelling and analysis of an IoT-enabled autonomous manufacturing system based on the performance metrics. Multi-agent interaction has been defined through a generalised control architecture, considering as an autonomous manufacturing system. The number of jobs entering and leaving the machines computes time-dependent performance metrics such as cycle time and work in process using Little's law. The mean arrival rate of each machine is calculated by evaluating the amount of work and a utilisation factor. The squared coefficient of inter-arrival rates based on successive substitution of job flow from machine x to machine y has also been obtained. This is preliminary work in the process of designing an IoT-enabled autonomous manufacturing system capable of self-learning and executing accurate decision faster.

Keywords Internet of Things · Cyber-physical systems · Factory modelling and simulation · Autonomous manufacturing systems

54.1 Introduction

According to Moore's law, the exponential growth of technology is forcing constant up-gradation in manufacturing the environment. Current manufacturing scenario lacks in adopting the changes, so an intelligent manufacturing system with a decentralised multi-agent approach shall be of a solution. In manufacturing systems, downtime can happen due to various disorders such as tool failure, workstation

K. B. Badri Narayanan · M. Sreekumar (✉)
Centre for AI, IoT, and Robotics, Department of Mechanical Engineering,
Indian Institute of Information Technology, Design and Manufacturing,
Kancheepuram, Vandalur–Kelambakkam Road, Chennai 600127, India
e-mail: msk@iiitdm.ac.in

R. G. Narayanan et al. (eds.), *Advances in Computational Methods in Manufacturing*, Lecture Notes on Multidisciplinary Industrial Engineering,
https://doi.org/10.1007/978-981-32-9072-3_54

failure, the breakdown of material handling devices like a robot, conveyor, and transfers system. This downtime issue generally increases the lead time and production cost. Immediate rescheduling of the process flow of the respective workpiece offers a solution for reducing systems downtime wisely and efficiently [1]. Also, reconfiguration changes the structure of the manufacturing layout that requires new scheduling [2–4]. Even state-of-the-art modern manufacturing facilities cannot be able to provide a complete solution for machine downtime.

For agent breakdown cases, present automation systems use a defined protocol. Hence for defining a new protocol, the system should be interrupted in the middle. This aspect increases the rigidness of the system. This can be avoided by using the machine-to-machine communication protocol called Process Field Net (PROFINET). Similarly, a cognitive approach can be implemented to adopt rapid changes in the manufacturing systems environment. Several unique paradigms of Intelligent Manufacturing Systems (IMS) such as holonomic systems [5, 6], cognitive systems [7], biological systems [8, 9], and reconfigurable manufacturing systems [10, 11] have already been established.

A manufacturing Petri Net (MPN)-based scheduling model for IoT-enabled hybrid flow shop manufacturing is discussed in [12]. IoT in the manufacturing system digitises the real-time production system and provides a faster smart decision. In the present work, an IoT enabled autonomous manufacturing system, as shown in Fig. 54.1, is designed by considering every machine as agents. Jobs are considered as encapsulated with radio-frequency identification (RFID) tags and represented as stable, smart objects (SSO). Every machine is attached with RFID readers and represented as dynamic smart objects (DSO). Similarly, every worker is

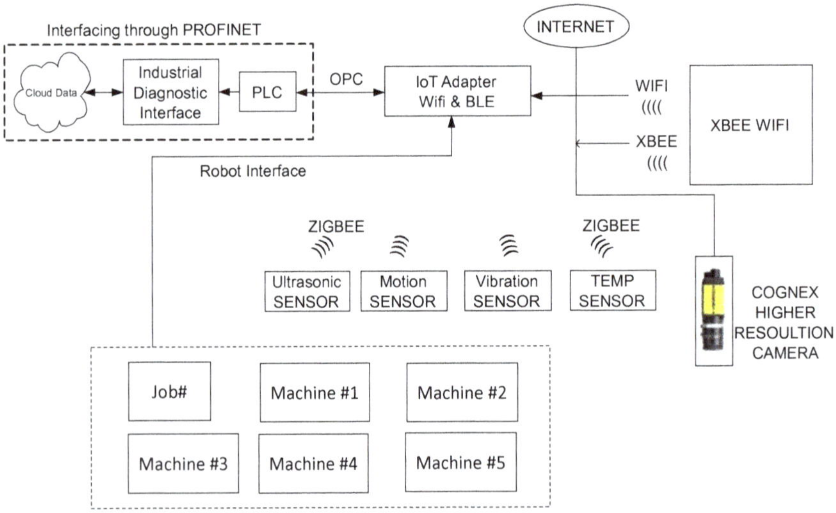

Fig. 54.1 Schematic arrangement of wireless communication in an IoT-enabled manufacturing system

provided with RFID smart ID card. Machining and buffering are recorded in the database by interactions between SSO, DSO, and smart ID card.

The rest of this article is constructed as follows. Section 54.2 discusses the core technologies such as swarm intelligence and control architecture of autonomous manufacturing systems. Section 54.3 is focused on modelling and analysis of the performance metrics of manufacturing systems. The conclusion is drawn in Sect. 54.4, followed by references.

54.2 Core Technologies

54.2.1 Swarm Intelligence

Swarm intelligence is a bio-inspired technique that mimics the localised interaction behaviour that exists amongst various species such as ant, insects, and other animals. Ant colony optimisation (ACO) is a swarm technique which mimics the food searching behaviour of ants. The interaction between ants is carried out basically by deposition of pheromone by a preceding ant. ACO can be implemented in manufacturing systems by considering machines as a group of autonomous and cooperative agents like ants. Such systems could be self-organised by a local interaction between SSO, DSO, and smart ID card. Every machine can calculate its pheromone value for a particular process. During machine failure, the agent interaction can be defined by comparing the machine specification and process information that are available in the database. The formulation of pheromone value can be accomplished by

$$P_X = \left[\alpha_{\mathrm{WIP}}(\mathrm{WIP}_O/\mathrm{WIP}) + \alpha_{\mathrm{CT}}(\mathrm{CT}_O/\mathrm{CT}) + \alpha_S\left(\frac{C_{SO}^2}{C_S^2}\right) + \alpha_I\left(\frac{C_{AO}^2}{C_A^2}\right) + \alpha_A\left(\frac{\delta_{xo}}{\delta_x}\right)\right]^{-1} \tag{54.1}$$

where P_X—pheromone value, WIP_O—maximum work in process, CT_O—maximum cycle time, C_{SO}^2 represents the maximum squared coefficient of variation for service time, C_{AO}^2—maximum squared coefficient of variation for the inter-arrival rate of jobs, δ_{xo}—maximum arrival rate of jobs to machine x. The factors $\alpha_{\mathrm{WIP}}, \alpha_{\mathrm{CT}}, \alpha_S, \alpha_I$ and α_A represent the weight of WIP, CT, C_{SO}^2, C_{AO}^2 and δ_{xo}, respectively. After generating the pheromone values of each process for all machines, the machine failures can be identified with the following cases

Case 1: If $\forall P_X < 0.2$
Then send a message from MES to Employee staff RFID /*all machines are good*/

Case 2: If $\exists ! P_X \geq 0.2$
Then send a message from MES to Employee staff RFID /*the machine x is selected for rectifying the failure*/
Case 3: If there are more than one $P_X \geq 0.2$
Then send a message from MES to Employee staff RFID /*the machine which is having the highest pheromone value selected for rectifying the first and followed by the subsequent highest value*/
Case 4: If $\forall P_X \geq 0.2$
Then

(i) send a message from MES to staff RFID /*All machines had pheromone value above 0.2*/
(ii) In response to the above message, send a message from staff to MES /*requesting the MES for rescheduling*/.

54.2.2 Control Architecture

The generalised control architecture of autonomous manufacturing systems is shown in Fig. 54.2. The core elements of the control architecture are agent interaction, agent behaviour, and database. The interaction between agents and manufacturing execution system (MES) occurs through Extensible Markup Language (XML). XML provides readable formats to both humans and machines. The industrial control system protocol Open Platform Communication (OPC) interfaces the machine with Programmable Logic Controller (PLC) and connects the physical devices like sensors integrated with machining centres. The database stores the information like job address, machine address, process details about jobs, sensor data, and tool failures. The agent cooperation is incorporated by calculating the pheromone value of each agent. Then the agents compare the pheromone value by interacting with nearest agents and finally, the agent which has the highest pheromone value is chosen to rectify the failure.

54.3 Performance Metrics

The mean of performance metrics considered for modelling and analysis of an IoT-enabled autonomous manufacturing system is cycle time CT, work in process WIP, arrival rate, squared coefficient of variation for inter-arrival, and service time. Using Little's law, the probability of seven different jobs flow across the manufacturing systems is formulated by equating the number of jobs entering and leaving the machine. This also computes the time-dependent variables such as cycle time and work in process. The mean arrival rate of each machine is calculated by

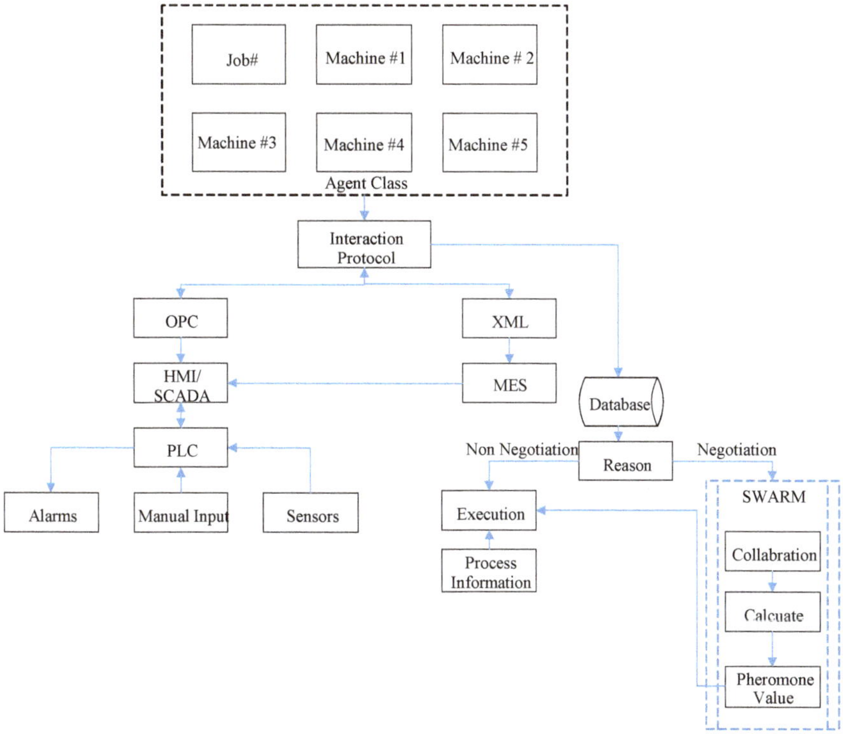

Fig. 54.2 A generalised control architecture of an autonomous manufacturing system

evaluating the amount of work and a utilisation factor. The squared coefficient of inter-arrival rates is obtained by successive substitution of job flow from machine x to machine y.

54.3.1 Workflow Modelling

A discrete workflow network, defining the arrival rate, is more complicated than the continuous systems. Consider machines, as shown in Fig. 54.3, the job is entering from an external source into the first machine with a mean rate of o. However, machine 1 also got rejected job with probability d. Hence, the flow into the machine 1 is not clear. Sequentially, the probability for machine 2 is also not clear since the job arrives from machine 1 has a mean rate of p_1. This quandary is a regular due to the discrete movement of jobs. These situations demand to calculate flow rates concurrently. For example, δ_x for $x = 1, 2\ldots n$ is used to indicate the total arrival of jobs for machine x. The mean flow rates are modelled mathematically using steady-state conditions as

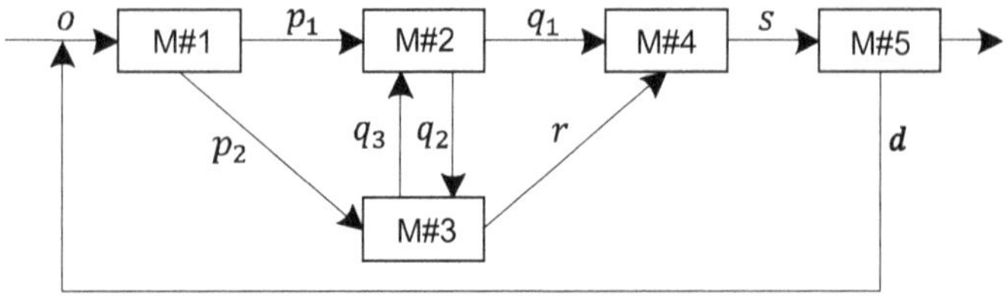

Fig. 54.3 Autonomous manufacturing system topology

$$\left.\begin{array}{l}\delta_1 = o + d\delta_5 \\ \delta_2 = p_1\delta_1 \\ \delta_3 = p_2\delta_1 + q_1\delta_2 \\ \delta_4 = \delta_3 + q_2\delta_2 \\ \delta_5 = s\delta_4\end{array}\right\} \quad (54.2)$$

The steps involved in the evaluation of autonomous manufacturing systems are detailed below.

Step 1. Compute the amount of work for each machine by considering n machines in a manufacturing system with i type jobs and each type flows is defined by transmitting matrix R_i and by assuming that summation of any row should be less than one. Let o_i denote mean arrival rate of type i jobs, and now the vector δ_i is given by [13]

$$\delta_i = \left[I - (R_i)^T\right]^{-1} o_i \quad (54.3)$$

Amount of work for each machine is calculated by

$$\mathrm{AW}_x = \sum_{i=1}^{m} \delta_{i,x} E[T_s(i,x)] \quad (54.4)$$

where m—maximum number of products,
$\delta_{i,x}$—the rate of product type i entering into machine x as shown in Table 54.1 and $E[T_s(i,x)]$ is mean service time of type i job at machine x.

The utilisation factor u_x, for machine x, is,

$$u_x = \frac{\mathrm{AW}_x}{c_x} = \frac{\sum_{i=1}^{m} E[T_s(i,x)]}{c_x} \quad (54.5)$$

Table 54.1 Arrival rate of each machine

Arrival rate of each machine x	Job types						
	J# 1	J# 2	J# 3	J# 4	J# 5	J# 6	J# 7
$\delta 1$	5.0813	5.4585	5.102	5.4348	5.102	5.0201	5.004
$\delta 2$	4.065	4.3668	0	0	0	0.2008	0.2002
$\delta 3$	0	3.7118	1.0204	0	1.0204	1.004	1.0008
$\delta 4$	0.813	4.5852	1.0204	4.3478	0	0	0.04
$\delta 5$	0.813	4.5852	1.0204	4.3478	1.0204	0.2008	0.04

Step 2. The meantime, squared coefficient of variation for the service time, and squared coefficient of variation for the arrival times at machine x are calculated by

$$E[(T_s(x)] = \sum_{i=1}^{m} \frac{\delta_{ix}}{\delta_x} E[T_s(i,x)] = \frac{\mathrm{AW}_x}{\delta_x} \tag{54.6}$$

$$C_s^2(x) = \frac{\left\{\sum_{i=1}^{m} \frac{\delta_{i,x}}{\delta_x} E[T_s(i,x)]^2 [1 + C_s^2(i,x)]\right\}}{\left\{\sum_{i=1}^{m} \left(\frac{\delta_{i,x}}{\delta_x}\right) E[T_s(i,x)]\right\}^2} - 1 \tag{54.7}$$

Step 3. The combined transmitting matrix $R = (r_{x,y})$ gives the transferring possibilities of a subjective job and is determined by

$$r_{x,y} = \frac{\sum_{i=1}^{m} \delta_{ix} r_{x,y}^{i}}{\delta_x} \quad \text{for } x, y = 1, \ldots, n \tag{54.8}$$

The flow of jobs from machine x that are routed directly to machine y will be called as x to y stream and the squared coefficient of variation of inter-arrival times to y from x is obtained by

$$C_a^2(y) = \frac{o_y}{\delta_y} C_a^2(0,y) + \sum_{x=1}^{n} \frac{\delta_x r_{x,y}}{\delta_y} \left[\begin{array}{l} r_{x,y}(1 - u_x^2) C_a^2(x) \\ + r_{x,y} u_x^2 \left(\frac{C_s^2(x) + \sqrt{c_x} - 1}{\sqrt{c_x}}\right) + 1 - r_{x,y} \end{array}\right] \tag{54.9}$$

Then, the mean cycle time spent within the manufacturing system by a type i job is given by

$$CT^{i} = \frac{\sum_{i=1}^{m} \delta_{i,x}\{CT_q(x) + E[T_s(i,x)]\}}{\sum_{y=1}^{n} o_{iy}} \tag{54.10}$$

54.3.2 Performance Analytics

The flow sequence of each product is shown in Table 54.2, and the processing time is shown in Table 54.3. The flow rate of each machine is assumed as [13] o = 5, p_1 = 0.8, p_2 = 0.2, q_1 = 0.6, q_2 = 0.2, q_3 = 0.2, r = 1, s = 1 and d = 0.1. The amount of work done is calculated by Eq. (54.4). The meantime and squared coefficient of variation for the service time at machine x can be calculated using the Eqs. (54.6) and (54.7).

Table 54.2 Sequence of each product type

Job sequence J	Job Types						
	J#1	J#2	J#3	J#4	J#5	J#6	J#7
Sequence 1	M# 1	M# 1	M# 1	M# 1	M# 1	M# 1	M# 1
Sequence 2	M# 2	M# 2	M# 3	M# 4	M# 3	M# 3	M# 3
Sequence 3	M# 4	M# 3	M# 4	M# 5	M# 5	M# 2	M# 2
Sequence 4	M# 5	M# 4	M# 5			M# 5	M# 4
Sequence 5		M# 5					M# 5

Table 54.3 Processing data of each machine for seven products

Machine x		M# 1	M# 2	M# 3	M# 4	M# 5
J# 1	$E[T_s(1,x)]$	0.071429	0.1	–	0.066667	0.06
	$C_s^2(1,x)$	0.8	1.2	–	1.5	0.75
J# 2	$E[T_s(2,x)]$	0.066667	0.055556	0.083333	0.06	0.0625
	$C_s^2(2,x)$	1.33	2	1.5	0.75	1.7
J# 3	$E[T_s(3,x)]$	0.065574	–	0.081633	0.070175	0.061538
	$C_s^2(3,x)$	1.43	–	1.56	1.5	1.72
J# 4	$E[T_s(4,x)]$	0.064725	–	–	0.069204	0.06079
	$C_s^2(4,x)$	1.53	–	–	1.6	1.8
J# 5	$E[T_s(5,x)]$	0.068259	–	0.079051	–	0.06006
	$C_s^2(5,x)$	1.527047	–	1.425532	–	1.76555
J# 6	$E[T_s(6,x)]$	0.06734	0.05305	0.077821	–	0.059347
	$C_s^2(6,x)$	1.526055	2.45	1.425532	–	1.751174
J# 7	$E[T_s(7,x)]$	0.066445	0.052493	0.076628	0.066445	0.058651
	$C_s^2(7,x)$	1.525533	2.65	1.377358	1.487805	1.737327
c_x		19	4	4	4	6

The multiple job study problems now shrink to a single job study. The transferring possibilities of a subjective job and is determined by Eq. (54.8). The mean time spent with the manufacturing system by a type i job is calculated using Eq. (54.10). The shop floor parameters like cycle time CT, work in process WIP, arrival rate, squared coefficient of variation for inter-arrival, and service time are shown in Fig. 54.4.

In Fig. 54.4, the performance metrics of each machine is monitored remotely. The database provides dynamics of real-time data for intelligent decision-making. Pheromone value of each machine is evaluated based on real-time data of performance metrics. Each machine has a different threshold value for different performance metrics. When the performance metrics getting deviated beyond the threshold value, messages are sent to employee's smart ID to overcome the deviation. The significance of this paper provides a solution to overcome the downtime of each machine using the cognitive ability of agents. Swarm intelligence improves the

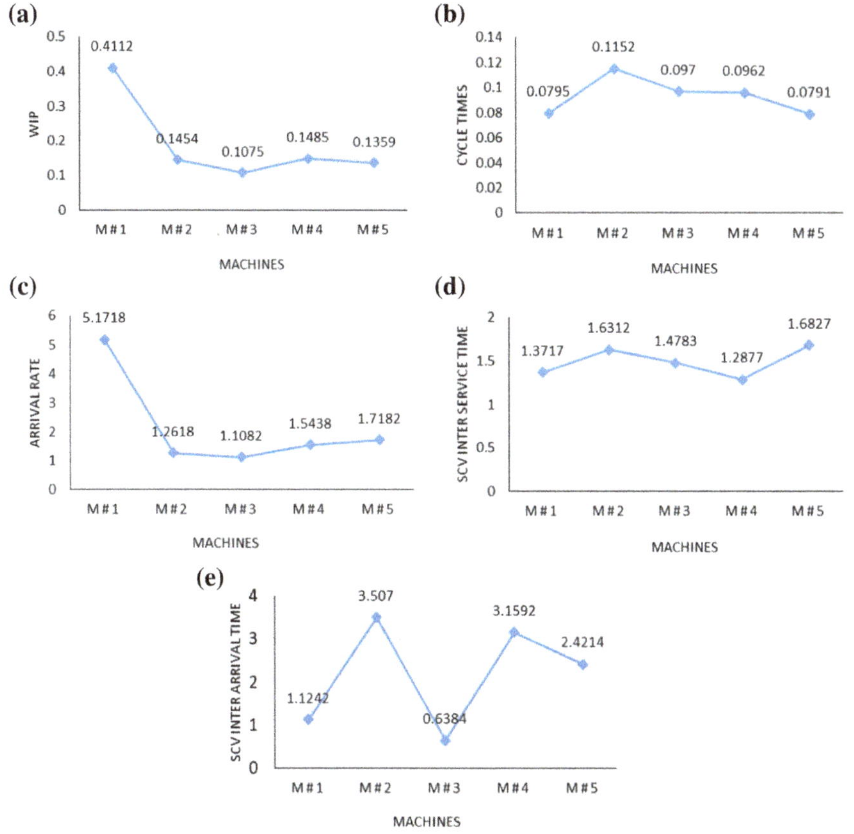

Fig. 54.4 Performance parameters **a** cycle time, **b** work in process, **c** arrival rate, **d** SCV inter-arrival time, and **e** SCV service time

autonomous behaviour of each agent to defend the disturbances without any upper-level aid and enhances the reliability of the system. The dashboard is transparently showing the dynamic status of machine health and the solution based on pheromone value priorities and addresses the failure immediately.

54.4 Conclusion

In this work, an attempt has been made to derive mathematical model and analyse a manufacturing system integrated with IoT based on the mean of performance metrics such as cycle time, work in process, arrival rate, squared coefficient of variation for service time, and inter-arrival rate. This model reflects the real-time operations of machining and buffering in a smart manufacturing system. The swarm intelligence of machining agents improves system robustness to face machine failures by multi-agent interaction, adaptable to changes, and reliable in product quality. The manufacturing system becomes rigid by decentralised control and autonomous behaviour in decision-making. The simulation results of all performance metrics provide dynamic states of each machine. The results also provide continual process adjustment in the manufacturing system to avoid failure of machines by evaluating the pheromone value. The manufacturing systems become intelligent to handle the disturbances in priority sequence using pheromone value and to reduce the downtime of the machine. Future research line extends to the development of the finite state of a machine for more precise modelling.

Acknowledgements Authors sincerely thank the Ministry of Electronics and Information Technology (MeitY), Government of India, for providing financial assistance under Visvesvaraya Ph.D. Scheme to carry out this research work at IIITDM Kancheepuram, Chennai, India.

References

1. Vieira, G.E., Hermann, J.W., Lin, E.: Rescheduling manufacturing systems: a framework of strategies, policies, and methods. J. Sched. 39–62 (2003)
2. Park, H.S., Choi, H.W.: Development of a modular structure-based changeable manufacturing system with high adaptability. Int. J. Precis. Eng. Manuf. 7–12 (2008)
3. Wang, Y.F., et al.: An integrated approach to reactive scheduling subject to machine breakdown. In: Proceedings of the IEEE International Conference on Automation and Logistics, pp. 542–547 (2008)
4. Shea, K., et al.: Design-to-fabrication automation for the cognitive machine shop. Adv. Eng. Inform. 251–268 (2010)
5. Leitão, P., Restivo, F.: ADACOR: a holonic architecture for agile and adaptive manufacturing control. Comput. Ind. 121–130 (2006)
6. Park, H.S., et al.: Agent-based shop control system under holonic manufacturing concept. In: Proceedings of the 4th Korea-Russia International Symposium, pp. 116–121 (2000)
7. Zäh, et al.: The cognitive factory. In: Changeable and Reconfigurable Manufacturing Systems, pp. 355–371. Springer, London (2009)

8. Bannat, A., et al.: Artificial cognition in production systems. IEEE Trans. Autom. Sci. Eng. 148–174 (2011)
9. Monostori, L.: AI and machine learning techniques for managing complexity, changes and uncertainties in manufacturing. Eng. Appl. Artif. Intell. 277–291 (2003)
10. Matsuda, M., Ishikawa, Y., Utsumi, S.: Configuration of machine tool agents for flexible manufacturing. In: 39th CIRP International Conference on Manufacturing Systems, pp. 351–357 (2006)
11. Wang, S., et al.: Towards smart factory for industry 4.0: a self-organized multi-agent system with big data based feedback and coordination. Comput. Netw. 158–168 (2016)
12. Wang, M., et al.: A MPN-based scheduling model for IoT-enabled hybrid flow shop manufacturing. Adv. Eng. Inform. 728–736 (2016)
13. Curry, G.L., Feldman, R.M.: Manufacturing Systems Modeling and Analytics. Springer Science & Business (2010)

Part VI
Material Fabrication

Chapter 55
Estimation of Distortion for CFRP Antenna Reflector During Autoclave Processing

Shrey Patel, Dhaval Shah, Shashikant Joshi and Kaushik Patel

Abstract Carbon fibre-reinforced polymer (CFRP) composites are broadly utilized in space applications due to their very high specific properties over typical metals and alloys. Autoclave curing is one of the finest composite producing techniques to achieve great dimensional accuracy of the advanced shaped component. The cure shrinkage, which is commonly referred to as warpage or spring-back distortion, is induced as a result of material thermo-elastic anisotropy during composite laminates manufacturing. In this paper, the distortion of CFRP reflectors is estimated using experimental and numerical approaches for autoclave processing. The antenna reflectors have been fabricated using HCU200/A45 prepreg and stainless steel AISI 430 as a part and tool material, respectively. The geometric modelling and finite element analysis (FEA) have been performed for parabolic reflectors using ABAQUS software along with COMPRO plug-in. Thermochemical and stress-deformation analysis has been performed to obtain the deformed shape of the antenna reflectors. The laminate thickness has been considered as a process parameter. The experimental and numerical results have been compared and located in sensible agreement with one another.

Keywords CFRP · Distortion · Finite element analysis · Autoclave

S. Patel · D. Shah (✉) · S. Joshi · K. Patel
Mechanical Engineering Department, Institute of Technology, Nirma University, Ahmedabad 382481, Gujarat, India
e-mail: dbshah@nirmauni.ac.in

S. Patel
e-mail: 17mmed05@nirmauni.ac.in

S. Joshi
e-mail: s.j.joshi@nirmauni.ac.in

K. Patel
e-mail: kaushik.patel@nirmauni.ac.in

R. G. Narayanan et al. (eds.), *Advances in Computational Methods in Manufacturing*, Lecture Notes on Multidisciplinary Industrial Engineering,
https://doi.org/10.1007/978-981-32-9072-3_55

55.1 Introduction

The use of CFRP prepreg composites in aerospace, automotive, and defence industries is growing mainly due to their high strength and stiffness to weight ratios. Although there is a wide scope of application of these materials, still there is a limited use because of quality reliability issues, high raw material, manufacturing cost, and lengthy manufacturing cycle. Autoclave process is the most popular for getting high-quality composites components compared to other composite manufacturing processes. This manufacturing process gives the minimum void content in manufactured composite components as compared with other vacuum infusion process, resin transfer moulding process, filament winding process, and hand layup process. The antenna reflectors used for transmitting the radio frequency signals between two devices in space applications, and therefore, the profile of the reflectors should be accurate. The COMPRO, a special purpose subroutine, can be used for simulation of autoclave manufacturing process along with ABAQUS software. The COMPRO plug-in consists of the material database for prepreg, fibre and resin system, and tool materials. The thermochemical, flow-compaction, and stress-deformation analysis modules are available for the simulation of composites.

Many researchers worked on the prediction of warpage and spring-back deformations for composite components. During the process of curing, Hahn and Pagano [1] found out that some stresses which developed during that process varied with respect to temperature development. Darrow and Smith [2] established three process parameters, i.e., mould expansion, thickness cure shrinkage, and fibre volume fraction gradients that effects on spring-in deformations. Twigg et al. [3] predicted warpage of flat laminates with considering the part length, tool–part interaction, the effect of release film, and release agent as process parameters. Stefniak et al. [4] performed experimental work to quantify the various mechanisms, i.e., property gradient mechanism, stress gradient mechanisms with respect to their relative contribution to composite part distortions. Kaushik et al. [5] studied the result of various process parameters, i.e., pressure, ramp rate, and the degree of cure on the coefficient of friction in the autoclave process. Fernlund and Poursartip [6] demonstrated the effect of the cure cycle and tool surface condition on spring-back deformation. Zeng and Raghavan [7] implemented three-dimensional process model through user subroutine interfaced with finite element software. Abusafieh et al. [8] described the effect of temperature and degree of cure by using experimental setup and FEA. Dong [9] derived the mathematical formula to calculate spring-in of flat laminates using the effective coefficient of thermal expansion. Mezeix et al. [10] simulated spring-back deformation of unidirectional carbon/epoxy flat laminates manufactured through the autoclave process.

In this research paper, experimental and numerical approaches have been used to estimate distortion for CFRP antenna reflectors. The stainless steel AISI 430 and HCU200/A45 prepreg tape have been identified as a tool and part material, respectively. The laminate thickness, i.e., 0.8, 1.6, 2, and 2.4 mm has been considered as a process parameter. In the experimental approach, CFRP antenna reflectors have been fabricated using the autoclave process. The geometrical modelling as well as FE analysis has been performed with the help of COMPRO plug-in using ABAQUS software. The thermochemical and stress-deformation analysis has been carried out for CFRP antenna reflectors to predict distortion. The comparison of distortions for fabricated reflectors using autoclave process and simulation has been carried out.

55.2 Experimental Approach

55.2.1 Material Selection

The AISI 430 grade of stainless steel has been selected as a tool material due to the ease of availability and machinability by comparing other grades of steel. The HCU200/A45 prepreg tape (tensile strength = 3450 MPa, fibre density = 1.8 g/ cm^3) has been identified, which consists of excellent surface finish and mechanical properties. The consolidation materials, i.e., release agent, vacuum bag, peel ply, and breather cloth have been required in the manufacturing of composite parts using the autoclave process. These consolidation materials have been sustained temperature up to 200 °C. The laboratory-scale autoclave machine of overall diameter 900 mm and length of 1000 mm has been used for the manufacturing process.

55.2.2 Fabrication of Antenna Reflectors

The 300 × 300 × 70 mm size block of stainless steel AISI 430 material has been procured. The 3D CAD model has been prepared using solid modelling software for paraboloid profile with a circumferential diameter of 240 mm at one face with the depth of 50 mm. The machining of stainless steel tool has been performed using NC code in the 5-axis VMC machine with the great surface finish as shown in Fig. 55.1. The three coats of release agent have been applied on the parabolic tool surface. The numbers of prepreg layers as per the required thickness of reflector have been cut into the circular shape of 300 mm diameter. Peel ply, breather cloth, and vacuum bag have been kept on layers of prepreg. The sealant tape has been applied over the entire assembly to prevent leakage of air and resin from the assembly. The vacuum valve has been placed at the centre of a vacuum bag to maintain pressure. The one end of the vacuum hose has been connected to a valve whereas another end at the outlet of the vacuum pump. The entire setup has been kept in the autoclave for curing as shown in Fig. 55.2. The manufacturer recommended curing cycle has been given to the autoclave. The demoulding process has been performed at atmospheric condition after the completion of the curing process. The finish cured and demoulded reflector is shown in Fig. 55.3.

Fig. 55.1 Stainless steel AISI 430

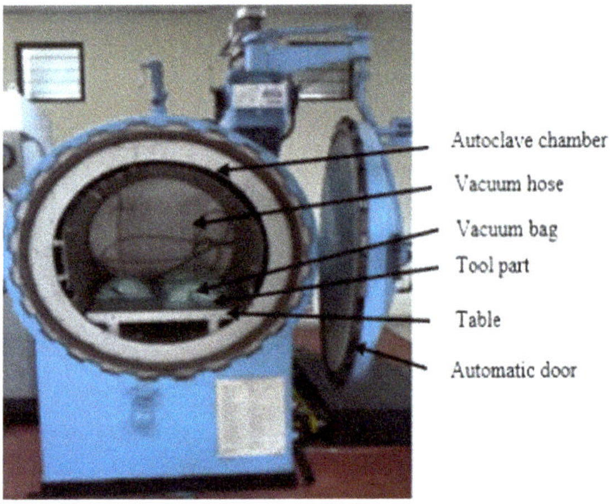

Fig. 55.2 Autoclave machine

Fig. 55.3 Manufactured CFRP antenna reflector

55.3 Numerical Approach

The geometry of the tool and part has been created as per actual dimensions in ABAQUS software. The thermochemical and stress-deformation analysis has been performed with the help of COMPRO plug-in for antenna reflectors to predict distortion [11]. The methodology followed in the numerical approach has been described as a flow chart as shown in Fig. 55.4.

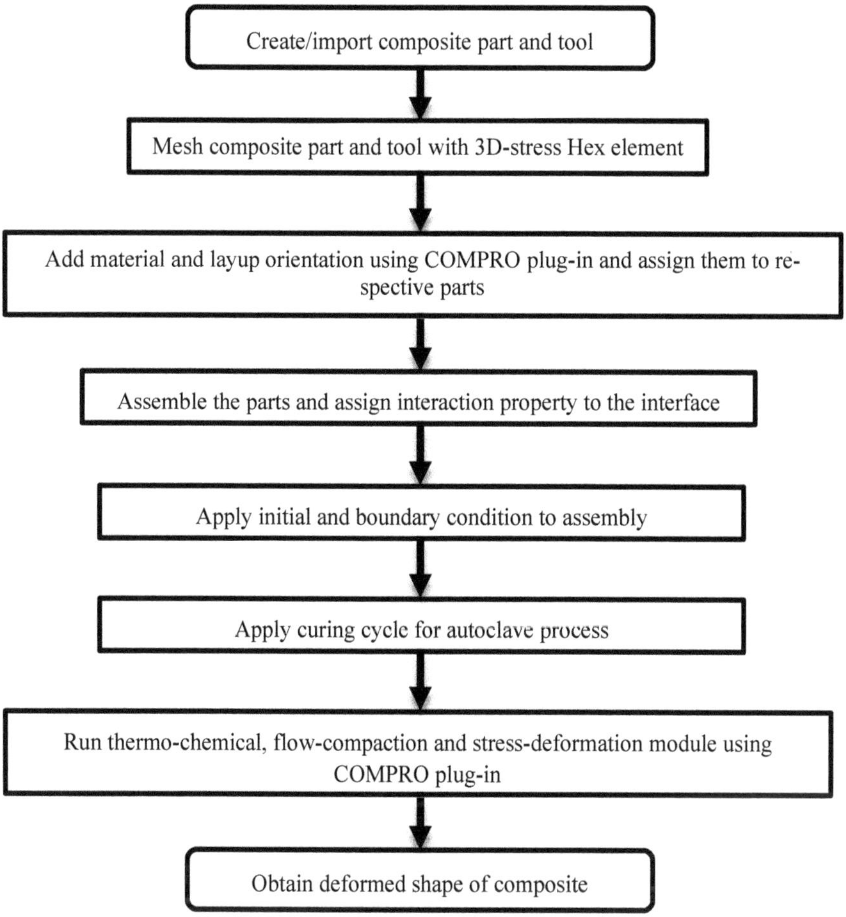

Fig. 55.4 Methodology for numerical simulation using ABAQUS with COMPRO plug-in

55.3.1 Simulation Procedure

The curing process of CFRP reflector has been simulated using ABAQUS along with COMPRO plug-in [12]. The paraboloid shape tool and reflector have been modelled using solid modelling software and imported in ABAQUS. The tool has been partitioned along two perpendicular planes passing through the centre of the model. The total of eight partitions has been created for the reflector in ABAQUS modeller. Each edge has been divided into five segments at a radius of 25, 50, 75, and 100 mm for the reflector. This will be facilitated for node generation at each junction and is further used to measure distortion at these locations. The thickness of the reflector should be modelled according to a number of prepreg layers as per thickness of each layer. The tool and reflector have been discretized using 3D Stress

20 node quadratic C3D20 elements. The meshed model for tool and reflector is shown in Fig. 55.5a, b, respectively. The stainless steel and carbon fibre prepreg material have been assigned for tool and reflector, respectively. These both materials have been imported using COMPRO plug-in. The through-thickness direction and in-plane direction has been assigned to the reflector as per stacking direction and fiber orientation to perform simulation using COMPRO. Assembly of tool and reflector has been performed by creating individual instances in the assembly module in ABAQUS. Frictional properties have been assigned to the interaction surfaces of tool and reflector. The tool top and reflector bottom surface have been selected as master and slave surface, respectively. The friction coefficient and limiting shear stress have been considered as 0.15 and 140,000 N/mm^{2,} respectively. Assembly of tool and reflector has been assigned an initial temperature of 20 °C. The fix boundary constraints have been applied for tool and reflector to restrict their free expansion during the cure cycle in *X*-, *Y*-, and *Z*-directions.

Thermochemical analysis has been performed using COMPRO plug-in. The heat transfer coefficient of 80 W/m^{2}k has been applied on all exterior surfaces of reflector and tool. The manufacturer recommended cure cycle has been given as temperature-time data for autoclave processing. The temperature profile and degree of cure plot have been obtained as a result of thermochemical analysis. The demoulding step has been defined to get the deformed shape of the reflector. The tool has been deactivated using model change command in the interaction module in tool removal step. The proper boundary conditions have been applied to the reflector during tool removal step. The residual stress profile and deformed shape have been obtained as a result of the stress-deformation analysis.

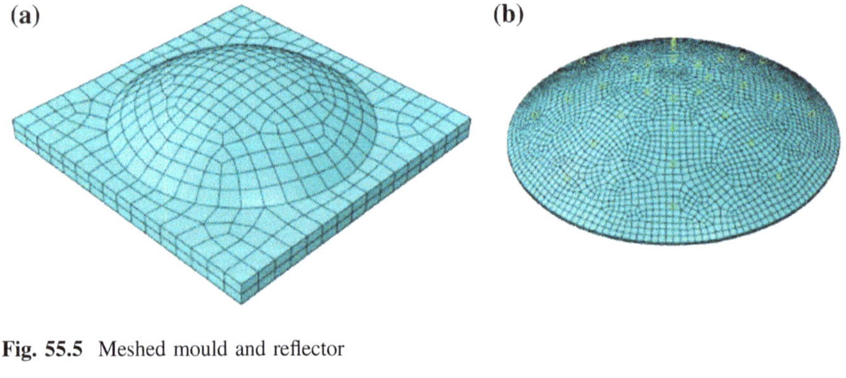

Fig. 55.5 Meshed mould and reflector

55.4 Results and Discussion

55.4.1 Distortion Measurement in Experimental Approach

The measurement of the specified coordinate points on the reflector surface has been performed using 3D scanning to confirm the accuracy of the parabolic shape obtained after curing. This has been accomplished by comparing the theoretical CAD model with 3D scan data. The 200 coordinate points have been obtained by projecting circles of diameter 50, 100, 150, 200, and 240 mm on the parabolic surface of the reflector. The deviation for scan data of reflector and CAD model has been measured for each coordinate point, and the same has been shown as a contour plot in Fig. 55.6. The distortion for each fabricated reflector has been calculated using root mean square (RMS) of deviations for all coordinate points. The predefined coordinate points on the CAD model and scanned data points on reflector along with deviation are given in Table 55.1. The final distortion, i.e., RMS value has been obtained as 0.37 mm.

55.4.2 Distortion Measurement in Numerical Approach

The deformation for total 41 coordinate points at the junction of division and partition on edge has been obtained as shown in Fig. 55.7. The original and deformed shape after demoulding for reflector is shown in Fig. 55.8. The deviation between coordinate points of original and deformed shape reflector are given in Table 55.2. The distortion for each fabricated reflector has been calculated using root mean square (RMS) of deviations for all coordinate points. The final distortion, i.e., RMS value has been obtained as 0.43 mm.

Fig. 55.6 Deviation of individual predefine coordinate

Table 55.1 Deviation of each coordinate point in autoclave curing process

Sr. no.	Predefined coordinate (Nominal) points (mm)			Scanned coordinate points (mm)			Deviation (mm)
1	0	−2.003	25	0	−2.051	24.992	0.04866
2	−3.911	−2.003	24.692	−3.911	−2.007	24.691	0.00412
3	−7.725	−2.003	23.776	−7.726	−1.978	23.78	0.02534
4	−11.35	−2.003	22.275	−11.353	−1.957	22.281	0.04649
5	−14.694	−2.003	20.225	−14.7	−1.947	20.232	0.05675
6	−17.677	−2.003	17.677	−17.683	−1.948	17.683	0.05565
.							
.							
.							
200	18.772	−46.083	118.521	18.728	−46.436	118.253	0.44524
Min	−120	−46.084	−120	−120.303	−47.054	−119.51	0.001
Max	120	0	120	120.454	0.0057	119.784	1.225
RMS							0.37

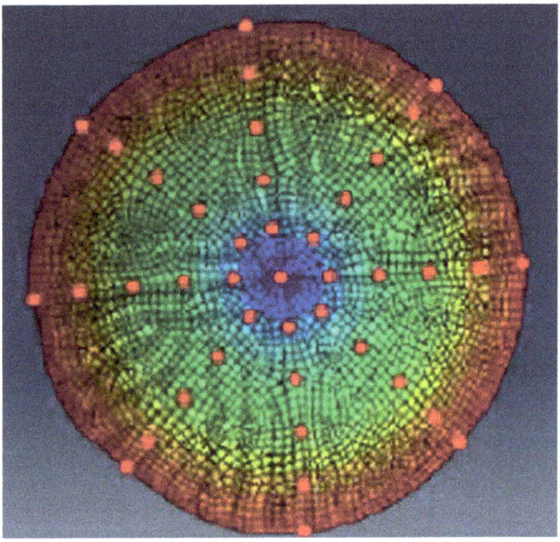

Fig. 55.7 Deformed coordinates of reflector

55.4.3 Comparison of Experimental and Numerical Results

Distortion of CFRP antenna reflectors has been determined using experimental and numerical approaches as given in Tables 55.1 and 55.2, respectively. Numerical results also compare quite well with experimental results. It has been shown that the difference between them in the range of 5–14%. The similar procedure has been repeated for different laminate thicknesses, i.e., 0.8, 1.6, 2, and 2.4 mm using both

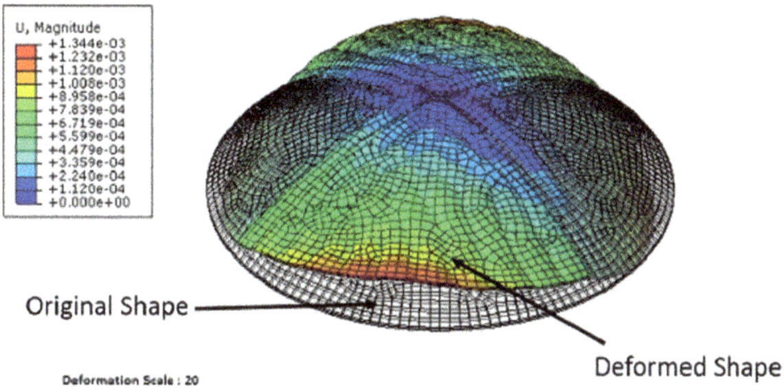

Fig. 55.8 Original and deformed shape of reflector

Table 55.2 Deviation of each coordinate point in the simulation process

Sr. no.	Predefined coordinate (Nominal) points (mm)			Deformed coordinate points (mm)			Deviation (mm)
1	0	0	0	0	0	0	0
2	25	−2	0	25.007	1.91	0.082	0.121
3	50	−8	0	50.010	7.91	0.160	0.187
4	75	−18	0	74.978	17.98	0.234	0.235
5	100	−32	0	99.885	32.15	0.302	0.356
6	120	−48	0	119.759	46.41	0.354	0.539
.							
.							
.							
41	84.8528	−48	−84.853	85.211	45.50	−84.290	0.886
Min	−120	−48	−120	−119.398	−47.39	−119.76	0
Max	120	0	120	119.759	0	119.648	1.225
RMS							0.43

approaches. The measured deformation using experimental and numerical approaches has been found as a good agreement with each other. However, for lower thickness components, the percentage difference is higher compared to high thickness components. The distortion for the corresponding reflector has been determined and compared with each other as shown in Fig. 55.9. The distortion decreases as laminate thickness increases due to the transverse contraction in the composite part. The close examination of the predicted final reflectors shape reveals that non-uniform thermal and cure shrinkage strains due to the coefficient of thermal expansion and coefficient of cure shrinkage anisotropies would result in changes in the shape of geometry. A sensitivity analysis revealed that the part thickness variation is the most important source of the predicted distortion.

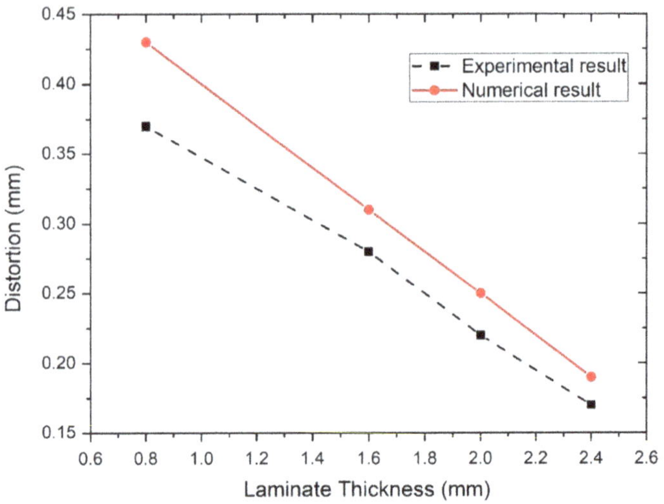

Fig. 55.9 Comparison of distortion for experimental and numerical result

55.5 Conclusions

Carbon fibre-reinforced polymer (CFRP) antenna reflectors have been fabricated using autoclave processing. The HCU200/A45 prepreg and stainless steel AISI 430 as a part and tool material, respectively, have been used in this study. The deviation in reflectors has been measured using 3D scanning technique. The geometric modelling and finite element simulation have been performed using ABAQUS software along with COMPRO plug-in. Thermochemical and stress-deformation analysis has been executed to obtain the distortion of the antenna reflectors. The distortion for reflectors has been determined using both approaches. The laminate thickness has been considered as a process parameter. The distortion decreases as laminate thickness increases. The measured deformation using experimental and numerical approach has been compared and found within 5–15% range.

Acknowledgements The authors express their gratefulness to the Convergent Manufacturing Technologies, CANADA for providing COMPRO software as well as for their technical support and help.

References

1. Hahn, H.T., Pagano, N.J.: Curing stresses in composite laminates. J. Compos. Mater. (1975)
2. Darrow, D.A., Smith, L.V.: Isolating components of processing induced warpage in laminated composites. J. Compos. Mater. (2002)

3. Twigg, G., Poursartip, A., Fernlund, G.: Tool–part interaction in composites processing. Part I: experimental investigation and analytical model. Compos. A **35**, 121–133 (2004)
4. Stefaniak, D., Kappel, E., Spröwitz, T., Hühne, C.: Experimental identification of process parameters inducing warpage of autoclave-processed CFRP parts. Compos. A **43**, 1081–1091 (2012)
5. Kaushik, V., Raghavan, J., Zeng, X.: Experimental study of tool–part interaction during autoclave processing of thermoset polymer composite structures. Compos. A **41**, 1210–1218 (2010)
6. Fernlund, G., Poursartip, A.: The effect of tooling material, cure cycle, and tool surface finish on spring-in of autoclave processed curved composite parts
7. Zeng, X., Raghavan, J.: Role of tool-part interaction in process-induced warpage of autoclave-manufactured composite structures. Compos. A **41**, 1174–1183 (2010)
8. Abusafieh, A., Federico, D., Connell, S., Cohen, E.: Dimensional stability of CFRP composite for space based reflectors
9. Dong, C.: Modeling the dimensional variations of composites using effective coefficients of thermal expansion. J. Compos. Mater. **43**(22) (2009)
10. Mezeix, L., Seman, A., Nasir, M.N., Aminanda, M., Rivai, Y., Castanie, A., Olivier, P., Ali, K.M.: Spring-back simulation of unidirectional carbon/epoxy flat laminate composite manufactured through autoclave process. Compos. Struct. **124**, 196–205 (2015)
11. Compro Model Documentation, Release 1.3. Convergent Manufacturing Technologies Inc. (2013)
12. COMPRO for ABAQUS Modelling Guidelines. Convergent Manufacturing Technologies Inc. (2014)

Chapter 56
Comparative Study of Hydroxyapatite Nanocomposites Reinforced with Zirconia and Titania Produced by Using HEBM

Vemulapalli Ajay Kumar, Penmetsa Rama Murty Raju, Nallu Ramanaiah and Siriyala Rajesh

Abstract Hydroxyapatite ($Ca_{10}(PO_4)_6(OH)_2$–HAp) conglomerate with various weight fractions (5, 10, 15, 20 and 25 wt%) is fabricated by reinforcing TiO_2 and $ZrO_2 + TiO_2$ nanoparticles (20–40 nm). In order to obtain homogeneous mixing of HAp, particles are reinforced using high-energy ball milling (HEBM) at 300 rpm for 1 h, compacted at 100 bar with holding time of 150 s and sintered at 1200 °C. X-ray diffraction (XRD) analysis, energy-dispersive spectroscopy (EDX) and microstructural analysis using field emission scanning electron microscopy (FESEM) have been carried out on the obtained samples. The mechanical characterization of the composites has been evaluated through flexural strength, compression strength, hardness, young modulus and fracture toughness. The mechanical properties are found to be improved with increasing content of TiO_2; however, with increasing content of $TiO_2 + ZrO_2$, mechanical properties are found to be improved significantly up to the addition of 20 wt % of reinforcement content. When 25 wt% of $TiO_2 + ZrO_2$ reinforcement is added to HAp, the mechanical properties are found to be decreased. The reduction could be due to the increase in dominant smaller particles of ZrO_2 and grain size. The improved mechanical properties are correlated with the observed microstructural features.

Keywords High-energy ball milling · ZrO_2 · TiO_2 · Mechanical properties

V. Ajay Kumar · P. Rama Murty Raju · S. Rajesh (✉)
Department of Mechanical Engineering, S.R.K.R. Engineering College, China Amiram, Bhimavaram 534204, India
e-mail: rajeshsiri.mech@gmail.com

V. Ajay Kumar
e-mail: ajayvijaya02@gmail.com

P. Rama Murty Raju
e-mail: dr.prmraju@gmail.com

N. Ramanaiah
Department of Mechanical Engineering, AU College of Engineering (A), Andhra University, Visakhapatnam 530003, India
e-mail: n.rchetty1@gmail.com

R. G. Narayanan et al. (eds.), *Advances in Computational Methods in Manufacturing*, Lecture Notes on Multidisciplinary Industrial Engineering,
https://doi.org/10.1007/978-981-32-9072-3_56

56.1 Introduction

The chemical composition or the structure similar to human hard tissues such as bone and teeth exhibits biocompatibility and bioactivity of hydroxyapatite (HAp). HAp, after implantation into human body as artificial material, can provide scaffolds for the formation of new bone tissues. It plays a role of osteoconduction because its calcium and phosphorous can be free from the material surface and absorption of the new tissues. In the field of medical sciences, it is considered to be a vital aspect with respect to human health, and the bioactive ceramics such as HAp are attractive candidates for body's hard tissues replacement [1–4] among different categories. The application of HAp has excellent biocompatibility properties, but the application is limited due to its low strength and brittle nature, but it has excellent biocompatibility [5]. Prokopiev et al. [6] reported microstructure and mechanical properties of sintered HAp on the sintering temperature at fixed compaction pressure (20 MPa) and also observed that Young's modulus and shear modulus (isotropic) increase with sintering temperature between 1200 and 1280 °C. Lee et al. [7] studied about Al_2O_3 and ZrO_2. Due to their excellent oxidation resistance, good biocompatibility and wear resistance, they are considered as a matrix as well as reinforcement phases. Que et al. [8] reported the major effect on the HAp structure and enhanced HAp properties on addition of titania to HAp. Due to the introduction of secondary phases, the phase changes in the composites at higher sintering temperatures are observed. Viswanath and Ravi Shankar [9] studied the interfacial reactions in hydroxyapatite/alumina nanocomposite in which alumina reacted completely with hydroxyapatite. Xihua et al. [10] reported the fabricating of hydroxyapatite matrix ceramic composites with low cost; alumina is introduced in hydroxyapatite.

Yu et al. [11] reported the implication of nano-ZrO_2 on the mechanical properties of reinforced HAp bone cement. Hongbo et al. [12] investigated first of its kind the fabrication of laminated and functionally graded HAp/Y-TZP composites by means of the spark plasma sintering (SPS). The functionally graded HAp/Y-TZP and laminated HAp/Y-TZP are designed strong Y-TZP core-bioactive HAp layer structure, a non-symmetric Y-TZP layer to HAp later structure or a bioactive HAp core-strong Y-TZ player structure. Aminzare et al. [13] studied the nanocomposite by HEBM in order to obtain highly dense objects with desired mechanical properties. The phase decomposition and mechanical properties have been investigated with the effects of alumina and titaniananoparticles on microstructure. Ahmed et al. [14] reported that the titanianano/hydroxyapatite composites produced by HEBM. The effect of addition of titania, sintering temperature, formed phases time and their influence regarding compressive strength of HAp were investigated. Khalil et al. [15] studied the mechanical and microstructure properties of hydroxyapatite and hydroxyapatite-yttria stabilized zirconia (3YSZ) with 20 vol%-(ZrO_2 + 3 mol% Y_2O_3) nanopowders developed by High Frequency Induction Heat Sintering (HFIHS) at different temperatures. The main objective of their study is to synthesize and then sinter the HAp–3YSZ powders with higher hardness, higher toughness, grain size in fine scale and constituent distribution homogeneously. Oktar [16] observed the influence of sintering process of the composite materials, made of hydroxyapatite, that were derived from the bovine bone, with 5 and 10 wt% of TiO_2, on their mechanical properties. The mechanical properties of the produced composite materials revealed the developed microstructure and the phases formed during sintering effecting the densification. Mobasherpour et al. [17] adopted a precipitation method to synthesize nanocrystalline HAp and investigated the effects of ZrO_2–Al_2O_3 on mechanical properties of nanocrystalline HAp powder. Improved bending and the tensile strength of HAp by using La_2O_3, ZrO_2, and TiO_2 as binders have been observed by Shin-Ike et al. [18].

In the present study, the application of HEBM technique is to produce HAp/TiO_2 and HAp/ZrO_2 + TiO_2 nanocomposites and the effect of the addition of TiO_2 and ZrO_2 + TiO_2 to HAp on the microstructure and mechanical strength of the composites was investigated. The correlation between the mechanical and microstructural properties was systematically established.

56.2 Materials and Methods

56.2.1 Sample Preparation

Hydroxyapatite (HAp) and its composite powders of particle size 20–40 nm with various wt% of TiO_2 and ZrO_2 + TiO_2 (0, 5, 10, 15, 20 and 25 wt%) were combined with HAp by HEBM. Powders were milled under dry condition using tungsten carbide hardened steel balls for 1 h with rotational speed 300 rpm. The balls to powders ratio considered were 20:1. The milling powder samples were compacted under pressure of 100 bar at room temperature for dwell time 150 s. The equipment used for compacting has two channels which can operate from 0.005 to 200 bars and can go from room temperature to 500 °C. These compacted powders are made into cylindrical specimens (compression strength, hardness) which comprises of 20 mm diameter and 5 mm height and a plain specimen (flexural strength and fracture toughness) about 40 mm length, 10 mm width and 3 mm thickness. To sum, sintering was performed in normal atmosphere using electric resistance box furnace at 1200 °C and the heating time 1 h. The specimens were fed into the furnace and heated up to desired temperature. The heating time was calculated from the point at which the desired temperature has been reached. All the sintered samples were furnace cooled.

56.2.2 Experimental Procedure

X-ray diffractometer (XRD) equipped with Cu–K_α radiation (wavelength λ = 1.5406 Å°) is used to distinguish HAp and its composites. This test is operated between 6 and 90° using a step size of 0.01° which follows the width (β) of their diffraction pattern; the average crystallite size (D) was estimated using Scherer's formula [14]. The microstructural information was known by field emission scanning electron microscopy (FESEM). The compositions of the elements Ca, P, O and Zr are evaluated using energy-dispersive analysis of X-rays (EDAX) attached with FESEM. The compressive strength of the nanocomposite ceramics of size 20 mm (diameter) and 5 mm (height) was conducted on UTM. The density of sinter samples was measured using distilled water by the Archimedes' method (relative density). This test was executed under a quasi-static loading range, where the initial strain rate (ε^*) is fixed at 0.001 s^{-1}; i.e., the machine crosshead speed ($\vartheta = \varepsilon * h_o$) is 0.005 mm/s. And h_o is the specimen's initial height, where the test is conducted on specimens which are loaded up to fracture [14].

Hardness test (Knoop hardness indenter) is evaluated on the basis of polished HAp composites which were analyzed using the Shimadzu HMV-2000 microhardness tester. The load on the specimen 4.905 N and holding time of 15 s are used in the measurements as per ASTM E384. The width-to-length ratio of the indenter is 1:7.11. The concavity or indention depth is around (1/30)th of the longer dimension, and its face angle for the rectilinear edge is 172° and for the breadth is 130°.

Knoop hardness 'H' can be determined by Eq. (56.1) as below [12]:

$$H = 14229 \frac{L}{d^2} \tag{56.1}$$

The load refers to 'L' (Newton), and the longer diagonal of the indentation refers to 'd' (mm).

Young's modulus along the composites was determined through the following Eq. (56.2) as given below [12]:

$$\frac{B'}{A'} = \frac{B}{A} - \alpha \frac{E}{H} \tag{56.2}$$

(B^1/A^1) gives the indent diagonal ratio, and (B/A) gives the ratio of the indenter dimensions. The term 'α' relates to a constant of a value of 0.45.

The three-point bending test is used to determine the flexural strength of the nanocomposite. The HAp composite specimens (40 mm length, 10 mm width and 3 mm thickness) were first prepared and refined to smooth out surface flaws before setting them on a three-point bending testing setup. The load (frame crosshead speed of 0.05 mm/min) was put from the top at middle of the bar. Through the indentation technique, the fracture toughness K_c (Eq. 56.3) was calculated, and the values were confirmed through the equation below [15]:

$$K_C = \xi \left[\frac{E}{H}\right]^{2/5} \frac{P}{d \times l^{1/2}} \tag{56.3}$$

'K_c' gives fracture toughness, 'E' refers Young's modulus (GPa), and 'H' gives the hardness (GPa). 'P' gives the applied load (N); 'ξ' shows the constant which is based on the geometry of the indenter ($\xi = 0.008$).

56.3 Results and Discussion

56.3.1 Microscopic Analysis

The morphologies of HAp and its composites are observed by a FESEM. HAp forms a crosslink with one and other during the growth process. This provides a conducive condition for the adhesion and growth of biological tissues. Figure 56.1 gives the XRD patterns of HAp and standard HAp (PdF card no. 74-0566). The diffraction pattern of HAp powder used for the current study is in good agreement with standard HAp pattern, and it is also observed that there are no reflections from impurities.

XRD patterns were used for the analysis of specimens which had undergone the process of pressing and then sintering at a temperature of 1200 °C. Figure 56.2 shows XRD patterns of HAp and its composites (HAp/TiO_2). When the TiO_2 powder was added, stronger peaks are detected along with HAp and TiO_2 when the compacting and the sintering processes were carried out under the same condition. With increase in TiO_2 (indicated with '$') content along with the HAp (indicated with '#'), TiO_2 phase reflections are present. Using Scherer's formula, all the XRD

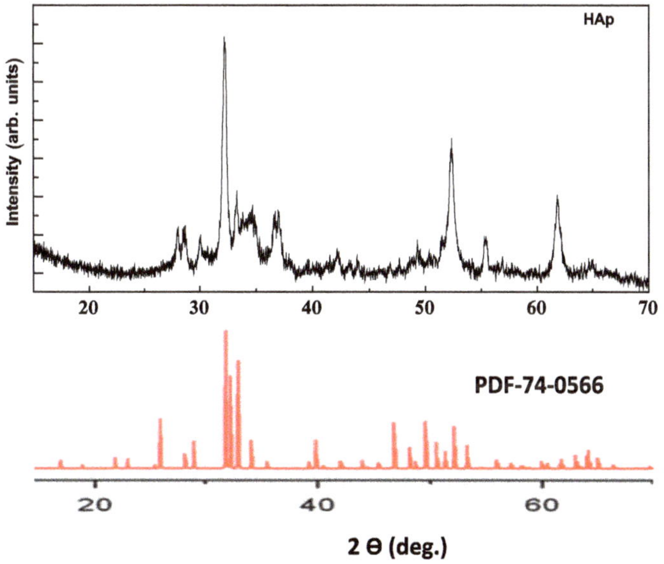

Fig. 56.1 XRD pattern of HAp sample and standard HAp

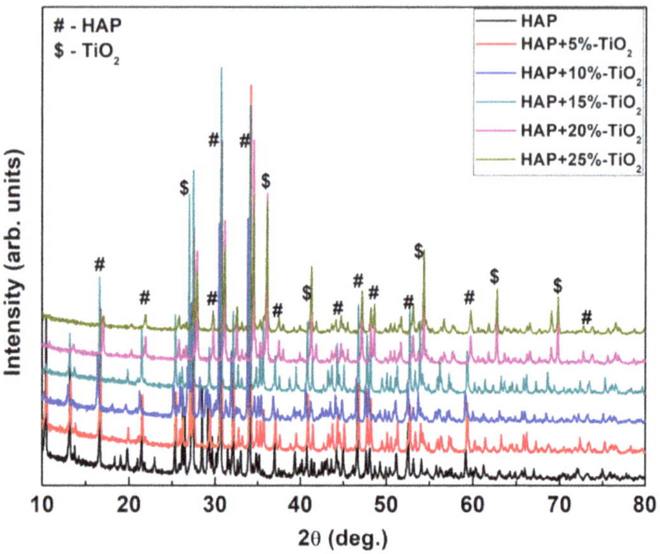

Fig. 56.2 XRD patterns of HAP/ZrO_2 composites after sintering

patterns crystallite sizes are calculated and observed, and they are in the range of 40.01–16.11 nm and 47.39–42.39 nm for HAp/TiO_2 and HAp/(TiO_2 + ZrO_2) composites, respectively. This shows clearly that there is a gradual decrease in the crystallite size when there is addition of TiO_2 in the process. It is analyzed with the increase of TiO_2 content; there will be increase in the reflections of TiO_2 phase. TiO_2 structural stability was studied with the improved TiO_2 content which in turn increased by intensities of diffraction patterns. There were no additional reflections due to the impurities.

Figure 56.3 depicts XRD patterns of HAp and its composites (TiO_2 + ZrO_2). The reflections observed were matched with the parent phase (HAp-marked with '#') and its composites (TiO_2 marked with '$' + ZrO_2 marked with '@'). It is observed that due to the impurities, there were no additional reflections. It is observed that ZrO_2 + TiO_2 attained more structural stability with increase in the diffraction patterns intensities due to the increasing of ZrO_2 + TiO_2 content. At 25 wt% of composite (TiO_2 + ZrO_2), strong tricalcium phosphate (TCP) peaks are detected along with zirconia and HAp. This formation of TCP may be attributed due to excessive reaction between the HAp and ZrO_2 with the improved wt% of ZrO_2 particles. The TCP structure is a typical biomaterial with excellent biocompatibility and osteoconductivity, by which the formation of new bone tissue is accelerated when these materials are implanted into the body which is observed by the Colon et al. [19]. TCP is fragile bioceramic and consequently cannot be used on their own as major load-bearing implants in the human body. From the identified phases as shown in Fig. 56.3, it can be seen that ZrO_2 has shown a dominant phase

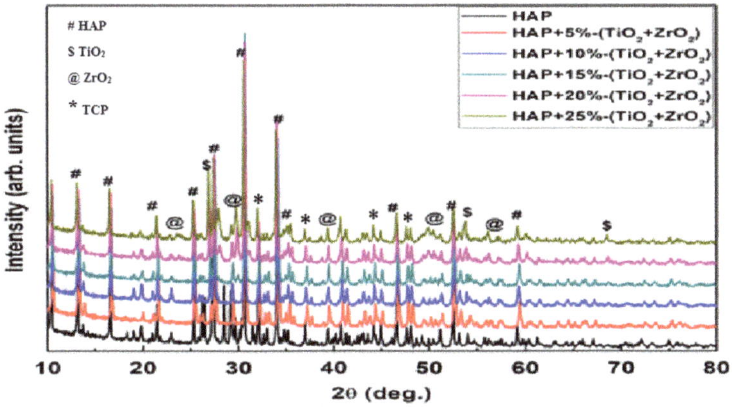

Fig. 56.3 XRD patterns of HAp/ZrO_2 + TiO_2 composites after sintering

over TiO_2 on addition of ZrO_2 + TiO_2 which is in good agreement with the studies conducted by Miao et al. [20].

Field emission scanning electron microscopy (FESEM) is used to analyze the morphology of the powders (HAp/5, 10, 15, 20 and 25 wt% TiO_2 and TiO_2 + ZrO_2) after HEBM. The morphology of the initial large particles is significantly changed due to fracture and de-agglomeration processes. After 1 h of HEB milling of the HAp/TiO_2 powders, sporadic-shaped morphologies are observed. These small particles were also observed after the cold welding phenomenon during ball milling.

FESEM micrographs of the HAp/TiO_2 composites which had undergone compacting and sintering at a temperature of 1200 °C with different wt% are shown in Fig. 56.4a–e. Based on the micrographs, it is observed that there is a considerable change in morphology due to the addition of TiO_2. This addition of different wt% may induce the nucleation which tends to the formation of leaf-like structures. The average size of leaf-like structure is observed around 18 μm, whereas a needle- and irregular shaped morphology of TiO_2 is observed for 25 wt% HAp/TiO_2 composite. Hence, it may be concluded that as the wt% of TiO_2 increases in the composite, the agglomerated irregular shape morphology is observed.

FESEM micrographs of the HAp/TiO_2 + ZrO_2 composites with different wt% are shown in Fig. 56.5a–e. A unique growth of rounded particle with small in size was exhibited by the composites with ZrO_2 + TiO_2 contents at various wt%. In Fig. 56.5a–e, the matrix is shown with smaller particles of ZrO_2 + TiO_2. From the micrographs, it is evident that there may be alteration in the composites microstructural properties due to the increase of ZrO_2 + TiO_2 content and homogeneous distribution of particles inside the grains which enhances mechanical properties. This may be attributed due to the interlocking of particles with the grain boundaries uniformly which can be observed from Fig. 56.5d and e.

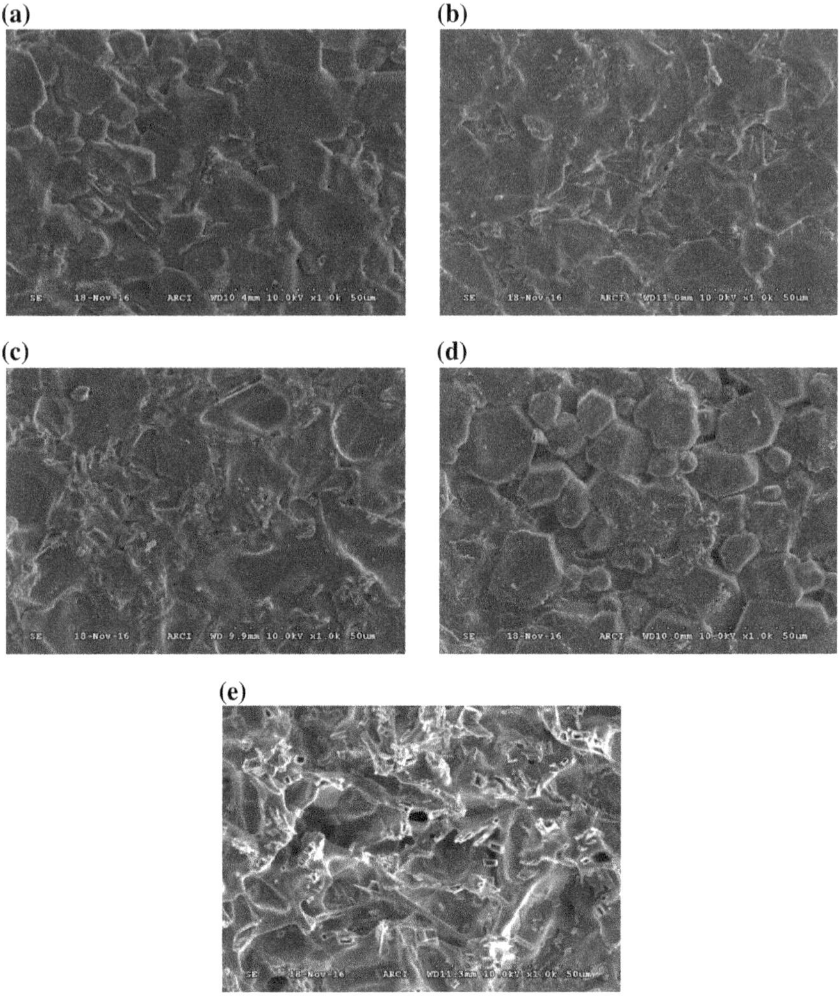

Fig. 56.4 FESEM micrographs of HAp/TiO_2 composites: **a** 5 wt%, **b** 10 wt%, **c** 15 wt%, **d** 20 wt% and **e** 25 wt%

Figure 56.6a shows energy-dispersive X-ray spectroscopy (EDX) analysis of HAp/TiO_2 composite. The presence Ca, P and Ti can be seen from the figure. Figure 56.6b shows the EDX analysis of HAp/TiO_2 + ZrO_2 composite. The EDX shows that Ca, P, Zr and Ti can be seen from Fig. 56.6b, whereas background grains exhibited strong Zr intensity but weak Ti intensity. The EDX results show that samples were free of metallic contaminants. However, contaminants could be found in very little percentages which are lower than the accuracy of the applied instrument.

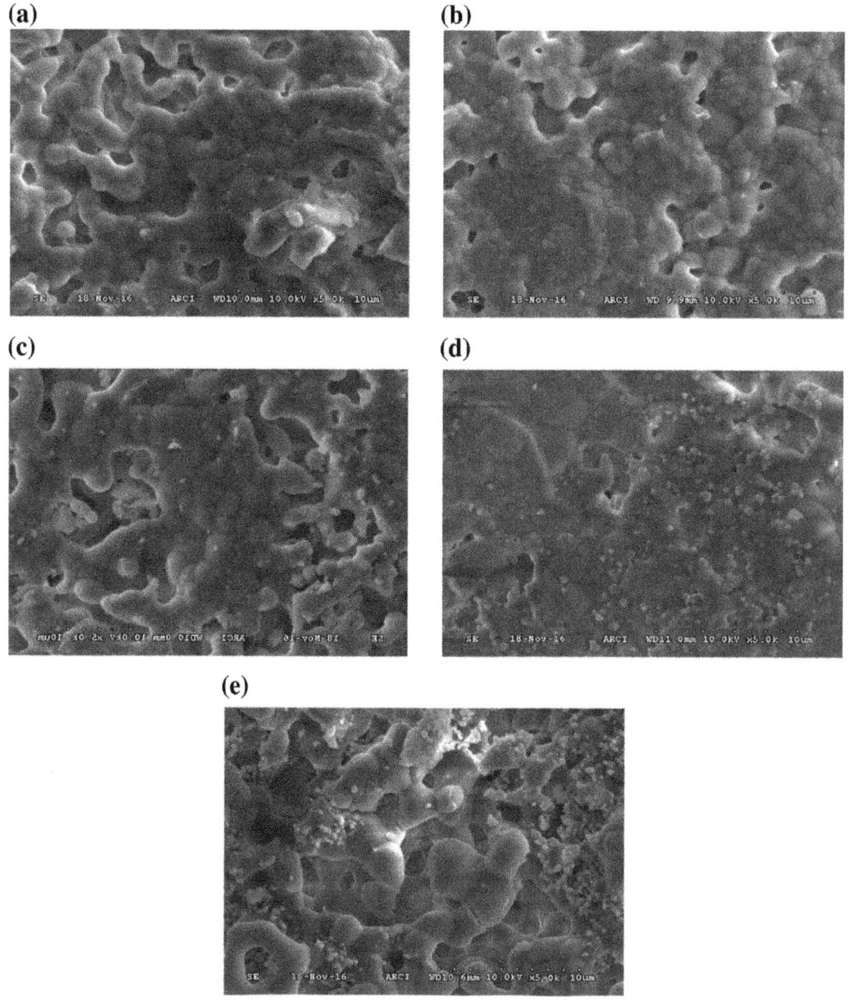

Fig. 56.5 FE-SEM micrographs of HAp/TiO_2 + ZrO_2 composites: **a** 5 wt%, **b** 10 wt%, **c** 15 wt%, **d** 20 wt% and **e** 25 wt%

Figure 56.7. shows the grain size of the HAp composites estimated from the ASTM line method [21] The grain size HAp/TiO_2 composite is observed to decrease gradually from 52 nm (5 wt% of TiO_2) to 20.1 nm with increase in TiO_2 content up to 25 wt%, whereas in HAp/ZrO_2 + TiO_2 composites, it is observed that the grain size is in the range of 42.1–19.09 nm.

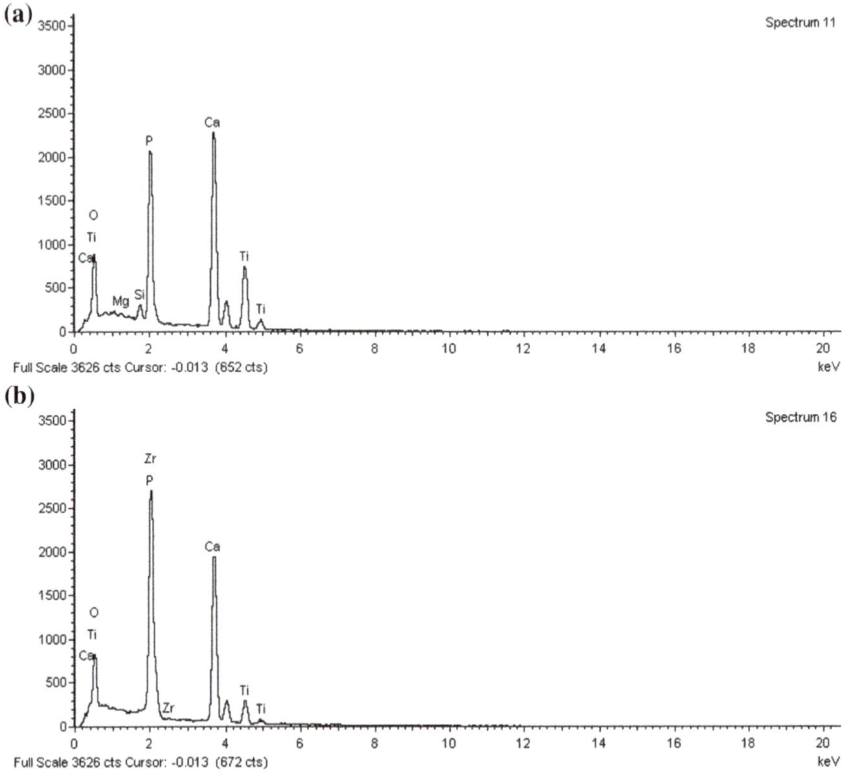

Fig. 56.6 EDAX composition of **a** HAp/25 wt% TiO_2 and **b** HAp/25 wt% ZrO_2 + TiO_2

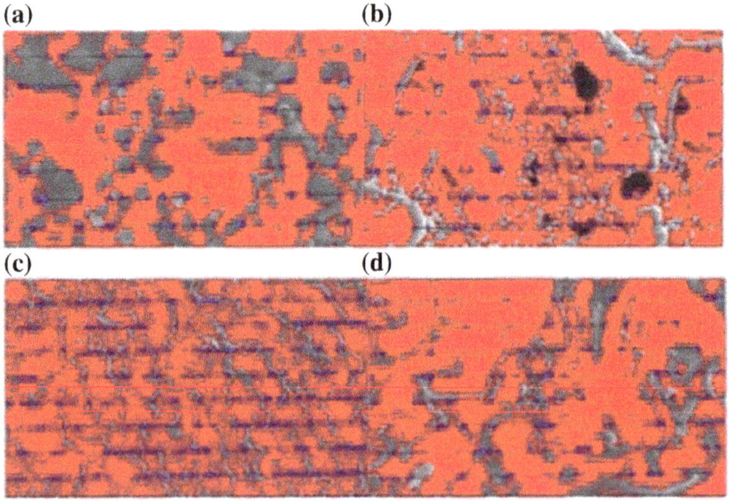

Fig. 56.7 Grain size of the composites (ASTM line method): **a** HAp + 10 wt% TiO_2, **b** HAp + 25 wt% TiO_2, **c** HAp/10 wt% ZrO_2 + TiO_2, **d** HAp/25 wt% ZrO_2 + TiO_2

56.3.2 Mechanical Properties

Based on the structural and morphological analysis, an attempt has been made to understand the formation mechanism of the composites TiO_2 and $ZrO_2 + TiO_2$. The measured density of HAp/TiO_2 and HAp/$ZrO_2 + TiO_2$ composites is shown in Fig. 56.8.

Calculated densities of HAp/TiO_2 and HAp/$TiO_2 + ZrO_2$ (5, 10, 15, 20 and 25 wt%) samples sintered at 1200 °C for 1 h are in close agreement with theoretical density. The density is found to be increasing gradually with TiO_2 content. With increase in TiO_2 content, it can be attributed that the grain size of HAp/TiO_2 composites decreases. The maximum density for HAp/$TiO_2 + ZrO_2$ composites has been obtained at 20 wt% reinforcement content. When 25 wt% of reinforcement is added to HAp, the density is found to be decreased, due to the background grains with strong Zr intensity but weak Ti intensity [20] of HAp nanocomposites, Fig. 56.5. Khalil et al. [15] has observed HAp and HAp-20 vol% 3YSZ composites at different sintered temperatures by HFIHS process.

Figure 56.9 shows the variation of hardness with reference to wt% of HAp composites. The effect of reinforcement on hardness of HAp composites can be studied as a function of density and grain size.

The increment of hardness is observed from 6.01 to 10.03 GPa. With the improvement in the TiO_2 content, the hardness of the HAp composites has been increased; at that time, grain size was observed a gradual decrease, whereas the density and the hardness increased. Among all the HAp composites (5, 10, 15, 20 and 25 wt%), maximum hardness value is 10.03 GPa, observed for the 25 wt% TiO_2 content. However, for HAp/$ZrO_2 + TiO_2$ composites, the hardness is increased from 5.56 to 14.01 GPa, with increasing content of $ZrO_2 + TiO_2$ from 5 to 20 wt%. The highest hardness (14.01 GPa) has been observed at 20 wt% $ZrO_2 + TiO_2$ content. Interestingly, a drop in hardness (9.98 GPa) has been observed for 25 wt% $ZrO_2 + TiO_2$ content. The similar work was conducted by Guo et al. [12] is observed to be in good agreement with Knoop hardness values (7.5 GPa at 20 wt% of Y-TZP) observed.

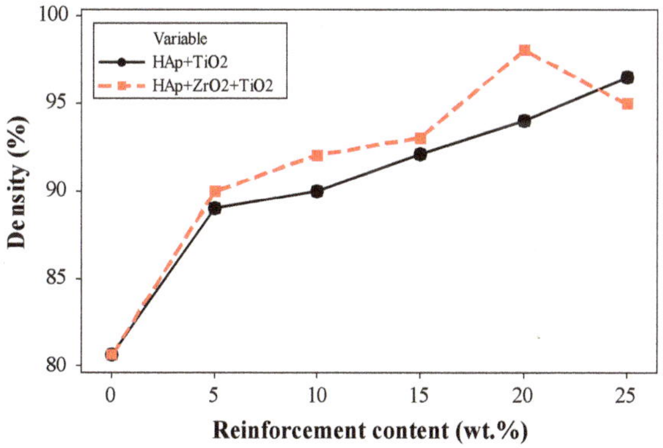

Fig. 56.8 Variation of density in HAp/TiO_2 and HAp/$ZrO_2 + TiO_2$ composites

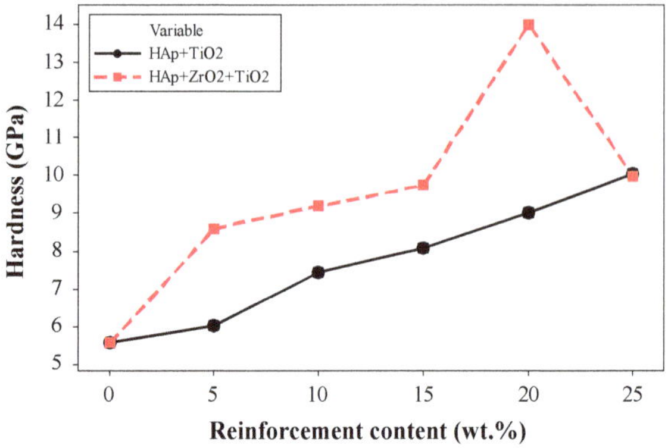

Fig. 56.9 Variation of hardness in HAp/TiO_2 and HAp/ZrO_2 + TiO_2 composites

Figure 56.10 shows Young's modulus of HAp/TiO_2 and HAp/ZrO_2 + TiO_2 sintered composite samples. With increase in wt% of TiO_2 content, Young's modulus of the HAp/TiO_2 composites is increasing from 91.2 to 128.5 GPa. The highest value of Young's modulus of 128.5 GPa has been obtained for 25 wt% of TiO_2 composite, whereas for HAP/ZrO_2 + TiO_2 composites, Young's modulus of the HAp composites was found to be increasing from 89.09 to 328.08 GPa. Young's modulus of sintered HAp composites samples increases with increasing ZrO_2 + TiO_2 content. For the HAP/ZrO_2 + TiO_2 composites, Young's modulus reaches a maximum at 20 wt%, reinforcement content followed by a slight decrease for 25 wt% composite. With increase in reinforcement content, grain size decreases and density increases; as a result, hardness is improved. The effect of the ceramic as additives is to improve the density and young's modulus values [8]. Due to the same reason, Young's modulus has also been improved with increase in reinforcement content.

Figure 56.11 shows the fracture toughness trend for HAp/TiO_2 and HAp/ZrO_2 + TiO_2 sintered composites with varying wt% of reinforcement. The fracture toughness is found to be varying from 0.742 to 3.05 MPa $m^{1/2}$. The highest value of fracture toughness of 3.05 MPa $m^{1/2}$ has been obtained for 25 wt% of TiO_2

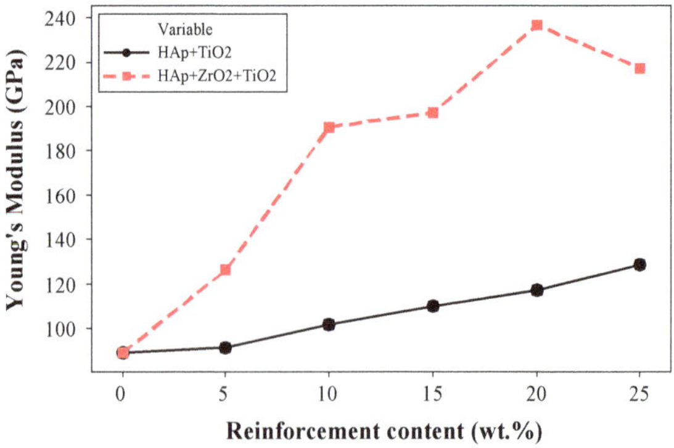

Fig. 56.10 Variation of Young's modulus in HAp/TiO_2 and HAp/ZrO_2 + TiO_2 composites

Fig. 56.11 Variation of fracture toughness in HAp/TiO_2 and HAp/ZrO_2 + TiO_2 composites

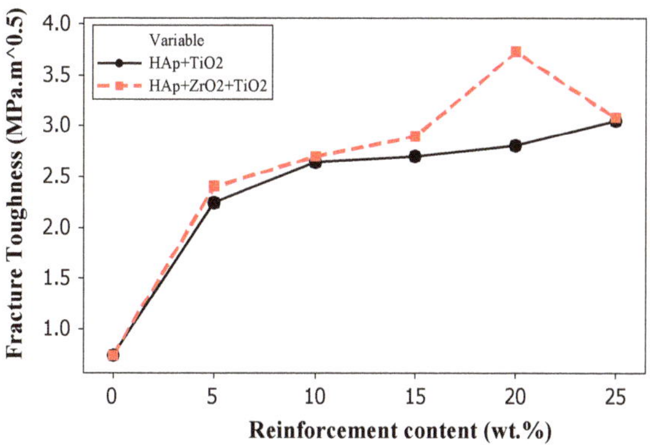

composite, whereas for HAp/ZrO_2 + TiO_2 composite, the values vary from 0.742 to 3.73 MPa $m^{1/2}$ and maximum fracture toughness 3.73 MPa $m^{1/2}$ has been obtained at 20 wt% of composite. Upon further addition of reinforcement content, i.e. 25 wt%, there is a drop in fracture toughness. It is true that hardness and Young's modulus depend on the density and grain size of composite. Further, it can also be concluded that fracture toughness in turn depends on hardness and Young's modulus.

Figure 56.12 shows the flexural strength of HAp/TiO_2 and HAp/TiO_2 + TiO_2 composites. With increasing wt% of TiO_2, the flexural strength varies from 50–206 MPa for HAp/TiO_2 composite. The highest value of fracture toughness of 206 MPa has been obtained for 25 wt% of TiO_2 composite, whereas for HAp/ZrO_2 + TiO_2 composites, the flexural strength varies from 50 to 236.8 MPa, and the maximum flexural strength is observed at 20 wt% of HAp/ZrO_2 + TiO_2 composite after which there is a dropin flexural strength for 25 wt% of reinforcement. The sintered samples with higher density improved mechanical properties, flexural strength, compression strength and hardness. Oktar [16] observed same results for bending and hardness at different temperatures, strength 86.2 MPa and hardness 143.5 HV (HAp-TiO_2 5 wt% at 1200 °C).

Fig. 56.12 Variation of flexural strength HAp/TiO_2 and HAp/ZrO_2 + TiO_2 composites

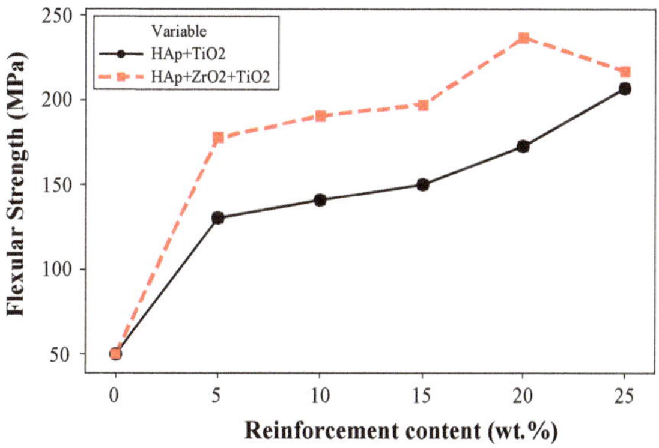

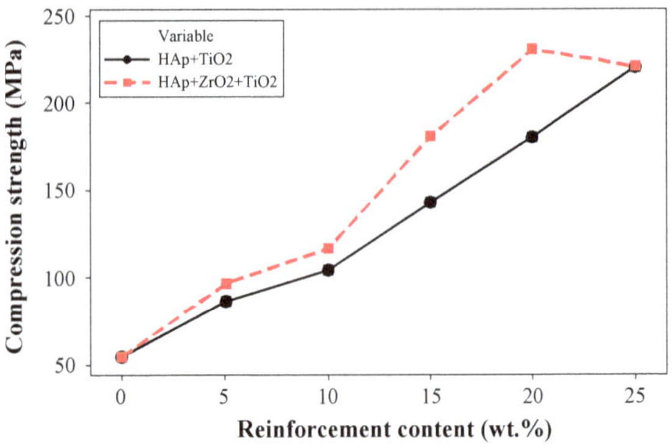

Fig. 56.13 Variation of compressive strength for HAp/TiO_2 and $HAp/ZrO_2 + TiO_2$ composites

Figure 56.13 can be observed that for HAp/TiO_2 composite, the compression strength varies from 50 to 220.06 MPa with increasing wt% of TiO_2. The maximum compression strength of 220.6 MPa is obtained at 25 wt% TiO_2, whereas the $HAp/(ZrO_2 + TiO_2)$ composite has compression strength values in the range of 50–230.67 MPa. The maximum compressive strength of 230.67 MPa was obtained for 20 wt%, after which it shows a drop in compression strength. The reason can be attributed to the increase in the dispersion of reinforcement particles in the matrix phase, Figs. 56.4 and 56.5. Que et al. [8] observed the hardness and young's modulus of the pure HAp samples and the HAp/TiO_2 (10 mol%) composites as a function of the sintering temperature. The highest hardness and Young's modulus, which are 2.81 and 45.33 GPa, respectively, are obtained at the sintering temperature of 1100 °C. Salman et al. [22] presented the fabrication and characterization of composite materials of hydroxyapatite and Ti. Hydroxyapatite (HA) powder was obtained from bovine bones (BHA) and human enamel (EHA) via calcination technique. Powder-compacts were sintered at different temperatures between 1000 and 1300 °C. The best mechanical properties for BHA–Ti composites were obtained after sintering in the range of 1200–1300 °C and for EHA–Ti composites in the range of 1100–1300 °C.

56.4 Conclusions

The HAp/TiO_2 and $HAp/ZrO_2 + TiO_2$ composites are successfully prepared using HEBM. All the mechanical properties are investigated with HAp and its composites. It is observed that with increase in reinforced content wt% of composite the mechanical properties are found to be improved.

The following major observations can be drawn from the present investigation:

i. All the composites of HAp and TiO_2 possess hardness values of 5.56–10.3 GPa, and the hardness values of $HAp/ZrO_2 + TiO_2$ composites lie in the range of 5.56–14.01 GPa.

ii. Young's modulus of the sintered HAp/TiO_2 composites varied from 89.09 to 128.5 GPa. Young's modulus of the sintered HAp/ZrO_2 + TiO_2 composites varied from 89.09 to 328.08 GPa which are high compared to the pure HAp.
iii. The composites of HAp and TiO_2 fracture toughness are found to be in the range of 0.742–3.05 MPa $m^{1/2}$, and HAp/ZrO_2 + TiO_2 composites fracture toughness was found to be in the range of 0.742–3.73 MPa $m^{1/2}$. Comparing the hardness, young's modulus and toughness of pure HAp, HAp/TiO_2 and HAp/(ZrO_2 + TiO_2), composites with 20 wt% are found to yield better.
iv. The flexural strength of HAp/TiO_2 composites increased from 50 to 206 MPa, whereas the flexural strength of HAp/ZrO_2 + TiO_2 composites was found to increase from 50 to 236.8 MPa.
v. The compressive strength of HAp/TiO_2 composites for 25 wt% addition is 220.67 MPa significantly higher than pure HAp, whereas the compressive strength of the 20 wt% HAp/ZrO_2 + TiO_2 composites is 230.67 MPa which is higher than pure HAp.
vi. By adding ZrO_2 + TiO_2, all the mechanical properties have been enhanced. HAp/ZrO_2 + TiO_2 composites had better mechanical properties than that HAp/TiO_2. The investigated biocomposite materials can be used for implant applications.

References

1. Pushpakanth, S., Srinivasan, B., Sreedhar, B., Sastry, T.P.: An in situ approach to prepare nanorods of titania-hydroxyapatite(TiO2-HAp) nanocomposite by micro wave hydrothermal technique. Mater. Chem. Phys. **107**(2–3), 492–498 (2008)
2. Xihua, Z., Changxia, L., Musen, L., Yunqiang, B., Junlong, S.: Fabrication of hydroxyapatite/diopside/alumina composites by hot press sintering process. Ceram. Int. **35**(5), 1969–1973 (2008)
3. Miao, X., Hu, Y., Liu, J., Huang, X.: Hydroxyapatite coating on porous zirconia. Mater. Sci. Eng., C **27**(2), 257–261 (2007)
4. Nath, S., Biswas, K., Basu, B.: Phase stability and micro structure development in hydroxyapatite-mullite system. Scr. Mater. **58**(12), 1054–1057 (2008)
5. He, L.-H., Standard, O.C., Huang, T.T.Y., Latella, B.A., Swain, M.V.: Mechanical behavior of porous hydroxyapatite. Acta Biomater. **4**(3), 577–586 (2008)
6. Prokopiev, O., Sevostianov, I.: Dependence of mechanical properties of sintered hydroxyapatite on the sintering temperature. Mater. Sci. Eng., A **431**(1–2), 218–227 (2006)
7. Lee, B.T., Jang, D.H., Kang, I.-C., Lee, C.-W.: Relation between microstructures and material properties of novel fibrous Al_2O_3-(m-ZrO_2)/t-ZrO_2 composites. J. Am. Ceram. Soc. **88**(10), 28974–2878 (2005)
8. Que, W., Khor, K.A., Xu, J.L., Yu, L.G.: Hydroxyapatite/titaniana—nanocomposites derived by combining high-energy ball milling with spark plasma sintering processes. J. Eur. Ceram. Soc. **28**(16), 3083–3090 (2008)
9. Viswanath, B., Ravishankar, N.: Interfacial reactions in hydroxyapatite/alumina nanocomposites. Scripta Mater. **55**(10), 863–866 (2006)

10. Xihua, Z., Changxia, L., Musen, L., Yunqiang, B., Junlong, S.: Fabrication of hydroxyapatite/dropsied/alumina composites by hot-press sintering process. Ceram. Int. **35**(5), 1969–1973 (2009)
11. Yu, W., Wang, X., Zhao, J., Tang, Q., Wang, M., Ning, X.: Preparation and mechanical properties of reinforced hydroxyapatite bone cement with nano-ZrO_2. Ceram. Int. **41**(9-A), 10600–10606 (2015)
12. Guo, H., AikKhor, K., Boey, Y.C., Miao, X.: Laminated and functionally graded hydroxyapatite/yttria stabilized tetragonal zirconia composites fabricated by spark plasma sintering. Biomaterials **24**(4), 667–675 (2003)
13. Aminzare, M., Eskandari, A., Baroonian, M.H., Berenov, A., Razavi Hesabi, Z., Taheri, M., Sadmezhaad, S.K.: Hydroxyapatite nanocompatites: synthesis, sintering and mechanical properties. Ceram. Int. **39**(3), 2197–2206 (2013)
14. Hannora, A.E., Ataya, S.: Structure and compression strength of hydroxyapatite/titania nanocomposites formed by high energy ball milling. J. Alloy. Compd. **658**, 222–233 (2016)
15. Khalil, K.A., Kim, S.W., Kim, H.Y.: Consolidation and mechanical properties of nanostructured hydroxyapatite-(ZrO_2 + 3 mol% Y_2O_3) bioceramics by high-frequency induction heat sintering. J. Mater. Sci. Eng. A **456**(1–2), 368–372 (2007)
16. Oktar, F.N.: Hydroxyapatite—TiO_2 composites. Mater. Lett. **60**(17–18), 2207–2210 (2006)
17. Mobasherpour, I., Solati Hashjin, M., Razavi Toosi, S.S., Darvishi Kamachali, R.: Effect of the addition ZrO_2-Al_2O_3 on nanocrystalline hydroxyapatite bending strength and fracture toughness. Ceram. Int. **35**(4), 1569–1574 (2009)
18. Shin-Ike, M., Tsutsui, J., Tanaka, A., Murayama, S., Fujita, A.: Attempts to improve the strength of sintered lanthanum-containing hydroxyapatite. J. Osaka Odontol. Soc. **52**, 854–861 (1989)
19. Colon, G., Aviles, M.A., Navio, J.A., Sanchez-Soto, P.J.: Thermal behaviour of a TiO_2–ZrO_2 microcomposite prepared by chemical coating. J. Therm. Anal. Calorim. **67**(1), 229–238 (2002)
20. Miao, X., Sun, D., Hoo, P.W., Lu, J., Hu, Y., Chen, Y.: Effect of titania addition on yttria-stabilised tetragonal zirconia ceramics sintered at high temperatures. Ceram. Int. **30**(6), 1041–1047 (2004)
21. Standard Test Methods for Determining Average Grain size Designation: E112–96 (ASTM International)
22. Salman, S., Gunduz, O., Yilmaz, S., Ovecoglu, Robert, M.L., Snyder, S. Agathapoulos, Oktar, F.N.: Sintering effect on mechanical properties of composites of natural hydroxyapatites and titanium. Ceram. Int. **35**(7), 2965–2971 (2009)

Chapter 57
Parametric Optimization of Corrosion Resistance of Electroless Ni–Co–P Coating

Subhasish Sarkar, Rishav Kumar Baranwal, Ishita Koley, Rupam Mandal, Tapendu Mandal and Gautam Majumdar

Abstract This research paper investigates the effect of cobalt on the corrosion resistance of electroless Ni–P coating by forming a ternary alloy with Ni–P. The corrosion resistance offered by the ternary coating was measured for coating formed by varying the composition of the bath. The varying bath parameters were the concentration of cobalt source (cobalt sulphate) (A), concentration of reducing agent (sodium hypophosphite) (B) and the temperature (C) of the bath. The corrosion resistance was measured using electrochemical impedance spectroscopy (EIS). Taguchi's method has been applied with corrosion resistance as response operator to find the optimum resistance to corrosion and the optimum bath parameters, and along with it, ANOVA is done to find significant interactions between factors. The scanning electron microscopy (SEM) and energy-dispersive X-ray spectroscopy (EDAX) were performed to find the surface morphology and surface composition of the optimized coating.

Keywords Taguchi · Corrosion resistance · ANOVA · Surface morphology

57.1 Introduction

Coatings in the recent time have gathered attention of many researchers as they enhance and introduce different surface properties. Their increased application in the industry [1] is one of the main reasons for it. Coatings can be done by various techniques like electrodeposition, physical vapour deposition, chemical vapour deposition and electroless coating. We have used the electroless technique as it is advantageous. Ni–P coatings are better known for their corrosion resistance [2]. The mechanical and tribological properties can be improved with the help of various heat

S. Sarkar · R. K. Baranwal (✉) · R. Mandal · G. Majumdar
Department of Mechanical Engineering, Jadavpur University, Kolkata 700032, India
e-mail: rishavbaranwal22@gmail.com

I. Koley · T. Mandal
Department of Materials and Mechanical Engineering, IIEST, Howrah 711103, India

R. G. Narayanan et al. (eds.), *Advances in Computational Methods in Manufacturing*, Lecture Notes on Multidisciplinary Industrial Engineering,
https://doi.org/10.1007/978-981-32-9072-3_57

treatment techniques. However, this can also be achieved with the help of using different elements and complexes. Corrosion is such a phenomenon which degrades the life of metals and causes unpredictable failure in various applications of such metals. Hence, it is very essential to improve the corrosion resistance of the metals or superimpose a coating which has ameliorated anticorrosion properties. This coating has a sacrificial nature. So, in order to increase the corrosion resistance, the introduction of a third element/complex has been preferred [3–5]. The use of cobalt in binary coatings has provided better microhardness compared to the binary coatings [12]. Cobalt introduces a new property of electromagnetic shielding [6, 7] in Ni–P coatings while increasing the corrosion resistance of the coating at the same time.

A number of optimization techniques have been used in the past to optimize the coating [8–11]. We have applied the Taguchi L_{27} orthogonal array design to optimize the corrosion resistance. The Taguchi analysis showed that the level 3 of A, level 3 of B and level 1 of C were the optimized parameters. Analysis of variance (ANOVA) has been applied to find the significant factors and the significant interactions which affect the corrosion resistance. It has been found that all the interactions between the factors are significant for the corrosion resistances. The corrosion resistance of the coatings was measured using EIS. The optimized corrosion resistances were 24.525 and 476.17 Ω cm^2. The EDAX analysis showed the weight percentage of cobalt at 10.90% which is just enough to improve the hardness and induce electromagnetic shielding. The nodular grains present in the coatings were shown by SEM analysis.

57.2 Experimental Methods

The coating took place over copper substrates of thickness 0.1 mm and dimensions 2×1.5 mm^2. The pre-treatment of the substrate surface was done by acid cleaning process in 25% HCl for 90 s. The surface of the substrate was activated by using palladium chloride solution at 55 °C for 10 s. It was then rinsed with distilled water, and then, it was immersed in the electroless bath.

The electroless bath consisted of nickel sulphate hexahydrate, cobalt sulphate, sodium hypophosphate, tri-sodium citrate dehydrate and ammonium sulphate. Their respective concentrations are given in Table 57.1.

Table 57.1 Bath composition

Bath constituents	Reagents	(g/L)
Nickel source	Nickel sulphate hexahydrate	5
Cobalt source	Cobalt sulphate	10/15/20
Reducing agent	Sodium hypophosphate	20/25/30
Complexing agent	Tri-sodium citrate dehydrate	15
Buffering agent	Ammonium sulphate	10

The temperature of the bath was varied between 80, 85 and 90 °C. The bath volume was 200 ml, and the deposition time was 60 min.

Cobalt sulphate and sodium hypophosphate control the percentage of cobalt and phosphorus in the coating, respectively, while temperature controls the rate of deposition of the coating.

To study the electrochemical behaviour or corrosion behaviour, standard electrochemical potentiodynamic polarization tests of the above specimens were performed using three electrode corrosion cells in 3.5% Na Cl solution (3.5 g of NaCl in 100 ml double-distilled water). Polarization experiments were carried out as per ASTM standard methods (ASTM G5: Potentiostatic and Potentiodynamic Anodic Polarization Measurements and ASTM G59: Polarization Resistance Measurements) using OrigaLys Potentiostat (software: OrigaMaster 5) with a voltage range from −900 to +600 mV and a scan rate of 1 mV/sec. Electrochemical impedance spectroscopy (EIS) was done to know the surface morphology of the effected region after potentiocyclic voltammetry testing. The frequency range for EIS was from 100 kHz to 100 Hz with initial voltage of 10 mV. The corrosion resistances of each coating are shown in Table 57.2. The corrosion resistance was calculated using the Randles circuit, which forms a double-layer capacitance at the interaction of the coatings'

Table 57.2 Three levels of three factors and their respective resistances

S. No.	$CoSO_4$ (g/L)	$NaHPO_2$ (g/L)	Temperature (°C)	Resistance (Ω cm^2)
1	10	20	80	442.65
2	10	20	85	223.24
3	10	20	90	977.51
4	10	25	80	1309.9
5	10	25	85	49.969
6	10	25	90	1198.5
7	10	30	80	476.17
8	10	30	85	180.31
9	10	30	90	10.115
10	15	20	80	873.60
11	15	20	85	49.079
12	15	20	90	576.79
13	15	25	80	236.67
14	15	25	85	159.60
15	15	25	90	6.74
16	15	30	80	246.49
17	15	30	85	655.58
18	15	30	90	664.24
19	20	20	80	265.12
20	20	20	85	322.48
21	20	20	90	21.9
22	20	25	80	196.72
23	20	25	85	140.15

(continued)

Table 57.2 (continued)

S. No.	$CoSO_4$ (g/L)	$NaHPO_2$ (g/L)	Temperature (°C)	Resistance (Ω cm^2)
24	20	25	90	133.87
25	20	30	80	233.98
26	20	30	85	876.79
27	20	30	90	344.27

surface and the corroding medium. The two layers are formed due to charge transfer across the two different phases at the interaction point.

Design of experiments is an effective tool to determine the optimum conditions. The Taguchi method uses 'larger the better' technique in the present case to optimize the coatings' response. The varying parameters in the bath are called factors, while the different values of a parameter are called levels. We used three factors, and each factor had three levels. The Taguchi method uses the signal-to-noise ratio for finding the optimum conditions for optimum resistance. More positive value of S/N ratio means lesser noise which in turn indicates the best level in that factor. With the help of ANOVA, we can identify the significant parameters and the significant interactions. This is done by interpreting the interaction plots between AxB, BxC and AxC.

The surface morphology and its composition were determined by SEM and EDAX, respectively.

57.3 Results

Taguchi Analysis:

The larger the better uses the following formula for calculating the S/N ratio:

$$\frac{S}{N} = -10\log_{10}\left[1/n\left[\sum \frac{1}{y^2}\right]\right]$$

The S/N ratio is the ratio of the desired signal to the undesired noise, which is calculated using the above-mentioned formula. If for a particular level of a factor, signal is high with lower noise; it will indicate a higher S/N ratio which is desired.

Using this concept, the plots of S/N ratio for different levels of each factor are shown in Fig. 57.1.

The average S/N ratio for each factor is shown by the dotted lines in Fig. 57.1.

From Fig. 57.1, we can see the most positive values are achieved for level 3 of A (cobalt sulphate), level 3 for B (sodium hypophosphate) and level 1 for C (temperature of bath). Thus, the optimized coating is A3B3C1, which is the 24th experiment having resistance as 233.98 Ω cm^2.

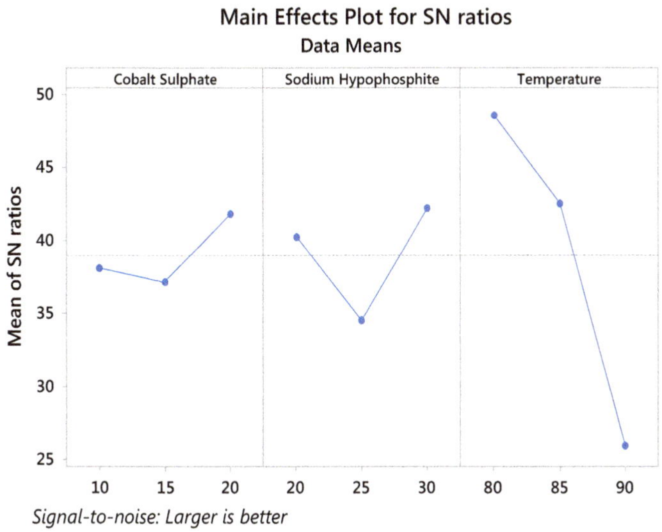

Fig. 57.1 S/N ratio of three levels of each parameter

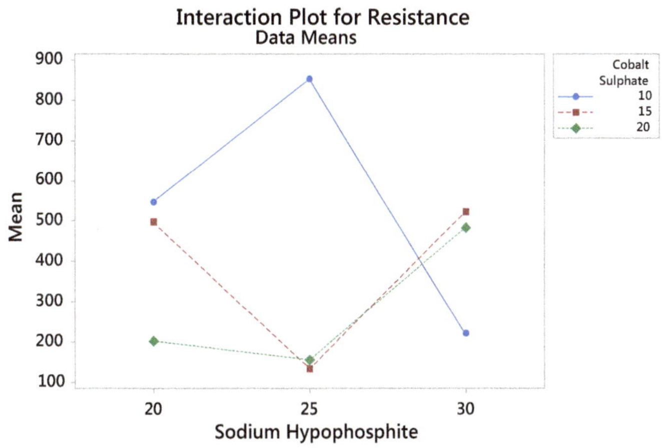

Fig. 57.2 Interactions between cobalt sulphate and sodium hypophosphite

The ANOVA results for R1 and R2 can be interpreted from Figs. 57.2, 57.3 and 57.4.

The blue, red and green lines represent the levels 1, 2 and 3, respectively, of the factors. If these lines remain horizontal or parallel to each other, then insignificant interactions between the factors have taken place. On the other hand, if the lines are inclined to each other and intersect at many points then good interactions have taken place between the factors. So for resistance offered to corrosion, we can say clearly, by observing the graphs that good interactions exist between A, B and C.

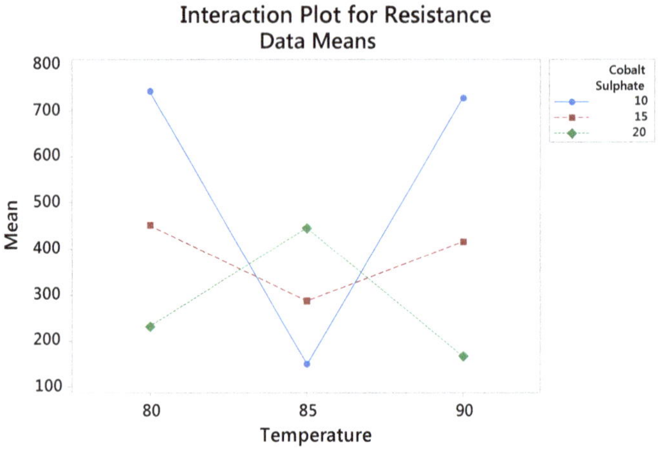

Fig. 57.3 Interactions between cobalt sulphate and temperature

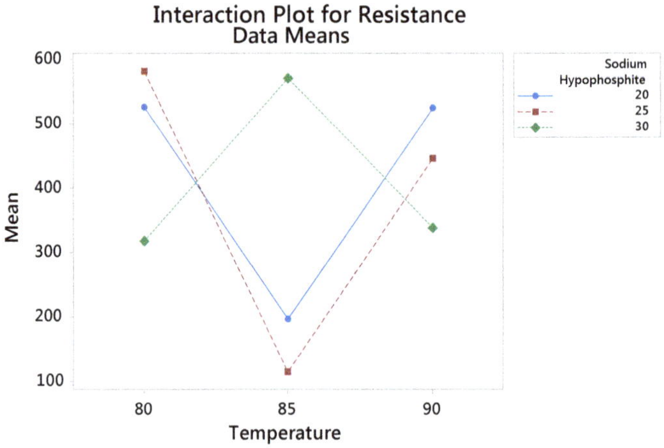

Fig. 57.4 Interactions between temperature and sodium hypophosphite

The energy-dispersive X-ray analysis shows the percentage of different elements on the surface of the coatings (Fig. 57.5). Its results are shown in Table 57.3. The cobalt % is 10.9% which is very influential in improving the various properties [12] and in this case corrosion resistance.

The composition is as shown.

Scanning electron microscopy was done to study the deposition on the surface of copper. Figure 57.6 shows the homogeneous deposition with very small agglomerated materials as globules. Since the surface is fully intact with coating. Hence, the porosity of the coating is very low. With low porosity, the corrosion agents which when gets deposited on the surface are unable to penetrate the surface of the coating which allows the coating to have higher corrosion resistance.

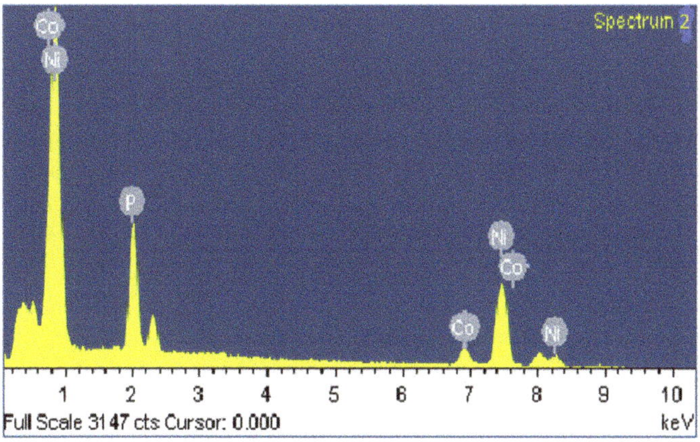

Fig. 57.5 EDAX image of the optimized sample

Table 57.3 EDAX result

Element	Weight %	Atomic %
Ni	76.89	69.34
Co	10.90	9.79
P	12.21	20.87

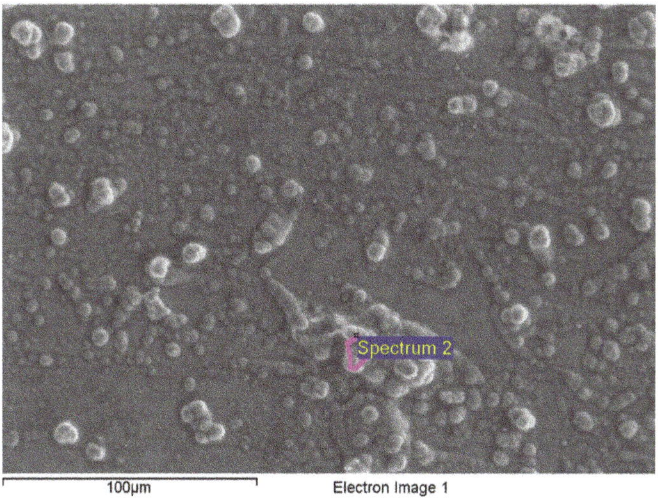

Fig. 57.6 SEM image of the coating

57.4 Conclusion

The electroless coating was optimized using Taguchi's method, and the optimized conditions are as follows: cobalt sulphate—10 g/L, sodium hypophosphite—30 g/L and temperature of bath at 80 °C. The increased resistance of the coating will have sacrificial behaviour. ANOVA results show that the interactions between each level (3 levels) of different factors (cobalt sulphate, nickel sulphate and temperature) are significant in determining the corrosion resistance of the coating. SEM results reveal intact coating taking place which allows the corrosion resistance to increase by hindering the movement of corroding materials into the substrate. The EDAX results show the weight percentage of cobalt at 10.9% which is enough to increase the corrosion resistance of the substrate material.

Acknowledgements We would like to thank Professor Deb Dulal Das of IIEST Shibpur, for the SEM and EDAX analysis.

References

1. Baudrand, D., Durkin, B.: Metal Finishing, pp. 20–24 (1998)
2. Narayanan, T.S.N.S., Baskaran, I., Krishnaveni, K., Parthiban, S.: Surf. Coat. Technol. **200**, 3438–3445 (2006)
3. Song, Y.-W., Shan, D.-Y., Chen, R.-S., Han, E.-H.: Surf. Eng. **23**(5), 334–338 (2007)
4. Rezagholizadeh, M., Ghaderi, M., Heidary, A., Vaghefi, S.M.M.: Prot. Metals Phys. Chem. Surf. **51**(2), 234–239 (2015)
5. Jiaqiang, G., Lei, L., Yating, W., Bin, S., Wenbin, H.: Surf. Coat. Technol. **200**, 5836–5842 (2006)
6. Narayanan, T.S.N.S., Selvakumar, S., Stephen, A.: Surf. Coat. Technol. **172**, 298–307 (2003)
7. Gao, Y., Huang, L., Zheng, Z.J., Li, H., Zhu, M.: Appl. Surf. Sci. **253**, 9470–9475 (2007)
8. Oraon, B., Majumdar, G., Ghosh, B.: Mater. Des. **29**, 1412–1418 (2008)
9. Oraon, B., Majumdar, G., Ghosh, B.: Mater. Des. **28**(7), 2139–2147 (2007)
10. Sahoo, P.: Mater. Des. **30**, 1341–1349 (2009)
11. Basavaraja, D.K., Skanda, M.G., Soumya, C., Ramesh, V.: Int. J. Mech. Aerosp. Ind. Mechatron. Manuf. Eng. **11**(4), 698–704 (2017)
12. Sarkar, S, Baranwal, R.K., Lamichaney, S., De, J., Majumdar, G.: Optimization of electroless Ni–Co–P coating with hardness as response parameter: a computational approach. Jurnal Tribologi **18**, 81–96 (1 Sept 2018)

Chapter 58
Investigation on Dimensional Accuracy for CFRP Antenna Reflectors Using Autoclave and VARTM Processes

Tushar Gajjar, Dhaval Shah, Shashikant Joshi and Kaushik Patel

Abstract The carbon fibre reinforcement polymer (CFRP) is widely used in space applications, automobile industries, aerospace and sports equipment because of good specific properties over conventional metals and alloys. Autoclave processing is one of the best manufacturing techniques to achieve high-dimensional accuracy for complex-shaped component. But, due to high initial and processing cost, vacuum-assisted resin transfer moulding (VARTM) is the alternative manufacturing method. In this research paper, CFRP antenna reflectors have been manufactured using autoclave process and VARTM manufacturing process to evaluate dimensional accuracy. The part thickness of laminate and layup orientation has been considered as a processing parameter in reflector manufacturing. The dimensions of the inner surface of the manufactured reflectors have been measured using 3D scanning techniques at the different locations. The deviation between theoretical CAD model and reflector dimensions at each location provides dimensional inaccuracy for the fabricated reflectors. The investigation on the dimensional inaccuracy of cured reflectors manufactured using the autoclave and VARTM manufacturing techniques has been carried out and the percentage difference has been found within the acceptable limit.

Keywords Dimensional accuracy · Autoclave · VARTM · Antenna reflector · Spring-back deformation

T. Gajjar
Mechanical Engineering Department, Sardar Vallabhbhai National Institute of Technology, Surat 395007, India
e-mail: tushar.gajjar79@gmail.com

D. Shah (✉) · S. Joshi · K. Patel
Mechanical Engineering Department, Institute of Technology, Nirma University, Ahmedabad 382481, India
e-mail: dbshah@nirmauni.ac.i

S. Joshi
e-mail: s.j.joshi@nirmauni.ac.in

K. Patel
e-mail: kaushik.patel@nirmauni.ac.in

R. G. Narayanan et al. (eds.), *Advances in Computational Methods in Manufacturing*, Lecture Notes on Multidisciplinary Industrial Engineering,
https://doi.org/10.1007/978-981-32-9072-3_58

58.1 Introduction

The composite material is a combination of two or more materials which are widely used in aerospace industries, space applications, automobiles and sports industries due to its high strength-to-weight ratio [1]. A prepreg is a pre-impregnated fibre and available as unidirectional, bi-directional and twill fabric tape [2]. The autoclave curing process is widely used for making complex-shaped composite components because of high-dimensional accuracy. But, due to high capital as well as processing cost, VARTM process is the alternative manufacturing method. The variation between cured laminated composite part dimensions and theoretical dimension is usually referred as spring-back deformation or warpage depending on the shape of components.

Many researchers worked on the prediction of the spring-back deformation and warpage for the composite products. Darrow and Smith [3] occupied three process parameters—thicknesses cure shrinkage, fibre volume fraction gradients and mould expansion that effects on spring-back deformation. Fernlund and Poursartip [4] exhibited the effect of the cure cycle and tool surface condition on spring-back deformation. Kaushik et al. [5] calculated the effectiveness of various parameters like pressure, degree of cure and ramp rate on the coefficient of friction in the autoclave process. Kappel et al. [6] measured the effect of crucial parameters such as part thickness, layup, part radius and scattering of occurring distortion during the autoclave manufacturing process. Correia et al. [7] developed an analytical formulation of governing equation for incompressible flow and this model successively used to measure the effect of process parameters such as inlet and outlet pressure, fibre architecture and layup orientation. Govignon et al. [8] performed experimentation on thickness variation during compaction to replicate reinforcement behaviour while pre-filling, filling and post-filling stage. Hammami et al. [9] studied the effects of the processing variables in vacuum infusion moulding process by developing the one-dimesional model. The compaction test was conducted with the maximum compaction pressure not beyond 0.1 MPa. It was established that for low compaction pressures (≤ 0.1 MPa), the number of layers did not have a significant effect on the compaction behaviour of the preform. Arulappan et al. [10] performed experiments on CFRP material with flat plate, *L*-shaped plate and the flat plate with a central circular hole. The pressure was varried for different shape. The flow of resin depends on flow medium. For large structure, high permeability medium (HPM) was selected for getting high flow rate and reduced filing time. Song et al. [11] developed an experimental model for incorporating resin flow through the preform as well as relaxation of the preform because resin pressure and part thickness variation were parameters in VARTM which play a vital role in manufacturing component.

In this research paper, autoclave and VARTM manufacturing processes have been used to predict dimensional accuracy of CFRP antenna reflectors. The AISI 430 grade stainless steel material has been chosen for manufacturing of parabolic mould. The thickness of the laminate and layup orientation has been considered as process parameters. The Hinpreg A45 CFRP prepreg is used in autoclave process, whereas carbon fibre dry fabric along with epoxy resin 520 and D-type hardener is used in the VARTM process. The parabolic-shaped reflectors have been cured using autoclave and VARTM processes. The dimensions at the inner surface of manufactured CFRP reflectors have been measured using 3D scanning techniques at different 200 locations. The deviation between the theoretical CAD model and reflector dimensions at each location gives dimensional inaccuracy or spring-back deformation for the manufactured CFRP reflectors. The comparison of spring-back deformation of cured reflectors manufactured using autoclave and VARTM manufacturing techniques has been carried out. The reflectors with different thickness and layup sequences have been fabricated using both manufacturing techniques, and dimensional accuracy has been matched.

58.2 Experimental Preparation

The stainless steel AISI 430 grade has been identified as mould material with 300 × 300 × 70 mm size. The 3D model of the parabolic mould has been prepared using solid modelling software. The dimension of the parabolic profile at one side is 240 mm circumferential diameter with 50 mm depth. The stainless steel mould block has been machined using NC code in 5–axis vertical milling centre with very high surface finish as shown in Fig. 58.1.

58.2.1 Reflectors Manufacturing Using Autoclave Process

The Hinpreg A45 unidirectional prepreg tape has been selected for the autoclave experimental process. Layers of prepreg are cut into the round shape of 300 mm diameter depending upon the required number of laminates and thickness as shown in Fig. 58.2. The consolidation materials like release agent, peel ply, breather cloth, vacuum bagging, etc. have been used in the autoclave process. In the autoclave process, the entire release agent has been applied on the parabolic mould surface.

Fig. 58.1 Manufactured stainless steel mould

Fig. 58.2 Unidirectional cut prepreg

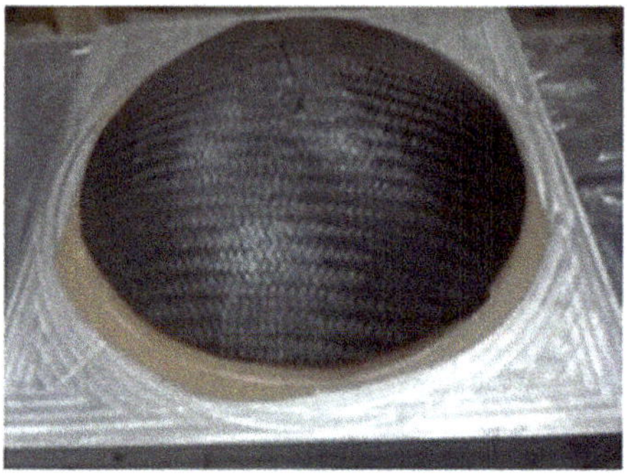

Fig. 58.3 Layers of CFRP prepreg on steel mould

The arrangement of prepreg layers with proper orientation is shown in Fig. 58.3. Peel ply permits free passage of volatiles and excess matrix while curing. To apply the vacuum and guide the removal of air and volatiles from the assembly, the breather fabric has been provided. Thicker breathers are required during great autoclave pressures. Sealant tape and vacuum bag are used for sealing the entire assembly. The whole set-up has been vacuum bagged with the help of vacuum film, vacuum valve and hose pipe as shown in Fig. 58.4.

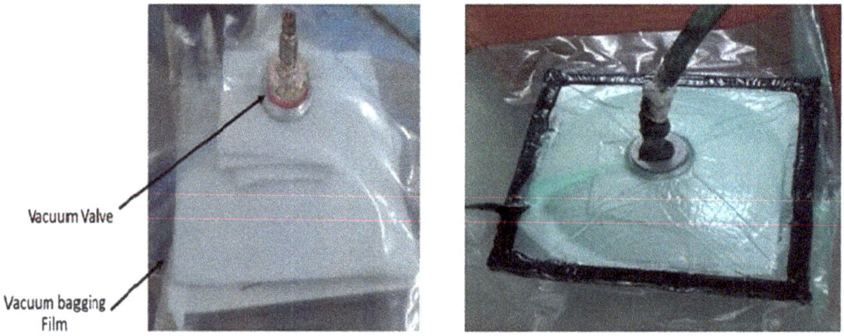

Fig. 58.4 Vacuum bagging and air suction process

Fig. 58.5 Four layers of carbon fibre

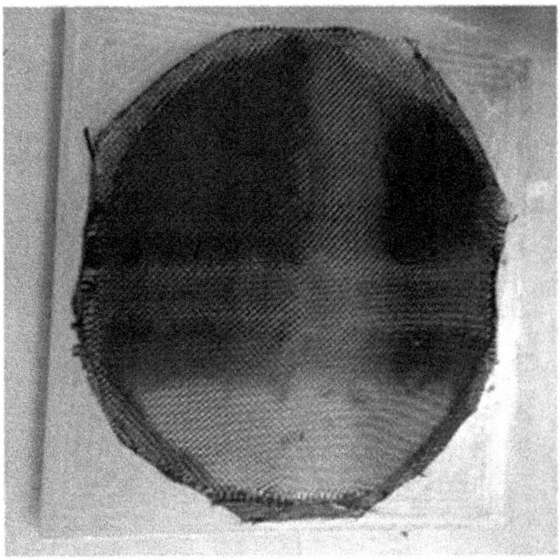

58.2.2 *Reflectors Manufacturing Using VARTM Process*

The parabolic shape reflectors are also manufactured using the VARTM process. The dry carbon fabric has been selected as a reinforcement and epoxy resin 520 and D-type hardener as a matrix material in the VARTM experimental process. The epoxy resin 520 having mix viscosity at 27 °C of 650 ± 100 MPas, the specific gravity of 1.15–1.18 g/cm^3 and pot life at 27 °C of 1.5–2 h has been considered. Similarly, D-type hardener consists of tensile strength of 70 MPa, the flexural strength of 120 MPa and impact strength of 80 N/m is mixed with one-fourth of resin ratio. The vacuum pump, infusion mesh, purported plastic, catch pot, etc. have been used for VARTM process. The layer of the release agent is applied on the paraboloid surface of the mould for ease of demoulding. The four layers of carbon fibre have been arranged on the mould surface in the proper sequence as shown in Fig. 58.5. Similarly, peel ply, purported plastic and infusion mesh are placed on fibre layers simultaneously and the whole assembly is sealed using sealant tape and vacuum bag as shown in Fig. 58.6. The PVC hose pipe is attached at resin inlet and another side for obtaining vacuum pressure. The resin is sucked into the system because of vacuum pressure as shown in Fig. 58.7.

58.2.3 *Curing of CFRP Reflectors*

In the autoclave process, the entire set-up—mould, composite prepreg along with consolidation materials, is kept in autoclave machine for curing purpose as shown in Fig. 58.8. The mould along with consolidate material has been cured for twenty-four hours at an atmospheric temperature in the VARTM process as shown in Fig. 58.9. The atmospheric pressure of 1 bar in VARTM, whereas 5 bar pressure in autoclave is retained machine throughout the curing cycle.

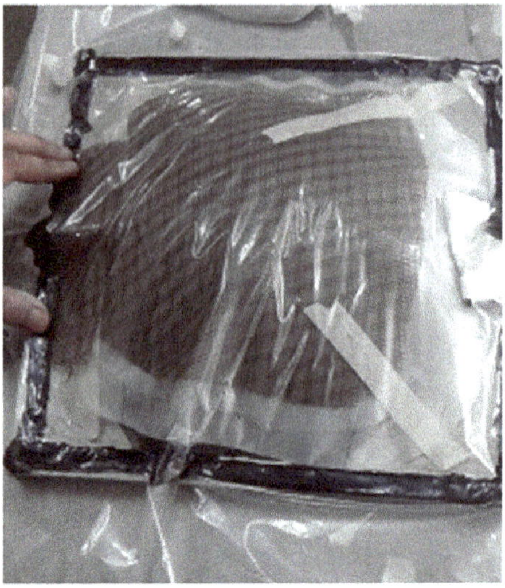

Fig. 58.6 Vacuum bag along with sealant tape

Fig. 58.7 Resin inlet hose and vacuum hose

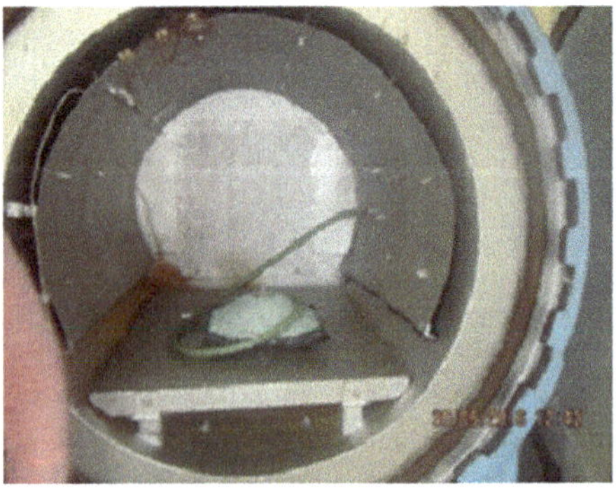

Fig. 58.8 Curing in autoclave machine

Fig. 58.9 Curing in VARTM process

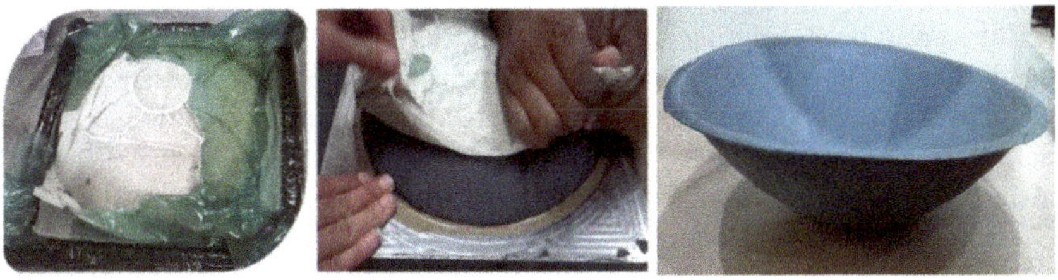

Fig. 58.10 Autoclave demoulding process and cured reflector

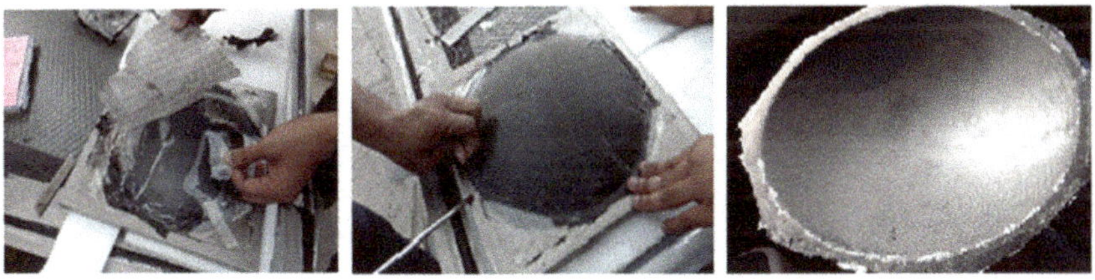

Fig. 58.11 VARTM demoulding process and cured reflector

The entire assembly is taken out from the autoclave machine after the curing process and consolidation materials are detached from the assembly at room temperature as shown in Fig. 58.10. The sequence of demoulding process for VARTM process and final cured reflector is shown in Fig. 58.11.

58.3 Dimensional Measurement Using 3D Scanning

The inner surface of reflectors is selected for dimension measurement and 200 number of coordinate points are identified on this surface. These points are obtained projecting circles of diameter 50, 100, 150 and 240 mm on parabolic surface of the reflector. Measurement of specified points is performed to ensure the accuracy of the parabolic reflector using the 3D scanner. This scanned model of the composite reflector is superimposed with the theoretical CAD model for the same to achieve deviation on each prescribed coordinate points. The deviation at each coordinate point is given using contour plot as shown in Fig. 58.12. The root means square of all 200 points deviations is referred as spring-back deformation for that reflector. The theoretical X, Y and Z coordinates and scanned X, Y and Z coordinates are given in Table 58.1 for autoclave curing, whereas Table 58.2 for VARTM curing process.

The spring-back deformation for the reflector manufactured using autoclave manufacturing process is 0.37 mm, whereas for the reflector manufactured using VARTM process is 0.35 mm. This value has been obtained for the same configuration reflector in each case. The difference between the autoclave manufacturing process and VARTM manufacturing process is around 5%. The different antenna reflectors have been manufactured with different laminate thickness and layup orientations. The 4, 8 and 12 number of layers and unidirectional, cross- and quasi-isotropic layup orientations have been considered. The deviation, i.e. dimensional inaccuracy, is observed and root mean square value is compared from VARTM and autoclave curing process for each case.

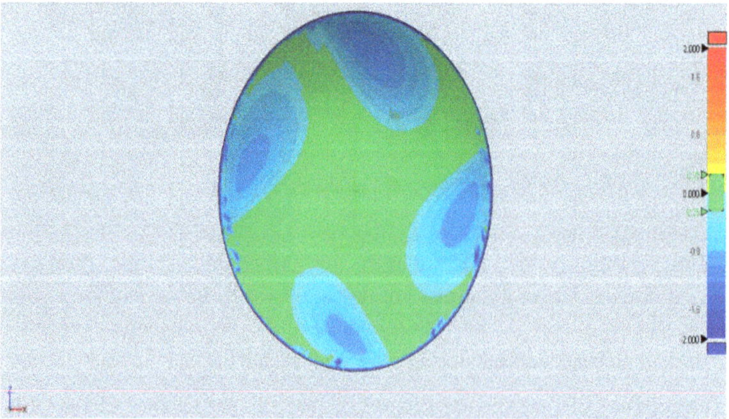

Fig. 58.12 Deviation contour of reflector at prescribed coordinate points

Table 58.1 Deviation of each coordinate point in autoclave curing process

Sr. No.	Predefined coordinate (Nominal) points (mm)			Scanned coordinate points (mm)			Deviation (mm)
1	0	−2.003	25	0	−2.051	24.992	0.04866
2	−3.911	−2.003	24.692	−3.911	−2.007	24.691	0.00412
3	−7.725	−2.003	23.776	−7.726	−1.978	23.78	0.02534
4	−11.35	−2.003	22.275	−11.353	−1.957	22.281	0.04649
5	−14.694	−2.003	20.225	−14.7	−1.947	20.232	0.05675
6	−17.677	−2.003	17.677	−17.683	−1.948	17.683	0.05565
.							
.							
.							
200	18.772	−46.083	118.521	18.728	−46.436	118.253	0.44524
Min	−120	−46.084	−120	−120.303	−47.054	−119.51	0.001
Max	120	0	120	120.454	0.0057	119.784	1.225
RMS							**0.37**

Table 58.2 Deviation of each coordinate point in VARTM curing process

Sr. No.	Predefined coordinate (Nominal) points (mm)			Scanned coordinate points (mm)			Deviation (mm)
1	25	−2.003	0	25	−1.998	0	−0.005
2	24.692	−2.003	3.911	24.693	−1.998	3.911	−0.005
3	23.776	−2.003	7.725	23.777	−1.998	7.726	−0.005
4	22.275	−2.003	11.35	22.277	−1.988	11.351	−0.015
5	20.225	−2.003	14.694	20.231	−1.957	14.699	−0.046
6	17.677	−2.003	17.677	17.683	−1.955	17.683	−0.049
.							
.							
.							
200	18.772	−46.083	118.521	18.831	−45.612	118.877	−0.594
Min	−119.99	−46.084	−119.99	−120.217	−46.265	−120.3	−1.168
Max	119.999	0	119.997	120.258	0.023	120.441	0.229
RMS							**0.35**

58.4 Conclusions

The CFRP antenna reflectors have been manufactured using autoclave and VARTM manufacturing processes with stainless steel AISI 430 mould. Hinpreg A45 uni-directional prepreg tape is used in the autoclave and bi-axial dry fabric with epoxy 520 resin along with hardener D has been used in the VARTM process. The dimension at the inner surface of CFRP manufactured reflectors has been measured using 3D scanning techniques at 200 different locations. The comparison of spring-back deformation of cured reflectors manufactured using autoclave manufacturing techniques and VARTM process has been carried out. The percentage deviations in spring-back deformation have been found within 5% range so the VARTM process gives more dimensional accuracy compared to autoclave process and VARTM is the cheapest process compare to autoclave process. The effect of laminate thickness and layup orientation has been considered as a process parameter. As laminate thickness increases, the spring-back deformation decreases. The quasi-isotropic layup orientation has minimum spring-back deformation.

References

1. Mazumdar, S.K.: Composites Manufacturing Materials, Product, and Process Engineering. CRC Press
2. White, S.R., Hahn, H.T.: Process modeling of composite materials: residual stress development during cure. Part I. Model formulation. J. Compos. Mater. (1992)
3. Darrow, D.A., Smith, L.V.: Isolating components of processing induced warpage in laminated composites. J. Compos. Mater. (2002)
4. Fernlund, G., Poursartip, A.: The effect of tooling material, cure cycle, and tool surface finish on spring-in of autoclave processed curved composite parts. In: Proceeding of ICCM (2003)
5. Kaushik, V., Raghavan, J., Zeng, X.: Experimental study of tool—part interaction during autoclave processing of thermoset polymer composite structures. Compos.: Part A **41**, 1210–1218 (2010)
6. Kappel, E., Stefaniak, D., Hühne, C.: Process distortions in prepreg manufacturing—an experimental study on CFRP L-profiles. Compos. Struct. **106**, 615–625 (2013)
7. Correia, N.C., Robitaille, F., Long, A.C., Rudd, C.D., Simacek, P., Advani, S.G.: Analysis of the vacuum infusion moulding process: I. Analytical formulation. Compos. Part A: Appl. Sci. Manuf. **36**(12), 1645–1656. (2005)
8. Govignon, Q., Simon, B., Kelly, P.A.: Simulation of the reinforcement compaction and resin flow during the complete resin infusion process. Compos. A Appl. Sci. Manuf. **41**(1), 45–57 (2010)
9. Hammami, A., Gebart, B.R.: Analysis of the vacuum infusion molding process. Polym. Compos. **21**(1), 28–40 (2000)
10. Arulappan, C., Duraisamy, A., Adhikari, D., Gururaja, S.: Investigations on pressure and thickness profiles in carbon fiber-reinforced polymers during vacuum assisted resin transfer moulding. J. Reinf. Plast. Compos. **34**(1), 3–18 (2015)
11. Song, X., Loos, A.C., Grimsley, B., Cano, R., Hubert, P.: Simulation of the vacuum assisted resin transfer moulding process. In: International Conference on Flo Process in Composite Materials, Newark, Delaware (2004)

Chapter 59
On Finite Element Analysis of Melting of Metallic Powders Using Microwave Energy at 2.45 GHz

Mohit Kumar, Gaurav Kumar and Vikas Kukshal

Abstract Use of microwave energy for processing of materials is gaining vast popularity due to its inherent superior characteristics like reduced processing time, better control on microstructure and enhanced mechanical properties. In this paper, the interaction of the microwave with metallic powder is analyzed using COMSOL Multiphysics software. Further, the effect of particle size on microwave absorption is also been analyzed. Microwave hybrid heating (MHH) technique is used to heat the metallic powder. Initially, metallic powder gets heated conventionally, i.e., through the heat transfer with separator plate and susceptor material. Once the metallic powder reaches the critical temperature, metallic powder starts to absorb the microwave and volumetric heating takes place. Also, the effects of input power on electric field strength and melting time of charge were analyzed. Further, it was found that as the size of the metallic powder particle is reduced, the time required for melting and casting of metallic powder reduces.

Keywords Microwave (MW) · Metallic powder · Temperature

Nomenclature

T	Temperature, K
t	Time of exposure, s
E	Time-harmonic electric field, V m^{-1}
H	Time-harmonic magnetic field, A m^{-1}

M. Kumar · G. Kumar (✉) · V. Kukshal
National Institute of Technology, Uttarakhand, Srinagar Garhwal 246174, Uttarakhand, India
e-mail: grv.kmr2015@gmail.com

M. Kumar
e-mail: mohitchauhan657@gmail.com

V. Kukshal
e-mail: vikaskukshal@nituk.ac.in

G. Kumar
Microwave Material Processing Laboratory, IIT Roorkee, Roorkee 247667, Uttarakhand, India

R. G. Narayanan et al. (eds.), *Advances in Computational Methods in Manufacturing*, Lecture Notes on Multidisciplinary Industrial Engineering,
https://doi.org/10.1007/978-981-32-9072-3_59

C_p Specific heat of material, $\mathrm{J\,kg^{-1}\,K^{-1}}$
h Convection heat transfer coefficient, $\mathrm{W\ m^{-2}\,k^{-1}}$
n Unit vector perpendicular to the surface of metal boundary
k Thermal conductivity, $\mathrm{W\ m^{-2}\,K^{-1}}$
ε Permittivity, $\mathrm{F\,m^{-1}}$
ε_r Relative permittivity
ε_0 Permittivity in free space, $\mathrm{F\,m^{-1}}$
ε^* Complex electric permittivity, $\mathrm{F\,m^{-1}}$
ε' Dielectric constant, $\mathrm{F\,m^{-1}}$
ε'' Dielectric loss factor, $\mathrm{F\,m^{-1}}$
μ Magnetic permeability, $\mathrm{H\,m^{-1}}$
μ_0 Magnetic permeability in free space, $\mathrm{H\,m^{-1}}$
μ_r Relative permeability, $\mathrm{H\,m^{-1}}$
μ'' Magnetic loss factor, $\mathrm{H\,m^{-1}}$
ω Angular frequency, $\mathrm{s^{-1}}$
σ Electrical conductivity, $\mathrm{S\,m^{-1}}$
σ_m Static electrical conductivity of metal, $\mathrm{S\,m^{-1}}$
c_0 Speed of light in free space, $\mathrm{m\,s^{-1}}$
γ Collision frequency of electrons, $\mathrm{s^{-1}}$
β Coupling parameter
ρ Density of material $\mathrm{kg\,m^{-3}}$
$\partial\Omega$ Small volume, $\mathrm{m^{-3}}$
ε_{rad} Emissivity
k_0 Wave number in free space

59.1 Introduction

The casting process is used to produce products with a wide range of variety in terms of complex shape, size and order quantity [1]. It is the most efficient manufacturing process to overcome complexity. Nowadays, cost, power and time saving are critical issues in the casting process. Microwave heating is a novel process which has a wide range of application in manufacturing processes like sintering, surface modification, casting, joining, drilling and cladding [2–7]. The past studies focused on the heating of metal powder [8, 9], melting [9] and composite casting with graphite mold [10] using microwave energy at 2.45 GHz. It is not possible to heat bulk metal directly using the microwave as bulk metal acts as a reflector to microwave at low temperature. But a recent study has shown that microwave hybrid heating (MHH) can be used to heat bulk metal and consequently, bulk metal can be melted and further cast in graphite mold [4].

Roy et al. [2] reported the microwave sintering process of metal powders (Fe–Ni and Fe–Cu) at a 2.45 GHz frequency and compared MW processed results with conventionally processed results. Properties like modulus of rupture (MOR), Rockwell hardness and sintered density of MW processed samples are substantially greater than conventional samples. It was reported that similar heating occurred with or without using susceptor, but the presence of susceptor provided rapid heating. Ghosh et al. [3] examined the characterization of microwave-heated aluminum powder (average particle size 25–30 μm with 99.7% purity) by XRD, SEM, TEM, EDX analysis and Fourier transform infrared (FTIR) spectroscopy. EDX analysis and FTIR studied revealed that Al powder transformed into Al–Al_2O_3 core–shell composite powder after the microwave heating process. SEM indicated that the morphology of microwave-processed Al powder was largely different from unprocessed Al powder [3]. Gupta and Sharma [7] developed nickel-based powder (EWAC, the average particle size of 40 μm) cladding on austenitic stainless steel (SS-316) substrate by using a domestic microwave system (1 kW and 2.45 GHz). The cladding was done successfully with an exposure time of 360 s. It was found that average Viker's micro-indentation hardness (304 ± 48 HV) of microwave clad is twice that of the SS-316 substrate (175 HV).

Chandrasekaran et al. [9] investigated the experimental and theoretical results of microwave melting of metals (Pb, Sn, Al and Cu) at 2.45 GHz and different power levels (520, 910 and 1300 W). Also, they compared the results with conventional heating and found that the energy required in microwave heating was half of the conventional heating (muffle furnace) and also processing time was very less as compared to the conventional heating. Singh et al. [10] reported the microwave melting and casting of metallic powders (EWAC or Ni-based powder + SiC) at 900 W and 2.45 GHz through MHH. It was observed that sample of EWAC + 10% SiC metal–ceramic composite casting is approximately 2.64 times harder than the sample of pure EWAC casting. Mishra and Sharma [4] reported the microwave casting of bulk aluminum alloy (AA-7039) by MHH at 1400 W and 2.45 GHz. The whole process was carried out inside the microwave oven cavity, and a new mechanism of in situ microwave casting was proposed. Four different temperature-dependent phases were also found during the study of time–temperature characteristics. Further, it was reported that the mode of heating in each phase was different. The time–temperature plot was validated using simulated data obtained from COMSOL Multiphysics software [11]. It was also reported that AA-7039 cast developed using in situ microwave casting through MHH showed superior property in the cast with less than 2% porosity and ultimate tensile strength of 148.46 ± 10 MPa [12].

The basic objective of the present work-study is to investigate the simulation result of the melting and casting of AA-7039 powder using MHH. The electromagnetic wave, frequency domain and heat transfer in solids were studied in COMSOL Multiphysics 5.2a software tool. The electric field distribution, temperature distribution and time–temperature profile of metallic powder were determined. The effect of input power on the electric field norm inside the microwave oven and melting time of charged or metallic powder were also discussed. In addition, the results showing the effect of particle size on melting time are also reported.

59.2 Modeling and Simulation

59.2.1 Model Development

A three-dimensional model of microwave (MW) casting of aluminum alloy 7039 powder was developed using COMSOL Multiphysics 5.2a software tool. The dimensions of the developed 3-D model of the microwave oven were taken from commercially available LG domestic microwave, and it operates at a power of 900 W with 2.45 GHz frequency source of a rectangular waveguide working in TE_{10} mode. The microwave oven cavity with dimensions of 335 × 335 × 215 mm^3 (W × D × H) and rectangular waveguide with dimensions of 50 × 43 86 mm^3 (W × D × H) were modeled to simulate the casting process. The powder material (particle radius 0.44 mm) was cast into a graphite mold cavity with dimensions of 24 × 45 × 10 mm^3 (W × D × H), and ceramic brick was used as a base of the setup. In the simulation, the MHH technique was used by taking SiC plate as a separator and SiC powder as a susceptor with a 4-mm thickness of both. The chemical composition of aluminum alloy (AA) 7039 used is shown in Table 59.1, and material properties used in the simulation analysis are shown in Table 59.2.

A simplified 3-D model of MW casting of metallic powder is shown in Fig. 59.1. The inset shows the zoomed view of metallic powder. A single sphere particle was generated with a preferred dimension, and an array operation was performed on the sphere particle. A number of particles were added in three dimensions (*X*-, *Y*- and *Z*-axes) by giving offset distance according to the mold dimensions using array operation.

59.2.2 Assumptions

In order to overcome the complexity and to simplify the model, the following assumptions were made:

(a) The ambient temperature and pressure of the system were assumed to be 27 °C and 1 atm, respectively.
(b) The oxidation of powder material and other chemical changes were not considered.
(c) Susceptor powder was taken as bulk metal (plate) to simplify the study.

Table 59.1 Chemical composition of AA 7039 [13]

Element	Zn	Mg	Mn	Fe	Cr	Si	Cu	Al
% weight	4.12	2.88	0.28	0.22	0.22	0.14	0.01	Balance

Table 59.2 Material properties used in the simulation study [11]

Materials	Density ($kg\ m^{-3}$)	Specific heat at constant pressure ($J\ kg^{-1}\ K^{-1}$)	Thermal conductivity ($W\ m^{-1}\ K^{-1}$)	Electric conductivity ($S\ m^{-1}$)	Permittivity	Magnetic permeability	Emissivity
AA-7039 (powder)	2700	$741.56 + 0.5071 \times T$	$238/(0.000011 \times T^{2.30} + (0.124/T))$	$3.774 \times 10^{7}/(1 + 0.0043 \times (300.15))$	1	1	0.19
Graphite	2100	830	470	1700	15	1	–
SiC	3210	1000	250	0.02	9.6	1	0.3
Ceramic brick	2000	900	0.5	0.001	1	1	–

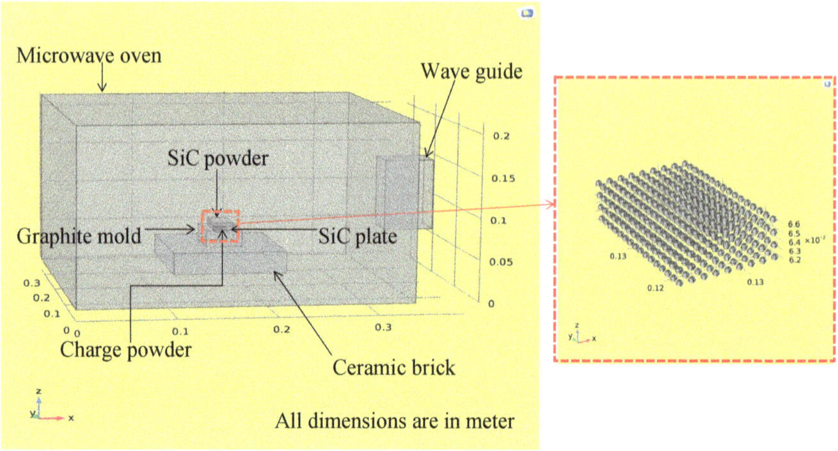

Fig. 59.1 A 3-D model of microwave casting of metallic powder (*inset* zoomed view of powder material)

(d) The wall of the microwave oven and waveguide were considered as copper metal.
(e) The ceramic bricks and air domain were not considered in the thermal analysis.
(f) The rectangular port was excited by transverse electric (TE_{10}) field in *Z*-direction only.

59.2.3 Governing Equations

A Hewlett-Packard (HP) work system (16 GB RAM) was used for simulation. Finite element-based COMSOL Multiphysics software was used to solve the current problem. The software couples the physics of electromagnetic wave in the frequency domain and heat transfer in the transient domain. Maxwell's equations were used to solve the electromagnetic wave of propagation. The electric field (E) inside the microwave oven is given by Eq. (59.1) [11]:

$$\nabla \times \mu_r^{-1}(\nabla \times E) - k_0^2\left(\varepsilon_r - \frac{j\sigma}{\omega\varepsilon_0}\right)E = 0 \quad (59.1)$$

where μ_r is the relative permeability, E is the time-harmonic electric field ($\mathrm{V\,m^{-1}}$), ε_r is the relative permittivity, $j = \sqrt{(-1)}$, σ is the electrical conductivity, ω is angular frequency, ε_0 is the permittivity of vacuum, i.e., 8.85×10^{-12} ($\mathrm{F\,m^{-1}}$), and k_0 is the wave number in free space which can be formulated in Eq. (59.2) [11]:

$$k_0 = \omega\sqrt{\varepsilon_0 \mu_0} = \frac{\omega}{c_0} \tag{59.2}$$

where c_0 is the speed of light in vacuum. The expression of complex electric permittivity (ε^*) of dielectric material is given by Eq. (59.3) [11]:

$$\varepsilon^* = \varepsilon' - j\varepsilon'' \tag{59.3}$$

where ε' is the dielectric constant ($\mathrm{F\,m^{-1}}$) and ε'' is the dielectric loss factor ($\mathrm{F\,m^{-1}}$). The complex electric permittivity ($\varepsilon^*_{\mathrm{m}}$) for metallic material can be expressed as in Eq. (59.4) [14]:

$$\varepsilon^*_{\mathrm{m}} = 1 - \frac{4\pi\sigma_{\mathrm{m}}}{\gamma\left[\left(\frac{\omega^2}{\gamma^2}\right)+1\right]} + j\frac{4\pi\sigma_{\mathrm{m}}}{\omega\left[\left(\frac{\omega^2}{\gamma^2}\right)+1\right]} \tag{59.4}$$

where σ_{m} is the static electrical conductivity of metals, its range for well-conducting metals is $10^{13}-10^{14}\,\mathrm{s^{-1}}$, and γ is the collision frequency of electrons, $\gamma = 8.5 \times 10^{14}\,\mathrm{s^{-1}}$ (for solid aluminum at room temperature) [15]. In Eq. (59.4), $\left(\frac{\omega^2}{\gamma^2}\right) \lll 1$; therefore, the second term approaches zero for the real part of $\varepsilon^*_{\mathrm{m}} \to 1$. For aluminum, Eq. (59.4) can be expressed as:

$$\varepsilon^*_{\mathrm{m}} = 1 + j\frac{4\pi\sigma_{\mathrm{m}}}{\omega} \tag{59.5}$$

As from the above equation, it is clear that heat dissipated (dielectric loss) in the metallic material depends on σ_{m}. Thus, the dependence of static electrical conductivity of metal (σ_{m}) on temperature (T) is as follows:

$$\sigma_{\mathrm{m}}(T) = \frac{\sigma_0}{1 + \beta(T - T_0)} \tag{59.6}$$

where β is the temperature coefficient of metal and σ_0 is the electrical conductivity of metal at a temperature T_0.

Fourier's energy balance Eq. (59.7) can be used to estimate the microwave (MW) energy dissipation as heat inside the powder material at the time of microwave exposure as [16, 17]:

$$\rho C_P \frac{\partial T}{\partial t} + Q_{\mathrm{Loss}} = Q(r, T) + \nabla.(k\nabla T) \tag{59.7}$$

where C_P is the specific heat capacity ($\mathrm{J\,kg^{-1}\,K^{-1}}$), ρ is the density ($\mathrm{kg\,m^{-3}}$), k is the thermal conductivity ($\mathrm{W\ m^{-2}\,K^{-1}}$), Q_{Loss} is the heat losses from material inside the microwave, and $Q(r, T)$ is the heat source or microwave (MW) energy absorbed by the powder material. For metal, it can be expressed as in Eq. (59.8) [16, 17]:

$$Q(r,T)=\frac{1}{2}\left[\omega\varepsilon_{\mathrm{m}}''(r,T)|E(r,T)|^2+\omega\mu_{\mathrm{m}}''(r,T)|H(r,T)|^2+\sigma_{\mathrm{m}}|E(r,T)|^2+\sigma_{\mathrm{i}}|E(r,T)|^2\right] \tag{59.8}$$

where $H(r,T)$ and $E(r,T)$ are the local electromagnetic field vectors at point r and temperature T in (K), μ_{m}'' is magnetic loss factor (H m^{-1}), $\varepsilon_{\mathrm{m}}''$ is dielectric loss factor (F m^{-1}) of metallic materials, and σ_{i} is ionic conductivity of metallic ions in the metal lattice.

59.2.4 Boundary Conditions

Boundary conditions used to approach the real experimental conditions are as follows:

- The impedance boundary condition was used to define the minimum penetration depth of microwave on the wall of the microwave oven and waveguide. This boundary condition is shown in Fig. 59.2a and can be expressed by Eq. (59.9) [11]:

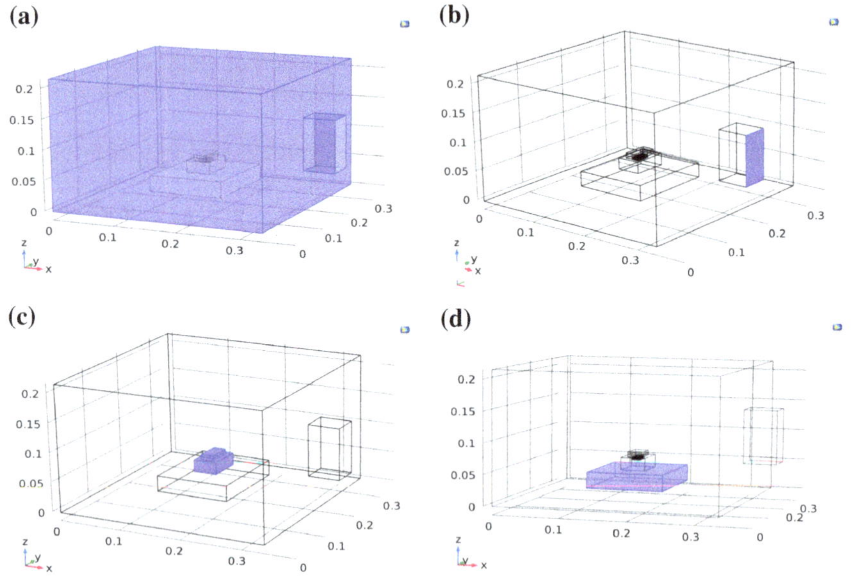

Fig. 59.2 Boundary conditions used in simulation: **a** impedance boundary condition, **b** port boundary condition, **c** heat transfer in solid and **d** thermal insulation

$$\sqrt{\frac{\mu_0\mu_r}{\varepsilon_0\varepsilon_r - j\frac{\sigma}{\omega}}}n \times H + E - (n.E)n = (n.E_s)n - E_S \tag{59.9}$$

where μ_0 is the permeability of vacuum, n is a unit vector which is perpendicular to the surface of the metal boundary, and E_S is the electric field component along the surface.

- A rectangular port boundary was used for wave excitation as the transverse electric (TE_{10}) field. The port boundary condition is shown in Fig. 59.2b. The port boundary condition at a small volume $\partial\Omega$ of edge element (S) is expressed by Eq. (59.10) [11]:

$$S = \frac{\int (E - E_1).E_1 \partial\Omega}{\int E_1.E_1 \partial\Omega} \tag{59.10}$$

- The heat transfer in solid condition is applied to the powder charge, mold, separator and susceptor as shown in Fig. 59.2c. The thermal insulation boundary condition (i.e., heat flux is zero) is shown in Fig. 59.2d and governed by Eq. (59.11) [11]:

$$-n \times (-k\nabla T) = 0 \tag{59.11}$$

- The heat flux and diffuse surface boundary conditions are applied on the surface of mold, separator and susceptor, which can be expressed by Eqs. (59.12, 59.13) [11]:

$$-n \times (-k\nabla T) = h(T_0 - T) \tag{59.12}$$

$$-n \times (-k\nabla T) = \varepsilon_{rad}\sigma_m\left(T_0^4 - T^4\right) \tag{59.13}$$

where h is the convective heat transfer coefficient and ε_{rad} is emissivity of the radiative surfaces.

59.2.5 Meshed Model

For simulating the above 3-D model, user-controlled mesh was used for meshing. Different types of free tetrahedral mesh like a finer, extra fine and extremely fine were used according to the suitable element size. Figure 59.3 shows mesh quality statistics (the average element quality and average growth rate) of 'finer' (Fig. 59.3a), 'extra fine' (Fig. 59.3b) and 'extremely fine' (Fig. 59.3c) mesh element sizes. The average element quality was good in case of extremely fined mesh with a low growth rate. The meshed model of microwave casting is shown in Fig. 59.4.

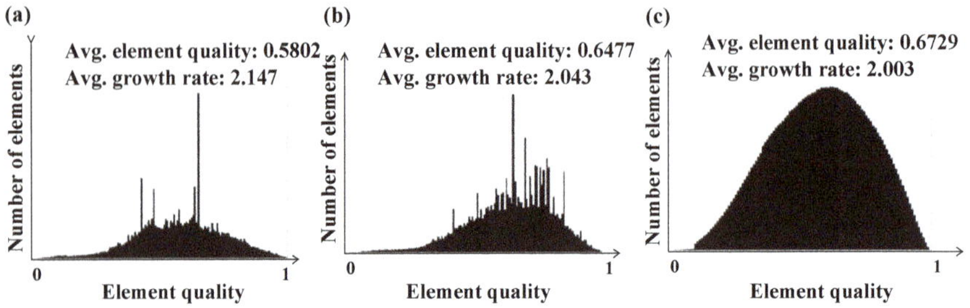

Fig. 59.3 Element quality comparison of **a** finer, **b** extra fine and **c** extremely fine element size

Fig. 59.4 Meshing of the experimental model

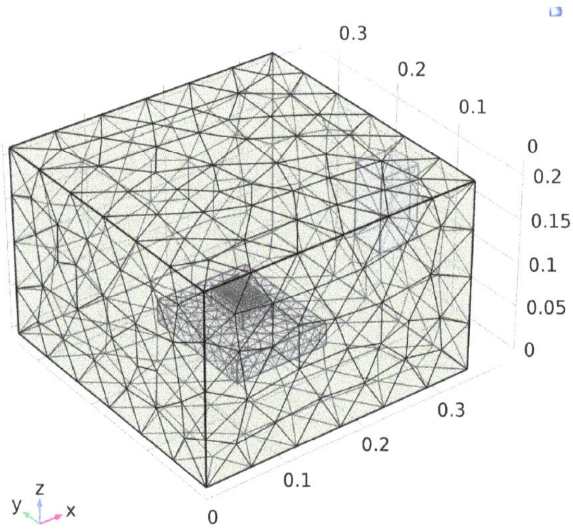

59.3 Results and Discussion

59.3.1 Electric Field Distribution and Temperature Distribution

The electric field distribution inside the microwave oven is shown in Fig. 59.5a. The energy is transmitted to the cavity at a power of 900 W and a frequency of 2.45 GHz. The distribution of the electric field is non-uniform inside the cavity because of the node and antinode formations due to the fluctuation of the electric field. So, the experimental setup should be kept precisely at a dense area of the electric field as it may affect the microwave absorption and hence, the energy required to melt the charged particle may be different.

As the electric field travels through dielectric materials like SiC plate and SiC powder, dielectric material starts to absorb it. This results in heating of dielectric

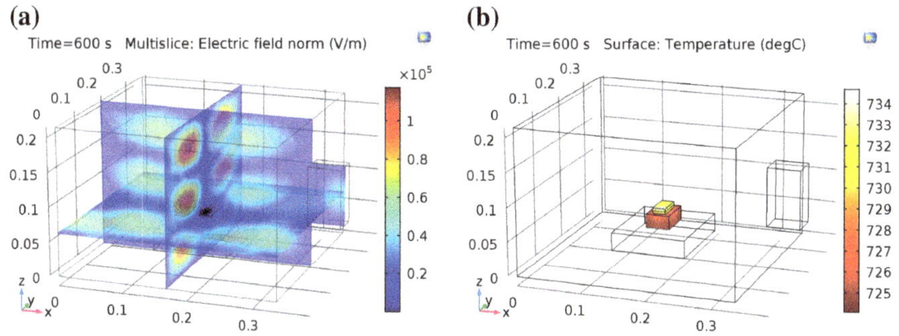

Fig. 59.5 **a** Electric field distribution inside the microwave oven and **b** temperature distribution

material. Initially, SiC powder couples with microwave radiation, starts heating up and transfers heat to SiC plate. SiC plate and graphite mold also start absorbing microwave and get heated rapidly, but charged powder absorbs very less microwave. Initially, the charged powder gets heated by conventional mode of heat transfer through SiC plate, susceptor and graphite mold. Once, the temperature of charged powder reaches up to a critical point, powder particles start absorbing microwave radiation and get heated by both conventional heating and microwave heating. The temperature increases rapidly with respect to the time, and melting of charged particles takes place. This heating phenomenon is known as microwave hybrid heating (MHH). This melted material is solidified and cast in a graphite mold.

The surface temperature distribution of heat transfer in the solid domain is shown in Fig. 59.5b. It is evident from the temperature distribution that temperature rise of SiC powder (susceptor) and SiC plate (separator) is more as compared to the metallic powder and graphite mold. This is due to the high dielectric constant, i.e., relative permittivity of SiC as compared to the metal as shown in Table 59.2. The time–temperature graph of charged powder is shown in Fig. 59.6. Initially, the heating rate of the charged powder is high but after reaching the melting point heating rate decreases. The melting of charged powder starts within 400 s, and after that temperature becomes constant. It can also be seen from Fig. 59.6 that slope of curve, i.e., rate of heating, increases after a certain time duration. This is the time duration after which powder particles start to absorb microwave radiation.

59.3.2 Effect of Input Power

The variation in input power affects the electric field norm inside the microwave oven and consequently affects the melting time of charged powder as shown in Fig. 59.7. The input power was varied 700–1300 W with step size 200 W with an exposure time of 600 s. It is observed from the figure that the maximum electric

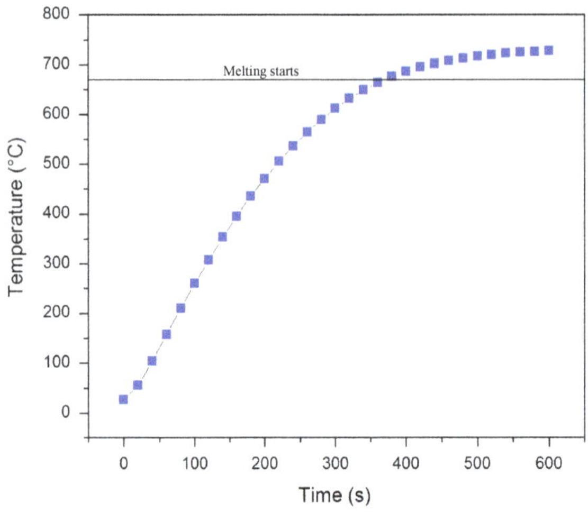

Fig. 59.6 Time–temperature graph at 900 W

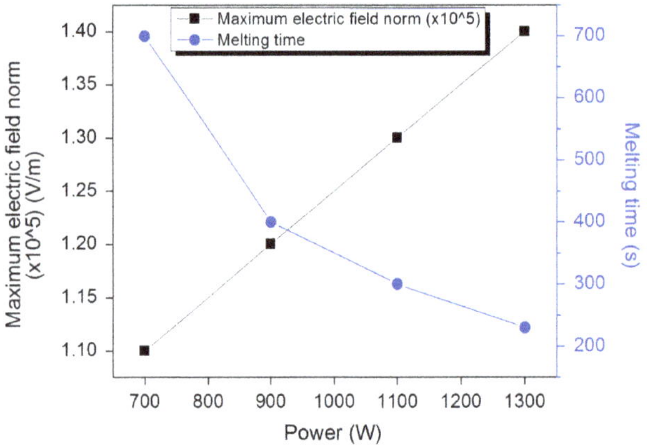

Fig. 59.7 Effect of power on electric field norm inside and the maximum temperature inside the charge

field norm inside the microwave increases linearly with input power. This happens because power absorbed inside the material is directly proportional to the electric field strength (Eq. 59.8). Also, the melting time of charged powder decreases with an increase in input power. The maximum electric field norm value varies from 1.1×10^5 to 1.4×10^5 V/m, and melting time decreased from 700 to 230 s by rising the input power from 700 to 1300 W, respectively.

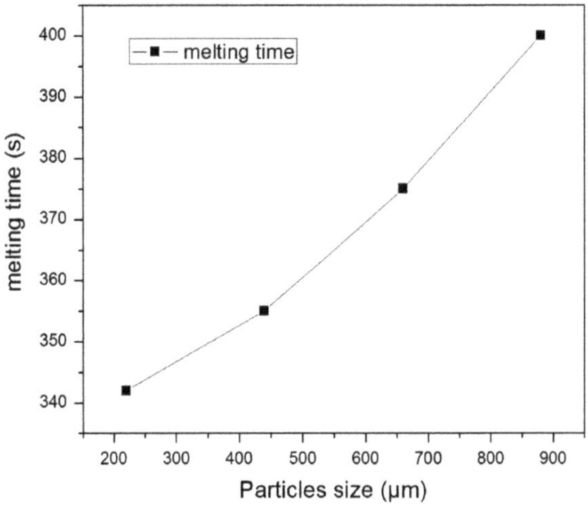

Fig. 59.8 Effect of powder particles in terms of melting time

59.3.3 Effect of Size of Powder Particles

Figure 59.8 shows the variation in melting time of charged powder with variation in particle size at 900 W. As the AA 7039 powder size reduces, the melting time of the charged powder reduces. Thus, it indicates that smaller-sized particles have higher microwave absorption capacity [18]. Microwave gets absorbed inside the material up to a particular depth called skin depth. As the particle size is reduced, the effective surface area which interacts with microwave increases and consequently, factors like Joule loss increase which increase the microwave absorption capacity [12, 19].

59.4 Conclusion

A 3-D microwave oven setup was modeled to find the precise location of experimental setup inside the cavity. COMSOL Multiphysics software tool was used successfully to simulate the microwave hybrid heating (MHH) of AA 7039 powder at a rated power of 900 W and 2.45 GHz frequency. Following are the conclusions based on the above study:

- It was observed that the experimental setup should be placed at the high magnitude or dense area of the electric field.
- It was found that the melting of the metallic powder achieved within 400 s and temperature rises to −730 °C in 600 s at the power of 900 W.
- It was observed that higher is the input power, more is the electric field norm, consequently reducing the melting time of charged powder.

- From the simulation results obtained, it can be concluded that decreasing the powder particle size results in more absorption of microwave energy and hence the time required for melting and casting of metallic powder reduces.

Acknowledgements The authors would like to express sincere gratitude to Prof. Apurbba Kumar Sharma for allowing and doing the simulation work using COMSOL Multiphysics in Microwave Material Processing Laboratory, IIT Roorkee.

References

1. Ravi, B.: Metal casting: computer-aided design and analysis. PHI Learning Pvt. Ltd. (2005)
2. Roy, R., Agrawal, D., Cheng, J., Gedevanishvili, S.: Full sintering of powdered-metal bodies in a microwave field. Nature **399**(6737), 668 (1999)
3. Ghosh, S., Pal, K.S., Dandapat, N., Mukhopadhyay, A.K., Datta, S., Basu, D.: Characterization of microwave processed aluminium powder. Ceram. Int. **37**(3), 1115–1119 (2011)
4. Mishra, R.R., Sharma, A.K.: On mechanism of in-situ microwave casting of aluminium alloy 7039 and cast microstructure. Mater. Des. **112**, 97–106 (2016)
5. Bansal, A., Sharma, A.K., Das, S.: Metallurgical and mechanical characterization of mild steel-mild steel joint formed by microwave hybrid heating process. Sadhana **38**(4), 679–686 (2013)
6. Kumar, G., Sharma, A.K.: Role of dielectric fluid and concentrator material in microwave drilling of borosilicate glass. J. Manufact. Processes **33**, 184–193 (2018)
7. Gupta, D., Sharma, A.K.: Development and microstructural characterization of microwave cladding on austenitic stainless steel. Surf. Coat. Technol. **205**(21–22), 5147–5155 (2011)
8. Yoshikawa, N.: Fundamentals and applications of microwave heating of metals. J. Microw. Power Electromagn. Energy **44**(1), 4–13 (2010)
9. Chandrasekaran, S., Basak, T., Ramanathan, S.: Experimental and theoretical investigation on microwave melting of metals. J. Mater. Process. Technol. **211**(3), 482–487 (2011)
10. Singh, S., Gupta, D., Jain, V.: Novel microwave composite casting process: theory, feasibility and characterization. Mater. Des. **111**, 51–59 (2016)
11. Mishra, R.R., Sharma, A.K.: On melting characteristics of bulk Al-7039 alloy during in-situ microwave casting. Appl. Therm. Eng. **111**, 660–675 (2017)
12. Mishra, R.R., Sharma, A.K.: Structure-property correlation in Al–Zn–Mg alloy cast developed through in-situ microwave casting. Mater. Sci. Eng. A **688**, 532–544 (2017)
13. Mishra, P., Upadhyaya, A., Sethi, G.: Modeling of microwave heating of particulate metals. Metall. Mater. Trans. B **37**(5), 839–845 (2006)
14. Buchelnikov, V.D., Louzguine-Luzgin, D.V., Anzulevich, A.P., Bychkov, I.V., Yoshikawa, N., Sato, M., Inoue, A.: Modeling of microwave heating of metallic powders. Physica B **403** (21–22), 4053–4058 (2008)
15. Eidmann, K., Meyer-ter-Vehn, J., Schlegel, T., Hüller, S.: Hydrodynamic simulation of subpicosecond laser interaction with solid-density matter. Phys. Rev. E **62**(1), 1202 (2000)
16. Mishra, R.R., Sharma, A.K.: A review of research trends in microwave processing of metal-based materials and opportunities in microwave metal casting. Crit. Rev. Solid State Mater. Sci. **41**(3), 217–255 (2016)
17. Maxim, I., Tanaka, M.: Numerical analysis of the microwave heating of compacted copper powders in single-mode cavity. Jpn. J. Appl. Phys. **50**(9R), 097302 (2011)

18. Mondal, A., Shukla, A., Upadhyaya, A., Agrawal, D.: Effect of porosity and particle size on microwave heating of copper. Sci. Sinter. **42**(2), 169–182 (2010)
19. Crane, C.A., Pantoya, M.L., Weeks, B.L., Saed, M.: The effects of particle size on microwave heating of metal and metal oxide powders. Powder Technol. **256**, 113–117 (2014)

Chapter 60
Tribological Behaviour of Nylon 6 Composites and Optimization of Process Parameters Using Hybrid Taguchi-GRA and CSA

Sudhir Kumar, Barnik Saha Roy and K. Panneerselvam

Abstract Friction and wear are the main causes observed by manufacturing industries which lead to the replacement of components. The objective of the present work is to investigate the tribological behaviours of Nylon 6 and glass-filled (GF) Nylon 6 composites using a pin-on-disc configuration. Pin materials were fabricated by injection moulding machine and tested for friction and wear against 320 grit size silicon carbide (SiC) sandpaper. The effect of the different load (N), glass-filled content (wt%) and sliding distance (m) were studied for the coefficient of friction (COF) and specific wear rate (SWR). Analysis of variance (ANOVA) procedure was used to find out the percentage contribution of process parameters (load, sliding distance and varying glass-filled content) with the COF and SWR. Process parameters were optimized by hybrid Taguchi-grey relational analysis (GRA) and cuckoo search algorithm (CSA). There is a close relationship between COF and SWR with wear mechanisms.

Keywords Polymer composites · COF · SWR · Taguchi · GRA · CSA

S. Kumar (✉) · B. S. Roy
Department of Mechanical Engineering, National Institute of Technology, Agartala 799046, India
e-mail: sudhir007gec@gmail.com

B. S. Roy
e-mail: barnik.me@nita.ac.in

K. Panneerselvam
Department of Production Engineering, National Institute of Technology, Tiruchirappalli 620015, India
e-mail: kps@nitt.edu

R. G. Narayanan et al. (eds.), *Advances in Computational Methods in Manufacturing*, Lecture Notes on Multidisciplinary Industrial Engineering,
https://doi.org/10.1007/978-981-32-9072-3_60

60.1 Introduction

Polymer matrix composites (PMCs) are replacing traditional metals and unfilled polymers due to their superior properties such as intrinsic lubrication property and wear resistance with low processing cost [1]. The application of polymer composites is mostly in sliding components in the food and chemical industry like gears and cams. This may be mainly attributed to the Nylon having inherent lubrication behaviour which avoided the use of oil lubrication [2]. Ramesh and Suresh carried out friction and wear test of polymer composites and observed grit size dominant factors followed by applied load and sliding distance [3]. Grey relational analysis (GRA) may be used to convert multiple objective optimization problems to a single objective problem [4–6]. Yang and Deb [7] investigated a new evolutionary and population-based algorithm cuckoo search for optimizing process parameters. Cuckoo search is based on the breeding behaviour of cuckoo birds. The study on hybrid Taguchi-GRA and CSA for multi-criteria optimization in tribological process parameters is very rare. The aim of present work is to investigate the friction and wear characteristic of Nylon 6 and glass-filled Nylon 6 composites against the abrasive paper. There is a close relationship between COF and SWR with wear mechanisms.

60.2 Materials and Methods

The material selected in this investigation for pin specimens was glass-filled Nylon 6 (0, 10, 20 and 30 wt%) in the form of granules as shown in Fig. 60.1. Injection moulding machine was used to fabricate the Nylon 6 composite specimens for mechanical (UTS, impact strength and hardness) and tribological test. The mould temperature was maintained at 240 °C. The size of stationary GF Nylon 6 pin specimen was 6 mm-diameter and 31 mm-length according to ASTM G99-05. Schematic illustration of the pin-on-disc set-up is shown in Fig. 60.2. Track diameter of the disc (320 grit size sandpaper) was 40 mm. The Taguchi technique is a powerful tool to optimize process parameters. The process parameters used for the experiment are presented in Table 60.1. Grey relational analysis is used to convert multi-objective (COF and SWR) into a single response optimization problem.

Cuckoo search (CS) is an evolutionary and population-based approach that mimics the breeding behaviour of cuckoo birds. The preliminary experiments show the promising results of this approach in the field of search and optimization. The idealized steps of CS can be stated as follows:

(1) Every cuckoo bird lays an egg on the randomly selected nest.
(2) The number of nests (agents) and eggs (individuals) in each nest is fixed.
(3) Nest with the high quality of eggs (elite members) is passed on to the next generation.

Fig. 60.1 Pure Nylon 6 and GF Nylon 6 (granule form) [1]

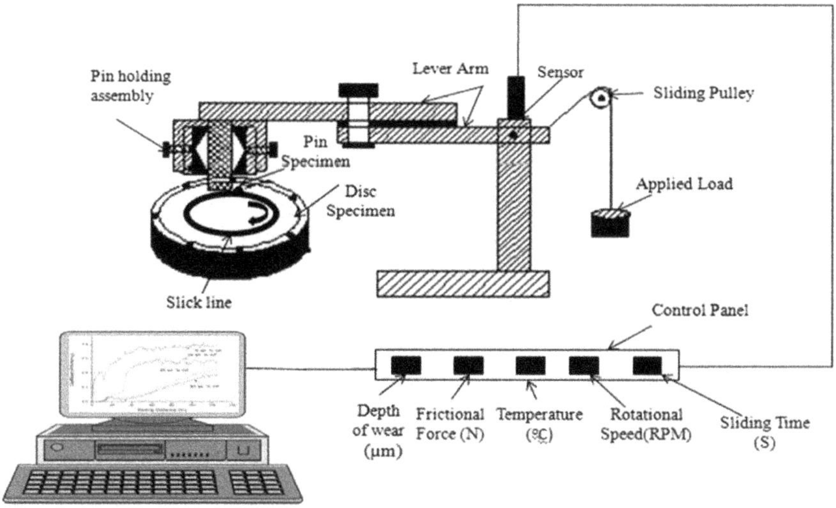

Fig. 60.2 Experimental set-up of pin-on-disc [8]

Table 60.1 Process parameters and their levels

Factor symbol	Control factors	Levels			
		1	2	3	4
A	Glass filled (wt%)	0	10	20	30
B	Load (N)	5	10	15	20
C	Sliding distance (m)	125	250	375	500

(4) If the host bird discovers an alien egg based on the external characteristics of an egg such as colour and spots, then host bird either throws the alien egg away or builds an entirely new nest at the new location.

For feature selection, the host nest *n* is defined as an agent containing multiple eggs (or individuals) called population. Each egg is represented as a binary string, with a 1 for inclusion of feature (or attribute) and a 0 for the exclusion of feature. Step 4 is implemented by maintaining the index of the worst nest and replacing it with better individuals. Levy flight is a random walk and named after French Mathematician Paul Levy, in which steps are defined in terms of step length having a probability distribution that is heavy-tailed. The direction of the steps in Levy flights can be isotropic or random. Various studies demonstrated the flight behaviour of many insects and animals based on Levy flights. In the past decade, these behaviours have been applied in optimization and optimal search and found to yield promising results [5–7].

60.3 Results and Discussion

The mechanical characterization includes evaluation of impact strength, UTS and hardness of the composites that are presented in Table 60.2. The UTS increased as to increase the GF wt% presented in Table 60.2. This may be attributed to the better distribution of GF and superior adhesion among the Nylon and GF. The hardness (shore D) increased as to increase the GF wt% as shown in Table 60.2. This may be attributed to brittleness character of GF. Nylon 6 material has high impact strength due to ductile nature and flexibility of the interface molecular chain resulting in absorbing more energy presented in Table 60.2. The impact strength increased as to increase the GF wt%. The improvement in impact strength of composites is due to good bonding strength between matrix and reinforcement. The experiments were conducted using L_{16} orthogonal array details which are shown in Table 60.3.

Table 60.2 Mechanical properties of various glass-filled contents (wt%) of Nylon 6 [1]

Glass-filled content (wt%)	Hardness (shore D) ASTM D2240-05	Notched impact strength (J/mm) ASTM D256-10	UTS (MPa) ASTM D638-14
0	64	0.137	51
10	65	0.075	54
20	73	0.125	71
30	74	0.125	86

Table 60.3 GRA for tribological properties of Nylon 6 composites

Exp. level	Weight loss (g)	COF (μ)	SWR (mm^3/Nm)	Normalized response		GRC		GRG	Rank
				COF	SWR	COF	SWR		
1($A_1B_1C_1$)	0.084	0.67	0.11	0	0	0.33	0.33	0.33	16
2($A_1B_2C_2$)	0.178	0.65	0.06	0.07	0.54	0.35	0.52	0.43	15
3($A_1B_3C_3$)	0.241	0.60	0.03	0.26	0.78	0.40	0.70	0.55	12
4($A_1B_4C_4$)	0.293	0.49	0.02	0.69	0.90	0.61	0.84	0.73	8
5($A_2B_1C_2$)	0.088	0.62	0.05	0.19	0.58	0.38	0.54	0.46	14
6($A_2B_2C_1$)	0.063	0.59	0.04	0.30	0.74	0.41	0.66	0.54	13
7($A_2B_3C_4$)	0.226	0.52	0.02	0.57	0.91	0.54	0.85	0.69	9
8($A_2B_4C_3$)	0.202	0.48	0.02	0.73	0.93	0.65	0.89	0.77	7
9($A_3B_1C_3$)	0.094	0.47	0.03	0.76	0.77	0.68	0.68	0.68	10
10($A_3B_2C_4$)	0.126	0.46	0.01	0.80	0.96	0.72	0.93	0.82	4
11($A_3B_3C_1$)	0.080	0.51	0.03	0.61	0.83	0.56	0.74	0.65	11
12($A_3B_4C_2$)	0.130	0.43	0.02	0.92	0.87	0.86	0.79	0.83	3
13($A_4B_1C_4$)	0.069	0.48	0.02	0.73	0.95	0.65	0.92	0.78	5
14($A_4B_2C_3$)	0.082	0.45	0.016	0.84	1	0.76	1	0.88	2
15($A_4B_3C_2$)	0.089	0.50	0.01	0.65	0.98	0.59	0.97	0.78	6
16($A_4B_4C_1$)	0.069	**0.41**	**0.02**	**1**	**0.96**	**1**	**0.93**	**0.96**	**1**

60.3.1 *Effect of GF Content, Load and Sliding Distance on COF*

The variation of COF w.r.t different sliding distances is depicted in Fig. 60.3. The load shared by both glass filled and Nylon 6 during sliding as depicted in Fig. 60.4, the major part of the load supported by GF resulting COF decreased. In the case of pure Nylon 6, the COF was observed more; however, further increase in glass-filled COF decreased due to rubbing of glass filled on the abrasive paper as shown in Fig. 60.5.

The COF of the Nylon 6 composites decreases as to increase load. This may be mainly due to the increase in heat generation of the surface of the specimen and contact zone of pin specimen. At 5 N load, the thermal motion of molecules of polymer observed weak, this may be because the heat generation at contact zone was low, then plastic deformation of polymer molecule cannot respond to load, and thus COF is relatively high. At 9.8 N loads, as the heat generation increases molecular segment of polymer movement gradually increases, then plastic deformation of polymer molecule can respond to load, and thus the corresponding COF gradually decreases. At 15 N loads, the heat generation increases molecular segments likely to move freely, the plastic deformed polymeric material at contact zone of pin and disc forms a low shear strength interfacial film which acts like as lubricant, and thus COF again decreases as shown in Fig. 60.5. The COF decreases

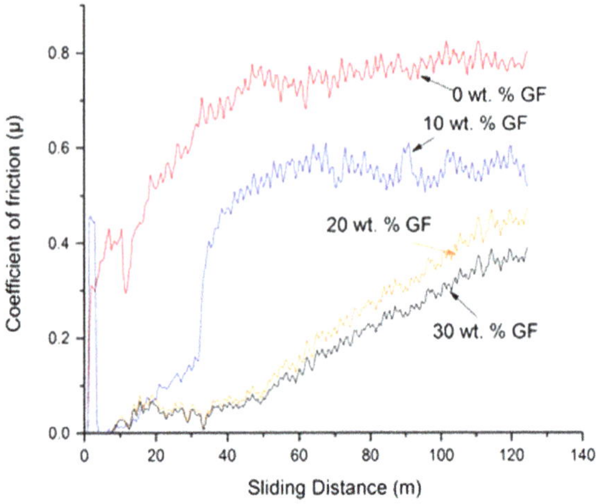

Fig. 60.3 Relation between sliding distance and coefficient of friction irrespective of load

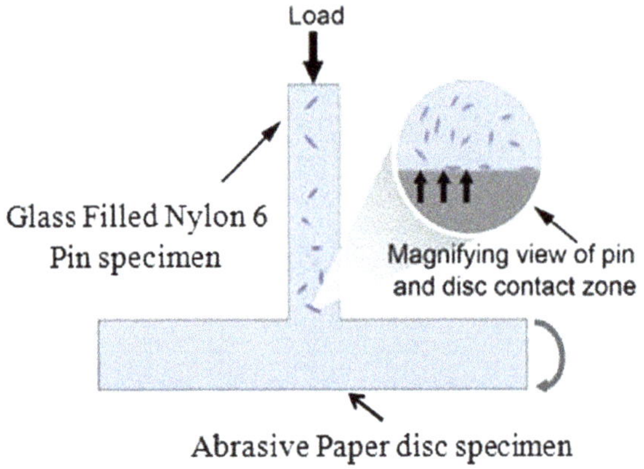

Fig. 60.4 Schematic representation of load sharing between GF and Nylon 6 [8]

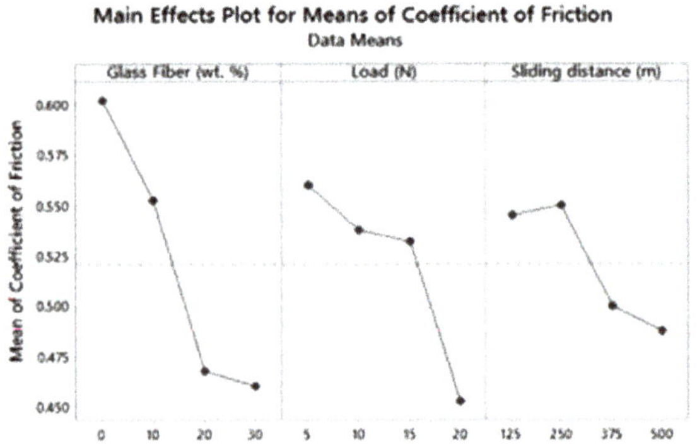

Fig. 60.5 Relation between process parameters and COF

with an increase in sliding distance for Nylon 6 composites. This may be mainly attributed to pin specimen slides over the same wear track, thus resulting in less effect of abrasiveness. Similar trends are found by other researchers [8, 9].

60.3.2 Effect of GF Content, Load and Sliding Distance on SWR

The values of SWR are observed from 0.0161 mm^3/Nm to 0.1189 mm^3/Nm. In the case of Nylon 6, SWR was very high and when glass filled increased then SWR decreased. When glass filled increases hardness and strength increases, so wear resistance improved. During sliding, load shared by glass filled and Nylon 6 is depicted in Fig. 60.4. This may be mainly attributed to improved wear resistance. The SWR decreases as an increase in load for Nylon 6 composites as shown in Fig. 60.6. Similar trends were found by other researchers [8, 9].

The variation of SWR w.r.t different sliding distances at a different GF is depicted in Fig. 60.8. The SWR of Nylon 6 composites decreased as an increase in the sliding distance. This may be mainly attributed to the repetition of the pin on same wear track gap on emery paper filled by debris.

60.3.3 Grey Relational Analysis and Implementation of Cuckoo Search Algorithm

The grey relational analysis results are shown in Table 60.3. The optimum parameter combination for minimum COF and SWR is shown in Fig. 60.7. ANOVA was done for grey relational grades, and the results are presented in Table 60.4. It was observed that glass filled has the dominant factor followed by load and sliding distance.

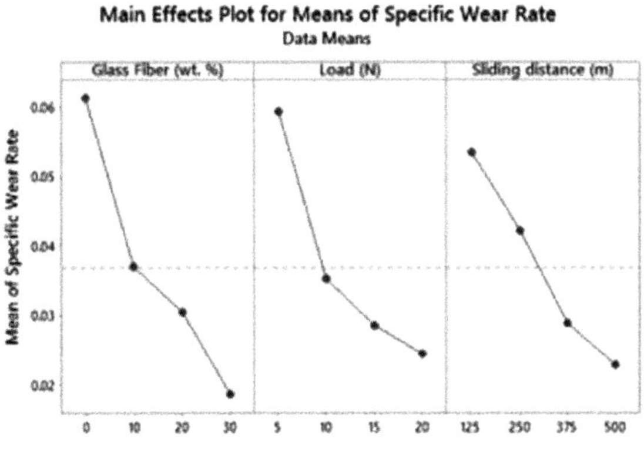

Fig. 60.6 Relation between process parameters and SWR

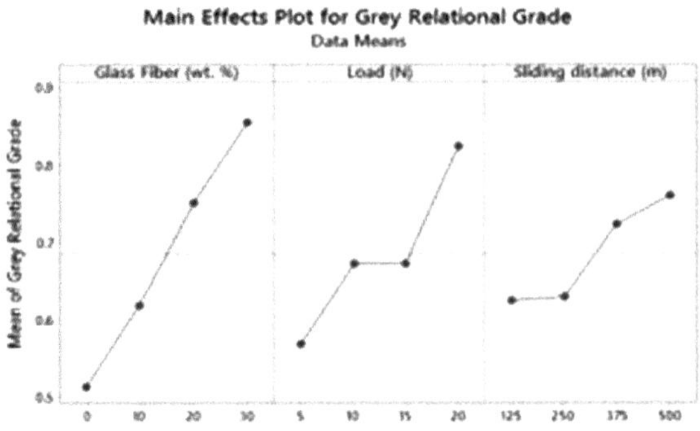

Fig. 60.7 Effect of control factors with their levels for GRG

Table 60.4 Analysis of variance of grey relational analysis

Source	DOF	Adj. SS	Adj. MS	F-cal.	P-value	% contribution
Glass filled (wt%)	3	0.26	0.08	129.85	0.000	57.92
Load (N)	3	0.13	0.04	65.07	0.000	29.02
Sliding distance (m)	3	0.05	0.01	27.25	0.001	11.74
Residual error	6	0.00	0.00	–	–	1.32
Total	15	0.46	–	–	–	–

Lower and upper limits of the parametric constraints are $0 \leq$ glass filled (wt%) ≤ 30, $125 \leq$ sliding distance (m) ≤ 500, $5 \leq$ load (N) ≤ 20. The cuckoo search parameter values are the number of nests $n = 25$, discover rate = 0.25, number of iteration = 100. Figure 60.8 shows the objective obtained during minimization for all iterations. It was revealed that firstly, the value of the objective function is at a higher value and as the iteration progresses, the objective function

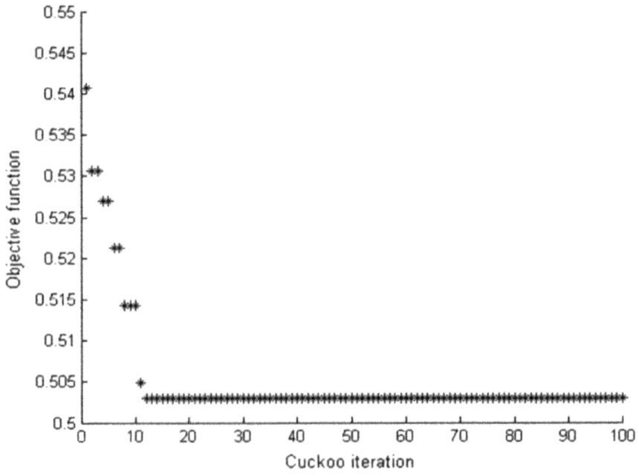

Fig. 60.8 Variation of objective function with iterations

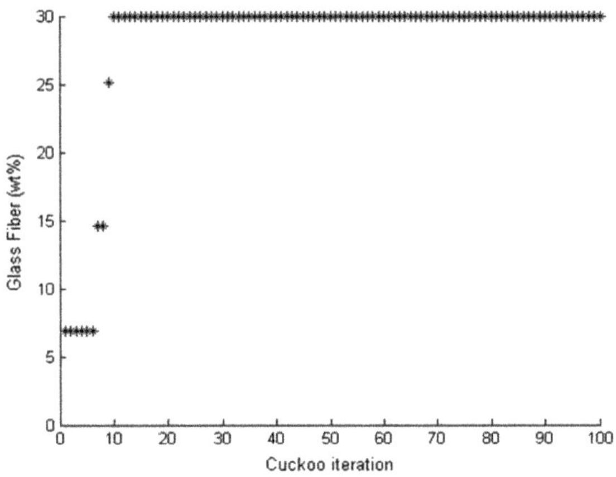

Fig. 60.9 Variation of glass filled with iterations

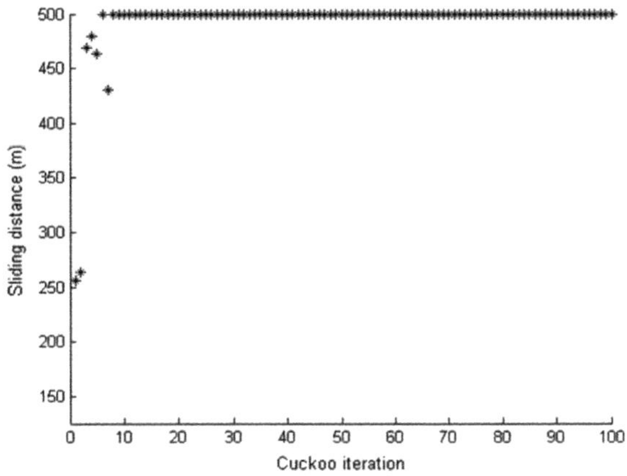

Fig. 60.10 Variation of sliding distance with iterations

values converge quickly. The best solution obtained is the objective function value of 0.5030 for glass-filled content of 30 wt%, sliding distance 500 m and load of 20 N for lower COF and SWR as shown in Figs. 60.9, 60.10 and 60.11.

60.3.4 Confirmation Tests for CSA

With the identified optimal process parameters of glass-filled content of 30 wt%, sliding distance of 500 m and load of 20 N, a confirmation experiment was conducted as shown in Table 60.5.

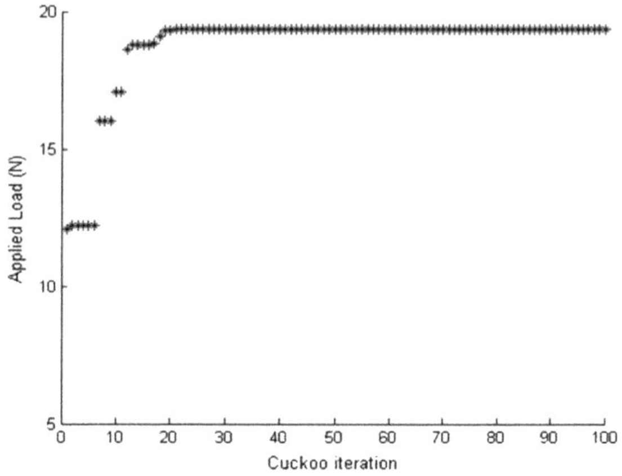

Fig. 60.11 Variation of load with iterations

Table 60.5 Measured responses of confirmation experiment for cuckoo search

	Optimal parameters	
	Prediction	Experiment
Parameter levels	$A_4B_4C_4$	$A_4B_4C_4$
COF (μ)		0.43
SWR (mm^3/Nm)		0.010
Grey relational grade (GRG)	0.988	0.967
Objective function = 1/(1 + GRG)	0.5030	0.5083

60.3.5 Worn Surface Morphology

Figure 60.12 shows a schematic illustration of abrasive wear mechanism. Figure 60.13a shows typical micrograph of unabraded counter-face of 320 grit size SiC abrasive paper. In case of low sliding distance, repetition of the pin over the wear track is less so clogging of abrasive wear debris is observed low as shown in Fig. 60.14a. However, in the case of 500-m sliding distance repetition of the pin over the wear track is more resulting in less effect of abrasiveness as shown in Figs. 60.13b and 60.14b. Consequently, decrease COF and SWR. Plastic deformation, micro-cracking and micro-cutting are the dominant wear mechanisms found in case of pure Nylon 6. This may be mainly attributed to more frictional heat generated at contact zone as shown in Fig. 60.15a. However, in the case of glass-filled Nylon 6, polymer softening decreased as depicted in Fig. 60.15b. This may be mainly attributed to the load shared by glass filled and Nylon 6 as shown in Fig. 60.4. Typical micrograph of 30 wt% GF Nylon 6 at 20 N shows more damage worn surface which results in de-bonding of the glass filled from the matrix. Consequently at 5 N, worn surface observed mild wear. Wear mechanism shows a good correlation with the trend of COF and SWR (Fig. 60.16).

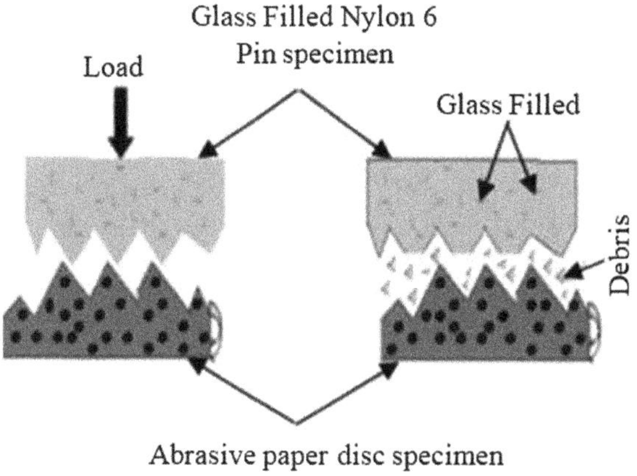

Fig. 60.12 Abrasive wear mechanism of GF Nylon 6 [8]

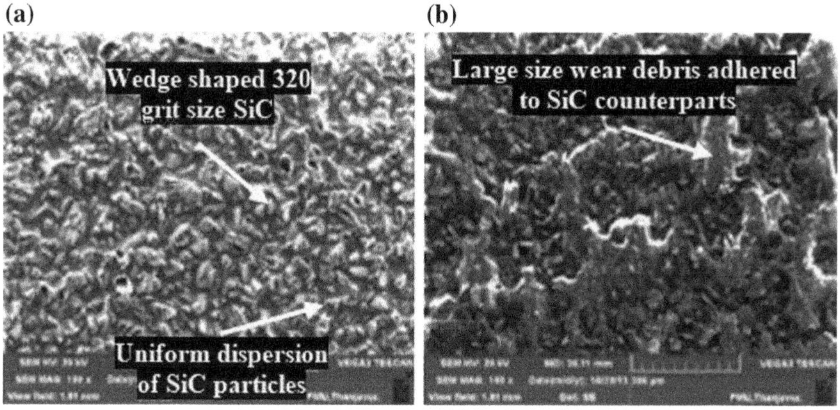

Fig. 60.13 Typical micrograph of **a** unabraded counter-face and **b** abraded counter-face at 500 m sliding distance

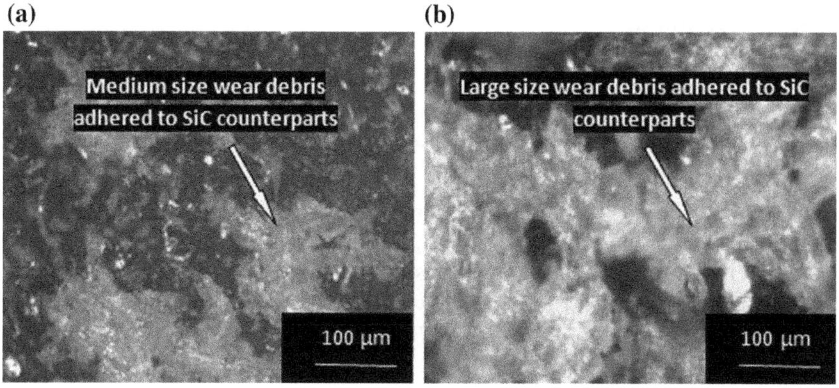

Fig. 60.14 Typical micrograph of abraded counter-face at sliding distance: **a** 125 m and **b** 500 m

(a) (b)

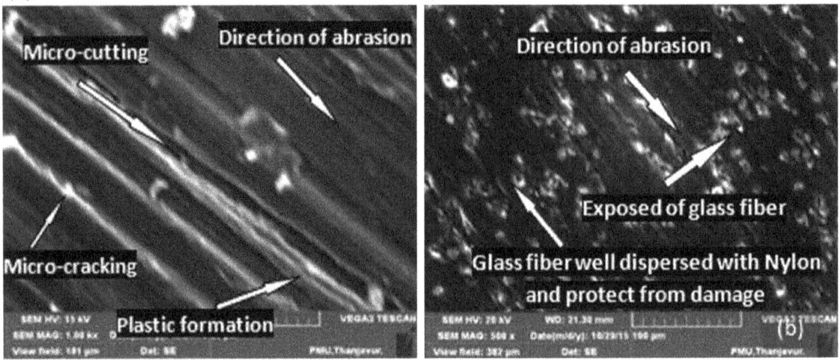

Fig. 60.15 Typical micrograph of worn surface at 20 N load and 500 sliding distance: **a** Nylon 6 and **b** 30 wt% GF Nylon 6

(a) (b)

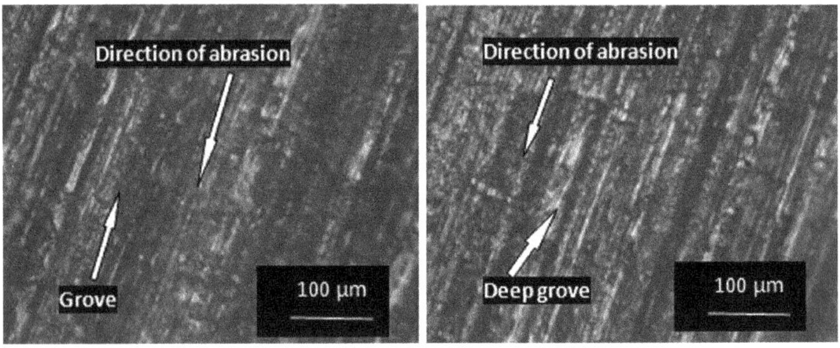

Fig. 60.16 Typical micrograph of worn surface of 30 wt% GF Nylon 6: **a** at 5 N load and **b** at 20 N load

60.4 Conclusions

From the experimental investigation, the following conclusions have been drawn:

- The present study demonstrated successful fabrication of glass-filled Nylon 6 composites using injection moulding.
- Glass-filled Nylon 6 improves the mechanical (hardness and tensile strength) and tribological properties (COF, weight loss and SWR).
- The weight loss of Nylon 6 composites increases as load increases; however, the addition of glass filled decreases the weight loss.
- The COF and SWR of glass filled Nylon 6 composites decreases as increase glass filled content and least achieved at 30 wt%.

- On the basis of ANOVA, glass filled content 57.92%, sliding distance 11.74% and load 29.02% exerts a significance influence on multiple responses.
- Cuckoo search shows that 30 wt% glass-filled content, 20 N load and 500 m sliding distance combinations of factor give the best solution.
- In case of Nylon 6 plastic deformation, micro-cracking and micro-cutting are dominant wear mechanisms; however, glass-filled Nylon 6 reduces these defects. The worn surface is characterized by the direction of abrasion.

References

1. Kumar, S., Panneerselvam, K.: Investigations on dry sliding wear behaviour of Nylon-6 and GFR Nylon-6 composites using Taguchi technique. Indian J. Eng. Mater. Sci. **25**, 89–97 (2018)
2. Zhang, S.W.: State-of-the-art of polymer tribology. Tribol. Int. **31**(1), 49–60 (1998)
3. Ramesh, B.N., Suresha, B.: Optimization of tribological parameters in abrasive wear mode of carbon-epoxy hybrid composites. Mater. Des. **59**, 38–49 (2014)
4. Kuo, Y., Yang, T., Huang, G.W.: The use of grey based Taguchi method for optimizing multi response optimization problems. Eng. Optim. **4**, 517–528 (2008)
5. Kumar, S., Panneerselvam, K.: Optimization of friction and wear of Nylon 6 and glass fiber reinforced (GFR) Nylon 6 composites against 30 wt.% GFR Nylon 6 Disc. J. Adv. Res. Mater. Sci. **19**, 14–32 (2016)
6. Senthilkumar, N., Tamizharasan, T., Anandakrishnan, V.: An hybrid Taguchi-grey relational technique and cuckoo search algorithm for multi-criteria optimization in hard turning of AISI D3 steel. J. Adv. Eng. Res. **1**(1), 16–31 (2014)
7. Yang, X.S., Deb, S.: Engineering optimisation by cuckoo search. Int. J. Math. Model. Numer. Optim. **1**, 330–343 (2010)
8. Kumar, S., Panneerselvam, K.: Two-body Abrasive wear behavior of Nylon 6 and glass fiber reinforced (GFR) Nylon 6 composite. Proc. Technol. **25**, 1129–1136 (2016)
9. Jafrey, D.D., Panneerselvam, K.: Abrasive wear of polypropylene/Cloisite 30B/Elvaloy AC 3427 nanocomposites. J. Compos. Mater. **52**(13), 1833–1843 (2018)

Chapter 61
Fabrication of Hexagonal Bar from Aluminum Alloy AA6063 Scrap by Frictional Stir Back Extrusion on Milling Machine

Gizaw Yohannes, Habtamu Beri and Perumalla Janaki Ramulu

Abstract Friction stir back extrusion (FSBE) is a novel and an efficient method of manufacturing sounded products by using specialized tool in one step (without melting of scrap). It can be saved about 40% of material, 26–31% of energy consumption, and 16–60% of labor costs compared with conventional extrusion. The main aim of this study is fabrication of hexagonal bar from aluminum alloy scrap by FSBE. The material used for this study is AA6063 as a scrap and H-13 tool steel as a die. Considered parameters were chamber rotational speed and vertical die transverse speed as major parameter on conventional milling machine. This work has proved the feasibility to extrude hexagonal-shaped bar by using FSBE by fixed die and rotates chamber. So far, none of the researchers carried out experiment by fixed die and rotate chamber in FSBE, and also no one applied FSBE to produce other than circular shape. Aluminum hexagonal-shaped bar is used as a supporter in medical and marine industry, as a raw material to the mass production of hexagonal head aluminum nut and bolt, Allen wrenches, etc. Extruded bar was defect-free, non-blurred grain boundary, and clear surface than the base material. The result showed that the compression strength of extruded sample was higher than the base material. The optimum rotational speed of chamber to produce hexagonal bar with the minimum twist angle is 250 rpm and transverse speed 6.25 mm/min. If the rotational speed increases, twist angle also increased.

G. Yohannes · H. Beri
Program of Mechanical Design and Manufacturing Engineering, School of Mechanical, Chemical and Materials Engineering, Adama Science and Technology University, Adama, Ethiopia
e-mail: gizawgotayohannes@gmail.com

H. Beri
e-mail: habtamu_beri@yahoo.com

P. J. Ramulu (✉)
Program of Mechanical Design and Manufacturing Engineering & Director of CoEAME, School of Mechanical, Chemical and Materials Engineering, Adama Science and Technology University, Adama, Ethiopia
e-mail: perumalla.janaki@astu.edu.et

R. G. Narayanan et al. (eds.), *Advances in Computational Methods in Manufacturing*, Lecture Notes on Multidisciplinary Industrial Engineering,
https://doi.org/10.1007/978-981-32-9072-3_61

Keywords Hexagonal bar · Friction stir back extrusion process · Aluminum alloy scrap · H-13 tool steel die and chamber

61.1 Introduction

Friction stir extrusion (FE) was developed by The Welding Institute (Cambridge, UK) in the early 1990s. It was derived from friction stir processing as a process of forming or reforming ceramic/metal composite material. Over the past two decades, significant amount of work had done on multiple aspect of friction stir process. Along with the rapid development of automotive, aerospace, and manufacturing industry in recent years, novel technologies like near-net shape manufacturing, additive manufacturing, solar energy jet, and fuel cell battery plays a critical role in producing safe, economical, and durable products. Friction stir extrusion process can meet this need as an innovative and economical material processing method as it directly converts metal wastes without melting to useful products. Compared with traditional manufacturing processes, it has more efficient, green, energy saving, and environmentally friendly technologies [1]. Friction stir extrusion relies on extensive plastic deformation and basic frictional heating, consolidates, and extrudes the small block, chips, and scrap directly into functional form. For example, direct recycling aluminum can save about 40% of material, 26–31% of energy consumption, and 16–60% of labor costs. The friction extrusion process falls into the category of direct recycling methods and it has all the advantages [2].

Aluminum and its alloys had received much attention in aerospace, automobile, and construction industries; due to their high strength-to-weight ratio, good formability, excellent corrosion resistance, low maintenance, good toughness, and high strength-to-weight ratio [3]. Tang and Reynolds [4] applied friction stir extrusion (FSE) technology and produced fully consolidated wires from aluminum alloy AA2050 and AA2195 machining chips. The fully equiaxed recrystallized microstructure in extruded wire was observed by optical microscopy. But, neither the trend of compression properties nor hardness test has been discovered. Abu-Farha [5] applied the derivative of FE which is called frictional stir back extrusion to produce sounded, defect-free aluminum alloy AA6063-T52 tube. A rotating tool is plunged into aluminum round bar specimens at a selected feed rate, forcing the processed material radially outward and thus forming tubes. Khorrami and Movahedi [6] fabricated the aluminum tubes by using a novel process of friction stir back extrusion and also investigated in detail the microstructural evolution as well as the mechanical properties of the formed tube. Behnagh et al. (2014), produced fully consolidated wires from aluminum alloy AA7277 machining chips by the friction stir extrusion. Components were stationary cartridge and a rotating plunger with a scroll-faced head. Hangai et al. [7] fabricated the AA1050 tubes filled with ADC12 foam by frictional stir back extrusion and foam able precursor was coated on the surface of the inner wall of the fabricated tube.

Babu et al. [8] carried out the hardness properties of friction stir processed AZ31 magnesium alloy wire at different processing conditions. The parameters were studied at the three different traveling speeds: 20, 30, and 40 mm/min with two different rotation speeds 600 and 800 rpm. The smallest precipitate size is obtained at spindle speed and feed rate combination of 800 rpm and 40 mm/min. Mechanical characterization and manufacturing of square bar was made of aluminum scraps through friction stir back extrusion process using a conventional lathe machine [9]. Baffari et al. [10] investigated the effect of process parameters and initial temper of the base metal on the wire quality by frictional stir extrusion. They considered rotational speed and vertical applied force. Extrusions were carried out with varying rotational speed (100, 200, and 300 rpm) and extrusion force (26.7, 35.6, 44.5, and 53.4 kN). Die and tool machined from H-13 tool steel and billet AA2050 chip and best sounded wire fabricated by 35.9 kN and 100 rpm are optimum. Although many researchers have done on frictional stir welding, frictional stir extrusion and channeling, some researchers experimentally proved frictional stir back extrusion by rotating the die and fixed chamber to produce tube, wire, rod, and c-channel. But no one applied frictional stir back extrusion to produce hexagonal-shaped bar which was proposed in this work. No work has been reported in the area of studying mechanical properties like hardness and compression strength of aluminum alloy AA6063 friction extruded bar, which was carried out in the present work.

61.2 Materials and Methodologies

The materials used to perform this work are as follows

- H-13 tool steel die and chamber
- Aluminum alloy billet
- 210 mm length M16 × 2 bolts
- Three-jaw chuck
- Rotary table
- T-headed bolt

The equipment to facilitate the friction stir back extrusion (FSBE) to produce hexagonal bar were adjustable wrench, chuck jaw Allen key, file and silicon carbide paper, small hammer, and box wrench. Table 61.1 shows the equipment and their specification used for the FSBE.

Universal Milling Machine

In the below Table 61.2, the maximum and minimum torque of milling machine was calculated by keeping other parameters constant. The specification of the milling machine was as indicated in Table 61.2.

Table 61.1 Machinery used for experiment and sample test

No.	Equipment	Tasks	Machine specification
1.	Universal milling machine	To fabricate hexagonal bar and produce key way in chamber	Excel vertical milling machine, FU-281
2.	CNC milling machine	Produce spiral in face of die	Fanuc series Oi Mate-MB
3.	Lathe machine	Turn, taper turn, face, threaded out, and bored of chamber and die	Excel XL-420
4.	Forging machine	Make hexagonal-shaped hole at center of die	Forge machine new 2010 version
5.	Drilling machine	To make hole in die	Universal drill TSAR-25
6.	Heat treatment machine	Heat treat both die and chamber up to HRC-55	Fours nagat 44550-montoir de Bretagne-France
7.	Universal test machine (UTM)	Compression strength of sample	GUNT WP 310 hydraulic
8.	Optical microscopy	Microstructural test	Magnification X50, X100, X200, X500, and X1000
9.	Vickers hardness tester	Hardness test	IS 1754/2002

Table 61.2 Specification of universal milling machine code F-281 (Excel machine manual)

No.	Spindle description	Units	Maximum range
1.	Main-driven range normally	RPM	40–2000
2.	Motor power	kW	5.5
3.	Maximum torque	Nm	1313.7
4.	Minimum torque	Nm	26.27
5.	Feed motor power	kW	1.5
6.	Feed motor rotational speed	RPM	920
Feed ranges			
	Longitudinal transverse	mm/min	12.5–630
		″	″
7.	Vertical, horizontal, and transverse.	″	5.2–262
8.	Gear transmission ratio	–	1:1

61.2.1 Design and Fabrication of Die and Chamber

Geometrical Design of Die and Chamber
The 3D modeling was made before manufacturing die and chamber. Figure 61.1a, b shows the surface of chamber and scroll-faced die with hexagonal-shaped center hole.

Chemical Composition of Aluminum Alloy Billet
The chemical composition of aluminum alloy billet was tested using spectrometer instrument. The obtained chemical composition of billet is shown in Table 61.3 and identified as AA 6063 alloy.

Figure 61.2 shows the scrap of aluminum alloy which was used for the FSBE to produce the hexagonal bars.

61.2.2 Process Parameters of Milling Machine to Fabricate Hexagonal-Shaped Bar

The process parameter like die rotational speed and plunge rate were varied by keeping die, chamber, plunge diameters, and contact land length of die parameters constant. The selected rotational speed was 200, 250, 315, 400, 500, and 630 rpm and with transverse speed 12.5 mm/min as shown Table 61.4. The range of rotational speed and plunge rate were considered based on previous research work [2]. The process parameters of frictional stir extrusion were optimized and decided the

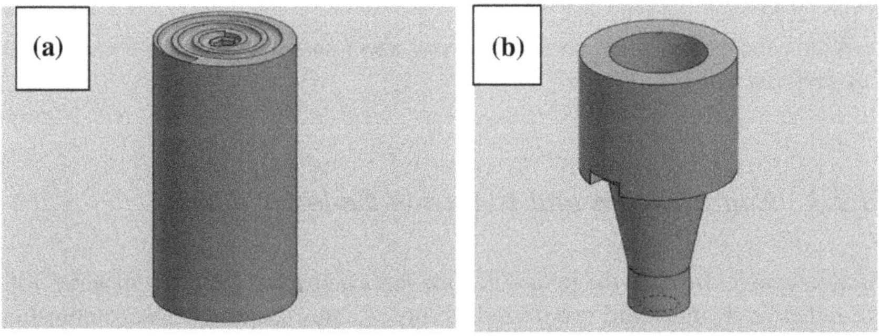

Fig. 61.1 Geometrical design of die (**a**) and chamber and (**b**) by solid work

Table 61.3 Chemical composition of base material (wt%)

Si	Fe	Cu	Mn	Mg	Zn	Sn	Cr	Pb	Ni	Al
0.58	1.35	0.017	0.014	0.7	0.092	0.039	0.033	0.015	0.019	97

Fig. 61.2 Aluminum alloy scrap chips

Table 61.4 Friction stir back extrusion process parameters

Test number	Die rotational speed (rpm)	Transverse speed of milling table (mm/min)
1	200	6.25
2	250	6.25
3	315	6.25
4	400	6.25
5	500	6.25
6	630	6.25

plunge rate 14 mm/min, hole size 5 mm, and initial contact land length was 2 mm to produced sounded wire [7].

Within the above given rotational speed and transverse speed, the experiment was performed.

61.2.3 Consolidation and Extrusion Period of Time

Friction consolidation is the period of time that compacted the scrap material with in cavities of chamber and maximized density of chips and scrap. The combination of high temperature and plastic deformation to be made the scrap material highly compacted. Friction extrusion is the whole process that includes consolidation, stirring, and extrusion by using scroll- or spiral-shaped die feed to the center plasticized material. The local temperature of extruded material increased up to 0.6–0.9 T_m [11]. In this experiment, the transverse feed was given by manually up to consolidate aluminum alloy scrap by above 0.6 T_m and engaged automatic feed after consolidate of material. The extrusion period was taken 1.28 min to each sample using automatic feed.

61.2.4 Fabrication of Hexagonal Bar

Specimen Fabrication Procedure

To produce sample bar, the prepared scrap was compacted within cavities of chamber using small-sized ball pen hammer. This pre-compact process was used to increase the length of output hexagonal bar. Then the compacted chips and scrap were taped by a circular disc 40 mm diameter and 5 mm thickness. The taper disc was made from the same aluminum alloy the chips made of to hold the scrap and increases friction between scroll-faced die up to consolidate. It was turned by lathe machine at high-speed steel cutter. To produce hexagonal sample bars, the compacted chamber with scrap was installed on the spindle of milling machine and also die was fixed milling machine table with proper alignment as indicated in Fig. 61.3. The consolidation time depends on rotational speed of chamber. Then after chips are plasticized automatic transverse gear engaged. In this process the machine table moved up 6.25 mm distance per one minute. Then after the sample bar extruded, the die was removed by engaging the clash downward to change feed direction.

Fig. 61.3 Setup of friction stir back extrusion process on milling machine

61.3 Results and Discussion

61.3.1 The Fabricated Die and Chamber

The scroll-faced die and one side cavity chamber are the basic equipment in friction stir back extrusion (FSBE). The function of this die and chamber is to consolidate, stir, and extrude aluminum scrap. To resist wear occurrence in the process of stirring and hexagonal-shaped sample bar production, it should be heat treated. The color looks like black after heat treatment as Fig. 61.4.

61.3.2 Extruded Samples at Different Speeds and Discussion

Figure 61.5 shows the samples fabricated at different rotational speed and their shapes at each rotational speed. For example, sample one is extruded at 630 rpm and 6.25 mm/min, sample two 500 rpm and 6.25 mm/min, etc.

Figure 61.5 shows the possibilities of FSBE to fabricate hexagonal-shaped bar and other shapes from aluminum alloy scrap. Most of extrudes twisted and circular shapes. The reason of twist is, in FSBE die shape, machine power and extruded material properties affects the shape, length, and mechanical properties of extrudes. Sample one is performed at the rotational speed of 630 rpm and transverse speed of 6.25 mm/min. In this process parameter, the extruded product is resulted with large twisting angle almost like a circular rod shaped. This sample is longer than the other and is about 280 mm in length and circular in shape due to higher rotational speed. The rotational speed played an important role and contributed 71.6% to the overall extrusion parameters. If the rotational speed increases, the possibility of consolidated aluminum to be changed into liquid form is high. In this time, the die land length must be done longer than the other because it takes few second to solidify for melted aluminum. To this work, the die center hole depth (hexagonal shape former) was 2 mm. the much-plasticized aluminum extrude out immediately by not formed hexagonal shape. It made circular shape due to high rotational speed and shortage of depth of hexagonal shape formed hole.

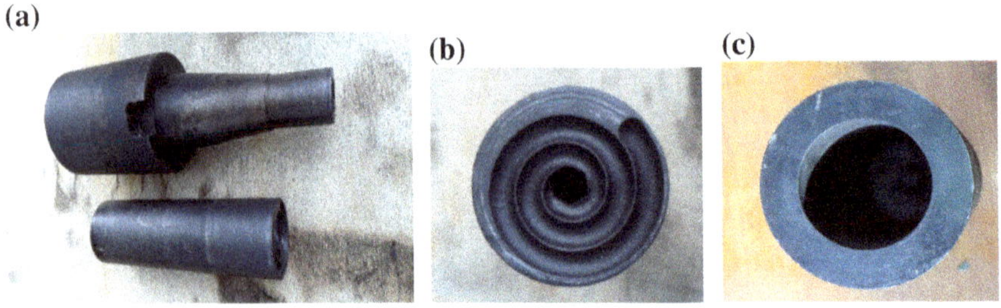

Fig. 61.4 Surface of die and chamber (**a**), spiral face of die, (**b**) and cavities of chamber (**c**)

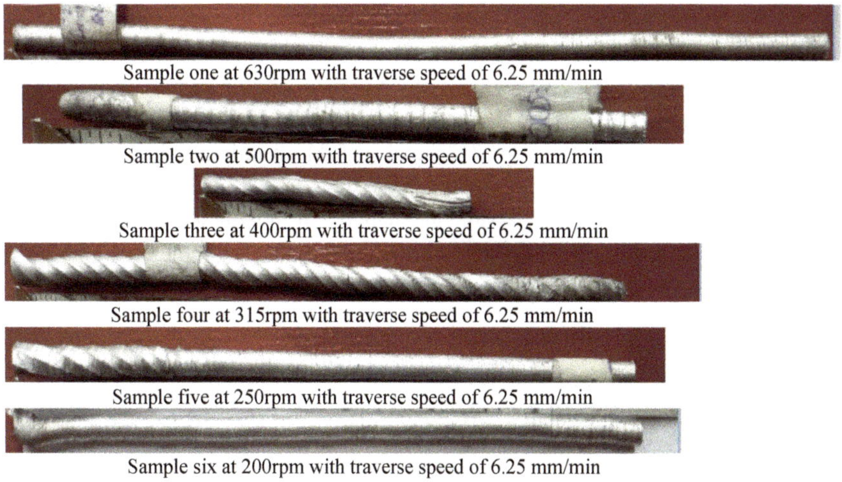

Fig. 61.5 Samples extruded by FSBE at different rotational speed

Sample two also almost the same shape to sample one, it is some hexagonal form related with sample one. It has been performed by rotational speed of 500 rpm and constant transverse speed 6.25 mm/min. The consolidate time took 5 min and extrusion time 1.28 min and the result is shown in Fig. 61.5 as 10.5 mm length and 6 mm diameter extrudes by friction stir extrusion. As explained above in sample one, rotational speed changed also resulted in shape change of the extruded products. The solution to this problem is design and manufacture die by deepened on each rotational speed of chamber. And also die center shape formed hole depth varied (means above 2 mm). Similarly, the other samples are manufactured at different process parameters as shown in Table 61.4.

Conclude to all the above discussion, the optimum rotational speed was 250 rotational speed and 6.25 mm/min transverse speed. In the case of surface smoothness, the optimum rotational speed was 325 rpm and the consolidation time also affects the physical properties of sample or hexa-bar (Table 61.5).

Table 61.5 Duration of scrap consolidation and extrusion time at each rotational speed

Sample No.	Rotational speed	Transverse speed (mm/min)	Consolidation time (min)	Extrusion period (min)
1.	630	6.25	5	1.28
2.	500	6.25	7	1.28
3.	400	6.25	7	1.28
4.	315	6.25	8	1.28
5.	250	6.25	9	1.28
6.	200	6.25	10	1.28

61.4 Conclusions

Defect-free sounded hexagonal-shaped bar has been fabricated by using FSBE process from aluminum alloy 6063 scrap. The considered process parameters are constant transverse speed 6.25 mm/min and variable rotational speed (200, 250, 315, 400, 500, and 630 rpm) of milling machine FU-281. From this interval of rotational speed, the accurate hexagonal-shaped bar obtained at 250, 315, and 400 rpm and all others are completely circle.

References

1. Li, X., Tang, W., Reynolds, A.P., Tayon, W.A., Brice, C.A.: Strain and texture in friction extrusion of aluminum wire. J. Mater. Process. Technol. **229**, 191–198 (2016)
2. Hosseini, A., Azarsa, E., Davoodi, B., Ardahani, Y.: Effect of process parameters on the physical properties of wires produced by friction extrusion method. Int. J. Adv. Eng. Technol. **3**(1), 592 (2012)
3. Behnagh, R.A., Mahdavinejad, R., Yavari, A., Abdollahi, M., Narvan, M.: Production of wire from AA7277 aluminum chips via friction-stir extrusion (FSE). Metall. Mater. Trans. B **45**(4), 1484–1489 (2014)
4. Tang, W., Reynolds, A.P.: Production of wire via friction extrusion of aluminum alloy machining chips. J. Mater. Process. Technol. **210**(15), 2231–2237 (2010)
5. Abu-Farha, F.: A preliminary study on the feasibility of friction stir back extrusion. Scr. Mater. **66**(9), 615–618 (2012)
6. Khorrami, M.S., Movahedi, M.: Microstructure evolutions and mechanical properties of tubular aluminum produced by friction stir back extrusion. Mater. Des. **1980–2015**(65), 74–79 (2015)
7. Ansari, M.A., Behnagh, R.A., Narvan, M., Naeini, E.S., Givi, M.K.B., Ding, H.: Optimization of friction stir extrusion (FSE) parameters through Taguchi technique. Trans. Indian Inst. Met. **69**(7), 1351–1357 (2016)
8. Babu, J., Anjaiah, M., Mathew, A.: Experimental studies on friction stir processing of AZ31 Magnesium alloy. Mater. Today Proc. **5**(2), 4515–4522 (2018)
9. Gelaw, M., Ramulu, P.J., Hailu, D., Desta, T.: Manufacturing and mechanical characterization of square bar made of aluminium scraps through friction stir back extrusion process. J. Eng. Des. Technol. **16**(4), 596–615 (2018)
10. Baffari, D., Reynolds, A.P., Li, X., Fratini, L.: Influence of processing parameters and initial temper on friction stir extrusion of 2050 aluminum alloy. J. Manufact. Proc. **28**, 319–325 (2017)
11. Khodabakhshi, F., Nosko, M., Gerlich, A.P.: Effects of graphene nano-platelets (GNPs) on the microstructural characteristics and textural development of an Al–Mg alloy during friction-stir processing. Surf. Coat. Technol. **335**, 288–305 (2018)

Chapter 62
Fabrication, Characterization and Evaluation of Mechanical Properties of Aluminium Hybrid Matrix Composite (Al6063/SiC-Bagasse Fly-Ash)

Getachew Gashaw, Perumalla Janaki Ramulu and Ch. Venkatesh

Abstract The fabrication and mechanical performance of Al-Mg-Si alloy matrix composites reinforced with silicon carbide (SiC) and bagasse fly-ash has been explored in this work. The fabrication and experimental investigation of mechanical properties of aluminium matrix (Al/SiC) composite with and without addition of bagasse fly-ash by varying wt% of reinforcements were investigated. Chemical composition of reinforcement, weighted required amount of matrix and reinforcements, powder mixing, stirring the slurry at 500 rpm speed and 5–10 min were the parameters used to develop the Al/SiC-bagasse fly-ash hybrid composite. Bagasse fly-ash particulates added with 7.5, 10, and 12 wt% SiC were utilized to prepare 10 and 12 wt% of the reinforcing phase with Al-Mg-Si alloy as matrix using stir casting method. The cast samples were machined for evaluating the properties by performing tensile testing, compression strength testing, and micro-hardness measurement. The results disclosed that the hardness of the hybrid composites increased with increase in bagasse fly-ash content with a maximum increment of 17% observed for the Al6063-7.5% SiC-10% bagasse ash composition (in comparison with the Al-10 wt% SiC-7.5 fly-ash reinforced composition). Tensile strength reduced of maximum value of 90–68 MPa at 7.5 SiC-7.5 bagasse ash compositions.

G. Gashaw · Ch.Venkatesh
Program of Mechanical Design and Manufacturing Engineering, School of Mechanical, Chemical and Materials Engineering, Adama Science and Technology University, Adama, Ethiopia
e-mail: gech21912@gmail.com

Ch.Venkatesh
e-mail: venkymech0607@gmail.com

P. J. Ramulu (✉)
Program of Mechanical Design and Manufacturing Engineering & Director of CoEAME, School of Mechanical, Chemical and Materials Engineering, Adama Science and Technology University, Adama, Ethiopia
e-mail: perumalla.janaki@astu.edu.et

R. G. Narayanan et al. (eds.), *Advances in Computational Methods in Manufacturing*, Lecture Notes on Multidisciplinary Industrial Engineering,
https://doi.org/10.1007/978-981-32-9072-3_62

Keywords SiC · Bagasse fly-ash · Stir casting · MMCs · Al/SiC

62.1 Introduction

Aluminium matrix composites (AMCs) have significant properties than other metal matrix composites with respect to high strength to weight ratio. The mechanical properties and wear resistance were improved by using SiC, TiC, TiB2 and fly-ash particulates to reinforce Al alloys [1]. Among all reinforcements, SiC was chemically compatible with aluminium and forms an adequate bond with the matrix without developing inter-metallic phase and had other advantages such as excellent thermal conductivity, good workability and low cost revealed in the literature [2]. In recent years, the use of fly-ash as reinforcement material in Al alloys was reported to be desirable from both environmental and economic significance due to its low cost. To overcome this, many research and development programmes were focused on Al metal matrix composites using low-cost industrial waste by-products as the reinforcement particulate [3]. Investigation of mechanical behaviour of aluminium alloy reinforced by nano- and micro-hard particles such as Al_2O_3 and SiC was a motivating area of research [4]. The present work, Aluminium hybrid matrix composite with reinforcements such as, SiC particle and waste of sugar cane bagasse fly-ash particle. Al/SiC-bagasse fly-ash composite provides more wear resistance, high coefficient of thermal expansion and high strength. Silicon carbide (SiC) is hard, wear resistance, high heat resistance ceramic material. With respect to low cost, less density and availability bagasse fly-ash is one of them which has inexpensive, low density reinforcement and available in large quantities by-product during combustion of coal in thermal power plants as solid waste. By which, composite material made with fly-ash as reinforcement are likely to overcome the cost barrier for extensive applications in automotive industries.

Metal matrix hybrid composites are the second-generation composite materials that have potential to substitute single-reinforced composite by improving properties [5, 6]. Vijayakumar et al. [7] studied the tensile strength and hardness of Al6061/SiC composite by varying the wt% SiC reinforcements 3–18%. The composite was manufactured by using stir casting of 600 rpm, 5 min stirring speed, and stirring time, respectively. The ultimate tensile strength and hardness results showed that the tensile strength and hardness of the developed metal matrix improved during increasing the wt% of SiC. Imran et al. [8] fabricated Al7075/bagasse ash–graphite of hybrid metal matrix composite (HMMC) by stir casting method. Reinforcements were added with three cases: in first case, 2% constant of bagasse ash with varying graphite 1, 3 and 5%; in second case, 4% bagasse ash and in third case, 6% constant with varying graphite percentage. Results showed that 6% bagasse ash with varying graphite 1, 3 and 5% more enhanced mechanical properties achieved as compared to case one and two of bagasse–ash combination. The hardness, ultimate tensile strength and yield strength of the developed composites were gradually increased with increasing the wt% of reinforcement, while ductility decreased. Similar studies were also carried out on MMCs and their characterizations with respect to their mechanical behaviour [9, 10]. Failure prediction of FRP laminates was studied for a new generation of materials to an aerospace application [11]. The present work significance is natural fibres have good mechanical property, but a finite resource. The usage of their wastage, such as fly-ash and plant wastes from industries as reinforcement, is recommendable. Bagasse fly-ash is a waste by-product of coal combustion in thermal power plants which is available in large quantities in Ethiopian sugar factories. The usage of this waste fly-ash particle reinforced into aluminium alloys, which can effectively enhance the various mechanical properties of the aluminium alloy. Automotive heavy components fabricated from metals, Due to an increase in the weight of components, consumption of fuel increases. Increasing volume percentage of SiC reinforcement the weight of the composite increase. Bagasse fly-ash leads to decrease in the density of the composite when using some percentage with silicon carbide as reinforcement for developing Al/SiC/fly-ash composite.

62.2 Experimental Methodology

62.2.1 Matrix and Reinforcement Materials

Base material-AA 6063 alloy

The Al6xxx alloys were chosen because of its wide use in automotive components and architectural materials, including pistons, disk rotors, cylinder liners and decorative parts, rivets, etc. AA 6063 was the most widely used because of its extrudability and used for the automotive space frame parts [12].

Reinforcement material–Silicon carbide (SiC)

Silicon carbide material was chosen as a reinforcement material because it is used in high hardness and high thermal resistance applications, such as automotive brakes, cylinders, clutches, valves and bullet proof vests, semiconductor electronics etc. It is also applicable in abrasive cutters, cutting tools, grinding wheels, nuclear reactors due to high temperature resistance, high thermal conductivity and low nuclear activation [13]. The grain size of silicon carbide varies micro to nano sizes. One way of improving mechanical properties of composite is reducing grain size of reinforcements depends on processing methods. The very small sizes of powder particle of SiC create agglomeration (mud) during casting of poor stirring. To overcome such problems, proper stirring action is used from 5–10 min. Figure 62.1 shows the micron-level silicon carbide powders which are used as a reinforcement ceramic material to fabricate Al/SiC-bagasse fly-ash hybrid composite.

Bagasse fly-ash

Bagasse is produced from sugarcane plant during processing in sugar production. For the present work, the bagasse ash was taken from Wonji Shewa Ethiopian

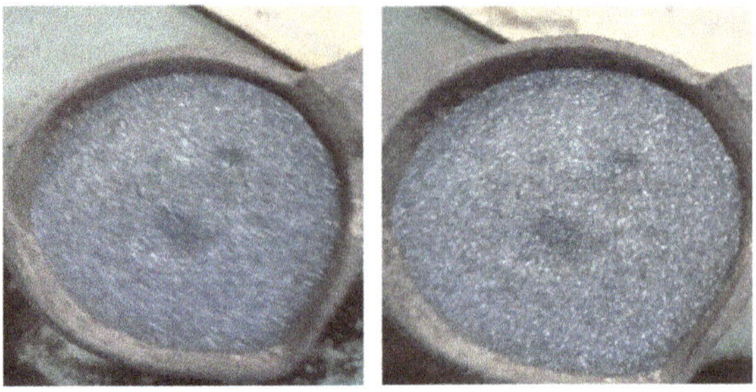

Fig. 62.1 Silicon carbide powders

Sugar Factory. Bagasse fly-ash was chosen due to low weight, density and increase in the strength. The ash was manually crushed and screened using sieves to make fine powder particles.

62.2.2 Methods

Physical Properties and chemical composition of bagasse fly-ash.

Density—1.96 g/cc and the specific gravity of fly-ash varies in the range of 0.6–2.8 gm/cc. Table 62.1 shows the chemical composition of the bagasse fly-ash by chemical analysis. The chemical composition analysis was performed on Dangote cement industries in Ethiopia.

The sugarcane bagasse ash is rich in SiO_2, which is good to replace SiC and used as reinforcement for metal matrix composites. In [14] research work, it was reported that alumina (Al_2O_3) also promise reinforcement material for production of metal matrix composite.

Powders mixing

Al-5% volume of SiC powders are mixed by stirrer on milling machine. The calculation for mass of aluminium, silicon carbide and bagasse fly-ash powder is needed for a sample of 130 mm length and 15 mm diameter. Pre-hearing of reinforcements from 600–700 °C of temperature before added in molten metal. Weight of powder added into matrix depends on the sample size. The blade of stirrer assembled on rotating part of milling machine speed of 500 rpm, 5–10 min stirring speed and time respectively, used to mix matrix and reinforcements.

Table 62.1 Chemical composition of bagasse ash in weight percentage

Composition wt%	SiO_2	Al_2O_3	Fe_2O_3	CaO	MgO
	%	%	%	%	%
	66.55	10.25	5.48	3.26	1.17

Mould preparation and fabrication the composite
To facilitate the easy removal of compacted specimen, tapered section has been used. The size of the mould depends on the size of samples which were used for tests. The required amount of reinforcements and matrix for each sample weighed and added in the crucible. Three samples were manufactured on one mould with the same composition of reinforcement. This is for the reputation of three experiments.

62.2.3 *Sampling and Sample Method*

In total, 8 number of experiments were conducted with three reputations for each type of mechanical testes. The wt% composition of reinforcements was taken as per available literatures, which has highest improvement in mechanical properties. The composition was 7.5, 10 wt% of SiC and 7.5, 10, 12 wt% of bagasse fly-ash and form eight combination of experiments. For specimen preparation with different percentage, composition of reinforcements manufactured was shown in Tables 62.2 and 62.3.

62.2.4 *Equipment Used for This Work*

Table 62.4 shows the equipment used for testing mechanical properties of the developed composite. Casting set-up was used to melt the matrix aluminium alloy during fabrication of Al/SiC–bagasse ash composite.

Table 62.2 Composite content in wt% of reinforcements

S. No	wt% of silicon carbide (SiC)	wt% of bagasse fly-ash	Al/SiC/bagasse fly-ash
1	7.5	7.5	Al6063-7.5% SiC-7.5% bagasse ash
2	7.5	10	Al6063-7.5% SiC-10% bagasse ash
3	7.5	12	Al6063-7.5% SiC-12% bagasse ash
4	10	7.5	Al6063-10% SiC-7.5% bagasse ash
5	10	10	Al6063-10% SiC-10% bagasse ash
6	10	12	Al6063-10% SiC-12% bagasse ash

Table 62.3 Composite content in wt% of silicon carbide only

S. No.	wt% of Al	wt% of SiC	Al/SiC
1	92.5	7.5	Al6063-7.5% SiC
2	90	10	Al6063-10% SiC

Table 62.4 Equipment used for this work

S. No.	Equipment	Task (Test name)	Machine model
1	Universal test machine (UTS)	Tensile test	WP 310 HYDROLIC UTM-Gunt Hamburg
2	Casting set-up	To melt and stir materials	

62.2.5 Mould Preparation and Fabrication

The sand mould was prepared for fabrication of Al/SiC/fly-ash hybrid composite based on the size of required specimens. 30 mm diameter and 150 mm length, 20 mm diameter and 60 mm length, specimen sizes were prepared for tensile test and compression test, respectively.

Fabrication

The matrix aluminium melted in a crucible which is placed inside furnace. It was melted at a temperature of 700 °C and reinforcements added and stirring with a molten metal for 5–10 min. The stirrer blade was assembled on hand drilling machine.

62.2.6 Tensile Test Specimen

The tensile test specimens of metal matrix composites were prepared based on the ASTM Standard D 3552. The flat and round configurations of metal matrix specimens are adopted. If the composite produced by diffusion bonding, flat shape of specimen would produce. In other case, metal matrix composite produced by various liquid infiltrations and other methods used for producing massive materials is better suited to circular cross-sectional shapes.

62.3 Results and Discussion

62.3.1 Fabrication of Al/SiC/Bagasse Ash Composite

Figure 62.2 shows the manufactured samples removed from moulds which were ready for machining and prepare samples.

Machining

Machining is performed on laze machine using carbide cutting tool. The specimens prepared based on ASME standard to conduct tensile, compression and hardness. Figure 62.3 shows the specimens machined and prepared for tensile tests.

Fig. 62.2 Manufacturing of Al/SiC-bagasse ash composite

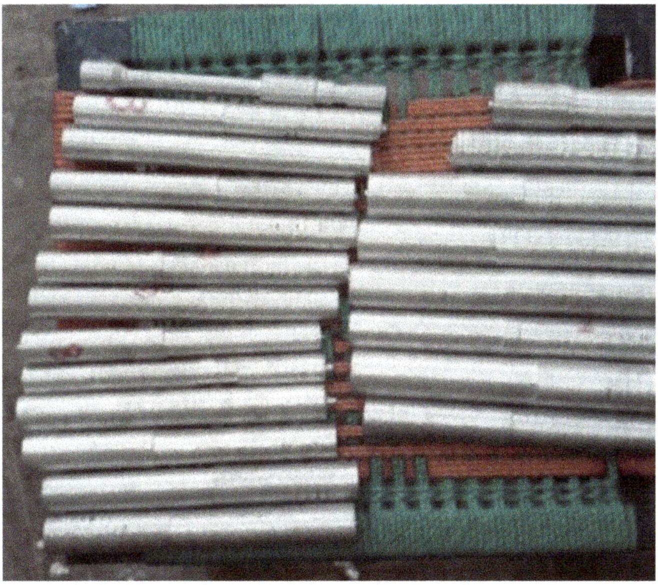

Fig. 62.3 Specimens for tensile and compression test

62.3.2 Mechanical Tests of the Al/SiC-Bagasse Ash Composite

Tensile test

Tensile strength of a material is the ability of resist loads which applied in the longitudinal direction of sample. Circular specimens were machined with a specific gage length, specimen diameter (*D*), notch diameter (*d*) and fillet radius (*r*) of 100,

Fig. 62.4 Failure of tensile test specimens

12, 7.5, and 9 mm, respectively. The tensile test was conducted on WP 310 computerized hydraulic universal testing machine (UTM) and the test specimens were shown in Fig. 62.4. The test was carried out on WP-300 hydraulic testing machine of 50 kN capacities with at a constant piston speed of 5 mm/min. Table 62.5 presented the complete test results of tensile data. The compared results are between single-reinforced composite (Al/SiC) and hybrid metal matrix composite (Al/SiC-bagasse ash) in addition to the base metal (Al6063). The variation of average tensile strength in each single and hybrid composite is presented in Table 62.6. Tensile test results showed that increasing percentage of reinforcement (SiC) only in the matrix reduces average tensile strength while increasing reinforcements, SiC and bagasse ash improves the tensile strength of Al/SiC-bagasse ash composite. At 10% SiC-10% bagasse ash hybrid composite have better improvement than other single and hybrid composites.

Table 62.5 Tensile strength of the developed hybrid metal matrix composites

S. No.	Al/SiC/fly-ash composite	Ultimate tensile strength (MPa)			Average ultimate tensile strength (MPa)
		Test 1	Test 2	Test 3	
1	Al6063-7.5% SiC-7.5% bagasse ash	52.49	81.71	70.4	68.2
2	Al6063-7.5% SiC-10% bagasse ash	88.5	76.43	60.7	75.2
3	Al6063-7.5% SiC-12% bagasse ash	74.5	61.12	78.98	71.6
4	Al6063-10% SiC-7.5% bagasse ash	76.8	68.8	76.32	73.89
5	Al6063-10% SiC-10%bagasse ash	84.71	87.53	76.62	82.95
6	Al6063-10% SiC-12% bagasse ash	87.5	79.00	85.09	83.86

Table 62.6 Tensile strength of the developed Al/SiC matrix composites

S. No.	Al/SiC composite	Ultimate tensile strength (MPa)			Average ultimate tensile strength (MPa)
		Test 1	Test 2	Test 3	
1	Al6063-7.5% SiC	78.46	45.4	64.87	62.91
2	Al6063-10% SiC	74.24	65.34	47.5	62.36

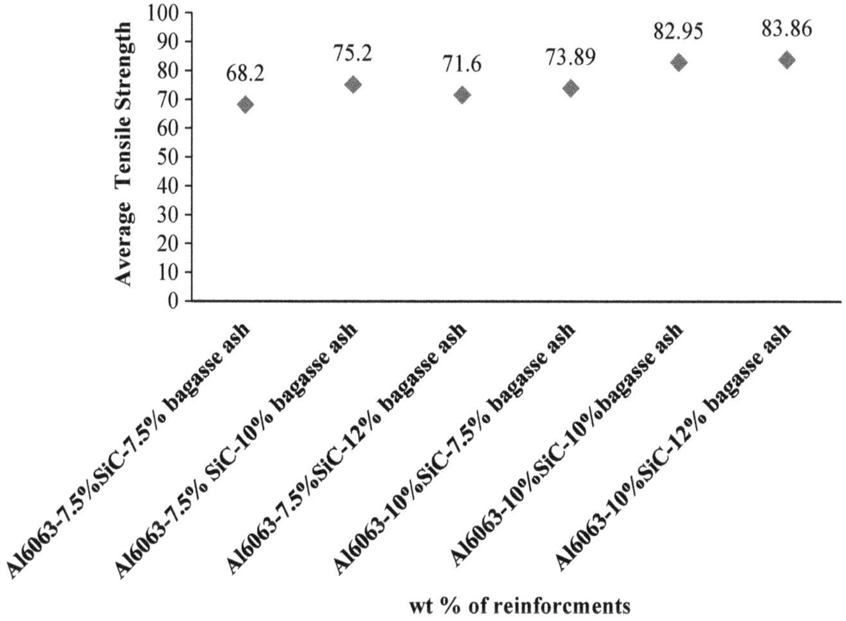

Fig. 62.5 Tensile strength and weight percentage of reinforcements

SiC and FA particles are very effective in improving the tensile strength of the composite. The addition of SiC and fly-ash particles in the matrix induces much strength to matrix alloy by offering more resistance to tensile stresses. It is well known that the thermal expansion coefficient of SiC and fly-ash particle is 3.25 × 106/°C and for aluminium alloy is 23 × 106/°C. The thermal mismatch between matrix and the reinforcement causes higher dislocation density in the matrix and load-bearing capacity of the hard particles which subsequently increases the composite strength. Figure 62.5 reveals the fracture surface of AA6063 10 wt% SiC 7.5 wt% fly-ash composite. It shows a network of dimples whose size is smaller compared to matrix alloy. The fly-ash and silicon carbide refined the grain size of matrix alloy and reduced the ductility which resulted in smaller size dimples.

62.4 Conclusions

The Al/SiC-bagasse fly-ash hybrid composite is successfully manufactured with different weight percentage of SiC and bagasse fly-ash reinforcements (7.5 and 10 wt% of SiC and 7.5, 10, and 12 wt% bagasse fly-ash). The mechanical properties of the developed composite are studied with conducting different tests.

The conclusions drawn from the present investigation are as follows:

- Bagasse fly-ash, the waste generated from sugarcane plant can be successfully used as a reinforcing material to produce metal matrix composite (MMC) component in aluminium matrix to be used in lightweight of automotive parts.
- The tensile strength, compression strength and hardness of the hybrid composite improved with increasing wt% of reinforcements.
- The hybrid Al/SiC-bagasse ash composite were better mechanical properties with compared single-reinforced Al/SiC composite.

References

1. Suresh, N., Venkateswaran, S., Seetharamu, S.: Influence of cenospheres of fly ash on the mechanical properties and wear of permanent moulded eutectic Al-Si alloys. Mater. Sci. Pol. **28**(1), 55–65 (2010)
2. James, S.J., et al.: Hybrid aluminium metal matrix composite reinforced with SiC and TiB2. Proced. Eng. **97**, 1018–1026 (2014)
3. Kala, H., Mer, K.K.S., Kumar, S.: A review on mechanical and tribological behaviors of stir cast aluminum matrix composites. Proced. Mater. Sci. **6,** 1951–1960 (2014)
4. Sajjadi, S.A., Ezatpour, H.R., Parizi, M.T.: Comparison of microstructure and mechanical properties of A356 aluminum alloy/Al_2O_3 composites fabricated by stir and compo-casting processes. Mater. Des. **34**, 106–111 (2012)
5. Singh, J., Chauhan, A.: Characterization of hybrid aluminum matrix composites for advanced applications—a review. J. Mater. Res. Technol. **5**(2), 159–169 (2016)
6. Bodunrin, M.O., Alaneme, k.k., Chown, L.H.: Aluminium matrix hybrid composites: a review of reinforcement philosophies; mechanical, corrosion and tribological characteristics. J. Mater. Res. Technol. **4**(4), 434–445 (2015)
7. Vijayakumar, S., et al.: Studies on mechanical properties of Al-SiC metal matrix composite. SSRG-IJMSE **2**(Nov–Dec), (2013)
8. Imran, M., et al.: Study of hardness and tensile strength of Aluminium-7075 percentage varying reinforced with graphite and bagasse-ash composites. Res. Effic. Technol. **2**(2), 81–88 (2016)
9. Alaneme, K.K., et al.: Fabrication characteristics and mechanical behaviour of rice husk ash—Alumina reinforced Al–Mg–Si alloy matrix hybrid composites. J. Mater. Res. Technol. **2**(1), 60–67 (2013)
10. Alaneme, K.K., Bodunrin, M.O., Awe, A.A.: Microstructure, mechanical and fracture properties of groundnut shell ash and silicon carbide dispersion strengthened aluminium matrix composites. J. K. Saud Univ. Eng. Sci. (2016)

11. Rao, H.J., Ramulu, P.J., Vardhan, M.V., Chandramouli, C.H.: Failure prediction in fiber metal laminates for next generation aero materials. In: IOP Conference Series: Materials Science and Engineering, vol. 149, No. 1, p. 012102. IOP Publishing, September 2016
12. Oyinbo, S.T., Ikumapayi, O.M., Ajiboye, J.S., Afolalu, S.A.: Numerical simulation of axisymmetric and asymmetric extrusion process using finite element method. Int. J. Sci. Eng. Res. **6**(6), 1246 (2015)
13. Lorrette, C., Réau, A., Briottet, L.: Mechanical properties of nanostructured silicon carbide consolidated by spark plasma sintering. J. Eur. Ceram. Soc. **33**(1), 147–156 (2013)
14. Baradeswaran, A., Perumal, A.E.: Study on mechanical and wear properties of Al 7075/ Al_2O_3/graphite hybrid composites. Compos. Part B Eng. **56**, 464–471 (2014)

Chapter 63
Fuzzy-EDAS (Evaluation Based on Distance from Average Solution) for Material Selection Problems

Divya Zindani, Saikat Ranjan Maity and Sumit Bhowmik

Abstract This present work proposes a novel fuzzy-based multi-criteria decision-making (MCDM) approach for addressing material selection problem under uncertain environment. The integrated framework integrates fuzzy decision making with evaluation based on distance from average solution (EDAS) MCDM method. Proposed framework fulfills the gap of restricted exploration of fuzzy decision-making tools in the domain of material design. The applicability of the proposed framework has been evidenced through a case study of a material selection of the casting plate used in sand casting. The selection was done considering five material alternatives and seven criteria. The presented fuzzy-based approach considers the objective weights while calculating the weights of the considered criteria.

Keywords Soft computing · Fuzzy theory · Material selection · Multi-criteria decision making

63.1 Introduction

Decision-making problems of the real world are complex and have to be solved under the influence of conflicting criteria. The decision-making problems when solved considering single criterion results in unrealistic solutions [1]. Therefore, in order to obtain suitable solutions, models for such problems are constructed considering multiple criteria. Systematic and scientific tools in the form of multi-criteria

D. Zindani (✉) · S. R. Maity · S. Bhowmik
Department of Mechanical Engineering, National Institute of Technology Silchar, Silchar 7888010, Assam, India
e-mail: divyazindani@gmail.com

S. R. Maity
e-mail: saikat.jumtech@gmail.com

S. Bhowmik
e-mail: bhowmiksumit04@yahoo.co.in

R. G. Narayanan et al. (eds.), *Advances in Computational Methods in Manufacturing*, Lecture Notes on Multidisciplinary Industrial Engineering,
https://doi.org/10.1007/978-981-32-9072-3_63

decision-making (MCDM) tools have been developed to deal with real-world problems considering multiple factors and criteria [2].

Material selection is a decision making problem that involves the selection of the best material from amongst the number of material alternatives. The rankings are obtained under the influence of multiple criteria that are often conflicting in nature. Therefore, material selection problems have been addressed with different MCDM tools. As for instance, the applicability of different preference ranking-based MCDM methods, i.e., operational competitiveness rating analysis (OCRA), organization, arrangement Et synthesize de donors relationnelles (ORESTE), grey system theory-based COPRAS (COPRAS-G) and extended preference ranking organization method for enrichment evaluations (EXPROM2) was evidenced by Chatterjee and Chakraborty [3] through the gear material selection problem. Rastogi et al. [4] employed the TOPSIS method for obtaining the final rankings of material for the phase change material. Material selection of brake friction material was addressed by Singh et al. [4] using combined VIKOR and AHP methodology. TODIM was applied by Zindani et al. [5] for the material selection problems of flywheel and spur gear. Evaluation based on distance from average solution (EDAS) with the design of experiments (DOE) was used recently by Chatterjee et al. [6] and the applicability was justified using gear material selection problem.

However, the problem becomes relatively more complex because of the associated uncertainty. The decision-making process then becomes relatively more challenging. In order to deal with qualitative criteria in material selection problems, fuzzy logic has been employed by the decisionmakers. Simplified fuzzy logic was used for dealing with qualitative properties of materials by Khabbaz et al. [7]. The applicability of the proposed methodology was demonstrated using material selection problems of the liquid nitrogen storage tank, the spar of an aircraft wing, and mast for a sailing boat. To deal with problems involving both the qualitative and quantitative properties, fuzzy-based MCDM models have been proposed and applied by the decisionmakers. Some of the examples include fuzzy-TOPSIS [8], fuzzy-VIKOR [9], etc.

EDAS method has been extended to solve under uncertain environments such as interval type-2 fuzzy sets [10], interval grey numbers [11], neutrosophic soft sets [12], interval-valued fuzzy soft sets [13], interval-valued neutrosophic sets [14], and intuitionistic fuzzy sets [15].

Although EDAS methodology has been evidenced for material selection problem, its applicability to address material selection problem under uncertain environment is yet to be explored. Therefore, the present work is a modest effort to fulfill this gap by proposing a fuzzy-EDAS methodology for material selection problems wherein triangular fuzzy numbers are used to represent linguistic information. The applicability of the proposed material selection framework has been demonstrated using material selection problem of casting plates used in sand casting. Rest of the paper has been organized into the following sections: basic concepts and arithmetic related to fuzzy set have been elaborated in Sect. 63.2. Section 63.3 presents a fuzzy-EDAS method to deal with material selection problems under an environment of uncertainty. An illustrative example of material selection problem has been demonstrated in Sect. 63.4 of the paper. Paper finally terminates with concluding remarks in Sect. 63.5.

63.2 Fuzzy Theory and Some Definitions

Zadeh [16] developed a fuzzy set theory to handle the problems comprising vague information. Some of the basic definitions pertaining to the fuzzy theory that will be essential in the present work have been delineated in this section.

63.2.1 Definition Number 1

Let $\tilde{b}$ denote a fuzzy set. A membership function $\mu_{\tilde{b}}(x)$ characterizes the fuzzy set which maps x to a real number lying in the interval [0, 1]. Value of the function $\mu_{\tilde{b}}(x)$ is known as the grade of membership of x in $\tilde{b}$.

63.2.2 Definition Number 2

A fuzzy set $\tilde{a}$ is a triangular fuzzy number when $\tilde{b} = (b_1, b_2, b_3)$. Membership function $\mu_{\tilde{b}}(x)$ for a triangular fuzzy number $\tilde{b}$ is denoted by Eq. (63.1):

$$\mu_{\tilde{b}}(x) = \begin{cases} 0, x \leq b_1, \\ \frac{x-b_1}{b_2-b_3}, b_1 \leq x \leq b_2, \\ \frac{b_3-x}{b_3-b_2}, b_2 \leq x \leq b_3, \\ 0, x > b_3, \end{cases} \tag{63.1}$$

where, b_1, b_2, and b_3 denote real numbers such that $b_1 < b_2 < b_3$.

63.2.3 Definition Number 3

Main operations between two triangular fuzzy numbers $\tilde{b} = (b_1, b_2, b_3)$ and $\tilde{c} = (c_1, c_2, c_3)$ can be expressed as follows:

Addition $\tilde{b}$ (+) $\tilde{c}$ = $(\mathrm{b}_1 + \mathrm{c}_1, \mathrm{b}_2 + \mathrm{c}_2, \mathrm{b}_3 + \mathrm{c}_3)$
Subtraction $\tilde{b}$ (−) $\tilde{c}$ = $(b_1 - c_3, b_2 - c_2, b_3 - c_1)$

Multiplication $\tilde{b}\ (\times)\ \tilde{c} = (b_1 \times c_1, b_2 \times c_2, b_3 \times c_3)$
Division $\tilde{b}\ (/)\ \tilde{c} = (b_1 \times c_1, b_2 \times c_2, b_3 \times c_3)$

63.2.4 Definition Number 4

Operations between a triangular fuzzy number and a real number k are described as below:

Multiplication $k\ (\times)\ \tilde{b} = (kb_1, kb_2, kb_3)$
Division $\tilde{b}/k = (b_1/k, b_2/k, b_3/k)$
Division $k/\tilde{b} = (k/b_1, k/b_2, k/b_3)$

63.2.5 Definition Number 5

The crisp value of a triangular fuzzy number $\tilde{b} = (b_1, b_2, b_3)$ is defined as by Eq. (63.2):

$$k(\tilde{b}) = \frac{1}{6}(b_1 + 4b_2 + b_3) \tag{63.2}$$

63.2.6 Definition Number 6

The maximum of zero and a triangular fuzzy number can be established using Eq. (63.3):

$$\psi(\tilde{b}) = \begin{cases} \tilde{b} \text{ if } k(\tilde{b}) > 0 \\ \tilde{0} \text{ if } k(\tilde{b}) \leq 0 \end{cases} \tag{63.3}$$

63.3 The Conceptual Framework of Fuzzy-EDAS Methodology

EDAS methodology was developed by Keshavarz Ghorabaee et al. [17] and the method ranks material on the basis of the average solution. The average solution is arrived at by calculation of the positive distance from average (PDA) and negative distance from average (PDA). The alternative that has higher PDA values and lower NDA values are the best-ranked material alternative.

In the case of fuzzy-EDAS methodology, alternatives are ranked in accordance with the decreasing value of defuzzified appraisal score. The steps involved in the fuzzy-EDAS method have been delineated in the ensuing discussion. Let $\tilde{A}$ denote the fuzzy decision matrix, i.e.,

$$\tilde{A} = \left[\tilde{a}_{ij}\right]_{n\times m} = \begin{bmatrix} \tilde{a}_{11} & \tilde{a}_{12} & \cdots & \tilde{a}_{1m} \\ \tilde{a}_{21} & \tilde{a}_{22} & \cdots & \tilde{a}_{2m} \\ \vdots & \vdots & \vdots & \vdots \\ \tilde{a}_{n1} & \tilde{a}_{n2} & \cdots & \tilde{a}_{nm} \end{bmatrix} \quad (63.4)$$

where j is the number of alternatives and i signify the criteria.

Following steps are traced in order to solve a decision-making problem using fuzzy-EDAS methodology.

63.3.1 Step 1

In this step of the proposed framework, objective weights are determined for each of the criteria Ci using Eqs. (63.5) and (63.6). Objective weights are determined for the decision matrix supplied by all the decisionmakers.

$$w_j^* = \frac{\sum_{i=1}^{m}\sum_{r=1}^{m}\left|p_{ij} - p_{rj}\right|}{\sqrt{\sum_{j=1}^{m}\left[\sum_{i=1}^{m}\sum_{r=1}^{m}\left|p_{ij} - p_{rj}\right|\right]^2}} \quad (63.5)$$

$$w_j^o = \frac{w_j^*}{\sum_{j=1}^{m} w_j^*} \quad (63.6)$$

63.3.2 Step 3

A fuzzy average decision matrix is developed with respect to all the considered criteria using Table 63.1 and Eq. (63.7):

$$\mathrm{AV}_j = \frac{\sum_{i=1}^{n}\tilde{a}_{ij}}{k} \quad (63.7)$$

Table 63.1 Linguistic terms for alternatives ratings

Linguistic term	Membership function
Very low (VL)	(0,0,0.1)
Medium low (ML)	(0, 0.1, 0.3)
Low (L)	(0.1,0.3, 0.5)
Medium (F)	(0.3, 0.5, 0.75)
High (H)	(0.5,0.75,0.9)
Medium High (MH)	(0.75,0.9,1)
Very high (VH)	(0.9, 1, 1)

63.3.3 Step 4

The distance of optimal solution from negative feasible solutions must be maximum whereas it should be minimum from the negative feasible solution. In this step of the fuzzy-EDAS methodology, matrices for fuzzy positive distance from average (PDA) and fuzzy negative distance from average (NDA) are calculated using Eqs. (63.8) and (63.9), respectively.

$$\mathrm{PDA} = \left[\mathrm{pda}_{ij}\right]_{n\times m} \tag{63.8}$$

$$\mathrm{NDA} = \left[\mathrm{nda}_{ij}\right]_{n\times m} \tag{63.9}$$

where for beneficial criteria

$$\mathrm{pda}_{ij} = \left\{ \frac{\psi\left(\tilde{a}_{ij} - \mathrm{AV}_j\right)}{k\left(\mathrm{AV}_j\right)} \right. \tag{63.10}$$

and

$$\mathrm{nda}_{ij} = \left\{ \frac{\psi\left(\mathrm{AV}_j - \tilde{a}_{ij}\right)}{k\left(\mathrm{AV}_j\right)} \right. \tag{63.11}$$

and for non-beneficial criteria

$$\mathrm{pda}_{ij} = \left\{ \frac{\psi\left(\mathrm{AV}_j - \tilde{a}_{ij}\right)}{k\left(\mathrm{AV}_j\right)} \right. \tag{63.12}$$

and

$$\mathrm{nda}_{ij} = \left\{ \frac{\psi\left(\tilde{a}_{ij} - \mathrm{AV}_j\right)}{k\left(\mathrm{AV}_j\right)} \right. \tag{63.13}$$

63.3.4 Step 5

In this step of the methodology, matrices are developed for fuzzy weighted positive and fuzzy weighted negative distances. Equations (63.14) and (63.15) are used for this purpose

$$\mathrm{sp}_i = \sum_{j=1}^{m} \left(\tilde{w}_j + \mathrm{pda}_{ij}\right) \tag{63.14}$$

and

$$\mathrm{sn}_i = \sum_{j=1}^{m} \left(\tilde{w}_j + \mathrm{nda}_{ij}\right) \tag{63.15}$$

63.3.5 Step 6

Fuzzy normalized values for fuzzy weighted positive and fuzzy weighted negative distances are computed using Eqs. (63.16) and (63.17)

$$\mathrm{nsp}_i = \frac{\mathrm{sp}_i}{\max_i(k(\mathrm{sp}_i))} \tag{63.16}$$

and

$$\mathrm{nsp}_i = 1 - \frac{\mathrm{sn}_i}{\max_i(k(\mathrm{sn}_i))} \tag{63.17}$$

63.3.6 Step 7

In the penultimate step, fuzzy appraisal score for different alternatives is calculated using Eq. (63.18)

$$\mathrm{as}_i = \frac{\mathrm{nsp}_i + \mathrm{nsn}_i}{2} \tag{63.18}$$

63.3.7 Step 8

In the last step, alternative materials are ranked in accordance with the decreasing value of defuzzified appraisal score. The best choice of the alternative material is the one with the highest value of appraisal score.

63.4 Illustrative Example

In this section, a case study on material selection of pattern used in sand casting has been solved using the above-proposed methodology. The material alternatives identified from literature survey [19] were wood (A_1), aluminum (A_2), steel (A_3), plastic (A_4), and cast iron (A_5). The criteria considered for the problem under consideration were machinability (C_1), wear resistance (C_2), strength (C_3), weight (C_4), reparability (C_5), resistance to corrosion (C_6), and resistance to swelling (C_7). Excluding weight, all the other criteria are benefited criteria. Three decisionmakers were asked to provide their performance rating. Table 63.2 depicts the performance rating for decisionmakers 1, 2, and 3. Each of the decision matrices was then converted to a triangular fuzzy number using information from Table 63.1. Table 63.3 depicts the fuzzy converted matrix for all the decisionmakers considered. Equations (63.4) and (63.5) were then used to obtain the objective weights for

Table 63.2 Performance ratings as given by decisionmakers

DM		C_1	C_2	C_3	C_4	C_5	C_6	C_7
1	A_1	VH	ML	M	VH	VH	VH	ML
	A_2	MH	MH	MH	MH	ML	M	VH
	A_3	M	VH	VH	ML	MH	ML	VH
	A_4	MH	M	MH	MH	M	VH	VH
	A_5	MH	VH	MH	ML	MH	ML	VH
2	A_1	VH	ML	M	VH	VH	VH	ML
	A_2	H	H	H	H	ML	M	VH
	A_3	M	VH	VH	ML	H	ML	VH
	A_4	H	M	H	H	M	VH	VH
	A_5	H	VH	H	ML	H	ML	VH
3	A_1	H	ML	M	H	VH	H	ML
	A_2	MH	MH	MH	MH	ML	M	VH
	A_3	M	VH	VH	ML	MH	ML	VH
	A_4	MH	M	MH	MH	M	VH	VH
	A_5	MH	H	MH	ML	MH	ML	H

Table 63.3 Triangular fuzzy numbers for performance ratings as obtained from different decisionmakers

		C_1	C_2	C_3	C_4	C_5	C_6	C_7
D_1	A_1	(0.9,1,1)	(0.1,0.3,0.5)	(0.3,0.5,0.75)	(0.9,1,1)	(0.9,1,1)	(0.9,1,1)	(0.1,0.3,0.5)
	A_2	(0.5,0.75,0.9)	(0.5,0.75,0.9)	(0.5,0.75,0.9)	(0.5,0.75,0.9)	(0.1,0.3,0.5)	(0.3,0.5,0.75)	(0.9,1,1)
	A_3	(0.3,0.5,0.75)	(0.9,1,1)	(0.9,1,1)	(0.1,0.3,0.5)	(0.5,0.75,0.9)	(0.1,0.3,0.5)	(0.9,1,1)
	A_4	(0.5,0.75,0.9)	(0.3,0.5,0.75)	(0.5,0.75,0.9)	(0.5,0.75,0.9)	(0.3,0.5,0.75)	(0.9,1,1)	(0.9,1,1)
	A_5	(0.5,0.75,0.9)	(0.9,1,1)	(0.5,0.75,0.9)	(0.1,0.3,0.5)	(0.5,0.75,0.9)	(0.1,0.3,0.5)	(0.9,1,1)
D_2	A_1	(0.9,1,1)	(0,0.1,0.3)	(0.3,0.5,0.75)	(0.9,1,1)	(0.9,1,1)	(0.9,1,1)	(0.1,0.3,0.5)
	A_2	(0.75,0.9,1)	(0.75,0.9,1)	(0.75,0.9,1)	(0.75,0.9,1)	(0,0.1,0.3)	(0.3,0.5,0.75)	(0.9,1,1)
	A_3	(0.3,0.5,0.75)	(0.9,1,1)	(0.9,1,1)	(0,0.1,0.3)	(0.75,0.9,1)	(0,0.1,0.3)	(0.9,1,1)
	A_4	(0.75,0.9,1)	(0.3,0.5,0.75)	(0.75,0.9,1)	(0.75,0.9,1)	(0.3,0.5,0.75)	(0.9,1,1)	(0.9,1,1)
	A_5	(0.75,0.9,1)	(0.9,1,1)	(0.75,0.9,1)	(0,0.1,0.3)	(0.75,0.9,1)	(0,0.1,0.3)	(0.9,1,1)
D_3	A_1	(0.9,1,1)	(0,0.1,0.3)	(0.3,0.5,0.75)	(0.9,1,1)	(0.9,1,1)	(0.9,1,1)	(0.1,0.3,0.5)
	A_2	(0.75,0.9,1)	(0.75,0.9,1)	(0.75,0.9,1)	(0.75,0.9,1)	(0,0.1,0.3)	(0.3,0.5,0.75)	(0.9,1,1)
	A_3	(0.3,0.5,0.75)	(0.9,1,1)	(0.9,1,1)	(0,0.1,0.3)	(0.75,0.9,1)	(0,0.1,0.3)	(0.9,1,1)
	A_4	(0.75,0.9,1)	(0.3,0.5,0.75)	(0.75,0.9,1)	(0.75,0.9,1)	(0.3,0.5,0.75)	(0.9,1,1)	(0.9,1,1)
	A_5	(0.75,0.9,1)	(0.9,1,1)	(0.75,0.9,1)	(0,0.1,0.3)	(0.75,0.9,1)	(0,0.1,0.3)	(0.9,1,1)

Table 63.4 Objective weights derived for different decisionmakers and the aggregated weights

	C_1	C_2	C_3	C_4	C_5	C_6	C_7
w_o (DM_1)	0.088	0.158	0.089	0.182	0.162	0.209	0.112
w_o (DM_2)	0.064	0.171	0.064	0.205	0.168	0.237	0.091
w_o (DM_3)	0.089	0.158	0.088	0.182	0.162	0.209	0.112
Aggregated weight	0.081	0.162	0.081	0.190	0.164	0.218	0.105

all the decision matrix obtained from the three decisionmakers. The aggregate weights were finally obtained by taking the average of all the objective weights for particular criteria. Table 63.4 depicts the individual objective weights and the aggregated objective weights. In the next step, the average decision matrix was formulated and has been tabulated in Table 63.5. Equation (63.7) was then used to arrive at the average solution which has been depicted in Table 63.6 which also shows the crisp value of the average solution obtained using Eq. (63.2). Positive distance from average (PDA) and negative distance from average (NDA) values were obtained using Eqs. (63.8–63.13) and Tables 63.7 and 63.8 represent these values. Fuzzy normalized values for fuzzy weighted positive and fuzzy weighted negative distances were computed using Eqs. (63.16) and (63.17). In the penultimate step, fuzzy appraisal score for different alternatives is calculated using Eq. (63.18). Finally, the alternatives were ranked on the basis of the defuzzified appraisal score. Table 63.9 portrays all these values. Aluminum was found to be the best alternative material.

63.5 Result and Discussion

Present work proposes a fuzzy-based integrated methodology in the form of fuzzy-EDAS for solving material selection problem. The proposed methodology aids the decisionmaker in addressing the linguistic performance rating of material alternatives with respect to the criteria. The presented material selection framework, therefore, aids in handling the complexity associated with the uncertain environment. However, the presented methodology only considers objective weights and does not take into consideration the subjective importance. To justify the applicability of the proposed method, material selection problem of casting pattern plate was solved. A total of five material alternatives and seven criteria were considered. Aluminum was found to be the best-ranked material alternative, whereas wood was the worst performing of all. As a future scope of the work, subjective weights could also be considered while arriving at the aggregated weights for each criterion. Further, quantitative properties in the decision matrix could also be given due consideration.

Table 63.5 Elements of the average decision matrix

	C_1	C_2	C_3	C_4	C_5	C_6	C_7
A_1	(0.85,0.967,1)	(0.067,0.233,0.433)	(0.3,0.5,0.75)	(0.85,0.967,1)	(0.9,1,1)	(0.850,0.967,1)	(1,0.3,0.5)
A_2	(0.583,0.8,0.933)	(0.583,0.80,0.933)	(0.583,0.80,0.933)	(0.583,0.80,0.933)	(0.067,0.233,0.433)	(0.3,0.5,0.75)	(0.9,1,1)
A_3	(0.3,0.5,0.75)	(0.90,1,1)	(0.9,1,1)	(0.067,0.233,0.433)	(0.583,0.783,0.933)	(0.067,0.233,0.433)	(0.9,1,1)
A_4	(0.583,0.80,0.933)	(0.3,0.5,0.75)	(0.583,0.80,0.933)	(0.583,0.8,0.933)	(0.3,0.5,0.75)	(0.9,1,1)	(0.9,1,1)
A_5	(0.583,0.80,0.933)	(0.85,0.967,1)	(0.583,0.80,0.933)	(0.067,0.233,0.433)	(0.583,0.783,0.933)	(0.067,0.233,0.433)	(0.833,0.967,1)

Table 63.6 Average solution matrix

	C_1	C_2	C_3	C_4	C_5	C_6	C_7
AV	(0.58,0.77,0.91)	(0.54,0.7,0.82)	(0.59,0.78,0.91)	(0.43,0.61,0.75)	(0.487,0.66,0.81)	(0.437, 0.59,0.72)	(0.72,0.85,0.9)
k (AV)	0.76	0.69	0.77	0.61	0.66	0.58	0.84

Table 63.7 Values of PDA

	C_1	C_2	C_3	C_4	C_5	C_6	C_7
A_1	(0.353,0.253,0.118)	(0,0,0)	(0,0,0)	(0.699,0.599,0.172)	(0.631,0.519,0.295)	(0.707,0.650,0.473)	(0.0,0)
A_2	(0.004,0.035,0.031)	(0.062,0.144,0.159)	(−0.009,0.026,0.030)	(1.415,1.610,1.415)	(0,0,0)	(0,0,0)	(0.206,0.175,0.119)
A_3	(0,0,0)	(0.519,0.432,0.255)	(0.403,0.286,0.117)	(0.415,1.610,1.415)	(0.147,0.188,0.193)	(0,0,0)	(0.206,0.175,0.119)
A_4	(0.004,0.035,0.031)	(0,0,0)	(−0.009,0.026,0.030)	(1.415,1.610,1.415)	(0,0,0)	(0.793,0.707,0.473)	(0.206,0.175,0.119)
A_5	(0.004,0.035,0.031)	(0.447,0.384,0.255)	(−0.009,0.026,0.030)	(1.415,1.610,1.415)	(0.147,0.188,0.193)	(0,0,0)	(0.127,0.135,0.119)

Table 63.8 Values of NDA

	C_1	C_2	C_3	C_4	C_5	C_6	C_7
A_1	(0,0,0)	(0.682,0.673,0.562)	(0.377,0.364,0.208)	(0.699,0.599,0.172)	(0,0,0)	(0,0,0)	(0.746.0.659,0.476)
A_2	(0,0,0)	(0,0,0)	(0,0,0)	(1.415,1.610,1.415)	(0.641,0.651,0.569)	(0.234,0.148,0.−046)	(0,0,0)
A_3	(0.367,00.358,0.209)	(0,0,0)	(0,0,0)	(0,0,0)	(0,0,0)	(0.633,0.605,0.496)	(0,0,0)
A_4	(0,0,0)	(0.346,0.288,0.106)	(0,0,0)	(1.415,1.610,1.415)	(0.285,0.244,0.112)	(0,0,0)	(0,0,0)
A_5	(0,0,0)	(0,0,0)	(0,0,0)	(0,0,0)	(0,0,0)	(0.633,0.605,0.496)	(0,0,0)

Table 63.9 Weighted sum of distances, their normalized values, and the appraisal scores

	sp_i	nsp_i	sn_i	nsn_i	as_i	k (as_i)	Rank
A_1	(0.42,0.36,0.19)	(1.69,1.46,0.78)	(0.35,0.32,0.19)	(−0.05, 0.04,0.43)	(0.82,0.75,0.61)	0.74	5
A_2	(0.30,0.35,0.31)	(1.21,1.42,1.26)	(0.42,0.20,0.14)	(−0.27,0.40,0.57)	(0.47,0.91,0.92)	1.17	**1**
A_3	(0.24, 0.45, 0.36)	(0.98,1.81,1.47)	(0.17,0.16,0.13)	(0.50,0.52,0.63)	(0.74,1.17,1.05)	1.08	3
A_4	(0.46, 0.48, 0.39)	(1.87,1.95,1.57)	(0.37,0.15,0.09)	(−0.11,0.56,0.72)	(0.88,1.26,1.15)	0.84	4
A_5	(0.38, 0.42, 0.36)	(1.53,1.69,1.45)	(0.14,0.13,0.11)	(0.59,0.61,0.68)	(1.06,1.15,1.06)	1.12	2

Acknowledgements The authors of the work would like to express their gratitude toward the anonymous reviewers in giving their valuable feedback that has aided in enhancing the overall quality of the present work. The authors would like to thank TEQIP-III at National Institute of Technology Silchar in providing financial assistance to attend the conference.

References

1. Chakraborty, S., Zavadskas, E.K., Antucheviciene, J.: Applications of WASPAS method as a multi-criteria decision-making tool. Econ. Comput. Econ. Cybern. Stud. Res. 1 **49**(1) (2015)
2. Kosareva, N., Zavadskas, E.K., Krylovas, A., Dadelo, S.: Personnel ranking and selection problem solution by application of KEMIRA method. Int. J. Comput. Commun. Control **11** (1), 51–66 (2016)
3. Chatterjee, P., Chakraborty, S.: Material selection using preferential ranking methods. Mater. Des. **35**, 384–393 (2012)
4. Rastogi, M., Chauhan, A., Vaish, R., Kishan, A.: Selection and performance assessment of Phase Change Materials for heating, ventilation and air-conditioning applications. Energy Convers. Manag. **89**, 260–269 (2015)
5. Zindani, D., Maity, S.R., Bhowmik, S., Chakraborty, S.: A material selection approach using the TODIM (TOmada de Decisao Interativa Multicriterio) method and its analysis. Int. J. Mater. Res. **108**(5), 345–354 (2017)
6. Chatterjee, P., Banerjee, A., Mondal, S., Boral, S., Chakraborty, S.: Development of a hybrid meta-model for material selection using design of experiments and EDAS method. Eng Trans (2018)
7. Khabbaz, R.S., Manshadi, B.D., Abedian, A., Mahmudi, R.: A simplified fuzzy logic approach for materials selection in mechanical engineering design. Mater. Des. **30**(3), 687–697 (2009)
8. Rathod, M.K., Kanzaria, H.V.: A methodological concept for phase change material selection based on multiple criteria decision analysis with and without fuzzy environment. Mater. Des. **32**(6), 3578–3585 (2011)
9. Girubha, R.J., Vinodh, S.: Application of fuzzy VIKOR and environmental impact analysis for material selection of an automotive component. Mater. Des. **37**, 478–486 (2012)
10. Ghorabaee, M.K., Amiri, M., Zavadskas, E.K., Turskis, Z.: Multi-criteria group decision-making using an extended EDAS method with interval type-2 fuzzy sets
11. Stanujkic, D., Zavadskas, E.K., Ghorabaee, M.K., Turskis, Z.: An extension of the EDAS method based on the use of interval grey numbers. Stud. Inf. Control **26**(1), 5–12 (2017)
12. Peng, X., Liu, C.: Algorithms for neutrosophic soft decision making based on EDAS, new similarity measure and level soft set. J. Intell. Fuzzy Syst. **32**(1), 955–968 (2017)
13. Peng, X., Dai, J., Yuan, H.: Interval-valued fuzzy soft decision making methods based on MABAC, similarity measure and EDAS. Fundamenta Informaticae **152**(4), 373–396 (2017)
14. Karaşan, A., Kahraman, C.: Interval-valued neutrosophic extension of EDAS method. In: Advances in Fuzzy Logic and Technology 2017, 13 Sept 2017, pp. 343–357. Springer, Cham
15. Stanujkic, D., Zavadskas, E.K.: A modified weighted sum method based on the decision-maker's preferred levels of performances. Stud. Inf. Control **24**(4), 461–470 (2015)
16. Kahraman, C., Keshavarz Ghorabaee, M., Zavadskas, E.K., Cevik Onar, S., Yazdani, M., Oztaysi, B.: Intuitionistic fuzzy EDAS method: an application to solid waste disposal site selection. J. Environ. Eng. Landscape Manag. **25**(1), 1–2 (2017)

17. Zadeh, L.A.: Fuzzy sets. Inf. Control **8**, 3 (1965)
18. Keshavarz Ghorabaee, M., Zavadskas, E.K., Olfat, L., Turskis, Z.: Multi-criteria inventory classification using a new method of evaluation based on distance from average solution (EDAS). Informatica **26**(3), 435–451 (2015)
19. Ekey, D.C., Winter, W.P.: Introduction to Foundry Technology. McGraw-Hill (1958)

Chapter 64
Tensile and Flexural Property Analyses of Glass/Carbon Embedded Hybrid Natural Fibre Composites

Jobin G. Jose, E. Sivakumar, Jibin T. Philip, Kore Mahesh and Basil Kuriachen

Abstract The composites developed from natural fibre have brought indisputable advantages over synthetic materials in terms of nontoxicity, efficient waste disposability, comparable strength, etc. Some automotive structures such as in aerospace and motor racing vehicles, strength-to-weight ratio is precisely imperative and fibre-reinforced composite materials are used to develop such structures with critical stiffness and weight sensitivities. The environmental effect and the subsequent failure of the fibre-reinforced polymers (FRP) have led to the study of its different mechanical properties. In this work, the fibre-reinforced composite material was fabricated with hybrid jute–coir fibre and polyester resin matrix embedded with glass/carbon fibres. Tensile and flexural strength of the developed natural fibre composites are tested using universal testing machine (UTM) as per the ASTM standards. The sample C2 [jute 75% + coir 25% and carbon 100% of 30% and 15% of total composite composition in weight percentage (wt%)] with yield, ultimate and flexural strength of 118, 122 and 81.04 MPa in-errantly proved to possess an upper edge over jute–coir–glass composites upon testing.

Keywords FRP · Natural Fibre · Composites

64.1 Introduction

Fibre-reinforced composite materials have made a drastic influence on the technology of part designing and manufacturing. Natural fibres are being used as reinforcements due to their good mechanical properties, cost-effectiveness,

J. G. Jose
Saintgits College of Engineering, Kottayam 686532, Kerala, India

E. Sivakumar
Centre for Bio-Polymer Science and Technology, Eloor 683501, Kerala, India

J. T. Philip (✉) · K. Mahesh · B. Kuriachen
National Institute of Technology Mizoram, Aizawl 796012, Mizoram, India
e-mail: philip.jibin07@gmail.com

R. G. Narayanan et al. (eds.), *Advances in Computational Methods in Manufacturing*, Lecture Notes on Multidisciplinary Industrial Engineering,
https://doi.org/10.1007/978-981-32-9072-3_64

biodegradability and lightweight [1, 2]. Moreover, the hybridization of natural fibre increases its mechanical, thermal and water absorption properties [3–5]. Fibre composites have various industrial as well as commercial applications. Nowadays, the metals/metal matrix composites are being replaced by polymers in various fields of their utilization. The orientation and stacking of fibres effect greatly on the properties of any developed composite [6]. Further, in continuous fibre composites, the reinforcements are arranged in the form of layers in different orientations to obtain the required properties with fibre volume up to 60–70% [7].

E-glass fibres have good properties such as high tensile strength high dimensional stability and high chemical resistance and thus it is preferred as the reinforcement material in most of the industries [8]. Carbon fibre is mostly used where lightness and high stiffness are required. Synthetic fibres are not eco-friendly and there comes the importance of using natural fibres to develop environment-friendly composite materials. Generally, these natural fibres do not possess any substantial properties to replace synthetic fibres when used as such. Hence, the natural fibres are hybridized with synthetic fibres to develop good property composites and reduce their impact on the environment. Mostly used natural fibres are jute, sisal, kenaf, hemp, etc. due to their desirable properties and availability in nature. E-glass fibres are often used for hybridization with jute fibre. In some cases, the natural fibre composites have properties better than glass fibre composites [9]. Coir natural fibres are found to have poor mechanical properties. Precisely, coir fibre composites showed only average values for tensile, flexural and impact strength which are poor when compared to glass fibre-reinforced polymer (GFRP) composites [10]. Nevertheless, natural fibres have good mechanical properties with polyester resin matrix [11].

The present work investigates the tensile and flexural strength of hybrid jute–coir fibre composite material compared with glass and carbon fibre at varying compositional ratios (wt%) with polyester resin as matrix. Methyl ethyl ketone peroxide and cobalt naphthenate are the catalyst and accelerator, respectively. Compression moulding technique was used to develop the samples and then subjected to property analyses on universal testing machine (UTM).

64.2 Experimental Methodology

64.2.1 Materials

In this work, the jute, coir, glass and carbon fibres are used as raw materials for the composite specimen preparation. Polyester resin was the matrix used, methyl ethyl ketone peroxide and cobalt naphthenate are the catalysts and accelerators, respectively. Jute fibre was imported from Jaipur, Rajasthan, India; coir fibre has been purchased from Alappuzha, Kerala, India; glass and carbon fibres are obtained from Sharon Enterprises, Ernakulam, Kerala, India. Polyester resin along with its catalysts and accelerators are also procured from Sharon Enterprises, Ernakulam, Kerala, India.

64.2.2 Composite Preparation

The specimens were prepared by compression moulding process. Stainless steel mould of dimension (200 × 200 × 3) mm was used for moulding the composite sheet specimens. Specimens were prepared according to proportion as described in Table 64.1. Calculation was done to find the required volume for moulding. The matrix and the reinforcement are taken in the ratio 55:45. A calculated amount of polyester resin along with its accelerator and catalyst was mixed thoroughly. Mixing of the matrix is done gently to avoid formation of voids and instantaneously poured onto the reinforcement in order to avoid curing in room temperature. Separator sheets were placed at the inner surface of the mould for quick and easy removal of the moulded sheet. A thin layer of matrix mixture was poured on the separator sheet and the fibres were kept on top. The fibres were arranged in a layer by layer manner in which natural fibres were kept outside and glass/carbon fibres were kept inside. Polyester resin mixture was poured on top of the fibres and spread all around. Separator sheet was placed on it and the mould was closed. Care was taken to avoid formation of air bubbles. Mould was kept in compression moulding machine and pressure was applied. The mould was allowed to cure at room temperature for 48 h. Further, the specimens were taken out of the mould and sides were shaped as shown in Fig. 64.1. The specimen was then cut into standard shapes required for tensile and flexural tests using counter cutting machine.

Table 64.1 Reinforced composite proportions

Sample	Jute/coir (30%)	Glass/carbon (15%)
G1	Jute 100%	Glass 100%
G2	Jute 75% Coir 25%	Glass 100%
G3	Jute 50% Coir 50%	Glass 100%
G4	Jute 25% Coir 75%	Glass 100%
G5	Coir 100%	Glass 100%
C1	Jute 100%	Carbon 100%
C2	Jute 75% Coir 25%	Carbon 100%
C3	Jute 50% Coir 50%	Carbon 100%
C4	Jute 25% Coir 75%	Carbon 100%
C5	Coir 100%	Carbon 100%

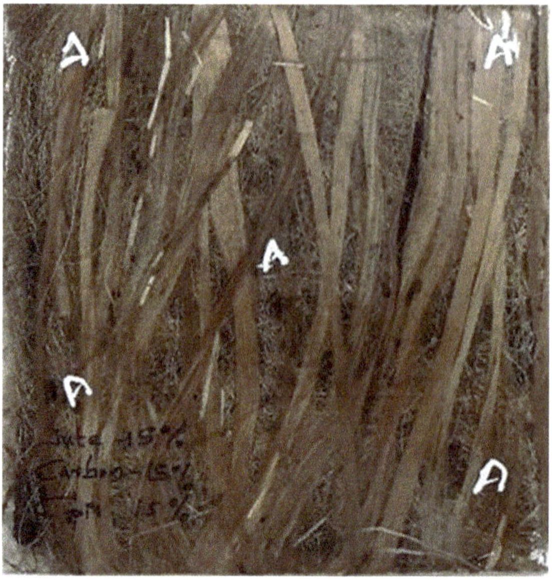

Fig. 64.1 Composite sheet specimen

64.2.3 Mechanical Property Testing

The tensile and flexure tests were carried out on the prepared specimens using standard instruments to study their mechanical properties. The data values such as displacement, force stress and strain at different time intervals were obtained from the system connected to the UTM, from which the mechanical property trend graphs of different specimens were plotted. The detailed methodology followed for various test methods are as follows.

Tensile Test. The tensile testing samples were cutted using a counter cutting machine according the ASTM standard D638 [12]. The edges of the samples were finished for tensile testing as shown in Fig. 64.2. The sample was fixed vertically on the UTM by clamping both ends on to the holder and load was applied gradually

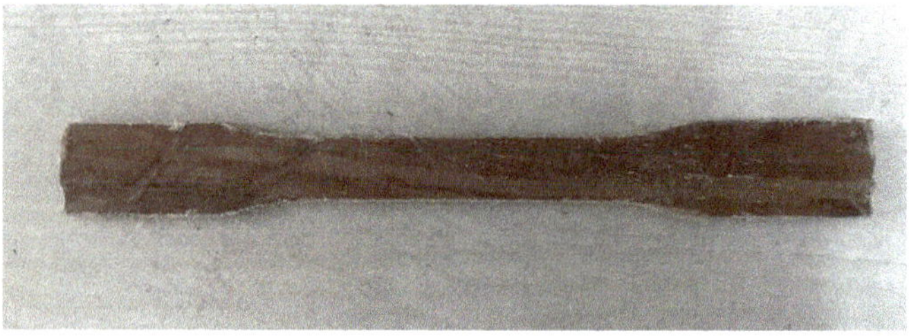

Fig. 64.2 Sample for tensile testing as per ASTM D638

until it fractures. The test was repeated for all the samples and its elongation value at the gauge section was recorded.

Flexural Test. Flexural test samples were prepared as per the ASTM standard D790 [13] as shown in Fig. 64.3. Three-point flexure test procedure was employed to carry out the entire sequence of methodology using UTM. The specimen was placed horizontally on the UTM as per standard procedural specifications by providing support at the end points. The force was applied gradually at the midsection of the sample until fracture occurs. The procedure was repeated for all the test specimens developed using different compositions of the fibres. The flexural strength values were recorded for all the test samples.

64.3 Results and Discussion

64.3.1 Tensile Properties

The tensile test results for all the samples are obtained and their yield stress and ultimate stress of glass and carbon fibre composite specimens are tabulated separately in Tables 64.2 and 64.3. It is observed that jute–coir composite samples are having less tensile strength when compared with jute composite samples. This is due to the hybridization of coir fibres in it. These samples show ductile behaviour. Comparing Tables 64.2 and 64.3, it is observed that the carbon composite specimens are having better tensile strength than glass composite specimens. It can also be observed that G3 and C3 samples are having less tensile properties. This is due to lack of proper adhesion between jute–coir–glass/carbon specimens as they are of equal fibre volume.

The force–displacement graph for jute–coir–glass and jute–coir–carbon fibre for tensile testing is as shown in Figs. 64.4 and 64.5, respectively. Comparing both the plots it can be interpreted that the samples G2 and C2 with jute 75% and coir 25% possess better properties than those samples with higher percentages of coir. Moreover, jute–coir–carbon fibre (C2) has greater withstanding capacity of 5990 N before breakdown than the jute–coir–glass fibre (G2) which has a value of 4160 N. So, the general trend is that the increase in percentage of coir with corresponding decrease in percentage of jute causes a reduction in tensile properties of the composite fibre. Hence, optimum value near to the sample compositions of G2 and C2 must be chosen for significant results.

Fig. 64.3 Sample for flexural testing as per ASTM D790

Table 64.2 Tensile strength of jute–coir–glass fibre composites

Sample	Yield stress (MPa)	Ultimate stress (MPa)
G1	90.5	100.5
G2	81	87.5
G3	45	47.1
G4	46.9	49.7
G5	48	50.5

Table 64.3 Tensile strength of jute–coir–carbon fibre composites

Sample	Yield stress (MPa)	Ultimate stress (MPa)
C1	133	140
C2	118	122
C3	21.1	33
C4	69.8	110
C5	71	80

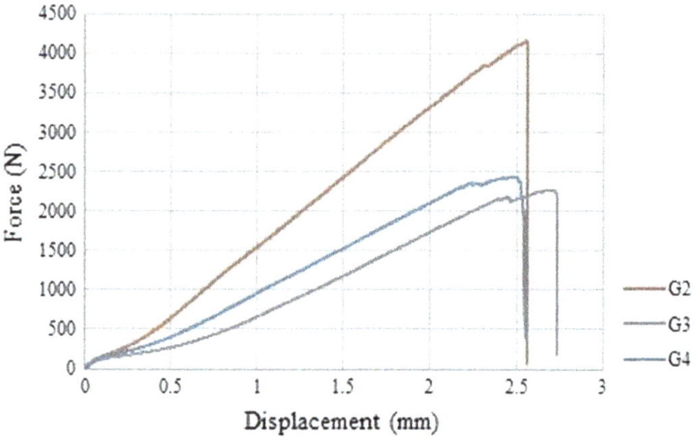

Fig. 64.4 Force–displacement plot for jute–coir–glass fibre for tensile test

64.3.2 *Flexural Properties*

The three-point flexural test values are obtained and are tabulated in Tables 64.4 and 64.5 separately for glass and carbon fibre composites. It is observed that flexural strength for jute–glass composite is less than jute–carbon composite. On investigation of hybridized jute–coir composite samples, it was observed that G3 is having greater flexural strength as 98.16 MPa and C3 is having the lowest as 54.05 MPa. Comparatively, jute-coir–carbon composite samples have good flexural strength than jute–coir–glass composite samples. The force–displacement graph for jute–coir–glass and jute–coir–carbon fibre for flexural test is as shown in Figs. 64.6

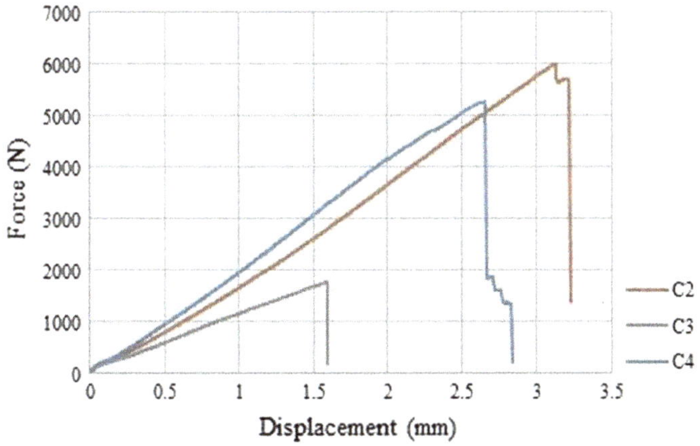

Fig. 64.5 Force–displacement plot for jute–coir–carbon fibre for tensile test

Table 64.4 Flexural strength of jute–coir–glass fibre composites

Sample	Flexural strength (MPa)
G1	100.22
G2	82.88
G3	98.16
G4	86.51
G5	77

Table 64.5 Flexural strength of jute–coir–carbon fibre composites

Sample	Flexural strength (MPa)
C1	125.8
C2	81.04
C3	54.05
C4	90.90
C5	80

and 64.7, respectively. Comparing both the plots, it can be noted that the samples G3, G4, C2 and C4 possess better withstanding capabilities to bending force. When considering the tensile properties discussed in the previous section along with the flexural properties, the samples G3 and G4 can be eliminated due to inconsistent behaviour in properties. Although these samples could withstand flexural loads of 170 and 198.2 N, their tensile load bearing capabilities are insignificant. Interestingly, the samples C2 and C4 sustained its consistency in both tensile and flexural property tests. On considering Fig. 64.6, it can be precisely noted that the samples C2 and C4 were able to withstand the flexural loads up to 145 N and 163.3 N, respectively. The depleting effect of the coir fibre with increase in its

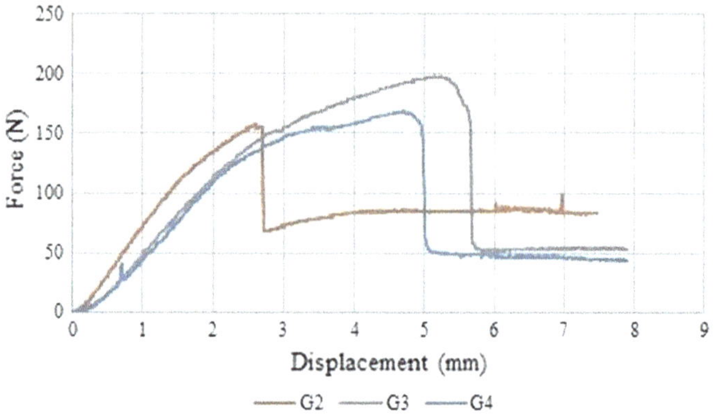

Fig. 64.6 Force–displacement graph for jute–coir–glass fibre for flexural test

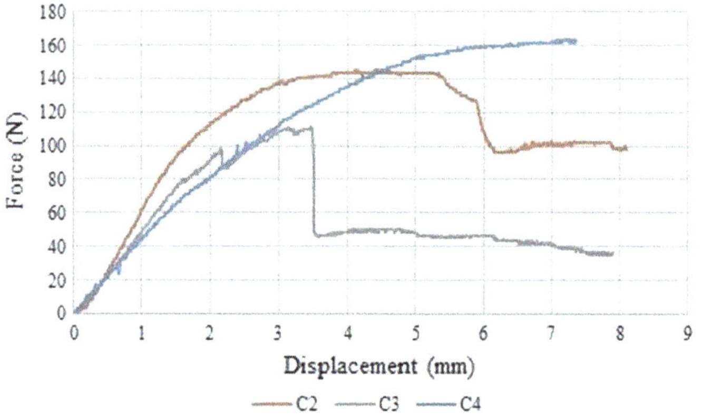

Fig. 64.7 Force–displacement graph for jute–coir–carbon fibre for flexural test

percentage is not much evident in the case of flexural tests. Hence, the jute–coir–carbon fibre is found to have an upper hand compared to the jute–coir–glass fibre.

64.4 Conclusions

The various proportions of jute–coir–glass fibre polyester resin composites and jute–coir–carbon fibre polyester resin composites were prepared and subjected to tensile and flexural tests for the comparative study. The following conclusions were derived from the experiment.

- Jute–carbon composites showed good tensile and flexural properties in comparison with jute–glass composites. Addition of coir into the composites decreased the mechanical properties of both jute–glass and jute–carbon fibre composites.

- Jute–coir–glass/carbon at equal volume proportion showed the lowest tensile strength due to lack of adhesion between fibres and the resin.
- The samples C2 and C3 showed the maximum and minimum tensile strengths among hybrid jute–coir fibre composites. Flexural strength was highest for G3 (98.16 MPa) and lowest for C3 (54.05 MPa) samples of the jute–coir fibre composites. Further, all the composites investigated showed ductile behaviour.
- The samples C2 and C4 of the jute–coir–carbon fibre composites possess better properties compared to the other samples tested for its mechanical properties. Specifically, the sample C2 was found to have consistent property of all the samples considered. The samples G1, G2 and G3 of the jute–coir–glass fibre composites were eliminated due to inconsistency in properties tested under various conditions.
- Finally, jute–coir–carbon fibre composites had greater tensile and flexural properties over jute–coir–glass fibre composites, comparatively.

Acknowledgements The authors would like to extend their sincere gratitude to Centre for Bio-Polymer Science and Technology (CBPST), Cochin, A unit of CIPET for providing the technical support for this initiative.

References

1. Ku, H., Wang, H., Pattarachaiyakoop, N., Trada, M.: A review on the tensile properties of natural fibre reinforced polymer composites. Compos. B Eng. **42**(4), 856–873 (2011)
2. Thakur, V.K., Thakur, M.K., Gupta, R.K.: Review: raw natural fibre–based polymer composites. Int. J. Polym. Anal. Charact. **19**(3), 256–271 (2014)
3. Lenda, T.A., Mridha, S.: Influence of moisture absorption on impact strength and failure behavior of hybrid jute-carbon/epoxy composite. In: Advanced Materials Research, vol. 264, pp. 457–462. Trans Tech Publications (2011)
4. Sreekala, M.S., George, J., Kumaran, M.G., Thomas, S.: The mechanical performance of hybrid phenol-formaldehyde-based composites reinforced with glass and oil palm fibres. Compos. Sci. Technol. **62**(3), 339–353 (2002)
5. Rajkumar, G., Srinivasan, J., Suvitha, L.: Natural protein fibre hybrid composites: effects of fibre content and fibre orientation on mechanical, thermal conductivity and water absorption properties. J. Ind. Text. **44**(5), 709–724 (2015)
6. Campbell, F.C.: Structural Composite Materials. ASM International (2010)
7. Sezgin, H., Berkalp, O.B.: The effect of hybridization on significant characteristics of jute/glass and jute/carbon-reinforced composites. J. Ind. Text. (2016). Doi: 1528083716644290
8. Bhagat, V.K., Biswas, S., Dehury, J.: Physical, mechanical, and water absorption behavior of coir/glass fibre reinforced epoxy-based hybrid composites. Polym. Compos. **35**(5), 925–930 (2014)
9. Wambua, P., Ivens, J., Verpoest, I.: Natural fibres: can they replace glass in fibre reinforced plastics. Compos. Sci. Technol. **63**(9), 1259–1264 (2003)
10. Harish, S., Michael, D.P., Bensely, A., Lal, D.M., Rajadurai, A.: Mechanical property evaluation of natural fibre coir composite. Mater. Charact. **60**(1), 44–49 (2009)

11. Hasan, K.F., Islam, M., Morshed, M.N., Sultana, M.Z., Mueeid, M.A., Wu, J.H.: Influence of naturally originated jute on the fabrication & mechanical properties of jute/polyester hybrid composite. Am. J. Polym. Sci. Eng. **4**(1), 82–90 (2016)
12. Standard, A.S.T.M. D638: Standard Test Method for Tensile Properties of Plastics. ASTM International, West Conshohocken (2010)
13. Standard, A.S.T.M. D790: Standard Test Methods for Flexural Properties of Unreinforced and Reinforced Plastics and Electrical Insulating Materials. ASTM International, West Conshohocken (PA) (2010)

Chapter 65
Mechanical Characterization of Hybrid Carbon–Glass-Reinforced Polypropylene Composites

Ansuman Padhi and Debiprasada Sahoo

Abstract In the present work, an attempt has been made to investigate the effect of varying wt% of short carbon fiber on the hybrid carbon–glass fiber/polypropylene (PP) composites. Composites of PP reinforced with short glass fibers and carbon fibers were prepared using two-roll milling machine and compression molding techniques with different vol. fraction and wt. percentage. The wt% and vol. fraction of glass fibers were kept constant, by varying the other two materials. Different mechanical properties such as tensile and impact properties were investigated by using suitable experimental setup. In tensile test, with the increase in carbon fiber wt% the strength (MPa) initially decreases but afterward it increases and is giving highest value for maximum wt% of carbon fiber. But for the impact test, the energy absorbed (J) gradually increases by increasing the carbon fiber wt%. From the test results, it shows that the material properties have been improved significantly compared to the polypropylene with only glass fiber, which encourages to use and opt for hybrid composite.

Keywords Hybrid composites · Carbon fiber · Glass fiber · Polypropylene · Mechanical properties

65.1 Introduction

Any single material may not be able to satisfy all the stringent requirements of properties to be used in high-end structural applications. This leads to the development and use of multi-phase integrated composite material systems. Composite materials are formed by macro-scale combination of two or more chemically distinct phases whose performances are superior to their constituent materials and often exhibit qualities that neither constituent possesses. FRP composites consist of

A. Padhi (✉) · D. Sahoo
Indira Gandhi Institute of Technology, Sarang, 759146 Dhenkanal, Odisha, India
e-mail: Anshmech25@gmail.com

R. G. Narayanan et al. (eds.), *Advances in Computational Methods in Manufacturing*, Lecture Notes on Multidisciplinary Industrial Engineering,
https://doi.org/10.1007/978-981-32-9072-3_65

fibers of high strength and modulus embedded or dispersed in matrix (polymer) with distinct interfaces between them.

In the past few years, the thermoplastic polymers have become most preferred materials for various applications due to several attractive properties including less weight, easy to process and cost-effectiveness. Hence, efforts have been made significantly to utilize polymers in various industrial applications, using different types of reinforcements including various kinds of fibers that are integrated into the polymers to increase their physical and mechanical properties. Thus, fiber-reinforced polymer matrix composites are comprehensively used due to their lightweight, high strength, biodegradability ability to be processed rapidly, high stiffness and good corrosion resistance. Among the synthetic fibers, carbon fiber is having high strength and stiffness with significantly light weight compared to same volume of glass fiber. But due to its cost, it can not be used in day-to-day applications. So in this paper, an attempt has been taken to manufacture a hybrid composite in which carbon fiber is added with different weight percentage ratios to the glass fiber polypropylene composite.

However, a number of studies have been carried out on hybrid composites with thermoplastic material (polypropylene) as matrix. Some of the significant findings are discussed below. Fu et al. [1] prepared and investigated the tensile properties of polypropylene (PP) composites reinforced with short carbon fibers (SCF) and short glass fibers (SGF) with extrusion compounding and injection molding techniques. Karsli et al. [2] studied the hybrid reinforcement effect of surface treated or untreated glass fiber (GF)/carbon nanotube (CNT) on the mechanical and electrical properties as well as morphology of polypropylene matrix composites. Gamstedt et al. [3] have studied the strength properties of polypropylene (PP) and polypropylene modified with maleic anhydride (MA-PP) reinforced by continuous longitudinal glass fibers. Ferreira et al. [4] have concerned studies of fatigue on polypropylene/glass fiber-thermoplastic composites produced from a bi-directional woven cloth of co-mingled E-glass fibers and polypropylene. Suresh et al. [5] discussed the influence of the forming pressure and coupler concentration on the mechanical behavior of polypropylene composite laminates reinforced with glass fiber. Ashori et al. [6] worked on improvement in the mechanical and thermo-mechanical properties of short carbon fiber (SCF)/polypropylene (PP) composite, resulting from coating of the SCFs with exfoliated grapheme nanoplatelets (xGnPs). The adhesion enhancement of xGnP–SCF/PP interface was clearly shown by scanning electron microscopy (SEM) images of the surfaces from tensile failure. Rezaei et al. [7] studied the thermal stability of short carbon fiber-reinforced polypropylene (SCF/PP) composite were prepared with melt blending and hot-pressing techniques. Swolfs et al. [8] prepared composite hybrids by taking bonding between carbon fiber prepregs and PP tapes. For a weak bonding, high penetration impacts resistance and the 20% ultimate tensile failure strain of self-reinforced PP were maintained. For a strong bonding, the ultimate tensile failure strain was reduced marginally, but the flexural performance was improved. Unterweger et al. [9] evaluated the mechanical properties of injection-molded CF-reinforced polypropylene (PP) composites under the influence of carbon fibers

(CF) surface properties and the amount of a coupling agent. Khan et al. [10] found that the damping ratio of the hybrid composites is enhanced with the addition of CNTs, which is attributed to sliding at the CNT–matrix interfaces. Arikan et al. [11] present the impact response of E-glass fiber-reinforced composites that have been manufactured with two types of resin, polypropylene and epoxy (thermoplastic and thermoset). Swolfs et al. [8] found that in the prepregs for a homopolymer PP matrix, the weak bonding among fiber and matrix affected the penetration impact resistance according to a linear rule-of-mixtures. Zhang et al. [12] investigated the influences of stacking sequence on the strength of hybrid composites comprising materials with differing stiffness and strength. Sullins et al. [13] worked on the effects of material treatment(s) on the mechanical behaviors of hemp fiber-reinforced polypropylene (PP) composites.

In this work, the composites have been prepared by taking polypropylene as matrix reinforced with short glass and carbon fiber, where the wt% of glass fiber kept constant in order to visualize the effect of CF. The above materials are prepared by using two-roll mill and compression molding techniques. This study mainly aimed to investigate the effect of carbon fiber hybridization on the glass-reinforced PP composites. The tensile properties as well as the energy absorption during impact were primarily considered in the experimentation.

65.2 Experimental Details

Materials used

The materials employed in this investigation were polypropylene, E-glass fiber roving and carbon fiber rovings. The mechanical and physical properties of these materials are listed in Table 65.1. The composition and volume fraction of each specimen are given in Table 65.2. The wt% of glass fiber was kept constant, while the wt% of other two materials is varied.

Specimen Preparation

The glass fiber and carbon fiber are cut into 3–5 mm in length by help of a scissor and make it to chopped fiber form. After that the polypropylene which is already procured is mixed with the above two fibers in the specified ratios with the help of a two-roll mill machine. These machines are extensively used for mixing and

Table 65.1 Mechanical properties of materials

Materials	Tensile strength (MPa)	Young's modulus (GPa)	Density (g/cm^3)
Glass fiber	2450	81.3	2.65
Carbon fiber	3665	230	1.75
Polypropylene	23.9	1.28	0.904

Table 65.2 Volume fraction and weight composition in different specimens

Specimen code	Polypropylene (PP) wt %/(Mass in gm.)	Carbon fiber (CF) wt%/ (Mass in gm.)	Glass fiber (GF) wt%/ (Mass in gm.)
A	0.8/400	0.05/25	0.15/75
B	0.75/375	0.1/50	0.15/75
C	0.7/350	0.15/75	0.15/75

kneading raw rubber, thermoplastics or EVA with chemicals, synthetic rubber into uniform materials. The uniform materials can be fed to calendar, hot presses or other processing machinery for preparing rubber or plastic products. Two-roller mills are the most simple in operation, where the material is crushed between two rollers before it continues on to its final destination. The spacing can be adjusted in between these two rollers by the operator. Smaller pieces are produced when the material is crushed in thinner spacing. The front roller in this study was set to 170 ° C, 15 rpm and rear roller to 175 °C, 10 rpm and a gap of 2 mm kept in between the two rollers, after that the mixing process will start (Fig. 65.1).

By setting the rollers quite tightly to each other, refined rubber in thin-sheet form (~0.10 mm) is produced. A smooth, uniform and free of grain or lumps sheet is produced. The impurities contained in reclaimed rubber are pressed by rolls to the both sides of rubber sheet which can be removed. Consequently, the purity of the reclaimed rubber is increased in refiner mills. Earlier, the finished sheet was pulled toward a wind-up attachment and layered few times to increase the thickness of the sheet (approx. 25 mm), after which they are cut using a hand knife. The cut sheets are then cleaned and stacked (Fig. 65.2).

Compression molding The final product is manufactured using the compression molding machine. Compression molding is a high-pressure and high-volume method suitable for molding intricate and high-strength fiberglass reinforcements. Advanced composite thermoplastics can also be compression molded with

Fig. 65.1 Mixing of fibers with polypropylene in two-roll mill

Fig. 65.2 Sheets obtained after two-roll mill operation

unidirectional tapes, woven fabrics, randomly oriented fiber mat or chopped strand. Compression molding is having ability to mold large and fairly complex parts. Also, it is one of the cost-effective molding methods compared with other methods such as transfer molding and injection molding; moreover, it trashes reasonably little material, giving it an advantage when working with expensive compounds.

The molding material is positioned in the mold cavity and the heated platens are closed by a hydraulic ram. The mold is subjected to pressure and heated until the curing reaction occurs. Temperature history during molding process is shown in Fig. 65.3. Holding time is set for 90 s for maximum temperature of the mold surface at 200 °C. To heat up from 50 to 200 °C, it took only 50 and 130 s for the cooling process from 200 to 50 °C. Altogether, molding was completed in 270 s. In this way, the three different compositions of composite sheets are prepared by compression molding as shown below. For quick and easy removal of composite sheets, polyester film applied at the inner surface of the mold. The dimension of the die is 160 × 160 × 3 mm^3. Care is taken to avoid formation of air bubbles. The die was allowed to cool at room temperature. Then the samples were taken out of the die, cut by hacksaw into different sizes for further experimentation as per standard (Table 65.3; Fig. 65.4).

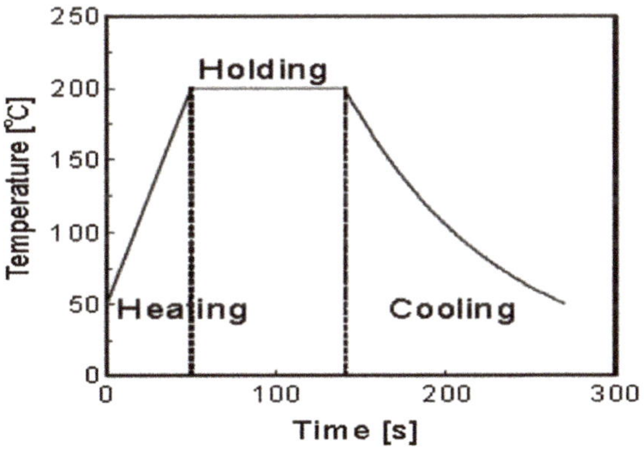

Fig. 65.3 Temperature versus time during a typical compression molding process

Table 65.3 Specimen dimensions for Tensile & Impact test

Test	Length (mm)	Width (mm)	Thickness (mm)
Tensile	165	12.7	3.2
Impact	63.5	12.7	3.2

Fig. 65.4 Compression molding machine

65.3 Results and Discussion

65.3.1 Tensile Test

Tensile tests were conducted on Instron universal testing machine (model 3382) in accordance with ASTM D638. The testing speed was 1 mm/min with at least three samples of each specimen were tested (Figs. 65.5 and 65.6; Table 65.4).

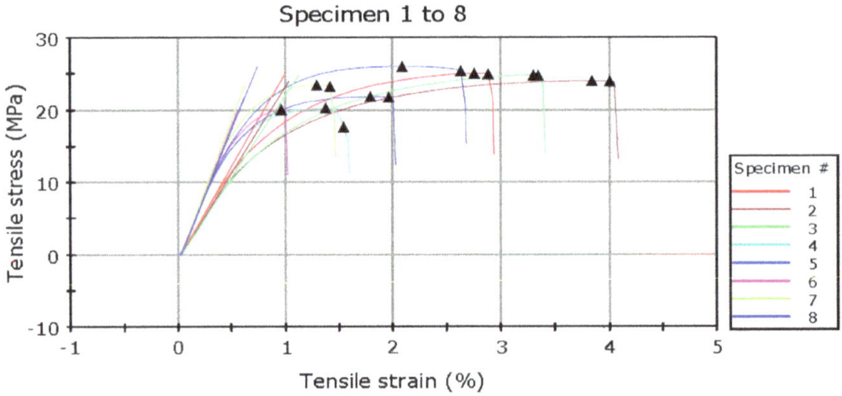

Fig. 65.5 Stress–strain curve of 8 specimens of three different compositions

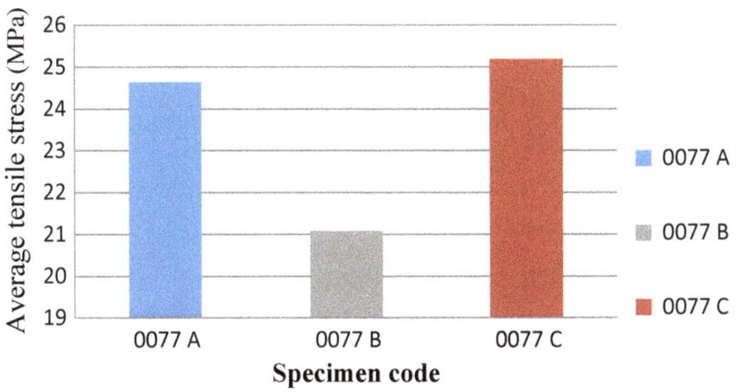

Fig. 65.6 Average ultimate tensile strength of three Specimens

65.3.2 Impact Test

The impact properties of the three specimens were determined using three samples for each composition with the dimensions as mentioned in Table 65.3. The Charpy impact test is a standardized high strain-rate test which measures the amount of energy absorbed by a material during fracture. This absorbed energy is a degree of a given material's notch toughness. The thickness of the specimen and the dimensions of the un-notched length were measured. The pendulum was raised to the left until it indicates the maximum energy range on the upper indicator unit. The specimen was placed horizontally across supports with the notch away from the pendulum. The pendulum was released to measure the energy absorbed during the impact. The energy absorbed during the impact test by different specimens was shown in Fig. 65.7. The test is done in accordance with the ASTM D 256.

It is observed that, the tensile test results shown in Fig. 65.6, specimen 8 have a better strength compared to all other seven specimens. The specimen 8 is having the composition of PP 70%, CF 15%, GF 15%. It has been noted that 5% CF content composition has more tensile strength compare to 10% CF content composition, but when 15% CF is added to PP and GF, it gives the maximum value of tensile strength and this value is 26.04 MPa.

Impact test results are shown in Fig. 65.7. It can be seen that impact strength of code "A" composition is 51.67 J/m. After that when CF% changes from 5 to 10%, the impact energy is increased to 66.67 J/m. But finally when the CF% changes from 10 to 15%, it gives a maximum value of impact energy and that value is 116.33 J/m.

Table 65.4 Tensile strength data sheet of all the 8 specimens

S. No.	Specimen code	Rate 1 (mm/min)	Width (mm)	Thickness (mm)	Maximum load (N)	Tensile stress at maximum load (MPa)	Average tensile stress (MPa)	Tensile strain at break (Standard) (%)	Tensile stress at break (MPa)
1	0077(A)	50.00	15.62	3.2	1254.2	25.09	**24.65**	2.88	24.99
2	0077(A)	50.00	15.54	3.16	1180.18	24.03		4	23.96
3	0077(A)	50.00	15.6	3.2	1239.94	24.84		3.33	24.81
4	0077(B)	50.00	15.2	3.28	1011.48	20.29	**21.08**	1.54	17.68
5	0077(B)	50.00	15.3	3.24	1084.28	21.87		1.96	21.78
6	0077(C)	50.00	15.3	3.2	981.74	24.05	**25.18**	0.96	20.05
7	0077(C)	50.00	15.3	3.2	1149.05	25.47		1.42	23.29
8	0077(C)	50.00	15.3	3.2	1274.99	26.04		2.62	25.43

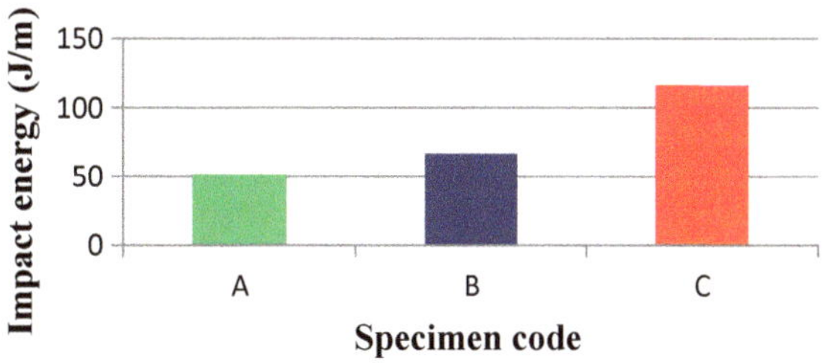

Fig. 65.7 Impact energy (J) of three different compositions

65.4 Conclusion

After conducting the tensile test and impact test of these three compositions, the following conclusions are derived.

- In tensile test the composition of 80% polypropylene, 5% carbon fiber and 15% glass fiber gives the maximum value as compared to 75% polypropylene, 10% carbon fiber and 15% glass fiber. But when the carbon fiber is changed from 10 to 15%, it gives a maximum value.
- In impact test, the value of the impact energy is gradually increased by increasing the carbon fiber %.
- So from the above two test, out of the three combinations the specimen code "C" showed the best result in terms of tensile strength and shock absorption properties.
- Hence, it is the best-suited mixture combination.
- The addition of carbon fiber plays a vital role in case of the glass fiber polypropylene composite. In order to make the specimen cost-friendly, less weight percentage of carbon fiber is taken over glass fiber.
- The test results show that the material properties have been improved significantly compared to the polypropylene with only glass fiber, which encourages to use and opt for hybrid composite.

References

1. Fu, S.Y., Lauke, B., Mäder, E., Yue, C.Y., Hu, X.: Tensile properties of short-glass-fiber-and short-carbon-fiber-reinforced polypropylene composites. Compos. A Appl. Sci. Manuf. **31** (10), 1117–1125 (2000)
2. Karsli, N.G., Yesil, S., Aytac, A.: Effect of hybrid carbon nanotube/short glass fiber reinforcement on the properties of polypropylene composites. Compos. B Eng. **63**, 154–160 (2014)
3. Gamstedt, E.K., Berglund, L.A., Peijs, T.: Fatigue mechanisms in unidirectional glass-fibre-reinforced polypropylene. Compos. Sci. Technol. **59**(5), 759–768 (1999)
4. Ferreira, J.A.M., Costa, J.D.M., Reis, P.N.B., Richardson, M.O.W.: Analysis of fatigue and damage in glass-fibre-reinforced polypropylene composite materials. Compos. Sci. Technol. **59**(10), 1461–1467 (1999)

5. Suresh, S., Kumar, V.S.: Experimental determination of the mechanical behavior of glass fiber reinforced polypropylene composites. Procedia Eng. **97**, 632–641 (2014)
6. Ashori, A., Menbari, S., Bahrami, R.: Mechanical and thermo-mechanical properties of short carbon fiber reinforced polypropylene composites using exfoliated graphene nanoplatelets coating. J. Ind. Eng. Chem. **38**, 37–42 (2016)
7. Rezaei, F., Yunus, R., Ibrahim, N.A.: Effect of fiber length on thermomechanical properties of short carbon fiber reinforced polypropylene composites. Mater. Des. **30**(2), 260–263 (2009)
8. Swolfs, Y., Shi, J., Meerten, Y., Hine, P., Ward, I., Verpoest, I., Gorbatikh, L.: The importance of bonding in intralayer carbon fibre/self-reinforced polypropylene hybrid composites. Compos. A Appl. Sci. Manuf. **76**, 299–308 (2015)
9. Unterweger, C., Duchoslav, J., Stifter, D., Fürst, C.: Characterization of carbon fiber surfaces and their impact on the mechanical properties of short carbon fiber reinforced polypropylene composites. Compos. Sci. Technol. **108**, 41–47 (2015)
10. Khan, S.U., Li, C.Y., Siddiqui, N.A., Kim, J.K.: Vibration damping characteristics of carbon fiber-reinforced composites containing multi-walled carbon nanotubes. Compos. Sci. Technol. **71**(12), 1486–1494 (2011)
11. Arikan, V., Sayman, O.: Comparative study on repeated impact response of E-glass fiber reinforced polypropylene & epoxy matrix composites. Compos. B Eng. **83**, 1–6 (2015)
12. Zhang, J., Chaisombat, K., He, S., Wang, C.H.: Hybrid composite laminates reinforced with glass/carbon woven fabrics for lightweight load bearing structures. Mat. Des. (1980–2015) **36**, 75–80 (2012)
13. Sullins, T., Pillay, S., Komus, A., Ning, H.: Hemp-fiber reinforced polypropylene composites: the effects of material treatments. Compos. B Eng. **114**, 15–22 (2017)

Chapter 66
Effect of Temperature on the Fracture Strength of Perfect and Defective MonoLayered Graphene

Kritesh Kumar Gupta and Sudip Dey

Abstract Graphene is one of the carbon allotropes which have one atom thickness. It is a monolayer of carbon atoms, which are bonded in a two-dimensional hexagonal lattice. It is a unit layer of graphite; stacking the monolayer graphene one over another forms the graphite. The exceptional electromechanical properties of graphene attracted the researchers and scientists to explore the wide application areas so that the properties of graphene could be harnessed. In this regard, the present paper analyses the perfect single-layer graphene sheet (SLGS) and defective graphene sheet by depicting the effect of temperature variations on the fracture strength of both of the sheets. To this end, the molecular dynamics (MD) simulations based on AIREBO interatomic potential field and Nose–Hoover thermostat technique are carried out. It is concluded by the study that with the increasing temperature, the fracture strength of the graphene reduces remarkably. It is also evident that introducing random vacancy of 2.5% in perfect graphene causes the yielding phenomenon at lower strains which remains absent in the perfect graphene. Also, the defective graphene shows the ductile fracture, confirming considerable yielding before complete fracture.

Keywords Fracture strength · Graphene · Molecular dynamics · AIREBO potential · LAMMPS

66.1 Introduction

The extraordinary characteristics of graphene have been used to develop nanoresonators [1], biosensors [2], nanocomposites [3] and many more nanoelectromechanical systems (NEMS). So far research has been conducted in order to determine the exact mechanical properties and strength of the monolayered graphene sheet.

K. K. Gupta (✉) · S. Dey
Department of Mechanical Engineering, National Institute of Technology Silchar, Silchar, India
e-mail: kriteshamie@gmail.com

R. G. Narayanan et al. (eds.), *Advances in Computational Methods in Manufacturing*, Lecture Notes on Multidisciplinary Industrial Engineering,
https://doi.org/10.1007/978-981-32-9072-3_66

Graphene is found to be one of the lightest (0.77 mgm^{-2}) and still stiffest material known. Its fracture strength and Young modulus are recorded as 130 GPa and 1 TPa, respectively. Finding the mechanical properties of graphene through experimental procedures are limited due to the fact that setting up the experimentation for the nanoscale material has its own complications and limitations. Experimentally, one of the common methods to extract graphene from graphite is chemical vapour decomposition (CVD), which is performed at elevated temperatures, it is obvious that during extraction some defects may be introduced to the graphene structure by vacancy, chemical reactions with foreign elements or by a change in the bond structure at elevated temperatures. This leads to getting incorrect desired results from the experimentation. The defect such as vacancies remarkably reduces the strength of graphene, which in return affects the performance of the NEMS.

In the past, researchers have carried out a vast study to accurately characterize the SLGS. To this end, Tsai et al. [4] explored the elastic properties of the graphene along with graphite flakes through molecular dynamics simulation and concluded that the single graphene layer has the higher value of Young's modulus of elasticity and shear modulus in comparison with graphite flakes. The effective characterization of graphene has been a challenge so far as there are a lot of discrepancies recorded regarding its exact characteristics, to address this Thomas and Ajith [5] performed the atomistic simulations using Tersoff interatomic potential and determined the mechanical, thermal and structural properties of the SLGS. The defects in graphene are prone to induced while its extraction, in this contrast Ansari et al. [6] performed the atomistic simulation using Tersoff–Brenner potential energy function to determine the mechanical properties of perfect and defective single-layered graphene, while Dewapriya and Rajapakse [7] investigated the rate of strain and temperature dependence of the failure criteria of single-layer graphene sheet and proposed a fracture mechanics model based on the continuum to characterize the toughness of defected graphene sheet. To explore the effect of Stone–Wales defect, Wang et al. [8] studied the fracture strength of defective graphene sheet using molecular dynamic simulation, the study was carried out at various temperature and it was concluded that the strength of SLGS is affected by the presence of defects, loading directions and temperature, whereas Wang et al. [9] investigated the mechanical responses of graphene membrane with various orientations of 5-8-5 defect using molecular dynamics simulation, and it was concluded from the study that the introduction of 5-8-5 defect in graphene sheet reduces its fracture strength and strain. To address the irregularities based on interatomic potential field, Anastasi et al. [10] presented the MD simulation of single-layer graphene under tensile loading. In order to carry out the simulation, several potential functions such as the bending angle, Morse, Lennard-Jones and torsion potential functions have been used from the mdFOAM library of openFOAM software. It was concluded from the study that graphene is anisotropic, and its fracture strength depends on the sheet size, while Singh et al. [11] predicted the elastic properties of graphene sheet at infinitesimal and finite strains by combining MM3 potential with the Cauchy–Born rule and compared the results with the outcome of a study carried out using

Tersoff–Brenner potential field and second-generation REBO potential field. The MD simulation utilizes different ensembles such as NPT, NVT and NVE for averaging the atomistic properties, and to differentiate the usage of NPT and NVT in the MD simulations, Clavier et al. [12] computed the elastic stiffness tensor at 1 bar and 60 K of FCC argon and compared the result obtained in both the cases. The graphene could be the potential reinforcement material for strengthening the nanocomposites in this context, and Papageorgio et al. [13] reviewed the mechanical properties of graphene and described the strategies to prepare the bulk graphene-based nanocomposites. It was concluded from the study that graphene as reinforcement in composites may improve the strength of the composite up to the next level.

Performing experimentation on graphene is a challenging task due to obvious limitations in experimental set-up for nanoscale analysis. Therefore, using MD simulation as the analysis tool gives the edge over understanding the complex nanomaterials in a better way. In the past, a lot of research is executed in order to determine the exact characteristics of pristine and defected SLGS. However, there is limited study available dealing with concentrations of defect present in the monolayer graphene. In this study, the mechanical properties and fracture strength of the perfect graphene and the graphene sheet with 2.5% vacancy concentration are determined using MD simulation. In order to observe the effect of temperature variation over the fracture strength of graphene, the analysis is carried out at various temperatures, i.e. at 1, 300, 900 and 1200 K. Young's modulus, fracture strength and failure strain are determined for both the directions (armchair and zigzag) of graphene, and results are compared.

66.2 Molecular Dynamics Simulation

66.2.1 Molecular Dynamics

Nanomechanical modelling is one of the best methods to analyse the properties of nanomaterials. Nanomechanics model may be differentiated into three main categories—the first principle approaches such as DFT, semi-empirical methods such as molecular dynamics simulation and the third is the modified continuum model.

The molecular dynamics is computationally affordable in comparison with the first principle approaches. Although it relies on several assumptions associated with the interatomic potential function. In the molecular simulations, materials are viewed as the collection of individual atoms which interact by exerting forces on each other and follow Newton's law of motion. In order to perform molecular dynamics simulations, three important ingredients are required; first, an initial system configuration such as position and velocities of the atoms [r (t = 0), v (t = 0)], interaction potential for the system and a way to integrate Newton's equation of motion. Simulation procedure in MD initiates by determining the forces acting on an individual atom by the neighbouring atoms. Force acting on an individual atom i may be represented as

$$f_i = -\frac{\partial E_i}{\partial r_i} \tag{66.1}$$

where E_i refers to the potential energy of the atom which is approximated by the interatomic potential field. After the determination of forces, the accelerations are determined using Newton's equation of motion

$$f_i = m_i \frac{\mathrm{d}^2 r_i}{\mathrm{d}t^2} = m_i a_i \tag{66.2}$$

Further atoms are allowed to accelerate with the determined acceleration for a certain duration known as a time step. At the end of each time step, the new positions and velocities are recorded and integrated using numerical integration algorithms such as velocity Verlet algorithm. The use of statistical ensembles such as NPT, NVT and NVE allows controlling the temperature and pressure of the system [14].

66.2.2 Modelling and Simulation

The graphene sheet of 5 nm × 5 nm with 1008 carbon atoms is modelled using visual molecular dynamics (VMD) [15], and the output file received from VMD is fed to Avogadro (an open-source atom manipulator) [16] in order to introduce a 2.5% vacancy in the perfect graphene sheet. Both the sheets (perfect and defected) were uniaxially deformed in x-direction using LAMMPS. LAMMPS [17] is an open-source code package by Sandia National Laboratories, USA. To avoid free edges, all three directions were applied with periodic boundary conditions. The sheets were firstly equilibrated to minimize the energy of the system and relax the graphene structure, over 30 ps prior to applying strain using a time step of 0.5 fs. In the relaxation period, the pressure component of the x- and y-directions was maintained zero using NPT ensemble. The NPT ensemble holds the no. of atoms, pressure and temperature of the system constant, and it uses Nose–Hoover thermostat to control the temperature of the system. Furthermore, to uniaxially deform the sheets a strain was imposed by elongation of the simulation box in both (armchair and zigzag) dimension and the applied strain rate was 10^9 s^{-1}. The stress components in y- and z-directions were kept at zero during simulating the tensile deformation of graphene by keeping in check the pressure component in the said directions using NPT ensemble. Figure 66.1 shows the topology of the perfect graphene sheet and graphene introduced to 2.5% vacancy concentration.

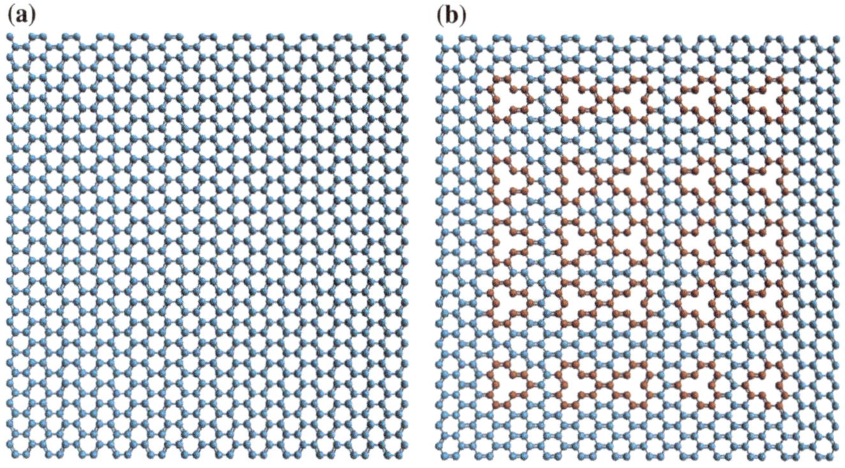

Fig. 66.1 Graphene sheet **a** perfect, **b** with 2.5% vacancy concentration

66.2.3 Potential Field

In molecular dynamics simulation, the potential field is an element which signifies the energy interactions between the atoms. The adaptive intermolecular reactive empirical bond order (AIREBO) potential is employed in the present study. The AIREBO potential is made by combining three different potential fields first, the original REBO potential, second the Lennard-Jones (LJ) and third the torsion potential. The original REBO term imparts the interatomic attractive and repulsive energy, the Lennard-Jones to account for the intermolecular interaction that includes dispersion and short-range repulsion and torsion component to signify the torsional interactions between the carbon–carbon bonds.

The general form to represent the AIREBO potential can be expressed as

$$E = \frac{1}{2}\sum_{i=1}\sum_{i\neq j}\left[E_{ij}^{\text{REBO}} + E_{ij}^{LJ} + \sum_{k\neq i}\sum_{l\neq i,j,k} E_{ijkl}^{\text{TORSION}}\right] \quad (66.3)$$

where E refers to the total potential energy with atoms i, j, k and l. The REBO term which refers the interatomic attraction and repulsion is represented as:

$$E_{ij}^{\text{REBO}} = V_{ij}^{R}(r_{ij}) + b_{ij}V_{ij}^{A}(r_{ij}) \quad (66.4)$$

where V^A and V^R refer the attractive and repulsive parts of the potential, respectively; r_{ij} signifies the distance between i and j atom up to the cut-off distance, and b_{ij} refers the bond order which allows for a change in the strength of the bond depending on the environment.

The LJ term of the potential can be described as:

$$E_{ij}^{\mathrm{LJ}} = f(r_{ij})V_{ij}^{\mathrm{LJ}} \tag{66.5}$$

$$V_{ij}^{\mathrm{LJ}} = 4\varepsilon_{ij}\left[\left(\frac{\sigma_{ij}}{r_{ij}}\right)^{12} - \left(\frac{\sigma_{ij}}{r_{ij}}\right)^{6}\right] \tag{66.6}$$

where $f(r_{ij})$ represents the set of cut-off functions σ_{ij}, and ε_{ij} are the distance at which interparticle potential is zero and the potential barrier, respectively.

Finally, the torsion term can be represented as:

$$E_{ijkl}^{\mathrm{TORSION}} = f(r_{ij}, r_{jk}, r_{kl})V^{\mathrm{TORSION}}(w_{ijkl}) \tag{66.7}$$

$$V^{\mathrm{TORSION}}(w_{ijkl}) = \varepsilon\left[\frac{256}{405}\cos^{10}\left(\frac{w}{2}\right) - \frac{1}{10}\right] \tag{66.8}$$

where r_{ij}, r_{jk} and r_{kl} are the bond lengths and w_{ijkl} is the dihedral angle and ε is the barrier height of the potential [18].

66.2.4 Calculation of Stress

In the molecular dynamics simulation, stress may be interpreted as Cauchy stress or virial stress. When it comes to the computational efficiency, the Cauchy stress is proved to be more considerable than virial stress. However, some amount of initial stress (when strain is zero) is induced by the Cauchy stress at higher temperatures. In this study, the computation of virial stress is carried out through atomistic simulation. Virial stress can be defined as:

$$\sigma_{ij} = \frac{1}{V}\sum_{\alpha}\left[\frac{1}{2}\sum_{\beta=1}^{N}\left(R_i^{\beta} - R_i^{\alpha}\right)F_j^{\alpha\beta} - m^{\alpha}v_i^{\alpha}v_j^{\alpha}\right] \tag{66.9}$$

where α and β refer to the serial number allocated to an atom and the number allocated to the neighbouring atom, respectively, while i and j signify the directional indices, and m^{α} and v_i^{α} are the mass and velocity of the atom α, respectively. R_i^{β} is the location of an atom β in the i direction; $F_j^{\alpha\beta}$ is the force due to an atom β on atom α along the j direction and V refers to the total volume of the material system. The virial stress represents a product of stress and volume, i.e. the stress value computed by LAMMPS would be in the units of pressure*volume, in order to extract the stress from the computed value it is essential to divide the value calculated by the LAMMPS with the total volume. Furthermore, in order to calculate the virial stress, the instantaneous volume is used, i.e. when the imposed strain is 'ε', the value of volume at the given instant would be $V_0(1+\varepsilon)$, where V_0 represents the initial volume of the system [19].

66.2.5 Constant Strain Method

In order to determine the elastic constants of graphene, it is subjected to a constant strain in the desired direction. As in this study, a strain rate of 10^{-9}/s is applied and the simulation box is allowed to deform in the applied direction. At the end of the time step, the resultant atomistic stresses are recorded in each direction. The relation between in-plane stress and stiffness matrix is written in terms of [4]

$$\begin{pmatrix} \sigma_{11} \\ \sigma_{22} \\ \tau_{12} \end{pmatrix} = \begin{pmatrix} C_{11} & C_{12} & 0 \\ C_{21} & C_{22} & 0 \\ 0 & 0 & C_{66} \end{pmatrix} \begin{pmatrix} \varepsilon_{11} \\ \varepsilon_{22} \\ \gamma_{12} \end{pmatrix} \tag{66.10}$$

Due to symmetry in the stiffness matrix, the values of elastic constants are as [4]

$$\begin{aligned} C_{11} &= C_{22}, \\ C_{12} &= C_{21}, \\ C_{66} &= \frac{C_{11} - C_{12}}{2}, \end{aligned} \tag{66.11}$$

As the applied strain is known and the atomistic stresses can be recorded, using these values the elastic constants can be determined. Once the elastic constants are known the value of Young's modulus and Poisson's ratio can be determined as [4]

$$E = \frac{C_{11}C_{22} - C_{12}C_{21}}{C_{22}} \tag{66.12}$$

$$\upsilon_{12} = \frac{C_{12}}{C_{22}} \tag{66.13}$$

66.3 Effect of Temperature Variation

The value of Young's modulus, elastics constants, Poisson's ratio and ultimate strength are determined, which shows a good agreement with the literature values. The said values are determined for the perfect graphene, which is as follows (Table 66.1)

Table 66.1 Validation of mechanical properties of graphene

Properties	Present study	Literature values
C_{11} (in GPa)	887.2	1094 (Thomas and Ajith [5])
C_{12} (in GPa)	153.1	136 (Thomas and Ajith [5])
Young's modulus (in TPa)	0.860	1.020 (Lee et al. [20])
Poisson's ratio	0.172	0.1248 (Thomas and Ajith [5])
fracture strength (in GPa)	145.4	130 (Lee et al. [20])

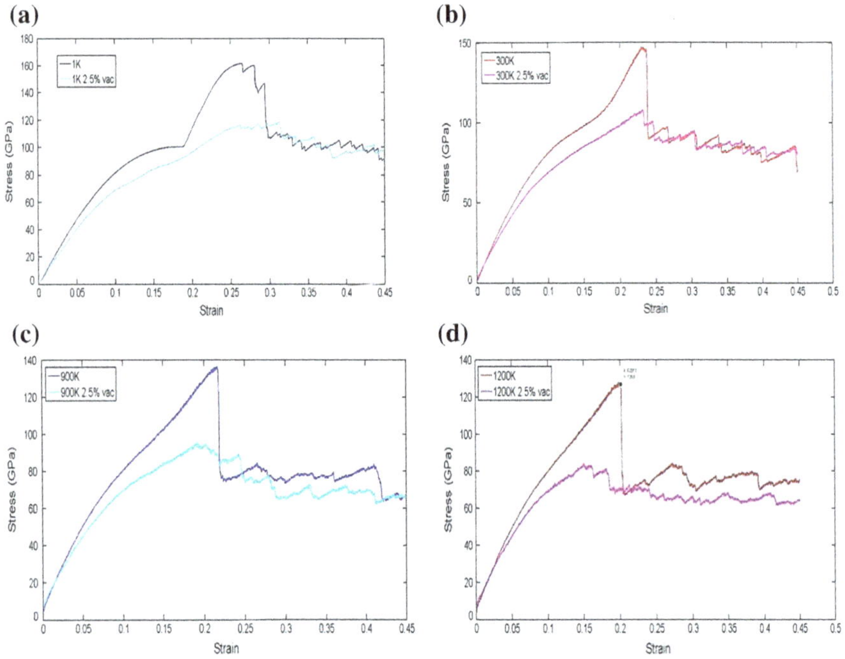

Fig. 66.2 Stress–strain behaviour of perfect and defective graphene loaded in an armchair direction **a** at 1 K, **b** at 300 K, **c** at 900 K, **d** at 1200 K. Defective sheets have 2.5% of randomly distributed vacancies

The stress–strain behaviour of perfect graphene and defected graphene at different temperatures, i.e. 1, 300, 900 and 1200 K is recorded in both armchair and zigzag directions, which is shown in Figs. 66.2 and 66.3, respectively. It is evident by the recorded values that the increment in temperature drastically reduces the fracture strength of the graphene in both the directions and failure occurs at fewer strain values. However, it is observed that the graphene depicts more resistance to failure in the transverse direction (zigzag) as compared to its longitudinal direction (armchair). It can also be seen that on uniaxially loading graphene in both the

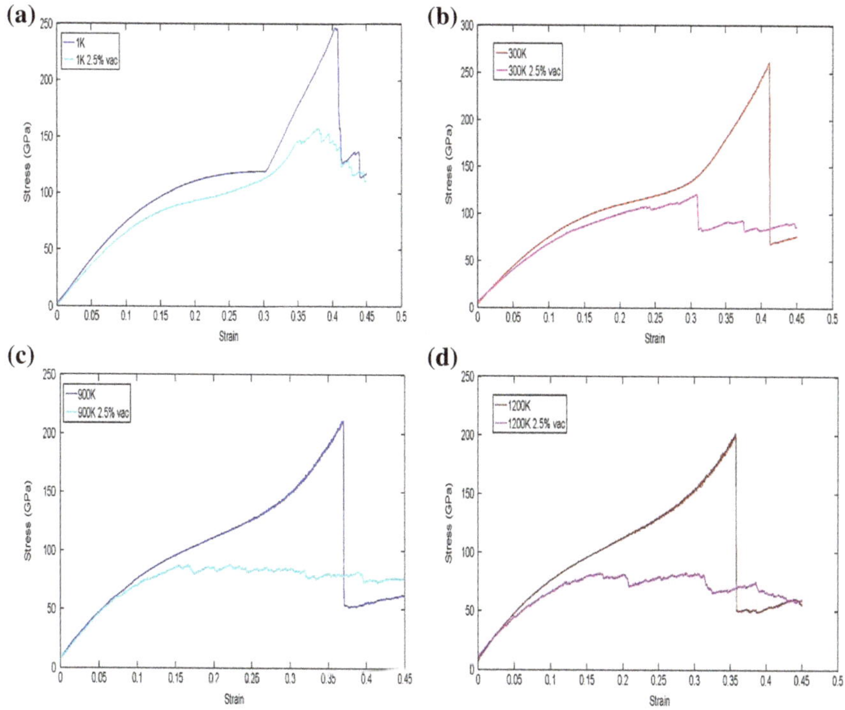

Fig. 66.3 Stress–strain behaviour of perfect and defective graphene loaded in a zigzag direction **a** at 1 K, **b** at 300 K, **c** at 900 K, **d** at 1200 K. Defective sheets have 2.5% of randomly distributed vacancies

directions (armchair and zigzag) at 1 K, it shows the sharp yielding point before fracture.

In both the cases, i.e. loading in an armchair and zigzag direction, the defected graphene does not show the instant fracture, and it is observed that on introducing the 2.5% vacancy to the graphene sheet, it shows the ductile fracture instead of brittle fracture. However, after some strain, it starts showing the yielding.

The given Tables 66.2 and 66.3 represent the recorded values of Young's modulus (*E*), fracture strength and failure strain of perfect graphene sheet at various temperatures in an armchair and zigzag directions, respectively.

Table 66.2 Young's modulus (*E*), fracture strength and failure strain of perfect graphene sheet loaded in an armchair direction due to variation in temperature

Temperature (K)	*E* (in TPa)	Fracture strength (in GPa)	Failure strain
1	0.814	146.9	0.293
300	0.812	145.4	0.236
900	0.809	135.7	0.217
1200	0.799	126.6	0.201

Table 66.3 Young's modulus (*E*), fracture strength and failure strain of perfect graphene sheet loaded in a zigzag direction due to variation in temperature

Temperature (K)	*E* (in TPa)	Fracture strength (in GPa)	Failure strain
1	0.745	246.3	0.407
300	0.756	261.9	0.411
900	0.764	211.4	0.370
1200	0.760	200.9	0.357

66.4 Conclusion

In summary, an MD simulation of uniaxial loading the perfect graphene sheet and defective graphene sheet (2.5% vacancy concentration) is carried out at various temperatures (1, 300, 900 and 1200 K) in both armchair and zigzag directions. The AIREBO potential field is used to incorporate the interatomic interactions between the constituent carbon atoms of the graphene sheet and to control the temperature of the system Nose–Hoover thermostat is employed, using NPT ensemble. The fracture strength of perfect and defective graphene sheet is observed at various temperatures, and it is concluded that the temperature has a profound impact over the fracture strength of the graphene sheet. It has been pointed in the study that how sometimes it is obvious that the extracted graphene sheet may have some defects, inherited by the way it is produced. The stress–strain behaviour of the perfect and defective graphene sheet is recorded at various temperatures by loading the graphene in both the directions, and it is concluded that graphene sheet possesses higher fracture strength in the zigzag direction as compared to its armchair direction. Furthermore, it is also been observed that the defective graphene sheet does not depict the brittle fracture, although the defected graphene starts yielding at lower strain than a perfect graphene sheet. The findings of the study provide the insights of variability in the fracture strength of graphene with respect to increasing temperature and with vacancy defects, using which the usability of graphene could be decided for the applications at elevated temperatures and where defects are unavoidable.

Acknowledgements The author(s) would like to acknowledge the financial support provided by the TEQIP-III to attend the conference ICCMM-2019 held at IIT Guwahati.

References

1. Dai, M.D., Kim, C.W., Eom, K.: Nonlinear vibration behaviour of graphene resonators and their applications in sensitive mass detection. Nanoscale Res. Lett. Springer Open J. **7**(1) (2012)
2. Kuila, T., Bose, S., Khanra, P., Mishra, A.K., Kim, N.H., Lee, J.H.: Recent advances in graphene-based biosensors. Biosens. Bioelectron. **26**, 4637–4648 (2011)
3. Wang, G., Wang, B., Wang, X., Park, J., Dou, S., Ahn, H., Kim, K.: Sn/graphene nanocomposite with 3D architecture for enhanced reversible lithium storage in lithium ion batteries. J. Mater. Chem. **19**, 8378–8384 (2009)
4. Tsai, J.L., Tu, J.F.: Characterizing mechanical properties of graphite using molecular dynamics simulation. Mater. Des. **31**, 194–199 (2010)
5. Thomas, S., Ajith, K.M.: Molecular dynamics simulation of the thermomechanical properties of the monolayer graphene sheet. Procedia Mater. Sci. **5**, 489–498 (2014)
6. Ansari, R., Ajori, S., Motevalli, B.: Mechanical properties of defective single-layered graphene sheets via molecular dynamics simulation. Superlattices Microstruct. **51**, 274–289 (2012)
7. Dewapriya, M.A.N., Rajapakse, R.K.N.D.: Molecular dynamics simulations and continuum modelling of temperature and strain rate dependent fracture strength of graphene with vacancy defects. J. Appl. Mech. **81** (2014)
8. Wang, M.C., Yan, C., Ma, L., Hu, N., Chen, M.W.: Effect of defects on the fracture strength of graphene sheets. Comput. Mater. Sci. **54**, 236–239 (2012)
9. Wang, S., Yang, B., Yuan, J., Si, Y., Chen, H.: Large-Scale molecular simulations on the mechanical response and failure behaviour of a defective graphene: cases of 5-8-5 defects. Sci. Rep. (2015)
10. Anastasi, A.A., Ritos, K., Cassar, G., Borg, M.K.: Mechanical properties of pristine and nanoporous graphene. Mol. Simul. (2016)
11. Singh, S., Patel, B.P.: Nonlinear elastic properties of graphene sheet using MM3 potential under finite deformation. Compos. B **136**, 81–91 (2018)
12. Clavier, G., Desbiens, N., Bourasseau, E., Lachet, V., Brussel-Dupend, N., Rousseau, B.: Computation of elastic constants of solids using molecular simulation: comparison of constant volume and constant pressure ensemble methods. Mol. Simul. **43**, 1413–1422 (2017)
13. Papageorgio, D.G., Kinloch, I.A., Young, R.J.: Mechanical properties of graphene and graphene-based nanocomposites. Prog. Mater. Sci. **90**, 75–127 (2017)
14. Cai, W.: Introduction to Molecular Simulations. Stanford University—Winter (2007)
15. Humphrey, W., Dalke, A., Schulten, K.: VMD—visual molecular dynamics. J. Molec. Graph. **14**, 33–38 (1996)
16. Hanwell, M.D., Curtis, D.E., Lonie, D.C., Vandermeersch, T., Zurek, E., Hutchison, G.R.: Avogadro: an advanced semantic chemical editor, visualization, and analysis platform. J. Cheminform. **4**, 17 (2012)
17. Plimpton, S.: Fast parallel algorithms for short-range molecular dynamics. J. Comput. Phys. **117**, 1–19 (1995)
18. Stuart, S.J., Tutein, A.B., Harrison, J.A.: A reactive potential for hydrocarbons with intermolecular interactions. J. Chem. Phys. **112**(14) (2000)

19. Thompson, A.P., Plimpton, S.J., Mattson, W.: General formulation of pressure and stress tensor for arbitrary many-body interaction potentials under periodic boundary conditions. J. Chem. Phys. **131** (2009)
20. Lee, C., Wei, X., Kysar, J.W., Hone, J.: Measurement of the elastic properties and intrinsic strength of monolayer graphene. Science **321**, 385–388 (2008)

Chapter 67
Effect of Process Parameters of Plasma Arc Preheating on Low Carbon Steel

Abhishek Kumar and Pankaj Biswas

Abstract In this study, both experimental and numerical investigations were performed for plasma preheating of AISI 1018 low carbon steel plates. The finite element (FE) software package ABAQUS 6.14 was used to perform the transient thermal analysis using Gaussian distribution as a heat source for plasma preheating. In this FE model, temperature-dependent thermal material properties of low carbon steel (AISI 1018) were used. The influence of different parameters, i.e., traverse speed, plasma arc power on thermal history was determined. Experimentally obtained transient temperature distribution and peak temperature compared fairly well with FE results with a maximum percentage error of around 6.8% and 7.53%, respectively.

Keywords Plasma heating · Finite element analysis · Transient thermal analysis · Heat generation · Process parameters

67.1 Introduction

Plasma heating has been come into the picture since the late 1800s. Around 30 years ago, NASA simulated the very high temperatures of reentry into the atmosphere of earth through plasma heating. The applications of plasma heating are widespread, especially in welding and preheating of the materials. The plasma can achieve 11,100 °C temperatures with very little combustion. The finite element method (FEM) has been the most common technique for the analysis of preheating

A. Kumar (✉)
Department of Mechanical Engineering, Jorhat Engineering College, Jorhat 785007, India
e-mail: abhisheksharma3344@gmail.com

P. Biswas
Department of Mechanical Engineering, IIT Guwahati, Guwahati 781039, India

R. G. Narayanan et al. (eds.), *Advances in Computational Methods in Manufacturing*, Lecture Notes on Multidisciplinary Industrial Engineering,
https://doi.org/10.1007/978-981-32-9072-3_67

problems which involve thermal, metallurgical and mechanical phenomena simultaneously. The plasma arc preheating system offers better temperature control, quicker reaction time, better control on process, efficient use of energy and lesser capital costs as compared to the other conventional methods.

Nandan et al. [1] observed the influence of a supplementary heat source in front of the friction stir welding (FSW) tool. This reduces the heat input for welding required from the FSW tool and thus the plunging loads of the FSW tool. Wu et al. [2] developed a modified three-dimensional conical heat source model and quasi-steady state for plasma arc heat source to reflect the thermomechanical process of plasma arc welding. Cabage et al. [3] implemented laser as a heat source in front of the FSW tool and reported 50% drops in the friction stir welding tool plunging forces. Kou et al. [4] patented a tungsten inert gas arc preheating system, which reduces FSW tool wear, extends the capability of FSW tool to weld harder materials and also helps weld dissimilar metals. Deshpande et al. [5] developed a conical heat source model to simulate the relationship between welding parameters and welding efficiency. They proposed that the relationship is useful for selecting a combination of weld parameters and keyhole welding.

From the above published literatures, it was observed that only a few number of researchers studied plasma arc preheating process. In this paper, a finite element (FE)-based 3-D heat transfer model was developed to calculate the effects of plasma preheating on temperature history over the AISI 1018 steel plates and validated it with the experimental results.

67.2 Experimental Details

The experiment was conducted for preheating AISI 1018 low carbon steel plate, and the temperature was measured during the preheating of the butt-joint configuration shown in Fig. 67.1. The dimension of the workpiece was 200 mm $\times$ 100 mm $\times$ 4 mm as shown in Fig. 67.2a. The edges of the workpieces were machined to obtain proper contact of butt joint and clamped to the bed without root opening. The plasma arc torch was clamped at 2 mm standoff distance in the designed fixture, and the steel plates were fixed on the horizontal moveable bed. Transient temperature was measured by using K-type thermocouples over the top surface of the plate at 10, 20, 30 and 40 mm far from the weld line as shown in Fig. 67.2b.

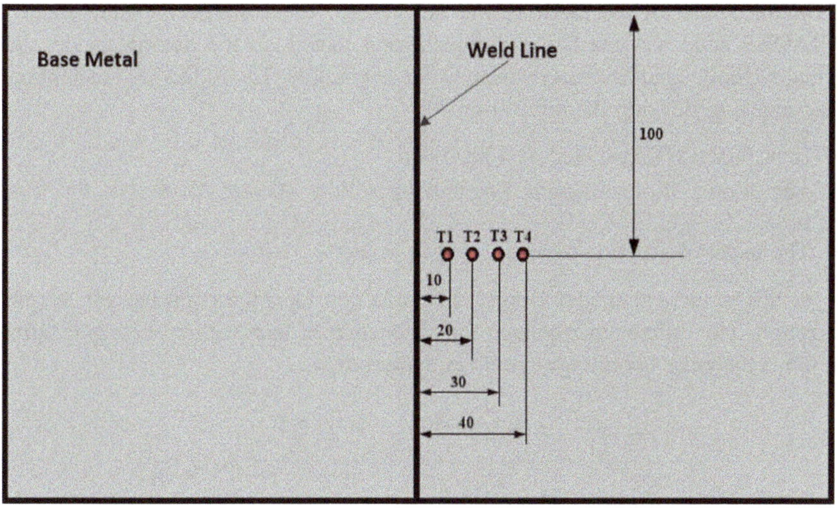

Fig. 67.1 Layout of the steel plate for plasma preheating

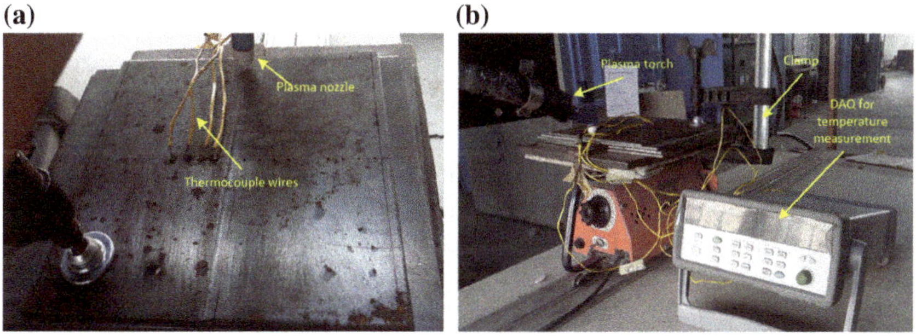

Fig. 67.2 **a** Butt-joint configuration for plasma heating; **b** experimental setup

67.3 Finite Element Analysis and Model Details

A numerical investigation by physical modeling gives a better understanding of physics involved in the process. A 3-D finite element transient thermal model was developed to analyze the heat flow and temperature distribution in preheating. The quality of the numerical model majorly depends on type, size and number of the element. A coarse mesh generally gives the wrong results; whereas, too much fine mesh increases the computational time. Therefore, to get reliable results, the optimum choice of the mesh should be present. The commercial finite element ABAQUS code was used to develop thermal model. In the thermal model, the actual welding conditions were taken as far as possible. The following assumptions were made to develop the thermal model:

- Heat flux was considered as a heat load.
- The plasma arc heat source was moving with a constant speed over the work piece.
- The material was homogeneous and isotropic.

In this numerical model x-coordinate was considered for plasma arc moving direction. The differential equation for conduction of heat without heat generation in the rectangular coordinate system is expressed as

$$\frac{\partial\left[K(T)\frac{\partial T}{\partial x}\right]}{\partial x}+\frac{\partial\left[K(T)\frac{\partial T}{\partial y}\right]}{\partial y}+\frac{\partial\left[K(T)\frac{\partial T}{\partial z}\right]}{\partial z}=\rho C(T)\frac{\partial T}{\partial t} \tag{1}$$

where ρ, C, K and T stands for density, specific heat, thermal conductivity and temperature of the material, respectively. The initial conditions for the FE model that covers all the elements of the workpieces are expressed as

$$T = T_{\mathrm{a}} \quad \text{for } t = 0 \tag{2}$$

where T_{a} is the atmospheric temperature for the present study. A convective heat transfer coefficient 25 W/m^2 °C for natural convection and ambient temperature 25 °C was used as a boundary condition for thermal analysis. At the bottom surface, a high value of heat transfer coefficient was assigned for considering the impact of the backing plate. The mathematical expression for combined convection and radiation heat loss is given as

$$k\frac{\partial T}{\partial n}=h(T-T_{\mathrm{a}})+\epsilon\sigma\left(T^4-T_{\mathrm{a}}^4\right) \tag{3}$$

where n, h, ε and σ are the normal direction vector of the boundary, convection coefficient, emissivity and Stefan-Boltzmann constant, respectively.

67.3.1 Heat Source Model

A three-dimensional thermal model for the preheating process was developed by using the temperature-dependent thermal properties and Newtonian convective heat loss for all surfaces open to the air. The plasma arc Gaussian distribution [6] heat flux is expressed as

$$q(r) = \frac{3Q_{\max}}{\pi R^2} e^{-3\left(\frac{r}{R}\right)^2} \tag{4}$$

where $q(r)$ indicates the heat flux at radius r, R indicates the radius over which maximum arc energy falls, and $Q_{\max}$ is the heat input to the plasma preheating source and given by

$$Q_{\max} = \eta \times V \times I \tag{5}$$

where η, V and I indicate power efficiency, voltage and heating current, respectively. In three-dimensional modeling, DC3D8 or 8-node linear heat transfer brick element was employed. The total number of elements and nodes used in this numerical model was 62,400 and 79,395, respectively. The workpiece was divided into three different mesh zones shown in Fig. 67.3. In the central region up to 15 mm away from the welding line, the element size was kept fine with 1 mm in size while out of this zone, gradually increasing mesh was assigned.

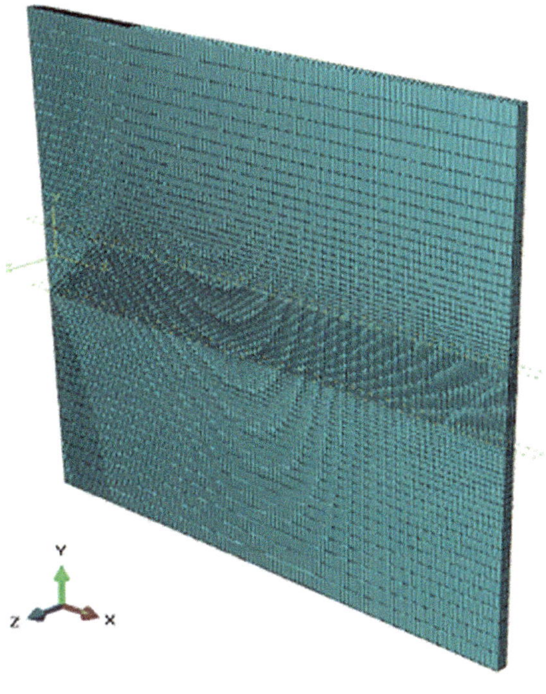

Fig. 67.3 Meshed zone for plasma heating

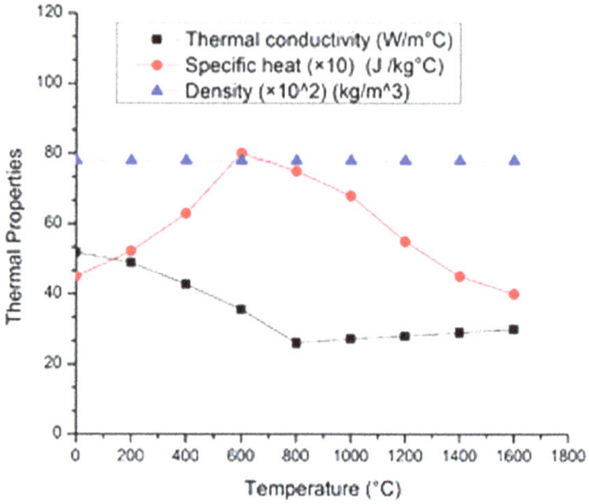

Fig. 67.4 Thermal properties of AISI 1018 low carbon steel

Table 67.1 Chemical composition [8] of AISI 1018 low carbon steel

C	Fe	Mn	P	S
0.14–0.20%	98.81–99.26%	0.60–0.90%	≤0.04%	≤0.05%

67.4 Material Properties

Thermal properties [7] of AISI 1018 low carbon steel used in the transient thermal analysis are shown in Fig. 67.4. Table 67.1 shows the Chemical composition [8] of AISI 1018 low carbon steel.

67.5 Results and Discussion

In this present FE model, experimental validation and effect of variation in parameters for plasma heating have been addressed. The 3-D numerical model results were developed and compared with the experimental data. The temperature distribution was computed using finite element simulation as shown in Fig. 67.5. The simulation was carried out on ABAQUS using DFLUX subroutine and validated it through experimental investigation. The parameters considered for performing numerical calculation and experimental investigation are presented in Table 67.2.

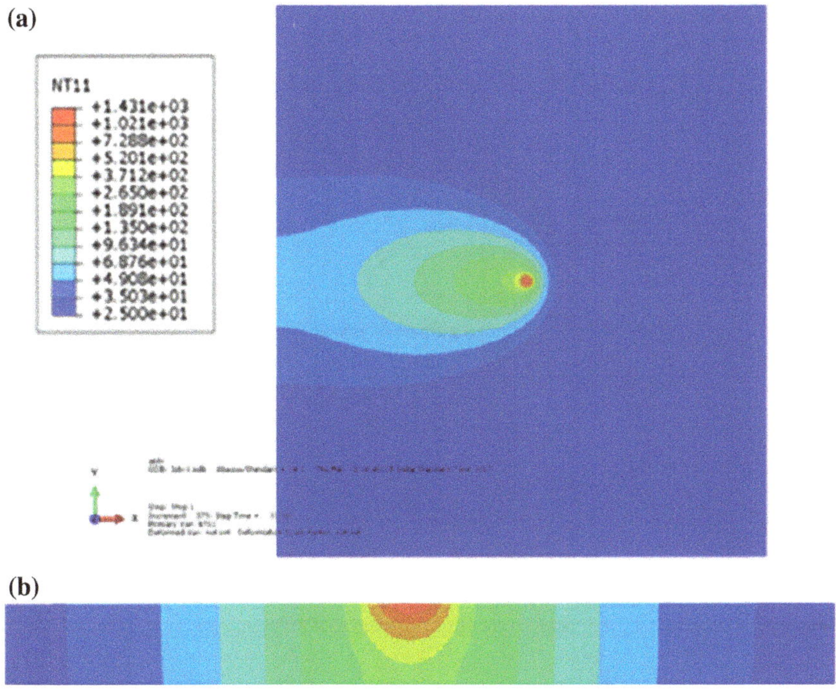

Fig. 67.5 **a** Simulated temperature distribution over the plate; **b** Temperature distribution along the thickness of the plate

Table 67.2 Process parameters for plasma heating

Workpiece length (x dir)	200 mm
Workpiece half-width (y dir)	100 mm
Voltage	13.2 V
Current	20 A
Arc radius	3 mm
Thermal efficiency [9]	0.7
Traverse speed	163.8 mm/min

67.5.1 Transient Temperature Profiles of Plasma

The comparison between the thermal model and data collected from K-type thermocouple for plasma preheating is shown in Fig. 67.6. From the combined numerical and experimental thermal profiles, it is observed that experimental data is much closer to the simulated data. The transient temperature curve plotted at different locations away from the center of weld line over the surface of the plates.

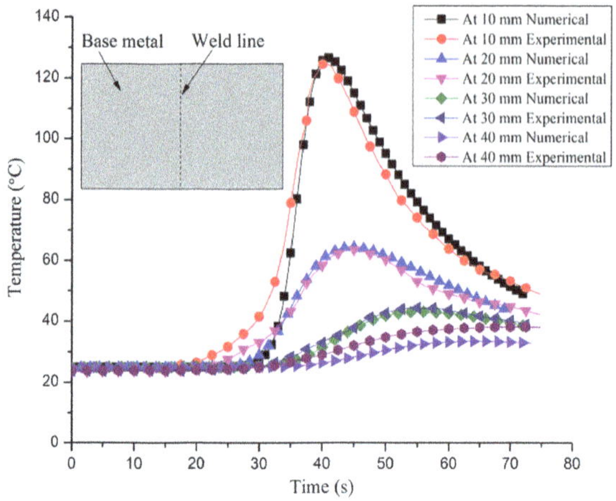

Fig. 67.6 Comparison between thermal model and experimental data at various distances from the weld line

Table 67.3 Percentage error between simulated and experimental data at various distances for preheating

Distance from the center of the plate (mm)	Average percentage error
10	10.82
20	4.77
30	3.39
40	8.35

The following Table 67.3 is representing the average percentage error between modeling results and experimental results for plasma preheating at 10, 20, 30 and 40 mm away from the weld line.

67.5.2 Peak Temperature–Distance Profiles on Top Surface

For comparing the results of temperature with respect to the perpendicular distance away from the center of weld line toward the traverse direction, the temperature is given by K-type thermocouple at four distinct locations 10, 20, 30 and 40 mm away from the weld line. Figure 67.7 shows the comparison between the experimental results and the results obtained from the numerical model for temperature distribution in the traverse direction of the weld line. The present numerical model is found to be in well agreement with the experimental data with an average error of 7.53% for peak temperature.

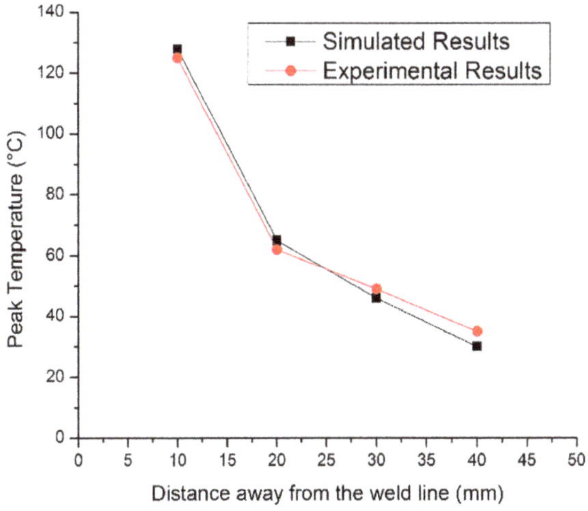

Fig. 67.7 Comparison of peak temperature profile for plasma preheating

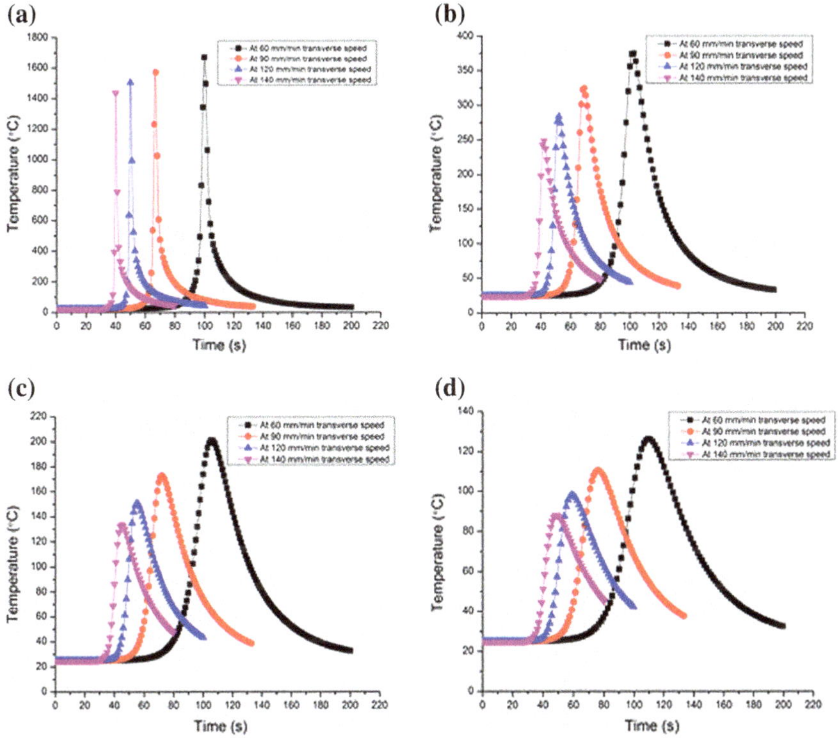

Fig. 67.8 Temperature profile at **a** center; **b** 5 mm; **c** 10 mm; **d** 15 mm away from the weld line at 30 A and 13.5 V

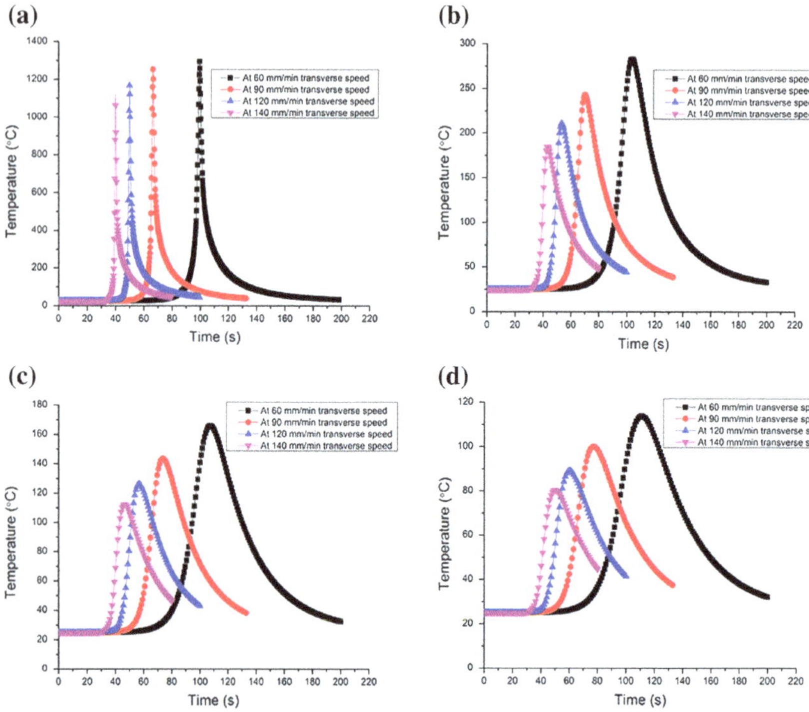

Fig. 67.9 Temperature profile at **a** center; **b** 5 mm; **c** 10 mm; **d** 15 mm away from the weld line at 20 A and 13.2 V

67.5.3 Effect of Variation in Traverse Speed on Transient Temperature Profile at Different Power Supply

The effect of variation in traverse speed on plasma at different power supply is shown in Figs. 67.8 and 67.9. The transient temperature curves were obtained from the numerically developed model at various locations showing that increase in traverse speed leads to a decrease in temperature of the plate.

67.6 Conclusion

In the present study, the plasma was used to preheat two AISI 1018 mild steel plates. The impact of plasma preheating on the thermal profile has been examined experimentally and numerically. The major conclusions are made from the analysis;

- A finite element thermal model was successfully developed for plasma preheating for AISI 1018 low carbon steel plate.
- It was found that finite element modeling of plasma preheating validated with the experimental results with around 6.8% and 7.53% error in transient profile and in peak temperature, respectively, which specifies that the developed thermal model can successfully implement in investigating the thermal histories during the plasma preheating.
- The influence of the power supply of plasma arc on temperature profile has been successfully observed over AISI 1018 low carbon steel plate. It was found that beyond 30 A current, the AISI 1018 steel plate starts melting. Hence, for preheating, the supply current should be less than 30 A.

References

1. Nandan, R., DebRoy, T., Bhadeshia, H.K.D.H.: Recent advances in friction-stir welding–process, weldment structure and properties. Prog. Mater Sci. **53**(6), 980–1023 (2008)
2. Wu, C.S., Wang, H.G., Zhang, Y.M.: A new heat source model for keyhole plasma arc welding in FEM analysis of the temperature profile. Weld. J.-New York **85**(12), 284 (2006)
3. Cabage, B.: New way to weld: ORNL team adds laser tech to expand the reach of a new welding technique. Oak Ridge National Laboratory Reporter (84), pp. 1–4 (2006)
4. Kou, S., Cao, G.: U.S. Patent No. 7,078,647. U.S. Patent and Trademark Office, Washington, DC (2006)
5. Deshpande, A.A., Short, A.B., Sun, W., McCartney, D.G., Xu, L., Hyde, T.H.: Finite element-based analysis of experimentally identified parametric envelopes for stable keyhole plasma arc welding of a titanium alloy. J Strain Anal. Eng. Des. **47**(5), 266–275 (2012)
6. Yaduwanshi, D.K., Bag, S., Pal, S.: Numerical modeling and experimental investigation on plasma-assisted hybrid friction stir welding of dissimilar materials. Mater. Des. **92**, 166–183 (2016)
7. Andrade, R.V.: Pipe Circularity Reformation Via Line Heating. Master's thesis, Massachusetts Institute of Technology (2001)
8. https://www.azom.com/article.aspx?ArticleID=6115
9. Teixeira, P.R.D.F., Araújo, D.B.D., Cunda, L.A.B.D.: Study of the gaussian distribution heat source model applied to numerical thermal simulations of TIG welding processes. Sci. Eng. J. **23**(1), 115–122 (2014)

Part VII
Process Simulation and Optimization

Chapter 68
Reduced-Order Finite Element Analysis of Rough Surface Contact Using Reduced Integration Elements

Ahad Noor, Saipraneeth Gouravaraju and Sachin Singh Gautam

Abstract This paper presents a reduced-order finite element analysis of rough surface contact. Frictionless, elastic contact between a rough deformable block with Gaussian roughness and a rigid flat plate is considered. An equivalent one-dimensional rough surface of a two-dimensional rough surface is generated using MATLAB. Generated rough surface is superimposed on ten layers of generated mesh of the block and analysis is done in commercial finite element software Abaqus. Four-noded quadrilateral elements with reduced-order integration are used to decrease the computational time. The rigid plate is placed close to the largest asperity of the rough surface and is then displaced in small steps. Contact force and real area of the contact are calculated. Results are compared with the results of the full-scale finite element analysis in the literature and are found to be in good agreement.

Keywords Finite element analysis · Rough surface · Gaussian roughness · Elastic contact · Abaqus

68.1 Introduction

Almost all the machines have contact interfaces which work under different types of loading conditions. It is a well-known fact that whatever be the manufacturing process, one cannot eliminate surface roughness. So roughness is important in determining the behaviour of the contacting surfaces. Roughness directly or indirectly affects friction, stress fluctuation, wear, leakage of liquid, and energy transfer

A. Noor · S. Gouravaraju · S. S. Gautam (✉)
Indian Institute of Technology Guwahati, Guwahati 781039, India
e-mail: ssg@iitg.ac.in

A. Noor
e-mail: ahad.noor@iitg.ac.in

S. Gouravaraju
e-mail: saipraneeth@iitg.ac.in

R. G. Narayanan et al. (eds.), *Advances in Computational Methods in Manufacturing*, Lecture Notes on Multidisciplinary Industrial Engineering,
https://doi.org/10.1007/978-981-32-9072-3_68

between the surfaces. The study of rough surface contact plays an important role in many industries like bearing, lens, household items, electronics, brake and clutch, medical equipment, and medical implants manufacturing industry. For better tribological design and performance, analysis of the contact between rough surfaces is very important. The high points on a rough surface are known as asperities/peak/summit/hill, and the low points are known as valleys. When two rough surfaces are brought into contact, the actual contact occurs at the peaks/asperities. Because of surface roughness, contact occurs only at these discrete contact locations. Then, the real area of contact can be calculated by summing all the discrete areas of contact at each asperity. Moreover, in many cases the actual area of contact is limited to only 20–25% of the nominal contact area [1]. As the contact area of these asperities is very small compared to the total or nominal contact area, very high stresses develop in the asperities causing the material to yield. In some applications where two rough surfaces are brought into contact repeatedly, the increase in interfacial stress leads to damage of the materials resulting in wear. So, roughness and complexity of the contact surface have emerged as an important factor in analysis of strength, fatigue, wear, damage, friction, and energy dissipation. Hence, analysing the contact behaviour of the rough surfaces is an interesting domain for researchers.

Hertz [2] has performed the first contact analysis to investigate the optical properties of lenses in contact. Hertz considered two elastic smooth spheres in contact. He considered the contact zone to be a circle and solved considering minimum deformation energy. He concluded that the contact area is proportional to the $(2/3)$rd power of the applied load. In analytical as well as numerical analysis, to create rough surface a distribution of height of asperities must be defined. Many researchers have considered Gaussian distribution. Moreover, Gaussian distribution of asperities is experimentally verified by Celedón et al. [3]. They measured surface roughness of a gold film and concluded that the height of asperities follows Gaussian distribution. Greenwood and Williamson [1] in their pioneering work on nominally flat rough surface contact developed one of the first elastic contact models. This model represents a rough surface by a number of hemispheres with the same curvature radius R, but the height of asperities is varied according to the Gaussian distribution. Each asperity was assumed to deform elastically according to the Hertz contact theory. The flat plate is considered to be at some distance from mean of rough surface asperities and using the probability theory calculated number of asperities in contact at that distance. Then, using the Hertz contact theory, the authors solved for the real area of contact and the contact pressure. They concluded that actual area of contact varies linearly with the applied load. Lu et al. [4] studied the initial contact between a rough surface and a flat plate using Greenwood-Williamson model (GW model) for low loads over a small area of contact. They considered an exponential distribution of rough surface. They proposed a modification for the initial position of the flat plate. They set it so that the flat plate is just in contact with the highest asperity, while the GW model considers the initial position of the flat plate somewhere in the middle. However, they considered initial contact so they limited their analysis to low loads only. Yastrebov et al. [5, 6] have done the analysis of self-affine fractal rough surface in elastic, frictionless, non-adhesive contact. They have created Gaussian rough surfaces numerically. They have used Fourier forward technique. They have applied pressure loading in small increments and have observed the evolution of real contact area under increasing pressure from zero up to full contact between rough surfaces. Later, they have done the analysis of contact between elasto-plastic rough surface and a rigid plane with two approaches: full-scale finite element analysis and to save computation time a reduced model with displacement boundary condition. In their reduced model, they considered single asperity, and to incorporate interaction between asperities, they have proposed a displacement boundary condition for the asperity outside the contact zone.

From the literature survey, it is clear that many researchers have extensively studied the contact between rough surface. Experimental calculation of actual area of contact is very difficult. So, researches mainly focused on analytical and numerical analyses. Due to the assumptions made in analytical studies, their accuracy is limited. Although the full-scale three-dimensional numerical simulations as in [5] give good results, the computational times are very high. So, in order to reduce the computational time without compromising on the accuracy in estimating the contact behaviour, in the present work, the method proposed by Scaraggi et al. [7] is used to generate an equivalent one-dimensional rough surface from a two-dimensional rough surface. A Gaussian, one-dimensional equivalent rough surface is created in MATLAB. This one-dimensional equivalent rough surface has all the information of two-dimensional rough surface. Finite element analysis is performed in commercial finite element software Abaqus. The rough surface is superimposed on the top surface of a two-dimensional block. To save the computational time, elements with reduced integration are considered. It is observed that the results obtained in the finite element analysis are consistent with the results of full-scale finite element analysis at low loads.

The remainder of the paper is structured as follows: Sect. 68.2 describes the representation and generation of rough surface. Section 68.3 presents the generation of the finite element mesh of the deformable block. Section 68.4 details the boundary conditions used and the methodology of the numerical simulations using Abaqus. Section 68.5 presents the obtained numerical results. Conclusions drawn from the study are presented in Sect. 68.6.

68.2 Description of the Rough surface Geometry

Any random rough surface can be represented by its power spectral density (PSD) denoted by $\Phi(k_x, k_y)$ of a profile in any direction [8] where k_x and k_y are the wavenumbers in x- and y-directions, respectively. In the present work, spectroscopic wavenumber which is the inverse of wavelength is used. It is normalized with the domain length L, i.e., $k = L/\lambda$. PSD contains all the statistical information of the surface topography and is unbiased by the particular scan size and pixel resolution used. Mathematically, PSD is the Fourier transform of the signal's autocorrelation function and is a function of wave vector. In the present work, isotropic self-affine fractal surfaces are used whose power spectral density is defined as [5]

$$\Phi\left(k_x, k_y\right) \sim \left(\sqrt{k_x^2 + k_y^2}\right)^{2(H+1)} \tag{68.1}$$

where H is Hurst exponent.

In the present work, the following form of PSD is considered [5]:

$$\Phi_{2\mathrm{D}}(|\mathbf{k}|) = \begin{cases} C_0 & \text{if } k_l < |\mathbf{k}| < k_r, \\ C_0 \frac{|\mathbf{k}|}{k_r} & \text{if } k_r \le |\mathbf{k}| \le k_s, \\ 0 & \text{otherwise} \end{cases} \tag{68.2}$$

where C_o is the constant that determines the roughness amplitude and $k = |\mathbf{k}|$ is the magnitude of the wave vector. In all the simulations, there is no plateau, so $k_l = k_r$.

Then, an equivalent one-dimensional rough surface profile is generated using the corrected equivalence condition between the mean square slopes of one-dimensional (1D) and two-dimensional (2D) rough surfaces. This is given by [7]

$$\Phi_{1\mathrm{D}}^{\mathrm{eqv}}(k) = \left(k_{1\mathrm{D}}/k_{2\mathrm{D}}\right)^2 \pi \Phi_{2\mathrm{D}}(k) \tag{68.3}$$

where $\left(k_{1\mathrm{D}}/k_{2\mathrm{D}}\right)^2 \approx 1.56$ is the correction factor. This equivalence increases the root mean square height of one-dimensional rough surface

$$\sqrt{(h^2)_{1\mathrm{D}}} = \left(k_{1\mathrm{D}}/k_{2\mathrm{D}}\right) \sqrt{(h^2)_{2\mathrm{D}}} \tag{68.4}$$

For given PSD and surface area A, the surface heights can be calculated as

$$h(\mathbf{x}) = \sum_{|\mathbf{k}} \tilde{h}(|\mathbf{k}) \mathrm{e}^{i(|\mathbf{k}|.\mathbf{x} + \phi(|\mathbf{k}))} \tag{68.5}$$

where $\left|\tilde{h}(|\mathbf{k})\right| = \frac{2\pi}{\sqrt{A}} \sqrt{\Phi(|\mathbf{k})}$.

Then, the two-dimensional and the equivalent one-dimensional rough surfaces are generated using the inbuilt fast Fourier transforms (FFTs) and inverse Fourier transforms (IFFTs) functions in MATLAB. These are shown in Figs. 68.1 and 68.2, respectively.

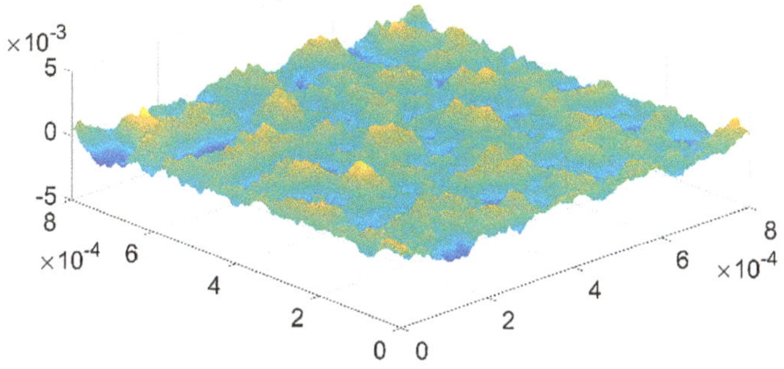

Fig. 68.1 Generated two-dimensional rough surface

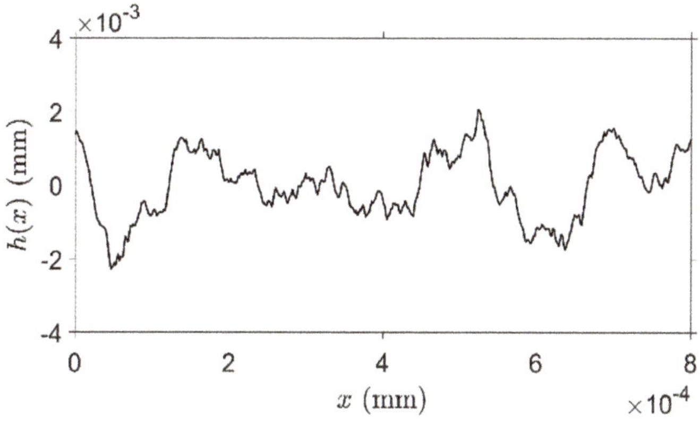

Fig. 68.2 Generated equivalent one-dimensional rough surface

68.3 Mesh Generation

To perform numerical analysis first, a mesh of the block having the dimensions 0.8 × 0.15 mm is considered. The finite element mesh is generated in Abaqus. It consists of 511 and 200 elements in the *x*- and *y*-directions, respectively. Four-noded plain stress quadrilateral element CPS4R with reduced integration is used for meshing. Since Abaqus does not have any option to create rough surface, it is generated in MATLAB. The height of the highest asperity is 0.0021 mm (2.1 μm). The generated rough surface heights $h(x)$ (see Fig. 68.2) are supposed to be superimposed onto the top layer of the initial mesh to create rough surface. But, this may produce badly distorted elements. In the worst case, it can produce negative Jacobians. Jacobian is the ratio of the area/volume of the deformed element to the undeformed element in a finite element mesh. So, if the Jacobian is negative, the

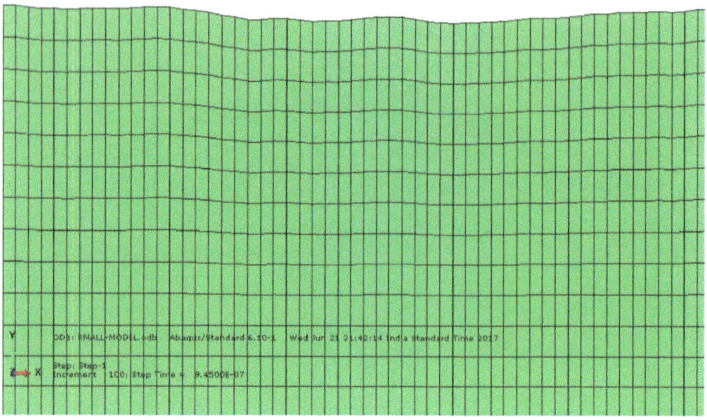

Fig. 68.3 Magnified view of generated roughness superimposed on ten layers of the block to avoid negative Jacobians

Table 68.1 Mechanical properties of steel

E (GPa)	ν	ρ (Kg/m^3)
200	0.3	7800

elements have undergone a non-physical movement. To avoid this, all the nodes in top 10 layers are moved by a fraction of the heights of the top surface and change in height is gradually brought to zero in tenth layer (see Fig. 68.3). This is done in following steps: Coordinates of calculated height in MATLAB is added on the nodes of top layer. Then, nodes of second layer are moved by adding 0.9 times height of rough surface; similarly, third layer is moved by 0.8 times, and this is done up to tenth layer nodes. The rigid surface is modelled as a straight horizontal line with its initial position at the 0.0001 mm from the highest point on the top surface. The resulting finite element mesh (zoomed view near the top surface) is shown in Fig. 68.3. The material for the rough surface is considered to be steel. Material properties of steel are given in Table 68.1. In the present work, only elastic contact is considered.

68.4 Boundary Conditions

In this simulation, fixed boundary conditions are applied on the bottom surface of the block in both the *x*- and *y*-directions. The vertical surfaces of the block are fixed in the *x*-direction. The top surface is in contact with a rigid flat plate which is placed at a distance of 0.0022 mm from the smooth surface (without roughness) and

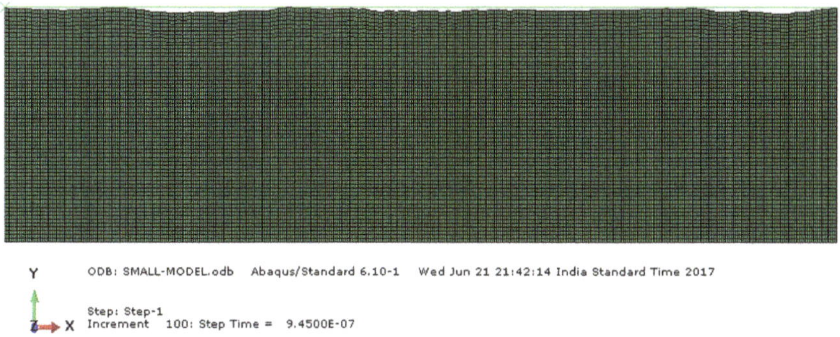

Fig. 68.4 Assembly of rigid plate and rough surface

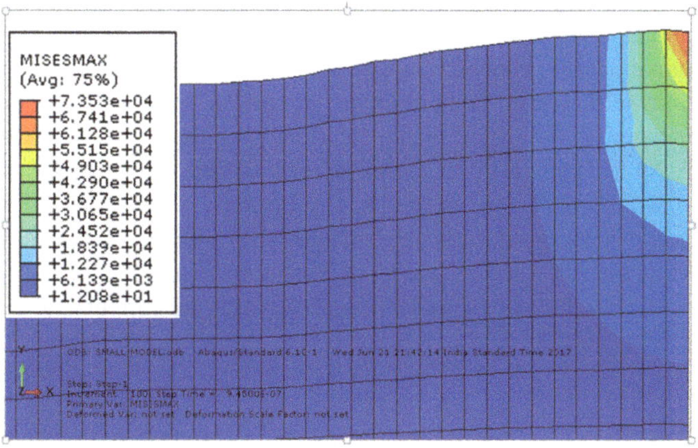

Fig. 68.5 Deformed surface with superimposed von Mises stress

0.0001 mm from the highest asperity as shown in Fig. 68.4. The rigid plate is given a total displacement of 0.002 mm in hundred time steps. However, it is worth mentioning that the Abaqus can reduce the time step to account for change in contact status. It is observed that initially the time step is nearly 10^{-8} s but as more and more asperities start to come in contact, Abaqus made the time step in the range of 9×10^{-9} s.

In Abaqus static analysis, the movement of upper rigid plate is guided by displacement boundary condition rather than velocity. Figure 68.5 shows the deformed surface at the end of simulation with superimposed von Mises stress. The problem is simulated on a Dell T620 workstation (32 GB RAM and 16 cores). The whole simulation is completed in approximately 4 min only.

68.5 Results and Discussion

The actual or real area of contact A_r and the corresponding contact force F are calculated at each load step. Total contact area is sum total of contact area of each asperity. Analytically, contact area is calculated by calculating the number of nodes in contact. Then, ratio of nodes in contact and total number of nodes on the top layer is calculated. This ratio is multiplied with nominal area to give actual area of contact. The same algorithm is used in numerical methods to calculate actual area of contact. Actual area is normalized with nominal area A_o to give dimensionless area of contact $A^* = A_r/A_o$. Contact force is normalized with the product of elastic modulus E and nominal area A_o to give dimensionless pressure $P^* = F/EA_o$. Figure 68.6 shows the evolution of normalized area of contact A^* with the normalized pressure P^*. The results obtained by Yastrebov et al. [6] for the contact of a two-dimensional rough surface with a rigid block are also included. It can be seen that the results from the present work match well with the numerical results of Yastrebov et al. [6] for low deformation regime while there is some discrepancy at higher deformation. It is important to note that the results of Yastrebov et al. [6] are obtained from a full-scale finite element simulation of two-dimensional rough surface which took 220 h, whereas in the current work the results are obtained from the analysis of an equivalent one-dimensional surface and at a much reduced computational time of 4 min. Although plot between actual area of contact and contact pressure is not linear in the present analysis, variation from linear behaviour as predicted by the full-scale FE model of Yastrebov et al. [6] is not very high for low loads, i.e., $P^* < 0.1$. However for loads $P^* > 0.1$, the current analysis predicts a nonlinear behaviour.

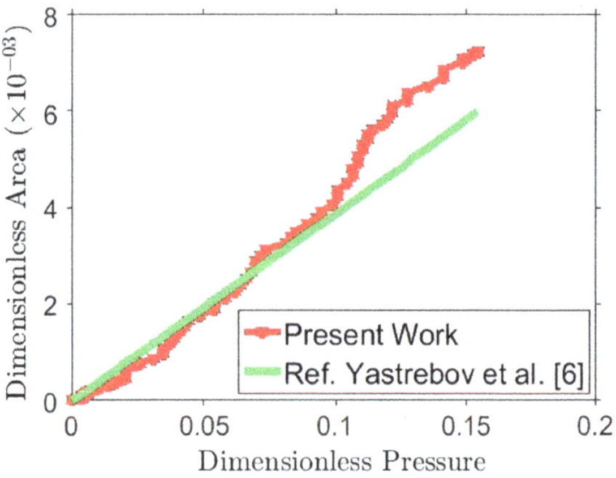

Fig. 68.6 Variation of dimensionless contact and dimensionless contact pressure

68.6 Conclusion

Numerical analysis of the contact between a linear elastic block and a self-affine rough surface is carried out using the finite element analysis software Abaqus. To save computational time, one-dimensional equivalent rough surface with reduced integration elements is used for analysis. Actual area of contact and contact force are calculated. It is observed that the variation of real area of contact with the applied contact pressure is close to linear in low load regime, and at higher loads, the variation is observed to be nonlinear. This behaviour can be explained in many ways. First, to decrease computational time, finite element approximation elements with reduced integration are considered which contain single Gauss point. So accuracy of result is compromised at the cost of calculation time. Full-scale analysis took 220 h [6] while this analysis of one-dimensional equivalent rough surface is completed in four minutes only. Second, equivalent one-dimensional rough surface used in this analysis can capture the essential features of its two-dimensional counterpart, and it is not exact. Although equivalent rough surface with reduced integration results showed some discrepancy at higher loads, computational time is reduced from 220 h to 4 min. Third, fixed boundary conditions are used at the vertical sides of the rough surface. This boundary condition is easy to apply. This can be applied by selecting "Encastre" option in Abaqus, but it can be used for approximation only because in real condition elements at the edges will not be fixed and show some deformation. For example, as rigid plate approaches the block, it will crush the rough surface and elements at edges will try to move outside. For exact results, more complex boundary conditions are needed which will be considered in the future work.

Acknowledgements The authors are grateful to the SERB, DST for supporting this research under project SR/FTP/ETA-0008/2014.

References

1. Greenwood, J.A., Williamson. J.B.P.: Contact of nominally flat rough surfaces. Proc. Royal Soc. London A Math. Phys. Eng. Sci. **295**(1442), 300–319 (1966)
2. Johnson, K.L.: Contact Mechanics. Cambridge University Press, Cambridge, UK (1985)
3. Celedon, C., Flores, M., Haberle, P., Valdes, J.: Surface roughness of thin gold films and its effects on the proton energy loss straggling. Braz. J. Phys. **36**(3B), 956–959 (2006)
4. Lu, P., O'Shea, S.: Mechanical contact between rough surfaces at low load. J Phys D Appl. Phys. **45**(47), 475303-1–475303-15 (2012)
5. Yastrebov, V.A., Anciaux, G., Molinari. J.-F.: From infinitesimal to full contact between rough surface: evolution of the contact area. Int. J. Solids Struct. **52**, 83–102 (2015)
6. Yastrebov, V.A., Durand, J., Proudhon, J., Cailletaud, G.: Rough surface contact analysis by means of finite element method and new reduced model. C.R. Mec. **339**(7–8), 473–490 (2011)
7. Scaraggi, M., Putignano, C., Carbone, G.: Elastic contact of rough surfaces: a simple criterion to make 2D isotropic roughness to equivalent 1D one. Wear **297**(1–2), 811–817 (2013)
8. Nayak, P.R.: Random process model of rough surfaces. ASME J. Lubr. Technol. **93**(3), 398–407 (1971)

Chapter 69
FEM-Based Modeling of Drilling-Induced Delamination in Laminated FRP Composites

Vishwas Divse, Deepak Marla and Suhas S. Joshi

Abstract In fiber-reinforced plastics (FRP) laminate, different laminae are adhesively bonded to each other. The interfaces between plies are weak because these are resin rich; therefore, there is a low-resistance path for crack propagation. As a result of this, delamination is a frequent mode of failure in the FRP laminate. The drilling of laminated FRP is always associated with delamination damage; therefore, delamination-free drilling is a major concern for aerospace industries. Experimental investigations of drilling-induced delamination have been widely studied in the literature, whereas very few studies are available related to its modeling. In this paper, the effect of the chisel-edge length of drill-bit on push-out delamination was studied. Finite element modeling (FEM) approach was used to model interfaces of laminae, and it was further extended to model delamination in drilling of FRP laminate. Cohesive zone method (CZM) based on traction–separation law was used in FEM to model the interfaces of laminae. It was observed that load-carrying capacity of beam in double cantilever beam (DCB) test reduced by 50% for '1 mm' initial crack length. Also, push-out delamination increased by around 10% with an increase in the chisel-edge length of drill bit from '0 mm' to '0.35 mm.' The FE models developed in this work were able to simulating delamination phenomenon with enough accuracy.

Keywords Delamination · Fiber-reinforced plastics · Drilling · Finite element modeling · Cohesive zone method

---

V. Divse · D. Marla · S. S. Joshi (✉)
Department of Mechanical Engineering, Indian Institute of Technology Bombay, Mumbai 400076, India
e-mail: ssjoshi@iitb.ac.in

R. G. Narayanan et al. (eds.), *Advances in Computational Methods in Manufacturing*, Lecture Notes on Multidisciplinary Industrial Engineering,
https://doi.org/10.1007/978-981-32-9072-3_69

69.1 Introduction

Composite materials are increasingly used as a structural material in aerospace and other engineering applications. Among different composites, FRP composites have wide applications in aerospace industries due to their high specific strength and stiffness. While assembling the FRP components, holes are required for fasteners like bolts or rivets. Mechanical drilling is widely used to create holes in these components [1]. However, this operation is always associated with damages such as fiber pullout or breakage, matrix cracking or thermal degradation, delamination due to its poor machinability [2]. Consequently, this could result in strength and stiffness degradation and may lead to failure of a component in service. Additionally, they also cause poor assembly tolerance and reduction in structural integrity. Delamination is a failure of an interface between plies. This is a highly undesirable problem because it affects the mechanical properties of FRP composites [3]. A number of methodologies have been adopted for the purpose of minimization of delamination. The quantification of delamination is done by the delamination factor (F_d). Several researchers developed empirical models to predict delamination factor in drilling [1, 4, 5]. They identified that drilling speed and feed have the highest effect on delamination. These models are applicable only for particular drilling conditions and materials. Artificial neural networks (ANNs) can predict delamination factor at different values of spindle speed, feed rate, drill diameter, drill point geometry, etc. [6–8]. It requires huge experimental data to train an ANN, and therefore, it is a tedious process to apply for different materials. Experimental techniques used to characterize delamination factor are challenging and costly. Hence, modeling of delamination is seen as an effective tool to predict delamination in drilling. FEM is a potential technique to model the delamination phenomenon in FRP composites [9]. FE models developed by some researchers can predict drilling-induced delamination [10–12]. These models are very complex in nature and have not considered the effect of drill-bit geometry. Most of the models presented in the literature to predict delamination are empirical and focused only on drilling parameters and neglected the geometry of the drill bit.

In this paper, a simple 2D FE model of drilling is presented to predict drilling-induced delamination. The main objective is to study the effect of chisel edge of drill bit on the push-out delamination. In Sect. 69.2, the modeling techniques of FRP laminate and their interfaces are discussed. Further, in Sect. 69.3, FE models of the double cantilever beam (DCB) and drilling of FRP laminate are presented. The results and conclusions based on simulations performed in this work are summarized in the last section.

69.2 Modeling Approach

FRP laminates are composed of a number of plies with different orientations. Each FRP ply is assumed to be homogeneous with orthotropic elastic properties. These plies are glued together with a resin-rich interface. A two-phase (FRP ply and interface) model (Fig. 69.1) is required to simulate the delamination process. The associated material and damage models of FRP ply and interface adhesive are described in the following subsections.

69.2.1 Fiber-Reinforced Plastics (FRP)

While modeling of FRP laminate, an individual ply in a laminate is assumed to be homogeneous, orthotropic, and in a state of plane stress [13]. Hashin's failure criterion [14] has been used to model the damage initiation in FRP material. The criterion considers the failure of both fibers and matrix under both tension and compression. Failure mechanisms of fibers are governed by the longitudinal stress with reference to the fiber orientation. Meanwhile, failure mechanisms of the matrix are governed by the transverse and tangential stresses to the fiber. Failure is said to occur if the failure criterion of that failure mode is equal to or greater than one.

69.2.2 Cohesive Zone Method (CZM)

The interface between the plies is resin rich; hence, it is weaker compared to the plies. Therefore, the delamination occurs only at the interface. Special elements such as cohesive elements are used in the FE model to simulate delamination. The

Fig. 69.1 Schematic representation of an FRP laminate

cohesive zone method (CZM) is based on traction–separation law [15]. The law has the capability to simulate the behavior of an interface from an initial linear elasticity to the final failure. Quadratic nominal stress criterion is used as a failure criterion [16]. The criterion assumes damage is initiated when a quadratic interaction function involving all nominal stress component ratios reaches a value of one. Damage evolution law based on fracture energy is incorporated in the model to simulate fracture at the interface. When the material is fully damaged, a crack will grow according to the quadratic fracture energy criterion.

69.3 Finite Element Modeling

2D FE models are developed to study the phenomenon of delamination based on the cohesive zone method. A commercial FE code Abaqus© is used for the simulations of the double cantilever beam test and the drilling of FRP laminate. DCB test is performed to understand delamination at interfaces of FRP laminate. Then, this approach is further extended to simulate push-out delamination in drilling process.

69.3.1 Double Cantilever Beam (DCB) Test

DCB test [17] is used to calculate the fracture energies of different modes of interface failure. Figure 69.2 shows an FE model of a DCB specimen. Two FRP layers are adhesively bonded together with an adhesive (here epoxy) layer. Each FRP layer is of 10 mm in length and 0.5 mm in thickness. Their elastic behavior is modeled with orthotropic elasticity, and damage initiation is modeled with Hashin's damage model [14]. An interface of 0.01 mm thickness between two

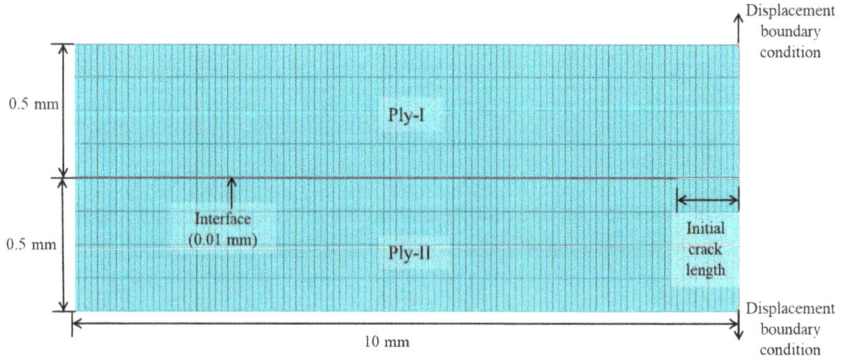

Fig. 69.2 FE model of double cantilever beam test

FRP layers is modeled based on CZM (see Sect. 69.2.2). Plain stress elements, 'CPS4R,' and cohesive elements 'COH2D4' are used to mesh FRP layers and interface layer, respectively.

The number of nodes on the adhesive layer and FRP layers is the same to achieve nodal connectivity at the interface. A displacement boundary condition is applied on the top and bottom corner of the nodes of FRP layers. Displacement of the nodes induces stresses in the elements which causes failure of the weakest section, i.e., interface. Quadratic damage criterion with an element deletion technique is used to simulate crack propagation along the interface. The load-carrying capacity of the interface in the mode-I type of delamination is predicted for different initial crack lengths.

69.3.2 Drilling of FRP Laminate

CZM can be effectively used to predict delamination of FRP laminate in a complex loading. Here, an attempt is made to predict delamination in drilling of FRP laminate. Drilling is a complex machining operation; hence, it is simplified by assuming no rotation of drill bit and a simple 2D FE model is developed. The main objective of the FE model is to predict the effects of thrust forces on the push-out delamination. Due to the simplification of the drilling process, this model does not take drilling torque into consideration. This model can predict delamination results with enough accuracy because push-out delamination is largely affected by thrust forces in drilling. The FE model of drilling (Fig. 69.3) shows a drill bit and four FRP plies that are adhesively bonded together with adhesive layers (epoxy). Drill bit is assumed to be rigid as it is 3–4 times harder than the FRP laminate. To minimize the computational time, whole geometry is scaled down. Each FRP layer is of 10 mm in length and 0.5 mm in thickness, and the diameter of a drill bit is taken as 1 mm. Their elastic behavior is modeled with orthotropic elasticity, and damage initiation is modeled with Hashin's damage model. Interfaces of 0.01 mm in thickness between the FRP layers are modeled based on CZM. Plain stress elements, 'CPS4R,' and cohesive elements 'COH2D4' are used to mesh the FRP layers and the interface layers, respectively. Element deletion technique is applied to delete failed elements based on damage criterion.

The interaction between the drill bit and workpiece is modeled using surface–surface contact available in Abaqus/Explicit. The contact is defined between the drill-bit surface and the FRP laminate nodes using sliding friction model with a coefficient of friction equal to 0.3. End nodes are fixed, and linear velocity in the vertical direction is applied to the drill bit. The effect of three different chisel-edge lengths on push-out delamination is predicted. The results obtained are discussed in the next section.

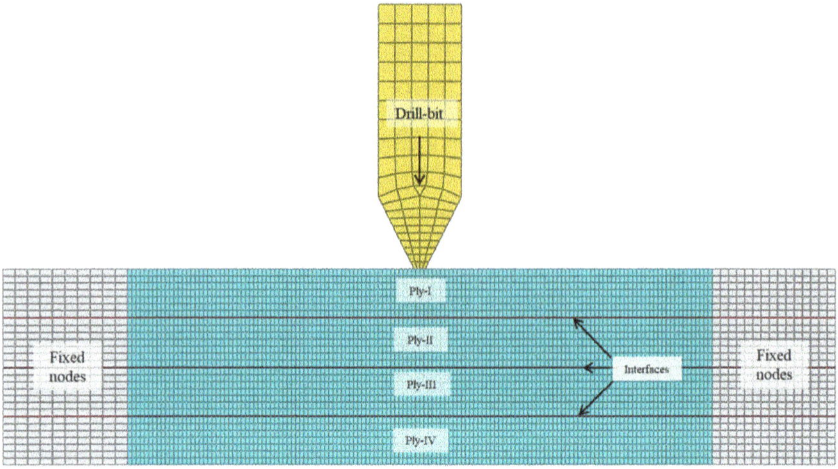

Fig. 69.3 FE model of drilling of FRP laminate

69.4 Results and Discussion

The results obtained in the simulation of double cantilever beam test and drilling of FRP laminate are discussed in the section.

69.4.1 Double Cantilever Beam (DCB) Test

These simulations were performed to predict the resistance offered by an interface for various initial crack lengths. Figure 69.4 shows a separation achieved at the interface by deleting the failed elements. Stresses in the *y*-direction (S22 in MPa) are the highest in the next critical elements at the interface. In this way, the crack propagates along the interface and delamination occurs. The simulation of DCB is run for five different initial crack lengths (0, 0.05, 0.25, 0.5, 1 mm). The initial crack helps in delamination; hence, the resistance offered by the interface layer reduces. This results in a reduction in load-carrying capacity with the size of initial crack length as shown in Fig. 69.5.

The shape of the load–displacement curve is triangular like a typical traction–separation curve that governs the behavior of an interface. The displacement indicates the separation between the nodes at the interface at the right edge of the beam (Fig. 69.4). The initial linear rise in the curves indicates the elastic behavior of the interface. The peak point is the maximum load-carrying capacity of the interface where the damage is initiated. Once the damage is initiated, the interface becomes weaker, its load-carrying capability goes down, and then it follows a linear softening law (Fig. 69.5). The load-carrying capacity is a strong function of an

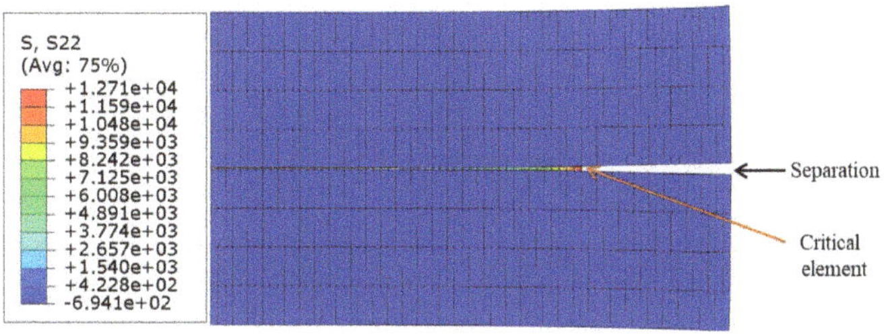

Fig. 69.4 Crack propagation along the interface during delamination

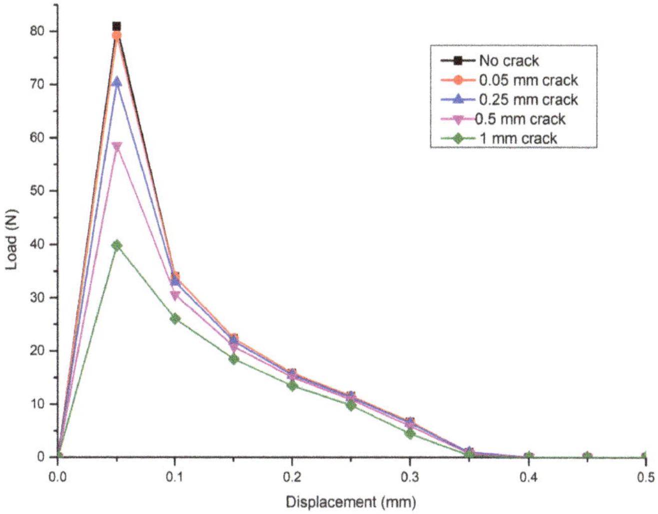

Fig. 69.5 DCB test results of load versus displacement for different initial crack lengths

initial crack length. The maximum resistance offered is 80 N with no initial crack, whereas 40 N was observed with a 1 mm initial crack.

69.4.2 *Drilling of FRP Laminate*

Simulation of drilling is performed for different chisel-edge lengths. For a given chisel edge, the drilling is achieved by deleting the failed elements (Fig. 69.6). As the drill bit moves in a vertically downward direction, the elements of the workpiece in contact with the drill-bit deform excessively. As the end nodes of the

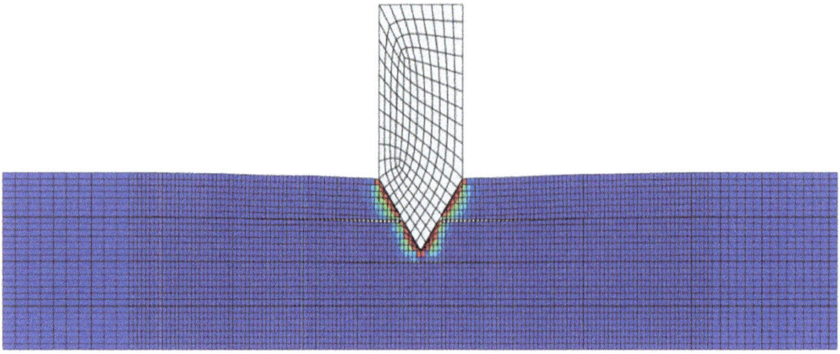

Fig. 69.6 Elements on the onset of failure in contact with drill bit

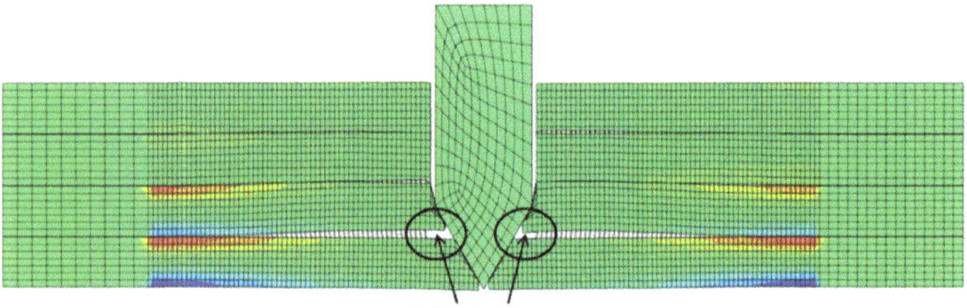

Fig. 69.7 Push-out delamination at the last interface in the drilling of FRP laminate

laminate are fixed, compressive and tensile stresses are induced in the laminate. When these stresses are enough to fail the FRP material, the damage is initiated based on Hashin's damage criterion.

Deletion of interface elements is based on quadratic damage criterion. In this way, simulation of a drilling process is accomplished in the 2D FE model. Drilling simulations are performed for three different chisel-edge lengths: 0, 0.15, and 0.35 mm, respectively. Figure 69.7 shows the delamination occurred at the last ply with a drill bit of '0 mm' chisel edge. The quantification of delamination is done using delamination factor $(F_d = D_{\max}/D)$, where $D_{\max}$ indicates the maximum diameter of the damaged zone and D is the actual hole diameter. As the length of the chisel-edge increases, there is an increase in the thrust force in drilling. Therefore, it is observed that delamination also increases with an increase in a chisel-edge length of a drill bit (Fig. 69.8).

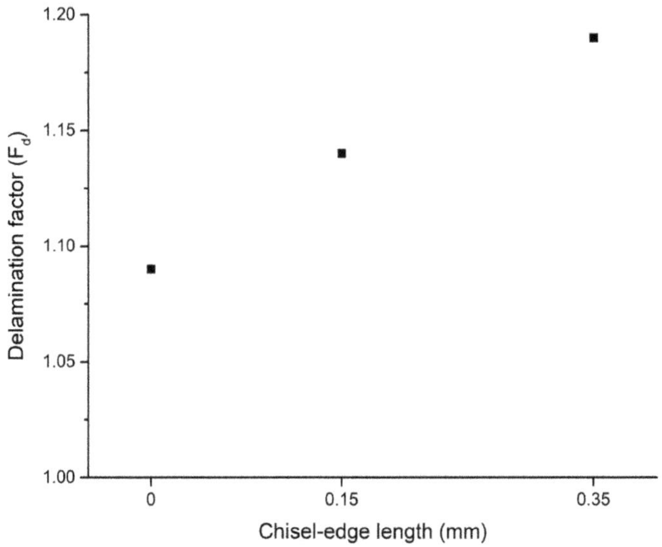

Fig. 69.8 Effect of the chisel-edge length on delamination

69.5 Conclusions

The finite element method can be effectively used to predict delamination in DCB test as well as in drilling of laminated composites. It was observed that load-carrying capacity of beam in double cantilever beam (DCB) test reduced by 50% for '1 mm' initial crack length. The available models to predict delamination in drilling are either analytical or the complex finite element models. These models have neglected the effect of chisel-edge length on the delamination. The present FE model is simple, takes less computational time, and produces results with enough accuracy. It takes a chisel-edge length effect into account. It was observed that push-out delamination increased by around 10% with an increase in the chisel-edge length from '0 mm' to '0.35 mm.' In addition, these models can be easily extended to predict the delamination for different drilling conditions and different FRP laminates.

References

1. Liu, D., Tang, Y., Cong, W.: A review of mechanical drilling for composite laminates. Compos. Struct. **94**(4), 1265–1279 (2012)
2. Davim, J.: Machinability of Fibre-Reinforced Plastics, vol. 4, pp. 1–70. DE Gruyter (2015)
3. Tao, C., Qiu, J., Yao, W., Ji, H.: The effect of drilling-induced delamination on tensile strength and prediction of residual strength of carbon fiber-reinforced polymer laminate. J. Compos. Mater. **50**(24), 3373–3384 (2016)

4. Gaitonde, V., Karnik, S., Rubio, J., Correia, A., Abrao, A., Davim, J.: Analysis of parametric influence on delamination in high-speed drilling of carbon fiber reinforced plastic composites. J. Mater. Process. Technol. **203**(1–3), 431–438 (2008)
5. Tsao, C., Hocheng, H.: Taguchi analysis of delamination associated with various drill bits in drilling of composite material. Int. J. Mach. Tools Manuf. **44**(10), 1085–1090 (2004)
6. Karnik, S., Gaitonde, V., Rubio, J., Correia, A., Abrão, A., Davim, J.: Delamination analysis in high speed drilling of carbon fiber reinforced plastics (CFRP) using artificial neural network model. Mater. Des. **29**(9), 1768–1776 (2008)
7. Mishra, R., Malik, J., Singh, I.: Prediction of drilling-induced damage in unidirectional glass-fiber-reinforced plastic laminates using an artificial neural network. Proc. Inst. Mech. Eng. Part B J. Eng. Manuf. **224**(5), 733–738 (2010)
8. Dhawan, V., Debnath, K., Singh, I., Singh, S.: A novel intelligent software-based approach to predict forces and delamination during drilling of fiber-reinforced plastics. Proc. Inst. Mech. Eng. Part L J. Mater. Des. Appl. **230**(2), 603–614 (2016)
9. Barbero, E.: Finite Element Analysis of Composite Materials using Abaqus, First edn. CRC press (2013)
10. Feito, N., Lopez-Puente, J., Santiuste, C., Miguélez, M.: Numerical prediction of delamination in CFRP drilling. Compos. Struct. **108**(1), 677–683 (2014)
11. Durao, L., Moura, M., Marques, A.: Numerical prediction of delamination onset in carbon/epoxy composites drilling. Eng. Fract. Mech. **75**(9), 2767–2778 (2008)
12. Phadnis, V., Makhdum, F., Roy, A., Silberschmidt, V.: Composites: Part A drilling in carbon/epoxy composites: experimental investigations and finite element implementation. Compos. Part A **47**, 41–51 (2013)
13. Mallick, P.: Fiber Reinforced Composites, Third edn. Taylor and Francis group (2007)
14. Hashin, Z.: Failure criteria for unidirectional fibre composites. J. Appl. Mech. **47**(June), 329–334 (1980)
15. Dassault Systèmes Simulia: Analysis User's Manual Volume 4. Abaqus 6.12, vol. IV, p. 29.3.8–10 (2012)
16. Abaqus Inc.: Analysis User's Guide, vol. III, p. 294 (2014)
17. Astm committee.: D5528-01: Standard test method for mode I interlaminar fracture toughness of unidirectional fiber-reinforced polymer matrix composites 1. Am. Stand. Test. Methods **3,** 1–12 (2014). (Reapproved 2007)

Chapter 70
Fuzzy Inference System-Based Neuro-Fuzzy Modeling of Electron-Beam Welding

Sanjib Jaypuria and Dilip Kumar Pratihar

Abstract Electron-beam welding, one of the high-energy density welding processes, is preferred for welding of nuclear, aerospace, and automotive materials. However, the input–output modeling of the welding is a challenging problem faced by many of the welding experts because of addition of new alloys to the industries. Soft computing techniques have been evolved in recent days to address this issue efficiently. This paper discusses two potential approaches of fuzzy inference system (FIS)-based neuro-fuzzy system for modeling of electron-beam welding response parameter. The following adaptive neuro-fuzzy inference systems (ANFIS): grid partition-based fuzzy interference system (G-ANFIS) and subtractive clustering-based fuzzy inference system (S-ANFIS) have been incorporated for the determination of depth of penetration in electron-beam welded copper plate in bead-on-plate configuration. In this study, the performances of above networks are compared on the basis of accuracy and computation time. Coefficient of correlation for S-ANFIS and G-ANFIS are found to be 0.98 in both approaches. Prediction of S-ANFIS is found to be 4 and 6% more accurate than results of G-ANFIS and regression model, respectively. Grid ANFIS consumes around 42 min to complete the computation, while it takes only 9.819 s for S-ANFIS. Based on the comparison, it is observed that the S-ANFIS model has the better prediction capability and less computational complexity than G-ANFIS and nonlinear regression model.

Keywords Electron-beam welding · ANFIS · Grid partitioning · Subtractive cluster · Fuzzy inference system

S. Jaypuria (✉) · D. K. Pratihar
Soft Computing Laboratory, Department of Mechanical Engineering, Indian Institute of Technology Kharagpur, 721302 Kharagpur, India
e-mail: sanjibjaypuria@iitkgp.ac.in

D. K. Pratihar
e-mail: dkpra@mech.iitkgp.ac.in

R. G. Narayanan et al. (eds.), *Advances in Computational Methods in Manufacturing*, Lecture Notes on Multidisciplinary Industrial Engineering,
https://doi.org/10.1007/978-981-32-9072-3_70

70.1 Introduction

Electron-beam welding (EBW) is an advanced fusion welding process, in which a high-speed electron beam is injected to the joining area of materials. The materials melt as the kinetic energy of the electrons transforms into heat upon impact. As a high-energy density process, EBW has inherent advantage of low-heat-affected zone and high aspect ratio, which yields relatively deep and narrow weld profile [1]. Electron-beam technologies are particularly useful in the industries like automotive, aviation, nuclear, medical, and railway. The primary input parameters, which affect the mechanical and metallurgical properties, are acceleration voltage, beam current, scanning velocity, focusing distance, and beam oscillation parameters [2]. Precipitated hardened Cu–Cr–Zr alloys have been selected as heat dissipation materials for the components of international experimental thermonuclear reactor (ITER) with high heat flux (HHF), such as the diverter, the limiter, and the first wall of the vacuum reactor [3].

Design of experiments, soft computing, and evolutionary computational networks have developed promptly and used for input–output modeling and optimization of welding [4]. Dutta and Pratihar [5] had modeled the dependence between input–output of gas tungsten welding (TIG) with three approaches, namely conventional regression, multilayer backpropagation neural network (BPNN), and genetic algorithm-tuned neural network (GANN). From the results, it was concluded that neural network is more robust in nature as compared to conventional regression. Moreover, GANN was found to outperform BPNN, as the former has fair chance of reaching the globally optimal solution. Jaypuria et al. [6] investigated the influence of EB welding input variables on mechanical properties and weld geometry of CuCrZr butt joint. A linear regression model had been developed to study the input and output relationship. Kim et al. [7] used a GA and a RSM simultaneously to predict the optimal welding conditions in the metal inert gas arc (MIG) welding process. Koleva [8] used statistical multiple regression techniques to estimate the depth and width of the welding schedule of EB welded stainless steels. Kanigalpula et al. [9] carried out electron-beam welding of copper plate and these bead-on-plate experiments were performed according to the central composite design. A statistical regression was performed to establish the input–output relationship of the EB joint. An optimization problem without constrains had been formulated to minimize the problem of spiking phenomena with maintaining the maximum average bead penetration. Results of genetic algorithm approach showed high accuracy with the real experiment results. Kanigalpula et al. [10] used genetic algorithm with penalty function approach to solve the complex constrained optimization problem associated with the EB welding of Cu–Cr–Zr alloy plates. They used four welding input parameters, namely voltage, beam current, scanning speed, and focusing distance to get suitable values of bead penetration, bead width, hardness of weld, and heat-affected zone.

Although many researches modeled input–output relation of EB welding for automation with respect to the accuracy of the model, there is no comparative study about the computational complexity. A few studies have been reported on fuzzy logic (FL)-based analysis related to welding process. Most of the articles concentrated their studies regarding fuzzy inference system (FIS) parameter optimization through genetic algorithm (GA) and particle swarm optimization (PSO) algorithm. However, no literature is available for different kinds of FIS-based ANFIS model for EB welding. Here, the modeling of EBW is done through two proposed neuro-fuzzy approaches to predict the response. The following systems: (i) grid partition-based fuzzy inference system, known as G-ANFIS and (ii) subtractive clustering-based fuzzy inference system, known as S-ANFIS, have implemented for establishing the nonlinear dependence between the EBW input and output parameters. Accelerating voltage, beam current, welding speed, and focusing distance are used as inputs to the FIS model to predict the bead penetration. The performance and computational complexity are compared for all the developed approaches.

70.2 Experimentation and Data Collection

Bead-on-plate (BOP) welding of Cu–Cr–Zr alloy plates of the dimensions 100 mm × 65 mm × 10 mm is taken in the present consideration. The input process variables which influence the bead penetration (BP) of EB welding are accelerating voltage (V), beam current (I), welding speed (S), and focusing distance (D). These experiments were conducted by using response surface central composite design (CCD) matrix with three center points. Hence, a total of $2^4 + 2\times4 + 3 = 27$ combinations of experimentation were conducted. Since the factors were expected to have quadratic effects, three central points were taken and three replicates were considered for each combination of input variables. The CCD matrix having 81 experimental data is utilized for this study. The weld geometry of the finished and etched samples was studied under optical microscope and macroscope. To test the performances of the developed input–output models, 12 random cases out of the 81 combinations were selected after maintaining their respective ranges unviolated. Training and test data set for verification of developed model was taken from Kanigalpula et al. [10]. Table 70.1 shows the test data matrix with different combinations of process parameters.

70.3 Adaptive Neuro-Fuzzy Inference System (ANFIS)—Based Modeling

Adaptive Neuro-Fuzzy Inference System (ANFIS) is a combination of artificial neural network (ANN) and fuzzy Takagi–Sugeno system [11]. It is a powerful tool, for carrying out various supervised learning tasks, such as regression and classification. The membership function parameters of initial FIS are fine-tuned by the use

Table 70.1 Test data matrix for testing proposed methods

Accelerating voltage (kV)	Beam current (mA)	Welding speed (mm/min)	Focusing distance (mm)	Experimental value (mm)
55	90	600	370	6.144
55	90	800	370	5.879
50	70	600	375	3.096
55	110	800	370	6.349
60	90	800	370	5.551
60	90	800	370	5.759
60	70	600	365	4.661
60	70	1000	375	3.797
55	90	800	370	5.423
60	90	800	370	5.181
55	90	800	370	5.121
50	110	1000	365	4.74

of backpropagation algorithm or in combination with the least square method. Then, the network learns and trains the FIS-based network on the data provided. This popular fuzzy system consists of following components: (1) a rule base, (2) a database, and (3) a decision-making mechanism [12]. The rule base consists of all possible rules from the number and levels of input variables. The membership function is decided form the database and the inference mechanism helps to establish the relation between input variables and rules. In ANFIS architecture, the initial FIS clusters the entire region and describes the behavior of the rules in the fuzzy embedded region.

70.3.1 Grid Partition-Based Fuzzy Inference System (G-ANFIS)

G-ANFIS is first reported by Abonyi et al. [13] in 1999. G-ANFIS works on the principle of generating single-output Sugeno-type fuzzy inference system using a grid partition on the data. The initial FIS generates the membership function parameters randomly just like any ANFIS network. G-ANFIS uses the grid partition (axis-paralleled partition) of fuzzy on the basis of predefined number of membership functions and their dimension. ANFIS starts from zero output during the training; it gradually learns and refines the fuzzy rules and functions parameters. The number of fuzzy rules increases exponentially, when the number of internal input variables increases, so with the increase of number of rules, it becomes computationally expensive and impractical for industries where a large number of variables are associated with the process. The grid partition might be useful for the system with small number of input variables mostly less than six and a large size of training data [14]. Here, four input variables are considered in this study. The proposed approaches have been developed with MATLAB 2015.

70.3.2 Subtractive Clustering-Based Fuzzy Inference System (S-ANFIS)

S-ANFIS [15] uses Sugeno-type FIS structure with subtractive grouping. S-ANFIS network is comparatively faster because of the faster estimation of cluster and cluster center in a given dataset. Here, each data point is assumed as cluster center and then defines the suitable cluster center of the dataset by calculating potential density of surrounding data points. In this manner, the first cluster center is chosen among the given data points and subsequently, the potential of data points near the cluster center is destroyed and second cluster center will be chosen based on potential density of surrounding data points. So, it becomes crucial to determine influence radius to construct one specific cluster and subsequent number of clusters with given data points. S-ANFIS returns a Sugeno-type FIS structure that models the data behavior and contains a set of fuzzy rules to cover the feature space [16].

70.3.3 ANFIS Models for Prediction of Depth of Penetration

In this work, both methods (G-ANFIS and S-ANFIS) are used for training of ANFIS and an attempt has been made to model and predict the response of EBW. Figure 70.1 shows the schematic view of developed ANFIS architecture for this study. A graphical user interface (GUI) of the updated rule base for a specific combination of input parameters is given in Fig. 70.2.

In the present study, out of the 81 input data pairs, 69 combinations are considered for training the FIS and 12 data pairs are used to test the accuracy and computation time. Takagi–Sugeno-type FIS structure of zero-order is used in both developed approaches. Gaussian and linear membership functions are considered in the current work for input and output layer, respectively. The reason behind choosing Gaussian function in input layer is the capability of the same to map the nonlinear process efficiently. The initial parameters associated with the membership function are randomly generated by ANFIS and tuned after subsequent training of FIS. A hybrid optimization algorithm, which is a mixture of backpropagation and least square method, is used to minimize the error (with error goal = 0) and a maximum iteration of 200. The number of membership functions is kept fixed at

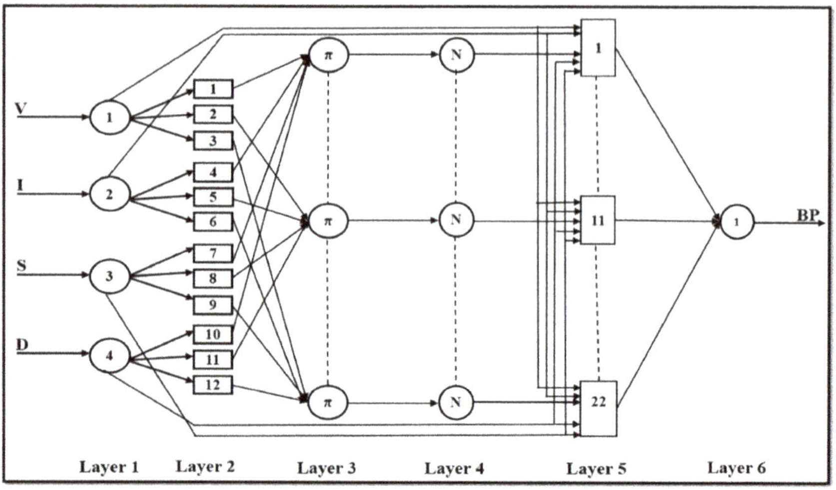

Fig. 70.1 ANFIS architecture

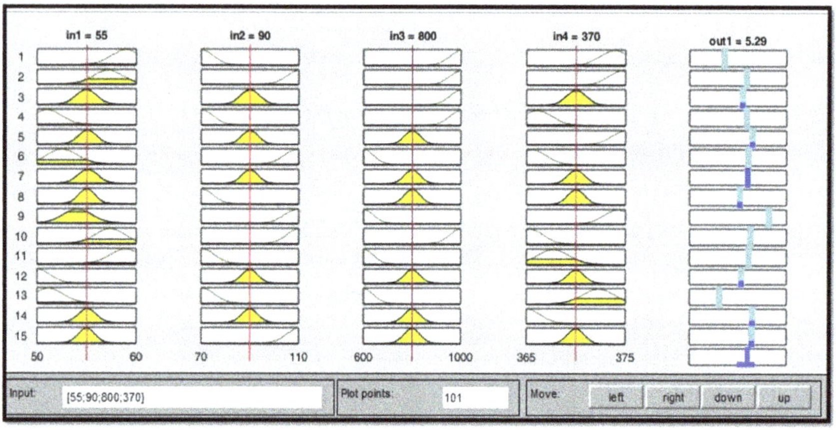

Fig. 70.2 Updated rule base for a specific input data combination [55, 90, 800, 370]

five, representing the linguistic variables like very low, low, medium, high, and very high. Table 70.2 represents the internal parameters associated with G-ANFIS and S-ANFIS.

Table 70.2 Internal parameters for proposed methods

Internal parameters	G-ANFIS	S-ANFIS
Number of nodes	1297	227
Number of linear parameters	3125	110
Number of nonlinear parameters	40	176
Total number of parameters	3165	286
Number of training data pairs	69	69
Number of checking data pairs	12	12
Number of fuzzy rules	625	22

70.4 Results and Discussion

70.4.1 Grid Partition-Based Fuzzy Inference System (G-ANFIS)

G-ANFIS model has been established to predict bead penetration as a function of welding input variables. The root-mean-square error (RMSE) values of G-ANFIS model are 0.15797 and 0.24419 for training and test data sets, respectively. The regression plots of the training data and test data sets are shown in Fig. 70.3 with the goodness of fit (R^2) values as 0.99635 and 0.98907, respectively. The total CPU time for execution of the program is 2510.9725 s. The computational complexity can be easily understood from 625 number of fuzzy (if-then) rules and 3165 number of internal parameters, which leads to curse of dimensionality [12, 14].

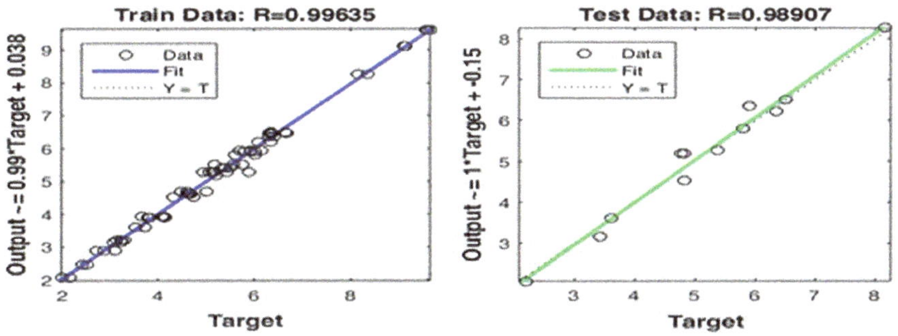

Fig. 70.3 Regression plots for training data and test data in G-ANFIS

70.4.2 Subtractive Clustering-Based Fuzzy Inference System (S-ANFIS)

The root-mean-square error (RMSE) values calculated for S-ANFIS architecture are 0.15507 and 0.23323 for the training and test data sets, respectively. The regression plots of the training data and test data sets are displayed in Fig. 70.4 with the (R^2) values as 0.99646 and 0.98915, respectively. The total CPU time for execution of the program is 9.819 s. Here, computation time is found to be quite less as compared to G-ANFIS because of elimination of unnecessary rules and dealing with only a few rules and internal parameters [15, 16].

70.4.3 Nonlinear Regression

A nonlinear statistical regression analysis is considered from available literature [10] to establish input–output relationships of the process and compare the results with other discussed ANFIS approaches. The generalized nonlinear regression expression is shown in below Eq. (70.1).

$$y = \beta_0 + \sum_{i=1}^{k} \beta_i x_i + \sum_{i=1}^{k} \beta_{ii} x_i^2 + \sum_{i=1}^{k} \sum_{i<j}^{k} \beta_{ij} x_i x_j + \varepsilon, \quad (70.1)$$

where y is the response parameter; $\beta_0, \beta_i, \beta_{ii}$ .... are the coefficients to be computed using a least-squares method. Here, ε stands for the error in fitting. Nonlinear regression equation for bead penetration is evaluated following Eq. (70.2), and given below.

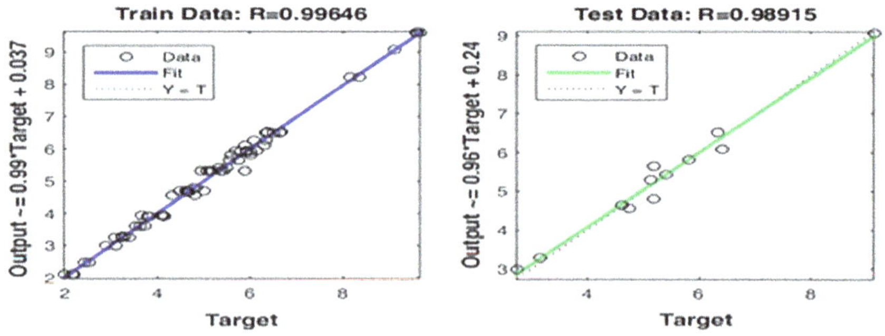

Fig. 70.4 Regression plots for training data and test data in S-ANFIS

$$
\begin{aligned}
BP_{uncoded} = {} & 5857.98 + 8.02798\,V + 0.0920434\,I + 0.0623438\,S - 33.0775\,D \\
& - 0.0313765\,V^2 - 0.00146853\,I^2 + 0.00000207134\,S^2 + 0.0456195\,D^2 + 0.0000113958\,VI \\
& - 0.0002530645\,VS - 0.0113766\,VD - 0.000145435\,IS + 0.00100352\,ID - 0.000114410\,SD
\end{aligned}
\tag{70.2}
$$

70.4.4 Discussion

The RMS values for both training and test dataset in various developed ANFIS approaches are represented in Table 70.3. The least RMS value of 0.15507 and 0.23323 for training and test cases of S-ANFIS signifies a better convergence rate because of efficient tuning of internal parameters. Here, the model uses subtractive clustering technique with Gaussian and linear membership function for input and output layers, respectively. It is very fast enough to extract the set of rules that models the input–output relationship of a given system by effectively calculating the number of clusters and cluster center. Figure 70.5 shows distinctly that there is a satisfying similarity between the experimental value and the forecasted assessment of depth of penetration. The percentage of deviation of predicted data with respect to the test cases is also seen to be minimum in S-ANFIS-based approach, which is shown in Fig. 70.6. The computation time involved in S-ANFIS approach is least of all approaches. It is due to the selected number of rules chosen for the training of a specific response, so it deals with less number of fuzzy rules and internal parameters.

Grid portioning or G-ANFIS is also showing comparable RMS error values for both the training and test cases. This may be due to the inclusion of all possible combinations of fuzzy rules in the FIS, which ensure all relevant rules are accounted in the training process. Although this approach is precise, the computational complexity involved in this FIS is complex [17]. So, here, both the performance criteria, i.e., RMSE and computation time of the FIS are opposing in nature. The observed trend is certainly due to the involvement of all possible rules and accumulation of internal parameters during training. In addition to this, the RMSE value of train and test cases of nonlinear regression model was found to be larger than train and test results of S-ANFIS and G-ANFIS, respectively. The prediction accuracy of nonlinear regression and G-ANFIS models has shown 6 and 4% inferior results than that of S-ANFIS model. The computation complexity of S-ANFIS is also found to be significant lower than that of G-ANFIS.

Table 70.3 Performance comparison of the proposed approaches

Performance criteria	G-ANFIS	S-ANFIS	Nonlinear regression
RMSE (training cases)	0.15797	0.15507	0.21936
RMSE (test cases)	0.24419	0.23323	0.24644
Computation time (s)	2510.972	9.819	

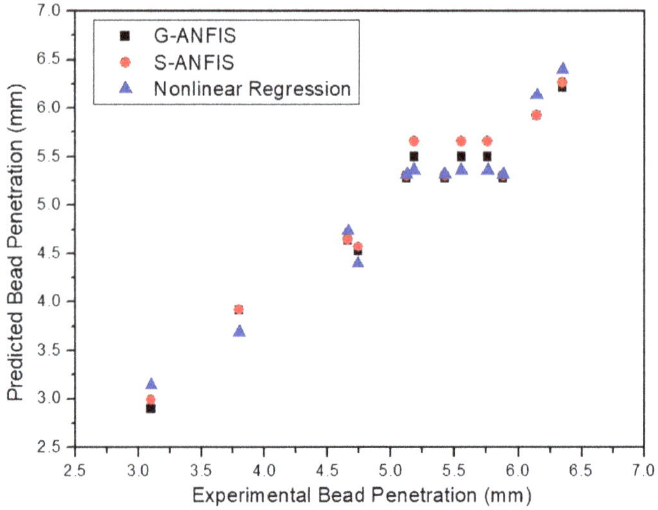

Fig. 70.5 Comparison of proposed approaches in terms of % deviation in prediction of bead penetration in test cases

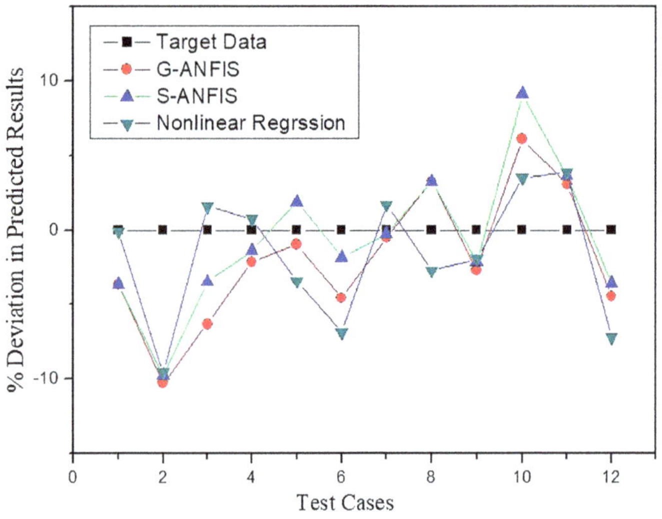

Fig. 70.6 Comparison between predicted and actual depth of penetration for the proposed approaches

Percentage deviation (Refer Fig. 70.6) result of S-ANFIS also signifies the good agreement with target data. Although the coefficient of correlation of both discussed approaches is nearly 0.98, the computational complexity of G-ANFIS creates difficulty for G-ANFIS model for real-time prediction. Therefore, it can be concluded that S-ANFIS architecture is the more suitable one to make predictions without hampering the accuracy and computational complexity. It can also be used for real-time prediction due to their less consumption of computational resources and relatively higher accuracy in prediction.

70.5 Conclusion

The prediction of bead penetration using input variables of EBW is approximated by the FIS architecture based on grid partition and subtractive clustering. The developed model is verified with the test data. From the results, the following conclusions have been drawn:

1. Two different kinds of FIS generated models, namely G-ANFIS and S-ANFIS, are used to train the ANFIS network. The developed models have a good agreement with the experimental results. It can be validated with high value of coefficient of correlation value, which is 0.98 in both cases. Root-mean-squared error of the test cases is found to be equal to 0.24419 and 0.23323 for G-ANFIS and S-ANFIS, respectively. The computation time for G-ANFIS and S-ANFIS is found to be equal to 2510 s and 10 s, respectively.
2. The performance of above-discussed ANFIS models is also compared with nonlinear regression model, which is taken from the literature. S-ANFIS prediction accuracy in terms of RMSE is found to be 6% more accurate than the nonlinear regression results.
3. S-ANFIS model is found to be suitable among discussed approaches in terms of prediction accuracy and computational complexity. S-ANFIS is also more preferred for real-time prediction due to less consumption of computational resources.

In future, the performances of proposed approaches can be compared with that of fuzzy C-means clustering (FCM) approaches and evolutionary-tuned ANFIS approaches to check the suitability of the algorithm for automation and online inspection.

These tools are very useful for automation and controller design of any manufacturing process, where it involves costly trial and error experiments to find the suitable parameters for a desired response. The application of this tool is not limited to manufacturing industries, it can also be widely used for all domains of industries, where real-time or online modeling is required.

References

1. Schultz, H.: Electron Beam Welding. Abington Publishing, England (2004)
2. Fabritsiev, S.A., Zinkle, S.J., Singh, B.N.: Evaluation of copper alloys for fusion reactor divertor and first wall components. J. Nucl. Mater. **233**, 127–137 (1996)
3. Lacki, P., Adamus, K., Wojsyk, K., Zawadzki, M., Nitkiewicz, Z.: Modeling of heat source based on parameters of electron beam welding process. Arch. Metall. Mater. **56**(2), 455–462 (2011)
4. Benyounis, K.Y., Olabi, A.G.: Optimization of different welding processes using statistical and numerical approaches—a reference guide. Adv. Eng. Softw. **39**(6), 483–496 (2008)
5. Dutta, P., Pratihar, D.K.: Modeling of TIG welding process using conventional regression analysis and neural network-based approaches. J. Mater. Process. Technol. **184**(1–3), 56–68 (2007)
6. Jaypuria, S., Doshi, N., Pratihar D.K.: Effects of welding parameters on mechanical properties in electron beam welded CuCrZr alloy plates. In: IOP Conference Series: Materials Science and Engineering, vol. 338, issue No. (1), pp. 012013 (2018)
7. Kim, D., Rhee, S., Park, H.: Modelling and optimization of a GMA welding process by genetic algorithm and response surface methodology. Int. J. Prod. Res. **40**(7), 1699–1711 (2002)
8. Koleva, E.: Statistical modelling and computer programs for optimisation of the electron beam welding of stainless steel. Vacuum **62**(2–3), 151–157 (2001)
9. Kanigalpula, P.K.C., Jaypuria, S., Pratihar, D.K., Jha, M.N.: Experimental investigations, input-output modeling, and optimization of spiking phenomenon in electron beam welding of ETP copper plates. Measurement **129**, 302–318 (2018)
10. Kanigalpula, P.K.C., Pratihar, D.K., Jha, M.N., Derose, J., Bapat, A.V., Pal, A.R.: Experimental investigations, input-output modeling and optimization for electron beam welding of Cu–Cr–Zr alloy plates. Int. J. Adv. Manuf. Technol. **85**(1–4), 711–726 (2016)
11. Jang, J.S.: ANFIS: adaptive-network-based fuzzy inference system. IEEE Trans. Syst. Man. Cybern. **23**(3), 665–685 (1993)
12. Gowtham, K.N., Vasudevan, M., Maduraimuthu, V., Jayakumar, T.: Intelligent modeling combining adaptive neuro fuzzy inference system and genetic algorithm for optimizing welding process parameters. Metall. Mater. Trans. B **42**(2), 385–392 (2011)
13. Abonyi, J., Andersen, H., Nagy, L., Szeifert, F.: Inverse fuzzy-process-model based direct adaptive control. Math. Comput. Simul. **51**(1–2), 119–132 (1999)
14. Jang, J.S.: Input selection for ANFIS learning. In: Proceedings of IEEE 5th International Fuzzy Systems, vol. 2, pp. 1493–1499. IEEE (1996)
15. Cobaner, M.: Evapotranspiration estimation by two different neuro-fuzzy inference systems. J. Hydrol. **398**(3–4), 292–302 (2011)
16. Admuthe, L.S., Apte, S.: Adaptive neuro-fuzzy inference system with subtractive clustering: a model to predict fiber and yarn relationship. Text. Res. J. **80**(9), 841–846 (2010)
17. Mathworks.: Fuzzy Reasoning Toolbox: User's Guide (r2015b) (2016)

Chapter 71
Numerical Analysis of Erosion of Rough Surface

Ahad Noor, Saipraneeth Gouravaraju and Sachin Singh Gautam

Abstract This paper presents a numerical analysis of solid particle erosion of a rough surface. A total of ten particles are modelled to impact the target surface. For validation, first a smooth surface is impacted with rigid particles and the erosion results are compared with experimental results in the literature. After validation, erosion of rough surface is simulated using the finite element software ABAQUS. Rough surfaces with different average surface roughness are generated using inbuilt fast Fourier transform and random distribution functions in MATLAB. The coordinates of the generated rough surfaces are then imported to and superimposed on the mesh of the target surface generated in ABAQUS and finite element analysis is performed. It is observed that surface roughness affects the erosion rate significantly. For impact angles of 30° and 40° erosion is found to be less than that of the smooth surface.

Keywords Finite element analysis · Rough surface · Gaussian roughness · Johnson-Cook model · ABAQUS

71.1 Introduction

Precise prediction of lifespan of machine elements and replacement (wherever necessary) can avoid industrial hazards. There are several reasons due to which premature failure of engineering machine elements take place. Solid particles moving at high velocity is considered to be one of the main reasons for premature failure of machine elements. Study of erosion is very important in engineering

A. Noor · S. Gouravaraju · S. S. Gautam (✉)
Indian Institute of Technology Guwahati, Guwahati 781039, India
e-mail: ssg@iitg.ac.in

A. Noor
e-mail: ahad.noor@iitg.ac.in

S. Gouravaraju
e-mail: saipraneeth@iitg.ac.in

R. G. Narayanan et al. (eds.), *Advances in Computational Methods in Manufacturing*, Lecture Notes on Multidisciplinary Industrial Engineering,
https://doi.org/10.1007/978-981-32-9072-3_71

applications where fluid travels at very high speed like oil and gas pipeline, rocket nozzles, compressor blades, drilling platforms, etc. When solid particles hit the surface repeatedly it causes material removal from the surface, this phenomenon is known as solid particle erosion (SPE). So, it would be beneficial to industries to predict life of machine elements for which the phenomena of erosion must be understood extensively. Understanding of erosion phenomena includes understanding loading condition, response of materials, materials properties, failure mechanism, and environmental factors like temperature and pressure. Erosion due to the impact of solid particle is one of the most extensively researched areas since 1960.

Finnie [1] performed analytical study of erosion. He considered a single non-spherical particle which has two sides, viz. flat side and cutting edge. He made an assumption that when particle strikes with flat side, erosion does not take place, and when it strikes with cutting edge, erosion takes place. He proposed a simple mathematical expression for prediction of erosion rate at different angles. When compared with experimental results, his equation was able to predict correct results at lower impact angles but not at normal impact. Later Bitter [2, 3] proposed that solid particle erosion cannot be explained with one model. He proposed that solid particle erosion takes place due to cutting action and plastic deformation. He solved energy equation for both the cases and provided mathematical expression to predict erosion rate at different impact angles. He developed concept of threshold velocity, according to which if a particle hits the surface with a velocity more than threshold velocity then only erosion takes place. His formula was simple but needed a constant, which was supposed to be determined experimentally.

By the era of Finnie and Bitter, it was well established that erosion is highly influenced by the impact angle. Although mechanism proposed by researchers mentioned above can predict erosion of ductile material, a generalized agreement could not be obtained even for simpler cases. So, to understand the erosion phenomena, some researchers concentrated on microscopic analysis by experiment. Hutching and Winter [4] attempted to explain the mechanism of erosion by scanning electron micrograph of impact zone. They used a 3 mm steel ball to impact surface of different samples made of aluminium, copper, and steel. They concluded that erosion of materials takes place due to shearing of material ultimately leading to lip formation.

Yerramareddy and Bahadur [5] performed experiments on Ti–6A–4V alloy substrate using silicon carbide particles at different impact angles. They studied the effect of heat treatment which includes effect of precipitation hardening microstructure. They reported that the temperature of heat treatment changes microstructure and so affects erosion rate. To understand mechanism of erosion, they studied scanning electron microscope image of impact zone. They reported that erosion rate depends on the particle's shape, i.e., blunt or sharp-edged. Depending on the shape, erosion takes place by different mechanism like cutting, ploughing, lip information, etc.

Oka et al. [6] performed an experimental analysis of solid particle erosion of 304 steel at different impact angles. They considered the effect of surface hardness also.

Based on the results of their experiment they proposed a predictive equation to estimate the erosion rate.

$$E_{90} = K(H_v)^{k_1}(v)^{k_2}(D)^{k_3} \tag{71.1}$$

where E_{90} is erosion in normal impact. H_v, v, and D are hardness, velocity, and diameter of particle, respectively. K accounts for the particle property factor, which includes angularity, hardness, etc., of the particle. k_1, k_2, and k_3 are exponents.

Erosion at a particular angle of impact $E(\alpha)$ is considered to be related to the erosion in normal impact E_{90} as

$$E(\alpha) = g(\alpha)E_{90} \tag{71.2}$$

where $g(\alpha)$ denotes the dependency of erosion on impact angle and is given as

$$g(\alpha) = (\sin\alpha)^{n_1}(1 + H_v(1 - \sin\alpha))^{n_2} \tag{71.3}$$

Although these equations as proposed by Oka et al. [6] predict the erosion rate correctly, they did not consider the effect of roughness.

Phenomena of solid particles erosion involve very high-velocity impact and very low interaction time, which makes it very difficult to study experimentally. So researchers were attracted by numerical methods to study erosion phenomena. Shimizu et al. [7] performed FE analysis of impact of a single particle at different angles. They concluded that for impact of spherical particle there is significant plastic deformation before failure. Before the development of numerical methods, it was very difficult to study the behaviour of erosion resistance coating and composite materials. Their thickness was determined with a trial and error basis. But numerical methods emerged as an important tool for analysis of coating and composite materials. Many researchers like Bielawski and Beres [8] performed a numerical analysis of composite materials.

Woytowitz and Richman [9] studied the effect of gap between impacting particles. Eltobgy et al. [10] performed finite element analysis of Ti–6A–4V composite and studied effect of impact angle and speed. Giriffin et al. [11] performed finite element analysis of steel coated with Al_2O_3. He used two different failure models for Al_2O_3 steel. Lui et al. [12] performed numerical analysis to investigate the effect of shape of the impacting particle.

Nelson et al. [13] performed erosion experiment at 30° impact angle with different rough sample made of mild steel. Yildierim and Muftu [14] have done analysis of particle impact on rough surface but they have not considered damage as they have performed their analysis in context of shot peening process. Anaraki et al. [15] has performed the analysis of erosion on rough surface and investigated effect of velocity, density, and diameter of particle but they have taken surface roughness value much high rather than in practical range and also they have not considered damage they assumed elements with 0.2% plastic strain as eroded volume.

From the literature survey, it was observed that extensive work has been done to analyse the effect of angel impact, size of particle, velocity of impact, etc. So it was decided to investigate the effect of surface roughness on erosion. We first validated our model which is described in the next section with experimental results of Oka et al. [6] Then different roughness were included in the model. Results were found to be consistent with the experimental results of Nelson et al. [13].

71.2 Model Description

In the present work, impact of ten particles striking at two closely spaced locations on the target material is simulated in ABAQUS/Explicit as shown in Fig. 71.1. The material considered here is 4340 steel. A stainless steel block of size 2 mm × 1.6 mm × 0.6 mm is considered as the target material. The bottom surface and all the vertical surfaces of the surface are assumed to be fixed (called the Encastre boundary condition in ABAQUS). The impacting particles are considered rigid. The diameter of particles is taken as 326 μm and mass of each particle is taken as 4.717×10^{-8} kg, which resembles with mass of silicon carbide particle of similar size. Rotary inertia of the particles is ignored. Particles are arranged such that each particle strikes the surface just after the impact of previous striking particle is complete, i.e., a rebound has occurred. Particles are placed at desired angles using rotation option in assembly module of ABAQUS. The particles are given an initial velocity of 104 m/s using predefined field option in ABAQUS. Particles are given velocity in *X*-and *Y*-direction depending on the angle of impact such that their resultant velocity comes to 104 m/s. Contact interaction between the particles and the rough surface is defined using surface to surface node contact algorithm, where the particle is considered as the master surface and all nodes of target are considered as slave surface. This contact algorithm provides an interaction between particle and target only, so no particle

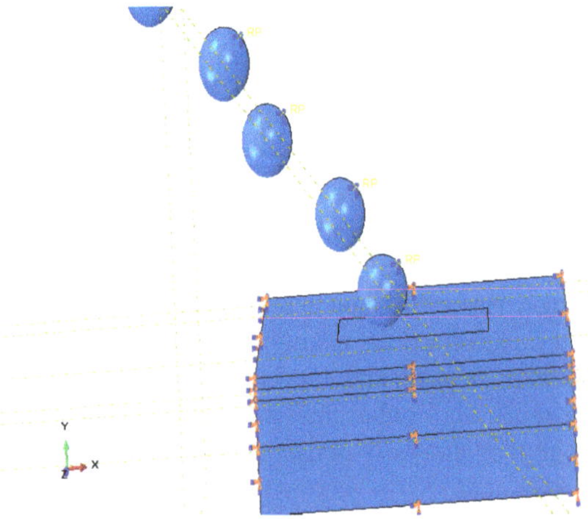

Fig. 71.1 Assembly of rigid particles and target material at an impact angle of 45° (only four particles are shown)

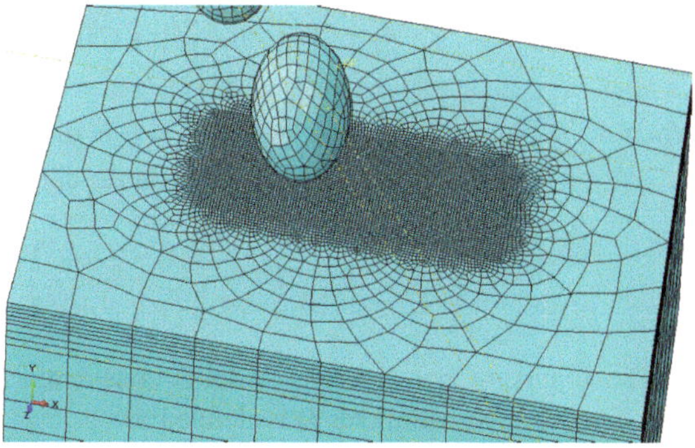

Fig. 71.2 Fine mesh near impact zone of 0.01 mm size while coarse mesh at other locations

collides during rebound of the previous particle. This helps us in placing particles closely and shortening the total simulation time. Friction between the surfaces is also considered with coefficient of friction value of 0.2, which was implemented using penalty, using interaction property option in ABAQUS. Heat generation due to friction and impact of particles were also considered. Eight nodded brick element (C3D8RT) with displacement degree of freedom is used for the analysis. Fine mesh of size 0.01 mm was used near impact zone while relatively coarse mesh was used in rest of the region as shown in Fig. 71.2.

The material data for 4340 steel is given in Table 71.1 [16].

Plastic properties are implemented using Johnson-Cook material model, which relates flow stress with equivalent plastic strain, temperature, and strain rate as shown in equation. Parameters of material model are given in Tables 71.2 and 71.3.

Table 71.1 Mechanical properties of 4340 steel

E (GPa)	v	ρ (kg/m^3)
200	0.3	7800

Table 71.2 Parameters for Johnson-Cook plasticity of 4340 steel

A (MPa)	B (MPa)	n	c	m	D_1	D_2	D_3	D_4	D_5
792	510	0.26	0.014	1.03	0.05	3.44	−2.12	0.002	0.61

Table 71.3 Composition of 304 and 4340 steel (%)

	C	Mn	Si	Ni	Cr	S	P
304 Steel	0.08	2.0	0.75	8–12	18–20	0.08	0.045
4340 Steel	0.42	0.70	0.28	1.83	0.79	0.0012	0.0009

$$\sigma_Y = \left(A + B\epsilon_p^n\right)(1 + C\ln\epsilon^*)(1 - T^*) \tag{71.4}$$

where ϵ_p is equivalent plastic strain, $\dot{\epsilon}^* = \frac{\dot{\epsilon}}{\dot{\epsilon}_0}$ is dimensionless plastic strain rate for $\dot{\epsilon}_0 = 1\ \text{s}^{-1}$ and T^* is homologous temperature defined as

$$T^* = \left(\frac{T - T_{\text{room}}}{T_m - T_{\text{room}}}\right) \tag{71.5}$$

The Johnson-Cook fracture model is considered in this work to simulate damage. The damage parameter D is defined as

$$D = \Sigma\frac{\Delta\epsilon}{\epsilon_f} \tag{71.6}$$

where $\Delta\epsilon$ is increment in equivalent plastic strain and ϵ_f is the equivalent strain to fracture. Fracture is assumed to occur when D reaches value of 1. The expression for equivalent strain to fracture ϵ_f is given by

$$\epsilon_f = (D_1 + D_2\exp(D_3\sigma^*))(1 + D_4\ln\epsilon^*)(1 + D_5T^*) \tag{71.7}$$

where $D_i(i = 1, 2, 3, 4, 5)$ are the five constant. Their values are given in Table 71.2 [16].

In their erosion experiments, Oka et al. [6] used 304 stainless steel. Reliable Johnson-Cook damage parameters for 304 steel are not available in the literature [17]. So, in the current work, the results of Oka et al. [6] are compared with the results in the current work obtained with 4340 steel in a qualitative manner. Liu et al. [12] also considered the properties of ARMCO Iron for their FE simulations while comparing with the results of Oka et al. [6]. Although both the steels are of different grade, trend of failure follows the same pattern. A comparison of composition of both materials is shown in Table 71.3.

71.3 Results and Discussion

In this section, first the erosion from a smooth surface is discussed for impact velocity of 104 m/s and at six different impact angles, viz. 10°, 20°, 30°, 40°, 60°, and 90°, respectively. Results are validated with the experimental results of Oka et al. [6]. Then the erosion from the rough surface is simulated.

Initially, up to the impact of four to five particles no erosion occurred as the plastic flow and plastic deformation of the material is observed. With further impact, damage to the material starts and erosion takes place. Eroded volume is calculated and normalized with eroded volume in case of normal impact, i.e., for 90°. For calculation of eroded volume, number of elements deleted after the impact

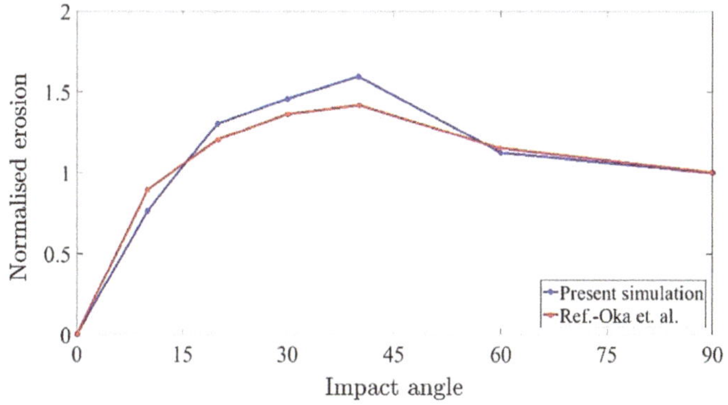

Fig. 71.3 Result for smooth surface and comparison to reference result

is counted. Then summing the volume of all the deleted elements gives us the total eroded volume. To convert eroded volume to dimensionless volume it is divided with eroded volume in case of normal impact. Results from the current finite element simulation are compared with results obtained by Oka et al. [6]. A comparison of present work and results obtained by Oka et al. [6] in their experiment is shown in Fig. 71.3. It can be seen that they are in good agreement. It is also observed that maximum depth occurs at normal impact but maximum volume occurs at 40° impact.

After validation, the effect of surfaces roughness on the erosion is studied. Four different rough surfaces with average surface roughness of 0.5 μm, 1 μm, 1.5 μm, and 2 μm are considered. Rough surface with autocorrelation length of 0.01 mm and mentioned average roughness values are generated using inbuilt Fourier transform and inbuilt random distribution function in MATLAB. Generated rough

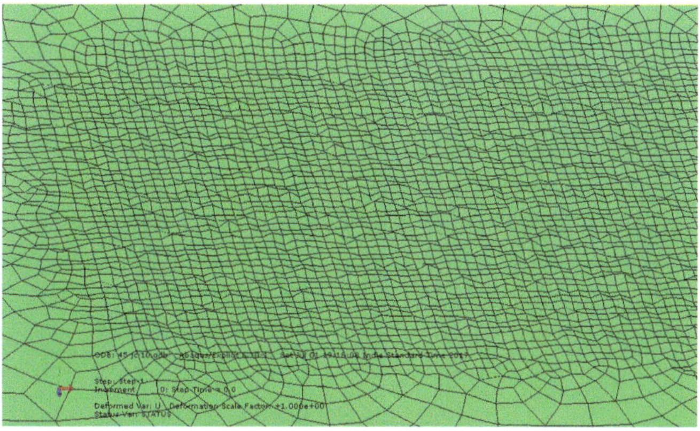

Fig. 71.4 Sample mesh of the target surface of the block after superimposing the coordinates of the generated rough surface

surface is superimposed on the top of the smooth surface discussed in the previous section (see Fig. 71.4). Then particle with a velocity of 104 m/s at different angles is impacted on the rough surface. Similar to smooth surface, eroded volume is calculated and normalized with volume of normal impact. Interestingly it is observed that at low impact angles, erosion decreased, and high angle is increased. Nelson et al. [13] also reported a decrease in erosion with increase in roughness for most of the cases but in one case they observed opposite trend also. At 30° impact angle, similar trend was observed in the current finite element simulation.

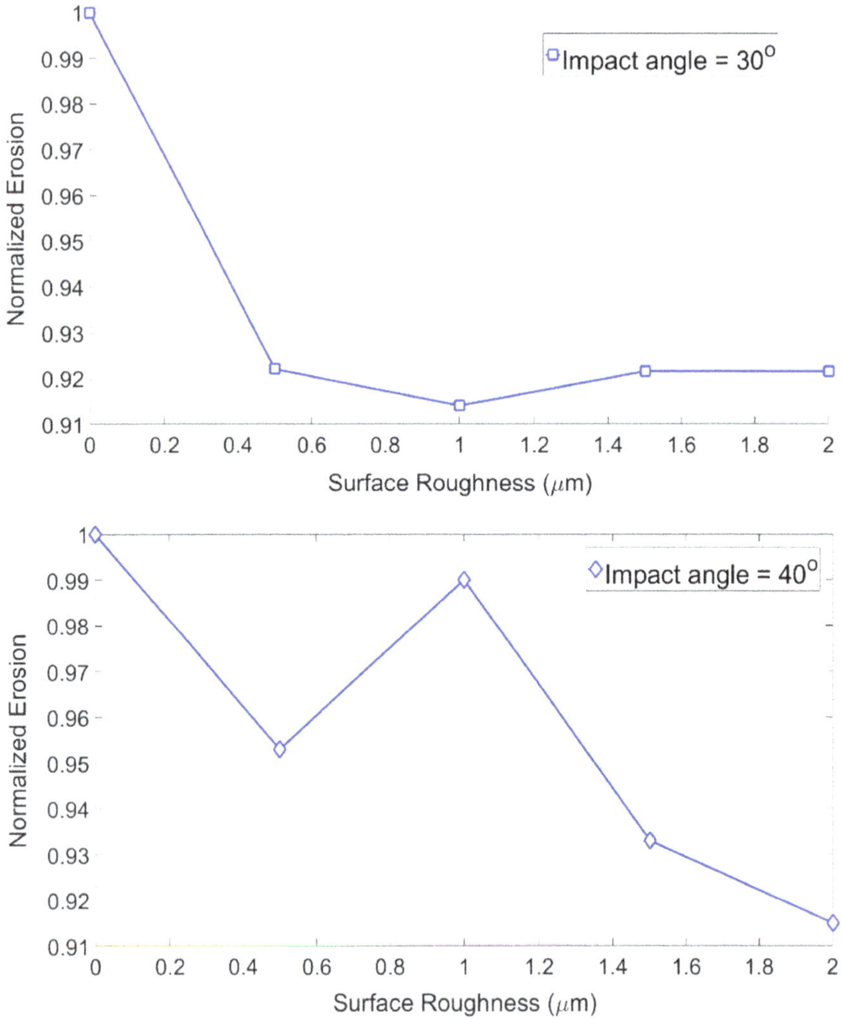

Fig. 71.5 Evolution of normalized erosion with surface roughness at impact angles of 30° and 40°

To quantify the effect of roughness, the eroded volume is normalized with the eroded volume of the smooth surface at each angle. Figure 71.4 shows the variation of normalized volume of erosion for rough surface with surface roughness for different impact angles. It can be seen that at lower impact angles, as the surface roughness increases, the erosion decreases. But at higher impact angles, erosion increases with increase in roughness. This observed behaviour is consistent with the observations of Nelson et al. [13] at an impact angle of 30°. (Figure 71.5)

71.4 Conclusion

A finite element analysis is carried out to study the erosion characteristics of a rough surface due to particle impact at different angles of impact. It is observed that surface roughness affects the erosion rate significantly. In the present analysis, two angles of impact are considered for the rough surface impact, viz. 30° and 40°. In both the cases, erosion rate is observed to be lesser than the erosion of smooth surface. This can be explained by Finnie's theory [1]. Finnie [1] argued that the surface roughness changes actual angle of contact for 30° impact. Results of the present numerical analysis follow the similar trend which was observed by Nelson et al. [13] experimentally.

Acknowledgements The authors are grateful to the SERB, DST for supporting this research under project SR/FTP/ETA-0008/2014.

References

1. Finnie, I.: Erosion of surface by solid particles. Wear **3**, 87–103 (1960)
2. Bitter, J.G.A.: A study of erosion phenomena part I. Wear **6**(1), 5–21 (1962)
3. Bitter, J.G.A.: A study of erosion phenomena part II. Wear **6**(3), 169–190 (1962)
4. Hutchings, I.M., Winter, R.E.: Particle erosion of ductile metals: a mechanism of material removal. Wear **27**(1), 121–128 (1974)
5. Yerramareddy, S., Bahadur, S.: Effect of operational variables, micro structure and mechanical properties on the erosion of Ti–6Al–4V. Wear **142**(2), 253–263 (1991)
6. Oka, Y.I., Okamura, K., Yoshida, T.: Practical estimation of erosion damage caused by solid particle impact: Part 1: effects of impact parameters on a predictive equation. Wear **259**(1–6), 95–101 (2005)
7. Shimizu, K., Noguchi, T., Seitoh, H., Muranaka, E.: FEM analysis of dependency of impact angle during erosive wear. Wear **233–235**, 157–159 (1999)
8. Bielawski, M., Beres, W.: FE modelling of surface stresses in erosion-resistant coatings under single particle impact. Wear **262**(1–2), 167–175 (2007)
9. Woytowitz, P., Richman, R.: Modeling of damage from multiple impacts by spherical particles. Wear **233–235**, 120–133 (1999)
10. ElTobgy, M.S., Ng, E., Elbestawi, M.A.: Finite element modeling of erosive wear. Int. J. Mach. Tools Manuf **45**(11), 1337–1346 (2005)

11. Griffin, D., Daadbin, A., Datta, S.: The development of a three dimensional finite element model for solid particle erosion on an alumina scale/MA956 substrate. Wear **256**(9–10), 900–906 (2004)
12. Liu, Z.G., Wan, S., Nguyen, V.B., Zhang, Y.W.: A numerical study on the effect of particle shape on the erosion of ductile materials. Wear **313**(1–2), 135–142 (2014)
13. Nelson, S.A., Baker, M.J., Deans, W.F.: The effect of surface roughness on the erosion of mild steel. In: International Pipeline Conference. American Society of Mechanical Engineers, pp. 803–810 (2004)
14. Yıldırım, B., Muftu, S.: Simulation and analysis of the impact of micron-scale particles onto a rough surface. Int. J. Solids Struct. **49**, 1375–1386 (2012)
15. Anaraki, A. P., Kadkhodapour, J., Taherkhani, B.: Simulation of erosion by particle impact on a rough surface. J. Fail. Anal. Prev. 784–789 (2014)
16. Johnson, G.R., Cook, W.H.: Fracture characteristics of three metals subjected to various strains, strain rates, temperatures and pressures. Eng. Fract. Mech. 31–48 (1985)
17. Bosetti, P., Maximiliano Giorgio Bort, C., Bruschi, S.: Identification of Johnson–Cook and Tresca's parameters for numerical modeling of AISI-304 machining processes. ASME. J. Manuf. Sci. Eng. **135**(5), 051021–051021-8 (2013)

Chapter 72
Numerical Simulation and Experimental Validation of Nanoindentation of Silicon Using Finite Element Method

Borad M. Barkachary and Shrikrishna N. Joshi

Abstract This paper reports finite element method-based numerical simulations of nanoindentation of Silicon to extract the mechanical properties, viz. Young's modulus and hardness. The simulations were carried out by using the Drucker–Prager constitutive model assuming a rigid spherical indenter of radius 300 nm to mimic the Hertzian contact which is typically observed underneath the Berkovich indenter. The dispalcement controlled boundary conditions comprising the depths of indentation as 45, 75, 95, 115, 130, and 140 nm were applied. Oliver–Pharr analytical method was used to obtain load-displacement plots. By using these plots, the values of Young's modulus and hardness were estimated. The results from the developed finite element simulation have also been validated by using the published experimental results. The estimated results from our simulations were found to be fairly matching with the experimental results. It was noticed that the FEM-based simulations provide an economical and fast option to the physical experiments in the estimation of mechanical properties of a material.

Keywords Nanoindentation · Silicon · Finite element method · Mechanical properties · Drucker-Prager material model

B. M. Barkachary
Department of Mechanical Engineering, Jorhat Institute of Science & Technology, Jorhat 781010, Assam, India
e-mail: borad.barkachary@gmail.com

S. N. Joshi (✉)
Department of Mechanical Engineering, Indian Institute of Technology Guwahati, Guwahati 781039, Assam, India
e-mail: snj@iitg.ac.in

R. G. Narayanan et al. (eds.), *Advances in Computational Methods in Manufacturing*, Lecture Notes on Multidisciplinary Industrial Engineering,
https://doi.org/10.1007/978-981-32-9072-3_72

72.1 Introduction

Nanoindentation testing is useful to obtain the unknown mechanical properties such as hardness and Young's modulus. This method comprises of pressing the substrate material (specimen) against a harder material (usually a diamond indenter), and the material properties are estimated based on the force–displacement plots obtained during the loading and unloading sequence. Over the decades, various techniques such as tensile/compressive test, shear test, triaxial, and scratch test, nanoindentation have been developed to measure the mechanical properties of materials, viz. Young's modulus, Poisson's ratio, yield strength, strain-hardening, shear strength, hardness, etc. Out of these techniques, the tensile/compressive are commonly used for characterization of ductile materials. However, these are not suitable for brittle materials such as Silicon, because the brittle materials have higher yield strength than the fracture strength. Thus, during these tests, brittle materials fail before yielding. The applied stress reaches the fracture strength first, thereby fracturing the specimen before any appreciable strain can form. Nanoindentation test is a simple method, which determines the mechanical properties of materials such as hardness and Young's modulus. It is also used to obtain strain-hardening exponent, creep, fracture toughness, viscoelastic properties, and stress-strain data.

While some experimental work on Silicon exists, little theoretical work, especially in the arena of numerical simulations, has been reported. Our motivation is to simulate the Nanoindentation of Silicon using the currently available material models and evaluate how well the results from such models compare to those obtained in experiments. In sum, the objectives of this study were (i) to explore nanomechanical response of Silicon to quasi-static nanoindentation, (ii) to demonstrate the feasibility of modeling a typical hardness test using finite element analysis, (iii) to show how basic mechanical properties can be extracted from such simulated results, and (iv) to compare and contrast the simulated results with the experimental results.

72.2 Theoretical Background

Nanoindentation is a robust technique used to investigate the nanomechanical properties of materials, such as hardness, Young's modulus, fracture toughness, and incipient plasticity, all of which are vital for designing structural/functional elements of MEMS/NEMS devices. Nanoindentation entails a simpler setup and specimen preparation is compared to other mechanical experimental characterization methods. Typical nanoindentation involves pressing a hard indenter into a substrate material whose properties are to be determined.

During a nanoindentation process, the indenter is pressed into the material as shown in Fig. 72.1a, due to which both elastic and plastic deformation processes

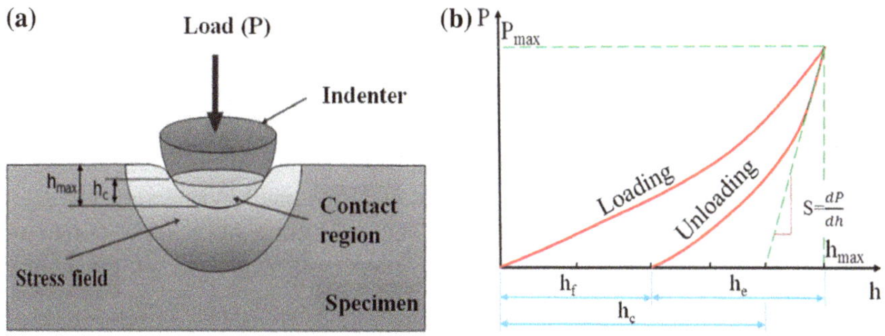

Fig. 72.1 Schematic of **a** nanoindentation process and **b** load–displacement plot

occur. This process produces an impression with a projected area A_p and surface area A_s of contact. This area depends on the geometry of the indenter and the contact depth, h_c. Basically, the nanoindentation process includes a loading and unloading cycle with or without dwell cycle. Figure 72.1b shows the indentation load (P) plotted against the displacement of the indenter tip (h) with respect to the top of the surface (before deformation) for one complete indentation cycle. The important parameters that have to be noted are the maximum depth ($h_{\max}$) of penetration at peak load, the peak load ($P_{\max}$) and the final residual depth after unloading (h_f). The slope S, also known as contact stiffness is determined by linearly fitting the upper 1/3th portion of the unloading curve. The contact stiffness, hardness, and Young's modulus are determined by using the Oliver–Pharr method [1, 2] as described in following Sects. 72.3.1 and 72.3.2 which is also mentioned in ISO 14577 and ASTM E2546.

72.3 FEM Modeling and Simulation of Nanoindentation of Silicon

With the advent of the computational techniques, it is now possible to use the FEA for analyzing complex engineering problems in a variety of engineering areas. Some of these problems include the study of heat transfer, contact mechanics, fluid dynamics, and electromagnetism [3]. In the present study, a commercial FE package ABAQUS™ was used to model the nanoindentation process of Silicon with a spherical rigid indenter of a known radius (300 nm). The use of a spherical indenter introduces a wider yet small stress field resulting in elasto-plastic deformation at shallow loads. Details of the model development are presented below.

72.3.1 Assumptions

As with any other engineering problem, there were several assumptions made while undertaking this study. First, it was assumed that no slippage occurs at the interface. Second, the contact between the indenter and the surface of the work material is assumed to be frictionless [4–6]. Workpiece and diamond indenter materials are assumed to be homogeneous and isotropic. The indentation process was carried out in quasi-static condition, so, temperature effect was neglected in the current simulation.

72.3.2 Geometric Modeling

The axisymmetric FEM model with its geometric dimensions is shown in Fig. 72.2. The dimensions of the work material were chosen in such a way that the deformation-induced stress waves reflected back from the fixed boundary can be avoided. Accordingly, the work material in this model was set to be an axisymmetric rectangular block of 2 μm × 2 μm. These dimensions are large enough to simulate a semi-infinite half-space for indentation with a maximum depth of 140 nm in this model. The rigid indenter was modeled with a radius of 0.3 μm.

72.3.3 Material Model

In this present simulation, a nanoindentation problem was considered to be based on continuum mechanics, where quasi-static loading of the indenter is applied. Since the von Mises yield criterion imply the dependence of material yielding solely on second deviatoric stress tensor and is independent of the first stress invariant, the yielding sensitivity to hydrostatic stress tensor is not incorporated for

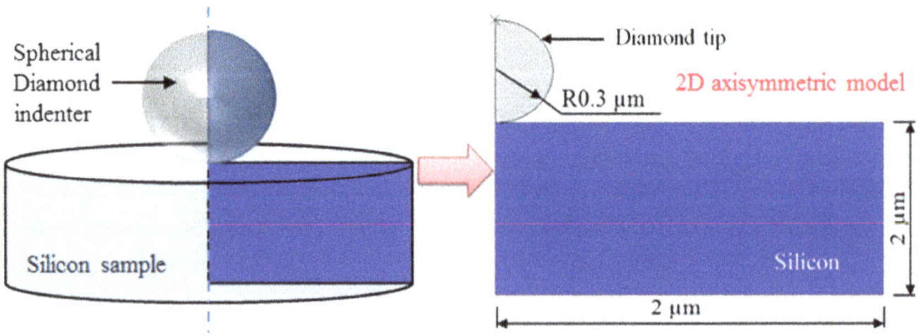

Fig. 72.2 Schematic of 2D axisymmetric indentation simulation model

pressure-sensitive materials. Drucker–Prager material model is used to model the material behavior of Silicon along with the elastic material properties. The Drucker–Prager yield criterion given by ref. [7] is a pressure-dependent model for determining whether the material has failed or undergone plastic yielding. The Drucker–Prager yield criterion can be described as a simple modification of the von Mises criterion, in which the hydrostatic stress component is also included to introduce pressure-sensitivity. The Drucker–Prager yield criterion is given by the following:

$$\sqrt{3 \cdot J_2} + I_1 \cdot \alpha - \kappa = 0 \tag{72.1}$$

where I_1 is the first invariant of the stress tensor, J_2 is the second invariant of the deviatoric stress tensor, α is the pressure-sensitivity coefficient, and κ is a material constant.

The quantity J_2 is given by the following:

$$J_2 = \frac{1}{6}\left[(\sigma_1 - \sigma_2)^2 + (\sigma_2 - \sigma_3)^2 + (\sigma_3 - \sigma_1)^2\right] \tag{72.2}$$

where σ_i are the principal stresses. The quantity κ is given by the following, which is equal to the yield stress in the case, when $\sigma_c = \sigma_t$, that is, no pressure dependency:

$$\kappa = \frac{2\sigma_c\sigma_t}{\sigma_c + \sigma_t} \tag{72.3}$$

where σ_c and σ_t are the yield stress in tension and compression, respectively.

The elastic materials properties for diamond and silicon are listed in Table 72.1 and Drucker–Prager material constants for silicon are listed in Table 72.2.

Table 72.1 Material properties of Diamond [7, 8] and Silicon [9]

Parameters	Unit	Diamond	Silicon
Density (ρ)	(kg/m^3)	3350	2328
Young's modulus (E)	(GPa)	1141	130
Shear modulus (G)	(GPa)	512	79.9
Poison's ratio (υ)		0.07	0.27
Fracture toughness (K_c)	(MPa $m^{1/2}$)	5.3 – 8	0.9
Thermal conductivity (k)	(W/m K)	700	150
Specific heat	(J/kg K)	520	700

Table 72.2 Constants for Drucker–Prager constitutive model for Silicon [10]

Parameters	Angle of friction (°)	Flow stress ratio	Dilation angle (°)	Yield stress (Nm^{-2})
Si	26	0.82	−20	15,867

72.3.4 Element and Meshing

Figure 72.3 shows the meshed models of the diamond tip and the workpiece. In the present work, two-dimensional, four-node quadrilateral element CAX4R was employed to discretize the model geometries. It is four-node quadrilateral element and provides reduced integration with hourglass control.

This element is chosen here because it is relatively inexpensive in terms of computation cost and time for problems involving non-linear constitutive behavior. For this type of element, the material calculations are only done at one point in each element by integrating all four nodes. Moreover, quadrangle element is one order higher than triangular element due to which it gives more accurate results. Also, by using this type of element, the symmetrical three-dimensional results can be viewed. The solution accuracy is dependent upon the type of mesh and its

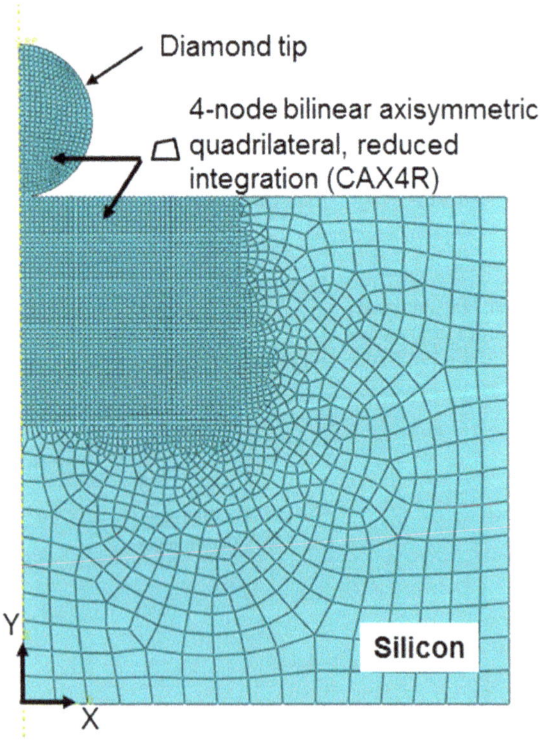

Fig. 72.3 Meshed geometries of the diamond tip and the workpiece

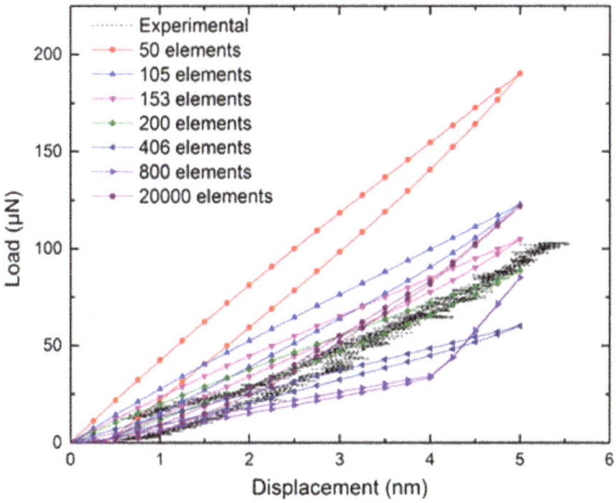

Fig. 72.4 Comparison of load–displacement plot with no of elements in indentation zone

refinement. Modification of mesh, known as *mesh refinement* is important to obtain a more accurate solution in any FEM simulation. The silicon block is discretized using fine mesh near the indentation zone, and coarse meshing is done away from the indenter to reduce the computational time without compromising the accuracy. The mesh sensitivity analysis has been carried out for a typical process condition: indenter tip radius—300 nm, max loading depth—45 nm, loading speed—0.1 mm/s. These process conditions are the same as that of the experimental conditions mentioned in the published literature by Rao et al. [11]. From the analysis, it was seen that the load–displacement plots at 20,000 elements are closer to the experimental results [11] in comparison with those obtained for the remaining cases as shown in Fig. 72.4. Therefore, the element size of 0.00001 mm, which generates 20,000 elements to discritze the geometries, was chosen for the discretization of the indentation zone.

72.3.5 Modeling of Surface Contact

The bottom surface of the indenter and the top surface of work materials formed a contact pair in the present numerical model. Surface-to-surface contact has been assigned between the contact pairs. Based on the suggestions from the literature ref. nos. [5, 6, 12], frictionless contact property is enabled between the two contact surfaces. It is because only a small portion of the indenter is considered to be in contact with the surface of the workpiece.

72.3.6 *Boundary Conditions*

Figure 72.5 shows the boundary conditions applied to the tool and workpiece. A fixed boundary condition is applied to the base of the work material. All the nodes on the base cannot move in any directions. The X-symmetry boundary condition is applied to the vertical surface of the workpiece as shown in Fig. 72.5.

It means that the workpiece geometry and diamond tip are symmetric along with the symmetry line and both form a revolving 3D geometry around the line. Rigid body constraint is applied to the diamond indenter at the reference point. First, it is restrained in the *X*-direction along with all the rotational degrees of freedoms. Since the process is very slow, i.e., quasi-static loading condition is assumed. Therefore, the load applied to the indenter is the displacement/velocity load of 0.1 mm/s in the current simulation. It is applied to the reference point of the diamond tip. As the contact area is very small and the loading speed is very slow, the effects of variation in temperature are not considered in the current simulation.

72.3.7 *Solution Methodology*

The numerical simulation of nanoindentation has been carried out in two steps: loading and unloading. The nanoindentation process includes complex interaction between the diamond tip and the workpiece along with large elastic-plastic

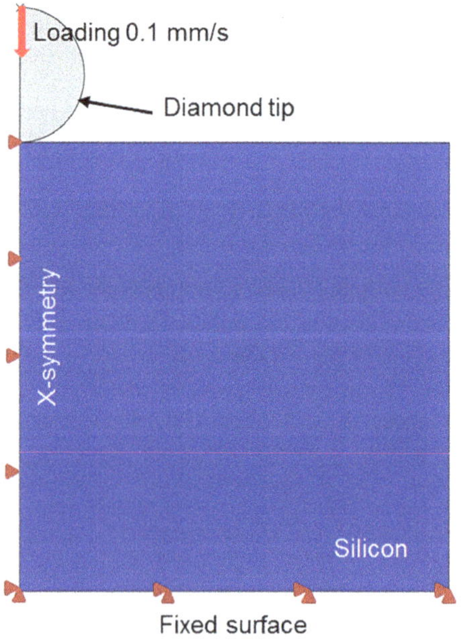

Fig. 72.5 Schematic of the boundary conditions applied

deformation of the workpiece during the nanoindentation process. Due to this complexity, both the steps have been solved by following the non-linear dynamic explicit solution step method with automatic (self-adaptive) time-step algorithm. In the dynamic explicit automatic time-step algorithm, the time increment is adjusted automatically to meet the corresponding convergence rate. The loading behavior of nanoindentation is simulated by increasing the indenter displacement in small increments. During the loading steps, the indenter is moved in the negative *Y*-direction, i.e., downward direction with a velocity of 0.1 mm/s and indents the work material to the maximum indentation depth. During the unloading step, the indenter returns to its original position incrementally. For each loading and unloading steps, 20 frames have been generated to reduce the data size.

72.4 Results and Discussion

The most common method for measuring hardness and Young's modulus using nanoindentation methods is the Oliver–Pharr analytical method. The underlying assumptions for the analytical model are:

(i) Deformation upon unloading is purely elastic
(ii) The compliance of the sample and the indenter tip is considered as springs in series
(iii) The process continuum is considered to be a homogeneous isotropic elastic half-space.

72.4.1 Measurement of Young's Modulus

From the extracted load–displacement plot, maximum load, corresponding maximum displacement at maximum load and the final displacement after the unloading was recorded. In the present case, the maximum load is 1275 μN, and maximum displacement is 45 nm. The displacement after unloading is 38.18 nm. It has been noted that there is no fixed rule for considering the unloading data points. The unloading data points are needed to be chosen by looking at the graphical data before deciding on the fitting parameters for analysis. Usually between 25 and 100% of the unload curve data are commonly used for the fitting function depending on the quality of the data [2, 13]. In the present case, we have found that 1/3th of the data points give good results while calculating Young's modulus and hardness. From the unloading curve, 1/3rd data points of the total unloading data points are linearly fitted, and the corresponding slope values were obtained as shown in Fig. 72.6.

Corresponding hardness and Young's modulus values were calculated for each depth of indentation following the Oliver–Pharr method. The stiffness of the contact

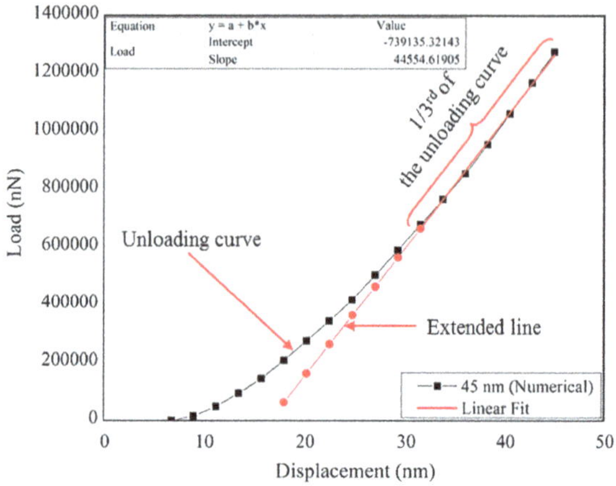

Fig. 72.6 Fitting the unloading curve and finding the slope

between the indenter and the material being tested is required to determine the mechanical properties of interest. The stiffness, S, is determined from the initial slope of the unloading curve.

$$\text{Stiffness}\,(S) = \frac{\mathrm{d}P}{\mathrm{d}h} \tag{72.4}$$

From the linearly fitted line, the slope is calculated using the linear line equation as shown in Fig. 72.6, which is nothing but the stiffness (S) = 44555 N/m^2. Once the stiffness is measured using (72.4), the reduced modulus E_r can be determined as:

$$E_r = S\frac{\sqrt{\pi}}{2\beta\sqrt{A}} \tag{72.5}$$

where S is the contact stiffness, A is the contact area, and E_r is the reduced modulus which accounts for the measured elastic displacement contributing both from the sample and the indenter tip. Depending on the indenter tip used, contact radius is calculated. In the present case, for spherical indenter, the contact radius is computed as

$$\text{Contact radius}, a = \sqrt{h(D-h)} = \sqrt{45(600-45)} = 158\,\text{nm} \tag{72.6}$$

where D is the diameter of the spherical indenter, and h is the maximum depth at maximum load. The contact area between the indenter tip and workpiece at maximum displacement is calculated as

$$\text{Contact area}, A = \pi a^2 = 7.85 \times 10^{-14}\,\text{m}^2 \tag{72.7}$$

After finding out the stiffness (S), contact area (A), and taking indenter geometry correction factor $\beta = 1$, the reduced modulus (E_r) is calculated as $1.41 \times 10^{11}\,\text{N/m}^2$

(by using the Eq. (72.5)). Next, the reduced modulus can be used to calculate the actual modulus of the sample, using the following relationship:

$$\frac{1}{E_r} = \frac{1 - \upsilon_i^2}{E_i} + \frac{1 - \upsilon_s^2}{E_s} \tag{72.8}$$

where $E_i = 1141$ GPa and $\upsilon_i = 0.07$ are the known values of the Young's modulus and Poisson's ratio for diamond indenter, respectively. Thus, by using Eq. (72.8), the actual modulus E_s for silicon sample is found to be $E_s = 153\,\text{GPa}$.

72.4.2 *Measurement of Hardness*

Hardness is the resistance to the permanent plastic deformation-induced in the material and is expressed as the ratio of the applied force (P_{max}) to the corresponding projected contact area (A) as shown in Eq. (72.9):

$$\text{Hardness}, H = \frac{P_{\max}}{\text{Contact area}} \tag{72.9}$$

By using Eq. (72.9), the hardness value of silicon were calculated as $H = 16.25$ GPa for $P_{max} = 1275$ μN and $A = 7.85 \times 10^{-14}$ m^2. After determining the required output parameters, the results were validated with the published experimental results. These are presented in the following section.

72.5 Experimental Validation of Nanoindentation Simulation

Nanoindentation simulations were carried out for silicon workpiece to validate the developed finite element-based nanoindentation model with experimental results [11]. All the process conditions in the simulations were set similar to the experimental process conditions mentioned in the Rao et al. [11] except the indenter geometry as the simulation was carried out in 2D. In numerical simulation spherical indenter tip with a diameter of 600 nm was considered, whereas, in the experiment, indentations were performed with a Berkovich indenter. The approximation of Berkovich indenter with a spherical indenter with nanometric radius is still valid as the tip of the Berkovich indenter is not completely sharp. Plots of experimental and numerical simulation results in terms of the load–displacement graph for indentation depths 45, 75, 95, 115, 130, and 140 nm are shown in Fig. 72.7. From the figure, it was observed that numerical results are in good agreement with the experimental results; however, the numerical results over predict the experimental results. Using Oliver–Pharr analytical procedure explained in previous Sects. 72.4.1

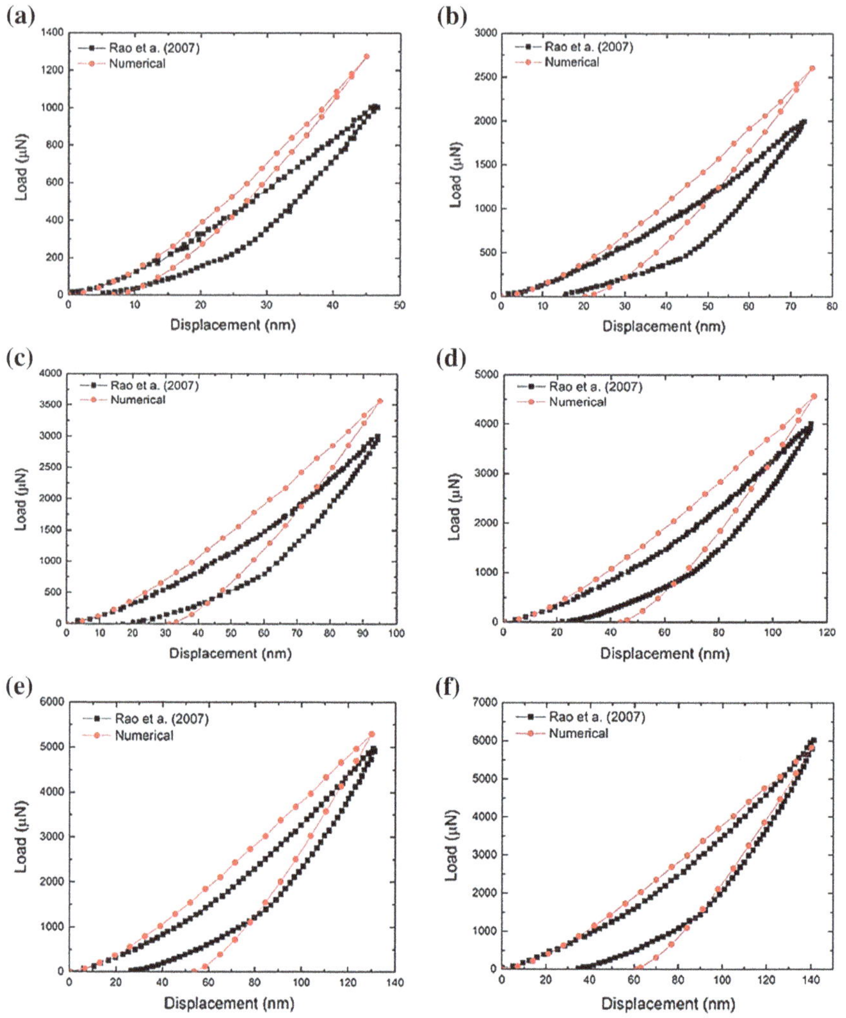

Fig. 72.7 Experimental [11] versus numerical comparison of P-h graph

and 72.4.2, Young's modulus and hardness values were calculated and shown in Table 72.3.

Table 72.3 also depicts the comparison of experimental results with numerical results, such as peak loads, projected area, calculated Young's modulus, and hardness values obtained for various levels of indentation depths. From the table, it is observed that peak load for simulations are higher than the experimental peak load. However, the projected area, Young's modulus, and hardness are found to be similar with minimal prediction error. The simulations are found to be predicting well with an overall prediction error of 10.9%. The projected area is found to be

Table 72.3 Comparison of Young's modulus and hardness at different indentation depth

Peak displacement	Peak load (μN)			Projected area (nm^2)			Young's modulus (GPa)			Hardness (GPa)		
	Experiment	Simulation	Prediction error (%)	Experiment	Simulation	Prediction error (%)	Experiment	Simulation	Prediction error (%)	Experiment	Simulation	Prediction error (%)
45 nm	1004.2	1275.2	27	81,157.7	78,492.9	−3	139	153	10	12	16	31
75 nm	2000.4	2604.4	30	121,056.3	123,750.0	2	176	167	−5	17	21	27
95 nm	3000.0	3566.8	19	149,824.4	150,778.6	1	196	176	−10	20	24	18
115 nm	4001.7	4568.5	14	173,974.9	175,292.9	1	193	185	−4	23	26	13
130 nm	4975.3	5296.1	6.4	192,637.2	192,028.6	−0	211	191	−9	26	28	7
140 nm	6026.0	5826.6	−3	203,447.6	202,400.0	−1	243	196	−19	30	29	−3
Average prediction error			16.6			1.3			9.2			16.5

similar to the experimental value with 1.3% prediction error. The average prediction error of Young's modulus and hardness were found to be 9.2 and 16.5%, respectively with the modified method.

From Fig. 72.6 and Table 72.3, it is observed that the calculated Young's modulus, hardness, and projected area obtained from the experiment and numerically simulated load–displacement graph are in good agreement. Furthermore, it is to be noted that the established values of Young's modulus and hardness values for silicon are 190 GPa and 12 GPa, respectively [14]. The slight variations in the results between the experiments and numerical simulations may be due to the consideration of rigid spherical indenter, isotropic and homogeneous material properties in the numerical model. Moreover, the presence of residual stresses strongly affects Young's modulus and hardness which may cause both to decrease [15–17]. Because, in the nanoindentation experiment, samples are prepared by mechanical means (such as machining), due to which residual stresses are left over the workpiece surface. Also, during the mechanical processing, brittle material undergoes amorphization, which generally has less Young's modulus and hardness as compared to pristine sample [18]. Further, Young's modulus calculation is largely dependent on an empirical factor presented by the Oliver–Pharr. However, despite all these limitations, the FEA fairly estimates the Young's modulus and hardness without conducting physical experiments, which saves a considerable time and cost.

72.6 Conclusions

In the present work, FEA-based simulations of displacement controlled quasi-static nanoindentation were performed on a silicon specimen using a spherical (300 nm tip radius) indenter. The simulations were primarily carried out as a testbed study to evaluate the mechanical properties, viz. Young's modulus and hardness of silicon. The computed values were found in good agreement with the experimental results. Calculated Young's modulus and hardness from the load–displacement plots were 153–196 GPa and 16–29 GPa, while the experimental values were noted to be 139–243 GPa and 12–30 GPa, respectively. The difference between the numerical and published experimental results may be due to (a) residual stresses (tensile or compressive) present in the experimental specimen, (b) experimental process may not follow the continuum assumption due to the presence of voids and dislocations, and (c) diamond indenter may not be that stiff as it is considered in the numerical model. Overall, it is found that finite element simulation provides an economical and quicker alternative to the costly and time-consuming physical experiments in obtaining the mechanical properties of the materials.

References

1. Oliver, W.C., Pharr, G.M.: An improved technique for determining hardness and elastic modulus using load and displacement sensing indentation experiments. J. Mater. Res. **7**(6), 1564–1583 (1992). https://doi.org/10.1557/JMR.1992.1564
2. Oliver, W.C., Pharr, G.M.: Measurement of hardness and elastic modulus by instrumented indentation: advances in understanding and refinements to methodology. J. Mater. Res. **19**(1), 3–20 (2004). https://doi.org/10.1557/jmr.2004.19.1.3
3. Zienkiewicz, O.C., Taylor, R.L.: The finite element method, vol. 36. McGraw-hill, London (1977)
4. Yu, N., Polycarpou, A.A., Conry, T.F.: Tip-radius effect in finite element modeling of sub-50 nm shallow nanoindentation. Thin Solid Films **450**(2), 295–303 (2004). https://doi.org/10.1016/j.tsf.2003.10.033
5. Bhattacharya, A.K., Nix, W.D.: Analysis of elastic and plastic deformation associated with indentation testing of thin films on substrates. Int. J. Solids Struct. **24**(12), 1287–1298 (1988). https://doi.org/10.1016/0020-7683(88),90091-1
6. Lu, C.J., Bogy, D.B.: The effect of tip radius on nano-indentation hardness tests. Int. J. Solids Struct. **32**(12), 1759–1770 (1995). https://doi.org/10.1016/0020-7683(94),00194-2
7. Mariayyah R (2007) Experimental and numerical studies on ductile regime machining of silicon carbide and silicon nitride. Doctoral dissertation, The University of North Carolina at Charlotte
8. Goel, S., Kovalchenko, A., Stukowski, A., Cross, G.: Influence of microstructure on the cutting behaviour of silicon. Acta Mater. **105**, 464–478 (2016). https://doi.org/10.1016/j.actamat.2015.11.046
9. Venkatachalam S (2007) Predictive modeling for ductile machining of brittle materials. Doctoral dissertation, Georgia Institute of Technology
10. Mir, A., Luo, X., Sun, J.: The investigation of influence of tool wear on ductile to brittle transition in single point diamond turning of silicon. Wear **364**, 233–243 (2016). https://doi.org/10.1016/j.wear.2016.08.003
11. Rao, R., Bradby, J.E., Ruffell, S., Williams, J.S.: Nanoindentation-induced phase transformation in crystalline silicon and relaxed amorphous silicon. Microelectron. J. **38**(6–7), 722–726 (2007). https://doi.org/10.1016/j.mejo.2007.04.006
12. Panich, N., Kraivichien, V., Yong, S.: Finite element simulation of nanoindentation of bulk materials. J. Sci. Res. Chulalongkorn Univ. **29**, 145–153 (2004)
13. Shuman, D.J., Costa, A.L., Andrade, M.S.: Calculating the elastic modulus from nanoindentation and microindentation reload curves. Mater. Charact. **58**(4), 380–389 (2007). https://doi.org/10.1016/j.matchar.2006.06.005
14. Reddy JD (2008) Mechanical properties of Silicon Carbide (SiC) thin films. Doctoral dissertation, University of South Florida
15. Tsui TY, Oliver WC, Pharr GM (1996) Influences of stress on the measurement of mechanical properties using nanoindentation: Part I. Experimental studies in an aluminum alloy. J. Mater. Res. **11**(3), 752–759. https://doi.org/10.1557/jmr.1996.0091
16. Kese, K.O., Li, Z.C., Bergman, B.: Influence of residual stress on elastic modulus and hardness of soda-lime glass measured by nanoindentation. J. Mater. Res. **19**(10), 3109–3119 (2004). https://doi.org/10.1557/JMR.2004.0404
17. Lee, W.S., Chang, S.L.: Nanoindentation behaviour and microstructural evolution of annealed single-crystal silicon. World academy of science, engineering and technology. Int. J. Chem. Mol. Nucl. Mater. Metall. Eng. **9**(7), 799–803 (2015). https://doi.org/10.1999/1307-6892/10001683
18. Yan J, Takahashi H, Tamaki JI, Gai X, Harada H, Patten J (2005) Nanoindentation tests on diamond-machined silicon wafers. Appl. Phys. Lett. **86**(18), 181913 (1–3). https://doi.org/10.1063/1.1924895

Chapter 73
Effect of Microchannel Pattern on Fluid Flow and Heat Transfer Characteristics

Dungali Sreehari and Apurbba Kumar Sharma

Abstract Three different patterns (rectangular-serpentine, U-serpentine and V-serpentine) of microchannels were modelled to investigate the fluid flow and heat transfer characteristics using the commercially available finite element analysis software—CFD Fluent. The microchannel models were based on 0.525-mm-thick silicon wafer substrate. The hydraulic diameter (0.4 mm) and channel length (114 mm) were kept constant in all the three different patterned microchannels to compare the performance of the microchannels. Three-dimensional (3D) simulations were performed to simulate near practical condition using the conjugate method in which heat transfer was considered in both solid and fluid zones. Reynolds number was varied in the range of 100–400, and uniform heat flux (30 kW/m^2) was applied at the bottom wall of the microchannels to investigate the effect of microchannel pattern on pressure drop, temperature, velocity profile and heat transfer coefficients. Results confirm that the higher heat transfer coefficient was obtained in U-serpentine microchannel, indicating better performance among the three.

Keywords Microchannel · Channel pattern · Heat transfer · Simulation

73.1 Introduction

With increasing the importance of microchannels in various fields of applications, especially in the electronics industry for cooling of microelectronic components, the research on the design, fabrication and performance evaluation of microchannels has increased. Due to the dense packing of the several microelectronic components in an electronic device, heat dissipation is a major concern. The microchannel is one

D. Sreehari (✉)
National Institute of Technology, Uttarakhand Srinagar (Garhwal) 246174, India
e-mail: sreehari@nituk.ac.in

D. Sreehari · A. K. Sharma
Indian Institute of Technology Roorkee, Roorkee, Haridwar 247667, India

R. G. Narayanan et al. (eds.), *Advances in Computational Methods in Manufacturing*, Lecture Notes on Multidisciplinary Industrial Engineering,
https://doi.org/10.1007/978-981-32-9072-3_73

of the best solutions to bring down these rising temperatures. Large surface-area-to-volume ratio and low weight and volume attribute the microchannel as an effective heat sink. The concept of the microchannel was introduced by Tuckerman and Pease [1] in the early 1980s. They fabricated microchannels on silicon by an orientation-dependent etching technique and concluded a maximum heat carried out was about 790 W/cm^2 from 1 cm^2 area. After their pioneering study, research on the microchannels application was performed by several researchers across the globe. Peiyi and Little [2] fabricated very fine microchannels on a silicon wafer by the photolithography technique and revealed that the width, depth and surface roughness of the microchannel affect the values of friction factor in the application of microminiature refrigerator. Weisberg et al. [3] numerically analysed the silicon microchannels using conjugate heat transfer method and presented some design formulations for the selection of channel dimensions. Experimental and numerical investigations were carried out by Qu et al. [4] on trapezoidal cross-sectional silicon microchannels. They predicted surface roughness of the microchannel could be the reason to obtain a significant difference between the experimental and numerically determined Nusselt number. Heat transfer characteristics such as temperature, Nusselt number and effect of Reynolds number were numerically analysed in three-dimensional (3D) rectangular silicon microchannels by Qu and Mudawar [5] using conjugate heat transfer method. Lee et al. [6] investigated rectangular microchannels of hydraulic diameter ranging from 318 to 903 μm by experimental and numerical analysis considering both thin wall model and 3D conjugate method. To investigate the geometrical effect of the microchannels on flow and heat transfer characteristics, Gunnasegaran et al. [7] numerically analysed three different cross sections (rectangular, trapezoidal and triangular) and concluded that rectangular cross-sectional microchannels have better heat transfer coefficients.

In general, the microchannel heat sinks are straight channels in which the streamlines of the coolant are nearly straight. Due to poor mixing of fluid in the straight microchannels, the heat transfer is ineffective. Therefore, an alternative wavy microchannel was proposed by Sui et al. [8]. They numerically investigated the performance of straight and wavy microchannels with the same rectangular cross sections using CFD simulations. From their results, it was observed that the performance of wavy microchannel is better than the straight microchannels. Lin et al. [9] further investigated the performance of wavy microchannels varying the wavelength and amplitude of the microchannels in the flow direction. Flow and heat transfer analysis was done in serpentine channels by Karale et al. [10]. They have compared the performance of general serpentine, sinusoidal serpentine and saw-tooth serpentine microchannels. Different channel configurations such as nonlinear, wavy, spiral, pin-finned and fractal-shaped were explored to enhance the heat transfer rate [11–14] further.

Although several serpentine microchannels were analysed in previous work, still it is not well understood which serpentine is more appropriate for high heat transfer with moderate pressure drop. Therefore, in the present work, three different patterns (rectangular-serpentine, U-serpentine and V-serpentine) of microchannels were modelled and analysed to compare the fluid flow and heat transfer characteristics.

73.2 Geometric Models of Microchannels

Silicon wafer (0.525-mm-thick) was selected as substrate material as it is commonly used to fabricate IC chips in the electronic devices. The schematic of three serpentine microchannels is shown in Fig. 73.1. The selection of dimension was based on the fabrication and substrate thickness constraints. Microchannels with a constant length of 114 mm were fabricated by micro-ultrasonic machining technique on the silicon wafer to a depth of 0.3 mm using 0.6-mm-diameter cylindrical tungsten carbide tool [15]. Hence, the same dimensions ($L \times W \times H$: 114 mm 0.3 mm × 0.6 mm) were used to model and analyse the fluid flow and heat transfer characteristics in all the three serpentine microchannels. However, it was observed, to maintain the constant length of the microchannel in all three different serpentines, the base areas of the microchannels vary significantly. The base areas of the rectangular-serpentine, U-serpentine and V-serpentine microchannels are 336, 244.5 and 384 mm^2, respectively. Therefore, a new dimensionless parameter

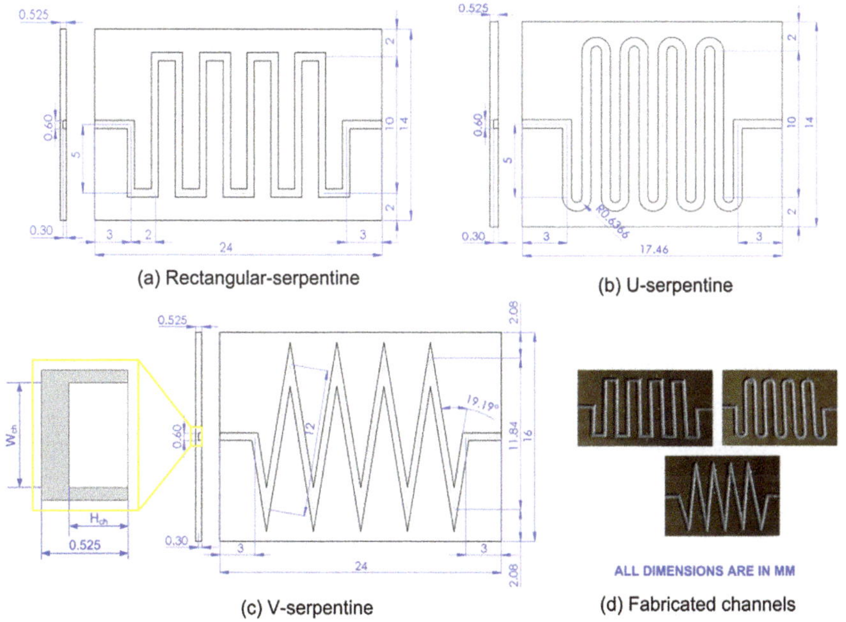

Fig. 73.1 Schematics of **a** rectangular-serpentine, **b** U-serpentine and **c** V-serpentine microchannels and **d** image of fabricated microchannels

Table 73.1 Sink area factors of microchannels

Pattern of microchannel	Total channel wall area (mm^2)	Base area (mm^2)	Sink area factor (%)
Rectangular-serpentine	136.8	336	40.71
U-serpentine	136.8	244.5	55.95
V-serpentine	136.8	384	35.63

—sink area factor (SAF) defined as the ratio of the total channel wall area to the total base area (Eq. 73.1)—was introduced to compare the performance of different serpentine microchannels with constant channel length and different base areas.

$$\text{Sink Area Factor (SAF)(\%)} = \frac{\text{Total channel wall area}}{\text{Base area}} \times 100\% \tag{73.1}$$

Table 73.1 gives the calculated values of sink area factors (SAFs) of the serpentine microchannels. The total channel wall area is the actual convective area responsible for heat transfer, i.e. the bottom wall area ($0.6 \times 114 = 68.4\ mm^2$) and the two side wall areas ($2 \times 0.3 \times 114 = 68.4\ mm^2$) of the microchannel.

73.3 Numerical Study

The fluid flow in the modelled microchannels was based on the continuity, momentum and energy equations. The flow was assumed to be steady, laminar and incompressible. Uniform heat flux ($30\ kW/m^2$) was applied at the bottom wall of the microchannels. The inlet velocity of the fluid was based on the Reynolds number which was varied in the range of 100–400.

73.3.1 Boundary Conditions

The boundary conditions used in the simulation are as follows:

(i) Inlet fluid temperature was constant ($T_{in} = 300$ K).
(ii) A constant heat flux ($30\ kW/m^2$) was given to the bottom wall, keeping the top and side walls as adiabatic.
(iii) Inlet velocity (V_{in}) was calculated using Eq. (73.2)

$$V_{in} = \frac{Re\mu}{\rho_f D_h} \tag{73.2}$$

$$D_h = \frac{4A}{P} = \frac{2W_{ch}H_{ch}}{W_{ch} + H_{ch}} \tag{73.3}$$

where Re—Reynolds number, μ—viscosity (Kg/m s), ρ_f—density of the fluid (kg/m^3), D_h—hydraulic diameter (μm), W_{ch} and H_{ch} are the width (μm) and depth (μm) of the microchannel.

(iv) Outlet pressure (P_{out}) was assigned a zero gauge pressure.

73.3.2 Grid Independence Test

The grid independence of the solution for rectangular serpentine was performed by considering the computational cells 2849658, 3289705 and 3951850. A variation of 1–2% was observed in the results of temperature and pressure using the computational cells of 3289705 and 3951850; while comparing the results obtained by the computational cells 2849658 and 3289705, a variation of 5–6% was observed. Therefore, 3951850 cells were considered for computation purpose in the numerical simulation study of rectangular-serpentine microchannels. Similar grid independence tests were performed for U-serpentine and V-serpentine microchannels. After obtaining the nearly identical solutions using the different number of cells, 5879476 and 6522096 cells were considered for computation purpose for U-serpentine and V-serpentine microchannels, respectively. It took nearly 60, 75 and 105 min to converge the solution in the rectangular-, U- and V-serpentine microchannels, respectively.

73.3.3 Model Validation

The model was validated with the work of Qu and Mudawar [5] to obtain the accurate results from the present numerical analysis. The boundary conditions, material—silicon substrate and fluid medium (liquid water) were considered to reproduce the results of Qu and Mudawar [5]. In order to compare the result, one of the operating conditions, i.e. wall heat flux of q = 50 W/cm^2 and Re = 140, was simulated. Figure 73.2a, b shows the comparative result of temperature distribution across a vertical plane at the middle of the substrate. It was observed that the present model showed similar results as of Qu and Mudawar [5].

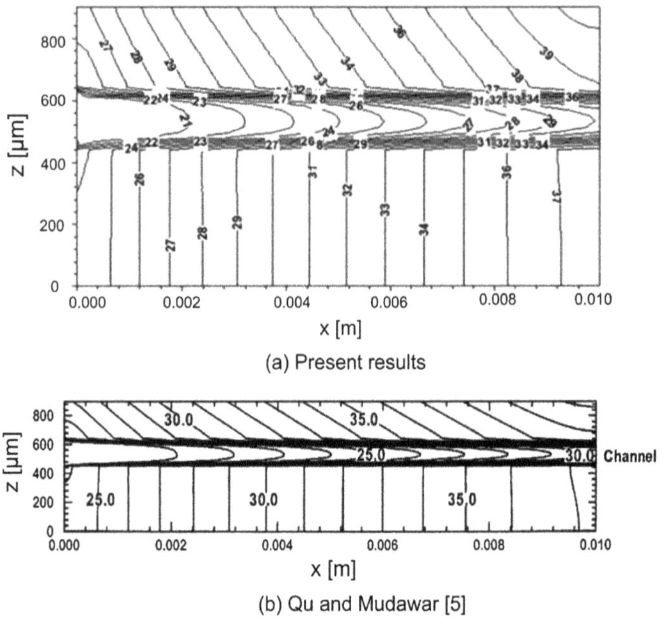

Fig. 73.2 Validation of the present model with Qu and Mudawar [5] results

73.4 Results and Discussion

The numerical simulations were carried out on the three patterns of microchannels considering a constant heat flux (30 kW/m^2) at the bottom wall and varying the Reynolds number from 100 to 400 in steps of 100. In the present section, comparative results in terms of pressure drop, temperature, velocity and heat transfer coefficient are presented.

73.4.1 Pressure Drop

Fig. 73.3 shows the variation of pressure drop (the difference between outlet pressure and inlet pressure) in different serpentine microchannels at different Reynolds numbers (*Re*). It was observed that the pressure drop increases with increase in the Reynolds number, irrespective of the serpentine. The reason is that as the Reynolds number increases, the velocity of the fluid flow increases which in turn increases the pressure drop as it is directly proportional to the square of the velocity. The results are in good agreement with the previous work of Gunnasegaran el al. [7]. Further, it was also observed that pressure drop in all the three serpentines is near similar at *Re*—100 and 200; highest in U—serpentine

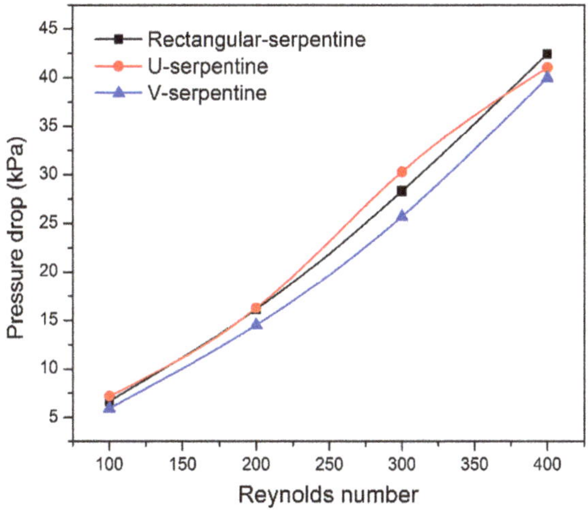

Fig. 73.3 Pressure drop variation with Reynolds number in different pattern microchannels

microchannels at *Re*—300 was observed compared to the other two serpentines. However, at higher Reynolds number (400), the pressure drop variation was marginal.

73.4.2 Temperature

The inlet temperature of the fluid was taken as one of the boundary conditions and was kept constant at 300 K in every simulation of the serpentines. The temperature at the outlet of the fluid was estimated from the simulation results. A difference of the inlet and outlet fluid was calculated in all the serpentine microchannels at different Reynolds numbers. A plot was drawn for the fluid temperature differences at different Reynolds numbers in rectangular-serpentine, U-serpentine and V-serpentine as shown in Fig. 73.4. It was observed that the fluid temperature difference decreases with increasing the Reynolds number, indicating that as the Reynolds number increases, the amount of heat carried away by the fluid was more. However, at higher Reynolds number (beyond *Re*—300), the rate of temperature difference was less. From Fig. 73.4, it was observed the V-serpentine microchannels have a larger fluid temperature difference compared to the remaining microchannels. The reason could be that due to the larger base area of V-serpentine, the amount of heat input was more and hence results in more outlet fluid temperature. Figure 73.5 indicates the contours of the temperature difference in the microchannel and substrate material.

The average base temperature of the substrate was considered from the simulation results, in order to investigate further. The same was plotted against the Reynolds number in Fig. 73.6. It was observed that the base temperature decreases with increase in the Reynolds number. The reason is that due to higher velocities at

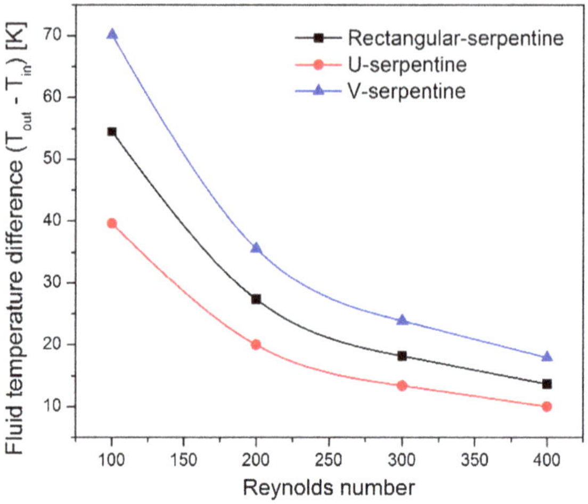

Fig. 73.4 Fluid temperature difference versus Reynolds number

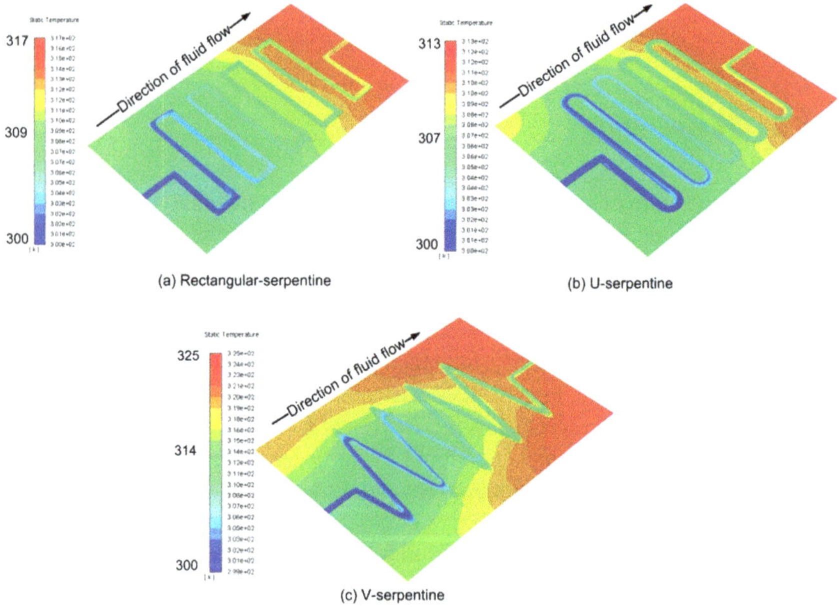

Fig. 73.5 Contours of temperature distribution

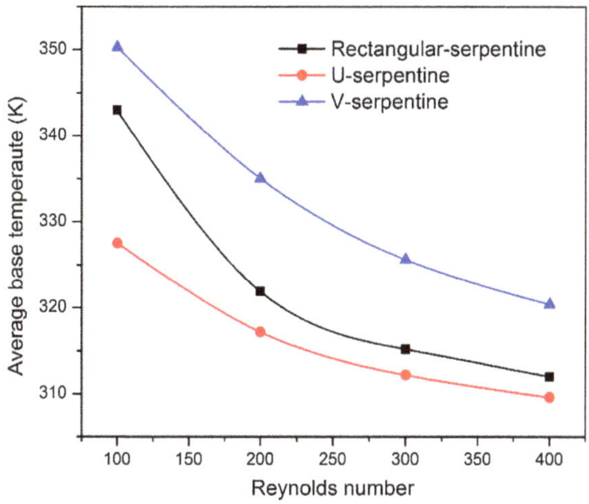

Fig. 73.6 Variation of the base temperature of the substrate with Reynolds number in different serpentine microchannels

higher Reynolds numbers, mixing of fluid was more to carry out more amount of heat from the substrate. However, at any particular Reynolds number, the base temperature of the V-serpentine microchannel was more. This attributes that the cooling effect on the V-serpentine was less compared to the other two serpentine channels. The reason could be that due to the presence of sharp corners in the V-serpentine, the possibility of fluid stagnation was more which results in higher base temperatures.

73.4.3 Velocity Profile

Figure 73.7 indicates the velocity profiles of all the three serpentine microchannel bends. From Fig. 73.7, it is clearly visible that the more amount of fluid was stagnated at the sharp corners of the V-serpentine followed by rectangular and least

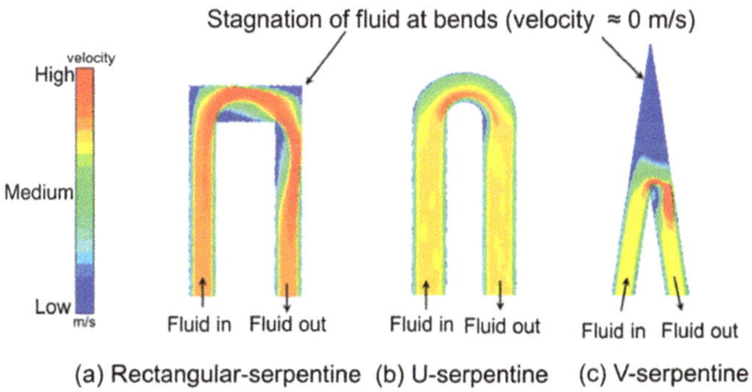

Fig. 73.7 Velocity profiles at bends of serpentine microchannels

in the U-serpentine microchannel. Therefore, the contribution of the fluid in these regions in heat transfer was very low. As the stagnant fluid region was high in V-serpentine, this leads to higher base temperatures. However, the stagnation region of fluid in U-serpentine was very less and hence lowers substrate base temperature (Fig. 73.6).

73.4.4 Heat Transfer Coefficient

The heat transfer coefficient of the fluid passing through the microchannels was calculated using Eq. (73.4).

$$h = \frac{q}{(T_{\mathrm{aw}} - T_{\mathrm{af}})} \tag{73.4}$$

where

T_{aw} Average bottom wall (base) temperature, K
T_{af} Average fluid temperature = $(T_{in} + T_{out})/2$, K.

In the calculation of heat transfer coefficient, both the substrate bottom wall (base) temperature and average fluid temperature were considered. Therefore, this parameter was used to compare the performance of the serpentine microchannels. Figure 73.8 shows the heat transfer coefficient of different serpentine microchannels at different Reynolds numbers. At higher Reynolds number, the velocity of the fluid is more which results in increased heat transfer coefficients as it maintains a

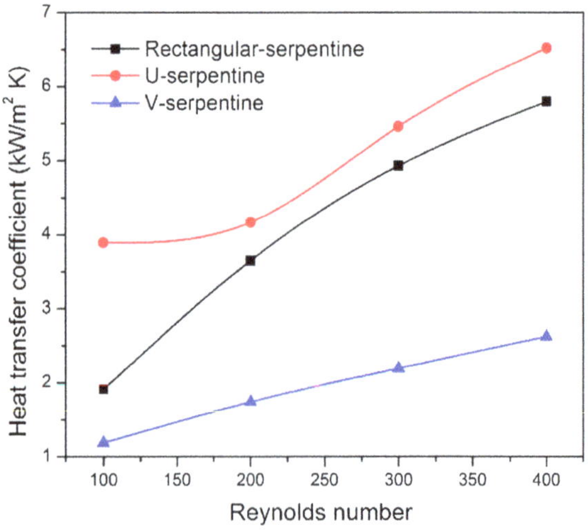

Fig. 73.8 Heat transfer coefficient variation with Reynolds number in different serpentine microchannels

high-temperature difference per unit time. At every *Re*, the value of heat transfer coefficient is more in U-serpentine microchannel, indicating better performing serpentine microchannel.

73.5 Conclusions

The characteristics of fluid flow and heat transfer such as pressure drop, temperature of fluid and substrate, velocity profile and heat transfer coefficient in three different serpentine microchannels at different Reynolds numbers were investigated numerically, and the results were discussed. The major conclusions drawn are as follows:

(a) Pressure drop was nearly similar in every microchannel. At *Re*—300, U-serpentine microchannel exhibits higher pressure drop compared to the remaining two microchannels.
(b) V-serpentine microchannel has the highest values of fluid temperature difference and average substrate temperatures.
(c) A larger amount of stagnation of fluid (velocity $\approx$ 0 m/s) was observed at the sharp bends of V-serpentine microchannels from the velocity profiles. However, smooth passage of the fluid was observed in U-serpentine microchannels.
(d) Among the three serpentine microchannels, the overall heat transfer coefficient in the U-serpentine microchannel has shown better performance compared to the other two.

References

1. Tuckerman, D.B., Pease, R.F.W.: High-performance heat sinking for VLSI. IEEE Electron Device Lett. **2**(5), 126–129 (1981)
2. Peiyi, W., Little, W.A.: Measurement of friction factors for the flow of gases in very fine channels used for microminiature Joule-Thomson refrigerators. Cryogenics **23**(5), 273–277 (1983)
3. Weisberg, A., Bau, H.H., Zemel, J.N.: Analysis of microchannels for integrated cooling. Int. J. Heat Mass Transf. **35**(10), 2465–2474 (1992)
4. Qu, W., Mala, G.M., Li, D.: Heat transfer for water flow in trapezoidal silicon microchannels. Int. J. Heat Mass Transf. **43**(21), 3925–3936 (2000)
5. Qu, W., Mudawar, I.: Analysis of three-dimensional heat transfer in micro-channel heat sinks. Int. J. Heat Mass Transf. **45**(19), 3973–3985 (2002)
6. Lee, P.S., Garimella, S.V., Liu, D.: Investigation of heat transfer in rectangular microchannels. Int. J. Heat Mass Transf. **48**(9), 1688–1704 (2005)
7. Gunnasegaran, P., Mohammed, H.A., Shuaib, N.H., Saidur, R.: The effect of geometrical parameters on heat transfer characteristics of microchannels heat sink with different shapes. Int. Commun. Heat Mass Transf. **37**(8), 1078–1086 (2010)

8. Sui, Y., Teo, C.J., Lee, P.S., Chew, Y.T., Shu, C.: Fluid flow and heat transfer in wavy microchannels. Int. J. Heat Mass Transf. **53**(13–14), 2760–2772 (2010)
9. Lin, L., Zhao, J., Lu, G., Wang, X.D., Yan, W.M.: Heat transfer enhancement in microchannel heat sink by wavy channel with changing wavelength/amplitude. Int. J. Therm. Sci. **118**, 423–434 (2017)
10. Karale, C.M., Bhagwat, S.S., Ranade, V.V.: Flow and heat transfer in serpentine channels. AIChE J. **59**(5), 1814–1827 (2013)
11. Zhang, J., Zhang, Y.F., Miao, M., Jin, Y.F., Bai, S.L., Chen, J.Q.: Simulation of fluid flow and heat transfer in microchannel cooling for LTCC electronic packages. In: 2009 International Conference on Electronic Packaging Technology and High Density Packaging, pp. 327–330 (2009)
12. Dehghandokht, M., Khan, M.G., Fartaj, A., Sanaye, S.: Numerical study of fluid flow and heat transfer in a multi-port serpentine meso-channel heat exchanger. Appl. Therm. Eng. **31**(10), 1588–1599 (2011)
13. Gong, L., Kota, K., Tao, W., Joshi, Y.: Parametric numerical study of flow and heat transfer in microchannels with wavy walls. J. Heat Transfer **133**(5), 051702 (2011)
14. Kandlikar, S.G.: History, advances, and challenges in liquid flow and flow boiling heat transfer in microchannels: a critical review. J. Heat Transfer **134**(3), 034001 (2012)
15. Sreehari, D., Sharma, A.K.: On form accuracy and surface roughness in micro-ultrasonic machining of silicon microchannels. Precis. Eng. **53**, 300–309 (2018)

Chapter 74
Comparative Analysis of GPU-Based Solver Libraries for a Sparse Linear System of Equations

Utpal Kiran, Subhajit Sanfui, Shashi Kant Ratnakar, Sachin Singh Gautam and Deepak Sharma

Abstract In this paper, a comparison of GPU-based linear solver libraries for the solution of sparse positive-definite matrices is presented. These large sparse matrices arise in a number of computational disciplines seeking a solution for partial differential equations. The solution of these matrices is often a time-consuming process that can be reduced by parallel computing. Since the development of GPU for general-purpose computing, a number of numerical solver libraries have evolved that can accelerate the solution procedure. The performance of three solver libraries has been evaluated in this paper for five different test matrices. These test matrices have been taken from different application domains with different sparsity patterns. Results demonstrate a higher speedup from the iterative solver over the direct solver on GPU and also over a multithreaded CPU implementation.

Keywords Finite element analysis · GPGPU · Sparse matrix · Linear solver

U. Kiran · S. Sanfui · S. K. Ratnakar · S. S. Gautam · D. Sharma (✉)
Department of Mechanical Engineering, Indian Institute of Technology Guwahati,
Guwahati, Assam 781039, India
e-mail: dsharma@iitg.ac.in

U. Kiran
e-mail: ukiran@iitg.ac.in

S. Sanfui
e-mail: s.sanfui@iitg.ac.in

S. K. Ratnakar
e-mail: r.shashi@iitg.ac.in

S. S. Gautam
e-mail: ssg@iitg.ac.in

R. G. Narayanan et al. (eds.), *Advances in Computational Methods in Manufacturing*, Lecture Notes on Multidisciplinary Industrial Engineering,
https://doi.org/10.1007/978-981-32-9072-3_74

74.1 Introduction

Most of the physical phenomena are governed by partial differential equations (PDEs). In various problems of engineering and science, it is not always possible to analytically solve the governing PDE to get the required solution. Most of the time simplifying approximations are made to reduce the PDE to a set of algebraic equations, using numerical methods [1].

Numerical simulations of manufacturing processes also use these numerical approximation techniques. Finite element analysis (FEA) is one of the most popular and widely used numerical methods. FEA facilitates construction, refinement, and optimization of a mechanical component before the start of manufacturing. This substantially reduces the cost and effort involved in production. The advantage of numerical simulation such as FEA is that it makes virtual testing possible in the early stages of production. Furthermore, the widespread availability of modern FEA software makes it very convenient to analyze the performance of the product and make the necessary improvements in the design.

The FEA formulation of the problem gives a system of algebraic equations. Solving this system of equations provides approximate values of the primary unknown, at a number of discrete points over the problem domain [2]. The resulting system of equations from FEA can be very conveniently represented in the form of matrices. Since FEA subdivides a large problem into smaller parts, due to local correlation in PDE discretization, the final matrix is sparse in nature [3].

There are two methods to solve the system of equations, the direct method and iterative method [1]. Direct methods are the procedures based on algebraic elimination. Gauss elimination and LU factorization are some of the most known algorithms in this category. Iterative methods start by assuming a trial solution or initial guess which is substituted in the system of equations, and then the difference between the obtained solution and the actual solution is determined, also known as an error or residual. Using the information from error value, the initial guess is improved iteration by iteration, so that it reaches closer to the actual solution. The conjugate gradient (CG) method is one of the most widely used iterative solvers [4].

In all practical FEA problems usually, the number of variables involved is very large, thus making the size of the sparse matrix huge. Solving a system of equations consisting of millions of variables becomes a very time-consuming exercise. In various studies, it has been observed that the solver may take anything between 70 and 90% of the total computation effort required (depending upon the problem size) [5]. Due to the facts discussed earlier, sparse matrix solver is an active research area.

74.1.1 Parallel Computing

One of the remedies to overcome a higher computational time of the solver is to use parallel computing, where a large problem is subdivided into small tasks and executed by multiple computing units (processors) simultaneously, with or without communicating with each other [6].

There are different types of parallel computing architecture. The choice of architecture is subjected to many factors, such as the requirement of the problem, availability of hardware, and cost. One of the most popular architectures is the distributed memory architecture, where each processor has its own memory system at its disposal, which it can use directly. Another is the shared memory architecture, where each processor has access to a centralized memory system from which they have to share. Usage of memory by each processor depends on its current requirement. These two architectures are generally referred to as traditional architectures.

A distributed memory architecture setup generally consists of a large number of CPU cores. This results in high setup and maintenance cost, making it somewhat impractical for common use. Shared memory architecture does not need any separate communication channel. All the CPU cores in shared memory architecture access a common global memory. Here, the changes made in memory by one processor are visible to all other processors. This may lead to inaccurate memory operation when more than one processor is trying to access the same memory location. This issue is commonly known as race condition [7].

In the literature, the parallelization of solver has been performed using traditional computing such as the shared memory architecture [5, 8, 9] and also distributed memory architecture [10–13]. Considering the inconvenience posed by the traditional architectures, in recent years, the novel architecture of graphics processing unit (GPU) has gained popularity among the researchers. After the launch of general-purpose GPU (GPGPU) by NVIDIA in 2007, the GPUs can be used for general-purpose computing. GPU is advantageous in the sense that they are relatively cheaper, maintenance-free, and portable. They can be used for parallel computing using the programming interface like Compute Unified Device Architecture (CUDA) [14] or OpenCL [15].

GPU has a highly parallel, multithreaded, many-core processor architecture, which is well suited for problems that can be expressed as data-parallel computations with high arithmetic intensity. Hence, GPU can be a viable choice for a sparse matrix solver. It has been used for the parallelization of solver in a few research works [16–19]. Moreover, some libraries compatible with GPU are also developed for a sparse matrix solver. However, these libraries have not been compared on a common platform on the problems of different domains. Therefore, the main contributions of this paper are (i) to compare the popular linear sparse solver libraries that are developed for GPU, with respect to the multithreaded CPU implementation, and (ii) to analyze the performance of the popular libraries by solving some sparse matrices evolved from different applications that have different matrix sizes, sparsity pattern, and nonzero entries.

The paper is organized into four sections. Section 74.2 presents the existing sparse matrix solver libraries developed for GPUs. In Sect. 74.3, the matrices evolved from different problems are discussed and the libraries are used to find the solution for a system of the linear system. The computation time is reported, and the speedup is shown. Section 74.4 concludes the paper.

74.2 Sparse Matrix Solver Libraries

Given the importance of the large sparse linear equation system solver, there are several libraries available in the literature, which takes the sparse matrix as an input and provides the solution as an output. The majority of works in the literature focus on CPU implementation, while a few of them have been developed to work on GPU hardware so that the solution can be obtained in a lesser amount of time. Many of these solver libraries can work with a user-defined matrix-vector product, which is the basic operation of any iterative solver.

Solver libraries provide highly optimized functions that perform many times faster than CPU-only solvers. They can be very useful for compute-intensive applications such as structural optimization, computational mechanics, heat flow problems, earthquake simulations to name a few. In the following subsections, a few of GPU sparse linear equation system solver libraries are discussed.

74.2.1 CUSP

CUSP is an open-source library written in C++ for sparse linear algebra on GPUs with CUDA architecture. It provides a flexible programming interface for operating on sparse matrices and solving sparse linear systems. CUSP uses iterative methods and also includes multiple pre-conditioners. This library supports many popular sparse storage formats such as COO, CSR, DIA, ELL, and HYB. The library is developed by NVIDIA [20].

74.2.2 cuSolver

Also developed by NVIDIA, the cuSolver library provides a collection of direct solvers for both dense and sparse matrices. This library comes in-built with the CUDA toolkit. For dense matrices, it has common matrix factorization and triangular solve routines, and for sparse matrices, a least square solver is used. This library also comes with an eigenvalue solver. It can be called from a C/C++ routine [21].

74.2.3 SuiteSparse

SuiteSparse is an open-source library, which has a Cholesky factorization accelerated by CUDA. It has a collection of a number of very popular sparse matrix packages such as UMFPACK, CHOLMOD, and SPQR [22].

74.2.4 PETSc

Portable, Extensible Toolkit for Scientific Computation (PETSc) is an open-source suite of data structures and routines for the parallel scalable solution of applications which are modeled by PDEs [23]. It has the capability of solving both linear and nonlinear systems of equations. It is available in C/C++ and Python. It uses MPI for message passing and supports GPU through CUDA or OpenCL. PETSc is the world's most widely used parallel numerical software library for partial differential equations and sparse matrix computations.

74.2.5 MAGMA

MAGMA is a collection of GPU-accelerated dense linear algebra libraries. It has been specifically developed to work with heterogeneous computing environment such as multicore CPUs and multi-GPUs [24].

74.3 Results and Discussion

In the present work, three libraries have been selected for solving the matrices using the direct and iterative methods on the GPU. The speedups of the GPU-based solvers are obtained with respect to a multithreaded implementation of Cholesky decomposition (CHOLMOD) provided by SuiteSparse library. The details of the three test libraries are provided in Table 74.1. The test matrices have been chosen

Table 74.1 Methods to solve a linear system of equations

	Library	Solution method	Architecture
Method 1	CUSP	Conjugate gradient (CG)	GPU
Method 2	cuSolver	Cholesky decomposition	GPU
Method 3	SuiteSparse	Cholesky decomposition	GPU
Ref. code	SuiteSparse	Cholesky decomposition	CPU multithreaded

from different applications of FEM and solved using libraries mentioned in Table 74.1. Five matrices with a different range of order and number of nonzeros have been selected from the Suitesparse matrix collection [3] for benchmarking. The details of these matrices are presented in Table 74.2. All the five test matrices are sparse and positive definite. The sparsity pattern of all test matrices is shown in Fig. 74.1. As can be seen from the figure, the sparsity pattern as well as the density of nonzeros can greatly vary for different applications. Since the focus of this paper is to analyze the performance of different solver libraries on matrices having different sparsity patterns, the problem formulation or mathematical modeling for the applications shown in Fig. 74.1 is not presented. However, the details of these applications can be found in [3], which is not included due to brevity. In the present work, the performance of solver libraries on matrices with significantly varying sparsity pattern is evaluated.

The hardware for the numerical experiment consists of Intel Xeon E5-2650 processor and NVIDIA Tesla K40 GPU. The CPU processor has 128 GB of memory and 12 physical cores clocked at 2.2 GHz. The GPU has 12 GB of memory, and the cores are clocked at 745 MHz.

The execution time for the matrices for different libraries is listed in Table 74.3. Execution time refers to elapsed GPU time for all three methods and CPU time for the reference code. The last column of Table 74.3 lists the time required by the reference multithreaded version. The speedups of the three solvers are computed with reference to this version and plotted in Fig. 74.2, using a semi-logarithmic scale. From the figure, it can be clearly seen that among all the libraries, the CG method in CUSP library takes the least amount of time for solving all the matrices. It can also be observed from the figure that the speedup for 'parabolic_fem' and 'thermal2' is significantly more pronounced than the other matrices. This can be

Table 74.2 Details of test matrices

Matrix	Application	Rows/columns	Nonzeros
thermal1	Steady-state thermal problem	82,654	574,458
thermal2	Steady-state thermal problem	1,228,045	8,580,313
nd6k	ND problem set	18,000	6,897,316
parabolic_fem	Diffusion–convection reaction	525,825	3,674,625
consph	FEM concentric spheres	83,334	6,010,480

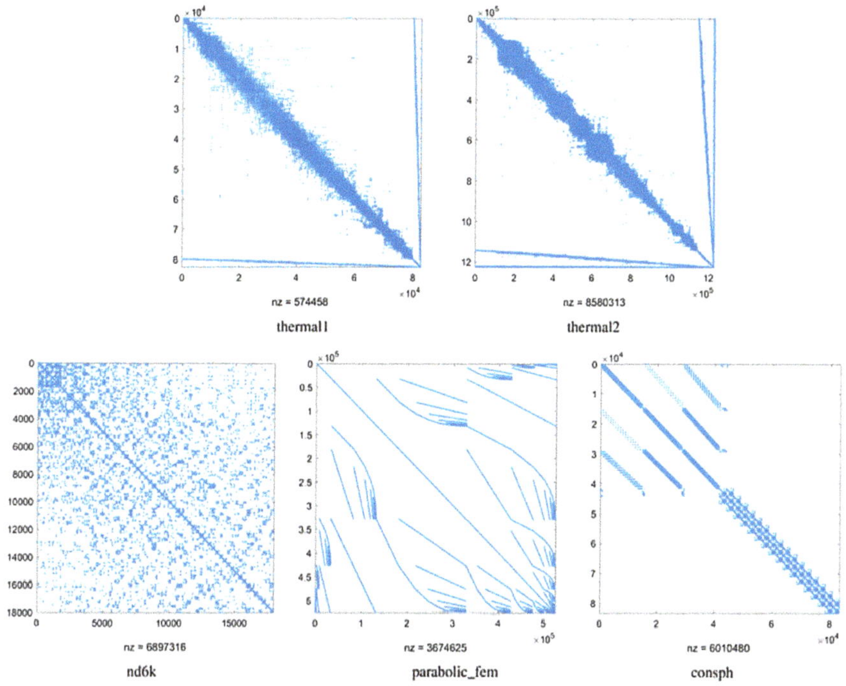

Fig. 74.1 Sparsity pattern of test matrices is represented by dots

Table 74.3 Execution time (in seconds)

Matrix	Method 1	Method 2	Method 3	Ref. code
thermal1	1.92	46.32	17.94	14.22
thermal2	18.11	–	195.92	262.74
nd6k	17.58	69.09	23.85	19.95
parabolic_fem	3.23	–	114.25	158.46
consph	27.84	–	153.85	177.08

attributed to the higher dimension of the said matrices. In the case of SuiteSparse library, the GPU implementation of CHOLMOD is seen to outperform the reference multithreaded CPU implementation for all matrices except 'thermal1' and 'nd6k'. This is due to the small size of the two matrices. It can be further concluded from the figures that the execution times of all the libraries are more dependent on the dimension of the matrix than the number of nonzeros in the matrices. For example, 'nd6k', despite having a high number of nonzeros as compared to the other matrices, suffers from a poor performance in all the methods as seen in Fig. 74.2.

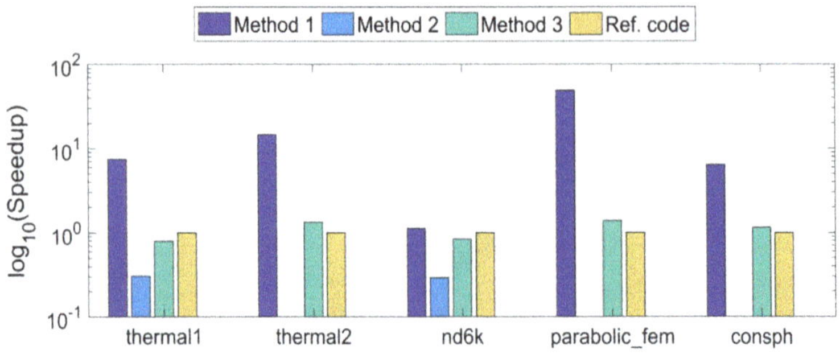

Fig. 74.2 Speedup of GPU solver libraries with respect to multithreaded ref. code

74.4 Conclusion

This paper presents a study of a solution of a sparse system of linear equations using a GPU. The performance of solver libraries, which use a GPU to accelerate the computation, is evaluated over a range of matrices. Five test matrices of varying sparsity pattern as well as a different number of nonzeros and dimension are taken. The matrices are solved by three popular libraries over GPU, and result is compared with a multithreaded CPU implementation. The GPU-based CUSP library achieves the least solution time for all the matrices. The other libraries that use GPU to accelerate computation also achieve lesser time than their CPU variants. It was observed that the effect of GPU acceleration becomes more apparent when matrix size increases. The highest speedup is achieved for the matrix of the largest dimension. The number of nonzeros is also found to be affecting the solution time.

References

1. Hoffman, J.D., Frankel, S.: Numerical methods for engineers and scientists. CRC Press (2001)
2. Zienkiewicz, O.C., Taylor, R.L.: The finite element method, vol. 36. McGraw-hill, London (1977)
3. Davis, T.A., Hu, Y.: The university of Florida sparse matrix collection. ACM Trans. Math. Softw. (TOMS) **38**(1), 1 (2011)
4. Shewchuk, J.R.: An Introduction to the Conjugate Gradient Method Without the Agonizing Pain. Technical Report. Carnegie Mellon University, Pittsburgh, PA, USA (1994)
5. Mahdavi, A., Balaji, R., Frecker, M., Mockensturm, E.M.: Topology optimization of 2D continua for minimum compliance using parallel computing. Struct. Multi. Optim. **32**(2), 121–132 (2006)
6. Pacheco, P.: Parallel Programming with MPI. Morgan Kaufmann (1997)
7. Farhat, C., Crivelli, L.: A general approach to nonlinear FE computations on shared-memory multiprocessors. Comput. Methods Appl. Mech. Eng. **72**(2), 153–171 (1989)
8. Farhat, C.: A simple and efficient automatic FEM domain decomposer. Comput. Struct. **28**(5), 579–602 (1988)

9. Babuška, I., Elman, H.C., Markley, K.: Parallel implementation of the hp-version of the finite element method on a shared-memory architecture. SIAM J. Sci. Stat. Comput. **13**(6), 1433–1459 (1992)
10. Borrvall, T., Petersson, J.: Large-scale topology optimization in 3D using parallel computing. Comput. Methods Appl. Mech. Eng. **190**(46–47), 6201–6229 (2001)
11. Masters, I., Usmani, A.S., Cross, J.T., Lewis, R.W.: Finite element analysis of solidification using object-oriented and parallel techniques. Int. J. Numer. Meth. Eng. **40**(15), 2891–2909 (1997)
12. Coelho, P.G., Cardoso, J.B., Fernandes, P.R., Rodrigues, H.C.: Parallel computing techniques applied to the simultaneous design of structure and material. Adv. Eng. Softw. **42**(5), 219–227 (2011)
13. Tezduyar, T., Aliabadi, S., Behr, M., Johnson, A., Mittal, S.: Parallel finite-element computation of 3D flows. Computer **26**(10), 27–36 (1993)
14. Sanders, J., Kandrot, E.: CUDA by Example: An Introduction to General-Purpose GPU Programming, 1st edn. Addison-Wesley Professional (2010)
15. Munshi, A., Gaster, B., Mattson, T.G., Ginsburg, D.: OpenCL Programming Guide. Pearson Education (2011)
16. Schmidt, S., Schulz, V.: A 2589-line topology optimization code written for the graphics card. Comput. Vis. Sci. **14**(6), 249–256 (2011)
17. Challis, V.J., Roberts, A.P., Grotowski, J.F.: High resolution topology optimization using graphics processing units (GPUs). Struct. Multi. Optim. **49**(2), 315–325 (2014)
18. Taylor, Z.A., Cheng, M., Ourselin, S.: High-speed nonlinear finite element analysis for surgical simulation using graphics processing units. IEEE Trans. Med. Imaging **27**(5), 650–663 (2008)
19. Goddeke, D., Buijssen, S.H., Wobker, H., Turek, S.: GPU acceleration of an unmodified parallel finite element Navier-Stokes solver. In: International Conference on High Performance Computing & Simulation, 2009. HPCS'09, pp. 12–21. IEEE (2009)
20. Bell, N., Garland, M.: Cusp: Generic parallel algorithms for sparse matrix and graph computations version 0.3. 0. Cusp-Library (2012)
21. NVIDIA Corporation, cuSOLVER Library User's Guide, https://docs.nvidia.com/cuda/pdf/CUSOLVER_Library.pdf
22. Davis, T.: Direct Methods for Sparse Linear Systems. SIAM Series on the Fundamentals of Algorithms, SIAM, Philadelphia (2006)
23. Balay, S., Abhyankar, S., Adams, M., Brown, J., Brune, P., Buschelman, K., Dalcin, L.D., Eijkhout, V., Gropp, W., Kaushik, D., Knepley, M.: Petsc user's manual revision 3.8 (No. ANL-95/11 Rev 3.8). Argonne National Lab. (ANL), Argonne, IL, United States (2017)
24. Tomov, S., Dongarra, J., Volkov, V., Demmel, J.: Magma Library. University of Tennessee and University of California, Knoxville, TN, and Berkeley, CA (2009)

Chapter 75
Hybridizing Evolutionary Multi-objective Algorithm Using Random Mutations and Local Searches

Riddhiman Saikia and Deepak Sharma

Abstract Multi-Objective Evolutionary Algorithms (MOEAs) have been successful in solving mathematical and real-world multi-objective optimization problems by evolving a set of optimal solutions, which are known as Pareto-optimal solutions. However, there are certain limitations with those algorithms such as slow convergence, lack of effective terminating condition to name a few. To address such challenges, hybrid MOEAs are being designed and studied where the global exploration power of MOEAs are combined with local exploitation modules of various numerical optimization techniques. However, hybridization itself brings new challenges in its implementation. In this work, a hybrid MOEA is presented in which random mutations are performed on the initial population to start with a better and diverse set of solutions. Moreover, a local search module is coupled to execute periodically on a least crowded non-dominated solution at a certain interval of generations. The proposed algorithm is tested on a set of benchmark multi-objective optimization problems and compared with the NSGA-II. The convergence plots demonstrate the superiority of the proposed algorithm over NSGA-II.

Keywords Hybrid multi-objective algorithm · Memetic algorithm · Mutation · Local search · NSGA-II

R. Saikia · D. Sharma (✉)
Department of Mechanical Engineering, Indian Institute of Technology Guwahati, Guwahati, Assam 781039, India
e-mail: dsharma@iitg.ac.in

R. Saikia
e-mail: saiki174103012@iitg.ac.in

R. G. Narayanan et al. (eds.), *Advances in Computational Methods in Manufacturing*, Lecture Notes on Multidisciplinary Industrial Engineering,
https://doi.org/10.1007/978-981-32-9072-3_75

75.1 Introduction

In Operations Research, multi-objective optimization is a sub-field of multi-criterion decision making in which optimal solutions are generated by simultaneous optimization of two or more conflicting objectives. Mathematically, the formulation of a multi-objective optimization problem is

$$\begin{gathered} \text{Min}\, F(x) = \{f_1(x), f_2(x), \ldots, f_M(x)\} \\ \text{subject to}\, x \in S \subset R^n \end{gathered} \tag{75.1}$$

Here, M is the number of conflicting objectives. Any optimal solution x^* is considered as Pareto-optimal, if there does not exist another $x \subset S$ that dominates x*. Meaning, a solution x is said to dominate x^* if it is not worse than x^* in any objective, and better in at least one [1]. Symbolically, it can be represented as $x \prec x^*$ (x dominates x^*), if $f_i(x) \leq f_i(x^*)$ for all $i = 1, \ldots, M$ and $f_j(x) < f_j(x^*)$ for at least one objective $j \in (1, \ldots, M)$ for the problem given in Eq. (75.1).

No solution to any multi-objective optimization problem can be considered as best since trade-offs between conflicting objectives are inevitable. Hence it is necessary to generate a set of solutions. The optimization problem in Eq. (75.1) can be solved using any numerical optimization technique by converting it into a single-objective composite function or converting all but one objective into a set of constraints. However, such approaches fail many a time to generate a well-distributed Pareto-optimal front. Evolutionary algorithms (EAs), however, have been an ideal candidate for solving such optimization problems because they work on a population of solutions. With appropriate operators for convergence and diversity, these EAs can well evolve a set of Pareto-optimal solutions.

In the literature, many multi-objective evolutionary algorithms (MOEAs) have been suggested, such as NSGA-II [2], SPEA-II [3], DoD [4], MOPSO [5], ASREA [6] to name a few. Despite excellent performances, MOEAs are being criticized for their slow convergence toward the Pareto-optimal front, lack of proof of convergence, lack of proper termination condition etc. To address some of these issues, many authors added various modifications to the MOEAs, commonly being modules for local search (LS), which is exploitation of an already-found solution in an attempt to improve its fitness. It is performed on one or many solutions at a time which can lead to a faster convergence if applied correctly [7].

While LS has many advantages, it is not a straightforward task to synchronize it with MOEAs. The first challenge with LS is that it is computationally exhaustive, since extra function evaluations consume extra resources. Too much of LS demands more computational resources and too less can lead to a sub-optimal solution [8]. Therefore, the challenges are when to execute LS and how much to execute? Another challenge is about a selection of solutions on which LS is applied. Since MOEAs are population-based optimization algorithms, the basis of choosing solution(s) for LS poses additional challenges [9].

There have been numerous attempts at answering such challenges. A remedy to overcome the difficulty of choosing weights for weighted scalar or composite function approach was proposed in [10], where the authors used three populations for generation update: the current population, the one after genetic operations and also after LS. If a solution after LS is degraded, then a better choice from other two populations is likely to be chosen over it for the next generation. Secondly, the authors used an acceptance rule for LS. Ishibuchi et al. [11] addressed the issue on how to choose solutions for LS. The authors used tournament selection based on fitness for choosing the best solutions of a population and performed LS on them. The authors noted improved performances and concluded that choosing better individuals for LS had a significantly positive impact. On the other hand, in some studies [12, 13], all non-dominated solutions have been considered for LS. Sindhya et al. [14] made an attempt to obtain good a balance between local and global search. The authors developed a concurrent hybrid MOEA where a probability based LS model was proposed. It has been argued that initial local searches can be wasted as GA operators would change the schema of solutions. But in later stages, it is desired that the solutions converge quickly towards the Pareto-optimal front. In this paper, an attempt is made in hybridizing NSGA-II with random mutations on the initial population and also coupled with local search during the generations. The initial mutations can help to further diversify the initial population and the local searches can help for faster convergence.

The rest of the paper is organized into four sections. Section 75.2 discusses the framework of hybrid MOEA. In Sect. 75.3, the performance of the proposed algorithm is presented after solving some benchmark multi-objective optimization problems. Section 75.4 concludes this paper with plans for future research.

75.2 Framework for Hybrid MOEA (HMOEA)

The framework for hybrid MOEA is shown in Fig. 75.1. The non-dominated sorting genetic algorithm, which is known as NSGA-II, has been chosen for hybridizing with the proposed ideas of random mutations and local searches. The hybrid NSGA-II is referred to as HMOEA in this paper. As can be seen from Fig. 75.1, an initial random population of size N is generated at the beginning. Since the initial population can affect the convergence of MOEAs [15], HMOEA is equipped with a module called 'initial mutation' which uses the polynomial mutation operator on every solution of the initial population with a high mutation probability of $(p_m = 0.5)$. The purpose of 'initial mutation' module is to bring diversity among the solutions of the initial population. After mutation, the total number of solutions is greater than N, which needs to be reduced to N. The environment selection of NSGA-II is chosen for selecting the best N solutions in which all solutions from the initial population and after random mutation are ranked using the non-dominated sorting operator which assigns a rank to every solution and store them in different fronts. Thereafter, the crowding distance operator is applied front-wise to choose less crowded solutions. Based on the rank and crowding distance, the best *N* solutions are selected. This is referred to as 'Evaluate population (P)' module in the HMOEA as shown in Fig. 75.1 after initialization.

At this stage, the termination condition is imposed on HMOEA in the decision box as shown in Fig. 75.1, which gets terminated after a fixed number of generations. It is noted that a sufficiently large number of generations are provided here so that the convergence of HMOEA can be observed with respect to the original NSGA-II. If the termination condition is not met, the population [which is now referred to as parent population (P)] is considered for 'selection' module. In the selection module, the binary tournament selection is used to choose better solutions for a mating pool. On this pool, the SBX crossover operator is applied to create new solutions in 'Crossover' module followed by polynomial mutation, which is referred to as 'Mutation' module in Fig. 75.1. The population consisting of new solutions is referred to as offspring population (Q). The parent population and offspring population are mixed together and the non-dominated sorting is applied to rank all solutions and store them into different fronts. The crowding distance is applied front-wise. The environment selection of HMOEA, which is same as NSGA-II, selects the best *N* solutions.

The random mutation brings diversity into the population while the environmental selection selects better fitness solutions in order to maintain a balance between diversity and selection pressure. The module for environment selection is shown as 'Environment Selection (P + Q)' in Fig. 75.1.

At this stage, a module of 'Local Search' is used which chooses a least crowded solution from the set of the non-dominated solution, that is, the front-1 solution.

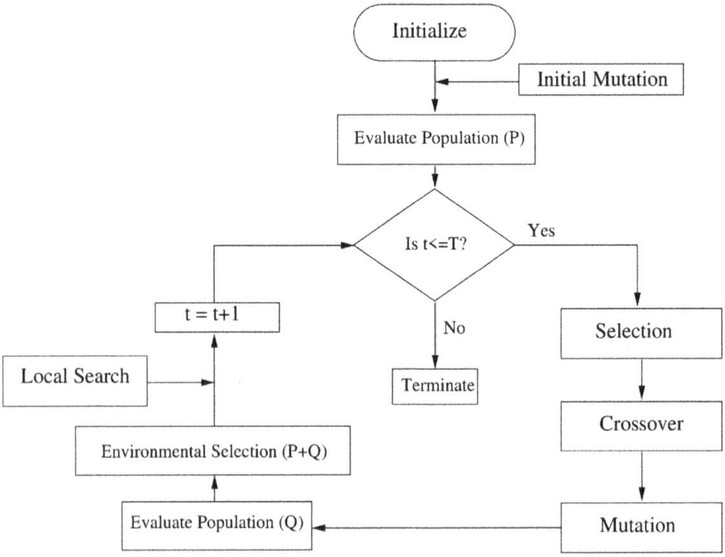

Fig. 75.1 Flowchart of hybrid multi-objective evolutionary algorithm

Thereafter, the multi-objective problem given in Eq. (75.1) is converted into a single-objective optimization problem using $\epsilon-$ constraint method [1]. The modified formulation for LS is given as

$$\begin{aligned} &\text{Minimize} f_i(x) \\ &\text{subject to } f_j \leq \epsilon_j, \forall j = (1, \cdots, M) \text{ and } j \neq i, \\ &x \in S \subset R^n. \end{aligned} \tag{75.2}$$

The single-objective problem of Eq. (75.2) is solved using the method of multipliers [16] in which the Cauchy's steepest descent method is used as the local search technique [16]. The solution after local search replaces its parent solution in the population. The challenge for the frequency of local searches is addressed by executing it once every $(T/5)$ generations, in which T refers to the fixed number of generations for which HMOEA is executed.

75.3 Results and Discussions

In this section, HMOEA is tested on five bi-objective benchmark problems, which are known as ZDT problems [17]. Among the ZDT problems, ZDT1, ZDT2, ZDT3, ZDT4, and ZDT6 problems are chosen. ZDT1, ZDT2, and ZDT3 have 30 number of variables and rest of ZDT problems have 10 number of variables. All variables are having the range from zero to one. ZDT5 problem is not considered because it has binary variables. The details of ZDT problems can be found in [17] which are not included in this paper due to brevity. HMOEA is also compared with the results of NSGA-II. Both MOEAs are run for the same population size of $N = 100$, number of generations $(T = 200)$, crossover probability of $p_c = 0.9$ and mutation probability of ($p_m = 0.1/$ number of variables). For local search, a maximum number of sequences for the method of multipliers is 40 and a maximum number of iterations for the Cauchy's steepest descent method is 20. The penalty parameter is kept 0.1 and the error for terminating the method of multiplier is 0.001.

75.3.1 Effect of Random Mutations

As part of HMOEA, random mutations were performed on the initial population for all test problems. Figure 75.2 shows the effect of initial mutation on the initial population. As can be seen, the population after random mutations is diverse and improved as compared to the randomly generated population after initialization. Due to the characteristics of ZDT6, many solutions are evolved at the extreme of 'f1' objective. As shown by [15], a better initial population helps MOEA for improved and faster convergence.

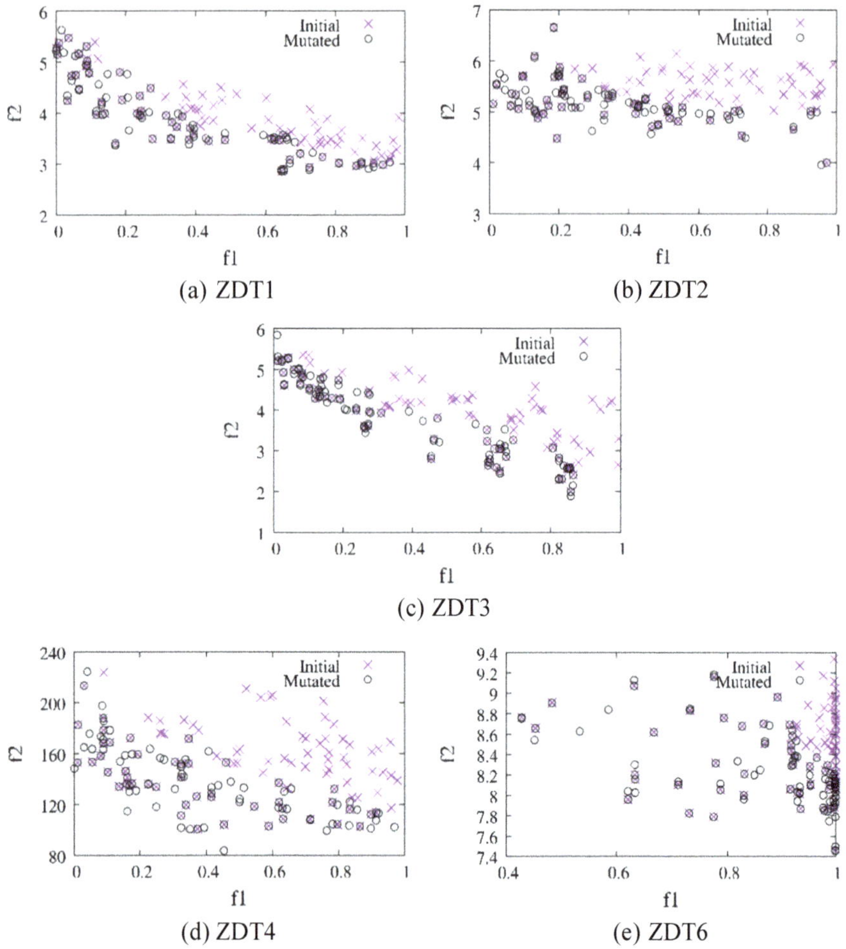

Fig. 75.2 Effect of random mutation on the initial population of HMOEA

75.3.2 Pareto-Optimal Solutions

The non-dominated solutions obtained by HMOEA for the chosen set of problems are shown in Fig. 75.3. It can be seen that HMOEA is able to generate non-dominated solutions that converge to the known Pareto-optimal fronts, for all the problems. It can be seen that ZDT1 has a non-convex Pareto-optimal front, ZDT2 has a convex Pareto-optimal front, ZDT3 has the discontinuous Pareto-optimal front, ZDT4 has a nonconvex Pareto-optimal front, and ZDT6 has nonlinearity along the convex Pareto-optimal front.

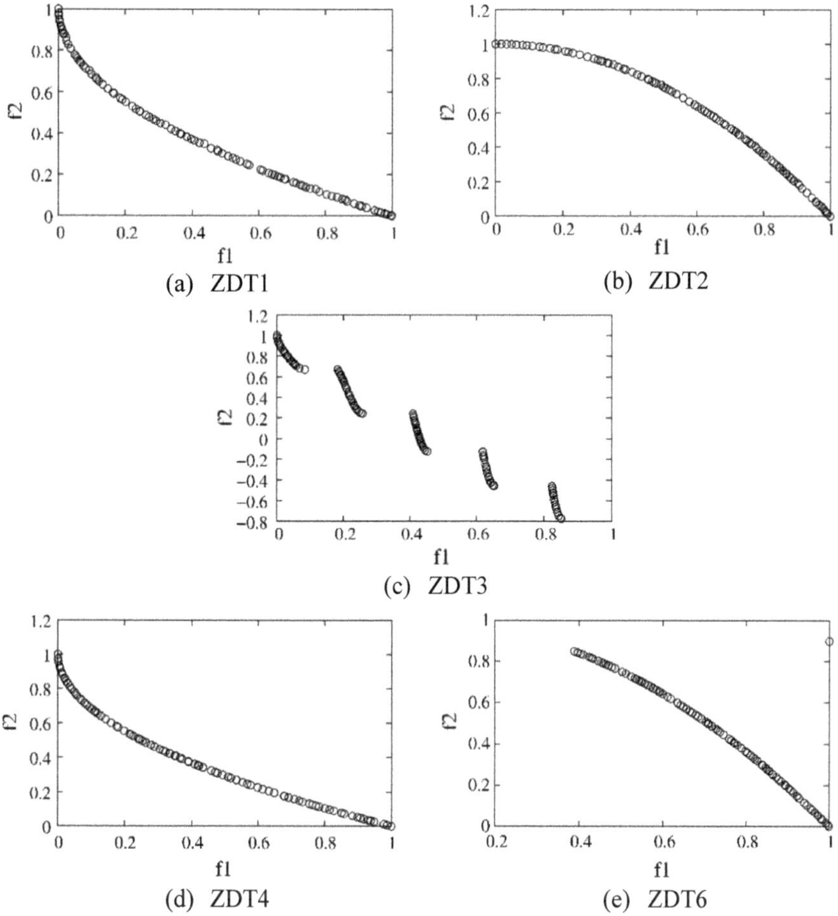

Fig. 75.3 The non-dominated solutions obtained from HMOEA

75.3.3 Convergence Plots

The convergence plots for HMOEA and NSGA-II for all five problems are shown in Fig. 75.4. These plots are made using the hypervolume (HV) indicator which signifies the volume occupied by the approximate set of non-dominated solutions generated by MOEAs and bounded by the reference point such that the reference point is weakly dominated by all the non-dominated solutions in the approximate set [1]. A semi-$\log_{10}$ scale is used for HV values over generations of MOEAs.

It can be seen from the figures that HMOEA has converged and stabilized faster than NSGA-II. The exact generation of convergence for both the algorithms is presented in Table 75.1, in which t_H and t_N denote generations required by

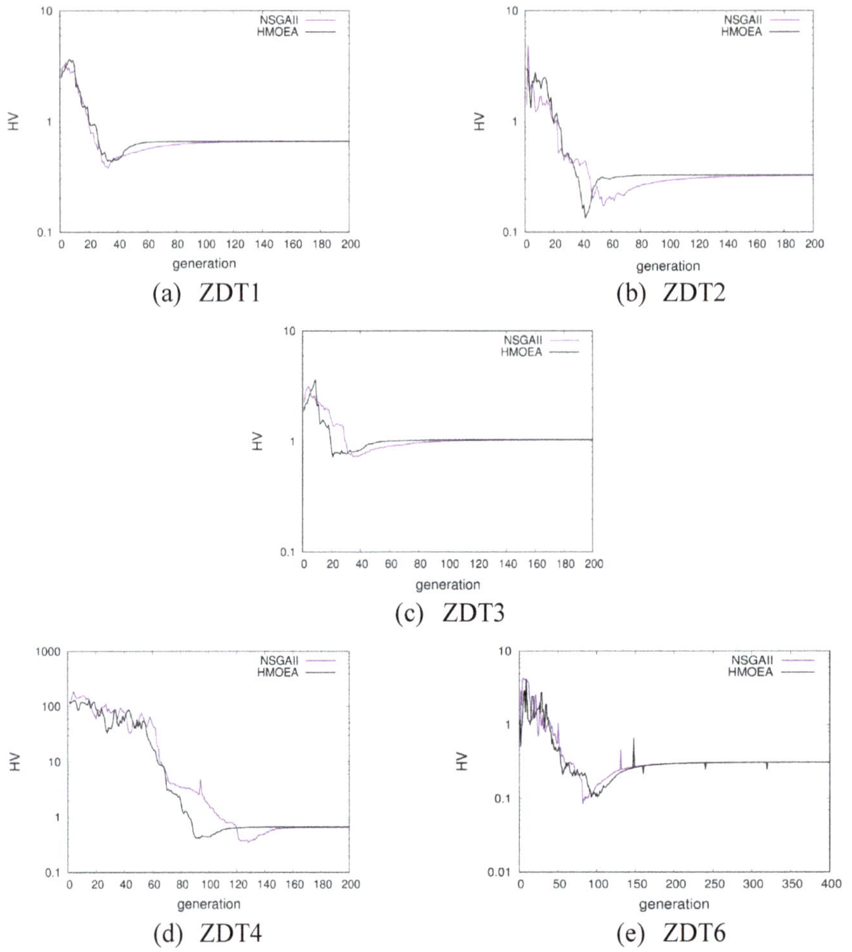

Fig. 75.4 Convergence plots of HMOEA over NSGA-II for all problems. HV values are plotted with $\log_{10}$ scale

Table 75.1 Number of generations required by HMOEA and NSGAII for solving ZDT problems

Problems	ZDT1	ZDT2	ZDT3	ZDT4	ZDT6
t_H	65	80	70	120	200
t_N	120	140	95	170	200
t_S	55	60	25	50	0

HMOEA and NSGA-II respectively, and t_S denotes the total generations saved by HMOEA over NSGA-II. Except for ZDT6, HMOEA is found to be superior over NSGA-II in saving a handful of generations.

75.4 Conclusions

This paper was an attempt to hybridize an existing MOEA with random mutations and local searches. From the results, it can be concluded that the random mutations with high mutation probability improved the diversity of the initial population. The convergence plots showed that local searches helped HMOEA to converge faster and saved a handful number of generations (or computations). Since the local search was performed on only one solution from the population and at the rate of $(T/5)$ generations, a well converged and nicely distributed set of non-dominated solutions was evolved that converged to known Pareto-optimal fronts. As a note on future work, HMOEA can be further improved by including an adaptive termination condition based on some performance indicators like HV. HMOEA also needs to be tested on other benchmark and engineering optimization problems and can be compared with other benchmark hybrid MOEAs.

References

1. Deb, K.: Multi-Objective Optimization Using Evolutionary Algorithms, 1st edn. Wiley, New York (2001)
2. Deb, K., Pratap, A., Agarwal, S., Meyarivan, T.: A fast and elitist multi-objective genetic algorithm: NSGA-II. IEEE Trans. Evol. Comput. **6**(2), 182–197 (2002)
3. Zitzler, E., Laumanns, M., Thiele, L.: SPEA2: Improving the strength pareto evolutionary algorithm for multiobjective optimization. In: Giannakoglou, K., et al. (eds) Evolutionary Methods for Design, Optimization and Control with Application to Industrial Problems (EUROGEN 2001), pp. 95–100. International Center for Numerical Methods in Engineering (CIMNE) (2002)
4. Sharma, D., Basha, S.Z., Kumar, S.A.: Diversity over dominance approach for many-objective optimization on reference-points-based framework. In: Deb et al. (eds) 10th International Conference Proceedings on Evolutionary Multi-Criterion Optimization. Michigan State University, East Lansing, USA, (2019) (to appear)
5. Agarwal, D., Sharma D.: Experimental study on bound handling techniques for multi-objective particle swarm optimization. In: Snášel, V., Abraham, A., Krömer, P., Pant, M., Muda, A. (eds) Innovations in Bio-Inspired Computing and Applications, vol 424, pp. 555–564. Advances in Intelligent Systems and Computing, Cham (2016)
6. Sharma, D., Collet, P.: An archived-based stochastic ranking evolutionary algorithm (asrea) for multi-objective optimization. In: Proceedings of The 12th Annual Conference on Genetic and Evolutionary Computation (GECCO'10), pp. 479–486. ACM, New York (2010)
7. Sindhya, K., Deb, K., Miettinen, K.: A local search based evolutionary multi-objective optimization approach for fast and accurate convergence. In: International Conference on Parallel Problem Solving from Nature, pp. 815–824 (2008)
8. Adrian, T.A., Hopgood, E.-M., Nolle, L., Battersby, A.: Hybrid genetic algorithms: a review. Eng. Lett. **13**(2), 124137 (2006)
9. Sindhya, K., Miettinen, K., Deb, K.: A hybrid framework for evolutionary multi-objective optimization. IEEE Trans. Evol. Comput. **17**(4), 495–511 (2013)
10. Ichibuchi, H., Narukawa, K.: Some issues on the implementation of local search in evolutionary multi-objective optimization. In: Genetic and Evolutionary Computation Conference, pp. 1246–1258 (2004)

11. Ishibuchi, H., Hitotsuyanagi, Y., Wakamatsu, Y., Nojima, Y.: How to choose solutions for local search in multi-objective combinatorial memetic algorithms. In: International Conference on Parallel Problem Solving from Nature, pp. 516–525 (2010)
12. Sharma, D., Kumar, A., Deb, K., Sindhya, K.: Hybridization of SBX based NSGA-II and sequential quadratic programming for solving multi-objective optimization problems. In: The proceedings of IEEE Congress on Evolutionary Computation (CEC), pp. 3003–3010, Singapore (2007)
13. Kumar, A., Sharma, D., Deb, K.: A hybrid multi-objective optimization procedure using PCX based NSGA-II and sequential quadratic programming. In: The Proceedings of IEEE Congress on Evolutionary Computation (CEC), pp. 3011–3018, Singapore (2007)
14. Sindhya, K., Deb, K., Miettinen, K.: Improving convergence of evolutionary multi-objective optimization with local search: a concurrent hybrid algorithm. Nat. Comput. **10**(4), 1407–1430 (2011)
15. Sharma, D., Deb, K., Kishore, N.N.: Domain-Specific initial population strategy for compliant mechanisms using customized genetic algorithm. Struct. Multi. Optim. **43**(4), 541–554 (2011)
16. Deb, K.: Optimization for Engineering Design: Algorithms and Examples, 2nd edn. PHI Learning Pvt. Ltd., New Delhi, India (2012)
17. Zitzler, E., Deb, K., Thiele, L.: Comparison of multiobjective evolutionary algorithms: empirical results. Evol. Comput. J. **8**(2), 125–148 (2000)

Chapter 76
Local Search Heuristics-Based Genetic Algorithm for Capacitated Vehicle Routing Problem

Deepak Sharma, Sukhomay Pal, Aayush Sahay, Puneet Kumar, Gaurav Agarwal and Katravath Vignesh

Abstract Vehicle routing problem (VRP) is an NP-hard problem, which looks for the optimal route for vehicles to deliver goods to the customers with minimum transportation cost. Since transportation and logistics incur a good percentage of cost on to the basic price of the products delivered, VRP has been an interest in many organizations and also for many researchers. Due to the limitations of exact algorithms, the heuristics/meta-heuristics have been a choice for evolving an optimal solution for VRP. In this paper, various local search heuristics are coupled with genetic algorithm (GA), and two more intense local search heuristics are proposed. GA is tested on six different capacitated VRP instances. Results demonstrate the efficacy of GA in obtaining the solutions close to the best-known solutions that too in less computation time.

Keywords Local search Heuristics · Vehicle routing problem · Genetic algorithm

---

D. Sharma (✉) · S. Pal · A. Sahay · P. Kumar · G. Agarwal · K. Vignesh
Department of Mechanical Engineering, Indian Institute of Technology Guwahati, Guwahati, Assam 781039, India
e-mail: dsharma@iitg.ac.in

S. Pal
e-mail: spal@iitg.ac.in

A. Sahay
e-mail: aayush.sahay@alumni.iitg.ac.in

P. Kumar
e-mail: puneet.kumar@alumni.iitg.ac.in

G. Agarwal
e-mail: gaurav.agarwal@iitg.ac.in

K. Vignesh
e-mail: v.katravath@iitg.ac.in

R. G. Narayanan et al. (eds.), *Advances in Computational Methods in Manufacturing*, Lecture Notes on Multidisciplinary Industrial Engineering,
https://doi.org/10.1007/978-981-32-9072-3_76

76.1 Introduction

Vehicle routing problem (VRP) is an NP-hard problem, which is difficult to solve in polynomial time [1]. It mainly finds optimal routes in transportation and logistics for people and organizations. In a typical supply chain, a requirement for VRP can be seen at different levels of the same organization ranging from transporting from its subsidiaries and manufacturing unit, delivering products to the users, or transporting employees from home to the organization and vice versa.

VRP has been solved by using sophisticated exact or heuristic algorithms in the literature. Exact algorithms aim to find a strict optimal solution, while heuristics are best in finding a good feasible solution. Another set of algorithms is hybrid approaches in which exact algorithm is coupled with heuristic to solve VRP. Due to the size of modern problems getting bigger day by day, heuristics and hybrid algorithms have been mostly used to solve VRP.

In specialized literature, the first GA to successfully solve the VRP was proposed by Prins [2] in which GA without trip-delimiters was hybridized with a local search procedure. GA showed competitive results with that of Tabu search (TS) and other heuristics. Nagata [3] proposed an edge-assembly crossover that was coupled with GA for solving capacitated VRP. Perboli et al. [4] proposed a hybrid GA in which the chain mutation was introduced with the *k*-chain local search. Masum et al. [5] in another effort represented trips of the route as chromosome of an individual. The authors used two types of crossover in which information from one individual was inserted into another to create a new child. Nazif and Lee [6] proposed a crossover operator, which was designed using a complete undirected bipartite graph. Vidal et al. [7] proposed a bi-criteria evaluation of individuals to maintain diversity in the population in order to avoid pre-mature convergence. The memetic algorithm proposed by Cattaruzza et al. [8] combined the split and local search procedure of Prins [2] along with the population diversity management of Vidal's approach [7]. Instead of single crossover and mutation, Contreras-Bolton and Parada [9] explored different combinations of crossover and mutation operators for better exploration and creation of new solutions. Pop and Horvat-Marc [10] considered theoretical aspects of seven neighborhoods motivated from VRP structure. These neighborhoods have been used in their local search approach such as intra-route and inter-route neighborhoods. For intra-route neighborhoods, the authors used the two-point move and three-opt move for improving single-vehicle routes. The shortest route was then selected. Similar steps were used for inter-route neighborhoods in which variations of opt moves to improve the solutions were performed. Kumar and Panneerselvam [11] studied different types of crossover operators used in GA in solving VRPs. Apart from that the authors also proposed 'sinusoidal motion crossover (SMC)' operator. Subramanian et al. [12] implemented a sequence of set partitioning in which an iterated local search-based heuristics and set partitioning were coupled. Wang and Lu [13] proposed a nearest addition method and sweep algorithm to solve the VRPs. Furthermore, an elitism conservation strategy was also used to hold superior chromosomes which can replace inferior chromosomes in the population. Chen et al. [14] proposed a hybrid heuristic method called 'iterated variable neighborhood descent (IVND)' algorithm which used perturbation strategy designed by a crossover operator that helped the algorithm to move out from the local minima.

From the above studies, it can be observed that many efforts have been made to make GA effective for solving VRP by introducing heuristics either at crossover and mutation or by introducing local search and moves. Since in the study [10], various local search and moves have been discussed but not implemented together, and the contribution of this paper is to implement those local search moves with GA along with two proposed intense local search heuristics and test on the known set of capacitated VRPs.

The paper is organized into five sections. Section 76.2 presents a formulation for capacitated VRP. Section 76.3 presents the description of GA which is coupled with various local search heuristics. Section 76.4 presents results and discussion. In Sect. 76.5, the paper is concluded with future work.

76.2 Problem Formulation for Capacitated VRP

VRP is represented as a set of nodes with the depot at index 0, a heterogeneous fleet of vehicles with load capacity *W*, and individual cost constraint *L*. Here, *L* can represent any cost, monetary, or otherwise. It is used to represent distance in our problem statement since cost incurred by a vehicle is proportional to the distance covered by it.

The mathematical model is defined on a complete undirected network $G = (N, E)$. *N* consists of node 0 and *C*, where node 0 is the depot and *C* represents the customers with a non-negative demand d_i. *E* represents the set of all connected edges where every edge (i,j) has a travel cost c_{ij} associated with it. The set of vehicles is represented by *V*. We define a decision variable X_{ij}^v, which is 1 if there is a vehicle in *V* going from node *i* to *j*, and 0 otherwise. Our main objective is to minimize the cost as represented in (76.1), and the constraints are defined in (76.2)–(76.6).

$$\text{Minimize} \sum_{v \in V} \sum_{(i,j) \in E} c_{ij} X_{ij}^v \tag{76.1}$$

$$\sum_{v \in V} \sum_{j \in N} X_{ij}^v = 1, \quad \forall i \in N \tag{76.2}$$

$$\sum_{i \in \mathrm{N}} d_i \sum_{j \in N} X_{ij}^v \leq W, \quad \forall v \in V \tag{76.3}$$

$$\sum_{j \in \mathrm{N}} X_{0j}^{v} = 1, \quad \forall v \in V \tag{76.4}$$

$$\sum_{i \in N} X_{ik}^{v} - \sum_{j \in N} X_{kj}^{v} = 0, \quad \forall v \in V, \quad \forall k \in \mathrm{N} \tag{76.5}$$

$$X_{ij}^{v} \in \{0, 1\}, \ \forall (i,j) \in E, \quad \forall v \in V. \tag{76.6}$$

Equation (76.2) represents that each customer is visited by only one vehicle. Equation (76.3) imposes constraints on the capacity of a vehicle. Equation (76.4) represents that a vehicle can leave the depot only once. Equation (76.5) is designed such that the number of vehicles entering the depot or customer k is equal to the number of vehicles leaving. Equation (76.6) represents binary variable X_{ij}^{v}.

76.3 Local Search Heuristics for Genetic Algorithm

GA has been chosen for implementing local search heuristics for solving VRPs. Following are the steps of GA:

Step 1: Initialize random population (π) without duplicate. Set generation counter $t = 1$ and a maximum allowed generations $T = 1000$;
Step 2: Calculate fitness of each chromosome in π by using the split algorithm [2];
Step 3: Select two random parent chromosomes by using binary tournament selection from π;
Step 4: Perform crossover on the selected parent chromosomes to create a new chromosome by using specialized single-point, multi-point, and orderly crossover operators;
Step 5: Perform local search heuristics on the new chromosome with a probability p_m by using randomized or comprehensive local search method;
Step 6: Select new chromosome if it is unique (no duplicate) and has better fitness value than the worst fitness chromosome in π. Replace the worst fitness chromosome by the new chromosome in π. Else, do not select the new chromosome;
Step 7: Perform intense local search after every $\tau = 500$ generations;
Step 8: Increase count $t = t + 1$. If $t \leq T$, go to **Step 3**. Else, terminate.

GA starts with an initial random population (π) of size (σ) in Step 1. A real-valued chromosome is chosen to represent a solution. An example is shown

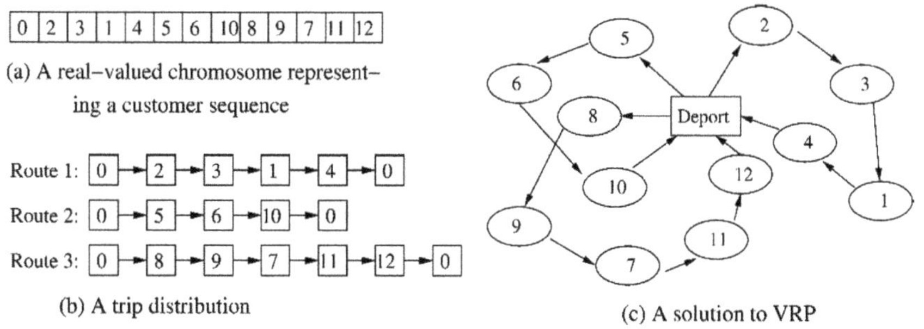

Fig. 76.1 A chromosome representation and trip distribution. Note that node 0 represents the depot

in Fig. 76.1a in which the customer sequence is represented the sequence in which vehicles are going to visit them. In Fig. 76.1b, the trips are distributed such that constraint on the capacity can be satisfied. Finally, a solution to VRP is shown in Fig. 76.1c.

The duplicate chromosomes are removed from the population for which an idea of spacing is used [2] in which the chromosomes should have the least similarity. Therefore, a split algorithm [2] is used that arranged the sequence of nodes in a chromosome such that least fitness of the chromosome can be found. An example is shown in Fig. 76.2 in which a chromosome, which is a solution to VRP, is shown before and after the split algorithm for making trips and calculating the least fitness for the given chromosome.

Once the initial random population is generated and the fitness is calculated for each chromosome, GA enters into its standard loop of generation at Step 3 at which two parent chromosomes are selected by using the binary tournament selection. The crossover operators are then applied to these two chosen chromosomes in order to create one new chromosome at Step 4. Three types of crossover operators are used with equal probability, which is referred to as single-point, multi-point, and orderly crossover operators. These operators are explained with the help of Fig. 76.3.

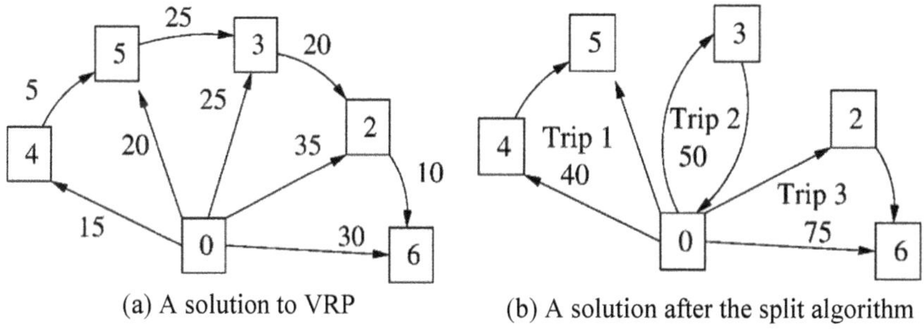

Fig. 76.2 Fitness evaluation of chromosome using the split algorithm [2]

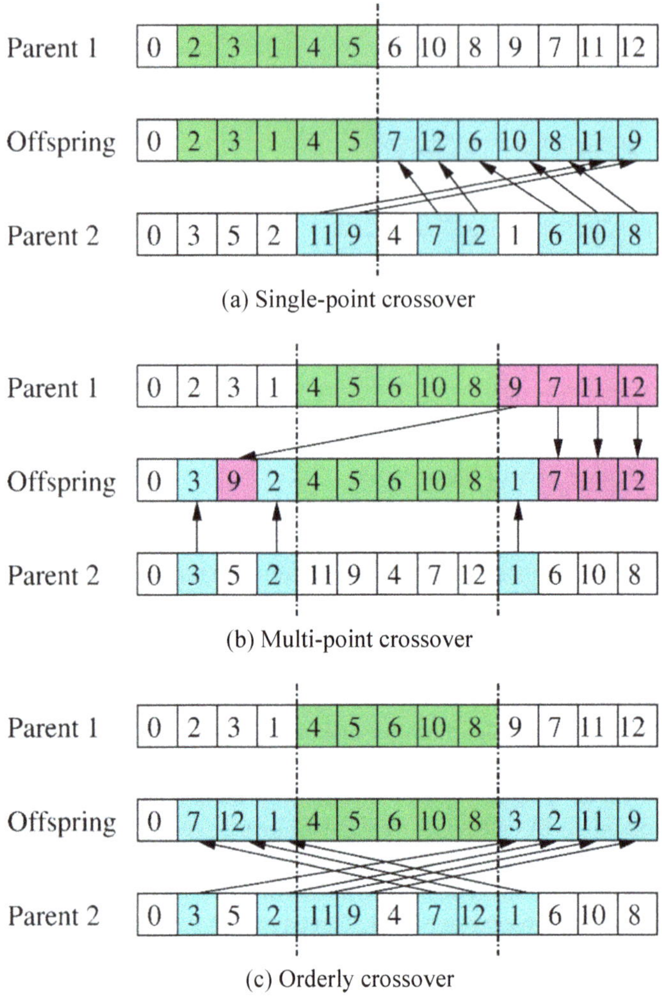

Fig. 76.3 Different crossover operators used with GA

In a single-point crossover, a mating site is chosen randomly, say at node 5. All the nodes of parent 1 copied into the new chromosome. Thereafter, the remaining nodes are copied from parent 2 such that there is no duplicate node in the new chromosome, but in the given sequence of parent 2. In multi-point crossover operator, two random sites are chosen so that a sequence between these sites can be copied to the new chromosome as shown in Fig. 76.3b. At the empty places of a new chromosome, the nodes are copied in the same places as of parent 2, but without repetition of nodes in the new chromosome. There can be a few places in the new chromosome are not filled, which can be filled from the sequence of parent 1 as shown in the figure. In orderly crossover operators, two sites are chosen randomly, and then the sequence from parent 1 is copied into the new chromosome at the places. Thereafter, the nodes, which do not appear in the new offspring, are copied in the given order/sequence as given in the parent 2.

After performing crossover, mutation is performed with a probability of p_m either by using randomized local search or comprehensive local search method at Step 5. Both of these random search methods are adopted from [2], in which the two nodes are selected randomly in the randomized local search. In the comprehensive local search method, two nodes are selected after scanning through all pairs of nodes. After selecting two nodes, six local search heuristics adopted from [10] are implemented which consist of (i) one-point move, (ii) two-point move, (iii) two-opt move, (iv) Or-opt move, (v) three-point move, and (vi) four-point move. Due to page restriction, the details for these moves are not given. However, the same can be found in [10].

At Step 6, the new chromosome is checked for insertion in the population. Its fitness is found by using the split algorithm. If the fitness of the new chromosome is better than the worst fitness chromosome in the population and also there is no duplicate, then it is selected. Otherwise, it is removed.

The main contribution of this paper is the proposed intense local search which is performed at Step 7. It has been observed through simulation that the local search performed at Step 6 gets stuck at the local optima. In order to escape from these local optima, intensive local search is done on all the chromosomes in the population after every $\tau = 500$ generation. First intense local search is 'k-nearest neighbors local search' in which the nodes, which are close by based on Euclidean distance, bring closer in the sequence as well. We loop through all nodes one by one, find that particular node's k-nearest neighbors, and apply local search between that node and the ith nearest neighbor, where $1 \leq i \leq k$. Each iteration of this local search stops at the first improving move. For a graphical representation, the k-nearest neighbors local search is shown in Fig. 76.4. Another intense local search is 'trip-wise improvement procedure.' It has been observed through simulations that the improvement in the solution fitness is very low and by minor tweaking the sequence of a chromosome, the solution can converge to the global optimal

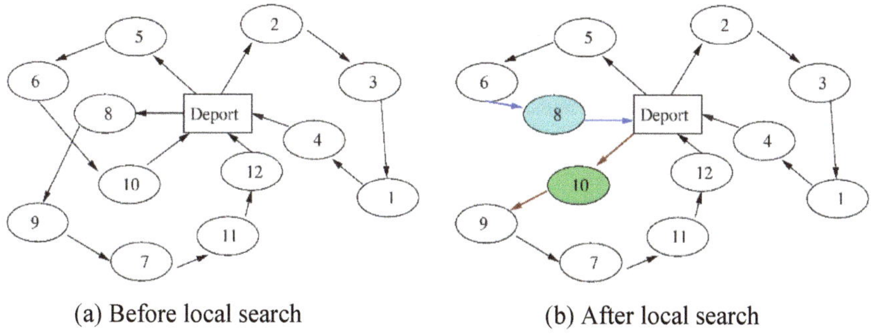

Fig. 76.4 An intense local search by using the k-nearest neighbors local search

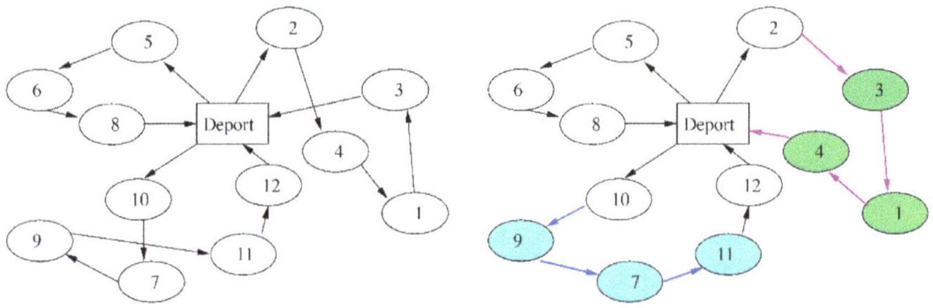

Fig. 76.5 Trip-wise improvement procedure for local search

solution. In this method, every solution is scanned in every trip and the nodes' sequence in the given trips is tweaked. A graphical presentation is shown in Fig. 76.5. At last, the generation counter is increased by one and the termination condition is check. If not terminated, GA then starts from Step 3 till it gets terminated.

76.4 Results and Discussion

GA with local search heuristics is now tested on a set of benchmark CVRPs from the literature. The chosen set of the problems belong to four Augerat instances [15], one Christofide instance [16], and one Fisher instance [17]. These instances have a different number of nodes, which are presented in Table 76.1. The best-known solution from the literature is also presented in the table to compare the outcome of the proposed GA with various local search heuristics.

For all instances, GA starts with an initial population of size 15, and it runs for 1000 generations. The probability of mutation is set to $p_m = 0.1$. GA runs on Intel core i7 with 1.8 GHz processor.

Table 76.1 Details of Augerat, Christofide, and Fisher instances. The letter 'C' refers to Christofide instance, the letter 'A' refers to Augerat instance, and the letter "F' refers to Fisher instance. The number signifies a number of nodes

Instances	A33	A48	A54	A61	C76	F72
Nodes	33	48	54	61	76	72
Best-known solution	661	1073	1167	1035	832	238

Table 76.2 Solutions obtained for different instances of CVRP

S. No.	Problem instance	Solution obtained	Number of vehicles	Average time (s)
1	A33	662	5	0.945
2	A48	1170	7	2.494
3	A54	1428	7	4.189
4	A61	1040	9	2.083
5	C76	846	10	2.876
6	F72	241	4	3.389

76.4.1 Obtained Solutions and Computational Time

The solutions obtained by solving various instances are presented in Table 76.2. It can be seen that the local heuristics coupled with GA can generate the solutions close to the best-known solutions given in Table 76.1. Since a homogenous fleet of vehicles is assumed for solving CVRP, the minimum of vehicles are also presented in the table. The average computation time is also reported which is calculated after running GA 30 times from different initial population. It can be seen that the developed GA is not expensive for solving CVRP having a maximum of 76 nodes. It is noted that the instances are different from each other; therefore, the computation time is not linear with the number of nodes in the problem instances.

76.4.2 Convergence Plots

The convergence plots for all test instances are shown in Fig. 76.6. It can be seen that the best fitness solution is improved drastically in very early generations and

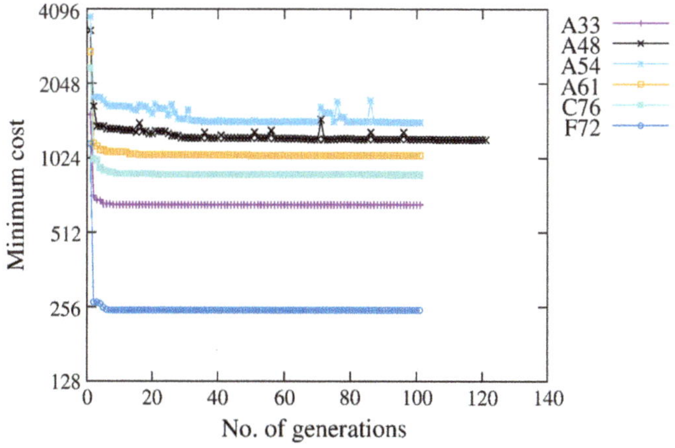

Fig. 76.6 Convergence plot of GA for different instances of CVRPs. The minimum cost values are plotted on a $\log_2$ scale

further their performance improved very slowly. The local search heuristics coupled with GA are able to make it efficient by improving chromosome drastically in the early generations.

76.5 Conclusions

In this paper, various local search heuristics were coupled with GA to improve its performance for solving various test instances of CVRP. From the results, it can be concluded that the hybrid GA can be used for generating a solution for CVRP as it was able to find the solution close the best-known solutions of the test instances. Since the computation time of GA was small, it can also be used for a large test instance and for application problems of CVRP. In the future, GA can be improved further so that it can generate the best-known solutions. Also, a large set of CVRP test instance needs to solve for assessing its performance.

References

1. Lenstra, J.K., Rinnooy Kan, A.H.G.: Complexity of vehicle routing and scheduling problems. Networks **11**, 221–227 (1981)
2. Prins, C.: A simple and effective evolutionary algorithm for the vehicle routing problem. Comput. Oper. Res. **31**(12), 1985–2002 (2004)
3. Nagata, Y.: Edge assembly crossover for the capacitated vehicle routing problem. In: European Conference on Evolutionary Computation in Combinatorial Optimization, pp. 142–153. Springer (2007)
4. Perboli, G., Pezzella, F., Tadei, R.: EVE-OPT: a hybrid algorithm for the capacitated vehicle routing problem. Math. Methods Oper. Res. **68**(2), 361–382 (2008)
5. Masum, A.K.M., Shahjalal, M., Faruque, F., Sarker, I.H.: Solving the vehicle routing problem using genetic algorithm. Int. J. Adv. Comput. Sci. Appl. **2**(7), 126–131 (2011)
6. Nazif, H., Lee, L.S.: Optimised crossover genetic algorithm for capacitated vehicle routing problem. Appl. Math. Model. **36**(5), 2110–2117 (2012)
7. Vidal, T., Crainic, T.G., Gendreau, M., Prins, C.: A hybrid genetic algorithm with adaptive diversity management for a large class of vehicle routing problems with time-windows. Comput. Oper. Res. **40**(1), 475–489 (2013)
8. Cattaruzza, D., Absi, N., Feillet, D., Vidal, T.: A memetic algorithm for the multi-trip vehicle routing problem. Eur. J. Oper. Res. **236**(3), 833–848 (2014)
9. Contreras-Bolton, C., Parada, V.: Automatic combination of operators in a genetic algorithm to solve the traveling salesman problem. PLoS ONE **10**(9), e0137724 (2015)
10. Pop, P.C., Horvat-Marc, A.: Local search heuristics for the generalized vehicle routing problem. In: 2012 International Conference on System Modeling and Optimization (ICSMO 2012), IPCSIT, vol. 23, pp 84–87 (2012)
11. Kumar, S.G.V., Panneerselvam, R.: A study of crossover operators for genetic algorithms to solve VRP and its variants and new sinusoidal motion crossover operator. Int. J. Comput. Intell. Res. **13**(7), 1717–1733 (2017)
12. Subramanian, A., Uchoa, E., Ochi, L.S.: A hybrid algorithm for a class of vehicle routing problems. Comput. Oper. Res. **40**(10), 2519–2531 (2013)

13. Wang, C., Lu, J.: A hybrid genetic algorithm that optimizes capacitated vehicle routing problems. Expert Syst. Appl. **36**(2), 2921–2936 (2009)
14. Chen, P., Huang, K., Dong, X.: Iterated variable neighborhood descent algorithm for the capacitated vehicle routing problem. Expert Syst. Appl. **37**(2), 1620–1627 (2010)
15. Augerat, P., Belenger, J.M., Benavent, E., Corberan, A., Naddef, D., Rinaldi, G.: Computational results with a branch and cut code for the capacitated vehicle routing problem. Technical Report RR 949-M, University Joseph Fourier, Grenoble, France (1995)
16. Christofides, N., Eilon, S.: An algorithm for the vehicle dispatching problem. J. Oper. Res. Soc. **20**, 309–318 (1969)
17. Fisher, M.L.: Optimal solution of vehicle routing problems using minimum k-trees. Oper. Res. **42**, 626–642 (1994)

Part VIII
Design

Chapter 77
Performance Evaluation of Adhesive Spur Gear with the Influence of Properties of Adhesive

R. Shailesh, M. Ramu, M. Govindaraju, K. Karthikeyan and V. Satheeshkumar

Abstract The effect of adhesive properties on performance of spur gear made of adhesive system referred to as adhesive spur gear is investigated. In addition, the repair ability and reusability of adhesive spur gear towards sustainability are also investigated. Two-part epoxy adhesive system constituting resin and hardener was used to fabricate spur gear samples. In order to modify adhesive properties, the hardener to resin ratio was varied. The repair ability and reusability of spur gear made by the adhesive system were checked by breaking a few teeth of the adhesive spur gears and repaired by adhesive. The repaired gears were tested and compared with the performance of actual spur gears made of adhesive. The gears were tested by a gear test rig developed indigenously. The performance of the gears with different hardener to resin ratios was evaluated by monitoring the number of cycles. The results show that the number of cycles increases with decrease in hardener to resin ratio. This is due to the better strength of the adhesive system with the resin-rich formulation. There is not much difference in performance between repaired gear and the corresponding actual gear. This ensures the sustainability of adhesive gears over conventional polymer gears.

Keywords Spur gear · Polymer gear · Epoxy adhesive · Gear performance · Hardener to resin ratio · Gear repair ability

---

R. Shailesh · M. Ramu · M. Govindaraju
Department of Mechanical Engineering, Amrita School of Engineering, Amrita Vishwa Vidyapeetham, Coimbatore, India

K. Karthikeyan
Department of Mechanical Engineering, Sri Ramakrishna Engineering College, Coimbatore 641022, India

V. Satheeshkumar (✉)
Department of Production Engineering, National Institute of Technology, Tiruchirappalli 620015, India
e-mail: satheeshv@nitt.edu

R. G. Narayanan et al. (eds.), *Advances in Computational Methods in Manufacturing*, Lecture Notes on Multidisciplinary Industrial Engineering,
https://doi.org/10.1007/978-981-32-9072-3_77

77.1 Introduction

Performance evaluation of the spur gears made of the adhesive system with the influence of adhesive properties and checking repair ability and reusability of the gears made of adhesive are the key activities of the present study. In this regard, White [1] invented a composite plastic metal gear in which different plastic materials were used for having gear mesh, and the gear hub was made of metal. In this case, the hybridization of plastic gear and metal gear hub together eliminates distortion and slip. Beddoe [2] invented a composite gear employing a set of deformable teeth disposed slightly in advance of the hard teeth to transfer load through hard teeth only when it exceeds a critical level. Santi [3] invented composite gears made from a rigid material like metal and from a thermoplastic material for the abnormal conditions like abrupt stop occurs. Kurokawa et al. [4] investigated the performance of injection-moulded spur gears with different types of composites made of poly-ether-ether-ketone with different types of carbon fibres (CFs). It was found that a composite gear reinforced with CF of the highest density indicated the highest load carrying capacity irrespective of the test conditions. Kim [5] proposed a method to evaluate the wear and durability of plastic spur gears made of acetal and nylon materials. In the results, acetal gears exhibited a large amount of wear with the development of plastic deformation and increased tooth stiffness due to brittle nature of the material. In the case of nylon gears, reduced amount of wear and prolonged service life with less gear temperature were observed due to the polymer softer in nature. Mao [6] proposed a design method for polymer composite gear by relating polymer gear surface temperature and wear rate. It was found that polymer gear wear rate increases with an increase in critical load for a specific geometry and then decreases. Nozawa et al. [7] investigated the tribological behaviour of polymer sheets adhered to steel gear teeth. It was found that noise suddenly increased when only one of the polymer sheets was failed from the gear surface due to low adhesive strength.

Kansal et al. [8] addressed the problem of evaluating the temperature distribution and thermal residual stresses that are developed due to the non-uniform cooling of the injection-moulded polystyrene gear. Senthilvelan and Gnanamoorthy [9, 10] investigated the influence of the gear rotational speed on performance of polymer composite gear tooth. It was found that gear rotational speed affects the performance of gears made of both the materials at different high running speeds and torques. The glass fibre reinforced nylon six gears showed better performance when compared with unreinforced gears due to its better mechanical strength and thermal deformation resistance. It was reported that the rise in the gear temperature under loading is due to hysteresis effects and friction. Kirupasankar et al. [11] reported the effect of applied torque on the transmission efficiency of polyamide nanocomposite spur gears. The increase in gear tooth temperature affects the tooth shape and thereby affects the gear performance. The improvement in mechanical properties of polyamide nanocomposite gears results in better power transmission efficiency. Anand Mohan and Senthilvelan [12] evaluated fatigue performance of polypropylene symmetric gears and asymmetric gears with and without the reinforcement of glass fibre. The results showed better load-carrying capacity of asymmetric gears when compared with symmetric gears due to the increased gear tooth width at the root region. Chaubey and Jain [13] conducted a detailed review on meso- and micro-gears used in various micro-systems and devices used for transmitting power and/or motion due to their reduced weight, dimensional accuracy, zero backlash, compact ability, better functional and operating characteristics, longer service life and sustainable performance. The review reveals about the manufacturing of meso- and micro-spur gear by various advanced manufacturing processes and the directions towards manufacturing quality gears. Hoskins et al. [14] generated a sound frequency spectrum with the influence of various polymeric gear materials and operating conditions. Results showed that there is a significant effect of surface characteristics, wear and temperature corresponding to the respective sound power levels. The effects of material, reinforcement and geometry show that the tribological properties of gear materials could be used to predict noise generated and the inaccuracy significance in polymer gear trains. Dearn et al. [15] described friction and wear by controlling the running temperatures by using a series of solid lubricant coatings like molybdenum disulphide, graphite flake, boron nitride (hexagonal) and poly-tetra-fluoro-ethylene (PTFE) deposited on flanks of the polymer gear teeth. Results showed a significant reduction in frictional forces with PTFE.

In line with adhesive materials, mostly the adhesive systems with two parts, namely resin and hardener are used for structural applications. In a compression testing of an adhesive system that is made of epoxy group resin and amine group hardener with different mixing ratios, the results show brittle behaviour with the epoxy-rich systems. This is due to the evolution of a rigid macromolecular structure. The adhesive mixture with hardener-rich formulation shows ductile behaviour [16]. Satheeshkumar and Ganesh Narayanan [17] demonstrated the mechanical performance of adhesive with two parts, namely resin—bisphenol A epichlorohydrin and hardener—polyamidoamine. It was postulated that resin-rich formulation turns the adhesive sample stronger, and hardener-rich formulation turns the adhesive sample more ductile.

In the above detailed literature review, it is observed that numerous investigations have been carried out on the manufacturing of polymer gears with different polymer materials. Different types of filler materials and reinforcements have been used for improving the gear performance. The polymer gears have been used predominantly, and efforts have been taken to achieve equivalent and better performance than metal gears. The polymer gear conventionally manufactured by the injection moulding process is replaced with a new gear once it is damaged. The damaged thermoplastic gears may be melted and reused. In the case of thermoplastic gears with typical fillers and reinforcements, repair ability is one of the major issues. The idea here is to develop polymer gears by using adhesive materials by choosing suitable properties for having better performance and to check the repair ability and reusability of adhesive gears in line with sustainable gear manufacturing. In the present preliminary study, the influence of hardener to resin ratio of adhesive on the performance of adhesive gear is investigated with actual and repaired gears.

77.2 Methodology

77.2.1 Materials Used and Mechanical Properties

Commercially available structural adhesive was used as a polymer matrix in this study. The adhesive system is a combination of two parts, namely resin (bisphenol A epichlorohydrin) and hardener (polyamidoamine). The suitable mixing ratio of the adhesive system is of 1:0.8 by weight as per supplier's instruction.

Gear Test Rig. Figure 77.1 shows the schematic of the test rig for evaluating the adhesive spur gear performance. It has a 0.5 HP DC shunt motor with maximum speed up to 1500 rpm. It is coupled with an alternator having the same capacity. The input to the motor and the field of the alternator is controlled externally by two separate autotransformers. The autotransformer is used to adjust speed and torque. There is an additional rectifier circuit consisting of diodes and capacitor for the AC to DC conversion. A 100 Ω rheostat is used as the load across the alternator. The testing gear is connected to the alternator shaft while the standard steel gear is connected to the motor shaft.

Evaluation of Adhesive Spur Gear Performance. Figure 77.2 shows the schematic of spur gear used for fabrication. Table 77.1 shows the parameters of the adhesive spur gear (Anand Mohan and Senthilvelan [12]). The adhesive mixture with different properties by varying hardener to resin (H/R) ratio like 0.6:1, 0.7:1, 0.8:1, 0.9:1 and 1:1 was cast into the suitable mould with a thickness of 6 mm (Fig. 77.3a). The cast adhesive mixture into the mould was allowed to cure for more than 24 h and removed from the mould carefully. Figure 77.3b shows fabricated spur gear made of epoxy adhesive system and a steel gear for a reference.

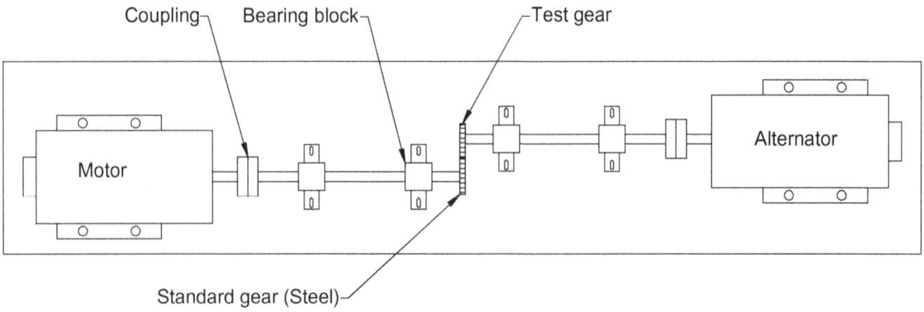

Fig. 77.1 Schematic of gear test rig

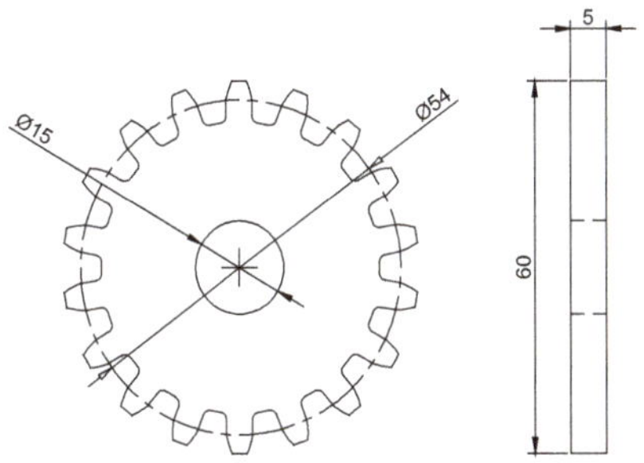

Fig. 77.2 Schematic of spur gear

Table 77.1 Test gear parameters

	Adhesive spur gear
Pitch circle diameter	54 mm
Addendum	3 mm
Module	3 mm
Pressure angle	25°
Number of teeth	18
Face width	6 mm
Root circle diameter	47 mm
Fillet radius	1.14 mm

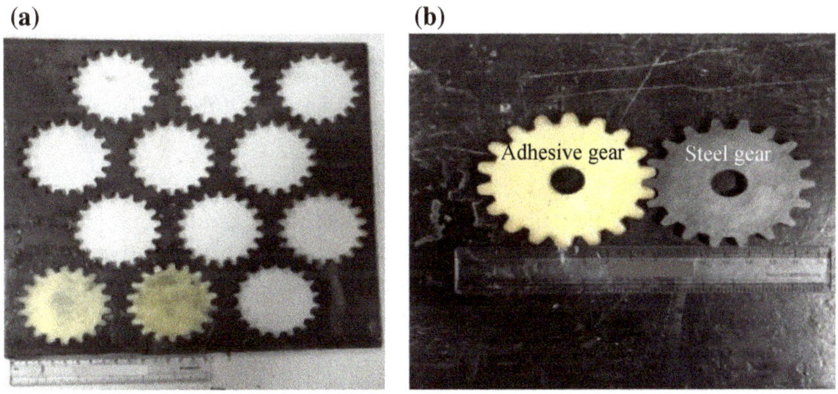

Fig. 77.3 **a** Making of adhesive gear using mould and **b** fabricated adhesive gear and steel gear

Further, a tooth in each fabricated adhesive gear with varying hardener to resin ratio was broken deliberately and inserted into the mould. The damaged location was repaired with the adhesive mixture maintaining the corresponding hardener to resin ratio. In the present work, the gear tested directly from the moulding process is referred to as actual gear, and the gear deliberately broken and repaired is referred to as repaired gear.

The adhesive gears were tested at a constant motor speed of 300 rpm and applied torque of 2 Nm. The platform of the testing setup was made rigid and heavy to minimize vibration. Initially, using a pre-test gear attached to the alternator, electrical equivalent of all torque conditions were calculated by measuring the voltage and current across the input of DC motor. The voltage and current were measured to be 35 V and 1.8 A, respectively. The voltage, current and speed transformation to equivalent torque were calculated by the Eqs. (77.1), (77.2).

$$P = V \times I \tag{77.1}$$

where *P*—power, *V*—voltage in V and *I*–current in A.

Also

$$P = \frac{2\pi NT}{60} \tag{77.2}$$

where *N*—Motor speed in rpm and *T*—torque in Nm.

During testing of gears, gear life was evaluated by monitoring the number of cycles. The failure of gear was ensured by sound difference and followed by tooth breakage. Three gears were tested in each case, and the fourth gear was tested in the case of not observing good repeatability.

77.3 Results and Discussion

77.3.1 Effect of Hardener to Resin Ratio of Adhesive on Failure Pattern of Gears

Figure 77.4a, b shows the failure pattern of actual gears and repaired gears with varying H/R ratios of adhesive. It is found that 1 or 2 teeth have failed in all cases. There is not much difference in failure pattern between actual gears and repaired gears. In the case of repaired gears, the failure of gear teeth is observed randomly, and not at the repaired location always.

Fig. 77.4 Failure pattern of adhesive gears: **a** actual gears and **b** repaired gears

77.3.2 Effect of Hardener to Resin Ratio of Adhesive on Gear Performance

Figure 77.5 shows a comparison of maximum number of cycles achieved by the adhesive gears with different hardener to resin ratios. The comparison of maximum number of cycles between actual and repaired adhesive spur gears is also shown. It is observed that the number of cycles of adhesive spur gear increases with decrease in H/R ratio of adhesive. The maximum number of cycles of about 40200 is observed for H/R ratio of 0.6:1. Based on the maximum number of cycles, the decrease in the percentage of maximum number of cycles is calculated as 19.4, 28.4, 44.8 and 60.4% with hardener to resin ratios of 0.7:1, 0.8:1, 0.9:1 and 1:1, respectively. This is due to the rich formulation of the resin increases the strength of the adhesive (Satheeshkumar and Ganesh Narayanan [17]) and thereby increases the gear performance. There is not much difference between actual and repaired adhesive spur gears.

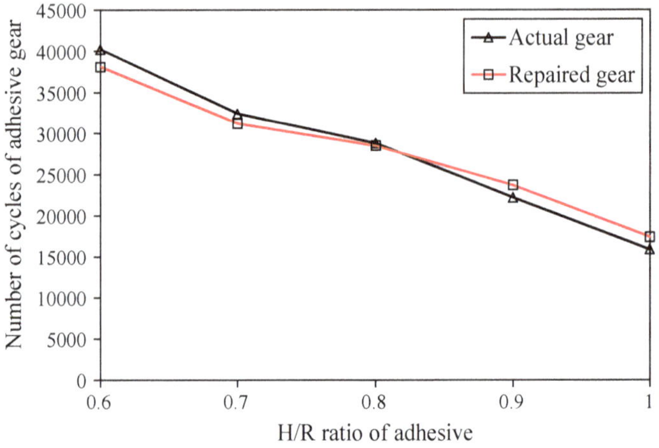

Fig. 77.5 Comparison of maximum number of cycles of adhesive spur gear between actual and repaired cases with different hardener to resin ratios (Data Variation: ±1200 cycles)

77.4 Conclusions

From the present study, the conclusions are summarized as follows:

- The performance of adhesive spur gears in terms of number of cycles increases with decrease in hardener to resin ratio of the adhesive system. This is due to the rich formulation of resin that improves the strength of adhesive. It is understood that improved strength of adhesive with a rich formulation of resin increases the strength of gear and thereby increases the number of cycles.
- There is not much difference in the performance of actual and repaired adhesive spur gears irrespective to hardener to resin ratio. This ensures that the adhesive gear has good repair ability and reusability. In the case of conventional thermoplastic gears made by injection moulding process, reusability is possible by melting the damaged gears and reinjection moulding process. But, the repair ability of those gears is difficult by injection moulding process and not suggestible as well.
- It is suggested that the adhesive with resin-rich formulation could be fabricated in sheet or plate form, and water jet cutting process could be used for spur gear cutting process in the case of mass production.

References

1. US3013440, White, J.T. [1961]
2. US3602058, Beddoe, B. [1971]
3. US005852951 A, Santi, J.D. [1998]
4. Kurokawa, M., Uchiyama, Y., Nagai, S.: Performance of plastic gear made of carbon fiber reinforced polyether-ether-ketone: Part 2. Tribol. Int. **33**, 715–721 (2000)
5. Kim, C.H.: Durability improvement method for plastic spur gears. Tribol. Int. **39**, 1454–1461 (2006)
6. Mao, K.: A new approach for polymer composite gear design. Wear **262**, 432–441 (2007)
7. Nozawa, J.I., Komoto, T., Kawai, T., Kumehara, H.: Tribological properties of polymer-sheet-adhered metal hybrid gear. Wear **266**, 893–897 (2009)
8. Kansal, G., Rao, P.N., Atreya, S.K.: Study: temperature and residual stress in an injection moulded gear J. Mater. Process. Technol. **108**, 328–337 (2001)
9. Senthilvelan, S., Gnanamoorthy, R.: Effect of gear tooth fillet radius on the performance of injection molded Nylon 6/6 gears. Mater. Des. **27**, 632–639 (2006)
10. Senthilvelan, S., Gnanamoorthy, R.: Effect of rotational speed on the performance of unreinforced and glass fiber reinforced Nylon 6 spur gears. Mater. Des. **28**, 765–772 (2007)
11. Kirupasankar, S., Gurunathan, C., Gnanamoorthy, R.: Transmission efficiency of polyamide nanocomposite spur gears. Mater. Des. **39**, 338–343 (2012)
12. Anand, Mohan N., Senthilvelan, S.: Preliminary bending fatigue performance evaluation of asymmetric composite gears. Mech. Mach. Theory **78**, 92–104 (2014)
13. Chaubey S.K., Jain, N.K.: State-of-art review of past research on manufacturing of meso and micro cylindrical gears. Precis. Eng. (2017) (In Press)
14. Hoskins, T.J., Dearn, K.D., Kukureka, S.N., Walton, D.: Acoustic noise from polymer gears —a tribological investigation. Mater. Des. **32**, 3509–3515 (2011)
15. Dearn, K.D., Hoskins, T.J., Petrov, D.G., Reynolds, S.C., Banks, R.: Applications of dry film lubricants for polymer gears. Wear **298–299**, 99–108 (2013)
16. d'Almeida, J.R.M., Monteiro, S.N.: The effect of the resin/hardener ratio on the compressive behaviour of an epoxy system. Polym. Test **15**, 329–339 (1996)
17. Satheeshkumar, V., Ganesh Narayanan, R.: Investigation on the influence of adhesive properties on the formability of adhesive bonded steel sheets. Proc. Inst. Mech. Eng. Part C: J. Mech. Eng. Sci. **228**(3), 405–425 (2014)

Chapter 78
Design Analysis and Modification of Sugarcane Fibrizer Hammer at Wonji Shoa Sugar Factory (WSSF) Ethiopia

Tesfaye Mathewos, Perumalla Janaki Ramulu and Esmael Adem Esleman

Abstract Wonji Shoa Sugar Factory (WSSF) is one of the top sugar factories in Ethiopia which produces sugar and electricity from sugarcane using three critical plants: mill plants, sugar processing plant, and a cogenerating power plant. Among three plants, the mill plant is the most important plant used to crush the cane for separating sugar juice from bagasse. In this plant, Fibrizer hammer device is used for cruising sugarcane. The Fibrizer performs the major share in milling process for fibrizing the chopped cane after set of cane knives using rotating knives. Most of the times, WSSF is facing breakages of Fibrizer hammer by which losses incurring to the factory. The main aim of the paper is to analyze the failure causes of existing Fibrizer hammer with respect to design and modify the necessary changes in design for Fibrizer hammer at WSSF. By which, get high preparation index (PI) and optimize in power consumption and cost. In addition, hardness of the existing Fibrizer hammer was tested to know the mechanical strength. For modification of existing Fibrizer hammer: comparison was made in terms of design, and load along with harmonic simulation analyses performed using finite element analysis. The results from finite element analysis revealed that the modified Fibrizer hammer

T. Mathewos
School of Mechanical, Chemical and Materials Engineering, Adama Science and Technology University, Adama, Ethiopia & Wonji Shoa Sugar Factory, Wonji, Adama, Ethiopia
e-mail: tesfamathewos@gmail.com

T. Mathewos
Wonji Shoa Sugar Factory, Wonji, Adama, Ethiopia

P. J. Ramulu (✉)
Centre of Excellence for Advanced Manufacturing Engineering, School of Mechanical, Chemical and Materials Engineering, Adama Science and Technology University, Adama, Ethiopia
e-mail: perumalla.janaki@astu.edu.et

E. A. Esleman
Program of Mechanical Design and Manufacturing Engineering, School of Mechanical, Chemical and Materials Engineering, Adama Science and Technology University, Adama, Ethiopia
e-mail: esm99a@gmail.com

R. G. Narayanan et al. (eds.), *Advances in Computational Methods in Manufacturing*, Lecture Notes on Multidisciplinary Industrial Engineering,
https://doi.org/10.1007/978-981-32-9072-3_78

having minimum deformation, minimum shear stress, and more centrifugal force has a relation to achieve better preparation index.

Keywords Fibrizer hammer · Harmonic analysis · Preparation index (PI) · FEM · Wonji Shoa Sugar Factory

78.1 Introduction

For completing the preparation of cane juice in the sugar industries, Fibrizer hammer is used after knifing. Formerly to break the cane, initial knifing was followed by heavy crusher. In this method, cane was not crushed properly in most of the time which leads to more power consumption and increase miscellaneous tasks. To overcome these issues, a few researchers studied and suggested the upgradation of technology to the sugarcane industries, in which Fibrizer hammer is the upgradated technology to all the sugarcane industries for crushing the cane. With this technology, betterment in level of preparation of index (PI) and reduction in the power consumption are attained during processing. For the best operations, less maintenance, low power consumption, and slip ring motors with high capacities are used, whereas steam turbines were used formerly for roller crushing instead of electric motors [1–4]. There were some of the studies carried out based on the Fibrizer hammer, i.e., design of hammer, modification, and optimization of Fibrizer hammer, etc. Podany et al. [5] dealt hammer construction for cane shredder, in which the hammer functions in shredder comprises crushing sugarcane stalks for further sugar production processing. For protection of hammer against wear and dynamic loading, hammer loading and other related calculations were made and suggested the weld deposit materials for strengthening the hammer. Ransing et al. [6, 7] studied about cane roller crusher for attaining good cane preparation for extraction of juice from cane in favor of cane Fibrizer. Cane knifing in chopper, cane crushing in Fibrizer, and extraction of juice in milling like three processes were performed on it. Based on the plant capacity, Fibrizer was changed in size and model, resulting in high PI. Nikam et al. [8] designed and analyzed a new flexible blade-type sugarcane leveler by numerical analysis. They concluded that the efficiency of cane leveler increased by changing design from fixed bitter to flexible blades. By modifying design cane lever, one can overcome the limitations of existing cane leveler. It also increases the productivity of milling section. Efficiency of cane leveler increases, and then it also increases the production.

From the above literature, it is observed that many of the researchers were carried out the research on the functional aspects of Fibrizer hammer, parameters effect, shredder design, power consumption, etc. No researchers discussed the Fibrizer hammers breakages and its consequences in the breakdown of the industry. In the present work, design, analyze and modification of the existing Fibrizer hammer for improvement of sugar production process at Wonji Shoa Sugar Factory (WSSF). In the case industry, i.e., at WSSF, frequent hammer breakage was observed at the hard-faced welded part and at the tail part of the Fibrizer hammer. Also, at hammer head tip wear, shearing of tip fixing bolt leads to huge crushing downtime. The recorded downtime during 2014–2017 period was 121 h in terms of cost 14,82,421.3 USD. The other problem in the factory was overtime payments during repair of Fibrizer hammer and electrical loss due to interruption of cane bagasse from mill plant to boiler. Based on these issues, the significance of this work is to improve the current Fibrizer hammer to increase juice extraction by improving the preparation index from 85 to 92%; to reduce overtime payments caused by Fibrizer hammer breakage; to supply continuous electrical power by continuous crushing; to save excess bagasse by continuous crushing which is used as a row material for pulp.

78.2 Experimental Methodology

To realize the stated objective, pertinent methodology was followed. The relevant data were collected from both WSSF and existing literature. The detailed specifications of the existing Fibrizer hammer at WSSF are shown in Table 78.1. The material used for Fibrizer hammer was structural steel. Fibrizer hammer bolt was made of mild steel with diameter, length, and thread length 22, 130 and 85 mm,

Table 78.1 Specifications of existing Fibrizer at Wonji Shoa Sugar Factory

S. No.	Description	Unit	Quantity	Remark
1	No's of Fibrizer hammer head	pcs	176	
2	Weight of single Fibrizer hammer head with tip	kg	29	
3	Speed of Fibrizer	RPM	750	
4	Type of Fibrizer			Swing type
5	Hub outer diameter	mm	1670	
6	Number of Hubs	pcs	23	
7	Hub inner diameter	mm	450	
7	Thickness of Hub	mm	50	
8	Material type of Hub			MS IS 2062
9	Design for preparation index	%	92	
10	Current preparation index	%	85	
11	Types of materials			Mild steel
12	Types of tip	kg	3.95 kg/pcs	Detachable
13	Types of bushing			Stainless steel
14	Materials of tip			Carbide tip

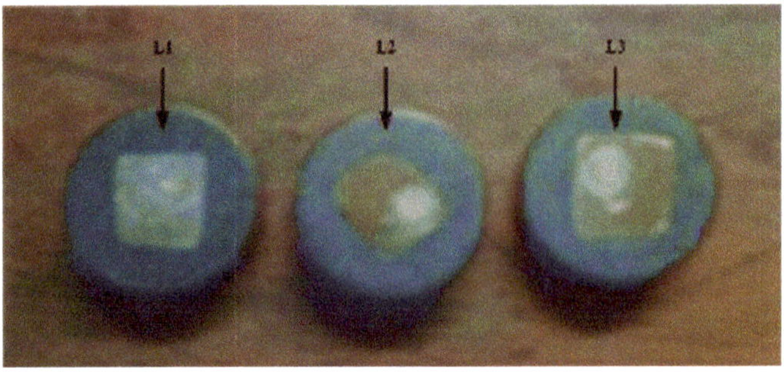

Fig. 78.1 Cut pieces from current Fibrizer hammer

respectively. The material of the Fibrizer hammer was tested at three different locations as L-1, L-2, and L-3 as shown in Fig. 78.1.

78.2.1 Experimental Analysis

Indian made Fibrizer hammer was submitted for failure cause analysis. The hammer was broken at the tail patron during operation. Failure causes were examined with following techniques.

78.2.1.1 Visual Examination

The part was broken and detached from rest of the part. The fracture surfaces were found corroded; the broken part was cleaned using a hydrochloric acid solution fetching from sugar processing department to enable them to be examined more closely. The nature of the failure parts was the same and similar. The welders were welding the worn out Fibrizer hammers and cane knife continuously to use like a spare part.

78.2.1.2 Stress Concentration Factor Analysis

In any machine, machine parts and structures' effect of stress concentration can arise from internal holes or voids formed from any manufacturing process [9].

Stress Concentration Factor Analysis of the Existing Fibrizer Hammer

Figure 78.2 is indicating the complete dimensions of the existing Fibrizer hammer. Center of gravity for load analysis was calculated at different zone of areas (A1, A2, A3, and A4 shown in Fig. 78.2). Parameters considered for existing Fibrizer

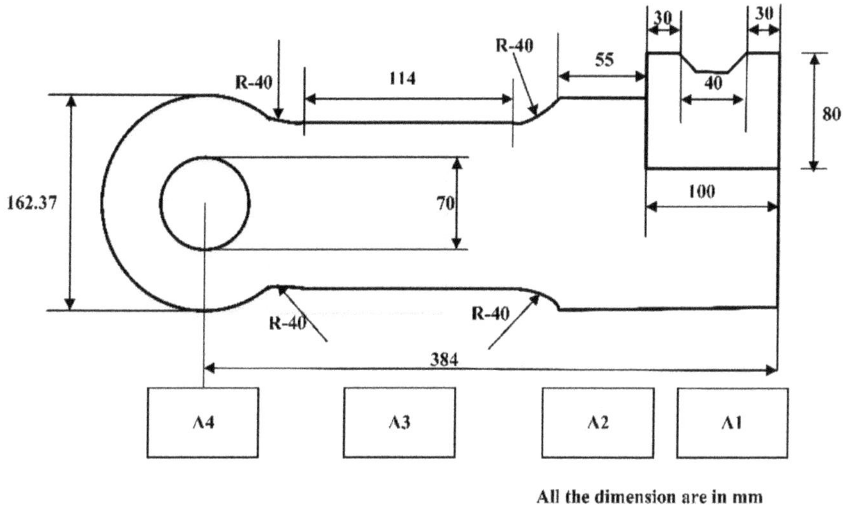

Fig. 78.2 2D schematic representation of current Fibrizer hammer

hammer design analysis were height at tail part of hammer $H = 180$ mm; web height $h = 110$ mm; and radius of the fillet of Fibrizer hammer $r = 40$ mm. From these parameters r/h; H/h; moment of inertia (I) and C which is distance from the neutral axis to the point of interest of hammer. By which, bending moment (M) and maximum shear stress $\left(\delta_{\max} = k_t \frac{MC}{I}\right)$ were evaluated considering a constant factor (k_t) value as 1.58.

78.2.1.3 Load Analysis of Current Fibrizer Hammer

Load analysis of current Fibrizer hammer was evaluated by using speed of Fibrizer (N) = 750 rpm and motor power = 2 MW. By calculating the center of gravity ($Y = (A1Y1 + A2Y2 + A3Y3 + A4Y4)/A1 + A2 + A3 + A4)$), angular speed ($\omega = 2\pi N/60$); linear velocity ($v = \omega \times Y$) and Torque ($T = 60P/2\pi N$).

Figure 78.3 indicates the forces which are acting on Fibrizer hub and hammer during the process. There were three major forces acting on the hub and hammers such as force created by the pin (F_p), centrifugal force (F_c), and force generated by cane being shredded (F_s). Centrifugal force (F_c) was calculated by $mr\omega^2$, where m is mass of hammer and r is distance of gravity center from rotation axis. Loading of one hammer during normal operation was value of tangential force, i.e., force created by pin (F_p) during conventional operation of Fibrizer. For this the force the following presumption were taken such as first hammer carries 75% of loading, second 20% and third 5% and evaluated the exact F_p.

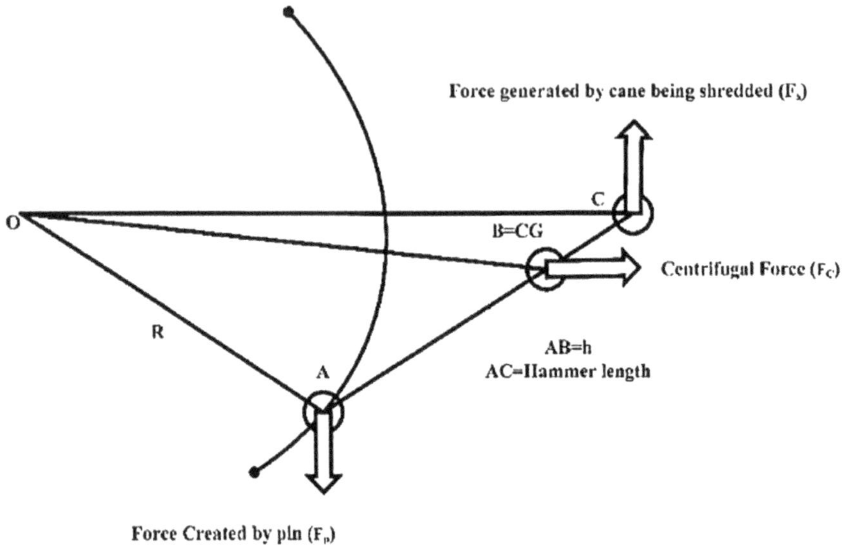

Fig. 78.3 Schematic of skeleton force diagram of Fibrizer hub and hammer

78.2.1.4 Hardness Measurement

The sample was taken from Fibrizer hammer at three places such as L-1, L-2, and L-3 indicated in Fig. 78.4. The hardness of the Fibrizer hammer was tested by Vickers hardness testing machine.

Fig. 78.4 Broken parts of Fibrizer hammer

78.2.1.5 Design Modification of Fibrizer Hammer

Necessity for Modification

The current Fibrizer hammer breakage problems at WSSF are happening frequently. The breakages locations were observed to be at the hard-faced welded part, at the tail part, and sometimes at the hole in which the Fibrizer hammer was fixed with rotatable shaft. The other problem is geometrical structure which exposed the Fibrizer hammer. This problem is not only at WSSF but also other sugar industries in Ethiopia and may be in the world sugar industries where this technology is under implementation. The details of the cost downtime losses were indicated in Table 78.2 in US dollars.

Load analysis of Modified Fibrizer Hammer

Figure 78.5 indicated the complete dimensions of the modified Fibrizer hammer. For this hammer also, the center of gravity for load analysis at different zone of areas such as A1, A2, A3, and A4 as shown in Fig. 78.5 was evaluated. Similar to existing Fibrizer hammer, Center of gravity (C.G), angular velocity (ω), linear velocity (v), torque (T), Centrifugal force (F_c), Work done in shredding per revolution, Force generated by pin (F_p), Force generated by cane being shredded (F_s) and Stress concentration analysis by which maximum shear stress were calculated for modified Fibrizer hammer. In addition to that, cost comparison was also made between existing and modified Fibrizer hammer.

Table 78.2 Loss through downtime

Year	Month	Nature of failure	Recorded downtime (h)	Loss due to downtime in Quintal	Cost of production in USD	Total in USD
2014	December–April	Fibrizer Hammer breakage	60	13,636.4	750,000	1,512,503
2015	April	Fibrizer hammer breakage & shearing of bolts	40.5	9204.6	506,253	
2016	May	Fibrizer hammer breakage & shearing of bolts	20.5	4659.1	256,250	

Note For 1 Qt = 55USD

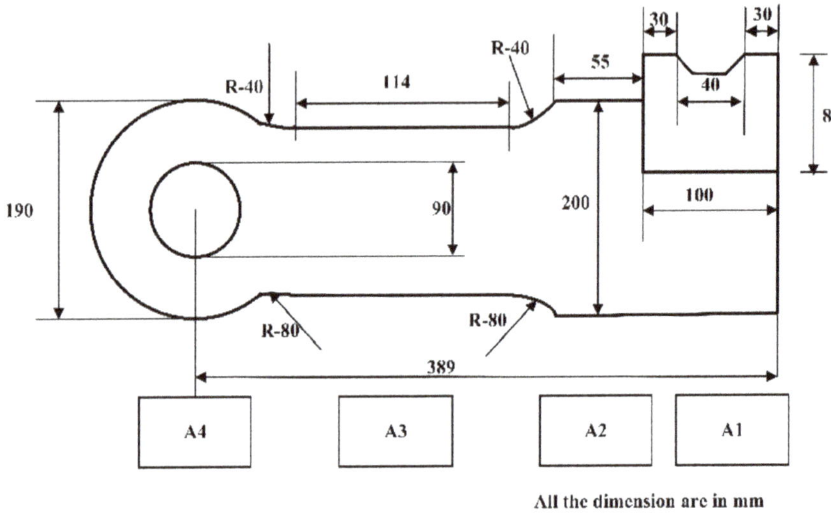

Fig. 78.5 2D schematic diagram of modified Fibrizer hammer

78.2.1.6 Harmonic Analysis Using Finite Element Simulation

A three-dimensional model of Fibrizer hammer was developed in a modeling software SOLDWORKS by comprehensively considering the structural feature, processing requirements, and solution calculation amount for current and modified Fibrizer hammers. This model was imported into ANSYS 18.0, a finite element software as shown in Fig. 78.6. The hammer part adopted the structural steel as per the records. Automatic mesh generation was activated and obtained that the number of nodes were 81,074, unit number was 48,701 for current Fibrizer, whereas 42,601 and 24,487 nodes and units were for modified Fibrizer.

One end of hammer where circular hole existed was fixed and other end of Fibrizer hammer was given the point load of 7136.6 N as boundary conditions. Load was applied along positive y-axis. Simulation results were recorded on harmonic response, total deformation, von Mises stresses, maximum shear stress, and

Fig. 78.6 3D model of from solid works to ANSYS current and modified with hard-facing Fibrizer Hammer

maximum amplitude. Young's modulus and passions ratios were 210 GPa and 0.3 considered.

78.3 Results and Discussion

78.3.1 *Stress Concentration Factor Analysis Result of the Existing and Proposed Fibrizer Hammer*

Stress concentration factor is evaluated with respect to the dimension of exiting Fibrizer hammer and proposed hammer. The final results are tabulated in Tables 78.3 and 78.4.

As the above results shows that the maximum shear stress of the new modified Fibrizer hammer is reduced compared to existing, it indicates that there is less shear stress and less stress concentration factor in modified Fibrizer hammer. The large *H*/*h* and the small *r*/*h* values give worse stress concentration (k_t).

78.3.2 *Load Analysis Result*

The result of load analysis for existing and proposed Fibrizer hammer results has been indicated in Table 78.5. The results show that maximum load can be derived with the modified Fibrizer hammer in all directions.

Table 78.3 Stress concentration factor analysis result for existing Fibrizer hammer

H (mm)	*h*	*r*	*r*/*h*	*H*/*h*	k_t	$I = 1/12bc$	$\delta_m = k_t MC/I$ (MPa)
180	100	25	0.25	1.9	1.58	0.0000042	22.6
180	110	40	0.27	1.7	1.34	0.0000055	16.1
180	120	50	0.33	1.58	1.28	0.0000072	13
180	130	60	0.38	1.46	1.23	0.0000092	10.9

Table 78.4 Stress concentration factor analysis for proposed

H (mm)	*h*	*r*	*r*/*h*	*H*/*h*	k_t	$I = 1/12bc$	$\delta_m = k_t MC/I$ (MPa)
190	150	80	0.53	1.27	1.2	0.0000114	8.8

Table 78.5 Load analysis comparison between existing and modified Fibrizer hammer

Hammer	F_c (N/Hammer)	F_p (N)	F_s (N)
Existing Fibrizer hammer	6623	406	314.2
Modified Fibrizer hammer	7170.6	529	342.5

In Table 78.5, the result of centrifugal force of the modified Fibrizer hammer is increased which does work. The higher the centrifugal force, the more would be the preparation index.

78.3.3 Vickers Hardness Results

The test is carried out on each sample at five different locations with three reputations. These results show that the hardness of the material is not the same throughout the hammer. At the hard-facing welded part, the hardness is high and less at the tail part. Due to this, the Fibrizer hammer frequently breaks at the welded part. The welding heat input is important welding parameter, which effects on the structure and properties of the weld metal. At location-1, sample harness result shows that the hardness varied between 110 and 124 VHN. At the location-2, Fibrizer hammer hardness is between 146 and 151 VHN, and at the location-3, hardness is between 152 and 167 VHN.

78.3.4 Cost Comparison Between Existing and Modified Fibrizer

Table 78.6 indicates the cost comparison between existing and modified Fibrizer. The difference in cost between them is $11,088.00.

78.3.5 Harmonic Analysis Results of Existing and Modified Fibrizer Hammers

ANSYS finite element software is used for the simulation to performing harmonic analysis. From the simulation results, total deformation, direction deformation, von Mises stresses, maximum shear stresses, and maximum amplitude are evaluated for existing and modified Fibrizer hammers.

Table 78.6 Cost Comparison

Hammer	No. of hammers	Unit cost in USD	Total cost in USD
Imported Fibrizer hammer	176 pcs	$167	$29,392.00
New fabricated Fibrizer hammer	176 pcs	$104	$18,304.00
	Difference	$63	$11,088.00

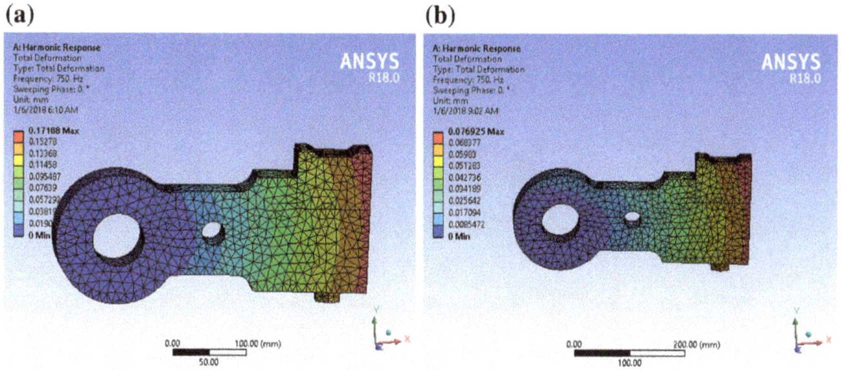

Fig. 78.7 Total deformation of **a** existing and **b** modified FH

From Figs. 78.7, 78.8, 78.9, and 78.10 analysis, results are tabulated in Table 78.7. From the results, it is understood that the total deformation in the modified Fibrizer hammer is lesser than the total deformation of the existing hammer. This shows that the modified Fibrizer hammer is more reliable than the current one. As the result shows that the maximum equivalent (von Mises) stress of current FH is increased by double, the higher the stress, the more the material will be exposed to be broken. The maximum shear stress result in the current Fibrizer hammer is more than the modified Fibrizer hammer. The smaller the radius (r) and web (h) in stepped plate of the Fibrizer hammer, the higher will be the stress.

In Table 78.7, analysis results show that there is less total and directional deformation in modified Fibrizer hammer. The equivalent (von Mises) stress and max. shear stress result also much less by half from the existing.

Similar manner frequency responses are verified for other surfaces also and tabulated in Table 78.8. From Table 78.8 data, the output of harmonic response has been taken in the form of amplitude which can be understood as mean stress development in any engineering components. Modified FH has more amplitude initially and gradually decreased compared to current FH. The overall behavior modified FH is taking less deformations and stresses by which life span of FH definitely will improve.

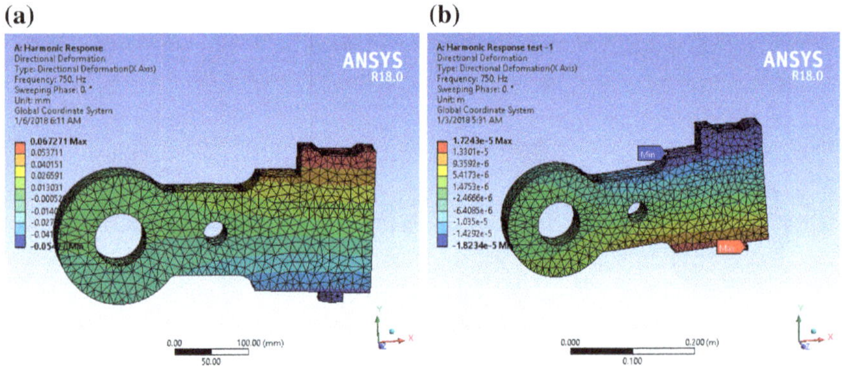

Fig. 78.8 Directional deformation of **a** existing and **b** modified FH

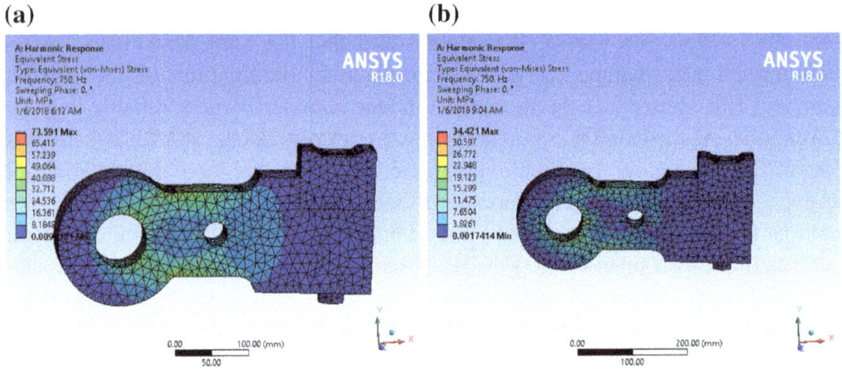

Fig. 78.9 Equivalent (von Mises) stresses of **a** existing and **b** modified FH

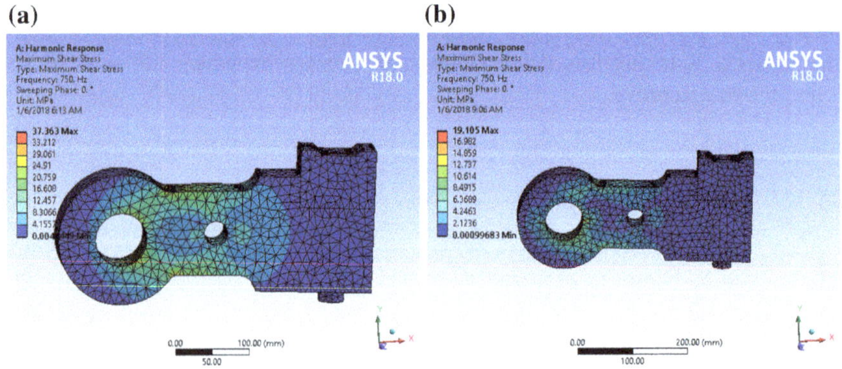

Fig. 78.10 Max. shear stresses of **a** existing and **b** modified FH

Table 78.7 Simulation results on current and modified FH

Parameter	Existing FH	Modified FH
Total deformation	0.1788 mm	0.076925 mm
Max. directional deformation	0.0672 mm	1.7243e − 5 = 0.0116 mm
Equivalent (von Mises) stress	73.591 MPa	34.421 MPa
Max. shear stress	37.363 MPa	19.105 MPa

Table 78.8 Frequency response of modified and current FH

Frequency responses at different surfaces	Modified FH max. amplitude (MPa)	Current FH max. amplitude (MPa)
1	114.9	127.91
2	216.1	184.94
3	28.1	37.25
4	4.5	5.97
5	0.29	0.44
6	0.26	0.24
7	0.59	0.28
8	0.59	0.44
9	0.2	12.07
10	0.38	43.24

78.4 Conclusions

Load and stress analysis of the existing Fibrizer hammer and the new modified Fibrizer hammer are performed. From the results, the stress concentration factor analysis revealed that there is high stress concentration factor on the fillet radius of the existing Fibrizer hammer and less at web thickness (h) that is the cause of the hammer to be broken at the tail part. It shows that the current Fibrizer hammer having design problem and stress concentration factor on the fillet radius of the modified designed Fibrizer hammer is reduced by half. The experimental result of hardness analysis at different locations reveals that the average Vickers hardness of existing Fibrizer is 140.76 VHN. From this experimental result, one can understand that the material property of the Fibrizer hammer is not uniform throughout the hammer. Total deformation, directional deformation, equivalent von Mises stress, max. shear stress and harmonic response analysis of current, and modified Fibrizer hammer were analyzed and compared using FEM software. The result confirmed that the modified Fibrizer hammer having less total deformation, directional deformation, equivalent von Mises stress, max. shear stress, and harmonic response analysis than the current Fibrizer hammer. Using the modified Fibrizer hammer will solve Wonji Shoa Sugar Factory Fibrizer hammer breakage.

References

1. Anderson, J.G., Camuglia, S.I.J., Trapp, N.: Enhancing the life of shredder hammer tungsten carbide tips. In: Proceedings of the Australian Society of Sugar Cane Technologists, pp. 508–518. Scribe Consulting (2005)
2. Anderson, T.L.: Fracture Mechanics: Fundamentals and Applications. CRC press (2005)
3. Hugot, E.: Handbook of Cane Sugar Engineering, 3rd edn. (1986–2014)
4. Rein, P.: Cane Sugar Engineering, 2nd ed. Verlag Dr. Albert Bartens KG (2016)
5. Podany, P., Džugan, J., Vacik, J.: Construction of hammer for sugarcane shredder. In: Advanced Materials Research, vol. 811, pp. 308–313. Trans Tech Publications (2013)
6. Ransing, K.A., Gajjal, D.S., Saidpatil, V.V.: Redesign and optimization of hub profile of Fibrizer in sugar cane industry. Int. Eng. Res. J. (IERJ), Special Issue **2**, 5790–5793 (2015a)
7. Ransing, K.A., Gajjal, D.S., Saidpatil, V.V.: Review of design and optimization of Fibrizer in sugar cane industry. Int. J. Innov. Sci. Eng. Technol. **2**(4) (2015b)
8. Nikam, P.P., Nimbalkar, A.A., Thorat, A.S., Nanaware, A.A., Taware, O.H.: Design modification and analysis of sugar cane leveler at sugar cane processing plant. Imperial J. Interdisc. Res. **3**(4) (2017)
9. Norton, R.L.: Machine Design: An Integrated Approach. Prentice-Hall Inc., NJ, USA (1996)

Chapter 79
FEA Validation of Experimental Results of First Ply Failure of Composite Structure

Ishan Sonawane and A. Muthuraja

Abstract Laminated composite has gained popularity (despite of their higher cost) in high-performance products that need to have high corrosion resistance, good thermal stability, excellent fatigue resistance, lightweight and high specific strength to take harsh loading conditions such as aerospace components. Based on the literature survey, it is observed that Tsai–Wu criterion is widely used because of its accuracy. In the current study, finite element analysis is being carried out by using MSC Nastran and MSC Patran to identify the critical ply location and strength of first ply failure within laminated composite material plates. Tsai–Wu failure criterion is used to predict the maximum failure indices on each ply of laminated composite material. The results obtained from FEA are validated by comparing with the experimental result.

Keywords Laminates · Composite materials · First ply failure · Finite element analysis

79.1 Introduction

Laminated composite materials have recently gained popularity for different applications in the construction of mechanical, aerospace, marine and automotive structures. In general, these structures require high reliability. Within the laminated structure, plies are oriented in a different direction so as to make the composite structure tailor able when multi-directional loads are applied. This makes the laminated structure highly anisotropic and structurally inhomogeneous. So for reliability conformance, we need to predict accurately that maximum load a ply can sustain before failure. During the design phase of aerospace composite structures,

I. Sonawane · A. Muthuraja (✉)
Department of Mechanical Engineering, Sandip University, 422213 Nashik, India
e-mail: a.muthuraja2010@gmail.com

I. Sonawane
e-mail: s.ishan.r@gmail.com

R. G. Narayanan et al. (eds.), *Advances in Computational Methods in Manufacturing*, Lecture Notes on Multidisciplinary Industrial Engineering,
https://doi.org/10.1007/978-981-32-9072-3_79

first ply failure is considered as critical phenomena in composite failure. However, after the first ply failure other ply will carry the loads till the complete failure of composites happens. First ply failure analysis of laminated composite has drawn close attention in recent years. Turvey [1] used analytical methods to study the linear and nonlinear first ply failure loads of simply supported symmetrically and anti-symmetrically laminated composite plates based on the classical lamination theory. Tsai–Hill failure criterion has been used to facilitate an initial flexural failure analysis [2, 3]. Initial flexural failure was analyzed under uniform or non-uniform loading [2]. It was obtained that the initial failure loads decreased with corresponding plate center deflections as plate patch size increases during loading at Z-direction. Tsai–Hill failure criterion was used for angle ply laminates of GFRP and CFRP [3]. It was concluded that the initial failure loads decrease rapidly as theta increases, and the simply supported GFRP strips were obtained to be insensitive to the number of constituent laminas under initial failure loads. Turvey [4] obtained that the initial failure loads were direction-dependent, and simply supported strips were shown superior and increase with increasing number of laminas. Kam and Jan [5] formulated a finite element model for a thick laminated composite plate on the basis of layer wise linear displacement theory. Finite element results were compared between experimental and analysis results for displacements and stresses. Reddy and Pandey [6] developed a finite element computational procedure for the first ply failure analysis of laminated composite based on the first-order shear deformation theory and for a tensor polynomial failure criterion like maximum stress, maximum strain, Tsai–Hill, Tsai–Wu and Hoffman. Kam et al. [7] used a nonlinear finite element method to study the deformation and first ply failure of thin laminated composite plates. This study was based on the principle of minimum total

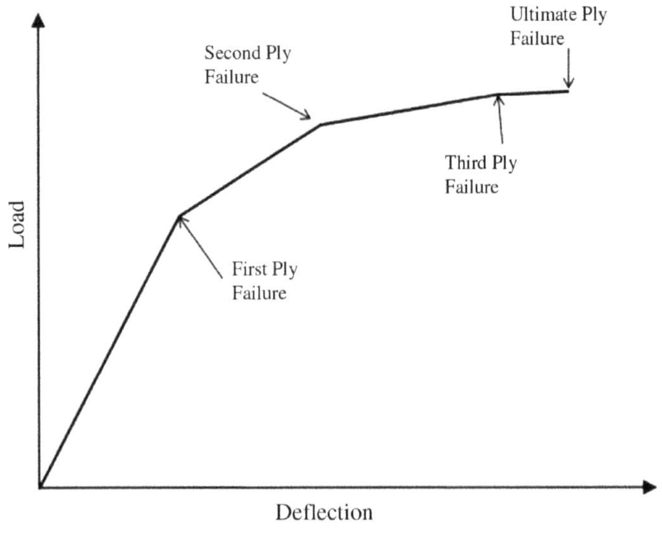

Fig. 79.1 Progressive ply failure

potential energy and von Karman–Mindlin plate theory. Reddy and Reddy [8] developed a finite element computational procedure to find first ply failure loads of composite laminates subjected to in-plane and transverse loads. This analysis was carried out for linear and nonlinear loading conditions, based on the first-order shear deformation theory and other failure criteria.

In the current study, finite element analysis software MSC Patran/Nastran is used to analyze the composite failure. Patran is used as pre-/post-processing, and Nastran is used as a solver. Nastran program was originally developed for NASA (Fig. 79.1).

79.2 Finite Element Method

79.2.1 Finite Element Modeling

In this paper, the finite element method is used to determine the first ply failure strength of laminated composite plate of 100 × 100 mm dimensions. The laminated composite plate is made with equal ply thickness of 0.121 mm and different orientations. Stacking sequence of $\left[45^0_3/0^0_3/-45^0_3/90^0_3\right]_S$ is used to construct the composite laminate. The lamina (graphite/epoxy) properties are considered from the experiments conducted [9] (Table 79.1).

The plate is modeled with 2D elements as the thickness is very less in comparison with other two dimensions. Also, the 2D elements have the ability to capture the bending stiffness. Elements are created on the mid surface and thickness is assigned. Vertically downward concentrated loads of 500, 1000 and 1335 N are applied independently at the center of the plate. It is observed from the result that the first ply failure occurs at a load of 1335 N. Thus, in this paper, a load of 1335 N

Table 79.1 Composite material properties

Material properties (linear elastic)	
Elastic modulus 1 (E11)	139,400 MPa
Elastic modulus 2 (E22)	7650 MPa
Shear modulus 12 and 13 (G12 and G13)	4350 MPa
Shear modulus 23 (G23)	1020 MPa
Poisson ratio (ϑ_{12})	0.29
Strength properties	
Tension stress limit (X_T)	1537.2 MPa
Tension stress limit (Y_T)	42.7 MPa
Compressive stress Limit (X_C)	1722.4 MPa
Compressive stress limit (Y_C)	213.9 MPa
Shear stress limit (S and T)	102.4 MPa
Interaction term	−3.215E − 6
Bonding shear stress limit	5000 MPa

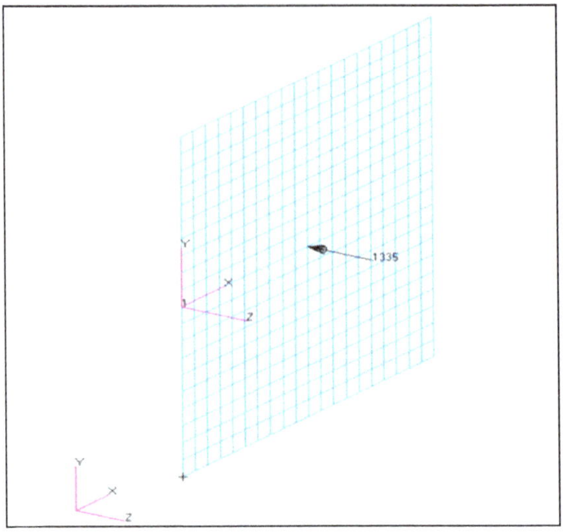

Fig. 79.2 Applied load

is applied as shown in Fig. 79.2, and all results are extracted at 1335 N applied load.

In accordance with experimental method carried out by [5], the model is constrained in all six directions at all four edges as shown in Fig. 79.3.

79.2.2 Results and Discussion

The strength of the composite plate is determined by using Tsai–Wu failure criteria. Figure 79.4 shows that the maximum failure index for non-layered plate occurs at the center of plate. Figure 79.5 shows that Tsai–Wu failure index of 1 is observed at the bottom ply location at the center location. During experimental tests, T. Y. Kam and F. M. Lai found that the first ply failure occurs at the bottom location with applied load of 1320 N. The FEM result is compared with the experimental result [5] and found that there is 1% difference of failure load between the experimental and FEA results. In both cases, first ply failure occurs at the bottom location.

79.3 Theoretical Approach

The constitutive model is considered in the linearly elastic region, and Tsai–Wu criterion is used to predict the first ply failure of the laminated plate using the maximum failure index. A number of failure criteria for composite laminate proposed over the years are maximum stress, maximum strain, Hill–Tsai, Tsai–Wu. Stress interaction between the laminates is ignored for maximum stress and

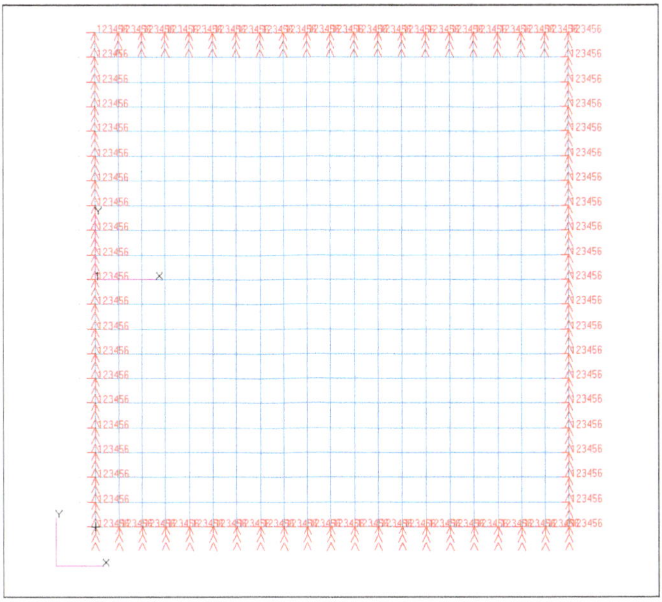

Fig. 79.3 Boundary conditions

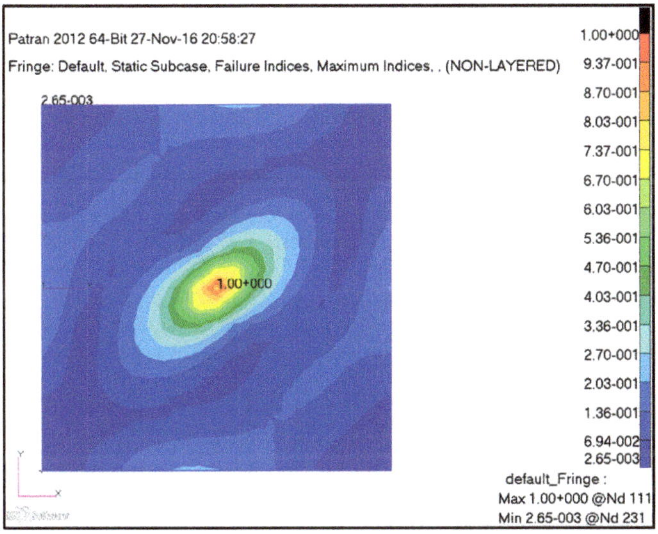

Fig. 79.4 Maximum failure index

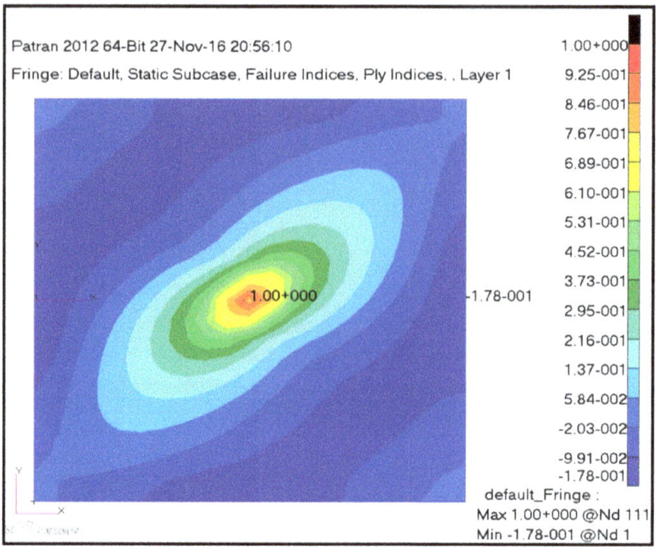

Fig. 79.5 First ply failure

maximum strain criteria; whereas, Hill–Tsai and Tsai–Wu criteria include full stress interaction.

The Tsai–Wu failure criterion is a macro-mechanical failure theory which is prominently used for anisotropic composite materials which have different strengths in tension and compression. Tsai–Wu uses strength tensors (F_i and F_{ij}) established through experiment. Theoretical predictions of Tsai–Wu criterion match very well compared with the experimental data.

Tsai–Wu theory predicts that the failure occurs when below criteria is satisfied

$$F_i\sigma_i + F_{ij}\sigma_i\sigma_j \geq 1 \text{ where } i,j = 1,2\ldots6$$

(Summation is done for repeated indices)

$$F_1 = \frac{1}{X_T} - \frac{1}{X_C} \quad F_2 = \frac{1}{Y_T} - \frac{1}{Y_C} \quad F_3 = \frac{1}{Z_T} - \frac{1}{Z_C} \quad F_{11} = \frac{1}{X_T X_C} \quad F_{22} = \frac{1}{Y_T Y_C}$$

$$F_{33} = \frac{1}{Z_T Z_C} \quad F_{44} = \frac{1}{R^2} \quad F_{55} = \frac{1}{S^2} \quad F_{66} = \frac{1}{S^2}$$

$$F_{12} = -\frac{1}{2\sqrt{X_T X_C Y_T Y_C}} \quad F_{13} = -\frac{1}{2\sqrt{X_T X_C Z_T Z_C}} \quad F_{23} = -\frac{1}{2\sqrt{Y_T Y_C Z_T Z_C}}$$

X_T, Y_T, Z_T (X_C, Y_C, Z_C) Lamina normal tensile (compressive) strength in 1, 2 and 3 directions.

R, S	Shear strengths in the 23 and 12 planes, respectively.
σ_1, σ_2, σ_3	Normal stress components; σ_4, σ_5, σ_6: shear stress components

For the analysis, shear strengths in the 12 and 13 planes are assumed to be same and $Y_T = Z_T$.

79.4 Conclusions

First ply failure strengths of laminated composite plates with different layups were determined using FEA, and the results were validated by using experimental data. The Tsai–Wu failure criterion is used to calculate the failure index.

1. It is observed that there is 1% difference of failure load between the experimental and FEA results. Bottom most layer shows the maximum failure index under 1335 N loading condition comparatively other layers. It shows the accuracy of first ply failure prediction from both the methods.
2. In composite laminate, the stacking sequence plays an important role in improving the strength with different loading conditions. During the design phase, stacking sequence changed frequently so as to obtain feasible design. By experimental methods, we can evaluate the first ply failure strength and location. But experimental setup and procedure results are in high cost. During design phase, it is difficult to go for experimental results every time, which results in

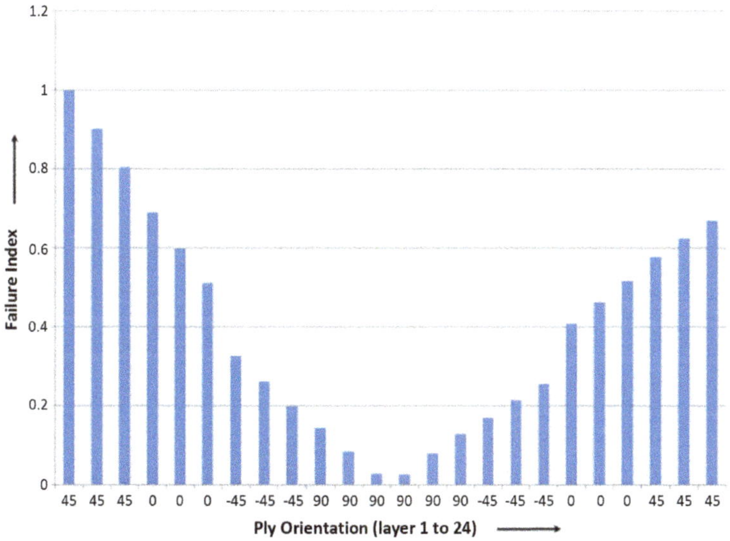

Fig. 79.6 Failure indices of different ply orientation

high cost. By FEA, we can predict the first ply failure strength and location. Also, we can predict the strength of other critical plies as shown in Fig. 79.6.

3. After the first ply failure, the composite will not fail suddenly, and the loads are transferred to other plies. It is important to detect the remaining strength in the structure till final failure. In many cases, composite structures are connected with metallic joints. Composites are weaker at the joint location. So, further study needs to be done on the first and progressive ply failure analysis at composite joint location.

Acknowledgements First and foremost, we praise and thank our almighty God whose blessings have bestowed on us the will power and confidence to carry out this paper. We would also like to thank all our friends and our parents who have prayed and helped us during the project work.

References

1. Turvey, G.J.: An initial flexural failure analysis of symmetrically laminated cross-ply rectangular plates. Int. J. Solids Struct. **16**, 451–463 (1980)
2. Turvey, G.J.: Initial flexural failure of square, simply supported, angle-ply plates. Fibre Sci. Technol. **15**, 47–63 (1981)
3. Turvey, G.J.: Flexural failure analysis of angle-ply laminates of GFRP and CFRP. J. Strain Anal. Eng. Des. **15**, 43–49 (1980)
4. Turvey, G.J.: A study of the onset of flexural failure in cross-ply laminated strips. Fibre Sci. Technol. **13**, 325–336 (1980)
5. Kam, T.Y., Jan, T.B.: First ply failure of laminated composite plates based on the layerwise linear displacement theory. Compos. Struct. **32**, 583–591 (1995)
6. Reddy, J.N., Pandey, A.K.: A first-ply failure analysis of composite laminates. Comput. Struct. **25**, 371–393 (1987)
7. Kam, T.Y., Chang, R.R., Sher, H.F., Chao, T.N.: Prediction of first-ply failure strengths of laminated composite plates using a finite element approach. Int. J. Solids Struct. **33**, 375–398 (1996)
8. Reddy, Y.S.N., Reddy, J.N.: Linear and nonlinear failure analysis of composite laminates with transverse shear. Compos. Sci. Technol. **44**, 227–255 (1992)
9. ASTM Standards and Literature References for Composite Material, 2nd ed. (1990)

Chapter 80
Triply Periodic Minimal Surface-Based Porous Scaffold Design and Analysis Subjected to Hard Tissue Reconstruction

Verma Rati, Nishant Singh, Sanjay Rai and Shekhar Kumta

Abstract Porous structures based on triply periodic minimal surfaces (TPMS) have occurred as appropriate candidates for scaffold design with high level of porosity and promising strength for bone replacement. This work describes a suitable procedure for design and modeling of 3D architecture of TPMS-based gyroid and primitive structures and identifying the optimal architecture for scaffold designing. Different models with varying porosity were reconstructed and analyzed to access the effective elastic moduli and compressive strength of each model. An optimal model that can be utilized in bone tissue replacement is identified eventually. Ti6Al4V which is considered as the best material for producing implants in conjunction with biocompatibility and strength is used in the study. The compression test performed by FEM revealed that the scaffold models with porosity level of 65 and 60% are best suited for cortical bone replacement and the model with 90% porosity can be used on the anatomical location, which are more inclined to cancellous bones.

Keywords Tissue engineering · Porous scaffolds · TPMS · Gyroid · Primitive · Finite element analysis

V. Rati · S. Rai (✉)
School of Biomedical Engineering, IIT BHU, Varanasi, India
e-mail: skrai.bme@iitbhu.ac.in

V. Rati
e-mail: rati.verma14@gmail.com

N. Singh
Department of Biomedical Engineering, NIT Raipur, Raipur, Chhattisgarh, India
e-mail: nksingh.bme@gmail.com

S. Kumta
Department of Orthopaedics and Traumatology, Prince of Wales Hospital, the Chinese University of Hong Kong, Shatin, New Territories, Hong Kong
e-mail: shekharkumta@gmail.com

R. G. Narayanan et al. (eds.), *Advances in Computational Methods in Manufacturing*, Lecture Notes on Multidisciplinary Industrial Engineering,
https://doi.org/10.1007/978-981-32-9072-3_80

80.1 Introduction

The field of tissue engineering aims to develop an efficient biological substitute in order to preserve, maintain, or enhance the function of tissue. Scaffolds play a significant part in deciding mechanical characteristics of desired tissue formed in it at a macroscopic or microscopic scale and also provides the vascularization needed by the newly forming tissues [1]. Scaffold must be biocompatible, bio-inert, bioactive, bioresorbable, and biodegradable so as to work efficiently to possess long term results [2]. Porous scaffold, therefore, provides necessary temporary atmosphere for adhesion, propagation, and differentiation of cell [3], together with the transport phenomena and degradation behavior [4]. Other advantages of these porous scaffolds include its flexibility in terms of mechanical properties, a large surface area that enables bio-functionalization and infection prevention and accelerates bone ingrowth as well as facilitates drug delivery in implants [5]. To establish a balance between the mechanical properties of scaffold and porosity is a challenge [1, 6]. Tissue regeneration efficiency is greatly affected by scaffold pore size [3].

Triply periodic minimal surface (TPMS) is an emerging area of interest for designing the scaffolds because of its unique properties like the ability to mimic the organic system and to build self-supportive structures for additive manufacturing [7, 8]. Another reason for TPMS-based scaffolds to draw the attention toward it is because of its capability to provide site-specific biological performance [9]. TPMS is defined in mathematical terms as surface curvatures which are infinite and periodic in 3D Euclidian space that owes to a surface capable of generating highly controllable and homogeneous scaffold design [10]. TPMS have implicit periodic surfaces in three independent Cartesian directions with mean curvature equal to zero [11]. These minimal surfaces play a very essential role in guiding chemical, cellular, and biochemical processes [10]. The TPMS-based porous structures have the potential to attain the lightweight bone models and medical requirements of bone tissue, thus providing a solution for the production of biomimetic bone in biomaterials and tissue engineering [12].

Observing the TPMS topology for Schoen's gyroid that perfectly suits the porous-solid structural configuration ideal for tissue engineering scaffolds [4, 13–16], this paper emphasizes on TPMS surface geometry and gyroid surface, particularly.

The present study aims to develop a porous scaffold based on TPMS geometry that possesses the mechanical properties (elastic modulus and compressive strength) proximate to bone. The study utilizes Schwarz's primitive (*P*) and Schoen's gyroid surface (*G*) structures to design bone scaffolds as their macroscopic structure is analogous to bone. Ti6Al4V (Grade 5) regarded as a biocompatible material is employed with its mechanical properties with scaffold porosities within the range of 60–90%. Unit cell length and pore size are varied with wall thickness kept constant at 102 μm for *G* and 133 μm for *P*, respectively, to achieve porosities within the desired range for both the structures (*P* and *G*). Finite element analysis was performed in ANSYS® software with static loading and boundary conditions for the determination of compressive strength based on von-Mises criteria to study the deformation behavior and to estimate the effective elastic moduli and compressive yield strength.

80.2 Materials and Method

80.2.1 TPMS-Based Porous Scaffold-Primitive (P) and Gyroid (G) Surface Modeling

Both the primitive (*P*) as well as gyroid surface are designed from the basic parametric equations with implicit functions described in Eqs. (80.1) and (80.2).

$$\text{Primitive}\,(P)\!:\ \cos(x) + \cos(y) + \cos(z) = 0 \tag{80.1}$$

$$\text{Gyroid:}\ \sin(x) * \cos(y) + \sin(y) * \cos(z) + \sin(z) * \cos(x) = 0 \tag{80.2}$$

The unit cell of TPMS is designed with a MATLAB code, executed in the MATLAB® (v. R2016a) software. "MESHGRID" command along with "ISOSURFACE" command was used to extract the isosurface 3D data from the volume 3D data generated by "MESHGRID" command.

The unit cell was extended periodically in x, y, and z directions by modulating the Eqs. (80.1) and (80.2) as described in Eqs. (80.3) and (80.4) [17].

$$\text{Primitive}\,(P)\!:\ F_P = \left[\cos\left(\frac{2\pi Nx}{L}\right) + \cos\left(\frac{2\pi Ny}{L}\right) + \cos\left(\frac{2\pi Nz}{L}\right)\right] - P \tag{80.3}$$

$$\begin{aligned}\text{Gyroid:}\ F_G(x,y,z) = {} & \left[\sin\left(\frac{2\pi Nx}{L}\right) * \cos\left(\frac{2\pi Ny}{L}\right) + \sin\left(\frac{2\pi Ny}{L}\right) * \cos\left(\frac{2\pi Nz}{L}\right)\right. \\ & \left. + \sin\left(\frac{2\pi Nz}{L}\right) * \cos\left(\frac{2\pi Nx}{L}\right)\right] - P\end{aligned} \tag{80.4}$$

where *N* defines the number of unit cells, L is length of the unit cell, and *P* is the strut thickness parameter that controls the volume fraction in a unit cell. For a scaffold with 3 × 3 × 3 unit cells (Fig. 80.1), value of *N* was fixed at $N = 3$, unit cell length L was varied due to which the pore diameter and total scaffold length were altered accordingly as listed in Table 80.1. Seven models for each surface (primitive: P_1–P_7 and gyroid: G_1–G_7) type were designed. As a result, a total of fourteen models were designed. P is kept constant at $P = 0$ to fix the volume fraction at 50%. "STLWRITE" command was added to the MATLAB code to export the geometry as stl file which can be imported to other software for finite element analysis (FEA).

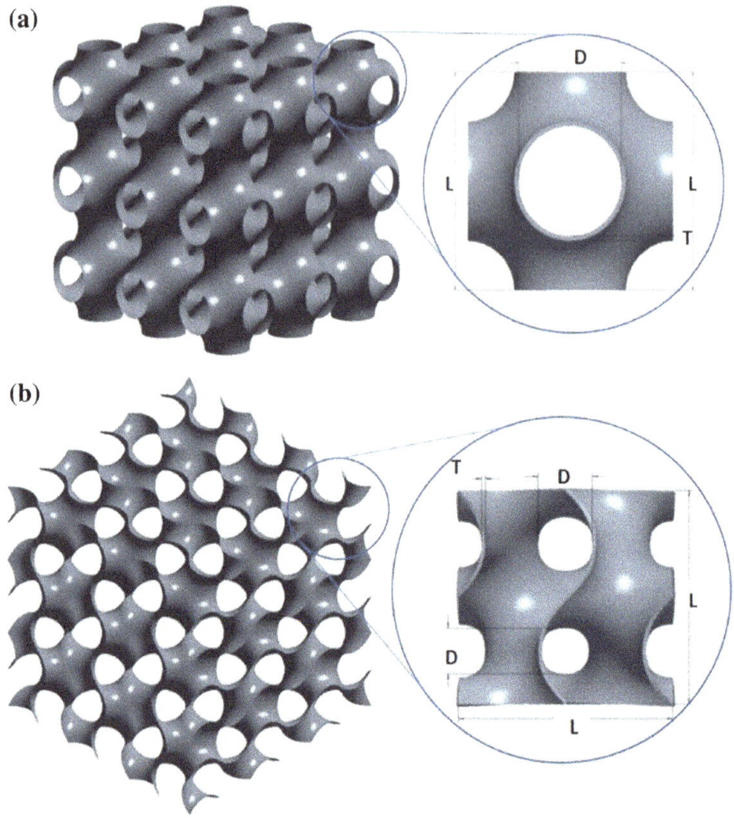

Fig. 80.1 Designed TPMS-based porous surface models. Isometric view with 3 × 3 × 3 unit cells for **a** primitive surface, **b** gyroid surface (Zoom in view inside the circle presents the unit cell length (*L*), pore diameter (*D*) and wall thickness (*T*) of each model)

Table 80.1 Parametric values of TPMS-based primitive (P_1–P_7) and gyroid (G_1–G_7) structures

	P_1, G_1	P_2, G_2	P_3, G_3	P_4, G_4	P_5, G_5	P_6, G_6	P_7, G_7
Porosity (%)	90	85	80	75	70	65	60
Unit cell length (*L*) (μm)	3141	2104	1570	1256	1055	911	791
Pore size (*D*) (μm)	1654	1074	776	600	488	407	340

80.2.2 Design and Formulation of Finite Element Model for FEA

The surface models of P surface and gyroid were imported in 3-matic software as stl files and fixed for any errors in the surface. Wall thickness was developed and the model was exported to ANSYS software for FEA after performing surface and volume meshing. The complete procedure is explained in the following subsections.

Developing wall thickness and preparing Finite Element Models

Stl files generated through MATLAB software were imported in 3-matic software for assigning wall thickness to the surface and carrying out surface and volume meshing. The surface of the models of gyroid and P were thickened with a uniform offset of 102 μm and 133 μm, respectively (Fig. 80.1). Since, offsetting the surface disturbed the surface geometry (mesh) of gyroid structures, surface meshing was performed with specific 3-matic tools to improve the quality of surface mesh. Additionally, to reduce the density of the triangles and improve the quality of mesh, quality preserving reduce triangles tool followed with uniform meshing with a target triangle edge length of 0.2 was performed in a manner that element quality is preserved. The shape-quality threshold and maximum geometrical error values were fixed to 0.5 and 0.05, respectively [7].

The generated models were checked for errors and fixed accordingly. Consequently, the number of inverted normal, noise shells, bad contours, planar holes, overlapping triangles, and intersecting triangles present in each model were fixed manually to obtain a single shell with no errors. Volume meshing was done to solidify the surface models with voxel size of Tet4 elements. The volumetric models of all scaffolds generated in 3-matic software were imported to Ansys software as stl files.

Assigning Material properties, Boundary conditions, and Load parameters

Ti6Al4V (Grade 5) material considered suitable in terms of a compatible biomaterial for implants is used for FEA [18]. The elastic isotropic material properties of Ti6Al4V adopted in the study include Young's modulus of 114 GPa, Poisson's ratio of 0.33, density of 4.43 g/cc, and compressive yield strength of 1070 MPa [19].

Remeshing with 4-node SHELL181 hexahedral element was performed for each model (Fig. 80.2). Two rigid plates are fixed at the top and bottom of each scaffold to assign the boundary conditions and loading. The bottom plate is constrained in all directions while a static compressive load of 75 N is applied in negative Y direction at the top plate (Fig. 80.3).

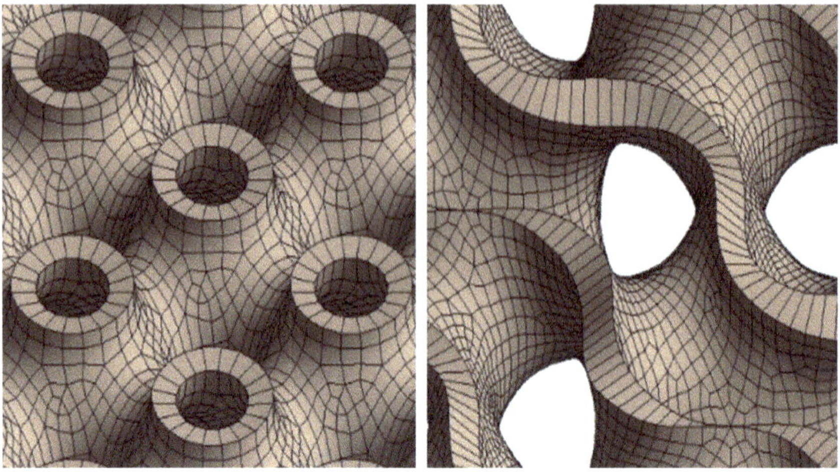

Fig. 80.2 Remeshed models with four-node hexahedral elements for primitive (left) and gyroid (right) surfaces

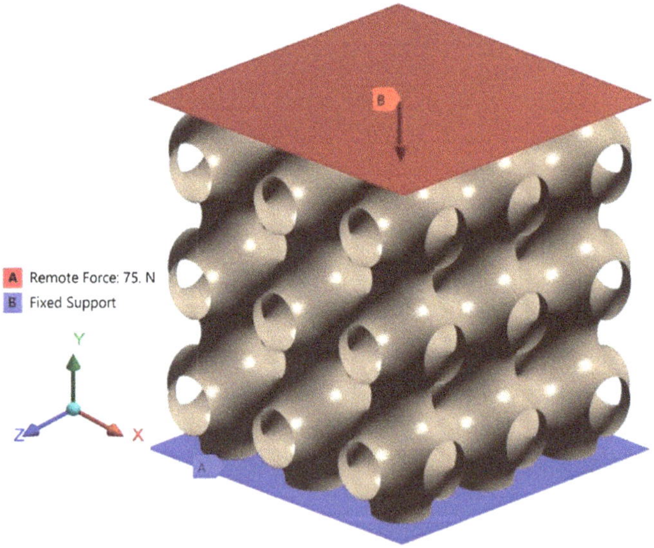

Fig. 80.3 Application of boundary conditions and load parameters. Two rigid plates, red and blue are fixed at top and bottom of designed scaffold structure and compressive load of 75 N is applied in negative Y direction

Table 80.2 Porous volume (V_P) and solid volume (V_S) of primitive (P_1–P_7) and gyroid (G_1–G_7) models

	P_1	P_2	P_3	P_4	P_5	P_6	P_7
(V_P) (μm^3)	2897.9	1287.9	763.62	495.08	348.26	260.81	195.89
(V_S) (μm^3)	28333.5	8395.1	3832.3	2000.8	1180.2	764.9	497.9

	G_1	G_2	G_3	G_4	G_5	G_6	G_7
(V_P) (μm^3)	3123.7	1402.2	780.92	499.79	352.65	262.7	198.37
(V_S) (μm^3)	31006490	9325.9	3875.8	1984.2	1176.2	756.207	496.191

80.3 Porosity

Porosity was calculated from the relation expressed in Eq. (80.3) [20].

$$N = 1 - \left(\frac{V_P}{V_S}\right) \qquad (3)$$

where V_S is the solid volume and V_P is the volume of porous construct. The solid volume constitutes the total volume of cubic construct. Solid volume (V_S) and porous volume (V_P) were calculated all primitive (P_1–P_7) and gyroid surface (G_1–G_7) models with different porosity levels (60–90%) as listed in Table 80.2.

80.4 Results

The effective elastic modulus as well as compressive strength was calculated for each model. Effective elastic moduli were determined from Hooke's law ($\sigma = \varepsilon E$) and compressive strength was obtained from von-Mises criteria. The performance characteristics of scaffold structure with 3 × 3 × 3 unit cells were observed under static compressive load of 75 N. Total von-Mises stress distribution and maximum deformation were measured for scaffolds with varying length and porosity (Fig. 80.4). Compressive strength for each model was determined by the load of 75 N that caused the maximum stress of 909.24 and 1102.5 MPa in *P* surface model and gyroid surface model, respectively, with 70% porosity (Fig. 80.4a, b). Maximum deformation of 28.542 and 22.875 μm was obtained for P surface model and gyroid surface model, respectively, with 70% porosity (Fig. 80.4c, d). Variation in elastic modulus and compressive strength with respect to porosity has been plotted (Fig. 80.5).

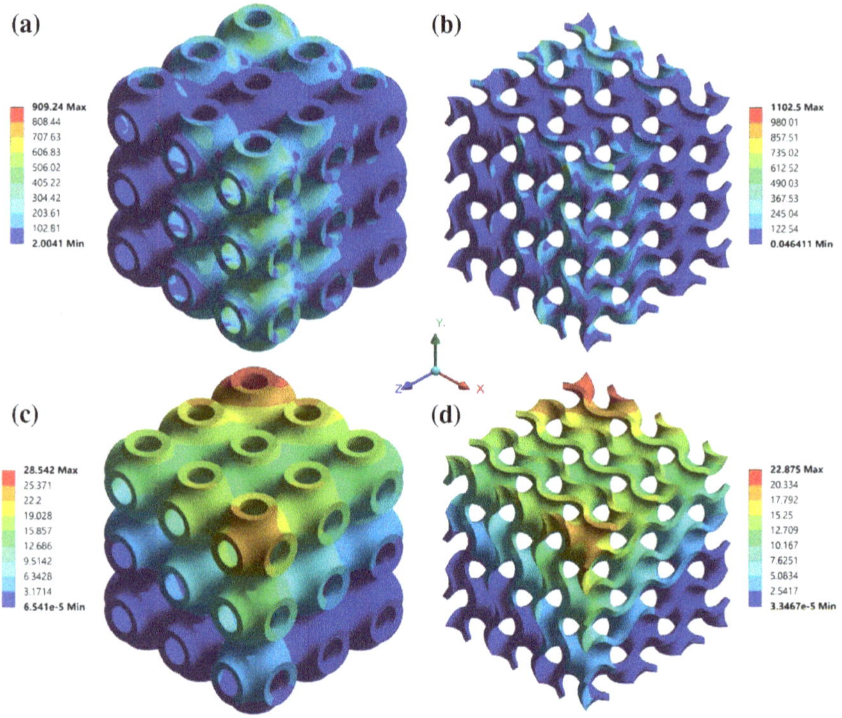

Fig. 80.4 Mechanical properties of scaffolds with 70% porosity obtained from FEA simulation. Von-Mises stress distribution (MPa) for **a** primitive structure: P_5, **b** gyroid structure: G_5 and total deformation (μm) for **c** primitive structure: P_5, **d** gyroid structure: G_5

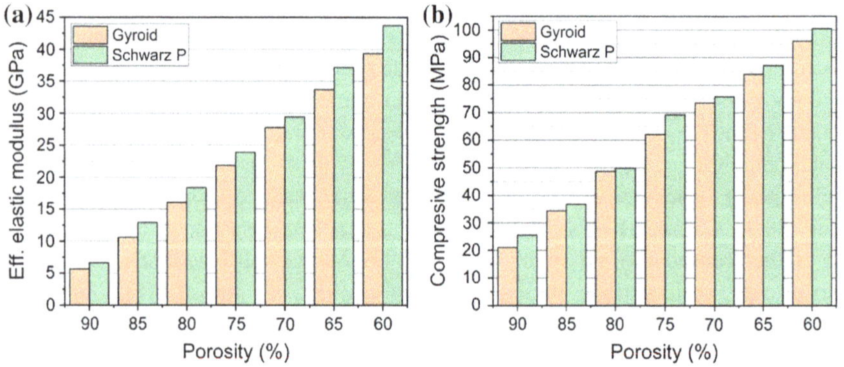

Fig. 80.5 Graphical representation of mechanical properties plotted against porosity. Plot of **a** effective elastic moduli and **b** compressive strength with respect to porosity for Schwarz *P* surface and gyroid surface

80.5 Discussion

Numerous studies have been carried out the design and manufacturing of TPMS-based porous scaffold materials because of the challenges imposed in this field [9, 20–22]. In order to obtain a scaffold that perfectly mimics mechanical properties similar to bone, the effective elastic moduli in addition to compressive strength of the scaffolds are obtained proximate to the bone [23, 24]. Elastic modulus for cortical bone has been noted in the range of 3–30 GPa, whereas for cancellous bone, it is in the range of 0.02–0.2 GPa. Also, the compressive strength for cortical and cancellous bone is reported as 100–230 MPa and 2–12 MPa, respectively [25]. Although, the mechanical characteristics of bone with compressive load show a discrepancy with according to various factors (age, gender, and anatomical locations) within a human body [26–28]. Therefore, a scaffold is designed in such a way so that a balance between the porosity level and mechanical properties can be maintained [29–31]. This prevents the bone from stress shielding as well as reabsorption due to bone–scaffold interaction [32, 33]. TPMS-based porous scaffolds also enhance the bone ingrowth and weight-bearing capacity due to their exceptional porous architecture that are lightweight in comparison with the conventional solid scaffolds [12, 33–35].

Figure 80.5 characterizes the mechanical properties comprising the effective elastic moduli along with stress distribution for primitive and gyroid surface models with varying porosity under a compressive load of 75 N in inverse Y direction. Effective elastic modulus and compressive strength for both structures (primitive and gyroid) was maximum for minimum porosity of 60% and followed a pattern to decrease with increasing porosity.

Observing the characteristics of effective elastic moduli (Fig. 80.5a), a maximum effective elastic modulus of 43.73 MPa for *P* surface and 39.36 MPa for gyroid surface was obtained in case of 60% porosity. However, a minimum value of 6.625 MPa for *P* surface and 5.628496 MPa for gyroid surface was obtained in case of maximum porosity of 90%.

Likewise, the compressive strength (Fig. 80.5b) was also observed to be maximum in case of 60% porosity for both primitive surface (100.519 MPa) and gyroid surface (96.029 MPa). A minimum value of von-Mises stress was obtained in case of 90% porosity for primitive surface (25.52178 MPa) and gyroid surface (21.06333 MPa). However, it can be noted that, for 80% porosity, the compressive strength was approximately equal for *P* surface (49.78 MPa) and gyroid surface (48.68 MPa).

As a result, it can be concluded that effective elastic moduli of gyroid surface models were lower than *P* surface models for all the porosity levels. Similar to the pattern of effective elastic moduli, compressive strength was also observed to be low for gyroid surface models as compared to *P* surface models for all the porosity levels ranging from 60 to 90%.

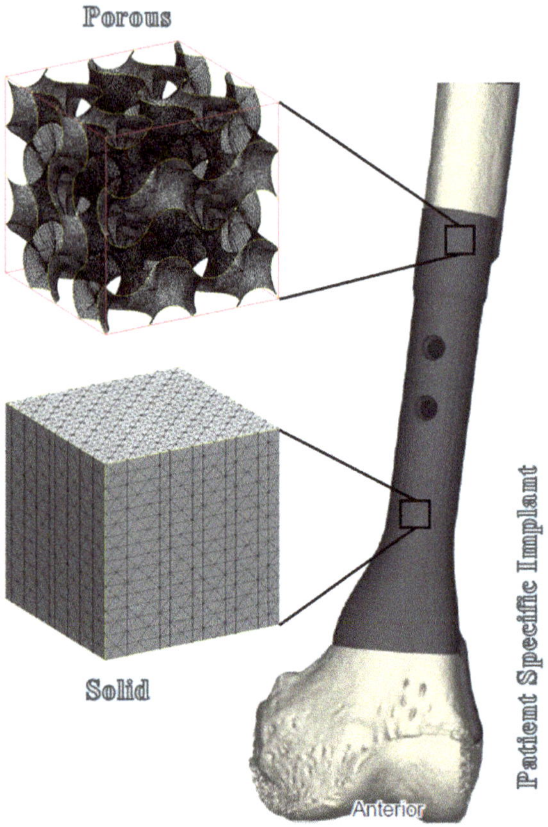

Fig. 80.6 Illustration of mapping of developed porous scaffold to treat large segmental bone defect

The conclusions of this pre-clinical analysis study elucidated the proficiency of design criterion for selecting an efficient scaffold structure in order to obtain the preferred value of the mechanical properties intended for patient-specific implant manufacturing that can be used for treating large segmental bone defects [36], where high porosity scaffolds are greatly needed (Fig. 80.6).

The overview of outcomes achieved in accordance with the mechanical properties of TPMS-based porous scaffold recommended their applicability at different anatomical locations according to the porosity level and suggested them to be appropriate structures for applications in bridging the bone gaps. However, the result is effective merely for desired scaffold structure requirements. The working fundamentals can be followed for other porous scaffold structures and biomechanical bone properties.

References

1. Almeida, H.A.: Design of tissue engineering scaffolds based on hyperbolic surfaces: structural numerical evaluation. Med. Eng. Phys. **36**(8), 1033–1040 (2014)
2. Zhang, X.Y.: Additively manufactured scaffolds for bone tissue engineering and the prediction of their mechanical behavior: a review. Materials **10**(1), 50 (2017)
3. Yang, N.: Mathematically defined gradient porous materials. Mater. Lett. **173**, 136–140 (2016)
4. Melchels, F.P.: Mathematically defined tissue engineering scaffold architectures prepared by stereolithography. Biomaterials **31**(27), 6909–6916 (2010)
5. Bobbert, F.S.L.: Additively manufactured metallic porous biomaterials based on minimal surfaces: a unique combination of topological, mechanical, and mass transport properties. Acta Biomater. **53**, 572–584 (2017)
6. Mehboob, H.: Finite element modelling and characterization of 3D cellular microstructures for the design of a cementless biomimetic porous hip stem. Mater. Des. **149**, 101–112 (2018)
7. Mohammed, M.I.: Design of three-dimensional, triply periodic unit cell scaffold structures for additive manufacturing. J. Mech. Des. **140**(7), 071701 (2018)
8. Mohammed, M.I.: Design and fabrication considerations for three dimensional scaffold structures. KnE Eng. **2**(2), 120–126 (2017)
9. Zheng, X.: Minimal surface designs for porous materials: from microstructures to mechanical properties. J. Mater. Sci. **53**(14), 10194–10208 (2018)
10. Blanquer, S.B.: Surface curvature in triply-periodic minimal surface architectures as a distinct design parameter in preparing advanced tissue engineering scaffolds. Biofabrication **9**(2), 025001 (2017)
11. Feng, J.: Porous scaffold design by solid T-splines and triply periodic minimal surfaces. Comput. Methods Appl. Mech. Eng. **336**, 333–352 (2018)
12. Wang, S.: Lightweight of artificial bone models utilizing porous structures and 3D printing. Int. J. Performability Eng. **13**(5), 633 (2017)
13. Olivares, A.L.: Finite element study of scaffold architecture design and culture conditions for tissue engineering. Biomaterials **30**(30), 6142–6149 (2009)
14. Melchels, F.P.: A poly (D, L-lactide) resin for the preparation of tissue engineering scaffolds by stereolithography. Biomaterials **30**(23–24), 3801–3809 (2009)
15. Gauvin, R.: Microfabrication of complex porous tissue engineering scaffolds using 3D projection stereolithography. Biomaterials **33**(15), 3824–3834 (2012)
16. Giannitelli, S.M.: Current trends in the design of scaffolds for computer-aided tissue engineering. Acta Biomater. **10**(2), 580–594 (2014)
17. Maskery, I.: Insights into the mechanical properties of several triply periodic minimal surface lattice structures made by polymer additive manufacturing. Polymer **152**, 62–71 (2018)
18. Li, X.: Fabrication and characterization of porous Ti6Al4V parts for biomedical applications using electron beam melting process. Mater. Lett. **63**(3–4), 403–405 (2009)
19. Welsch, G., Boyer, R., Collings, E.W. (eds.): Materials properties handbook: titanium alloys. ASM international (1993)
20. Walker, J.M.: Design and mechanical characterization of solid and highly porous 3D printed poly (propylene fumarate) scaffolds. Prog. Addit. Manuf. **2**(1–2), 99–108 (2017)
21. Dias, M.R.: Optimization of scaffold design for bone tissue engineering: a computational and experimental study. Med. Eng. Phys. **36**(4), 448–457 (2014)
22. Murr, L.E.: Next-generation biomedical implants using additive manufacturing of complex, cellular and functional mesh arrays. Philos. Trans. R. Soc. Lond. A: Math. Phys. Eng. Sci. **368** (1917), 1999–2032 (2010)
23. Al-Ketan, O.: Mechanical properties of a new type of architected interpenetrating phase composite materials. Adv. Mater. Technol. **2**(2), 1600235 (2017)
24. Montazerian, H.: Porous scaffold internal architecture design based on minimal surfaces: a compromise between permeability and elastic properties. Mater. Des. **126**, 98–114 (2017)

25. Wang, X.: Topological design and additive manufacturing of porous metals for bone scaffolds and orthopaedic implants: a review. Biomaterials **83**, 127–141 (2016)
26. Choi, K.: The elastic moduli of human subchondral, trabecular, and cortical bone tissue and the size-dependency of cortical bone modulus. J. Biomech. **23**(11), 1103–1113 (1990)
27. Rho, J.Y.: Mechanical properties and the hierarchical structure of bone. Med. Eng. Phys. **20**(2), 92–102 (1998)
28. Rho, J.Y.: Young's modulus of trabecular and cortical bone material: ultrasonic and microtensile measurements. J. Biomech. **26**(2), 111–119 (1993)
29. Li, D.: Optimal design and modeling of gyroid-based functionally graded cellular structures for additive manufacturing. Comput. Aided Des. **104**, 87–99 (2018)
30. Ataee, A.: Anisotropic Ti-6Al-4V gyroid scaffolds manufactured by electron beam melting (EBM) for bone implant applications. Mater. Des. **137**, 345–354 (2018)
31. Kadkhodapour, J.: Failure mechanisms of additively manufactured porous biomaterials: effects of porosity and type of unit cell. J. Mech. Behav. Biomed. Mater. **50**, 180–191 (2015)
32. Arabnejad, S.: Fully porous 3D printed titanium femoral stem to reduce stress-shielding following total hip arthroplasty. J. Orthop. Res. **35**(8), 1774–1783 (2017)
33. Arabnejad, S.: High-strength porous biomaterials for bone replacement: a strategy to assess the interplay between cell morphology, mechanical properties, bone ingrowth and manufacturing constraints. Acta Biomater. **30**, 345–356 (2016)
34. Kadkhodapour, J.: Investigating internal architecture effect in plastic deformation and failure for TPMS-based scaffolds using simulation methods and experimental procedure. Mater. Sci. Eng., C **43**, 587–597 (2014)
35. Taniguchi, N.: Effect of pore size on bone ingrowth into porous titanium implants fabricated by additive manufacturing: an in vivo experiment. Mater. Sci. Eng., C **59**, 690–701 (2016)
36. Reichert, J.C.: The challenge of establishing preclinical models for segmental bone defect research. Biomaterials **30**(12):2149–2163 (2009)

Chapter 81
Design of Medical Device Product Using Multiphysics Simulations

Rajesh Kapuganti

Abstract The primary function of kidneys is to remove waste and other impurities from the blood of a human being on a daily basis. When the kidneys stop functioning properly, the toxins from the blood have to be removed through an external device, dialyser. The dialyser mimics the kidney and draws the blood out of the patient, purifies the blood and infuses them back into the patient's body. When the blood is infused back to the human body, it has to be at 37.6 °C failing which would lead to hypothermia, a fatal condition. To avoid this fatal condition, the body is heated to human body blood temperature 37.6 °C. The blood flow rate in the bloodline can vary over a wide range but still the blood has to be heated to 37.6 °C which is human body blood temperature. To raise the human body temperature to 37.6 °C, a sub-component in the dialyser machine called blood warmer takes care of heating. Since it is a medical device which interacts with the patient, the precision of 0.5 °C is expected over a range of flow rates based on patient's medical condition. The scope of this work is the design of a precise blood warmer which is effective than existing. The multiphysics simulations helped the engineers to visualize, observe and understand the physics involved during the design stage.

Keywords Multiphysics · Blood warmer · Product design · Flow rate and heat supply

81.1 Introduction

Dialysis is a process through which waste products artificially removed from the blood to compensate for the presence of an abnormally functioning kidney. In a healthy kidney, excess fluids and electrolytes such as sodium and potassium are removed from the body in the form of urine. In the case of kidney disease, the loss

R. Kapuganti (✉)
HCL Technologies, Chennai 600101, India
e-mail: rajesh-kapuganti@hcl.com

R. G. Narayanan et al. (eds.), *Advances in Computational Methods in Manufacturing*, Lecture Notes on Multidisciplinary Industrial Engineering,
https://doi.org/10.1007/978-981-32-9072-3_81

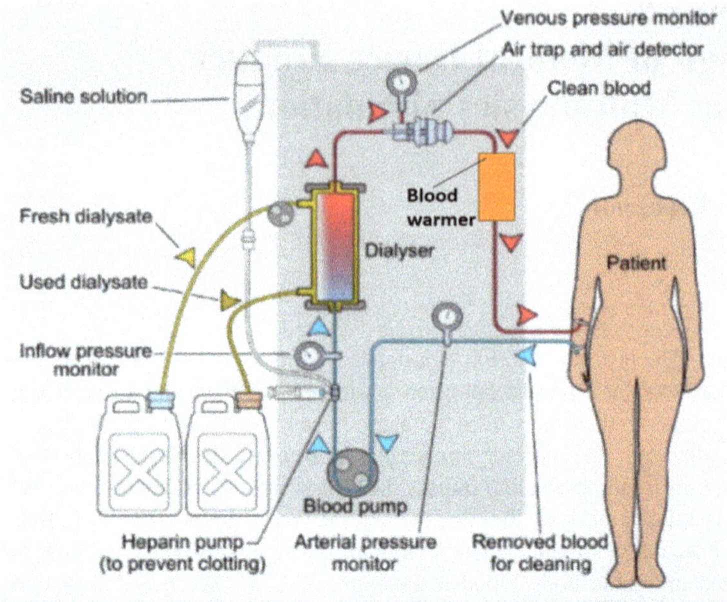

Fig. 81.1 Blood warmer

of kidney function results in the accumulation of excess waste, toxins or fluids in the body thereby causing other complications such as hypertension, heart failure, metabolic acidosis and hypokalemia [1]. Dialysis is performed in the situations of acute kidney failure and to remove toxins in blood from the body (Fig. 81.1).

81.1.1 Blood Warmer

Blood warmer is a sub-component in dialyser machine which is intended to deliver the blood or fluids before to transfusion to a human body at 37 °C which is the set temperature. Malfunctioning of the blood warmer device could lead to hypothermia. The blood warmer heats the blood to the desired temperature (37 °C) that is non-toxic for infusion.

The resistance heating technology which has a silicone heater patch and aluminium plates between which the fluid tubing passes was selected based on the thermal efficiency results. In designing the heating plate in the blood warmer, the data needed are type of plate material, plate dimensions, the effective area of the heating plate, the upper and lower temperature limits of the heating plate and the discharge of blood flow [1].

Though current blood warmers can heat blood to the human body blood temperature, the challenges are temperature management, heat mapping across the plate, optimum sensor location and the determination of the right compensation factor for dynamic control algorithm. This study was performed to improve the performance and functional aspects of the blood warmer design.

The device manufacturer supplied the complete list of product requirement specifications which had all the details of constraints, functionalities and geometric details. The design and development of blood warmer subassembly involved a multidisciplinary team which consisted of simulation, software, hardware, CAD, electrical, testing, industrial design and materials engineers.

However, in this paper, only the electro-thermo-fluid interactions of the multiphysics simulations are discussed. The multiphysics simulation procedure and the results are discussed.

81.2 Multiphysics Simulation Procedure and Results

81.2.1 Simulation Procedure

The multiphysics simulations were performed to determine the power input required for eight different flow rates to reach a desired set temperature of 37.6 °C from the inlet temperature of 30 °C. The range of flow rates used in the simulation are 100, 150, 200, 250, 300, 350, 400 and 450 ml/min. By varying the power input applied on the heating coils of heater plates, the power required to achieve the desired set temperature was determined. A convective heat transfer coefficient of 10 W/m^2 K was assumed considering that the heating unit is placed in an enclosure. The operating fluid was assumed as water instead of blood to avoid the nonlinearities. Humidity effects were not considered in the simulation.

The voltage difference was applied on the coil path of the top plate and the bottom plate (to generate joule heating in the plates). A perfect thermal contact was assumed between the disposable and heater plates.

The blood warmer assembly used in the dialysis machine consists of four components, the fluid domain, blood flow path, the heater coil and the two aluminium heater plates. The material properties of the different components used in the simulations are as listed in Table 81.1 (Fig. 81.2).

81.2.2 Simulation Results

The power required for the flow rates 100, 150, 200, 250, 300, 350, 400, 450 (ml/min) are 60, 89, 118, 148, 177, 206, 234, 264 (W), respectively, as listed in the Table 81.2.

Table 81.1 Material properties of the blood warmer assembly

S. No.	Component	Material	Property	Values	Unit
1	Fluid	Water	Specific heat	4181.8	J/kg K
			Density	1000	kg/m^3
			Dynamic viscosity	1.002	(Pa-s, N-s/m^2) $\times$ 10^{-3}
			Kinematic viscosity	1.004	(m^2/s) $\times$ 10^{-6}
			Thermal conductivity	0.58	W/(m K)
2	Blood flow path	Polyethylene	Specific heat	1800	J/kg * K
			Density	1500	kg/m^3
			Thermal conductivity	0.209	W/(m K)
3	Heater plate	Aluminium	Specific heat	956	J/kg * K
			Density	2699	Kg/m^3
			Thermal conductivity	165	W/(m K)

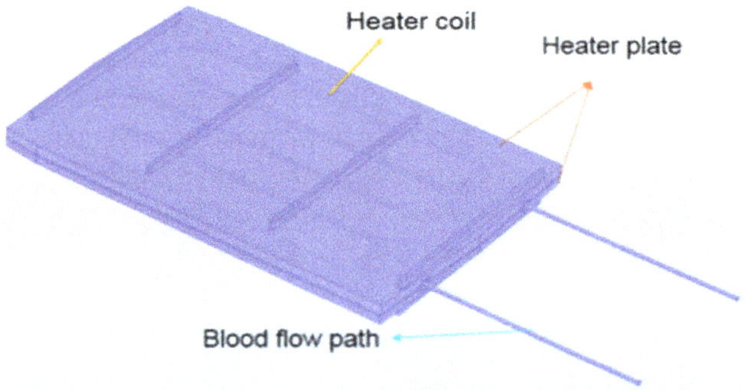

Fig. 81.2 Blood warmer assembly

Table 81.2 Heat energy required versus mass flow rate

S. No.	Mass flow rate (ml/min)	Heat energy required (W)
1	100	60
2	150	89
3	200	118
4	250	148
5	300	177
6	350	206
7	400	234
8	450	264

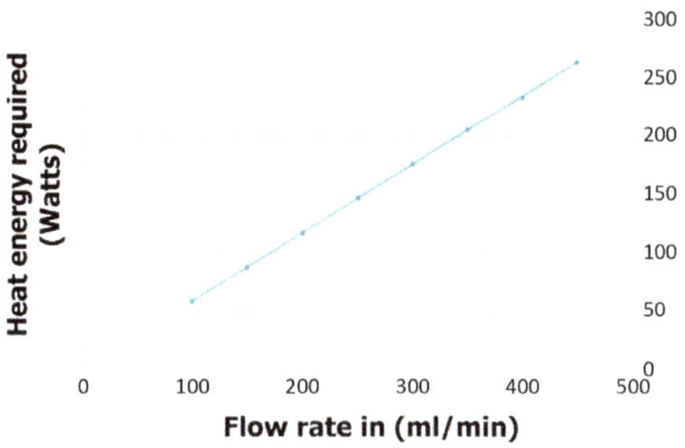

Fig. 81.3 Relationship between heat energy and flow rate

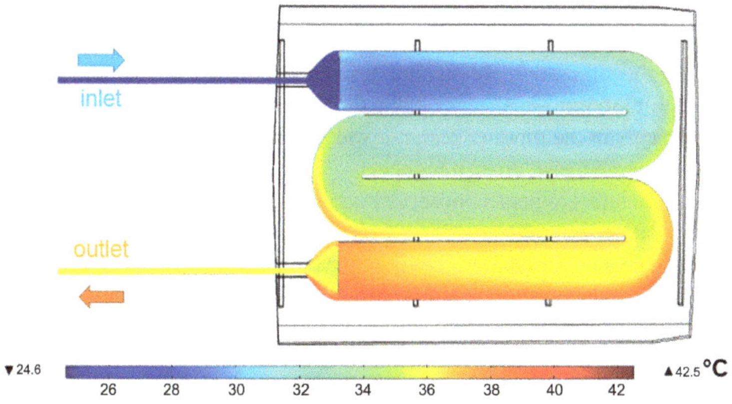

Fig. 81.4 Temperature distribution along the blood flow passage

There is a linear directly proportional relationship between the mass flow rate and heat energy required as depicted in the Fig. 81.3. The spatial thermal distribution of the blood flow path is depicted in the Fig. 81.4. The temperature of the fluid increases along the blood flow passage.

The temperature distribution is higher in the exit region of the blood warmer and lower at the entry region of the blood warmer. The spatial thermal variations on the heater plate are depicted in Fig. 81.5.

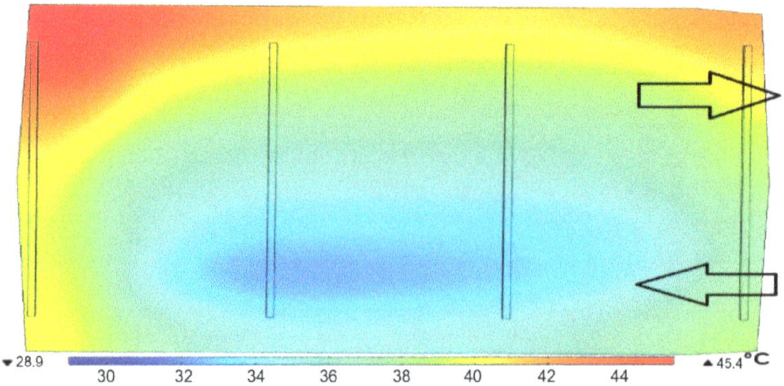

Fig. 81.5 Temperature distribution in the heater plate

81.3 Conclusion

1. Based on the simulation results, the heat energy required at different flow rates to heat the blood to 37 °C was obtained. This information was useful for further developments in the product design cycle. There is a linear relationship between the flow rate and heat energy required.
2. The temperature gradient is observed across flow passage as it gets heated up.
3. The correlation between testing and multiphysics simulation results was good. The multiphysics simulations were coupled with MATLAB for creating the algorithms, which was possible because of the virtual prototyping capability of COMSOL.
4. The product development process has been shortened in the concept and prototyping phase by effectively using multiphysics simulations.

References

1. Hendrarsakti, J., Ichsan, Y.: Experimental study of isothermal plate uniformity for blood warmer development using geothermal energy. IOP Conf. Seri. Earth Environ. Sci. **42** (2016)
2. http://e-pao.net/epSubPageExtractor.asp?src=education.Health_Issue.Dialysis_Everything_you_need_to_know_By_Sanjeev_Gulati

Chapter 82
Finite Element Analysis of a Disc Brake Mounted on the Axle of a Train

E. Madhusudhan Raju, Somayajula Kavya Geetika and Devireddy Jitesh Krishna

Abstract Axle-mounted disc brakes are generally employed in high-speed trains owing to the lesser weight and sustainable thermal properties. Most widely used material for manufacturing of the disc brake is grey cast iron (GCI), because of its availability. However, it being a brittle material has certain limitations. Hence, an aluminium composite material (AlSiC) is considered for analysis under different loading conditions. Three models of brake discs are developed on which both structural analysis and coupled analysis were done and results are compared.

Keywords Brake disc · Finite element analysis · Stress analysis

82.1 Introduction

In the olden days, the braking system of railway coaches consisted of wheel thread braking system. But they are not applicable to high-speed trains, as they damage the wheel thread. Hence, axle-mounted disc brakes are being employed in coaches. The braking system of a train involves a very complex process and is of great importance by making an essential contribution to the passenger's safety. This complexity results from numerous mechanical, thermal, pneumatic and electrical phenomena occurring during the process of braking. The actions of the above processes act on the different parts of the train with variable intensities. Also, the thermal deformation and overheating of the disc brake system during braking are turning out to be crucial. It is difficult to determine beforehand and analyse discs and linings separately, due to the amount of heat transferred to the discs, which may depend on their design and the train speed. In this paper, the effects of thermal and structural loads coupled together on the brake disc are analysed with different materials.

E. Madhusudhan Raju · S. Kavya Geetika · D. Jitesh Krishna (✉)
MED, UCE, OU, Hyderabad, India
e-mail: jiteshd.be19@uceou.edu

R. G. Narayanan et al. (eds.), *Advances in Computational Methods in Manufacturing*, Lecture Notes on Multidisciplinary Industrial Engineering,
https://doi.org/10.1007/978-981-32-9072-3_82

82.2 Literature Review

Reibenschuh et al. [1] used two brake discs (new and old) for both structural and thermal analysis. The material used for both the discs was spheroidal graphite cast iron. They found that the stresses developed in the new disc were more than worn disc, but stress values for both the discs were within the limits.

Maleque et al. [2] has developed a method of selection for selecting a better material in terms of optimization for the brake disc system, highlighting on substitution of cast iron by any lightweight material. Requirements for the material performance were scrutinized, and alternate solutions were assessed among different materials like cast iron, aluminium alloy, titanium alloy, ceramics and composites. The analysis led to a conclusion of usage of an aluminium metal matrix composite as the most appropriate and optimum material for brake disc system.

Yildiz and Duzgun et al. [3] had studied the stress analysis of ventilated brake discs by means of the finite element method. During this study, three different designs of ventilated brake discs were manufactured, and their braking system performances were examined experimentally together with a solid disc. These assessments indicate that the application of variable force along the brake pads leads to a reduction of the stress on ventilated discs from 8.8% to 19.1%.

Jung et al. [4] had studied shape optimization and the thermal characteristics of a ventilated disc. In this study, an analysis technique was developed to estimate the temperature rise and thermal deformation of the ventilated disc. Also, the braking condition and properties of the disc and pad could be estimated by this technique. Mathematical explanations for the analytical process of the braking power generation were developed.

82.3 Problem Definition

82.3.1 Aim of the Project

The primary objective of this paper is to present the various stresses and deformations induced in the brake disc models made up of grey cast iron (GCI) and aluminium composite materials (AlSiC) under static structural and coupled (structural and thermal) loading conditions. Three models were developed for the estimation of stresses, one of which is similar to the existing design, whereas the other two are new designs suggested to overcome the limitations of the existing one. The existing model is of the bolted joint type, and the other models are of push fit type. Model 2 is the push fit type of the existing model, and Model 3 is push fit type with cross-drills to reduce the weight. To determine the boundary conditions for the problem, an emergency braking situation of 160 kmph to stop on a flat horizontal track was utilized. The input loads, pressure, rotational velocity and heat flux induced were calculated and applied on the disc, and the output in terms of von

Mises stresses, total deformation and temperature induced in the brake disc was recorded and used for the finite element analysis.

82.3.2 2D Drawings of Models

See Figs. 82.1, 82.2 and 82.3 and Tables 81.1 and 82.2.

82.3.3 Software Used

1. Modelling—SOLIDWORKS 2016
2. Analysis—ANSYS WORKBENCH R16.

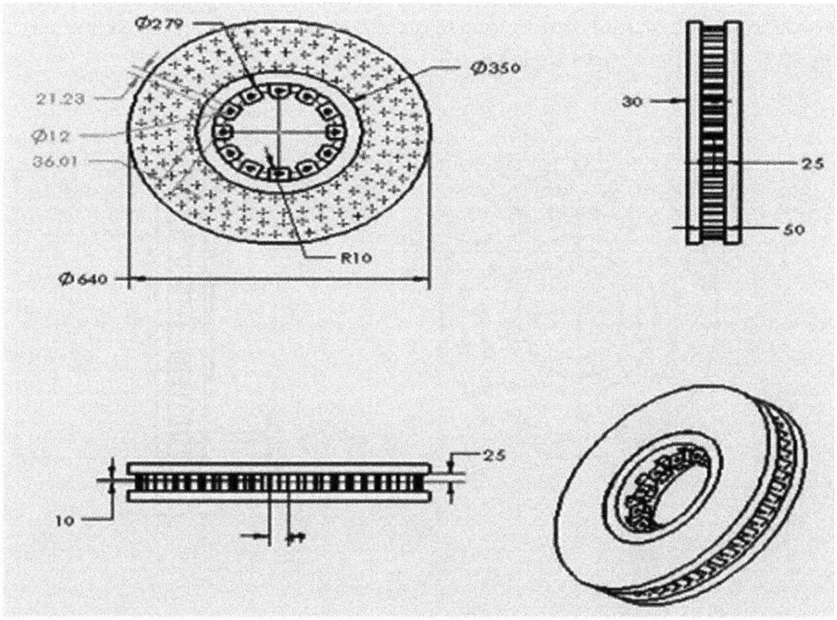

Fig. 82.1 2D model drawing of Model 1 (existing) [5]

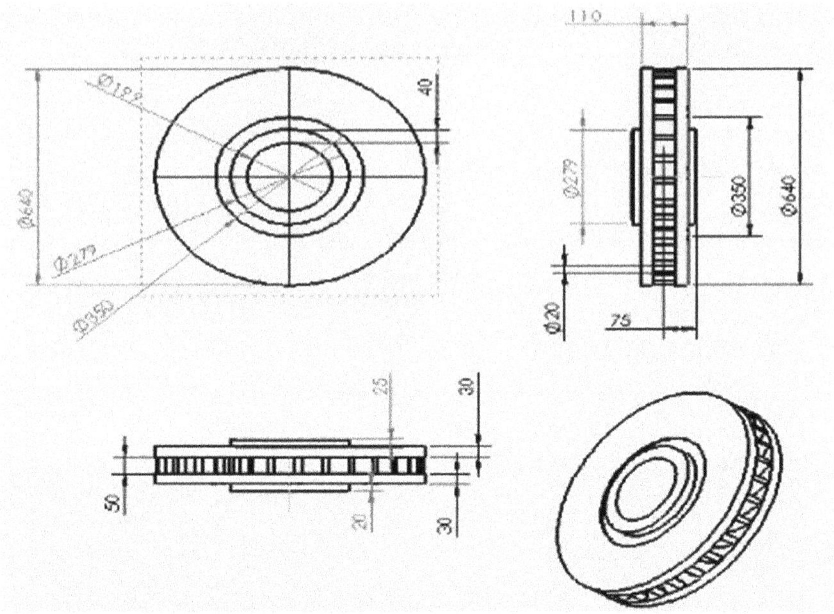

Fig. 82.2 2D model of Model 2 (push fit type)

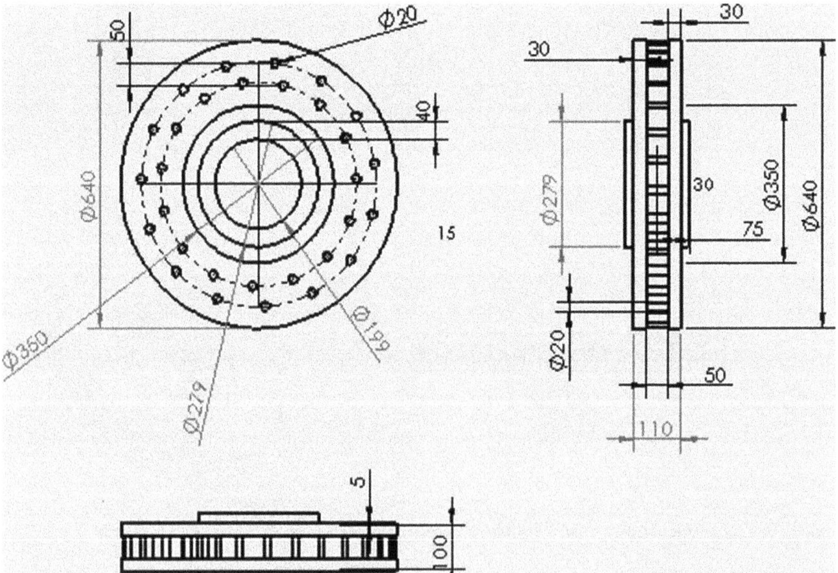

Fig. 82.3 2D model drawing of Model 2 (push fit type with cross-drills)

Table 82.1 Dimensions of disc brake [5]

S. No.	Description	Dimension
1	Dimensions	640 × 110 mm
2	Outer diameter of friction ring	$640^{+0/-1}$ mm
3	Inner diameter of friction ring	350 mm
4	Width of friction ring	$110^{+0/-0.3}$ mm
5	Inner bore	119 mm (H6)
6	Width of hub	150 mm

Table 82.2 Disc brake material properties

S. No.	Parameter	Units	GCI	AlSiC
1	Density	kg/m^3	7200	2950
2	Yield strength	MPa	–	2400
3	Ultimate strength	MPa	240	340
4	Young's modulus	GPa	43	93
5	Poisson's ratio	–	0.28	0.24
6	Shear modulus	GPa	43	93
7	Bulk modulus	GPa	83	147
8	Thermal conductivity	W/m°C	52	197
9	Specific heat	J/kg-C	447	808
10	Coefficient of thermal expansion	$*10^{-6}$	0.11	6.5

82.4 Research Methodology

82.4.1 Theoretical Background

Introduction. Due to the complex nature of the loading on the disc brake, the von Mises stresses induced due to the loading conditions are examined to determine the failure conditions of the model. The von Mises yield criterion (also known as the maximum distortion energy theory of failure) suggests that the yielding of a ductile material begins when the second deviatoric stress invariant reaches a critical value. This theory works well for most cases, especially when the material is ductile in nature behaviour.

Structural load calculations. Refer Appendix (Table 82.3).

82.4.2 Solution Methodology

The solution methodology to determine the better material and better geometry for railway brake disc is given by the following steps:

Table 82.3 Boundary conditions

S. No.	Analysis	Boundary conditions
1	Static structural	Pressure on disc—1.5 MPa Rotational velocity—97.12 rad/s
2	Coupled (static structural + thermal)	Temperature, as derived from transient thermal analysis for different materials and above-mentioned structural load

1. Dimensions, as given in Table 82.1, are taken from Railway Specification No. RDSO/2015/CG-03 'Brake disc for LHB-type coaches' [5] for modelling of brake discs.
2. Brake discs are modelled using SOLIDWORKS 2016 software and saved as an IGES file.
3. Import 3D model of brake discs into ANSYS WORKBENCH 16.
4. Analysis of brake discussing ANSYS WORKBENCH 16 by applying the boundary conditions as given above.
5. Static structural analysis is done by applying the pressure on each side of the disc surface and rotational velocity to investigate the von Mises and total deformation developed in the models.
6. The coupled analysis is carried out at all time points with respective temperatures by applying structural loads to obtain the stresses, deformation and strain under the coupled load of thermal and structural.
7. The obtained results are tabulated and compared. The material and geometry with the least values are considered as better material and better geometry.

82.5 Results and Discussions

82.5.1 Static Structural Analysis

The models developed in SOLIDWORKS 2016 were converted into IGES file and imported back into ANSYS WORKBENCH R16 for the finite element analysis. Boundary conditions as calculated in Sect. 82.3.1 were applied on the brake disc models. The constant pressure of 1.5 MPa was applied on both sides normal to the frictional surface of the discs. A rotational velocity of 97.12 rad/sec (ramped) was given at the axis of the disc. Fixed support was provided in the eye of fixing bolt. Both disc temperature and surrounding environment temperature are taken as 45 °C. The analysis was run for 36 s, i.e. the time taken by the vehicle to stop due to the application of the emergency brake. The output results of von Mises stress and total deformation developed in the model were recorded (Figs. 82.4, 82.5 and 82.6).

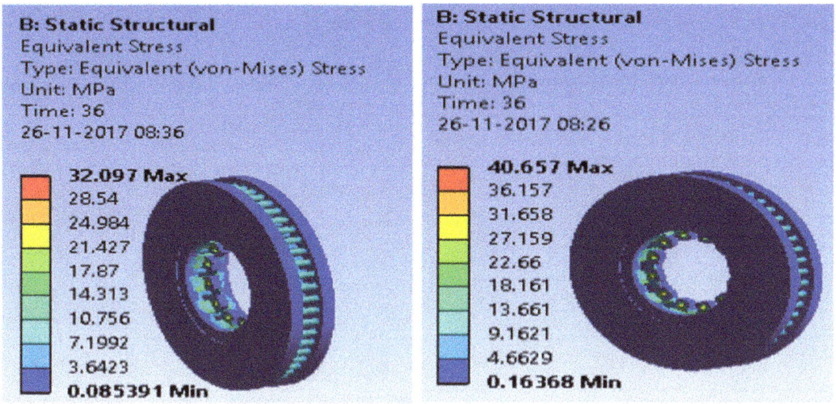

Fig. 82.4 SS GCI and AlSiC profile of von Mises stress (Model 1)

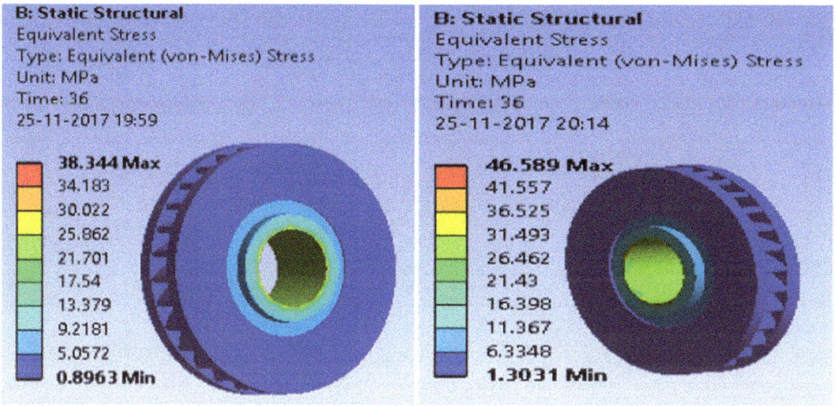

Fig. 82.5 SS GCI and AlSiC profile of Von Mises stress (Model 2)

82.5.2 Coupled Analysis

In the coupled analysis, the thermal load was coupled with the structural load to find out the combined effect on brake disc models. The temperature induced at various time points was imported into static structural, and then structural loads were applied. The analysis was run for 36 s, i.e. the time taken by the vehicle to stop due to the application of the emergency brake. The output results of von Mises stress and total deformation developed in the model were recorded (Figs. 82.7, 82.8 and 82.9).

Some important points that can be drawn from the analysis are:

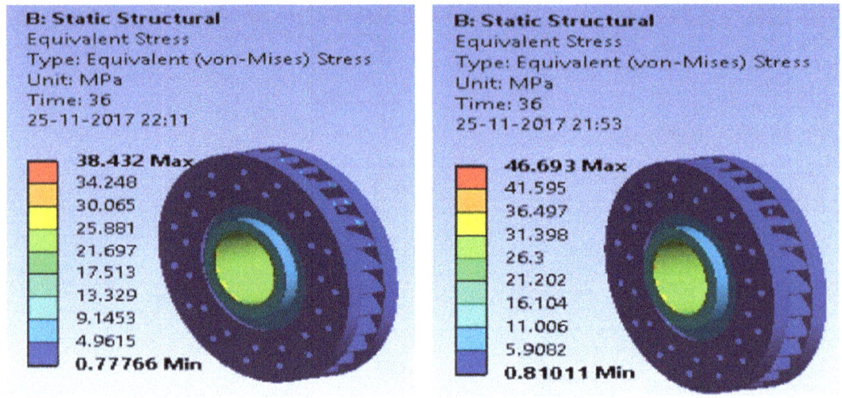

Fig. 82.6 SS GCI and AlSiC profile of von Mises stress (Model 3)

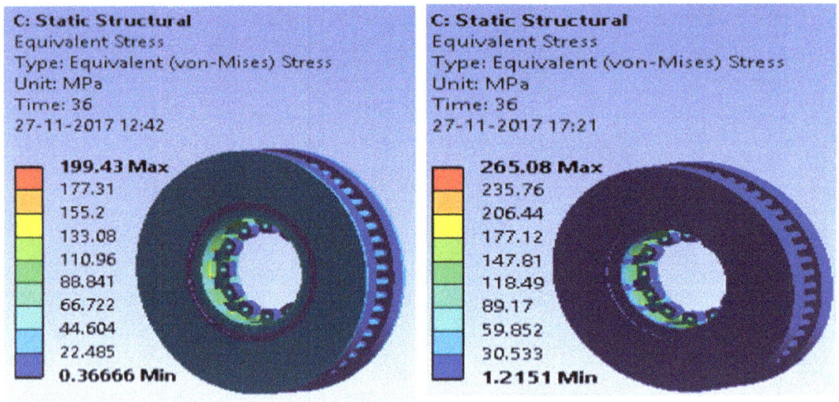

Fig. 82.7 CA GCI and AlSiC profile of von Mises stress (Model 1)

- When compared to the Factor of Safety offered by the GCI models, the AlSiC models offer higher Factor of Safety.
- For the same applied load, the AlSiC models have lower thermal stresses than the GCI models, as AlSiC material has greater thermal conductivity and heat dissipation capability.
- The weight of AlSiC models is lesser when compared to the GCI models (GCI model having a weight of about 134 kg gets reduced to 54 kg in case of AlSiC material).

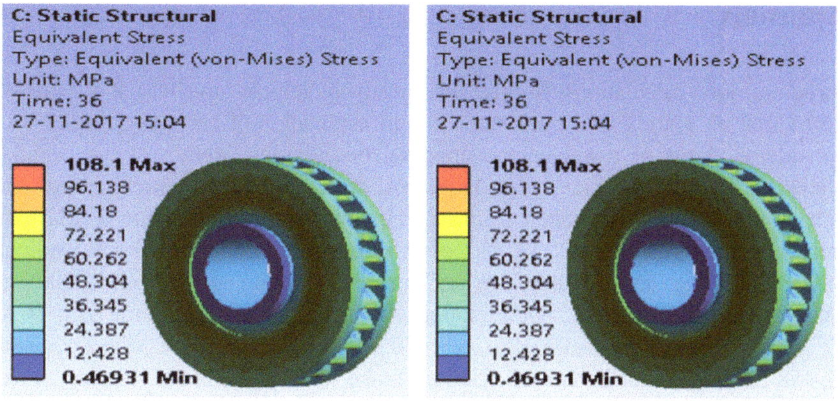

Fig. 82.8 CA GCI and AlSiC profile of von Mises stress (Model 2)

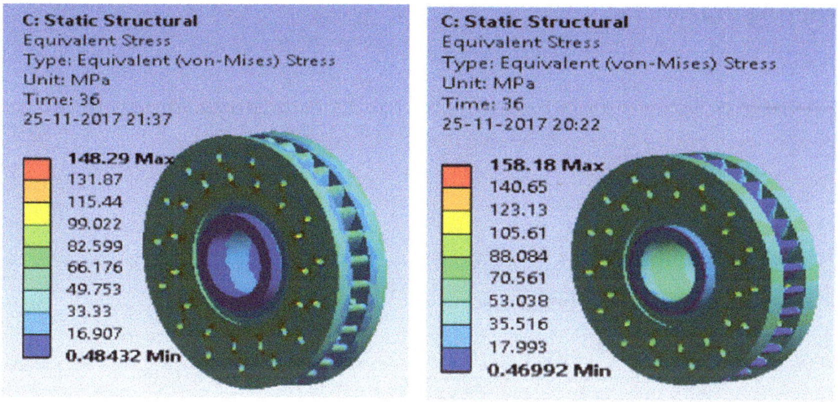

Fig. 82.9 CA GCI and AlSiC profile of von Mises stress (Model 3)

82.6 Future Scope

- Suggested models of brake discs are of a solid design which needs to be pressed on to the axle before pressing of the wheels. For replacement of these brake discs, wheels are necessary to be pressed out, which is not required if the models are of a split type. Hence, further experiments may be done with split-type brake disc models.
- Carbon matrix composites may be used to reduce the sound barrier and higher resistance to temperatures.
- Brake pad analysis with various materials suitable to the disc material may be carried out to optimize pad material.

Appendix

Structural load calculations: In this case, a railway vehicle travelling at a speed of 160 kmph on a horizontal track stops due to application of emergency brake was considered. Time of travel before stopping, deceleration, weight of the vehicle, clamping force on the brake disc, brake pad area, coefficient of friction, etc., for calculating the loads are taken from the railway specification. Value of Mass of railway vehicle—M = 64000 kg, No. of axles per vehicle = 4, Maximum load per axle = 16000 kg, no. of brake discs per axle = 2, Load on each wheel = 8000 kg, Start speed v_0 = 44.4 m/s, Deceleration a = 1.2 m/s^2, Braking time t_a = 36 s, Effective radius of the brake disc r_{disc} = 0.247 m, Radius of the wheel r_{wheel} = 0.458 m, Mean coefficient of friction brake pad μ = 0.35, Clamping force F_c = 42.1 kN, Surface area of brake pads A_c = 400 cm^2, Maximum temperature under sun = 70 °C, Maximum temperature under shade = 45 °C, Factor of Safety = 1.5 (Fig. 82.10).

Stopping distance

$$S = v_o t_s - \frac{1}{2} a t_s^2 = 822.24m \tag{82.1}$$

Determination of pressure on disc

Pressure acting on the brake disc,

$$P = \frac{F_c}{A_c \times \mu} = \frac{42.1 \times 1000}{800 \times (10)^{-4} \times 0.35} = 1.5\,\text{Mpa on each side} \tag{82.2}$$

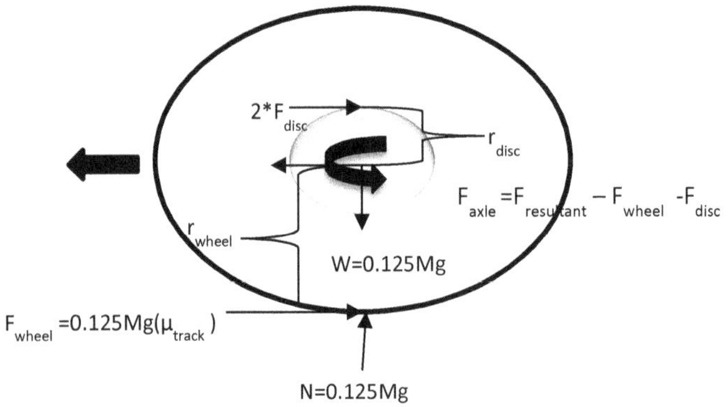

Fig. 82.10 Representation of forces acting on wheel and disc brake

where

F_c Clamping force (i.e. 42 kN)
A_c Contact area of brake pad on each side (i.e. 400 cm^2)
μ Coefficient of friction (i.e. 0.35).

Angular velocity

$$\omega = \frac{\text{Velocity}}{\text{radius}} = \frac{v_0}{r_{\text{wheel}}} = \frac{44.44}{0.458} = 97.12\,\text{rad/s} \tag{82.3}$$

Thermal load

The kinetic energy for one wheel (disc brake) is equivalent to the energy balance

$$\begin{aligned} 0.125 * \frac{1}{2} * M * v^2 &= \int_0^{t_s} P(t)\mathrm{d}t \\ &= 2 * F_{\text{disc}} \int_0^{t_s} v_{\text{disc}}(t)\mathrm{d}t \end{aligned} \tag{82.4}$$

The energy change at the moment is equal to the heat flux on the surface of the disc.

Equation (82.4) is valid in the case of constant braking deceleration. The braking force on the disc is equal to Eq. (82.7)

$$F_{\text{disc}} = \frac{0.125 * \frac{1}{2} * M * v_0^2}{2 * \frac{r_{\text{disc}}}{r_{\text{wheel}}}\left(v_0 * t_s - \frac{1}{2} * a * t_s^2\right)} = 8940\,\text{N} \tag{82.5}$$

The heat flux at the moment, which affects one half of the disc, is calculated according to the

$$\begin{aligned} Q(t) &= F_{\text{disc}} * v_{\text{disc}}(t) = F_{\text{disc}} * \frac{r_{\text{disc}}}{r_{\text{wheel}}}(v_0 - a * t) \\ &= 8940 * \frac{0.247}{0.458}(44.44 - 1.2 * t) = 214261 - 5786 * t\,\text{Watts} \end{aligned} \tag{82.6}$$

$$\text{Area of friction surface} = \frac{\pi}{4}\left(0.64^2 - 0.35^2\right) * 2 = 0.45\,\text{m}^2$$

$$Q(t) = \frac{214261 - 5786 * \mathrm{t}}{0.45}\ \mathrm{W/m^2} \tag{82.7}$$

For the case of emergency braking on horizontal track from 160 kmph to stop, the analysis was carried out in 36 steps, each step being 1 s long.

References

1. Reibenschuh, M., Oder, G., Čuš, F., Potrč, I.: Modelling and analysis of thermal and stress loads in train disc brakes—braking from 250 km/h to standstill. J. Mech. Eng. **55**(7–8), 494–502 (2009)
2. Maleque, M., Dyuti, S., Rahman, M.: Material selection method in design of automotive brake disc. In: Proceedings of the World Congress on Engineering, London (2010)
3. Yildiz, Y., Duzgun, M.: Stress analysis of ventilated brake discs using the finite-element-method. Int. J. Autom. Technol. **11**(1), 133–138 (2010)
4. Jung, S.P., Young, G.K., Park, T.W.: A study on thermal characteristic analysis and shape optimization of a ventilated disc. Int. J. Precis. Eng. Manuf. **13**(1), 57–63 (2012)
5. 'Brake disc for LHB type coaches', Indian Railways specification RDSO/2015/CG-03

Chapter 83
Modelling and Analysis of Composite Conical Space Adapter

Rajesh Addanki, Pranay Narinder Kumar, Ayush Mahajan and Manoj Kumar Buragohain

Abstract This paper is concerned with the development of model for the finite element analysis of the conical space adapter which is subjected to axial compression. It also provides the information about the parameters required for generating the input for modelling the structure. Analysis is performed using Ansys, i.e. software for finite element analysis. The results of analysis include the behaviour of the adapter under axial compression, e.g. stresses in the helical and circumferential ribs, displacement of the space adapter, critical buckling load and global buckling. Modal analysis has been carried out to study the modes of vibrations and obtain the frequency of vibration of the structure.

Keywords Composite lattice structures · Finite element analysis · Buckling analysis · Critical buckling loads · Modal analysis

Nomenclature

β	Semi-cone angle
α_0	Geodesic angle at the start (at larger radius)
α_f	Geodesic angle at the end (at smaller radius)
θ	Angle of rotation of fibre
D	Larger diameter of the adapter
d	Smaller diameter of the adapter
b	Width of the fibre
t	Thickness of the fibre
H	Height of the space adapter
W	Width of the fibre

R. Addanki (✉) · P. N. Kumar · A. Mahajan · M. K. Buragohain
Composite Product Development Centre, Advanced Systems Laboratory, DRDO, Hyderabad 500069, India
e-mail: rajeshaddanki@rediffmail.com

P. N. Kumar
e-mail: pranaynkumar@gmail.com

R. G. Narayanan et al. (eds.), *Advances in Computational Methods in Manufacturing*, Lecture Notes on Multidisciplinary Industrial Engineering,
https://doi.org/10.1007/978-981-32-9072-3_83

83.1 History of Lattice Structures

The development of grid-stiffened structures dates back to WWII English bomber Wellington. The application of these structures is not limited to space applications. Earlier, grid-stiffened structures were made by machining metals, e.g. aluminium isogrid structures by machining aluminium slabs. But today the material has changed from metal to carbon fibre composites. Currently, the fabrication methods of a lattice structure are done by grooving the fibre path over the mandrel of required dimensions, and fibres coated in epoxy resin are wound over it. After curing it at a required temperature, the desired shape is obtained.

One of the earliest lattice structures was developed by Russian-based laboratory. Central Institute of Special Machine Building (CRISMB) was Moscow radio tower [1] designed by Shukov in 1921. It stands 148.3 m tall and marks the advent of the development of lattice structures for Russians. It looks quite similar to modern-day lattice structures.

In 1985, US navy developed a launch vehicle adapter fitting for their satellite named as [2] Petite Amateur Navy Satellite (PANSAT) for communication purposes. The study conducted on this adapter fitting provides a basis for structural design and analysis for the adapter. The thesis gives an in-depth knowledge about the finite element analysis of such structures.

Since the inception of space adapters, they were manufactured in various shapes and sizes [3–6] offering flexibility in design to the user, such as using a cylindrical and conical adapter. A conical configuration of the shaped adapter has been modelled out for easiness of mounting the payload with available space and thereby reducing the additional stage. Even though the manufacturing of such structures is limited to axisymmetric shapes, the use of these structures remains dynamic.

83.2 Introduction

A typical space adapter consists of helical and circumferential ribs as shown in Fig. 83.1 wherein the impregnated fibres are laid along the grooves for circumferential and longitudinal pattern to form a lattice structure. When the structure is subjected to axial compression, the helical ribs transfer the load to circumferential ribs as tensile force. Now, this tensile force is transferred to a member to the adjacent structure which stabilizes its shape and increases the load carrying capacity. The load-carrying capacity is greatly influenced by dimensions of lattice structure. Therefore, keeping the space constraints in mind, the roadmap of designing process is laid down.

The space adapter is used to install payload over the last stage of the composite rocket motor casing. The conical-shaped adapter is made of carbon fibre with epoxy resin which provides higher strength to weight ratio than conventional metallic

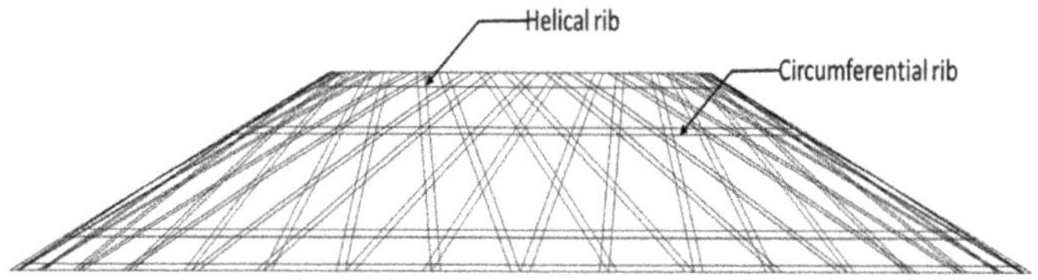

Fig. 83.1 Schematic of space adapter

structure made of steel or aluminium. Further, handling and installation of the adapter becomes easy due to its low weight.

In the present context, a space adapter is designed and modelled for finite element analysis [7]. The analysis consists of static analysis followed by buckling analysis. For calculating the natural frequency, modal analysis is carried out. The boundary conditions and load cases are explained in further sections. The results obtained from the analysis are also discussed in the paper. Further, prospects and research areas are also discussed which can be explored in the future for further research.

83.3 Configuration

The conical adapter is designed with two circumferential stiffeners in addition to metallic stiffeners at top and bottom and 20 crossed axial stiffeners with varying helical angle. A total of 40 layers with 0.5 mm thickness are considered for analysis. The helical angles with fibre path variation cater for a height of about 300 mm with outer diameter to inner diameter ratio of three. The helical ribs of cross section 8 × 8 to 20 × 20 in multiples of four have worked out to cater the axial payload of about 50 kg with 10 g axial force. A typical 2D view of space adapter is shown in Fig. 83.2.

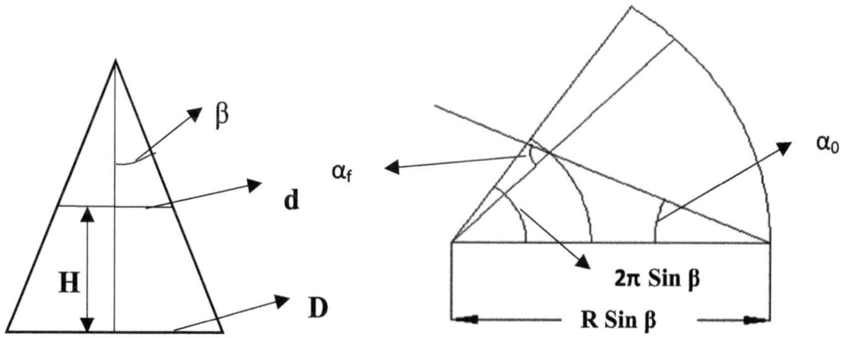

Fig. 83.2 Development of the cone

The empirical relations for calculating the path, angle and radii are as given below (from Eqs. 83.1 to 83.6).

$$\frac{r\varnothing}{\sin\beta} = r\theta \tag{83.1}$$

$$\theta = \frac{\varnothing}{\sin\beta} \tag{83.2}$$

$$\varnothing = \alpha - \alpha_0 \tag{83.3}$$

Also, as per Clairut's equation

$$R\sin\alpha_0 = r\sin\alpha \tag{83.4}$$

$$\alpha = \sin^{-1}\left(\frac{R\sin\alpha_0}{r}\right) \tag{83.5}$$

Substituting Eqs. 83.2 and 83.5 in Eq. 83.3, we get,

$$\theta = \frac{1}{\sin\beta}\left[\sin^{-1}\left(\frac{R\sin\alpha_0}{r}\right) - \alpha_0\right] \tag{83.6}$$

83.4 Input Generation

The geometric parameters used for the design and modelling the conical space adapters serve as the inputs. Equations which are used for modelling are stated in the previous section. Various cross sections were designed and analysed to arrive at a final configuration. The parameters used are as follows:

- Radius of the frustum of the cone (both smaller and larger)
- No. of helical ribs
- No. of circumferential ribs
- Winding angle at start and end
- Width of fibre
- Cross section of the rib.

Once the above parameters are finalized then model becomes ready for analysis. The values of above parameters are used in the governing equations refered in Sect. 83.2 to obtain the set of coordinates. The properties of composites, which were experimentally evaluated from the sample test as per ASTM standards, are used for the analysis by testing the small specimens constructed in the scale down model of the full-scale prototype. After the characterization of the high-strength carbon fibre with epoxy resin, the following properties are obtained which are given

Table 83.1 Carbon epoxy composite properties

Parameter	E_{xx} (MPa)	E_{xy} (MPa)	E_{yz} (MPa)	μ_{xx}	μ_{xy}	μ_{yz}	G_{xx} (MPa)	G_{xy} (MPa)	G_{yz} (MPa)
Value	117,000	6000	6000	0.3	0.29	0.29	3000	2000	2000

Note Where symbols have their usual meaning like elastic modulus, shear modulus and Poisson's ratio in respective directions of the laminate

Table 83.2 Properties of steel

Properties	Value (MPa)
E	201,000
μ	0.3

below in Table 83.1. Also, the material input properties for the steel are listed in Table 83.2.

83.5 Modelling

The helical path with circumferential ribs has been modelled in ANSYS which is software based on finite element analysis. The relations mentioned in the previous sections are used to create a set of coordinates of fibre path. With +θ coordinates and −θ coordinates have been developed the path is fed to software. In the similar lines, 90° path (circumferential ribs) was also modelled. The nodal points and the element generation were done by brick node 185. In total, 20 helical ribs with an angular difference of 18° between any two +θ helical ribs are considered. The nodes are merged to eliminate any discrepancy at the crossovers. While modelling, a special attention must be given to crossovers (where ribs are meeting each other) because crossover points act as single unit not different one, otherwise results obtained may be erroneous or not at all accurate. The adapter model with helical and circumferential ribs is shown in Fig. 83.3. Figure 83.4 shows the element detail depicting the nodal merging at the crossover points.

83.6 Boundary Conditions and Load Cases

The space adapter is modelled in global Cartesian system in such a way that the bottom surface lies in X-Y plane and positive z-axis is aligned with axis of the cylinder. Following are the boundary conditions applied on the mode for the analysis:

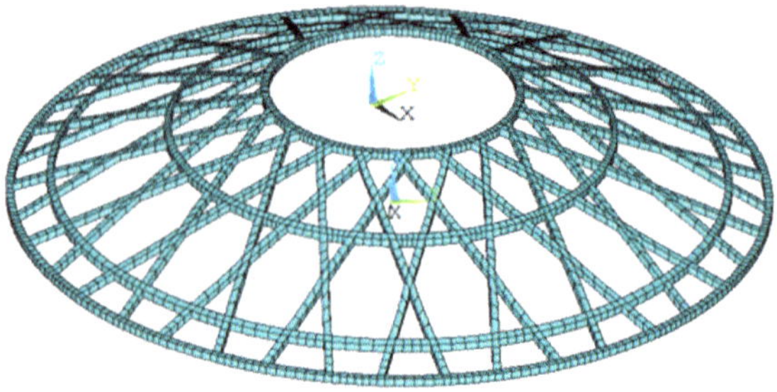

Fig. 83.3 FEA model of space adapter

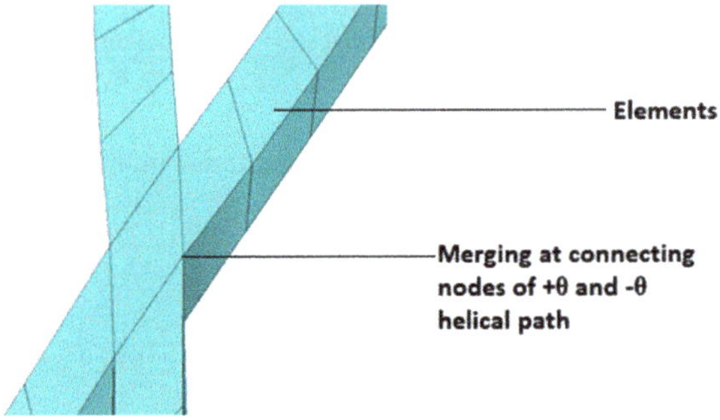

Fig. 83.4 Elements in the model

- Displacement of the bottom surface has been arrested and at the top of it is kept free where the load is applied, i.e. at $z = 0$ $U_x = 0$, $U_y = 0$, $U_z = 0$ and at $z = H$ U_x, U_y, $U_z \neq 0$.
- A uniform load is applied on the top nodes at $z = H$.
- To have the interface, a metallic bulkhead at top and bottom of the adapter (lower and higher Diameter) steel is considered with properties as follows.

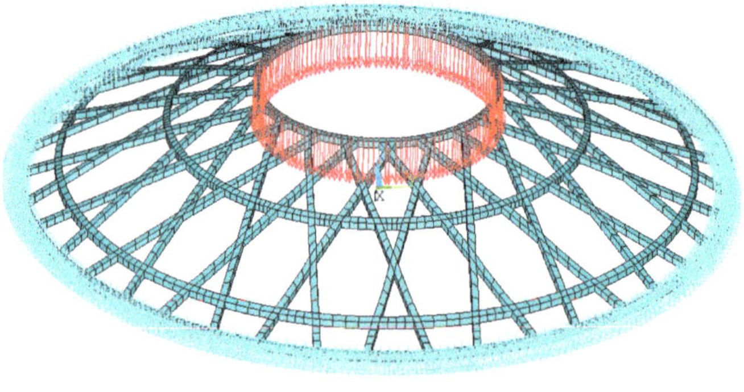

Fig. 83.5 Load application and boundary condition for the adapter

Design load, in this model, is applied for two types of analysis, i.e. static and buckling. Axial compressive load used for analysing the structure is 10 tons and the loading pattern is shown in Fig. 83.5.

The model is fixed at the bottom and load is applied at the top. A load of 10 tons is applied on 280 nodes at the top. Then load per node comes out to be 357.14 N.

In addition to static analysis, modal analysis was also conducted on the space adapter model (20 × 20) to study the vibrations and calculate frequency of the adapter.

Modal analysis is used to calculate the natural frequency and mode shapes of a continuous structure. A continuous structure has infinite degrees of freedom, but finite element approximates with a finite number of degrees of freedom.

Block Lanczos method is used for performing the modal analysis, which uses an algorithm for finding the null space of a matrix over a finite field. The same method is used in the present study for the matrix functions and in calculating buckling factor. The eigen buckling analysis is performed and three modes are extracted for the buckling factor.

The results of modal analysis serve as the basis for harmonic analysis. Harmonic analysis solves time-dependent equations of motion for linear structures undergoing steady state vibrations.

83.7 Results and Discussions

Several models were configured with different cross sections ($w \times t$) varying as 8 × 8, 16 × 16, 18 × 16, 20 × 20. Few results are represented to validate the strength and configuration of the space adapter, and the buckling factors variations are tabulated in Table 83.3.

A typical buckling mode is also represented in Fig. 83.6.

It is observed that, as the rib cross section increases, the load carrying capacity of the adapter also increases. The buckling factor of 5.62 indicates that critical load is 5.62 times of applied load. Higher buckling factor ensures that design will not fail under the action of axial compressive loads. The stresses are developed in the layers of the model.

The graphical plot in Fig. 83.7 demonstrates the behaviour of stress in 40 layers of the model at three different heights, i.e. top, middle and bottom of the adapter.

Table 83.3 Buckling factor variations

SI No.	Cross section of the ribs (w × t)	Buckling factor
1.	8 × 8	0.2295
2.	16 × 16	0.9122
3.	18 × 16	1.1011
4.	20 × 20	5.6200

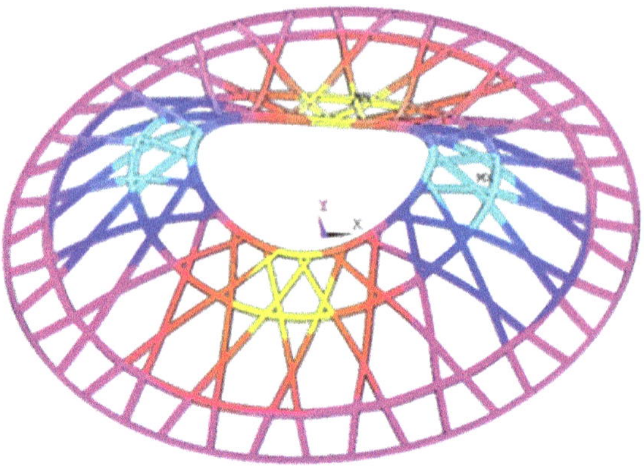

Fig. 83.6 Buckling mode of the adapter

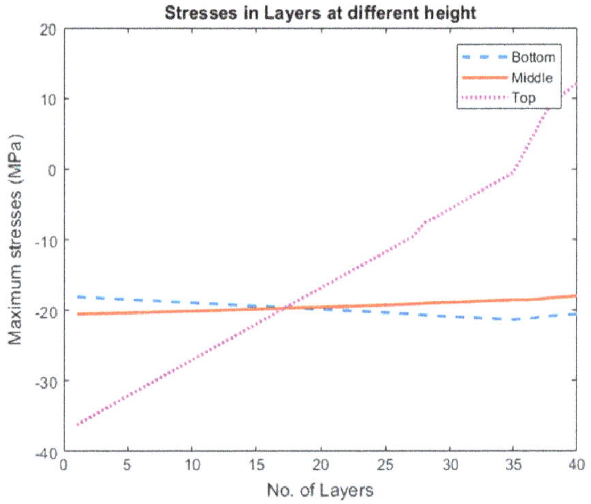

Fig. 83.7 Stress pattern in the helical plies at various height

The behaviour of stress shown in the graph indicates that the layer between 15 and 20 experiences same stress and behaves as the neutral layer.

The graphical plot in Fig. 83.8 depicts the stress in layers at the crossovers of the helical rib. The plot indicates that stresses start as compressive but as we move along the transverse direction to the layers, the nature of stress starts changing. The topmost layer, i.e. layer 40, experiences lesser compressive force compared to lowermost one. Up to layer no. 35, the reduction in compressive stress is gradual, but later there is drastic reduction in the compressive stress value.

The graphical plot in Fig. 83.9 shows that stresses in layers of circumferential rib are largely tensile attributing to the fact that helical ribs transfer compressive stress to the circumferential ribs which change it into tensile. There is gradual decrease in the stress in the transverse direction indicating that the load is distributed uniformly.

As we noted earlier, that nature of stress is different in helical and circumferential rib, the stress pattern changes from compressive to tensile is clearly seen in

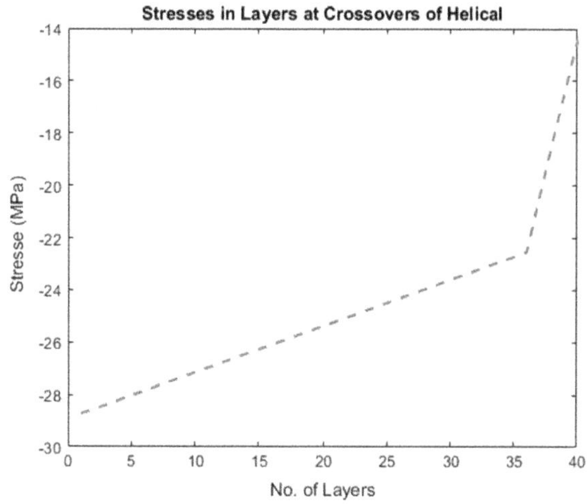

Fig. 83.8 Stress pattern in the helical plies at crossover

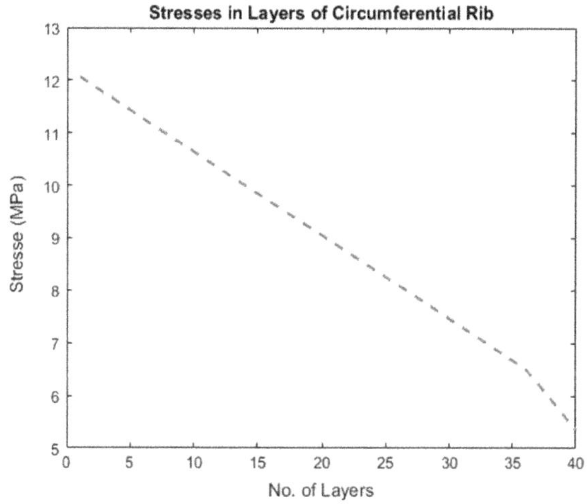

Fig. 83.9 Stress pattern in the circumferential ribs

Fig. 83.10. It can be seen, at the crossover, the stresses change from compressive to tensile indicating the transfer of the loading and is evident from the plot that the load is transferred from lowermost ply to topmost ply. Since, the value as well as the nature of the stress changes gradually up to layer 35 and the slope is high.

Figure 83.11. shows that as load is increased, the displacement also increases in the negative *z*-direction, i.e. along the height of the adapter and also follows a linear relationship. The deformed shape of the adapter superimposing onto the undeformed configuration is shown in Fig. 83.12.

The results obtained from static analysis have been saved and further utilized for extraction of mode shapes in modal analysis. In modal analysis, three modes were

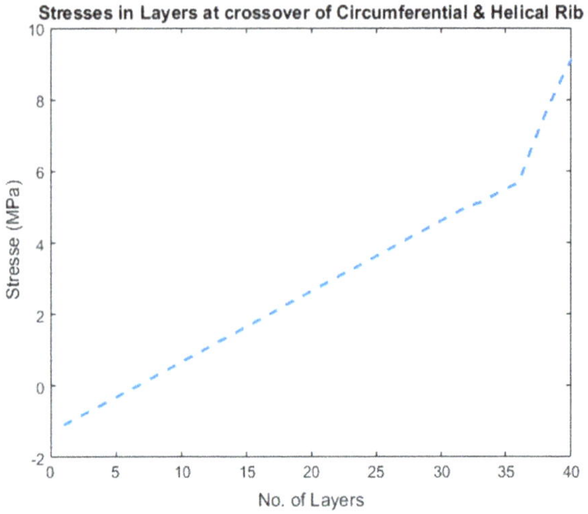

Fig. 83.10 Stress pattern in the circumferential and helical crossovers

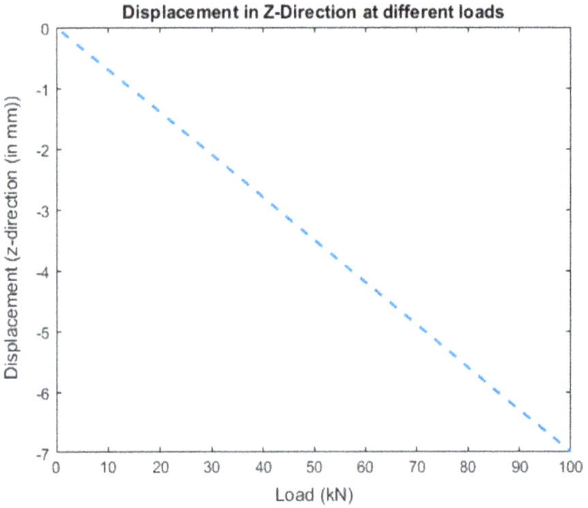

Fig. 83.11 Displacement in z-direction at different loads

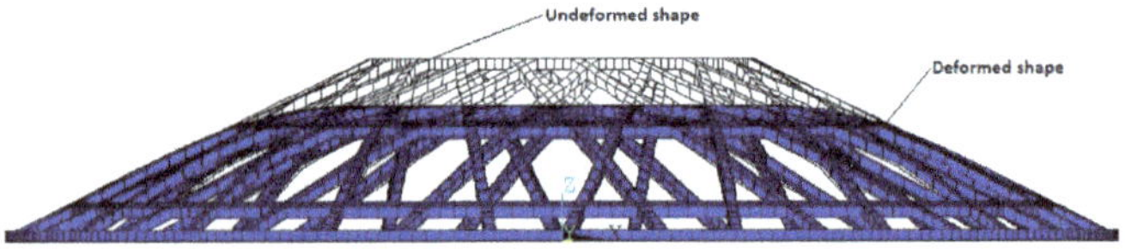

Fig. 83.12 Deformed shape versus original shape

extracted and the results obtained in modal analysis are as follows (refer Fig. 83.13):

The modal frequencies are listed in Table 83.4.

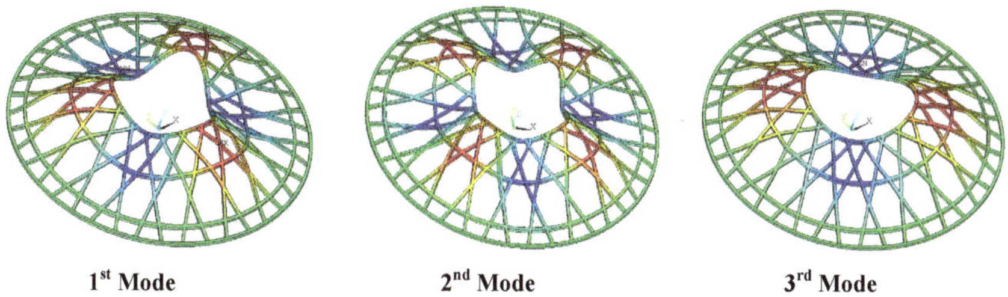

Fig. 83.13 Three mode shapes of the adapter

Table 83.4 Frequency of the mode shapes

Modes	First	Second	Third
Frequency (Hz)	5.09	5.15	6.56

83.8 Conclusion

The results from different models are studied and found that if there is increase in the cross section of the helical groove of carbon fibre, the buckling factors increase. Also, if two circumferential ribs are used, it contributes in increasing buckling factor. It can also be noted that, the bottommost layer in the ply sequence is the critical layer in the helical ply as it is experiencing the highest value of stress (in both the ribs circumferential and helical plies).

The modal analysis shows the structures that have three modes. Since the value of natural frequency is low it can be concluded that the stiffness of the structure is high enough to tolerate the load fluctuations during the actual application as stiffness is inversely proportional to frequency.

Further studies can be made considering the change in winding angle (helical rib angle), varying circumferential rib location, total no. of helical ribs and total no. of helical plies to find out buckling load capacity. The present structure designed is having a weight of approximately 15 kg and weight optimisation studies with higher modulus carbon fibre can be studied.

Acknowledgements A sincere thanks to the Director, ASL and CRMC team for their valuable inputs in the process.

References

1. Vasiliev, V.V., Barynin, V.A., Razin, A.F.: Anisogrid lattice structure-survey of development and application. Compos. Struct. **54**, 361–370 (2001)
2. Gannon, B.B.: Design and analysis of launch vehicle adapter fitting for petite amateur navy satellite (PANSAT)

3. Vasiliev, V.V., Razin, A.F.: Anisogrid conical adapter for commercial space applications. In: AIAA, p. 3440 (2005)
4. Marichela, D.: Advanced grid stiffened structure. A thesis Louisiana State University and Agricultural and Mechanical College
5. Vasiliev, V.V., Barynin, V.A., Razin, A.F.: Anisogrid composite lattice structures-development and aerospace applications
6. Vasiliev, V.V., Razin, A.F.: Anisogrid composite lattice structures for spacecraft and aircraft application. Compos. Struct. **76**, 182–189 (2006)
7. Ansys Products Ltd., version 17.0

Chapter 84
Parametric Study of Axisymmetric Circular Disk Subjected to Rotational Autofrettage

S. M. Kamal and R. Kulsum

Abstract Rotational autofrettage is a potential strengthening technique for thick-walled circular disks/cylinders used in many engineering applications. In particular, the rotating disks used as a rotor in an aircraft engine, rotating impeller wheel of a compressor and fasteners in high pressure pipelines are the important application areas, where rotational autofrettage is useful in mitigating the detrimental tensile stresses during operation. The process induces compressive residual stresses in the disk on unloading a previously applied plastically deforming centrifugal load. The amount of residual stresses generated during the process depends upon different influencing parameters. The geometrical dimension of the disk, rotational speed and material properties are the important parameters to be considered for assessing the effectiveness of rotational autofrettage of a disk. In this paper, a parametric study of the different influencing parameters during the rotational autofrettage of a thick-walled axisymmetric circular disk is carried out.

Keywords Rotational autofrettage · Hydraulic autofrettage · Rotational speed · Residual stress

84.1 Introduction

Usually, autofrettage is performed in thick-walled cylinders to induce compressive residual stresses in the vicinity of the bore surface by a process of loading and unloading of a plastically deforming load applied at the inner surface of the cylinder. The process may also be useful in strengthening thick-walled hollow circular disks such as fastener holes and rotating disks used in automobile and aerospace industries. During the process, the material at the neighborhood of the inner bore surface is subjected to plastic deformation and the material at the outer portion of the cylinder is in the elastic state. The early known autofrettage processes

S. M. Kamal (✉) · R. Kulsum
Department of Mechanical Engineering, Tezpur University, Tezpur 784028, Assam, India
e-mail: smkmech@tezu.ernet.in

R. G. Narayanan et al. (eds.), *Advances in Computational Methods in Manufacturing*, Lecture Notes on Multidisciplinary Industrial Engineering,
https://doi.org/10.1007/978-981-32-9072-3_84

are hydraulic [1], swage [2] and explosive autofrettage [3]. The hydraulic autofrettage employs a very high internal pressure, whilst the swage autofrettage employs an oversized solid mandrel to be pushed through the bore of the cylinder to achieve the desired plastic deformation in the cylinder wall. The explosive autofrettage utilizes explosive charge to achieve the beneficial effect of autofrettage. The hydraulic and swage autofrettage processes have received considerable attention of the researchers attempting theoretical and experimental studies on the processes [4–11]. The explosive autofrettage process has been less investigated. Recently, researchers have developed some new methods of achieving autofrettage, viz. thermal autofrettage [12] and rotational autofrettage [13] to circumvent certain disadvantages associated with the earlier processes. In thermal autofrettage, a high radial thermal gradient is used for achieving plastic deformation at the inner side of the vessel. Detailed theoretical and experimental studies of the process have been carried out in Refs. [12, 14, 15]. Some recent works on thermal autofrettage have been reported in [16, 17].

Rotational autofrettage is a process where the beneficial compressive residual stresses are generated in the vicinity of a thick-walled disk/cylinder by rotating the component at sufficiently high speed and then gradually reducing its angular velocity to zero [13, 18]. The high angular velocity initiated to the disk/cylinder produces a large centrifugal force plasticizing the inner bore region keeping the region near the outer surface in the elastic state. This phase is called the loading phase. In a later stage when the centrifugal force is made zero by ceasing the rotation of the disk/cylinder, the material near the bore surface is subjected to compressive residual stresses. This phase is called the unloading phase. When the disk/cylinder with compressive residual stresses is put in the actual working environment, it mitigates the magnitude of tensile stresses induced in the component due to service load. This eventually enhances the load-bearing capacity of the component. The rotational autofrettage was first conceptualized by Zare and Darijani [13] in 2016. They carried out a theoretical analysis of the process for thick-walled cylinders assuming plane strain condition. The analysis was based on Tresca yield criterion and its associated flow rule, elastic-perfectly plastic material model with the incorporation of Bauschinger effect. The authors further extended their analysis to incorporate linear kinematic hardening [19]. Most recently, the rotational autofrettage has been studied for thick-walled hollow circular disks by Kamal [18]. He assumed plane stress condition and Tresca yield criterion and its associated flow rule for the analysis. The effect of strain hardening was taken into account using Ludwik's hardening law. The author carried out the numerical simulation of the model considering aluminum and SS304 disks. It was shown that the rotationally autofrettaged disk could enhance its working internal pressure, temperature difference and centrifugal load individually during service. Kamal [18] also carried out a three-dimensional finite element method (FEM) analysis of the rotational autofrettage of a disk. The FEM results were compared with the analytical models, and it was found that the FEM stresses were in close agreement with the theoretical stresses.

It has been observed that there are only a few research papers [13, 18, 19] available in the open literature so far in the area of rotational autofrettage. All the literatures deal with the theoretical analysis of the process revealing it as a potential process. In any process, the process parameters play an important role in satisfying certain design requirements. For example, in an autofrettage process, the ultimate design requirement is to achieve the compressive residual stress at the inner surface of the component and the corresponding increase in the load carrying capacity as per the actual working condition. In order to achieve the required design objectives in rotational autofrettage process, the process parameters such as the geometrical dimension of the component, rotational speed and material properties should be selected appropriately. No literature is available studying the effect of process parameters on rotational autofrettage. Thus, in this work, a parametric study of the various process parameters in the rotational autofrettage of a thick-walled disk is carried out. The effect of different parameters on the residual stress and on different load-carrying capacity is studied. The study is based on the work of Kamal [18]. Before carrying out detailed parametric study of the process, the plane stress model of Kamal [18] is briefly described in Sect. 84.2.

84.2 Plane Stress Model of Rotational Autofrettage Proposed by Kamal [18]

An axisymmetric disk with inner radius a and outer radius b is considered. Under plane stress condition, the wall of the circular hollow disk splits into two deformed zones, viz. the inner plastic zone and the outer elastic zone when subjected to a plastically deforming centrifugal force by rotating the disk at sufficiently high rotational speed. The details are described in [18]. The radial and hoop stresses induced in the disk due to rotation, when the centrifugal force is not sufficient to deform the disk plastically, are given by

$$\sigma_r = (3+v)\frac{\rho\omega^2}{8}\left(a^2+b^2-\frac{a^2b^2}{r^2}-r^2\right) \tag{84.1}$$

$$\sigma_\theta = \frac{\rho\omega^2}{8}\left\{(3+v)\left(a^2+b^2+\frac{a^2b^2}{r^2}\right)-(1+3v)r^2\right\}, \tag{84.2}$$

where ρ is the density of the disk material, ω is the rotational speed, ν is the Poisson's ratio, a is the inner radius and b is the outer radius of the disk. The yield onset speed of the disk is given by

$$\omega_i = \left[\frac{4\sigma_Y}{\rho\{(1-v)a^2+(3+v)b^2\}}\right]^{1/2} \tag{84.3}$$

where σ_Y is the yield strength of the material. On increasing the rotational speed beyond ω_i, the material at the inner surface and some portion beneath it deforms plastically, whereas the outer region remains in the elastic state. Thus, an inner plastic zone, $a \leq r \leq c$ and an outer elastic zone, $c \leq r \leq b$ are formed within the wall of the disk. The radial position, c is known as the radius of elastic–plastic interface. The stresses in the plastic and elastic zones are given by the following equations:

Plastic zone, $a \leq r \leq c$

$$\sigma_r = \frac{c}{r}\left(\frac{c^2 - b^2}{b^2 + c^2}\right)\left\{\sigma_Y + (1+3\nu)\frac{\rho\omega^2 c^2}{8}\right\} + \sigma_Y\left(1 - \frac{c}{r}\right) + (3+\nu)\left(\frac{\rho\omega^2 b^2}{8}\right)\left(\frac{c}{r}\right)\left(\frac{2b^2}{b^2+c^2} - \frac{c^2}{b^2}\right) + \frac{\rho\omega^2}{3}\left(\frac{c^3}{r} - r^2\right) - \frac{K}{r}\int_r^c \left(\varepsilon_{eq}^p\right)^n \mathrm{d}r_1 \tag{84.4}$$

$$\sigma_\theta = \sigma_{eq} = \sigma_Y + K\left(\varepsilon_{eq}^p\right)^n \tag{84.5}$$

Elastic zone, $c \leq r \leq b$

$$\sigma_r = \left(\frac{1+3\nu}{b^2+c^2}\right)\frac{\rho\omega^2 c^2}{8}\left(c^2 - \frac{b^2c^2}{r^2}\right) + (3+\nu)\frac{\rho\omega^2 b^2}{8}\left\{1 - \frac{c^2}{b^2+c^2} + \frac{b^2c^2}{(b^2+c^2)r^2} - \frac{r^2}{b^2}\right\} + \left(\frac{c^2}{b^2+c^2}\right)\sigma_Y\left(1 - \frac{b^2}{r^2}\right) \tag{84.6}$$

$$\sigma_\theta = \left(\frac{1+3\nu}{b^2+c^2}\right)\frac{\rho\omega^2 c^2}{8}\left\{c^2 + \frac{b^2c^2}{r^2} - \frac{(b^2+c^2)r^2}{c^2}\right\} + (3+\nu)\frac{\rho\omega^2 b^2}{8}\left\{1 - \frac{c^2}{b^2+c^2} - \frac{b^2c^2}{(b^2+c^2)r^2}\right\} + \left(\frac{c^2}{b^2+c^2}\right)\sigma_Y\left(1 + \frac{b^2}{r^2}\right) \tag{84.7}$$

The residual stresses induced in the disk are obtained by subtracting the elastic stresses (Eqs. 84.1 and 84.2) from the respective stresses in the plastic and elastic zones. It is assumed that the unloading process is purely elastic devoid of Bauschinger effect. The unknown radius of elastic–plastic interface c can be obtained from Eq. (84.4) by using the boundary condition of vanishing radial stress at the inner radius. A numerical solution procedure such as bisection method may be employed for this. The present parametric study is based on the equations presented in this section.

84.3 Parametric Study

In this section, a parametric study of different process parameters influencing rotational autofrettage of thick-walled disk is carried out. The geometrical parameter such as wall thickness ratio, rotational speed, density and yield stress of the material influences the performance of rotational autofrettage of a thick-walled disk. In the following subsections, the influence of these parameters on the rotational autofrettage of typical thick-walled disks is studied in detail.

84.3.1 Influence of the Wall Thickness Ratio (b/a)

In any autofrettage process, the geometric parameter, wall thickness ratio (*b*/*a*) has significant influence on the load requirement, residual stresses and increase in service load-carrying capacity. In the following subsections, the effect of wall thickness ratio on different factors is discussed in detail.

Effect of Wall Thickness Ratio on Rotational Speed In rotational autofrettage, the disk is subjected to a sufficiently high rotational speed in the loading phase. Yielding onsets at a rotational speed ω_i at the inner radius of the disk according to Eq. (84.3). Expressing Eq. (84.3) in non-dimensional form, one obtains

$$\Omega_i = \frac{2}{\left\{(1-\nu) + (3+\nu)\left(\frac{b}{a}\right)^2\right\}^{1/2}} \tag{84.8}$$

where Ω_i is the non-dimensional yield onset rotational speed given by

$$\Omega_i = \omega_i a\left(\frac{\rho}{\sigma_Y}\right)^{1/2} \tag{84.9}$$

Figure 84.1 shows the variation of normalized yield onset rotational speed Ω_i with the wall thickness ratio (*b*/*a*). It is observed that the initial normalized rotational speed required to cause yielding decreases with the increase of wall thickness ratio. This indicates that for higher *b*/*a* ratios, the disk requires comparatively lesser rotational speed for partially plasticizing the wall of the disk during loading phase for attaining the maximum residual stress at the inner side after unloading. For wall thickness ratios up to five, there is a significant reduction in the autofrettage rotational speed requirement. The subsequent increase of wall thickness ratio beyond five provides less than 5% reduction in speed.

Effect of Wall Thickness Ratio on Residual Stress The wall thickness ratio also influences the residual stress induced at the inner surface of the disk. In order to assess the effect of wall thickness ratio on the residual stress, a typical SS304 disk is

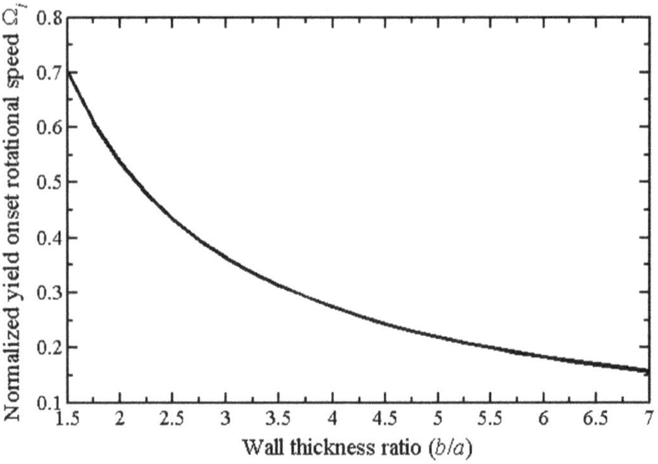

Fig. 84.1 Variation of normalized yield onset speed with wall thickness ratio

considered with the material properties given as follows: Young's modulus of elasticity, $E = 193$ GPa, yield stress, $\sigma_Y = 205$ MPa, Poisson's ratio, $\nu = 0.3$, density, $\rho = 8000$ kg/m^3. The numerical simulation of the residual stresses is carried out for the disk as described in Sect. 84.2. The detailed expressions are available in Ref. [18]. In Ref. [18], the author has discussed the effect of strain hardening and found that its effect is not very significant. Hence, in the present study, the effect of strain hardening is neglected. Consequently, for numerical simulation, the hardening parameter K is taken as zero in Eqs. (84.4) and (84.5) in Sect. 84.2. The simulated results are presented in Table 84.1 showing the variation of normalized residual hoop stress at the inner and outer radii with the wall thickness ratio (*b*/*a*). The residual hoop stress distribution along the positive radial direction of the disk for all (*b*/*a*) ratios is also shown in Fig. 84.2.

It is observed from Table 84.1 that as the wall thickness ratio is increased, the normalized compressive residual hoop stress increases at the inner radius. In all cases, the maximum rotational speed of the disk during loading phase is also shown. The maximum rotational speed is considered in such a way that it causes maximum overstrain level (ε)[1] in the disk, and at the same time the residual stresses should not initiate reyielding after unloading. The maximum speed in the loading phase follows an increasing trend with the wall thickness ratio as expected. The normalized tensile residual hoop stress induced at the outer surface of the disk is also presented in Table 84.1 for different wall thickness ratios. It is observed that the normalized tensile residual stress at the outer surface increases up to $b/a = 4.5$. This indicates that the sensitivity to stress corrosion cracking of the outer surface of

[1]The overstrain level, ε is defined as the percentage of disk's wall thickness that undergoes plastic deformation.

Table 84.1 Variation of normalized residual hoop stress at the inner and outer radii with wall thickness ratio

Wall thickness ratio (b/a)	Normalized maximum rotational speed in the loading phase $\Omega = \omega a\left(\frac{\rho}{\sigma_Y}\right)^{1/2}$	Normalized residual hoop stress $\left(\frac{\sigma_\theta^{Res}}{\sigma_Y}\right)$	
		At the inner surface	At the outer surface
1.5	0.794	−0.2806	0.1906
2	0.654	−0.4863	0.2935
2.5	0.554	−0.6362	0.3432
3	0.480	−0.7510	0.3891
3.5	0.423	−0.8414	0.4212
4	0.377	−0.9101	0.4532
4.5	0.341	−0.9630	0.4219
4.8	0.322	−0.9900	0.4047
5	0.309	−0.9969	0.3218

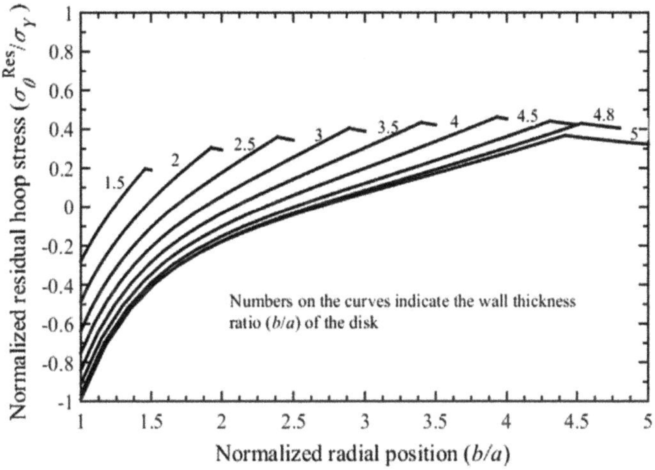

Fig. 84.2 Normalized residual hoop stress distribution in the disk for different wall thickness ratios corresponding to maximum rotational speed

the disk increases with the wall thickness ratio of the disk up to 4.5 for the maximum rotational speed. However, from b/a = 4.8, the maximum normalized tensile hoop stress at the outer surface of the disk starts decreasing. This may be due to the decrease of relative percentage overstrain in the disk. Thus, for the disks with b/a = 4.8 and above, it appears that the maximum compressive residual stress at the inner surface is almost the order of the yield stress of the material. It is observed that the tensile residual hoop stress at the outer surface increases gradually up to

$b/a = 4$, and thereafter it follows a decreasing trend. The lowering of the tensile hoop stress at the outer surface may be considered as relatively advantageous along with the maximum possible compressive residual stress at the inner surface. The residual hoop stress distribution in Fig. 84.2 shows that the maximum magnitude of the compressive residual hoop stress appears at the inner surface of the disk. It is to be noted that the maximum tensile residual hoop stress occurs at the radius of elastic–plastic interface for all (b/a) ratios. Figure 84.2 shows that the variation of the maximum tensile residual stress follows an increasing trend up to $b/a = 4.5$, and thereafter it starts decreasing with the increase of b/a.

Effect of Wall Thickness Ratio on Different Load Carrying Capacities The rotationally autofrettaged disk can bear significantly higher loads in service when it is subjected to internal pressure, temperature difference and rotational speed individually [18]. The effect of b/a on the different load-carrying capacities of the disk is also studied corresponding to the cases presented in Table 84.1. In Fig. 84.3, the maximum percentage increase in the service load for different wall thickness ratios is shown when the disk is subjected to internal pressure, radial temperature difference and rotational speed individually. The percentage increase in all service loads is estimated with reference to a corresponding non-autofrettaged disk. The maximum load-carrying capacity of the non-autofrettaged disk is decided based on the respective yield onset load following Tresca yield criterion. For internal pressure and radial temperature difference, the yield onset loads in non-dimensional form are given by [20].

$$\bar{P}_Y = \frac{P_Y}{\sigma_Y} = \frac{1}{2}\left(1 - \frac{a^2}{b^2}\right) \tag{84.10}$$

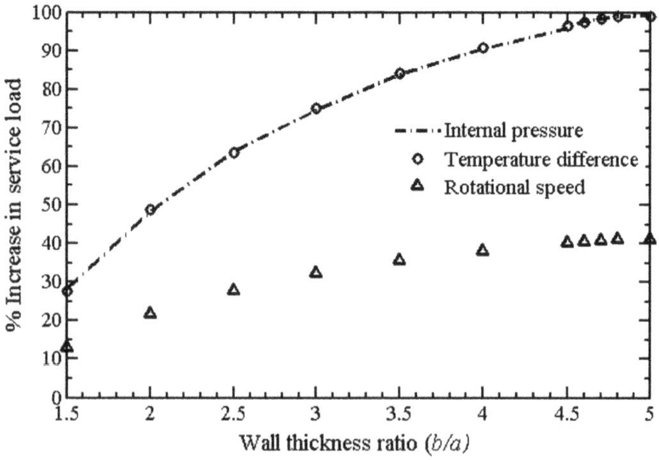

Fig. 84.3 Variation of percentage increase in the maximum different service loads as a function of wall thickness ratio

$$\theta_{Y}=\frac{E\alpha\Delta T_{Y}}{\sigma_{Y}}=\frac{2}{\left\{\frac{\ln\left(\frac{b}{a}\right)-1}{\ln\left(\frac{b}{a}\right)}+\frac{1}{\left\{\left(\frac{b}{a}\right)^{2}-1\right\}}\left\{\left(\frac{b}{a}\right)^{2}+1\right\}\right\}} \tag{84.11}$$

where P_Y is the yield onset pressure, ΔT_Y is the yield onset temperature difference and α is the coefficient of thermal expansion. The yield onset rotational speed is given by Eq. (84.8). The maximum load during service in the autofrettaged disk is obtained by superposing the loading stresses with the rotational residual stresses as discussed in Ref [18] such that the autofrettaged disk does not yield.

It is observed from Fig. 84.3 that the percentage increase in the internal pressure loading and temperature difference of the autofrettaged disk compared to the non-autofrettaged one is of the same order for all wall thickness ratios. The percentage increase in the rotational speed of the disk in service is comparatively lesser than the increase in the other two loads. The percentage increase in all the loads is increased significantly up to the wall thickness ratio 4.6. There is no significant change in the percentage increase of the relative load-carrying capacities as the wall thickness ratio is increased beyond b/a = 4.6. Thus, for attaining the maximum load carrying capacity in a disk, the maximum wall thickness ratio can be considered as b/a = 4.6. The maximum increase in the internal pressure and the radial temperature difference is about 97% and that in the rotational speed is about 41% during operation corresponding to the wall thickness ratio b/a = 4.6.

84.3.2 *Influence of Autofrettage Rotational Speed*

In rotational autofrettage, the plastically deforming load is in the form of a centrifugal force due to a sufficiently high rotational speed. Thus, the rotational speed plays an important role during rotational autofrettage of a disk. In Table 84.1, it is shown that the maximum rotational speed requirement for rotational autofrettage increases with the increase of wall thickness ratio. Consequently, the residual hoop stress generated at the inner and outer surfaces of the disk is also estimated for all wall thickness ratios up to five. Here, in this section, keeping the wall thickness ratio constant, the effect of rotational speed on the residual stress and the corresponding load-carrying capacity when subjected to different individual loads are assessed. For instance, the wall thickness ratio is taken as three for a typical SS304 disk. The normalized rotational speed for initiation of yielding at the inner side for b/a = 3 is 0.363, and the normalized rotational speed for completely plasticizing the disk wall is 0.481. Thus, for all simulations the rotational speed is taken in the range, $0.363 \leq \Omega \leq 0.481$. Within this range, the residual stress at the inner and outer surfaces of the disk is presented in Table 84.2 along with the percentage overstrain (ε) and the percentage increase in different individual loads during operation.

Table 84.2 Effect of rotational speed on percentage overstrain, residual stress and different load-carrying capacities

Normalized rotational speed (Ω)	Percentage overstrain $\varepsilon = \frac{c-a}{b-a} \times 100\%$	Normalized residual stress $\left(\frac{\sigma_\theta^{Res}}{\sigma_Y}\right)$		% increase in the maximum service load		
		At the inner surface	At the outer surface	Pressure	Temperature difference	Rotational speed
0.381	5.50	−0.1036	0.001453	10.30	10.86	5.05
0.399	12.79	−0.2148	0.007219	21.28	21.74	10.22
0.418	22.37	−0.3314	0.02043	33.13	33.69	15.38
0.437	35.07	−0.4533	0.04743	45.32	45.65	20.55
0.456	52.11	−0.5805	0.1033	58.05	58.69	25.72
0.475	78.19	−0.7131	0.2488	71.22	71.73	30.88
0.480	98.79	−0.7537	0.4367	74.51	75	32.43

It is observed from Table 84.2 that as the rotational speed in the loading phase is increased for a constant wall thickness ratio, the percentage overstrain, level of normalized residual stresses at the inner and outer surfaces and the percentage increase in the individual pressure loading, temperature difference across the wall thickness and the rotational speed during working increases. The maximum percentage increase in the respective individual service load is achieved for $\Omega = 0.480$. In order to attain a high rotational speed of the disk, high power/torque motor will be needed in practice. This may increase the equipment cost. However, one can choose the autofrettage speed depending upon the strength requirement of the disk in the actual working environment and accordingly select a motor of required capacity. It is also observed from Table 84.2 that at lower rotational speed, the outer surface of the disk is subjected to very low tensile residual stresses. Thus, at lower rotational speed, the outer surface of the disk is less sensitive to the stress corrosion cracking.

84.3.3 Influence of Material Properties

In rotational autofrettage of thick-walled disk, the properties of the disk material plays a very important role on the rotational speed during loading phase and residual stresses. The yield stress (σ_Y) of the material and density (ρ) are the two dominating material properties which influences the rotational autofrettage of a disk. It follows from Eq. (84.3) that for materials with high density and low yield stress predicts yielding at low rotational speed. In order to assess the effect of ρ and σ_Y, disks of different materials with different combinations of ρ and σ_Y are considered as listed in Table 84.3. The radial dimensions for all disks are kept constant

Table 84.3 Materials with different combinations of ρ and σ_Y

Materials	ρ (kg/m^3)	σ_Y (MPa)
Aluminum	2700	50.3
SS304	8000	205
Ti-0.3Mo-0.8Ni	4510	380
Titanium unalloyed	4510	170
ASTM-A723	7850	1050

and is taken as a = 0.1 m and b = 0.3 m for instance. The Poisson's ratio for all materials is taken as 0.3.

Now, the rotational autofrettage for all disks is numerically simulated such that the inner surface is subjected to the maximum possible compressive residual stress. The results are presented in Table 84.4. It is observed that in order to generate the maximum possible compressive residual stress ($-0.7537\sigma_Y$ in all disks), the rotational speed increases proportionately for higher yield stress. For the low values of yield stress and density, e.g., aluminum, the rotational speed required to achieve the maximum possible compressive residual stress is lower. Thus, for such materials the speed required for rotational autofrettage can be attained using a low power/low torque electrical motor. This makes the process comparatively inexpensive. However, for high values of yield stress and density, e.g., ASTM-A723, makes inconvenience in achieving the rotational autofrettage in terms of the high speed requirement. For ASTM-A723 disk, the rotational speed required to achieve the maximum possible compressive residual stress is 1756.86 rad/s (16776.78 rpm), which is very high. This requires a high power motor to induce high speed in the disk. Also, in such cases, extra care needs to be taken for dynamic balancing of the rotating system. Materials possessing same density (e.g., Ti-0.3Mo-0.8Ni and titanium unalloyed) require higher rotational speed with the higher yield stress. The materials with medium density and lower yield stress, e.g., SS304 requires moderate speed, which is also convenient for rotational autofrettage.

Table 84.4 Effect of material properties on rotational autofrettage

Materials	ω_i (rad/s)	ω (rad/s)	Compressive residual stress at the inner surface
Aluminum	495.1027	655.66	$-0.7537\sigma_Y$
SS304	580.6643	768.96	$-0.7537\sigma_Y$
Ti-0.3Mo-0.8Ni	1052.9	1394.39	$-0.7538\sigma_Y$
Titanium unalloyed	704.2541	932.64	$-0.7538\sigma_Y$
ASTM-A723	1326.6	1756.86	$-0.7538\sigma_Y$

84.4 Conclusions

In this work, a parametric study of a new method of rotational autofrettage for a thick-walled axisymmetric circular disk is carried out. The study is based on the theoretical analysis of Kamal [18]. The influence of different process parameters on the performance of rotational autofrettage of thick-walled disk is carried out. First, the effect of wall thickness ratio on the rotational speed is studied followed by its effect on the residual stresses and different load-carrying capacities. Then, the effect of rotational speed, material density and yield stress on the residual stress and load-carrying capacities is considered. It is found that as the wall thickness ratio is increased the rotational speed required to cause plastic deformation for achieving rotational autofrettage in the disk is decreased. The rotational speed requirement is the minimum for wall thickness ratio of five with less than 5% reduction in speed upon subsequent increase of wall thickness ratio. It is also found that with the increase of wall thickness ratio, the compressive residual stress induced at the inner surface increases corresponding to the maximum rotational speed. For $b/a = 4.8$ and above, the maximum compressive residual stress induced in the disk is -0.99 σ_Y. The study of the influence of wall thickness ratios on the different load-carrying capacities shows that with the increase of wall thickness ratio, the percentage increase in the individual pressure loading, temperature difference and rotational speed of the disk in service increases. It is found that the maximum load-carrying capacities in all the three individual loads is achieved for $b/a = 4.6$ with an error of less than 1%. The percentage in the working pressure loading and temperature difference are of the same order for all wall thickness ratios. However, the percentage increase in the centrifugal load (rotational speed) is lesser compared to the increase in working pressure and temperature difference for wall thickness ratios. The effect of rotational speed for constant wall thickness ratio of the disk shows that on increasing the rotational speed, the residual stresses at the inner and outer surfaces and the percentage increase in the different individual loads during working increases for a typical disk. The yield stress of the material and density also influence the rotational autofrettage of a disk. It is observed that the lower the yield stress of the material, the lower the rotational speed necessary to achieve a certain level of autofrettage. Materials with high yield stress and high density require very high rotational speed to achieve rotational speed. The necessary high speed may be produced by utilizing a high power/high torque motor in such cases. However, this will make the process comparatively expensive. High rotational speed may also cause dynamic imbalance of the system. Thus, rotational autofrettage is convenient for low yield stress and low density materials. Finally, it may be concluded that for the applications of the disks as a rotating component in aircraft and automobile industries or as fastener holes subjected to high pressure/temperature, rotational autofrettage is a very useful strengthening method. For this purpose, the process parameters in the rotational autofrettage of the disk should be selected as per the strength requirement of the disk against the working load during operation.

References

1. Jacob, L.: La Résistance et L'équilibre Elastique des Tubes Frettés. Mémoire de L'artillerie Navale **1**(5), 43–155 (1907). (in French)
2. Davidson, T.E., Barton, C.S., Reiner, A.N., Kendall, D.P.: New approach to the autofrettage of high-strength cylinders. Exp. Mech. **2**, 33–40 (1962)
3. Mote, J.D., Ching, L.K.W., Knight, R.E., Fay, R.J., Kaplan, M.A.: Explosive autofrettage of cannon barrels. Army Materials and Mechanics Research Center, Watertown, Massachusetts, p.02172 Report No. AMMRC CR 70-25. (1971)
4. Rees, D.W.A.: Autofrettage theory and fatigue life of open-ended cylinders. J. Strain Anal. Eng. Des. **25**(2), 109–121 (1990)
5. Gao, X.L.: An exact elasto-plastic solution for an open-ended thick-walled cylinder of a strain-hardening material. Int. J. Press. Vessel Pip. **52**, 129–144 (1992)
6. Parker, A.P., Underwood, J.H., Kendall, D.P.: Bauschinger effect design procedures for autofrettaged tubes including material removal and Sachs' method. ASME J. Press. Vessel Technol. **121**(4), 430–437 (1999)
7. Huang, X.P., Moan, T.: Residual stress in an autofrettaged tube taking Bauschinger effect as a function of the prior plastic strain. ASME J. Press. Vessel Technol **131**(2), 021207-1–021207-7 (2009)
8. Alexandrov, S., Jeong, W., Chung, K.: Descriptions of reversed yielding in internally pressurized tubes. ASME J. Press. Vessel Technol. **138**(1), 011204-1–011204-10 (2016)
9. Iremonger, M.J., Kalsi, G.S.: A numerical study of swage autofrettage. ASME J. Press. Vessel Technol. **125**, 347–351 (2003)
10. Bihamta, R., Movahhedy, M.R., Mashreghi, A.R.: A numerical study of swage autofrettage of thick-walled tubes. Mater. Des. **28**, 804–815 (2007)
11. Rees, D.W.A.: A theory for swaging of discs and lugs. Meccanica **46**, 1213–1237 (2011)
12. Kamal, S.M., Dixit, U.S.: Feasibility study of thermal autofrettage of thick-walled cylinders. ASME J. Press. Vessel Technol. **137**(6), 061207-1–061207-18 (2015)
13. Zare, H.R., Darijani, H.: A novel autofrettage method for strengthening and design of thick-walled cylinders. Mater. Des. **105**, 366–374 (2016)
14. Kamal, S.M., Dixit, U.S.: Feasibility study of thermal autofrettage process. In: Narayanan, R. G., Dixit, U.S. (eds.) Advances in Material Forming and Joining, pp. 81–107. Springer, New Delhi (2015)
15. Kamal, S.M., Borsaikia, A.C., Dixit, U.S.: Experimental assessment of residual stresses induced by the thermal autofrettage of thick-walled cylinders. J. Strain Anal. Eng. Des. **51**(2), 144–160 (2016)
16. Shufen, R., Dixit, U.S.: A finite element method study of combined hydraulic and thermal autofrettage process. ASME J. Press. Vessel Technol. **139**(4), 041204-1–041204-9 (2017)
17. Shufen, R., Dixit, U.S.: An analysis of thermal autofrettage process with heat treatment. Int. J. Mech. Sci. **144**, 134–145 (2018)
18. Kamal, S.M.: Analysis of residual stress in the rotational autofrettage of thick-walled disks. ASME J. Press. Vessel Technol. **140**(6), 061402-1–061402-10 (2018)
19. Zare, H.R., Darijani, H.: Strengthening and design of the linear hardening thick-walled cylinders using the new method of rotational autofrettage. Int. J. Mech. Sci. **124–125**, 1–8 (2017)
20. Kamal, S.M., Dixit, U.S.: A comparative study of thermal and hydraulic autofrettage. J. Mech. Sci. Technol. **30**(6), 2483–2496 (2016)

Chapter 85
Influence of Varying Fully Constrained Circular Boundary Condition Area on Damage Mechanism of GFRP Laminate Under Low Velocity Impact Loading

Mahesh and K. K. Singh

Abstract Modern aircrafts, automobiles, space crafts, etc., need light weight and high-performing structures or materials to reduce the total weight of the structure without compromising in strength which highly influences the efficiency of that structure and fuel cost associated with it. Thus, fiber-reinforced polymer (FRP) laminates are the most suitable materials for these kinds of applications due to their high specific strength to weight ratio which is better than many other conventional metal or alloy materials. But these materials are highly prone to catastrophic and sudden failure because of their unpredictable and undetectable crack growth and formation in the laminate structure. These undetectable cracks, also known as barely visible impact damage (BVID), which are induced into the laminate under low velocity impact. Example includes dropping of a tool in an assembly line. Low velocity impact (LVI) is a highly complex phenomenon as it consists of subsurface micro cracks (fiber and matrix cracking), debonding, delamination, and fiber breakage. Due to anisotropic nature of the material, prediction of damage induced becomes difficult. Hence, in this numerical analysis, LS-DYNA is used to carry out the LVI loading onto a bidirectional plain woven glass fiber-reinforced polymer (GFRP) laminate using a hemispherical striker of mass 20 kg with a nose radius of 5 mm impacted at a velocity of 2 m/s. During the analysis, fully constrained circular boundary condition (FCCBC) is used and the radius of the circular boundary condition (CBC) is varied from 15 to 35 mm with an increment of 10 mm. Effect of variation in boundary condition diameter on damage area is examined using von Mises stresses, energy versus time, force versus time, and displacement versus time graphs.

Mahesh (✉) · K. K. Singh
Indian Institute of Technology (ISM), Dhanbad 826004, Jharkhand, India
e-mail: itsshindhe@gmail.com

K. K. Singh
e-mail: kksingh@iitism.ac.in

R. G. Narayanan et al. (eds.), *Advances in Computational Methods in Manufacturing*, Lecture Notes on Multidisciplinary Industrial Engineering,
https://doi.org/10.1007/978-981-32-9072-3_85

Keywords GFRP · Low velocity impact (LVI) · Boundary condition · Damage mechanism

85.1 Introduction

FRP composites gained widespread application during the World War-II due to their high strength to weight ratio. Since then, it is used in various industries. Impact is one of the critical loading conditions for FRP composites particularly low velocity impact is fatal because it creates subsurface or barely visible impact damages or nonvisible cracks to the naked eyes. These cracks, over a prolonged period of time grows throughout the laminate and reduces strength of the laminate or structure which leads to sudden or catastrophic failure of the structure. Even though experimental investigation can be carried out to study the damage mechanism, understanding the damage behavior and evolution becomes complex because these materials are highly anisotropic in nature. Moreover, damage mechanism involves various failure mechanisms such as fiber breakage, matrix cracking, delamination, etc. Thus, numerical analysis helps in better understanding of damage mechanism. Damage mechanism and its pattern are influenced by various aspects such as impactor shape, impacting angle, etc., which are all considered as impact loading parameters. One among them is boundary condition because this determines the material space available for absorbing energy from the impactor to the laminate particularly if the boundary condition (BC) is fully constrained clamped type. Although wave propagation in FRP laminate influences the damage pattern and deformation of the material but only at high velocity impact, and this effect is less observed for low velocity impact (LVI).

An extensive literature review was conducted by [1–3] on low velocity impact behavior of FRP laminates and covered various aspects of laminates as well as impact parameters. Moreover, these review works covered the various finite element analysis tools used particularly LS-DYNA. Li et al. [4] studied the GFRP laminates behavior under LVI loading using drop-weight impact machine and simulated the experimental results using ABAQUS FEA tool with edge constrained boundary condition. Both experimental and numerical results showed that central deflection and extent of damage at front as well as back surface behaved linearly with respect to the impact energy. Both numerical and experimental results were reasonably in good agreement. Rawat et al. [5] used LS-DYNA to study the behavior of symmetric and asymmetric GFRP laminates with and without pre-cracks under low velocity impact. From results, it was clear that without pre-crack symmetric laminate performed better than symmetric with pre-crack and asymmetric laminate with and without pre-crack. Author's used circular boundary condition for both symmetric and asymmetric laminate design. Finally, concluded that energy absorption is influenced by symmetric nature of the plies arranged in the laminate about the mid-plane. Rawat et al. [6] investigated the influence of impactor shape on damage area and damage shape under LVI. Eight ply symmetric GFRP laminate was impacted with flat, hemispherical, spherical, and oval-shaped impactor. LS-DYNA numerical simulation showed that flat-headed impactor caused maximum damage area while oval-shaped impactor caused minimum damage area but it fractured maximum fibers in the laminate. It was concluded that contact surface area with the laminate primarily influenced the damage extent and pattern. Ansari and Chakrabarti [7] examined the influence of boundary condition and impactor nose angle on GFRP laminates under impact loading. Experimental work is carried out on pneumatic gun with flat-and conical-headed impactor and ANSYS was used for numerical simulation. Clamped and simply supported boundary conditions were considered during the analysis. From experimental and numerical analysis, it was concluded that the blunt impactor showed more resistance to penetration under simply supported boundary condition. Liu et al. [8] investigated the satin weave and unidirectional hybrid carbon fiber laminates damage resistance under LVI. Drop-weight impact tester was used for experimentation while ABAQUS was used for numerical validation. From experimental and numerical analysis, it was concluded that damage resistance was affected by lay-up sequence. Delamination was found to be the major failure mode and observed a good correlation between numerical and experimental results. Hashin failure criteria was used by Kim et al. [9] to develop a three-dimensional progressive damage model for GFRP laminate. Developed model was simulated using LS-DYNA and experimental validation was done by conducting impact and flexural testing. Author's concluded that to predict the behavior of impact, interlaminar delamination, and damage progression, three-dimensional progressive failure can be used. Constant Menna et al. [10] investigated GFRP laminates made of different thickness under low velocity impact using LS-DYNA numerical analysis tool. From analysis, it was found that delamination was the major damage mode of failure and majorly observed at mid-plane. Both experimental and numerical analysis found to be in good agreement.

Many available literature is related to behavior of FRP laminate under different boundary conditions, impactor nose shapes, laminate ply orientation, symmetric nature of laminate about the mid-plane, etc. But there is no work that correlates the damage area or energy absorption by the laminate with the boundary condition area. Thus, in this simulation work bidirectional woven glass fibers-reinforced with epoxy is considered. Laminates are made of eight plies with symmetric configuration [11] and impacted with a hemispherical impactor made of mild steel. Circular boundary condition with varying area is considered and LS-DYNA is used to carry out the numerical simulation. To author's knowledge more work is done on different boundary conditions and impactor shapes but a very less work is done on influence of varying boundary condition area and its effect on damage mechanism of FRP laminates under LVI.

Table 85.1 Properties of material used [5]

Young's modulus (GPa)			Poisson's ratio			Tensile strength (GPa)			Compressive strength (GPa)			Shear strength (GPa)		
Laminate properties														
E_{11}	E_{22}	E_{33}	ν_{21}	ν_{31}	ν_{32}	T_X	T_Y	T_Z	C_X	C_Y	C_Z	S_{ba}	S_{ca}	S_{cb}
26	26	8	0.1	0.25	0.25	0.85	0.85	0.12	0.72	0.72	0.5	0.105	0.065	0.065
Impactor properties														
210			0.3											

1, 2 and 3 indicates the material properties with respect to that direction or plane
X, *Y* and *Z* indicates the material properties with respect to that direction or plane
a, b and c indicates the material properties with respect to that direction or plane

85.2 Numerical Simulation

85.2.1 Materials Used

Bidirectional plain woven glass fiber and epoxy laminate are considered for laminates. All the materials properties are provided in Table 85.1 and orientation of the plies is $[(0/90)/(+45/-45)/(+45/-45)/(0/90)]_S$. A mild steel hemispherical-shaped impactor of diameter 10 mm is considered.

85.2.2 Numerical Modeling

MAT_59_SOLID_COMPOSITE_FAILURE_SOLID_MODEL is used to model the GFRP laminate. This card considers the material as orthotropic material and is modeled for enhanced damage modeling. MAT_20_Rigid is used to model the mild steel impactor. Contact between plies and impactor is modeled using AUTOMATIC_SURFACE_TO_SURFACE while TIED_SURFACE_TO_SURFAC- E is used to model the contact between the plies. BOUNDARY_ CONDITION_ SPC is used to model the circular boundary condition with varying radius of 15, 25, and 35 mm. Area of boundary conditions is 706.85, 1963.49, and 3848.45 mm^2. These boundary condition from here onwards are referred as boundary condition 15 (BC 15), boundary condition 25 (BC 25), and boundary condition 35 (BC 35), respectively.

85.3 Results and Discussion

As the boundary condition area is varied, the response time and von Mises stresses generated during the impact are also varied. Effective von Mises stresses (Fig. 85.1) are high for boundary condition 15 at midplies when compared to other two boundary conditions because less material space or region is available to absorb the

Fig. 85.1 Effective von Mises stresses at time $T = 1$ ms and $T = 3.5$ ms

impacted energy. Whereas, at boundary condition 35, the increase in von Mises stresses is gradual because sufficient material space is available for the material to react with the impacted energy. Table 85.2 provides the maximum effective von misses stresses generated in each plies at time $T = 1$ ms and it is observed that at boundary condition area 35, there is gradual drop in the effective von Misses stresses from top ply to the mid ply and then gradual increase from midply to the bottom ply while exactly opposite is observed in case of boundary condition 25, while a sudden increase in effective von Mises stresses and reaches maximum value at mid ply and then starts declining gradually from mid ply to the bottom ply

Table 85.2 Maximum von Mises stresses under varying boundary condition at time $T = 1$ ms

Ply position in laminate from top to bottom	Boundary condition = 15	Boundary condition = 25	Boundary condition = 35
1	2.941e−1	3.744e−1	4.732e−1
2	3.367e−1	5.749e−1	3.916e−1
3	5.035e−1	3.141e−1	3.806e−1
4	5.999e−1	2.914e−1	3.236e−1
5	5.199e−1	3.205e−1	4.093e−1
6	4.692e−1	3.633e−1	4.148e−1
7	4.048e−1	4.389e−1	3.924e−1
8	3.579e−1	2.810e−1	4.095e−1

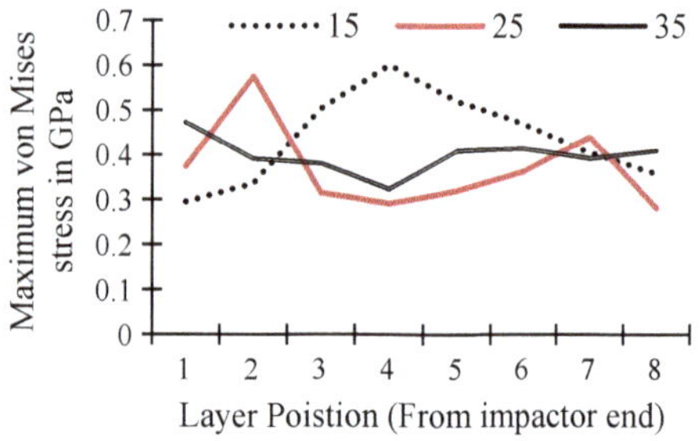

Fig. 85.2 Effective von Mises stresses generated at time $T = 1$ ms

(Fig. 85.2). This kind of behavior is mainly influenced by the boundary condition area and this transaction in behavior is observed from boundary condition 25 which bridges between boundary condition 15 and 35.

Complete fiber fracture is observed in case of boundary condition 15 at time $T = 3.5$ ms, while no such fracture is observed for boundary condition 35 whereas, boundary condition 25 showed mixed behavior as shown in Fig. 85.1. Effective von Mises stresses is high for boundary condition 35 in each ply when compared to boundary condition 15 (Table 85.3). Here, gradual increase and then decrease trend is observed for boundary condition 15, while exactly opposite characteristic is exhibited by boundary condition 35 as seen in Fig. 85.3.

Almost square damage is observed in all boundary conditions. Initially, at time $T = 1$ ms, damage area for boundary condition 15 is high when compared to other boundary conditions, while at time $T = 3.5$ ms damage area is high for boundary condition 35 then boundary condition 25 and boundary condition 15 (Fig. 85.4) because boundary condition 15 at initial stage reacted to impactor with all available material space. But over a period of time, damage area gets saturated for boundary condition 15 while for BC 35, there is enough material space available to react with the impactor to absorb the energy over a period of time through deformation. But boundary condition 15, after saturation absorbed energy is dissipated in terms of damage height, and maximum damage height is also observed for boundary condition 15 (Fig. 85.5 and Tables 85.4, 85.5).

Table 85.3 Maximum von Mises stresses under varying boundary condition at time T = 3.5 ms

Ply position in laminate from top to bottom	Boundary condition = 15	Boundary condition = 25	Boundary condition = 35
1	1.845e−1	1.831e−1	3.097e−1
2	2.674e−1	1.541e−1	2.018e−1
3	2.749e−1	1.183e−1	3.390e−1
4	1.900e−1	1.244e−4	2.507e−1
5	2.318e−1	1.566e−1	5.072e−1
6	1.326e−1	1.043e−1	1.863e−1
7	1.257e−1	8.455e−1	2.266e−1
8	1.363e−1	1.169e−1	1.817e−1

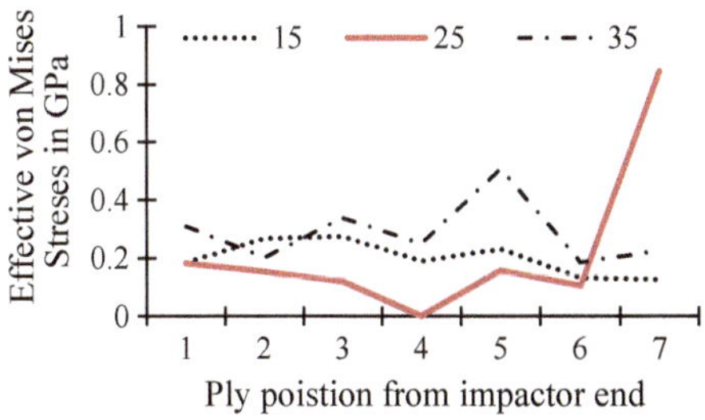

Fig. 85.3 Effective von Mises stresses at time T = 3.5 ms

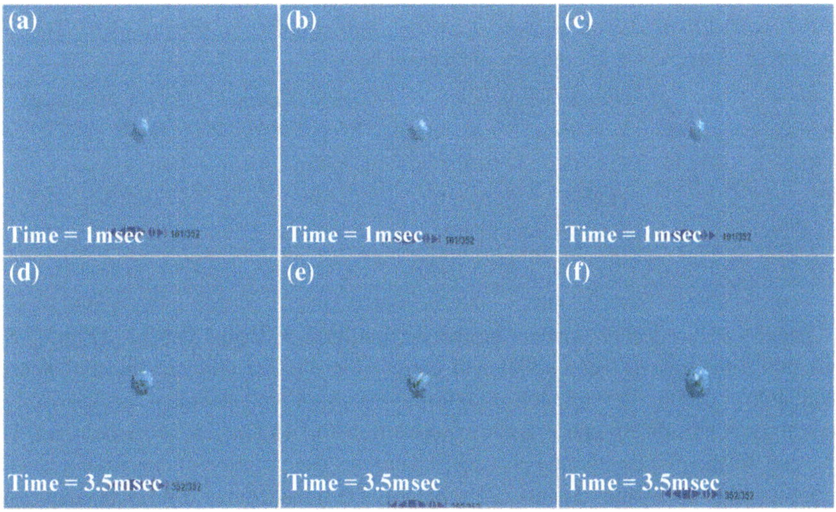

Fig. 85.4 Damage along X-and Y-direction at time T = 1 ms **a** and **d** boundary condition 15, **b** and **e** boundary condition 25, and **c** and **f** boundary condition 35

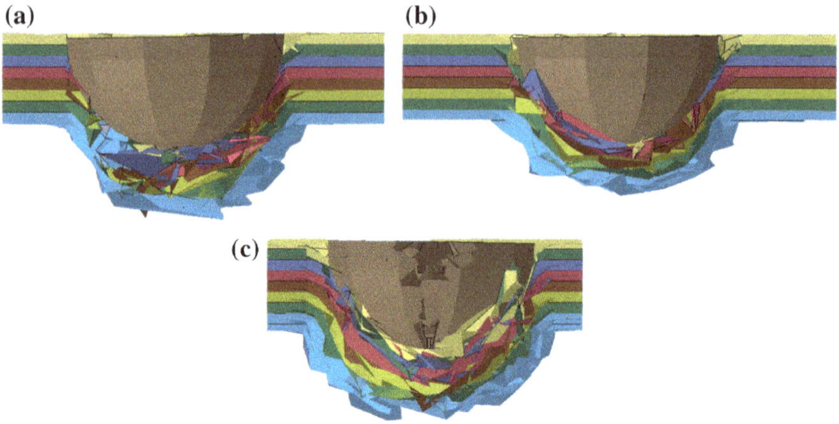

Fig. 85.5 Damage height at time $T = 3.5$ ms **a** BC 15 **b** BC 25 and **c** BC 35

Table 85.4 Approximate damage area

Time = 1 ms	Damage length in mm		Approx. damage area at base mm^2
Boundary condition	*X*-direction	*Y*-direction	
15	10.135	10.0464	101.8202
25	10.0703	8.2277	82.8554
35	8.2384	8.1574	67.2039

Table 85.5 Approximate damage area and damage height

Time = 3.5 ms	Damage length in mm		Approx. damage area at base in mm^2	Approx. damage height in mm^2
Boundary condition	*X*-direction	*Y*-direction		
15	11.1415	10.167	113.2746	8.3503
25	11.9445	10.2682	122.6485	7.4180
35	13.0267	12.663	164.9571	8.3008

Energy absorption is more for boundary condition 35 than 15 and 25 (Fig. 85.6) but this difference is very small and causes for this variation is material space availability during the impact. Maximum resultant displacement is shown by boundary condition 35 and is more compared to BC 15 and 25. Moreover, BC 15 showed linear displacement because there is no deformation in material instead involves fiber breakage. But BC 25 indicates the transaction form linear displacement to more elongate or bell-shaped deformation (Fig. 85.7). More fiber breakage

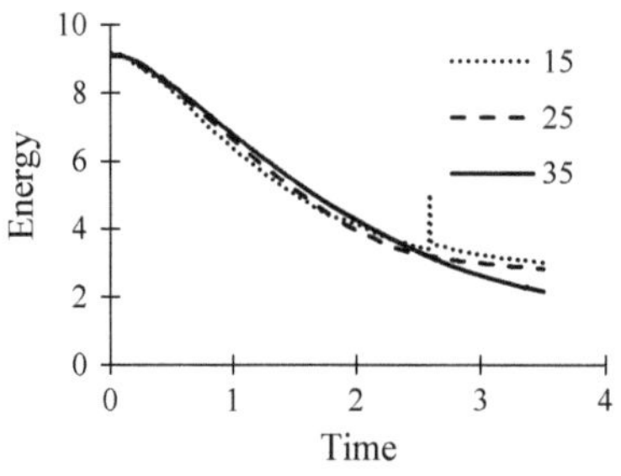

Fig. 85.6 Impact energy (J) versus time (ms)

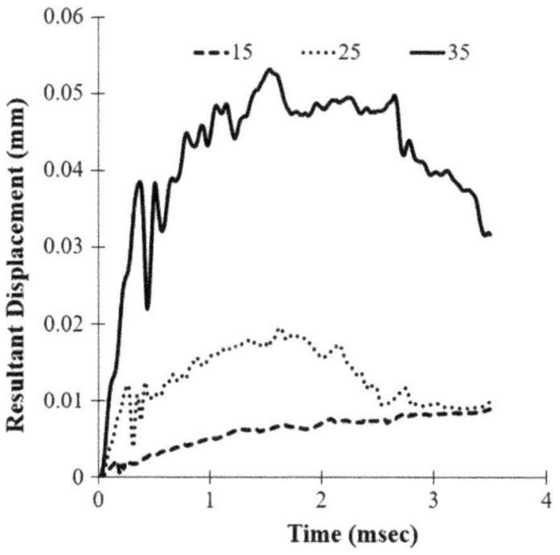

Fig. 85.7 Resultant force (N) versus time (ms)

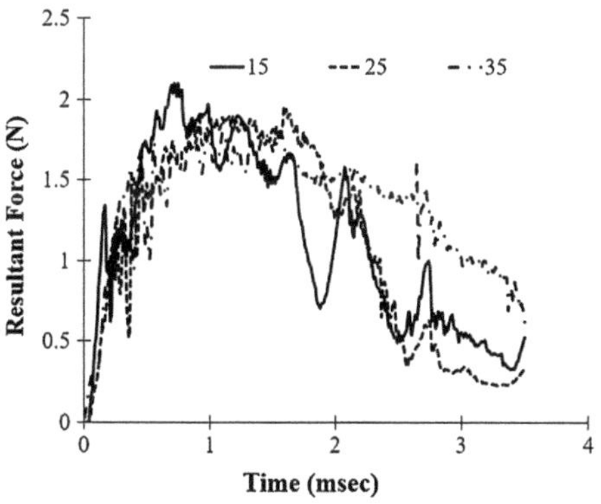

Fig. 85.8 Resultant displacement (mm) versus time (ms)

is observed in case of boundary condition 15 than other boundary conditions (Fig. 85.8). because there is more space available at boundary condition 35 for material to undergo sufficient deformation.

85.4 Conclusion

Influence of boundary condition area is studied for GFRP laminate under LVI using LS-DYNA. Three circular boundary conditions 15, 25, and 35 mm in radius are considered for simulation. Damage length in *X*-and *Y*-direction, damage height, effective von Mises stresses, resultant displacement, energy absorption, and resultant force are evaluated. From numerical simulation results and observation following points were concluded.

- As the boundary condition area reduces, the damage area reduces by 31% at 3.5 ms for boundary condition 15 when compared to boundary condition 35 but fiber breakage and damage height increases in case of boundary condition 15.
- As the boundary condition area reduces, the deformation associated with the material also reduces. Reduction in deformation for boundary condition 15 is found to be 69% less when compared to boundary condition 35 at time $T = 3.5$ ms.
- As the boundary condition reduces, the energy absorption capacity reduces but this reduction is less when compared to higher boundary condition. In case of boundary condition 35 absorbed 9% more energy compared to boundary condition 15.
- As the boundary condition area is reduced from 35 to 15, stress generation is also increased because in case of boundary condition area 15, the material space availability is less compared to 35. Moreover, maximum stress generation is also due to the fiber breakage which is maximum in case of boundary condition area 15 than 35.
- Stresses generated in the material show drastic and sudden increase for small boundary condition area.

References

1. Agarwal, S., Singh, K.K., Sarkar, P.K.: Impact damage on fibre-reinforced polymer matrix composite—a review. J. Compos. Mater. **48**, 317–332 (2014)
2. Singh, N.K., Singh, K.K.: Review on impact analysis of FRP composites validated by LS-DYNA. Polym. Compos. **36**, 1786–1798 (2015)
3. Rawat, P., Singh, K.K.: An impact behavior analysis of CNT-Based fiber reinforced composites validated by LS-DYNA: a review. Polym. Compos **38**, 175–184 (2017)
4. Li, Z., Khennane, A., Hazell, P.J., Brown, A.D.: Impact behavior of pultruded GFRP composites under low-velocity impact loading. Compos. Struct. **168**, 360–371 (2017)
5. Rawat, P., Singh, K.K., Singh, N.K.: Numerical investigation of low velocity impact in symmetric and asymmetric GFRP laminate with and without precrack. Adv. Mater. Proc. **2** (3), 152–155 (2017)
6. Rawat, P., Singh, K.K., Singh, N.K.: Numerical investigation of damage area due to different shape of impactors at low velocity impact of GFRP laminate. Mater. Today Proc. **4**, 8731–8738 (2017)

7. Ansari, M.M., Chakrabarti, A.: Effect of boundary condition and impactor nose angle on impact behavior of FRP composite: experimental and FE analyses. Mater. Proc. **4**(9), 9645–9649 (2017)
8. Liu, H., Falzona, B.G., Tan, W.: Experimental and numerical studies on the impact response of damage tolerant hybrid unidirectional/woven carbon-fiber reinforced composite laminates. Compos. B **136**, 101–118 (2018)
9. Kim, D.H., Jung, K.H., Lee, I.G., Kim, H.J., Kim, H.S.: Three-dimensional progressive failure modeling of glass fiber reinforced thermoplastic composites for impact simulation. Compos. Struct. **176**, 757–767 (2017)
10. Menna, C., Asprone, D., Caprino, G., Lopresto, V., Prota, A.: Numerical simulation of impact test on GFRP composites laminates. Int. J. Impact. Eng. **38**, 677–685 (2017)
11. Singh, K.K., Singh, N.K., Jha, R.: Analysis of symmetric and asymmetric glass fiber reinforced plastic laminates subjected to low-velocity impact. J. od Compos. Mater. **50**(14), 1853–1863 (2015)

Chapter 86
Process Capability Index for Geometrically Distributed Quality Characteristics

N. K. Singh

Abstract Process capability indices play a very important role in the control and the improvement of quality characteristics of products during the production process. Various process capability indices have been developed for quality characteristics which follow normal, Weibull, exponential and other distributions. Many researchers have also developed process capability indices for quality characteristics which follow bivariate normal and exponential distributions, respectively. Some of these indices are being used by industries for the quality improvement of products during the production process for competing in global markets. Some quality characteristics of the products should be treated as discrete random variables following distributions like geometric distribution rather than widely known continuous distribution like normal. Such situations are encountered while dealing with quality characteristics like number of operations before the failure of electrical, electronic or mechanical switches and components. The control and the improvement of such quality characteristics of such products during the production process, in order to ensure the conformity to the specification (usually one-sided), are quite important. No much work has been done so far to develop process capability index applicable to such quality characteristics following the geometric distribution. In this paper, the author proposes a process capability index for a number of successes or operations before the first failure of a component or product assumed to follow a geometric distribution and derives the expectation and variance of the estimated index. The optimal choice of process capability interval related to the said index has also been discussed. The said process capability index may be low for a highly capable process and for the minimum variation of the index, the percentage area lying outside natural tolerance limit may be low as 0.02, and the probability of successes or operations before the first failure may be high as 0.99.

Keywords Process capability index · Geometric distribution · Quality characteristics · Sampling properties

---

N. K. Singh (✉)
Indian Institute of Technology (Indian School of Mines), Dhanbad 826004, India
e-mail: nirmal@iitism.ac.in

R. G. Narayanan et al. (eds.), *Advances in Computational Methods in Manufacturing*, Lecture Notes on Multidisciplinary Industrial Engineering,
https://doi.org/10.1007/978-981-32-9072-3_86

86.1 Introduction

Process capability index C_p is a simple relative number or numerical value comparing the value of required process variability (required tolerance interval) to natural process variability (natural tolerance interval). Kane [1] introduced the process capability index C_p, also called process potential index, which considers spread only.

$$C_p = \frac{(U - L)}{6\sigma} \tag{86.1}$$

$$C_p \geq 1 \quad \text{For a capable process}$$

where U and L are upper and lower specification limits, respectively.

However, C_p focuses on the dispersion of the studied process and does not take into account the centering of the process. To overcome this problem, Kane [1] introduced the process performance index, C_{pk} as,

$$C_{pk} = \min\left\{\frac{\text{USL} - \mu}{3\sigma}, \frac{\mu - \text{LSL}}{3\sigma}\right\} \tag{86.2}$$

where μ is the process mean and σ is the process standard deviation of the in-control process.

The reason for six in the denominator in Eq. (86.1) is that under the assumption of normality, 6σ spread covers 99.73% of the data, leaving only 0.27% in non-conforming range. Larger values of C_p indicate high process capability while smaller values (less than 1.0) indicate lower process capability and incapable process.

The indices C_p and C_{pk} do not take into account the target value. Chan et al [2] point out that the index while ensuring that the process spread is within the specification limits, fails to take into account the target value, T of the process. To correct this, the authors suggested a modified version of C_p, Thus, this index is defined as

$$C_{pm} = \frac{U - L}{6\tau} = \frac{C_p}{\sqrt{1 + \left(\frac{\mu - T}{\sigma}\right)^2}} \tag{86.3}$$

where $\tau^2 = \sigma^2 + (\mu - T)^2$

It is obvious that C_{pm} is smaller than C_p.

The index C_{pm} was also proposed independently by Taguchi and Hsiang [3]. C_{pm} is, sometimes, called Taguchi capability index.

A capability index, which is more sensitive than C_{pk} and C_{pm} with regards to departures from the process mean to the target value, was introduced by Pearn et al. [4] as

$$C_{pmk} = \min\left\{ \frac{\mathrm{USL} - \mu}{3\sqrt{\sigma^2 + (\mu - T)^2}}, \frac{\mu - \mathrm{LSL}}{3\sqrt{\sigma^2 + (\mu - T)^2}} \right\} \quad (86.4)$$

A superstructure or family of capability indices, containing C_p, C_{pk}, C_{pm} and C_{pmk} involving two parameters, is introduced by Vannman [5] as,

$$C_p(u, v) = \frac{d - u|\mu - m|}{3\sqrt{\{\sigma^2 + v(\mu - T)^2\}}} \quad (86.5)$$

where $d = \frac{U-L}{2}$ and $m = \frac{U+L}{2}$

By letting $u = 0$ or 1 and $v = 0$ or 1 in the above equation, we can obtain four basic indices, i.e., $C_p(0,0) = C_p$, $C_p(1,0) = C_{pk}$, $C_p(0,1) = C_{pm}$ and $C_p(1,1) = C_{pmk}$.

Wang and Hubele [6] discussed geometric distance approach to reduce the dimensionality of the multivariate (normal and non-normal) data to univariate data and used established univariate process capability index (PCI) for process performance analysis.

Castagliola [7] extended the paper of Castagliola and Castellanos [8] for the estimation of bivariate PCIs in the case of non-normal bivariate distributions. The proposed method is based on the use of Johnson's distribution/transformation in order to transform the bivariate non-normal distribution into an approximate bivariate normal distribution. Numerical examples have been presented and some criteria are given in order to choose the appropriate Johnson's distribution.

Raut and Attar [9] provided a case study based on process capability in the vehicle industry. In this case study, area of analysis was the measurement section for which they selected the body shop production. The main highlights of the case study were to go through a comprehensive explanation of the entire body shop and the departments covered in it.

Senvar and Sennaroglu [10] examined Clements' Approach (CA), Box-Cox transformation (BCT) and Johnson transformation (JT) methods for process capability evaluations through Weibull-distributed data with dissimilar parameters to figure out the effects of the tail behaviors on process capability and matched their performances in terms of accuracy and precision.

In Sect. 86.2 of this paper, the author proposes a process capability measure for the number of success or operations before the failure assumed to follow a geometric distribution and derives the expectation and sampling variance of the estimated index. The optimum choice of process capability interval has been discussed in Sect. 86.3.

86.2 Process Capability Index for Geometric Distribution

86.2.1 Process Capability Index (Ig)

The probability density function (pdf) for geometric distribution is given by

$$f(x) = qp^x \tag{86.6}$$

$$\text{Let} \sum_{x=0}^{y} f(x) = 1 - p^{y+1} = \alpha \tag{86.7}$$

where x is the number of operations/successes before the first failure of the equipment or component, and p and q are probabilities of success and failures, respectively. α is the percentage area lying outside natural tolerance limit.

The natural process interval may be taken as (y, α) with the probability content $1 - \alpha$.

From Eq. (86.7), we get

$$y = \frac{\log\left(\frac{1-\alpha}{p}\right)}{\log p}$$

Then the process capability index (Ig) can be proposed as

$$Ig = \frac{L}{y} = \frac{L.\log p}{\log(\frac{1-\alpha}{p})} \tag{86.8}$$

Assuming y, the number of successes/operations follows a geometric distribution, L is the lower specification limit and U, the upper specification limit is ignored. It is also assumed that there is no measurement error.

It may be noted that process capability indices, e.g., C_p, C_{pk}, C_{pm} *and* C_{pmk}, mentioned in Sect. 86.1 should always be greater than unity for a capable process, but in Eq. (86.8) Ig may be taken less than unity for a capable process.

The sample estimate of Ig comes out to be

$$\hat{Ig} = \frac{L}{\hat{y}} = \frac{L.\log \hat{p}}{\log\left(\frac{1-\alpha}{\hat{p}}\right)} = LG(\hat{p}), \quad \text{say} \tag{86.9}$$

where $\hat{y}$ = Sample quantile of order α

86.2.2 Estimation of Îg

We have (from Johnson and Kotz [11]).

$$E(x) = \frac{p}{q}$$

From Eq. (86.6)

Maximum likely estimate $m.1.e$ of $\hat{p}$ comes out to be,

$$\hat{p} = \frac{\bar{x}}{\bar{x}+1},$$

where $\bar{x}$ = Average number of successes before the first failure.

Then putting this value of $\hat{p}$ in Eq. (86.9), it is found that,

$$\hat{I}g = \frac{L.\log(\frac{\bar{x}}{\bar{x}+1})}{\log\left\{\frac{(1-\alpha)(\bar{x}+1)}{\bar{x}}\right\}} = L\psi(\bar{x}), \quad \text{say} \tag{86.10}$$

86.2.3 Sampling Properties of Îg

Bias of $\hat{I}g$

From standard results in large sample theory (see, e.g., Kendall and Stuart [12]).

$$E(\hat{I}g) = L.E\{\psi(\hat{p})\} \approx L.\psi(p) = Ig$$

Thus, $\hat{I}g$ is asymptotically unbiased for Ig.

Sampling variance of $\hat{I}g$

It is known that

$$\text{Var}(\hat{I}g) = L\left(\frac{\delta\psi}{\delta\bar{x}}\right)^2 \text{Var}(\bar{x}) \tag{86.11}$$

$$\bar{x} = \frac{p}{q}$$

where Var $(\bar{x}) = \frac{\text{Var}x}{n} = \frac{1}{n}\cdot\frac{p}{q^2}$

On differentiation of $\psi(\bar{x})$ in Eq. (86.10) with respect to $\bar{x}$, it is found that

$$\frac{\delta\psi}{\delta\bar{x}} = \frac{L.\log(1-\alpha)}{\bar{x}(\bar{x}+1)\left[\log\left\{\frac{(1-\alpha)(\bar{x}+1)}{\bar{x}}\right\}\right]^2}$$

On putting this value in Eq. (86.11) and on simplification, it is found that

$$\mathrm{Var}(\hat{I}g) = \frac{L^2}{n}\frac{(1-p)^2}{p}\frac{\{\log(1-\alpha)\}^2}{\left\{\log\left(\frac{1-\alpha}{p}\right)\right\}^4} \quad (86.12)$$

86.3 Optimum Choice of α

Usually, the percentage area α lying outside natural tolerance limit will be given from assurance (confidence) requirement. However, one may, even for the sake of the theoretical interest, try to choose the value of α such that, for a given process parameter p, will result in the most stable estimate of process capability index *Ig* (in repeated sampling).

On differentiation of Var ($\hat{I}g$) in Eq. (86.12) with respect to α and equating it to zero, it is found that for minimum variation of Var ($\hat{I}g$).

$\alpha = 0$ or 1 which are not feasible solutions.

So, the variance minimizing choice of α depends on p and vice versa.

From Eq. (86.12) and Table 86.1, on numerical analysis, it is observed that when α increases, Var ($\hat{I}g$) decreases, p remaining constant. It means that the variation in the process capability may be less when the percentage area lying outside natural tolerance is more resulting in less Nos. of successes or operations of products. It is also observed from Table 86.2 that when p increases Var ($\hat{I}g$) increases, α remaining constant. It means that the variation in the process capability may be more in order to achieve a high probability of successes or operations. More efforts may be needed for the improvement of the process to achieve high Nos. of successes or operations of products.

So, for the minimum variation of process capability index *Ig*, α may be chosen as low as possible, corresponding to a fixed p. For example, choice of α may be chosen as 0.02 if p (or its estimate) comes out to be 0.99.

Table 86.1 Values of Var ($\hat{I}g$) corresponding to α at constant p

Values of α at $p = 0.99$	(n/L^2) x Var ($\hat{I}g$)
0.02	19.787
0.03	2.864
0.04	0.983
0.05	0.487

Table 86.2 Values of variance Var ($\hat{I}g$) corresponding to p at constant α

Values of p at $\alpha = 0.02$	$(n/L^2) \times$ Var ($\hat{I}g$)
0.99	19.787
0.985	736.95
0.97	181.43
0.96	19.95
0.95	6.099
0.94	2.747

86.4 Typical Applications

The process capability index (Ig) may be applied in finding out capability or variability of process or system having Nos. of flights or Nos. of landings before the first failure after planned repair and preventive maintenance of an aeroplane. This may also prove to be helpful in taking decision on the variability of the process for the production of electrical, electronics and computer components and hardware in order to increase their probability of success (p) resulting in their reliability as required in reliability-centered maintenance in special case.

86.5 Conclusion

The process capability index (Ig) may be less than unity and positive for a capable process for the quality characteristics following the geometric distribution and for the minimum variation of Ig, the percentage area (α) lying outside natural tolerance limit may be low as 0.02 and probability of successes or operations before the first failure may high as 0.99.

Acknowledgements I am very much thankful to Prof. (Retd.) S. P. Mukherjee former Centenary Professor of Statistics Department of Statistics and former Dean Faculty of Science Calcutta University Kolkata and presently Mentor of Indian Association for Productivity, Quality and Reliability Kolkata for constant inspiration for research.

References

1. Kane, V.: Process capability indices. J. Qual. Technol. **18**, 41–52 (1986)
2. Chan, L.K., Cheng, S.W., Spiring, F.A.: A new measure of process capability C_{pm}. J. Qual. Technol. **30**, 162–175 (1988)
3. Taguchi, G., Hsiang, T.C.: A tutorial on quality control and assurance—the Taguchi methods. Presentation given at the Annual Meeting of the Americal Statistical Association, Las Vegas, LV (1985)
4. Pearn, L.J., Kotz, S., Johnson N.L.: Distributional and inferential properties of process capability indices. J. Qual. Technol. **24**, 216–223 (1992)

5. Vannman, K.: A unified approach to capability indices. Research report, division of quality technology and statistics. Luleo University, Sweden. Stat. Sin. **3**(2) (1993)
6. Wang, F., Hubele, N.: Quality evaluation using geometric distance approach. Int. J. Reliab. Qual. Saf. Eng. **6**, 139–153 (1999)
7. Castagliola, P.: Evaluation of non-normal process capability indices using Burr's distributions. Qual. Eng. **8**, 587–593 (1996)
8. Castagliola, P., Castellanos, J.: Capability indices dedicated to the two quality characteristics case. Qual. Technol. Quant. Manage. **2**(2), 201–220 (2005)
9. Raut, R., Attar, P.R.: Optimization of process capability in an automobile industry: a case study. Int. Res. J. Eng. Technol. **3**(7) (2016)
10. Senvar, O., Sennaroglu, B.: Comparing performances of Clements, box-cox, Johnson methods with Weibull distributions for assessing process capability. J. Ind. Eng. Manag. **9**(3), 634–656 (2016)
11. Johnson, N. L., Kotz, S.: Distributions in Statistics, vol. I (Discrete Distribution). Wiley, New York, NY (1969)
12. Kendall, M.G., Stuart, A.: The Advanced Theory of Statistics, vol. 1, pp. 228–245. Charles Griffin, London (1973)

Chapter 87
Thermomechanical Analyses of Single Sided Single Pass Submerged Arc Welding of AISI 304 Austenitic Stainless Steel

P. V. S. S. Sridhar, Vishnu Nair, Pankaj Biswas and Pinakeswar Mahanta

Abstract The submerged arc welding (SAW) is an automated process that is capable to weld thick and thin plates of steel and non-ferrous materials due to its ease of operation, less spatter and high deposition rates. SAW is extensively used to fabricate bridges, pipe lines, pressure vessel construction, ship building and surface cladding. In present study numerical simulation on effect of welding speed on residual stresses for SAW with square butt geometry of austenitic stainless steel is explored. A moving heat source model considering heat generation is used in this present study. The element birth and death technique is used to incorporate the effects of filler metal deposition. The peak temperatures observed in experimentation were correlated with that of numerical results and the variation was observed to be 4.77%. At higher welding speeds the low peak temperatures were observed which is due to less heat penetrating into the joint. It was observed in the present study at higher welding speeds the residual stress and von Mises stress were in increasing trend whereas normal and transverse stresses are negligible along the weld line.

Keywords Submerged arc welding · Residual stress · Austenitic stainless steel

---

P. V. S. S. Sridhar (✉) · P. Biswas · P. Mahanta
Department of Mechanical Engineering, Indian Institute of Technology Guwahati, Assam 781039, India
e-mail: p.sridhar@iitg.ac.in

P. Biswas
e-mail: pankaj.biswas@iitg.ac.in

P. Mahanta
e-mail: pinak@iitg.ac.in

V. Nair
Engineering Research Centre, TATA Motors Ltd., Jamshedpur 831004, Jharkhand, India
e-mail: vsnunair@gmail.com

R. G. Narayanan et al. (eds.), *Advances in Computational Methods in Manufacturing*, Lecture Notes on Multidisciplinary Industrial Engineering,
https://doi.org/10.1007/978-981-32-9072-3_87

87.1 Introduction

The submerged arc welding is a popular production process due to its ease of automation, high deposition, deeper penetration, no spatter and fumes and ability to produce sound quality welds in thin and thicker section of materials. The process is extensively used to weld ferrous metals including all varieties of stainless steels and nonferrous alloys like copper and titanium [1]. In fusion welding heat is applied introduced through a heat source which is responsible for fusion between the plates to be joined, in addition to that several phenomena like distortion, residual stresses and metallurgical structural changes are inevitable as a result of applied heat. In order to get a sound weld, it is necessary to understand the effect of heat flow in a welding process. The residual stresses, distortion, microstructure and HAZ hardness introduced into the weld have direct bearing on the applied welding power. The local heating and cooling occurring during the welding process affects the structural changes in the material. Hence, the knowledge of heat flow and its resultant effects need to be predicted for establishing optimum welding parameters.

The heat source from which heat flows into the material is different for various welding processes. The analytical model and its solution for heat flow during welding was first, introduced by Rosenthal. In his famous paper, he assumed that conduction is the predominant mechanism of heat transfer, while assuming the shape of the weld pool in 2D and 3D [2, 3]. However, the Rosenthal model's yields poor predictions. The model developed by Ohring and Lugt [4] for gas metal arc welding (GMAW) considered all possible physical phenomena like evaporation, heat transfer by convection, radiation and viscous deformable free-surface conditions. The main drawback of his model is they assumed constant material properties. Brown and Song [5] considered 3D model of heat source for better understanding of the underlying phenomena. The essential parameters that are to be incorporated in the simulation, a moving heat source, welding parameters like speed, voltage and current, deposition of filler material into the joint. Goldak et al. [6] proposed double ellipsoidal heat source and incorporated the size and shape of heat source for shallow and deeper penetrations welds. Biswas and Mandal [7] formulated a numerical model and macros to simulate a distributed moving heat source, and also considered for deposition of filler metal in submerged arc welding for predicting weld distortions and residual stresses. Podder et al. [8] developed a generalized solution for a heat source model for submerged arc welding of low carbon and M.S. plates. The physical phenomena like forces of buoyancy, Lorentz and Marangoni were taken into account while formulating the model. The principal objective of modeling of welding heat source is to represent the true source of heat, which is the welding arc. The heat is dissipated from the weld pool by conduction, convection and radiation. In the present work effect of process variable i.e., welding speed on residual stress in the austenitic stainless steel is explored.

87.2 Experimental Details

The temperature measurement during was carried out using *K*-type thermocouples. This method was used during the submerged arc welding of butt welded single pass single sided joints from 304 austenitic stainless steel and the results were compared with those of finite element method. The thermocouples were located on the top surface of the plate using a thermocouple attachment setup. Temperatures were measured at two distances from the weld melt line on one side of the plates. The location of thermocouple was 17 mm from weld center line. Figure 87.1 Schematic of test specimen with location of thermocouple. The dimensions of weld sample are 200 mm × 100 mm × 8 mm. A data acquisition and controller (Make: AGILENT 34972A) is used to collect the signals, which was interfaced to a PC. It was set to collect temperature data for interval of 500 ms. Figure 87.1 shows Schematic of test specimen and actual SA welding.

The work pieces were welded in square butt condition, in a single pass with reverse polarity (DCEP). After conducting several trial runs, the welding parameters were decided. The voltage (*V*), current (*I*) and travel speed (*S*) in each joint were given in Table 87.1. The arc efficiency (η) was taken as 0.85 [9, 10] and was used to calculate the heat input per mm Q_w.

$$Q_w = \frac{\eta VI}{S} \tag{87.1}$$

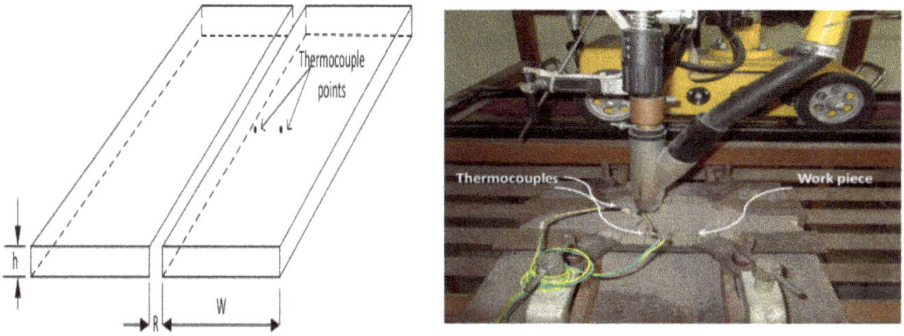

Fig. 87.1 Schematic of test specimen with location of thermocouple and SAW with thermocouples attached to the test specimen

Table 87.1 Parameters considered for analysis of variation of speed

Exp.	Welding arc power (W)	Thickness (mm)	Speed (m/hr)
1.	24 V × 430 A = 10,320	8	20
2.	24 V × 430 A = 10,320	8	22
3.	24 V × 430 A = 10,320	8	24

87.3 Thermomechanical Analysis

The present study involves transient thermal analysis to predict the temperature distribution in a welded sample and mechanical analysis to study structural impacts on the weld caused due to thermal load. A three dimensional finite element model was developed to study thermos-mechanical behavior of weld.

87.3.1 Assumptions for Thermo-Mechanical Analysis [7]

Following are the assumptions made to develop the model for welding process:

- All the thermal properties are considered as a function of temperature.
- No forced convection is considered unless mentioned.
- Linear Newtonian convection cooling is considered on all the surfaces except the weld zone.
- Radiation loss being small, the same is neglected.
- Arc efficiency is incorporated to account for other losses.
- Phase change is not considered.

87.3.2 3D Finite Element Model

A three dimensional finite element model was developed using ANSYS 18.0 to study the temperature distribution and heat transfer in a SAW process. In the thermal model actual conditions were incorporated as far as possible. The governing differential equation for heat conduction for a homogenous, isotropic solid in the rectangular coordinate system (x, y, z) is:

$$\frac{\partial}{\partial x}\left[K\frac{\partial T}{\partial x}\right]+\frac{\partial}{\partial y}\left[K\frac{\partial T}{\partial y}\right]+\frac{\partial}{\partial z}\left[K\frac{\partial T}{\partial z}\right]=\rho C\frac{\partial T}{\partial t} \tag{87.2}$$

where ρ = density of plate material (kg/m^3), C = specific heat (J/kg °C) and K = thermal conductivity (W/m °C). The initial condition and boundary conditions applied on the model are mention below.

Initial Condition
An initial temperature was assumed for the entire plate equivalent to ambient temperature.

$$T = T_{\infty} \quad \text{for } t = 0$$

where T_{∞} is the ambient temperature.

Boundary Condition

As the heat flows over the weld surface, the same is conducted to the rest of the plate. The following equation represents the same:

$$q_n = -q_{\text{sup}} \tag{87.3}$$

Here q_{sup} represents the heat supplied from an external welding arc. And q_n represents the component of the conduction heat flux vector normal to the work surface.

As the heat losses due to convection from the plate surface. Considering h_f as convection coefficient, the heat loss q_{conv} is evaluated as:

$$q_{\text{conv}} = h_f(T - T_\infty) \tag{87.4}$$

Structural Analysis

The residual stress and plastic strain were evaluated post heat transfer analysis where in temperature distribution in heat affected zones was studied. Tack weld condition was applied and using the results of thermal analysis the structural analysis was carried out. The following are the formulations for stress-strain evaluation used in the current study:

$$\nabla\sigma + F = 0 \tag{87.5}$$

where

∇ Differential operator
σ Stress
F Body force

The stress strain relationship is as follows:

$$\{\sigma\} = [D]\{\varepsilon\} \tag{87.6}$$

where

$\{\sigma\}$ Stress tensor
$[D]$ Stiffness matrix
$\{\varepsilon\}$ Strain tensor

The von Mises yield criterion is used to analyze the stress and strain behavior post thermal analysis.

87.4 Modelling Details

The model was developed in ANSYS 18.0. For transient thermal analysis an eight-node SOLID70 element was considered. As the results of thermal analysis was further used to determine the residual stress and strain induced in the plate, thus

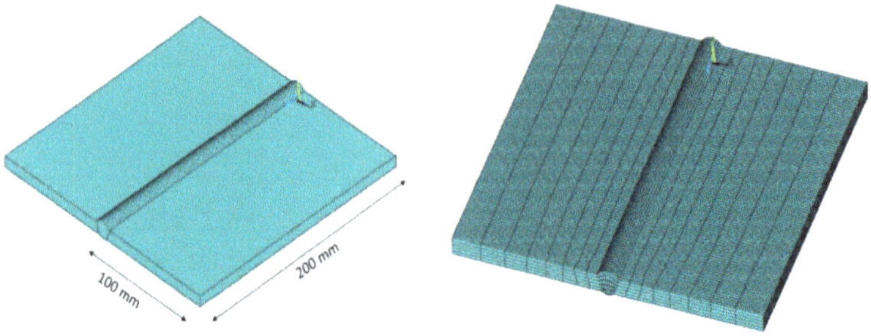

Fig. 87.2 3D solid model and FE Meshed model

the eight- node SOLID70 element was later replaced to eight-node SOLID185 element to perform structural analysis. The weld beads obtained from experiments were analyzed using Nikon SMZ25 microscope with magnification of 0.5× and 30 W scanning power. The bead geometry thus obtained was used for developing FE model. The modelling details single sided SAW is shown in Fig. 87.2.

The solid model was built based on actual experimental bead cross section, further the model was meshed using ANSYS 18.0. Fine mesh was carried out along the weld region to obtain better temperature gradient along weld zone and to reduce computation time coarse mesh was used away from the bead zone. The 3D solid model and FE mesh model was shown in Fig. 87.2. The temperature dependent material properties [11] of austenitic steel is mentioned in Figs. 87.3 and 87.4.

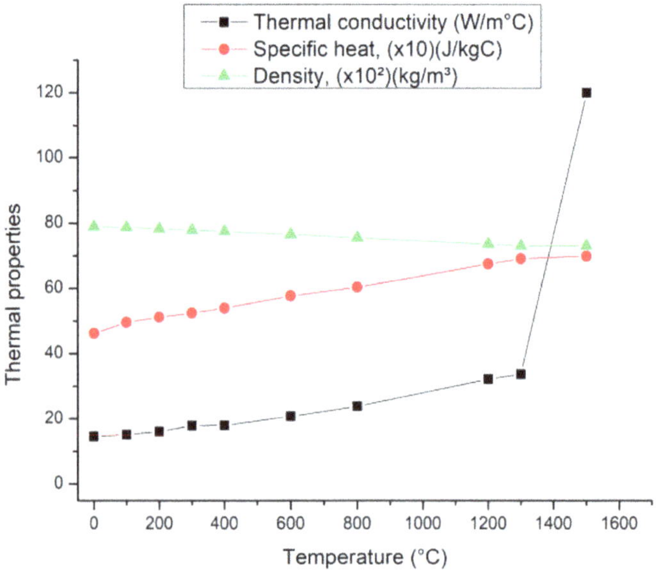

Fig. 87.3 Transient thermal properties of 304 stainless steel [11]

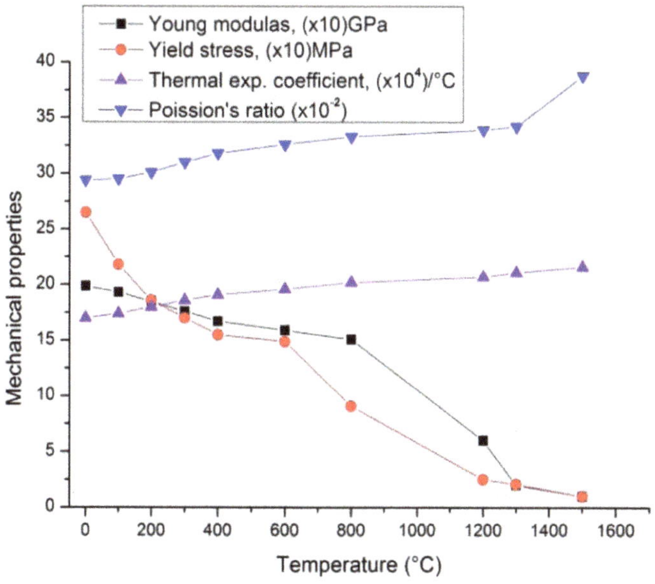

Fig. 87.4 Transient mechanical properties of 304 stainless steel [11]

87.4.1 Modelling the Moving Heat Source

The modelling of a transient thermal analysis with arc moving from one location to another with respect to time in order to generate uniform weld bead, the element birth and death technique was used. The entire weld pool was divided into multiple elemental volumes and subsequently with respect to time, heat was added and removed consecutively from one bead to another, thus attaining a moving heat source model. The quantity of heat added and removed was determined by following equation:

$$Q = \frac{\eta W}{v} \tag{87.7}$$

where Q is the quantity of heat added and removed, η is the efficiency of welding process, v is the volume of the individual weld bead considered, W is the welding power.

87.5 Results and Discussions

The heat applied to create a weld joint is dissipated by conduction to the base metal and by convection to the environment. Thus incorporating all possible conditions, and by varying the process parameter the numerical analysis was carried out in order to study the variation in thermal and mechanical behavior of the weld. The same process parameters used to develop weld during experiments were considered for numerical analysis. The model is cooled till 500 secs after application of thermal load.

87.5.1 Moving Heat Source and Thermo-Mechanical Analysis

The simulated moving heat source model obtained numerically is shown in Fig. 87.5 for the welding variable voltage of 24 V, current of 430 A and speed of 20 m/h.

The tack weld boundary condition was applied and the results from thermal analysis were imposed to analyze the mechanical behavior of the model. Figures 87.6 and 87.7 shows the contour of transverse residual stress and residual stress along thickness; longitudinal residual stress and von Mises stress respectively.

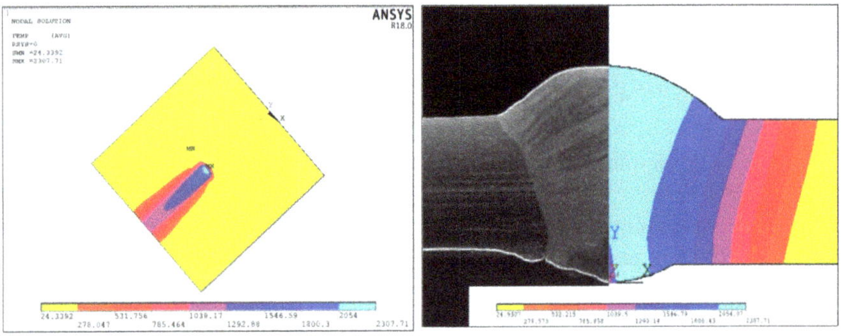

Fig. 87.5 FE model moving heat source and Experimental and numerical bead cross section

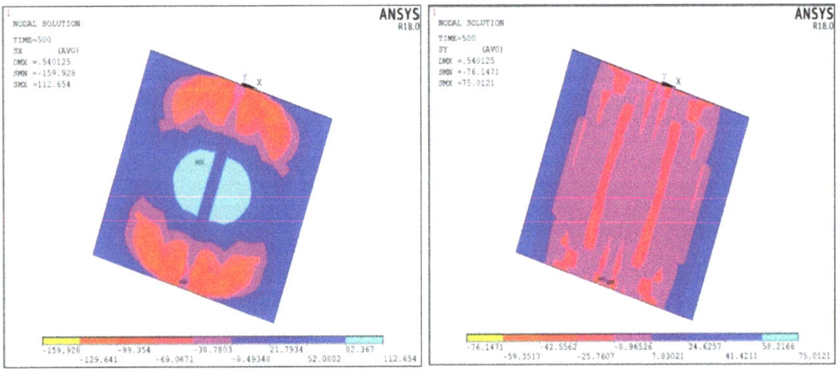

Fig. 87.6 Contour plots of transverse residual stress and residual stress across thickness

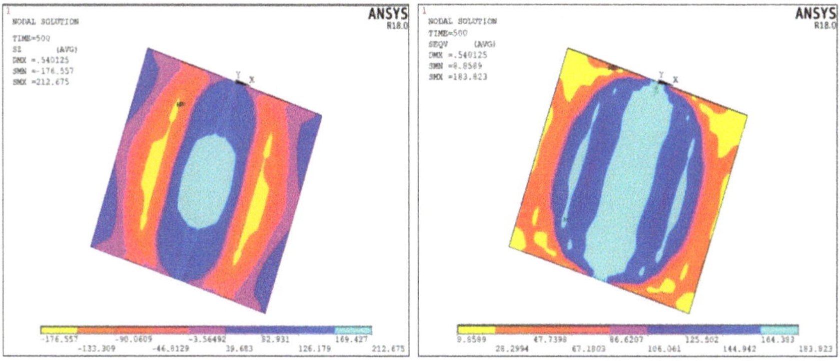

Fig. 87.7 Contour plots of longitudinal residual stress and von Mises stress

Due to variation in temperature gradient along the width direction (away from the weld), there are variation in shrinkage pattern which lead to development of tensile stress contour in the weld zone and its vicinity, and to balance the same compressive stresses are generated in rest of the portion of plate [12]. The stress reversal takes place between molten zone and adjoining zone due to drastic change in physical and mechanical properties of molten material and solid material [13]. Also at the start and end of a weld, large shrinkage pattern causes the variation in stress contour, as compressive stresses are generated in the ends and tensile stress are generated along the weld. This variation is evident in transversal and longitudinal stress contour [13]. As, the model is cooled after thermal loading cycle, phase transformation occurs in the welded plate establishing plastic deformation which is measured in terms of plastic strain in this study [16]. The material closer to weld path faces tensile strain, as a result of contraction due thermal loading, which is later balanced by compressive strain away from the weld zone [14]. As material is deformed plastically, significant intergranular residual strains develop in the loading direction owing to strain redistribution due to slip along the preferred systems. The repeated expansion and contraction during a thermal loading cycle along with consequent cooling effect increases the dislocation density along the weld [15]. Since heat transfer is more turbulent along the thickness direction than along the width, this affects the plastic strain contour along the thickness direction [16].

87.5.2 Effect of Welding Speed

The major process parameters that influenced the shape and quality of weld bead are arc power and welding speed. In the present study effect of welding speed on residual stress is considered for analysis. Three samples were considered to analyse effect of residual stresses on varying of welding speed. The results were analysed based on observations from peak temperature distribution, residual stress on the

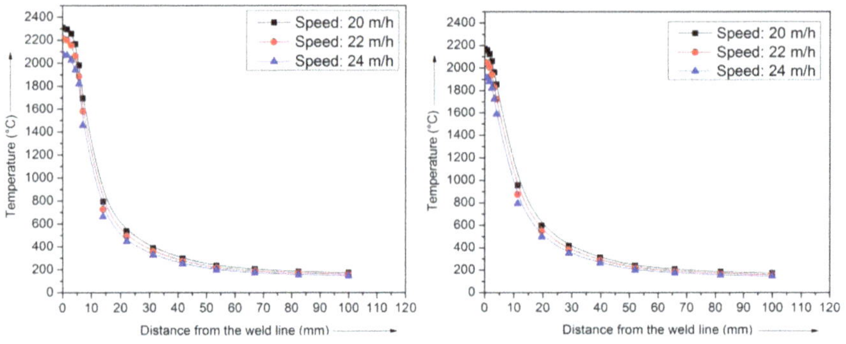

Fig. 87.8 Peak temperature perpendicular to weld line top and bottom surface

plate. The residual stress and plastic strain were observed along the weld centre line and perpendicular to the weld centre line. Residual stresses were considered for all three directions: transversal, longitudinal and along thickness direction of the plate. Also, von Mises stresses were calculated.

The various parameters used to study the effect of variation of welding speed are tabulated in Table 87.1.

The variation of peak temperature perpendicular to the weld line for three welding speeds for top surface and bottom surface of the plate is plotted in Fig. 87.8.

It was observed from the above temperature pattern that peak temperature reduces with increase in welding speed on both top and bottom side of the plate. Further analysis is carried out to determine the pattern of the residual stresses for top and bottom portion of the welded plates along and perpendicular the weld line.

Figure 87.9 shows the variation in residual stress for both top and bottom side of the plate along the weld line. Transversal residual stress (S_x), residual stress along

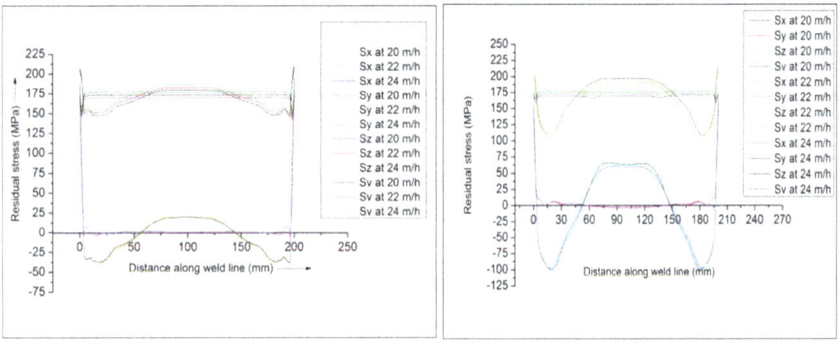

Fig. 87.9 Residual stress in *X*, *Y*, *Z* direction and von Mises stress along weld line for top side of plate and for bottom side of plate respectively

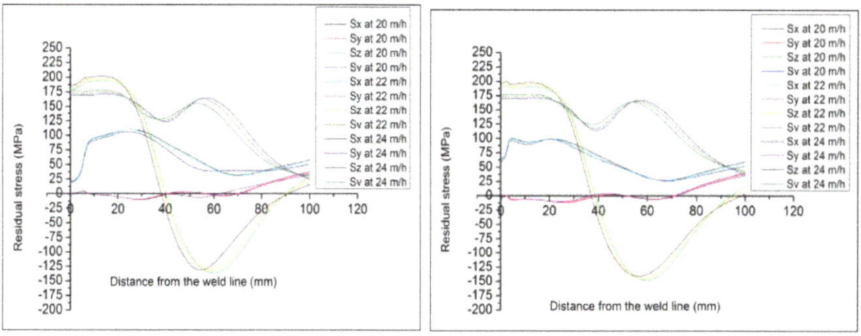

Fig. 87.10 Residual stress in *X*, *Y*, *Z* direction and von Mises stress perpendicular to weld line on top surface on bottom surface respectively

thickness direction (S_y), longitudinal residual stress (S_z) and von Mises stress (S_v) are considered for study. Figure 87.10 shows the residual stress in *X*, *Y*, *Z* direction and von Mises stress perpendicular to weld line on top surface on bottom surface respectively.

87.5.3 Analysis of Residual Stress on Top Surface of Plate

With increase in speed, longitudinal tensile stress and von Mises tensile stress increases, whereas, the variation in normal stress and transverse stress were negligible along the weld line. Both longitudinal tensile stress and von Mises tensile stress increases with increase in welding speed in the weld zone and in its near vicinity but reduces with increase in welding speed in the portion away from weld indicating increase in longitudinal compressive stress, but maximum value of longitudinal compressive stress was observed to decrease with increase in welding speed. Whereas transversal tensile stress reduces with increase in speed away from the weld center line, and negligible variation was noticed in normal stress away from the weld center line on the top surface. It is also noticed that the magnitude of normal stress both along the weld and away from the weld is not as evident as that of other components of stresses.

87.5.4 Analysis of Residual Stress on Bottom Surface of Plate

With increase in speed, longitudinal tensile stress and von Mises tensile stress increases, whereas, the variation in normal stress was negligible along the weld line. And transversal stress (both tensile and compressive) decreases with increase in welding speed along the weld center line. Both longitudinal tensile stress and von Mises tensile stress increases with increase in welding speed in the weld zone and in its near vicinity but reduces with increase in welding speed in the portion away from weld. The maximum longitudinal compressive stress decreases with increase in welding speed which is located away from the weld vicinity. The Transverse residual tensile stress diminishes away from the weld center line with increase in welding speed, and negligible variation in normal stress was noticed away from the weld center line. It is also noticed that the magnitude of normal stress both along the weld and away from the weld is not as evident as that of other components of stresses.

87.6 Validation of the Model

The temperature profile was recorded for all welding parameters. For purpose of validating the numerical model, Experiment No. 1 thermal couple profile was matched with numerically obtained result for distance of 17 mm from the weld centre line. Figure 87.11 depicts that close agreement of experimental observation with FE model.

The peak temperature obtained at 17 mm from weld line was co-related with numerical values and is in good agreement with a maximum deviation of 4.77%. Hence the developed heat source model is validated with the experimental result.

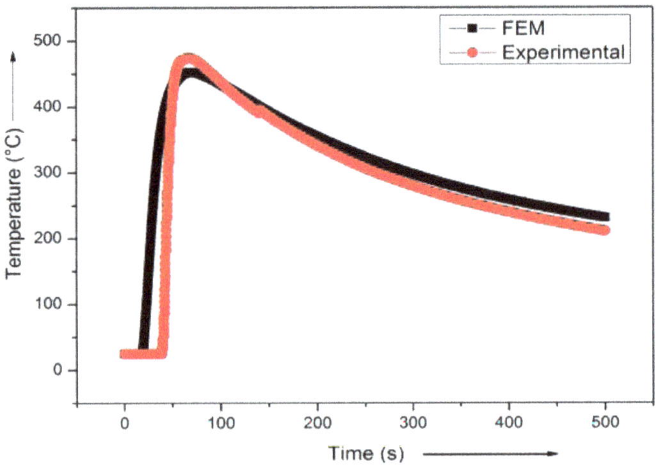

Fig. 87.11 Validation of experimental and numerical temperatures at 17 mm from weld line

87.7 Conclusions

1. The SA welding successfully conducted on AISI 304 stainless steel with a square butt configuration.
2. The temperature distribution obtained from the experiments is matched with that of developed heat source model and variation in peak temperatures is around 4.77%.
3. It was observed that with increase in welding speed the peak temperatures observed were low as higher speeds reduce the heat input into the material.
4. Thermomechanical analysis was done to determine the residual stress in *X*, *Y*, *Z* direction and von Mises stress for top and bottom of the plate.
5. With increase in speed, longitudinal tensile stress and von Mises tensile stress increases, whereas, the variation in normal stress and transversal stress were negligible along the weld line.

References

1. Houldcorft, P.T.: Submerged arc welding, 2nd edn. Abington Publishing, Cambridge UK (1989)
2. Rosenthal, D.: Mathematical theory of heat distribution during welding and cutting. Weld. J. **21**(5), 220-s–234-s (1941)
3. Rosenthal, D.: Theory of moving sources of heat and its application to metal treatments. Trans. AIME **43**(11), 849–866 (1946)
4. Ohring, S., Lugt, H J.: Numerical simulation of a time dependent 3-D GMA weld pool due to a moving arc. Weld. J. **79**(12), 416-s–424-s (1999)
5. Brown, S., Song, H.: Implication of three-dimensional numerical simulation of welding of large structures. Weld. J. 55-s–62-s (1992)
6. Goldak, J., Bibby, M., Moore, J., House, R., Patel, B.: Computer modeling of heat flow in welds. Metall. Trans. **17B**, 587–600 (1986)
7. Biswas, P., Mandal, N.R. Thermomechanical finite element analysis and experimental investigation of single pass single sided submerged arc welding of C-Mn steel plates. Proc. IMechE Part B J. Eng. Manuf. (224), 627–639 (2009)
8. Podder, D., Mandal, N R., Das, S.: Heat source modelling and analysis of submerged arc welding. Weld. J. (93), 183-s–192-s (2014)
9. Pathak, A.K., Datta, G.L.: Three dimensional finite element analysis to predict the different zones of microstructures in submerged arc welding. Proc. IMechE Part B **218**, 269–280 (2004)
10. Friedman, E.: Thermomechanical analysis of welding process using finite element method. Trans. ASME J. Press. Vessel Technol. **97**(3), 206–213 (1975)
11. Farid, V.H., Asl, A.Z.: Numerical and experimental investigation of T-shape fillet welding of AISI 304 stainless steel plates. Mater. Des. **47**, 615–623 (2013)
12. Velaga, S.K., Ravisankar, A.: Finite element based parametric study on the characterization of weld process moving heat source parameters in austenitic stainless steel. Int. J. Press. Vessels Pip. **157**, 63–73 (2017)

13. Ravisankar, A., Velaga, S.K., Rajput, G., Venugopal, S.: Influence of welding speed and power on residual stress during gas tungsten arc welding (GTAW) of thin sections with constant heat input: A study using numerical simulation and experimental validation. J. Manuf. Process. **16**, 200–211 (2014)
14. Dean, D., Wei, L., Hidekazu, M.: Determination of welding deformation in fillet-welded joint by means of numerical simulation and comparison with experimental measurements. J. Mater. Process. Technol. **183**, 219–225 (2007)
15. Murat, O.A., Fitzpatrick, M.E.: Determination of plasticity following deformation and welding of austenitic stainless steel. Mater. Sci. Eng., A **701**, 203–213 (2017)
16. Yadaiah, N., Bag, S.: Effect of heat source parameters in thermal and mechanical analysis of linear GTA welding process. ISIJ Int. **52**, 2069–2075 (2012)

Chapter 88
Stochastic Longitudinal Vibration of Single Walled Carbon Nanorods—A Non Local Elasticity Approach

Tanmoy Loha and Sudip Dey

Abstract The present study deals with stochastic nonlocal non-dimensional natural frequencies of single walled carbon nanorods and rotating cantilever beam. In this study, the chordwise bending–vibration behavior of nanocantilever is analyzed. The nonlocal parameter focused on the small-size effects when dealing with nanosize structures such as single-walled carbon nanotubes (SWCNTs). The Eringen's nonlocal elasticity theory in conjunction to the governing differential equations are utilized to solve the problem. The stochastic nonlocal natural frequencies for rotating nano cantilever are observed by employing the differential quadrature method (DQM). The effects of the scale is computationally analyzed based on Monte Carlo simulation (MCS). In another case free longitudinal frequency of a nanorods considering two types of boundary condition is observed, namely, clamped-clamped and clamped-free. The longitudinal vibration of the system are described by a set of partial differential equations, derived by using D'Alembert's principle and using by employing Fourier infinite series in conjunction to separation of variables. The nanoscale effects implement a important role on the frequency response of nanorods subjected to rotation. The statistical analysis are carried out based on stochastic input parameters such as material properties (elastic modulus, density) and geometric properties (length).

Keywords Nanorods · Stochasticity · Non local effect · Fundamental frequency · Scale effect

88.1 Introduction

Due to introduction of graphene and carbon nanotubes [1], nanostructure have gained immerse popularity in the field of the scientific community. Such materials can be employed to obtain the nanodevices made of complex structures. Due to

T. Loha (✉) · S. Dey
Mechanical Engineering Department, National Institute of Technology Silchar, Silchar, India
e-mail: tanmoyloha5@gmail.com

R. G. Narayanan et al. (eds.), *Advances in Computational Methods in Manufacturing*, Lecture Notes on Multidisciplinary Industrial Engineering,
https://doi.org/10.1007/978-981-32-9072-3_88

eminent effectiveness of usability in nanodevices also its eminent characteristics draws attention on the improvement of nanodevices and nanomachines. These nanomaterials contains superior mechanical, thermal, electrical, chemical and electronic properties. In classical continuum mechanics strain field of distant point to reference point is eliminated but whereas in non local elasticity approach, nonlocal elasticity and the differential quadrature method (DQM) application in order to determine flapwise bending vibration of a rotating nanocantilever consider these distance effect between surroundings atoms. Therefore stress field at a reference point depends not only that particular point but also on the strain at every other points at that corresponding domain. An experimental approach at nano scale is quite difficult and time consuming because handling all influencing parameters of a nanostructure simultaneously and obtained result by considering all parameters together is critical task and trouble worthy task. Furthermore employing molecular dynamics (MD) simulation needs large time and high computational efficacy.

In the present work the size effect and surroundings effect of molecule, strain field effect of nearby atoms on a structural element familiar as nonlocal theory is taken into consideration. The nonlocal elasticity theory was first introduced by Eringen's [2] to study nanomaterials behaviour and utilization of nanomaterials. Vibrational response of various nano or micro level structures like as single and multiple beam, rod, nanoplate and nanodisc in conjunction with elastic, visco elastic or other varieties of sheet thickness has been explored by various authors [3–5]. The actual significance of nonlocal elasticity theory stimulated us to study various features of nano/micro structure and implements that specific properties into nanodevices. The viability of nonlocal continuum theory in the field of nanotechnology was first reported by Peddieson et al. [6]. Various efforts related to nonlocal elasticity theory have been reported on several well established literature e.g. [7–9]. Recently Lee and Chang [10] have analysed vibrational response of single-walled carbon nanotube (SWCNTs) filled with fluid embedded in an elastic medium using winkler-type model. Further Murmu and Pradhan [11] established a single non local beam model to evaluate flapwise bending vibration behaviours of a rotating cantilever. They observed that vibration responses of a rotating nano cantilever is strongly effected in small scale. Recently Pradhan [12] has carried out finite element analysis for getting various mechanical response buckling, bending and vibration for non local beam under various boundary conditions like hinged-hinged, clamped-hinged, and clamped-free are extended their results to taper beams. According to Wang [13] a modified non local beam model was developed and they have found that nano scale non local parameter and critical flow velocity has a great impact on vibration and stability response of a nanotube carrying fluid. Danesh et al. illustrated the scale effect on tapered nanorod subjected to longitudinal vibration using the nonlocal theory. The authors have used the differential quadrature method to solve motion equations for three different boundary conditions [14]. Aydogdu, was modelled Carbon nanotube immersed into elastic medium i.e. a nanorod encompassed by elastic layers using the Eringen's nonlocal theory. The author distinguished the longitudinal frequencies between the nonlocal and classical continuum models [15]. Abdollahi [16] implement the EBFs on bounded domains, the effects of the boundary conditions are handled by truncating the kernel/attenuation function in the constitutive relations. This helps to found residuals in the governing equations, also to minimize the residuals near the boundary region by using weighted residual approach. Danilo Karličić et al. applied nonlocal Kirchhoff-Love plate theory to illustrate mechanical behaviours of single-layered and multi layered graphene sheets used as an orthotropic material. Here they have derived the governing equation of motion of a simply supported orthotropic nanoplate. They also had showed that magnetic field could be successfully implemented to enhance sensitivity of nanomechanical sensors [17]. Bahaadini [18] defined the instability behaviour of cantilever nanotubes on dynamic structural fluid and they have incorporated end concentrated follower force and distributed tangential load to improve accuracy of the results like surface elasticity and residual stresses. The frequency response function is an important structural parameter which is defined as the ratio of output response of a structure due to an applied force. Such modal response can be measured in the form of displacement, velocity, and acceleration. In spite of recent development the model which considered in published work [11], stochasticity effect based on parametric study has not yet been considered. Uncertainty quantification on various properties of the above said model has not yet been considered because geometric and material uncertainties leading to unpredictable system response on natural frequencies. Reliability based failure analysis of single walled carbon nanorods structure still not observed. Based on carbon nanostructure various design configuration of nanomachines are found but for that proper understanding of mechanical behaviour is needed. One such nano machines is nano turbine whose performance depends upon dynamics properties of beam like nanocantilever elements. From design point of view study of bending vibration on nanocantilever, nanorod are essential for complexity in its lattice structure. Due to interdependency of large number of parameters in complicated fabrication processes, the system properties are inevitably random in nature resulting in uncertain response of the nano structures. Many researchers worked on nano structures in past but mainly focused on deterministic analysis [1–4]. The stochastic natural frequency responses are solved by random eigenvalue problem with an improved nonlocal elasticity based approach. Eringen's nonlocal elasticity theory is idealized as nanocantilever beam to study the flapwise bending vibrational frequencies of rotating nanocantilever. The longitudinal frequency responses of nanorod is determined by using methods of separation of variables on ordinary differential governing equation. The various effects of nonlocal parameters, small scale effects and nondimensional angular velocities on nanostructure are investigated in the present study.

88.2 Nonlocal Parameter and Nonlocal Elastic Model

The basic parameter in non local model e_o remains constant for each nanostructure. The parameter e_o should be taken in such way so such non local model yields certain approximation when dispersion curve due to lattice effect on plane waves are considered. The term $e_o a$, $e_o l_c / l_a$ are the nonlocal parameters or the scale coefficient. Non-local elasticity theory has the capability to envisage the statistical behaviours on longitudinal vibration response and to predict buckling characteristics of CNTs. However, main distinguishable features of nonlocal theory is to optimize nonlocal parameter's value to get better effective outcomes from any systems. In this theory nanoscale effects are taken into consideration when lattice dimensions of atoms considered as material parameter. The most generalized form of the constitutive relations for nonlocal elasticity involves an integral over the whole body. The nonlocal linear homogeneous equations are explained by Pradhan and Murmu [19]. Based on Euler–Bernoulli hypothesis nonlocal constitutive relations for general one dimensional beam can be expressed as

$$\sigma(x) - (e_o a)^2 \frac{\delta^2 \sigma}{\delta^2 x} = E\varepsilon(x) \tag{88.1}$$

where σ and ε denote the normal stress and strain due to nonlocality respectively. E denotes Young's modulus of the element. The term $(e_o a)$ is nothing but nonlocal parameter or the scale-coefficient as depicted in a well-established literature [19].

88.3 Materials

In general, during synthesis of nanostructures, high stiff and light-weight materials are employed so that the structure can sustain large external load without failure. In the present study, single walled carbon nanotube is used because of its extraordinary material properties. It is made of graphene sheet which is idealized to construct the present model. Carbon nanotubes (CNTs) being the allotrope of carbon have a cylindrical closed packed structures. These CNTs have unusual properties, which are valuable for nanotechnology, optics, electronics industry and other fields of materials science and technology. Owing to the material's exceptional strength and stiffness, nanotubes are constructed with length-to-diameter ratio of up to 132,000,000:1 significantly larger than for any other material.

88.4 Methodology

88.4.1 Governing Equation of Motion of Rotating Nanocantilever

Let us assume a nano cantilever whose length is L and one end fixed at point O to a rigid hub of radius r rotates at a constant angular speed Ω rad/s in counter clockwise direction shown in Fig. 88.1. From different literature survey it is observed that the classical Euler-elastic beam theory offers a most reliable model for mechanical deformation due to axial loading in one-dimensional nanostructures, such as carbon nanotubes, nanowire, nonobeam, nanoplate. Further it can also be assumed that plane of cross section normal to longitudinal axis remains plane before and after deformation and deformation due to bending is considered as very small. Thus rotating nanocantilever model is considered as an Euler–Bernoulli beam model. In the present theory the plane cross sections of the beam remain plane during flexure and that the radius of curvature of a bent beam is larger compared to the beam's depth. Here W denote deflection along flapwise deformation direction.

The moment stress resultant relation can be expressed as $M(x,t) = \int_A z\sigma_x \mathrm{d}A$

Using Eq. (88.1) and from moment relation, nonlocal moment relation can be expressed as

$$M - (e_o a)^2 \frac{\delta^2 M}{\delta x^2} = -EI \frac{\delta^2 W}{\delta x^2} \tag{88.2}$$

Now based on Euler-Bernoulli's hypothesis is applied in basic differential form of Euler Lagrange's equation as follows

$$\frac{\delta^2 M}{\partial x^2} + \frac{\delta}{\delta x}\left(\Gamma \frac{\delta}{\delta x}\right) - \rho A \frac{\delta^2 W}{\delta t^2} = 0 \tag{88.3}$$

where Γ denotes the tensile force due to centrifugal action of rotation of cantilever beam. The force $\Gamma(x) = \int_x^L \rho A \Omega^2 (r + x) \mathrm{d}x$. Here r is taken as the hub radius which denote length of beam from centre point of rotation to edge of cantilever beam, Ω denotes angular velocity with which cantilever rotates. Now The second order derivative of M is determined from Eq. (88.2) by substituting into Eq. (88.1) it is found that

$$M = -EI \frac{\delta^2 W}{\delta x^2} + (e_o a)^2 \left[-\frac{\delta}{\delta x}(\Gamma \frac{\delta W}{\delta x}) + \rho A \frac{\delta^2 W}{\delta t^2} \right] \tag{88.4}$$

Now substituting the expression for M from Eq. (88.3) into Eq. (88.4), the governing partial differential expression for vibration of rotating cantilever are obtained as

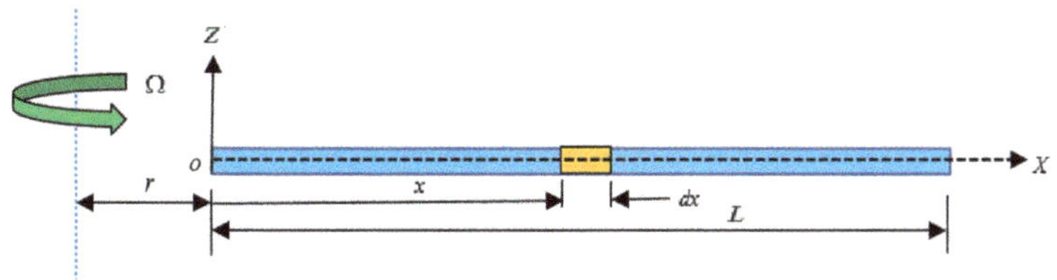

Fig. 88.1 Basic layout of a rotating nanocantilever based on Euler–Bernoulli approach

$$\frac{\delta^2}{\delta x^2}\left(EI\frac{\delta^2 W}{\delta x^2}\right) - \frac{\delta}{\delta x}\left(T\frac{\delta W}{\delta x}\right) + \rho A\frac{\delta^2 W}{\delta t^2} + (e_o a)^2\frac{\delta^2}{\delta x^2}\left[\frac{\delta}{\delta x}\left(T\frac{\delta W}{\delta x}\right) - \rho A\frac{\delta^4 W}{\delta t^4}\right] = 0 \tag{88.5}$$

The four boundary conditions for a nanocantilever beam at two ends are expressed as

when $x = 0$, $W = \frac{\delta W}{\delta x} = 0$ and $\frac{\delta^2 W}{\delta x^2} = \frac{\delta^3 W}{\delta x^3} = 0$ at $x = L$,

After non-dimensionalization, the principal parameter employing differential quadrature (DQ) method [19], the Eigen value equation can be obtained as $K_{\text{total}}\{W\} = \lambda^2\{W\}$, where K_{total} denotes the cumulative DQ matrix. From the above Eigen matrix, the Eigenvalues are solved and the fundamental frequencies are calculated for longitudinal vibration of nanocantilever based on nonlocal elasticity approach [20].

88.4.2 Characteristics Equations of Nanorod

In present study, a nanorod having length L and arbitrarily selected area of cross section A is considered as illustrated in Fig. 88.2a. It is assumed that cross-sectional area along X-direction remains constant. The materials of nanorod are considered to be homogeneous and isotropic in nature. To find out fundamental frequencies of nanorods vibrating longitudinally, it is assumed to be vibrated along X-direction. For analysing of present model, it is considered as an infinitely small element having length dx at location x and the equilibrium condition is based on D'Alembert's principle as furnished in Fig. 88.2b. A normal resultant force N is applied due to normal stress σ_x in axial direction. In addition, the continuously distributed axial loads $f\ (x,\ t)$ are neglected. On the right side of infinitesimal element elementary force $+$dN is considered.

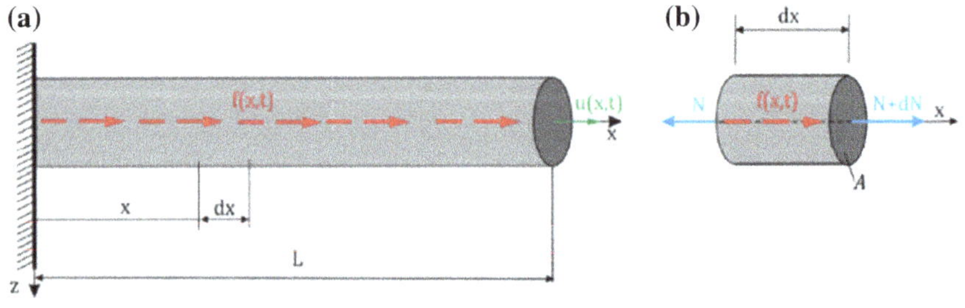

Fig. 88.2 **a** Model of a nanorod vibrating longitudinally; **b** corresponding infinitesimal element subjected to resultant stresses in axial direction [21]

By considering D'Alembert's equilibrium equation, equating sum of all forces acting along x axes is equal to zero which states

$$-N+(N+\mathrm{d}N)+f(x,\,t)\mathrm{d}x=\rho A\mathrm{d}x\frac{\partial^2 u}{\partial t^2} \tag{88.6}$$

where $dN=\frac{\partial y}{\partial x}dx$ is the resultant stress N of the small infinitesimal element, $u = u(x,\ t)$ is the axial deformation along x-direction. After some transformations the following equilibrium equation is obtained in the differential form,

$$\frac{\partial N}{\partial x}+f(x,t)=\rho A\frac{\partial^2 u}{\partial t^2} \tag{88.7}$$

where, N denotes the resultant axial stress over whole cross section area defined by $N(x,t)=\int_A \sigma_x \mathrm{d}A$, where σ_x represent normal stress in infinitesimal element.

According to Eringen's nonlocal elasticity model the constitutive equations in differential form are specified as follows

$$\sigma_x-(e_o a)^2\frac{\partial^2\sigma_x}{\partial x^2}=E\varepsilon_x \tag{88.8}$$

$$N-(e_o a)^2\frac{\partial^2\sigma_x}{\partial x^2}=E\varepsilon_x. \tag{88.9}$$

By substituting (88.7) into (88.9) the following governing differential equation of motion based on D'Alembert's principle and Eringen's model of nonlocal elasticity in terms of axial displacement is obtained

$$\rho A\frac{\partial^2 u}{\partial t^2}-f(x,t)-EA\frac{\partial^2 u}{\partial x^2}=(e_o a)^2\frac{\partial^2}{\partial x^2}\left(\rho A\frac{\partial^2 u}{\partial t^2}-f(x,t)\right) \tag{88.10}$$

In the present study two types of boundary conditions are considered.

(a) -Clamped-Clamped- $U(0,\ t) = U(L,\ t) = 0$
(b) -Clamed-Free- $U(0,\ t) = N(L,\ t) = 0$

where $N(L,\ t)$ is nonlocal axial resultant force acting at the end of a nanorod. From (88.7) and (88.9) the following equation can be derived

$$N(x,t)=(e_o a)^2\frac{\partial^3 u}{\partial x\partial t^2}+EA\frac{\partial u}{\partial x}. \tag{88.11}$$

Considering oscillatory motion of nanorod with time and using separation of variables methodology, the expression of governing equation is assumed as $\theta(x,t)=\sum_{n=1}^{\infty}X_n(x)T_n(t)$, where $T_n(t)=e^{i\omega_n t}$ is the time harmonic function

$X_n(x)$ is the mode shape function corresponding to longitudinal vibration which is governed by some end boundary condition of the system. Neglecting continuously distributed axial load $f(x, t)$ the following ordinary differential expression with corresponding mode shape function $X_n(x)$ can be obtained as

$$\frac{\mathrm{d}^2 X_n(x)}{\mathrm{d}x^2} + \alpha_n^2 X_n(x) = 0 \tag{88.12}$$

where α_n is the characteristics value of the mode shape function corresponding to longitudinal vibration, which depends on corresponding boundary condition

$$\alpha_n = \frac{\omega_n^2}{\frac{E}{\rho} - (e_o a)^2 \omega_n^2}. \tag{88.13}$$

The nonlocal frequencies of a nanorod subjected to the longitudinal vibration can be determined by using boundary condition given below (Figure 88.3).

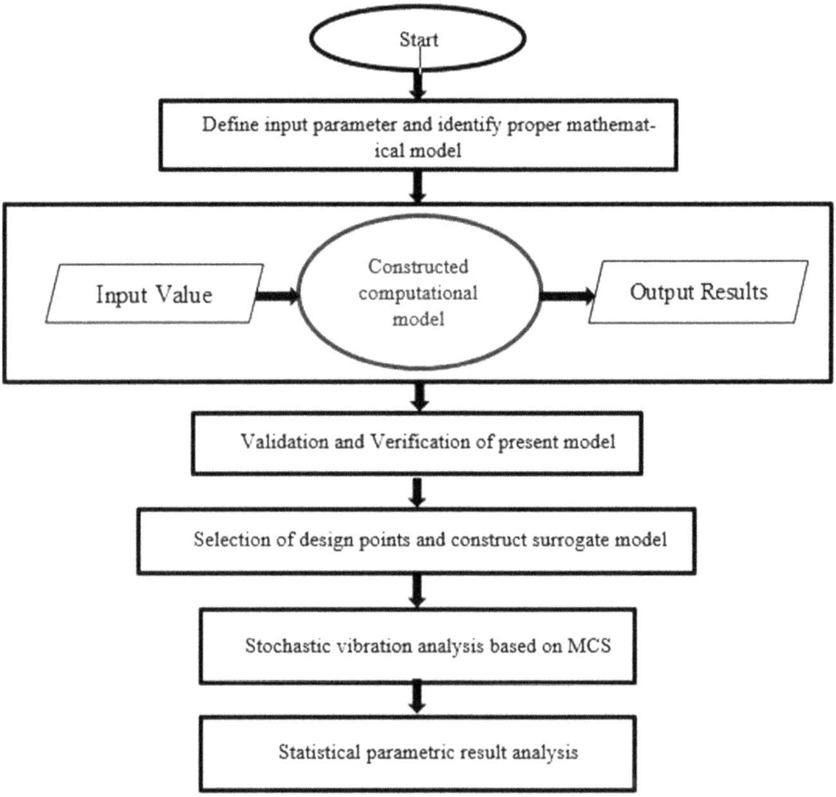

Fig. 88.3 Flow diagram of present methodology

$$\omega_n = \sqrt{\frac{E\alpha_n^2}{\rho\left(1 + (e_o a)^2 \alpha_n^2\right)}}. \quad (88.14)$$

88.5 Result and Discussion

In the present study, the following assumptions are considered such as, body force is neglected and employed the D'Alembert's Principle for equilibrium equation. Any contribution of shear force in cross section to transverse deformation is ignored i.e. transverse shear force is neglected, only deformation due to force applied in longitudinal is taken into consideration in this study. The method of separation of variable is considered using Fourier infinite series to solve ordinary differential equation (ODE). The fundamental mode is considered for validation and verification while the scale effect is considered to be caused by small internal molecular distance. Moreover, single walled carbon nanotubes formed by rolling graphene sheet is considered having material properties such as, Young's Modulus (*E*) 6.85 TPa and density (ρ) 9517 kg/m^3 [21].

Nonlocal elasticity theory is used to illustrate small scale effect on rotating nano cantilever beam. In present study, nonlocal elasticity model based on Eringen's theory is solved by differential quadrature method and the fundamental natural frequencies are obtained by considering the nonlocal parameter. After solving governing differential equation, the solution of governing differential equation is analysed by MATLAB code. The results are validated and verified based on Murmu and Pradhan [11] wherein the stochasticity analysis is carried out. The accuracy of fundamental natural frequencies and mode shapes depend on mode and material properties. The small scale effects on angular velocity and mode number is also investigated. For local model nonlocal parameter is assumed to be zero. But when non local parameter is taken into consideration then fundamental frequencies value corresponding to each curve becomes very close together. It is also observed that as the nondimensional angular velocity raised (see Fig. 88.4), non-dimensional fundamental frequencies are also found to increase due to effect of centrifugal force.

Moreover it can be observed that in nonlocal model of nano cantilever beam considered has been greatly influenced by angular velocity on the fundamental frequencies of nonobeam. Moreover local and nonlocal models is considered then more significantly small scale effect on mechanical vibration and buckling response are raised when cantilever rotates at high speed angular speed. It is also observed that higher frequency at high angular velocity zone is because of the simultaneous effect of both nonlocal parameter and angular velocity due to rotation. Fatigue effect on mechanical structure are also dependent on number of reversal of cyclic frequency applied on that particular structure. This investigation provides good information for the design of carbon nanotubes blade of nano turbine. It has been observed that when the fact of non local parameters are disregard longitudinal

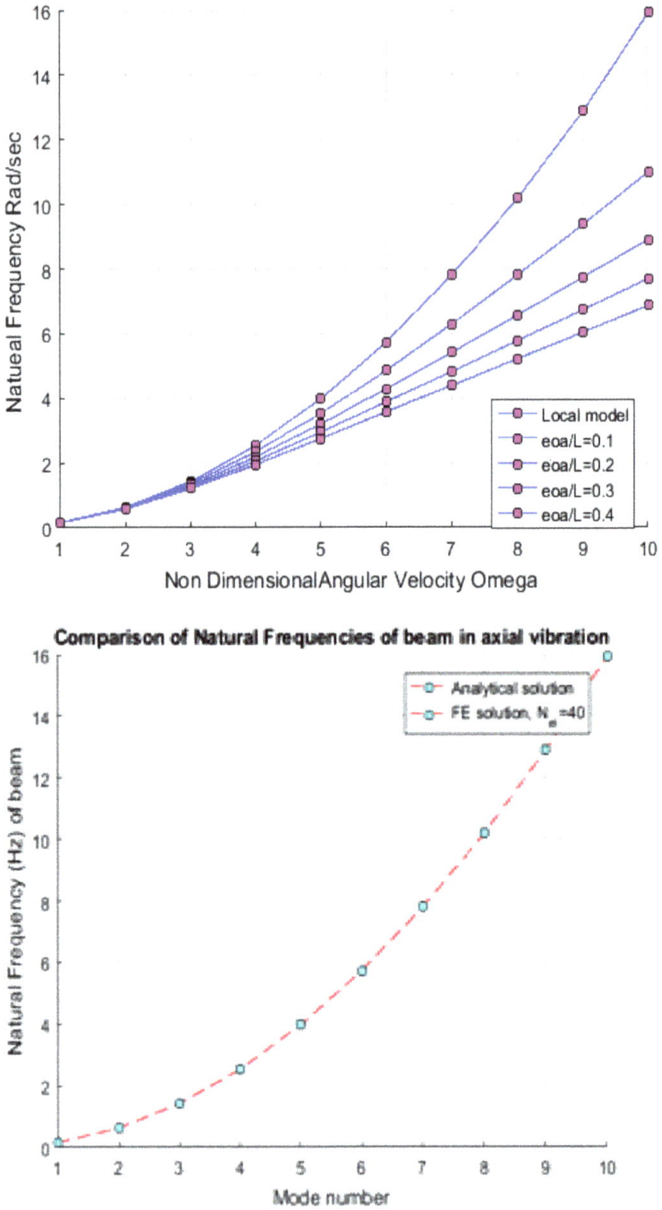

Fig. 88.4 Validation of non-dimensional natural frequency with corresponding angular velocity for different nonlocal parameters of a rotating nanocantilever beam and mode number with published literature [19]

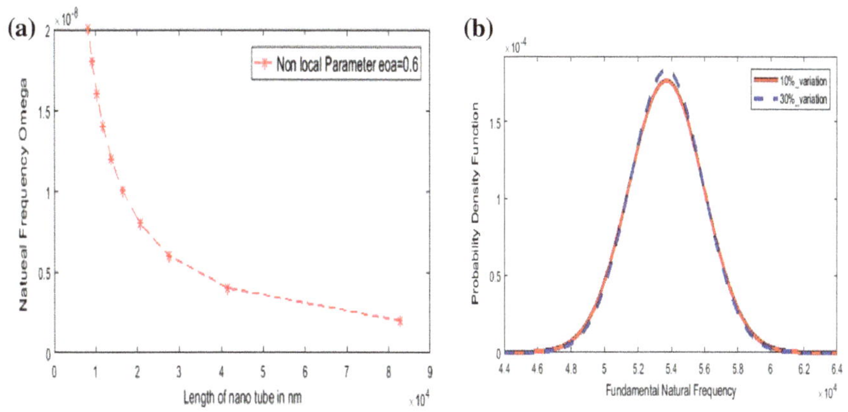

Fig. 88.5 **a** Validation of fundamental frequency changes with variation in length of nanorod with clamped free condition [21]. **b** Stochasticity analysis of fundamental natural frequencies of carbon nanorod based on randomness on input parameter (Non local parameter and Mode number)

vibrational frequencies are overestimated. Unlike transverse vibration of nanorod, natural frequency has got decreased with an increase in nonlocal parameter (see Fig. 88.5). Non local effect on longitudinal frequency of a nanorod has more prominent and distinguishable effect when any higher vibrational modes are considered. In the following table comparison and validation of the fundamental natural frequency of nanorod obtained from non local theory and those obtained from lattice dynamics based on MD simulation.

The following parameters and dimensions have been employed to validate the results as in given Table 88.1: $\rho = 9517$ kg/m^3, $E = 6.85$ TPa, $L = 12.2$ nm. Here fundamental frequencies are compared with the MD simulation for single walled carbon nanorod after obtaining results for clamped-free vibration using nonlocal elasticity theory. More consistent results are obtained when non local parameter goes to unity (see Table 88.1).

Since the value of internal length parameter a is fixed for SWCNT so in order to fit the result natural frequency optimization of parameter e_o is required for each mode of vibration corresponds to variation in geometrical properties of nanorod. However molecular dynamics (MD) simulation has been restricted to smaller scale because of higher time and cost consumption therefore nonlocal theory is used for more complex nanostructures analysis.

Above Fig. 88.5 indicates the effect of length non local parameter and of nanorod on fundamental frequency and it also shows characteristics of natural frequency of carbon nanorod based on random variation of material property. In present study each outcomes i.e. natural frequency on 10, 20, 30% are varied by using normal distribution and it can be observed that for small variation in material property within a certain limit natural frequency follows an normal distribution with a mean value constant and certain standard deviation. This parametric study helps us to observe uncertainty determination due to variation in material property due to

Table 88.1 Comparison and validation of nonlocal parameter influences on natural frequency of longitudinal vibration of nanorod

Non Local Parameter (e_0a)	Mode Number (n)	Present Analysis of Fundamental Frequency using MATLAB	Fundamental Frequency on Murmu and Adhikari published book [21]	Fundamental frequency using molecular dynamics on G. CAO et al. literature [22]	Percentage of error with Murmu and Adhikari book [21]	Percentage of error with molecular dynamics simulation on G. CAO et al. Literature [22]
0	1	0.488	0.549	0.544	11.11	10.29
0.5	1	0.536	0.548	0.544	2.19	1.47

structural inhomogeneity and anisotropic behaviour of structure in nanoscale. A notable decrease in fundamental frequencies with increase in length is observed (see Fig. 88.5). However, the fundamental nondimensional frequency has a decreasing in nature exponentially with a raises in the non-local parameter but this effect becomes insignificant with a greater length of the nanorod.

88.6 Conclusion

Present work shows the non local longitudinal frequency analysis of a single walled carbon nanorod and rotating nanocantilever, where first a non local mathematical model based on Eringen's non local stress theory has been used, then uncertainty effect based on parametric study of randomness on material property has been observed. It can be observed that nonlocal parameter has an important role when scale effect is considered on nanostructure. It has also been well established that when first vibrational mode consider, the fundamental non dimensional frequencies evaluated with nonlocal based models have higher magnitude than those obtained by local model. With increase in magnitude of angular velocity natural frequency on cantilever beam increased and also when length increases natural frequency of nanorod decreases exponentially.

The novelty of present study includes the computational investigation on stochastic frequency response analysis using computational code of a nanorod with uncertainty in the material properties i.e. young's modulus, density and non local parameter, mode number. Computational code is very accurate as compared with published work. Also when randomness is incorporated in material property natural frequency responses like a normal distribution curve then central tendency of frequency for continues variation of various material property can easily be evaluated. Stochasticity has also been investigated based on randomness in input parameter which shows uniform variation in fundamental frequency curves with very effective manner.

Present work is helpful in manufacturing point of view because Present study deals a significant contribution on stochastic behaviour of fundamental frequencies for a nanoscale structure so that structure become safe when that particular structure used in some specific domain of a machine element basically Nano Electro Mechanical System (NEMS) systems manufacturing like turbine blade, centrifugal pump blade etc. As manufacturing of any systems requires first fully structural analysis like bending, buckling, vibrational behaviours of any static or dynamic structures then according to structural analysis report manufacturings are being carry forwarded. From the physical point of view, it can be summarized that the non-local parameter dampened away fundamental natural frequencies. Fatigue response of a nanorod as well overall stiffness also depends on frequency response of the nanorod. To the best of authors' knowledge, this is the first attempt for applying Nonlocal elasticity in uncertainty quantification of nanostructure. Finally it is concluded that the present methodology can also be extended to analyse statistical behaviour of nanoplates as well as buckling behaviour analysis of nanoplate.

References

1. Lijimia, S.: Helical microtubules of graphitic carbon. Nature **354** (1991)
2. Eringen, A.C.: Differential equation of nonlocal elasticity and solution of screw dislocation and surface waves. J. Appl. Phys. **54** (1983)
3. Kozi, P., Pavlovic, R., Karlic, D.: The flexural vibration and buckling of the elastically connected parallel beams with a Kerr-type layer. Mech. Res. Commun. **56** (2014)
4. Ansari, R., Rajabiehfard, R., Arash, B.: Nonlocal finite element model for vibrations of embedded multi-layered graphene sheets. Comput. Mater. Sci. **494**, 831–838 (2010)
5. Oniszczuk, Z.: Free transverse vibrations of an elastically connected complex beam-string system. J. Sound Vib. **24**(4), 703–715 (2002)
6. Peddieson, J., Buchanan, G.G., McNitt, R.P.: Application of nonlocal continuum models to nanotechnology. Int. J. Eng. Sci. **41**, 305–312 (2003)
7. Zhang, Y.Q., Liu, G.R., Xie, X.Y.: Free transverse vibration of doubled walled carbon nanotubes using a theory of nonlocal elasticity. Phys. Rev. B **71** (2005)
8. Lu, P., Lee, H.P., Lu, C., Zhang, P.Q.: Dynamic properties of flexural beams using a non local elasticity model. J. Appl. Phys. **99** (2006)
9. Reddy, J.N., Roque, C.M.C.: Analysis of Timoshenko nanobeams with a non local formulation and meshless method. Int. J. Eng. Sci. **49**, 976–984 (2011)
10. Lee, H.L., Chang, W.J.: Surface effect on frequency analysis of nanotubes using nonlocal Timoshenko beam theory. J. Appl. Phys. **108** (2010)
11. Pradhan, S.C., Murmu, T.: Application of nonlocal elasticity and DQM in the flapwise bending vibration of a rotating nanocantilever. Physica E **42** (2010)
12. Pradhan, S.C.: Non local finite element analysis and small scale effects of CNTs with Timoshenko beam theory. Finite Elem. Anal. Des. **50**, 8–20 (2012)
13. Wang, C.M., Tan, V.B.C., Zhang, Y.Y.: Timoshenko beam model for vibration analysis of multi-walled carbon nanotubes. J. Sound Vib. **294**(4-5), 1060–1072 (2006)
14. Danesh, M., Farajpour, A., Mohammadi, M.: Axial vibration analysis of a tapered nanorod based on nonlocal elasticity theory and differential quadrature method. Mech. Res. Commun. **39**, 23–27 (2012)

15. Aydogdu, M.: Axial vibration analysis of nanorods (carbon nanotubes) embedded in an elastic medium using nonlocal elasticity. Mech. Res. Commun. **43**, 34–40 (2012)
16. Abdollahi, R., Boroomand, B.: Nonlocal elasticity defined by Eringen's integral model - Introduction of a boundary layer method. Int. J. Solids Struct. **51** (2014)
17. Karličić, D., Kozić, P., Adhikari, S., Cajić, M., Murmu, T.: Nonlocal mass-nanosensor model based on the damped vibration of single-layer graphene sheet influenced by in-plane magnetic field. Int. J. Mech. Sci. **96–97**, 132–142 (2015)
18. Bahaadini, R., Hosseini, M., Jamalpoo, A.: Nonlocal and surface effects on the flutter instability of cantilevered nanotubes conveying fluid subjected to follower forces. Physica B **509**, 55–61 (2017)
19. Murmu, T., Pradhan, S.C.: Small scale effect on buckling of single layered graphene sheet under biaxial compression via nonlocal continuum mechanics. Comput. Mater. Sci. **47**(1), 268–274 (2009)
20. Pradhan, S.C., Murmu, T.: Vibration analysis of nano single layered graphene sheets embedded in a elastic medium based on non local elasticity theory. J. Appl. Phys. **105** (2009)
21. Murmu, T., Adhikari, S., McCarthy, M.: Non-Local Structural Mechanics, 1st edn. Danilo Karličić, © ISTE Ltd. (2016)
22. Cao, G., Chen, X., Kysar, J.W.: Thermal vibration and apparent thermal contraction of single-walled carbon nanotubes. J. Mech. Phys. Solids **54** (2000)

Chapter 89
Finite Element Analysis and Three-Point Bend Testing of Indian Rice Grains with Fissure Effects

P. V. Gharat, D. K. Shinde and D. N. Raut

Abstract In today's world, one of the major problems faced by the rice-milling industries is breakage of rice during processes like de-husking/de-hulling, whitening, and polishing. For designing of milling machinery and for reducing the breakage of different rice varieties while processing, it is important to understand the rupture force of rice kernels of different varieties under the three-point breaking test. In this study, the different varieties of rice grains with different fissure conditions were investigated. The rupture force investigated by using three-point breaking test which is conducted on Instron UTM with specially designed adjustable support fixture, in which span can be adjusted for different varieties of rice grains. Five brown rice varieties—Rashi Poonam, Kolam, Komal, Basmati, and Daptari tested at a rate of 1 mm/min. The finite element analysis has helped to validate the result of the flexural stresses in the rice grain induced during test and then the crack propagation is simulated in the analysis. It was seen that out of the five rice varieties under tests, Rashi Poonam has the highest average rupture force of value 32.29 N. The results obtained from this test and computational analysis can help for developing methods and types of equipment used for whitening and polishing processing of rice with different varieties.

Keywords Brown rice · Three-point breaking test · Fissure detection

P. V. Gharat
Department of Production Engineering, Veermata Jijabai Technological Institute, Mumbai 400031, India

D. K. Shinde (✉) · D. N. Raut
Faculty Department of Production Engineering, Veermata Jijabai Technological Institute, Mumbai 400031, India
e-mail: dkshinde@vjti.org.in

R. G. Narayanan et al. (eds.), *Advances in Computational Methods in Manufacturing*, Lecture Notes on Multidisciplinary Industrial Engineering,
https://doi.org/10.1007/978-981-32-9072-3_89

89.1 Introduction

Rice is the seed of the grass species Oryza sativa, and it is one of the most important cereals in the world. The worldwide production of rice in 2014 was 741 million tonnes. India is one of the world's largest producers of rice. India's rice production is 20% of all worldwide rice production. In India, rice is pre-eminent crop and the staple food of the people of the various parts of the country. For the year 2013–2014, India's rice production was 106.54 million tonnes. In India, rice is a major crop and staple food for more than 70% of the population, thus plays a vital role in the national economy [1]. Rice passes through various processing like harvesting, de-husking, whitening, and polishing which may lead to breakage of rice grains. Rice is preferably consumed as whole grains. The percentage of unbroken rice grains gives good economical value to farmers and millers. Therefore for accurate design of the de-husking, whitening and polishing machines to minimize the grains breakage, information on physical properties such as length, width, and thickness and mechanical properties of rice grains such as breaking force and graph of force versus deformation must be provided [2].

The various physical properties like length, thickness, and mass of rice variety PR-106 were measured by Gadge. The moisture content of rice was 13.34% (wet basis). The average length, width, and thickness were 6.61, 1.75, and 1.40 mm, and unit mass volume of rice kernels were 0.017 g, and 0.051 cm^3, respectively. The mean diameter and surface area and aspect ratio were 2.52 mm, 20.1 mm^2, and 26.58%, respectively [3]. The study by Zareiforoush et al., the rupture force of paddy grains was measured under quasi-static compressive loading. The Alikazemi and Hashemi varieties of paddy rice were tested in horizontal and vertical orientations at two different loading rates of 5 and 10 mm/min. For Alikazemi variety of paddy rupture force values decreases as the loading rate increased from 5 to 10 mm/min. The rupture force values for horizontal orientation were 125.69 and 33.51 N at a loading rate of 5 mm/min and 10 mm/min, respectively. Similarly, the rupture force values for vertical orientation were 117.38 and 33.51 and the loading rate of 5 mm/min and 10 mm/min respectively [2]. In another study, Lu et al., the compressive force and bending force of rough rice, brown rice, and white rice were determined by conducting compression and three-point bending tests. It was observed that thinner kernels required low breaking force and therefore lead to more breakage during milling [4].

I. Bagheri et al., investigated the mechanical strength of 12 varieties of brown rice at four levels of moisture content 8, 10, 12, and 14 wet basis. Brown rice grains were quasi-statically loaded at two rates 10 and 15 mm/min. The results show that the breaking force decreases with increasing moisture content and also with an increase in a loading rate. The breaking of brown rice with a lower level of moisture content was a sudden failure with less deformation and breaking of brown rice with a higher level of moisture content was gradually crushing with more deformation [5]. For three different levels of amylase ratio, the mechanical properties of different varieties of rice grains were invested by Y. J. Shu et al. The mechanical properties were measured using compression and three-point breaking test. According to their study, the rice variety with the high amylose ratio has the highest compression strength. Their study can help to develop methods and devices used for milling processing of rice with different levels of amylose to reduce the milling losses [6]. The another study carried out by Ponlakrit Kumklam, in which the strength of rice grains measured by breaking force and study done on the correlation of breaking force and head rice yield. Bending test was done by keeping fixed base support distance of 3.5 mm with flat loading head of thickness 1.5 mm and deformation rate was 1.5 mm/min. From the test, the bending force was closely related to the head rice yield [7].

The main objective of this research was to measure physical properties such as length, width, and thickness and to determine the mechanical property of five different varieties of brown rice grains in terms of rupture force under three-point breaking test and examining the effect of rice variety and number of fissures on rupture force.

89.2 Materials and Methods

The brown rice varieties Rashi Poonam, Kolam, Komal, Basmati, and Daptari used for this test were obtained from one of the rice mills from Palghar, Maharashtra, India. The hot-air oven-drying method was used to determine initial moisture content of the rice samples. The initial moisture contents of grains were 5.35, 5.69, 4.98, 6.13, and 5.42% (wet basis) for Rashi Poonam, Kolam, Komal, Basmati, and Daptari varieties, respectively.

89.2.1 Detection of Fissures in Rice Grains

This study investigates the rupture force for five different rice varieties with the number of fissures in rice grains. To detect fissures in rice grains, fissure detection technique is newly developed. Imaging of the fissures in rice kernels was done with the help of DSLR camera. Camera used was Nikon D5200 with 18–55 primary lens and close up lenses/secondary lenses (numex 52 mm close up +1, +2, +4 and +10) were additionally attached for better zooming. For better zooming without the use of secondary lenses can be achieved by attaching a primary lens in reverse direction to the camera. For better results, the camera was on a tripod stand.

The light source was white LED bulb (Wipro LED, 9 W). All samples were placed in plain brown-colored glass, and images were captured using DSLR camera. After capturing the images of rice grains, all are edited in Lightroom CC image-editing software (version 3.4) for better clarity of fissures. While editing, clarity was increased to full value, highlights, shadow, and whites reduced to zero value and exposure, the contrast was adjusted according to images and finally, the

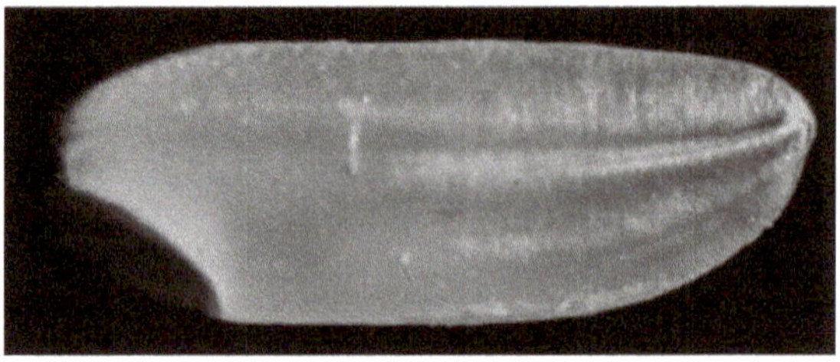

Fig. 89.1 Image of brown rice with no fissure

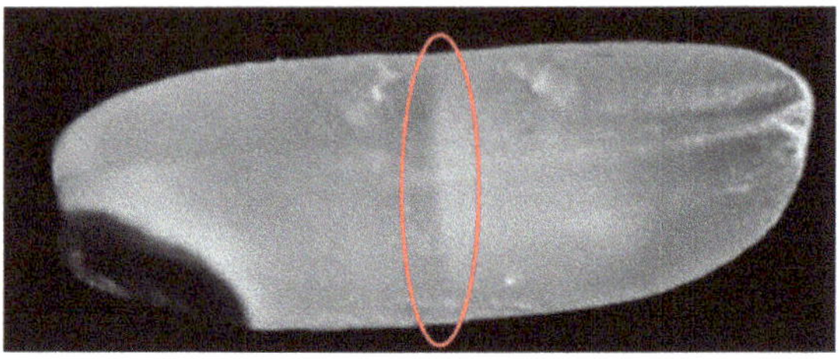

Fig. 89.2 Image of brown rice with one fissure

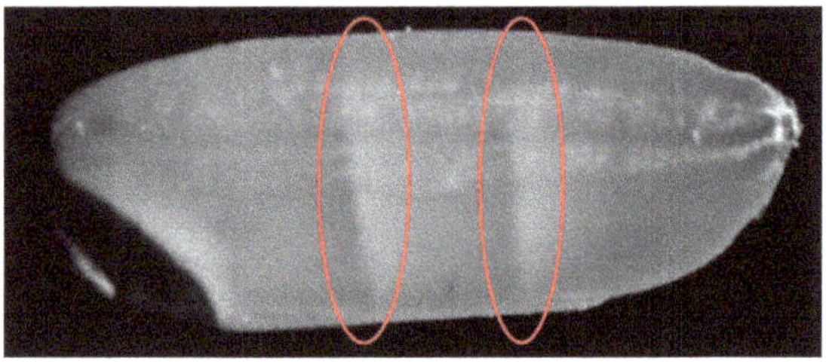

Fig. 89.3 Image of brown rice with two fissures

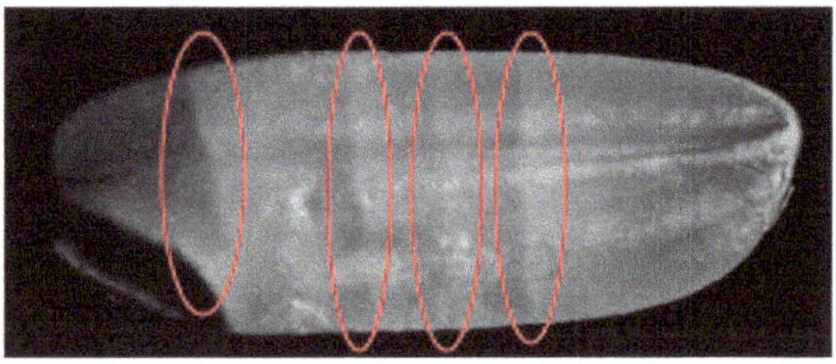

Fig. 89.4 Image of brown rice with more than two fissures

black and white effect was given to images. Figure 89.1 shows the images of brown rice with a different number of fissures. Figures 89.1, 89.2, 89.3, and 89.4 show the brown rice with 0, 1, 2, and more than 2 fissures, respectively.

89.2.2 Types of Fissures Observed in Rice Grains

Various types of fissures are mentioned in Figs. 89.5 and 89.6 show the image of brown rice grains (Basmati rice) with different types fissures.

From Fig. 89.5,

Type A fissures—Fissures perpendicular to longitudinal axis of grain—Non-terminating type: The fissures present in a rice grain which are perpendicular to the longitudinal axis of grain and ended somewhere in rice grain.

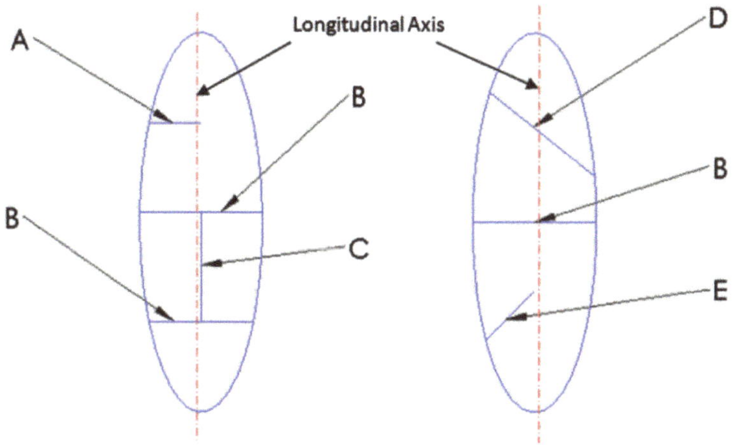

Fig. 89.5 Types of fissures in rice grains

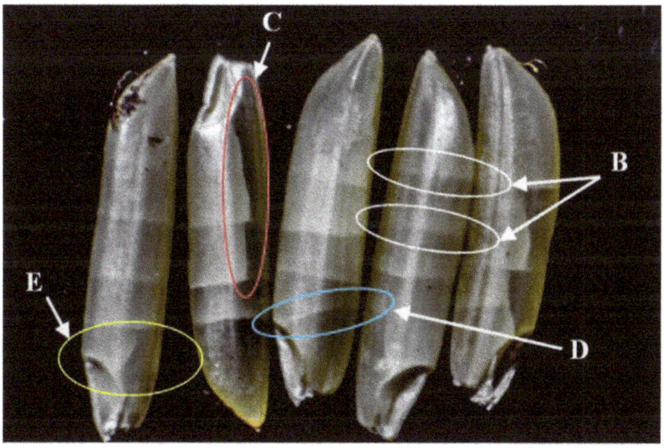

Fig. 89.6 Image of detection of different types of fissures in rice grains

Type B fissures—Fissures perpendicular to longitudinal axis of grain—Terminating type: The fissures present in a rice grain which are perpendicular to the longitudinal axis of grain and passed throughout the rice grain along that axis.
Type C fissures—Fissures parallel to longitudinal axis of grain—Terminating type: The fissures present in a rice grain which are parallel to the longitudinal axis of grain and passed throughout the rice grain along that axis.
Type D fissures—Cross fissure—Terminating type: The fissures present in a rice grain which are neither parallel nor perpendicular to the longitudinal axis of grain and ended somewhere in rice grain.
Type E fissures—Cross fissure—Non-terminating type: The fissures present in a rice grain which are neither parallel nor perpendicular to the longitudinal axis of grain and passed throughout the rice grain.

It was observed that in the case of fissured kernels the percentage of fissures perpendicular to longitudinal axis of grains (A and B-types) was more than any other types of fissures. Each and every fissured kernel had A and B type of fissures present in it. Percentage of fissures parallel to longitudinal axis of grains(C type) were more in Basmati rice variety than any other rice variety.

89.2.3 Physical Characteristics of Rice Grains

The shape of the rice kernel was found to be ellipsoid shape with three dimensions —length, width, and thickness. The physical dimensions of 20 kernels of each variety were measured using micrometer screw gauge (Mitutoyo Corporation) having least count of 0.01 mm.

The measured lengths of Rashi Poonam, Kolam, Komal, Basmati, and Daptari rice grains were found 5.79 ± 0.2 mm, 5.41 ± 0.19 mm, 6.64 ± 0.21 mm, 7.66 ± 0.33 mm, 5.45 ± 0.18 mm, respectively. From the dimensions of rice grains, it can be seen that the length of Basmati rice variety was more than other types of rice varieties. So, Basmati was long rice grain while others were short/ medium rice grains. Physical dimensions of five types of rice varieties are given in Table 89.1.

Table 89.1 Physical dimensions of five types of rice for three-point breaking test

S. No.	Type of rice	Length (mm)	Width (mm)	Thickness (mm)
1	Rashi Poonam	5.79 ± 0.2	2.0 ± 0.08	1.57 ± 0.04
2	Kolam	5.41 ± 0.19	1.65 ± 0.07	1.52 ± 0.05
3	Komal	6.64 ± 0.21	2.08 ± 0.07	1.57 ± 0.04
4	Basmati	7.66 ± 0.33	1.97 ± 0.06	1.75 ± 0.06
5	Daptari	5.45 ± 0.18	1.62 ± 0.15	1.54 ± 0.05

89.2.4 Three-Point Breaking Test

The mechanical property of brown rice grains was determined in terms of average rupture force in three-point breaking test for five different varieties Rashi Poonam, Kolam, Komal, Basmati, and Daptari. The experiment was performed at a loading rate of 1 mm/min. Three-point breaking test was performed using a Universal testing machine (Instron Universal Testing Machine-3345, Fiest Laboratory, MEMS Department, IIT Bombay).

For testing, all rice grains were de-husk by hand. In each batch, 20 rice samples were taken for testing. Rice samples were sorted out according to the number of fissures present in it. Fissured were observed by fissured detection technique which was newly developed with the use of DSLR camera. For three-point bending tests, the whole rice grain was kept on an adjustable base support in which span was adjustable. The diameter of the end of support and compression tip was 1.45 mm. As rice shape is narrow toward the ends, the 15% of overhang was considered for three-point breaking test so that rice grains cannot break at their end due to the small area as compared to the middle portion. The adjustable base supports for three-point test apparatus for the breaking of a rice grain is shown in Fig. 89.7.

Different parts of the adjustable fixture for three-point breaking test are:

1. Base of fixture.
2. Two guiding shafts: for guiding movable supports.
3. Two movable supports: for support the rice grain while testing.
4. Circular tip on each support: for reducing stress concentration.
5. Main threaded shaft: half anticlockwise and another half clockwise threading.
6. Handle for rotation of the main threaded shaft.

The position of three points for breaking test:

The position of three contact points while loading is shown in Fig. 89.8. Compression Pouch is always in the middle of rice grain.

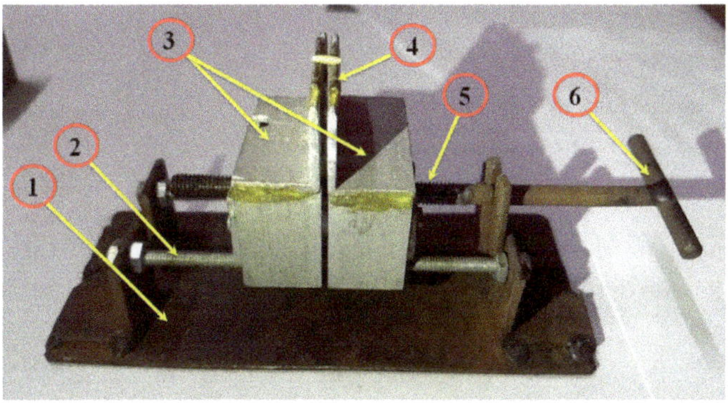

Fig. 89.7 Adjustable base supports for three-point breaking test

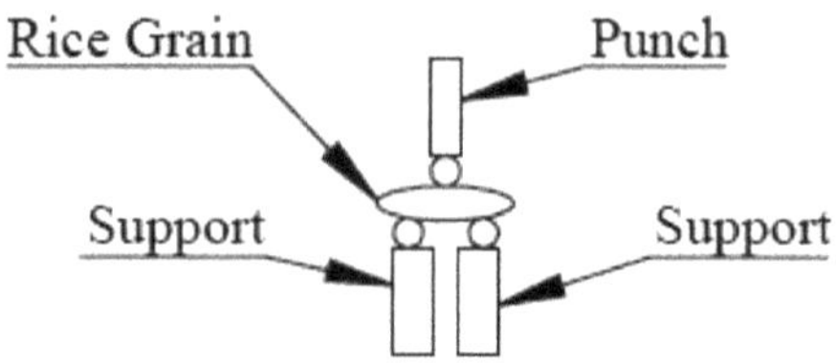

Fig. 89.8 Position of three contact points

89.3 Results

Table 89.2 shows the rupture force value for different varieties of brown rice grains. Data shows that variety and number of fissures affect the rupture force. In the case of Basmati variety, the average force to rupture the grain was obtained as 18.69 ± 6.98 N, 9.56 ± 5.63 N, 6.32 ± 3.36 N, and 3.49 ± 1.38 N for grains without fissures, with 1 fissure, 2 fissures, and more than 2 fissures, respectively. Similarly, rupture force values for Kolam, Komal, Daptari, and Rashi Poonam variety are given in Table 89.2. From data, it can be seen that rupture force value decreases as fissures increase in rice grains. Rashi Poonam variety has the highest average rupture force of value 32.29 N for grains without fissures. While Komal variety, the rupture force value 28.52 N which is much closer to the value of rupture force for Rashi Poonam variety.

89.3.1 Finite Element Analysis of Rice Grain (Three-point Breaking Test)

Finite element analysis of rice grain under three-point breaking test was done in ANSYS Workbench Software. As rice grain is having complex geometry, for analysis, rice grain is considered as a cylindrical object with diameter equal to equivalent diameter of rice grain calculated using formula:

Table 89.2 Rupture force by three-point breaking test

S. No.	Type of rice	Breaking force (N)			
		0 fissure	1 fissure	2 fissures	More than 2 fissures
1.	Basmati	18.69 ± 6.98	9.56 ± 5.63	6.32±3.36	3.49 ± 1.38
2.	Kolam	15.44 ± 6.57	10.30 ± 4.72	5.21 ± 4.69	3.81 ± 2.08
3.	Komal	28.52 ± 4.41	16.83 ± 4.39	11.29 ± 3.56	6.34 ± 2.21
4.	Daptari	16.23 ± 5.98	12.36 ± 3.87	7.12 ± 2.28	4.51 ± 1.13
5.	Rashi Poonam	32.29 ± 6.55	14.68 ± 5.89	9.90 ± 2.70	6.96 ± 3.06

$$\mathrm{De} = \left(L * \frac{(W + T)^2}{4} \right)^{1/3}$$

where L = Length, W = Width of rice grain, T = Thickness of rice grain [8]. Mechanical properties of rice grains considered for analysis were: Young Modulus = 417MPA, Poisson's ratio = 0.4476, density 2700 kg/m^3 [9]. Physical dimensions of Rashi Poonam rice variety were taken for analysis. Figure 89.13 shows the finite element analysis of rice grain under three-point breaking test, and Fig. 89.9a, b shows the analysis results for the equivalent stress and equivalent strain, respectively.

From analysis result, it can be seen that for 7% of strain, the equivalent stress in rice was 322 MPa which at the contact of rice bottom surface and support. The crack may occur at contact of support and bottom surface of rice grain where the stress intensity is maximum. In experimental results for Rashi Poonam rice variety without any fissure, the maximum stress was found was 430 MPa and maximum strain was 9%. In three-point bending test, the crack started at contact of bottom surface of rice grain and support and then propagate toward the contact of punch to rice grain as shown in Fig. 89.10. High-speed camera used for recording the crack propagation was Phantom MIRO M110 with supporting software—Phantom Camera Control Application for controlling operation of camera. Illumination is done by fiber optic dual light illuminator (White light source).

89.3.2 A Typical Graph of Load (N) Versus Deformation (mm) Obtained After Three-point Breaking Test of Rice Grains

From load (N) versus deformation graph as shown in Fig. 89.11, the testing starts at point A, i.e., punch moves downward with a set speed (1 mm/min). The region from point A to point B is almost parallel to *X*-axis. This region shows that up to point B, there is no contact of punch with rice grain. (This is due to we touch punch to rice grain manually, so for every sample, it may vary). From point B, actual loading gets started. The region from point B to point C shows the linear relationship between load and extension. At point C, the breaking of rice grains occurs.

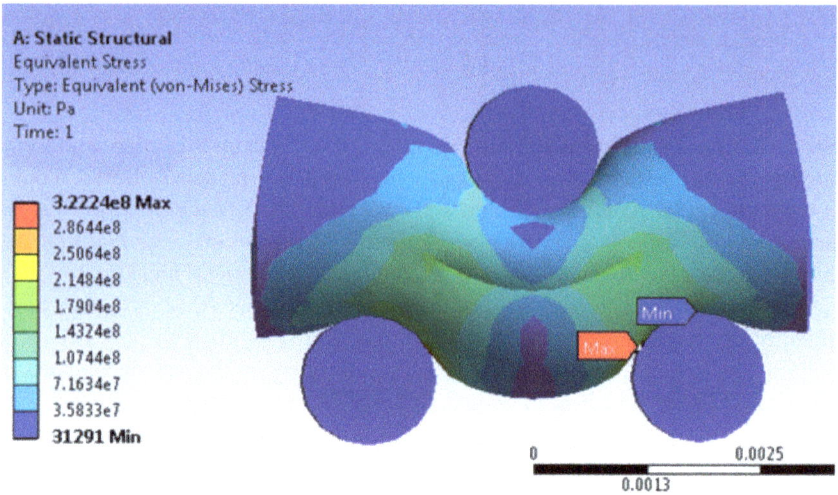

(a) Analysis for Equivalent Stress

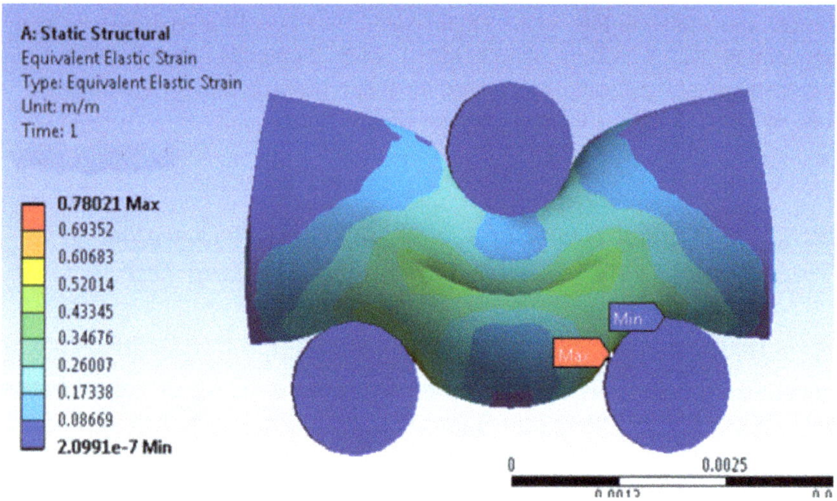

(b) Analysis for Equivalent Strain

Fig. 89.9 FEM analysis of the rice grain with failure area of fissure in the three-point bending loading. **a** Analysis for equivalent stress, **b** analysis for equivalent strain

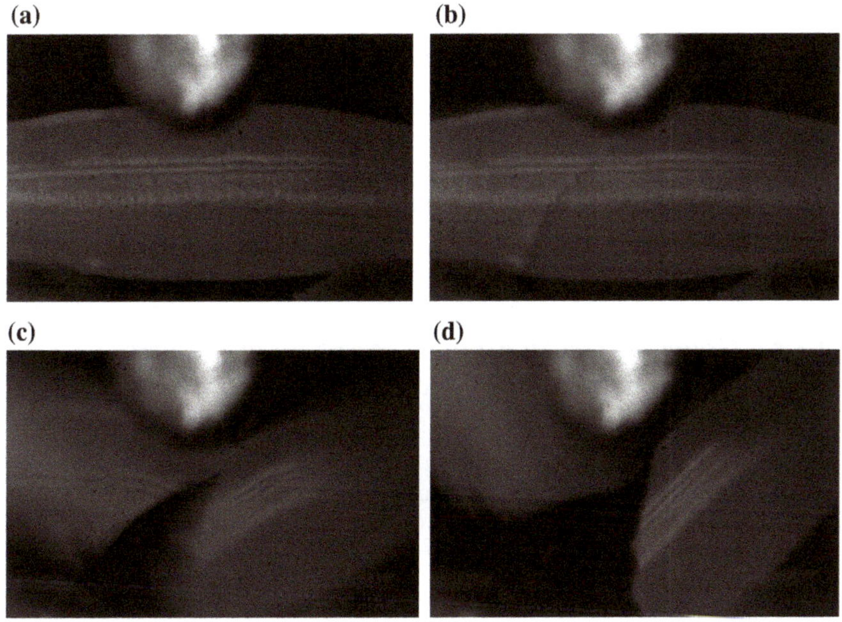

Fig. 89.10 Images of the crack propagation in rice grain while applying the load

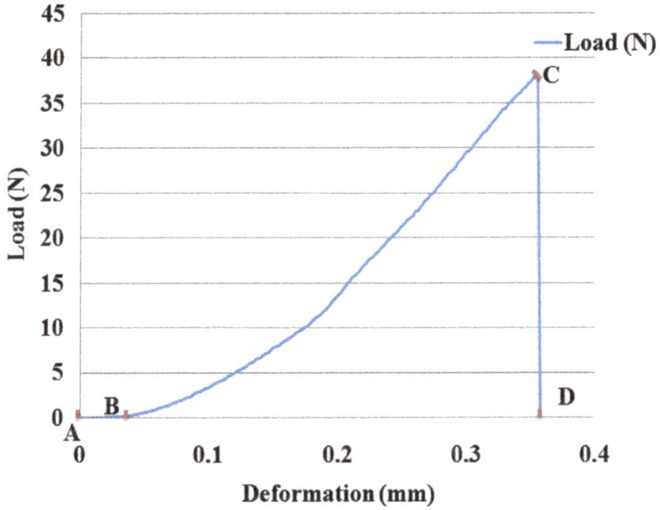

Fig. 89.11 Typical graph of load (N) versus deformation (mm)

89.3.3 Modified Graph of Load (N) Versus Deformation (mm) Obtained After Three-Point Breaking Test of Rice Grains

This graph is obtained by removing Region A-B and Region C-D is from the typical graph. This graph shows the linear relationship between load and deformation (Fig. 89.12).

89.3.4 Graph of Breaking Force Versus no of Fissures for Five Varieties of Rice Grains

Figure 89.13 shows the graph of breaking force value for five different types of rice varieties with a different number of fissures present in the rice kernel. From the

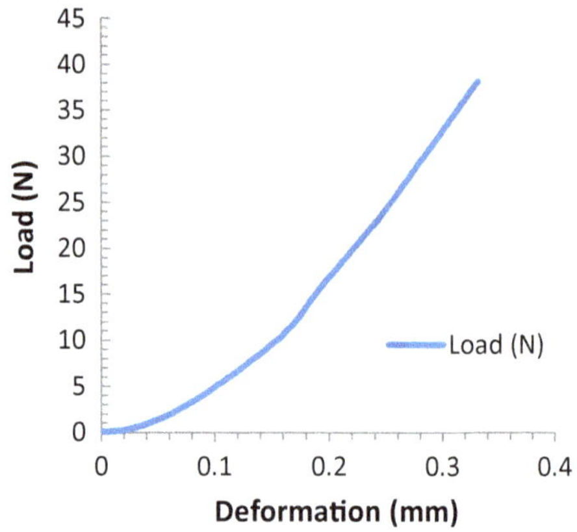

Fig. 89.12 Graph of load (N) versus deformation (mm)

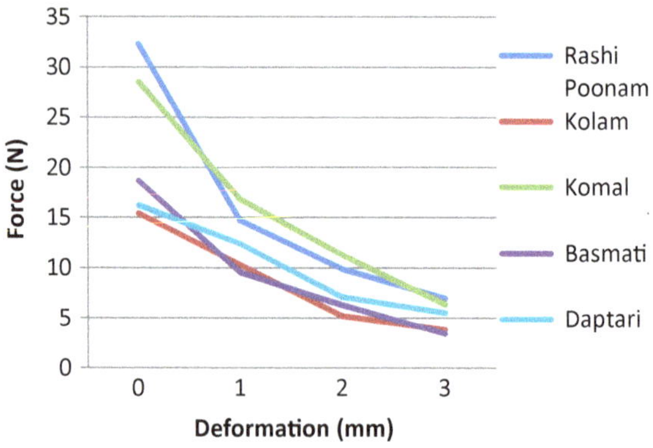

Fig. 89.13 Graph of breaking force versus number of fissures (*On X-axis: Point 3 indicates more than two fissures)

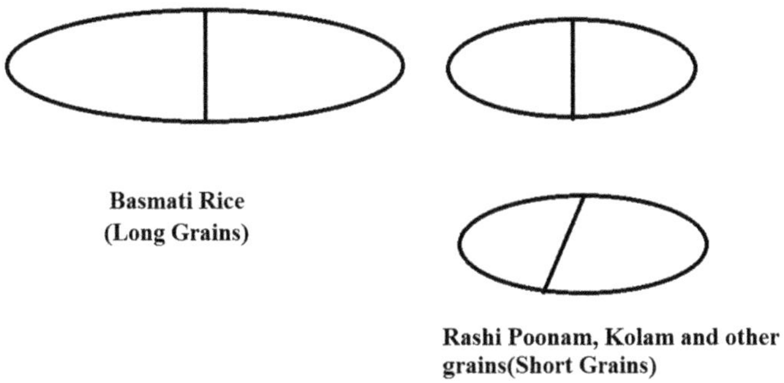

Fig. 89.14 Breaking patterns of rice grains while three-point breaking test

graph, it can be clearly seen that the breaking force required decreases as the number of fissures increases in rice grain. Rice grains become so weak if there are more than two fissures in a grain and for those rice grains breaking force values are much closed irrespective of rice varieties.

89.3.5 Different Breaking Patterns of Rice Grains While Three-Point Breaking Test

After testing and imaging of rice grain it is oberved that rice kernels with fissures subjected to 3-point breaking test, crack initiated at big fissure. When there was no fissure in rice grain, then crack is initiated from the contact point of one of the base support and propagates upward toward the point of contact between rice grain and punch. Each case of the rice grain, crack propagated from bottom surface to top surface of rice grain while loading because of bottom surface always in tension while 3-point breaking tests (Figure 89.14).

89.4 Conclusions

The three-point breaking test has been performed on five varieties of brown rice grains with different types of fissures present in brown rice grains. The values of the rupture force of brown rice grains Rashi Poonam and Komal rice varieties were taken higher than rupture force value of brown rice grains. So, it can be concluded that the Rashi Poonam and Komal varieties have more resistant to breakage than the Basmati, Kolam and Daptari varieties during the various operations such as de-husking, milling, and polishing. The results obtained from three-point breaking test of brown rice grains can help for developing of methods and types of equipment used for whitening and polishing processing of rice with different varieties.

Finite element analyis is result have validated the cracks are likely to occur due to the applied the rupture forces on the rice grain in the three point bending test. If rice millers able to find breaking force value for particular variety of rice grains before processing, then it may easy for rice millers to adjust correct pressure for processing that particular batch of rice grains so further breaking can be avoided. The testing of rice grain can be done on the low load capaicty carrying universal testing machine shall be recommeded for better rupture analysis of rice grain in the experimental testing.

Acknowledgements We thank Prof. Uday Sheorey and Prof. Rakesh Mote (Professor, Mechanical Department, Indian Institute of Technology, Bombay) for their excellent guidance throughout the project. We also thank Metallurgical Engineering and Material Science Department, IIT Bombay, for support to carry out various testing.

References

1. Mahajan, G., Kumar, V., Chauhan, B.S.: Rice production worldwide. Springer International Publishing AG, Switzerland (2017)
2. Zareiforoush, H., Komarizadeh, M.H., Alizadeh, M.R.: Mechanical properties of paddy grains under quasi-static compressive loading. N. Y. Sci. J. **3**(7), 40–46 (2010)
3. Ghadge, P., Prasad, K.: Some physical properties of rice kernels: variety PR-106. J. Food Process Technol. **3**(8), 1–5 (2012)
4. Lu, R., Siebenmorgen, T.J.: Correlation of head rice yield to selected physical and mechanical properties of rice kernels. Am. Soc. Agric Eng. **38**(3), 889–894 (1998)
5. Bagheri, I., Dehpour, M.B., Payman, S.H., Zareiforoush, H.: Rupture strength of brown rice varieties as affected by moisture content and loading rate. Aust. J. Crop Sci. **5**(10), 1239–1246 (2010)
6. Shu, Y.J., Liou, N.-S., Moonpa, N., Topaiboul, S.: Investigating damage properties of rice grain under compression load. In: International Conference on Experimental Mechanics and Twelfth Asian Conference on Experimental Mechanics, Proceedings of SPIE 923402, pp. 1–5, Bangkok, Thailand (2014)
7. Kumklam, P.: Predict head rice yield with mechanical strength of kernels. J. Appl. Sci. Res. **9** (12), 6107–6110 (2013)
8. Ravi, P., Venkatachalam, T.: Important engineering properties of paddy. Sci. J. Agric. Eng. **4**, 73–83 (2014)
9. Chattopadhyay, P.K.: The rheological properties of rice grain. J. Food Process. Eng. **17**, 1–17 (1994)

Chapter 90
Finite Element Analysis of Potential Liner Failures During Operation in Spherical Pressure Vessel

R. Pramod, N. Siva Shanmugam, C. K. Krishnadasan and K. Sankaranarayanasamy

Abstract Composite overwrapped pressure vessel (COPV) is widely used in aerospace industry to contain pressurized gases. The behaviour of pressure vessel components (liner and composite overwrap) at applied pressure and potential failures during its operation requires adequate study. The finite element (FE) analysis is performed for potential failures occurring in a 420 mm diameter aluminium alloy (AA) 6061-T6 lined spherical pressure vessel. In the present study, the design of spherical liner with apt manufacturing techniques is presented. The liner contribution towards sustaining the minimum burst load is determined and liner burst test is conducted to check the location of failure. Further, typical liner failures like buckling and bond failure occurring due to inefficient contact between the liner and composite overwrap, which eventually forms a debond region are also studied. The liner burst pressurization results predicted the location of failure in liner at transition region as the liner is undergoing thickness changes in that particular region. The mode of liner buckling, buckling load is predicted when liner is subjected to compression during depressurization cycle. Further, at debond region, the liner is unable to transfer pressure load effectively to composite overwrap experiencing severe deformation.

Keywords Finite element analysis · Liner · Buckle · Debonding

R. Pramod · N. S. Shanmugam (✉) · K. Sankaranarayanasamy
Department of Mechanical Engineering, National Institute of Technology, Tiruchirappalli 620 015, Tamil Nadu, India
e-mail: nsiva@nitt.edu

C. K. Krishnadasan
Structural Design and Analysis Division, Vikram Sarabhai Space Centre, Thiruvananthapuram 695 022, Kerala, India

R. G. Narayanan et al. (eds.), *Advances in Computational Methods in Manufacturing*, Lecture Notes on Multidisciplinary Industrial Engineering,
https://doi.org/10.1007/978-981-32-9072-3_90

90.1 Introduction

Launch vehicle and satellite require pressure vessels during operation for managing fluids and propulsion systems. Composite overwrap pressure vessel (COPV) with inclusion of a metal liner offers less vessel weight in comparison to conventional metal pressure tanks [1]. Aluminium and titanium alloys are generally used materials for liner because of less density and corrosion-resistant properties exhibited by these materials, respectively [2]. Non-load sharing and load sharing are the two types of liner. In the first case, liner is having minimum thickness and weight designed primarily to provide permeation barrier for preventing escape of gases. In the latter case, the liner contributes towards sharing the applied pressure load with higher thickness and weight [3]. Lee et al. [4] designed a spherical pressure tank using the Ti-15 V-3Al-3Cr-3SN material. Hemispheres were made by hot spin forming and blow forming process followed by TIG welding to join the two hemispheres. Pressurization test was conducted; the strain and acoustic emission signals were obtained to examine the outcome of selected manufacturing technique on the modes of failure and working performance of the tank. Chandran et al. [5] performed FE analysis by designing and analysing the COPV using titanium liner and T-300 carbon fibre composite. The effect of debonding location and length of debond in the liner-composite interface for weld and transition region was investigated. Almeida Jr et al. [6] investigated the effect of load sharing ability of aluminium and stainless steel liner material with various thicknesses in cylindrical liner geometry. Further, analyzed the stress and strain distribution at various regions of liner and composite structure. Many research works have been carried out in developing pressure vessels using AA 6061-T6 liner. The numerical analysis performed mainly focused on the modelling techniques and analysis of composite overwrap, failure analysis of composite overwrap and methods to analyse filament wound COPV [7–9]. From the literature review, it is noted that the research work has not focused on the analysis of failure and its modes in liner under applied internal pressure. Studying the behaviour of liner by physically testing the pressure vessel by subjecting it to repeated testing shall be economically high and time-consuming as it comprises of conducting several trials to obtain in-depth information generally involved in a design process. Instead, numerical analysis is conducted using the FE method, as an alternative to experimental testing [10]. In the present case, it is emphasized on cost reduction by replacing titanium with aluminium liner material and will be helpful in meeting quality conditions effectively. Hence, in the present analysis using ABAQUS and FE analysis software, liner analysis is conducted to study its behaviour, when subjected to internal burst pressure. Also, contact failure issues such as buckling and debonding, which lead to loss of COPV are also analyzed.

90.2 Materials and Design

The selected liner material in COPV is aluminium alloy (AA) 6061-T6. Assigned physical [11] and mechanical material properties are shown in Tables 90.1 and 90.2, respectively. The spherical liner design is shown in Fig. 90.1, as being a non-contributory type liner, the thickness in dome region is kept minimum to achieve weight savings. The formed part is fabricated by forming a 3.5 mm thick

Table 90.1 Physical properties of AA 6061-T6

Young's modulus, MPa	Poisson's ratio	Density, kg/mm^3
73,100	0.34	2.76×10^{-6}

Table 90.2 Mechanical properties for base and weld metal AA 6061-T6

Part	Yield stress, MPa	Yield strain, %	Tensile stress, MPa	Tensile strain, %
Base	240	7	290	19
Weld	140	4	180	8

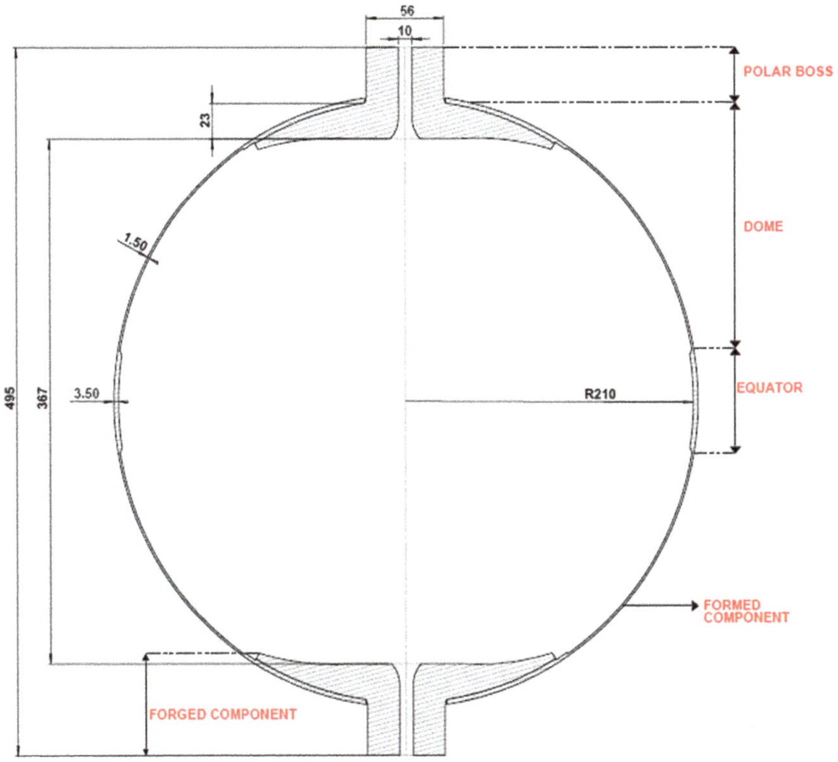

Fig. 90.1 Pressure vessel liner configuration (dimensions in mm)

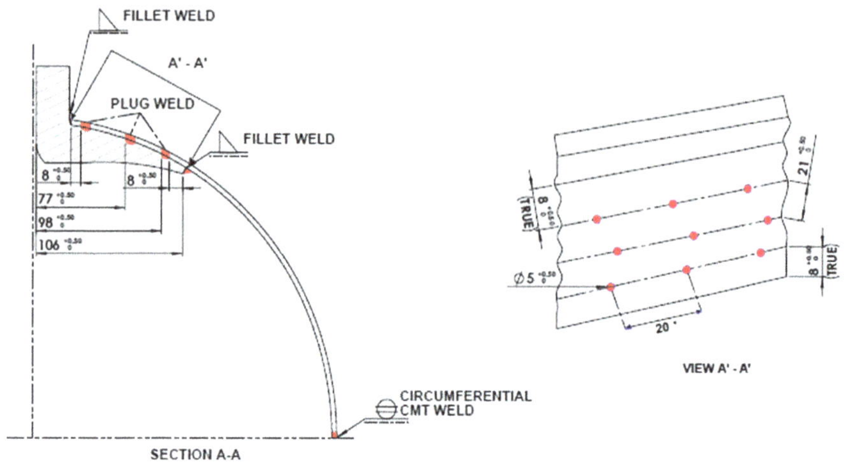

Fig. 90.2 Representation of plug, fillet and CMT welding locations in the liner

sheet using die and punch system, and forged part for end boss is obtained by machining a solid block of AA 6061-T6. The parts are further joined by plug welding technique, and welding the hemispheres circumferentially using cold metal transfer welding (CMT) technique as shown in Fig. 90.2. The true distance between the points of laying plug weld to edge of fillet weld is maintained at 8 mm with each plug having radius of 5 mm. Between each plug weld, a distance of 21 mm is maintained and positioned at three stages with pitch angle of 20° as shown in Fig. 90.2 (section A′–A′).

90.3 Finite Element Modelling

The liner is modelled as solid part. The effect of welding process changes the inherent base material properties in aluminium. Hence, by considering the effects due to welding, weld metal properties are assigned at pole and equator region, and length to be assigned is determined from an experimental trial. The length for assigning weld metal properties in liner is determined from Fig. 90.3. A miniature level AA 6061-T6 tube test specimen (replicating the large-scale liner structure) having a thickness of 3.5 mm were subjected to CMT welding to determine the weld bead width (BW) and heat-affected zone (HAZ). It was observed that BW and HAZ comprised of 15 mm (refer Fig. 90.3a) and this was measured using Struers welding expert with magnification of 20X. Due to less heat input of CMT process, various researchers have observed narrower heat-affected zone in AA 6061

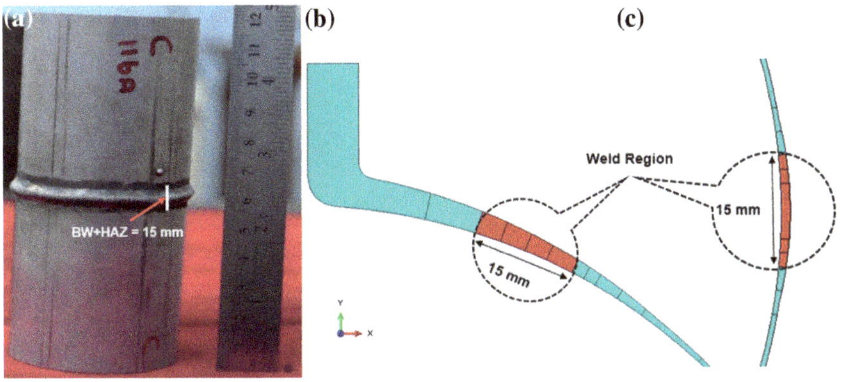

Fig. 90.3 Weld bead and HAZ measurement in CMT weld AA 6061-T6 pipe specimen (Left) and Weld region assignment in the FE modelled liner (Right)

weldment using CMT [12, 13]. In the present experimental trial, (refer Fig. 90.3a) the HAZ is found negligible as measured along with weld bead. Hence, weld bead + HAZ is considered as a single region and measured to be 15 mm in length. In the present analysis, weld material properties are assigned for a length of 15 mm as shown in Fig. 90.3b in the pole region and Fig. 90.3c in the equator region. The symmetric boundary condition is applied on side faces of liner by restricting *Z*-axis movement (U3 = 0). Also, bottom face of liner is restricted by *Y*-axis movement (U2 = 0) and inlet face movement is restricted by *X*-axis movement (U1 = 0), respectively as shown in Fig. 90.4. The internal pressure load is applied on the interior face of liner (refer Fig. 90.4). Liner burst test is conducted to determine its pressure load contribution and check the failure location due to applied load. The applied burst pressure load on the liner is determined using the Eq. 90.1.

$$P = \frac{2 * \sigma * t}{R}$$
$$P = \frac{2 * 290 * 1.5}{210} = 4.15\,\text{MPa} \tag{90.1}$$

where σ = Ultimate tensile strength of AA 6061-T6, t = Minimum thickness of liner and R = Radius of liner. Dynamic explicit analysis is conducted with non-linear geometric effects. For prediction of failure, Johnson–Cook (JC) damage material model is utilized. The failure is expected when parameter $D = 1$. The mathematical expression is shown in Eqs. 90.2 and 90.3. The constants for the JC model used in the evaluations for AA 6061-T6 is shown in Table 90.3 [14]. Load angle (ξ) effect is considered as $\xi = 0$.

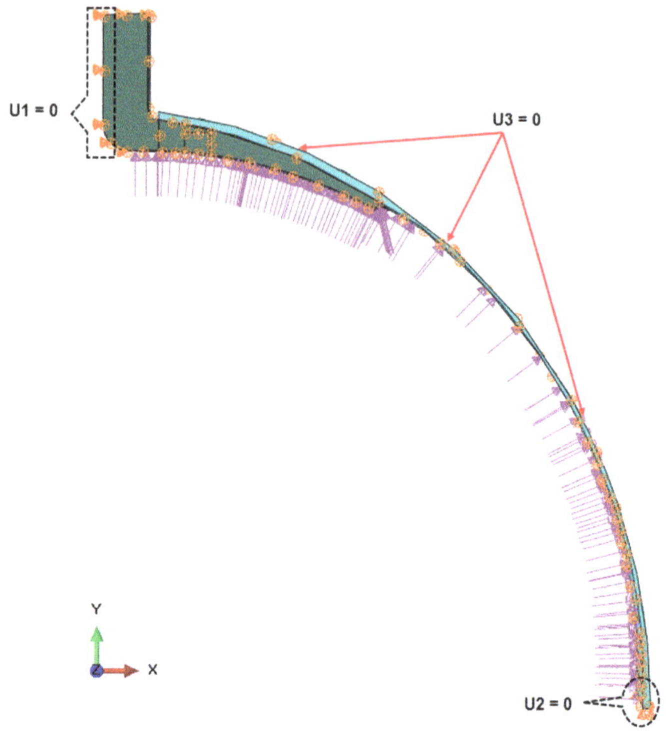

Fig. 90.4 Load and boundary conditions assigned on liner

Table 90.3 JC damage parameters assigned for AA 6061-T6

D_1	D_2	D_3	D_4	D_5
−0.77	1.45	−0.47	0.00	1.60

$$D = \sum \frac{\Delta \varepsilon}{\varepsilon_f} \quad (90.2)$$

$$\bar{\varepsilon}_f = [D_1 + D_2 \exp D_3\, \sigma^*][1 + D_4 \ln \dot{\varepsilon}^*][1 + D_5 T^*] \quad (90.3)$$

where

$\bar{\varepsilon}_f =$ Strain at fracture
$\dot{\varepsilon}^* =$ Strain rate
$T^* =$ Temperature
D_1, D_2, D_3, D_4 and D_5 are material constants.

Table 90.4 Description of meshing parameters

Part	Shape of element	Meshing technique	Elements
Plugs	Hex-dominated	Sweep	5891
Formed part	Hex-dominated	Free	7158
Forged part	Hex-dominated	Sweep	1038
Fillet	Hexagonal	Sweep	168

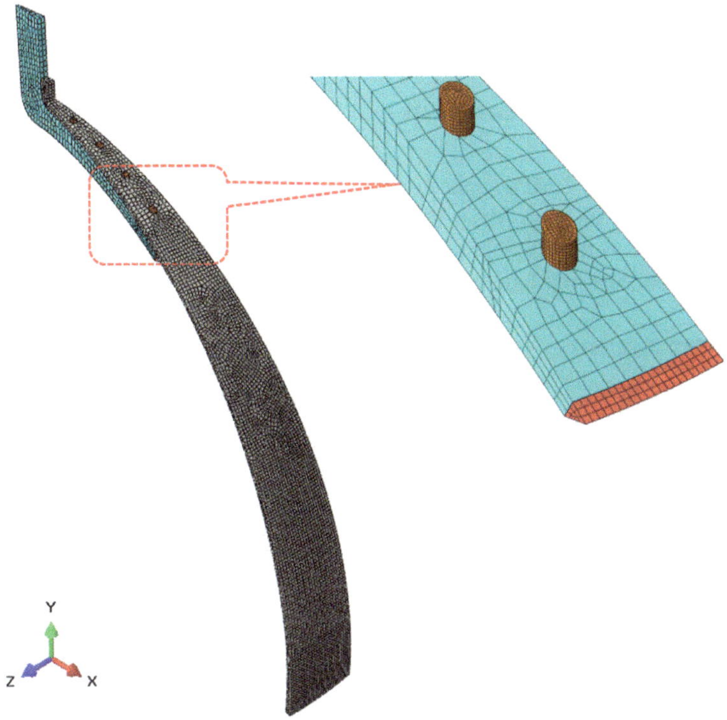

Fig. 90.5 Meshed model of liner and mesh profile in plug, fillet weld (inset)

The meshing of liner is performed by maintaining fine mesh near the plug weld region and on plug welds. The primary element type used for meshing parts is C3D8R (An eight node linear brick, reduced integration with hourglass control).

Table 90.4 shows a brief description of meshing parameters utilized in meshing of parts and Fig. 90.5 shows the meshed model of liner and mesh profile in plug and fillet weld.

The stable bond between liner and composite overwrap shall fail in case of design irregularities, complex curvatures in liner, thus creating a debond region. In the present analysis, a debond region is created by removing the contact interaction between liner and composite overwrap to study the ill-effects on liner. The eigenvalue linear analysis is performed for determining the load at which liner buckles during depressurization with subsequent buckling modes.

90.4 Results and Discussion

The burst load of 4.32 MPa applied on inner surface of liner at plastic stage showed severe yielding with Von Mises stress of 309 MPa (refer Fig. 90.6a) and plastic strain of 0.76% (refer Fig. 90.6b) at the transition regions near to pole and equator region. Further, the first failure location of liner is seen at transition region near the

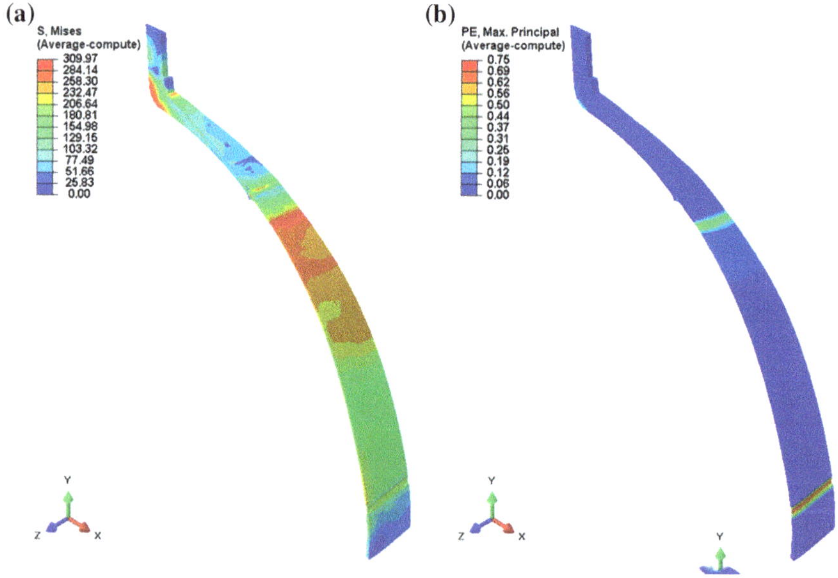

Fig. 90.6 Distribution of Von Mises stress **a** and plastic strain **b** at plasticity stage in liner

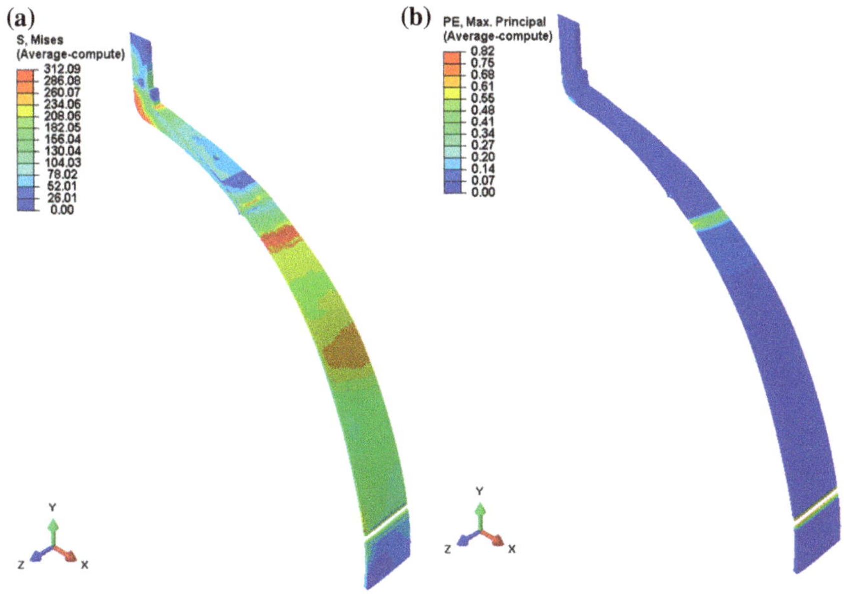

Fig. 90.7 Distribution of Von Mises stress **a** and plastic strain **b** at first failure stage in liner

equator section. Failure is observed in terms of crack appearing in liner and complete rupture is observed as equator region gets completely detached from the liner section at Von Mises stress of 312 MPa (refer Fig. 90.7a) and plastic strain of 0.82% (refer Fig. 90.7b). From the analysis, it was also required to know, if the failure is occurring in weld zones of liner as weld specimen has lower strength (refer Table 90.2). Observation showed that second failure of liner in terms of crack propagated at the transition region in dome section at Von Mises stress of 358 MPa (refer Fig. 90.8a) and plastic strain of 0.99% (refer Fig. 90.9a). Also Figs. 90.8c, d and 90.9c, d shows the enlarged view of the failure regions with complete separation of ruptured parts. It is also noted from Figs. 90.8b and 90.9b the Von Mises

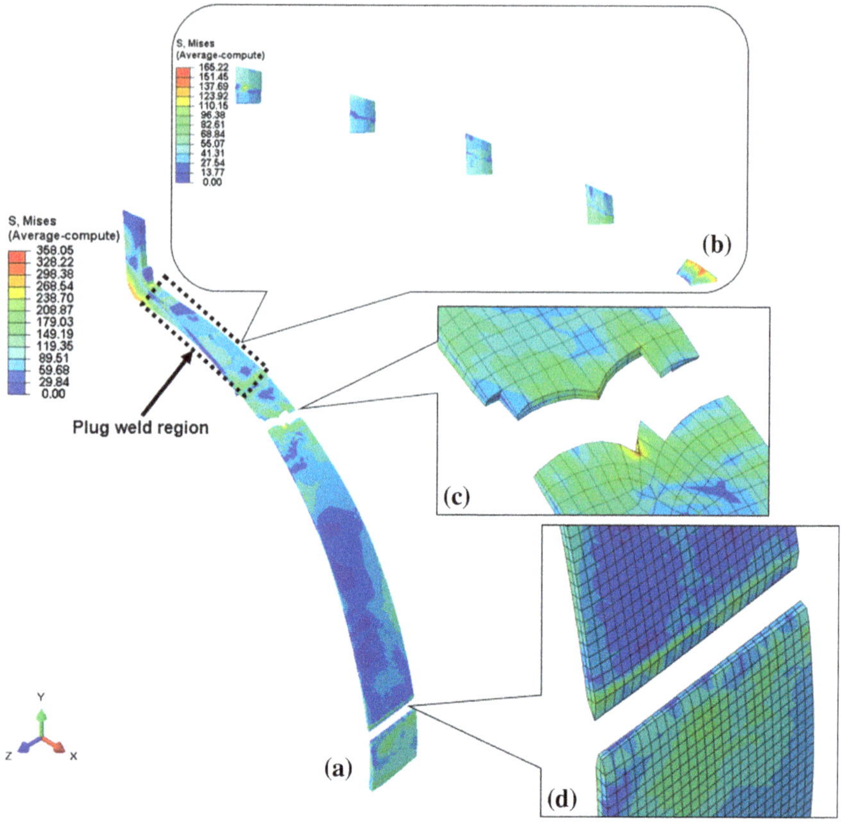

Fig. 90.8 Distribution of Von Mises stress in liner **a**, plug and fillet weld **b**, ruptured locations near dome **c** and equator **d** regions in liner

stress (165 MPa), plastic strain (0.056%) at the plug and fillet weld at the time of liner failure proved that it was intact and these welds can successfully take the impact of burst load and provide effective welding medium to join forged and formed part.

At debond regions, the internal pressure load applied on liner will not be effectively transferred to composite overwrap. This occurs because of improper bonding and insufficient bond pressure between liner-composite overwraps. Hence

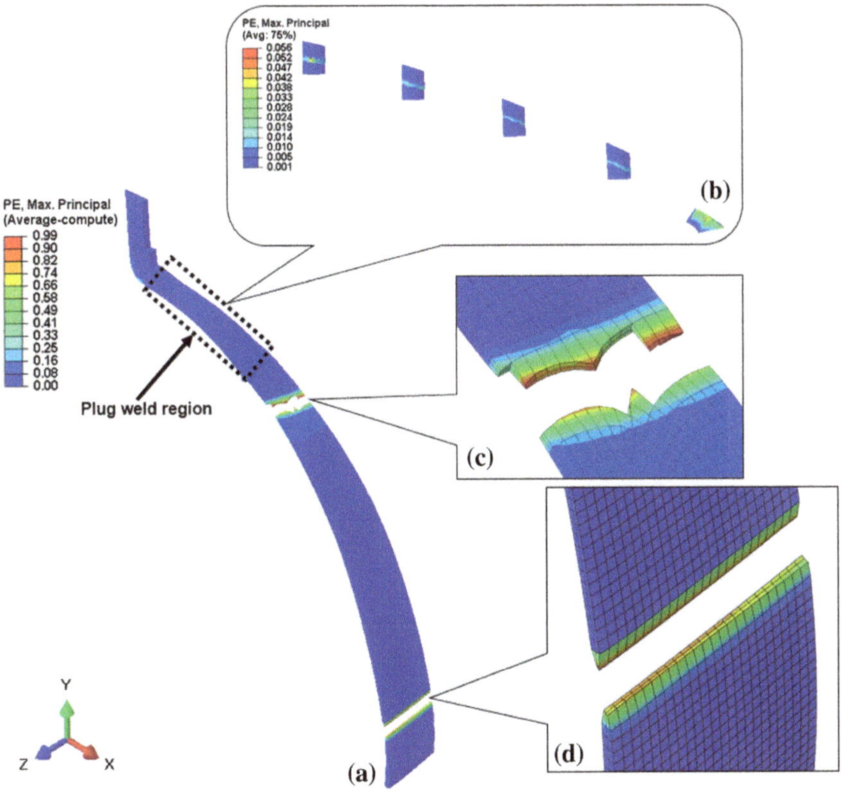

Fig. 90.9 Distribution of plastic strain in liner **a**, plug and fillet weld **b**, ruptured locations near dome **c** and equator **d** regions in liner

a void created in-between will allow liner to consume pressure load more than its capability, which will allow it to deform excessively as it cannot effectively transfer the load to composite overwrap. Peak Von Mises stress of 293 MPa (refer Fig. 90.10) is seen at the debond locations. The obtained stress of 293 MPa is 3 MPa more than the ultimate tensile strength of AA 6061-T6 proving that liner will fail in presence of debond. Graphical representation of Von Mises stress along the axial length of liner is shown in Fig. 90.11. Figure 90.12 shows the maximum deformation (7.93 mm) in liner at debond location. This plastic deformation causes

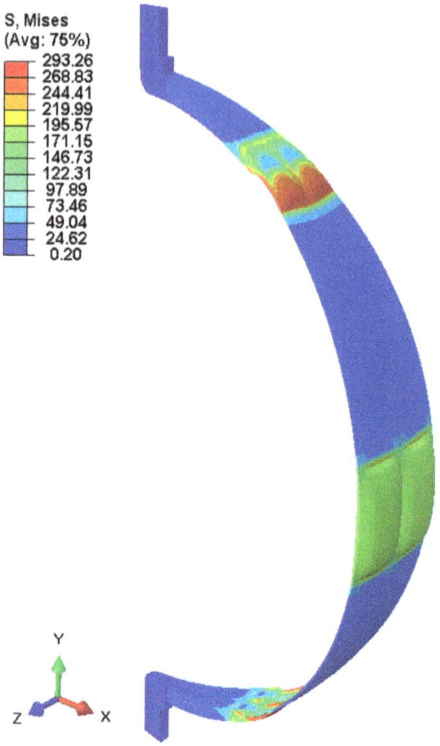

Fig. 90.10 Von Mises stress distribution in the debond model

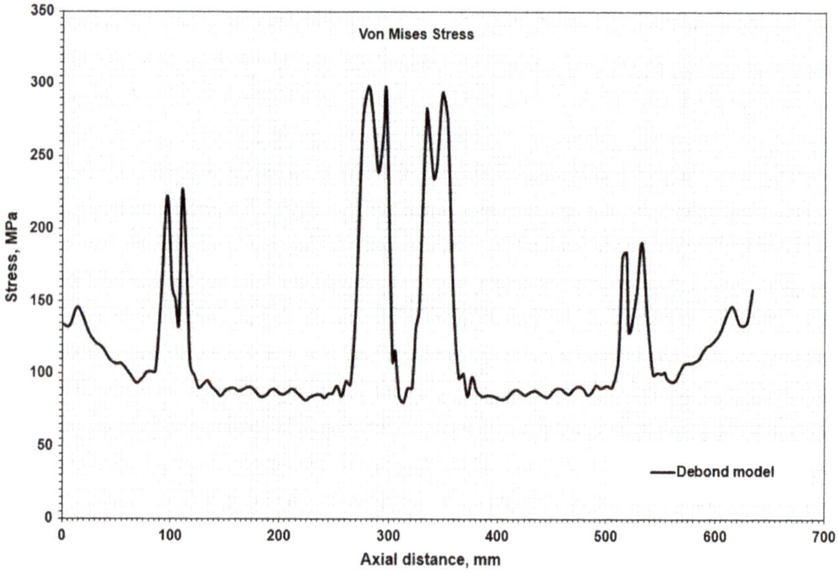

Fig. 90.11 Graphical representation of Von Mises stress distribution in the debond model

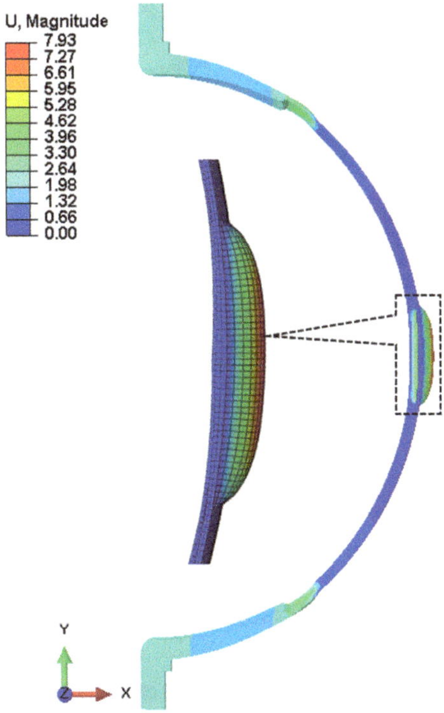

Fig. 90.12 Displacement plot in the debond model

additional problem, as during consecutive pressurization-depressurization cycles, it may cause liner to buckle; hence, bringing down the efficiency of liner and possible failure during operation. The eigenvalue linear buckling analysis is conducted to observe the buckling of liner (plastic condition) when a compressive load of 1 MPa is assigned to obtain load at which the liner buckles. During the depressurization phase, if the external load on the liner exceeds the critical buckling load of F_{cr} = 5.04 MPa the liner will buckle as seen in Fig. 90.13.

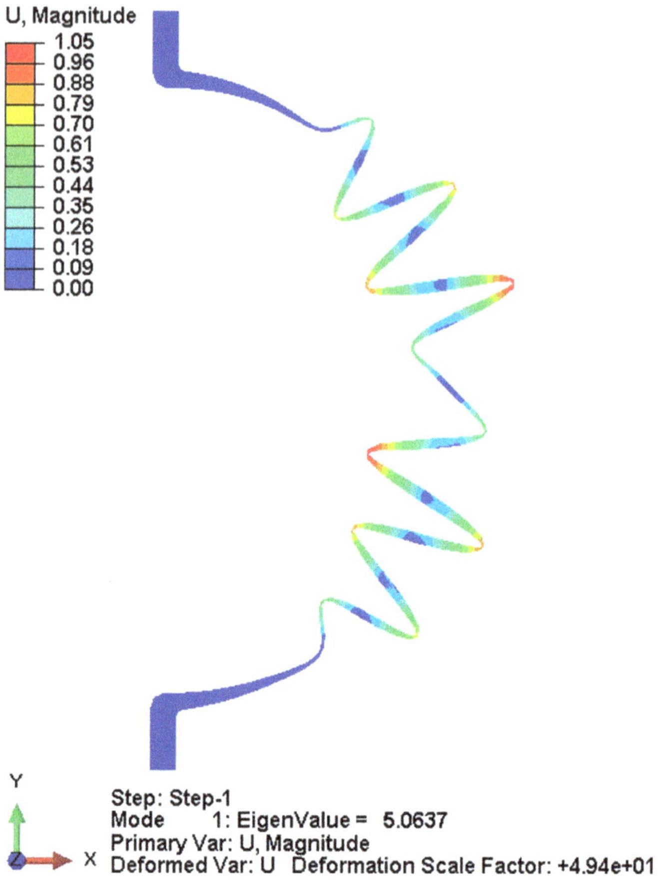

Fig. 90.13 Buckling of liner during depressurization

90.5 Conclusions

The three types of failures occurring in liner during operation are analyzed and the following conclusions are drawn:

- The presence of thickness variation in liner forms a transition region. Burst pressure load of 4.32 MPa, applied on liner showed that failure of liner occurred at these transition regions, i.e. equator region (312 MPa Von Mises stress and 0.82% plastic strain) and dome region (358 MPa Von Mises stress and 0.99% plastic strain). Further, results revealed that plug and fillet weld regions are safe from failure (165 MPa Von Mises stress and 0.056% plastic strain) due to applied burst pressure load.

- The presence of debond region had a major effect on liner as the observed Von Mises stress (293 MPa) exceed the ultimate tensile strength (290 MPa) of AA 6061-T6 proving that liner fails in presence of debond region.
- The predicted eigenvalue buckling load shows if the external pressure acting on the outer surface of liner exceeds of 5.04 MPa (depressurization phase) then the liner will buckle.

References

1. McLaughlan, P.B., Grimes-Ledesma, K.: Composite overwrapped pressure vessels. Natl. Aeronaut. Space Adm. 1–20 (2011)
2. Kawahara, G., McCleskey, S.: Titanium lined carbon composite overwrapped pressure vessel. In: 32 AIAA/ASME/SAE/ASEE Joint Propulsion Conference. Lake Buena Vista, pp. 3–12 (1996)
3. Lark, R.F.: Recent advances in lightweight filament wound composite pressure vessel technology. Natl. Aeronaut. Space Adm. 1–33 (1977)
4. Lee, H.-S., Yoon, J.-H., Park J.-S., Yi, Y.-M.: A study on failure characteristic of spherical pressure vessel. J. Mater. Process. Technol. **164–165**, 882–888 (2005)
5. Varun Chandran, B.V., Tiwari, S.B., Suresh, R, Krishnadasan, C.K., Sivasubramonian, B., Anoop Kumar, S.: Design and analysis of composite overwrapped pressure vessel. In: Proceedings of International Conference on Materials for the Future—Innovative Materials, Processes, Products and Applications, 109–114 (2013)
6. Almeida, J.H.S. Jr., Faria, H., Marques, A.T., Amico, S.C.: Load sharing ability of the liner in type III composite pressure vessels under internal pressure. J. Reinf. Plast. Compos. 2274–2286 (2014)
7. Son, D.-S., Chang, S.-H.: Evaluation of modeling techniques for a type III hydrogen pressure vessel (70 MPa) made of an aluminum liner and a thick carbon/epoxy composite for fuel cell vehicles. Int. J. Hydrogen Energy **37**, 2353–2369 (2012)
8. Liu, P.F., Xing, L.J., Zheng, J.Y.: Failure analysis of carbon fiber/epoxy composite cylindrical laminates using explicit finite element method. Compos. B **56**, 54–61 (2014)
9. Gray, D.L., Moser, D.J.: Finite element analysis of a composite overwrapped pressure vessel. In: 40th AIAA/ASME/SAE/ASEE Joint Propulsion Conference and Exhibit, pp. 1–15 (2004)
10. Akula, V.M., Shubert, M.: Analysis of debonding of filament wound composite pressure vessels. In: American Society for Composites—28th Annual Technical Conference, pp 341–354 (2013)
11. Kaufmann, C., Williams, K.: Optimization of a numerical simulation involving the impact of an AP-T C44 12.7 mm projectile on a semi-infinite monolithic 6061-T6 aluminium target, pp. 1–17. Defence Research and Development Canada (2004)
12. Nalajam Pavan Kumar, Arungalai Vendan, S., Siva Shanmugam, N.: Investigations on the parametric effects of cold metal transfer process on the microstructural aspects in AA6061. J. Alloys. Compd. 255–264 (2016)
13. Liang, Y., Hu, S., Shen, J., Zhang, H., Wang, J.: Geometrical and microstructural characteristics of the TIG-CMT hybrid welding in 6061 aluminum alloy cladding. J. Mater. Process. Technol. 18–30 (2017)
14. Lesuer, D.R., Kay, G.J., LeBlanc, M.M.: Modelling large-strain, high-rate deformation in metals. In: Third Biennial Tri-Laboratory Engineering Conference Modelling and Simulation, Pleasanton, pp. 1–12 (1999)

Index

R. G. Narayanan et al. (eds.), *Advances in Computational Methods in Manufacturing*, Lecture Notes on Multidisciplinary Industrial Engineering,
https://doi.org/10.1007/978-981-32-9072-3

www.ingramcontent.com/pod-product-compliance
Ingram Content Group UK Ltd.
Pitfield, Milton Keynes, MK11 3LW, UK
UKHW020317270926
475829UK00012B/106
* 9 7 8 9 8 1 3 2 9 0 7 1 6 *